PROCEEDINGS OF THE SECOND INTERNATIONAL SYMPOSIUM
HYDRAULICS / HONG KONG / CHINA / 16-18 DECEMBER 1998

Environmental Hydraulics

Edited by
J.H.W.Lee
The University of Hong Kong, China
A.W.Jayawardena
The University of Hong Kong, China
Z.Y.Wang
International Research and Training Centre on Erosion and Sedimentation (IRTCES), China

A.A.BALKEMA / ROTTERDAM / BROOKFIELD / 1999

Cover photo: Satellite images courtesy of Geocarto International Centre, Hong Kong; Internet site: www.geocarto.com © 1998 Geocarto/RSGS

The texts of the various papers in this volume were set individually by typists under the supervision of each of the authors concerned.

Published by
A.A.Balkema, P.O.Box 1675, 3000 BR Rotterdam, Netherlands
Fax: +31.10.413.5947; E-mail: balkema@balkema.nl; Internet site: www.balkema.nl
A.A.Balkema Publishers, Old Post Road, Brookfield, VT 05036-9704, USA
Fax: 802.276.3837; E-mail: info@ashgate.com

ISBN 90 5809 035 3
© 1999 A.A.Balkema, Rotterdam
Printed in the Netherlands

Environmental Hydraulics, Lee, Jayawardena & Wang (eds) © 1999 Balkema, Rotterdam, ISBN 90 5809 035 3

Table of contents

1.3 *Jets and plumes* – I

1.4 *Jets and plumes* – II

1.5 *Outfall modelling*

2 Transport and mixing processes
2.1 Turbulent shear flows

2.2 Turbulent mixing in open channel flow

2.3 Stratified flows

2.4 Air-water interaction

3.2 *Cooling water studies*

3.3 *Environmental impact studies/Three Gorges Project*

4.6 Coastal hydraulics

4.7 Miscellaneous

Environmental Hydraulics, Lee, Jayawardena & Wang (eds) © 1999 Balkema, Rotterdam, ISBN 90 5809 035 3

Editor's Foreword

The Second International Symposium on Environmental Hydraulics (ISEH-98) was organized by the Department of Civil Engineering, The University of Hong Kong. It was jointly held in parallel with The Seventh International Symposium on River Sedimentation (ISRS-98) at the Sheraton Hotel, Hong Kong, during December 16-18, 1998.

The objective of ISEH-98 is to bring together engineers, scientists, and practitioners with a common interest in hydraulic problems related to the environment. The conference aims to provide a forum for the exchange of ideas and experiences among hydraulics researchers, modellers, and practising engineers.

In response to the call for papers, about 230 abstracts were received by the secretariat. Prospective authors were invited to submit six-page manuscripts for final review. The proceedings were prepared from camera-ready manuscripts submitted by the authors with minor editing in some cases. Due to time and technical constraints, we have not been able to include every submitted paper, but we would like to thank those who have shown an interest in ISEH-98.

This Volume contains 3 keynote lectures, 7 invited lectures and some 130 contributions from 20 countries. The papers cover a broad spectrum of solution methods and research approaches – ranging from basic study of turbulent structures in open channel flow to case studies of quantitative environmental impact assessment. A rich body of the latest advances in hydrodynamic and water quality modelling, turbulent transport and mixing processes, and stratified flow can be found herein – as well as examples of sediment-environment interactions and notable engineering applications. The latter include in particular problems related to sustainable development in the Pearl River Delta including the Hong Kong waters, and on the Yangtze River. The collection of papers reflects the vibrant research activity on environmental hydraulics, as well as the complementary role of numerical modelling and experimental/field studies in the solution of practical engineering problems.

We would like to thank the valuable assistance of members of the International Scientific Committee, the Advisory Committee, and the Organizing Committee, and in particular the panel of referees. The generous support of the financial sponsors, in particular the Croucher Foundation and the K.C.Wong Educational Foundation, is gratefully acknowledged. Finally, the advice and expert handling of the publication of the Proceedings by Balkema are well appreciated.

J.H.W.Lee
A.W.Jayawardena
Z.Y.Wang

December 1998

Organization

Organized by
Department of Civil Engineering
The University of Hong Kong

Sponsored by
Hong Kong Institution of Engineers
Chinese Hydraulic Engineering Society
International Research and Training Centre on Erosion and Sedimentation (IRTCES)
International Association for Hydraulic Research
American Society of Civil Engineers
Japan Society of Civil Engineers
International Association for Hydrological Sciences
Hong Kong Polytechnic University
Hong Kong University of Science and Technology
American Society of Civil Engineers (ASCE) – Water Resources Engineering Division
Chinese Hydraulic Engineering Society (CHES)
Hong Kong Institution of Engineers (HKIE)
International Association for Hydraulic Research (IAHR)
Japan Society of Civil Engineers (JSCE)

Financially supported by
The Croucher Foundation
K.C.Wong Education Foundation
Hydraulics Research Wallingford, UK
Delft Hydraulics, Netherlands
ATAL Engineering Ltd.
Penta-Ocean Construction Co. Ltd
Environmental Hydraulics Visiting Fellowship, The University of Hong Kong
William M.W.Mong Engineering Research Fund
Dr Kai-kit Wong

COMMITTEES

1   Hydrodynamic and water quality modelling

1.1 Hydrodynamics and water quality of Hong Kong water

Environmental Hydraulics, Lee, Jayawardena & Wang (eds) © 1999 Balkema, Rotterdam, ISBN 90 5809 035 3

Keynote lecture: The CBS (Characteristic Based Split) algorithm in hydraulic and shallow water flow

O.C.Zienkiewicz
Institute for Numerical Methods in Engineering, Department of Civil Engineering, University of Wales Swansea, UK

P.Ortiz
CEDEX, Centro de Estudios de Técnicas Aplicadas, Computational Engineering Division, Madrid, Spain

ABSTRACT: An efficient finite element model for the computation of a wide range of shallow water problems is introduced. The algorithm presented can be used in an explicit, semi-explicit and in a nearly and fully implicit forms. The semi-explicit version shows robustness and economy in problems characterised by low Froude numbers, even for large timesteps. Also, the optimal diffusion properties of the method makes it suitable for very demanding high speed flows such as supercritical flows in hydraulic structures. In this case, an explicit version is recommended. The so-called 'nearly' implicit form considers implicitly the diffusion term. Here this possibility is illustrated for a scalar transport equation coupled with the Shallow Water Equations.

1. INTRODUCTION

Finite element modelling of the depth-integrated Shallow Water equations is a useful predictive tool for a wide range of problems in coastal, harbour and hydraulic engineering. A new finite element methodology is here proposed by the authors for the solution of the Shallow Water equations, called 'Characteristic Based Split' algorithm. The split procedure introduced has its basis on the fractional step method for the Navier Stokes equations for incompressible flows (Chorin, 1968). The extension of this technique for compressible flows and shallow water flows (Zienkiewicz et al, 1998, Zienkiewicz et al, 1995), allows a single characteristic velocity and, therefore, the application of the Characteristic-Galerkin method.

The simplest form of the procedure is represented by its explicit form. This case is currently adopted when high speed problems are studied (represented by Froude numbers close to or higher than one).

The semiexplicit form of the model is defined by the implicit computation of the pressure. This option provides a critical timestep (for pure convection) in terms of the flow velocity instead of the wave celerity as in the explicit form. This is a relevant property when low Froude number flows must be considered, such as, in general, tidal currents. Important savings in computation can be reached in this situation, obtaining, sometimes, up to 20 times the critical (explicit) time-step, without affecting considerably the accuracy of the results. This advantage is illustrated in this paper for a typical tide propagation problem in the Severn Estuary, where errors in amplitude and phases are computed for different timestep computations.

Finally, another form of the algorithm can be obtained when diffusion terms are not neglected. In this situation practical horizontal viscosity ranges (and diffusivity in the case of transport problems) can produce limiting timesteps much lower than the limit by convection. To circumvent this restraint, an implicit computation of the diffusion terms is required.

2. SHALLOW WATER EQUATIONS

Shallow water equations in their depth integrated form can be written, using the summation convention, as:

$$\frac{\partial h}{\partial t} + \frac{\partial U_i}{\partial x_i} = 0$$

$$\frac{\partial U_i}{\partial t} + \frac{\partial F_{ij}}{\partial x_j} + \frac{\partial p}{\partial x_i} + Q_i = 0 \tag{1,2}$$

where $(i,j=1,2)$ and $U_i = hu_i$ (depth-integrated horizontal velocities) and h (total height of water) are the unknowns. In above $F_{ij} = hu_iu_j$ is the (i) component of the (j) flux vector and the pressure p is:

$$p = \frac{1}{2}g(h^2 - H^2) \tag{3}$$

where H is the depth of water. The total depth h can be written as: $h = H + \eta$, where η is the surface elevation respect to the mean water level (see Figure 1). Q_i represents the (i) component of the source vector, here defined as:

$$Q_i = -g(h - H)\frac{\partial H}{\partial x_i} + g\frac{u|u|}{C^2h} + r_i - \tau_i \tag{4}$$

The terms of the RHS represent, respectively, the source term produced by the bottom slope variation, friction forces (Chezy-Manning formula), Coriolis force and wind tractions. Here, for simplicity, viscous terms are not considered.
For long waves, the wave celerity is related with the height of water as:

$$c^2 = \frac{dp}{dh} = gh \tag{5}$$

3. THE CHARACTERISTIC BASED ALGORITHM

The numerical procedure for the discretization of the equations (1) and (2) can be summarised following a sequence similar to the fractional step method:

a) Explicit computation of an intermediate variable ΔU^*_i considering the momentum equations ommiting the pressure gradient terms, by means of the Characteristic-Galerkin method (Zienkiewicz et al, 1995). This step gives:

$$\frac{\Delta U^*_i}{\Delta t} = -\left[\frac{\partial F_{ij}}{\partial x_j} + Q_i\right]^n + \frac{\Delta t}{2}\left[u_k\frac{\partial}{\partial x_k}\left(\frac{\partial F_{ij}}{\partial x_j} + Q_i + (1-\theta_2)\frac{\partial p}{\partial x_j}\right)\right]^n \tag{6}$$

where $(i,j,k=1,2)$. It should be remarked that all the terms in the RHS were computed at time : $t = n\Delta t$. This expression is modified if diffusion terms are added. Its form is going to be described in the next section.

b) Computation of the pressure as:

$$\frac{1}{c^2}\frac{\Delta p}{\Delta t} - \theta_1\theta_2\frac{\partial}{\partial x_i}\frac{\partial(\Delta p)}{\partial x_i} = -\frac{\partial}{\partial x_i}\left(U_i^n + \theta_1\Delta U_i^*\right) + \theta_1\Delta t\frac{\partial}{\partial x_i}\frac{\partial p^n}{\partial x_i} \tag{7}$$

4

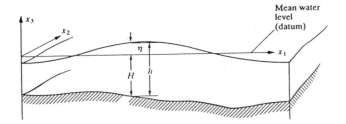

Figure 1 : Representation of the variables

(i=1,2); $(0 \leq \theta_1, \theta_2 \leq 1)$.

and the computation of the new surface elevation is carried out by using eq. (5).

c) Computation of the final velocity as:

$$\frac{\Delta U_i}{\Delta t} = \frac{\Delta U_i^*}{\Delta t} - \frac{\partial p^n}{\partial x_i} - \theta_2 \frac{\partial (\Delta p)}{\partial x_i} \tag{8}$$

where the correction coming from the pressure term is calculated again based in the discretisation along the characteristic lines (Zienkiewicz et al , 1995).

The final matrix form of the algorithm is, after the spatial discretization:

For the first step:

$$M \Delta U^* = - \Delta t \left[C U^n + M Q^n \right] - \frac{\Delta t^2}{2} \left[K_u U^n + K_p p^n + f_Q \right] + bt1 \tag{9}$$

where:

$$M = \int_\Omega N^T N \, d\Omega$$

$$C = \int_\Omega N^T u_j \frac{\partial N}{\partial x_j} \, d\Omega$$

$$K_u = \int_\Omega \frac{\partial}{\partial x_k} (N^T u_k) \frac{\partial}{\partial x_j} (N u_j) \, d\Omega$$

$$K_p = \int_\Omega \frac{\partial}{\partial x_k} (N^T u_k) \frac{\partial N}{\partial x_j} \, d\Omega$$

and

$$f_Q = \int_\Omega \frac{\partial}{\partial x_k} (N^T u_k) N \, d\Omega \, Q_i \tag{10}$$

and bt1 represents the boundary term:

5

$$bt1 = \frac{\Delta t}{2} \int_{\partial\Omega} N^T u_k \left(\frac{\partial F_{ij}}{\partial x_j} + Q_i \right)^n \cdot n_k \, d\Gamma$$

For the pressure step (step b)) the final form is:

$$\left[M + \theta_1 \theta_2 \Delta t^2 H \right] \Delta p = -\Delta t \, Q \left[U^n + \theta_1 \Delta U^* \right] - \Delta t^2 H p^n + bt2 \tag{11}$$

where now:

$$M = \int_{\Omega} \frac{1}{c^2} N^T N \, d\Omega$$

$$H = \int_{\Omega} \frac{\partial N^T}{\partial x_i} \frac{\partial N}{\partial x_i} \, d\Omega$$
and

$$Q = \int_{\Omega} N^T \frac{\partial N}{\partial x_i} \, d\Omega \tag{12}$$

The boundary terms bt2 are now:

$$bt2 = \Delta t^2 \int_{\partial\Omega} N^T \left(\frac{\partial p^n}{\partial x_i} \cdot n_i \right) d\Gamma + \theta_1 \theta_2 \Delta t^2 \int_{\partial\Omega} N^T \left(\frac{\partial (\Delta p)}{\partial x_i} \cdot n_i \right) d\Gamma \tag{13}$$

If integration by parts is applied for the second term of the left hand side of (11) and for the second and third term of the right hand side of (11), the last boundary integral represents (as an aproximation of order Δt) the normal component of the momentum equation on the boundary. If normal velocities are prescribed (or related with the pressure as in an open boundary condition), this is included at the pressure correction stage in a weak form.

Remark 1: The discretised form of the source term: $-g(h - H) \frac{\partial H}{\partial x_i}$ is such that for the particular solution of the Shallow Water equations: $h = H + \beta; \quad U_i = 0$, where β is a constant value, the numerical solution gives this particular solution. This condition (called by some authors as a conservative property(e.g. Bermudez et. al.(1994)) is satisfied if $H(x_1, x_2)$ is defined at nodal points and its discretisation is given by: $H = N^j H^j$.

Remark 2: Assuming the values of the time integration parameters such that:

$$\tfrac{1}{2} \leq \theta_1, \theta_2 \leq 1$$

the equation (11) is solved implicitly and the stability limit of the whole solution is governed by the critical time step defined by the computation of ΔU^*:

$$\Delta t \leq \Delta t_{crit} = \alpha \frac{l}{|u|}$$

where l is the element size and $\alpha = 1/\sqrt{3}$ for consistent mass matrix or $\alpha = 1$ if mass lumping is used. Taking $\theta_2 = 0$, the equation (11) becomes explicit and the global stability limit is nearly the same as that for the Taylor-Galerkin method (Zienkiewicz et al, 1995).

4.TRANSPORT EQUATIONS

The application of the characteristic based split algorithm for any scalar transport equation is straightforward, because of the absence of the pressure gradient term.. Then, the second and third step are not necessary.
The computation of the scalar s is analogous of the equation (6), but now, diffusion term is added:

$$\frac{\Delta S}{\Delta t} = -\left[\frac{\partial(u_i S)}{\partial x_i} + R\right]^n + \frac{\Delta t}{2}\left[u_k \frac{\partial}{\partial x_k}\left(\frac{\partial(u_i S)}{\partial x_i} + R + \frac{\partial(h\Phi_i)}{\partial x_i}\right)\right]^n + \left[\frac{\partial(h\Phi_i)}{\partial x_i}\right]^{n+\theta_3} \tag{15}$$

where Φ represents the fluxes of the transported quantity, R is the source, $S = hs$, and $0 \le \theta_3 \le 1$.

The simplest closure is adopted for these fluxes, defined by an eddy diffusivity "k" and the gradient of the transport quantity.
Now the final matrix form, neglecting terms higher that second order, is written as (preserving the notation defined in (10)):

$$(M + \theta_3 \Delta t\, D)\,\Delta S = -\Delta t\left[C S^n + M R^n\right] - \frac{\Delta t^2}{2}\left[K_u\, S^n + f_R\right] - \Delta t D\, S^n + bt \tag{16}$$

where now:

$$D = \int_\Omega \frac{\partial N^T}{\partial x_i} k \frac{\partial N}{\partial x_i} d\Omega$$

$$f_R = \frac{\partial}{\partial x_k}\left(N^T u_k\right) N\, d\Omega \cdot R_i^n$$

and

$$b.t. = \Delta t \int_{\partial\Omega} Nk\left(\frac{\partial s}{\partial x_i}\right)^{n+\theta_3} \cdot n_i\, d\Gamma \tag{17}$$

As a practical illustration, the parameters involved in the study of transport of salinity in a real case of a river are here considered. The region studied was of approximately 55 kilometers long and the mean value of the eddy diffusivity was of k=40 $\frac{m^2}{s}$. The limiting timestep for convection (considering 3 components of tides) was of 3.9 s.. This limit was severely reduced to 0.1 s. if the diffusion term is active and solved explicitly. The convective limit was recovered assuming an implicit solution with $\theta_3 = 0.5$.
The comparisons of diffusion error between computations with 0.1 s. and 3.9 s. had a maximum diffusion error of 3.2 % for the 3.9 s. calculation, showing the accuracy required for engineering purposes, taking

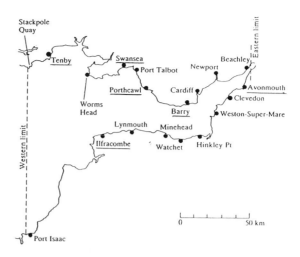

Figure 2 a):Bristol Channel: Location Map

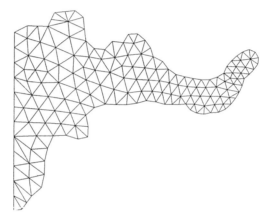

Figure 2..b):Bristol Channel: Mesh

into account that the timestepping was increased 40 times. The cost of computation is drastically reduced with the implicit computation.

This reduction is very important when, in practical applications, the behaviour of the transported quantity must be computed for long-term periods of time, as was the case of the example commented, where the evolution of the salinity needed to be calculated for more than 60 periods of equivalent M_2 tide.

5. APPLICATIONS

As an illustration of industrial applications of the model, two very different kind of problems are here presented.

In the range of low Froude numbers, an example of tide propagation problem is commented. Secondly, two examples of supercritical flows are shown.

5.1 Bristol Channel. Tide Propagation

The propagation of a M_2 tide on the Bristol Channel (see location map on Figure 2,a) was studied and the numerical results were compared with measurements obtained by the IOS (Institute of Oceanographic Sciences) (HRS, 1981).
The test was carried out with an approximation of the real bathymetry interpolated in a mesh of 256 elements and 172 nodes (Figure 2,b).

The semi-explicit form of the model was adopted, with the integration parameters : $\theta_1 = \dfrac{1}{2}, \theta_2 = \dfrac{1}{2}$.

A constant real friction coefficient (Manning) of 0.038 was adopted for all the estuary. Coriolis forces was included. The effect of the Coriolis forces in this problem was important, because the results shows a better agreement with the measurements in terms of phase.
The table 1 represents the comparisons between observations and computations in terms of amplitudes and phases for 7 different points which are represented in the location map of the Figure 2,a. This comparison were done for three different timesteps. The first timestep (50 s.) is approximately 0.8 the critical timestep for an explicit computation of the heights, (computed with the sum of the wave celerity and current velocity). The second timestep used was of 200s. and the third timestep was of 400s. (approximately 8 times the explicit limit mentioned above).
The maximum error in amplitude only increases in 1.4 % when the timestep of 400 s. is used respect to the timestep of 50 s., while the absolute error in phases (-13) is two degrees more than the case of 50 s. (-11).

Table 1 :Amplitudes (cm.) and phases (degrees) of M2 tide in the Bristol Channel:
Observations and computations for timesteps: 50 s.,200 s.,400 s..
(Absolute errors in phases and relative error in amplitudes are represnted in parenthesis)

Station	Observations (HRS)		Model (dt=50 s.)		
	a(cm.)	g(d)	a(cm.)	g(d)	% (a)
Boundary(mid)	235.	160.	238.	160(0)	1.2
Tenby	262.	170.	254.	170(0)	3.
Swansea	315.	173.	307.	176(+3)	2.5
Ilfracombe	308.	162.	300.	165(+5)	2.6
Barry	382.	182.	387.	186(+4)	1.3
Porthcawl	317.	173.	327.	178(+5)	3.
Avonmouth	422.	202.	432.	191(-11)	2.3

Station	Model (dt=200 s.)			Model (dt=400s.)		
	a(cm.)	g(d)	% (a)	a(cm.)	g(d)	% (a)
Boundary(mid)	238.	159(-1)	1.2	238.	162(+2)	1.2
Tenby	255.	170(0)	2.6	252.	170(0)	3.8
Swansea	307.	176(+3)	2.5	307.	176(+3)	2.5
Ilfracombe	299.	165(+3)	2.9	295.	166(+4)	4.
Barry	387.	185(+3)	1.3	387.	186(+4)	1.4
Porthcawl	325.	178(+5)	2.5	327.	178(+5)	3.1
Avonmouth	431.	189(-13)	1.9	430.	189(-13)	1.9

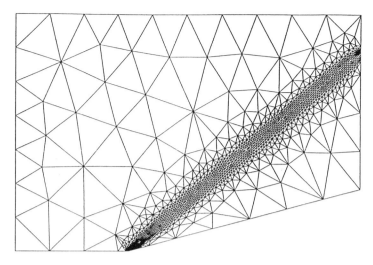

Figure 3..a): Supercritical flow in a channel with a wall constriction. Final mesh
Nodes= 1512. Elements = 2963. One remeshing

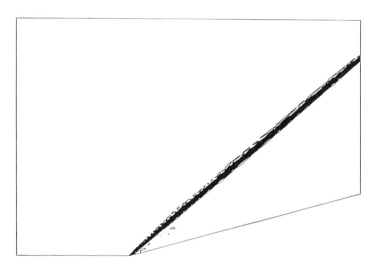

Figure 3..b): Supercritical flow in a channel with a wall constriction. Heights contours

5.2 Supercritical flow in a channel with a wall constriction

A regular triangular mesh of 2066 nodes and 3955 elements was firstly used for this case. One remeshing
was performed, giving a mesh with 1512 nodes and 2963 elements (see Figure 3,a) .

With only one remeshing an accurate capture of the shock-type solution is reached.The inflow Froude
number was 2.5.

The boundary conditions imposed are height and velocities prescribed at the inflow boundary (left
boundary), slip boundary condition at the wall (lower boundary) , free variables at the outflow boundary
(right boundary) and symmetric boundary condition on the upper boundary.

The solution adopted was explicit with an interior damping increased by using a limiting value of timestep
(local).

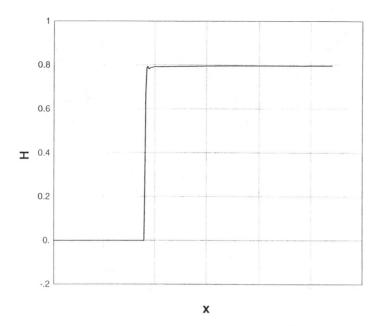

Figure 3..c):.Supercritical flow in a channel with a wall constriction.
Heights along the constricted wall.
Theoretical Solution (Ippen,1951): $\eta = 0$; $\eta = 0.807$

Otherwise, neither artificial diffusion nor friction were included.
The Figure 3,b shows the contours of heights calculated, while the Figure 3,c shows the heights along the constricted wall. It can be seen in the Figure 3,c the very good agreement with the theoretical inviscid solution of the problem (Ippen, 1951).

5.3 Symmetric channel of variable width

For a supercritical flow in a rectangular channel with a symmetric transition on both sides, a combination of a 'positive' jump, as in the previous case, and 'negative' waves, causing a decrease in depth, appears. The profile of the latter is gradual and an approximate solution can be obtained by assuming no energy losses and that the flow near the wall turns without separation.

The constriction and enlargement here analised was of $15°$, and the initial regular mesh generated have 9790 nodes and 19151 elements.
The supercritical flow has an inflow Froude number of 2.5 and the boundary conditions are as follow:
 height and velocities prescribed in inflow boundary (left boundary of Figure 4), slip boundary condition on walls (upper and lower boundaries in Figure 4) and free variables on the outflow boundary (right side of Figure 4).
Here the explicit version with local timestep was again adopted as in the previous case.
The computation has been carried out with only one remeshing. The final mesh has been reduced from the original one to 13652 elements and 6979 nodes. The Figure 4 represents the contours of heights, were 'cross' waves and 'negative' waves are contained. One can observe the 'gradual' change in the behaviour of the negative wave created at the origin of the wall enlargement.

11

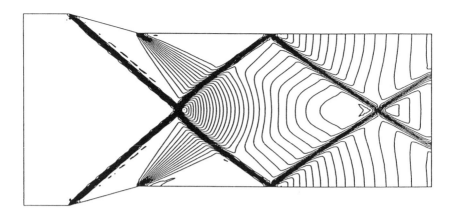

**Figure 4. : Symmetric Channel of variable width. Contours of h. Inflow Froude number:2.5
elements: 13652. Nodes: 6979. Constriction: 15 degrees.
Left boundary: inflow. Right boundary: outflow.**

6.CONCLUSIONS

The algorithm proposed shows remarkable properties for all the range of problems which can be treated with the Shallow Water approximation: economy and efficiency for low Froude number type of problems, nearly optimal diffusion properties, necessary for very demanding problems such as high speed flows (including shock formation in the solution), and simplicity in terms of the inclusion of transport equations. The Characteristic-Galerkin method can be, obviously, fully exploited in the case of Transport equations, but the Characteristic Based Split algorithm permits to do this also when the momentum equations must be treated. The implicit solution for the diffusion terms allows to retain the timestep in terms of the velocity, giving an efficient solution for long term transport problems.

7.REFERENCES

Bermudez, A. and Vazquez, M (1994). "Upwind methods for hyperbolic conservation laws with source terms". *Computers and Fluids*, 23, 8, 1049-1071.

Chorin, A. J. (1968). "Numerical Solution of the Navier Stokes Equations", *Math. Comput.*, 22, 745-762.

Hydraulic Research Station (1981). "Severn Tidal Power". *Report EX 985*.

Ippen, A. (1951). "High velocity flow in open channels: symposium", *Trans. ASCE*, V, 116.

Zienkiewicz, O.C. and Ortiz P. (1995). "A split-characteristic based finite element model for the Shallow Water equations", *Int. Journ. Num. Meth. Fluids*, 20, 1061-1080.

Zienkiewicz, O.C., Nithiarasu, P., Codina R., Vazquez, M. and Ortiz, P. (1998) "An efficient and accurate algorithm for Fluid Mechanics problems. The Characteristic based split procedure". (To be published). *Int. Journ.. Num. Meth. Fluids*

Environmental Hydraulics, Lee, Jayawardena & Wang (eds) © 1999 Balkema, Rotterdam, ISBN 90 5809 035 3

Keynote lecture: Recent advances in computational hydraulics

T.J.Weare
HR Wallingford Group Limited, UK

ABSTRACT: The current programme of European research underpinning recent advances in computational hydraulics is reviewed. The emphasis of the review is on the extensive programme of publicly funded pre-competitive research.

1. INTRODUCTION

There is a perplexing range of powerful computational hydraulics tools available to civil engineers today. For a recent review see Weare, 1996. In the following paper I attempt to summarise the current programme of research underway in Europe that is helping to develop and extend these computational hydraulic tools. HR Wallingford is actively engaged in much of this research and I have drawn heavily from our experience. Most of the major research projects referred to are the result of pan-European collaboration involving the leading national hydraulics laboratories and many of the leading academic institutions. Space does not permit me to give proper credit to all of the contributors involved. Fuller credit is contained in the reference papers.

2. RIVERS AND RIVER BASINS

Computational tools for 1D modelling of river and urban catchments have been commercially available for many years now. Software products such as ISIS or Mike 11 for river modelling and MOUSE or HydroWorks for urban drainage now have a wide following. These tools and others are developed and supported on a wholly commercial basis. Great effort is going into the integration of functions to provide appropriate tools for today's holistic needs. ISIS and HydroWorks, for example, starting from a basic hydraulic engine have evolved to embrace also the hydrology, real time control, sediment transport, and water quality. The data requirements for some of the larger commercial studies is considerable and much of the recent development of these tools has concentrated on the data/information management and project management tasks. Typical of the current state of the art in this field is InfoWorks. In developing InfoWorks, Wallingford Software have combined a Relational Database with a Geographical Analysis to provide a single environment that integrates asset and business planning with modelling. The result is a powerful desktop tool for predicting events, consolidating data, communicating information and improving the decision making progress.

In the main the development of these commercial products will continue as a self-sustaining business. However, underpinning this activity there needs to be a programme of generic research. The widespread and frequent severe flooding that has been a feature of this decade has stimulated a number of research projects into the cause and mitigation of river flood risks. Many of these involve the use or development of computational models. Notable areas of research are described in the following.

2.1 The Use of 3-D Computational Models for River Flow Simulation

In a recent study for the UK Government a number of commercially available 3D CFD codes were applied to a variety of river flow test cases. Flows studied include, secondary flow in river bends, bifurcations, flow control structures, intake structures and groynes. Key points to come out of the research are:

- no one model is good for all applications,
- the codes can reproduce the detailed velocity distributions measured in the lab and the field, and
- the most probable area of application for 3D modelling in the near future is for the design of hydraulic structures.

Areas of difficulty are:

- recirculating flows where the results are sensitive to the choice of model parameters and
- the ability of the modeller to assimilate all the information available from the software and make effective judgements on the model results.

Further details of the research can be found in Samuels (1998). The authors conclude that although CFD codes are effective, particularly for local modelling around structures and for testing alternative plan layouts, they cannot yet be considered as a complete replacement for the traditional scale models for large areas or particularly complex situation which would require excessive computer resources. Also the proper use of these codes requires specialist skills, more so than with 1D packages. Hence their use will probably be restricted in the short to medium term.

2.2 Flood modelling.

Considerable advances have been made in the analysis and modelling of riverine flooding through a number of European Commission funded projects. An excellent review of this field is contained in the paper by Casale and Samuels (Casale and Samuels, 1998). Advances are noted in six areas:

- *Weather radar and other remotely sensed data.* Radar imagery potentially provides a powerful means of measuring the rainfall distribution over large areas in real time. Investigations are in progress on how a river catchment model reacts to the measurement errors that are inherent to this source of data (**DARTH,** **D**evelopment of **A**dvanced **R**adar **T**echnology for Application to **H**ydrometeorology).
- *Hydro-meteorological modelling.* There are marked differences in spatial and time scale between the models developed by meteorologists and hydrologists. A recurring theme in the European concerted action research has been the need for better communication between these two groups (**RIBAMOD,** **Ri**ver **Ba**sin **Mod**elling, Management and Flood Mitigation). High resolution limited area meteorological modelling now has grid scale down to 5km (**TELFLOOD,** Forecasting floods in urban areas downstream of steep catchments). Short-range quantitative precipitation forecasting is being improved through detailed modelling of the atmospheric column (**HYDROMET,** The development of active on-line **hyd**rological and **met**eorological models to minimise impact of flooding).
- *Flood forecasting.* There is a need to balance the degree of sophistication of the forecasting models with the forecast requirements and the quality of information available on precipitation and on the catchment state. Under the project **AFORISM** (A comprehensive forecasting system for flood risk mitigation and control) progress has been made in the inter-comparison of different approaches to rainfall-runoff modelling, and in then integration of innovative technologies in an operational decision support tool for flood forecasting and flood impact analysis.
- *River basin modelling.* The focus of EC funded research has been on the use of simulation models as research tools rather than on the fundamental model development. Conclusions from **RIBAMOD** include: the need to integrate models across the disciplines involved, the risk of end-user requirements moving beyond the fundamental capability of the calculation "engine", and the recurrent theme of the disparate spatial and temporal scales at play.
- *Debris and sediments.* Flooding in mountainous areas is often compounded with the additional hazard of

debris flow. Modelling the occurrence of landslides was an important part of the recently completed **FRIMAR** project (**F**lood **ri**sks in **m**ountain **ar**eas). The **DEBRIS FLOW** project (Debris flow management in the alpine region) completed in 1996 achieved progress in the modelling of initiation mechanisms, propagation and deposition phenomena. Special attention was paid to the flow through channels and to the extension of the deposition areas of lobes or sheets of debris. The dynamics of unsteady debris flows and the problems of shock front impacts on structures was investigated, with promising results for the design of civil engineering structures.

- *Flood risk assessment.* The EC has been funding research on methods for assessing, mapping and communicating flood risk. Within the **FLOODAWARE** project (Applied researches on a transferable methodology, devoted to **flood aware**ness and mitigation, helping the decision negotiation processes, adapted to a changing environment, and respecting the water resources.!) a framework is being developed that integrates hydrological modelling, river flow simulation, digital terrain modelling, land use assessment and risk mapping. The methodology is being tested in several countries.

Key areas for future research and development of computational tools have been identified and include:

- Improve the coupling of meteorological and hydrological forecasting, in particular the forecasting of rainfall from current radar analysis.
- For the investigation of the effects of climate change on flood risk, a key research issue if the generation of precipitation fields at the appropriate spatial and temporal scale from the results of GCM (Global Climate Model) simulations of future climate scenarios.
- Combining models to provide decision-makers with tools which address the practical management of river systems. A particular challenge is the linking of models of water movement and riverine ecology.

A major collaborative project underway at present, which addresses many of these needs, is **EUROTAS** (**Eu**ropean **R**iver Flood **O**ccurrence and **T**otal Risk **A**ssessment **S**ystem). The objective is to develop and demonstrate tools and procedures for the assessment and management of flood risk, including the effects of climate change. The research has three main components:

- The development of an integrated framework for whole catchment modelling based upon an "open-systems" approach.
- The demonstration of the feasibility and benefits of integrated modelling to answer real scientific and practical issues on the changing nature of flood risk in five different European river catchments.
- The development of procedures to determine the impact of river engineering works and environmental change on flooding and the assessment of flood risk.

The principal output of the research will be a prototype integrated catchment modelling system which will include decision support for the procedures developed in the course of the research. This project and a number of others that HR Wallingford is leading or co-ordinating can be tracked through the HR web page (http://www.hrwallingford.co.uk).

2.3 Dambreak analysis

Another collaborative research project that is receiving European funding is **CADAM**, **C**oncerted **A**ction on **Dam**-break **M**odelling. Having officially started on 1st February 1998 **CADAM** has already generated interest world-wide. The principal aims are:

- to exchange dam-break modelling information between participants:
  Universities ←→ Research Organisations ←→ Industry
- to promote the comparison of numerical dam-break models and modelling procedures with analytical, experimental and field data
- to promote the comparison and validation of software packages developed or used by the participants
- to define and promote co-operative research

CADAM continues the work started by an IAHR working group, itself established and led by Electricté de France (EDF) following the IAHR Congress in 1995. HR Wallingford is acting as *Co-ordinator* for this project which has participants from over 10 different countries across Europe. Links with other experts around the world are being sought to ensure that state-of-the-art techniques and practices are considered.

Early priorities for research include:

- Define and disseminate tests for breach modelling
- Tests to be applied to real geometry's (using physical models) to investigate how models resolve features in real valleys
- Compile a database of actual dam failures plus available data
- Investigate the classical flow formulae – are they applicable for dam-break?
- Are the Shallow Water Equations applicable to dam-break?
- Identify the important processes within the whole dam-break event (i.e. hydrodynamics / breaching etc.)
- Define dam failure modes and identify reasons for them
- Further targeted modelling – valley confluences / structures / 1D 2D models

Like many of the European collaborative research programmes CADAM's progress will be marked and measured by a series of workshops. The first was held at HR Wallingford in March 1998. The second is scheduled for October 8th/9th, 1998 and will be hosted by the University of the Federal Armed Forces (Institute of Hydrosciences) in Munich. The Munich meeting is to be an 'open' workshop at which state of the art papers will be presented on a number of topics including,

- Breach formation: history, field data, case studies, formation processes and modelling
- Dam-break sediment processes: process identification, sediment / debris transport and modelling
- Dissemination of test cases for comparison with 'real valley' models

The meeting will be open to non CADAM members. Full details of the meeting programmes and further information on the research programme may be found on the CADAM website: www.hrwallingford.co.uk/projects/cadam

3. COASTAL WATERS

3.1 Commercial tools for tidal flow and wave propagation

A wide range of computational tools is now in routine commercial use to describe the propagation of tides and waves in coastal waters. The computing power now available permits the use of 2 and 3-D finite element models for commercial studies, with all the flexibility and user friendliness that this implies. A good example of the latest commercial tool is **TELEMAC 3-D** developed by Electricite de France, see Hervouet, 1996. The 3-D Navier Stokes equations with the hydrostatic pressure hypothesis are solved on an unstructured triangular mesh. The numerical scheme for discretisation in time uses an operator splitting method so that each term in the equation can be processed optimally, allowing high performance, robustness and accuracy. The finite element solution used in systems such as TELEMAC supplants the finite difference solutions commonly used hitherto. HR Wallingford has recently successfully enhanced the system to include a water quality module in **TELEMAC** and applied it to a number of studies in Hong Kong including the Green Island reclamation.

A good example of a modern wave-modelling tool is the **ARTEMIS** software, also out of the EDF stable. This is now in regular use for commercial design studies of coastal ports and harbours, see for example Tozer (1998). **ARTEMIS** is used to solve the mild slope equation and has the ability to handle resonance effects and to include the effects of wave breaking. The model can be run in various modes, from monochromatic unidirectional waves to short crest random wave spectra. The unstructured finite element mesh solution (identical to that in **TELEMAC**) allows irregular boundaries to be fitted accurately. However

16

the ultra fine mesh required for accurate resolution of the waves can lead to high running costs. It is common practice therefore to use the less complete but computationally more efficient models based on ray tracing techniques to provide a coarse screen to identify the critical wave conditions. It is also usual to test final designs of any major port or harbour scheme in a large-scale physical model.

Most of these computational tools for the modelling of waves and currents are now being developed as entirely commercial products by the major hydraulics laboratories. Nationally and internationally funded research tends to focus on other processes in the coastal zone. In particular, considerable effort is being directed towards achieving a better understanding of sediment dynamics (including interaction with coastal structures), and long-term coastal morphodynamics. This in turn will lead to improved computational tools. Major projects are summarised in the following.

3.2 Coastal morphology – sand and shingle

Considerable strides have been made in the past 5 years in the development and application of computational models for coastal morphodynamics. Prediction of coastal morphology over yearly and decadal time-scales forms an important part of many coastal engineering design studies. In 1993 the UK ministry responsible for flood protection started a 5 year project, called **CAMELOT** (**C**oastal **A**rea **M**odelling for **E**ngineering in the **Long T**erm), to investigate methods suitable for engineering use to predict long-term morphology of exposed sand or shingle/gravel coastlines. The **CAMELOT** project has also linked with a number of EU funded research projects under the **MAST** programme (**Ma**rine **S**cience and **T**echnology).

CAMELOT models are based, to a greater or lesser extent, on theoretical representations of the main wave, current and sediment transport processes. No single model type can cover all engineering applications involving coastal morphology. HR Wallingford has developed a range of models for different purposes:

MODEL	TIME-SCALE	TYPE	SPEED
COSMOS-2D	Hours to Weeks	Profile model	Fast
PISCES	Hours to Weeks	Area model	Slow
BEACHPLAN	Months to Decades	Plan model	Fast

Mostly these use well-established theories, but in a few cases some theoretical development was necessary. An example is the development of a shingle transport formula which has been incorporated in the **COSMOS-2D** model (a beach profile model describing cross-shore processes only) and in the **BEACHPLAN** model (bulk long-shore processes only). **COSMOS-2D** can be used to predict the evolution of a cross-shore slice of a uniform beach over a period of hours or weeks, for storm events. **BEACHPLAN** is used to predict the evolution of a representative coastline contour over a period of months to decades. Both are computationally fast and can therefore be used to investigate the limits of predictability due to chaotic behaviour by "ensemble" modelling (see below).

PISCES is an interactive coastal area model, capable of simulating the various processes of wave propagation, current distribution, the resulting sediment flux, and the associated response of the coastline. At the heart of PISCES is a streamwise sediment flux algorithm based on the work of Soulsby, 1998. An interesting research application is described by Chesher, 1995 in which PISCES was applied to the study of a breach through a coastal sand strip, connecting a large lagoon with the Gulf of Guinea at Keta in Ghana. Comparisons were made with data supplied by Delft Hydraulics from a physical model study of the area based on data collected locally. Successful morphological evolution over a period of approximately one year was obtained following the development of an algorithm to describe the erosion of the sides of the breach.

One of the early findings of the **CAMELOT** research was the realisation that morphodynamic models based on detailed representations of the main physical processes of waves, tides and sediment transport were doomed to fail when applied to long time-spans, over years and decades. The lack of long-term data for validation, the impossibility of knowing future sequences of waves and other forcing conditions, and the possibility of "chaotic" types of morphological behaviour all militate against the straightforward deterministic approach. Different methodologies have been developed for long-term predictions including:

- Process/input filtering to allow models to operate at long time-scales
- Probabilistic predictions rather than deterministic
- Multiple model runs to quantify uncertainty due to sequencing of forcing conditions
- Sensitivity analysis to quantify uncertainty due to input data errors
- Ensemble modelling to establish limits of predictability due to chaotic behaviour

For further description of the **CAMELOT** project see Southgate, 1998.

3.3 Intertidal mudflats

A large proportion of estuary coastlines in the UK (and elsewhere) consists of intertidal mudflats, particularly in areas sheltered from severe waves. Mudflats play an important role in coastal protection, acting together with saltmarshes to dissipate the tidal and wave energy acting on the coast. They also provide habitats for a large range of species of flora and fauna. A European Commission funded project **INTRMUD** (The Morphological Development of **Inter**-tidal **Mud**flats) has brought together a large number of European experts on mudflats, from both physical and biological backgrounds to develop statistical/mathematical descriptions of the processes controlling mudflat behaviour. In a recent account of the HR Wallingford contribution to **INTRMUD** Roberts et al (1998) summarise the challenge as follows.

The difficulty of modelling the morphodynamics of mudflats, as with other water and sediment systems, arises principally because of the large range of time scales and spatial scales at which important processes occur. However, there are a number of aspects of the behaviour of muddy sediments that present different or additional problems to those experienced when considering non-cohesive sediments. One of the most complex issues specific to mudflats is the way the bed properties change with time, through consolidation and biological processes. The resistance to erosion of muddy sediments varies with the state of consolidation, on a time scale of hours to days. This property of muds leads to two (or more) layer situations where a mobile, transient layer of soft mud overlies an overconsolidated, stronger geological deposit. Biological processes can also contribute to the range of time scales to be considered. For example, there is evidence of an annual cycle of varying resistance to erosion on many mudflats of temperate latitudes, where increased activity in the summer months provides protection from erosion through the medium of extra-cellular polysaccharides.

The other principal difference between muddy and sandy sediments is the much lower settling velocity of mud than sand which introduces a separation between sources and sinks of sediment. Both the settling velocity and consolidation issues mean that ideas that have been found to work well with medium-term modelling of sandy shore morphodynamics, such as the "morphological tide" as a means of input filtering, may be much more difficult to apply to mudflats.

The key feature of mudflats that Roberts et al seek to predict is the topography of the mudflat. This is certainly linked to a number of secondary aspects, which in many situations may be of the greatest ultimate interest, such as the suitability of an intertidal area as an ecological habitat. The topography of the mudflat arises from the balance of erosion and accretion of sediments at each point of the mudflat. However, there are numerous practical and theoretical difficulties in making on accurate long-term prediction of changes in mudflat morphology by integrating models of short-term erosion/accretion. In the research underway at HR Wallingford the approach to developing a method for long-term prediction of mudflat morphology is a hybrid. An equilibrium hypothesis is linked with process models to evaluate the form of the equilibrium for specific combinations of forcing, geology and sediment supply, and to calculate the rate of change of out-of-equilibrium mudflats. Various equilibrium models have been proposed in the past, either for muddy shores in particular or for estuaries in general. Models include: estuary regime relationships, uniform peak bed shear stress, uniform dissipation of wave energy, etc. These and other possible equilibrium criteria are being evaluated together with the development of appropriate parameterisations of process models of deposition, erosion and sediment transport, examined in the context of analysis of the field measurements arising from the **INTRMUD** project as a whole. The findings of **INTRMUD** are due to be reported in 1999.

3.4 Scour at coastal structures

The majority of marine structures will at some point during their design lives be vulnerable to the erosion of sediment at their base due to the scouring action of waves and tidal currents. Recent research at HR Wallingford has provided new insights into the mechanisms of scouring at coastal structures, and has lead to the development of improved methods for predicting the onset and extent of scour holes for a wide range of structures. The research is presented in Whitehouse, 1998, and Powell, 1998. These studies have considered both sand and shingle sized sediments and utilised a combination of physical model tests for the shingle and numerical model tests for the finer sand. For the latter case, the **COSMOS** model was used.

Based on the 2-dimensional physical model tests, scour at the base of seawalls was found to be a function of the following primary variables:

- Initial water depth at the toe of the seawall
- Wave height and period
- Number of waves
- Sediment size
- Seawall geometry and type

The results of the studies are presented in convenient iso-parametric plots for both shingle and sand sediments. The plots, which relate to scour after 3000 waves, clearly illustrate the complexity of the relationship between scour depth, water depth and incident wave conditions. Although the research concentrates mainly on vertical seawalls, some physical model results were collected for sloping and rough, permeable walls.

An additional consideration for breakwaters is the scour development at the head. This has been examined in recent research at the Technical University of Denmark, reported in Fredsoe, 1997 and Sumer, 1997. Information on scour prediction techniques for a wide range of other structures can be found in Hoffmans and Verheij, 1997 and Whitehouse, 1998. The first of these covers scour downstream of barrages and sills, at abutments and spur dykes, piled structures and other coastal and offshore structures. The second reference covers scour at piled structures, pipelines, large caisson or gravity structures, seawalls and breakwaters. Combined they form an invaluable reference source of practical information on scouring.

A major research programme **SCARCOST** (**S**cour **A**round **Co**astal **St**ructures), funded by the EC, is now underway to study the potential risk of scour in the vicinity of coastal structures. One of the main aims is to prepare and disseminate practical guidelines. The work plan is summarised in Sumer et al 1998. The project is an excellent example of the scale of international and interdisciplinary co-operation required to advance our knowledge in this field. The 9 partner institutes represent both coastal engineers and geotechnical engineers. The work is subdivided into two tasks (1) the flow processes in the water column and the resulting sediment transport/scouring and (2) the behaviour of the sediment bed due to the wave-induced pressures and flows within the bed sediment. A combination of laboratory and field measurements is planned as well as some computational modelling. The project runs through to the year 2000.

3.5 Ship manoeuvring and mooring studies

Steady progress has been made in recent years in developing computational tools to assist port engineers optimise the design of approach channels and berths. One example is real time navigation simulation which is a commercial tool that has been used for some years to enable port designers to evaluate proposed changes with regard to safe ship manoeuvring. Exploiting the latest advances in virtual reality engines enables highly realistic simulations to be achieved at affordable costs. An example of a system that has been used in many commercial applications is the **HR Mardyn Navigation Simulator**, McBride, 1998. This uses a Silicon Graphics Onyx Reality Engine to generate a very high quality, fully anti-aliased and texture mapped view from the bridge of the ship updating 30 times a second. Any time of day or night, and any range of visibility can be reproduced. The underlying mathematical model is of a standard type, describing the three important degrees of freedom, surge, sway and yaw. The modular structure of the software allows the modelling of different ship types in all manoeuvring conditions including ship-ship interactions. A flexible approach ensures that the modelling methodology is suitable for the site, the design vessels and the expected environmental conditions making this a powerful tool to be used by port engineers at the feasibility stage of any port modification or design. This innovation has been achieved primarily as a commercial development. A second area of innovation is in the modelling of the movement of moored ships. Excessive horizontal motions of moored ships are a problem, in general making cargo handling more difficult or posing safety hazards. With large modern ships the largest amplitude movements are generally long period slow drift type motions. These are often generated by various second-order effects. For this reason large scale physical modelling has long been used as a design tool to investigate berth tenability. However, for the early stages of design studies, time and budgetary constraints may preclude the use of a physical model. In these situations there is often a need to investigate the viability of a number of possible design options, and a good, well-validated computational model can provide an attractive and versatile alternative. In the UK a nationally funded programme of research has provided a solution in the form of a suite of models, **UNDERKEEL, QUAYSHIP, DRIFTKEEL** and **SHIPMOOR**, see McBride, 1997. The hydrodynamic wave forces on a ship, the ship motions and mooring forces are calculated using computational models in two stages. Initially UNDERKEEL and QUAYSHIP are used to determine linear wave forces. Linear forces, however, are less significant in many cases than the non-linear long period effects, which cause ranging and other slow drift motions of moored ships. Therefore, DRIFTKEEL is used to compute these slow drift forces. The incident wave forcing on the ship is determined, allowing for diffraction effects, hydrodynamic damping and added mass effects. All of these wave-related computations are made initially in the frequency domain, for unit amplitude incident waves and ship movements. They are only later applied in the time domain, employing principles of superposition to generate wave-forcing sequences for any desired prototype wave spectrum and wave direction. This procedure is used for computational efficiency.

Although relatively small, compared to other hydrodynamic effects, the long period second-order forces are important because they occur at frequencies close to resonant frequencies for moored ship motions in surge, sway and yaw. In addition, ship motion at these low frequencies is, in most cases, only lightly damped and consequently, motion amplitudes can be large. It is usual for these to be the largest horizontal motions of a moored ship. They are therefore potentially damaging to moorings and disruptive to cargo handling operations. Once the environmental forces on the ship are established, the second stage of the modelling process consists of their application, with the constraining effects of the mooring system, to solve for the vessel motion. The computational model used for this process is SHIPMOOR.

SHIPMOOR is a time domain model. It operates by computing the ship's accelerations from the forces acting on it. Accelerations are then integrated to compute velocities which are, in turn, integrated to compute positions. Thus, the computation proceeds thorough time, giving the ship's position, velocity and mooring forces as they evolve over a series of discrete time steps. Non-linear characteristics of mooring lines and fenders can be reproduced. The model reproduces all six degrees of freedom of ship motion of which the horizontal motions in surge, sway and yaw are usually the most significant for generating forces in fenders and hawsers. However, all of the six motions can affect cargo handling operations.

The models have been extensively verified against both field data and physical model results. The models have also been used successfully in feasibility studies involving moored LNG carriers, container ships, general cargo vessels and oil tankers. For a more complete description see Spencer, 1995 and 1996.

4. CONCLUDING REMARKS

Development of computational tools for the hydraulic engineer continues apace and on a number of fronts. An enormous investment programme is underway by each of the major commercial hydraulics labs in an effort to exploit the latest advances in information technology. The results of this development tend to be well enough promoted commercially so that there has been little need to refer to it in depth here. Instead, in the examples quoted here, I have attempted to illustrate the considerable body of publicly funded research that is currently in progress and which is vital in underpinning these tools. The examples are drawn largely from UK and European research projects. A striking feature of much of this research is the extent of the multi-disciplinary, multi-national collaboration required to achieve progress in this field.

5. REFERENCES

Casale, R. and Samuels, P.G. (1998). "Hydrological Risks: Analysis of recent results from EC research and technological development actions". European Commission Environment and Climate Programme.

Chesher, T.J. (1995). "Numerical Morphodynamic Modelling of Keta Lagoon", Coastal Dynamics '95, 4-8th September, 1995, Gdansk, Poland.

Fredsoe, J and Sumer, B.M. (1997). "Scour at the round head of a rubble-mound breakwater." Coastal Engineering, 29, 231-262.

Hervouet, J.-M. (1996). "Introduction to the TELEMAC system". Report HE-43/96/073/A. Departement Laboratoire d'Hydraulique, Electricite de France.

Hoffmans, G.J.M.C and Verheij, H.J. (1997). "Scour Manual." Balkema, Rotterdam.

McBride, M.W., Spencer, J.M.A. and Smallman. J.V. (1997). "Modelling of Moored Ship Motions in India." Proceedings of the 2nd Indian National Conference on Harbour and Ocean Engineering, Trivandrum, India, 7th-11th December 1997.

McBride, M.W., Smallman. J.V. and McCallum, I.R. (1998). "Safety Assessment for Ship Manoeuvring inn Port Areas." Proceedings of the 29th Navigation Congress, The Hague, The Nederlands, 6th-11th September 1998.

Powell, K.A. and Whitehouse, R.J.S. (1998). "The occurrence and prediction of scour at coastal and estuarine structures." UK Ministry of Agriculture Fisheries and Food, 33rd Conference of River and Coastal Engineers, Keele University, 1-3 July 1998.

Roberts, W and Whitehouse, R.J.S. (1998). "Long-term modelling of mudflat morphology", INTERCOH 98.

Samuels, P.G., May, R.W.P and Spalivero, P. (1998). "The use of 3-D computational models for river flow simulation". UK Ministry of Agriculture Fisheries and Food, 33rd Conference of River and Coastal Engineers, Keele University, 1-3 July 1998.

Soulsby, R.L. (1998). "Dynamics of Marine Sands – a manual for practical applications", published by Thomas Telford.

Southgate, H.N. (1998). "CAMELOT, Final Project report", HR Wallingford report SR 523, May 1998.

Spencer, J.M.A. (1995). "Time Domain Models of Ships: Second-order Mathematical Model." HR Wallingford report SR 428, March 1995.

Spencer, J.M.A. and Beresford, P.J. (1996). "A second-order computational model of wave-induced movements of moored ships." 11th International Harbour Congress, Antwerp, Belgium, 1996.

Sumer, B.M. and Fredsoe, J (1997). "Scour at the head of a vertical-wall breakwater." Coastal Engineering, 29, 201-230.

Sumer. M.S., Whitehouse, R.J.S. and Torum, A. (1998). "Scour Around Coastal Structures (SCARCOST)." 3rd European Marine Science and Technology Conference, Lisbon, 23-27 May 1998.

Tozer, N.P. (1998). "ARTEMIS –Wave Disturbance Model, Evaluation and Guidelines on its use". HR Wallingford report IT 450, January 1998.

Weare, T.J. (1996). "Computational hydraulics in 1997". Proceedings of the Second International Conference on Hydrodynamics/ Hong Kong/ 16-19 December 1996.

Whitehouse, R.J.S (1998). "Scour at Marine Structures. A Manual for Practical Applications." Thomas Telford, 194 pp.

Environmental Hydraulics, Lee, Jayawardena & Wang (eds) © 1999 Balkema, Rotterdam, ISBN 90 5809 035 3

3D water quality modelling of Deep Bay

T.A. Nauta – *WL/Delft Hydraulics, Netherlands*

H.S. Lee – *Hong Kong Environmental Protection Department, Water Policy and Planning Group, China*

A. Kwok – *Montgomery Watson Limited, Hong Kong, China*

E. Gmitrowicz – *Hyder Environmental Limited, Hong Kong, China*

ABSTRACT: Deep Bay, a large shallow bay on the east bank of the Pearl Estuary, was designated by the Hong Kong - Guangdong Environmental Protection Liaison Group (HKGEPLG) in 1990 as the highest priority study area requiring protective conservation action. To protect the environment of Deep Bay and control the impact arising from the rapid developments on both sides of the bay, the HKGEPLG endorsed in December 1993 a recommendation from its Technical Subgroup (TSG) to carry out a study on Deep Bay's capacity to assimilate and disperse pollutants.

This paper focuses on the set up and application of the computer modelling system Delft3D, which was used to improve understanding of the dispersive and assimilative capacity of the bay and to evaluate strategic management options to protect the water quality in the bay from further deterioration.

1. INTRODUCTION

The Deep Bay catchment (Fig. 1) has been important for agriculture and fisheries. Intensive pig rearing in the catchment and increasing amounts of untreated sewage from the growing urban areas has resulted in large organic pollution loads entering Deep Bay. In the past few decades, brackish and fresh water ponds that have been dug for aquaculture since the 1920s have increasingly been filled for urban development. Recent development in the Shenzhen Special Economic Zone has even been more rapid than on the Hong Kong side of

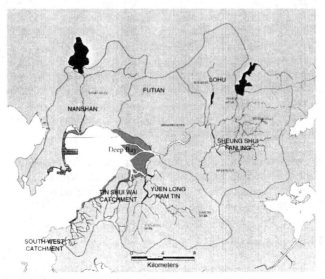

Fig. 1 Deep Bay Catchment

the catchment. Rapid development is resulting in pollution that is putting the ecosystem under stress. As a result of the wide range of land and water uses in and around Deep Bay, certain conflicts arise as each use has its own specific set of environmental quality criteria which are required for effective and safe use of the resources.

Traditionally assessment of environmental quality in Deep Bay was based upon monitoring results. Since 1986, the Environmental Protection Department of Hong Kong (HKEPD) has been monitoring Deep Bay as part of the long term water quality monitoring programme for Hong Kong waters. HKEPD also keeps records of monitoring data from other government programmes. In addition, various projects in the area included monitoring programmes covering different locations, time periods and parameters. Given the size of the study area and the costs involved, however, monitoring programs will always be limited in their extent of coverage, both spatially and temporally. Despite existing monitoring programmes in Deep Bay, knowledge of the spatial behaviour of substances and significant processes was considered to be still limited. The application of models can help by providing a means of integrating existing knowledge on physical, chemical and biological processes with monitoring data in both space and time, and by so doing strengthen capacity to manage the environment in a sustainable way.

The majority of substances entering the water system, regardless of source, are transported through the system by relative 'large scale' circulation patterns and may be subject to various water quality processes. By incorporating transport and water quality processes, models can be used to systematically review effects on the environment and make the relationship between the upstream sources of pollution (emission) and the downstream concentration levels and subsequent ecological effects. A common difficulty in assessing environmental changes in such a shallow system is the high level of natural variability that often exists in the observed concentrations. Many substances exhibit significant variations over a longer time scale, not only due to for instance winds, temperature, light conditions and other natural effects, but also varying loads and open sea inflows. A modelling framework can give a structured approach to making causal assessments for observed conditions. The present Deep Bay modelling study made use of the state-of-the-art and widely applied Delft3D model to obtain an improved understanding of the different hydrodynamic, transport, water quality and ecological processes in Deep Bay.

2. STUDY AREA

Deep Bay is a large shallow bay on the east bank of the Pearl Estuary, adjacent to the relatively deep flow channel of Urmston Road. The bay has a surface area of approximately 112 km^2, with a length of about 15 km and an average depth of 3 m. The total catchment area of the bay is about 535 km^2.

The tide in Deep Bay is mixed with a strong diurnal component imposed on a semi-diurnal cycle. The hydrodynamics of the bay strongly depend on the season. In the dry season, the bay is usually comparatively well mixed. The bay receives runoff and freshwater input from several main rivers in the catchment, including Shenzhen River, Dasha River, Shan Pui River (receiving flow from the Kam Tin River and Yuen Long Creek) and Tin Shui Wai drainage Channel, as well as some minor streams in the Northwest and Southwest of the catchment. Due to the strong influence of the Pearl River runoff, the water of Deep Bay is primarily estuarine with relatively low salinity levels during summer months.

With a subtropical monsoon climate, the Deep Bay region has mild weather, abundant rainfall and long durations of sunshine. The weather changes with the summer and winter monsoons, resulting in seasonal variation in temperature and rainfall. The characteristic dry season period is from December to March and the characteristic wet season period is from June to August.

3. THE MODELLING FRAMEWORK

Available data, obtained from a vast variety of sources and conveying a large quantity of information, were insufficient to calibrate and validate the detailed three-dimensional hydrodynamic and water quality modelling framework. To obtain an improved understanding of the functioning of Deep Bay a comprehensive monitoring programme was carried out, including the different physical, chemical and biological conditions for two periods in time: one survey in March 1996 illustrative of dry season conditions and one survey in August illustrative of wet season conditions. The first detailed set of data was used for calibrating the Delft3D modelling framework and the second set of data was used for validation.

The data collected and used for the calibration and validation of the model included the following components:

- a 26-hour tidal cycle hydrodynamic and water quality survey in Deep Bay (including surface, mid-depth and near bottom sampling);
- a limited sediment survey;
- a river flux survey, covering most important freshwater inputs; and
- supplementary data from different sources describing major meteorological conditions, and Pearl Estuary boundary conditions.

Two different Delft3D modules were used to represent the hydrodynamic and water quality phenomena within Deep Bay: Delft3D-Flow and Delft3D-Waq. Delft3D-Flow solves the shallow water equations. For the Pearl Estuary and Deep Bay a set of nested models was applied on a curvilinear grid. A calibrated Pearl River model covering the whole of Hong Kong waters and the Pearl Estuary was used to determine the inflowing boundary conditions for the inner Deep Bay model. The curvilinear grid, with ten equidistant vertical layers (sigma-co-ordinates), as applied to Deep Bay included in total 70*85 grid points of which approximately 2700 are active points. Grid sizes vary between 25m in the small channels in the inner part of Deep Bay to approximately 800m near the open sea boundaries (Fig. 2).
The water quality module Delft3D-Waq solves the advection diffusion equation (i.e. the transport equation) and is applied on the same grid and 5 equidistant layers (aggregation over the vertical).

Following data collection, the water quality modelling of Deep Bay was focused upon the nutrient and oxygen balances, phytoplankton, BOD, suspended solids and bacterial pollution, due to their potential to generate the greatest environmental impacts. Copper was also modelled to add support to the calibration of water circulation and dispersion in the Bay. Due to the limited data set, it was also necessary to assume that the loads of all rivers (and the partitioning in fractions) except the Shenzhen River and its tributaries are time constant for the model calibration, with time variability of loads over the last 3 days, when available, for the

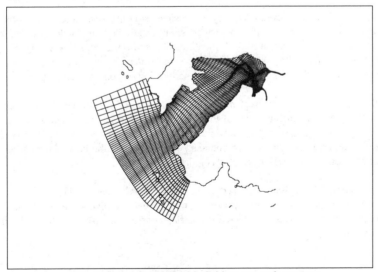

Fig. 2   Deep Bay Model – Grid set-up and coverage

verification period. Although the lack of long term time series data on loads limits the ability of the model to simulate the full degree of water quality variability in Deep Bay, a comprehensive data set was collected over a limited period of time and this data is considered suitable for model calibration and verification of the water quality model.

For efficiency reasons (with respect to computing and analysis time) both hydrodynamic and water quality modules were operated in 2D and 3D modes. Once the 2D model simulations were in a dynamic equilibrium regarding the horizontal distribution, the modules were operated in 3D mode to redistribute over the vertical. Before doing so it was assured that the horizontal distribution of substances between the 2D and the 3D modelling approaches were essentially in agreement.

By repeatedly simulating, in two-dimensional mode, a period of approximately 14 days preceding the days of the intensive monitoring and using restart files, and by using an estimation of the initial sediment concentrations based on field observations, the historical build up of concentration levels in water and sediment were obtained. Hereafter the three-dimensional model was run for two days. The last day, with repeated hydrodynamic input of the first day, was used for presentation and interpretation.

All water quality simulations have been performed on a HP 735 workstation. The 2D model uses an alternate direction implicit scheme, similar to the hydrodynamic modelling approach. Runtime of one simulation (18 days, 15 active and 7 inactive substances, considerable output specifications) is approximately 8 hours (computational time step is 90 seconds). The 3D model uses horizontally a flux corrected and vertically an implicit (backward) scheme. Runtime of one simulation (2 days) is approximately 20 hours (computational time step is 30 seconds).

4. CALIBRATION AND VALIDATION RESULTS

The ultimate test of any mathematical model is its ability to reproduce reality. The governing philosophy is that a correct description of the processes in the model enhances the prospect for successful application of the model to a new period / system and for extrapolation of the model beyond the bounds of previous applications. The approach for the determination of the process coefficients was to use a judgement in selecting parameters that are consistent with available data and local expertise as well as with tabulations of accepted values and expert judgement. The parameters were selected in a sequential manner and the model was finally judged adequate based on the comparison of predictions and observations.

In the calibration process it was believed necessary to consider carefully where the assumptions made for the model and the setting for the process coefficients depart significantly from available data and local expertise, and where the observations are misleading. Both model results and observations reveal a strong influence of the small channels in inner Deep Bay on the transport of the water quality parameters. For this reason the selection and representation of the exact locations of the monitoring stations was crucial.

It was considered inappropriate to take great measures to refine specific features of the model if other ones remain crude. For this reason a starting point of the model set up and application was to maintain a certain robustness and to avoid fine-tuning of processes while more dominant processes were only globally understood or described. Despite the great uncertainties in boundary conditions and the lack of time-series for the loads from the key point sources, the calibration result was judged satisfactory.

The dry season conditions for 1996 were used in the calibration process. The model was validated by demonstrating that it reproduces the wet season conditions for 1996. Applying the new boundary conditions, meteorological forcing and river loads and the same coefficient setting as used for the calibration, the model validation was carried out.

4.1 Hydrodynamic results

Calibration using the measured and modelled 3-D salinity results revealed a good degree of similarity regarding water levels and current velocities. The influence of smaller channels on spatial salinity distribution was well reproduced (Fig. 3). Simulated water levels at Chiwan and Tsim Bei Tsui were satisfactory, although there was a difference between measured and simulated mean level results at Chiwan which required water levels to be adjusted slightly during the wet season.

With respect to salinity verification against wet season data, the simulated horizontal distribution agreed well with actual measured data, with highly stratified field conditions in the outer Bay and almost well mixed

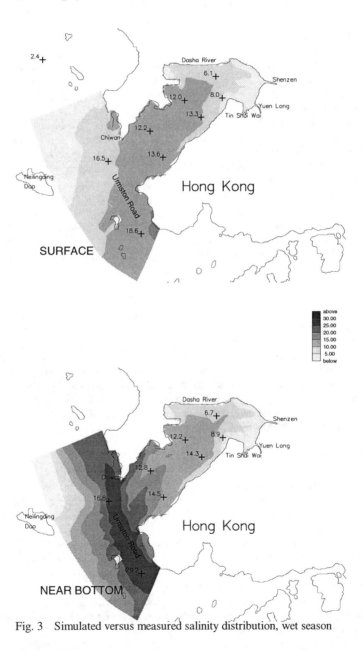

Fig. 3   Simulated versus measured salinity distribution, wet season

conditions in the inner Bay. Within Urmston Road the simulated surface salinities were up to 4-5 ppt greater than field measured concentrations due to the coarse vertical resolution of the model in relation to the steep measured salinity gradient. While field measurements revealed the salinity distribution to be complex, the simulated verification results were considered acceptable.

Measured and simulated velocities were generally in good agreement, with the largest discrepancies being near the model boundary, outside the Bay. The duration of ebb flow is underestimated in this area, a feature attributable to tidal boundary conditions as the same effect was seen in the overall Pearl Estuary Model. The overall result of the calibration and verification of the Deep Bay model is that tidal flow fields, water levels and the relatively complex salinity structure were correctly reproduced by the model and that the hydrodynamic model predictions provided a good basis for water quality modelling.

4.2 Water Quality results

The results of water quality monitoring within Deep Bay displayed the influence of variable estuarine conditions and pollution sources on the spatial distribution and measurements of parameters. The influence of increased fresh water flow during the wet season produced a lower salinity range and higher suspended solids range compared to those recorded during the dry season survey. Levels of chlorophyll-a were also highest during the wet season, while the nutrients concentrations, which were consistent with analysis of historical EPD routine monitoring, were not limiting the growth of phytoplankton in either season. Similarly, the proportion of the phytoplankton component within the total organic carbon increased during the wet season despite a reduction in the absolute values measured.

The water quality model was calibrated against dry season data and the results of the 2-D calibration simulations produced results which were generally similar to field measurements. The 3-D simulations subsequently undertaken showed no significant vertical layering during the dry season conditions that were simulated, and the calibration results produced a similar overall good correspondence with field data. This conclusion holds for all substances studied: dissolved oxygen, nutrients, phytoplankton, BOD, suspended solids, bacteria and copper.

The water quality verification runs were undertaken for the wet season and the simulations produced results which were consistent with the field data (Fig. 3), though suspended solids were slightly higher than the measured field data for both 2-D and 3-D simulations. The 3-D simulations displayed slightly more detail in their spatial distribution patterns, although no pronounced vertical layering was observed in the inner and middle bay as might have been expected during the wet season. There was a small degree of layering in the outer bay due to the influence of Urmston Road. Within the inner bay concentration patterns were most pronounced in the smaller channels due to their role in distributing water. This role can be seen with respect to the distribution of anaerobic water from inflowing rivers. It is therefore believed that once the levels of oxygen in these rivers improves (through a reduction of the organic pollution loads), oxygen levels in the Inner Bay will increase from its present low levels.

5. CONCLUSIONS

The overall approach taken in the calibration and verification of the Deep Bay modelling framework produces results which are consistent with the field data, resulting in a tool which was fit for the purpose of assessing the dispersive and assimilative capacity of Deep Bay.

The approach to reconstruct a certain period in time could only be made more successful if reliable, accurate and long term records of field observations of boundary conditions (horizontal and vertical detailed description), river inflows, sediment accumulation (spatial concentration patterns in the upper layer of the sediment and accretion rates), meteorological forcing, etc. could be used. In such a case the model needs to be run for a long period of time to take account of the historical build up of water and sediment concentrations. However, it would be far from pragmatic (with respect to computation times) and beyond any allowable project budget, to adopt such an approach.

6. ACKNOWLEDGEMENT

The authors would like to thank the Director of Environmental Protection of the Hong Kong Special Administrative Region Government for his permission to publish this paper. The views expressed are those of the authors and do not necessarily reflect those of the HKSAR.

Environmental Hydraulics, Lee, Jayawardena & Wang (eds) © 1999 Balkema, Rotterdam, ISBN 90 5809 035 3

3-dimensional water quality and hydrodynamic modelling in Hong Kong
I. Concepts and model set-up

A.K.M.Ng
Environmental Protection Department, Hong Kong Special Administrative Region, China

A.Roelfzema & L.J.M.Hulsen
WL/Delft Hydraulics, Netherlands

ABSTRACT: The Hong Kong Special Administrative Region (HKSAR), PRC is located on the south-eastern side of the Pearl River Estuary. The river discharges 336 billion m^3 of freshwater runoff a year into the estuary carrying large quantities of sediment and nutrients which characterise the hydrographic and water quality condition in the region, especially during the wet season. A suite of 3-dimensional hydrodynamic and water quality models has been set up for the Hong Kong SAR Government to assess the water quality impact of sewage disposal schemes and engineering developments. This paper discusses the hydrographic and water quality characteristics in Hong Kong and the nearby region, model features required to incorporate the important characteristics and complicated coastline of Hong Kong, field surveys conducted and model set-up for calibration and verification of the 3-dimensional models.

1. INTRODUCTION

Hong Kong is one of the most densely populated cities in the world with about 4 million people living on either side of Victoria Harbour (Figure 1) where much of the commercial activities in this busy city are held. However, Hong Kong's hilly terrain is not conducive to urban development and large tracts of the city were actually reclaimed from the sea. With the rapid growth in population and economic activities over the last few decades, sewage treatment infrastructure has lagged behind and the water quality in Victoria Harbour has noticeably deteriorated with time. The Hong Kong government first applied mathematical water quality and hydrodynamic models to assess the water quality effect of land reclamation and pollution control schemes in the early 1980s. Since then, the spending on infrastructural developments by the government has increased tremendously and it is necessary to ensure that the government is well advised on water quality effect of these developments with the most appropriate models. In view of the rapid advancement in software and hardware technology over the last decade, the Hong Kong government decided to upgrade its suite of mathematical models to improve the modelling capability to today's standards.

To initiate the model upgrade process, the Hong Kong government, with the advice of three local academics, set up a model review panel to review the local hydrographic and water quality characteristics and to define the modelling requirements for Hong Kong, taking due considerations of the state-of-the-art technology. Model features in terms of accuracy, computational efficiency, water quality processes library, utility functions and user-friendliness were considered in the model selection process with the objective to select the most appropriate models for Hong Kong's conditions.

After the detailed requirements were defined, WL | delft hydraulics was commissioned to undertake the assignment to upgrade the modelling capabilities of the Hong Kong government (WL|delft hydraulics, 1998). This paper describes the details of the hydrographic and water quality characteristics around Hong Kong waters, the features and set-up of the new upgraded mathematical models which cater for these major characteristics, and field data used to calibrate and validate the models. The new model suite will serve as an important tool to assess the water quality effects of sewage disposal schemes and engineering developments in Hong Kong in the future.

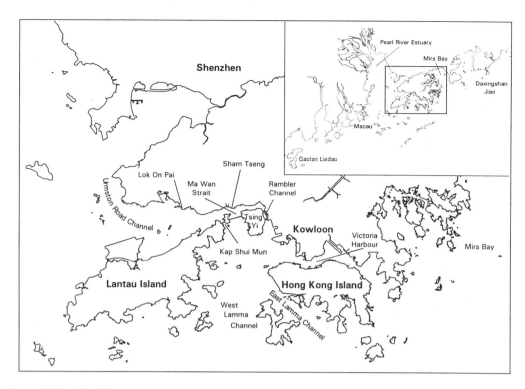

Figure 1 Location Plan

2. HYDROLOGICAL AND WATER QUALITY CHARACTERISTICS

2.1 Meteorology

Hong Kong is located on the south-eastern coast of China; north of the South China Sea (SCS) and the meteorology is dominated by the north-east monsoon in the winter and south-west monsoon in the summer with average wind speeds of 9 m/s and 6 m/s respectively. As a result of the monsoon winds, a residual current of about 0.2 m/s is observed offshore in the northeast and southwest directions during the summer and winter respectively. This strongly affects the dispersion pattern of the freshwater discharge from the Pearl River, whose estuary is located west of Hong Kong, and hence the quality of Hong Kong waters.

The annual average rainfall in Hong Kong is about 2,200 mm with more than 75% of it precipitated from May to September. The temperature is lowest in January with an average air temperature of 16 °C and highest in July with an average temperature of 29 °C. Mean daily global solar radiation varies from 10.7 MJ/m^2 in February to 19.2 MJ/m^2 in July.

2.2 Tidal Flows

The tides in Hong Kong are mixed and mainly semi-diurnal. The K1 and O1 diurnal constituents and M2 and S2 semi-diurnal constituents are the principal tidal components for Hong Kong. The tidal oscillations in the SCS are mainly co-oscillating tides, spreading from tidal waves of the Pacific Ocean. The tide enters the SCS from the northeast and propagates along the south-coast of China, passing Hong Kong in the southwest direction. A time delay of about 50 minutes can be observed between Daxingshan Jiao on the east and Gaolan Liedao on the west (Figure 1). The tidal wave propagates from the northeast, curving around the southwest part

32

of the Lantau Island and continuing, while curving, towards the Urmston Road and Ma Wan Strait. The Ma Wan Strait forms such a restriction for tidal wave that it induces a time delay of some 20 minutes between Tsing Yi Island and Sham Tseng, located west and east of the Ma Wan Strait. The tide-induced residual currents are in the order of several cm/s and mainly directed towards the southwest. These currents are negligible when compared to the monsoon wind-driven currents.

2.3 Hydrography

The Pearl River is the largest river system in south China. It has a total catchment area of 453,700 km^2 with an average annual freshwater runoff of 336 billion m^3 (from PRWRC). About 80 % of the freshwater runoff passes through the estuary during the wet season from April to September. The runoff is discharged into the estuary through eight major outlets at a rate of about 20,000 m^3/s in the wet season and reduces to some 4,000 m^3/s in the dry season. The large runoff creates vertical as well as lateral gradients in salinity, especially during the wet season. The effect of the Pearl River on the salinity and quality of Hong Kong waters decreases from west to east with vertical salinity stratification of 24 ppt and 10 ppt at Urmston Road Channel and Mirs Bay respectively during the wet season.

The Pearl River runoff contains high concentration of nutrients and sediment. The runoff carries about 71 million tons of sediment into the estuary each year (Zhao, 1983). The turbid water in the estuary inhibits phytoplankton growth and has an important bearing on the eutrophication processes in the estuary and further offshore. When the sediment is settled out in the estuary, the nutrient can sustain a rapid growth of phytoplankton offshore in the surface layer and supersaturated concentration of dissolved oxygen is often observed.

3. MODEL FEATURES AND SET-UP

3.1 Hydrodynamic Model

The large volume of Pearl River runoff requires a fully 3-dimensional model to be applied in this region in order to resolve the vertical density and flow stratification. During the wet season the brackish water flows out of the estuary above and in opposite direction to the incoming oceanic water. To handle the vertical variations, the hydrodynamic model divides the water column into 10 equal layers using sigma transformation. The eddy viscosity and diffusivity are modelled with the k-ε turbulence model.

The eight major outlets of the Pearl River system spread over a distance of 110 km. While the area of interest is within Hong Kong waters its hydrography and water quality are strongly affected by the Pearl River runoff. The model boundary must be far enough to cover all these outlets to enable robust simulation of salinity and water quality distribution in the region. This consideration determines that the model's western boundary be extended some 120 km west of Victoria Harbour. The model's southern boundary was extended 75 km offshore in order to contain most of the Pearl River freshwater plume advected north-easterly during the wet season. On the other hand, the model's resolution must be reasonably fine to accurately simulate the flow through major channels in Hong Kong waters and for assessment of the effect of land reclamation within the harbour. The model uses a boundary fitted, curvilinear grid system to optimise the extensive model coverage with different level of details and reasonable computational time as shown in Figure 2. The model covers an area of about 250 km by 200 km with grid sizes of about 7.5 km at the model's ocean boundaries and 70 m within Victoria Harbour. The CPU time for a HP model C180 workstation to simulate a 25-hour tidal cycle takes about 4.9 hrs.

The grid direction at the open sea is aligned in the north-east to south-west direction parallel to the residual flow field. At the open sea boundary the water level and salinity are prescribed as boundary conditions. The water level is prescribed by means of the amplitudes and phases of the nine most important tidal constituents. Tilting of the mean water surface is applied to represent the known residual oceanic current close to Hong Kong (Uittenbogaard and Hulsen, 1998). Wind conditions for the larger open sea area that are derived from offshore wind stations are applied to realistically reflect the oceanic monsoon winds.

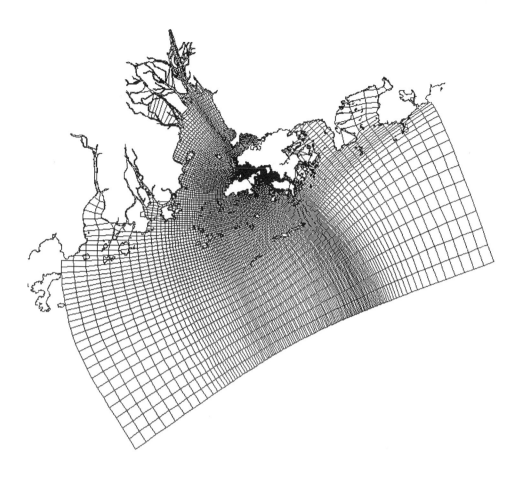

Figure 2. Layout of the new Hong Kong base model

3.2 Water Quality Model

Water quality modelling is often conducted to assess the level of treatment and the discharge location of water pollution control schemes required to achieve the designated water quality objectives of a water body in Hong Kong. The model must therefore be able to simulate all the major water quality kinetic processes so that the concentrations of pollutants may be accurately determined. There is an annual cycle of water quality changes due to the seasonal variations in the Pearl River runoff, temperature and solar radiation. The water quality model has different grid size and simulation period requirements from the hydrodynamic model. Since most water quality reaction rates are slow and the lateral variation in concentration is less rapid than tidal flow, the water quality model is usually required to run for a longer period of time with a coarser grid.

The water quality model was set up to run on a yearly period using tidal flow results. It is however not practical to provide a whole year of the underlying hydrodynamics with the hydrodynamic model due to limitations on computer speed and storage requirements. Net propagation of substances takes place by residual flows. Tidal flows are mainly important to enhance horizon and vertical mixing. Representative tides in the wet and dry season with averaged residual flow were run repeatedly to make up a year-cycle water quality simulation. The water quality model is run within the tide with a time-step of one hour and 5 aggregated vertical layers. This enables intra-tidal dispersion to be accounted for without the need to use a very large dispersion coefficient with reduced accuracy and computational efficiency. The yearly period of water quality simulation is composed of 149 days of representative dry season tide followed by 76 days of representative wet season tide and the year is complete with another 140 days of representative dry season tide at the end.

The demands on the water quality modelling grid come from the fact that no artificial mixing must result from a schematisation that is too coarse. Various sensitivity tests of grid aggregation schemes and numerical techniques were conducted to ensure that the original salinity pattern as computed by the fine grid, short time-step hydrodynamic model is reproduced by the water quality model. The CPU time to simulate a one year water quality simulation with a 4 by 4 aggregation scheme and 5 vertical layers on a HP model C180 workstation is about 8 hrs.

Eutrophication processes in Hong Kong play an important role on achievement of some important water quality objectives such as dissolved oxygen and ammonia. The water quality model simulates some 16 state variables, including salinity, dissolved oxygen, BOD, suspended solids, 2 algal species, nutrients (dissolved and detritus components) and coliform bacteria. Detritus components and sediment are subjected to transport, sedimentation, erosion and burial in the bottom sediment layers.

The concentration of parameters of water quality in the Pearl River tributaries is a major unknown. Estimates were derived from measured concentrations at nearby EPD's monitoring stations (EPD, 1997) and the results of the Deep Bay Water Quality Regional Control Strategy Study (Axis-CES, 1997). As no measurements near the model seaward boundaries were available, constant boundary concentrations were used. Within the present study, all discharges have been considered to be continuous and have been kept constant over the course of a year.

4. DATA FOR MODEL SET-UP, CALIBRATION AND VALIDATION

The hydrodynamic model was calibrated and validated with three data sets collected in 1987, 1990 and 1992 for calibrating and validating the former modelling suite. Data collection was focused on tidal variation in hydrodynamic and water quality data during representative spring and neap tides in the wet and dry seasons in these field surveys. The field surveys conducted for these studies provided valuable information at different locations for calibration and verification of the hydrodynamic model. In general, the comparison between model results and field measurements for the hydrodynamic model is satisfactory (Postma et al. 1998).

Though the water quality data collected in the previous field surveys provide useful information on the spatial variation of water quality parameters, they are too sparse temporally to provide seasonal variation in water quality for calibration of the water quality model which runs on a year-long cycle. Routine EPD multi-year monitoring data collected monthly or bi-monthly at 82 stations all over Hong Kong waters were analysed to provide long-term averaged water quality distribution for comparison with the model results (Postma et al. 1998). Monthly averaged solar radiation based on Earth Radiation Experiment Satellite measurements for the period 1985 to 1989 were used as model input.

Data gaps for model calibration and verification were identified during data analysis and model calibration processes. Recommendations for further field surveys for additional model calibration and verification were made. The strategy for the recommended surveys is to provide systematic spatial and long-term variations in hydrography and water quality constituents. The surveys would also collect data at the month of the Pearl River and seaward boundaries of the model. Field surveys are scheduled to start in July 1998 for completion in December 1998 and these are expected to provide the necessary field data for the verification and further improvement of the water quality and hydrodynamic models.

5. CONCLUSIONS

Through this project a major improvement has been achieved not only on the quality of the upgraded models used by the Hong Kong government but also that insights into and knowledge of the characteristics of the Hong Kong and surrounding waters have been gained. These so called "base" models have now been set up, calibrated and validated against existing data and therefore constitute a set of state-of-the-art, three dimensional modelling tool. From the data analysis and modelling work, data gaps were identified and recommendations for additional data gathering and survey strategies were made. These survey activities are to be performed in the framework of other projects, but their results will be utilised to further improve the new base models in the future. The models will be used to assess the impacts of possible future developments in the Hong Kong coastal areas. The software has been implemented on new hardware, dedicated to the requirements of this new suite of models. Through the flexibility of the underlying modelling software, the Hong Kong government can adapt and modify the models in accordance with the specific project requirements, allowing for optimal model applications from the efficiency as well as accuracy points of view.

6. ACKNOWLEDGEMENT

Permission by the Directors of Civil Engineering and Environmental Protection of HKSAR government to publish the work in this paper is gratefully acknowledged.

7. REFERENCES

Axis– CES (Asia) Joint Venture (1997). Deep Bay Water Quality Regional Control Strategy Study.

Environmental Protection Department, Government of HKSAR (1997). Marine Water Quality in Hong Kong for 1996.

Pearl River Water Resources Commission, PRC. Pearl River.

Postma, L., Stelling, G.S. and Boon, J.G. (1998). 3-Dimensional Water Quality and Hydrodynamic Modelling in Hong Kong, III. Stratification and Water Quality. To appear in proceedings of *7th International Symp. on River Sedimentation and the 2nd International Symp. on Env. Hydraulics.* 16-18 Dec., 1998.

Uittenbogaard, R.E. and Hulsen, L.J.M. (1998). 3-Dimensional Water Quality and Hydrodynamic Modelling in Hong Kong, II. Forcing Residual Currents. To appear in proceedings of *7th International Symp. on River Sedimentation and the 2nd International Symp. on Env. Hydraulics.* 16-18 Dec., 1998.

WL | delft hydraulics (1998). Upgrading of the Water Quality and Hydraulic Mathematical Models – Executive Summary.

Zhao Huanting (1983). Hydrological Characteristics of the Zhujiang (PEARL River) Delta. *Tropic Oceanology.* Vol.2 No.2 May 1983.

Environmental Hydraulics, Lee, Jayawardena & Wang (eds) © 1999 Balkema, Rotterdam, ISBN 90 5809 035 3

3-dimensional water quality and hydrodynamic modelling in Hong Kong
II. Forcing residual currents

R.E.Uittenbogaard & L.J.M.Hulsen
WL/Delft Hydraulics, Netherlands

ABSTRACT: This paper shows that the simulated residual flow, affecting significantly the salt distribution in Hong Kong waters, complies with extensive and excellent long-term ADCP observations only if the monsoon-dependent residual currents along the North coast of the South China Sea are correctly imposed through the open-sea boundary conditions.

1. INTRODUCTION

The proper simulation of flow and transport in estuaries includes a part of the adjacent sea. The sea is then truncated numerically by using so-called open boundaries which require additional conditions, here designated as open-sea boundary conditions.

Experience in this and other projects shows that the quality of the estuarine simulation delicately depends on the correctness of open-boundary conditions which must satisfy a proper balance with wind forcing inside, as well as communicating the external tidal motions and residual currents from outside the simulation domain. For the latter, we applied the following model, for details see (WL | DELFT HYDRAULICS, 1998).

2. MODEL FOR WIND-DRIVEN FLOW IN SOUTH CHINA SEA

Wyrtki (1961) shows that the South China Sea (SCS) exhibits a permanent thermocline. The monsoon winds drive the surface layer and induce return currents near the thermocline. Therefore we assume that the surface layer is a neutrally-stratified unidirectional flow with constant thickness 2D. Depth-averaging of the stationary momentum equations over half the surface layer then yields:

$$D\left(f\,\underline{\hat{e}}_z * \overline{\underline{u}} + g\vec{\nabla}\zeta\right) = \underline{\underline{\sigma}} \cdot \hat{\underline{e}}_z\Big|_{surf} - \underline{\underline{\sigma}} \cdot \hat{\underline{e}}_z\Big|_{ml} \tag{2.1}$$

with upward unit vector $\hat{\underline{e}}_z$ and $\underline{\underline{\sigma}}$ the stress tensor due to averaging of momentum advection and the tensor includes internal friction, the overbar indicates depth-averaging.

The SCS is approximated as a long channel with its longest axis inclined 33^0 from the north direction. A fortunate coincidence is that the dominant monsoon wind directions are roughly parallel to this direction. In this drastic schematisation, the y-coordinate and v-velocity component are parallel to the longest channel axis. The x-direction and the corresponding u-velocity component cross the channel width from roughly NW to SE direction. The channel is about 500 km wide and 3,000 km long. The SW boundary of the channel is at the connection of the SCS to the Java Sea. The distance between Hong Kong and the SW boundary is about 2,000 km. The NE boundary ends near Taiwan. The channel width is based on the distance between Macao or Hong-

Kong to Port Uson on the Philippines (see Figure 1). Wyrtki (1961, fig. 7.4) reports the seasonal variations in the difference in water level between Macao and Port Uson.

Both the channel's geometry as well as observations (Wyrtki, 1961) allow for neglecting the flow in u-direction so that the depth-averaged version of the incompressibility condition demands: $\partial \bar{v} / \partial y = 0$. From this and (2.1) follow the strongly-simplified equations, in x- and y-direction, for the wind-driven surface-layer flow in geostrophic balance:

$$g\frac{\partial \zeta}{\partial x} - f\bar{v} = 0 \quad \text{and} \quad gD\frac{\partial \zeta}{\partial y} + \tau_{ml} - \tau_y^{wind} = 0 \;, \qquad\qquad (2.2) \text{ and } (2.3)$$

with τ_y^{wind} the wind stress of the typical wet season winds in positive y-direction and

$$\rho_o \tau_y^{wind} \equiv \rho_0 |u_*| u_* = \rho_a c_D |U_w| U_w \;, \qquad\qquad (2.4)$$

holds with air density ρ_a, drag coefficient c_D, wind speed U_w and wind-shear stress velocity u_*.

Subsequently, we must consider the shear stress τ_{ml} at the level $z = D$ defined by zero residual velocity. Considering the large depth of the SCS, the turbulence in the surface layer is mainly determined by wind forcing and not by the tidal motions nor by the residual current. Analysis and numerical simulations (Svensson, 1978) indicate the following parabolic vertical distribution of the eddy viscosity of vertical exchange of horizontal momentum inside the neutrally-stratified surface layer:

$$\upsilon_T = \kappa u_* z(1 - z/2D) \;;\; \kappa \approx 0.41 \;, \qquad\qquad (2.5)$$

with u_* defined by (2.4). For the internal friction of the residual current we apply the eddy viscosity concept which yields for the shear stress τ_{ml}:

$$\tau_{ml} = \upsilon_T(D)\frac{\partial v}{\partial z}\bigg|_{ml} \quad \text{with} \quad \upsilon_T(D) \approx 0.2 u_* D \;. \qquad\qquad (2.6)$$

The velocity profile for a wind-driven flow in neutrally-stratified flow is logarithmic as if it were the flow along a rigid wall with shear-stress velocity u_* and roughness length z_s (Craig, 1996), thus

$$\tau_{ml} = \upsilon_T \frac{\partial v}{\partial z}\bigg|_D \equiv r\bar{v} \quad \text{with} \quad r \approx \frac{0.2\,u_*}{\log(D/z_s)-1} \;. \qquad\qquad (2.7)$$

Here the shear rate $\partial v/\partial z$ at level $z = D$ is expressed conveniently in the velocity $\bar{v}$ which is $v(z)$ depth-averaged over half the surface layer with thickness 2D. Notice that τ_{ml} depends linearly on $\bar{v}$ because the eddy viscosity is determined by the wind and not by $\bar{v}$ proper, $\bar{v}$ is determined solely by the wind-shear velocity u_*.

Due to the proximity of the equator, the variability of the Coriolis parameter must be taken into account and the β-plane approximation (Pedlosky, 1979) yields

$$\frac{\partial f}{\partial x} = \frac{\beta_x}{R} \;;\; \frac{\partial f}{\partial y} = \frac{\beta_y}{R} \;;\; \beta_x = \frac{1}{R}\frac{\partial f}{\partial \theta}\sin\phi \;;\; \beta_y = \frac{1}{R}\frac{\partial f}{\partial \theta}\cos\phi \qquad\qquad (2.8)$$

with θ the azimuth angle, θ=0 on the equator, radius R of the Earth and φ the inclination of the y-axis relative to North. For the application to the SCS, φ is small and the ratio between channel width and length is also small and therefore the x-dependence of f is neglected i.e. $\beta_x \approx 0$.

Substitution of (2.6), (2.7) and (2.8) into (2.3) yields the drastically simplified problem of solving:

$$g\frac{\partial \zeta}{\partial x} - \beta_y \frac{y}{R}\bar{v} = 0 \; ; \quad gD\frac{\partial \zeta}{\partial y} + r\bar{v} - \tau_{surf}/\rho_o = 0 \; . \tag{2.9}$$

Differentiation of the second equation with respect to y and using $\partial \bar{v}/\partial y = 0$ shows that ζ must be linear in y and this suggest the separation of variables $\zeta = yh(x)$ with $h(x)$ a non-dimensional function. The solution for $\zeta(x,y)$ as well as for the residual current $\bar{v}$ then reads:

$$\zeta = \hat{\zeta}\frac{y}{y_{max}}\left\{1 - (1-\eta_o)e^{-x/L}\right\} \; ; \quad \hat{\zeta} = \frac{\tau_{surf}\, y_{max}}{gD} \quad ; \quad L = \frac{rR}{\beta_y D} \tag{2.10a}$$

$$\bar{v} = \hat{v}(1-\eta_o)\exp(-x/L) \; ; \quad \hat{v} = \frac{\tau_{surf}}{r} \approx 5u_* \log\left(\frac{D}{ez_s}\right) . \tag{2.10b}$$

The free parameter η_o allows for a linear profile of ζ along the North SCS coast (x=0). We set y_{max}=2,000 km being the position of Hong Kong relative to the SW boundary of the SCS at the Java Sea. For clarity of its dependence, we express the residual current magnitude $\hat{v}$ into physical parameters:

$$\hat{v} \approx 5\left(\frac{\rho_a c_D}{\rho_w}\right)^{1/2} U_w \log\left(\frac{D}{ez_s}\right) . \tag{2.11}$$

This expression shows a linear relationship between residual current and the wind velocity. The surface elevation along the North coast (x=0) is determined by:

$$\zeta(x = 0) = y\,\eta_o \frac{\rho_a c_D}{\rho_o gD} U_w^2 \quad . \tag{2.12}$$

In case of zero surface elevation along the North coast η_o=0 holds. The latter condition matches approximately the observed sea-level topography in August, see (Wyrtki, 1961, plate 4c).

Figure 1 presents (2.10) for a 140 m thick surface layer i.e. D=70 m with U_w=5 m/s ; z_s=0.1 m ; c_D=0.001 ; ρ_a=1.2 kg/m^3 ; ρ_o=1035 kg/m^3. The drag coefficient c_D follows from Smith and Banke (1975) for the given moderate wind speed.

3. FORCING RESIDUAL FLOW IN THE 3D HYDRODYNAMIC MODEL

For objective simulation of salt patterns in Hong Kong waters, the grid of our 3D hydrodynamic model, including Coriolis force, has an extensive sea section, see Figure 2, having three so-called open-sea boundaries (OSB). For wet-season conditions, on the surface of this sea section a 5 m/s SW wind is applied. Further, on the corners of the OSB, the observed tidal elevation is imposed through the amplitudes and phases of astronomical constituents. The zero-frequency elevation, say $\zeta_0(x,y)$, is then the remaining degree of freedom. In case of $\zeta_0(x,y)$=0 along the OSB, Figure 2 shows the surface-velocity field at some instant in time and for reference are added 5 salt iso-lines. The oscillatory orientation of the surface velocity near the SE open-sea boundary are artificially generated baroclinic Rossby waves due to deviations from geostrophic balance. Further, the simulated residual current in the Lema channel, south of Lantau Island, is nil whereas long-term ADCP clearly show a 200 mm/s NE residual current. In the wet season, the latter current in

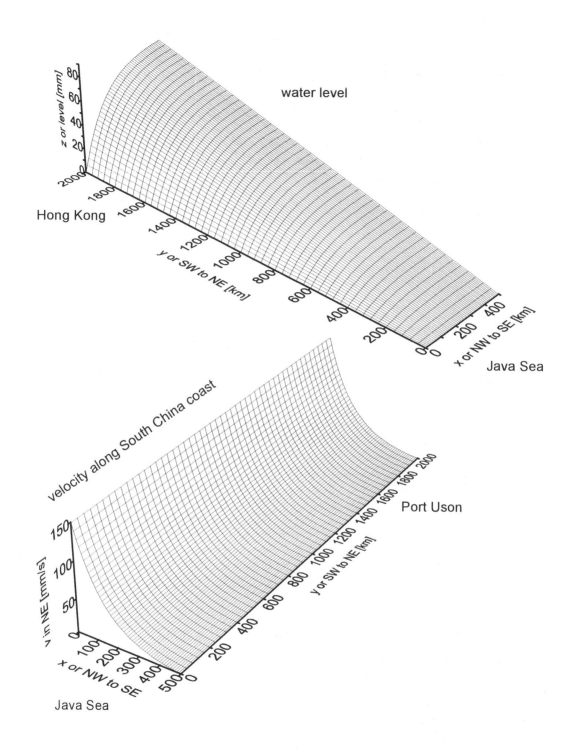

Figure 1. Analytic solution for wind-driven flow in South China Sea.
SW wind of 5 m/s, thermocline depth 140 m.

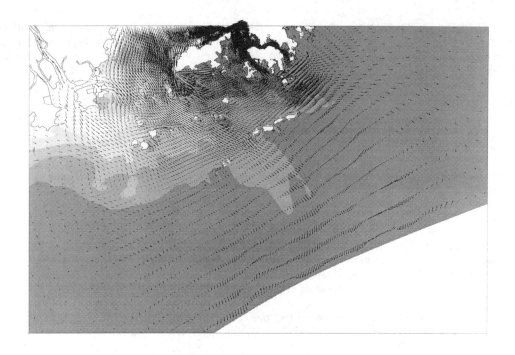

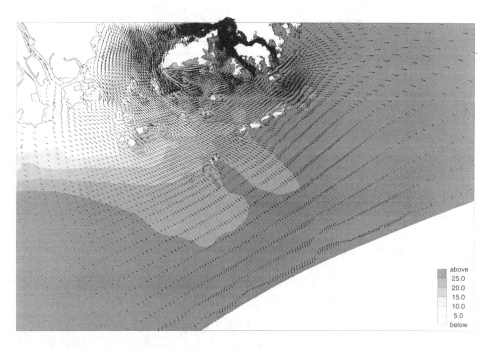

above
25.0
20.0
15.0
10.0
5.0
below

Figure 2. Wet season simulation with 3D Upgraded HK model
Surface velocity field with salinity contours
Zero-frequency elevation (upper) and tilting (lower)

conjunction with the prevailing SW wind, is responsible for the NE transport of Pearl River water south of Lantau Island. The Rossby waves are absent when the tilting of Section 2 and Figure 1 is imposed on $\zeta_0(x,y)$ along all OSB. Further, the salt pattern is now in good agreement with observations. For details of the latter we refer to WL | DELFT HYDRAULICS (1998) and the accompanying paper (Postma et al., 1998).

REFERENCES

Craig, P.D. (1996). Velocity profiles and surface roughness under breaking waves. *J. Geoph. Res.*, 101, pp. 1265-1277.

Pedlosky, J. (1979). *Geophysical fluid dynamics*. Springer Verlag.

Postma, L., Stelling, G.S. and Boon, J.G. (1998). 3-Dimensional Water Quality and Hydrodynamic Modelling in Hong Kong, III. Stratification and Water Quality. To appear in proceedings of *7th International Symp. on River Sedimentation and the 2nd International Symp. on Env. Hydraulics.* 16-18 Dec., 1998.

Smith, S.D. and Banke, E.G. (1975). Variations of the sea surface drag coefficient with wind speed. *Quart. J. Roy. Meteor. Soc.*, vol. 101, pp. 665-673.

Svensson, U. (1978). *A mathematical model for the seasonal thermocline*. Ph.D.Thesis, Dept. Water Resources Eng., Un. Lund, Sweden, report no. 1002.

WL | DELFT HYDRAULICS (1998). Upgrading of the Water Quality and Hydraulic Mathematical Models - Final Report.

Wyrtki, K. (1961). *Scientific results of marine investigations of the South China Sea and the Gulf of Thailand*. NAGA report vol. 2, Scripps Institute of Oceanography, Las Jolla, California.

Environmental Hydraulics, Lee, Jayawardena & Wang (eds) © 1999 Balkema, Rotterdam, ISBN 90 5809 035 3

3-dimensional water quality and hydrodynamic modelling in Hong Kong III. Stratification and water quality

L. Postma, G. S. Stelling & J. Boon
WL/Delft Hydraulics, Netherlands

ABSTRACT: The Hong Kong Special Administrative Region (SAR) government has commissioned the setting up and delivery of an Upgraded model for water movement, stratification and water quality of Hong Kong waters. Using the Delft3D modelling suite hydrodynamic and water quality models have been set up (Ng et al., 1998). The hydrodynamic model has proven to reproduce spatial and temporal patterns of water levels, velocities and stratification correctly for typical dry season and wet season situations. The water quality model was set up to reproduce bacteria, BOD-DO, Suspended sediment, N-, P- and Si-fractions and Algae throughout the year.

1. INTRODUCTION

The residual flow South of Hong Kong Island is driven by either the SW monsoon for the wet season or the NE monsoon for the dry season. It is enhanced by the Coriolis driven circulation in the South China Sea (Uittenbogaard and Hulsen, 1998). This causes the Pearl River outflow to generate a stable stratification around Hong Kong Island in the wet season, making a 3-dimensional approach necessary. From the Delft3D modelling suite the application of the FLOW module and the WAQ (water quality) module are described in this paper. The Delft3D-FLOW module contains an efficient solver of the 3-dimensional hydrodynamic equations under the hydrostatic pressure assumption. It simultaneously solves equations for the salinity and temperature and the equations of the k-ε turbulence model. All equations are solved on a curvilinear grid both horizontally and vertically (sigma co-ordinates). A specially accurate treatment of the horizontal advection terms enhances the performance (Stelling and van Kester, 1994). In this way all physics associated with turbulent viscosity and diffusivity and their relation to density differences are included in the model and hardly any calibration coefficient is left. The Delft3D-WAQ module uses the computed hydrodynamics (levels, flow and turbulent diffusion) to solve the advection diffusion equation for the many substances involved (WL | delft hydraulics, 1998).

2. STRATIFICATION MODELLING SET-UP

The degree of stratification in the wet season and the shape of the fresh water plume is partly determined by the residual flows in the fresh and salt water layers in Ma Wan Strait. For a model to address the effects of reclamations on residual circulation patterns it is necessary that this stratification is predicted reliably. For this reason the model was equipped with an adjustable number of vertical layers using Sigma co-ordinates. Runs with 10 and 20 layers have been used to address model sensitivity to the number of layers. The 10 layer model had equivalent results to the 20 layer model, so this number of layers could be used furthermore. For the horizontal grid it is important to cover the main channels and straits around Hong Kong Islands with a sufficient degree of detail. On the other hand it is important that the whole fresh water Pearl River outflow plume is contained in the model during wet-season simulations. The grid extends 50-100 km away from Hong Kong waters up to the 70m depth contour. This is perhaps not the depth contour that is generally accepted as where the continental shelf ends (Gerritsen et al., 1995), but it was considered far away enough to apply tidal constituents and constant open ocean concentrations as boundary conditions. This made the model independent of specifically measured time series of water levels and/or concentrations for its functioning. The tidal constituents for the water level at the open boundaries were derived in two ways. Firstly by translating nearest long-term tidal stations towards the open boundary using co-tidal charts. Secondly the boundary conditions have been derived using inverse modelling (ten Brummelhuis et al., 1993). The latter resulted in about the same model accuracy, but with a more smooth and realistic translation function. The resulting grid is shown in (Ng et al., 1998) with some 10,560 'wet' computational elements horizontally, using a time step size of 60 seconds. Sensitivity analysis has shown that applying horizontally a 2x2 finer grid or using half the time step size is not affecting model results to a significant degree, so the numerical truncation error is within acceptable limits. The open ocean salinity concentration was set at 35 permille.

The bottom roughness is the only adjustable parameter left. It is represented by Mannings coefficient and it is coarsely adjusted to 0.020 for the parts more shallow than 25 meters and 0.026 for the deeper parts. The most uncertain factor in the forcing of the model was the absence of actual daily or monthly flows for the Pearl River tributaries. A total flow of 19405 m³/s for the 8 tributaries is used for the wet season and 4116 m³/s for the dry season, as derived from long term monthly averages. For the subdivision along the 8 branches results from a dedicated study were used. Sensitivity analysis has been carried out indicating that major changes in modelled salinity results are to be expected for major changes in the river flow only. This is supported by the good model performance even though the actual values for the river flow were unknown.

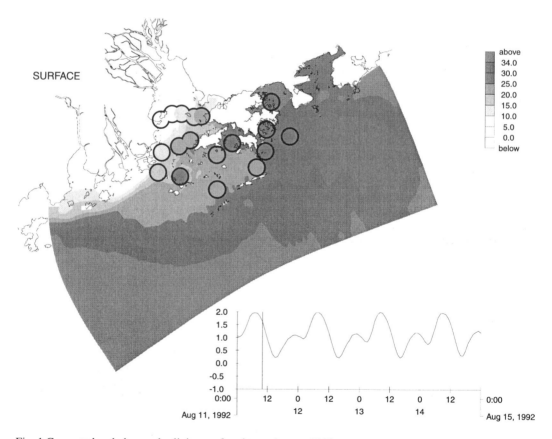

Fig. 1 Computed and observed salinity, surface layer, August 1992

3. STRATIFICATION MODELLING RESULTS

The 1987, 1990 and 1992 data available for calibration and validation (Ng et al., 1998) had about the same structure. At isolated points, on several depths, a diurnal cycle of concentrations, temperature and salinity was measured. These measurements were available for the 4 combinations of dry-wet season, spring-neap tide conditions. The 1987 Victoria Harbour study had its 18 stations around Hong Kong Island; the 1990 PADS study covered the whole of Hong Kong waters with 13 stations and the 1992 SSDS monitoring was the first to have some 12 stations in Mainland China waters as well. Because of the physical impossibility to monitor all stations at the same time, time shifts of hours up to days could be noted for the different stations. This is why the spatial model results have been compared with locally averaged values. The model results themselves have not been averaged over the tidal cycle to preserve the more pronounced features of the direct result. This means that the comparison is indicative. Figure 1 shows the wet season surface layer salinities together with the 1992 measurements as processed this way. De adaptation time as used for this model result spanned over 40 days. The open boundary conditions were derived from tidal constituents only and spatially constant monsoon winds were used rather than measured time series for wind.

During the process of model set-up a somewhat less favourable comparison was found. Further inspection of the data revealed that the monsoon winds had turned form SW to NE about 1.5 days before the measurements where conducted. The model was forced however with a constant SW monsoon wind of 5 m/s. Once the turning wind was applied, the much better result of Figure 1 was obtained. Figure 2 shows the detail surface layer North of Lantau Island without and with turning monsoon winds.

Figure 3 gives the measured and modelled salinity profiles at station WF2 (North of Wanchan Qundao). It shows both the good performance of the stratification modelling and the influence of the horizontal tidal excursion. Station 15 (Figure 4) of the Victoria Harbour programme is located west of Hong Kong Island. The submerged locations of maximum velocity are clearly visible at 12 and mid-night. They reflect the own movement of the saline under-layer both in the measurements and in the model results.

Although the dry season covers the larger part of the year, it is less dynamic with respect to Hong Kong Waters. The Pearl River flow diminishes, it does not reach Hong Kong waters and flows out along the SW shoreline of the Mainland China (Figure 5). The dry season simulation nevertheless gave good opportunity to validate the transport through Ma Wan Strait. The vertical shear in the horizontal velocity is the most important horizontal transport here. It causes a transport of somewhat fresher water from the Pearl River area through Ma Wan Strait, resulting in a week but distinguishable horizontal concentration gradient. This transport is compensated for by the residual flow through the Strait. The flow is 650 m^3/s in westerly direction as a model spring-neap cycle average. Sensitivity analysis has shown that deviations from this residual flow rate, would give a distinctly different horizontal salinity gradient around Lantau Island.

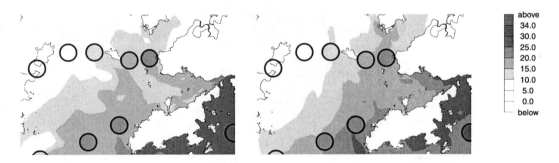

Fig. 2 Computed and observed salinity, Pearl Estuary surface layer, without(l) and with(r) turning monsoon

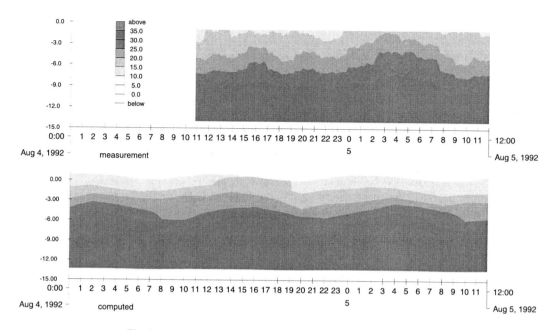

Fig. 3 Computed and observed salinity profiles at station WF2

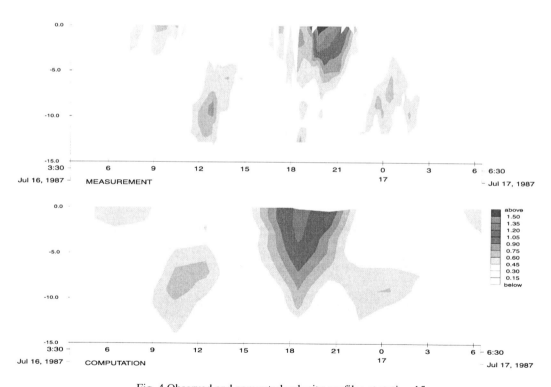

Fig. 4 Observed and computed velocity profiles at station 15

4. WATER QUALITY MODELLING

The advection and turbulent diffusion (and reduction of it vertically by stratification) at computed by the hydrodynamical model applies to the other substances. Superimposed on this are the physical and bio-chemical water quality processes. For Hong Kong waters a selection is made for the water quality processes to be addressed. Bacteria, the BOD-DO processes, Nitrogen-, Phosphorus- and Silica cycles, Algae and suspended sediment have been modelled with all associated fractions and processes (Thomann and Mueller, 1987). Water temperature was used as forcing function based on measurements.

Water quality modelling for estuaries means dealing with different time scales. Water quality processes show a year cycle, determined by seasonal changes in forcing like flow, temperature, sunlight etc. Usage can be made of the fact that horizontal transport of substances generally has velocities that are an order of magnitude smaller than the propagation of the tidal wave. This allows for a coarser grid. Furthermore, once stratification is computed by the hydrodynamic model, the number of layers for the water quality model can be relaxed somewhat. Not too much however otherwise the physics of horizontal shear dispersion in the model must be replaced by a calibrated dispersion coefficient. Finally advanced accurate and stable solvers (Saad and Schultz, 1986) allow for a longer time step. This gave a water quality modelling set-up with 4x4 aggregation horizontally for calibration purposes, and 2x2 aggregation for the final computations with 5 layers vertically. A time step of one hour could be applied. This provided a speed factor of 500 to 2000. Salinity computations have been performed with the water quality model to show that the aggregation in space and time would produce the same results as those of the 3-dimensional stratification model for both wet and dry season. A problem that remains is the construction of a year cycle of hydrodynamics. This

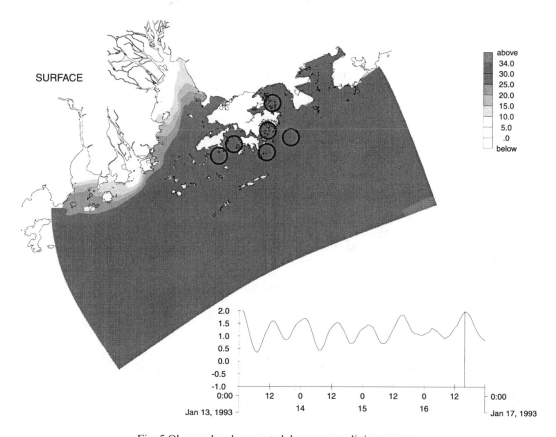

Fig. 5 Observed and computed dry season salinity patterns

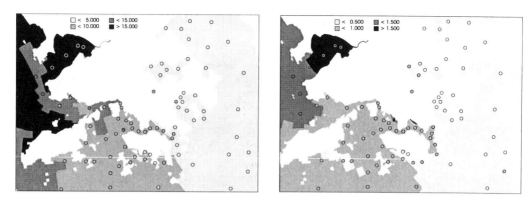

Fig. 6 Comp. and obs., wet-season, surface-layer, long term averaged susp. sediment (l) and total-N (r)

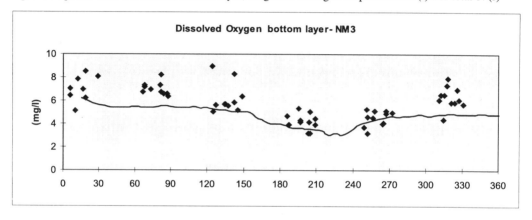

Fig 7. Computed and measured oxygen values at NM3 (Urmston Road) bottom layer.

would take a month computation time. Instead a representative dry season period has been used repeatedly from January-May. A wet season period for June to mid August and the dry season period again for the end of the year. Other forcing like solar radiation and temperature have been specified as time functions with values obtained by monthly averaging over many years.

Although the Routine Monitoring Programme of Hong Kong EPD has never been designed to support model calibration, it was the data set with the longest time record and most spreading over the year (monthly and bi-monthly values) and with most spatial coverage for Hong Kong waters. This is why measured wet-season and dry-season averaged values have been derived from this data set and also time functions have been taken from these stations. Figure 6 shows the wet-season top layer values for suspended solids and total Nitrogen. A time series for Dissolved Oxygen is given in Figure 7 with measurements from some 10 years.

5. CONCLUSIONS

A major step is set in mathematical modelling of stratification and water quality of Hong Kong waters. The good reproduction of measured velocities and salinities, the independence of measured boundary conditions and the intrinsic physics give confidence that the model is a valuable assessment tool. The water quality model has more aspects of calibration. So additional measurements are needed and presently conducted.

6. ACKNOWLEDGEMENT

Permission by the Directors of Civil Engineering and Environmental Protection of HKSAR government to publish the work in this paper is gratefully acknowledged.

7. REFERENCES

Brummelhuis, P.G.J. ten, Heemink, A.W. and Boogaard, H.F.P. van den (1993). "Identification of shallow sea models", *Int. J. for Num. Meth. In Fluids,* vol. 17 no 8.

Gerritsen, H., Vries, H.de, and Philippart, M. (1995). "The Dutch Continental Shelf Model", *Coastal and Estuarine Studies,* Lynch D.R. and Davies, A. M. (Eds.), Am. Geoph. Union.

Ng, A.K.M., Roelfzema, A. and Hulsen L.J.M. (1998). "3-Dimensional water quality and hydrodynamic modelling in Hong Kong - Concepts and Set-Up", To appear in *Proceedings of 7th Int. Symp. on River Sed. and the 2nd Int. Symp. on Env. Hydraulics.* 16-18 Dec., 1998.

Saad, Y and Schultz M.H. (1986). "GMRES: a generalized minimal residual algorithm solving nonsymmetric linear systems", *Siam J. Sci. Stat. Comp.* 7, p856-869.

Stelling, G.S. and Kester, J.A.T.M. van (1994). "On the approximation of horizontal gradients in sigma co-ordinates for bathymetry with steep bottom slopes", *Int. J. Num. Meth. Fluids,* Vol 18, 915-955.

Thomann, R.V. and Mueller, J.A. (1987). *Principles of surface water quality modelling and control,* Harper & Row, New York, N.Y.

Uittenbogaard, R.E. and Hulsen, L.J.M. (1998). "3-Dimensional water quality and hydrodynamic modelling in Hong Kong - Forcing residual currents", To appear in *Proceedings of 7th Int. Symp. on River Sed. and the 2nd Int. Symp. on Env. Hydraulics.* 16-18 Dec., 1998.

WL | delft hydraulics (1997). "User-Manual Delft3D-WAQ".

1.2 Numerical modelling

Environmental Hydraulics, Lee, Jayawardena & Wang (eds) © 1999 Balkema, Rotterdam, ISBN 90 5809 035 3

Invited lecture: Strategy for coupling hydrodynamic and water quality models for addressing long time scale environmental impacts

Alan F. Blumberg & James J. Fitzpatrick
HydroQual Incorporated, Mahwah, N.J., USA

ABSTRACT: The cost of implementing remediation strategies to halt and reverse the degradation of the marine environment has led to the development of sophisticated water quality models. These models are three-dimensional and time-varying and are dependent on high spatial resolution, three-dimensional, time-varying hydrodynamic models for their transport and mixing information. To provide for an effective use of water quality models for problems involving annual and longer time scales, procedures for coupling them to hydrodynamic models have been developed. In this paper, methods for indirect coupling, whereby the hydrodynamic model is run first and its transport information stored off-line for later use by the water quality model, time averaging and spatial aggregation are introduced and applied to a case study in Massachusetts Bay. The results from this case study (and others) are used to confirm the validity of the coupling procedures.

1. INTRODUCTION

The coastal zones of the world are being subjected to increasing environmental stresses and degradation caused by the discharge of nutrients and potentially toxic wastes because of population growth and industrial development. Strategies are being formulated to halt and reverse the degradation of the marine environment. These include the construction of advanced wastewater treatment facilities, pollutant load relocation and the construction of tidal barriers. However, the costs of implementing such remediation strategies can be formidable and we have become increasingly dependent on water quality models, used in conjunction with field monitoring efforts, to provide a cost effective means of studying various alternatives.

Today water quality models, and, in particular, eutrophication and pollutant fate models, exist which are three-dimensional and time-varying and which are coupled, in one form or another, to three-dimensional, time-varying hydrodynamic models (for example, Chapelle, et al. 1994, HydroQual 1991, HydroQual 1995). These models are formulated to simulate conditions over multi-year periods. This is particularly necessary when evaluating the effects of nutrient reduction strategies on water quality where the simulation of the annual cycle of phytoplankton primary productivity becomes critical. Because of the differences between the usage requirements of these models, a key issue to be considered in modern hydrodynamic/water quality modeling is how to interface or couple them together in an efficient manner. This paper will address some of the approaches we have used to successfully couple hydrodynamic and water quality models in the marine environment. We will present the procedures we have used in the coupling process, together with an example of their application to the relocation of the Boston Harbor sewage outfall into Massachusetts Bay.

2. COUPLING STRATEGY

There are basically two methods for coupling hydrodynamic and water quality models: direct and indirect coupling. Direct coupling can refer to either the incorporation of a water quality submodel within the same code as the hydrodynamic model or having separate hydrodynamic and water quality model codes, but with the water quality model using the same spatial grid and time-steps as the hydrodynamic model. Indirect coupling involves temporal averaging and/or spatial aggregation of the hydrodynamic model output, storing this information and subsequently using it as input to drive the water quality model.

Generally, problem settings which require only one or two spatial dimensions and/or only a few water quality state-variables carry sufficiently low computational burden to allow direct coupling. However, direct coupling may be computationally prohibitive for problem settings which require all three spatial dimensions and/or numerous water quality state-variables. For example, Table 1 presents the computational time-steps and grid sizes, number of state-variables, and computational runtime for a number of coupled hydrodynamic/water quality modeling studies we have performed. As can be seen, the runtimes for the water quality models are quite lengthy, even when using advanced super-workstation computers.

Considering these lengthy water quality model runtimes, it is our practice in modeling studies to use indirect coupling, whereby the hydrodynamic model is developed first and run separately from the water quality model. The transport information generated by the hydrodynamic model is stored off-line for later use by the water quality model. The use of indirect coupling provides an opportunity to achieve extraordinary computational benefits. First is that a water quality model that is based on off-line transport often can be integrated with a time-step of an order of magnitude larger than required for a hydrodynamic model. This supercycling (Hecht, et al., 1998) of water quality model state variables is possible because the time-step in a hydrodynamic model is limited by rapid (external and internal) wave speeds whereas water quality model time-steps are constrained by slower advective currents. Additional benefits are obtained through both the temporal averaging and/or through the spatial averaging or grid aggregation of the hydrodynamic model output. These two averaging techniques will be discussed in turn.

2.1 Temporal Averaging

Temporal or time-averaging of hydrodynamic model output can benefit water quality model computations in two basic ways. Foremost is that large amounts of disk space can be easily consumed when storing time-varying, three-dimensional velocity and diffusion fields. Therefore, any form of time averaging can reduce disk storage requirements and can also reduce the computational overhead associated with reading hydrodynamic model output during the water quality model simulation. In addition, time averaging the

Table 1. Computational Characteristics of Several Studies Involving Coupled Hydrodynamic and Water Quality Models

	Hydrodynamic Model		Water Quality Model			
Location	Number of Grid Cells	Runtime[2] (hrs/yr)	Number of State-Variables	Number of Grid Cells	Original Runtime[2] (hrs/yr)	Aggregated Runtime[3] (hrs/yr)
Mamala Bay	25,000	70	9	6,000	270	80
Massachusetts Bay	60,000	53	25	5,000	120	11
Delaware Estuary	620	8	22	620	32	8
Mississippi River	14,790	44[1]	31	1,488	95	10
Escambia Bay	11,640	53	25	11,640	108	36

[1] including a sediment transport submodel
[2] the time-step varies from study to study
[3] based on time and space aggregation

varying hydrodynamic model output produces currents that are smaller than the instantaneous currents. The use of smaller currents can increase the size of the computational time-step required to maintain stability for numerical codes which use explicit time integration schemes.

The extent to which time averaging can be used depends on the nature of the hydrodynamic characteristics of the region. To understand the nature of the averaging period, consider an instantaneous variable ϕ and its decomposition into time-averaged and time-varying components $\overline{\phi}$ and ϕ', respectively,

$$\phi = \overline{\phi} + \phi' \tag{1}$$

where

$$\overline{\phi} = \frac{1}{T} \int_{t_0}^{t_0+T} \phi \, dt \tag{2}$$

$$\overline{\phi'} = 0 \tag{3}$$

Here, T is the averaging period. The time-averaged transport equation for a water quality constituent, c, can be written as

$$\frac{\partial \overline{c}}{\partial t} + \frac{\partial \overline{u}_i \overline{c}}{\partial x_i} + \frac{\partial \overline{u'_i c'}}{\partial x_i} = \frac{\partial}{\partial x_i}\left(\overline{D}_{ij} \frac{\partial \overline{c}}{\partial x_i}\right) \tag{4}$$

where $\overline{D}_{ij}$ is the time-averaged diffusion coefficient. The time-averaged velocity $\overline{u}_i$ is smaller than the instantaneous one. Therefore, the time-step can be enhanced in the water quality model. The effect of time averaging as can be seen in Equation (4) is to introduce an eddy correlation term. Its magnitude is a function of the averaging period. It becomes important then to determine just how large the averaging period (T) can be before the eddy correlation term in Equation (4) begins to dominate. Once the eddy correlation term begins to dominate, the water quality constituents calculated with the time-averaged hydrodynamic model output will no longer match those calculated with instantaneous hydrodynamic model output. The water quality model results will degrade and not compare well against observations.

The effect of time averaging can be illustrated by considering a series of model computations using a hydrodynamic model of the New York Harbor system (Blumberg et al., 1998). Panels A through C of Figure 1 present comparisons of salinities over a fifteen-day period computed by the hydrodynamic and water quality models, wherein the hydrodynamic flow and dispersion fields have been averaged over different time intervals before being used by the water quality model. Panel A presents a comparison wherein the hydrodynamic output was averaged over periods of thirty minutes before being used by the water quality model. By the third day small differences in the surface salinities computed by the two models can be observed. These differences between the two model computations continue to increase slightly over approximately the next nine days before leveling off. Maximum differences in surface salinities between the two models is approximately 1 ppt or about a five to eight percent difference between the two models. Interestingly, there is little difference between the bottom salinities computed by the two models. Panel B presents comparisons, wherein the hydrodynamic model output has been averaged over a one hour period. Over the first six days of the simulation the differences between the two models are similar to those observed in panel C for the 30-minute averaging period. However, by day 10 of the simulation these differences have increased to approximately 2-2.5 ppt or between a ten to fifteen percent difference between the two model

55

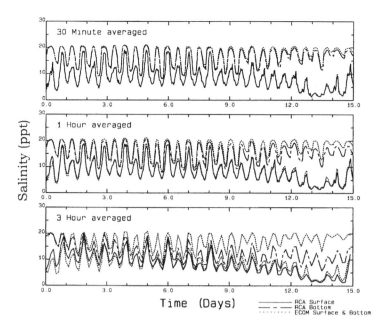

FIG. 1. Comparisons of salinity over a fifteen day period computed by the hydrodynamic (ECOM) and
water quality (RCA) models at a stratified station in New York Harbor for three different
averaging periods

computations. Finally, panel C presents a comparison for a three-hour averaging period. As can be seen, there are now significant discrepancies between the surface and bottom salinities as computed by the two models. Indeed by day 10 of the simulation, the differences in surface salinities computed by the two models are on the order of 50 percent. It appears for New York Harbor that, for averaging periods greater than one hour, information relative to the essential hydrodynamic processes is being lost as a consequence of the averaging period.

Procedures have been developed (Hamrick 1990) for using tidal cycle averaging periods. The approach is based on the idea of tidal dispersion analogous with Taylor's shear dispersion theory, where the term $\overline{u_i'c'}$ can be defined as $-D_{ij}^T \frac{\partial \bar{c}}{\partial x_j}$ and where D_{ij}^T is a dispersion coefficient. Dortch et al. (1992) successfully applied the approach to the Chesapeake Bay system. However, it has not appeared to work elsewhere. The development of alternative approaches to extending the averaging period is an active area of research (Tang and Adams, 1998).

2.2 Spatial Aggregation

Spatial aggregation of hydrodynamic grids and model output for use in water quality model computations can also significantly reduce computational burden during an environmental investigation. The savings in disk storage requirements and computational burden that can be achieved are obvious. For example, if it is possible to combine two hydrodynamic model grid cells in both the X- and Y-axes of the computational grid into one water quality grid cell, the number of computational cells required for water quality model computations is reduced by four. In addition to this direct computational savings, another factor of two is usually achieved in computer codes which utilize explicit time schemes. This is due to the fact that the stability criteria, which determines the maximum computational time-step, will increase by a factor of two for a 2x2 grid aggregation. For this aggregation, the spatially averaged flows and mixing coefficients are just the sum of the flows and the arithmetic average of the mixing coefficients, along the appropriate X- and Y-axes, respectively. The aggregated horizontal and vertical mixing coefficients can be computed as the harmonic mean of the individual diffusivities, using a reciprocal averaging technique (Atkinson et al., 1998). While these spatial aggregation concepts seem to work, it appears that a rigorous theoretical basis for spatial aggregation of fine grid hydrodynamic transports to drive a coarser resolution water quality model will be difficult at best. This is due in part to the natural spatial variability of the turbulence diffusion coefficients and the large gradients in currents which exist near coastlines. Until the difficulties are addressed, the best manner in which to proceed is not to be overly ambitious in the amount of aggregation applied.

3. CASE STUDY

Both temporal averaging and spatial aggregation have been used successfully in a water quality modeling study to investigate the impacts on water quality in Boston Harbor and Massachusetts and Cape Cod Bays resulting from upgrades to the City of Boston's wastewater treatment facilities and the relocation of the effluent outfall from Boston Harbor into Massachusetts Bay (HydroQual, 1995). The analysis involved simulations spanning an eighteen-month period. Figure 2 presents both the fine-grid hydrodynamic model used to develop the circulation for Massachusetts and Cape Cod Bays, as well as, the spatially aggregated grid used for water quality modeling. The details of the hydrodynamic modeling component of the study may be found in Blumberg et al. (1993). The water quality grid is both a subset and a spatially aggregated version of the fine-grid hydrodynamic model. The model parameters are provided in Table 1. It was necessary to include a portion of the Gulf of Maine in the hydrodynamic model in order to properly represent the circulation within Massachusetts Bay and Cape Cod Bay that results from freshwater discharges, originating outside of the area of interest. For the purposes of the water quality investigation it was not necessary to include the Gulf of Maine portion of the hydrodynamic model, since an analysis of the hydrodynamic model output suggested that exchange between the Gulf of Maine and the Massachusetts Bays system would be limited.

In order to ensure that the grid aggregation does not result in the inability of the water quality model to compute the transport of pollutants properly, the water quality model includes salinity as a state-variable and comparisons are made between the salinity computed by the water

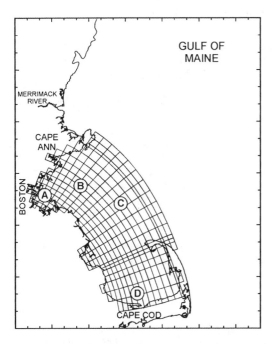

FIG. 2. The grid segmentations used in Massachusetts Bay for the hydrodynamic model (top) and the water quality model (bottom). Stations where salinity comparisons are made are noted.

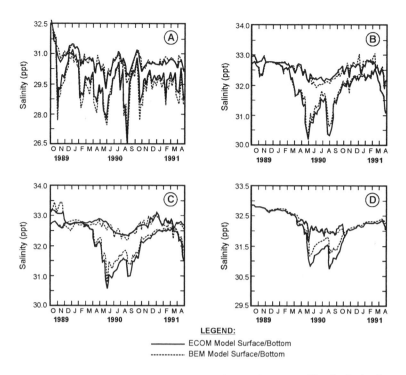

LEGEND:
——— ECOM Model Surface/Bottom
------------ BEM Model Surface/Bottom

FIG. 3. Comparisons of salinity over an eighteen month period computed by the hydrodynamic (ECOM) and Bays eutrophication model (RCA-BEM) at three stations using an hour averaging period

quality model versus that computed by the hydrodynamic model. Figure 3 presents such a comparison for three stations within the model domain (Figure 2). In this figure the annual cycle of surface and bottom salinities computed by the water quality model compare favorably with those computed by the hydrodynamic model, indicating that neither the time-averaging (i.e., the water quality model used one-hour time-averaged hydrodynamic flows and mixing coefficients) nor the spatial aggregation resulted in a significant loss of ability to reproduce the transport computed by the hydrodynamic model.

4. CONCLUSIONS

A series of procedures have been developed that permit an effective coupling between three-dimensional, time-varying water quality and hydrodynamic models. The procedures use indirect coupling wherein the hydrodynamic model is run first and its transport information stored off-line for later use by the water quality model. Further, they involve temporal averaging and spatial aggregation of the transport necessary for water quality simulations. A case study focusing upon the relocation of the Boston Harbor sewage outfall into Massachusetts Bay suggests that the coupling procedures are valid. More research is needed, however, to extend the range of the averaging and aggregation used in that case.

5. ACKNOWLEDGMENTS

Discussions over many years with Dominic Di Toro have greatly benefitted the coupling issues developed here. The authors gratefully acknowledge Kai-Yuan Yang, Richard Isleib and Sandy Faraday for their efforts in support of the methods presented here.

6. REFERENCES

Atkinson, J. E., Gupta, S. K., DePinto, J. V. and Rumer, R. R. (1998). "Linking hydrodynamic and water quality models with different scales", *J. Envir.Engr.*, ASCE 124:399-408.

Blumberg, A. F., Khan, L. A. and St. John, J. P. (1998). "Three-Dimensional Hydrodynamic Simulations of the New York Harbor, Long Island Sound and the New York Bight", Accepted to *J. Hydraulic Engin.*

Blumberg A. F., Signell, R. P. and Jenter, H. L. (1993). "Modeling Transport Processes in the Coastal Ocean", *J. of Mar. Env. Engr.,* 1:31-52.

Chapelle, A., Lazure, P. and Menesguen, A. (1994). "Modelling eutrophication events in a coastal ecosystem. Sensitivity analysis", *Est. Coast. Shelf. Sci.,* 39: 529-548.

Dortch, M. S., Chapman, R. S. and Abt, S. R. (1992). "Application of Three-Dimensional Lagrangian Residual Transport", *J. Hydraulic Engin.,* 118: 831-849.

Hamrick, J. M. (1990). "The Dynamics of Long-Term Mass Transport in Estuaries. Residual Currents and Long-Term Transport", R. T. Cheng (Ed.), *Coastal and Estuarine Studies*, Vol. 38, Springer-Verlag, New York.

Hecht, M. W., Bryan, F.O. and Holland, W.R. (1998). "A consideration of tracer advection schemes in a primitive equation ocean model", *J. of Geophysical Res.*, 103:C2:3301-3321.

HydroQual, Inc. (1991). "Water Quality Modeling Analysis of Hypoxia in Long Island Sound", Report N-139.

HydroQual, Inc., 1995. "A water quality model for Massachusetts Bay and Cape Cod Bay: The Bays eutrophication model calibration". Prepared for the Massachusetts Water Resources Authority. Mahwah, NJ.

Tang, L. and Adams, E. E. (1998). "Interfacing hydrodynamic and water quality models with the Eulerian-Lagrangian Method", Estuarine and Coastal Modeling, Proceedings of the 5th International Conference, ASCE, 153-165.

Environmental Hydraulics, Lee, Jayawardena & Wang (eds) © 1999 Balkema, Rotterdam, ISBN 90 5809 035 3

Impact of a tide-overtopping barrage on water quality and subsequent management

J. M. Maskell
HR Wallingford Limited, UK

Betty Ng
The Environment Agency, Wales, UK

ABSTRACT: This paper describes and examines the water quality problems caused by a tide-overtopping barrage, using the Tawe Barrage in Wales, UK as an example. A three dimensional water movement and water quality model (WQFLOW-3DSL) of the Tawe Impoundment has been developed and calibrated. It is used to simulate the water quality processes and predict the effectiveness of remedial measures by the regulator (The Environment Agency) and the operator (City and County of Swansea), as a management tool for decision making.

1. INTRODUCTION

Although the construction of barrages offers a number of economic benefits, such engineering structures also have significant impacts on the water quality in the impoundment. The Tawe Barrage in Wales, UK provides such an example.

The River Tawe discharges into Swansea Bay in South Wales. Prior to impoundment there was a tidal reach of some 6km with a tidal range of up to 10m and areas of mud-flats exposed at low water. Swansea City Council wished to promote urban regeneration of the area and in 1978 commissioned a feasibility study for a barrage across the lower reaches of the estuary.

The feasibility study investigated the impact of an overtopping weir. A number of options were considered with weir heights of -2.1 to 0.1 mOD (N) (Atkins, 1983 and Broyd et al 1984). Stratification was not thought to be important as flushing by spring tides was considered to prevent full development of stratification and a one-dimensional model was used to predict the impact on water quality which was concluded to be unlikely to be affected significantly. The final design was different from that tested and incorporated a fish pass, navigational lock, a primary weir and a secondary weir at 3.05m OD(N) and 3.35m OD(N) respectively.

The Tawe Barrage is only overtopped during spring tides and the Tawe impoundment is stratified during most of the spring-neap cycle, with brackish water overlying a saline layer. Since the completion of the Tawe Barrage in July 1992, water quality in the impounded estuary has been regularly monitored. Deteriorations in water quality appeared to occur after prolonged saline stratification of the water column (Rogers, 1992; Rogers and Bryson, 1994). The saline stratification is only eroded by significant river spates. During summer months, dissolved oxygen levels in the bottom saline layer often fall to 1mg/l or less. Details of the role of The Environment Agency in relation to barrage developments in England and Wales, and in particular in protecting the aquatic environment of the Tawe impoundment, can be found in Jones *et al* (1996).

2. CAUSES OF WATER QUALITY PROBLEMS

Poor water quality is often found in the impoundment during summer months when the river flows are low. A number of different factors, such as saline stratification, discharges from combined sewer overflows (CSO), sediment oxygen demand and algal bloom, can all contribute to the problem. In order to be able to provide effective management of the water quality problem, it was necessary to understand the relative contribution of these factors. A three-dimensional segmented and layered hydrodynamic and water quality model of the Tawe impoundment (WQFLOW-3DSL) has been constructed and calibrated by HR Wallingford to study the various processes.

2.1 The mathematical model

The model was used in a two-dimensional vertical mode as the impoundment is relatively narrow. It extends from Morriston Road Bridge in the north to a point 500m seaward of the LW mark in Swansea Bay (see Figure 1). The tidal limit in the impoundment is Beaufort Weir, 5.6 km upstream of the barrage.

The longitudinal resolution varies from 500m in the Approach channel in Swansea Bay to 100m immediately downstream of the barrage up to Steps, increasing to about 200m at the landward boundary giving 46 in-plan segments each of which is divided into a maximum of 24 layers. The vertical resolution is 0.5m over the tidal range (-5 to 5m ODN). Below -5ODN the layer thickness increases so that the water column below -5m ODN is represented by a maximum of 3 layers.

2.2 Model calibration

The Tawe model was calibrated and validated against salinity and water quality data for June 1994, when particularly poor water quality and algal blooms occurred during the low spring and neap tides in the first half of the month and a rainfall event on the 21st June resulted in flushing out of much of the saline water. Figure 2 shows the Tawe river flows and tide levels in the impoundment during June 1994. A number of remedial measures were employed by City and County of Swansea between 11th and 20th June to improve water quality in the impoundment. These measures included the operation of penstocks to lower the impoundment level, the opening of lock gates to release water, physical mixing of the water column and injection of compressed air into the water column.

Very little information is available on the operational details of the remedial measures deployed by City and County of Swansea during the period from 11th to 20th June. The water quality survey carried out during the period of remediation showed that there was only very localised mixing of the water column by air injection and no benefit was gained from the recirculating pump (Rogers and Mee, 1994). The use of penstocks was also found to have very limited improvement in the lower reaches of the impoundment. However, the release of water through the navigation locks on 14th and 15th June 1994 was found to produce a bigger impact. A maximum draw-down of 1.3-1.4m was achieved, the impoundment being refilled during the subsequent flood tide. No information on the actual operation of the lock is available and the only recorded data were the water level variations. It is apparent that the lock gates were only partially open. A number of sensitivity tests were undertaken to examine the effect of different opening strategies which may have been used. The extreme effects being to concentrate flushing in the bottom metre or so of the water column above the upstream sill of the lock and to have more uniform flushing through the depth. These tests indicated that the different assumptions about the method of opening the lock gates resulted in different water quality in the impoundment. Hence the lack of information on drawdown using the lock precludes the use of the period 11th June to 20th June from model calibration.

The model reproduced the main features of observed stratification and dissolved oxygen in the

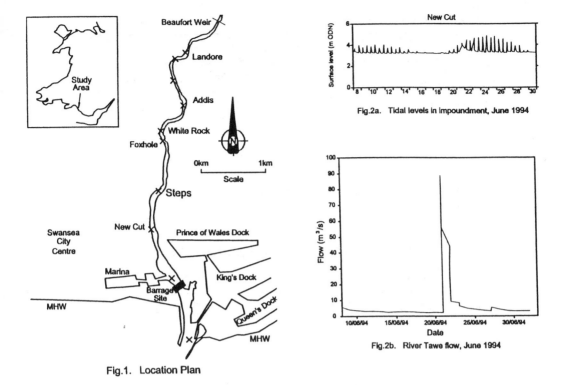

Fig.2a. Tidal levels in impoundment, June 1994

Fig.2b. River Tawe flow, June 1994

Fig.1. Location Plan

impoundment. Figure 3, 4 and 5 show the comparison of observed and predicted dissolved oxygen and salinity in the impoundment on 21st and 29th June 1994. These figures demonstrate that the model reproduced the observed flushing of most of the saline water during the rainfall event of 21 June 94. The observed rapid re-establishment of saline stratification during the next few spring tides in June 94 was also simulated successfully.

2.3 Main contributions

Sensitivity tests using the calibrated model were carried out to investigate the relative impact of various processes on water quality in the Tawe impoundment. A range of tests were undertaken to investigate the impact of river flow, CSO discharges, algal blooms and sediment oxygen demand.

Model results indicated that river flows of more than $40m^3/s$ were required to flush the majority of saline water out of the impoundment leaving some saline water below 1m ODN. This agreed with Environment Agency experience (Rogers and Mee, 1994).

Tests undertaken to determine sensitivity to CSO discharges indicated that CSO discharges have a relatively small impact on the water quality in the impoundment. In May-June 1994 only 8% of the BOD from freshwater sources came from CSO discharges despite several significant rainfall events in May. There are at present no measurements of sediment oxygen demand (SOD) in the impoundment. WDFLOW-3DSL simulates SOD by depositing particulate organic matter onto the bed where it decays exerting an oxygen demand on the bed. Particulate matter enters the impoundment from the river, several small streams, CSO discharges and from Swansea Bay. During low flow conditions in May-June 1994 the total input of BOD from freshwater sources was about 1 tonne/d of which it was estimated 20% was particulate.

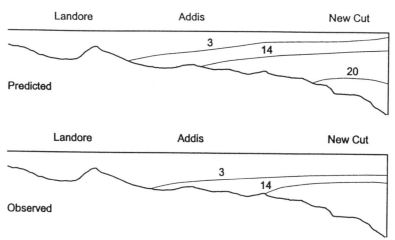

Fig.3. Predicted and observed salinity, 21st June 1994

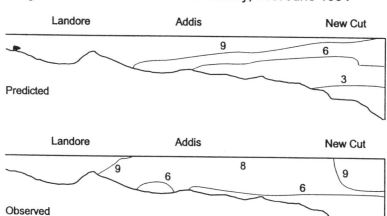

Fig.4. Predicted and observed disolved oxygen, 21st June 1994

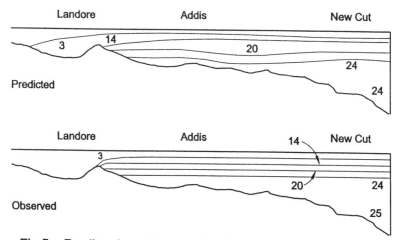

Fig.5. Predicted and observed salinity, 29th June 1994

The total input of BOD from Swansea Bay on a spring tide is about 0.8 tonnes. Stratification within the impoundment ensures that a significant proportion is retained in the lower layer. Various sensitivity tests were undertaken varying the particulate to dissolved ratios of BOD in each of the inputs. The results indicated that even with high estimates of particulate BOD the calculated SOD was only about $1g/m^2/d$.

The dominant processes controlling dissolved oxygen levels in the impoundment are the strong stratification and the significant BOD load contained in the lower saline layer.

3. REMEDIAL MEASURES

As discussed in Section 2, opening the lock gates on the ebb tide allowing water to flow out and sluicing water through the penstocks have been used to improve water quality in the Tawe Impoundment during poor water quality periods. At present, the City and County of Swansea, in collaboration with the Environment Agency, are considering the effectiveness of aerator blocks and propellers to improve the water quality of the Tawe impoundment. The aim of these devices is to improve the water quality by increasing the vertical mixing thus reducing the stratification. A field trial is planned for August 1998.

A number of bubble diffusers will be deployed on the river bed. A propeller will be placed near the halocline behind the barrage producing a horizontal jet and a larger propeller will be located lower in the water column to produce a vertical jet. These devices have been incorporated into the WQFLOW-3DSL hydrodynamic and water quality model. Literature values are used for the model parameters of the remedial devices.

3.1 Bubble mixers

The bubble diffusers operate by producing a column of air bubbles that entrain water thus increasing mixing. The bubble diffusers can potentially improve water quality in two ways. Firstly, by increasing vertical mixing and breaking down the stratification, and secondly by direct transfer of oxygen from the air bubbles to the water column. Bubble diffusers are most often used in reservoirs where weak thermal stratification is more readily overcome than saline stratification. In a reservoir situation, the benefits of bubble diffusers are mainly due to mixing. However, in situations of saline stratification, reaeration becomes more important as mixing is reduced by the stronger density difference.

The current trial proposal is to deploy up to 20 diffusers around Addis. Model runs of using 20 bubblers at Addis (for 44 hours and with an oxygen mass rate per diffuser of 37.5kg/day) were carried out to demonstrate the impact of the mixing alone and the other impact of both the mixing and the direct oxygenation. The model results suggest that the use of bubble diffusers will produce a local improvement in mixing giving rise to slightly elevated oxygen concentrations below the halocline.

3.2 Horizontal jet

It is proposed to use a propeller mixer at about 1.2m below the water surface and next to the barrage. This propeller will produce a horizontal, which is intended to increase mixing. Model runs show that the turbulent kinetic energy jet produced by the propeller could significantly reduce the saline stratification and increase the dissolved oxygen below the halocline, immediately behind the barrage. There may also be noticeable effects as far as Addis.

3.3 Vertical jet

It is proposed to use a banana-blade propeller at depth below the halocline, immediately behind the barrage to produce a vertical jet, with the aim of lifting salt water to the surface layer so that it flows over the fish pass. In the model, the propeller acts effectively as a pump. Water is removed from the model at the level of the propeller at a user-defined rate. The power input and the size of the propeller are used to determine whether the jet will have sufficient energy to carry salt water to the surface layer.

Model runs show that locally the salinity is decreased above and below the halocline by 3-4ppt. There also appears to be a slight increase in dissolved oxygen throughout the water column. The reduction in salinity below the halocline is also seen at Addis although there is no significant increase in salinity above the halocline. There is also an increase in dissolved oxygen below the halocline.

4. CONCLUSIONS

The case study of Tawe Barrage illustrates the type of water quality problems that can be experienced in the impoundment of a tide-overtopping barrage. Effective management of the post-impoundment water quality problems require both regular monitoring and deployment of remedial measures if necessary. The calibrated water quality model of Tawe enabled the investigation of the relative contributions of different causes of poor water quality and the relative effectiveness of various remedial measures/options. The field trials of remedial measures in August 1998 may provide improved model parameter values for the bubble diffusers and propellers. The model will be used to aid management of the Tawe impoundment.

5. ACKNOWLEDGEMENTS

The authors are grateful to the Environment Agency for permission to publish this paper. The views and opinions expressed here are those of the authors and not necessarily those of the Agency. The authors would also like to express their gratitude to Elfed Jones, Nicholas Odd, Keith Davies, Niall Reynolds, Simon Clark and Ken O'Hara for their assistance in this project.

6. REFERENCES

Atkins Research and Development (1983). "Tawe barrage feasibility study". W.S. Atkins report submitted to Swansea City Council.

Broyd, T. W, Hooper, A. G. and Kingsbury, R. S. W. M. (1984) "The effects of constructing a barrage across the Tawe Estuary". *Wat. Sci. Tech.* Vol 16. Pp 463-475.

Jones, F. H., Gough, P. J. and Axford S. (1996) "Barrage Developments in England and Wales - the role of the Environment Agency in protecting the aquatic environment", *Barrages: Engineering design & environmental impacts,* edited by Burt, N and Watts J., John Wiley & Sons Ltd.

Rogers, A. P. (1992) *Tawe Barrage: Water Quality changes following impoundment,* NRA (Welsh Region) Technical memo. ref: EQ/P1/2/7, 25.9.92.

Rogers, A. P. and Bryson, P. (1994) *An assessment of the impact of the Tawe Barrage on water quality in the Tawe estuary,* NRA (Welsh Region), Report ref. PL/EAW/94/2.

Rogers, A. P. and Mee, D. M. (1994) *A report on the deterioration in water quality of the Tawe Impoundment, including observations of the impact of salmonid fish migration during June 1994,* NRA report no. PL/EAW/94/8.

Environmental Hydraulics, Lee, Jayawardena & Wang (eds) © 1999 Balkema, Rotterdam, ISBN 90 5809 035 3

Salinity simulations in San Francisco Bay

Edward S. Gross, Jeffrey R. Koseff & Stephen G. Monismith
Environmental Fluid Mechanics Laboratory, Stanford University, Calif., USA

ABSTRACT: A three-dimensional hydrodynamic and scalar transport model is applied to simulate salinity in South San Francisco Bay. All model parameters are identical to those used in a model calibration in which the model was shown to reproduce current meter data accurately. The 64 day time period studied is characterized by low freshwater input. For this period the salinity data is reproduced well by the model. Because no model parameters are adjusted, the salinity simulation is considered to be a validation of the model. The validated model is used to investigate the effect of wind forcing on the model results.

1. INTRODUCTION

Mathematical transport models are frequently used to study natural systems. In order to provide confidence in these models a model validation is frequently performed. Successful validation implies that the important physical transport processes are resolved accurately and that mixing induced by numerical error is small relative to the physical mixing. Once the model is validated it can be applied with confidence to simulate scalar transport in situations where field data is not available to compare with model results.

2. GOVERNING EQUATIONS

The governing equations used in our numerical simulations are the Navier-Stokes equations after Reynolds averaging, assuming hydrostatic flow and applying the Boussinesq approximation. The momentum equations are:

$$\frac{\partial u}{\partial t} + u\frac{\partial u}{\partial x} + v\frac{\partial u}{\partial y} + w\frac{\partial u}{\partial z} = -g\frac{\partial \zeta}{\partial x} - \frac{g}{\rho_o}\frac{\partial}{\partial x}\left(\int_z^\zeta \rho' dz\right) + \nu_h\left(\frac{\partial^2 u}{\partial x^2} + \frac{\partial^2 u}{\partial y^2}\right) + \frac{\partial}{\partial z}(\nu_v\frac{\partial u}{\partial z}) + fv, \quad (1)$$

$$\frac{\partial v}{\partial t} + u\frac{\partial v}{\partial x} + v\frac{\partial v}{\partial y} + w\frac{\partial v}{\partial z} = -g\frac{\partial \zeta}{\partial y} - \frac{g}{\rho_o}\frac{\partial}{\partial y}\left(\int_z^\zeta \rho' dz\right) + \nu_h\left(\frac{\partial^2 v}{\partial x^2} + \frac{\partial^2 v}{\partial y^2}\right) + \frac{\partial}{\partial z}(\nu_v\frac{\partial v}{\partial z}) - fu, \quad (2)$$

and the continuity equation is:

$$\frac{\partial u}{\partial x} + \frac{\partial v}{\partial y} + \frac{\partial w}{\partial z} = 0, \quad (3)$$

where t is time, $u(x,y,z,t)$, $v(x,y,z,t)$, $w(x,y,z,t)$ are the velocity components in the horizontal x,y directions and vertical z direction, respectively, $\zeta(x,y,t)$ is the water surface elevation above an undisturbed water level, f is the Coriolis parameter, ν_h is the horizontal eddy viscosity, $\nu_v(x, y, z, t)$ is the vertical eddy viscosity, ρ_o is a constant reference density, $\rho'(x, y, z, t)$ is the local variation from the reference density, and g is gravitational acceleration.

The governing equation for each scalar constituent is:

$$\frac{\partial s}{\partial t} + \frac{\partial(us)}{\partial x} + \frac{\partial(vs)}{\partial y} + \frac{\partial(ws)}{\partial z} = \frac{\partial}{\partial x}\left(\epsilon_h \frac{\partial s}{\partial x}\right) + \frac{\partial}{\partial y}\left(\epsilon_h \frac{\partial s}{\partial y}\right) + \frac{\partial}{\partial z}\left(\epsilon_v \frac{\partial s}{\partial z}\right) \tag{4}$$

where $s(x, y, z, t)$ is the scalar concentration, $\epsilon_h(x, y, z, t)$ is the horizontal diffusion coefficient, $\epsilon_v(x, y, z, t)$ is the vertical diffusion coefficient and $u(x, y, z, t)$, $v(x, y, z, t)$, $w(x, y, z, t)$ are the velocity components in the horizontal x, y directions and vertical z direction, respectively. In the salinity simulations the horizontal diffusion coefficient, ϵ_h, is assumed to be zero.

3. NUMERICAL METHOD

The numerical method used to solve the governing equations is the TRIM method (Casulli and Cattani, 1994). In our applications z_0 coefficients are used instead of Manning's n coefficients. Otherwise no changes are made to the published TRIM method.

The method as published by Casulli and Cattani (1994) did not include a turbulence closure method or a method for scalar transport. In these simulations we use a variation of the Mellor Yamada level 2.5 turbulence closure model (Mellor and Yamada, 1982) known as the QETE turbulence closure (Galperin et al., 1988). The numerical implementation is described in detail by Gross et al. (1998c).

Details of the horizontal transport method, the vertical transport method and a combination of these methods using a directional splitting approach are described by Gross (1998) and Gross et al. (1998a,b).

These extensions to TRIM do not introduce additional limitations on the time step chosen in simulations. Thus the efficiency of TRIM is maintained in these simulations. A complete discussion of stability conditions for the 3-D approach is given in Gross et al., (1998a).

4. SALINITY SIMULATIONS

The model was calibrated against water level and current meter data collected by the USGS (e.g. Cheng and Gartner, 1984) by adjusting the roughness coefficients (Gross, 1998). Because the model parameters used in the salinity simulations are identical to those used in the calibration the following results comprise a validation of the model.

The time period chosen for the salinity simulations is a period of fairly well-mixed conditions which was used previously in depth-averaged simulations (Gross et al., 1998b). Due to a lack of salinity data in the shoals, the model results are compared only with data collected in the channel.

A major assumption inherent in these simulations is that tidal dispersion is adequately resolved using the 200 meter grid. Because we believe that the scales which cause tidal dispersion are adequately resolved the horizontal diffusion/dispersion coefficient used in the simulations is zero. Vertical shear dispersion resolved by the numerical method produces small scale mixing while tidal dispersion produces the large scale stirring.

4.1. Salinity Data

The water surface elevation and salinity time series data used to specify the northern boundary condition was collected at the west end of the Bay Bridge. The sensors are at 2.7 and 12.0 meters below mean lower low water (MLLW) and the data interval is 15 minutes. A similar salinity time series is used for comparison with model results at the San Mateo Bridge. The sensors are at 1.7 and 13.9 meters below MLLW.

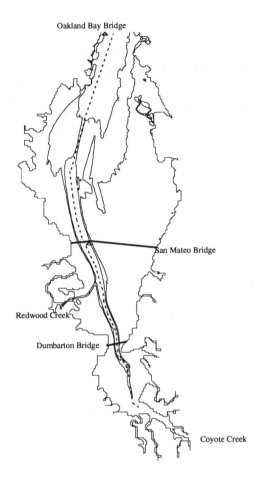

FIG. 1. Geographical features and bathymetry of South San Francisco Bay with contours at 5 and 10 meters. The Δ symbols indicate the location of the time series data and the dashed line indicates the location of the transect data

Vertical profiles of salinity are collected in the channel of South San Francisco Bay (Edmunds *et al.*, 1997) from the Oakland Bay Bridge to the region South of the Dumbarton Bridge (SDB) excluding Coyote Creek, as shown in Fig. 1. The data was collected over a three hour period on each of the dates used in this study as the research vessel traveled from SDB to the Oakland Bay Bridge. These profiles were collected in 10 locations in SSFB on June 13, 1995, July 18, 1995 and August 16, 1995 (Julian day 164, 199 and 228). The data on Julian day 164 is used to specify the initial conditions and the data collected on the other two occasions is compared to model results.

There are several limitations of the data set which affect the degree of agreement that can be expected between the data and model results. First, salinity time series are available only at one horizontal location at the northern boundary. Since data is available at this location, salinity along the entire open boundary is assumed to be uniform horizontally. Second, salinity data for the initial conditions is available for the channel but not the shoals. Finally, the data used to specify initial conditions was collected over approximately 3 hours, and thus does not give a synoptic specification of salinity.

4.2. Model Setup

The 200 meter horizontal grid spacing and 1 meter vertical grid spacing result in a total of 13,949 water columns and 116,244 active grid points. Quiescent initial conditions are used for model hydrodynamics and the hydrodynamics are allowed to "spin up" for 34 hours before the initial salinity conditions are specified. The initial conditions incorporate the vertical variability of salinity using ten vertical profiles of salinity. The data used to specify the salinity initial conditions was collected on June 13, 1995 starting at 6:38 AM SDB and ending at 10:12 AM at the Oakland Bay Bridge (Edmunds *et al.*, 1997).

The largest source of freshwater input to SSFB during dry summer months is the discharge from wastewater treatment plants. However, the evaporation rate is on the same order of magnitude as the freshwater input rate (Selleck *et al.*, 1966). As both are relatively small they are omitted from the simulations.

The measured surface salinity plotted in Fig. 2 indicates that, during the time period modeled, the salinity field in SSFB changed from winter conditions of large freshwater input towards summer conditions of oceanic salinity. The total length of the domain is greater than 50 kilometers while typical tidal excursions in the channel are about 10 kilometers (Cheng *et al.*, 1993). Thus the salinity in SDB is not affected instantly by changes in salinity at the boundary, but instead after many tidal cycles of tidal stirring.

A time variable, but spatially uniform, wind forcing is specified using data collected in Redwood City during the period of the salinity simulations. Summer winds are generally from the Northwest with a typical amplitude of about $4m/s$ during the afternoon with weaker winds during the night and morning. The wind induces a mean flow up-estuary (southward) in the shoals and down-estuary (northward) in the channel.

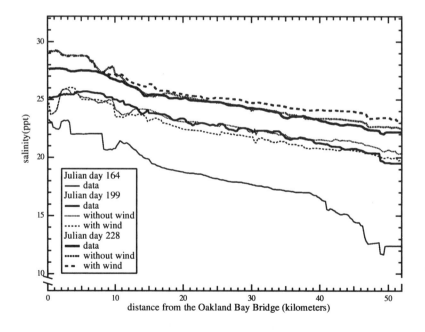

FIG. 2. Longitudinal transects of measured and computed surface salinity

4.3. Model Results

The field data was collected on Julian day 199 and Julian day 228 during relatively weak ebb tides. Fig. 2 shows the comparison of simulated surface salinity with field data. Fig. 3 shows the measured salinity field and the computed salinity field in the channel on Julian day 228. The salinity field for the simulation which did not include wind forcing is also plotted. Not only is the top salinity reproduced well by the three-dimensional model, there is also good qualitative agreement between the predicted and measured stratification. The stratification reaches a maximum of roughly 1 *ppt* in both the data and the simulations.

The time series data collected at the San Mateo Bridge is also used in the model validation. The salinity measured at the top sensor is reproduced well by the model. However, the bottom salinity is overestimated by the model, indicating that the stratification predicted by the model is greater than the actual stratification at the San Mateo Bridge.

The model tends to slightly overpredict stratification and intertidal salt transport, suggesting that the predicted longitudinal baroclinic exchange is stronger than the actual longitudinal baroclinic exchange. The longitudinal baroclinic exchange flow would be overpredicted if the eddy diffusivities predicted by the QETE turbulence closure model are too small, as has been observed by Stacey *et al.* (1998).

4.4. Effect of Wind on Salt Transport

During summer the wind usually blows from the Northwest in SSFB. The residual circulation caused by the wind forcing during this period is up-estuary (southward) in the shoals and down-estuary (northward) in the channel, which is opposite to the depth-averaged baroclinic circulation (Gross, 1998). Thus the effect of wind is to decrease the transport of salt into SSFB.

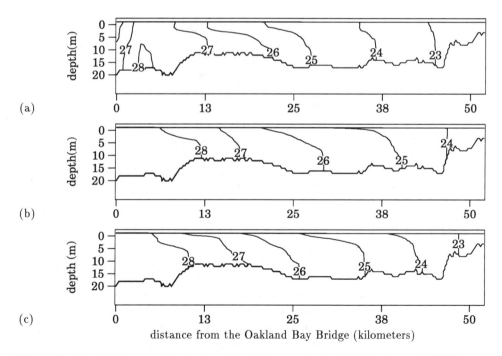

FIG. 3. Distribution of measured and computed salinity over the depth of the channel in SSFB on Julian day 228; (a) data, (b) model without wind (c) model with wind

Table 1. l_1 error of the three-dimensional salinity simulations with and without wind. Units are *ppt*

simulation	transect − 199	transect − 228	time series
with wind	0.025	0.015	0.019
without wind	0.016	0.029	0.036

Table 1 shows that the simulation which includes wind forcing reproduce the surface salinity transect on Julian day 228 and the salinity time series data more accurately than the previous simulation, and that on Julian day 199 the simulations which include wind forcing were less accurate. As seen in Fig. 3 the salinity predicted by the model decreases when wind forcing is considered. However, the degree of stratification is not strongly altered by the wind forcing. In general, the three-dimensional simulations which include wind forcing are more accurate than the three-dimensional simulations which neglect wind forcing.

5. SUMMARY

The three-dimensional TRIM model (Casulli and Cattani, 1994) has been modified to include the QETE turbulence closure model (Galperin *et al.*, 1988) and conservative a scalar transport scheme. The resulting method was applied to study South San Francisco Bay in a 64 day simulation of salinity. Because of the large time step (3 minutes) used, simulations using 116,244 active grid points can be completed in less than 1 day of CPU time on an Unix workstation using a MIPS R10000 processor. The model accurately reproduced salinity data during the period studied.

The salinity data was reproduced accurately without considering horizontal turbulent mixing, evaporation or local input of freshwater in the simulations, suggesting that these processes have less influence on the longitudinal salinity variation in SSFB than the processes that were resolved by the model. However field data used was only available in the channel. Wind forcing has a small, but significant, effect on model results and the salinity data is reproduced more accurately by the model when wind forcing is considered. The wind has little effect on the calculated stratification.

6. Acknowledgements

This study was funded by the United States Environmental Protection Agency San Francisco Estuary Project contract number 22-1509-1001-01, the California Regional Water Quality Control Boards (San Francisco Region) and the Leavell Family Faculty Scholarship. The salinity data was provided by Dr. J. Thompson of the United States Geological Survey. The TRIM model was provided by Professor Vincenzo Casulli.

7. REFERENCES

Casulli, V. & Cattani, E. 1994. Stability, accuracy and efficiency of a semi-implicit method for three-dimensional shallow water flow. *Computers and Mathematics with Applications*, 27, 4, 99–112.

Cheng, R. T., Casulli, V., & Gartner, J. W. 1993. Tidal, residual, intertidal mudflat (TRIM) model and its applications to San Francisco Bay, California. *Estuar., Coastal Shelf Sci.*, 369, 235–280.

Cheng, R. T. & Gartner, J. W. 1984. *Tidal and Residual Currents in San Francisco Bay, California. Results of Measurements 1979-80.* Open-File Report 84-4339, U. S. Geological Survey, Menlo Park, CA.

Edmunds, J. L., Cole, B. E., Cloern, J. E., & Dufford, R. G. 1997. *Studies of the San Francisco*

Bay, California, Estuarine Ecosystem: Pilot Regional Monitoring Results, 1995. Open-File Report 97-15, U. S. Geological Survey, Menlo Park, CA.

Galperin, B., Kantha, L. H., Hassid, S., & Rosati, A. 1988. A quasi-equilibrium turbulent energy model for geophysical flows. *J. Atm. Sci.*, 45, 55–62.

Gross, E. S. 1998. *Numerical Modeling of Hydrodynamics and Scalar Transport in an Estuary.* PhD thesis, Stanford University.

Gross, E. S., Casulli, V., Bonaventura, L., & Koseff, J. K. 1998a. A semi-implicit method for vertical transport in multidimensional models. *Int. j. numer. methods fluids*, 190, 802–817.

Gross, E. S., Koseff, J. K., & Monismith, S. G. 1998b. An evaluation of advective transport schemes or simulation of salinity in a shallow estuary. *J. Hydr. Engrg., ASCE*, accepted for publication.

Gross, E. S., Koseff, J. K., & Monismith, S. G. 1998c. Three-dimensional salinity simulations in South San Francisco Bay. *J. Hydr. Engrg., ASCE*, in progress.

Mellor, G. L. & Yamada, T. 1982. Development of a turbulence closure model for geophysical fluid problems. *Reviews of Geophysics and Space Physics*, 20, 851–875.

Selleck, R. E., Pearson, E. A., Glenne, Bard, & Storrs, P. N. 1966. *Physical and hydrological characteristics of San Francisco Bay.* Report 65-10, University of California at Berkeley, Sanitary Engineering Research Laboratory.

Stacey, M. T., Monismith, S. G., & Burau, J. R. 1998. Observations of turbulence in a partially stratified estuary. *J. Phys. Oceanography*, in progress.

Environmental Hydraulics, Lee, Jayawardena & Wang (eds) © 1999 Balkema, Rotterdam, ISBN 90 5809 035 3

On the use of laboratory observations to validate numerical models

D.L. Boyer, N. Pérenne, D.C. Smith & C. Robichaud

Enviornmental Fluid Dynamics Program, Department of Mechanical and Aerospace Engineering, Arizona State University, Tempe, Ariz., USA

ABSTRACT: In order to develop data sets which can be used as benchmarks for coastal circulation numerical models, laboratory experiments are being conducted in a cylindrical tank in which a continuous continental shelf model, interrupted only by a single smooth canyon, is placed along the periphery of the test cell. Prior to experimentation, the tank is filled with a linearly stratified fluid and the tank is then slowly brought up to solid body rotation with Coriolis parameter f. To initiate the experiments, the turntable rotation rate is then either (i) impulsively changed by $\pm\Delta\omega$ or (ii) modulated sinusoidally about the background rotation rate f/2. The objectives of the experiments are to (i) observe and better understand the motion field in the vicinity of a submarine canyon and (ii) provide benchmarks for the development of numerical coastal circulation models. The present communication reports on some preliminary laboratory and numerical experiments directed toward these goals.

1. INTRODUCTION

The continental shelf serves as a transition zone between the coastline and the open ocean. The coastal ocean is used extensively for transportation and, owing to the high population density along coastal regions, is associated with significant levels of pollution from a variety of sources. The shelf is also the conduit for the transport of material (e.g., nutrients and particulate matter of natural or man-made origin) from regions near the coastline to the open ocean and the reverse. The continental slope/shelf break/shelf region is interrupted at irregular intervals by submarine canyons. It is generally agreed that the motion fields in such canyons are characterized by (i) enhanced turbulence, upwelling and internal wave generation, (ii) increased across shelf/slope material transport and (iii) coastal-trapped wave modification and generation. Because of the difficulties of making measurements in submarine canyons, especially on their steep slopes, it is not surprising that the numbers of field observations in canyons is quite limited; Hickey (1997). While some knowledge of the effects of canyons on coastal current systems has been gained by field observations, analytical and numerical models, and laboratory experiments, the general problem is still poorly understood. It is clear that the development of a prognostic capability for the general flow patterns within and in the vicinity of canyons will have to depend on increasingly realistic numerical models coupled with adequate field data. One nagging issue related to this numerical model development is the lack of suitable benchmarks to serve as validation tools for the models. The thesis of this communication is that laboratory models can play this role.

2. PHYSICAL SYSTEM AND PARAMETERS

The physical system considered is given schematically in Figure 1. An axially symmetric shelf (horizontal), shelf-break, continental slope topography, interrupted only by a single smooth canyon, is placed in the center of a circular test cell of radius R_T; the inner shelf ends at a vertical coastline at a radius R_c and the shelf break occurs at the radius r_s. The canyon is characterized by its length L and width W (measured along the shelf break). The system is filled with a linearly stratified fluid of depth h_d in the deep ocean and h_s over the shelf. Prior to initiating an experiment, the system is brought to solid body rotation with Coriolis parameter $f = 2\Omega_0$, where Ω_0 is the turntable rotation rate. Experiments are initiated by an impulsive change in the rotation rate (i.e. $\pm \Delta\omega$) to produce an approximately steady flow relative to an observer fixed to the canyon. The resulting time-dependent and mean motion field at various depths can thus be measured.

The steady forcing case has been studied by Klinck (1996) who found that the flow direction is a critical parameter in determining the flow response. For the coast on the left facing downstream, he found the long-time behavior being upwelling favorable and, for the coast on the right, downwelling in nature.

To observe the flow, neutrally buoyant particles of nominal density $\rho_0 = 1.02 \pm .005$ gmcm^{-3} are released into the flow. A horizontal light sheet illuminates the particles whose motions are recorded by a super VHS video system. These video images are then analyzed to yield the horizontal velocity, vertical vorticity and horizontal divergence field at selected levels using the DigImage software developed at Cambridge University. Conductivity probes are used to measure the background stratification in the tank as well as the temporal development of the stratification at selected horizontal locations.

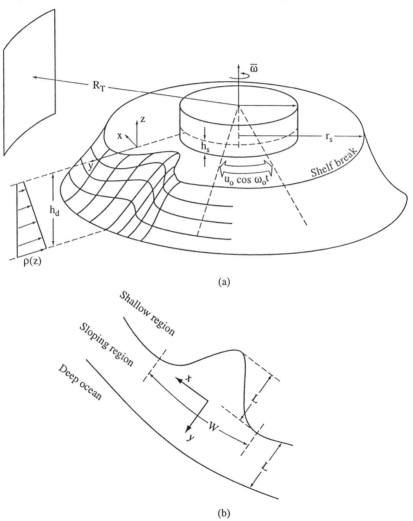

(a)

(b)

Figure 1. Physical system (a) and top view (b).

76

Table 1. System parameters.

Symbol	Parameter	Lab	Ocean
h_s	Water depth on the shelf	2.5 cm	200 m
h_d	Water depth in the deep ocean	12.5 cm	1500 m
r_s	Radius to shelf break	55 cm	—
R_c	Radius to coastline	35 cm	—
R_T	Radius of test cells	90 cm	—
N	Brunt-Väisälä frequency	$1.5 \ s^{-1}$	$3(10)^{-3} \ s^{-1}$
W	Width of the canyon	20 cm	20 km
L	Length of the canyon	15 cm	20 km
f	Coriolis parameter	$0.5 \ s^{-1}$	$(10)^{-4} \ s^{-1}$
u_0	Excursion of background flow	$0.5 \sim 2 \ cm \ s^{-1}$	$20 \ cm \ s^{-1}$
ν	Viscosity	$0.01 \ cm^2 s^{-1}$	$0.001 \ m^2 s^{-1}$
$Ro = u_0 / (fW)$	Rossby number	$0.02 \sim 0.2$	$0.1 \sim 0.2$
$E = \dfrac{\nu}{fh_s^2}$	Ekman number	$3(10)^{-3}$	$2.5(10)^{-4}$
$Bu = \dfrac{N^2 h_d^2}{f^2 W^2}$	Burger number	3.8	0.4

Table 1 provides a list of the laboratory parameters investigated as well as typical values of the parameters one might expect for ocean canyons.

3. SOME PRELIMINARY RESULTS

3.1 Laboratory

Figure 2 depicts time series measurements of the normalized isopycinal elevation $z^* = z/h_d$ against the dimensionless time $t^* = ft/2\pi$ for a horizontal probe location at the center of the canyon $x = y = 0$ (see Figure 1) at the vertical depths indicated; these are (i) $z = -(h_s - 1 \ cm)$ just above the rim of the canyon, (ii) $z = -h_s$ at the level of the canyon rim, (iii) $z = -(h_s + 1 \ cm)$ just below the rim, (iv) $z = -(h_s + 3 \ cm)$ deep in the canyon and finally (v) $z = -(h_s + 5 \ cm)$ which is 1 cm above the slope at that particular horizontal location, respectively. Each time series corresponds to a single experiment, but we waited 2-4 hours before moving the probe and performing another deceleration/acceleration of the turntable, so that the fluid was at rest before each measurement. The repeatability of these measurements proved to be good.

A number of observations are apparent on this figure. One obvious aspect of the plots is the relatively strong oscillation superimposed on the mean having the same frequency as the background rotation of the turntable. Our analysis has shown that this owes to a slight misalignment between the gravitational vector and the background rotation. Steps are being taken to improve the turntable alignment.

Owing to the homogenization of the fluid layer above the shelf (not shown) by evaporation and boundary mixing the upper region of the fluid is little affected by the presence of the canyon, as evidenced by the first plot of Figure 2.

The second and third plots nicely indicate the initial downwelling that occurs in the center of the canyon upon initiation of the impulsive motion. These plots, however, then evolve into upwelling regions with the mean flow being steady (except for the high frequency barotropic gravity mode discussed above. Note the phase lag of the initial downwelling at the upper elevations which owes to the time required for a fluid parcel to advect from the "upstream" canyon boundary to the canyon center (probe) location.

3.2 Numerical Model

The numerical model employed in these experiments is that of Haidvogel et al. (1991) (SPEM, version 3.5). The flow is forced in a periodic channel with a gradient in streamfunction applied between the onshore and offshore boundaries. This results in alongshore mean flows with magnitude 0 (10 cm/s). The initial stratification ($N^2 = 2 \times 10^{-3}$ s^{-1}) is adjusted to make the Burger number agree in the model and the laboratory experiment. The domain size in the numerical experiments is 32 km x 16 km with a resolution of 0.5 km. The resulting canyon width scale is approximately 7 km, in rough agreement with that of natural ocean canyons. Figures 3a and 3b show the flow fields at the surface after four days of integration for upwelling (Figure 3a) and downwelling (Figure 3b) favorable forcing. Model bathymetry is shown in figure 3c. The resulting flow fields are obviously different, thus supporting Klinck's (1996) result. In both cases cyclonic vorticity is predominant, with the downwelling case reflecting a double cyclone system while the upwelling case has a single cyclone. Numerical issues currently being addressed are i) the relative proximity of the periodic boundaries to the canyon, ii) the use of periodic boundary conditions, and iii) the form of turbulence closure in the model and its relation to the laminar nature of the laboratory flow.

3.3 Comparisons

Attempts were made to compare the temporal evolution of the normalized isopycinal deflections for the upwelling favorable case of Figure 2. Figure 4 depicts the laboratory (------) and numerical model (solid)

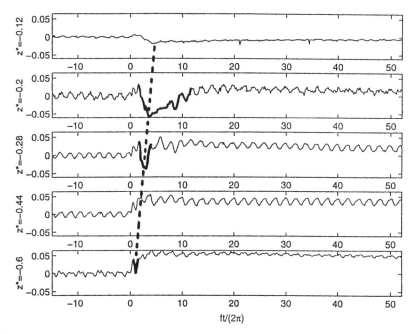

Figure 2. Temporal evolution of normalized, relative, isopycinal heights for various dimensionless depths z* = z/h$_d$ for parameter values Ro = 0.10 and Bn = 4. The bold portion of the plot is the period of downwelling as associated with the spin-up phase of the motion.

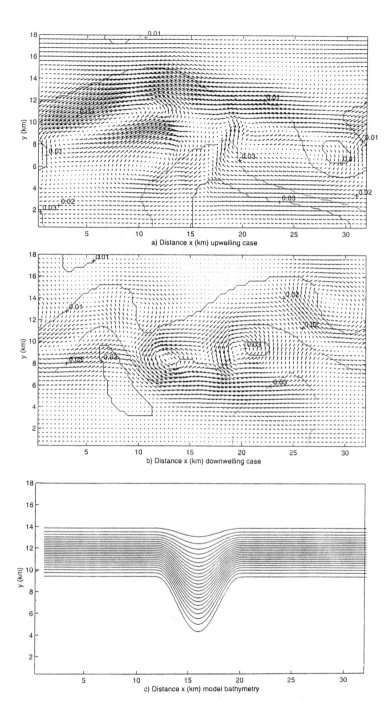

Figure 3-
(a) Upwelling case. Velocity and salinity field at model surface after 4 days of constant forcing. Maximum velocity magnitude is 0 (10 cm/s). Salinity values are contoured as the deviation from 26.0 ppt and a scale factor of 10 has been applied making maximum salinity deviation values .004 ppt.
(b) Same as (a) for the downwelling case.
(c) Canyon bathymetry, depth in m. Shelf width is 9 km, canyon width W is 7 km, continental shelf depth is 200 m and offshore depth is 800 m.

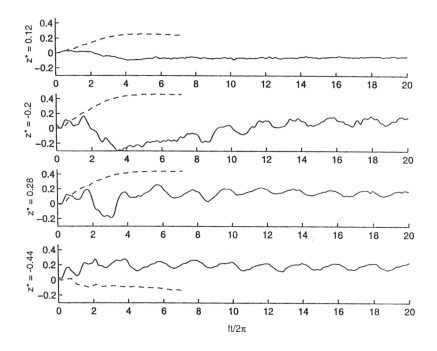

Figure 4. Upwelling favorable case (Ro = 0.1, Bu = 4).

runs for the initial stages of the upwelling case of Figure 2. One notes that the numerical model, while predicting an evolution to an upwelling mode at longer times, does not adequately predict the initial downwelling phase.

4. CONCLUSION

Our ability in the long term to better predict the nature of coastal current systems will necessarily rely on numerical models. These models will not only require accurate initial (from observations) and boundary conditions, but also a proper parameterization of small-scale processes. Synoptic data in space and time are not typically available from field measurements to test numerical models. The present study aims to investigate the feasibility of substituting laboratory data for these field observations.

This project to date has illustrated some of the difficulties of achieving this goal but nevertheless has shown promise that good data sets (velocity and density) can be obtained in the laboratory and that careful analyses will have to be made concerning the differences between the laboratory and numerical models and the ocean.

5. REFERENCES

Haidvogel, D.B., Wilkin, J. and Young, R. (1991). "A semi-spectral primitive equation ocean circulation model using vertical sigma and orthogonal curvilinear horizontal coordinates," *J. Geophys. Res.*, **101**, 151-185.
Hickey, B.M. (1997). "The response of a steep-sided, narrow canyon to time-variable wind forcing," *J. Phys. Ocean.*, **27(5)**, 697-726.
Klinck, J.M. (1996). "Circulation near submarine canyons: A modeling study," *J. Geophys. Res.*, **101**, 1211-1223.

Environmental Hydraulics, Lee, Jayawardena & Wang (eds) © 1999 Balkema, Rotterdam, ISBN 90 5809 035 3

Prediction of the behaviour of marine oil spills: Applications based on random walk techniques

Z. Li
Department of Environmental Science and Technology, Dalian University of Technology, China

C.T. Mead
HR Wallingford, UK

ABSTRACT: A numerical model has been developed to simulate the transport and fate of oil spilled at sea. The model combines the transport and fate processes of spilled oil with the random walk technique. Oil movement under the influence of tidal currents, wind-driven currents, and turbulent eddies is simulated by the PLUME-RW dispersion model developed by HR Wallingford. The weathering processes in the model represent physical and chemical changes of oil slicks with time, and comprise mechanical spreading, dispersion, evaporation and emulsification. Shoreline stranding is determined approximately using a capacity method for different shoreline types. This paper presents details of the model, and describes the results of various sensitivity tests. The model is suitable for oil spill contingency planning.

1. INTRODUCTION

The frequency of accidental oil spills, and the associated aquatic pollution, are growing concerns. There is a need for model systems that can be used in both rapid spill response and contingency planning. Many oil spill models have been constructed to represent oil transport and fate processes (Mackay *et al* 1980; Spaulding *et al* 1992; Humphery *et al* 1993, ASCE Task Committee 1996). These models include transport calculations which determine the oil movement in space and time, and fate models which estimate oil partition between various environmental compartments and changes of oil properties. Most models use a mass balance approach to track the amount of oil in compartments, which include the sea surface, the atmosphere, the water column, the shoreline, the sea bed, biodegraded, etc.

For some years, random walk models have been used as planning tools for assessing the probable impacts of waste discharges on the marine environment. This paper describes an extension of this modelling method to simulate the behaviour of oil spilt at sea. The concepts upon which the new model, OIL-RW, is based form the basis of other oil spill models, but the intention here is to focus on the combination of these concepts with the random walk technique.

2. MODEL FORMULATION

In OIL-RW, spilled oil is assumed to consist of a large number of model particles, with each particle representing a defined quantity of oil. Effectively, model particles are treated as 'mass points', with their transport determined by tidal currents, wind-driven currents, turbulent eddies (represented using the random walk method), gravitational spreading and buoyancy. Each particle on the water surface is treated as an individual oil slick when simulating weathering, so that its volume and properties change with time. Dispersion and initial oil distributions are determined using stochastic methods. Model particles generally move in the horizontal plane on the water surface, but some move in three dimensions within the water column after entrainment by breaking waves.

2.1 Random Walk Method

Vertical Structure of Tidal and Wind-driven Currents: In order to compute the movement of oil droplets in the sea, the oil spill model fits an analytic current profile to tidal currents computed by two-dimensional, depth-averaged flow models, and superimposes a wind-driven current profile. The profiles are based on the equations (Mead 1991):

$$u(z) = (u_0)_T / k_0 \ln(30.1z / k_s) \tag{1}$$

$$U_w(z) = U_s(3(1 - z/d)^2 - 4(1 - z/d) + 1) \tag{2}$$

where u is the tidal current speed (ms^{-1}), $(u_0)_T$ is the tidal current friction velocity (ms^{-1}); k_0 is von Karman's constant, z is the height above the sea bed (m), k_s is the roughness length (m); U_w is the wind-driven current speed (ms^{-1}) and U_s is the surface wind-driven current speed (ms^{-1}) and d is water depth (m). Usually, U_s is 3-3.5 percent of the wind speed at a height of 10m above the sea surface (ASCE Task Committee 1996).

Turbulent Diffusion: Horizontal and vertical diffusive displacements are used to simulate the effects of turbulent eddies on oil slick movement, and these are added to the ordered movements that represent advection by mean currents. The eddy diffusivity is a specified constant for horizontal diffusion, but is determined from the following equation for vertical diffusion:

$$D = 0.16 \, z^2 (1 - z/d) \left| \frac{U_*}{k_0 z} \right| \tag{3}$$

where D is the eddy diffusivity ($m^2 s^{-1}$), and U_* is the total friction velocity due to wind-driven and tidal currents (ms^{-1}). The directions of the displacements are random.

2.2 Oil Weathering

Evaporation: Evaporative loss of oil from the sea surface is determined by a mass-transfer equation derived by Stiver and Mackay (1984):

$$\frac{dF}{dt} = \frac{K_E}{h} \exp \left[k_1 + k_2 (C_1 + C_2 F) \right] \tag{4}$$

where F is the fraction of oil evaporated, and k_1 and k_2 specify the temperature dependence of the vapour pressure: $k_1 = 48.5 - 0.1147 T_0$, $k_2 = 4.5 \times 10^{-4} T_0 - 0.1921$ (T_0 is the ambient temperature (K)). K_E is a mass transfer coefficient for evaporation (m/s), $K_E = 2.5 \times 10^{-3} U_{10}^{0.78}$, h is oil slick thickness (m), and C_1 and C_2 are constants derived from crude oil distillation data (Leech 1992)

Emulsification: Emulsification is modelled using the algorithm of Mackay *et al* (1980). The rate of water incorporation into oil slicks is given by:

$$\frac{dY_w}{dt} = K_A (1 + U_{10})^2 (1 - K_B Y_w) \tag{5}$$

where Y_w is the fractional water content of the oil, K_A is a water incorporation rate constant for emulsifying oils ($\sim 2 \times 10^{-6}$) and K_B is an oil dependent constant; 0.7 for crude oils and heavy fuel oil.

Weathering: The weathering of oil can be estimated from the equations of Mackay *et al* (1980):

$$\mu = \mu_0 \exp(C_4 F) \cdot \exp\left(\frac{2.5 Y_w}{1 - C_3 Y_w} \right) \tag{6}$$

$$\rho = Y_w \rho_w + (1 - Y_w)(\rho_{crude} + C_{dn} F) \tag{7}$$

where μ is the emulsion viscosity (mPa s) and μ_0 is the crude oil viscosity (mPa s), C_4 is an oil dependent constant (1 for gasoline, light diesel; 10 for crude oil), C_3 is the mousse viscosity constant, ~ 0.65; ρ is the density of emulsified oil (kg/m^3), ρ_{crude} is the density of crude oil (kg/m^3), ρ_w is the density of seawater (kg/m^3), and C_{dn} is a constant based on density-composition data,.

2.3 Oil Transfer

Spreading: In OIL-RW, oil spreading on the sea surface is represented by enhancing the horizontal diffusion for the surface particles. A spreading diffusivity is derived on the basis of the thick slick formulation of Mackay *et al* (1980). Each model particle undergoes additional random walk diffusion given by:

$$ D_{sp} = \frac{K_1 A^{0.33}}{8\pi} \left(\frac{V}{A}\right)^{1.33} \tag{8}$$

where D_{sp} is the spreading diffusivity (m^{2}s^{-1}), A is the surface area of the oil slick (m^2), V is the volume of the oil slick (m^3), and K_1 is a constant with a default value of 150 s^{-1} (Mackay et al 1980). This specification of particle displacements causes patches of model particles to simulate the gravitational spreading of oil slicks.

Dispersion: The dispersion of oil from the sea surface into the water column is calculated using the method of Delvigne and Sweeney (1988). The dispersion rate is expressed as:

$$ Q_d = C_0 \cdot D_d^{0.57} S_{cov} \cdot F_{wc} \cdot d_0^{0.7} \Delta d \tag{9}$$

where Q_d is the dispersion rate of oil droplets with diameters in the range Δd centred on d_0 (kg/m^2 s), C_0 is an empirical dispersion coefficient dependent on the oil type and weathered state, D_d is the dissipated breaking wave energy per unit surface area (J/m^2): $D_d = 0.0034\ \rho_w \cdot g \cdot H_{rms}^2$ (H_{rms} is the rms wave height (m), ρ_w is the density of water (kg/m^3), g is the acceleration due to gravity (ms^{-2})), S_{cov} is the fraction of the sea surface covered by oil ($0 < S_{cov} < 1$), F_{wc} is the fraction of the sea surface hit by breaking waves per unit time (s^{-1}): $F_{wc} = 0.032\ (U_{10} - U_T)/T_w$ (U_T is the threshold wind speed for wave breaking (5ms^{-1}), T_w is the breaking wave period (s)), and d_0 is the oil droplet diameter (m, the oil droplet size range is typically 10-500μm). Marine observations indicate that breaking wave events rapidly distribute the oil in a near-surface zone, typically restricted to a depth: $Z_m = (1.5 \pm 0.35)H_b$ (H_b is the breaking wave height (m)).

OIL-RW assigns each model particle a droplet diameter in the range 10-500μm at random. The probability of a model particle dispersing from the sea surface into the water column is calculated using eqution (9) and the process is applied stochastically. The dispersion coefficient C_0 is calculated in OIL-RW using formulae based on the experimental data of Delvigne and Hulsen (1994): when μ is less than 100cP, $C_0 = 1500$; whilst when $\mu \geq 100$cP, $C_0 = 1500 \times 100/\mu$.

Buoyancy: Vertical particle movements result from buoyancy effects, as well as random turbulence caused by tidal flow and wind-driven currents. Buoyancy is a function of oil droplet diameter and density. The vertical buoyant velocity is given by Stoke's law.

Stranding: When oil spills occur during onshore winds, oil may impinge on nearby shorelines. However, field observations of large spills indicate that the capability of beaches to hold oil is limited. Once the shoreline capacity is reached, oil will be exposed to longshore transport processes. According to Humphrey et al (1993), the maximum capacity of a beach for oil can be expressed as:

$$ C_{max} = L \cdot W \cdot D \cdot \eta_{eff} \tag{10}$$

where C_{max} is the maximum capacity of a beach for stranding (m^3), L, W and D are the length, width, and depth of sediments on the beach respectively (m), and η_{eff} is the effective porosity of the sediments on the beach (0.12~0.46).

3   RESULTS AND DISCUSSION

3.1 Sensitivity Studies

As the oil properties summarised in Section 2 determine the distribution of oil between different phases, it is necessary to validate the model parameters using experimental data. Buchanan and Hurford (1987) have reported measurements of an experimental spill of 20 tons of crude oil (mainly Forties Crude) at 52°10′N, 2°23′E in the North Sea between 14 and 17 July 1987. Measurements of the wind conditions and oil evaporative loss, viscosity, water content and density were performed over a 75-hour period.

Figure1 shows the effects of changes in the value of the model's emulsification constant on the evaporation rate, water content and viscosity of spilled oil, together with comparisons with the experimental data. The figure demonstrates that the value of the emulsification constant, K_A, has little effect on the evaporation rate of oil, but has significant effects on the water content and viscosity. A value of 1×10^{-6} was selected for use in the model, as this gave optimum agreement between the model results and the observations.

Figure 2 shows the effect of viscosity on the dispersion coefficient. It can be seen that the dispersion coefficient, C_0, can be regarded as constant when the oil viscosity is less than 100cP, but decreases rapidly when the viscosity exceeds 100cP. This indicates that weathering tends to reduce dispersion rates. The algorithm used to calculate C_0 agrees well with the experimental data.

The effects of the oil-dependant constant, C_4, on viscosity are shown in Figure 3, which demonstrates that oil type has significant effects on weathering processes. Viscosity changes with time are larger for crude oils with high C_4 values than for lighter oils.

3.2 Overall mass balance

A model test run was carried out to examine the overall mass balance of a hypothetical oil spill 6km off the coast of south east Dorset, United Kingdom. In the test, $43{,}270 m^3$ of Arabian Heavy Crude were released over a 48-hour period. The wind was constant at $7 ms^{-1}$ from 120°N. Oil release began at high water.

The mass balance as a function of time is shown in Figure 4. About 20% of the oil evaporates rapidly, on first exposure to the ambient water. At the same time, the percentage of entrained oil also increases relatively quickly, but tends to decrease subsequently, due to the increase in viscosity associated with emulsification. The fluctuations in the quantity of oil entrained occur because resurfacing is affected by vertical diffusion, which varies with the changing tidal current speeds. Around 21 hours after the start of the spill, part of the oil reaches the shoreline and becomes stranded. The quantity of stranded oil increases with time until the shoreline becomes saturated.

The simulated slick locations at intervals over a two-day period are shown in Figure 5. Initially, the size of the slick increases, and it moves periodically with the tidal current. After the cessation of the spill at 48

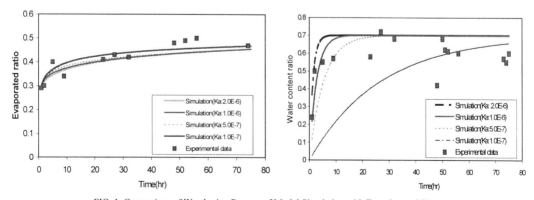

FIG. 1. Comparison of Weathering Process of Model Simulation with Experimental Data

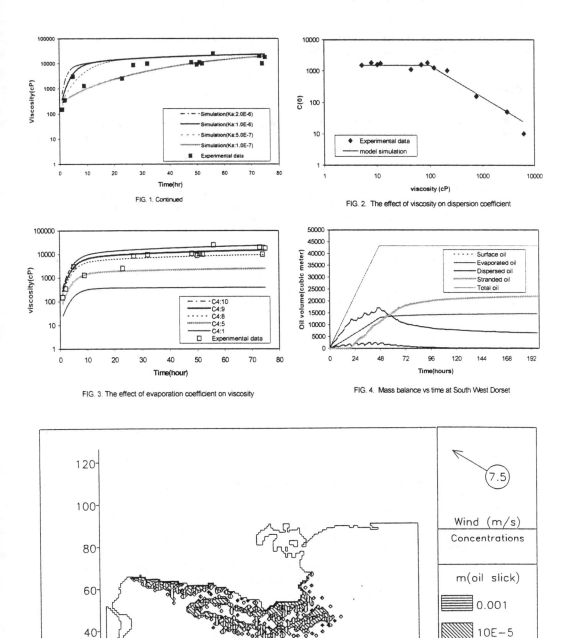

FIG. 1. Continued

FIG. 2. The effect of viscosity on dispersion coefficient

FIG. 3. The effect of evaporation coefficient on viscosity

FIG. 4. Mass balance vs time at South West Dorset

FIG. 5. Trajectories of spilled oil of two days after spilling

hours, the proportions of surface and entrained oil decrease as increasing quantities of oil become stranded on the shoreline north west of the source under the influence of the imposed wind.

4 CONCLUSION

A comprehensive oil spill model has been developed, primarily for use in contingency planning. The model combines oil spill trajectory computations, based on the random walk technique, with weathering calculations. As well as simulating surface oil movement, it includes vertical exchanges and partitions between various phases.

The oil spill model has been validated by comparison with observational data. Evaporation, emulsification and weathering are simulated well. A functional form was selected for the dispersion coefficient, which resulted in good agreement between the model and experimental data.

5 ACKNOWLEDGEMENTS

This work was carried out as part of a placement under the Academic Links with China Scheme, supported by the British Council. The authors would like to thank W J Foley and A A Adewuyi for their assistance during the development of the model.

6 REFERENCES

ASCE Task Committee on modelling of oil spills of the water resources engineering division, 1996. "State-of-the-art review of modelling transport and fate of oil spills", *J. of Hydraulic Engineering,* November. pp. 594-609

Buchanan, I., N. Hurford, 1987. *Results of the forties fate trial, July 1987.* Stevenage: Warren Spring Laboratory. Report No. LR 671 (OP)M

Delvigne, G.A.L. and C.E. Sweeney, 1988. "Natural dispersion of oil". *Oil and Chemical Pollution* 4: 281-310

Delvigne, G.A.L. and L.J.M. Hulsen, 1994. "Simplified laboratory measurement of oil dispersion coefficient-Application in computations of natural oil dispersion", *Proc., 17th arctic marine oilspill program tech.* Seminar, Environment Canada, Ottawa, Ontario, 173-187

Humphery, B., E. Owens, and G. Sergy, 1993. "Development of a stranded oil in course sediment (SOCS) model", *Proc, 1993 oil spill conf.,* Am. Petr. Inst., Washington, D.C. 573-582

Leech, M. V., 1992. *Development of an oil and chemical spill simulation model for the north west European continental shelf and the western Mediterranean/Adriatic Seas-Final report for the European Commission.* LR 810 (MPBM), Warren Spring Laboratory

Mackay, D., S. Paterson and K. Trudel, 1980. *A mathematical model of oil spill behaviour.* University of Toronto, Prepared for Environmental Protection Service, Fisheries and Environment, Canada.

Mead, C. T., 1991. "Random Walk simulations of the dispersal of sewage effluent and dredged spoil". *Proceedings of the 10th Australasian conference on coastal and ocean engineering,* Auckland, New Zealand, pp477-480

Spaulding, M. L., E Howlett, E Anderson and K Jayko, 1992. "OILMAP: A Global Approach to Spill Modelling". *Technical Seminar, Edmonton,* Canada, June 9-11

Stiver, W. and D. Mackay, 1984. "Evaporation rates of spills of hydrocarbons and petroleum mixtures". *Environmental Science and Technology,* V 18, PP834-840

Environmental Hydraulics, Lee, Jayawardena & Wang (eds) © 1999 Balkema, Rotterdam, ISBN 90 5809 035 3

Modelling of sediment plumes caused by dredging operations

P.A. Mackinnon
Department of Civil Engineering, The Queen's University of Belfast, UK

Y. Chen
University of Bradford, UK

G. Thompson
Binnie Black and Veatch, Redhill, UK

ABSTRACT: While attempting to improve their efficiency in abstracting marine sediments, trailing suction hopper ('trailer') dredgers release large quantities of fine sediment into the waters in which they operate. This procedure causes a substantial local increase in suspended sediment concentration, particularly in the area close the water surface, where visible plumes are created. If operations are carried out without adequate control, these plumes can result in breaches of water quality standards and the associated detrimental impacts on marine life, on commercial and industrial uses (such as fisheries and cooling water intakes), and the aesthetic quality of the water. This paper documents the development and application of two mathematical models, an advection-dispersion model and a surface plume model, which were designed to predict the impact of trailer dredgers operating in a marine environment. The models were first used in an assessment of dredging to extract marine sand in Hong Kong's coastal waters.

1. INTRODUCTION

In optimising the abstraction of marine sediments, trailing suction hopper ('trailer') dredgers release large quantities of fine sediment into the waters in which they operate. The process is known as 'overflowing' and is carried out deliberately in order to maximise the volume of material loaded onto the vessel during each journey from the dredging site to the point of offloading. Whilst the procedure improves the efficiency of operation, it also causes a substantial local increase in suspended sediment concentration in the water and can therefore generate a significant impact on the surrounding environment. As the discharge is usually at or close to the water surface, the sediment discharged forms a visible plume. Fig. 1 shows the surface plume generated by a trailer dredger extracting marine sediment in Hong Kong's coastal waters.

In order to minimise the risk of the adverse environmental effects that can occur, developers and water quality regulators are making increased use of mathematical models to predict the environmental impacts of dredging operations. However, the physical processes that take place during overflow differ from the natural sequence of erosion, transport and deposition, therefore standard sediment transport models would normally be inadequate for this particular application.

FIG. 1. Surface plume generated by trailer dredger during 'overflow'

This paper documents the development and application of mathematical models for assessing the impact of trailer dredgers operation in a marine environment. Two models are described:

• a two-layer advection-dispersion model, which was used to assess the far field effects of dredging over a prolonged period;
• a surface plume model, which was used to predict transient local effects each time the dredger completed its dredging cycle.

The first of these models is based on the Bradford University 'DIVAST' software, but has been adapted in order to simulate the specific sediment transport processes that take place when sediment is released near the water surface. The second model uses an analytical plume development simulation to predict the peak sediment concentrations in the immediate vicinity of the dredger and to examine the potential impact of wind on plume movement during the early stages of dispersion. The use of two separate models is related to the specific processes which take place during overflow. Neither model was capable of simulating all of the key processes and therefore the combined use of two models was preferred. The most significant sedimentation processes generated by the operation of a trailer dredger are described in the Section 2 below.

2. PROCESSES ENCOUNTERED DURING DREDGER OVERFLOW

Observations of the effects caused by trailer dredgers show that, even with a keel discharge instead of a surface overflow, a significant quantity of sediment is carried to the surface by air entrained in the discharge. Immediately after release, sediment concentrations near the water surface are relatively high, resulting in a clearly visible plume in the wake of the dredger. However, the quantity of sediment in the surface plume tends to diminish rapidly, due to dispersion and gravitational effects.

There is evidence to suggest that the vast majority (up to 90%) of sediment released by the dredger falls directly to the sea bed at a rate which is much higher that the settling velocity of the individual particles. This is caused by the formation of a 'density current'. When the sediment reaches the sea bed, it forms an unconsolidated 'fluid mud' layer, in which the particles are much more susceptible to erosion than similar particles in undisturbed bed sediments.

The relatively small percentage of sediment that is not drawn into the density current remains in suspension, but sediment concentrations close the water surface reduce rapidly as the particles descend under the influence of gravity and are distributed through the water column due to turbulent mixing. Observers of plumes generated during extensive dredging operations in Hong Kong in the early 1990s indicate that, within one hour of the dredger passing a specific location, the suspended sediment concentration close to the water surface at that location had typically returned to a near-ambient value. Any sediment that remains in suspension is dispersed under the influence of tidal and oceanic currents. If the currents are sufficiently weak, this material will eventually settle on the sea bed. However, due to the transport that takes place in the interim period, suspended material may be deposited at significant distance from the dredging site.

To summarise, the main effects of a sediment overflow from a trailer dredger are: the generation of short term local surface plumes; the creation of an unconsolidated fluid mud layer in the dredged area; the production of increased concentrations of suspended sediment throughout the water column; and the deposition of sediment at locations remote from the dredging site.

3. ADVECTION-DISPERSION MODEL

3.1 Background Information

The first of two models employed to assess the impacts of dredging was based on the Bradford University 'DIVAST' (Depth Integrated Velocity And Solute Transport) software (Falconer and Chen, 1996). The original DIVAST software permits two-dimensional advection-dispersion modelling. It relies on the division of the flow domain into individual cells and solves the equations of motion using the finite difference method.

3.2 Adaptation of the Advection-Dispersion Model

As the first application of the model was the assessment of dredging operations in Hong Kong's coastal waters, the software (which is normally used to model flow which does not vary with depth) was amended to incorporate layered flow, a phenomenon encountered in this region when the nearby Pearl River is in flood. For this particular application, the model was adapted to simulate the specific sediment transport processes that take place when sediment is released near the water surface and the exchange of sediment particles between different layers of water.

The amended DIVAST model was used to quantify three of the four main impacts identified at the end of Section 2. Due to the nature of the model, it was unsuitable for predicting the characteristics of surface plumes.

3.3 Equations Governing Sediment Behaviour

The equations governing sediment transport and dispersion were the standard DIVAST pollutant transport and dispersion equations. Based on methods used in previous related work (Binnie & Partners, 1989a), the processes of erosion and deposition were defined as follows:

$$\text{Deposition} \qquad \frac{dm}{dt} = w_s c \left(1 - \frac{\tau_b}{\tau_d}\right) \text{ when } \tau_b \leq \tau_d \qquad (1)$$

$$\text{Erosion} \qquad \frac{dm}{dt} = M\left(\tau_b - \tau_e\right) \text{ when } \tau_b \geq \tau_e \qquad (2)$$

where
m = mass of sediment (kg)
t = time (s)
c = concentration of suspended sediment (kg/m^3) } values computed within model
τ_b = bed shear stress (N/m^2) }
τ_d = critical stress for deposition (N/m^2)
τ_e = critical stress for erosion (N/m^2)
M = empirical erosion constant (kg/N/s)
w_s = settling velocity (m/s)

Deposition of suspended sediment: Within the model, deposition of suspended sediment normally occurs from the lower layer of water (the 'bed' layer), but in areas where the overall depth is too shallow for two distinct layers to exist, deposition may occur from the upper (or 'surface') layer.

Erosion of fluid mud: For sediment particles such as those in the unconsolidated fluid mud layer, the relationship between the erosion shear stress and the dry density may be written as (Delo 1988a; Delo 1988b):

$$\tau_e = 0.0005 \, \rho_h^{1.4} \qquad (3)$$

where ρ_h = dry density (kg/m^3)

When the shear stress exceeds τ_e, the fluid mud particles at the sea bed are re-suspended and they behave in a manner identical to all other suspended material.

Re-erosion of material deposited after suspension: Particles resting on the sea bed following suspension are assumed to have a significantly higher erosion shear stress than the particles within the fluid mud layer. The shear stress values used in a specific dredging study are provided later in this document.

3.4 Model Input

Sediment load: Within the model, the mass of sediment, m, generated as a result of dredger overflow is specified in two portions. The first of these is the 'bed load' or the quantity of sediment which returns directly to the sea bed due to the density current. The bed load is therefore used to define the location and quantity of fluid mud generated as a result of dredging. The load is specified as a source of sediment in each model cell within the dredged area. The sediment within each cell is continuously replenished during the period when dredging is under way. For the purpose of modelling, the bed load is assumed to comprise 90% of the total sediment released.

The second portion of the released sediment is the 'suspended load' or the relatively small proportion of material that remains in the water column, dispersing under the influence of currents before deposition. This is normally specified as a continuous load in each affected cell, but intermittent loading may be used if required. During modelling, the suspended load is quantified as 10% of the total sediment released.

4. SURFACE PLUME MODEL

4.1 Background information

The PLUME surface plume model was developed to provide a means for examining the transient conditions of high sediment concentration which occur in the vicinity of the dredger immediately after dredging takes place. The advection of the plume is based on particle paths generated by the advection-dispersion model. Using Brooks' method, in which longitudinal dispersion is assumed to be insignificant, the lateral dispersion of sediment is computed. The model includes the settlement of particles due to gravity, and thus accounts for the rapid decay of the surface sediment plume. By incorporating the effect of wind on the path of the plume, an estimate of the effect of wind on plume dispersion may be made.

4.2 Principles of the Plume Model

The plume model assumes a semi-infinite space. The discharge is represented as a cross-sectional source, the area of which is determined largely by the physical characteristics of the sediment release. The model is used to calculate the progressive expansion of the plume as it travels away from the discharge point.

The governing equations used in the plume model include the equation for the continuity of mass and the equation for pollutant transport. The transport processes modelled include the advection due to movement of the water body, the turbulent diffusion and the shear dispersion perpendicular to the path of the plume. The dispersion-diffusion coefficient is computed on the basis of eddy viscosity, shear velocity and the depth of water. Options are available for using constant dispersion or a value which is related to the plume width. The formulae for the sediment concentration on the plume centre line, S_m, and the plume width, B_i, at time t_i are as follows:

$$S_m(x_i, y_i, t_i) = S_0 \exp(-kt_i)\,\mathrm{erf}\left(\frac{b}{2\sigma_y}\right) \tag{4}$$

$$\frac{B_i}{b} = \left\{\left[1 + \frac{24}{b^2}\Sigma_{j=1}^{i-1}Dy_j\tau_j\right]^{1-\frac{m}{2}} + \frac{12(2-m)}{b^2}Dy_i\Delta t_i\right\}^{\frac{1}{2-m}} \tag{5}$$

where
i	time increment ($i=1,2,3,\ldots$)
S_0	initial sediment concentration in plume at $x=y=0$
k	rate of reduction of sediment concentration due to particle settling
b	initial plume width
t_i	total time elapsed since the start of the simulation
Δt_i	time step at $t=t_i$
x_i	co-ordinate of plume centre line in the x-direction
y_i	co-ordinate of plume centre line in the y-direction
u_i	plume velocity in the x-direction at time t_i
v_i	plume velocity in the y-direction at time t_i
$Dy_i{}^*$	$=\alpha U_* H_i$, the dispersion-diffusion coefficient
U_*	shear velocity
H_i	total water depth
m	coefficient for constant, linear or other dispersion-diffusion relationship with plume width
C_i	Chezy coefficient ($=H_i^{1/6}/n$)
n	Manning's roughness coefficient
α	eddy viscosity coefficient
τ_j	variable timestep

$$\sigma_y = \sqrt{4\Sigma_{j=1}^{i}Dy_j\Delta t_j} \tag{6}$$

$$Dy_i\tau_i = \left\{[1 + \Sigma_{j=1}^{i-1}Dy_j\tau_j]^{1-\frac{m}{2}} + \frac{12(2-m)}{b^2}Dy_i\Delta t_i\right\}^{\frac{2}{2-m}} - 1 - \Sigma_{j=1}^{i-1}Dy_j\tau_j \tag{7}$$

4.3 Model Input

The co-ordinates, x_i and y_i, of the plume centre line are derived from output from the advection-dispersion model. Values for the initial sediment concentration, S_0, the rate of reduction of sediment concentration due to particle settling, k, the initial plume width, b, and the total water depth, H_i, must be specified as input for each particular test. The coefficient, m, which defines the relationship between dispersion-diffusion and plume width, must also be specified. For applications in coastal waters, the Manning's roughness coefficient, n, is generally assumed to be around 0.022 and the eddy viscosity coefficient, α, is set at 0.6 (Fischer, 1979). Examples of the other values used in an application of the model are provided in Section 5 of this document. All of the other parameters listed in Section 4.2 are computed within the model.

5. APPLICATION OF MODELS

5.1 Use of Models to Assess Impacts of Dredging in Hong Kong

The first application of the models was in the assessment of the impact of dredging operations in Hong Kong. The models were used to indicate the extent, magnitude and duration of impacts from different dredging scenarios in specific offshore areas.

5.2 Advection-Dispersion Model

The mass of sediment released was calculated using data compiled during from previous similar operations and provided by the client's dredging consultants. The sediment erosion and deposition characteristics were based on values observed during previous dredging operations in Hong Kong's coastal waters. A critical shear stress, τ_d, of 0.1 N/m^2 was used as the threshold for deposition of suspended material (Binnie & Partners, 1989a).

A dry density, ρ_h, of 20-25 kg/m^3 (Binnie & Partners, 1989b) was assumed for the sediment released during overflow. On this basis and using Equation (3), a value of 0.04 N/m^2 was calculated for the minimum shear stress, τ_e, for the erosion of unconsolidated fluid mud. The results of modelling showed that, with this relatively low minimum shear stress, the fluid mud generated by the density current was readily re-suspended. A shear stress, τ_e, of 0.3 N/m^2 was assumed as the threshold for re-erosion of material deposited on the sea bed as opposed to the fluid mud layer (Binnie & Partners, 1989a).

Based on information from previous studies in the same region (Binnie & Partners, 1989a), a value of 0.0015 Kg/N/s was adopted for the empirical constant, M, in the erosion equation. This value was used for erosion of both fluid mud and other sediment deposits arising from dredging operations. Using overflow sediment particle diameters between 0.013mm and 0.063mm (provided by the dredging consultants), and using a recognised procedure (Vanoni, 1977), values of 0.002 to 0.038 m/s were calculated for the particle settling velocity, w_s.

5.3 Plume Model

For surface plume modelling, an initial plume width, b, of 50m was used. The total water depth was based on the typical water depth at the particular dredging area. The rate of reduction in sediment concentration in the surface plume due to particle settling, k, was set at values between 0.35 and 0.40 per hour, depending on the characteristics of the material being dredged.

5.4 Scenario Tests

The different scenarios investigated included:

- dredging in two different designated 'marine borrow areas';
- simultaneous operation of up to four dredgers;
- different dredging, transport and off-loading sequences;
- abstraction of material with different settlement characteristics;
- operation during different seasons when the water is mixed or layered and when different oceanic currents dominate flow patterns;
- the effect of different wind conditions on the movement of surface plumes.

The results from scenario tests were presented in four separate forms: concentration and distribution of sediment temporarily suspended within the surface plume (top 2m); concentration of sediment suspended in the surface layer of water (top 8m approximately); concentration of sediment suspended in the lower layer of water (below 8m depth); mass of sediment deposited on the sea bed. Typical results are shown in Fig. 2.
The results were compared with the Water Quality Objectives for the dredging area and the adjacent Water Control Zones (as defined by Hong Kong Government Environmental Protection Department, 1991). The

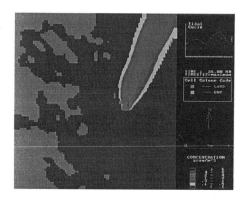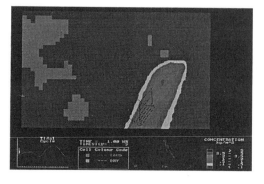

FIG. 2. Typical results from scenario tests using the models

output allowed an assessment of the likely impacts on designated mariculture zones and bathing areas, coral reefs and water intakes.

6. CONCLUSIONS

Through amendments to recognised hydrodynamic and water quality models, techniques have been established for modelling the specific sediment transport processes that occur when a trailer dredger operating in coastal waters releases fine sediment close to the water surface. By using two different models, it was possible to predict short- and long-term changes in suspended sediment concentrations. The results were used in an assessment of the impacts of trailer dredger operations on the marine environment.

7. ACKNOWLEDGEMENTS

This paper is published with the permission of the Director of Civil Engineering, Government of the Hong Kong Special Administrative Region.

8. REFERENCES

Binnie & Partners (Hong Kong) Ltd (1989a). "Two layer Mathematical Model Simulation of Mud and Particulate Effluent Transport Calibration and Validation", *Hydraulic and Water Quality Studies in Victoria Harbour*, Mathematical Model Report, Part II, Volume 9A.

Binnie & Partners (Hong Kong) Ltd. (1989b). "Properties of Hong Kong Mud", *Hydraulic and Water Quality Studies in Victoria Harbour*, Mathematical Model Report, Part II, Volume 9B.

Delo, E.A. (1988a). "The behaviour of estuarine muds during tidal cycles", Hydraulics Research Ltd, Report SR 138.

Delo, E.A. (1988b). "Siltation and stability of cohesive dredged slopes manual", Hydraulics Research Ltd, Report SR 180.

Falconer, R. A. and Chen, Y. (1996). "Modelling sediment transport and water quality processes on tidal floodplains", *Floodplain Processes*, Chapter 11, pp. 361-398, John Wiley & Sons Ltd., Chichester, England,.

Fischer, H.B., List, E.J., Koh, R.C.Y., Imberger J., Brooks, N.H. (1979). *Mixing in inland and coastal waters*, Academic Press, New York.

Hong Kong Government Environmental Protection Department, (1991). *Marine Water Quality in Hong Kong - 1991.*

Vanoni, V.A. (1977). *Sedimentation Engineering*, ASCE, New York.

Environmental Hydraulics, Lee, Jayawardena & Wang (eds)© 1999 Balkema, Rotterdam, ISBN 90 5809 035 3

Three-dimensional mathematical modeling of pollutant transport in stratified estuaries

Der-Liang Young, Bor-Chou Her & Yeng-Fung Wang
Department of Civil Engineering and Hydraulic Research Laboratory, National Taiwan University, Taipei, Taiwan, China

ABSTRACT: This study, in the flow computation, uses a three-dimensional shallow water BEM(Boundary Element Method) model. However, to investigate the temperature distribution, we adopt a three-dimensional advection-diffusion ELBEM(Eulerian-Lagrangian Boundary Element Method) model. The advection part of ELBEM model is treated by the concept of Eulerian-Lagrangian scheme, which overcomes the limitation of traditional BEM scheme which is incapable to deal with the arbitrary velocity field. The linkage of three-dimensional shallow water BEM model and advection-diffusion ELBEM model enables us to deal with the fluctuation of free surface and density stratification in a stratified estuary.The simulation results are satisfactory through the comparison with analytical solutions and the results from other numerical schemes. Furthermore, this model is applied to simulate conditions which flow field is effected under different boundary conditions such as surface wind shear or heat flux. The simulation results, as comparing with those of homogeneous flow field, show the phenomenon of overturn under the condition of unstable density stratification. As far as stably stratified flow fields are concerned, the phenomena of multiple circulating gyres and the thermocline are also captured.

1.INTRODUCTION

With the advent of computer technology, such as the efficient computation of parallel computing and high-performance computers of main frames, work stations or even PC, the numerical modelling of large scale three-dimensional flow fields becomes more feasible nowadays. The purpose of this study is to establish a three-dimensional boundary element model to analyze the pollutant transport in the stratified estuaries, bays, or coastal waters. Under the assumption of shallow water theory, an Ekman boundary layer model is developed with the consideration of dersity stratification due to temperature and salinity or even the pollutant concentrations. The wind shear effect will also be taken into account as far as flow mixing mechanism is concerned.

We have successfully developed a 3-D boundary element model to investigate the pollutant transport in a homogeneous (non-stratified) estuary, Young and Wang (1995a), Young and Wang (1995b), Young and Wang (1995c). With the framework of the previous studies, we are able to extend the homogeneous bodies of water into the more complex and realistic stratified basins. However, with the consideration of stratified effects, the phenomena of circulating gyres and the associated formation, maintenance of thermocline in a stably stratified environment, as well as the overturn and the hydrodynamic instability in a unstably stratified estuary will highly enhance the degree of difficulty of numerical simulations.

2.GOVERNING EQUATIONS

Under the assumptions of (1) shallow water approximation, that is, the vertical scale is much less then the horizontal scale ,and hydrostatic pressure distribution in vertical direction, (2) small Rossby number, (3) ignoring the horizontal diffusion in the momentum equations, (4) constant eddy viscosity and eddy diffusivity, (5) constant Coriolis parameter, and (6) Boussinesq approximation, we obtain the following governing equations

$$\frac{\partial u}{\partial x} + \frac{\partial v}{\partial y} + \frac{\partial w}{\partial z} = 0 \tag{1}$$

$$\frac{\partial u}{\partial t} - fv = -\frac{1}{\rho_0}\frac{\partial p}{\partial x} - \frac{1}{\rho_0}\frac{\partial p'}{\partial x} + \upsilon\frac{\partial^2 u}{\partial z^2} \tag{2}$$

$$\frac{\partial v}{\partial t} + fu = -\frac{1}{\rho_0}\frac{\partial p}{\partial y} - \frac{1}{\rho_0}\frac{\partial p'}{\partial y} + \upsilon\frac{\partial^2 v}{\partial z^2} \tag{3}$$

$$0 = -\frac{1}{\rho}(\frac{\partial p}{\partial z} + \frac{\partial p'}{\partial z}) - g \tag{4}$$

$$\rho = \rho_0 + \rho', \rho' << \rho \tag{5}$$

$$\frac{\partial T}{\partial t} + u\frac{\partial T}{\partial x} + v\frac{\partial T}{\partial y} + w\frac{\partial T}{\partial z} = \frac{\partial}{\partial x}(K_x^T\frac{\partial T}{\partial x}) + \frac{\partial}{\partial y}(K_y^T\frac{\partial T}{\partial y}) + \frac{\partial}{\partial z}(K_z^T\frac{\partial T}{\partial z}) + S^T(x,y,z,t) \tag{6}$$

$$\rho = \rho_0\{1 - [7(T - T_0)^2 - 750\phi] \times 10^{-6}\} \tag{7}$$

Herein, υ =eddy viscosity, $f = 2\omega s \sin\theta$ =Coriolis parameter, ω =angular speed of earth rotation, θ =the latitude, g =gravitational acceleration, ρ =density, ρ' =density deviation form the standard water density, ρ_0 =water density at $4°c$ =$999.972\,Kg/m^3$, T =temperature in degree Celsius, p' =pressure variation due to ρ', ϕ =salinity , in parts per thousand, K_x^T, K_y^T, K_z^T =eddy diffsivity for temperature.

3. NUMERICAL MODEL

After the non-dimensional procedure, and the discretization of difference scheme in time, together with the boundary element method, we obtain the following boundary integral equation for the hydrodynamic equations for the flow field.

$$u^{n+1}(\vec{x}) = \int_{\Omega_0}\left\{\hat{u}_x\left[-\left(\frac{\partial\eta}{\partial x}\right)^{n+1} - \left(\frac{\partial p'}{\partial x}\right)^{n+1} + Su^n\right] + \hat{v}_x\left[-\left(\frac{\partial\eta}{\partial y}\right)^{n+1} - \left(\frac{\partial p'}{\partial x}\right)^{n+1} + Sv^n\right]\right\}d\Omega_0$$

$$- \int_{\Gamma_0}\left\{u^{n+1}\left[E_k\frac{\partial\hat{u}_k}{\partial z}\vec{k}\right]\cdot\vec{n} + v^{n+1}\left[E_k\frac{\partial\hat{v}_y}{\partial z}\vec{k}\right]\cdot\vec{n}\right\}d\Gamma_0 + \int_{\Gamma_0}\left\{\hat{u}_x\left[E_k\frac{\partial u^{n+1}}{\partial z}\vec{k}\right]\cdot\vec{n} + \hat{v}_x\left[E_K\frac{\partial v^{n+1}}{\partial z}\vec{k}\right]\cdot\vec{n}\right\}d\Gamma_0 \tag{8}$$

$$v^{n+1}(\vec{x}) = \int_{\Omega_0}\left\{\hat{u}_y\left[-\left(\frac{\partial\eta}{\partial x}\right)^{n+1} - \left(\frac{\partial p'}{\partial x}\right)^{n+1} + Su^n\right] + \hat{v}_y\left[-\left(\frac{\partial\eta}{\partial y}\right)^{n+1} - \left(\frac{\partial p'}{\partial x}\right)^{n+1} + Sv^n\right]\right\}d\Omega_0$$

$$- \int_{\Omega_0}\left\{u^{n+1}\left[E_k\frac{\partial\hat{u}_k}{\partial z}\vec{k}\right]\cdot\vec{n} + v^{n+1}\left[E_k\frac{\partial\hat{v}_y}{\partial z}\vec{k}\right]\cdot\vec{n}\right\}d\Gamma_0 + \int_{\Gamma_0}\left\{\hat{u}_y\left[E_k\frac{\partial u^{n+1}}{\partial z}\vec{k}\right]\cdot\vec{n} + \hat{v}_y\left[E_K\frac{\partial v^{N+1}}{\partial z}\vec{k}\right]\cdot\vec{n}\right\}d\Gamma_0 \tag{9}$$

where

$$p' = \int_z^\eta \rho^* \, dz \quad , \quad \rho^* = \frac{\rho'}{\rho_0}$$

After the introduction of Eulerian-Lagrangian boundary element method (ELBEM), the corresponding non-dimensional-diffusion equation for temperature is written as follows:

$$\alpha(\bar{x}_i, t^{n+1}) T(\bar{x}_i, t^{n+1}) = \int_{\Omega_E} \hat{T}(\bar{x}_i, t^{n+1}; \bar{x}, t^n) T(\bar{x}, t^n) d\Omega_E - \int_{t^n}^{t^{n+1}} \int_{\Gamma_E} \hat{T}(\bar{x}_i, t^n; \bar{x}, t) q(\bar{x}, t) d\Gamma_E dt + \int_{t^n}^{t^{n+1}} \int_{\Gamma_E} \hat{q}(\bar{x}_i, t^n; \bar{x}, t) T(\bar{x}, t) d\Gamma_E dt$$

(10)

Furthermore, we use the constant element to deal with the temperature domain, and obtain the following equation

$$\alpha(\bar{x}_i, t^{n+1}) T(\bar{x}_i, t^{n+1}) = \int_{\Omega_E} \hat{T}(\bar{x}_i, t^{n+1}; \bar{x}, t^n) T(\bar{x}, t^n) d\Omega_E - \int_{\Gamma_E} \hat{T}^*(\bar{x}_i, t^{n+1}; \bar{x}, t^n) q(\bar{x}, t^n) d\Gamma_e + \int_{\Gamma_E} \hat{q}^*(\bar{x}_i, t^{n+1}; \bar{x}, t^n) T(\bar{x}, t^n) d\Gamma_E$$

(11)

where,

$$\hat{T}^* = \frac{1}{[4\pi K r]} erfc(\eta^*)$$

(12)

$$\hat{q}^* = \frac{1}{[4\pi r^2]} \frac{\partial r}{\partial n} \left[-\frac{2\eta^*}{\sqrt{\pi}} \exp(-\eta^{*2}) - erfc(\eta^*) \right]$$

(13)

$$\eta^* = \frac{r}{2\sqrt{K(t^{n+1} - t^n)}}$$

(14)

The boundary element methods will not be presented here due to the limitation of space, and the previous papers of the writers are recommended for the details.

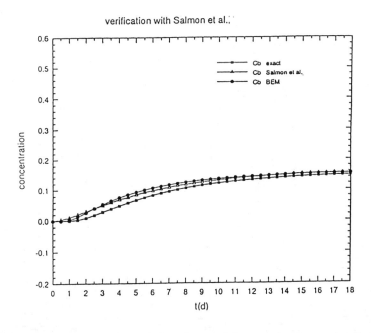

Fig.1:Comparison of Concentration evolution of exact, Salmon et al and present study (BEM)

4. NUMERICAL RESULTS

We first calibrate a simple large and shallow lake to test the performance of the numerical models. A square lake with the geometric sizes of 62.5km × 62.5km × 80m is undertalsen. The horizontal and vertical eddy diffusivities are 20m²/s and 0.02 m²/s respectively. For the concentration of pollutant at the position of (45.9km × 32.3km × 80m), the comparison of the exact solution, Salmon et al (1980) and present model is shown in Fig.1. The three results are all very agreeable.

We now consider another case study as performed by Davies (1983). This is a rectangular shallow basin with the dimensions of 400km × 400km × 65m. The following physical parameters are chosen: $\nu = 650 \, \text{cm}^2/\text{s}$,f=0.00012222 rad/s , $\tau_{yy} = -14.6341463 \, \text{cm}^2/\text{s}$ =shear stress in the y-direction, $\Delta t = 409.1s$. As far as the boundary conditions are concerned, except there is heat transfer on the air-sea interface, the other sides are all insulated. For the non-stratified (homogeneous) lake, the velocity distributions at the profiles of x=0.4,y=1.0 and z=-0.5 are depicted at Fig.2a for the time-step of 50. The corresponding figures for the time-step of 100 is also described in Fig.2b. it is concluded from the numerical results that for the homogeneous lake, there is only one circulating gyre in the whole water body after the long action of wind shear stress. The flow field is very simple. However, where there are both wind shear and heat transfer on the water surface, the flow and temperature fields are more complicated, as shown in Fig.3. This graph reveals the distribution of density and flow at the time-step of 100. The density stratification as well as the multiple circulating gyres in the lake are very dominant. We can see very clearly there are two circulating gyres and also the phenomenon of overturn the physics of the stratified basin in general is more interesting and more challanging in the modeling of bodies of water. Since the present model is able to treat the constant eddy viscosity and diffusivity only, a quantitative and vigorous comparison of Davies model (1983) is not possible. However, the qualitative features of both models all render fundamental fluid dynamics for the homogeneous and stratified basins.

5. CONCLUSION

We draw the following conclusions:
(1) A three-dimensional pollutant transport model in the stratied bodies of water is established after the

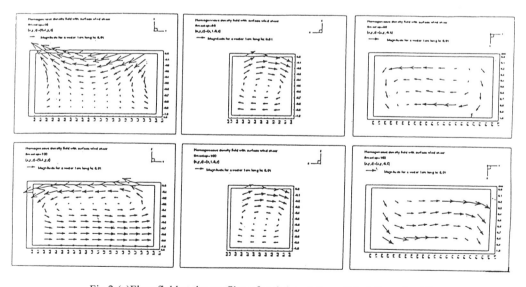

Fig.2:(a)Flow field at the profiles of x=0.4, y=1.0, z=-0.5 at time-step 50

(b)Same as (a) at time-step 100

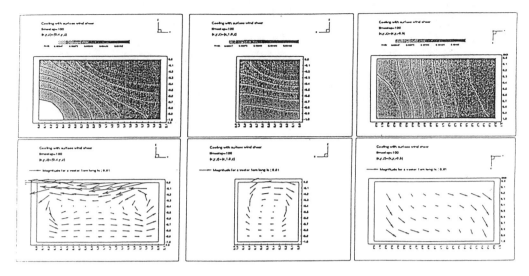

Fig.3:Density (top) and flow (bottom) fields at the profiles of x=0.4, y=1.0, z=-0.5 at time-step 100

calibration and verification of the existing literature. The model consists of two submodels: the flow equation is solved by the three-dimensional shallow water theory using the boundary element method, while the density, temperature, or pollutant concentrations are solved by the three-dimensional advection-diffusion equation using the Eulerian-Lagrangian boundary element method. The overall performance of the present model is satisfactory and comparable to the other numerical models.

(2) The present model is able to handle both the homogeneous and stratified environments under the air-sea interaction at the water surface to consider the mass, momentum and heat transfers. The homogeneous circulation of water is the simplest case, however, we have obtained the physical characteristics of thermocline for stably stratified basins and overturn for the unstably stratified environments.

This study is supported by the NSC 85-2611-E-002-005 grant of the National Science Council of Taiwan, it is greatly acknowledged.

6.REFERENCES

Davies, A.M. (1983) "Formualtion of a Linear Three-Dimensional Hydrodynamic Sea Model Using a A Galerkin-Eignfunction Method", *Int. J. num. Meth. Fluids* ,3,33-60.

Salmon,J.R., Liggett, J.A. and Gallagher,R.H.,(1980). "Dispersion Analysis in Homogeneous Lakes", *Int. J. num. Meth. Engrg*, 15, 1627-1642.

Young,D. L., and Wang, Y. F.,(1995a). "Fundamental Solutions and Boundary Integral Method of 3D Shallow Waters, *Proceedings of the First Asian Computational Fluid Dynamics Conference,* Hong Kong, 623-628.

Young, D. L., and Wang, Y. F.,(1995b). "Boundary Integral Equation Simulations of 3D Shallow Water With Moving Free Surface Boundary", *Proceedings of Hing-Performance Computing(HPC') ASIA* , 1995,Taipei, Taiwan, EN047, 1-16.

Young, D. L.,and Wang,Y. F.,(1995c). "Boundary Element Computations of Transient 3D Shallow Water Circulation," *Water Resources Engineering*, ed by W.H. Espey, Jr. and P.G. Combs, 1,1303-1307.

1.3 Jets and plumes – I

Environmental Hydraulics, Lee, Jayawardena & Wang (eds) © 1999 Balkema, Rotterdam, ISBN 90 5809 035 3

Invited lecture: The effects of wind stress on the spread of a buoyant surface layer

B.C.Wallace
NSW Department of Public Works Hydraulics Laboratory, Manly Vale, N.S.W., Australia

D.L.Wilkinson
Department of Civil Engineering, University of Canterbury, Christchurch, New Zealand

ABSTRACT: The effects of wind stress on the spread and structure of a buoyant surface layer of finite volume are investigated.

1.   INTRODUCTION

Density fronts are found in a variety of geophysical flows, freshwater outflows from estuaries and river mouths, exchange flows in ocean straits and on a much smaller scale urban runoff into harbours may all exhibit strong frontal structures. All of these phenomena can have a significant impact on the local environment and prediction of their behaviour is of some importance. Buoyancy plays a dominating role in the dynamics of these flows. However, the effects of wind stress can significantly affect their motion and structure.

The early work on the spread of buoyant surface plumes subjected to wind stress was largely based on the assumption that the surface plumes were simply advected by the current due to the surface wind stress. Hoult (1972) used this assumption to predict the trajectory of oil spills. More recently, Pearson et al (1992) made similar assumptions to predict the spreading of surface plumes produced from the Sydney deep water outflows. Wilkinson and Wood (1987), in a qualitative laboratory study in which a constant flux of buoyant fluid was released below a free surface subjected to wind stress, found that the simple advection hypothesis was invalid during the early stages of the motion. The presence of wind stress significantly affected the structure and motion of the buoyant layer. The upwind flow was deeper and its motion retarded with respect to the down wind front which was elongated and moved faster than did a front in the absence of wind. Figure 1 shows a photograph from these experiments and the asymmetry of the propagating fronts caused by the effects of wind stress is very apparent.

Figure 1.   Propagation of a surface plume subjected to wind stress (wind is blowing from left to right)

Experiments by Simpson and Britter (1980) in which dense gravity currents advancing into a counter flow were subjected to varying bottom boundary shear produced by a moving floor showed pronounced changes in the steepness of the front with the sign of the shear in the bottom boundary layer. The effects on the shape of the front were exactly similar to those produced by wind stress acting on a spreading buoyant layer at the water surface.

Jirka and Arita (1987) in a study of the structure of arrested density currents used perturbation methods to show that the presence of vorticity in the flow approaching an arrested dense layer determined whether the frontal structure would take the form of a salt water wedge or be similar to that of a propagating gravity current in stationary ambient surroundings. For inviscid flows they found that the internal angle of the flow was $\pi/3$ corresponding to the theoretical model first proposed by von Karman (1940). The imposition of a uniform shear in the approach flow amounting to only 10% of the mean ambient velocity over the thickness of the gravity current resulted in a flattening of the frontal angle by approximately one third. Conversely, when the shear was such that the velocities at the bottom boundary were greater than that of the mean flow, the frontal angle was found to steepen by a similar amount. This is in accord with experimental observations.

In the experiments described in this paper a finite volume of buoyant fluid was released at the surface of immiscible fluid in which there was no mean motion. Gravity caused the buoyant fluid to spread laterally in both directions and in doing so it developed well defined fronts. In the first series of experiments which were conducted without wind, the geometry of the front and its rate of spread were recorded and related to the characterising parameters of the buoyant release.

This was followed by a second series of experiments in which the surface of the ambient fluid was subjected to wind stress and the effect of this on the form and propagation of the front was investigated.

2. EXPERIMENTAL FACILITY

The experiments were performed in a glass walled wind-water tunnel 8 m long, with an internal width of 295 mm and a total depth of 500 mm. The mean water depth was approximately 230 mm during these experiments. The tank could be tilted and this was used to control the longitudinal pressure gradient in the air flow by means of adjusting the divergence between the water surface which remained horizontal, and the roof of the air tunnel which tilted with the flume. The tilt adjustment was employed to obtain a zero pressure gradient in the air flow above the working section of the tunnel. The water in the tank could be recirculated to create a current however this facility was not used in the experiments described here.

The air flow was produced by a suction fan at the upwind end of the tunnel and flow straightners downstream from the fan ensured uniform velocities and low levels of turbulence in the tunnel. The wind speed in the tunnel could be continuously varied from zero to 3 m/s. Figure 2a shows a schematic view of the tank.

The buoyant layers were formed from an aqueous solution of methanol. Sodium fluorescein was added for flow visualisation and sodium chloride was added to act as a tracer for conductivity measurements. The unsteady surface plumes were generated by releasing a reservoir of the buoyant

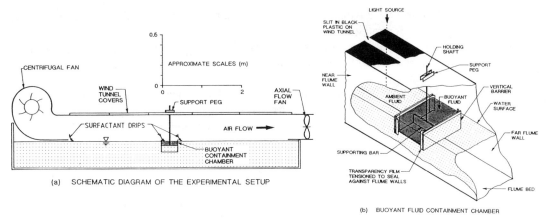

Figure 2. **Schematic diagram of the experimental facility**

fluid which was floating on the denser quiescent fluid below. The containment chamber is shown schematically in Figure 2b. It consisted of two thin vertical barriers which spanned the tunnel; the barriers were supported by a central connecting bar. Sealing strips at the ends of the barriers prevented any leakage of the buoyant fluid prior to the commencement of the experiment. Chambers with lengths of 11.8, 23.5 and 35.5 mm were used to provide a range of length to depth ratios for the initial release. The buoyant fluid was careful floated on top of the quiescent fluid in the chamber and the small amount of mixing which occurred at the interface during filling was removed using a disc siphon. The experiments were commenced by rapidly raising the chamber allowing the buoyant fluid to spread unimpeded.

Flow visualisation was the principle means of recording the experiments. A 35 mm camera equipped with a remotely controlled motor drive was used to record the spread of the buoyant layer. The tunnel was illuminated with a light sheet from above. A pulsed hydrogen bubble system was used to obtain vertical velocity profiles through the spreading buoyant layer. Measurements were taken along the centre-line of the tunnel to minimise the boundary layer effect propagating from the tank walls.

Wind speed profiles were measured using pitot-static tubes. Mean velocities measured along the centre-line of the tunnel and at the quarter points from the sides differed by less than 3%. The shear velocity at the water surface U_* was evaluated to the measured velocity profile in the air flow above the surface. Care was taken to ensure that measurements were taken within the turbulent boundary layer.

Velocity profiles were measured in the wind drift layer using the hydrogen bubble technique. The shear velocity and momentum flux in the drift layer was evaluated by fitting logarithmic curves to these profiles. Momentum fluxes were found to functions of the mean wind speed and the fetch length.

3.   SPREAD OF THE BUOYANT LAYER IN THE ABSENCE OF WIND

The behaviour of two dimensional surface plumes formed when a finite volume of buoyant fluid is released at the surface of a deep and a much larger volume of immiscible fluid may be described in terms of the following parameters: d_0 the initial depth of the buoyant fluid, L_0 initial length of the buoyant fluid, g the effective gravitational acceleration of the buoyant fluid, the kinematic viscosity of the buoyant fluid. It is assumed that the ratio Δ of the density difference to the density of either fluid is small and that differences in density has a negligible effect on inertial response of either fluid. Furthermore, it is assumed that the viscosities of the two fluids are similar.

Rottman & Simpson (1983) conducted similar experiments in which a finite volume of fluid was released at the bottom of a tank into a much larger volume of fluid of slightly lesser density. The experiments were designed to simulate a dense gas spill. The initial depth of the denser fluid was much less than that of the surrounding fluid. The bottom fronts which formed were similar to the surface fronts which are described in this section and were similar to those observed in lock exchange experiments. In the initial phase of motion which they termed the slumping phase, the dominant force balance was that between gravity and inertia and they found that during this phase of the motion the velocity of the front was nearly constant. The non-dimensional length of the layer L/d_o was found to satisfy the relationship:

$$\frac{L}{d_o} \approx t \left(\frac{\Delta g}{d_o} \right)^{1/2} \qquad (1)$$

As the dense layer continued to spread, viscous forces became increasing important. During this buoyancy-viscous phase of the motion the buoyant force F_B scales as: $F_B \approx \Delta \rho g d^2$ where d is the depth of the dense layer.

The viscous force F_V due to the development of the boundary layer at the interface of the buoyant fluid and the ambient fluid scales as: $F_v \approx \rho v \frac{c}{\delta} L$

where c is the celerity of the front and δ is the thickness of the developing boundary layer.

At low Reynolds numbers δ may be approximated by: $\delta \approx (vt)^{1/2}$

Equating these two forces leads to the following expression for the spread of the buoyant layer in the buoyancy viscous regime:

$$L \approx \left(\frac{\Delta g A^2}{v^{1/2}} \right)^{1/4} t^{3/8} \qquad (2)$$

where A is the 2 dimensional volume of the buoyant layer.

The transition between the buoyancy-inertial regime and buoyancy-viscous regime occurs at a length scale:

$$L_t = \left[(\Delta g)^{1/2} \frac{L_0^4 d_0^{5/2}}{v} \right]^{1/5} \qquad (3)$$

and a time scale:

$$t_t = \left[\frac{L_0^4}{(Ag)^2 v} \right]^{1/5} \qquad (4)$$

and these are used to normalise the spread of the buoyant layer in Figure 3.

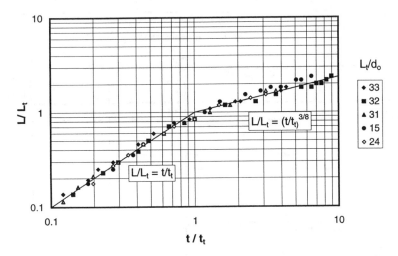

Figure 3. Normalised spread of the buoyant layer as a function of time

Experimental data showing the variation of normalised spread of the buoyant layer as a function of the normalised time is shown in Figure 3. The aspect ratio L_o/d_o in these experiments ranged from 0.6 to 5.2 and it is evident that the transition length and time scales collapse the experimental data into a single relationship.

There are two distinct phases of spread and the transition between these regimes occurs at characteristic length and time scales of approximately unity. The spread law for the buoyancy-inertial phase conforms with the linear power relationship reported by Rottman & Simpson (1983) in Equation 1, while the spread during the buoyancy viscous phase conforms with the 3/8 power law in Equation 2.

It follows from the linear relationship between spread and time in the buoyancy-inertial regime that the normalised celerity of the front is constant and can be expressed in terms of a frontal densimetric Froude Number (F_f) given by:

$$F_f = \frac{c_f}{(\Delta g d_o)^{1/2}} \qquad (5)$$

where c_f is the celerity of the front. F_f was found to have a weak 1/8 power dependency on the initial aspect ratio of the buoyant release L_o / d_o which leads to the following expression for the celerity of the front:

$$\frac{c_f}{(\Delta g A^{1/2})^{1/2}} = 0.60 \pm 0.08 \qquad (6)$$

4. SPREAD OF THE BUOYANT LAYER IN THE PRESENCE OF WIND

4.1 Upwind propagation

In the presence of wind shear the spread of the buoyant layer is asymmetric as was seen in Figure 1. The upwind front deepens and moves more slowly while the downwind front is thinner and moves more rapidly. Furthermore as the buoyant layer elongates and its depth reduces, the buoyant force

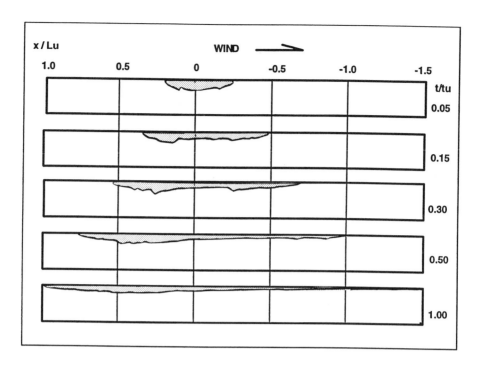

Figure 4. Form of the buoyant layer following release into a low stress wind

propelling it upwind also diminishes causing the upwind front to slow as the momentum of the wind drift layer become increasingly dominant component of the force balance. Finally the upwind front comes to rest and then slowly commences to retreat as the entire buoyant layer begins to drift downwind driven by the wind induced stress at the surface. Figure 3 shows the form of the buoyant volume at different times after its release. The times in this figure are expressed as fractions of the time taken for the upwind front to come to rest (t_u).

The initial depth of the buoyant layer d_o was appreciably greater than the depth of the wind drift layer and consequently the initial motion of the front was similar to that with no wind and its speed of propagation was given by Equation 1. The upwind motion of the layer was arrested when the kinematic momentum of the wind drift layer m was comparable with the buoyancy force of the layer Δgd^2 yielding m $/\Delta gd^2$ as the characterising variable where d is the depth of the spreading buoyant layer. If L_u is the upwind extent of the buoyant layer when is finally comes to rest then continuity dictates that d ~ A / L_u so that m ~ Δg (A / L_u)^2 and the limiting upwind extent of the buoyant front is given by:

$$\frac{L_u}{A^{1/2}} = C\left[\frac{m}{\Delta gA}\right]^{1/2} \qquad (7)$$

where C is a constant.

Figure 5 shows the non-dimensional maximum upwind extent of the buoyant layer measured in the experiments as a function of the wind drift momentum. The data conform well to the relationship in Equation 7 and the value of the constant C in the fitted curve is 3.8.

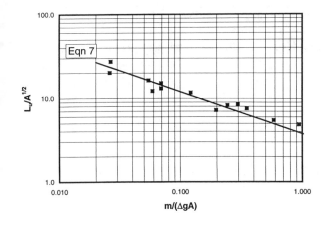

Figure 5. Arrested length of the upwind front as a function of the normalised momentum of the wind drift

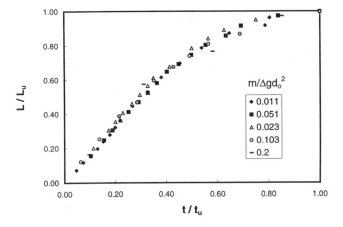

Figure 6. Normalised upwind extent of the buoyant layer as a function of time

The time scale of the upwind spread during the buoyancy-inertial phase was shown to be $t_0 = (d_0/\Delta g)^{1/2}$ and it follows that upwind extent of the layer normalised with respect to its ultimate extent should be determined by the time scale t / t_0. Data from a representative range of experiments are shown in Figure 6 and it is evident that these buoyancy-inertial length and time scales result in a collapse of the data.

It should be noted that the momentum of the wind drift in these experiments was such that the layer did not attain the buoyancy-viscous regime of spread. Different behaviour could have been expected had this been the case.

4.2 Downwind propagation

The behaviour of the downwind front was found to depend on the shear velocity of the wind-drift layer U^*. If the initial speed of the front [characterised by $(gd_0)^{1/2}$] exceeded the surface drift velocity [characterised by U^*] then the downwind front initially appeared similar to a front with no wind. If on the other hand the velocity of the wind-drift layer exceed the buoyancy induced propagation velocity of the front, then buoyant fluid was advected away in the wind-drift layer from the main body of buoyant fluid. The parameter characterising this mode of behaviour was the shear Froude Number of buoyant volume $F^* = U^* / (gd_0)^{1/2}$. For values of $F^* < 1$ a distinct front was initially evident on the downdrift front, however for $F^* > 1$ the downdrift layer took the form of a tapering wedge. For the experiment shown in Figure 4, F^* was less than 1.

Ultimately the elongation of the buoyant release reduced the buoyancy induced spread and even when F* < 1, wind drift finally determined the downwind spread of the buoyant fluid. A distinct front on the down drift side could only persisted for a limited time in the presence of wind shear.

5.   CONCLUSIONS

Experiments conducted into the effects of wind stress on the spread of a buoyant layer of finite volume revealed that relatively low momentum in the wind drift layer compared with the buoyant forces causing the spread of the layer have a significant affect on its form and rate of spread. Wilkinson (1982) found that the front of air cavities intruding into ducts containing water showed a similar sensitivity to relatively minor forcing at the front; in that case due to surface tension. The upwind advance of the front was ultimately arrested and its extent was determine by the momentum in the wind drift layer and the buoyancy of the layer.

The form of the downwind front depended on the celerity of the wind drift layer relative to the buoyancy-inertial speed of the front. When the speed of the wind drift exceeded the buoyancy speed of the front then buoyant fluid was advected in a thin surface wisp from the bulk of the layer. If the buoyancy speed if the front exceeded the speed of the wind drift layer then a wedge like front was formed. Ultimately however as the depth of the buoyant layer reduced with continued spread, its speed fell below that of the wind drift and buoyant fluid commenced to drain into the wind drift.

6.   REFERENCES

Hoult, D.P. (1972). Oil spreading on the sea. *Ann. Rev. Fluid Mech.* 4, 341-368.

Jirka, G.H. & Arita, M. (1987). Density currents or density wedges: boundary layer influence and control methods. *J. Fluid Mech.* 177, 187-206.

Peirson, W.L. , Wilson, J., Constantinides, G., Miller, B.M., Lim, B.B, & Jayewardene, L. (1992). Sydney Deepwater Outfalls Environmental Monitoring Program, Post-Commissioning Phase, An Integrated Ocean Outfall Simulation Sytem. *Australian Water and Coastal Studies Pty Ltd. Interim Rep. 92/01/06.*

Rottman, J.W. & Simpson, J.E. (1983). Gravity currents produced by instantaneous releases of a heavy fluid in a rectangular channel. *J. Fluid Mech.* 135, 95-110.

Simpson, J.E. & Britter, R.E. (1980). A laboratory model of an atmospheric mesofront. *Quart. J.R. Met. Soc.* 16, 485-500.

von Karman, T. (1940). The engineer grapples with nonlinear problems. *Bull. Am. Math. Soc.* 46, 615-683.

Wilkinson, D.L. (1982). Motion of air cavities in long horizontal ducts. *J. Fluid Mech.* 118, 109-122.

Wilkinson, D.L. & Wood, I.R. (1987). Blocking of layered flows in channels of gradually varying geometry. *Proc. 3rd Int. Symp. On Stratified Flows, Pasadena,* 133-145.

Environmental Hydraulics, Lee, Jayawardena & Wang (eds) © 1999 Balkema, Rotterdam, ISBN 90 5809 035 3

Numerical simulation on turbulent buoyant jets in a cross-flow of stratified fluids

Robert R. Hwang
Institute of Physics, Academia Sinica, Taipei, Taiwan, China

T. P. Chiang
Institute of Naval Architecture and Ocean Engineering, National Taiwan University, Taipei, Taiwan, China

ABSTRACT: In this study, an investigation using a 3-D numerical model, which treats conservation of mass, momentum and salinity simultaneously, was carried out to study the detailed flow field of a buoyant jet in a uniform cross-stream of stable linearly stratified environment. A k-ε turbulence model was used to simulate the turbulent phenomena and close the solving problem. Results indicate that the numerical computation simulates satisfactorily the plume behavior in a stratified crossflow. The ambient stratification tends to inhibit the flow development by inducing pairs of vortex as the plume flowing downstream and results in causing the jet-flow oscillation from its maximum height-of-rise.

1. INTRODUCTION

Wastes are often introduced into the environment as buoyant jets. The turbulent mixing of a buoyant jet in a density-stratified crossflow is of significant complexity in understanding. The turbulent shear generated by the discharge results in efficient mixing, and this reduces the concentration of pollutants. When compared with a similar flow in unstratified fluid, the major effects of density stratification are to limit the vertical rise of a buoyant jet and to restrict the mixing with the surrounding ambient fluid. Hence, understanding the mixing and dilution processes, and establishing the capability to estimate the rise and width of a buoyant jet in stratified fluid, are necessary for designing and controlling a waste disposal system.

A number of approximate predictive methods for the plume flow in stratified surroundings have been developed in the literature of Abràham (1965), Schwartz and Tulin (1972), Sneck and Brown (1974), Wright (1984), and Hwang and Chiang (1986,1995). All these are basically integral methods based on the simplifying assumptions that the entrainment rate and the similarity profiles for both the velocity and temperature are made.

Along with recent developments in numerical computations, there has been considerable progress in the construction of turbulence model of wide applicability. It is therefore now possible to use an acceptable turbulence model for the differential method to obtain numerical solutions for jets in a stratified cross flow. The aim of this study is to calculate by a finite-difference method the flow field of a round turbulent buoyant jet issuing vertically into a uniform free stream of density-stratified fluid. In solving the three-dimensional time-averaged governing equations, the two-equation model of standard k-ε model is adopted to simulate the turbulent transported quantities.

2. FORMULATION OF FLOW PROBLEMS

We consider the three-dimensional flow field of a round buoyant jet of diameter D_j with constant injection velocity W_j issuing vertically into a uniform velocity cross flow of stably linear stratified fluid. Figure 1 shows the geometry and coordinates of the flow problem. With the Boussinesq approximation, the density variation is accounted for only in the gravitational term, the governing differential equations for the mean velocity, salinity and concentration are :

Continuity:

$$\frac{\partial U_i}{\partial x_i} = 0 \qquad (1)$$

Momentum:

$$U_j \frac{\partial U_i}{\partial x_j} = -\frac{1}{\rho}\frac{\partial P}{\partial x_i} - \beta g(T - T_a)\delta_{3i} + \frac{\partial}{\partial x_j}\left(-\overline{u_i u_j}\right) \qquad (2)$$

Salinity:

$$U_j \frac{\partial T}{\partial x_j} = \frac{\partial}{\partial x_j}\left(-\overline{u_j T'}\right) \qquad (3)$$

Concentration:

$$U_j \frac{\partial C}{\partial x_j} = \frac{\partial}{\partial x_j}\left(-\overline{u_j C'}\right) \qquad (4)$$

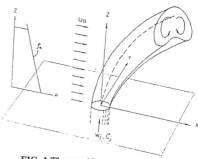

FIG. 1 The considered flow problem

where U_i is the time averaged velocity components in the i coordinate direction, ρ is density, P is the mean pressure, T is the salinity, T_a is the local ambient salinity, C is the mean concentration of tracer injecting from the jet, and $-\overline{u_i u_j}$ and $-\overline{u_j T}$ are the turbulent transport rates of momentum and mass fluxes respectively. Far upstream of the jet exit, the local ambient salinity, or the ambient density, is assumed to be linear in the vertical direction. In order to solve equations (1)-(4), a turbulence model for the turbulent stresses and mass flux has to be specified. The turbulence model adopted in this study is the standard $k-\varepsilon$ model (Rodi,1984) in which an isotropic eddy diffusivity is assumed.

It can be seen that the time-averaged flow field is symmetrical about the xz plane passing through the center of the jet. The calculation domain was chosen that, in the y direction, it extended from the symmetry plane to a location where y/D_j is 60. In the z direction, one boundary of the domain considered was the bottom plane, and the other was placed sufficiently far away (say, H) so that uniform crossstream conditions could be approximately assumed there. This location of the outer boundary was found by computational experiments. For a jet-to-crossstream velocity ratio (R) of 8, for example, the outer boundary was located at a z/D_j of 80.

In the x direction, the upstream boundary was placed six jet diameters upstream of the jet and the position of the downstream boundary was located sufficiently far downstream that the flow velocity became almost parallel to the x direction. For example, in the case of the velocity ratio R=8, the downstream boundary was positioned at $x/D_j=125$.

The boundary conditions for the calculation domain prescribed above were specified as follows:
(1) On the upstream yz plane $x=X_1$,

$$U=U_a, \ V=W=C=0, \ T_a = T_{ao}+\frac{dT_a}{dz}z, \ k = 0.04\,U_a^2, \ \varepsilon = k^{3/2}/(0.06H) \qquad (5)$$

(2) At the downstream yz plane $x=X_2$, $\partial(V,W,T,k,\varepsilon,C)/\partial x = 0$, $U_B^n = U_{B-1}^n$, where n represents the iteration level, and B the boundary point.
(3) On the jet exit,

$$W=W_j, \ U=V=0, \ T=T_j, \ C=C_j, \ k = 0.001W_j^2, \ \varepsilon = k^{3/2}/(0.5D_j)$$

(4) The xz plane through the origin (y=0),

$$V=0, \ \partial(V,W,T,k,\varepsilon,C)/\partial y = 0 \qquad (6)$$

(5)The other xz plane $y=Y_2$ and the top xy plane $z=Z_2$ were considered to be far from the jet and had the

Table 1   The jet and ambient flow conditions conducted in this study

case	Jet				Ambient flow		Flow parameter		
	D_j (cm)	W_j (cm/s)	$\Delta\rho/\rho_{ao}$ $\times10^{-2}$	Re_j $\times10^3$	U_a (cm/s)	$\alpha\times10^3$ (cm^{-1})	R	F	G (s^{-2})
1	0.40	199.7	0.0	8.0	24.9	0.0	8.0		0.0
2	0.82	82.5	0.76	6.8	10.3	0.36	8.0	33	0.35
3	0.82	82.5	0.76	6.8	4.6	0.36	18.0	33	0.35
4	0.82	82.5	0.76	6.8	10.3	0.72	8.0	33	0.70

same flow conditions as the upstream yz plane condition of (1).

(6) On the bottom xy plane z=0, the wall-function treatment in the near-wall region is adopted.

3. NUMERICAL METHOD

In this study, the numerical computations were performed on a nonuniform and staggered MAC (marker and cell) grid system (Spalding,1972). In the staggered grid system, the pressure and other dependent scalar variables such as k and ε, are calculated at nodal points between those of the mean velocity.

To solve the partial differential equations (2), (3), (4) and the transport equations of k and ε which are of elliptic type, a computer code based on the power law difference (PLD) method (Patankar and Splalding, 1972) is constructed for use in the present study. Briefly, in a control volume finite-difference method, the convection together with diffusion terms of transport equation for variable ϕ ($\phi = U,V,k,W,T,C,\varepsilon$) of

$$\frac{\partial}{\partial x_j}(\rho U_j\phi) = \frac{\partial}{\partial x_j}(\Gamma\frac{\partial\phi}{\partial x_j}) + S_\phi \tag{7}$$

is discretized by PLD, source term by second order central difference, and then integrated within a control volume element to obtain an algebraic equation. The pressure field P is solved with SIMPLEC algorithm of Von-Doormaal (1984). The system of linear algebraic equations is solved by the alternating direction line by line iteration method. The convergence criterion is specified as the relative difference of these dependent variables, i.e. U, V, W, k and ε, at all nodal points being smaller than 10^{-4}.

4. Presentation and discussion of results

The computed jet-flow development are supported by extensive experimented results obtained by Hwang and Chiang (1986). Figures 2 and 3 present the comparison of the computed trajectories of jet flows with the experimental trajectories in both the unstratified and stratified ambient environments. It notices that the agreement between the calculation and the experiment on jet trajectories is reasonably well. In the present study, numerical computations of four flow cases are conducted to study the effect of ambient fluid stratification on a jet flow subject to a uniform velocity cross flow. The jet and the ambient flow conditions of these flows are listed in Table 1.

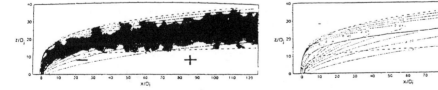

FIG. 2 Comparison of computed jet Trajectory for R=8, F=∞, and G=0 with experimental photography [— is computed isoconcentration contours of present study; …indicates jet boundaries and trajectory obtained from integral approach (Hwang and Chiang, 1986)

Figures 4, 5 and 6 show the U-W velocity vectors, W-velocity contours, and iso-concentration contours of the jet flow interacted with the crossflow for varied jet configurations and ambient stratifications on the plane of symmetry (y=0). Figure 4 shows the computed results for the pure momentum jet in the crossflow of unstratified fluid where R=8, F=∞ and G=0 case. Above the exit, the jet driven by its initial momentum, deflects promptly then continues to move upwards. It can be seen from Fig. 4(b) that the W contours of vertical velocity vary monotonically. The turbulent shear generated by the discharge results in efficient mixing which rapidly reduces tracer concentrations. Figure 4(c) shows the computed concentrations of the dyed jet to be reduced down to 10% of the undiluted jet concentration in the vicinity of the jet exit.

The other three cases considered show the influences of ambient density stratification and main-stream velocity on the jet flow development. A major effect of density stratification is to limit the rise of a buoyant jet and to restrict the dilution compared to a similar flow in an unstratified ambient fluid. Figures 5(a) and 5(b) show the effects of ambient stratification on the jet flow development at y=0 plane for flow configuration of R=8, F=33.0, and G=0.35sec^{-2}. Here $F = W_j / \left[\left(\Delta \rho / \rho_{ao} \right) g D_j \right]^{1/2}$, is the densimetric Froude number, where $\Delta \rho = \rho_{ao} - \rho_j$ is the density difference between the ambient density at the level of jet exit and the jet density, and G=αg is the square of Brunt-Vaisala frequency of the stratified environment. In a stably stratified environment, the plume first behaves like a buoyant jet. The initial momentum and the buoyancy of the plume causes the jet flow to move upward, bent over in the crossstream direction and mix with the heavy bottom fluid. The density deficit of the plume then decreases continuously and becomes zero at a certain level. From here on the buoyancy force is negative and the flow decelerates and turns down after reaching a maximum height. The jet trajectory is illustrated by the iso-concentration contours shown in Fig. 5(c).

Increasing the ratio of the jet velocity to the main-stream velocity will increase the initial momentum of jet so that the buoyant jet rises higher and mixes more rapidly with dense bottom fluid. Figure 6 shows the

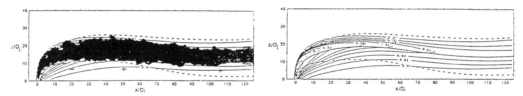

FIG. 3 Comparison of computed jet Trajectory for R=8, F=∞, and G=0.35 s^{-2} with experimental photo-graphy [— is computed isoconcentration contours of present study; ...indicates jet boundaries and tra-jectory obtained from integral approach (Hwang and Chiang, 1986)

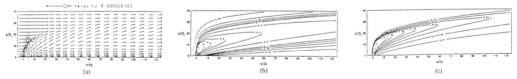

FIG. 4 The computed results on the y=0 plane for R=8, F=∞ and G=0 (unstratified case) (a) The U-W velocity vectors; (b) The W velocity contours; (c) The iso-concentration contours

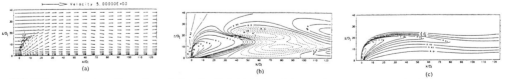

FIG. 5 The computed results on the y=0 plane for R=8, F=33.0 and G=0.35s^{-2}, (a) The U-W velocity vectors; (b) The W velocity contours; (c)The iso-concentration contours

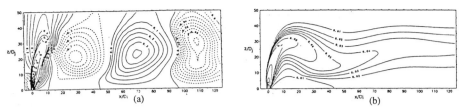

FIG. 6 The computed results on the y=0 plane for R=18, F=33.0 and G=0.35s^{-2}, (a)The W velocity contours; (b)The iso-concentration contours

computed results on the y=0 plane for the case of R=18, F=33.0 and G=0.35 sec^{-2}. The upward and downward motion of the jet flow is clearly evident in the W-velocity contours of the Fig.6(a). The wavelength of the oscillating motion of a buoyant jet in a stratified crossflow from its maximum height-of-rise is inversely proportional to the velocity ratio of R. The larger the jet exit ratio, the smaller the wavelength of the oscillating motion. In the correspondence with the wave-like motion of the jet flow, the jet trajectory also exhibits an oscillating trajectory. This can be seen from the iso-concentration contours shown in Fig. 6(b).

As described previously, the horizontal momentum of the cross flow causes the jet to be deflected along the flow direction. Since the jet shear layer is permeable and deformable, the surrounding cross-flow passes and penetrates through the jet shear layer resulting in the formation of a pair of vortices on the jet cross-section. Figure 7 indicates the calculated velocity contours of the jet cross section at various heights, z/D$_j$, along the jet near the emitting exist. The pair of vortices on the cross section is counter clockwise on the left and clockwise on the right. The vortex pair moves upward along the jet-flow development. Figure 8(a) depicts the vortex pair on the cross section moves upward and reduces its magnitude of the peak value of the vorticity along with its diffusion outward. This can be seen more clearly in the various transverse sections of the streamwise vorticity field shown in Fig. 8(b). The streamwise-component of vorticity is

defined as $\varpi_x = \dfrac{\partial W}{\partial y} - \dfrac{\partial V}{\partial z}$. The deformation of the jet cross-section can also be seen from the iso-concentration contours of Fig. 8(c).

In a stably stratified environment, one effect of density stratification is to prevent the vertical rise of the buoyant jet. The retardation of the flow development influenced by the stratified environment then leads to the formation of a secondary pair and a third pair of vortices above and below the primary vortex pair with a reverse direction. It can be noticed at Fig. 9 for the case of R=8, F=33.0, and G=0.35 sec^{-2}. The various transverse sections of the streamwise-component vorticity field of Fig.9(b) depict the evolution process more clearly. The effect of the ambient stratification is to produce the vorticity of opposite sign to the main vorticity. As the stratification of the ambient fluid is increased, the retardation of the jet flow from its source efflux becomes more significant. Influenced by the same mechanism, the development of iso-concentration contours, which have the same shape as the jet cross-section, is also altered by the effects of density stratification, as illustrated in Fig. 9(c). Both the iso-concentration contours and the shape of jet cross-section have deformed from a symmetric kidney-shape to an oblate ellipse. The decaying rate of the trace concentration is also retarded and decreased due to the effect of the ambient stratification, as show in Figs. 4 and 5.

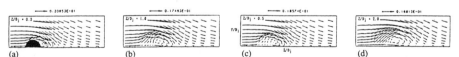

FIG. 7 Computed U-V velocity vectors of jet deflected by main-stream near emitting exit

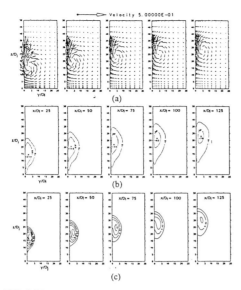

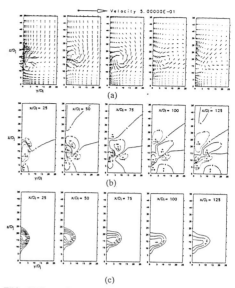

FIG. 8 The evolution of vortex motion on the jet cross-section for R=8, F=∞ and G=0 (unstratified case) (a)The V-W velocity vectors (b) The streamwise-component vorticity contours. Contour magnitudes are amplified by 10^3. (c) The iso-concentration contours

FIG. 9 The evolution of vortex motion on the jet cross-section for R=8, F=33.0 and G=0.35 s^{-2} (stratified) (a) The V-W velocity vectors (b) The streamwise-component vorticity contours. Contour magnitudes are amplified by 10^3. (c) The iso-concentration contours

5. CONCLUSION

The detailed structure of the flow development for a buoyant jet discharged in a cross flow of stably linear stratified environment is not clear at present. In this study, a three-dimensional numerical model based on a two-equation turbulence model is applied to predict the plume behavior in stratified cross flow. Computations of the model have given results which are, in general, in satisfactory agreement with available experimental data. The formation of the secondary and third pairs of vortices which are not induced   in the unstratified environment causes the jet flow oscillation from its maximum height-of-rise and alteration to the entrainment mechanism in the stratified cross flow. The ambient stratification, prohibits the development of the plume radius and reduces the mixing rate as well the plume rise and a wave-like trajectory is formed for the case of higher density stratification and higher jet-to-crossstream velocity ratio.

6. REFERENCES

Abraham, G. (1965) "Entrainment principle and its restriction to solve jet problem", *Journal of Hydraulic Research*, 31, 1-23.

Hwang, R.R., Chiang, T.P. (1986) "Buoyant jets in a crossflow of stably stratified fluid", *Atmospheric Environment,* 20, 1887-1890.

Hwang, R.R., Chiang, T.P. (1995) "Numerical simulation of vertical forced plume in a crossflow of stably stratified fluid", *ASME, Journal of Fluids Engineering*, 117, 696-705.

Patankar, S.V., Spalding, D.B. (1972) "A calculation procedure for heat, mass, and momentum transfer in three-dimensional parabolic flows", *International Journal of Heat and Mass Transfer*, 5, 1878-1906.

Rodi, W. (1984) "Turbulence model and their applications in hydralics", *International Association for Hydraulics Research*, Delft.

Schwartz, J., Tulin, M.P. (1972) "On the mean path of buoyant bent-over chimney plumes", *Atmospheric Environment*, 6, 19-36.

Sneck, H.J., Brown, D.H. (1974) "Plume rise from large thermal sources such as cooling tower", *Journal of the Heat Transfer*, 96, 232-238.

Spalding, D.B. (1972) "A novel finite differential expression involving both first and second derivatives", *International Journal of Numerical Methods in Engineering*, 4, 551-559.

Van-Doormaal, J.P., Raithby, G.D. (1984) "Enhancements of the SIMPLE method for predicting incompressible fluid flows", *Numerical Heat Transfer*, 7, 147-163.

Wright, S.J. (1984) "Buoyant jets in density-stratified crossflow", *Journal of Hydraulic Division, ASCE*, 110, 643-656.

Environmental Hydraulics, Lee, Jayawardena & Wang (eds) © 1999 Balkema, Rotterdam, ISBN 90 5809 035 3

The axisymmetric jet in a coflow

S.J.Gaskin
Department of Civil Engineering and Applied Mechanics, McGill University, Montreal, Que., Canada
I.R.Wood
Department of Civil Engineering, University of Canterbury, Christchurch, New Zealand

ABSTRACT: The axisymmetric jet in a coflow has been modeled using the excess momentum equation, the continuity equation and an entrainment function to determine the approximate variation of the mean jet properties. The entrainment into the jet changes as the jet changes from a strong jet to a weak jet. The entrainment function allows for the additional entrainment occurring in the weak jet indicated by the more convoluted jet boundary. In the limits the model agrees with dimensional analysis.

1. INTRODUCTION

In environmental fluid mechanics the study of buoyant jets is relevant to the mixing and dilution of pollutants in water bodies. Buoyant jets in both stationary and flowing ambients have been much studied. Their mean properties have been obtained using integral models in which the momentum equation and the continuity equation are combined with a closure equation. The closure equation can be an entrainment function, in which the entrainment coefficient is proportional to the mean center line velocity (Morton, Taylor & Turner 1956). An alternative closure equation is the spread function, which assumes a constant rate of spread of the turbulent jet (Wright 1994, Wood et al. 1993).

For buoyant jet flows in the oceans, the crossflow (or current) is a small proportion of the initial jet velocity. In this case, the entrainment into the buoyant jet can be modeled in the irrotational region outside the jet flow by superimposing the crossflow velocity (Gaskin 1996) with the sink flow due to a jet in a still ambient (Taylor 1958). This is the motive for further exploring the entrainment function.

The buoyant jet flow can be divided into several regions in which different parameters dominate, dimensional analysis allows the form of the mean properties to be determined for each regime. A numerical model should in the limit satisfy all the regimes. The jet in a coflow, which is one of the simpler regimes, will be considered here.

2. PROPERTIES OF AXISYMMETRIC AND PLANE JETS

The velocity and turbulence profiles for plane and axisymmetric jets are self preserving in a still fluid (Hussein, J.H., Capp, S.T. and George, W.K 1994, Papanicolaou and List 1988 and others). For plane and axisymmetric jets in a coflow the excess mean velocities are self-preserving but the product of the turbulence velocities are not (Nickels and Perry 1996 and others). This suggests that the entrainment changes as the flow changes from a strong jet to a weak jet but is self preserving in these limits. A useful solution can be obtained by assuming self preservation of the mean excess velocities over the range and an approximate form of the entrainment function.

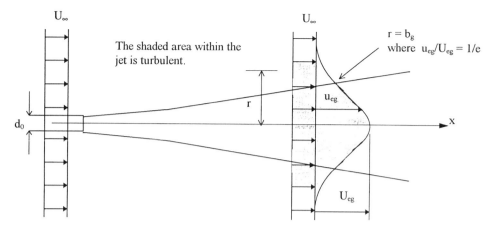

U_∞ is the coflow velocity
u_{eg} is the local time averaged excess velocity measured at r and is approximately Gaussian
U_{eg} is the maximum of the local time averaged excess velocity
b_g is the value of r where u_{eg}/U_{eg} equals 1/e

FIGURE 1. The Axisymmetric Jet in a Coflow.

3. THE AXISYMMETRIC JET IN A COFLOW

An axisymmetric jet in a coflow is a single jet released parallel and in the direction of the flowing ambient. The most basic case of the jet in large duct and with no turbulence in the coflow is considered here. The flow consists of a non-turbulent flow outside the jet boundaries and a turbulent region inside the jet boundaries, see Figure 1. The turbulent region consists of the turbulent jet flow and the turbulent portion of the coflow which is within the jet boundaries. This turbulent flow is carried by a velocity equal to the sum of the coflow velocity and the excess jet velocity.

The flow is long and narrow and the duct is large (area A) so the boundary layer assumption can be made. This implies that the excess velocity (u_{eg}) and tracer distributions are self similar and they are assumed to have a Gaussian distribution given by

$$u_{eg} = U_{eg} \exp-\left(\frac{r}{b_g}\right)^2 \tag{1}$$

The continuity equation between the port exit ($_o$) and a point downstream ($_x$) is

$$\int_0^A u_{eg} 2\pi r dr + \int_0^A U_\infty 2\pi r dr = \frac{\pi}{4} d_o^2 U_{eo} + A U_{\infty o} \tag{2}$$

Conservation of momentum in the x-direction is given by

$$\int_0^A (u_{eg} + U_\infty)^2 2\pi r dr + \int_0^A \frac{p_x}{\rho} 2\pi r dr = \int_0^A (U_{eo} + U_\infty)^2 2\pi r dr + \int_0^A \frac{p_o}{\rho} 2\pi r dr \tag{3}$$

where p_o and p_x are the pressures at o and x. These two equations are combined by subtracting U_∞ times the continuity equation from the conservation of momentum equation. Dividing by the maximum average velocity (U_{eg}) and the width (b_g) allows each term to be expressed in terms of the dimensionless self similar profile shape. Assuming p_o is zero gives,

$$U_{eg}^2 b_g^2 \int_0^\infty \left[\left(\frac{u_{eg}}{U_{eg}} \right)^2 + \frac{p_x}{\rho U_{eg}^2} \right] 2\pi \frac{r}{b_g} d\frac{r}{b_g} + U_\infty U_{eg} b_g^2 \int_0^\infty \frac{u_{eg}}{U_{eg}} 2\pi \frac{r}{b_g} d\frac{r}{b_g} = \frac{\pi}{4} d^2 \left(U_{eo}^2 + U_\infty U_{eo} \right) \qquad (4)$$

$$\underbrace{\hspace{5.5cm}}_{\text{term I}} \qquad \underbrace{\hspace{4cm}}_{\text{term II}}$$

The velocity term of the dimensionless part of I, after integration, yields the shape constant I_m of $\pi/2$. The data of Hussein et al. (1994) suggests that the pressure term of I is a function of the dimensional turbulence velocities and has a value of approximately 10% of the velocity term. These two terms will be combined in the shape constant I_m as 1.1 times $\pi/2$ which is equal to 1.72. The dimensionless part of II after integration yields the shape constant I_q of π.

The excess velocity (U_{eg}) and the width (b_g) are made dimensionless (') by respectively dividing by U_∞ and the length scale for the transition from the strong jet to the weak jet, $l_{J,WJ}$, where

$$l_{J,WJ} = \left(\frac{\pi}{4} \frac{\left(U_{eo}^2 d^2 + U_\infty U_{eo} d^2 \right)}{U_\infty^2} \right)^{\frac{1}{2}} \qquad (5)$$

Equation (4) can then be written as

$$
\begin{array}{rcl}
I_m U'^2 b'^2 \quad + \quad I_q U' b'^2 & = & 1 \\
f \quad + \quad e & = & 1 \\
I_q U' b'^2 \quad + \quad \left(\dfrac{I_q^2}{I_m} b'^2 \right) \dfrac{I_m}{I_q} U' & = & 1 \\
q' \qquad\qquad \dfrac{I_m}{I_q} U' & = & 1
\end{array}
\qquad (6)
$$

where $f = I_m U'^2 b'^2$ is the momentum due to the jet excess velocity and $e = I_q U' b'^2$ is the momentum due to the coflow velocity. The volume flux, q', is the turbulent flow within the jet boundaries, which has a term related to the excess velocity ($I_q U' b'^2$) and a term related to the turbulent coflow ($I_q^2/I_m \, b'^2$). Defining U' and b' in terms of e and f and subtituting gives q' equal to e/(1-e). The velocity carried by q' is $I_m/I_q \, U'$ and this is approximately equal to the top hat velocity, $0.5U'$, used by Morton et al. (1956) for jets in a still fluid.

The entrainment into the turbulent flow, q', is driven by the excess velocity U' and is given by

$$\frac{d}{ds'} q' = \frac{de}{dx'} \left(\frac{1}{1-e} \right)^2 = 2\pi b' \alpha_c U' = 2\pi \alpha_c \frac{(1-e)^{\frac{1}{2}}}{I_m^{\frac{1}{2}}} \qquad (7)$$

where α_c is the entrainment constant for a jet in a coflow. When the coflow is small (e is small) α_c tends to the value for a jet in a still fluid, α_j, which is equal to 0.057 (Hussein et al. 1994). This gives solutions for b' and U' for a jet in a still fluid of

$$b' = 2\pi \frac{\alpha_j}{I_q} x' = 0.11 x'$$

$$U'_{eg} = \frac{I_q}{I_m^{\frac{1}{2}} 2\pi\alpha_j x'} = \frac{6.7}{x'} \tag{8}$$

Nickels and Perry's (1996) data for the strong jet in a coflow give a constant of 6.9 for the excess velocity. Observations indicate that the boundary between the jet flow and the irrotational ambient is more convoluted in the weak jet than in the strong jet. Townsend (1976) has suggested that the variation in the entrainment between strong jets and weak jets or wakes can be explained by assuming that a set of ordinary eddies common to all jets causes the basic entrainment and an additional set of eddies causes additional folding that is found only in the weak jets. To provide for the additional entrainment, when there is a more convoluted boundary, in the weak jet case (large e) and to satisfy the strong jet case, the entrainment function is written as,

$$\alpha_c = \alpha_j \left(1 + k_e e\right) \tag{9}$$

The entrainment into the turbulent flow can then be rewritten as,

$$\frac{d}{ds'} q' = \frac{de}{dx'}\left(\frac{1}{1-e}\right)^2 = 2\pi b' \alpha_c U' = 2\pi\alpha_j \left(1 + k_e e\right)\frac{\left(1-e\right)^{\frac{1}{2}}}{I_m^{\frac{1}{2}}} \tag{10}$$

The limiting behaviour of a weak jet is checked by letting f tend to zero and substituting for e. This gives the velocity decay as varying with $x^{-2/3}$ and the width growth varying as $x^{1/3}$, which are the values obtained by simple dimensional analysis (Wood et al. 1993).

A complete solution can be obtained by integrating the equations starting from the initial conditions determined at the end of the zone of flow establishment.

4. CONCLUSIONS

The jet in a coflow has been modeled using the continuity and conservation of momentum equations, and a modified entrainment function for the closure equation. The turbulent flow within the jet boundaries has a component due to the excess velocity and a component due to the coflow and it is carried by a velocity equal to the sum of the coflow and excess velocities. The excess velocity drives the entrainment into the turbulent jet. The modified entrainment function allows for the additional entrainment occurring in the weak jet, which is due to additional eddies indicated by the increasingly distorted weak jet boundary. The modified equations agree with dimensional analysis in the limits of the strong jet and the weak jet.

5. REFERENCES

Gaskin, S. (1996). "Single buoyant jets in a crossflow and the advected line thermal", *Ph.D. Thesis* Civil Engineering, University of Canterbury, Christchurch, N.Z.

Hussein, H,J., Capp, S.P. and George,W.K. (1994). "Velocity measurements in a high-Reynolds- number, momentum conserving, axisymmetric, turbulent jet", *J. Fluid Mech.*, 258, 31-75.

Papanicolaou, P.N. and List, E.J. (1988). "Measurements of round vertical axisymmetric buoyant jets", *J. Fluid Mech.*, 195, 341-391.

Morton, B.R., Taylor, G.I. and Turner, J.S. (1956). "Turbulent gravitational convection from maintained and instantaneous sources", *Proc. Royal Soc. London*, Series A, 234, 1-23.

Nickeis, T.B. and Perry, A.E. (1996). "The turbulent coflowing jet", *J. Fluid Mech.*, 309, 157-182.

Taylor, G.I. (1958). "Flow induced by jets", *J. Aero/Space Sciences*, XXV, 456-457.

Townsend, A.A., (1976). *The structure of turbulent shear flow*, Cambridge University Press, Cambridge.

Wood, I.R., Bell, R.G. and Wilkinson, D.L. (1993). *Ocean Disposal of Wastewater*, World Scientific Publ., Singapore.

Wright, S.J. (1994). "The effect of ambient turbulence on jet mixing", *Recent Research advances in the fluid Mechanics of Turbulent Jets and Plumes*, NATO ASI Series E: Applied Science, 255, 13-27 .

Environmental Hydraulics, Lee, Jayawardena & Wang (eds) © 1999 Balkema, Rotterdam, ISBN 90 5809 035 3

A semi-analytical self-similar solution of a bent over jet in crossflow

Li Lin, J.H.W.Lee & V.Cheung
Department of Civil Engineering, The University of Hong Kong, China

ABSTRACT: The vorticity dynamics of the bent-over buoyant jet in the far field have been investigated by two-dimensional numerical solutions with similarity transformation. By introducing Prandtl's free shear layer model to enclosure turbulent governing equations, the semi-analytical computed streamlines and vorticity clearly indicate a vortex-pair flow. The distribution of scalar field in the advected flow grows up from single cell to kidney-shaped double peak structure. The calculated jet flow field agrees well with experimental results.

1. INTRODUCTION

The discharge of a turbulent jet horizontally into a current is a focus in hydraulic researches. Many experiments (Keffer 1963; Moussa,Trischka & Eskinazi 1977; Cheung 1991) have revealed a transition from an initially vertical jet through a bent-over phase during which the jet becomes almost parallel with the main streamwise and a secondary vortex pair flow in the transverse section of the jet. Chu (1974) have finished the investigation of related scalar mixing, visual trajectory and limited tracer concentration in the bent-over phase of a momentum-dominated buoyant jet in crossflow by experiments, and a striking double peak of scalar concentration distribution is found in the transverse section for the bent-over phase of the jet. On the other hand, numerical studies (Patankar 1977; Chu 1985; Lee 1991) were carried out with intergral methods by Euler or Lagrangian models.

From experimental results, there is no universal concentration distribution for a buoyant jet in the bent over phase of a round buoyant jet into crossflow. It may vary from a radially symmetry shape to a double peak kidney shape. But there is a clear relationship between the distribution and the momentum, either the initial momentum or that generated by the buoyancy force. For a given discharge and current, the concentration distributions at successive cross-sections in the bent-over phase are self-similar.

By its nature, integral methods can not reveal any information on the jet structure, therefore an understanding of the vorticity dynamics in the bent-over phase of a buoyant jet into crossflow is necessary to explain the observed concentration structures. A similarity solution is employed for this purpose (Yih 1981, Cheung 1991). Employing Prandtl's free shear layer model for turbulent closure, an analytically self-similar solution was derived for the asymptotic stage; a three term approximation of the series solution was given. Although the mixing process was not discussed, the flow field was illustrated for a value of dimensionless eddy viscosity for which the approximation is valid. But from the latest research, the applicable value λ in their analytical solution are too small (around 3) to give the real description of its behavior, it is necessary to introduce numerical solution to analyze the realistic far field of a round buoyant jet in crossflow.

In this paper, a semi-analytical method is used for study the cross-sectional mean flow and passive scalar field of bent-over jet in a current. Based on the hypothesis that the x-component of the jet velocity equals to the ambient velocity in the bent-over phase, the vorticity dynamics in the far field have been investigated by 2-D numerical solutions with similarity transformation. The computed streamlines and vorticity clearly indicate a vortex-pair flow, and the shape of the concentration field in the advected flow varies graduately from symmetrical form to kidney-shaped double peak structure.

2. THEORY

2.1 Far Field of A Jet into Crossflow

Consider a turbulent round jet, of initial momentum flux M_o, discharging into a steady uniform crossflow with velocity U_a in the horizontal direction (Fig.1). From the concept of basic flow regimes (Scorer 1978, Fischer 1979), the bent over phase exists in *Momentum Dominated Far Field* ($z \gg l_m$), where l_m represents the characteristic length at which the momentum-induced velocity ($M_o^{1/2} z^{-1}$) decay to that of the ambient velocity value. In the far field, the jet is significantly deflected, and flow behavior is analogous to an advected line puff (ALP) translating at U_a. For jets with 3-D trajectories, the asymptotic relations for the far-field remain valid. In the absence of body forces, the jet excess momentum flux is conserved. With no initial excess x-momentum, the averaged excess x-velocity of the jet is zero, and the jet may be assumed to be globally advected with a horizontal velocity of U_a.

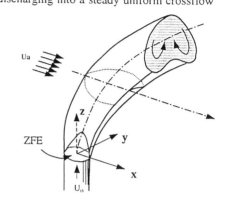

Fig.1 Buoyant jet in Crossflow

2.2 Governing Equations

The governing equations can be formulated for the conservation of mass, momentum and tracer mass in any vertical section with a constant streamwise velocity ($u=U_a$) as followed:
Continuity:

$$\frac{\partial v}{\partial y} + \frac{\partial v}{\partial z} = 0 \tag{1}$$

Momentum:

$$U_a \frac{\partial v}{\partial x} + v \frac{\partial v}{\partial y} + w \frac{\partial v}{\partial z} = -\frac{1}{\rho}\frac{\partial p}{\partial y} + \varepsilon(\frac{\partial^2 v}{\partial y^2} + \frac{\partial^2 v}{\partial z^2}) \tag{2}$$

$$U_a \frac{\partial w}{\partial x} + v \frac{\partial w}{\partial y} + w \frac{\partial w}{\partial z} = -\frac{1}{\rho}\frac{\partial p}{\partial z} + \varepsilon(\frac{\partial^2 w}{\partial y^2} + \frac{\partial^2 w}{\partial z^2}) \tag{3}$$

Tracer mass:

$$U_a \frac{\partial \theta}{\partial x} + v \frac{\partial \theta}{\partial y} + w \frac{\partial \theta}{\partial z} = \frac{\varepsilon}{Pr}(\frac{\partial^2 \theta}{\partial y^2} + \frac{\partial^2 \theta}{\partial z^2}) \tag{4}$$

where (U_a,v,w) are the mean velocity components in the respective (x,y,z) directions; θ is the mean concentration of a tracer; ε is the kinematic eddy viscosity, which is assumed to be constant over any vertical section. *Pr* is Prandtl number, which equals to the ratio of eddy visicosity to eddy diffusivity.

Conservation of vertical momentum flux and mass flux are concepted as:

$$M_v = U\iint w\,dydz = M, \quad Q_o\theta_o = U\iint \theta\,dydz \tag{5}$$

For this 2-D flow field, introducing stream function and vorticity as:

$$v = \frac{\partial \varphi}{\partial z}, \quad w = -\frac{\partial \varphi}{\partial y} \; ; \qquad \hat{\xi} = \frac{\partial w}{\partial y} - \frac{\partial v}{\partial z} \tag{6}$$

124

Then, the momentum equations can be written as x-Vorticity:

$$U_a \frac{\partial \hat{\xi}}{\partial x} + v \frac{\partial \hat{\xi}}{\partial y} + w \frac{\partial \hat{\xi}}{\partial z} = \varepsilon (\frac{\partial^2 \hat{\xi}}{\partial y^2} + \frac{\partial^2 \hat{\xi}}{\partial z^2}) \tag{7}$$

2.3 Transformation of the Governing Equations

As the flow in the system is entirely driven by the introduced jet advected by the crossflow, the variation of jet characteristics depends upon M, U and the spatial coordinate x. By dimensional analysis, length, velocity and eddy viscosity scales in the jet crossflow can be deduced as:

$$1 = k(\frac{Mx}{U^2})^{1/3} \qquad V^* = K(\frac{MU}{x^2})^{1/3} \qquad \varepsilon = \alpha(\frac{M^2}{Ux})^{1/3} \tag{8}$$

where K, k, α are constants.

Non-dimensionalize as: $(\eta, \varsigma) = (y, z)/1$, $(V, W) = (v, w)/V^*$, $h = \theta/\theta^*$. Then, governing equations will be transformed into new format within a non-inertial reference frame in which the solution approaches self-similar solution in steady state. Especially for vorticity equation, after transformation with characteristic length scales, as follow:

$$-\frac{k}{3K}\left(\eta \frac{\partial \xi}{\partial \eta} + \varsigma \frac{\partial \xi}{\partial \varsigma}\right) - \frac{k}{K}\xi + \left(V\frac{\partial \xi}{\partial \eta} + W\frac{\partial \xi}{\partial \varsigma}\right) = \frac{\alpha}{Kk}\left(\frac{\partial^2 \xi}{\partial \eta^2} + \frac{\partial^2 \xi}{\partial \varsigma^2}\right) \tag{9}$$

With vertical momentum flux, $k^2 K = 1$, if $k = A\alpha^n \Rightarrow 3n = n+1 \Rightarrow n = \frac{1}{2}$ from above equation's coefficient (in magnitude). When $A^3 = 6A$, or $A = \sqrt{6}$, $k = \sqrt{6\alpha}$ and $\lambda = (6\alpha^3)^{-1/2} = 6K/k$, the governing equation can be simplied into:

$$\frac{1}{Pr}\left(\frac{\partial^2 h}{\partial \eta^2} + \frac{\partial^2 h}{\partial \varsigma^2}\right) + 2\left(\eta \frac{\partial h}{\partial \eta} + \varsigma \frac{\partial h}{\partial \varsigma}\right) + 4h = \lambda\left(V\frac{\partial h}{\partial \eta} + W\frac{\partial h}{\partial \varsigma}\right) \tag{10}$$

$$\frac{\partial^2 \xi}{\partial \eta^2} + \frac{\partial^2 \xi}{\partial \varsigma^2} + 2\left(\eta \frac{\partial \xi}{\partial \eta} + \varsigma \frac{\partial \xi}{\partial \varsigma}\right) + 6\xi = \lambda\left(V\frac{\partial \xi}{\partial \eta} + W\frac{\partial \xi}{\partial \varsigma}\right) \tag{11}$$

$$\frac{\partial^2 \psi}{\partial \eta^2} + \frac{\partial^2 \psi}{\partial \varsigma^2} = -\xi \tag{12}$$

3. NUMERICAL SOLUTION

3.1 Computational Parameter and Boundary Conditions

In all the results reported herein, $Pr=1$, $\lambda \in [0,200]$. The domain for computation is chosen to be Ω: $(\eta_1, \eta_2) \times (\varsigma_1, \varsigma_2) = (-8,8) \times (-8,8)$. Numerical experiments show that the domain is big enough for cross-sectional scope of the bent-over jet in crossflow, so the boundary can be treated as infinite boundary. To obtain accurate solution, 3000 time steps suffice for all cases. Convergence is declared when the normalized residual R is less than 2.4×10^{-4} for every equation. By virtue of symmetry assumption, the numerical solution is sought for half of the flow domain, $\eta \geq 0$. On the line of symmetry, $\eta = 0$, $\partial h/\partial \eta = 0$, $\xi = 0$, $\psi = 0$, $\partial^2 \psi/\partial \eta^2 = 0$. At the infinite boundary, all of variables are set to zero.

3.2 Computational Scheme

Alternating-direction implicit (ADI) method is adopted to discretize equation of stream function. For concentration and vorticitity, FTCS method is adopted by applying forward-time and centered-space differences to Pseudo-transient governing equations. The pseudo-time step is an additional parameter in the scheme that can be varied in order to accelerate convergence. Upon convergence, the pseudo-time term

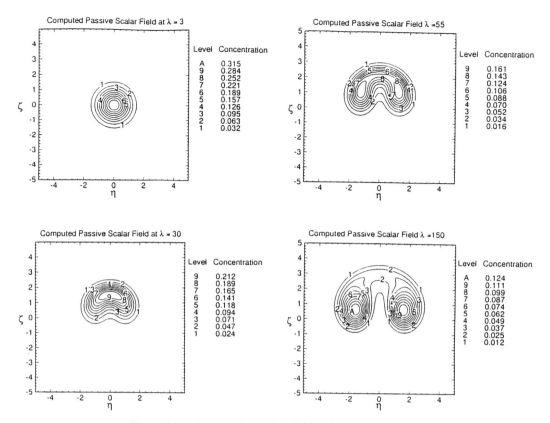

Fig.2 Numerical passive scalar field at λ=3, 30, 55, 150

vanishes, and steady state equations can be satisfied. With Fourier analysis method, stability condition for vorticity & passive scalar equations are found Δt as:

$$6\Delta t + 2\Delta t(\frac{1}{(\Delta x_i)^2} + \frac{1}{(\Delta y_j)^2}) \le 1, \qquad \frac{|2x_{i,j} - \lambda V|}{\Delta x_i}\Delta t + \frac{|2y_{i,j} - \lambda W|}{\Delta y_j}\Delta t \le 1 \qquad (13)$$

Extensive tests have confirmed that the general features and characteristics reported in this paper are unrelated to the numerical procedure adopted. Two simulation with different resolution on a coarser 43×68 and finer 75×138 grid were made to validate the numerical accuracy. All the contour plots are very similar to the high-resolution results, and differences for all the non-dimensional parameters presented are limited to at most 3 percent.

4. NUMERICAL RESULTS

4.1 Scalar Field

In order to find the best λ value to show basic characteristic of bent-over jet in crossflow with minimum relative error, numerical experiments have been designed within range of λ∈[0,200]. Tab.1 and Fig.2 show vertical cross-sectional scalar field at different λ values. As a salient feature of bent-over jet, double peak concentration distribution can generally be simulated in the kidney-shaped sectional scalar field, which become self-similarity after λ>45. Generally, at smaller λ, scalar field is similar to analytical solution of Yih (1981) and Cheung (1991). As increasing λ value, the concentration shape gradually changes from circular distribution to flat, then to kidney-shaped with symmetry maxima. However, when λ>150, the distance of maxima in scalar field intends to increase and to bifurcate the kidney shape with alone two cells distribution.

In order to quantity scalar field with length scales, definition of characteristic lengths of cross section is shown in Fig.3. Corresponding coefficient relationship are found out for each λ..

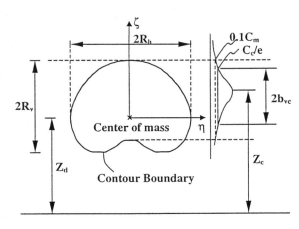

Fig.3 Definition of characteristic lengths of cross section

i) As a basic parameter to characterize the sectional shape, scalar field aspect ratio is defined as the ratio of horizontal half-width R_h over vertical half-width R_v inside contour $C=0.1C_m$, which C_m stands for the maximum concentration over the section. Numerical results show that the aspect ratio is roughly equal to 1.3 for 45<λ<200, but gradually reduce, with a minimum value of 1.0, for small λ. The ratio is essentially consistent with experimental results of 1.27 by Cheung (1991) and comparable to the value of 1.3 suggested by Pratte & Baines (1967) for the developed phase from an analysis of flow visualization. ii) Centerline trajectory coefficient C_{2p} is equal to 1.8965 for λ=55, with an average value of 1.8504 for 50<λ<100. This value is close to experimental value of 1.56 and 1.63 in Wong (1991) and Chu (1996), respectively; and comparable to 1.77 from the buoyant jet experiment of Ayoub (1971). iii) Scalar dilution C_{3p}=0.483 for λ=55. The value is comparable to the experimental value of 0.47 in Wong (1991), but much higher than the computed value of 0.283 for line puffs in Lee (1996). iv) Coefficient of center of mass is 0.7717 for λ=55, it is very close to computed value of 0.75 for line puffs in Lee (1996), and also comparable well with 0.71 of experimental result of Chu (1996).

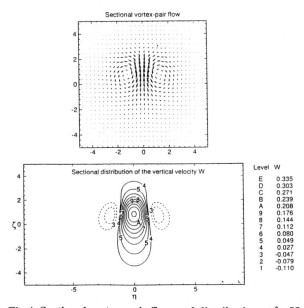

Fig.4 Sectional vortex-pair flow and distribution at λ=55

Table 1. Similarity relations derived from numerical solution

Jet characteristics	Coefficient	λ=30	λ=55	λ=70	λ=150	Experiments
Ratio of jet width	R_h / R_v	1.3684	1.3224	1.3139	1.3371	1.25 (Chu 1996) 1.3 (Pratte 1967)
Vertical velocity	$W_m = C_v M_o^{1/3} t^{-2/3}$	0.9708	1.3269	1.4653	2.0623	1.21 (Chu 1996)
Trajectory	$z_d = C_{2pp} M_o^{1/3} t^{1/3}$	1.6581	1.4337	1.3074	0.9574	1.55 (Chu 1996)
Centerline trajectory	$z_c = C_{2p} M_o^{1/3} t^{1/3}$	1.8710	1.8965	1.8631	1.6643	1.77 (Ayoub 1991)
Center of mass	$L \sim z_d$	0.8601	0.7717	0.7155	0.5869	0.75 (Lee 1996)
Scalar concentration	$C_o / C_m \sim M_o^{2/3} t^{2/3} / V_o$	0.4946	0.4830	0.4511	0.3799	0.47 (Wong 1991)
Centerline half-width	$b_{vc} = C_{1p} z_c$	0.2408	0.1779	0.1975	0.1846	0.27 (Lee 1996)
Max/average velocity	$W_m / \overline{W}$	2.7676	3.2183	3.0812	2.7114	3.17 (Chu 1996)
Turbulent viscosity	$\varepsilon / W_m L$	0.0368	0.0187	0.0146	0.0070	0.018 (Lee 1996)
Maximum concentration	C_m/C_c	1.0	1.2496	1.6951	4.0048	1.2 (Chu 1996)

Note: Contour boundary of this study is on $0.1 C_m$; Lee 1996 is numerical results of turbulent line puff by J.H.W.Lee, Rodi.W & Wong C.F., its contour boundary on $0.01\, C_m$; Paul 1996 is experimental measurements of advected line puff in crossflow by Paul Chu, its contour boundary on $0.25\, C_m$; Wong 1991 is experimental measurements of line puff in crossflow by Wong C.F., its contour boundary on C_m /e.

4.2 Velocity Field

The resulted secondary flow pattern, which is very similar to that in advected line puff (Lee & Chen 1997), is shown in Fig.4, to illustrate the production of the vortex-pair system in the bent-over phase, as another major feature of a jet in a crossflow. The shape of flow field is well-preserved for λ>50, the maximu value of the normalized stream function ψ*=-0.2,-0.2028,-0.2146,-0.2326 at λ=3,10,30,55, respectively. On the other hand, the vorticity field is qualitatively similar to the stream function in the main body of the jet. For λ=55, the vortex eyes are located at y/l=±0.24, $(z-z_d)$/l=-0.06.

5. CONCLUSIONS

With self-similarity laws & Prandtl's free shear layer model, a non-dimensional governing equation is developed with semi-analytical method. Based on the basic equation, a numerical study of mixing behavior of cross-sectional of a bent over jet in current is developed. The salient feature of the predicted flow and scalar mixing rates are supported well by experiment and others numerical results. The main conclusions are i) Computed double-vortex flow is similar to the observed bent-over jet in crossflow and analogy ALP. The scalar field is sectional characterized by a kidney-shaped outline containing a double peak of concentration maxima at the high λ values. ii) Comparing with experimental data and others computed results, only when λ=55, in our numerical range [0,200], is the best value to give the typical scalar field and flow field of a bent-over jet in current. iii) The expression of eddy viscosity as a function of the product of characteristic vertical velocity and length scale, based on self-similarity law, to transform complicated problem of bent-over jet in crossflow to 2-D cross-sectional study, and to avoid unnecessary complex governing equations and surplus parameters.

REFERENCES

Cheung V. (1991). "Mixing of a round buoyant jet in a current", *PhD thesis*, Univ. of Hong kong.
Chu C.K.P (1996). "Mixing of a turbulent advected line puffs", *PhD thesis* , Univ. of Hong kong.
Chu V.H. (1985). "Oblique turbulent jets in a crossflow", *J.Engg. Mech.*, ASCE, 111, 1343-1360.
Lee J.H.W., Rodi W. and Wong C.F. (1996). "Turbulent line momentum puffs", *J. Engg. Mech.*, ASCE, 122(1),19-29.
Moussa Z.M. etc (1977). "The near field in the mixing of a round jet with a cross-stream", *J.Fluid Mech.*, 80, 49-80.
Patankar S.V. etc (1977). "Prediction of the three-dimensional velocity field of a deflected turbulent jet", *J.Fluids Engg.*, ASME, 99,758-762.
Wong C.F (1991). "Advedted line thermals and puffs", *M.Phil thesis*, Univ. of Hong kong
Yih C.S. (1981). "Similarity Solutions for Turbulent Jets and Plumes", *J. Engg. Mech.*, ASCE, 107(3), 455-477.

Environmental Hydraulics, Lee, Jayawardena & Wang (eds) © 1999 Balkema, Rotterdam, ISBN 90 5809 035 3

The spreading rate of a single coflowing axisymmetric jet

H.J.Wang & M.J.Davidson
Department of Civil Engineering, Hong Kong University of Science and Technology, China

ABSTRACT: The preliminary results of an experimental investigation into the behavior of a single jet in a coflowing ambient fluid are presented. The experimental results are compared with the predictions from two numerical models, one of which uses the centerline-velocity to define the spreading rate while the other uses the cross-sectional average-velocity for this task. The comparison of spread data with the model predictions indicates that the centerline velocity is the appropriate characteristic velocity in defining coflowing jet spread. However the dilution data departs from the centerline velocity model, which suggests that the turbulent fluxes may become more significant when approaching the transition region. Detailed experimental investigations into this problem are continuing.

1. INTRODUCTION

Studies of the behavior of single coflowing jets often provide the basis for understanding more complex buoyancy driven flows. Improvements in our understanding of these more fundamental problems is central to our ability to deal with many engineering problems, such as estimating the initial dilution of wastewater released from outfall diffusers.

For an axisymmetric jet issuing into a coflowing ambient fluid, there exists two distinct regions. The strong jet region occurs when the jet behavior is dominated by the jet excess momentum and the weak jet region occurs when the jet behavior is strongly affected by the coflowing ambient flow. In between these regions is the transition region where the flow transforms from strong jet behavior to weak jet behavior.

Although the behavior of single coflowing jet has been studied for many years (Antonia & Bilger 1973, Knudsen 1988, Nickels and Perry 1997, and others), there are still some important unanswered questions. The definition of the spreading rate of the discharge in the weak jet region is one of them. In the past, experimental investigations into the behaviour of single coflowing axisymmetric jets have provided insights into the discharge behaviour in the strong jet and transition regions, but there is no data available from the weak jet region. Data in the transition region shows significant scatter and does not provide any reliable indication as to the appropriate definition of the spread function in the weak jet region. The scatter in the data may result from the presence of ambient turbulence, which can have a strong influence on the discharge behaviour in the transition and weak jet regions (Wang 1996).

In this paper we report on the preliminary results of an experimental investigation into the behaviour of an axisymmetric coflowing jet. The aim of the investigation is to obtain detailed information about the nature of the discharge in the weak jet region, in an essentially non-turbulent ambient current. Initially however we focus on the application of the laser-induced fluorescence experimental technique in the strong jet and transition regions, and this data is presented here.

2. MODELING A SINGLE COFLOWING AXISYMMETRIC JET

A common method for dealing with a jet problem is to assume the excess velocity and tracer distributions are Gaussian and self-similar (Forstall & Shapiro, 1950; Knudsen, 1988). The governing equations based on the conservation of momentum and tracer fluxes are then developed. For a single coflowing jet the conservation of momentum flux equation can be written as:

$$I_m U_e^2 b^2 + I_q U_e U_a b^2 = M_{e0} \tag{1}$$

while the conservation of tracer flux has the form:

$$I_c U_a C b^2 + I_{qc} U_e C b^2 = C_0 Q_0 \tag{2}$$

where
U_e is the centerline excess velocity,
U_a is the ambient velocity.
M_{e0} is the initial jet excess momentum,
C_0 is the initial tracer concentration,
C is the centerline tracer concentration,
Q_0 is the initial volume flux of the flow,
b is a characteristic jet radius based on the Gaussian velocity distribution,

and I_m, I_q, I_c, I_{qc} are the shape factors, the values of which can be determined by integrating the velocity and concentration distributions. It is worth noting that when determining the values of the shape factors I_m and I_{qc}, contributions from the turbulent fluxes must also be taken into account. In addition a second characteristic radius can be defined based on the Gaussian concentration distribution and this is represented by λb, where λ is a constant and it is assumed to have a value of 1.19 (Davidson and Pun 1998).

To close the system of equations, an assumption about the jet spread rate is made. For a single axisymmetric jet issuing in a stagnant ambient fluid, the spread function is given as (Abraham, 1965)

$$\frac{db}{dx} = k \tag{3}$$

where k is constant.

To extend the spread function to coflowing ambient fluids, it must be modified to account for the impact of the ambient current. Based on the work of Patel (1971), in the presence of an ambient current, the spread function becomes:

$$\frac{db}{dx} = k \frac{u}{u + U_a} \tag{4}$$

Where u is a characteristic jet velocity

Traditional models of single coflowing jets employ the centerline excess velocity (U_e) to define the rate of spread of the discharge (Patel, 1971; Antonia & Bilger, 1973; Knudsen, 1988). The resulting spread equation is:

$$\frac{db}{dx} = k \frac{U_e}{U_e + U_a} \tag{5}$$

However, recently Chu (1994) and Wright (1994) have suggested that the average or top-hat velocity ($U_e/2$) is a more appropriate velocity scale for the jet spread function. Replacing the center-line velocity by the average velocity in the spread function results in the following relationship:

$$\frac{db}{dx} = k\frac{U_e}{U_e + 2U_a} \qquad (6)$$

However, Chu's study focused on an axisymmetric jet in a stagnant environment and in this still jet (and the strong jet case where $U_e \gg U_a$) the rate of spread is a constant. Therefore the choice of characteristic velocity has little impact. Conversely in the weak jet (strongly-advected) region, where $U_e \ll U_a$, the choice of characteristic velocity alters the rate of spread by a factor of 2. Wright studied the coflowing jet problem, but the experimental results analyzed show considerable scatter and it is difficult to draw concrete conclusions about the appropriate spreading rate in the weak jet region.

Two numerical models for a single coflowing jet have been generated. One employs a spread assumption with the centerline velocity as the characteristic velocity (Eq. (5)), while the other uses a spread function with the average velocity as the characteristic velocity (Eq. (6)). Combined with the conservation of momentum and tracer flux, the two models provide predictions of the variation of the mean velocity, spread and dilution with downstream distance. Predictions of spread variations along the jet axis from the two models are shown in Figure 1 and these are compared with available experimental data.

In Figure 1 it can be seen that there is no difference in the spread predictions from the two models in the strong jet region. From the transition region, however, the two curves diverge and the jet spread growth rates are different by a factor of two in the weak jet region. A comparison of the models' predictions with data from the weak jet region will give a clear indication of the appropriate velocity scale for defining the spreading rate of a coflowing jet. However, as mentioned previously existing data sets (Knudsen 1988 and Nickel and Perry 1997) are inconclusive because of a lack of information in the weak jet region and scatter

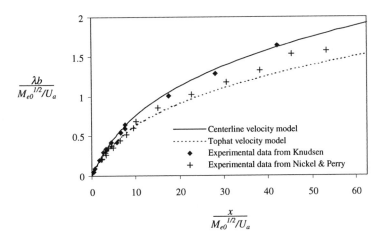

Figure 1   Spread predictions and existing experimental data for a single coflowing jet

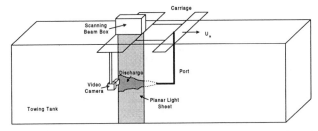

Figure 2   A schematic diagram of the experimental system used to study single axisymmetric jets in a coflowing ambient fluid

131

in the data in the transition region, which may in part be due to the presence of ambient turbulence. In the present investigation the experiments are designed to reduce the effects of ambient turbulence to a negligible level.

3. EXPERIMENTAL EQUIPMENT

A comprehensive experimental investigation is currently being carried out into the coflowing axisymmetric jet problem. The experiments are being conducted in a towing tank which is 1.6m deep, 15m long and 2m wide. Laser Induced Fluorescence (LIF) techniques are employed to accurately measure concentration profiles in the strong jet, transition and weak jet regions. The experimental system used in this study is shown schematically in Figure 2.

During the experiments, fluorescent dye water is pumped from a storage tank to a horizontally positioned port (diameter = 1.67mm). The water is then discharged from the port as a single axisymmetric jet into the receiving water in the towing tank. The discharge port is mounted on a carriage, which can be towed along the tank at a constant velocity. When the discharge port moves (in the stationary ambient fluid) in the opposite direction to that in which it is discharging, relatively the ambient water will have a velocity in the same direction as the discharge velocity and thus a coflowing axisymmetric jet is simulated.

A parallel planar laser light sheet is generated by a laser beam, from a 6 Watt Argon Laser, which passes through a beam scanning box. The laser light sheet cuts a vertical observation plane through the discharge centerline. The beam scanning box is towed with the discharge source, so that the light sheet stays in the same position relative to the source. The fluorescent dye in the discharge water fluoresces at the laser light wavelength (514nm), so that the jet behavior can be observed when it passes through the light sheet. The concentration of the tracer is related to the light intensity of the fluorescence through a calibration procedure and corrections are made for light sheet variations and attenuation. A video camera, mounted on and moving with the carriage, records the behavior of the jet in the light sheet.

This setup effectively reduces the level of ambient turbulence, because of the very small discharge structure and the essentially stationary ambient fluid. In addition the discharge is controlled such that it is initially in the strong jet region, where any turbulence generated is entrained into and digested by the jet, hence in the weak jet region the effects of any turbulence produced by the discharge structure are avoided. The coflowing jet in this experimental investigation can therefore be considered as a single axisymmetric jet issuing in a non-turbulent coflowing ambient fluid.

4. RESULTS AND DISCUSSION

Dimensional analysis yields a length scale (l_{sw}) which provides an estimate of the location of the transition from strong jet to weak jet behavior (Knudsen, 1988). This length scale has the following form:

$$l_{sw} \sim \frac{M_{e0}^{\frac{1}{2}}}{U_a} \sim \frac{\sqrt{(1-U_{ar})\pi/4}}{U_{ar}} d_p \qquad (7)$$

Where M_{e0} is the initial jet excess momentum,
               U_{ar} is the initial velocity ratio ($U_{ar} = U_a/U_0$).

Previous experimental results indicate that this transition occurs over a significant distance, starting when the distance from source is approximately equal to l_{sw} and the transition is not yet complete when the distance from the source approaches 100 l_{sw}. It is clear from Eq. (7) that the length scale for the strong jet to weak jet transition is governed by the initial velocity ratio and port diameter. The jet behavior in the strong jet, transition and weak jet regions can therefore be studied by changing the initial velocity ratio.

To date the single coflowing jet behavior has been studied at 5 different initial velocity ratios (U_{ar} = 0.00344, 0.00681, 0.0135, 0.0265, and 0.0527). In all of these experiments the initial discharge flowrate

(Q_0) was constant at 200ml/min. The initial velocity ratios were varied by changing the ambient velocity. During the experiments, the distance between the discharge port and laser light sheet was 500mm, and the laser light sheet was approximately 500mm wide and 5mm deep.

An image processing program is employed to digitize, process and analyze the images from the experiments. The data analysis provides information on the centerline (minimum) dilution and jet spread along the jet axis. The spread data from these initial experiments is plotted against downstream distance in Figure 3, where this data is compared with predictions from the two models.

The experimental results in Figure 3 are obtained in the strong jet and transition regions, and as yet no data is available in the weak jet region. However the experimental data obtained to date is consistently in agreement with predictions from the model employing the centerline velocity, as the appropriate velocity scale in defining the spread of the discharge. This indicates that predictions from the centerline velocity model are more accurate than those from the cross-sectional average velocity model.

The centerline dilution ($C_r = C_o/C$) experimental data is plotted against downstream distance in Figure 4 and the predictions from the two models are also plotted for comparison.

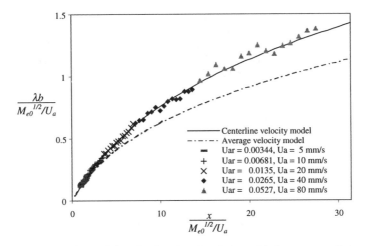

Figure 3   Experimental data for the spread variation with downstream distance

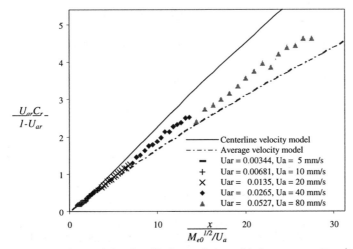

Figure 4. Experimental data for dilution variation with downstream distance

133

Interestingly the experimental data is initially consistent with the centerline velocity model predictions, but the data diverges from the curve and falls between the centerline and the average velocity models as the downstream distance increases. An analysis of the fluctuation statistics obtained during the experiments indicates the root-mean-square of the concentration fluctuations increases dramatically in the transition region. Previous experimental investigations have also shown that the root-mean-square of the velocity fluctuations increases in the transition region (Nickel & Perry, 1997). These results indicate that the portions of tracer and momentum transported by the turbulent fluxes change as the flow enters the transition region. This will have an impact the values of some of the shape factors in the conservation equations (I_m and I_{qc} in equations (5) and (6)). It is possible that changes in the values of the shape factors are responsible for the observed differences between the experimental data and the dilution predictions in the transition region. This issue is currently being investigated in more detail. It is interesting to note that as the flow approaches the weak jet region the shape factors I_m and I_{qc} are no longer significant and one would expect the experimental data and the centerline velocity model predictions to once again show good agreement.

5. CONCLUSIONS

The preliminary results of an experimental investigation into the behaviour of single coflowing jet have been presented. These results indicate that the centerline velocity is the appropriate characteristic velocity in defining the spreading rate of a coflowing jet. However, the mean dilution data deviates from this model in the transition region and it is possible that this is the result of increases in the portions of tracer and momentum that are transported by the turbulent fluxes. The experimental investigation is continuing, so that experimental data can be obtained in the weak jet region.

6. ACKNOWLEDGMENTS

The authors would like to acknowledge the financial support of the Research Grants Council of Hong Kong.

7. REFERENCES

Abraham, G. (1965) Horizontal Jets in Stagnant Fluid of Other Density, *Journal of Hydraulics Division, ASCE*, 91, No. HY4, Proc. Paper 4411, 139-154.

Antonia, T. A. and Bilger, R. W. (1973) An experimental investigation of an axisymmetric jet in a coflowing air stream, *J. Fluid Mech.* 61, 805-822.

Chu, V. H. (1994), Lagrangian scaling of turbulent jets and plumes with dominant eddies, in Davies, P.A. and Valente Neves, M. J. (eds.), Recent Research Advances in the Fluid Mechanics of Turbulent Jets and Plumes, 45-72, Kluwer Academic Publishers, Netherlands.

Davidson, M. J. (1989) The behavior of single and multiple , horizontally discharged, buoyant flows in a nonturbulent coflowing ambient fluid. Ph.D. Thesis, Rep. 89-3, Dept. of Civil Engineering, University of Canterbury, Christchurch, New Zealand.

Davidson M. J., and Pun K. L., (1998). Weakly-advected jets in a cross-flow. Journal of Hydraulic Engineering, ASCE. In press.

Fischer, H. B., List, E. J., Koh, R. C. H., Imberger, J., and Brooks, N. H. (1979) Mixing in Inland and Coastal Waters, Academic Press, New York, USA.

Knudsen, M. (1988) Buoyant horizontal jets in an ambient flow. Ph.D. Thesis, Rep. 88-7, Dept. of Civil Engineering, University of Canterbury, Christchurch, New Zealand.

Nickels, T. B. and Perry, A. E. (1996) An experimental and theoretical study of the turbulent coflowing jet. *J. Fluid. Mech.* 309, 157-182.

Patel, R. P. (1971) Turbulent Jets and Wall Jets in Uniform Streaming Flow. *Aero. Quart.* XXII, 311-326.

Wood, I. R., Bell, R. G., and Wilkinson, D. L. (1993) Ocean Disposal of Wastewater, World Scientific Publishing, Singapore.

Wright, S. J. (1994), The effect of ambient turbulence on jet mixing, in Davies, P.A. and Valente Neves, M. J. (eds.), Recent Research Advances in the Fluid Mechanics of Turbulent Jets and Plumes, 13-27, Kluwer Academic Publishers, Netherlands.

Environmental Hydraulics, Lee, Jayawardena & Wang (eds) © 1999 Balkema, Rotterdam, ISBN 90 5809 035 3

Momentum-based solutions for buoyant jet trajectories in still and flowing ambient fluids

K.L.Pun & M.J.Davidson
Department of Civil Engineering, Hong Kong University of Science and Technology, China

I.R.Wood
Department of Civil Engineering, University of Canterbury, Christchurch, New Zealand

ABSTRACT: A perturbation analysis based on the relevant components of momentum is employed to locate the two-dimensional centreline trajectories of buoyant jets in still and flowing ambient fluids. The use of this approach is conceptually different from a traditional analysis, commonly found in Eulerian integral models, which makes use of the relevant components of the centreline velocity for locating discharges. Comparisons with available data indicate that the momentum based trajectory equations can accurately predict the flow trajectories and are in good agreement with the trajectory predictions from a Lagrangian integral model - JETLAG. Lagrangian models, such as JETLAG, employ a cross-sectional average velocity to locate the flow trajectory and this is equivalent to the momentum based method described here.

1. INTRODUCTION

The ability to accurately locate wastewater released from an outfall diffuser can be critical when, for example, determining whether or not interference between neighbouring discharges is possible. Complex flows, representative of such wastewater discharges, can be subdivided into regions (limiting cases) where the behaviour of the discharged fluid is governed by a single parameter (see for example Davidson and Pun 1998a), which can be a single initial discharge parameter or some combination of the initial and ambient parameters. This governing parameter is indicative of the dominant components of momentum within the region it governs. Solutions based on simplified forms of the momentum and mass flux equations are able to accurately predict the growth and dilution of the released wastewater within each flow region. However, the flow trajectory in these regions can be significantly altered by relatively small components of momentum, originating from either the initial discharge, buoyancy forces or the ambient current. Neglecting these secondary influences within the regional models can result in erroneous trajectory predictions. This is particularly important when these relatively small components of momentum are acting perpendicular to dominant momentum component(s).

Traditionally these flows are located using a perturbation analysis, which is based on the relevant components of the centreline velocity. Evidence of this can be found in Wright (1977), Doneker and Jirka (1990), Wood (1993) and others. However, it is also possible to define the trajectory relationships based on the relevant components of momentum. This approach assumes that the location of the flow centreline is a function of the total flow momentum in the cross-section, rather than the momentum local to the central regions of the flow. In this paper we employ a momentum ratio method to incorporate secondary influences on the trajectory of the released fluid in two flow regions. The behaviour in these regions is discussed in the context of a horizontally discharged buoyant jet in a still ambient fluid and a vertically discharged buoyant jet in a cross-flow.

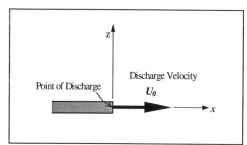

FIG. 1 Configuration of a horizontally
discharged buoyant jet in a still ambient fluid

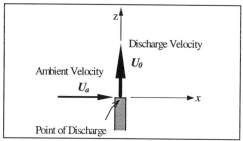

FIG. 2 Configuration of a vertically discharged
buoyant jet in a cross-flowing ambient current

2.   HORIZONTALLY DISCHARGED BUOYANT JET IN A STILL AMBIENT FLUID

A schematic diagram of the discharge configuration for a horizontally discharged buoyant jet in a still
ambient fluid is presented in Fig. 1. The jet and plume flow regions form with this type of discharge and
within these regions secondary effects due to the initial discharge momentum or buoyancy forces can
deflect the flow trajectory.

2.1 Buoyancy in the Jet Region

The flow in the jet region is driven by the constant initial momentum (M_0) in the horizontal discharge
direction. The horizontal momentum (M_H) can therefore be written as:

$$M_H = M_0 = \frac{I_m}{I_Q} UQ \tag{1}$$

where Q is the local volume flux ($= I_Q Ub^2$), U is the centreline mean velocity in the direction of the
dominant momentum component (the discharge direction in this case), b is the spread of the flow, and I_m
and I_Q are shape constants with values of 1.74 and 3.14 respectively.

Initially the discharge has no vertical momentum, but this increases with distance from the source because
of the buoyancy forces acting on the flow. We write the vertical momentum (M_V) of the discharge as:

$$M_V = U_D Q \tag{2}$$

where U_D is the average deflection velocity (generated by the buoyancy forces in this case). The rate of
change of the buoyancy-induced vertical momentum is then written in the following form:

$$\frac{d}{dx}[QU_D] = I_\Delta \Delta b^2 \tag{3}$$

where I_Δ is a shape constant with a value of 3.6, Δ is the centreline density deficit ($=[(\rho_a - \rho)/\rho_a]g$), ρ_a
and ρ are the densities of the ambient fluid and the local jet fluid, and g is the gravitational acceleration.

In the jet region we know that $M_V << M_H$ and hence $U_D << U$. We therefore assume that the only significant
impact of the buoyancy-induced velocity is to deflect the flow. It is appropriate then to substitute the density
deficit flux, spread and velocity equations for a jet (see for example Davidson and Pun 1998a) into eq. 3 and
carry out an integration to get:

$$\frac{U_D}{U_0} = \frac{1}{2} \frac{I_\Delta I_m}{I_{Q\Delta} I_Q} \frac{1}{F_{r0}^2} \left[\frac{x}{d_p} \right] \tag{4}$$

where $I_{Q\Delta}$ is a shape constant which has a value of 2.0, d_p is the port diameter, U_0 is the initial discharge
velocity, and F_{r0} is the initial densimetric Froude number.

The differential trajectory relationship obtained from a perturbation analysis based on the relevant

136

momentum components is then:

$$\frac{dz}{dx} = \frac{\langle Buoyancy\ Induced\ Momentum\rangle_{Deflection}}{Jet\ Momentum} = \frac{I_Q U_D}{I_m U}$$ (5)

A momentum ratio solution for the jet trajectory under the influence of buoyancy forces is then obtained as:

$$\frac{z}{d_p} = \frac{I_\Delta I_m^{1/2} k_G}{3 I_{Q\Delta} \pi^{1/2}} \frac{1}{F_{r0}^2} \left[\frac{x}{d_p}\right]^3$$ (6)

where k_G is the spread constant and has a value of 0.11.

2.2 Initial Momentum in the Plume Region

In the plume region, the flow is driven by the buoyancy-induced vertical momentum and this can be written as:

$$M_V = UQ$$ (7)

The governing equation in this region is then:

$$\frac{d}{dz}[QU] = I_\Delta \Delta b^2$$ (8)

However, the initial momentum of the discharge continues to have an influence on the flow trajectory in the direction of the initial discharge. The initial momentum which deflects the plume trajectory in the discharge direction is written as:

$$M_0 = QU_D$$ (9)

In this region $M_V \gg M_H$ and hence the deflection velocity is very small when compared to the component of the velocity in the direction of the dominant momentum component (U). We therefore introduce the spread and velocity solutions for a plume (Pun and Davidson 1998) into eq. 9 and solve for the deflection velocity:

$$\frac{U_D}{U_0} = \left[\frac{\pi^2 I_{Q\Delta} I_m}{12 k_G^4 I_Q^3 I_\Delta}\right]^{1/3} F_{r0}^{2/3} \left[\frac{1}{z/d_p}\right]^{5/3}$$ (10)

The differential trajectory relationship for the plume in terms of the relevant components of momentum is then:

$$\frac{dz}{dx} = \frac{Plume\ Momentum}{\langle Initial\ Momentum\rangle_{Deflection}} = \frac{I_m}{I_Q} \frac{U}{U_D}$$ (11)

The deflection of the plume trajectory resulting from the initial momentum is then:

$$\frac{x}{d_p} = \left[\frac{Z_{JP}}{d_p}\right] - \left[\frac{12\pi I_{Q\Delta}^2}{I_m I_\Delta^2 k_G^2}\right]^{1/3} F_{r0}^{4/3} \left\{ \left[\frac{1}{z/d_p}\right]^{1/3} - \left[\frac{1}{Z_{JP}/d_p}\right]^{1/3} \right\}$$ (12)

where $\frac{Z_{JP}}{d_p} \left(= C_{JP}\left[\frac{\pi}{4}\right]^{1/4} F_{r0}\right)$ is the distance to the transition from a jet to a plume. The constant C_{JP} has a value of 2.3.

3.   VERTICALLY DISCHARGED BUOYANT JET IN A CROSS-FLOW

A vertical buoyant discharge in a weak cross-flow normally forms a jet and then becomes a plume due to the buoyancy effects. A schematic diagram of this type of discharge is presented in Fig. 2. The ambient cross-flow, which is perpendicular to the discharge, deflects the flow trajectory in the horizontal direction.

137

The influence of the ambient current can be dealt with in the jet and plume regions as follows:

3.1 A Weak Ambient Current in the Jet Region

The flow in the jet region is driven by the initial momentum of the discharge and this is in the vertical direction, so:

$$M_V \approx \frac{I_m}{I_Q} UQ \approx M_0 \qquad (13)$$

In this situation the relatively small buoyancy-induced vertical momentum is in the discharge direction and hence it has no significant influence on the flow trajectory. In contrast the relatively small ambient velocity is perpendicular to the discharge direction and the entrained ambient momentum is given by:

$$M_H = QU_a \qquad (14)$$

where U_a is the ambient velocity. Expressing the differential trajectory relationship in terms of the relevant components of momentum we get:

$$\frac{dz}{dx} = \frac{Jet\ Momentum}{\left\langle Ambient\ Momentum \right\rangle_{Deflection}} = \frac{I_m}{I_Q}\frac{U}{U_a} \qquad (15)$$

Recognising that while $M_V >> M_H$ the flow behaves in essentially the same manner as jet released into a still ambient fluid (Davidson and Pun 1998b) we can substitute for the jet velocity and integrate to obtain:

$$\frac{x}{M_0^{1/2}/U_a} = \frac{I_Q}{I_m}\frac{k_G I_m^{1/2}}{2}\left[\frac{z}{M_0^{1/2}/U_a}\right]^2 \qquad (16)$$

It is worth noting that a similar jet trajectory equation can be derived by employing the relevant centreline velocity components in the above analysis. In this situation the analysis yields $\frac{dz}{dx} = \frac{U}{U_a}$ which gives the trajectory relationship as:

$$\frac{x}{M_0^{1/2}/U_a} = \frac{k_G I_m^{1/2}}{2}\left[\frac{z}{M_0^{1/2}/U_a}\right]^2 \qquad (17)$$

Equation 17 is similar to the equation based on the momentum ratio approach (eq. 16). The only difference is the ratio of two shape factors ($I_Q/I_m \approx 2$). This difference can be eliminated if a cross-sectional average velocity is selected as the characteristic velocity for locating the discharge trajectory, where the average velocity is based on the ratio of the relevant momentum flux to the mass flux, that is, $U_{avg}=U\ I_m/I_Q$. The use of the centreline velocity ratio method is based on the assumption that the central regions of the flow are transported by the central velocities of the flow. In contrast, the momentum ratio method considers the total momentum in the flow cross-section to be responsible for locating the discharge. The latter implies that there are significant motions perpendicular to the mean flow direction. Recent experiments have shown that this is indeed the case and the importance of the large-scale eddies in providing this type of motion (Davidson and Pun 1998b).

3.2 A Weak Ambient Current in the Plume Region

In the plume region the buoyancy-induced vertical momentum is dominant over the initial momentum and the ambient current deflects the flow trajectory. Following the same procedure the trajectory relationship in this case can be expressed as:

$$\frac{dz}{dx} = \frac{Plume\ Momentum}{\left\langle Ambient\ Momentum \right\rangle_{Deflection}} = \frac{I_m}{I_Q}\frac{U}{U_a} \qquad (18)$$

Substituting the velocity solution for a plume (Pun and Davidson 1998) into eq. 18 and integrating yields:

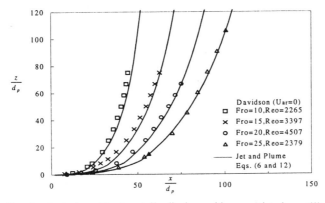

FIG. 3 Predicted trajectories of horizontally discharged buoyant jets in a still ambient fluid

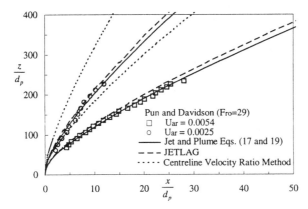

FIG. 4 Predicted trajectories of vertically discharged buoyant jets in a cross-flow

$$\frac{x}{Q_{\Delta 0}\big/U_a^3} = \frac{I_Q}{I_m}\left[\frac{9I_{Q\Delta}I_m k_G^2}{16I_\Delta}\right]^{1/3}\left[\frac{z}{Q_{\Delta 0}\big/U_a^3}\right]^{4/3} \tag{19}$$

where $Q_{\Delta 0}\,(=Q_0\Delta_0)$ is the initial density deficit flux of the flow. A centreline velocity ratio approach yields a similar equation, but again there is the difference of the shape factor ratio I_Q/I_M. This difference can also be resolved through an appropriate selection of the characteristic velocity when defining the flow trajectory.

4. APPLICATION AND VERIFICATION

In this section, the momentum ratio solutions for the flow trajectories in the jet and plume regions are compared with experimental data. Transitions between flow regions are dealt with through the use of length-scales and virtual sources (Davidson and Pun 1998a). The trajectory predictions of a Lagrangian-integral model (JETLAG - Lee and Cheung 1991) and a representative Eulerian-integral (centreline velocity) model are also included for the discharge case of a vertical buoyant jet in a cross-flow. Fig. 3 shows the trajectory predictions for horizontally discharged buoyant jets in a still ambient fluid. Good agreement is obtained in the jet and plume regions when compared with the data of Davidson (1989). In Fig. 4, the momentum-ratio solutions for the trajectories of vertically discharged buoyant jets in a weak cross-flow compare well with the data of Pun and Davidson (1998) and are consistent with the predictions of the

JETLAG model. As discussed, the momentum ratio approach is equivalent to the use of an appropriately defined cross-sectional average velocity for locating the flow centreline. Lagrangian models employ this appropriate cross-sectional average velocity for locating the flows and these models progress in a stepwise manner along the flow trajectory obtaining the average (top-hat) properties of the flow. However, the present method can be employed to provide explicit analytical solutions for the expected behaviour in any particular flow region. The formulation provides physical insight into those factors influencing the flow trajectory within the region and the solution then defines the expected limiting behaviour. Perhaps most importantly, there is no arbitrary selection of the appropriate characteristic velocity for locating the discharge.

5. CONCLUSIONS

Momentum-ratio based solutions for the flow trajectories of a horizontally discharged buoyant jet in a still ambient fluid and a vertical buoyant jet in a weak cross-flow have been derived. The secondary effects of the relatively weak momentum components originating from the initial discharge, buoyancy forces and ambient current have been taken into account in locating the flow trajectories. The momentum based approach differs from the traditional method of formulating trajectory equations, which is based on the relevant centreline velocity components. The derived trajectory equations have been shown to provide accurate trajectory predictions for the cases considered. In addition the use of the total cross-sectional momentum flux when locating the flow centreline, is shown to be equivalent to the use of an appropriate cross-sectional average. The momentum based method requires that there be significant motions perpendicular to the mean flow direction in the cross-section, such that a fluid particle moves with a range of cross-sectional velocities as it progresses downstream. In this way the total momentum in any given cross-section contributes to the overall progress of the discharged fluid. The large scale eddies, present in these flows, are responsible for generating the required significant perpendicular motions on the scale of the flow.

6. ACKNOWLEDGEMENT

The authors acknowledge the financial support of the Research Grants Council of Hong Kong.

7. REFERENCES

1. Davidson, M. J. (1989), The behaviour of single and multiple, horizontally discharged, buoyant flows in a non-turbulent coflowing ambient fluid, Ph.D Thesis, Department of Civil Engineering, University of Canterbury, Christchurch, New Zealand, Report 89-3.
2. Davidson M. J., and Pun K. L., (1998a). A hybrid model for the prediction of initial dilutions from outfall discharges. Journal of Hydraulic Engineering, ASCE. In press.
3. Davidson M. J., and Pun K. L., (1998b). Weakly-advected jets in a cross-flow. Journal of Hydraulic Engineering, ASCE. In press.
4. Doneker, R. L., and Jirka, G. H. (1990), Expert system for hydrodynamic mixing zone analysis of conventional and toxic submerged single port discharges (CORMIX1), EPA/600/3-90/012, Feb. 1990, USEPA.Hung, W. K. (1998), A jet at an oblique angle to a cross-flow, Final Year Project Report, Civil and Structural Engineering Department, The Hong Kong University of Science and Technology.
5. Lee, J. H. W., and Cheung, V. (1991), Generalized Lagrangian model for buoyant jets in current, Journal of Environmental Engineering, ASCE, 116(6), 1085-1106.
6. Pun, K. L., and Davidson M. J., (1998). On the behaviour of advected plumes and thermals. Journal of Hydraulic Research, International Association of Hydraulic Research (IAHR). In press.
7. Wood, I. R. (1993), Asymptotic solutions and behaviour of outfall plumes, Journal of Hydraulic Engineering, ASCE, 119(5), 555-580.
8. Wright, S. J. (1977), Effects of ambient cross-flow and density stratification on the characteristic behaviour of round turbulent buoyant jets, Report No. KH-R-36, W. M. Keck Lab. of Hydr. and Water Resour., California Inst. of Tech., Pasadena, California.

Anisotropic buoyant turbulence model and its application

Zhou Xueyi, Chen Yongcan, Yang Kun & Li Yuliang
Department of Hydraulic and Hydropower Engineering, Tsinghua University, Beijing, China

ABSTRACT: By analyzing the components of Reynolds stresses of implicit algebraic stress model(IASM), this paper assumes that Reynolds stresses in buoyant turbulent flows are induced by both strain and buoyancy. Thereby, a nonlinear anisotropy buoyant turbulence model is developed by applying equilibrium hypothesis to Reynolds stress transports. The model has taken account of buoyancy and avoided numerical singularity. The model is applied to simulate vertical buoyant jets from the bottoms of tidal flows. The numerical results are reasonable and state that the model can predict the important role of buoyancy in tidal flows.

1. INTRODUCTION

Buoyant flow is one of the fundamental flows. The difference of density between discharged   fluid and the ambient fluid can cause buoyant turbulent flows , such as sewage discharge in estuaries or coastal zones, and heated effluents from thermal power plants into rivers. Turbulence is intensified by buoyancy, and its accurate simulation becomes more difficult.

At present , implicit algebraic stress model (IASM) is often used to simulate turbulent flows. It takes account of turbulent anisotropy and buoyancy , but it could be failed because of numerical singularity if local strain is large(Gatski & Speziale,1993). To improve accuracy and to avoid numerical singularity , turbulent stresses are modeled with different ways(Pope,1975, Rubinstein & Barton, 1990, Shih, et al 1995).   But these models don't take account of buoyancy . The key problem in the paper is to develop a new model which can not only take account   of   anisotropy and buoyancy to properly describe turbulent flow, but also avoid numerical singularity .

2. ANISOTROPIC BUOYANT TURBULENCE MODEL

The governing equations of 2D planar flows are

$$\frac{\partial \rho}{\partial t} + \frac{\partial \rho U}{\partial x} + \frac{\partial \rho V}{\partial y} = 0 \tag{1}$$

$$\frac{\partial \rho U}{\partial t} + \frac{\partial}{\partial x}\left(\rho UU - \mu\frac{\partial U}{\partial x}\right) + \frac{\partial}{\partial y}\left(\rho UV - \mu\frac{\partial U}{\partial y}\right) = -\frac{\partial P}{\partial x} - \frac{\partial \overline{\rho uu}}{\partial x} - \frac{\partial \overline{\rho uv}}{\partial y} \tag{2}$$

$$\frac{\partial \rho V}{\partial t} + \frac{\partial}{\partial x}\left(\rho UV - \mu\frac{\partial V}{\partial x}\right) + \frac{\partial}{\partial y}\left(\rho VV - \mu\frac{\partial V}{\partial y}\right) = -\frac{\partial P}{\partial y} - \frac{\partial \overline{\rho uv}}{\partial x} - \frac{\partial \overline{\rho vv}}{\partial y} - \alpha\Delta C\rho_a\, g \tag{3}$$

$$\frac{\partial \rho C}{\partial t} + \frac{\partial}{\partial x}\left(\rho UC - D\frac{\partial C}{\partial x}\right) + \frac{\partial}{\partial y}\left(\rho VC - D\frac{\partial C}{\partial y}\right) = -\frac{\partial \overline{\rho uc}}{\partial x} - \frac{\partial \overline{\rho vc}}{\partial y} \tag{4}$$

$$\frac{\partial \rho k}{\partial t} + \frac{\partial}{\partial x}\left[\rho U k\right] + \frac{\partial}{\partial y}\left[\rho V k\right] = \frac{\partial}{\partial x_l}\left[\rho C_k \frac{k^2}{\varepsilon}\frac{\partial k}{\partial x_l}\right] + \rho P_k + \rho G_k - \rho \varepsilon \tag{5}$$

$$\frac{\partial \rho \varepsilon}{\partial t} + \frac{\partial}{\partial x}\left[\rho U \varepsilon\right] + \frac{\partial}{\partial y}\left[\rho V \varepsilon\right] = \frac{\partial}{\partial x_l}\left[\rho C_\varepsilon \frac{k^2}{\varepsilon}\frac{\partial \varepsilon}{\partial x_i}\right] + C_{\varepsilon 1}\rho \frac{\varepsilon}{k}\left(P_k + G_k\right)\left(1 + C_{\varepsilon 3}R_f\right) - C_{\varepsilon 2}\rho \frac{\varepsilon^2}{k} \tag{6}$$

$$\rho = \rho_a(1 - \alpha \Delta C) \tag{7}$$

Where $U, V, P, C, \rho, k, \varepsilon$ are, respectively, horizontal velocity, vertical velocity, dynamical pressure, concentration, density, turbulent kinetical energy and dissipation rate. D is the molecule diffusion coefficient of pollutants (often be neglected). μ is the dynamical viscosity. $\Delta C = C - C_a$, C_a, ρ_a are the ambient concentration and density respectively. $\alpha = \frac{1}{\rho_a}\left(\frac{\partial \rho}{\partial C}\right)$, $\alpha > 0$, effluent driven by negative buoyancy ; $\alpha < 0$, effluent driven by positive buoyancy ;

R_f is the flux Richardson number defined as $R_f = -\frac{1}{2}\frac{G_\perp}{P_k + G_k}$, where $G_\perp$ is buoyant production term of the lateral turbulent component. For vertical jets, $G_\perp = 0$; for horizontal jets, $G_\perp = 2G_k$.

Reynolds stresses $\overline{u_i u_j}$ are calculated by the non-linear anisotropic buoyant model .It is given by:

$$\overline{u_i u_j} = \left\{\overline{u_i u_j}\right\}_1 + \left\{\overline{u_i u_j}\right\}_2$$

$$= \left\{\frac{2}{3}k\delta_{ij} - 2C_\mu \frac{k^2}{\varepsilon}\overline{S}_{ij} - C_D C_\mu^2 \frac{k^3}{\varepsilon^2}\left(\dot{\overline{S}}_{ij} - \frac{1}{3}\dot{\overline{S}}_{mm}\delta_{ij}\right) - C_D C_\mu^2 \frac{k^3}{\varepsilon^2}\left(\overline{S}_{ik}\overline{S}_{kj} - \frac{1}{3}\overline{S}_{mn}\overline{S}_{mn}\delta_{ij}\right)\right\}$$

$$+ \left\{k\frac{\left(1 - C_3\right)\left(G_{ij} - \frac{2}{3}\delta_{ij}G_k\right)}{C_1\varepsilon}\right\} \tag{8}$$

Where

$$P_k = -\overline{u_i u_l}\frac{\partial U_i}{\partial x_l}, \qquad P_{ij} = -\left(\overline{u_i u_l}\frac{\partial U_j}{\partial x_l} + \overline{u_j u_l}\frac{\partial U_i}{\partial x_l}\right),$$

$$G_k = \alpha g_i \overline{u_i c}, \qquad G_{ij} = \alpha g_j \overline{u_i c} + \alpha g_i \overline{u_j c},$$

$$g_i = \begin{cases} 0, & i=1 \quad \text{horizontal coordinate} \\ -g, & i=2 \quad \text{vertical coordinate} \end{cases}, \qquad \dot{\overline{S}}_{ij} = \frac{D\overline{S}_{ij}}{Dt} - \frac{\partial U_i}{\partial x_k}\overline{S}_{kj} - \frac{\partial U_j}{\partial x_k}\overline{S}_{ki},$$

$$\overline{S}_{ij} = \frac{1}{2}\left(\frac{\partial U_i}{\partial x_j} + \frac{\partial U_j}{\partial x_i}\right). \qquad \frac{D\overline{S}_{ij}}{Dt} \text{ is often be neglected .}$$

Indeed, $\overline{u_i u_j}$ can be divided into two parts: $\left(\overline{u_i u_j}\right)_1$ and $\left(\overline{u_i u_j}\right)_2$, as shown in formula (8). In the paper; NLSM of Speziale(Speziale,1987) is selected to calculate stress $\left(\overline{u_i u_j}\right)_1$, Stress $\left(\overline{u_i u_j}\right)_2$ is calculated by Rodi's equilibrium hypothesis(Zhou, 1991).

The model can not only take account of anisotropy and buoyancy but also avoid numerical singularity.

Tab.1   Constants of model

C_μ	C_k	C_ε	$C_{\varepsilon 1}$	$C_{\varepsilon 2}$	$C_{\varepsilon 3}$	C_D	C_1	C_3
0.09	0.09	0.07	1.44	1.92	0.8	1.68	2.2	0.55

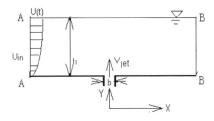

Fig.1   Jet   in   tidal   Flow

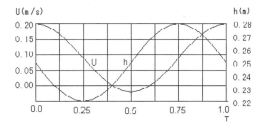

Fig.2   Depth and velocity of tidal flow

Fluxes $\overline{u_i c}$ are considered with the hypothesis of isotropical gradient diffusion, i.e.

$$\overline{u_i c} = -\frac{v_t}{\sigma_c}\frac{\partial C}{\partial x_i} \qquad (9)$$

where σ_c is the Prandle number. σ_c is 0.5 for plane free jet, or 0.85-0.9 for near wall jet.

The constants in the equations are listed in Tab. 1.

The reliability and accuracy of the model are verified by the comparisons between predictions and experimental data(Yakhot & Orszag, 1992, Chen & Singh, 1986, Yang, 1997).

3. SIMULATION TO BUOYANT JET IN TIDAL FLOW

3.1 Boundary Conditions

Before describe boundary conditions, we must make computational domain clear. Due to back and forth movement of tidal flows, jets will influence velocity and concentration distribution from upstream to downstream. The inlet and outlet alternate along with change of flow direction. Empirically, inlet section $A-A$ and outlet section $B-B$ are determined to locate at $x=-20h$ and $x=20h$ respectively. h is the water depth , as shown in Fig.1. Here, boundary conditions will be introduced as follows.

3.1.1 Inlet section A-A

We assume inlet velocity distributes exponentially along vertical direction, that is

$$U_{in} = U(t)\left(\frac{y}{h(t)}\right)^{\frac{1}{n}}, \qquad V_{in} = 0, \qquad k_{in} = U_*^2\left(1-\frac{y}{h(t)}\right)/\sqrt{C_\mu}, \qquad \varepsilon_{in} = k_{in}^{1.5}C_\mu^{0.75}/(K_a y)$$

Where U is the surface velocity at the inlet at time t; h is the water depth at time t, as shown in Fig.2; T is period of tides, n is a constant varying from 8 to 10; K_a =0.40 ∼ 0.435.

Without experimental data, it is difficult to determine the inlet concentration. However, since the mass transport at inlet satisfies Eq.(4), if diffusion and vertical convection are neglected, the inlet boundary condition can be simplified as $\dfrac{\partial C}{\partial t}+\dfrac{\partial(UC)}{\partial x}=0$

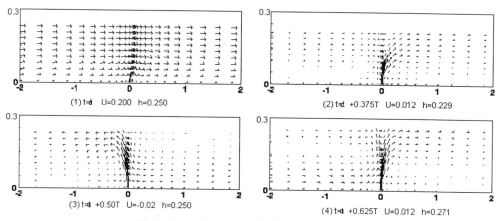

Fig.3   Velocity vectograph of buoyant jet in tidal flow

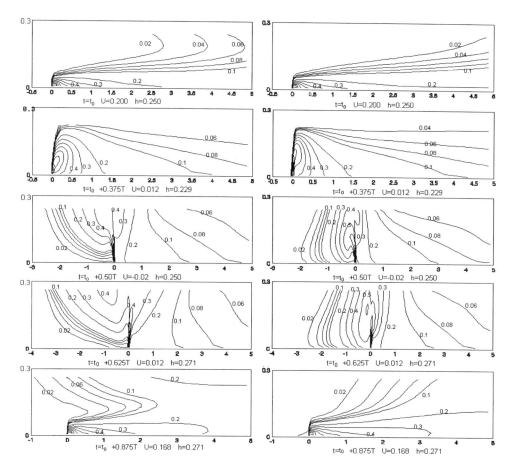

Fig.4   Concentration contour map(buoyant jet)   Fig.5   Concentration   contour   map(pure   jet)

3.1.2 Jet conditions

The conditions of the jet are given as $U_{jet} = 0$, $V_{jet} = V_{jet}^0$, $C_{jet} = C_{jet}^0$

Empirically, turbulent kinetical energy and dissipation rate are calculated from $k_{jet} = 0.005\left(V_{jet}^0\right)^2$, $\varepsilon_{jet} = C_\mu k_{jet}^{1.5} / (0.04b)$.

Where V_{jet}^0、C_{jet}^0 are, respectively, inlet velocity and inlet concentration, b is jet outlet width.

3.1.3 Outlet section B-B

Because the distance from outlet section to jet outlet is so long that we can assume fluid state at outlet has been fully developed, that is, boundary conditions can be written as $\dfrac{\partial \phi}{\partial x} = 0$, for individual variables except $V = 0$.

3.1.4 Wall conditions

Non-slid condition is adopted on wall, i.e. $U = V = 0$; Due to no mass exchange on wall concentration satisfies $\dfrac{\partial C}{\partial x} = 0$; k_1、ε_1 are obtained through wall functions $k_1 = U_*^2 / C_\mu$, $\varepsilon_1 = U_*^3 / (K_a y_1)$
Where U_* is friction velocity; U_1 is the velocity along wall ; y_1 is the distance from wall to the center of velocity control element close to wall.

3.1.5 Free water surface A-B

In order to simplify the conditions, we assume water surface locates on a horizontal line which can vertically move along with time. Vertical velocity: $V_s = 0$; Dissipation rate is computed from following empirical formula $\varepsilon_s = {k_s^{2/3}}\Big/{(0.07 K_a h)}$ Other variables are determined with symmetry condition: $\dfrac{\partial \phi}{\partial y} = 0$. In addition, pressure satisfies $\dfrac{\partial P}{\partial n} = 0$ on all boundaries except at water surface point of the downstream, where the pressure is set to be zero.

3.2 Results and Analysis

Parameters for buoyant flow are listed as follows:

Water depth: $h = 0.25 - 0.03\sin(2\pi t / T)$

Tidal velocity: $u = 0.09 + 0.11\cos(2\pi t / T)$

Tidal period: $T = 240s$

Jet velocity: $V_j = 0.1m / s$

Jet outlet width: $b = 2cm$

Jet concentration: $C_{jet} = 1.0$

Density difference : $\Delta\rho / \rho = 0.005$ (Correspondingly, $\alpha = -0.005$)

145

Tab.2 The comparisons of two cases

Case		Pure jet	Buoyant jet
Width of water surface with $S < 5$		5m	5.3m
Interval of water Surface with $S < 5$	Toward upstream	1m	3m
	Toward downstream	2.7m	3.8m
Concentration	Near jet outlet	high	low
	Near surface	low	high
Contour lines	Near jet outlet	plumpy	thin
	Near surface	closed	trumpet-shaped

In order to analyze the buoyancy effect, non-buoyant jet with $\Delta\rho / \rho = 0.0$ and other parameters same as the above is also computed. Fig.3 shows the velocity field of the buoyant flow. Obviously, the velocity field is in accordance with that of classic tidal flows. Fig.4 shows the concentration field of the buoyant flow, meanwhile, Fig.5 shows the concentration field of the non-buoyant flow. Comparing the two cases, we can obtain the buoyancy effect , as listed in Tab.2,in which S denotes dilution, i.e. $S = C_{jet}/C$. As shown in Tab.2, because buoyancy enhances turbulence, mass diffusion at jet outlet becomes quick and the pollutant cloud at the surface becomes large. Therefore, buoyancy affects significantly concentration distribution in the near-zone, so buoyancy effect to near-zone cannot be neglected.

4   SUMMARY

The advantage and disadvantage of several typical turbulence models are analyzed in this paper. On the basis of IASM , the production of stress is attributed to strain and buoyancy . According to this idea, by adding stress term induced by buoyancy to ameliorate NLSM, the non-linear anisotropic buoyant turbulence model is put forward. The presented model avoids numerical singularity and it's time-saving for the simulation. When applied to planar buoyant jets, this model has a higher precision than LSM and its program code can be easily revised from LSM. Finally. the model is applied to simulate velocity field and concentration field of buoyant jets in tidal flows. It is shown that the velocity field is reasonable, and the concentration field can reflect buoyancy effect to concentration distribution in near-zone of jets in tidal flows.

5   REFERENCES

Chen, C.J. & Singh, K.(1986),   "Development of a Two-Scale Turbulence Model and Its Applications",IIHR Report No. 299.

Gatski, T.B. & Speziale, C.G.(1993),   "On explicit algebraic stress models for complex turbulent flows", J.Fluid Mech., 254, 59-78.

Pope, S.B.(1975),   "A more general effective-viscosity hypothesis", J. Fluid Mech., 72(2), 331-340.

Rubinstein, R. & Barton, J.M.(1990),   "Nonlinear Reynolds stress models and the renormalization group", Phys. Fluids A2(8).

Shih, T.H., Liou, W.W. & Shabbir, A., et.al(1995),   "A New $k - \varepsilon$ eddy viscosity model for high Reynolds number turbulent flows", Computers Fluids 24(3), 227-238.

Speziale, C.G.(1987),   "On nonlinear $k - l$ and $k - \varepsilon$ models of turbulence", J.Fluid Mech.,78,450-475.

Yakhot, V. & Orszag, S.A.(1992), "Development of turbulence models for shear flows by a double expansion technique", Phys. Fluids A4(7).

Yang Kun(1997), "Buoyant turbulence model and its application of computation of discharge from the bottom of tidal flow", Doctoral dissertation, Department of Hydraulic and Hydropower Engineering, Tsinghua University,Beijing,P.R.China

Zhou Lixing(1991), "Numerical simulation of two-phrase turbulent flow and combustion", Tsinghua University Press,China.

Environmental Hydraulics, Lee, Jayawardena & Wang (eds) © 1999 Balkema, Rotterdam, ISBN 90 5809 035 3

Discharge and mixing of artificially upwelled deep ocean water

Clark C.K. Liu & Huashan Lin
Department of Civil Engineering, University of Hawaii at Manoa, Honolulu, Hawaii, USA

ABSTRACT: Artificial Upwelling and MIXing (AUMIX) is a new technology for enhancing open ocean mariculture using nutrient-rich deep ocean water (DOW). The researchers at the University of Hawaii have conducted AUMIX research in three phases. The first phase involves the development of a wave-driven artificial upwelling device that cost effectively brings nutrient-rich deep ocean water to the surface. The second phase, as reported by this paper, investigates the formation of a nutrient-rich plume in the ocean surface. The third phase will investigate DOW enhanced marine biology and fishery.

1. INTRODUCTION

The ocean water, from depths of 300 meters and below, is rich in nutrients and relatively free of pathological organisms and other man-made pollutants. Natural upwelling provides a steady supply of nutrients; consequently, high fish production has been observed in these areas. If some nutrient-rich DOW could be artificially distributed in the surface waters, it would open a potentially vast new source of food by dramatically increasing fish and other marine organism populations.

The challenge of bringing up DOW cost effectively has been accomplished by the Phase I Artificial Upwelling and MIXing (AUMIX) research in which a wave-driven artificial upwelling device was developed and tested (Liu and Jin 1994). The Phase II research investigates the ways to enhance the mixing characteristics of the DOW discharge.

2. A BRIEF REVIEW

Near-field mixing of a DOW effluent takes place immediately after it is discharged, where the effluent behaves like a turbulent buoyant jet and is subject to the action of both momentum and buoyancy. Further away from the discharge point is a far field, where the momentum and buoyancy of the effluent are completely dissipated and the effluent plume is transported by ocean currents and eddies. Near-field mixing of a turbulent buoyant jet was studied extensively mainly concerning the ocean disposal of municipal and industrial wastewater. Empirical equations were formulated by dimensional analysis and hydraulic experiments (Wright, 1977; Fischer, *et al.* 1979; Roberts, *et al.* 1989). Mathematical models were developed based on the Lagrangian integral method and entrainment hypothesis (Baumgartner, *et al.*, 1994; Lee and Cheung 1990). The entrainment hypothesis was introduced to overcome the closure problem associated with turbulent flows. More recently, turbulent models of jet mixing were developed which do not require the entrainment hypothesis (Hossain and Rodi 1982; Papanicolaou and List 1988;Martynenko and Korovkin 1994).

Modeling analyses conducted specifically for DOW discharge are rather limited. The near-field mixing of discharge from ocean thermal energy conversion (OTEC) plants into a stagnant and stratified ocean were investigated (Jirka et al.,1977; Coxe et al.,1981). Objectives of these studies were mainly to evaluate the environmental impact of large cold water discharges. Jirka et al. (1977) found that OTEC effluent would not re-circulate in the near-field to create an OTEC operational problem. Paddock and Ditmas (1983) developed nomograms for estimating the near-field trajectory and average dilution of OTEC effluent plumes.

Little research work has been done on wave effects on turbulent jet mixing. Ismail and Wiegel (1983) indicated that turbulent jets expand at a greater rate in the presence of waves due to the interaction of waves with jet driven flow. Several recent physical and mathematical modeling studies have shown that waves have significant effect on the mixing process of turbulent buoyant jet (Ger, 1979; Chin 1987, 1988; Hwung and Chyan ,1990; and Hwang et al.,1996). Chin (1987) found the additional mixing caused by surface waves is proportional to wave induced velocity and derived an empirical relation of initial dilution of a buoyant jet.

3. MATHEMATICAL FORMULATION

3.1 Dilution of a Turbulent Buoyant Jet

A turbulent buoyant jet has both jet and plume characteristics. Based on a dimensional analysis, Fischer *et al* (1979) defined dimensionless volume flux $\overline{\mu}$ and dimensionless distance along the axis of jet ζ as:

$$\overline{\mu} = S(\frac{R_o}{R_p}) \tag{1}$$

$$\zeta = 0.25(\frac{z}{l_Q})(\frac{R_o}{R_p}). \tag{2}$$

where R_o is jet Richardson number *or* $R_o = QB^{1/2}/M^{5/4}$; R_p is plume Richardson number and takes a value of about 0.557; z is the distance measured along the centerline of a turbulent jet. Fischer *et al* (1979) further showed when ζ is much smaller than 1 a turbulent buoyant jet is dominated by its initial momentum flux and $\overline{\mu} = \zeta$, or

$$S = 0.25\frac{z}{l_Q} \tag{3}$$

When ζ is much greater than 1, a turbulent buoyant jet behaves like a pure plume and $\overline{\mu} = \zeta^{5/3}$, or

$$S = 0.1(\frac{z}{l_Q})^{5/3}(\frac{R_o}{R_p})^{2/3} \tag{4}$$

3.2 Near-field Mixing of a Turbulent Buoyant Jet with Wave Actions

A dimensionless length scale z/Z_m was introduce by Chin (1987) to study wave effects on jet mixing. Z_m is a length scale that measures the distance along a jet centerline where the wave-induced momentum is comparable to the source momentum. Z_m is small when wave-induced velocity is large relative to the initial jet velocity.

As a surface wave passes, it causes water particles to move – elliptic movements when the waves are shallow or intermediate, circular movements when the waves are deep. In a similar fashion, effluent tracers in a wave will experience same kinds of elliptic or circular movements and promote their mixing with ambient water. Another dimensionless length scale r_o/l_Q can be introduced to measure wave effects on jet mixing. r_o is the effective radius of these elliptic or circular movements.

The wave effects can be determined by a functional relationship,

$$S_w - S = f(\frac{r_o}{l_Q}, \frac{z}{Z_m})$$
(5)

where S_w is the dilution at the presence of waves; and S is the dilution without wave.

4. HYDRAULIC EXPERIMENTS

Experiments were conducted in a 12 m × 1.2 m × 0.9 m glass wave tank (Figure 1). During the experiments, still water depth in the channel was kept at 0.65 m. The diameter of the injection nozzle orifice is 0.45 mm. The size of jet nozzle orifice was selected so that the dimensionless distance z/l_Q is large enough to allow a detailed investigation of the effects of various discharge and environmental parameters. The experiments were conducted with jet discharges larger than 14 cm³/sec in order to assure that the jet is turbulent. An electronic weight scale was used to measure the average flow rate from the injection nozzle.

Wave height ranging from 1 cm to 4 cm and wave period from 0.7 second to 2 seconds were used in the experiments.

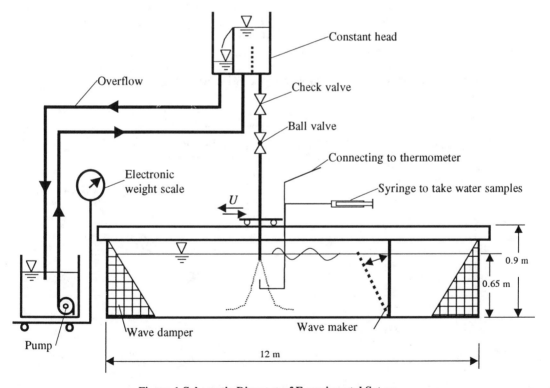

Figure 1 Schematic Diagram of Experimental Set-up

The Rhodamine dye tracer was used to simulate the DOW mixing. Dye tracer samples were taken by a syringe to avoid disturbing experimental flow conditions and to provide stable average readings. Negative buoyancy was created by adding sodium chloride to the injection water.

A series of experiments were conducted to investigate the dilution of a turbulent buoyant jet with and without wave actions. A few typical results are shown in Figure 2. The closed agreement of the experimental results with the prediction of Equation (4) indicates that the experimental setup and measuring devices are quite reliable (Figure 2a).

Wave effects on jet mixing were investigated experimentally in this study by using different jet discharges under same wave conditions and by using different waves at the same jet discharge. Based on experimental results and by following an asymptotic approach of dimensional analysis (Fischer, et al., 1979), an empirical equation was derived to estimate wave influence on the dilution along jet centerline,

$$S_w - S = \frac{r_o}{l_Q} \frac{z}{Z_m} \qquad (6)$$

This simple equation is rather general and can be used to predict the dilution of a turbulent buoyant jet with and without wave actions. It is valid when $r_o = 0$ or no wave action and when $z = 0$ or near the injection point where the action of wave velocity is overshadowed by the initial jet velocity.

5. DOW EFFLUENT DISCHARGE

Open ocean mariculture with nutrient-rich DOW requires that the effluent plume stays in the upper trophic zone without significant dilution. The equation derived above was used to investigate the mixing of DOW effluent brought up by an artificial upwelling device. The following environmental conditions used in the artificial upwelling study was adopted (Feng, 1995): Discharge nozzle diameter = 1.2 meters; Effluent flow rate = 0.95 m³/s; Wave Height = 1.9 meters; Wave Period = 12 seconds. A typical tropical ambient temperature profile (Paddock and Ditmas, 1983) was applied. The effluent temperature was assumed to be 15°C. The salinity of both the ambient and effluent was 3.5%. The investigation was conducted by locating the injection nozzle at the depth 0, 5, 10, 20, 30, 40 and 50 meters.

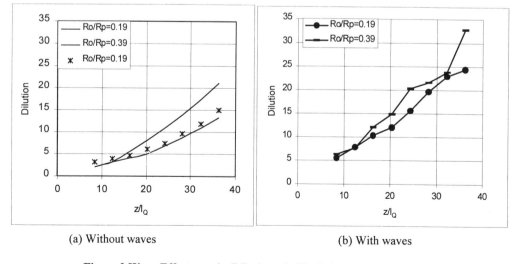

(a) Without waves   (b) With waves

Figure 2 Wave Effects on the Dilution of a Turbulent Buoyant Jet

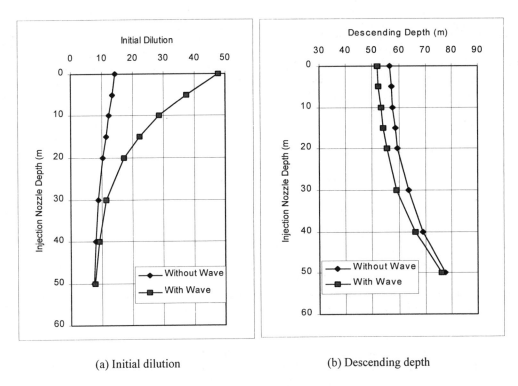

(a) Initial dilution   (b) Descending depth

Figure 3 Wave Effects on DOW Effluent Plume

The equilibrium depth and initial dilution were calculated by using the interpolation method formulated by Paddock and Ditmas (1983). The effluent dilution was calculated by Equation (4) without wave actions and by Equation (6) with wave actions. The results are shown in Figure 3.

The calculated results show that wave effects on jet mixing are insignificant when the injection nozzle is located below 30 meters from the ocean surface. Surface waves have smaller effects on the descending depth than on the initial dilution.

6. CONCLUDING REMARKS

Dilution of a turbulent buoyant jet in the ambient ocean is larger at the presence of surface waves. However, the wave effect on jet mixing reduces rapidly with water depth. This study indicates that, in order to sustain a nutrient-rich DOW plume in tropical open ocean, the effluent diffuser should be located at about 30 m below the ocean surface.

7. REFERENCES

Baumgartner, D.J., Frick, W.E.and Roberts,P.J.W.(1994). "Dilution models for effluent discharges", *EPA/600/R-94/086*.

Chin, David A. (1987). "Influence of surface waves on outfall dilution", *J. Hydr. Engrg. Div.*, ASCE,

113(8): 1006-1017.

Chin, David A. (1988). "Model Of Buoyant Jet Surface Wave Interaction", *J. Waterway, Port, Coastal and Ocean Engineering*, ASCE, 114(3): 331-345.

Coxe, D.H., Fry, D.J. and Adams, E.E.(1981). "Research on the external fluid mechanics of OTEC plants: Reports covering experiments in a current", MIT-EL 81-049, Energy Lab. Cambridge, Mass. 277 pp.

Feng,G. (1995). "Experimental and theoretical study of wave-driven artificial upwelling", Master of Science Thesis, Department of Civil Engineering, University of Hawaii at Manoa, Honolulu, HI.

Fisher, H.B., List,E.J.,Koh,R.C.Y.,Imberger,J and Brooks,N.H. (1979). "*Mixing in inland and coastal waters*", Academic Press, New York, N.Y.

Ger, A. M. (1979). "Wave Effects On Submerged Buoyant Jets", *Hydraulic Engineering in Water Resources Development and Management*, IAHR, p.295-300.

Hossain,M.S. and Rodi,W. (1982). "A turbulent model for buoyant flow and its application to vertical buoyant jets", *Turbulent Jets and Plumes* (Edited by W. Rodi), p121-178. Pergamon Press, New York.

Hwang, Robert, and Chiang, Te-Pu (1986). "The Flow of Buoyant Jets in Density-Stratified Crossflow", *Proc. Natl. Sci. Counc. ROC(A)*, Vol. 10, No.3, 1986, p290-297.

Hwang, Robert R. and Yang, W. C. & Chiang, T. P.(1996). "Effect of Surface Waves on A Buoyant Jet", *J. Marine Env. Engg.* Vol.3, p63-84.

Hwung, H. H. and Chyan, J. M. (1990). "The Vortex Structure Of Round Jets In Water Waves", *Proceeding of International Conference on Physical Modeling of Transport and Dispersion in Conjunction with the GARBISH. KEULEGAN centenial Sysp.* IAHR, pp.10a.19-10a.24.

Ismail, N.M and R.L. Wiegel,R.J. (1983). "Opposing wave effect on momentum jets spreading rate", *J. Waterway, Port, Coastal and Ocean Engineering,* 109: 465-483.

Jirka, G.H., Johnson, R.P., Fry, D.J. and Harleman, D.R.F.(1977). "Ocean Thermal Energy Conversion Plants: Experimental and analytical study of Mixing and Recirculation", Rep. 231, R.M. Parsons Lab., M.I.T., Cambridge, Mass.

Lee, J.H.W. and Cheung,V. (1990). "Generalized Lagrangian model for buoyant jets in current", *J. Environ. Engrg Div.,ASCE,*116(6): 1085-1106.

Liu, C.C.K. and Jin, Q. (1995). "Artificial upwelling in regular and random waves", *J. Ocean Engng.,*22(4):337-350.

Martynenko, O.G. and Korovkin, V.N.(1994). "Flow and heat transfer in round vertical buoyant jets", *Int. J. Heat Mass Transfer*, 37(1): 51-58.

Paddock, R. A. and Ditmars, J. D. (1983). "Initial Screening of License Applications for Ocean Thermal Energy Conversion (OTEC) Plants with regards to their interaction with the environment", Agronne National Laboratory.

Panicolaou, P.N. and List, E.J. (1988). "Investigations of round vertical turbulent buoyant jets", *J. Fluid Mech.* 195:341-191.

Roberts, P.J.W., Snyder, W.H., and Baumgartner, D.J. (1989). "Ocean outfalls: I: submerged wastefield formulation", J. Hydr. Div., ASCE, 115(1): 1-25.

Wright, S.J. (1977). "Mean Behavior of Buoyant Jets in a Crossflow", *J. Hydr. Div.,* ASCE, 103(5): 499-513.

Environmental Hydraulics, Lee, Jayawardena & Wang (eds) © 1999 Balkema, Rotterdam, ISBN 90 5809 035 3

Experiments on 2-D submerged vertical jets with progressive water surface waves

J. Kuang & C.T. Hsu
Department of Mechanical Engineering, Hong Kong University of Science and Technology, Clear Water Bay, China

ABSTRACT: The effect of progressive water surface waves on two-dimensional vertical impinging jets is studied experimentally. The submerged turbulent jets were generated by discharging water through a slot orifice into a water ambient. The velocity profiles across and along the jet center plane at x=0 were measured with a laser Doppler velocimeter (LDV) for cases with and without surface waves. The experimental results show that for the case with waves: (1) the mean velocity profile of jet remains Gaussian and self-similar in the zone of established flow (ZEF); (2) the waves enhance the jet mixing by approximately 30%, as by comparing the spreading constant of $\alpha=1.6$ for jet with waves to that of $\alpha=1.2$ for jet without waves; and (3) the surface waves will cause the jet maximum centerline to deflect from the center plane of x=0.

1. INTRODUCTION

Round and 2-D jets have been widely studied theoretically and experimentally for decades because of their significance in engineering applications. Jets were studied by Heskestad (1965), Gutmark and Wygnanski (1976), Jirka and Harleman (1979), Andreopoulos *et al.* (1986) and many others in a stagnant environment, and to some extent were better understood. Jets with steady cross-flows were also investigated by Andreopoulos and Rodi (1984), Kelso *et al.* (1996) and many others. Their measurement and flow-visualization results showed higher turbulent intensity in mixing layer and the deflection of jets due to cross-flows.

Relatively, few investigations on jets were carried out in an environment with unsteady flows, particularly with wave motion. Chyan and Hwung (1993) investigated the interaction between a vertically discharging turbulent round jet and progressive waves with laser measurement techniques. The jet deflection in the action of waves was observed. They also observed that the mean velocity profile of the jet is still Gaussian distribution in the case of weak waves. Koole and Swan (1994) investigated experimentally a horizontally discharging 2-D jet with propagation waves, and concluded that velocity distribution is no longer Gaussian in the presence of waves. Zhou *et al.* (1996) measured the velocity fluctuations in a plane turbulent wall jet excited by a sinusoidal acoustic pressure and found that the acoustic excitation will increase the turbulent intensity in lateral and vertical direction. Their results also showed the self-similarity in mean velocity and fluctuations. Hsu *et al.* (1996) studied the interaction of round jets with standing waves. The velocity profiles and tracer concentration were measured with LDV and LIF, respectively. They found the mean velocity distribution remains Gaussian in a standing wave environment.

As a sequel to the work of Hsu *et al.* (1996), this paper reports some measurements on a planar turbulent jet discharging vertically into a water ambient with progressive surface waves. The flow field of jet was measured with a laser Doppler velocimeter. The mean velocity profiles across and along the center plane at x=0 are presented in this paper. The experiments were carried out at different wave frequencies and amplitude, however, only the results at frequency of f=1.6Hz and amplitude of a=3.5mm were presented here. The experimental facility and setup are described in section 2, and the experimental results and discussions are presented in section 3. Finally in section 4, conclusions are summarized.

2. EXPERIMENTAL SETUP

Experiments were carried out in a wave tank of 6m long, 400mm wide and 400mm high. Progressive water surface waves were generated with a paddle-type wavemaker installed at one end of the tank, and eliminated with an absorber at the opposite end. The reflection coefficient of the wave absorber is about 5%. Two-dimensional turbulent jets were generated with a jet generator by discharging water vertically through an orifice of 2mm wide and 395mm long. The discharged water was recirculated with a recirculation system to maintain the water level during experiments. Figure 1 shows the schematic of the wave tank, the jet generator and the water recirculation system. The distance between the jet orifice and the water surface is 26.5cm.

Velocities in the x direction of wave propagation and vertical z direction were measured with a laser Doppler velocimetry system (LDV), operated in a back-scattering mode. A laser probe in the LDV system is mounted on a three dimensional traverse which scans the measurement volume of focused laser beams through the flow field to be measured. The position accuracy of the traverse is ±0.01mm. To improve the LDV signal quality the flow is seeded with fine titanium oxide (TiO_2) particles of 1~5μm in diameter.

3. RESULTS AND DISCUSSIONS

3.1 Self-Similar Velocity Profile

It is well known that the mean velocity profile and the fluctuation distribution of a 2-D turbulent free jet in the zone of established flow (ZEF) is self-similar. The self-similar profile of the mean velocity W as normalized by the centerline velocity W_c (maximum velocity) can be described by a Gaussian distribution given by

$$\frac{W}{W_c} = exp(-\kappa\eta^2) \tag{1}$$

where κ is a constant and $\eta=x/(z-z_0)$ a dimensionless lateral distance with z_0 being the virtual origin of the jet. Andreopoulos *et al.* (1986) suggested a constant of $\kappa=50$ for free jets. Turbulent jets spread linearly with

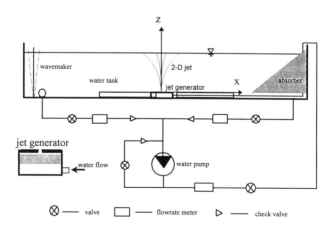

Fig. 1. The schematic diagram of experimental setup

downstream distance z as described by

$$x_{1/2} = \alpha(z - z_0)$$ (2)

where $x_{1/2}$ is the location where $W = W_c/2$ and α a constant to characterize the jet spreading. From Eqs. (1) and (2) we have $\alpha = \sqrt{ln2/\kappa}$. Heskestad (1965) reported a constant of $\alpha=0.11$ and $z_0=6d$, and Gutmark and Wygnanski (1976) a constant of $\alpha=0.1$ and $z_0=2d$, where d is the width of jet orifice; the corresponding values of κ are 57 and 69, respectively.

The self-similar mean velocity profiles as measured in the present experiments at different z-locations in the zone of established flow for the jet without surface waves are shown in Figure 2. The symbols represent the measurement data. The dimensionless water depth H/d is 130 and the Reynolds number Re defined by jet exit velocity, W_0, and orifice width, d, is 2340. The fitting by the Gaussian distribution of Eq.(1) with

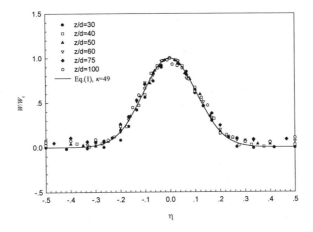

Fig. 2. Self-similarity of mean velocity profiles for a 2-D free jet
in the zone of established flow (H/d=130, Re=2340)

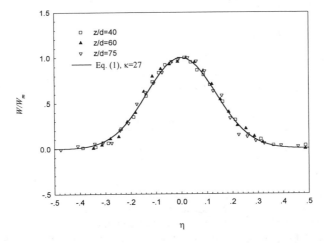

Fig. 3. Mean velocity profiles of a 2-D jet in the zone of established flow
with surface wave (H/d=130, Re=2340, f=1.6Hz and a=3.5mm)

157

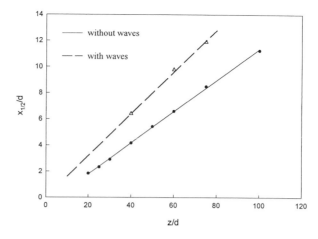

Fig. 4. The linear spreading of a 2-D turbulent jet in environment with
and without waves (H/d=130, Re=2340, f=1.6Hz, a=3.5mm)

constant κ=49 as plotted in Fig. 2 shows an excellent agreement. For the same jet, surface waves having a frequency of f=1.6Hz and an amplitude of a=3.5mm were imposed and the velocity profiles were measured. With waves, the location of the maximum velocity is not aligned with the center plane of x=0 because of the Stokes drift current associated with the progressive waves. This phenomena of jet deflection was observed and reported by Chyan and Hwung (1993). To show the similarity feature, the measured velocity profiles were shifted to the maximum at $x=x_0$ and plotted in Fig 3 against $\eta=(x-x_0)/(z-z_0)$. Figure 3 shows no twin-peaks in velocity profile as reported by Chyan and Hwung (1993), this is because the wave is relatively weak in the present experiments. As shown in Fig. 3, the Gaussian curve with a constant κ=27 in Eq.(1) fits excellently with the experimental data. From these results, it is concluded that the velocity profile of jet in ZEF remains self-similar and Gaussian in the environment of weak progressive waves. The jet spreading for the case with waves is greater than the case without waves by comparing Fig. 3 with Fig. 2. This is also evident from the difference in the values of κ.

The spreading of jet is plotted in Fig. 4 for both cases. A linear spreading as described in Eq.(2) is also found for the jet in the ambient with surface waves. Our present experiments give α=0.12, z_0=5d for the case of no surface waves and α=0.16, z_0=0 for the case with surface waves.

3.2 Vertical Velocity along x=0

The jet centerline is aligned with the center plane of x=0 for a 2-D free jet, but it is deflected under the action of surface waves as demonstrated above. This deflection can also be detected by measuring the velocity along the center plane. Theoretically, the centerline velocity of a 2-D free jet in ZEF is inversely proportional to the root square of vertical distance from jet orifice (Schlichting, 1979), i.e., in dimensionless form of

$$\frac{W_c}{W_0} = k\left(\frac{z-z_0}{d}\right)^{-1/2} \tag{3}$$

where k is a constant equal to 2.3 as suggested by List (1982). Figure 5 shows the decay of the centerline velocity for jet impinging onto a free surface in a stagnant ambient as measured in the present experiments for three different flowrates at Q=40 l/min, 60 l/min and 80 l/min, corresponding to Re=1560, 2340 and 3120, respectively. The solid line in Fig. 5 represents the fitting to Eq.(3) with k=2.1 in the range of z/d=10 to 90. It confirms the decay law in the zone of established flow. In the impinging zone of z/d>90, the velocity collapses rapidly to zero. This impinging behavior was also observed by Gutmark, et al. (1978).

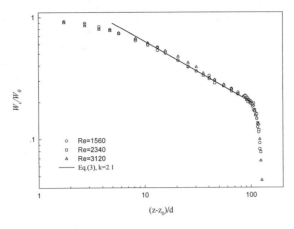

Fig. 5. The decay of centerline velocity for 2-D jets in a stagnant
ambient and its impinging onto a free surface (H/d=130)

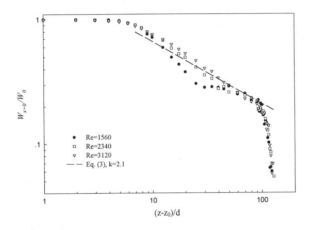

Fig. 6. The decay and the deflection of jets at different exit velocities
for the case with surface waves (H/d=130, f=1.6Hz, a=3.5mm)

Figure 6 shows the measurement results for the same flowrates but imposed with the surface waves at frequency of 1.6Hz and amplitude of 3.5mm. As showed in Fig. 6, the velocity at x=0 plane no longer satisfies the Eq.(3) but decreases in the range of z/d=10 to 60. The present data also show that the wave effect on the jet depends on the jet exit velocity. For a weaker jet such as Re=1560, a lower velocity is found at the center plane in the range of z/d=10 to 50. But for a relatively stronger jet such as Re=3120, the velocity decrease is much smaller. This indicates that a weaker jet is much easier to be deflected by the surface waves. Chyan and Hwung (1993) divided the jet affected by waves into three regions: the deflection region, the transition region and the developed region. By examining the case of Re=1560 in Fig. 6, it seems that the deflection region is located in z/d=10 to 30 and the transition region in z/d=30 to 60. The curve of Eq. (3) with k=2.1 is also plotted in Fig. 6 for comparison.

4. CONCLUSIONS

Mean velocities of jet flow field along and across the center plane of x=0 are measured in the environment with and without waves in this experiment. Similar to the case without wave, the jet mean velocity profile remains self-similar and Gaussian distribution in ZEF in a weak wave environment. The jet spreading is larger in the case with waves. The present measurement give the spreading constant of $\alpha=0.12$ for the case without waves and $\alpha=0.16$ with waves. The drift current associated with progressive waves causes the jet deflection. The wave action is serious for weak jet and to the contrary for strong jet. It appears that the ratio of the wave orbital velocity to the jet exit velocity, i.e., $2\pi fa/W_0$, will be an important parameter for the jet-wave interaction. Further investigations are necessary.

ACKNOWLEDGMENT

This work is supported by the Hong Kong Government under the RGC Grant No. HKUST 708/95E and under the RIG Grant No. RI95/96. EG15.

REFERENCES

Andreopoulos, J., Praturi, A. and Rodi, W. (1986). "Experiments on vertical plane buoyant jets in shallow water", *J. Fluid Mech.* 168, 305-336

Andreopoulos, J. and Rodi, W. (1984). "Experimental investigation of jets in a crossflow", *J. Fluid Mech.* 138, 93-127

Chyan, J. M. and Hwung, H. H. (1993), "On the interaction of a turbulent jet with waves", *J. Hyd. Res.* 31, 791-810

Kelso, R. M., Lim, T. T. and Perry, A. E. (1996). "An experimental study of round jets in cross-flow", *J. Fluid Mech.* 306, 111-144

Gutmark, E., Wolfshtein, M. & Wygnanski, I. (1978). "The plane turbulent impinging jet", *J. Fluid Mech.* 88, 737-756

Gutmark, E. & Wygnanski, I. (1976), "The planar turbulent jet", *J. Fluid Mech.* 73, 465-495

Heskestad, G. (1965), "Hot wire measurement in a plane turbulent jet", *J. Appl. Mech.* 32, 721-734

Hsu, C. T., Chen, X & Qiu, H. (1996), "LDV measurement of turbulent jets interacting with standing water waves", *Proceedings of the second international conference on hydrodynamic*, Hong Kong, 761-766

Jirka, G. H. & Harleman, D. R. R. (1979), "Stability and mixing of a vertical plane buoyant jet in confined depth", *J. Fluid Mech.* 94, 275-304

Koole, R. & Swan, C. (1994), "Measurements of a 2-D non-buoyant jet in a wave environment", *Coastal Engg.*.24, 151-169

List, E. J. (1982), "Mechanics of turbulent buoyant jets and plumes", *Turbulent buoyant jets and plumes*, Pergamaon, 1-68, Ed. Rodi, W.

Schlichting, H. (1979), *Boundary layer theory*. McGraw-Hill, New York, 1942, 7th ed.

Zhou, M. D., Heine, C. & Wygnanski, I. (1996), "The effect of excitation on the coherent and random motion in a plane wall jet", *J. Fluid Mech.* 310, 1-37

Turbulence of a non-buoyant jet in a wave environment

M. Mossa
Dipartimento di ingegneria delle Acque, Politecnico di Bari, Italy

ABSTRACT: This paper presents experimental results of a turbulent non-buoyant jet vertically discharged in a stagnant ambient and of the same jet discharged in a flow field of regular waves. The study was carried out in the wave channel of the laboratory of the Department of Water Engineering of Bari Polytechnic (Italy). Jet velocities were measured with a backscatter four-beam two-component fiber-optic LDA system. Results indicate that oscillating velocity components (obtained by phase-averaging the horizontal and vertical velocity components acquired by LDA system) cannot be described by classic wave motion theories. Amplitudes, particularly for the vertical oscillating velocity components, are larger than those obtainable through these theories. This result may be justified by both the effect of pressure variation on the nozzle due to wave motion, and nonlinear jet-wave interaction. Comparison of the root-mean square of turbulent velocity components indicates the effect of wave presence. For cross sections further from the nozzle, the experimental values of shear turbulent Reynolds stresses indicate that for configurations with the presence of wave motion, an inversion occurs of the sign compared to the jet issued in quiescent ambient. The cross sectional profiles of the shear wave Reynolds stresses are similar to those of the shear turbulent Reynolds stresses, but their magnitude is smaller. Comparison of the wave Reynolds stress in each cross section indicates that the lower the stress the smaller the wave period.

1. INTRODUCTION

Environmental problems have assumed an increasingly pivotal role in recent years. One of the problems which causes particular concern and is still subject to study is wastewater ocean outfall. While there are several studies in literature on non-buoyant and buoyant jets and their interaction with currents (Rajaratnam, 1976), few deal with jet-wave interaction, with the majority emphasizing the importance of a wave flow field in diffusion processes (Ger, 1979; Chyan and Hwung, 1993; Koole and Swan, 1994; Calabrese and Di Natale, 1994; Mossa, 1996a and 1996b; Mossa and Petrillo, 1997) and the necessity of experimental tests to explain jet-wave interaction dynamics and possibly confirm the validity of mathematical models present in literature (Chin, 1988). Studies of jets discharged in a stagnant ambient have in fact been carefully carried out by researchers. Although stagnant ambient conditions are of great interest, they are almost never present in real coastal environmental problems, where the presence of waves or currents is frequent. As a result, they cannot be analyzed without considering the surrounding environment, which is only rarely under stagnant conditions. In particular, the study of jet-wave interaction still lacks experimental results.

This paper deals with the situation previously described, and shows the experimental results of a turbulent non-buoyant jet vertically discharged in a stagnant ambient and in the presence of a wave flow field in order to compare both situations.

2. EXPERIMENTAL SET-UP

Experiments were carried out in a wave channel at the Water Engineering Department of Bari Polytechnic laboratory. The channel is about 45 m long and 1 m wide; its wall made up of crystal glass sheets 1.2 m high, supported by iron frames with a center-to-center distance of about 0.44 m, where resistance probes for wave profile measurements may be placed. During testing, mean water depth near the paddle was h=0.8 m. The velocity field was measured by using a backscatter, two-component four-beam fiber-optic LDA system. A 5 W water-cooled argon-ion laser, a transmitter, a 85 mm probe (focal length 310 mm, beam spacing 60 mm) and Dantec 58N40 FVA Enhanced signal processor were used. The accuracy of velocity measurements is ±2%. For jet-wave interaction configurations, the measurement system allows us to assess, at the same time as the velocity components, the wave elevation profile, by use of a resistance probe placed in a transversal section of the channel crossing the laser measurement volume. The accuracy of measurements made with the resistance probe is ±0.5%. The entire system is assisted by a process calculator.

This study was carried out for a jet discharged in a stagnant ambient and for the same jet interacting with progressive wave flow fields. The vertical non-buoyant jet was introduced through a 2.01 mm in-diameter nozzle, with volume flow rate equal to 22.22 cm³/s, discharge velocity U_0 equal to 6.42 m/s and Reynolds number equal to 13482. The round nozzle was located at about 11 m from the wavemaker at 16.7 cm from the bottom of the channel. As for wave motion, regular wave trains were reproduced in the channel characterized by periods for each configuration of 2.00 s, 1.43 s and 1.00 s respectively. Table 1 reports wave height (H), wave length (L), wave period (T), H/L and h/L parameters, relative to jet configurations with wave motion. The wave flow field in the channel can be described with Stokes II order theory according to the classic criteria. The reflection coefficient in the channel is not greater than 9%.

For each configuration we measured the vertical and horizontal velocity components at points of the longitudinal section of the channel crosses the nozzle. The cross sections of the jet, along which measurements were carried out, were 5, 10, 60, 110, 160, 210 and 260 mm from the nozzle, with the exception of configurations 2 and 3 in Table 1, for which measurements were carried out in sections 5, 60 and 110 mm from the nozzle. For some jet cross sections, velocity measurements were carried out on both sides in order to verify symmetry and/or antisymmetry conditions of the flow field with respect to the jet axis.

3. THEORETICAL BACKGROUND

Fundamental equations governing the problem can be derived from the Navier-Stokes equation. Any physical quantity is split into the steady mean flow component, the fluctuation component due to the statistical contribution of the wave (oscillating component) and the fluctuation component of turbulence (see, for example, Hussain and Reynolds, 1970). Therefore, the u_i component of velocity can be expressed as follows:

$$u_i(x_i,t) = \langle u_i \rangle(x_i,t) + u_i'(x_i,t) = \overline{u_i}(x_i) + \tilde{u}_i(x_i,t) + u_i'(x_i,t) \qquad (1)$$

where the angular brackets $\langle \rangle$ are an operator to take an ensemble average, the tilde symbol indicates fluctuations due to waves (or oscillating components), the prime symbol indicates turbulent fluctuations and

Table 1. Main characteristics of the wave motion fields

Case	H [cm]	L [m]	T [s]	H/L	h/L
1	4.20	5.10	2.00	0.0082	0.1569
2	4.40	3.05	1.43	0.014	0.2623
3	4.13	1.56	1.00	0.027	0.5128

the over-bar indicate steady mean flow (time-averaged components). In addition, t is time quantity, and x_i ($i=1,2,3$; $x_1 = x$, $x_2 = y$, $x_3 = z$) the coordinates of a Cartesian frame with z extending positive upwards from the nozzle, x the longitudinal axis extending positive offshore and y normal to the former two. The velocity components that were measured in the present study are $u_1=v$ in the x direction (cross velocity component, conventionally established as positive if oriented toward the shore) and $u_3=u$ in the z direction (longitudinal velocity component, conventionally established as positive if upwardly oriented).

Some useful properties between two signals $f(t)$ and $g(t)$ that follow from the basic definition are

$$\langle f' \rangle = 0, \ \overline{\tilde{f}} = 0, \ \overline{f'} = 0, \ \overline{\langle f \rangle} = \langle \overline{f} \rangle = \overline{f}, \ \langle \tilde{f}g \rangle = \tilde{f}\langle g \rangle, \ \langle fg \rangle = \overline{f}\langle g \rangle, \ \overline{\langle \tilde{f}g' \rangle} = \overline{\tilde{f}g'} = 0 \qquad (2)$$

where the latter states that, on average, background turbulence and organized motion are not correlated.

4. EXPERIMENTAL RESULTS

For the sake of brevity only the main quantities of the cross section 60 mm from the nozzle (configuration 1 of Table 1) are reported in Figs. 1a-f.

The analysis of the root-mean square of turbulent velocity components relative to the section 5 mm from the nozzle shows the existence of the core in the case of the jet issued in the stagnant ambient. Indeed, the diagram presents a local minimum on the center line. As for configurations of jet-wave interaction it is observable that the profile of the root-mean square presents a local minimum on the jet axis which is less evident than in the case of jets discharged in the quiescent ambient. Furthermore, we observed that the profile shows lower values as the wave period is reduced. At the cross section 10 mm from the nozzle the rms profiles show the absence in the jet with the presence of wave flow motion of the local minimum in the center line and lower values than those of the same jet discharged in quiescent ambient. Indeed, for the jet cross section 10 mm from the nozzle and, generally speaking, for sections nearer to the nozzle, the wave flow field causes the jet to oscillate even though it maintains many of its hydrodynamic characteristics, i.e. preserves its identity. The result of this type of interaction is a mixing of the characteristics typical of the jet in stagnant ambient and, therefore, a flattening of these characteristics on the jet cross section. In sections 60 mm (Figs. 1a-b) and 110 mm from the nozzle we observed that the root-mean square of turbulent velocity components of jets in wave environment increases proceeding from the jet axis toward the external region. In general, we noted that the root-mean squares of the turbulent velocity components of jets issued in wave environment are smaller than those of the same jet discharged in a stagnant ambient near the axis. On the contrary, they are greater in measurement points of the cross sections further from the jet axis. In addition, the root-mean squares of turbulent velocities of jet-wave interaction, especially for the longitudinal velocity components, tend to increase with the distance from the axis. The root-mean square increases as the wave period is reduced. Although still valid, these conclusions are not as evident for cross turbulent velocity components. As for the cross sections of the *developed jet region* (see Chyan and Hwung, 1993), we observed that root-mean squares of both longitudinal and cross turbulent velocity components are greater than for the same jet discharged in a stagnant ambient.

The analysis of cross-sectional profiles of crosscorrelations $\overline{u'v'}$, which represent the ratio between the time-averaged shear turbulent Reynolds stresses $\overline{\tau_{xz}}$ and the water density ρ, shows that, for sections 10 mm and 60 mm from the nozzle the absolute values of the shear turbulent Reynolds stresses are smaller for jets in wave environment than those of the same jet issued in stagnant water. Furthermore, the analysis pointed out that $\overline{u'v'}$ decreases with the decrease of the wave period (Fig. 1e). It is worth noting that jets in wave environment, starting from the cross section 210 mm from the nozzle, show an inversion of the sign of the time-averaged shear turbulent Reynolds stresses when compared to the same jet discharged in stagnant water. Indeed, starting from the jet cross section 210 mm from the nozzle, the profile of $\overline{u}(x)$ is flat near the

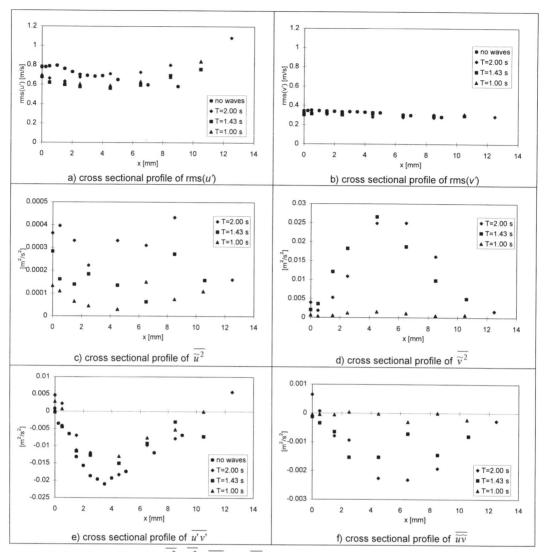

Fig. 1. Profiles of rms(u'), rms(v'), $\widetilde{u^2}$, $\widetilde{v^2}$, $\overline{u'v'}$ and $\widetilde{uv}$ at the cross section 60 mm from the nozzle (conf. 1 Table 1)

axis (Chyan and Hwung, 1993; Mossa, 1996a), where therefore the contribution of $\partial\overline{u}/\partial x$ in the equation of Boussinesq for the time-averaged shear turbulent Reynolds stresses becomes smaller than that of the same jet discharged in still water. Furthermore, in the sections in which the profile of $\overline{u}(x)$ presents twin peaks (see also Sharp, 1986 and Koole and Swan, 1994), $\partial\overline{u}/\partial x$ for measurement points near the axis has an opposite sign with respect to points further from the jet axis. At points in which the time-averaged shear turbulent Reynolds stresses of jet-wave interaction flow assume an opposite sign compared to those of the same jet discharged in a stagnant ambient, fluid flow further from the axis has an upwardly dragging effect on fluid near the axis, an observation confirmed by $\overline{u}(x)$ profiles, which show local minimum values in the jet center line.

Figures 1c-d show cross-sectional profiles of $\widetilde{u^2}$ and $\widetilde{v^2}$ (i.e. ratio between normal Reynolds wave stresses and the water density). Both $\widetilde{u^2}$ and $\widetilde{v^2}$ present smaller values in the case of waves with smaller periods.

We observed that $\overline{\tilde{v}^2}$ vanishes in the jet center line and presents a peak in intermediate positions between the jet axis and its external area. Figure 1f report cross sectional profiles of $\overline{\tilde{u}\tilde{v}}$ (i.e. ratio between shear wave Reynold stresses and the water density).

Although the cross sectional profiles of the shear wave Reynolds stresses are similar to those of the shear turbulent Reynolds stresses, their magnitude is smaller. Comparison of the wave Reynolds stress in each cross section indicates that the lower the stress the smaller the wave period. From analysis of the previous figures we observed that the oscillating horizontal and vertical velocity components (linked to the statistical contribution of the wave motion field) are not always uncorrelated, as would be the case, on the contrary, in the hypothesis that they were described through Airy or Stokes II order theories. Experimental analysis carried out showed that turbulent velocity components and those linked to (statistical) contribution of wave motion (oscillating velocity components) are not correlated, as foreseen by the latter in eqs. (2).

It is worth noting that the oscillating vertical velocity components of all measurement points present maxima and minima for which the absolute value is far greater than that obtainable from application of classic wave motion theory. This behaviour can be ascribed to two causes: the first, linked to the surface profile variation for the presence of the wave motion and, as a consequence, of the pressure value at the nozzle; the second, linked to jet-wave interaction. Although this topic merits further investigation, the study of Skop (1987) may be considered. In this regard, Figs. 2a-d show power spectra of horizontal and vertical velocity components in measurement points of cross section 160 mm from the nozzle. Along with the u_i velocity components spectra, we also report those of u'_i turbulent components, obtained by subtracting from the former the phase-averaged velocity components. In order to highlight the existence of peaks in u_i components power spectra, we have shown them in linear scale diagrams. From the analysis of figures previously referred to, it is observable that for measurement points closer to the jet axis a peak at double frequency compared to that of the wave flow field is present in vertical velocity component spectra (see at this proposal Mossa, 1996b). The power spectra of turbulent longitudinal and cross velocity components, presented in logarithmic scales, stress the absence of the peak at frequencies equal to or multiple that of the wave flow motion and the presence of the typical −5/3 slope of the inertial subrange.

5. CONCLUSIONS

The present study analyzes turbulent non-buoyant jets vertically discharged in a stagnant ambient and in the presence of wave motion. In particular, we have split the velocity acquired with LDA system in a time-averaged component, an oscillating component (statistical contribution of the wave) and a turbulent random component. Main results indicate the following:

1) Oscillating velocity components cannot be described by classic wave motion theories. Amplitudes, particularly for the vertical oscillating velocity components, are larger than those obtainable through these theories. This result may be justified by both the effect of pressure variation on the nozzle due to wave motion, and nonlinear jet-wave interaction.

2) Comparison of the root-mean square of turbulent velocity components indicates the effect of wave presence. Indeed, we observe that the profiles of the turbulent velocity components rms of the sections close to the nozzle no longer present a local minimum on the jet axis and those profiles tend to have lower values when the wave period diminishes. The root-mean squares of turbulent velocity components for sections further from the nozzle are consistently greater than those of the same jet discharged in stagnant ambient.

3) For cross sections further from the nozzle, the experimental values of shear turbulent Reynolds stresses indicate that for configurations with the presence of wave motion, an inversion occurs of the sign compared to the jet issued in quiescent ambient. Although the cross sectional profiles of the shear wave Reynolds stresses are similar to those of the shear turbulent Reynolds stresses, their magnitude is smaller. Comparison of the wave Reynolds stress in each cross section indicates that the lower the stress the smaller the wave period.

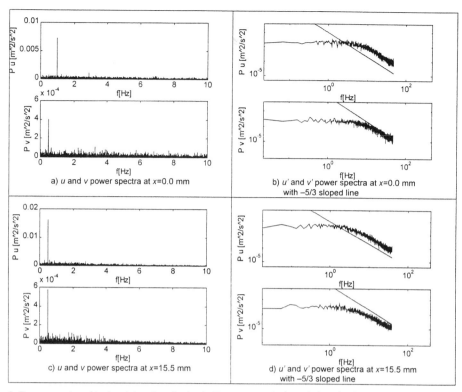

Fig. 2. Power spectra of horizontal and vertical velocity components (u and v) and turbulent horizontal and vertical velocity components (u' and v') at the cross section 160 mm from the nozzle (configuration 1 of Table 1)

6. REFERENCES

Calabrese, M. and Di Natale, M. (1994). "Diffusione di un getto liquido sommerso in presenza di un moto ondoso stazionario", *XXIV Conv. di Idr. e Costr. Idr.*, vol. I, Napoli, Italy, pp. (T1)241-254 (in Italian).

Chin, D.A. (1988). "Model of buoyant-jet-surface-wave interaction", *Jour. of Waterway, Port, Coastal and Ocean Eng., ASCE*, vol. 114, no. 3, pp. 331-345.

Chyan, J.-M. and Hwung, H.-H. (1993). "On the interaction of a turbulent jet with waves", *Jour. of Hydr. Res., IAHR*, vol. 31, no. 6, pp. 791-810.

Ger, A.M. (1979). "Wave effects on submerged buoyant jets", *Proc. 8th Congress Int. Ass. for Hydraul. Res.*, New Delhi, Part C, pp. 295-300.

Hussain, A.K.M.F. and Reynolds, W.C. (1970). "The mechanics of an organized wave in turbulent shear flow", *Journal of Fluid Mechanics*, vol. 41, part 2, pp. 241-258.

Koole, R. and Swan, C. (1994). "Dispersion of pollution in a wave environment", *Proc. 24th Coastal Eng. Conf.*, part 5, pp. 3071-3085.

Mossa, M. (1996a). *Diffusione di un getto in un campo di liquido in quiete o in moto ondoso*, PhD thesis, Università degli Studi della Calabria, Cosenza, Italy (in Italian).

Mossa, M. (1996b). "Spettri di potenza delle componenti di velocità di un getto interagente con un campo di moto ondoso", *IV Congress of A.I.VE.LA.*, University of Ancona, Italy, pp. 79-109 (in Italian).

Mossa, M. and Petrillo, A. (1997). "Turbulent energy transport of a jet in stagnant or wave environment", *Proc. 27th Congress Int. Ass. for Hydraul. Res.*, San Francisco, Theme B, vol. 1, pp. 173-178.

Rajaratnam, N. (1976). *Turbulent jets*, Elsevier, Scientific Publishing Comp., Amsterdam.

Sharp, J.J. (1986). "The effects of waves on buoyant jets", *Proc. Inst. Civ. Eng.*, part 2, 81, pp. 471-475.

Skop, R.A. (1987). "An approach to the analysis of the interaction of surface waves with current fields", *Appl. Math. Modelling*, vol. 11, pp. 432-437.

Environmental Hydraulics, Lee, Jayawardena & Wang (eds) © 1999 Balkema, Rotterdam, ISBN 90 5809 035 3

Laboratory measurements of a jet discharged into waves

C. Swan
Department of Civil and Environmental Engineering, Imperial College, London, UK

S. H. Kwan
Department of Civil and Environmental Engineering, Imperial College, London, UK (Presently: Ove Arup and Partners (HK) Limited)

ABSTRACT: This paper concerns the near-field characteristics of a jet discharged into a regular wave field. A new laboratory investigation is presented and the measured data ensemble-averaged with respect to the phase of the wave motion. Comparisons with data describing an identical jet discharged into a steady cross-flow suggest that under some circumstances the wave-induced ambient fluid flow may be approximated by a sequence of quasi-steady currents. Further comparisons with a modified integral solution, based on a formulation proposed by Lee and Cheung (1990), are shown to be in reasonable agreement. However, if the orientation of the jet (relative to the wave motion) is such that at some phases of the wave cycle the jet experiences an opposing or counter flow, there is evidence to suggest that a new mixing mechanism becomes significant. This leads to asymmetric jet profiles which cannot be modelled by a simplistic integral model.

1. INTRODUCTION

The first study of a jet discharged into waves was undertaken by Shuto and Ti (1974). They considered the case of a vertical jet discharged into standing waves, and showed that the time-averaged surface dilution is higher than that for a jet discharged into a stagnant ambient. The behaviour of a buoyant jet discharged into waves was considered qualitatively by Sharp (1986). In respect of deep water waves, Sharp suggested that in the area close to the nozzle the jet was not significantly disturbed by the waves. In contrast, in shallow water conditions he noted that an enormous pollutant dilution occurs immediately downstream of the nozzle. These results clearly suggest that the strength of the jet relative to the ambient wave motion is significant. This point was further confirmed by Chin (1987) who undertook a series of experiments to measure the time-averaged surface dilution of horizontal jets discharged into waves. Furthermore, Chin also noted that a large contribution to the overall dilution arises in the near-field, close to the nozzle.

More recently, Chyan and Hwung (1993) carried out a series of experiments in which a vertical non-buoyant jet was discharged into waves. A flow visualisation method based on Laser-Induced Fluorescence (LIF) was employed, and their results (largely expressed in terms of time-averaged data) appear to be very different from those that relate to a jet discharged in a stagnant ambient. In particular, the centreline profiles of both the velocity and the concentration were found to decay in a multi-staged manner. This decay involves a large initial decrease, a zone of near uniformity, and finally a more gradual decay. Similar results have also been presented by Koole and Swan (1994) in respect of a two-dimensional slot-jet discharged into waves. To explain these results, they developed an integral model similar to that proposed by Chin (1988) to simulate a jet-in-waves. Comparisons with this model demonstrated that the wave-induced displacement of the jet was significant, and provided a plausible explanation (at least in qualitative terms) for both the non-Gaussian time-averaged radial distributions and the apparent multi-stage centre-line decay. However, to achieve these results a large increase in the coefficient of radial entrainment was required, with little by way of physical justification.

Kwan and Swan (1997) considered this point and concluded that, in several respects, the analysis of time-averaged data is misleading, and that it is more informative to consider laboratory data ensemble-averaged with respect to the phase of the wave-cycle. Furthermore, it was also noted that much previous research has concentrated on a qualitative description of a jet discharged in waves, and that as a result there was no clear evidence as to the nature of the entrainment mechanisms, or the extent to which the entrainment coefficients will be modified by the oscillatory wave motion. The present paper will address these points.

2. EXPERIMENTAL INVESTIGATION

2.1 Experimental Set-up

A new series of laboratory observations has been undertaken in which a hot water jet was discharged beneath a regular wave train. Several different discharge orientations (measured relative to the ambient wave motion) were considered. In particular, the present paper will present laboratory data relating to three very different cases: a vertical jet-in-waves; a horizontal jet-in-opposing-waves; and a horizontal jet-in-cross-waves. A schematic diagram illustrating these discharge orientations is presented in figure 1. In addition to these cases, comparable data defining the near-field characteristics of a vertical jet discharged into a steady cross-flow (or current) were also recorded. This latter case acts as an effective bench mark with which to check the formulation of the integral model (see below) and to determine the relative importance of any additional wave-induced mixing.

The experimental observations were undertaken in the hydraulics laboratory within the Department of Civil & Environmental Engineering at Imperial College, London. In total, three different test facilities were employed. Firstly, the vertical jet in a steady cross-flow was generated within a large open channel flume which is 1.5m wide and 6m long. After the installation of 'honeycomb' sheets at the upstream end of the channel, preliminary measurements confirmed that the current was to all intents and purposes uniform with depth, and the root-mean square turbulence intensity equal to 3% of the mean velocity. This was considered adequate for the purpose of the present tests. The second set of tests, concerning a vertical jet-in-waves (figure 1a), were undertaken in a 30m long wave flume in which the water depth was maintained at 0.6m. In contrast, those cases involving a horizontal jet-in-opposing-waves (figure 1b) and a horizontal jet-in-cross-waves (figure 1c) were carried out in a small wave basin. This facility has a plan area of 8m×6m, and allows a constant water depth of 0.46m throughout the working section. In these latter cases special attention was paid to the downstream conditions so as to reduce the effects of wave reflection. In the case of the wave flume a large passive absorber, consisting of poly-ether foam, reduced the reflection coefficient to below 3%. In the case of the wave basin a sloping beach with a gradient 1:6.5, together with additional passive absorbers, reduced the reflection coefficient to approximately 8%.

In each of the cases noted above the hot water was supplied from a large heating tank which is equipped with two large thermostatically controlled heating elements. The water within this tank is continuously circulated to ensure that the temperature of the water is both constant and uniform. Once the required temperature has been achieved, the water was pumped, via an insulated pipe, to a constant head tank prior to its discharge into either the open channel, the wave fume, or the wave basin. The same apparatus, and indeed the same exit nozzle, was

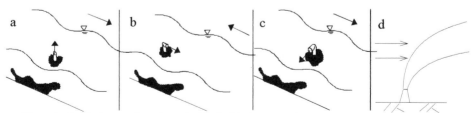

Figure 1   Jet discharge orientations (a) Vertical jet-in-waves (b) Horizontal jet-in-opposing-waves
(c) Horizontal jet-in-cross-waves and (d) Vertical jet-in-cross-flow

used in each experiment. Furthermore, to avoid any difficulties associated with the bottom boundary layer the nozzle was positioned at least 75mm above the bed.

The instrumentation used in the present tests consisted of surface-piercing wave gauges to measure the time-history of the water surface elevation, $\eta(t)$. These gauges were estimated to have an accuracy of ± 1mm, and were primarily used to define the phasing of the wave cycle. In addition, near-field temperature data were recorded using a purpose built K-type thermocouple, the response of which was sufficient to resolve the temperature fluctuations arising throughout a wave period. In all cases data were recorded for 96 wave-cycles with a sampling frequency of 200 Hz. Repeated control experiments confirmed that records of this length were sufficient to define the phase-averaged temperature profiles with both a high level of accuracy ($\pm 0.01^\circ$C) and repeatable.

2.2 Experimental Parameters

Table 1 provides details of the experimental parameters applicable to each of the four test cases. Within this table U_{wmax} represents the maximum wave-induced ambient velocity in a direction perpendicular to the discharge velocity. This value was calculated at the position of the nozzle (ie. the discharge point) using a linear wave theory. The ratio of the jet discharge velocity (or exit velocity) to this maximum wave-induced ambient velocity, U_o/U_{wmax}, is used to quantify the strength of jet relative to the ambient wave motion. In effect, this provides an equivalent to the current ratio K, widely used to characterise a jet discharged in a steady current. In each of the test cases noted on table 1, the nozzle had an exit diameter of 6mm, and the regular wave train (applicable to cases 1-3) had a period of 1 second.

2.3 Method of Analysis

To generate the phase-averaged results discussed in section 3, an ensemble-averaging technique similar to that outlined by Lam (1995) was employed to remove the incoherent turbulent temperature fluctuations. In the discussion that follows the phase angle was determined from the water surface elevation measured above the location of the jet discharge. Furthermore, this phase angle has been normalised by 2π to produce a phase number, ϕ_I, which lies within the range $0 \leq \phi_I \leq 1.0$; where $\phi_I = 0$ corresponds to the occurrence of a zero up-crossing (figure 2a); $\phi_I = 0.25$ the arrival of a wave crest (figure 2b); $\phi_I = 0.5$ the occurrence of a zero down-crossing (figure 2c); and $\phi_I = 0.75$ the arrival of a wave trough (figure 2d).

3. EXPERIMENTAL RESULTS AND DISCUSSION

3.1 Variation of the phase-averaged jet-axial temperatures over a wave cycle

Due to the unsteadiness of the ambient wave motions, the phase-averaged jet-axial temperature varies over a wave cycle. In order to demonstrate this variation, the phased-averaged jet-axial temperatures are plotted against the phase number, ϕ_I, in figure 3. These jet-axial temperatures were recorded in experiment 3 at a distance of 10 diameters downstream from the nozzle. Qualitatively, an explanation for this temperature variation can be arrived at by approximating the wave motion by a series of quasi-steady cross-flows. In the case of experiment 3 the strongest cross-flows occur at phases $\phi_I = 0.25$ and $\phi_I = 0.75$; whereas, the weakest cross-flows occur at phases $\phi_I = 0.0$ and $\phi_I = 0.5$. These values are clearly related to the occurrence of the maximum and the minimum phased-averaged temperatures. For example, the maximum phase-averaged temperature occur shortly after phases $\phi_I = 0.0$ and $\phi_I = 0.5$ and are therefore related to the occurrence of the weakest cross-flows; while the minimum phased-averaged temperatures occur shortly after phases $\phi_I = 0.25$ and $\phi_I = 0.75$, and are therefore related to the strongest cross-flows. Results of this type were found in each case and suggest that the occurrence of the maximum wave-induced mixing and jet-dilution is strongly dependent upon the instantaneous local ambient velocity, and may therefore be approximated by a sequence of quasi-steady currents.

Table 1   Experimental Parameters

Exp. No.	Type of jet flow	Discharge Temperature, T_o (°C)	Discharge Velocity W_o (or U_o) (m/s)	Level of discharge (m)	Ambient Wave Conditions	
					Wave Amplitude (m)	U_{wmax} (m/s) and U_o/U_{wmax}
1	Vertical jet-in-waves	17	0.62	0.075	0.030	0.033 (18.8)
2	Horizontal jet-in-opposing-wave	17	0.76	0.010	0.030	0.030 (25.3)
3	Horizontal jet-in-cross-wave	17	0.76	0.015	0.032	0.068 (10.9)
4	Jet-in-cross-flow	17	0.76	0.075	----	Ambient Current Velocity = 0.066 m/s Current Ratio = 11.5

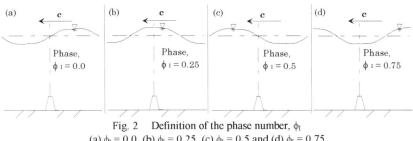

Fig. 2   Definition of the phase number, ϕ_I
(a) $\phi_I = 0.0$, (b) $\phi_I = 0.25$, (c) $\phi_I = 0.5$ and (d) $\phi_I = 0.75$

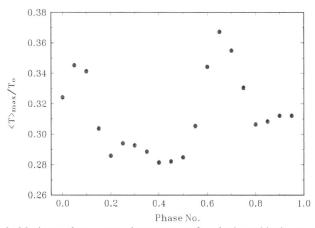

Figure 3   Maximum phase-averaged temperatures for a horizontal jet-in-cross-waves
(experiment 3, data obtained 10 diameters downstream of the nozzle)

To further emphasise this point table 2 contrasts the maximum and minimum phase-averaged temperatures recorded in wave case 3, with similar data relating to two separate cases of a jet discharged in a steady ambient cross-flow. These comparisons clearly suggest that provided the jet to ambient flow strength ratio are consistent, the maximum and minimum phase-averaged temperatures can be explained in terms of a quasi-steady current.

Table 2 Comparisons of the maximum and minimum of the phase-averaged jet-axial temperatures for a horizontal jet-in-cross-waves with temperatures measured in corresponding jets discharged in cross-flows.

Event	$<T>/T_o$ in the jet-in-wave	T/T_o in jet-in-cross-flow	Details of the jet-in-cross-flows
Maximum of the phase-averaged jet-axial temperature variation	0.37	0.38	After Patrick (1963) $U_0/U_a = 20.2$
Minimum of the phase-averaged jet-axial temperature variation	0.28	0.30	Exp. 4 of the present study $U_0/U_a = 11.5$

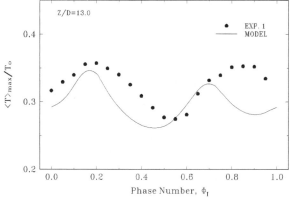

Figure. 4   Phase-averaged jet-axial temperature measured in Experiment 2
at the elevation of 13 diameters above the nozzle

3.2 Integral Modelling

In an attempt to explain and / or predict the variation in the phase-averaged jet-axial temperatures, comparisons were made between the present data and a new integral model based on the formulation outlined by Lee and Cheung (1990). This approach is based upon a semi-Lagrangian formulation and thus it is relatively straight forward to incorporate the unsteadiness of the wave-induced ambient flow field. The integral model was first compared to the laboratory data describing a jet discharged in a steady cross-flow (case 4, table 1). In this comparison the widely accepted entrainment coefficients were applied and the model shown to be in good agreement with the measured data. On this basis similar comparisons were made between the model and the various jet-in-wave cases. Figure 4 concerns experiment 2 and contrasts the measured and predicted data recorded 13 diameters above the nozzle. Although this agreement is clearly not perfect, the simple integral model reproduces the general trend of the data, and in particular provides a reasonable estimate of the maximum and minimum values with no change in the entrainment coefficients. This is in marked contrast to Koole and Swan (1994).

3.3 Asymmetric Profiles

Figure 5 concerns the horizontal jet-in-opposing-waves considered in experiment 2 and describes the phase-averaged temperature profile at $\phi_1 = 0.4$, 10 diameters downstream of the nozzle. The arrow given on this figure indicates the direction of the wave-induced ambient fluid velocity. In this case the radial asymmetry of the temperature profile is clearly apparent, and perhaps looks similar to that which might be expected for a jet discharged into a steady cross-flow. However, at subsequent phases of the wave cycle, the direction of the ambient flow changes. This is particular important since the present result suggests that where a jet is first subject to a cross-flow, and then subject to an opposing or counter-flow, a small volume of the jet fluid is separated from the main body of the jet. This represents a new mixing mechanism which is clearly apparent in

171

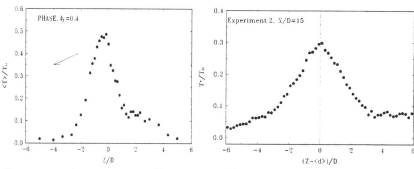

Figure 5 Phase-averaged temperature profile, (experiment 2, 10 diameters downstream of nozzle)

Figure 6 Asymmetric average temperature profile (based on the shifted phase-averaged profiles)

figure 6. This figure again concerns experiment 2, but provides a time-averaged description of the radial temperature profile in which the effect of the jet-displacement has been removed. The data presented relates to conditions 15 diameters downstream of the discharge, and clearly defines the distinct asymmetry caused by the additional mixing (or separation of the jet) that only occurs on one side of the profile, depending on the relative orientation of the wave motion.

4. CONCLUSIONS

A new series of experimental observations concerning a buoyant jet discharged into a regular wave field has been presented. Several cases include a vertical jet-in-waves, a horizontal jet-in-opposing-waves and a-horizontal jet-in-cross-waves have been considered. In each case the measured data has been averaged with respect to the phase of the ambient wave motion. These results show that in the near-field region the behaviour of a jet discharged into waves is similar to a jet discharged into a quasi-steady current. However, if at any stage during the wave cycle, the jet is subject to a cross-flow followed by a counter or opposing flow, small volumes of the jet fluid will be separated from the main flow. This represents a new mixing mechanism which can lead to significant asymmetry of the jet profile, and which cannot be modelled by the existing integral solutions.

5. REFERENCES

Chin, D. A. (1988). "Model of buoyant-jet-surface-waves interaction." J. Waterway, Port, Coastal and Ocean Eng., 114(3), 331-345.

Chyan. J. M. and Hwung, H. H. (1993). "On the interaction of a turbulent jet with waves." J. Hyd. Res., 31(6), 791-809.

Kwan, S. H. (1997). "A study of the near-field characteristics of turbulent jets discharged into waves." PhD thesis, University of London, UK.

Kwan and Swan (1997) "Near-field measurements of a horizontal buoyant jet in waves and currents." Proc. 25th. Inter. Conf. Coastal Engng. 4, 4569-4582.

Koole, R. and Swan, C. (1994a). "Measurements of a 2-D non-buoyant jet in a wave environment." Coastal Eng., 24, 151-169.

Lam, K. M. (1995). "Phase-locked eduction of vortex shedding in flow past an inclined flat plate." Phys. Fluid, 8(5), 1-10.

Lee, J. H. W. and Cheung, V. (1990). "Generalised Lagrangian model for buoyant jets in current." J. Envir. Eng., A.S.C.E., 116(6), 1085-1106.

Patrick, M.A. (1967). "Experimental investigation of the mixing and penetration of a round turbulent jet injected perpendicularly into a transverse stream" Trans. Inst. Chem. Engers. 45, 16-31.

Shape, J. J. (1986). "The effects of waves on buoyant jets." Proc. Inst. Civ. Eng., Part 2, 81, 471-475.

Shuto, N. and Ti, L. H. (1974). "Wave effects on buoyant plumes." Proc. 14th Conf. Coastal Eng., Copenhagen. A.S.C.E., New York, 2199-2209.

Environmental Hydraulics, Lee, Jayawardena & Wang (eds) © 1999 Balkema, Rotterdam, ISBN 90 5809 035 3

Numerical simulations of a jet in a wave-induced oscillatory flow

S.H. Kwan
Department of Civil and Environmental Engineering, Imperial College, London, UK

C. Swan
Department of Civil and Environmental Engineering, Imperial College, London, UK (Presently: Ove Arup and Partners (HK) Limited)

ABSTRACT: This paper outlines a full three-dimensional numerical simulation of a round turbulent jet discharged into a wave-induced oscillatory flow field. These calculations have been achieved within a commercially available computational fluid dynamics code (FLUENT), and are based upon a standard k-ε turbulence model in which the empirical coefficients were identical to those originally proposed by Rodi (1991). The numerical scheme employs a finite volume formulation, and the results are shown to be in good general agreement with laboratory data. In particular, the near-field mixing processes and the importance of the vortex structure are clearly identified.

1. INTRODUCTION

Previous work concerning jets discharged beneath a series of progressive gravity waves has either sought to provide experimental data describing the velocities and/or concentrations within the near-field region, or has sought to model these flows using an integral solution based on the principle of self-similarity with empirically derived entrainment coefficients. Although these approaches are essential if one is to achieve a fundamental understanding of the jet characteristics, they are limited in two important respects. Firstly, for a fully 3-D flow field the provision of detailed laboratory data is both time consuming and restrictive. Secondly, the integral model provides no physical insight into the nature of the entrainment process, and as such cannot be used to examine the details of the flow field. To overcome this difficulty the present paper describes a numerical simulation of a jet discharged in waves, and makes direct comparisons with similar calculations involving a jet discharged into a steady cross-flow.

2. NUMERICAL SIMULATIONS

2.1 Test cases

The present scheme employs a finite volume formulation in which the equations of fluid motion are solved using a commercially available computational fluid dynamics code, FLUENT. The calculations are based upon a standard k-ε turbulence model in which the empirical coefficients are identical to those proposed by Rodi (1991). Two different cases of a jet discharged in waves have been considered. The first corresponds to a vertical jet-in-waves; while the second a horizontal jet-in-cross-waves. In both these cases the wave period was 1 second and the water depth was 0.46m. These values were chosen so as to model the experimental data described by Swan and Kwan (1998). In addition, to provide comparable numerical data a jet in a steady cross-flow has also been considered. Details of these three test cases are given in table 1.

2.2 Boundary conditions

Each of the numerical simulations has been performed within a three-dimensional rectangular domain, positioned around the jet nozzle. A general diagrammatic description of this domain is given on figure 1. To simplify the explanation of this domain, and in particular the boundary conditions, the vertices of the domain are numbered 1-8; the eight faces (or boundaries) of the domain are referred to as faces 1234, 1256,..., etc; while the edges of the domain are described as line 12, line 34,..., etc. Overall, the dimensions of the box are defined by x_s, y_s, z_s. The boundary conditions applied on the various surfaces and the dimensions of the domain used in each of the simulations are summarised in table 2.

Table 1   Test conditions: jet discharge and ambient wave conditions

Simulation	Type of jet flow	Discharge Temperature, T_o ($^\circ$C)	Discharge Velocity W_o (or U_o) (m/s)	Level of discharge (m)	Wave Amplitude (m)
1	Vertical jet-in-waves	17	0.62	0.075	0.030
2	Horizontal jet-in-cross-wave	17	0.76	0.015	0.032
3	Jet-in-cross-flow	17	0.62	N.A.	Ambient current velocity = 0.038 m/s

Table 2   Boundary conditions and domain sizes

	Simulation 1	Simulation 2	Simulation 3
Face 1234	Outlet condition[1]	Wave velocity applied[2]	Outlet condition
Face 1458	Wave velocity applied	Wave velocity applied	Ambient current velocity and temperature
Face 1256	Condition of symmetry[3]	Outlet condition	Outlet condition
Face 3478	Condition of symmetry	Wave velocity applied (except near the nozzle)	Condition of symmetry
Face 2367	Wave velocity applied	Wave velocity applied	Outlet condition
Face 5678	Wave velocity applied	Wave velocity applied	Wall condition
x_s, y_s and z_s	30D, 10D and 22D	30D, 15D and 30D	20D, 8D and 20D
Location of Discharge	In the middle of Line 87	In the middle of Face 3478	On Line 87 and 4D from Line 48
Direction of Discharge	In z direction	In y direction	In z direction

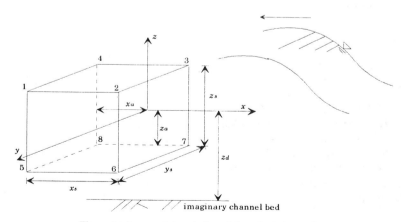

Figure 1   A general description of the calculation domain

Notes:
1. <u>Outlet conditions</u>: On those faces which are described as having an outlet condition applied, the rate of change of all quantities (ie. velocity and temperature) in a direction perpendicular to the face is zero. In addition, the velocity perpendicular to the face is also zero.
2. <u>Wave velocities applied</u>: On these faces the appropriate wave-induced velocities are calculated using linear wave theory and prescribed at all grid points.
3. <u>Condition of symmetry</u>: This condition requires that the rate of change of <u>all</u> quantities in a direction which is perpendicular to the face is zero.

2.3 Mesh generation.

Within the calculation domain the mesh generation was achieved using PREBFC. This is a commercially available mesh generator used in conjunction with the CFD package. Within the computational domain variable grid spacing was employed to optimise both the efficiency and the accuracy of the solution. In particular, fine grids were placed close to the nozzle exit, while coarser grids were placed close to the boundaries of the calculation domain. Depending on the geometry of the three cases, the number of cells within the computational domain varies from some 53,000 to 115,000. This corresponds to an averaged cell size of $0.1D^3$, where **D** is the exit diameter of the nozzle. In all cases the calculations were undertaken until both the phase-averaged velocities and the phase-averaged concentrations achieved an equilibrium state. At this point, the calculations were said to be 'dynamically steady'.

3. DISCUSSION OF RESULTS

3.1 Vortex Structure

The results of simulation 3, relating to a jet discharged in a steady cross-flow, confirm that the vortex-structure forms at a relatively small height above the discharge point. Figure 2 presents the simulated velocity field on a horizontal plane which is 7 diameters above the nozzle. Within this figure, the simulated temperature and vertical-momentum flux contours are provided, together with the velocity vectors showing the distortion of the ambient flow about the jet. Patterns of this type have been discussed previously by Chan and Kennedy (1972). Indeed, they note that this kind of velocity field is responsible for the so-called forced entrainment, both on the upstream face of the jet and, to a lesser extent, on the downstream face.

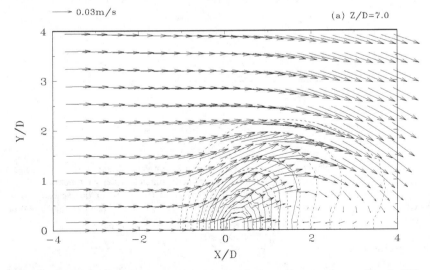

Figure 2. Simulated velocity field describing a jet discharged in a steady cross-flow (z=7D)
--- temperature contours, — vertical momentum flux contours, → velocity vector

175

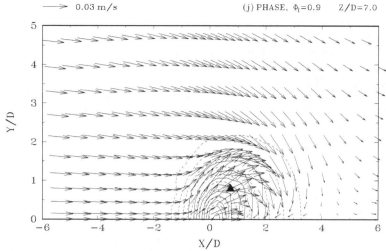

Figure 3. Simulated velocity field describing a vertical jet discharged in waves (z=7D)
--- temperature contours, —— vertical momentum flux contours, → velocity vector.

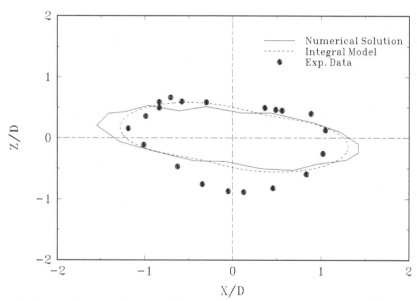

Figure 4   Predicted and measured locations of the instantaneous jet-centres for a horizontal jet-in-cross-waves

Figure 3 shows a similar sequence of results relating to simulation 1, or a vertical jet in waves. Once again the results correspond to the calculated velocity field arising on a plane located 7 diameters above the nozzle. The data provided on figure 3 describes the flow field 0.15s after the jet is subject to the strongest quasi-steady cross-flow. The similarities between figures 2 and 3 are clear, and suggest that the ambient fluid is again entrained into the jet via the forced entrainment mechanism.

3.2 Simulated jet displacement

Simulation 2 corresponds to a horizontal jet discharged into a series of cross-waves (ie. the phase velocity of the waves is perpendicular to the initial jet axis). In this case the jet is effectively subjected to a range of

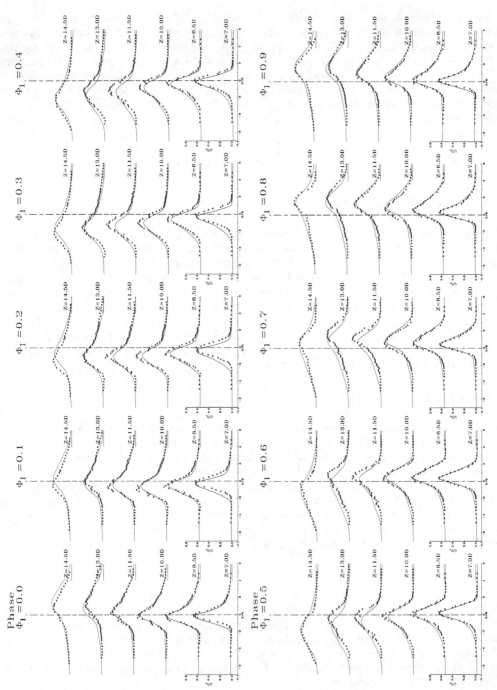

Figure 5   Comparisons between the measured and simulated phase-averaged vertical velocity profiles at various elevations

177

quasi-steady cross-flows. As a result, the displacement of the jet is such that it moves around the nozzle centreline. To confirm this behaviour a series of laboratory observations was undertaken within the hydraulics laboratory at Imperial College (see Kwan and Swan, 1998). A comparison between the present numerical predictions and this laboratory data is provided on figure 4. In addition, the results of an integral model based upon the formulation proposed by Lee and Cheung (1990) are also included. Within this latter solution the entrainment of the ambient fluid is based upon the forced entrainment hypothesis, with the value of the entrainment coefficient identical to that proposed by Lee and Cheung (1990). Further details concerning this model are provided by Kwan (1997). At each particular phase of the wave cycle the jet centre (presented in terms of z,x and non-dimensionalised with respect to the exit diameter, D) is defined as the centroid of the highest 5%of the phase-averaged temperatures recorded on the plane. The comparisons provided on figure 4 show that both the integral solution and the numerical model are in reasonable agreement with the laboratory data.

3.3 Phase-averaged velocity profile

To further investigate the capability of the numerical model, the calculated phase-averaged vertical velocity profiles are compared with a corresponding set of experimental data. In respect of these experiments, the laboratory apparatus, the measuring instrumentation, and the method of data analysis are identical to that described by Swan and Kwan (1998). Figure 5 considers 6 vertical elevations above the nozzle and contrasts the measured and predicted data at 10 equally spaced phases of the wave cycle, defined in terms of the phase number, ϕ_I. This parameter lies within the range $0 \leq \phi_I \leq 1.0$; where $\phi_I = 0$ corresponds to the occurrence of a zero up-crossing; $\phi_I = 0.25$ the arrival of a wave crest; $\phi_I = 0.5$ the occurrence of a zero down-crossing; and $\phi_I = 0.75$ the arrival of a wave tough. At all stages of the flow the numerical model provides a reasonably good description of the measured data. In particular, the deflection of the jet away from the nozzle centreline is well predicted, as is the development of asymmetric jet profiles.

4. CONCLUSIONS

The present paper has shown that a commercially available CFD package (FLUENT), coupled with a standard k-ε turbulence model (Rodi, 1991), can be used to provide a full three-dimensional simulation of a round turbulent jet discharged into an unsteady wave field. Three separate cases have been considered: a vertical jet in waves; a horizontal jet in cross-waves; and a jet in a steady cross-flow. In all cases a 'converged' solution, in which both the time-averaged velocities and the concentrations maintained an equilibrium state, were achieved. Comparisons between these results and new experimental data show good agreement. In particular, the wave-induced displacement of the jet axis is well modelled, as is the development of the asymmetric jet profiles. This latter point is particularly important since a typical integral solution, which generally forms the basis of most near-field calculations, does not and cannot predict the occurrence of jet asymmetry.

5. REFERENCES

Chan, T-L. and Kennedy, J.F. (1972). "Turbulent non-buoyant or buoyant jets discharged into flowing or .quiescent fluids." IIHR report no. 140, Iowa Institute of Hyd. Res., The University of Iowa.

Kwan, S. H. (1997). "A study of the near-field characteristics of turbulent jets discharged into waves." PhD Thesis, University of London, UK.

Lee, J. H. W. and Cheung, V. (1990). "Generalised Lagrangian model for buoyant jets in current." J. Envir. Eng., A.S.C.E., 116(6), 1085-1106

Rodi, W. (1991). Turbulence models and their application in hydraulics - a state-of-the-art review. IAHR Monograph.

Swan, C. and Kwan, S. H. (1998). "Laboratory measurements of a jet discharged into waves." 2nd International Symposium on Environmental Hydraulics. Hong Kong.

1.4 Jets and plumes – II

Environmental Hydraulics, Lee, Jayawardena & Wang (eds) © 1999 Balkema, Rotterdam, ISBN 90 5809 035 3

Radially spreading surface flow

Michael MacLatchy & Gregory Lawrence
Department of Civil Engineering, University of British Columbia, Vancouver, B.C., Canada

ABSTRACT: When discharged in shallow water a vertical buoyant jet will experience relatively limited dilution before it surfaces. In this case, dilution occurring in the radially spreading surface layer emanating from the vertical jet will become of importance if regulatory requirements for mixing are to be achieved. The present study was undertaken to identify the details of the structure of the radially spreading surface layer associated with the discharge of a vertical buoyant jet in shallow water. Flow visualization was used to examine the surface flow for a variety of initial jet conditions. Large scale vortex structures propagating outward from the region of surface impingement were observed to occupy the full extent of the surface flow. Significant entrainment into the radial flow occurred as a result of these structures. The radial surface flow was found to be highly intermittent and discontinuous in nature, a well defined interface between the outward flowing upper layer and inward flowing lower layer did not exist at any given time. No internal hydraulic jumps were observed.

1. INTRODUCTION

As a result of their importance in effluent discharges and other situations, round buoyant jets have received considerable attention (Fischer et al. (1979)). The most fundamental case for a round buoyant jet is that of a vertical jet in a quiescent or near-quiescent water body. While a reasonable understanding exists of the mechanisms of entrainment and mixing in a vertical buoyant jet (Fischer et al. (1979), List (1982)), consideration of the case where the jet is discharged in shallow water leads to some complications. Since the vertical extent of the buoyant jet is limited when discharged in shallow water, the degree of entrainment, and the behaviour of the jet, after it surfaces takes on greater importance if environmental regulations are to be met. While several researchers have examined the problem of the radially spreading surface layer, the nature of entrainment and mixing into this surface layer remains elusive.

In investigations of a radial surface buoyant jet in a circular tank, Chen (1980) found that entrainment occurred into the surface flow via the action of large scale instabilities or coherent structures. At some distance from the jet exit, buoyancy caused these entrainment structures to collapse, marking the transition from momentum dominated conditions (near-field) to buoyancy dominated conditions (far-field), and the radial surface jet became a plume. Koh (1971) reported similar behaviour while studying the discharge of a two-dimensional horizontal surface buoyant jet. As with Chen (1980), Koh observed the existence of an entrainment region where the surface flow increased in thickness due to entrainment of ambient water.

In contrast to these investigations, Lee & Jirka (1981) and Wright et al. (1991) have both presented models of vertical buoyant jets in shallow water accompanied by laboratory investigations to validate their models. In Lee & Jirka's case their model of the radially spreading upper layer was based on the assumption of the existence of an internal hydraulic jump. Wright et al. (1991) based their model of the transition from near-field to far-field in the radially spreading surface flow on the assumption of the existence of a density jump of the maximum entraining type, based on the definitions developed by Wilkinson & Wood (1971).

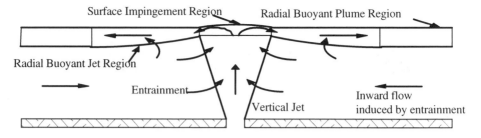

Figure 1: Generalized schematic of flow regions of a vertical buoyant jet in shallow water

Generally most researchers, including Lee & Jirka (1981) and Wright et al. (1991) have divided the flow of a vertical buoyant jet in shallow water into four regions. These regions are illustrated in figure 1. The first region is the vertical jet itself. This is followed by the surface impingement region, in which the vertical flow of the jet is redirected into an outward spreading radial flow, as it is bounded by jet fluid on all sides entrainment into this region is considered to be negligible. The third region, in which the flow is still momentum dominated, is the region of interest in this study. In this radial buoyant jet region there is some uncertainty as to the form which the flow takes, whether it is an entraining shear layer or some variant of internal hydraulic jump. The final region is the far-field region where the buoyancy becomes dominant and the flow more plume like in nature, and hence shall be referred to as the radial buoyant plume region here.

The objective of the present study is to determine the nature of the radial buoyant jet region. We use flow visualization to determine whether this region contains an internal hydraulic jump, an entraining shear layer, or some combination of the two.

2. EXPERIMENTS

Experiments to investigate the radial spreading of a vertical buoyant jet after it surfaces were conducted in a circular tank of 1.8m diameter and 0.3m depth. This circular tank was contained in a square tank of 2m by 2m and 0.4m depth to facilitate flow visualization. The vertical jet was discharged in the center of the tank, and the surface flow produced by the vertical jet was radially symmetric within the tank. The vertical

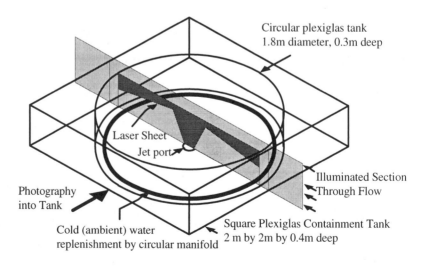

Figure 2: Schematic of Experimental Apparatus

jet discharge was equipped with interchangeable nozzles of 2, 4 and 6 cm diameter, thus allowing the depth to port diameter ratio to be varied. A schematic of the experimental tank is provided in figure 2.

The experimental apparatus was designed to simulate an infinite ambient by allowing the upper layer flow to spill out of the circular tank into the square containment tank, over the walls of the circular tank. The duration of experiments was increased and the build up of jet water and dye in the circular tank was reduced by replacing the ambient water as it was entrained by the jet. This ambient water replacement was accomplished by means of an injection manifold installed in the bottom outside edge of the circular tank, encompassing the whole circumference of the tank. The injection manifold has a series of small holes, covered with air filter material, along it to evenly distribute the replacement water. In this manner the inward flow of water in the lower layer induced by entrainment into the upper layer flow could be simulated as would occur in an infinite ambient environment.

Flow visualization was accomplished with sodium fluorescene dye injected into the jet discharge line by a small peristaltic dosing pump. Flow illumination was provided by a 4 Watt argon ion laser, with a laser sheet produced by a resonant scanning mirror controlled by a function generator. This laser sheet was projected in a plane through the center of the circular tank and parallel to one wall of the square tank. Image recording was done both with a 35 mm still camera and a video camera, under dark room conditions. Prior to the commencement of an experiment a scale was placed in the same plane that the laser sheet would be illuminating, and photographed to allow later measurement from the recorded images. Once the scale had been photographed, the position and focal length of the cameras were not changed. Experiments were of approximately 10 minutes duration.

Both Lee & Jirka (1981) and Wright et al. (1991) indicate that the parameters of importance governing the behaviour of the vertical buoyant jet are the densimetric Froude number, regular Froude number of the jet, and the depth to diameter ratio (hereafter depth ratio). These quantities are defined as follows:

densimetric Froude Number: $\qquad$ $F_o = u/\sqrt{g'_o D}$ $\qquad$ (1)

regular Froude Number: $\qquad$ $F = u/\sqrt{gD}$ $\qquad$ (2)

depth ratio $\qquad$ H/D $\qquad$ (3)

Where: u = jet exit velocity.

$\qquad$ $g' = (\Delta\rho/\rho)g$, modified gravitational constant.

and: $\qquad$ $\Delta\rho$ $\qquad$ = density difference between jet and ambient water.

$\qquad$ ρ $\qquad$ = density of ambient water, i.e. cold water.

$\qquad$ g $\qquad$ = gravitational constant.

$\qquad$ H $\qquad$ = depth of ambient water.

$\qquad$ D $\qquad$ = jet nozzle diameter.

Table 1. Experiments performed

Exp. No.	D (cm)	H/D	F	F_o	Q (l/s)	g'_o (cm/s^2)	Exp. No.	D (cm)	H/D	F	F_o	Q (l/s)	g'_o (cm/s^2)
1	6	5	0.21	2	0.47	10.9	14	4	7.5	1.0	50	0.79	0.40
2	6	5	0.42	4	0.94	10.9	15	4	7.5	1.0	100	0.79	0.10
3	6	5	0.84	8	1.80	10.9	16	4	7.5	2.0	20	1.18	9.8
4	6	5	0.25	5	0.54	2.45	17	4	7.5	2.0	50	1.18	1.6
5	6	5	0.25	10	0.54	0.6	18	4	7.5	2.0	100	1.18	0.4
6	6	5	0.25	20	0.54	0.15	19	2	15	0.5	5	0.08	10.9
7	6	5	0.5	5	1.08	9.8	20	2	15	1.0	10	0.15	10.9
8	6	5	0.5	10	1.08	2.45	21	2	15	2.0	20	0.28	9.8
9	6	5	0.5	20	1.08	0.61	22	2	15	2.0	50	0.28	1.6
10	4	7.5	0.53	5	0.43	10.9	23	2	15	2.0	100	0.28	0.4
11	4	7.5	1.05	10	0.86	10.9	24	2	15	5.0	20	0.70	61
12	4	7.5	2.11	20	1.70	10.9	25	2	15	5.0	50	0.70	9.8
13	4	7.5	1.0	20	0.79	2.45	26	2	15	5.0	100	0.70	2.45

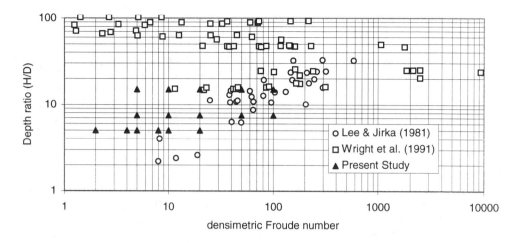

Figure 3: comparison of the experiments of Lee & Jirka (1981), Wright et al. (1991) and the present study by densimetric Froude number and depth ratio.

For each depth ratio, a series of three different densimetric Froude numbers, F_o, and two different Froude numbers, F, were run to investigate the behaviour of the flow. Variation of the densimetric Froude number was accomplished by varying the proportions of hot and cold water in the total jet discharge required to achieve the desired Froude number. Details of the experiments conducted are provided in Table 1.

The experiments conducted in this study range in depth ratios from 5 to 15 and densimetric Froude numbers from 2 to 100. The range of experimental parameters investigated in this study is compared to those of Lee & Jirka (1981) and Wright et al. (1991) in figure 3. With Wright et al. (1991), the experiments in this study are only directly comparable at a depth ratio of 15, the lowest that Wright et al. used, as they tended to favor higher depth ratios. Lee & Jirka (1981) has similar experiments in the middle of their range, where depth ratios vary between 7.5 and 15 and densimetric Froude numbers range from 10 to 100.

Ambient recharge flow rates were determined by careful observation of the flow in the circular tank. The ambient recharge flow was carefully adjusted until the upper layer was just spilling over the edge of the circular tank with minimal recirculation within the circular tank. Care was taken to ensure that the ambient recharge flow was not larger than the flow being entrained into the jet, to avoid having the upper layer flow washed or carried out of the circular tank by the replenishment flow. For this reason it was considered desirable to have a minor degree of recirculation occurring at the wall of the circular tank since this indicated that the replenishment flow was not too high.

3. RESULTS

No radial internal hydraulic jumps were observed in any experiments. Figure 4 is typical of the radially spreading flow observed in the upper layer for all experiments. Generally, immediately outside the surface impingement zone large instabilities or vortices appeared causing readily apparent entrainment and mixing. These structures were of the same scale as the depth of the surface layer, and in fact comprised the surface layer, and would tend to grow as they traveled radially outward from the surface impingement zone. Gradually as these instabilities moved outward they became less energetic and harder to distinguish from one another.

Photographs of the surface flow from experiment 21 ($F_o = 20$, $F = 2.0$, $H/D = 15$) are presented in figure 4. Large scale interfacial instabilities, vortex cells, visible as the brightest areas, are apparent starting from the exit from the surface impingement zone at the left. Ambient (darker) water can be seen intruding

Figure 4: Flow visualization of radially spreading surface layer. Fo=50, F =2 H/D=15, approximate scale 1.7 cm = 10 cm, time interval 0.4s

almost to the free surface between the vortex cells. As the cells move outward (to the right), they grow in size. These cells can be seen in the movement of features **a** and **b** in the sequential photographs. Feature **c** is a group of wisps of upper layer water caught between the surface flow and the underlying inward flow, and as a result, they do not move significantly over the sequence of photos.

As can be seen from the photographs the radial surface flow was highly intermittent and lacked a continuous, distinct interface. The thickness of the upper layer varied constantly as vortices would form, grow while moving outward, and subside. In addition, examination of the video recordings revealed that the vortices or billows tended to merge with increasing radial distance. This pattern of behaviour is very similar to that reported in mixing layers by Koop & Browand (1979).

First hand observations during the course of experiments, and review of the video recordings revealed that there was a continuation of vorticity from the vertical jet into the radially spreading region. The most energetic of the vortices in the vertical jet could be clearly seen to travel upward into the surface impingement region, to then emerge in the radial flow. Thus the larger instabilities were continuous in nature from the vertical to the radial flow. Unfortunately this phenomena is most readily apparent when the actual motion can be observed and is not visible in the still photographs.

During some experiments where very high jet flow rates were used, it was found that a large proportion of the surface layer would be recirculated back into the tank at the circular wall. When such a large degree of re-circulation did occur at the wall, it was due to the inadequacy of the ambient water replenishment flows from the ring diffuser. The entrainment demands of the radially spreading upper layer were satisfied by drawing upper layer water down into the lower layer at the circular weir, to eventually be re-entrained back into the upper layer flow. This effect was not one of an internal control imposed by the circular weir, but one that arose because there was not an infinite ambient from which large entrainment flows could be drawn.

4. COMMENTS

In all experiments a radial form of entraining shear layer was observed. In the near-field region, significant mixing and entrainment occurred due to the formation and propagation of large instabilities or vortices that penetrated the full depth of the upper layer. These vortices were continuations of the instabilities formed in the vertical buoyant jet region. As the flow progressed radially outward this energetic entrainment and mixing was observed to gradually subside though it did not completely die off within the confines of the experimental tank.

It should especially be noted that there was not a continuous distinct interface in the two layer flow in the near-field. Instead the upper layer flow took the form of a series of highly intermittent instabilities which propagated outward from the surface impingement region. The nature of the flow in the near-field means that conventional internal hydraulic principles do not apply to the flow. As a continuous upper layer did not exist, this implies that interfacial long waves are not a possibility, as there was not an interface for them to act along. Since interfacial long waves cannot exist in the near-field, and cannot govern the flow, the concepts of internally supercritical and internally subcritical flow are therefore hard to apply, this effectively precludes the existence of internal hydraulic jumps. The application of standard hydraulic equations, such as continuity and momentum, while possible on a time-averaged basis, are not necessarily representative of the flow conditions in the upper layer at any given instant, due to its discontinuous, intermittent, nature.

5. CONCLUSIONS

When a vertical buoyant jet enters shallow water there are four distinct regions: the vertical buoyant jet region, the surface impingement region, the momentum dominated near-field surface flow and the buoyancy dominated far-field surface flow. As the flow exits the surface impingement region there will be entrainment of ambient water, and an increase in the thickness of the radial surface flow as it spreads outward. An internal hydraulic jump does not occur as part of this flow. The radially spreading surface layer exists as a series of ring vortices or instabilities, which grow in size and slow as they move radially outward. The upper layer flow in the near-field is an intermittent dynamic flow, lacking a continuous distinct interface, which makes the occurrence of internal hydraulic jumps impossible.

6. REFERENCES

Chen, J. C. 1980. "*Studies on Gravitational Spreading Currents*", W. H. Keck Laboratory Report No. KH-R-40, California Institute of Technology, Pasadena, California, 436 pp.

Fischer, H. B. , List, E. J. , Imberger, J. , Koh, R. C. Y, Brooks, N. H. 1979. "*Mixing in Inland and Coastal Waters*", Academic Press, New York, 483 pp. .

Koh, R. C. Y. 1971. "Two-dimensional surface warm jets", *Journal of the Hydraulics Division, ASCE*, Vol. 97, HY6, pp. 819-836.

Koop, C.G., Browand, F.K. 1979. "Instability and Turbulence in a Stratified Fluid with Shear", *Journal of Fluid Mechanics*, Vol. 93, pp.135-159.

Lee, J. H. W, Jirka, G. H. 1981. "Vertical round buoyant jet in shallow water", *Journal of The Hydraulics Division, ASCE*, Vol. 107, pp. 1651-1675.

List, E.J. 1982. "Turbulent Jets and Plumes", *Annual Review of Fluid Mechanics*, Vol. 14, pp. 189-212.

Roberts, P.J.W. and Wright, S.J. 1983, "Discussion of vertical round buoyant jet in shallow water", *Journal of The Hydraulics Division, ASCE*, Vol. 109, pp. 490-494.

Wilkinson, D. L. , Wood, I. R. 1971. "A rapidly varied flow phenomena in a two layer flow", *Journal of Water Mechanics,* Vol. 47, part 2, pp. 241-256.

Wright, S. J. and Buhler, J. 1986. "Control of Buoyant Jet Mixing by Far Field Spreading " Proceedings of Symposium on Advancements in Aerodynamics, Water Mechanics, and Hydraulics, Minneapolis, Minnesota, pp. 736-743.

Wright, S. J. , Roberts, P. J. W. , Zhongmin, Y. , Bradley, N. E. 1991. "Surface dilution of round submerged buoyant jets", *Journal of Hydraulic Research*, Vol. 29, No. 1, pp. 67-89.

Environmental Hydraulics, Lee, Jayawardena & Wang (eds) © 1999 Balkema, Rotterdam, ISBN 90 5809 035 3

Laboratory experiments on the impact of a buoyant jet with a solid boundary

A. Cavalletti & P.A. Davies
Department of Civil Engineering, The University, Dundee, UK

ABSTRACT: Model experiments are presented to show and quantify the distortions to the structures of the velocity field of a plane, negatively-buoyant jet caused by the impact of the jet with a bottom boundary. The techniques of shadowgraph and particle tracking are used to monitor continuously the behaviour of the perturbed flow and the two-dimensional velocity field within the jet, for a range of momentum and buoyancy fluxes and fluid depths. The shape of the disturbance field close to impact is characterised by a pair of vortices displaced symmetrically from the centreline of the incident jet. The height reached by the associated vortex pairs after the impact with the solid boundary is shown to have clear dependence upon the Richardson number of the buoyant jet. The distortion of the jet centerline velocity decay within the impact zone is discussed.

1. INTRODUCTION

The motivation for the study is the phenomenon of seafloor impact and bottom spreading of so-called produced water discharged from marine petroleum exploration platforms, though there is relevance also to the behaviour of microbursts and downdraughts in the atmosphere. Produced water is formation water entrained into oil flow during production and it contains a wide variety of dissolved inorganic salts and organic compounds. Usually, it has not only a high salt concentration but also an elevated temperature relative to the seawater into which it is discharged. In the present study, the produced water problem has been modelled in laboratory experiments by a plane, negatively-buoyant turbulent jet discharging into a receiving flow of quiescent, homogeneous water. Data from North Sea exploration operations have been used to guide the parameter ranges of the model experiments, though, for simplicity, salinity differences alone are utilised to generate the driving buoyancy flux of the jet.

Two sets of experiments, having a fixed jet depth H (defined as the distance between the jet source and the bottom boundary) have been carried out. The flow rate Q and the density value ρ_0 of the buoyant jet at the source have been changed in the experiments of the same set, so that cases of momentum-dominated and buoyancy-dominated flows have been investigated. The shape of the descending jet before the impact, the maximum height of the flow reached after the impingement with the flat boundary and the spreading of the current along the bottom have been analysed using the shadowgraph technique. The two dimensional velocity field of the buoyant jet has been determined by analysing data collected on the centreline vertical section of the jet with the *DigImage* video-based, particle-tracking software. Particular attention has been focussed on the decay of the centreline axial velocity close to the impact region at the bottom boundary. Turbulent buoyant jets have been studied by many investigators (see, for example, List, 1982). Experimental and theoretical works have studied the free turbulent buoyant jet, giving particular importance to the centreline velocity decay (see, for example, Kotsovinos, 1975; Kotsovinos and List, 1977; Yannopoulos and Noutsopoulos, 1990). Beltaos and Rajaratnam (1972) have studied the flow behaviour of a pure impinging jet.

2. BASIC CONSIDERATIONS

The Reynolds number Re plays an important role in determining the turbulent nature of the flow and, for the case of a plane buoyant jet, it may be conveniently defined as $Re = V_0 b/\nu_0$, where V_0 is the velocity at the source, b is the source width of the jet and ν_0 is the kinematic viscosity of the source fluid. In the present experiments, only values of the Reynolds number greater than 4000 have been used in order to have a fully turbulent flow. A further dimensionless number characterising the behaviour of turbulent buoyant jets is the Froude number Fd, defined here as $Fd = V_0/(g'b)^{1/2}$, where $g' = g(\Delta\rho)/\rho_0$ is the modified gravitational acceleration and $\Delta\rho$ is the density difference between the source density ρ_0 and the ambient receiving water density ρ_a. The value of the Froude number determines whether the behaviour of the discharged fluid at a given distance from the source is jet-like (relatively high values of Fd) or plume-like (relatively low values of Fd). As is well-known, a vertical buoyant jet behaves like a pure jet near the source, but like a pure plume in the far field, even if the initial Froude number is large (Chen and Rodi, 1980). For purposes of comparison, it is useful to introduce at this stage a further dimensionless dynamical parameter, the Richardson number Ri defined by $Ri = Fd^{-4/3}$ and a dimensionless length scale Z (Chen and Rodi, 1980) $Z = (z/b)(Ri)$.

3. EXPERIMENTAL PROCEDURE

A two-dimensional vertical buoyant jet is realised by introducing a constant downwards discharge Q of saline water of prescribed density ρ_0 into a rectangular tank (3.5 m x 0.3 m x 0.9 m) filled with fresh water of density ρ_a. The receiving water is quiescent and the discharged water is introduced through a rectangular slot of thickness $b = 0.5$ cm and length $h = 26.5$ cm. The geometrical dimensions of the slot are chosen in order to maintain the two-dimensionality of the buoyant jet (Kotsovinos, 1975). The slot is positioned below the free water surface, so that the buoyant jet is submerged. The jet depth (*i.e* the distance H between the jet source and the solid boundary is adjustable and two different values $H = 0.74$ m and $H = 0.64$ m have been employed. Parametric experiments have been performed in terms of discharge rate q (= Q/b) and density difference $\Delta\rho$, in order to cover a large range of Richardson number. Two techniques have been used, namely, shadowgraph flow visualisation and particle tracking. In the first, a small amount of red vegetable dye is introduced in the jet to enhance the flow visibility and all the experiments are recorded by means of a camcorder. The same sets of experiments are run a second time using an experimental set up in which a vertical section of the buoyant jet is illuminated by a two dimensional beam light positioned underneath the tank. Small neutrally-buoyant particles are added to the introduced water at the source (with no dye) and images of the buoyant jet interaction with the bottom boundary are recorded by a CCD video camera placed in front of the (glass-sided) tank. The automated particle tracking function of *DigImage* software has been used to follow the motion of the neutrally-buoyant tracer particles and to determine the velocity field in the buoyant jet. Values of the centreline axial velocity in the buoyant jet in the impingement region have also been determined. A sketch of the experimental setup is shown in Fig.1.

4. EXPERIMENTAL RESULTS

4.1 Qualitative Observations

The starting buoyant jet immediately before the impingement is characterised by the presence of two vortices of scale comparable with the local half width of the jet and displaced symmetrically from the centreline. Fig 2 shows a typical set of particle streaks for the buoyant jet in the impinging region. This behaviour has been found by previous studies; see, for example, Noh *et al* (1992) and Ching *et al* (1992). While the buoyant jet is descending, fluid from the receiving water is entrained into the buoyant jet, with the entrainment being more enhanced for high Richardson number flows. After the impact with the solid boundary, each of the vortices detaches from the centreline (see, for example, Fig 2c) and starts decaying following the gravitational collapse of the fluid within it. Consequently, the flow is characterised by a gravity current flowing along the bottom surface in a direction away from the point of impact.

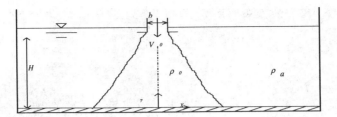

Fig 1. Schematic sketch of physical system

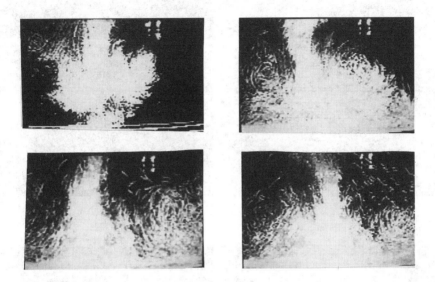

Fig 2.   Sequence of particle streak images illustrating the impact of the buoyant jet with the bottom
              solid boundary.

Figure 3 shows the evolution of the descending buoyant jet in the two limiting cases having respectively Ri = 0.055 and Ri = 0.016.

4.2 Height of the vortex

In order to estimate the maximum height z^* reached by the vortex as a result of the impact, an equivalence between the potential and the kinetic energy of a fluid particle impinging on a bottom boundary can be assumed.   That is, $z^* g' \approx v_p^2$, from which, expressing the particle velocity v_p as $v_p \approx v_0 (Ri)^{1/2}$ (a valid approximation for a plume-like flow), the relation $z^* g'/v_0^2 \propto Ri$ can be easily derived.   The dependence of the height of the vortex on the Richardson number predicted by is shown by the experiments and it is well represented in Fig 4 where a dimensionless graph has been plotted.   The two sets of experiments having $H = 0.74$ m (solid symbols) and $H = 0.64$ m (open symbols) show good accordance with the prediction and collapse well in a single curve.

4.3 Centreline axial velocity decay

The decay of the centreline velocity has been studied by many authors (see, for example, Chen and Rodi (1980) for a review).   Although this behaviour has been understood in free turbulent buoyant jets, few investigations have been carried out on the velocity decay of a buoyant jet impinging on a solid boundary. Fig 5 shows the normalised centreline velocity decay v_m/v_0 with dimensionless distance z/b from the solid

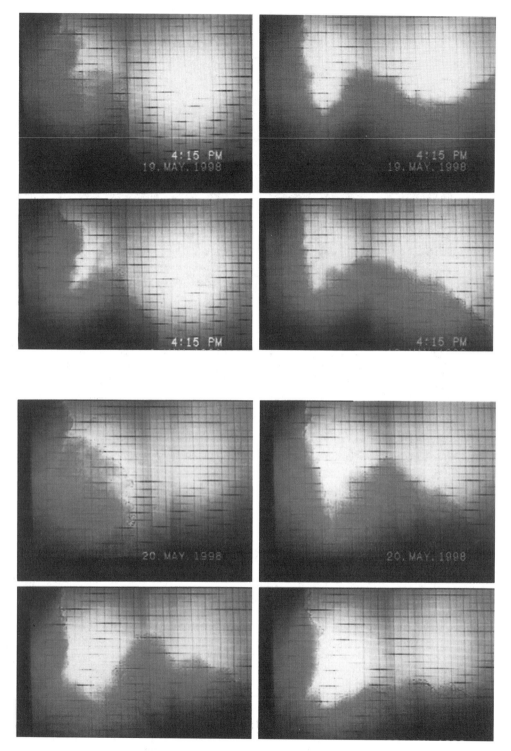

Fig 3   Shadowgraph sequences of descending buoyant plume for Ri = (i) 0.055 and (ii) 0.016.

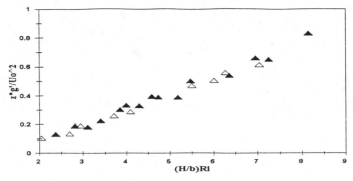

Fig 4: Dimensionless plot of height z^* of vortex versus $(H/b)(Ri)$ for $H = 0.74$ m (solid symbols) and 0.64 m (open).

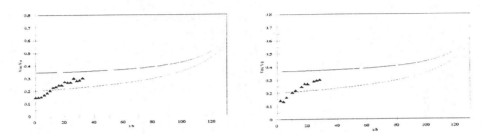

Fig 5   Data (solid symbols) showing decay of normalised centreline velocity v_m/v_0 with dimensionless distance z/b from solid boundary ($z = 0$) for $Ri = 0.022$ (left) and 0.026 (right). Solid and dotted lines represent respectively the free buoyant jet and the free jet-like flow approximations, as calculated by Yannopoulos and Noutsopoulos (1990).

boundary, as calculated for two different experiments having the same total depth $H = 0.74$ m. In these graphs, the data (solid symbols) have been compared with the centreline axial velocity distribution as calculated by Yannopoulos and Noutsopoulos (1990). Solid and dotted lines indicate respectively the approximation for a free turbulent buoyant jet and for a free turbulent pure jet. As the flow approaches the boundary ($z = 0$) of the impinging region, the axial velocity v_m decreases faster to the zero value.

Fig 6 shows a composite plot of v_m/v_0 versus z/b, for different initial Ri values and an example of the time-mean velocity distribution is shown in Fig 7.

5. CONCLUSIONS

The studies have looked at cases in which the impacting fluid is behaving in a plume-like or jet-like manner, in the sense that the transition distance (from the source) characterising the change from the latter to the former is both greater than and less than the fluid depth. The impact of the buoyant jet with the rigid bottom surface is seen to be accompanied by distortion of the eddy field in the incident jet. As a result, the incident vortices are distorted significantly upon impact, with the vortices containing erupted dense fluid extending vertically over a distance that is dependent primarily upon the Richardson number Ri of the flow. Specifically, the data on height of penetration show good scaling with Ri when using the normalising length scale g'/u_0^2, for a full range of Ri and Re and for two different depths of receiving water. No evidence for a sharp transition between jet and plume values is seen in the shadowgraph data.

Following impact and vertical eruption in the side vortices, fluid is seen to flow away from the impact region as a density current. Measurements of the velocity decay along the centreline of the plume show a behaviour

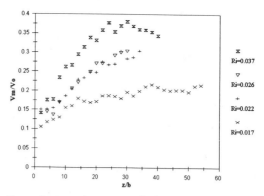

Fig 6   Plots of centreline velocity v_m/v_0 versus z/b, for different initial Ri values indicated.

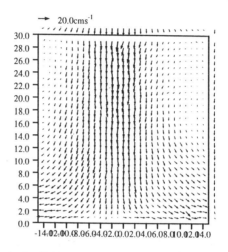

Fig 7.   Typical time-mean velocity field acquired from *DigImage* particle tracking in the impact zone.

that is consistent with the modifications by the boundary of the free buoyant jet data presented by Yannopoulos and Noutsopoulos (1990).

6. REFERENCES

Beltaos, S. & Rajaratnam, N. (1972). "Plane turbulent impinging jets", *J. Hydr. Res.*, 11(1), 29-59.

Chen, J. C., & Rodi, W. (1980). **Turbulent Buoyant Jets - a Review of Experimental Data**, HMT, 4, Pergamon Press.

Ching, C. Y., Fernando, H.J.S & Noh, Y. (1993). "Interaction of a negatively buoyant line plume with a density interface", *Dyn. Atmos. Oceans*, 19, 367-388.

Kotsovinos, N. E. (1975). "A study of the entrainment and turbulence in a plane buoyant jet", PhD Thesis, Report No KH-R-32, California Institute of Technology, Pasadena, USA.

Kotsovinos, N. E. and List, E. J. (1977). "Plane turbulent buoyant jets, Part I: Integral properties", *J. Fluid Mech.*, 81, 25-44.

List, E.J, (1982), "Turbulent jets and plumes". *Ann. Rev. Fluid Mech.*, 14, 189-212.

Noh, Y., Fernando, H. J. S. & Ching, C. Y., (1992). "Flow induced by the impingement of a two-dimensional thermal on a density interface", *J. Phys. Oceanogr.*, 22(10), 1207-1220.

Yannopoulos, P. & Noutsopoulos, G. (1990). "The plane vertical turbulent buoyant jet", *J. Hydr.Res.*, 28(5), 565-580.

Environmental Hydraulics, Lee, Jayawardena & Wang (eds) © 1999 Balkema, Rotterdam, ISBN 90 5809 035 3

Barge dumping of rubble in deep water

J. Bühler
Institut für Hydromechanik und Wasserwirtschaft ETH, Zürich, Switzerland

D.A. Papantoniou
Performance Technologies S.A., Agios Dimitrios, Greece

ABSTRACT: When a load of coarse material is dumped from a barge in deep water, it descends in the water body as an irregularly shaped particle cloud, which is called a suspension thermal. As such a thermal keeps sinking it slows down and undergoes a transition to a smooth, bowl-shaped cluster. An important difference betweeen these two flow regimes is that in the thermal stage the fluid inside the cloud moves in unison with the particles, while it remains nearly motionless in the cluster stage as the particles fall through it. In this paper results of experiments on clusters are presented, and a description of the flow is proposed which covers both the thermal and the cluster stages, i.e. the entire range of flows which are of interest for barge dumping.

1. INTRODUCTION

Rubble from tunnel construction or dredged material is often disposed of in designated areas of lakes and coastal waters (Fig. 1). In order to assess the environmental impact on benthic organisms at some distance from the disposal site, or the lateral extent of the sediment deposition, information on the dynamics of the fluid mass set in motion by the falling solid material is required. Slack (1963) investigated the release of large numbers of particles in air, and his visualisations show smooth, bowl - shaped clusters which were formed when the settling rate of individual particles was greater than one-third of the sedimentation rate of the whole cluster. Boothroyd (1971) pointed out that two different types of particle clouds exist. One type is similar to a thermal of warm air, and the slip velocity of any particle relative to the interstitial fluid surrounding it is much smaller than the velocity of the cloud. In contrast, the two velocities are nearly equal for the second, clusterlike, type, such that the sinking velocity of the cloud is nearly equal to the individual settling velocity of the particles in calm fluid. Nakatsuji et al. (1990) reported that for plane particle clouds a constant front velocity was approached after some time.

Bühler and Papantoniou (1991) suggested that suspension thermals eventually evolve into sinking particle clusters (they called them particle swarms, which now seems less appropriate for highly organized structures than the term clusters used by Slack and Boothroyd). They concluded that the transition from the thermal to the cluster stage occurs when the front velocity according to the similarity theory for thermals has decayed to a value which is about equal to the terminal sinking velocity u_t of individual particles in calm fluid. By assuming near self-preservation of the flow these authors also derived power laws for the widening rate of clusters. One option was that the width of 3-d clusters would grow with the cube root of the distance from the source, and that of plane clusters with the square root. They also carried out some experiments with isometric sand of 1.5 - 2 mm sieve size, and found the widening rate to be consistent with the cube root law.

More systematic measurements of the velocity and growth rate in the thermal and cluster stages were carried out by Rahimipour and Wilkinson (1992), who confirmed the scaling of the transition from a thermal to a cluster. Tamai et al. (1991) as well as Noh and Fernando (1993) investigated plane releases. Noh and Fernando proposed a different concept for the transition length scale which they determined on the basis of

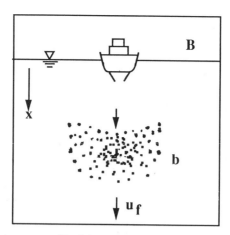

Fig.1 A particle cluster

local properties of the flow, and which depends on the fluid viscosity as well as on u_t. Their measurements of the front velocity and the width of the clouds, however, are also consistent with the transition concept by Bühler and Papantoniou (1991), and with a growth with the square root of the travel distance (their Figs. 5 and 7). Li (1997) performed numerical simulations and carried out experiments for particle clouds up to the transition to the cluster stage. Both of his approaches confirm that the front velocity approaches the settling velocity far enough from the source. In the present paper conservation equations based on those for thermals are proposed for clusters, and experimental data are presented which are used to determine the relevant flow constants.

2. ANALYSIS

When a mass M of solid particles of density ρ_s is submerged in a liquid of uniform density ρ_l, it gives rise to an excess mass $M_e = (\rho_s-\rho_l)M/\rho_s$ and a resulting buoyancy force $B = g\,M_e$, where g is the gravitational acceleration. Both of there quantities remain preserved during the descent.

In the thermallike initial stage of the flow the cloud grows in size due to the turbulent entrainment of ambient fluid through its interface. As long as the sinking velocity of the cloud is much larger than the terminal settling velocity u_t, the average downward velocity u of fluid within the cloud can be considered equal to that of the solid phase, and the cloud motion is adequately described by conventional thermal theory (see e.g. Baines and Hopfinger, 1984). This theory supposes that the entrainment rate per unit interfacial area is proportional to u, that the entrained fluid contains no net forward momentum, and that pressure outside the thermal is hydrostatic. For a roughly spherical suspension thermal of width b descending in an unstratified liquid of density ρ_l the integrated conservation equations for mass and momentum and density can then be written as

$$\frac{\pi}{6}\frac{d}{dt}(\rho b^3) = \alpha'(\rho_1\rho^2)^{1/3}\pi b^2 u \tag{1}$$

$$\frac{\pi}{6}\frac{d}{dt}\left((\rho + k_v\rho_1)b^3 u\right) = gM_e \tag{2}$$

$$\rho = \rho_1 + \frac{6M_e}{\pi b^3} \tag{3}$$

194

α' is an entrainment coefficient, and k_v an added mass coefficient which accounts for the fact that part of the mass containing forward momentum is outside the visible boundaries of the cloud, i.e. that it contains no buoyancy. The modification of the entrainment coefficient with the bulk density ρ of the fluid within the cloud is adopted from Baines and Hopfinger (1984) who quote values of $\alpha' = 1/3$ for thermals with small excess density, and $k_v = 0.5$. The entrainment coefficient for such thermals can be obtained by plotting b^2 vs. t, but it is usually based on measurements of the growth rate $db/dx_f = 2\alpha$ of the clouds, where $x_f = x+b/2$ is the position of their front.

An essential feature of the momentum equation (2) is that the entire moving mass is considered to remain associated with the thermal, i.e. that the thermal has no wake. This assumption is not justified for clusters as most of the fluid entering from below is not permanently entrained, but leaves again through the upper fringes of the cloud, giving raise to a wakelike flow in its lee. Similarly, the velocity u in clusters is associated with the fluid phase only. It can be inferred that the local fluid velocity increases from zero at the lower rim of the cluster to to a larger value at its upper rim, as more and more particle wakes are contributing to the downward motion. Variations of the particle velocity over the extent of a cluster are of order u and can be neglected in comparison to u_t, so that the cloud can be taken to move at a velocity $dx/dt \cong u+u_t$ over the entire range from thermals to clusters.

Provided that the growth of a cluster can also be described as being due to interfacial shear and turbulent entrainment, the continuity equation (1) can be retained for these flows, possibly with a different value α'_c of the entrainment coefficient. Conversely, the throughflow of fluid must be considered in the momentum equation. While the fluid entering a cluster from below is still at rest and has no momentum in the frame of reference fixed in undisturbed ambient fluid, the outflow of momentum through the upper boundary needs to be accounted for. In the cluster stage the visible cloud sinks essentially with the settling velocity u_t, and the fluid phase has a much smaller local velocity of order u. The mass of water leaving through the upper rim per unit time can then be estimated as $c_c\rho_l u_t b^2 \pi/4$, and the momentum associated with it as $c_c\rho_l u u_t b^2 \pi/4$. The coefficient c_c is supposed to refine the geometric and kinematic approximations used here and needs to be determined from experiments. A simple description for the transition from a thermal to a cluster can then be formulated by adding this term to the momentum equation for thermals, or

$$\frac{\pi}{6}\frac{d}{dt}\left((\rho + k_v\rho_1)b^3u\right) + \frac{\pi}{4}c_c\rho_1 u\, u_t\, b^2 = gM_e = B \tag{4}$$

After a suspension is released from a barge at a velocity u_0 much larger than u_t, the first term in the momentum equation is the dominant one, while the second term takes over later in the cluster stage. The approach proposed here suggests that the transition from the thermal to the cluster stage is gradual as far as the motion of the entire cloud is concerned. Noh and Fernando's (1993) visualizations also show more local processes, and suggest that the particles drop out of thermals quite suddenly. This agrees with our own observations that, at some point in time, the particles near the outer rim fail to make a full ascent to the top of the large toroidal vortex in which they are embedded, and drop out of the cloud.

For clusters we shall simply neglect the first term of (4) in comparison with the second one. A substitution into (1) then leads to

$$\frac{d(b^3)}{dt} = \frac{24\alpha'_c B}{\pi c_c \rho_1 u_t} \tag{5}$$

integration of the result by setting $b = 0$ at the virtual origin $t = 0$, leads to

$$b^3 = \frac{24\alpha'_c Bt}{\pi c_c \rho_1 u_t} \tag{6}$$

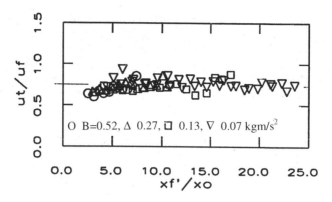

Fig. 2 Decay of the front velociy u_f

as $u_t \cong x_f/t$, we also find

$$b^3 \cong \frac{24\alpha'_c B x_f}{\pi c_c \rho_1 u_t^2} \tag{7}$$

Bühler and Papantoniou (1991) started out from a balance of the second and third terms of eq. (4), which also describes the excess momentum in the far-wake of a body of buoyancy B sinking in a fluid. The scale b then corresponds to the wake width at a given downstream distance, and u_t is the velocity of the body. They argued that this equation should also be valid just above the falling body if is sufficiently porous, as it is in case of clusters. The one-third power law was derived by assuming that the growth rate db/dt is proportional to the velocity scale u of the fluid. For wakes, this similarity assumption leads to a width which increases with the cube root of the distance from the falling body. Experiments for axisymmetric wakes do not clearly support this growth law. Cannon and Champagne (1991) showed that plots of the width with the square root of the downstream distance (the similarity law for plane wakes) also lead to satisfactory results. Similarly, Wen and Nacamuli (1996) found that a growth of the width of 3-d thermals and clusters with the square root of the travel time (the similarity law for plane clusters) is consistent with their experimental data for a large range of particle sizes and densities. Bühler and Papantoniou (1995) pointed out that the cluster stage may come to an end when the particles have drifted so far apart that their wakes no longer interact within the cluster. In this final stage the cloud width could also be expected to grow with the square root of the travel distance due to the random, self-induced oscillations which the particles perform during their descent.

3. EXPERIMENTAL DATA

A number of experiments on 3-d clusters were carried out by the authors in a tank of 1.5x3m size and 1.1m depth, and results on coarse sand of 1.5-2 mm sieve size were reported by Bühler and Papantoniou (1991). The material was released at the surface in a corner of the glass-walled tank, 0.5 m from two of the side walls. Further details about the experimental procedure are given in that paper. Here we examine the corresponding results for tests with coarser sand, with an isometric sieve size of 2-3 mm. A mean diameter of the particles was estimated as 2.62 mm by making use of their known density $\rho_s' = 2.6$ g/ccm, and by measuring the weight of a predetermined number of particles. For a given size distribution this method leads to an equivalent particle size which is somewhat larger than the mean diameter, as large particles contribute much more to the weight of the sample than small ones. The sinking velocity for this diameter at 20C in water was determined from Raudkivi (1976) as $u_t = 21.8$ cm/s. This value represents an average for quartz sand, and is less than the value for spherical particles. The decay of the velocity of the clouds during their descent is shown in Fig. 2. Five tests were carried out, three for values of B corresponding to 0.54, 0.27, and 0.13 kgm/s^2 respectively, and two more for 0.07 kgm/s^2 which are not distinguished in Fig. 2. The distance

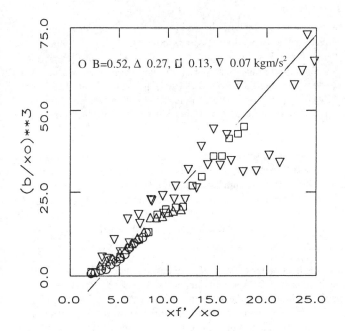

Fig. 3 Cube root growth law of cluster width b with frontal distance x_f'

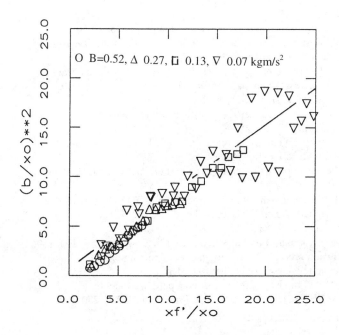

Fig. 4 Square root growth law of cluster width

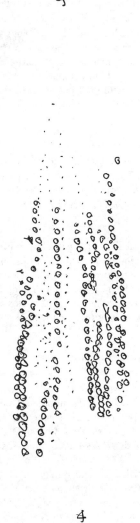

Fig. 5 Particle positions at 0.04s intervals, B = 0.13 kg m/s^2

x_f' of the front from the real source of the flow is normalized with the scale $x_0 = (B/\rho_1)^{1/2}/u_t$. An asymptote at $u_f \cong 4/3 \, u_t$ is clearly recognizable, and the ratio of the two velocities correponds closely to the value of 1.4 reported by Bühler and Papantoniou (1991) for 1.5 - 2mm sand. Nakasuji et al. (1990) carried out experiments with glass spheres of uniform size and found the front velocity to be in agreement with u_t. This suggests that the largest particles in our clusters were near the front. Li (1997) also carried out experiments with sand but computed the settling velocity for spherical particles, the resulting value was also found to agree with the front velocity. The ratio u_f/u_t thus appears to depend rather significantly on the sphericicity and size distribution of the particles.

Fig. 3 represents the growth of particle clouds based on the cube root law. By fitting a straight line to the region far from the source one obtains a value of 1.5 for $b/(x_f x_0^2)^{1/3}$. This is consistent with the values of 1.3 to 1.5 reported by Bühler and Papantoniou (1995) for 1.5-2 mm sand, and corresponds to a value α'_c/c_c of 0.4. The transition from the initial region to the cluster region occurs at an abscissa value of about 3, and the virtual origins of the flows ($x_f = 0$) in the cluster stage is downstream from the real source. In Fig. 4 the square of the width is plotted against the travel distance as suggested by the similarity law for plane clusters. The slope corresponds to $b/x_0 = 0.8 \, (xf/x_0)^{1/2}$.

Fig. 5 shows the particles trajectories in one of the smaller clusters. The distance between the end marks is 95 cm and particle positions are shown at intervals of 0.04 s. Dots denote particles near the front of the cluster, open circles near its sides and back. Variations of the velocity of about 20% are observed, with particles near the front moving faster than the rest. The gently curved trajectories show some indication of the oscillations of the particles about their mean path. These could become responsible for the self-induced part of the growth in the final stage of the flow, when the particle wakes no longer interact within the cloud.

4. REFERENCES

Boothroyd, R.G. (1971). *Flowing gas-solids and suspensions*. Chapman and Hall, London.

Bühler, J., and Papantoniou, D. A. (1991). Swarms of coarse particles falling through a fluid. *Environmental Hydraulics*, Lee & Cheung (eds),135, Balkema Rotterdam.

Bühler, J., and Papantoniou, D. A. (1995). The growth rate of particle clusters sinking through a fluid body.*HYDRA 2000*, Proc. XXVI IAHR Congress, London, V.2, 196-201.

Li, C. W. (1997). Convection of particle thermals. *Jour. Hydr. Res*., 35(3), 355-376.

Luketina, D., and Wilkinson, D. (1994). Particle clouds in density stratified environments. V. 2, Session B4, *Proc. Int. Symp. on Strat. Flow*, Grenoble, June 29-July 2.

Nakatsuji, K., Tamai, M., and Murota, A. (1990). Dynamic behaviours of sand clouds in water. *Int. Conf. on Phys. Modelling of Transport and Dispersion*, M.I.T. Boston, 8C.1

Noh, Y., and Fernando, H. J. S. (1993). The transition in the sediment pattern of a particle cloud. *Phys. Fluids* A 5, 3049.

Rahimipour, H., and Wilkinson, D. (1992). Dynamic behaviour of particle clouds. 743, *11th Australasian Fluid Mech. Conf.*, Univ. of Tasmania, Hobart, Australia, Dec. 14-1.

Raudkivi, A.J. (1976). *Loose boundary hydraulics*. 2nd ed., p.11, Pergamon Press.

Slack, G. W. (1963). Sedimentation ofa large number of particles as a cluster in air. *Nature*, V. 200, 1306

Tamai, M. , and Muraoka, K. (1991). Diffusion process of turbidity in direct dumping of soil. *Environmental Hydraulics*, Lee & Cheung (eds), 147, Balkema Rotterdam.

Wen, F., and Nacamuli, A. (1996). The effect of the Rayleigh number on a particle cloud, in *Hydrodynamics*, Chwang, A.T. and Lee J.H.W., eds., V.2, Balkema, Rotterdam, 1275-1280.

Environmental Hydraulics, Lee, Jayawardena & Wang (eds) © 1999 Balkema, Rotterdam, ISBN 90 5809 035 3

Effects of boundary and buoyancy on jet behavior in cold water

Ruochuan Gu
Department of Civil and Construction Engineering, Iowa State University, Ames, Iowa, USA

ABSTRACT: The problem of jet-boundary and jet-buoyancy interactions is theoretically and numerically investigated. A 2-D numerical simulation model is applied to a plane buoyant jet discharged horizontally into a bounded, stationary, and temperature-stratified cold waterbody with an ice cover. Numerical simulations are conducted to capture the key processes, including ambient water entrainment, jet deflection, recirculation, attachment to a boundary, impingement on a wall, and wall-jet flow. Effects of boundaries and their symmetry, external forces, buoyancy, and stratification on jet behavior are analyzed. The peculiar hydrodynamic and hydrothermal feature, i.e. buoyancy reversal, is identified and illustrated through numerical modeling. Flow and mixing processes and circulation patterns, e.g. spreading caused by the adjacent boundary and driven by buoyant forces, are examined and described using the simulation results. The impact of boundaries perpendicular and parallel to the flow direction is compared with the offset jet phenomenon.

1. INTRODUCTION

A submerged water discharge into a cold and stratified waterbody is often encountered in environmental and hydraulic engineering problems (Ashton, 1986; Ellis, et al. 1991). After an ice cover forms, water temperature typically varies from 0 °C at the ice-water interface to 2-4 °C at the bottom (water-sediment interface), producing a weakly stable stratification. The behavior of the jet is governed by buoyancy and stratification as well as (vertical and horizontal) boundaries. When the density of discharged water is different from that of ambient water, a buoyant jet is formed and driven to rise or fall by a buoyant force (Gu and Stefan, 1988). When water is discharged into a waterbody near a horizontal boundary parallel to the initial jet direction (offset from a vertical wall), the jet may be deflected towards the boundary and eventually attach to it (Gu,1996). Temperatures of natural waters in cold regions during winter usually cool to 4 °C or below. Submerged warm water discharges (4-20 °C) into cold waterbodies with ice covers have peculiar flow and mixing features. In the nearfield within a very short distance from the point of discharge, the jet is a momentum-dominated flow having the properties of a free jet. Attachment may occur when the jet is driven by buoyancy or deflected by an adjacent solid wall or a free surface and tends to flow along the boundary (Fig. 1). In the region around the attachment point, i.e. the impingement region and part of the pre-attachment region, the jet has some characteristics an impinging jet has. The offset jet becomes a wall jet--its limiting case in the farfield. Gu (1996) performed 2-D simulations of non-buoyant offset jets in uniform ambients and compared with laboratory measurements.

This paper investigates the roles of boundary and buoyancy in controlling jet behavior and mixing processes in a ice-covered water body. Effects of boundaries perpendicular and parallel to flow and of buoyancy on jet attachment and impingement are compared. Symmetric flow configuration is used to de-emphasize the Coanda effect caused by offset in the analysis of buoyancy and stratification effects, while non-buoyant jets and uniform ambients are considered for investigating boundary effect. This 2-D modeling study provides a detailed description of the flow characteristics and offer significant insight into the dynamic flow processes caused by the adjacent boundary and ambient stratification and buoyancy. The results of present work should aid in better physical understanding and modeling of complex jet flows and provide information useful in design of environmental water discharges.

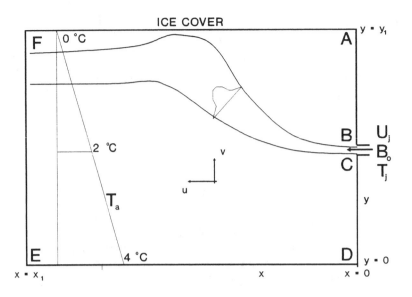

Fig. 1. Schematic of a buoyant jet in a stratified cold waterbody.

2. MODEL

2.1 Governing Equations

The 2-D model employed in this study was based on an earlier one introduced by Patankar (1980; 1981). The earlier model was modified with incorporation of buoyancy, a numerical solution scheme for unsteady and nonlinear problem, and appropriate boundary conditions by Gu and Stefan (1991). The present model is capable of directly simulating incompressible turbulent plane jets discharged vertically or horizontally. The governing equations include four mean flow equations (continuity, horizontal momentum, vertical momentum, and temperature), two κ-ε turbulent transport equations(Hossain and Rodi, 1986), and an equation of state in the Cartesian coordinate system.

2.2 Boundary Conditions

Boundary conditions need to be specified on all surfaces of the computational domain. Boundaries presented in this study include free boundary (line AB and CD), inflow (BC) and outflow (FE), and solid wall (AF and ED) as shown in Fig. 1. A free boundary is defined as the location where velocity and a scalar quantity are nearly equal to its free-stream ambient value (like a liquid wall). At a solid surface, the no-slip condition is applied, i.e. both mean and fluctuating velocities are zero. The dependent variables at the near-wall point are connected to the wall conditions, i.e. wall shear stress, heat fluxes, and wall temperature, by applying a linear function for the viscous sublayer and the log-law of the wall just outside the region.

2.3 Characterizing Parameters

The major parameters characterizing the discharge of an buoyant jet are jet nozzle Reynolds number defined as Re $= U_j B_0/\nu$ and densimetric Froude number defined as $Fr_d = U_j/|g'B_0|^{1/2}$ characterizing the ratio of momentum to buoyancy. Richardson number defined as Ri $= g'B_0/U_j{}^2$ describes the strength and sign of buoyancy (Ri $= 1/Fr_d{}^2$). The sign of jet buoyancy (positive or negative) is determined by $g' = g(\rho_{a,j}-\rho_j)/\rho_{a,j}$, where $\rho_{a,j}$ and ρ_j are densities of ambient water and injected water, respectively, at the jet origin. The injected water temperature, T_j, determines the initial direction of the buoyant force through the density-temperature relation with the maximum

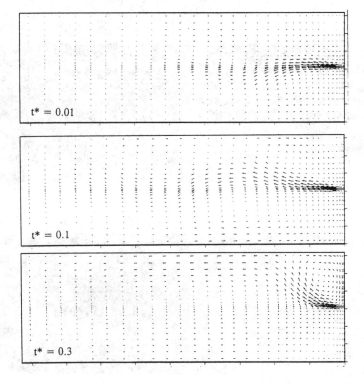

Fig. 2. Simulated velocity fields for a jet of 8 °C (Re = 35000 and Fr_d = 13)

density at 4 °C. A dimensionless time is defined as the real time t multiplied by the ratio of flow rate per unit length of the slot jet to the volume of water body per unit width, i.e. $t^* = tU_jB_o/(y_1x_1)$. At $t^* = 1$, t is equivalent to the time required to completely withdraw the water at a flowrate of $Q_j = U_jB_o$.

3. RESULTS AND DISCUSSION

The evolution of flow and thermal fields of a submerged jet in an ice-covered lake and effects of buoyancy and stratification on flow pattern were examined through numerical simulations using the 2-D model. Cases with different injection temperatures, 4 °C ≤ T_j ≤ 20 °C, were simulated and analyzed. Distributions of velocity and temperature and details in flow pattern and mixing process were obtained for each set of conditions. Comparison of jet behavior was made by changing any one of the parameters t^*, T_j and Fr_d (or Ri) and fixing the other two.

At the discharge point a jet can be positively or negatively buoyant depending on ambient temperature relative to its own temperature and density. The jet will rise if the initial buoyancy is positive and sink if negative. Fig. 2 presents flow fields in the form of velocity vectors at various times (t^* = 0.01, 0.1 and 0.3) for T_j = 8 °C with Fr_d = 13. These time points were selected to represent three mixing stages, i.e., initial, intermediate, and long-term. As the jets are initially cooled quickly down to 4 °C (maximum density) by dilution due to entrainment, a reversal in buoyancy (from positive to negative) turns the rising jets into falling ones (t^* = 0.01). As time progresses more mixing between the jet and the ambient occurs. The ambient becomes warmer and more uniform. Rapid cooling of the jet is slowed down. A reversal in buoyancy from negative to positive drives the jet upwards to the ice cover (t^* = 0.1), impinging on the top surface and becoming a wall jet. The jet impinges on the ice cover at t^* = 0.3.

Flow patterns and trajectories of jets with different discharge temperatures under a same buoyancy condition (Fr_d) at a specific time are compared. Presented in Fig. 3 is the simulated isotherms of four buoyant jets with T_j = 4, 8, 10, and 20 °C at t^* = 0.5, in which Fr_d = 6 and Re = 8000-70000. The 4 °C discharge is driven to the bottom by

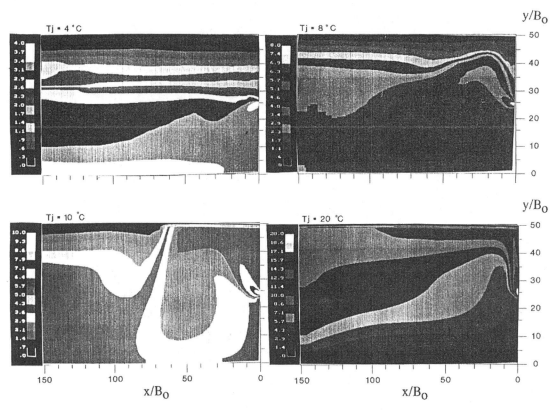

Fig. 3. Simulated thermal fields at t* = 0.05 for jet temperatures of 4, 8, 10, and 20 °C

negative buoyant force all the time. As cooled down to 4 °C by the ambient water, the leading edge of a warm jet is always sinking until the ambient water warms up to 4 °C or higher in all cases of $T_j > 4$ °C. Strong spatial and temporal buoyancy reversals occur at $T_j = 8$-10 °C. Negative buoyancy dominates a jet of low discharge temperature (4-6 °C). Positive buoyancy plays a very important role in a high discharge temperature situation ($\geq$ 15 °C). Higher Fr_d (lower buoyancy) results in straighter jet trajectory and longer attachment distance.

Table 1 summarizes discharge conditions (T_j, Fr_d, Re, and Ri), jet behavior, and flow patterns in the nearfield (0< x/x_1 < 0.5) at various time points. The initial jet buoyancy strength is represented by Richardson number (Ri). It is shown that the flow pattern and mixing process at a specific time point largely depends on the discharged temperature, T_j, and buoyancy, $g'B_0/U_j^2$. If the water body has a finite volume, as is often the case, the ambient cold water will eventually be replaced by warmer water recirculating from the jet. If all of the ambient water warmed to 4 °C after certain period of time, the jet discharge ($T_j > 4$ °C) would become positively buoyant everywhere and rise continuously. After the ambient water is completely replaced by the discharged water, the jet becomes neutral or non-buoyant in all cases, reaching a equilibrium state and forming a horizontal jet. However, it would take a extremely long time for a large natural lake to be completely filled with the discharged water under a small jet flow rate. It is also possible that only a part of the lake or semi-open waterbody is involved in the circulation induced by the jet.

Similar flow patterns can be observed if one compares the processes of jet attachment, boundary impingement and wall-jet flow after the jet is deflected towards the boundary by buoyant force with that of a non-buoyant offset jet close to a boundary and deflected by the Coanda effect (Gu and Stefan, 1991; Gu, 1996). Attachment to either the bottom or the top boundary may occur after a buoyant jet is driven up or down by positive or negative buoyant force, followed by impingement and wall-jet flow. However, a non-buoyant offset jet remains attached to the boundary once it is deflected by reduced pressure. The driving force for the deflection of a buoyant jet results from density difference, which turns the jet towards its boundary. Buoyancy reversal and non-uniform ambient

Table 1. Summary of near-field jet characteristics and flow processes

T_j (°C)	Re	Ri	t*							
			0	0.005	0.010	0.025	0.050	0.1	0.3	∞
4	8000	-0.0278	S	S	S	S, A	S, A			N
6	35000	0.0003	R≈N	S	S	S	S, A	S, A	S, A	N
8	35000	0.0061	R	R→S	R→S	S	S	R→S	R, A	N
10	35000	0.0156	R	R→S	R→S	S	R→S	R, A	R, A	N
15	35000	0.0541	R	R	R→S	R→S	R, A			N
20	70000	0.0278	R	R	R	R, A	R, A			N

"R", "S", "R→S", "A", and "N" denote Rise, Sink, change from Rise to Sink along a jet trajectory due to spatial reversal of buoyancy, Attachment to and impingement on a boundary, and Neutral or non-buoyant, respectively.

water density together make jet attachment and boundary impingement unsteady. Attachment length is inversely related to the buoyancy strength ($Ri = g'B_0/U_j^2$). Since all jets studied are highly turbulent at Re = 8000-70000, flow patterns are not considerably sensitive to Reynolds number. Jet detachment from the boundary may occurs because of buoyancy reversal. Therefore, wall-jet flow following jet impingement may not appear in the buoyant jet case. In contrast, jet separation from boundary does not exist in the non-symmetric and non-buoyant offset jet cases. In a symmetric and buoyant situation, jet attachment to and impingement on either the top or the bottom boundary can be caused by buoyant force or by Coanda effect or by the combination of two. Jet attachment and impingement eventually disappear if the jet is located sufficiently far from the boundary and if no disturbances appear. The jet becomes a non-buoyant one, i.e., its equilibrium state after the ambient water is completely mixed by continuous discharge, entrainment, and dilution.

4. CONCLUSIONS

Simulation results showed that both boundary and buoyancy are the important factors controlling jet behavior and mixing processes. Peculiar hydrodynamic and hydrothermal features of the jets caused by the temporal and spatial reversals in buoyancy were identified. The flow pattern is very sensitive to the temperature at the discharge due to jet buoyancy and ambient stratification. A warm water jet can be sinking, or it can permanently rise to the surface where it may melt the ice cover depending on the state of mixing achieved in the ambient water. For $T_j > 4$ °C the leading edge of a jet is always sinking due to the cooling effect of the ambient water until the ambient warms up by the jet to 4 °C or higher. Strong negative buoyancy drives the 4-6 °C jets to sink and attach to the bottom, while 15-20 °C jets are driven to rise and attach to the ice cover by strong positive buoyant forces. Strong spatial and temporal buoyancy reversals appear in the situations of $T_j = 8-10$ °C. Dilution of a warm discharge by entrainment of cold ambient water causes the reversal in buoyancy, which alters the jet's flow direction. There are no attachment until t* = 0.025 in all cases. The time of first attachment depends on T_j and Ri. Higher jet densimetric Froude number results in straighter jet trajectory and longer attachment distance. T_j also influences attachment distance. Attachment disappears once the jet becomes neutral and after the finite volume of ambient water is completely replaced by discharged water as t* → ∞.

The jet was simulated as an offset jet by assuming a liquid wall, i.e. the free boundary on the inflow side of the flow domain. The effect of buoyancy on jet behavior was studied by using a symmetric geometry to partially decouple it from that of Coanda effect caused by boundary. The flow characteristics shown by the processes of deflection, attachment, impingement and wall-jet decay exhibit some similarities of the buoyant jet with the boundary-affected jet. However, some specific deviations are expected because of different driving forces, buoyant force in the former and reduced pressure in the latter. Attachment may not occur in the case of symmetric and non-buoyant offset jet in a uniform ambient. Detachment of a buoyant jet in stratified cold waters from a boundary may occur due to spatial buoyancy reversal. Attachment of a non-symmetric and non-buoyant offset jet is persistent, while intermittent attachment can result from temporal buoyancy reversal in the buoyant and symmetric case. Both buoyancy and Coanda effect should be considered in any attempts to analyze buoyant offset jets.

5. REFERENCES

Ashton, G. D. (1986). *River and Lake Ice Engineering*, Water Resources Publications, Littleton, CO.

Ellis, C., Stefan, H. G., and Gu, R. (1991). "Water temperature dynamics and heat transfer beneath the ice cover of a lake", *Limnology and Oceanography,* 36, 324-335.

Gu, R., and Stefan, H. G. (1988). "Analysis of turbulent buoyant jet in density-stratified water", *J. of Environmental Engineering,* ASCE, 114, 878-897.

Gu, R., and Stefan, H. G. (1991). *Buoyant jet flows and mixing in stratified lakes, reservoirs or ponds,* St. Anthony Falls Hydraulic Laboratory Project Report No. 318, University of Minnesota, p. 214.

Gu, R. (1996). "Modeling two-dimensional turbulent offset jets", *J. Hydr. Eng.,* ASCE, 133, 617-624.

Hossain, M. S., and Rodi, W. (1986). "A turbulent model for buoyant flows and its application to vertical buoyant jets", In: W. Rodi (Editor), *Turbulent Buoyant Jets and Plumes,* Pergamon Press, Oxford.

Patankar, S. V. (1980). *Numerical Heat Transfer and Fluid Flow,* Hemisphere Publishing Corporation, McGraw-Hill, New York, N.Y.

Patankar, S. V. (1981). *A calculation procedure for two-dimensional elliptic simulations,* Numerical Heat Transfer, 4, 409-442.

Environmental Hydraulics, Lee, Jayawardena & Wang (eds) © 1999 Balkema, Rotterdam, ISBN 90 5809 035 3

A numerical study on the stability of a vertical plane buoyant jet in confined depth

C.P. Kuang & J.H.W. Lee
Department of Civil Engineering, The University of Hong Kong, China

ABSTRACT: A plane turbulent buoyant jet discharging vertically into confined depth is studied using the Renormalization Group(RNG) $k - \epsilon$ model. The steady two-dimensional turbulent flow and temperature field are computed using the finite volume method on an unstructured triangular mesh. The numerical predictions demonstrate three generic flow patterns for different jet discharge and environmental parameters: i) a flow with circulation cells of alternate rotation for non-buoyant discharge; ii) a stable buoyant discharge with the mixed fluid leaving the near-field in a surface warm water layer; and iii) an unstable buoyant discharge with flow recirculation and re-entrainment of heated water. A stratified counterflow region always appears in the far-field for both stable and unstable buoyant discharges. The near field interaction and hence discharge stability is governed by only two dimensionless parameters - the discharge densimetric Froude number and the relative depth. The computed velocity and temperature fields agree well with the laboratory flow-visualization and temperature measurements, and the theoretical predictions of Jirka & Harleman (1979).

1. INTRODUCTION

Many industrial processes generate wastewater which is often discharged into natural water bodies, such as rivers, lakes or seas. A case of particular significance is the discharge of thermal effluent into a shallow waterbody from a line source placed at the bottom, where the vertical confinement of the flow domain has a significant influence on the flow development. An extensive review of such thermal diffusers for waste heat disposal has been given by Jirka (1982a). The present study is concerned with the case of a two-dimensional vertical discharge from a line source without considering end effects. Fig.1 shows a 2D vertical buoyant jet with initial velocity V_j, jet width B, excess temperature θ_o, discharging into an otherwise stagnant ambient fluid of depth H and uniform density ρ_a. The governing dimensionless parameters are the jet densimetric Froude number $F_o = V_j[(\Delta\rho_0/\rho_a)gB]^{-1/2}$ and the relative submergence H/B, where $\Delta\rho_0/\rho_a$ is the relative density difference between the source and ambient fluid. Whether buoyant or non-buoyant, a vertical jet flow develops which spreads and entrains ambient fluid. The jet impinges on the free surface, thereby causing a small superelevation which in turn leads to an excess hydrostatic pressure that drives the fluid sideways away from the impingement region. The flow development beyond the impingement region, and indeed the interaction between the jet and the adjacent flow, depends strongly on the parameters considered; a wide range of flow patterns can develop in the shallow water. The rate of entrainment depends on the buoyancy of the jet but also on the development of the ambient flow. A theoretical and experimental investigation of this problem has previously been reported (Jirka & Harleman 1979; Jirka 1982b). Their study (the JH study) was based on the coupling of integral zonal models, and the measurements were restricted to visualization and temperature measurements. Andreopoulos *et al.* (1986) performed detailed velocity and turbulence measurements for several cases in a relatively short channel with strong downstream control; most of the flow data were however limited to the vertical buoyant jet zone. All the previous

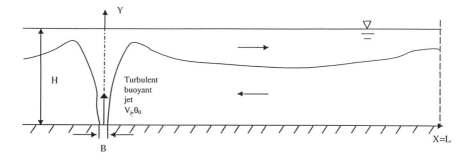

Fig.1 Schematic diagram of a vertical plane buoyant jet in shallow fluid (stable discharge configuration)

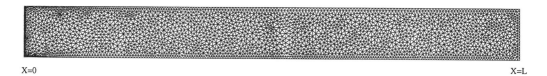

X=0   X=L

Fig.2 Unstructured triangular mesh for Case 1 (stable discharge)

experiments were insufficient to give a detailed physical picture of the jet-ambient flow interaction. As far as we are aware, numerical modelling of this turbulent buoyant flow has also not been reported. We present herein a numerical study of the flow and temperature fields of a vertical plane buoyant jet in confined depth.

2. THEORY AND FORMULATION

By virtue of the small density differences, i.e. $(\rho_a - \rho_0)/\rho_a \ll 1$, we adopt the Boussinesq approximation and the following Reynolds-averaged equations for steady incompressible flow:

$$\frac{\partial U_i}{\partial x_i} = 0 \tag{1}$$

$$U_j \frac{\partial U_i}{\partial x_j} = -\frac{1}{\rho}\frac{\partial p}{\partial x_i} + \frac{\partial}{\partial x_j}\left(\nu_{eff}\left(\frac{\partial U_j}{\partial x_i} + \frac{\partial U_i}{\partial x_j}\right) - \frac{2}{3}k\delta_{ij}\right) + \frac{\rho - \rho_a}{\rho_a}g_i \tag{2}$$

$$U_j \frac{\partial h}{\partial x_j} = \frac{\partial}{\partial x_i}\left(\alpha_h \nu_{eff}\frac{\partial T}{\partial x_i}\right) + U_i\frac{\partial p}{\partial x_i} + \left(\nu_{eff}\left(\frac{\partial U_j}{\partial x_i} + \frac{\partial U_i}{\partial x_j}\right) - \frac{2}{3}k\delta_{ij}\right)\frac{\partial U_i}{\partial x_j} \tag{3}$$

where U_i = fluid velocity in x and y directions, p = dynamic pressure, and $h = C_pT$ is the static enthalpy of the fluid; C_p and T are the specific heat and temperature of the fluid respectively, ρ=density, and the gravitational acceleration vector $\vec{g} = (0, -g, 0)$. The effective viscosity ν_{eff} is the sum of the molecular viscosity ν and the turbulent viscosity ν_t. α_h is the inverse turbulent Prandtl number for energy. Water properties are taken as temperature dependent. The Renormalization Group (RNG) $k - \epsilon$ model (Yakhot 1986) is adopted for turbulence closure:

206

$$\nu_{eff} = \nu \left[1 + \sqrt{\frac{C_\mu}{\nu}} \frac{k}{\sqrt{\epsilon}} \right]^2 \tag{4}$$

$$U_i \frac{\partial k}{\partial x_i} = \frac{\partial}{\partial x_i}(\alpha_k \nu_{eff} \frac{\partial k}{\partial x_i}) + \nu_t S^2 - g_i \frac{\alpha_h \nu_t}{\rho} \frac{\partial \rho}{\partial x_i} - \epsilon \tag{5}$$

$$U_i \frac{\partial \epsilon}{\partial x_i} = \frac{\partial}{\partial x_i}(\alpha_\epsilon \nu_{eff} \frac{\partial \epsilon}{\partial x_i}) + C_{1\epsilon} \frac{\epsilon}{k}(\nu_t S^2 + (1 - C_{3\epsilon})(-g_i \frac{\alpha_h \nu_t}{\rho} \frac{\partial \rho}{\partial x_i})) - C_{2\epsilon} \frac{\epsilon^2}{k} - R \tag{6}$$

where k is the turbulence kinetic energy (TKE), and ϵ is referred to as dissipation rate(DR) of k. α_k and α_ϵ are the inverse Prandtl numbers for k and ϵ, $S^2 = 2 S_{ij} S_{ij}$ is the modulus of the mean rate of strain tensor expressed as $S_{ij} = \frac{1}{2}(\frac{\partial U_i}{\partial x_j} + \frac{\partial U_j}{\partial x_i})$, and R is a rate-of-strain term that can be evaluated from S, k, ϵ. The RNG theory gives values of the constants $C_\mu = 0.0845$, $C_{1\epsilon} = 1.42$, $C_{2\epsilon} = 1.68$, as well as a means of evaluating $\alpha_k, \alpha_\epsilon$ in terms of ν/ν_{eff}. Low-Reynolds-number effects are included in the evaluation of viscosity, permitting laminar-like behavior to be predicted; the model is hence more applicable to the partly-turbulent characteristics of the flow induced by a laboratory jet in confined depth.

Our objective is to numerically simulate the laboratory experiments of Jirka and Harleman (1979). In the JH study, a slot jet with $V_j = 0.1 - 1.6 \; m/s$, $B = 0.5 - 6.4$ mm, $\theta_o = 6 - 16^o C$ was used; the jet discharges into a two-dimensional channel of length $L = 3.7 - 6.1$m and depth $H = 0.3 - 0.6$m. The end of the channel, $x = L$, is connected to a larger reservoir and hence provides a control section. In our numerical calculations, the discharge stability and mixing are studied for a total of 25 cases, covering broadly the same range of F_o and H/B as in the JH experiments. The salient features of the numerical experiments are summarised.

3. NUMERICAL SOLUTION AND DISCUSSION

The governing equations are solved using the finite volume method and the SIMPLEC algorithm for pressure-velocity coupling, as embodied in the FLUENT/UNS code. The equations are discretized on a two-dimensional, unstructured mesh composed of triangular cells(Fig.2) on which Cartesian velocities and other variables are defined at the cell centre. Cell-face dependent variables are interpolated from the cell centre values using a second order scheme. No under-relaxation of pressure is required, while a factor of 0.1 is adopted for velocities, 0.8 for temperature and 0.5 for other variables. By virtue of symmetry, the numerical solution is sought for a half of the flow domain, $x \geq 0$. On the plane of symmetry ($x = 0$), the symmetrical boundary conditions are imposed; in addition, the free surface ($y = H$) is modelled as a symmetry plane. At the outlet boundary ($x = L$) where reverse flow is expected, the zero pressure condition is prescribed. All other variables are extrapolated from the interior of the domain using zero gradient condition. At the wall boundary, the no-slip condition and zero heat flux are imposed. In the region close to the walls, the standard wall function is applied. At $t = 0$, zero initial velocity and pressure, and constant temperature of 293K are prescribed. The jet discharge and ambient conditions for all the cases are given in Table 1. Convergence is declared when the maximum scaled residual is less than 1×10^{-6} for the energy equation and 1×10^{-3} for other equations. About 5000 iterations are sufficient for convergence in all cases.

The computational results show that the flow pattern changes radically with the jet buoyancy. Fig.4 shows the velocity field of three representative cases - corresponding to a strongly buoyant, weakly buoyant and non-buoyant discharge respectively. For a strongly buoyant discharge (Case 1; same as Experiment 6 in JH study), a stratified two-layer counterflow develops which does not communicate with the vertical buoyant-jet zone. The flow in the lower layer provides all the fluid entrained into the jet and a small backflow exists in the near bottom layer in the near-field. This is classified as a stable

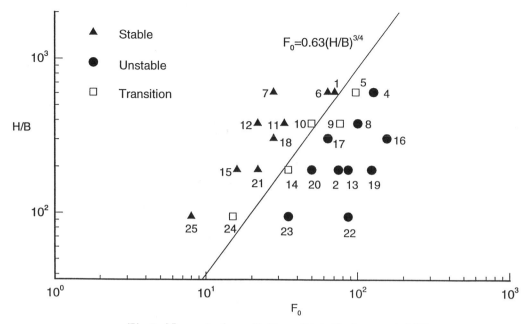

Fig.3 Numerical prediction of jet discharge stability

situation by Jirka & Harleman (1979). In the region of $x/H < 2.5$, a rapid increase in layer thickness occurs. The temperature distribution in the stratified counterflow region, $x/H > 2.5$, shows a strong density interface and horizontal and vertical uniformities in each layer (Fig.5). Thus there are negligible mixing across the interface and negligible heat loss in the free surface. The computed temperature profiles show that the interface slopes upwards towards the outlet. For a weakly buoyant discharge (Case 2 - Expt 12 of JH study), a recirculating cell exists in the discharge vicinity ($x/H \leq 2.5$) and a stratified two-layer counterflow exists in the region $x/H > 2.5$. The re-entrainment of heated fluid grossly reduces the dilution ability of the discharge system. This is classified as an unstable discharge. The counterflow region is similar to that of the stable discharge mode. For both cases, the computed interfacial shape and layer depth of the counterflow region are in good agreement with the experiments and theoretical prediction given in the JH study (Fig.5). For non-buoyant discharge (Case 3), the flow structure consists of circulation cells of alternate rotation. The length scale of the primary cell is about 2H; the primary recirculation zone forms a secondary counter-rotating zone by momentum transfer at its outer boundary and this forms a third one and so on. It can be seen (Fig.4) that the secondary cells are considerably shorter and weaker than the primary cell.

The numerical predictions of discharge stability are compared in Fig.3 with the theoretical criterion $F_0 = 0.63(H/B)^{3/4}$ of the JH study (which was validated by their experiments). In our calculations, the discharge is defined as unstable when an obvious temperature rise at the bottom boundary ($\theta/\theta_0 \geq 1.5\%$) and an obvious recirculating cell are computed in the near field. The discharge is defined as stable when no temperature rise and no recirculating cell are computed in the near-field. Other intermediate flow configurations are categorised as transition discharges. The excellent agreement with the JH study gives support to the present numerical model, and the theory for behaviour of buoyant jets in confined depth.

4. CONCLUSION

A numerical study on the stability of vertical plane turbulent buoyant jets discharging in confined

Table 1   **Summary of Jet discharge and Ambient Parameters (This Study)**

CASE	H (m)	B (mm)	V_j (m/s)	L (m)	θ_0 (^0C)	R_{ej}	L/H	F_0	H/B	$0.63(H/B)^{3/4}$	Stability category
1	.610	1.0	.461	6.1	16.4	921	10	70	610	77	stable
2	.610	3.2	.650	6.1	8.9	4159	10	75	191	32	unstable
3	.610	1.0	.461	6.1		921	10	∞	610	77	unstable
4	.610	1.0	.650	6.1	10.0	1300	10	128	610	77	unstable
5	.610	1.0	.500	6.1	10.0	999	10	98	610	77	transition
6	.610	1.0	.400	6.1	15.0	799	10	64	610	77	stable
7	.610	1.0	.200	6.1	20.0	399	10	28	610	77	stable
8	.610	1.6	.650	6.1	10.0	2080	10	101	381	54	unstable
9	.610	1.6	.500	6.1	10.0	1599	10	78	381	54	transition
10	.610	1.6	.400	6.1	15.0	1279	10	51	381	54	transition
11	.610	1.6	.300	6.1	20.0	960	10	33	381	54	stable
12	.610	1.6	.200	6.1	20.0	639	10	22	381	54	stable
13	.610	3.2	.800	6.1	10.0	5119	10	88	191	32	unstable
14	.610	3.2	.400	6.1	15.0	2559	10	36	191	32	transition
15	.610	3.2	.200	6.1	20.0	1279	10	16	191	32	stable
16	.305	1.0	.800	6.1	10.0	1599	20	157	305	46	unstable
17	.305	1.0	.400	6.1	15.0	799	20	64	305	46	unstable
18	.305	1.0	.200	6.1	20.0	399	20	28	305	46	stable
19	.305	1.6	.800	6.1	10.0	2559	20	124	191	32	unstable
20	.305	1.6	.400	6.1	15.0	1279	20	51	191	32	unstable
21	.305	1.6	.200	6.1	20.0	639	20	22	191	32	stable
22	.305	3.2	.800	6.1	10.0	5119	20	88	95	19	unstable
23	.305	3.2	.400	6.1	15.0	2559	20	36	95	19	unstable
24	.305	3.2	.200	6.1	20.0	1279	20	16	95	19	transition
25	.305	3.2	.100	6.1	20.0	640	20	8	95	19	stable

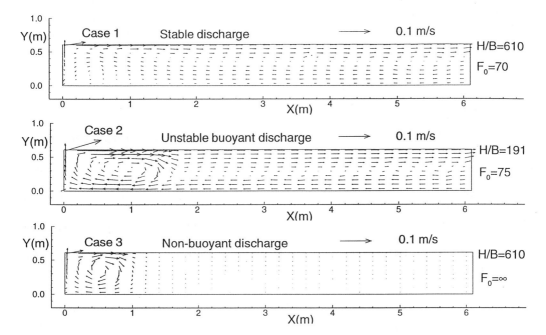

Fig.4 Computed velocity field for three representative situations: i) stable and ii) unstable buoyant discharge, and iii) non-buoyant discharge

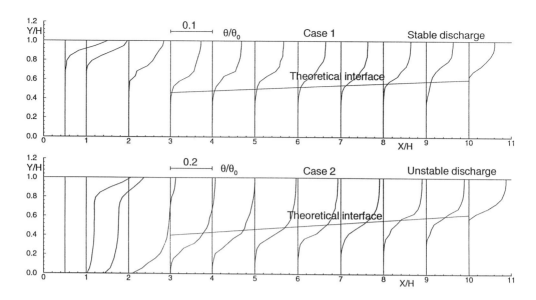

Fig.5 Computed vertical temperature profiles for stable and unstable discharge (theoretical interface from Jirka and Harleman 1979)

depth has been carried out. For non-buoyant discharges, an unstable flow develops with a primary recirculation cell of about 2.0 to 2.5H length which drives a secondary cell further away from the discharge. For buoyant discharges, the numerical computations demonstrate that discharge stability depends solely on the near behaviour of the jet - i.e. the jet densimetric Froude number and the relative submergence. Unstable discharge configurations are characterized by vertical recirculation cells leading to re-entrainment of mixed fluid into the jet. The numerical results give further insight into the flow and temperature fields; the computed stability criterion is in excellent agreement with previous experimential and theoretical results.

5. REFERENCE

Andreopoulos, J., Praturi, A. and Rodi W. (1986). "Experiments on vertical plane buoyant jets in shallow water." *J. Fluid Mech.* 168, 305-336.

Jirka, G.H. (1982a) "Multiport diffusers for heat disposal: a summary." *J. Hydraul. Div.* ASCE 108, 1425-1468.

Jirka, G.H. (1982b) "Turbulent buoyant jets in shallow fluid layers." In *Turbulent Buoyant Jets and Plumes*(ed. W.Rodi), HMT Series, Vol.6. Pergamon.

Jirka, G.H. and Harleman, D.R.F. (1979). "Stability and mixing of a vertical plane buoyant jet in confined depth." *J.Fluid Mech.* 94, 275-304.

Yakhot et al. (1986). "Renormalization group analysis of turbulence, I. Basic Theory." *J. Sci. Comput.* 1, 1-51.

Environmental Hydraulics, Lee, Jayawardena & Wang (eds) © 1999 Balkema, Rotterdam, ISBN 90 5809 035 3

Simultaneous velocity and concentration measurements of buoyant jet discharges with combined DPIV and PLIF

Adrian Wing-Keung Law & Hongwei Wang
School of Civil and Structural Engineering, Nanyang Technological University, Singapore

ABSTRACT: An experimental system is developed that allows the simultaneous measurements of velocities and concentrations within a study area using the techniques of Digital Particle Image Velocimetry (DPIV) and Planar Laser Induced Fluorescence (PLIF). The hardware components for the system are described in this paper. It was found that the simultaneous measurements are possible under appropriate combination of the amount of the two different tracers. For verification purposes, the system was used to measure the characteristics of a round jet discharging into stagnant environment. Comparison between the measurement results and the large body of existing information for pure jets is satisfactory.

1. INTRODUCTION

In recent years, the measuring techniques of Digital Particle Image Velocimetry (DPIV) and Planar Laser Induced Fluorescence (PLIF) are extensively used to investigate mixing processes related to buoyant jets phenomenon (e.g. Haven and Kurosaka, 1997; Weisgraber and Liepmann, 1998). The principles of DPIV and PLIF are well known and can be found in numerous articles. The experimental setup for the two techniques is rather similar. DPIV requires the projection of a laser sheet onto the flow field at successive time intervals and the subsequent capturing of the images detailing the position of seeding particles that reflect the laser light. Analysis of the difference in positions of the particles reveals the Lagrangian velocity distribution of the flow field. Similarly, PLIF requires the projection of a laser sheet onto the flow field to excite the dissolved fluorescence. Either digital cameras or video cameras are then used to capture the images of the fluorescence for qualitative assessment of the flow structure, and/or quantitative analysis of the concentration distribution.

With the similarity in the experimental setup, it is not surprising that the two techniques are sometimes applied to the same experiments for the measurements of both the velocity and concentration fields (e.g. Haven and Kurosaka, 1997). Thus far, in almost all cases the application of the two techniques is performed separately without synchronisation. Although valuable information can already be gained by analysing independently the velocity and concentration data, the interaction between the flow structure and the concentration distribution cannot be readily revealed without the synchronicity of the two records.

There are many advantages if the two techniques can be applied simultaneously. First, there will be significant saving in experimentation time. Experiments have to be repeated if the DPIV and PLIF are performed separately, thus doubling the number of experiments necessary. In addition, there may be inaccuracies involved in repeating the same experiment for the two measurements, such as changes in ambient conditions and the typical inaccuracies in controlling the discharge characteristics. These can be eliminated with the coupled measurements. Perhaps the most important advantage of the coupling is that it provides synchronicity of the two databases allowing the computation of both the time mean and turbulent transport quantities in the mixing processes. This is in additional to the planar information for the velocities and concentrations individually. Thus a complete analysis of the transport processes within the study area can be obtained. Investigation of the coupling of these two techniques as a measurement approach is the primary objective of the present study.

As far as the authors are aware, the only previous attempt to couple the two techniques is by Simoens and Ayrault (1994) who performed measurements of concentration flux in turbulent flows through the combination of PIV and PLIF. Their approach was primarily based on photographic film in contrast with the fully digital approach present in this paper. Photographic processes are undesirable for PLIF measurement because the relationship between the PLIF image greyscale and the species concentration may not be linear. As a result, the dynamic range of photographic PLIF is limited. Directional ambiguity associated with conventional PIV that requires complicated techniques such as image shifting and colour coding to resolve was also not addressed in their setup. The moderate precision of the techniques used by Simoens and Ayrault (1994) resulted in a significant degree of uncertainty in the second order quantities such as turbulent mass transport. Moreover, even if photographic cameras can be synchronised to record images continuously, the rate of sampling should be very low and the amount of recordings is quite limited due to the mechanical nature of the photographic camera.

The present setup improves on the accuracy of the measurements by adopting fully digital approaches. The DPIV setup in this study can measure velocity at sub-pixel accuracy (typically 0.1 pixel) using method similar to Willert and Gharib (1991). Through cross-correlation of two successive images, the velocity polarity is readily determined with DPIV. A double pulsed mini Nd:Yag laser is also employed that can "freeze" the flow field within a few nanoseconds as compared the typical temporal resolution of a chopped argon laser that is in the order of milli-second. The pulsed laser has an instantaneous power density much higher than a normal argon laser, thus the signal-to-noise ratio of images can be significantly enhanced.

On the combined application of the DPIV and PLIF techniques, the key issues are to achieve the simultaneous data capturing and to minimise the interfering effect between them. DPIV requires markers such as neutrally buoyant particles to provide the signal within a captured image for data processing. These seeding particles reflect the laser light onto a camera that is typically fitted with a filter for the laser frequency. Their positions are then captured in the images. Higher concentration of these particles improves the accuracy of the DPIV for auto- and cross-correlation analysis. However, blockage of the laser light by these particles will induce a diffracted illumination pattern with varying light intensity in their immediate vicinity and in their shadow. This unevenness in light intensity is highly undesirable to the PLIF. In the application of PLIF, the concentration measurements are taken by recording in images the fluorescent light intensity distribution in the flow field excited by a consistent emission of an illuminating laser light source. The degree of fluorescence would depend on the dye concentration and the illuminating laser intensity. Hence, if the laser intensity is consistent, the dye concentration can then be analysed by the fluorescent light pattern. However, with uneven intensity distribution in the laser illumination sheet created by the high concentration of the marker particles embedded in random positions within the flow, the concentration may no longer bear a direct relationship with the degree of fluorescence. The significance of the interference effect will of course depend on the concentration of the seeding particles.

In the following sections, the combined DPIV and PLIF system developed in this study is first presented. Subsequently the verification experiments with jet discharging into stagnant environment are described. It will be shown that the comparison is satisfactory, demonstrating the coupled system as a useful experimental tool for the measurements of turbulent mixing processes.

2. COMBINED DPIV AND PLIF SYSTEM

Figure 1 shows a schematic diagram of the setup. The DPIV and PLIF components will be described separately followed by a discussion on the synchronisation of the two components.

2.1 DPIV System

An enhanced Dantec FlowMap DPIV system was employed as an instantaneous whole-field velocimetry tool. The basic system provides five stages of DPIV data acquisition process: seeding, illuminating, recording, processing and analysing the flow field (Dantec, 1997). The light source was a dual-cavity frequency-doubled Q-switched pulsed mini Nd:YAG laser. The energy level was 25mJ per pulse and pulse

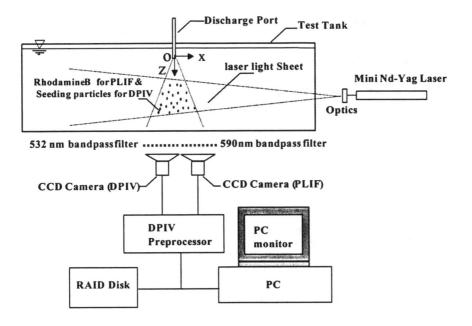

Figure 1. Schematic diagram of the experimental setup.

duration was about 7ns (by using Q switch). Near neutral buoyant polyamid particles with nominal diameter of 50μm were used as seeding particles. The emitted light from the mini laser was green at a wavelength of 532nm, which was frequency-doubled from infrared by a built-in harmonic generator. The residual infrared was removed by a harmonic separator. The maximum pulse repetition rate of each laser cavity was 15Hz. The obtained velocity field was integrated in a time interval much smaller than the time scale of the flow and thus the velocity field can be regarded as an instantaneous measurement.

The illumination system incorporated light sheet optics mounted at the front of the laser head. The typical thickness of the light sheet was 3 mm. The divergence angle was 32°. The two laser beams were co-linearly aligned to ensure that the two light sheets illuminated the same spatial area. Two Kodak Megaplus ES1.0 Charge Coupled Device (CCD) cameras were configured in this system, one for the DPIV measurement and the other for PLIF. The camera has a CCD chip that comprises both 1008×1018 light-sensitive pixels and an equal number of storage cells. The active area of the CCD chip is 9.1 × 9.2 mm with pixel pitch 9 × 9 μm, pixel size 3.4 ×7 μm and fill factor 60% (using microlens array). The camera is black-and-white with an intensity resolution of 8 bits or 256 greyscales. However, it is difficult to employ the full resolution range for the PLIF measurement due to the fact that the camera was primarily configured for DPIV purpose.

A laser pulse pair is needed for one DPIV measurement. The first laser pulse is timed to expose the first frame, which is transferred from the light-sensitive cells to the storage cells immediately after the laser pulse. The second laser pulse is then fired to expose the second frame of the frame pair. These two frames are then transferred to the DPIV processor for cross-correlation function computation. The readout of a frame lasts 66 ms, so the frame pair is read out after 132 ms, giving a maximum DPIV measurement rate of around 7.5 Hz. The minimum time interval between the pulse pair is 2μs. This determines the upper limit of velocity measuring range of the cross-correlation DPIV.

The integration time for the first frame of the frame pair is very short (5~255μs). On the contrary, the second frame is integrated for almost 65 ms. Therefore, the second frame is exposed to both the signal laser light and the ambient or background light. If the ambient light level is high, there will be poor contrast between the seeding images and the background. The second frame in the frame pair will appear brighter and have less contrast than the first one. This problem was prevented by placing an optical filter on the

213

DPIV camera, thus allowing only a narrow band of wavelengths around 532 nm that was scattered by the seeding particles to pass.

In order to use DPIV for turbulence measurements, the system must be able to capture the flow field with high temporally frequency for a sufficient duration. The present DPIV system possesses enhancement specifically tailored for this purpose. The coupled DPIV and PLIF measurements can be performed up to 7.5 Hz for a duration up to 2 minutes that should be sufficiently long to establish the mean flow properties for most laboratory experiments involving turbulent mixing. The relatively long measuring duration is made possible through the utilisation of an 18 GB RAID cabinet as an immediate transfer buffer before the images are slowly written to the hard disk.

2.2 PLIF System

Fundamentally for PLIF, the water solution of Rhodamine B was excited by the pulsing laser light sheet at a wavelength of 532 nm, that resulted in the emission of fluorescent light at longer wave lengths around 590 nm. The two-dimensional fluorescent light field was then captured by the PLIF camera. An optical filter with a small bandpass around 590 nm was placed before the aperture of this camera, blocking out the scattered 532 nm laser light and all ambient light except the fluorescent emission. This arrangement made it possible to simultaneously measure the two dimensional velocity and concentration fields.

The energy distribution of the laser light emission characteristics was carefully examined including the two individual laser beams and between consecutive pulses from the same cavity to ensure that the Nd:Yag laser was satisfactory for PLIF application. The aqueous solution of Rhodamine B fluoresces almost immediately when illuminated by the laser sheet. The fluorescence still lasts for a few nanoseconds after the pulse duration. The specification of the timing diagram of the PLIF camera was verified to ensure that the fluorescence images could be captured by the present set-up.

Calibration procedures were carefully planned before the experiments to ensure the accuracy of the PLIF results. The procedures were numerous and could not be listed here due to space constraint.

2.3 Coupled DPIV and PLIF

The PLIF camera was set to Single-frame Mode while the DPIV camera worked in Double-frame Mode. Each PLIF image was taken from a pulse pair that was the same as the DPIV pulse pair. Thus the PLIF image was double exposed by the two consecutive laser pulses (each pulse had a duration of 7ns) from both cavities. The purpose of the double exposure for PLIF images was to improve the sensitivity of the PLIF measurement. For simultaneous DPIV and PLIF measurements, the time interval between pulses in a pulse pair was determined by the DPIV requirement and was typically set from 100μs to 20ms. It was found that the time interval between the two pulses which the PLIF images were exposed did not affect the image intensity.

The Dantec DPIV software can only provide the instantaneous velocity fields and the raw greyscale images. For PLIF alone and for coupled DPIV and PLIF analysis, a 32-bit Windows program was developed in-house based on Visual C++ 5.0. The program analyses from thousands of captured images the turbulent velocity and concentration characteristics to compute the mean and turbulent transport in the flow field.

The interference effect between the two tracers, namely the effect of the fluorescent light on DPIV and the presence of seeding particles on PLIF, was investigated by applying the system to a calibration tank and to the following experiments of jet discharging into stagnant water with different concentrations of particle tracer and fluorescent dye. A satisfactory combination of the two tracers was then selected based on the results obtained.

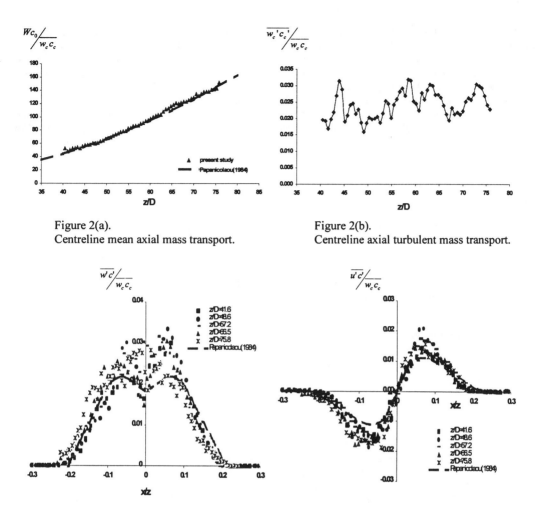

Figure 2(a).
Centreline mean axial mass transport.

Figure 2(b).
Centreline axial turbulent mass transport.

Figure 2(c).
Cross-sectional variation of axial turbulent
mass transport.

Figure 2(d).
Cross-sectional variation of radial turbulent
mass transport.

3. VERIFICATION EXPERIMENTS

Experiments with a jet issuing into stagnant environment were performed to verify the combined system and to quantify the degree of accuracy in the measurements of the induced mixing. The setup for the experiments is also shown in Figure 1. The subject of a pure round jet discharging into a stagnant environment has been well studied experimentally in the past. The large body of existing experimental results provides good references for verification purposes.

Only the results of one test are briefly presented here due to space limitation. The test conditions consisted of a jet diameter of 0.45 cm and discharge velocity of 2.52 m/s. The jet was discharged into a transparent glass tank with dimension 3m long x 1m wide x 0.8m water depth. The jet fluid was seeded with seeding particle concentration of 52 g/m^3 and Rhodamine B concentration of 573 µg/l. The combined DPIV and PLIF system was applied with a sampling frequency of 5Hz and a sampling duration of 60 seconds.

A large amount of information of the mixing processes is collected with the measurements. Here only results of the mean and turbulent mass transport are shown. Generally the results demonstrate satisfactory agreement with information obtained from previous studies. Figure 2(a) illustrates the changes in the mean axial mass transport along the centreline of the jet. The dotted curve in the figure is computed based on the results by Papanicolaou (1984) with centreline decay constants of 6.71 and 6.37 for mean axial velocity and concentration respectively. The agreement is excellent as shown in the figure. The centreline turbulent mass transport plotted in Figure 2(b) was measured to fluctuate between 0.016 to 0.032. Figures 2(c) and 2(d) show the experimental results of the cross-sectional profiles of the axial and radial turbulent mass transport respectively. The mean cross-sectional profiles from Papanicolaou (1984) measured using combined LDA and LIF are also plotted for comparison. The agreement is not as good as the mean transport quantities but still appears to be satisfactory.

4. SUMMARY

A coupled DPIV and PLIF system is developed in this study for the experimental measurements of mixing processes. The different issues on the application of the techniques of DPIV and PLIF alone as well as the coupling of the two were examined. The verification results for the case of a round jet into stagnant environment agreed well with existing data. With the ability to capture both qualitatively (with the images) as well as quantitatively (with DPIV and PLIF) the mixing characteristics, the system appears to be extremely useful for future experimental studies of pollutant transport.

ACKNOWLEDGMENT

Funding support from the Academic Research Fund of the Nanyang Technological University for this project is gratefully acknowledged. The assistance of Edmund Lee in various aspects of the experimental setup is also appreciated.

REFERENCES

DANTEC Measurement Technology (1997). *FlowMap Installation & User's guide.*
Haven, B. A., and Kurosaka, M. (1997). "Kidney and anti-kidney vortices in crossflow jets." *J. Fluid Mech.*, Vol. 352, pp.27-64.
Papanicolaou, P.N. (1984). *Mass and momentum transport in a turbulent buoyant vertical axisymmetric jet.* Ph.D thesis, California Institute of Technology.
Simoens, S. and Ayrault, M. (1994). "Concentration flux measurements of a scalar quantity in turbulent flows." *Experiments in Fluids*, Vol.16, pp.273-281.
Weisgraber, T.H., and Liepmann, D. (1998). "Turbulent structure during transition to self-similarity in a round jet." *Experiments in Fluids*, Vol.24, pp.210-224.
Willert, C.E. and Gharib, M. (1991). "Digital particle image velocimetry." *Experiments in Fluids*, Vol.10, pp.181-193.

Environmental Hydraulics, Lee, Jayawardena & Wang (eds) © 1999 Balkema, Rotterdam, ISBN 90 5809 035 3

A study of plane wall jet for destroying density current

Zhang Zhengquan
Yangtze River Scientific Research Institute, China

Duan Wenzhong
Wuhan University of Hydraulic and Electric Engineering, China

ABSTRACT : This paper proposes a new way to use a plane wall jet for destroying density current. Through theoretical analyses, a relation of mean velocity in cross section of the plane wall jet is developed and a expression of jet velocity is derived necessary for wall jet destroying density current under different water-sand conditions. By means of flume experiment, a detailed study is made on the phenomena, mechanism, and sand-intercepting and sand flushing effects of using wall jet destroying density current. And an exammation of these formulas is made with experiment data.

1. INTRODICTION

Within upper and lower approach channels of ship lock in a water control project, there often exist deposition of density current which causes inadequate depth of navigation and endangers safety of navigation. Problem of solving sediment deposition of a approach channel in a project has become one of important technique problem of a completed project. In the past, this problem was generally solved by mechanical clearage(dredging) and "flushing of sediment by water flow" or elutriation by water. For a navigation project of such high head as The Three Gorges, due to long approach channel and large deposit volume , the mechanical clearage is not only expensive, but it is also of influence on navigation in removing sediment; the elutriation by water requires installing sand sluice or outlet which not only increases cost of construction but also give rise to a series of such important technique problem as stability of high side slope; cavitation and vibration of   high-velocity flow; and large volume of water consumption. A study was made by some experts on methods of introducing outside water (guest water) and pneumatic (water) barrier destroying density current. The reason why these enqineering measures are difficult to implement is that water source problem exists and deposit-reducing efficiency is low. For this reason, author proposes a new method of using plane wall jet for destroying density current. Through theoretical analyses and experiment study, results show that it is a efficient approach to preventing and reducing sediment deposition of approach channel.

2. STUDY OF 2-D PLANE WALL JET

By the 2-D plane wall jet we mean a arrangement of a plane slot jet setup at the bottom of a approach channel.   On the basis of shooting flow theory, 2-D Plane wall jet submerged can be written, negleting viscous shear stress, as the following differential equation of boundary layer:

$$\bar{u}\frac{\partial \bar{u}}{\partial x}+\bar{v}\frac{\partial \bar{u}}{\partial y}=-\frac{1}{\rho}\frac{\partial \bar{p}}{\partial x}+\frac{\partial}{\partial y}(-\overline{u'v'}); \qquad \frac{\partial \bar{u}}{\partial x}+\frac{\partial \bar{v}}{\partial y}=0 \qquad (1a,b)$$

Boundary conditions of 2-D plane wall jet is that when y=0 、 u=0 、 v=0 ; when y → ∞, u=0
Adopting Prandtl relation of tubulent shear stress:

$$\tau_t = -\overline{\rho u'v'} = \rho \nu_t \frac{\partial u}{\partial y} \; ; \qquad\qquad \nu_t = Kb(u_m - u_c) \qquad\qquad (2a,b)$$

By (1)、(2) and boundary conditions, conservation relationship of momentum flux of wall jet may be obtained (Zhang etal, 1997)

$$\int_0^b \rho u^2 dy = J_0 - \int_0^x \tau_0 dx \qquad\qquad (3)$$

Upon autho's study (Zhang, 1997), wall shear stress in initial and mainbody sections of plane wall jet can be formulated as:

$$\tau_{0c} = k_1 (\frac{D_0}{x})^{\frac{1}{2}} f(\text{Re}_0)(\frac{1}{2}\rho U_0^2) \; ; \qquad\qquad \tau_{0m} = k_2 (\frac{D_0}{x}) \; f(\text{Re}_0)(\frac{1}{2}\rho U_0^2) \qquad\qquad (4a,b)$$

By substituting (4) into (3), by integration and arrangement, we have

$$\int_0^b \rho u^2 dy = (1-\beta)J_0 \qquad\qquad (5)$$

where $\qquad \beta = f(\text{Re}_0)\left[k_1 (\frac{x_c}{D_0})^{\frac{1}{2}} + \frac{1}{2}k_2 (\ln \frac{x}{x_c}) \right] \qquad\qquad (6)$

By letting U is mean velocity of any cross section in jet flow; U_0 is mean velocity of jet slot; D_0 is thickness of jet slot; b is thickness of jet flow in any cross-section. Adopting hydraulic metyod, then, Equation (5) may be simplified to the following form $\qquad \rho U^2 b \doteq (1-\beta)J_0$. By $\quad J_0 = \rho U_0^2 D_0 \quad$ and $\quad \eta = \sqrt{1-\beta}$, we get

$$U = \eta\sqrt{D_0 / b}\, U_0 \qquad\qquad (7)$$

Equation (7) is mean velocity in cross section of 2-D plane wall jet in static environment. On the basis of author's data of jet experiment (Zhang, 1997), $0.707 < \eta < 1$, that is , $\eta \doteq 0.85$, then

$$\frac{U}{U_0} = 0.85(\frac{D_0}{b})^{\frac{1}{2}} \qquad\qquad (8)$$

where $\quad b = 0.14X + D_0$

3. THEORETICAL ANALYSES OF PLANE WALL JET DESTROYING DENSITY CURRENT

Study made by our predecessors (Xie,1981; Chen,1990) showed that head velocity of density current in static water is

$$U_f = C_1 \sqrt{\frac{\Delta\rho}{\rho} g h_f} \qquad\qquad (9)$$

Let $\quad h_f = C_2 H$, then, head velocity of density current is

$$U_f = C_3 \sqrt{\frac{\Delta\rho}{\rho} gH} \qquad\qquad (10)$$

218

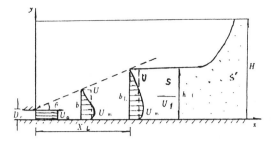

FIG.1   Mechanism of plane Wall Jet Destroying Density Current

Previous study also showed that when velocity of environment flow $U' = \dfrac{q}{H} \geq U_f$, density current stops moving

forwards,   let $U' = \dfrac{q}{H} = C_4 U_f$. If thickness of jet flow is equal to that of density current, that is, $b = h_f = \dfrac{H}{2}$

By (7), mean velocity in cross section of jet flow at this moment may be obtained:

$$U = \eta \sqrt{\frac{D_0}{b}} U_0 = \eta \sqrt{\frac{2D_0}{H}} U_0 \tag{11}$$

By $b = \varepsilon x + D_0$ and $b = h_f = \dfrac{H}{2}$, we may obtain a distance from jet slot to a point at which thickness of jet

flow reaches that of density current, as shown in Fig.1, then,

$$X_L = \frac{H - 2D_0}{2\varepsilon} \tag{12}$$

When jet flow passes through distance X_L its thickness equals that of density current; when its mean velocity (U) in cross section equals or is greater than mean velocity required by stopping surrounding flow from producing density current, density current may be prevented in the same way from going upstrem, i.e. satisfying

$$U \geq U' = C_4 U_f \tag{13}$$

By (13)、 (11) and (10), we may obtain jet velocity required in preventing density current from going upstream:

$$U_0 \geq \frac{C_3 C_4}{\eta} \sqrt{\frac{\Delta\rho}{\rho} gH^2 / (2D_0)} \tag{14}$$

If muddy water concentration is defined as the following form $\rho' = \rho + (1 - \dfrac{\rho}{\rho_s}) S'$, where S' is sediment

content of density current; ρ and ρ_s are density of water and sediment,   then

$$\frac{\Delta\rho}{\rho} = \frac{\rho_s - \rho}{\rho_s \rho} S' \tag{15}$$

By (15) and (14) we find

$$U_0 \geq \frac{C_3 C_4}{\eta} \sqrt{\frac{\rho_s - \rho}{\rho_s \rho} S' g H^2 / (2D_0)}$$ (16)

And assuming that there is a linear relation between sediment confent of density current and that of main flow $P < 0.02mm$, that is

$$S' = C_5 P < 0.02mmS$$ (17)

where $P < 0.02mm$ is percentage making up by $d < 0.02mm$; C_5 is proportional coefficient. By substituting (17) into (16), relation between U_0 and hydraulic sediment factors may be obtained:

$$U_0 \geq \frac{C_3 C_4}{\eta} (C_5 P < 0.02)^{\frac{1}{2}} \left[\frac{\rho_s - \rho}{\rho_s \rho} g H^2 S / (2D_0) \right]^{\frac{1}{2}}$$ (18)

Assuming that critical velocity of jet slot required by plane wall jet destroging is U_{0k} and let $C = \frac{C_3 C_4}{\eta} (C_5 P < 0.02)^{\frac{1}{2}}$, by (18) we arrive at

$$U_{0k} = C \sqrt{\frac{\rho_s - \rho}{\rho_s \rho} g H^2 S / (2D_0)}$$ (19)

From above equation, it may be seen that jet velocity required in destroging density current is directly proportional to depth H and sediment content $S^{\frac{1}{2}}$ of main flow and varies inversely as thickness of jet slot $D_0^{\frac{1}{2}}$.

4. EXPERIMENTAL STUDY

Using refined coal powder, by analogue to the water-sand conditions of sediment model in dam region of the Three Gorges Project, a experiment of wall jet destroying density current is made in a glass flume.

4.1 Flow Field of clear Water and phenomena of wall Jet Destroying Density Current

After gushing or emitting from wall slot and admixing with surrounding static waters, jet flow moves forwards; jet boundary spreading outwards constantly, its active cross section enlarging by degrees, water discharge incrasing downstream, and flow velocity decreasing gradually. When jet boundary approaches a definite distance from water surface, boundary line rises suddenly. Due to momentum entrainment and friction action of dynamic and static waters, in navigation current within upper part of jet flow, there is a transverse circulation; on its bottom, the circulation is in the same direction as jet flow and on its top, direction of two is in the very reverse. Extent and strength of the circulation are related to water depth H and initial velocith U_0. The more the water depth, the more its extent, it is the same the other way round; the more the initial velocity, the greater its strength, the reverse is true. In addition, between jet slot and dead-end reach (caecum reach), there is a weak transverse circulation, as shown in Fig.2.

Phenomena of wall Jet Destroying Density Current: At adefinite distance within a flume, a wall jet setup is installed. when density current approaches a distance in the front of jet slot or clearance, the jet slot begins gushing flow from it. Destroying phenomena of density current varies with velocity or discharge of jet flow from jet slor.

a.   When jet velocity is small and head of density current moves over to the influence range of jet flow, its speed dereases distinctly and its thickness weakens some what; When it lies close to jet slot, for relative velocity of jet flow is greater on bottom and smaller on top, head of density current rises upwards and aross the jet slot continues going upstream, but thickness of density current decreases evidently.

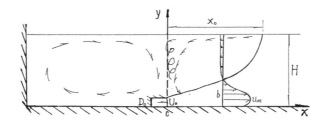

FIG.2 Sketch of Flow Field

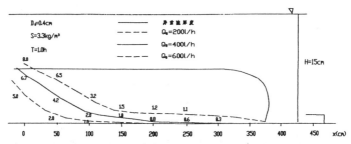

FIG.3 Effect of Jet Discharge on Thickness of Density Current

b.   When jet velocity is great, head of density current moves forwards a distance and due to decrease in velocity, it ceases to move by degrees. Its head cut down considerably. With increasing time, entire thickness of density current slashes progressively and retreat phenomena takes place.

From jet flow to nearwater surface, since direction of main flow is opposite to that of jet flow, sediment in suspension transports towards navigation channel in which a part of sediment is drawn into jet flow and it spurts outside the channel. With the passage of time, sediment carried on top of jet region reduces and also circulation in flume carries suspended sand outside the channel gradually. And finally, clear water has all emerged in the entire navigation channel. It follows that effect of wall jet destroying density current is very ideal for preventing sediment from depositing.

4.2   Effect of Jet Discharge and its Time on Head Thickness of Density Current

On condition that D_0=0.4cm, depth H=15cm, sediment content S=3.3kg/m^3, jet discharge are 200 、 400 、 700 l/h   respectively, and jet time is one hour, the thickness of density current is observed, as shown in Fig.3.
It may be seen from this figure that the more the jet discharge, the more is the thickness of density current reduces. Also, experiment shows that under a definite condition of jet flow, jet time will have distinct influence on thickness of density current, the longer is the jet time,the more is head thickness of density current reduces   until it reaches balance state appropriate to jet flow.

4.3   Relation between Critical Discharge or Velocity of Destroying Density Current and Hydraulic Sediment Factors

Jet discharge required for destroying density current under conditions of six different depths of the same discharge and seven sediment contents of the same depth is observed in this experiment. Now that under a definite discharge, flow velocity varies with jet distance.

Experiment holds that according to different water-sand conditions, it may be controlled with calculating X_L value. when head of density current reaches a distance X_L from jet slot, thickness of density current is equal just to that of jet flow. If density current stops moving forwards at this moment, this discharge is called critical discharge destroying density current or jet discharge Q_{0k} required for destroying density current.

221

Experiment study shows that when sediment content S of main flow is constant, jet discharge Q_{0k} increases with increase in depth H; in the case of the same depth H, Q_{0k} increases with increase in jet thickness D_0; when H and D_0 is constant, the more the sediment content, the more the required Q_{0k}, the reverse is true. by experiment data, Let c=0.5 in (19), we get

$$U_{0k} = 0.50 \sqrt{\frac{\rho_s - \rho}{\rho_s \rho} gH^2 S / (2D_0)}$$ (20)

4.4 Study of Sand Retention Efficiency

In order to study the sand refention effect of wall jet, jet eductor is moved over to about 50cm from jet slot. As soon as stage is adjusted to stability, the eductor is opened to spout until jet flow reaches Q_0 value and partition for separation clear water from muddy one is again opened, after about 8 hours, sediment coutent, grain size, and deposit thickness are measured respectively in all cross section. Analyses of observation data obtains the following knowledge:

a. The more the jet discharge, the move distinct the sediment content reduces progressively, the reverse is true; jet discharge does not have distinct influence on grain size. For the main, grain size in flume reduces distinctly as compared with grain size at jet slot and tends to become finer progressively.

b. After gushing from jet slot, volume of deposition in its flume reduces distinctly as compared with failure to gush from it. Degree of its reduction is related to jet discharge Q_0. The more the jet discharge Q_0 the more the deposit volume reduces, the reverse is true. In the same discharge Q_0 and thick jet slot, its volume of deposition is slightly greater.

c. Rate of sand retention is related to depth H, sediment content (S) of main flow, thickness (D_0) of jet slot, and different jet discharge Q_0. When sediment content is great, the rate of sand retention is low, the reverse is true. When S and H are kept constant, the more the jet discharge, the more the rate of sand retention, the reverse is true; when Q_0 is kept constant, the more the jet thickness D_0 the less the rate of sand retention, the reverse is true, for the main, when H is 15cm, S is 3~4.1kg/m³, jet discharge Q_0 is 200~600l/h, rate of sand retention is 71%~97%. Meanwhile, it can be again observed from experiment that in 30-40cm ditance on the front of jet slot, deposit elevation reduces considerably; nothing hapens in flume botton near jet slot, Showing that plane wall jet installed in approach channel have better sluicing (retending) action on befor (behind) jet slot.

5. CONCLUSION

Plane wall jet is effective measures of destroying density current in dead-end reach. Critical velocity at jet slot

$$U_0 = 0.50 \sqrt{\frac{\rho_s - \rho}{\rho_s \rho} gH^2 S / (2D_0)}$$

may give a reference to design of practical project destroying density current.

6. REFERENCES

Cheng Guifu (1990). "Relation between velocities of density current movement and environment flow", Sedimentation Research. (in Chinese)

Xie Jianheng(1981). "Engineering of river sedimentation" Volume I, Water Conservancy Publishing House. (in Chinese)

Zhang Zhengquan and Duan Wenzhong(1997). "Theoretical Analyses of using plane wall jet for destroying density current", Journal of Yangtze River Scientific Research Institute. (in Chinese)

Zhang Zhengquan(1997). "Plane wall jet and its theory and experiment study", Degree Paper, Wuhan University of Hydralic and Electric Engineering. (in Chinese)

Environmental Hydraulics, Lee, Jayawardena & Wang (eds) © 1999 Balkema, Rotterdam, ISBN 90 5809 035 3

The velocity field of a circular jet in a counterflow

C.H.C.Chan & K.M.Lam
Department of Civil Engineering, The University of Hong Kong, China

ABSTRACT: This paper describes an analysis of the velocity field of a circular jet in a counterflow. The treatment is based on the relative motion and momentum criteria. Starting from our previously derived Chan-Lam Equation for the centreline velocity decay, the spreading of the jet in the counterflow is computed with the assumption of conserved momentum and Gaussian distribution in the radial direction of mean flow velocities. Expressions for the jet widths and radial velocity profiles are obtained. The mean velocity fields at a number of jet-to-current velocity ratios are measured in the laboratory with laser-doppler anemometry (LDA) and the results compare well with the analytical expression.

1. INTRODUCTION

The turbulent mixing and dispersion of jets in a stagnant ambient is quite well understood. However, a main flow stream is present in many real-life situations as a result of river flows and ocean currents. It has been shown that presence of an ambient current modifies the initial mixing of a submerged jet (Lee and Neville-Jones, 1987). Through the decades of experimental and theoretical studies, the mixing behaviour of a circular jet in a coflow and crossflow is well understood. However, there have been comparatively few studies on the mixing characteristics of a circular jet in a counterflow. Recently, Lam and Chan (1998) and Chan (1998) reported complicated three dimensional behaviours of that type of flow using laser-induced fluorescence (LIF). They also analysed the centreline velocity decay of a jet in a counterflow (Chan and Lam, 1998). This paper aims to extend the analysis to the whole velocity field. The results from the model are compared with laboratory measurements with LDA.

2. EXPERIMENT

In order to validate the analytical velocity profiles which are derived in the next section, experiments were carried out in a 10 m X 0.45 m X 0.3 m wide laboratory flume. A counter-flowing jet was formed by issuing water from a circular nozzle against the main flow stream of the flume. The horizontal nozzle had an exit diameter of $D = 10$ mm and was fed from a constant head tank. The nozzle was located horizontally at the centre of the flume and at the mid-depth of the main flow. The jet velocities U_j ranged from 3 to 10 times the magnitude of the counter-flowing current. The jet Reynolds number thus ranged from 3,000 to 10,000. The ambient flow velocity in the flume was kept constant at a fixed value of $U_o = 10$ cm/s, while the jet exit velocity U_j was adjusted to give a range of jet-to-current velocity ratios U_j/U_o. Velocities along the radial direction were measured with a DANTEC two-colour fibre-optic laser-doppler anemometer. Measurements were performed in backscatter mode with a 3-watt Argon-ion laser and FVA enhanced processors (DANTEC 58N40) with frequency shift. The flow was seeded using pollycrystalline powders which were neutrally buoyant having a nominal diameter of 10 μm.

3. ANALYSIS OF THE VELOCITY FIELD

Unlike a jet in a stagnant ambient, the entrained fluid of the jet in the counterflow carries negative momentum. The forward momentum flux of the jet in a counterflow will therefore drop downstream. However, if the jet in counterflow is referred to an observer moving with the counterflow, or in the towing jet situation where a jet with exit velocity (U_j+U_o) is towed forward with a velocity U_o in a stagnant ambient, the entrained fluid does not carry any momentum and will not have any effect on the momentum flux along the axial direction of the towed jet. We define the momentum flux with respect to the stagnant ambient as M, which should be conserved since the entrained fluid is stagnant. Thus, the momentum flux should be equal to the initial momentum flux (noting that the jet exit velocity is (U_j+U_o)):

$$M=\frac{\pi D^2}{4}(U_j+U_o)^2 \tag{1}$$

At any section, the velocity profile along the radial direction is assumed as:

$$U=(U_c+U_o)\ e^{-(\frac{r}{b})^2} \tag{2}$$

The jet width $b(x)$ is defined as the radius r where the axial velocity is e^{-1} times the centreline velocity, (U_j+U_o), which is also known as the excess velocity of the jet relative to the counterflow and U_c is the centreline velocity respect to the nozzle. The momentum flux in the towed jet (with respect to the counter current) can be obtained as:

$$M=\int_0^\infty (U_c+U_o)^2\ e^{-2(\frac{r}{b})^2}\ 2\pi r dr \tag{3}$$

which can be evaluated as:

$$M=\frac{\pi}{2}b^2(U_c+U_o)^2 \tag{4}$$

Equating Eq.(1) and Eq.(4), the spreading width b can be obtained as,

$$b=\frac{D}{\sqrt{2}}\ \frac{U_j+U_o}{U_c+U_o} \tag{5}$$

which is dependent on U_c. An expression for the centreline velocity has been derived in Chan and Lam (1998) as:

$$U_c=\frac{U_j l_M}{x^*}+U_o\log(\frac{U_j l_M}{x^*}-U_o)-U_o\log(U_o-U_j), \tag{6}$$

$$x^*=x+d_v$$

where $d_v = U_o/(U_j+U_o)$ 6.2 D and $l_M = 6.2$ D $U_j/(U_j+U_o)$.

The analytical jet width as calculated from Eq.(5) is shown in Fig. 1 together with the experimental results. The experimental jet width is obtained by locating the radial position at which the excess velocity is e^{-1} of the excess centreline velocity. A modified analytical jet width $b^*(x)$ is introduced in order to match with the experimental results and is defined as follows:
For the upstream half of the jet field, it is the same as the original analytical jet width in Eq. (5),

$$b^*(x)=b(x) \quad for \quad x<\frac{l_p}{2} \tag{7}$$

For $x>l_p/2$, the derivative of $b^*(x)$ respect to x is the same as the value at the half of the penetration length,

$$\frac{db^*}{dx}=[\frac{db}{dx}]_{x=\frac{l_p}{2}} \quad for \quad x \geq \frac{l_p}{2} \tag{8}$$

For simplicity, let the constant slope there be m. $b^*(x)$ can be expressed as;

$$b^*(x)= \int_{x=\frac{l_p}{2}}^{x} m \ dx \tag{9}$$

With the initial condition, $b^*=b(l_p/2)$ at $x=l_p/2$;

$$b^*(x)=m(x-\frac{l_p}{2}) + b(\frac{l_p}{2}) \quad for \quad x>\frac{l_p}{2} \tag{10}$$

To better match the jet widths observed in the experiments, an x-offset needs to be included. The final modified analytical jet width, denoted by $b_c(x)$, is thus obtained as:

$$b_c(x)=b^*(x+x_o) \tag{11}$$

The value of x_o which best fits the experimental data at all velocity ratios is found to be $l_c/2$. The variation of $b_c(x)$ with x/D are shown in Fig. 1.

For the radial profile of the axial velocity, it is modelled by considering that the spreading composes of a forward spreading as well as a reversed flow due to the counterflow. In the near field of this flow, say $x/l_p<0.7$ as suggested by Yoda and Fiedler, 1996, these profiles are superposition of two flow regions; the inner forward flow region where the jet dominates and the outer reverse flow region. The forward spreading is defined as $b_f(x)$ and it is found to be described as follow;

$$b_f(x)=b^*(x+x_o) \tag{12}$$

where x_o is found to be $1D$.

It is assumed that the width of the jet still tends to increase linearly when the jet flow is reversed, thus the reverse spreading, $b_r(x)$ is taken as;

$$b_r(x)=b_f(2l_p-x) \tag{13}$$

An example of Eq.(13) at $U_j/U_o=3$ is shown in Fig. 2(a). Since the flow basically composes of a

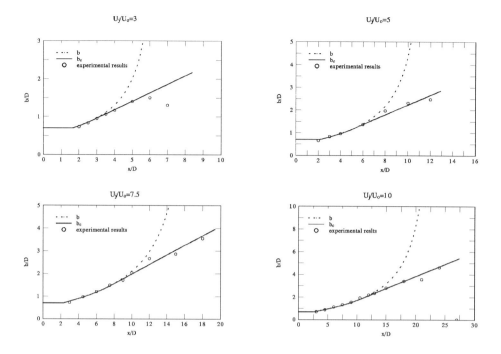

Fig. 1 Comparison of the theoretical and experimental jet width

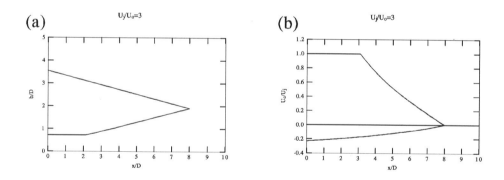

Fig. 2 (a) The reverse spreading of $U_j/U_o=3$
(b) The imaginary reverse centerline velocity of $U_j/U_o=3$

forward and a reverse flow, the magnitude of centreline velocity in the reverse flow region, U_{rc} can be obtained from the momentum criteria as expressed in Eq.(5) with initial condition $U_c=0$ at $x=l_p$;

$$U_{rc}(x)=\frac{D}{\sqrt{2}}(\frac{1}{b_j(l_p)}-\frac{1}{b_r})(U_j+U_o) \qquad (14)$$

An example of the imaginary reversed centreline velocity at $U_j/U_o=3$ as expressed by Eq.(14) is shown in Fig. 2(b). To be consistent with the overall centreline velocity obtained in Eq.(6). The

226

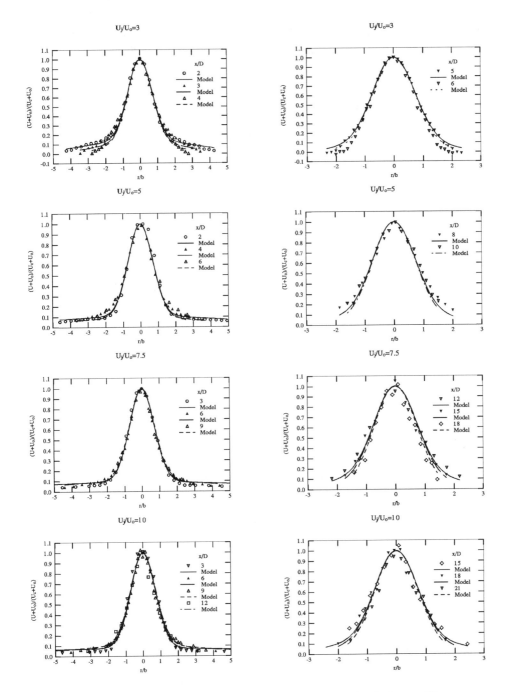

Fig. 3 Comparison of the theoretical and experimental velocity profile

centreline forward velocity component should be subtracted from the overall centreline velocity as:

$$U_{fc} = U_c - U_{rc} \tag{15}$$

Now that the centreline velocities and the jet widths of the forward and reverse flows have been defined, the profile of excess axial velocity in the combined flow can be calculated from the Gaussian distribution as:

$$U(x,r)+U_o=(U_{fc}+U_o)e^{-(\frac{r}{b_f})^2}+U_{rc}e^{-(\frac{r}{b_r})^2} \qquad (16)$$

In Fig. 3, the radial distributions of $U(x,r)$ in Eq.(16) are compared with LDA measurements. The comparison is made at the four velocity ratios ranging from 3 to 10 and at different streamwise stations beyond the potential core region. The figures are plotted as U/U_c against r/D. On the whole, the analytical expression agrees well with the experimental results. At the velocity ratio of 3, Fig. 3 shows both the analyzed and experimental values do not follow a single Gussian curve clearly. As expected, the non-similar characteristics is due to the fact that the radial mean velocity profiles are composed of a jet-like core region which is surrounded by a relatively broad region of a reverse flow.

4. REFERENCES

Chan, H.C. (1998), "Investigation of a round jet into a counterflow." Ph.D. Thesis, The University of Hong Kong.

Chan, C.H.C. and Lam, K.M. (1998), "Centreline velocity decay of a circular jet in a counterflowing stream." Phys. Fluids. Vol.10, No.3, 637-644.

Lam, K.M. and Chan, H.C. (1997), "A round jet in an ambient counter-flowing stream." J. Hydr. Engrg., ASCE, Vol. 123, No. 10, 895-903.

Lee, J.H.W. and Neville-Jones, P.(1987), "Initial dilution of horizontal jet in a crossflow", J. Hydr. Div., ASCE, Vol. 113, 615-629.

Yoda M. and Fiedler H.E. (1996), "The round jet in a uniform counterflow: flow visualization and mean concentration measurements", Exp. Fluids 21, 427-436.

Environmental Hydraulics, Lee, Jayawardena & Wang (eds) © 1999 Balkema, Rotterdam, ISBN 90 5809 035 3

On the penetration of a round jet into a counterflow at different velocity ratios

C.H.C.Chan & K.M.Lam
Department of Civil Engineering, The University of Hong Kong, China

S.Bernero
Hermann-Fottinger-Institut für Stromungsmechanik, Technical University of Berlin, Germany

ABSTRACT: This paper investigates the penetration of a round jet into a counterflow under a wide range of jet-to-current velocity ratios. Experimental data of centreline velocities at velocity ratios ranging from 1.3 to 60 obtained by LDA in two laboratories, HKU and TUB, under different facility conditions are reviewed. It is shown that the drop of jet centreline velocity in counterflow can be described by some universal decay curves with various degrees of sophistication. Similarity is also found on the longitudinal and radial turbulence intensities along the jet centreline.

1. INTRODUCTION

Jet flow is one of the basic phenomena in environmental hydraulics, and its interactions with an external stream are of great importance in innumerable practical applications, both in nature and in technology. Although jets in a coflow or crossflow have been widely investigated over the past years (Rajaratnam, 1976), relatively few studies have been carried out on the jet in counterflow, since a jet flowing into a uniform stream, coaxial to it but in the opposite direction, presents additional experimental and theoretical difficulties, related to flow reversal and to the pronounced instability.

In the last few years, two research groups, one at the University of Hong Kong (HKU) and the other at the Hermann-Föttinger Institute of Fluid Mechanics at the Technical University of Berlin (TUB), have been investigating independently on jet in a counterflow (Yoda and Fiedler, 1996; Lam and Chan, 1997; Chan and Lam, 1998; Bernero and Fiedler, 1998). Velocity and concentration data in the jet have been obtained at both laboratories and theoretical models have been independently derived for the centreline velocity decay. However, differences exist between the two laboratories in the experimental facilities, the background flow conditions, the jet-to-current velocity ratios under investigation, and the approach to the theoretical models. It is thus felt that a deeper understanding of the flow problem can be achieved by gathering the results of the two laboratories. In addition, some experiments were recently preformed in HKU with the jet appartus of TUB. This paper presents a collection of the LDA data from the two laboratories and the collaboration experiments.

2. EXPERIMENTAL FACILITIES

At TUB, a water tunnel has been especially built for the study of the jet in counterflow. It had a vertical, 30 cm by 30 cm test section and could produce a uniform counterflow with velocities up to 13 cm/s. At this velocity, which was the value at which most of the experiments were carried out, the turbulence level was about 1.6 % and in the central 70% of the cross section the velocities varied within less than 1%. The jet was formed by issuing water through one of the three nozzles with fifth order polynomial inner contours. The nozzles had round outer profiles and were mounted on a pipe of inner diameter 40 mm. The exit diameters of the nozzles were 10, 5, and 2 mm and the contraction ratios were 16, 64, and 400 respectively. Tests were run at velocity ratios from 1.3 up to 60. Flow velocities were measured with a one-component Dantec laser-Doppler anemometry (LDA) system employing a 60 mW He-Ne laser. Forward scatter measurements were made with a Dantec counter processor and a frequency shifter.

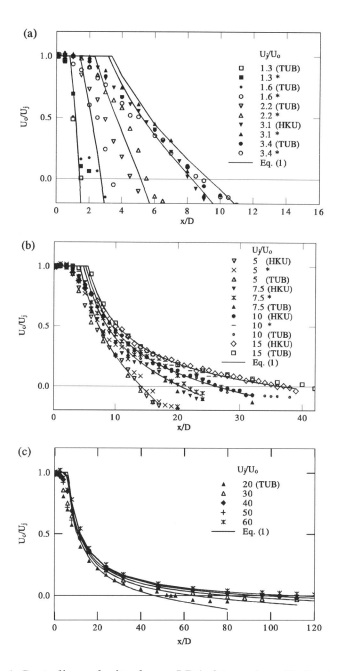

Fig.1 Centreline velocity decay: LDA data and prediction by Eq. (1)
(a)Low velocity ratios (b)Median velocity ratios (c)High velocity ratios
velocity ratios marked by * are data from collaboration experiments

At HKU, jet experiments were carried out in a 10 m long by 0.3 m wide laboratory flume. The jet in counterflow was formed by discharging water from a circular nozzle fed from a constant overhead tank against the main flow in the flume. The counterflow velocity in the flume was usually set at 10 cm/s and the water depth was about 40 cm with the nozzle placed at mid-depth. The variation of flow velocity within the test cross section of the flume was less than 5% and the background turbulence intensity in the flume was about 4%. The horizontal nozzle had an exit diameter of 10 mm and was mounted on a pipe of inner diameter 20 mm, thus resulting in a contraction ratio of 4. The velocity ratios ranged from 3 to 15. Velocities were measured with a Dantec two-colour fibre-optic laser-doppler anemometer (LDA). Measurements were performed in the backscatter mode with a 3-watt Argon-ion laser and FVA enhanced processors (Dantec 58N40) with frequency shift. The flow was seeded with neutrally buoyant polycrystalline powders with a nominal diameter of 10 μm.

The collaboration experiments were performed in the water flume at HKU but with the jet pipe and the 10 mm diameter nozzle of TUB mounted horizontally. Velocities along the jet centreline have been measured with LDA at velocity ratios from 1.3 to 10.

3. MODELS FOR CENTRELINE VELOCITY DECAY

At HKU, an expression for the centreline velocity U_c has been derived in Chan and Lam (1998) based on an advection effect of the counterflow on fluid elements issued from the jet nozzle. For a jet of exit velocity U_j and diameter D in a counterflow of velocity U_o, the centreline velocity is given by:

$$U_c = \frac{U_j l_M}{x^*} + U_o \log\left(\frac{U_j l_M}{x^*} - U_o\right) - U_o \log\left(U_o - U_j\right)$$ (1)

where the distance x^* from the actual jet origin is obtained from the distance x from the nozzle corrected with an offset d_v by $x^* = x + d_v$ with $d_v = U_o/(U_j+U_o)\ 6.2\ D$; and where the characteristic length scale is given by $l_M = 6.2\ D\ U_j/(U_j+U_o)$. The model has been shown to predict well the LDA velocity data as well as the penetration distance x_p from either velocity or LIF (laser-induced florescence) concentration measurements.

Based on the considerations of Yoda and Fiedler (1998), a simple model has been derived at TUB from the superposition of a jet flow and a uniform flow. The centreline velocity decay in the non-dimensional form suggested by Beltaos and Rajaratnam (1973) is predicted by the following relationship:

$$\left(\frac{U_c + U_o}{U_j}\right)\frac{x_p}{D} = \left(\frac{k}{2}\right)\frac{1 + \sqrt{1 - x/x_p}}{x/x_p}$$ (2)

As in Beltaos and Rajaratnam (1973), a value of $k = 5.83$ is used. In this model, the penetration distance can be predicted by $x_p = (k/2)U_j/U_o$.

4. RESULTS AND DISCUSSION

The drop of jet velocity along the centreline has been measured with LDA at HKU, at TUB, and during the collaboration experiments. Data are available at many velocity ratios covering a wide range from $U_j/U_o = 1.3$ to 60. Figs. 1a to 1c show the LDA data of velocity decay with axial distance in form of U_c/U_o against x/D at low, median and high velocity ratios. There are a number of velocity ratios at which measurements have been performed in both facilities and in the collaboration experiments. In general, the LDA data from different sources agree with one another and give the same decay of U_c with x. Larger discrepancies are observed at the few lowest velocity ratios at which, however, there are relatively few measurement points so that an evaluation cannot be very accurate. Another reason for the discrepancies may be the different degrees of distortion of the counterflow near the jet nozzle. In both facilities, the jet pipe is much larger than the jet nozzle exit and the counterflow needs to flow around the nozzle and jet pipe when approaching the jet exit, therefore, the issuing jet will not see parallel counterflow streamlines. The distortion is much pronounced at small velocity ratios and is dependent on the blockage of the jet pipe which is different between the two facilities. Also shown in Fig. 1 are the predictions by the HKU model, Eq. (1). The curves predict the experimental data quite well at all velocity ratios, but again except at the few lowest velocity ratios.

Experimental determination of penetration distance x_p at different velocity ratios can be made from the centreline

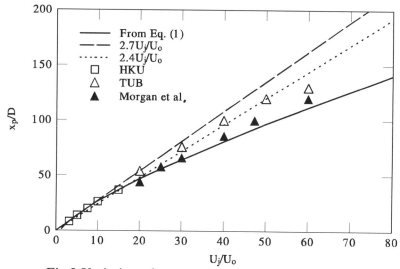

Fig.2 Variation of penetration distance with velocity ratio

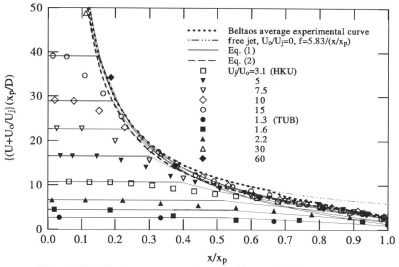

Fig. 3 Similarity curves of centreline velocity decay

velocity decay as the axial distance where the local velocity becomes zero. The data are shown in Fig. 2, together with the prediction from Eq. (1) and the linear relationship reviewed in Rajaratnam (1976). The model of TUB also assumes a linear relation between x_p and U_j/U_o but as can be observed in Fig. 2, the penetration distances predicted by Eq. (1) increase with velocity ratios in a manner slightly-departing from a linear relation. The penetration of a turbulent jet into a counterflowing turbulent pipe flow and the confinement effect of the enclosing pipe were investigated in Morgan et al. (1976). In the low jet momentum regime (defined in that reference as the one for $U_jD/U_oB < 0.5$, where B is the pipe diameter), the confinement effect was shown to be negligible and the relationship between x_p and U_j/U_o was found to be linear. Data of x_p from that reference are included in Fig. 2, which are well fitted by the HKU prediction. The data from Morgan et al. (1976) shown in Fig. 2 correspond to the case of $B/D = 83$, therefore they lie in the low jet momentum regime until $U_j/U_o = 40$. The data of TUB, which were measured for the high velocity ratios with $B/D = 150$, satisfy instead the condition of low jet momentum regime up to $U_j/U_o \approx 75$ and, accordingly, depart later from the linear relationship.

(a)

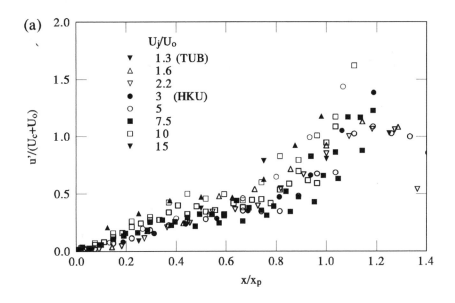

(b)

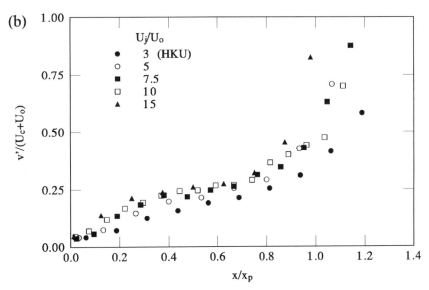

Fig. 4 Turbulence intensities along jet centreline
(a) Longitudinal u'   (b) Lateral v'

Beltaos and Rajaratnam (1973) suggests a non-dimensional representation for which the centreline velocity decays for several velocity ratios collapse on a single curve which is approximated by the hyperbolic decay for a free jet. Fig. 3 shows that curve, the present models of Eqs. (1) and (2), and the LDA data plotted in that non-dimensional form. The model of TUB, Eq. (2), predicts the velocities only in the zone of established flow, that is at locations downstream of the potential core and upstream of the penetration distance. It shows good agreement with LDA data but the range of validity is limited, since at low velocity ratios the potential core occupies a significant part of the distance x_p. The model of HKU, Eq. (1), show instead good agreement with the experimental data over a wider range of x values, especially at low velocity ratios. It is noted that the HKU predictions at different velocity ratios do not exactly collapse into a single curve in Fig. 3, though very close to one another. The difference becomes more obvious at low velocity ratios for $U_f/U_o < 10$.

In addition to mean flow velocities at the centreline, turbulence intensities in the axial u velocity component as well as in the lateral v velocity component are available from the LDA data. Fig. 4 shows their variations along the jet centreline. When plotted in the reduced form of $u'/(U_c+U_o)$ against x/x_p, the data are observed to collapse roughly on a similarity curve. This observation that velocities can be scaled by local velocities relative to the counterflow velocity and axial distance can be scaled by the penetration distance supports indirectly the assumption that the flow can be viewed as a superimposition of a simple jet and a uniform counterflow.

ACKNOWLEDGEMENT

The collaboration experiments and the stay of the second author at HKU are supported by a grant under the Germany/Hong Kong Joint Research Scheme awarded jointed by the German Academic Exchange Service (DAAD) and the Research Grants Council of Hong Kong. The authors are also grateful to the other investigators of the grant, Prof. H.E. Fiedler of TUB and Prof. N.W.M. Ko of HKU for their advice and support.

REFERENCES

Beltaos, S. and Rajaratnam, N. (1973), "Circular turbulent jet in an opposing infinite stream", *Proc. 1st Can. Hydr. Conf.*, Edmonton, 220-237.

Bernero S. and Fiedler H.E. (1998), "Experimental investigations of a jet in counterflow", *Advances in Turbulence VII (Ed.: U. Frisch)*, Kluwer, 35-38.

Chan, C.H.C. and Lam, K.M. (1998), "Centreline velocity decay of a circular jet in a counterflowing stream", *Phys. Fluids*, Vol. 10(3), 637-644.

Lam, K.M. and Chan, H.C. (1997), "A round jet in an ambient counter-flowing stream" *J. Hydr. Engrg., ASCE*, Vol. 123(10), 895-903.

Morgan, W.D., Brinkworth, B.J. and Evans, G.V. (1976), "Upstream penetration of an enclosed counterflowing jet", *Ind. Eng. Chem., Fundam.*, Vol. 15(2), 125-127.

Rajaratnam, N. (1976), "*Turbulent Jet*", Elsevier Scientific Publishing Co.

Yoda, M. and Fiedler, H.E. (1996), "The round jet in a uniform counterflow: flow visualization and mean concentration measurements", *Exp. Fluids*, Vol. 21, 427-436.

1.5 Outfall modelling

Environmental Hydraulics, Lee, Jayawardena & Wang (eds) © 1999 Balkema, Rotterdam, ISBN 90 5809 035 3

Round plumes at plane surfaces

Walter E. Frick
Ecosystems Research Division, USA EPA, Athens, Ga., USA

ABSTRACT: Plume models are widely used to support regulatory discharge permitting decisions and to help design outfalls. With thousands of projects worldwide, these models must be as accurate as possible. However, recent studies show that these models are subject to inaccuracies arising from the round plume assumption. For example, use of this assumption inadequately describes processes ranging from gravitational collapse to plume-surface interaction. In sharply-bending plumes, it also inadvertently introduces negative mass into the formulation. The onset of this condition is reported by EPA's PLUMES UM model (a statement that dilution is over-estimated), but is not corrected. The round plume assumption can be modified to remove the negative mass anomaly, and to include the center-of-mass correction. Predictions strongly suggest the overlap condition is a criterion for plume lateral spreading. The successful modification reported here suggests the round plume assumption is viable for modeling behavior at unconstrained boundaries.

1. INTRODUCTION

The round plume assumption is basic to most integral flux and Lagrangian plume models (Fan, 1967; Winiarski and Frick, 1976; Schatzmann, 1979; Teeter and Baumgartner, 1979; Lee et al., 1987; and Baumgartner et al., 1994). It combines utility and rigor. Since there is no universally recognized closure of the plume equations based on first principles, plume modeling relies on the use of entrainment hypotheses. In essence, the mass, or volume, of the plume element or control volume is known before the dimensions are computed. What can be easier than computing the radius from the well known cylinder equation? Furthermore, since plumes often represent high velocity flows that convert their kinetic and potential energies to turbulence, they expand radially over much of their initial mixing trajectories.

On the other hand, it is easy to find instances in which the assumption is, apparently, woefully inadequate. For example, any plume intersecting the free surface or the bottom of a water body will be bounded by planes. One of the strengths of the CORMIX models (Doneker and Jirka, 1989) is that they account for plume interaction with boundaries. PLUMES UM (Baumgartner et al., 1994) estimates the effect of plume-surface interaction in shallow streams by using the reflection technique (Davis, 1990). However, in general, all these models deal with this complication in less than rigorous ways.

Other cross-sections are even more problematic. For example, an elliptical cross-section has been proposed to deal with gravitational deformation (Frick et al., 1990). In stratified ambient flow the plume cross-section deforms from a force couple directed towards the center depth of the plume element. However, compared to the round plume cross-section, the elliptical cross-section introduces enormous complications that may prove to be insurmountable when other effects, such as surface interaction and plume merging, must be considered. Rectangular cross-sections also fare poorly. Unconstrained plume surfaces, or free boundaries, are not well represented by planar surfaces. There remains the reality that plume boundaries expand radially over much of the initial dilution trajectory and questions arise concerning the continuity of mass. Generally, how would one redistribute the mass of two merging rectangular plume cross-sections?

Consequently, in the on-going modification of the PLUMES UM model to produce the three-dimensional vector model (VM) within a Windows paradigm (Windows Interface for Simulating Plumes (Wisp), Frick et al., 1998), the round plume assumption is retained. Also considered are the Projected Area Entrainment (PAE), or forced entrainment, and Taylor aspiration hypotheses, but here the similarity with UM ends. The new model is vector based, hence the name Vector Model, making it fully three-dimensional (Frick and Roberts, 1998).

The main emphasis of this paper is on the modification of the round plume assumption to accommodate intersections with plane surfaces, focusing on the negative volume anomaly (Frick et al., 1994), and the self-intersection of the plume element's two cross-sections. The general approach propounded by Frick et al. (1994) is used as a basis for further development of the theory. For greater generality, the earlier parameterization of forced entrainment (Winiarski and Frick 1976; Frick, 1984), is replaced by an accurate construction of the plume surface by a series of facets that encircle the plume and serve as flux surfaces. This approach can better describe the properties of the entrained flow. In addition, the center-of-mass correction introduced by Frick et al. (1990) is also further developed and implemented. The improvements allow a physically consistent solution in the overlapped region and a model that gives robust results over a range of time steps. Results indicate that further progress using the approach is contingent on a further generalization of the round plume assumption to establish an element consisting of a composite of polyhedra and rounded sectors to account for the deformation of plumes.

2. THEORY

VM is a Lagrangian model with strong similarities to other Lagrangian models, particularly, the three-dimensional Lagrangian plume model, JETLAG (Lee et al., 1987, and, Cheung, 1990) and UM (Frick et al., 1994). Thus, the governing equations are now widely published and need not be repeated here. (In any case, for the sake of brevity, a full description is not possible.) Also, the traditional round plume assumption is most pertinent; it is that for the mass contained in a right cylinder of fluid:

$$m = \rho \pi b^2 h_c \qquad (1)$$

where m is the mass of the plume element, ρ is its average density, b is the radius, and h_c is the length of the plume element measured along the locus of the centers of the plume's round cross-sections. However, the Lagrangian models predict the trajectory of the center-of-mass of the plume element. Heretofore, the fact that the center-of-mass trajectory is not generally coincident with the trajectory of the center of the round plume cross-section has not been generally acknowledged. This oversight also is corrected here.

As has been stated, in Lagrangian plume models the radius is a dependent variable. In terms closely resembling modeling practice, the new radius may be expressed

$$b_{t+\Delta t} = \sqrt{\frac{m_{t+\Delta t}}{\rho \pi h_{c,t+\Delta t}}} \qquad (2)$$

where t is time measured from the instant the element is discharged to the present position, and, Δt is the program time-step size, which is optimized internally.

Frick et al. (1991) show that there are two radii that satisfy the computed mass and, when the element is overlapped, Equation 2 gives the larger of the two. Under these circumstances, Equation 2 yields an artificially large radius because implied is an integrating height that extends from one cross-section to the other. In Figure 1 this height is perpendicular to the plane of the figure.

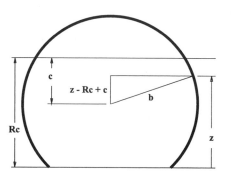

Figure 1. Some cross-sectional variables.

238

It is zero when $z = 0$ and h when $z = R_c$. In traditional treatments, the limits of integration extend to the lower radius and the height, which is a difference, changes sign. In other words, it is allowed to become negative. Consequently, negative and positive volumes are added (integrated) and a larger radius is deduced. As a consequence, since radius is consistently overestimated, entrainment is overestimated as well.

The answer to the so-called negative volume anomaly is, conceptually, simple: the plume element must be limited to the region between the radius opposite the trajectory center-of-curvature and the line of overlap, in other words, to the positive part of the volume (Frick et al., 1994). The element volume and center-of-mass correction, the latter to assure a physically consistent distribution of mass, are derived here.

2.1 Correcting for Overlap and Center-of-Mass

In regions of overlap, the positive, and only physically real, volume of the plume element can be found using a theorem due to Pappus (Symon, 1953), stating that the volume of revolution is the product of the cross-sectional area of the element, the centroid and the angle of rotation.

$$vol = area \frac{h}{R_c} \left(\frac{1}{area} \int_0^{R_c - c + b} 2\sqrt{b^2 - (R_c - z - c)^2} \, z \, dz \right) \tag{3}$$

where vol is the element volume, $area$ is the cross-sectional area of the un-overlapped part of the element, c is the distance from the center-of-mass to the center of the circular cross-section, R_c is the radius-of-curvature of the plume center-of-mass trajectory, and z is the variable of integration. Unlike h_c, h is measured along the center-of-mass. This is the volume in terms relative to the line of overlap, i.e., $z = 0$ at the center-of-curvature, which lies within the round cross-section since the element is overlapped.

Modern mathematical software (e.g., Mathematica (Wolfram, S., 1996)) simplifies solving complex integrals. The solution of Equation 3 is

$$vol = \frac{h}{R_c} \frac{\left(\frac{b^2 \pi}{2} + \sqrt{b^2 - (R_c - c)^2} (R_c - c) + b^2 \sin^{-1}\left(\frac{R_c - c}{b} \right) \right) \left(2\sqrt{b^2 - (R_c - c)^2} (R_c - c)^2 + b^2 \left(-3c\pi + 4\sqrt{b^2 - (R_c - c)^2} + 3\pi R_c \right) + 6b^2 (R_c - c)\sin^{-1}\left(\frac{R_c - c}{b} \right) \right)}{\left(3b^2 \pi + 6\sqrt{b^2 - (R_c - c)^2} (R_c - c) + 6b^2 \sin^{-1}\left(\frac{R_c - c}{b} \right) \right)} \tag{4}$$

For small angles, the distance between the center-of-mass and center of the round cross-section of the plume element, c, is

$$R_c = \frac{1}{vol} \int_0^{R_c - c + b} 2\sqrt{b^2 - (z + c - R_c)^2} \, h \frac{z^2}{R_c} \, dz \tag{5}$$

The solution to Equation 5 is

$$Rc = \frac{3b^4 \pi + \sqrt{b^2 - (R_c - c)^2} (R_c - c)^3 + 2b^2 (R_c - c)\left(-6c\pi + 13\sqrt{b^2 - (R_c - c)^2} + 6\pi r \right) + 6b^2 \left(b^2 + (R_c - c)^2 \right) \sin^{-1}\left(\frac{R_c - c}{b} \right)}{4\left(2\sqrt{b^2 - (R_c - c)^2} (R_c - c)^2 + b^2 \left(-3c\pi + 4\sqrt{b^2 - (R_c - c)^2} + 3\pi r \right) + 6b^2 (R_c - c)\sin^{-1}\left(\frac{R_c - c}{b} \right) \right)} \tag{6}$$

In VM, Equations 4 and 6 are solved simultaneously by successive approximation for b and c. Equation 6 yields the correct limiting value of $1.25b$ at the point of overlap and yields $3\pi b/16$ when the line of overlap is through the center of the round cross-section, at which point $R_c = c$.

2.2 An Instability in UM

It is worthwhile to discuss the instability reported in connection with a previous attempt to eliminate the negative volume anomaly (Frick et al., 1994). In UM, the forced entrainment terms defined relative to a local coordinate system where the z-axis points in the direction of the element's velocity vector and the y-axis points toward the center-of-curvature. This makes it easy to express the sum of the corresponding "growth," cylinder," and "curvature" forced-entrainment terms. However, as the radius shrinks in high curvature regions, the growth term reverses sign and must be set to zero. The sudden removal of the term from the sum reduces the entrainment rate, causing a decrease in curvature. With curvature reduced, the growth term reappears and the cycle repeats, an apparent numerical instability related to the entrainment formulation. This instability disappears in VM. In VM, the boundary between plume and ambient fluid is modeled by covering the outside boundary by a series of webbed, triangular facets. At each time step, the plume element velocity and radii vectors are used to create a "spoked-wheel" depicting the cross-section. One might imagine it looking like a bicycle wheel. Corresponding points on the rim of each wheel are connected to form facets. The vector cross-product of the vectors defining the facets neatly provide a measure of the facet area and direction into or out of the plume element. In the un-overlapped part of the element, the facet vector points outward, a fact that can be used to test for entrainment. This approach provides the depth from which the entrained fluid is withdrawn (Frick and Roberts, 1998), which is called "distributed entrainment."

3. RESULTS

3.1 Model Stability

A time forward-stepping model like VM should be reasonably insensitive to the magnitude of the time step. In UM and VM the time step is controlled by a mass criterion. The previous time step is used to estimate entrainment into the plume element during the current time step, and then is compared to the "mass criterion." The criterion is simply a small percentage of the mass at any time. In UM, with special exceptions, it is chosen to approximately double the mass every 100 time steps and is numerically equal to 0.006955.

Unlike UM, VM uses two growth criteria. The second criterion limits changes per time step in the curvature of the plume trajectory. This procedure allows the mass criterion to be increased to 0.02. Table 1 lists the results of the limited quality assurance tests conducted at the time of this writing. Input values are listed in Table 2. The distributed entrainment model is used through the level of overlap. To establish model convergence, a simplified entrainment assumption is used beyond that level—the mass increase per time step is simply held constant. This is necessary because the model, using the PAE hypothesis, predicts that the diameter grows without limit, indicating upstream intrusion, or "anvil" formation.

As can be seen, in a broad range of values of the curvature criterion the solutions compare closely, comparable to variations experienced using UM, in which variations of about two percent for a similar range is typical. A value of 0.01 appears about optimal for the curvature criterion. A step by step examination of the output shows that the predictions change smoothly in all regions.

3.2 Comparison of UM and VM outputs

While it is necessary to offer a model that is computationally stable, stability in itself is insufficient to assure the accuracy of the model. Considering limitations on time, one way to gain confidence in the new model is to compare it with established models, especially one as conceptually similar to VM as UM is. A limited comparison is given in Table 2, showing both input and UM and VM predictions. The discharge and ambient conditions correspond somewhat to a laboratory setting, but have been deliberately chosen to cause plume overlap. The discharge depth is arbitrarily chosen. The estimated trapping level UM dilution is 128.1. At the point of overlap, the predicted UM dilution is 205.3. The user is warned: "dilution overestimated," indicating that dilution will be overestimated beyond that point; it ultimately attains a value of 380.5, an 85%

Table 1. Predicted dilution sensitivity to changes to the curvature criterion.

Curvature Criterion	Trap Level Dilution	Overlap Dilution	Maximum Rise Dilution	Rise at Overlap (cm)
0.0025	168.3	181.0	748.5	32.9
0.0050	169.8	219.2	494.4	35.5
0.0075	170.0	223.2	494.7	35.8
0.0100	171.0	227.1	498.0	36.0
0.0150	171.0	230.2	492.0	36.1

Table 2. UM and VM input and predictions.

Port parameters

total flow $(m^3\text{-}s^{-1})$	salinity (psu)	temperature (C)	density $(kg\text{-}m^{-3})$	depth (m)	diameter (m)	velocity $(m\text{-}s^{-1})$	vertical angle (deg.)	aspiration coefficient
3.142E-06	3.6	20.0	1001.00551	10.0	0.0025	0.64	90	0.10

Ambient parameters

depth(m)	current$(m\text{-}s^{-1})$	density $(kg\text{-}m^{-3})$	salinity(psu)	temp(C)
0.0	0.008	1013.4174	20.0	20.0
10.0	0.008	1028.5991	40.0	20.0

UM prediction

plume depth(m)	plume diameter(m)	dilution	x-distance(m)
10.00	0.0025	1.000	0.000
9.632	0.1631	**128.1**	0.08383 -> trap level
9.565	0.2620	**205.3**	0.1243 -> overlap -> **dilution overestimated**
9.537	0.4366	**380.5**	0.1871 -> maximum rise

VM prediction

plume depth(m)	plume diameter(m)	dilution	R_c/radius	Center Offset/radius
10.00	0.0025	1.000		
9.677	0.2110	**171.0**	3.655	0.0697 -> trap level
9.640	0.3351	**227.1**	0.984	0.0363 -> overlap
9.633	0.4885	250.6	0.5844	0.5964
9.631	0.5260	254.5	0.6186	0.5429
9.599	**5.4875**	**498.0**	0.042	0.9686 -> maximum rise

increase above the overlap point dilution.

The VM trapping level dilution, 171.0, is much larger than the corresponding UM estimate, partly due to the distributed mass entrainment used in VM and the more complex superposition principle used in UM. By the time the plume overlaps, the center-of-mass correction helps to reduce the dilution rate so that they converge somewhat, cf. 205.3 to 227.1. The next two lines show where R_c/b and c/b cross: 0.589, or $3\pi/16$. The entrainment (dilution = 498.0) after overlap is proportionally even greater than for the incorrect UM. However, the latter result is based on an ad hoc entrainment assumption used for demonstration purposes.

The removal of negative mass from the model simulation is clearly important to the initial dilution calculation, as is the correction for the distribution of mass about the center-of-mass. However, other model details also influence the differences in prediction. Noteworthy is the distributed entrainment concept. At maximum rise the element is profoundly overlapped: the ratio of R_c to the radius is only 0.042. This strongly suggests plume anvil formation and the necessity to reformulate the models in this region.

4. CONCLUSION

To most modelers, the round plume assumption may seem deficient in addressing plume-surface interactions and other planar effects in model simulations. One such interaction is plume element self-intersection, created when the plume bends sharply so that its cross-sections overlap. The overlap problem, or negative volume anomaly (Frick et al., 1994), identifies a conceptual artifact in integral flux and Lagrangian models that inadvertently introduces negative mass into plume prediction. However, previously the effect of the anomaly was unquantified because the only model to correct for it became numerically unstable as a result. Also, that effort did not include a correction for the center-of-mass. The VM model includes a correction of the negative volume anomaly, a rigorous method for distributing mass about the center-of-mass of the plume element, and a distributed entrainment function. The new entrainment formulation abandons the parameterization of forced entrainment found in UM in favor of constructing the surface boundary of the element by a series of contiguous triangular facets. The facet approach is superior in that it draws ambient fluid into each facet from its level in the ambient. The approach produces a stable model which shows, in a test case, that the use of the projected area entrainment hypothesis implies unlimited growth in the overlapped region. In other words, it supports the conclusion that the overlap criterion is an indicator of commonly observed lateral plume spreading and upstream intrusion. However, since the solution rapidly diverges, it is not useful for demonstrating the correctness of the model in the overlapped region. A simplified plume assumption developed simply to test the consistency of the model in the overlapped region shows the solution is stable and accurately reproduces two known endpoints. In general, the predicted plume dimensions are consistent with the geometry of lateral spreading. This application shows that the round plume assumption is still viable, and can be modified to accommodate plume-surface interaction. More work is recommended to develop a composite element consisting of polyhedra adjacent to plane boundaries and rounded sectors at free boundaries. Also, since the projected area entrainment hypothesis only indicates the existence of lateral spreading, more theoretical and experimental work is needed to better define entrainment in the overlapped region.

5. ACKNOWLEDGEMENTS

I thank Dr. Wu-Seng "Winston" Lung for some inspirational discussions that contributed to the general conception of this work, Dr. Luis Suarez for his guidance in the use of Mathematica and other mathematical aspects of this work, and Drs. Craig Barber, Anne Sigleo, Bob Swank, and Jim Weaver for their comments.

6. REFERENCES

Baumgartner, D.J., Frick, W.E., and Roberts, P.J.W. (1994). Dilution models for effluent discharges (3rd Ed.) EPA/600/R-94/086, U.S. EPA, Newport, OR, USA.

Cheung, V. (1991). Mixing of a round buoyant jet in a current. Ph.D. Thesis, Dept. of Civil and Structural Engrg., Univ. of Hong Kong, Sep. 1991.

Davis, L.R. (1990). A review of buoyant plume modeling. Presented at AIAA/ASME Thermophysics and Heat Transfer Conference, 18-20 Jun. 1990, Seattle, WA, USA.

Doneker, R.L., Jirka, G.H. (1990). Expert system for hydrodynamic mixing zone analysis of conventional and toxic submerged single port discharges (CORMIX1). EPA/600/3-90/012, U.S. EPA Athens, GA.

Fan, L.N. (1967). Turbulent buoyant jets into stratified or flowing ambient fluids. Report No. KH-R-15, W.M. Keck Lab. Of Hydraulics and Water Resources, California Inst. of Technology, Pasadena, CA.

Frick, W.E. (1984). Non-empirical closure of the plume equations. *Atmospheric Environment*, **18, 4**.

Frick, W.E. and Roberts, P.J.W. (1998). Environmental Protection Agency (USA) program Wisp: Windows Interface for Simulating Plumes. 3rd International Conference on Hydroscience and Engineering, 31 Aug - 3 Sep 1998, Cottbus, Germany.

Frick, W.E., Baumgartner, D.J., and Fox, C.G. (1994). Improved prediction of bending plumes. *J. of Hydraulic Research*, **32, 6**.

Frick, W.E., Baumgartner, D.J., and Roberts, P.J.W. (1998). Dilution models for effluent discharges (4th Ed.) introducing Windows interface for simulating plumes (Wisp) (draft), ERL-Athens, NERL,ORD, U.S. EPA, Athens, GA, USA.

Frick, W.E., Bodeen, C.A., Baumgartner, D.J., and Fox, C.G. (1990). Empirical energy transfer function for dynamically collapsing plumes. Proceedings of Int'l Conference on Physical Modeling of Transport and Dispersion, MIT, 7-10 Aug 1990. Boston MA, USA.

Lee, J.H.W., and Cheung, V. (1989). Generalized Lagrangian model for buoyant jets in a current. ASCE *J. of Environmental Engineering*, **116, 6**, pp. 1085-1106.

Lee, J.H.W., Cheung, Y.K., and Cheung, V. (1987). Mathematical modeling of a round buoyant jet in a current: an assessment. Proceedings of International Symposium on River Pollution Control and Management, Shanghai, China, Oct 1987.

Schatzmann, M. (1979). An integral model of plume rise. *Atmospheric Environment*, **13**, pp. 721-731.

Symon, K.R. (1964). Mechanics (Third Printing). Addison-Wesley Publishing Co., Inc., Reading, MA, USA.

Teeter, A.M., and Baumgartner, D.J. (1979). Prediction of initial mixing for municipal ocean discharges. CERL Pub. 043, U.S. EPA Environmental Research Laboratory, Corvallis, OR, USA.

Winiarski, L.D., and Frick, W.E. (1976). Cooling tower plume model. EPA/600/3-76-100, ERL-Corvallis, ORD, U.S. EPA, OR, USA.

Wolfram, S., 1996. Mathematica (software). Wolfram Research, Champaign, IL, 61820, USA.

Environmental Hydraulics, Lee, Jayawardena & Wang (eds) © 1999 Balkema, Rotterdam, ISBN 90 5809 035 3

3-D numerical simulation on dilution behavior of near field for river diffuser

Wenxin Huai & Wei Li
Department of River Engineering, Wuhan University of Hydraulic and Electric Engineering, China

T. Komatsu
Department of Civil Engineering, Kyushu University, Fukuoka, Japan

ABSTRACT: The 3-dimensional turbulence modelling and hybrid finite analytic method are used to predict the dilution behavior of near field for multiport buoyant jets in river. The predict dilution are good by comparison with available laboratory measurements. A empirical formula for temperature dilution in near field for this kinds flow is given.Comparison of the dilution between three-port and one-port, it is found that multiport discharge is better than one-port one. There are bifurcation and Coanda effect in the near field for river diffuser.

1.INTRODUCTION

The submerged multiport diffuser is a very effective outfall structure for discharging heated effluents from stream-electric generating stations when rapid mixing is desired. Generally speaking,a multiport diffuser consists of several discharge ports emanating from a large manifold pipe located near the bottom of a water body. The flow can be divided into there regions,i.e. the individual-jet region,the transition region where the individual jets merge each other,and the two-dimensional region following the merging of the jets. To date, most research efforts related to multiport diffusers in shallow, flowing environments have concentrated on the two-dimensional region. For the information on the near field of multiport diffusers. J.H.W.Lee (1980,1984, 1985),A.D. Parr et al (1981) and E.E.Adams(1981) given the good results by theory analysis and test study. A.D.Parr et al(1979)presented the results of an experimental study of the behavior of heated jets in shallow,flowing, laterally-confined ambient. The test conditions correspond to the flow regions classified as supercritical and subcritical by Cederwall(1971).The parameters used to describe jet behavior in the initial regions for a specific vertical discharge angle,θ,and a relatively high jet densimetric Froude number ($Fr=u_j/\sqrt{(\Delta\rho/\rho)gD}$,say greater than 10) are relative depth (H/D),or relative submergence ((H-h)/D); the velocity ratio ($K=u_j/u_a$) and the relative distance downstream from port (x/D),in which the terms are indicated in the Fig.1. In many cases,diffuser performance at low ambient flow is of principal interest. When the parameter R_m($R_m=(T_E-T_a)/(T_m-T_a)=(q_a+q_j)/q_j=V+1=$mixing ratio) becomes extremely important,in which T_E=initial jet temperature;T_a =ambient temperature;T_m =fully mixed temperature;$q_j=u_j(\pi D^2/4S)$ =jet discharge per unit width,where S is the port spacing; and $V=q_a/q_j$=volume flux ratio.The mixing ratio represents the excess temperature dilution that would exist if complete mixing occurred and is definition sketch of Fig.1.

In this paper,the 3-dimensional turbulence modelling and hybrid finite analytic method(HFAM) with non-uniform staggered grids are used to predict the dilution behavior of near field for river diffuser.

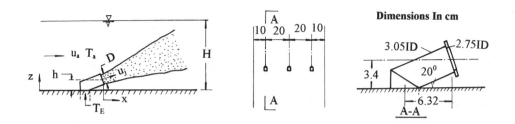

Fig.1 Definition Sketch and Physical Model Size

2. MATHEMATICAL MODELLING

The coordinate system and physical model size in Fig.1 is considered. The time-averaged, three-dimensional, steady state equations governing the turbulent flow may be expressed in Cartesian tensor notation as

$$\frac{\partial}{\partial x}(\rho u\phi)+\frac{\partial}{\partial y}(\rho v\phi)+\frac{\partial}{\partial z}(\rho w\phi)=\frac{\partial}{\partial x}\left(\Gamma_{eff,\phi}\frac{\partial\phi}{\partial x}\right)+\frac{\partial}{\partial y}\left(\Gamma_{eff,\phi}\frac{\partial\phi}{\partial y}\right)+\frac{\partial}{\partial z}\left(\Gamma_{eff,\phi}\frac{\partial\phi}{\partial z}\right)+S_{\phi} \tag{1}$$

in which u,v and w are velocity components in x,y and z direction respectively, ϕ may stand for constant 1 (for continuity equation),any of velocity components u,v,w(for momentum equation),the temperature T (for energy equation) or a species concentration C (for concentration equation). S_{ϕ} is the corresponding source (for example,the pressure gradient terms appears in momentum equation),Γ_{eff} is the effective viscosity must be determined by the turbulent kinetic energy k and its rate of dissipation ε .The equation of k and ε can be written as

$$\frac{\partial}{\partial x}(\rho uk)+\frac{\partial}{\partial y}(\rho vk)+\frac{\partial}{\partial z}(\rho wk)=\frac{\partial}{\partial x}\left(\Gamma_{eff,k}\frac{\partial k}{\partial x}\right)+\frac{\partial}{\partial y}\left(\Gamma_{eff,k}\frac{\partial k}{\partial y}\right)+\frac{\partial}{\partial z}\left(\Gamma_{eff,k}\frac{\partial k}{\partial z}\right)+S_{k}, \tag{2}$$

$$\frac{\partial}{\partial x}(\rho u\varepsilon)+\frac{\partial}{\partial y}(\rho v\varepsilon)+\frac{\partial}{\partial z}(\rho w\varepsilon)=\frac{\partial}{\partial x}\left(\Gamma_{eff,\varepsilon}\frac{\partial\varepsilon}{\partial x}\right)+\frac{\partial}{\partial y}\left(\Gamma_{eff,\varepsilon}\frac{\partial\varepsilon}{\partial y}\right)+\frac{\partial}{\partial z}\left(\Gamma_{eff,\varepsilon}\frac{\partial\varepsilon}{\partial z}\right)+S_{\varepsilon}, \tag{3}$$

$$S_{k}=G-\rho\varepsilon \ , \ S_{\varepsilon}=C_{1}\frac{\varepsilon}{k}G-C_{2}\rho\frac{\varepsilon^{2}}{k}, \quad G=\mu_{turb}\frac{\partial u_{j}}{\partial x_{j}}\left(\frac{\partial u_{i}}{\partial x_{j}}+\frac{\partial u_{j}}{\partial x_{i}}\right), \tag{4),(5),(6}$$

$$\Gamma_{eff,T}=\mu_{eff}\big/\sigma_{T} \ , \ \Gamma_{eff,k}=\mu_{eff}\big/\sigma_{k} \ , \ \Gamma_{eff,\varepsilon}=\mu_{eff}\big/\sigma_{\varepsilon}$$

The model constant are $C_{\mu}=0.09,C_{1}=1.44,C_{2}=1.92,\sigma_{k}=1.0,\sigma_{\varepsilon}=1.3,\sigma_{T}=0.7$.

For the problem of heat, the buoyancy should be considered. The Boussinesq approximation is taken.The buoyancy terms in S_{ϕ} in x and y direction is zero,in z direction can be express as

$$S_{z,buoyancy}=-g(\rho_{local}-\rho_{ref}) \tag{7}$$

if the ambient temperature is taken as reference temperature,and the coefficient of expansion, β ,is introduced,equation(7) becomes $S_{z,buoyancy}=\rho\beta g(T_{local}-T_{amb})$.

Table 1 Flow Parameter

No	H(cm)	u_a	T_a	T_E	u_j	D(cm)	Fr	K	R_m
1	22.75	13.45	30.0	40.7	50.63	2.79	15.9	3.8	21.27
2	22.75	4.48	23.72	36.0	50.63	2.79	15.9	11.3	7.75
3	26.87	4.48	21.83	25.7	50.63	2.79	31.8	11.3	8.93
4	14.48	4.48	22.0	34.8	50.63	2.79	15.9	11.3	5.29
5	14.48	14.45	30.0	40.7	50.63	2.79	15.9	3.8	13.89

in which u_a, u_j is in cm/s, T_a, T_E is in ^0C.

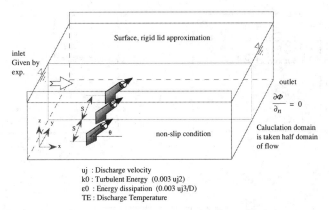

uj : Discharge velocity
k0 : Turbulent Energy (0.003 uj2)
ε0 : Energy dissipation (0.003 uj3/D)
TE : Discharge Temperature

Fig.2 Calculated Domain and Boundary Condition

3. BOUNDARY CONDITIONS AND NUMERICAL METHOD

The calculated domain and boundary conditions are shown in Fig.2.In order to compare the results with experimental data of Parr and Sayre(1979).We use their experimental parameters to calculate.(see table 1)
Hybrid finite analytic method (HFAM) has been proved to be an effective method in numerous numerical computations (Li and Huai,1997). For the content of HFAM can be seen reference(Huai,1998).

4.RESULTS AND DISCUSSION

4.1 Center Line Dilution

The points of minimum excess temperature dilution at any section in the point in the section with the maximum excess temperature, ΔT_{max}. These points occur along the center line of the jet. Fig.3 shows the center line excess temperature dilution. $\Delta T_E/\Delta T_{max}$, where ΔT_E, is the initial excess temperature,for the three-port runs at most of downstream sections. In Fig.4,the solid points stand for the data of Parr and Sayre, the hollow points stand for the results of simulation,It can been obtained that the results agree with the ones of experimental data. In mean while,the dilutions show a linear increases at low values of x/D.
We rearrange with the results in Fig.3, i.e. ,the minimum excess temperature dilution for the runs are presented in Fig.4 in generalized form at downstream sections for runs. The points are the results of simulation,the curves are the ones of experimental data. The results of No.4 and No.5 "peeled off" from the solid curve,the reason is

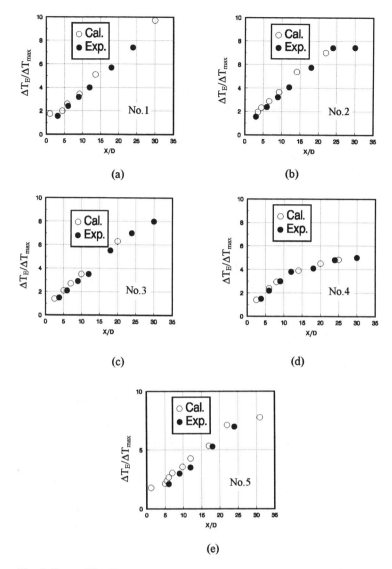

(a)

(b)

(c)

(d)

(e)

Fig. 3 Center Line Excess Temperature Dilution Verses Distance Downstream

these runs having shallow receiving water depths and low values of the fully-mixing excess temperature dilution. Meanwhile,the curves show a linear increase at low values of x/D,followed by a gradual leveling off as the mixing ratio,R_m ,is approached. In the range $x/D<2R_m$,the curves in Fig.5 can be represented by the linear relationship

$$\Delta T_E/\Delta T_{max}=0.28(x/D+2.5) \qquad (7)$$

From the Fig.3 and Fig.4,we can obtain that the mathematical modeling and numerical method is effective and feasible.Based on the parameters in table 1, we calculated the dilution field for one-port for this kind flow (Huai,1998). The minimum excess temperature dilution are present in Fig. 4 in generalized form at downstream section for one-port in dotted line. The results of three-port are given in same Fig. also. The difference of dilution between one-port and three-port is very clear.

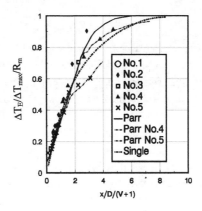

Fig. 4 Generalized Minimum Excess Temperature Dilution

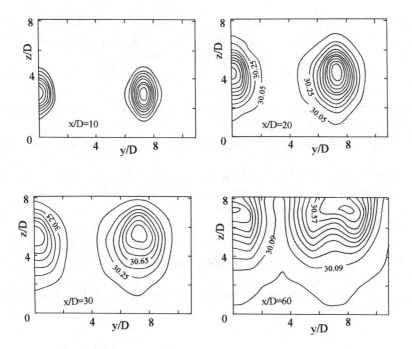

Fig. 5 Temperature Contours at x/D=10,20,30,60 for Run No.1

4.2 Bifurcation and Coanda Effect

Fig.5 present the temperature contours at x/D=10,20,30 and 60 for Run No.1, it is found that there are three high temperature region in section x/D=60 with the development of jet. this is the bifurcation shown by temperature contours. On the other hand, from Fig.5, it can be seen that adjacent two jet interact and deflect about the finger to a new path at an angle to the original stream. It is known as the Coanda effect after the inventor Henri Coanda.

5. CONCLUSIONS

Numerical simulation were performed to investigate jet behavior of multiple submerged heated jets in laterally confined,shallow, flowing ambient. In order to provide data and numerical method that can be used to predict the performance of thermal multiport diffusers on rivers. The principal results of this study are as follows:

a) The mathematical modelling and numerical method are effective,it can be used to simulate the behavior of diffuser in rivers.

b) For this kind flow considered in this paper,the dilution of excess temperature dilution along jet centerline varied linearly with x/D in the region $x/D < 2R_m$. and it can be described by equation (10). Beyond this region,the excess temperature dilution asymptotically approached the value of the mixing ratio,R_m.

c) There are bifurcation and Coanda effect in the near field for river diffuser.

6. REFERENCE

Adams,E.E.(1982). "Dilution analysis for unidirectional diffusers",*J.Hydr.Div.*,ASCE,108(3),327-342.

Cederwall,K.(1971). "Buoyant slot jets into stagnant or flowing environments",W.M. Keck Laboratory,*Report No.KH-R-25*,California Institute of Technology, Pasadena, Calif.Apr.

Huai,W.X.(1998), "Behavior of near field for 3-d single buoyant jets in shallow coflowing receiving waters",*J. of Hydrodynamics*,Ser.A,13(1),79-87.(in Chinese)

Lee,J.H.W.(1980). "Multiport diffuser as line source of momentum in shallow water",*Water Resources Research*,16(4),695-708.

Lee,J.H.W.(1983). "Multiple shallow water jets in coflowing current",*Proceedings 20th IAHR Congress*,3,141 -149.

Lee,J.H.W.(1984). "Boundary effects on a submerged jet group",*J. of Hydr. Research*,22(4),199-216.

Lee.J.H.W.(1985). "Comparison of two river diffuser models",*J. Hydr. Enging*,ASCE,111(7),1069- 1079.

Li,W.and Huai, W.(1995). "Calculation of whole field for round buoyant jet in linearly stratified environment",*J. of Hydr. Research*,33(6),865-876.

Li, W., and Huai, W.(1997).*The theory and application of buoyant jets*,Science Press of China,Beijing,China.

Parr,A.D. and Sayre,W.,(1979). "Multiple jets in shallow flowing receiving waters", *J. Hydr. Division.*,Proc. ASCE,105(11),1357-1374.

Parr,A.D. and Melvile,J.G.(1981), "Nearfield performance of river diffusers",*J.of Environmental Engineering Div.*,ASCE,107(5),995-1008.

Environmental Hydraulics, Lee, Jayawardena & Wang (eds) © 1999 Balkema, Rotterdam, ISBN 90 5809 035 3

Evaluation of mixing zone models: CORMIX, PLUMES and OMZA with field data from two Florida ocean outfalls

Hening Huang & Robert E. Fergen
Hazen and Sawyer, Raleigh, N.C., USA

John J. Tsai & John R. Proni
Atlantic Oceanographic and Meteorological Laboratory, National Oceanic and Atmospheric Administration, Miami, Fla., USA

ABSTRACT: This paper presents an evaluation of three mixing zone models: CORMIX, PLUMES, and OMZA using field data from two Florida ocean outfalls: Hollywood and Miami-Central outfalls. The Hollywood outfall has a single port outlet and the Miami-Central outfall has a multiport diffuser. Both outfalls discharge secondary effluent. For the nearfield, all of the three models predict realistic initial dilutions for the tests at the Hollywood outfall. The three models predict realistic initial dilutions too for the tests at the Miami-Central outfall except three cases in CORMIX predictions and two cases in PLUMES predictions (out of 20 cases). For the nearfield and farfield combined, CORMIX significantly overestimates dye concentrations for the tests at the Hollywood outfall but underestimates dye concentrations within the 300 m to 400 m range for the tests at the Miami-Central outfall. PLUMES predictions agree reasonably well with the field data for the range from 300 m to 800 m but do not agree well within the 300 m range for the tests at both outfalls. OMZA predictions agree well with the field data within the 800 m range for the tests at both outfalls.

1. INTRODUCTION

Three mixing zone models: CORMIX, PLUMES, and OMZA are currently available and have been used in the outfall mixing zone analysis. However, the evaluation of these models with field data are quite limited. Especially, the farfield predictions of these models were not compared and evaluated.

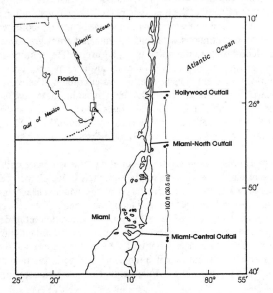

FIG. 1. Locations of Three Southeast Florida Ocean Outfalls

Table 1 Characteristics of Hollywood and Miami-Central Outfalls

Parameter	Hollywood	Miami-Central
Average discharge (m³/s)	1.32	5.26
Discharge depth (m)	27.0	28.2
Discharge off shore (m)	3,100	5,700
Diffuser length (m)	0	39
Number of ports	1	5
Spacing of ports (m)	0	9.8
Diameter of ports (m)	1.52	1.22
Port orientation	Horizontal	Vertical

The purpose of this paper is to evaluate these three mixing zone models using field data obtained at the Hollywood and Miami-Central outfalls during the Southeast Florida Ocean Outfall Experiment II (SEFLOE II) project (Hazen and Sawyer, 1994). Both nearfield and farfield (up to 800 m range from the outfall) dilutions are considered.

2. OUTFALL CHARACTERISTICS

The Hollywood and Miami-Central outfalls are located off the east coast of south Florida and in the western boundary region of the Florida Current (Fig.1). These two outfalls discharge secondarily treated domestic sewage whose average density is about 0.998 g/cm³. The Hollywood outfall has a single port outlet and the Miami-Central outfall has a multiport diffuser. Table 1 summarizes the characteristics of these two outfalls.

3. FIELD TESTS

Nearfield initial dilution was derived from two types of measurements, dye and salinity. In the dye studies, Rhodamine-WT was injected into effluent at the treatment plants. Dye concentrations in effluent were measured continuously using a fluorometer. Dye concentrations in the ocean surface boils were measured using an on-board fluorometer connected to a continuous data logging unit and from grab water samples. In the salinity studies, temperature and conductivity were measured using a towed CTD; salinity deficit was taken as a tracer to determine initial dilutions. During the tests at both outfalls, surfacing plumes occurred because of the positive buoyancy of effluent and uniform ambient density over the water column. Therefore, the field measured initial dilution is the minimum surface or near-surface dilution. In addition to dye and salinity measurements, parameters affecting dilution, including effluent discharge rates, ocean currents, and ambient densities were also measured. A total of 14 data sets were obtained at the Hollywood outfall and 20 were obtained at the Miami-Central outfall. Details on the tests and data can be found in Proni et al. (1994).

In the dye tests, in addition to sampling at outfall boils, sampling was also made within about 800 m range from the boil(s). During a test, the research vessel made transacts in the surface plume at different distances from the boil(s). Along the track, grab water samples were taken at the water surface using sampling bottles; dye concentrations of these samples were later measured using a fluorometer. Dye concentrations along ship tracks were also continuously measured using an on-board fluorometer which had an underwater sampler towed and kept at approximately 2 m beneath the water surface. However, some instrumentation problems were found for the continuous measurements. Therefore, data reported here are solely from surface grab samples. Note that measured dye concentrations from grab samples may be considered to be "instantaneous" concentrations, because no time averaging was involved in these samples. Five field tests were conducted at each outfall. Tables 2 and 3 summarize the test conditions.

250

Table 2 Summary of Field Test Conditions at Hollywood outfall (February 9-10, 1992)

No.	Test span	Effluent dye Concentration (ppb)	Effluent flow rate mgd (m³/s)	Current speed at 16.5 m (cm/s)	Current direction (degree)
1	11:31-13:04	590	33.0(1.446)	16.2	180
2	19:40-21:13	595	32.5(1.424)	12.9	200
3	23:00-23:59	600	30.5(1.336)	17.1	190
4	03:06-04:20	610	24.0(1.051)	15.4	190
5	05:55-07:04	645	23.1(1.012)	15.5	190

Table 3 Summary of Field Test Conditions at Miami-Central outfall (February 7-8, 1992)

No.	Test span	Effluent dye concentration (ppb)	Effluent flow rate mgd (m³/s)	Current speed at 16.8 m (cm/s)	Current direction (degree)
1	10:08-11:15	555	143.4(6.282)	16.2	190
2	13:51-14:58	585	154.5(6.768)	16.5	135
3	18:07-19:18	575	140.7(6.164)	19.2	300
4	22:55-23:59	550	137.9(6.041)	20.0	0
5	02:55-04:41	500	105.9(4.639)	15.5	0

4. MIXING ZONE MODELS

Three mixing zone models: CORMIX, PLUMES, and OMZA are compared and evaluated. CORMIX is developed by Cornell University under the support or USEPA. CORMIX consists of three sub-models: CORMIX1, CORMIX2, and CORMIX3, which are for single port, multiport diffuser, and surface discharges, respectively (Doneker and Jirka 1990; Akar and Jirka, 1991). The methodology of CORMIX for nearfield buoyant jet mixing process is based on asymptotic analysis to classify nearfield flow-patterns (e.g., Jirka and Doneker, 1991) and the use of asymptotic solutions. In the farfield, CORMIX uses a buoyant spreading or a turbulent diffusion model. In the transition between nearfield and farfield, CORMIX uses a control volume model to connect nearfield and farfield models. CORMIX Version 3.1 was used in the evaluation.

PLUMES is developed by USEPA (Baumgartner et al., 1994). It consists of three sub-models: UM, RSB, and Brooks model. UM is a nearfield buoyant jet model based on the Lagrangian formulation of conservation of mass (continuity), momenta, and energy, and the Projected Area Entrainment (PAE) hypothesis (Frick, 1984) and the traditional Taylor entrainment hypothesis. The Lagrangian plume equations are solved numerically at each time steps to give dilutions along plume trajectories. RSB also is a nearfield model but based on experimental studies of Roberts et al. (1989a, b, and c). Brooks model is a farfield turbulent diffusion model for surface plumes (Brooks, 1960). Buoyant spreading, another important farfield dilution mechanism, is not considered in PLUMES. Also, PLUMES does not contain a model for the transition between the nearfield and farfield mixing processes. The UM model of PLUMES was used for initial dilution calculation.

OMZA is developed by Hazen and Sawyer P.C. (Huang and Fergen, 1996).   For the nearfield buoyant jet mixing process, OMZA uses a three-rank jet-classification concept and an all-regime prediction method to predict jet behaviors. For the farfield plume mixing processes, OMZA uses a model that includes both buoyant spreading and turbulent diffusion to predict farfield dilution (Huang and Fergen, 1997). For the transition

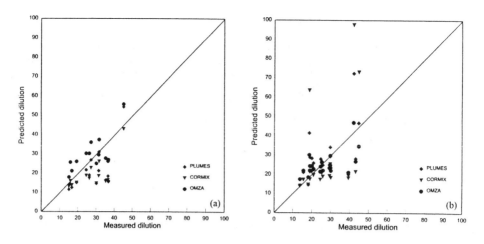

FIG. 2. Comparison of Model Predictions for Initial Dilution with Field Data

between the nearfield and farfield, OMZA uses a modified control volume model to connect the nearfield and farfield models.

5. RESULTS AND DISCUSSION

Figures 2a and 2b show the comparison of initial dilution predicted by the three models with field data at the Hollywood and Miami-Central outfalls respectively. Note that the initial dilution predicted by CORMIX and OMZA is the centerline or minimum dilution at the water surface and therefore can be directly compared with the field measured dilution. However, the initial dilution predicted by PLUMES/UM is the flux-averaged dilution. In order to compare with field initial dilution data, a factor of 1.77 for single port discharge and 1.41 for multiport diffuser discharge are applied to PLUMES/UM nearfield predictions.

It can be seen from Fig. 2a that for the Hollywood outfall, all of the three models provide realistic initial dilution predictions. However, CORMIX and PLUMES/UM somewhat underestimate initial dilutions. It can be seen from Fig. 2b that for the Miami-Central outfall, the three models provide realistic initial dilution predictions again except three cases in CORMIX predictions and two cases in PLUMES predictions (out of 20 cases). The reason for the spikes in CORMIX predictions may be due to the discontinuity in asymptotic flow regimes. However, the reason for the spikes in PLUMES/UM predictions is not clear. It should be mentioned that CORMIX Version 2.1 was used to compare with the same initial dilution data before (Fergen and Huang, 1994). It was found that CORMIX Version 2.1 somewhat overestimated initial dilution at the Hollywood outfall and produced 15 unrealistic predictions out of 20 cases at the Miami-Central outfall. CORMIX Version 3.1 is evidently improved on Version 2.1.

The measured dye concentrations in the surface plume are presented as a function of range from the boil (Fig.3). An envelope of the data points may represent the upper bound of dye concentrations within the surface plume. Such an envelope is assumed to be the plume centerline concentration and is used to compare with the predictions from the mixing zone models. Note that the farfield dilution predicted by PLUMES and OMZA is the centerline dilution while that predicted by CORMIX is the average dilution. In order to compare with farfield dye concentration data, a factor of 1.41 is applied to the CORMIX farfield predictions.

Figure 3 shows the measured dye concentrations in the surface plumes along with the model predictions for plume centerline dye concentrations. Note that results for Tests 3, 4, and 5 at the Hollywood outfall are not presented because the peak concentrations at the boils were not caught during these tests. Results for Test 3 at the Miami-Central outfall were not presented too because the field sampling did not cover the whole plume.

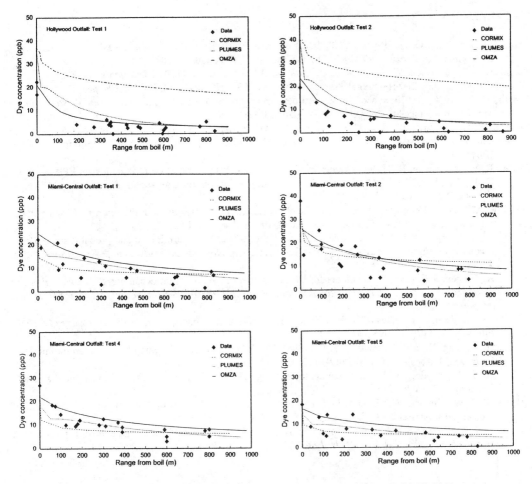

FIG. 3. Comparison of Model Predictions for Dye Concentrations with Field Data

It can be seen from Fig. 3 that CORMIX significantly overestimates dye concentrations for the tests at the Hollywood outfall but underestimates dye concentrations within the 300 m to 400 m range for the tests at the Miami-Central outfall. PLUMES predictions agree reasonably well with the field data for the range from 300 m to 800 m but do not agree well within the 300 m range for the tests at both outfalls. OMZA predictions agree well with the field data within the 800 m range for the tests at both outfalls.

6. CONCLUSIONS

Three mixing zone models: CORMIX, PLUMES, and OMZA are evaluated using field data from two Florida ocean outfalls. Consistency and reasonable agreement with field data are found for initial dilution predictions from these three models. However, farfield predictions for dye concentrations (which are also affected by the initial dilution prediction) from the three models are not consistent and only OMZA predictions show good agreement with the field data. It should be recognized that the conclusions regarding the predictive ability of these three models are based on these two Florida ocean outfalls under the tested conditions. It is possible that under different conditions and at different outfalls, these three models might produce results different from those presented here. Further evaluation of these models is desirable.

7. REFERENCES

Akar, P. J., and Jirka, G. H. (1991). CORMIX2: *An expert system for hydrodynamic mixing zone analysis of conventional and toxic multiport diffuser discharges*, EPA/600/3-91/073.

Baumgartner, D. J., Frick, W. E., Roberts, P. J., W., and Bodeen, C. A. (1993). *Dilution models for effluent discharge*, U.S. Environmental Protection Agency, ERL-N (Newport), N-210, 191 pp.

Brooks, N. H. (1960). "Diffusion of sewage effluent in an ocean current", *Proc. Int. Conf. Waste Disposal Mar.Environ.*, 1st, 246-267, Oxford: Pergamon Press.

Doneker, R. L., and Jirka, G. H. (1990). *Expert system for hydrodynamic mixing zone analysis of conventional and toxic submerged single port discharges* (CORMIX1), DeFrees Hydraulics Laboratory, Cornell University.

Fergen, R. E., and Huang, H. (1994). "Comparison of SEFLOE II field initial dilution data with EPA models - CORMIX and UM", *Water Environment Federation 67th Annual Conference & Exposition*, Chicago, Illinois, October 16-20, 1994.

Frick, W. E. (1984). "Non-empirical closure of the plume equations", *Atmospheric Environment*, 18(4), 653-662.

Hazen and Sawyer, P.C. (1994). SEFLOE II *Final Report*, Hollywood, FL.

Huang, H., and Fergen, R.E. (1996). "OMZA: A computer model for deterministic and probabilistic outfall mixing zone analysis", *Water Environmental Federation 69th Annual Conference & Exposition*, Dallas, Texas, October 5-9, 1996.

Huang, H., Fergen, R.E. (1997). "A model for surface plume dispersion in an ocean current", *Proceedings of Theme B, Water for A Changing Global Community, The 27th Congress of the Int'l Assoc. for Hydraulic Research*, 370-375.

Jirka, G. H., Doneker, R. L. (1991). "Hydrodynamic classification of submerged single-port discharges', *J. Hydr. Eng.* ASCE, 117(9), 1095-1112.

Proni, J. R., Huang, H., and Dammann, W. P. (1994). "Initial dilution of Southeast Florida ocean outfalls", *J. Hydr. Eng.*, ASCE, 120(12), 1409-1425.

Roberts, P. J. W., Snyder, W. H., and Baumgartner, D. J. (1989a). "Ocean outfalls. I: Submerged wastefield formation", *J. Hydr. Eng.*, ASCE, 115(1), 1-25.

Roberts, P. J. W., Snyder, W. H., and Baumgartner, D. J. (1989b). "Ocean outfalls. II: Spatial evolution of submerged wastefield", *J. Hydr. Eng.*, ASCE, 115(1), 25-48.

Roberts, P. J. W., Snyder, W. H., and Baumgartner, D. J. (1989c). "Ocean outfalls. III: Effects of diffuser design on submerged wastefield", *J. Hydr. Eng.*, ASCE, 115(1), 49-70.

Environmental Hydraulics, Lee, Jayawardena & Wang (eds) © 1999 Balkema, Rotterdam, ISBN 90 5809 035 3

Jet scouring of ambient sediment in the outfall of stage 2 Shanghai Sewerage Project

Wei Heping & Liu Cheng
School of Environmental Engineering, Tongji University, Shanghai, China

J.H.W.Lee
Department of Civil Engineering, The University of Hong Kong, China

ABSTRACT: The outfall of the Stage 2 Shanghai Sewerage Project will be constructed in the Bailong Gang zone of Yangtze Estuary where the sediment content is high and the water depth is shallow. Lowering the port height above river floor as far as possible to fully utilize the depth of receiving water is one way to increase the near field dilution, but doing so will increase the possibility of the sediment deposition over the diffuser. Based on the theories of model similarity and sand movement, experiments on the jet scouring of ambient sediment are conducted under both the conditions that there is sediment deposited over diffuser and the ambient current is sediment-laden. The ability of scouring ambient sediment for every feasible scheme is analyzed when the port height is lowered.

1. INTRODUCTION

In the Stage 2 Shanghai Sewerage Project, the wastewater of Wujing, Minhang, Pudong new region and some of the wastewater intercepted from Puhuan, Xuhui and Luwan will be transported to Bailong Gang zone and discharged into the Yangtze Estuary via a submerged multi-port diffuser. The Bailong Gang region has the following characteristics (Tongji University, 1998): 1) the volume of sewage flow is large and the design dry weather flow is 1.7 million m^3/day (19.7 m^3/s); 2) the water depth along the diffuser only ranges from 4.7 to 7.0m at its low level; 3) the landscape conservation is very important, as a tourist center is planned downstream and Pudong International Airport is being constructed nearby; 4) the diffuser length is restricted to about 200 m due to the presence of the main navigation channel; and 5) Yangtzi River is laden with sediment and the sediment content in Bailong Gang area is rather high; the depth-averaged value can reach a level of around 1.10 kg/m^3.

As there is little room for changing the diffuser length and the water depth, lowering the port height above river floor to increase the effective submerged depth becomes one of the effective ways to increase the near field dilution. Theory and field data have shown that the centerline dilution increases significantly with the square of the elevation (Lee and Neville-Jones 1987); Xu (1991) also showed that the dilution increases with the water depth, but lowering the port height increases the possibility of ambient sediment deposition over the diffuser ports. The present study is to find the optimal port height regarding the ability of jet scouring ambient sediment. It is necessary to obtain the dynamic equilibrium between the sediment scouring and deposition around the diffuser ports, and avoid any unfavourable sediment deposition.

2. MODEL DESIGN

The model is undistorted and based on the Froude similarity. The linear dimension ratio of prototype to model is $\lambda_L=50$. Other ratios are as follows (Zuo, 1984):

flow rate ratio $\lambda_Q = \lambda_L^{5/2} = 17678$

flow velocity ratio $\lambda_V = \lambda_L^{1/2} = 7.07$

A total of 70 experiments were conducted in a 3.5m long glass flume, using a rectangular cross-section 0.40m wide and 0.7m deep (figure 1).

The selection of model sand is one of the key issues in the model design to study the regulation of sediment deposition and scouring. If the prototype sand is coarse enough, sand of same density but smaller size can be used as the model sand. In the present study, the prototype sand is rather fine; if particles of the same density are used as the model sand, their size will be so small that they will behave quite differently from those in the prototype as flocculation and coagulation may occur. In this study, an artificial phenolic plastic sand with density smaller than the prototype sediment is used in order to achieve the model similarity for coarser model sand.

The median particle diameter of bed sediment is 0.008mm and that of suspended sediment is 0.005mm in the vicinity of the diffuser. The model sand particle size is determined according to the similarity of sediment incipient velocity between the model sand and nature sediment (Wuhan Institute of Hydraulics and Power, 1981). The median size of model sand adopted in the test of sediment deposited over ports is 0.3mm, and the grain size distribution is similar to that of bed sediment. The median size of model sand adopted in the test of sediment-laden water is 0.1mm, and the grain size distribution is similar to that of suspended sediment.

3. JET SCOURING WITH SEDIMENT DEPOSITED OVER DIFFUSER PORTS

3.1 Phenomenon of Jet Scouring Deposited Sediment

After the deposition of sediment over the diffuser ports, the processes of jet scouring ambient sediment can be divided into three stages: (1) initial fluidization—for very low flow rates, the flow seeps through the deposited sediment. As the flow rate slowly increases incrementally, isolated pockets of disrupted sand migrate up through the bed to form a localized boil on the sand surface; (2) full fluidization—As the flow rate continues to increase, the areas of boiling sand enlarge until the bed above the port is fluidized. In this stage, the eroded sand particles form berms or remain in suspension within the fluidized region; (3) stable and clear scouring pit—as the jet flow rate and energy further increases, the fluidized zone is widened. When the jet velocity reaches a critical value, the sand in the fluidized zone is carried away from it by the jet flow, and the scouring pit is widened rapidly until an equilibrium state is reached when there are few suspended

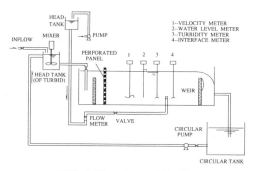

FIG. 1. Experimental Set Up

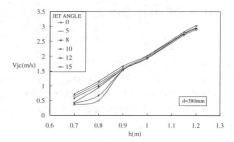

FIG. 2. Third Stage of the Jet Scouring	FIG.3. Influence of jet angle θ and deposited
Ambient Sediment Processes	sediment thickness h to critical sediment scouring
	velocity v_{jc}

sands in the pit (see figure 2). For easy reference, the jet velocity when the third stage is just reached is defined as the critical sediment scouring velocity (v_{jc}).

3.2 The Influence of the Jet Angle and the Deposited Sediment Thickness upon the Critical Sediment Scouring Velocity

The critical sediment scouring velocities are measured with the jet angles changing between 0° to 15° and the deposited sediment thickness over ports varying from 0.7 to 1.5m. (Note: values of all variables are expressed in prototype dimensions in this paper). The experimental results are shown in figure 3. It can be seen that the critical sediment scouring velocities increase approximately linearly with the deposited sediment thickness, and decrease as the jet angle increases.

3.3 The Influence of the Jet Angle upon the Size of the Scouring Pit

The size of scouring pits is measured when the jet velocities increase step by step. In the stage of initial fluidization, the area of boil on the sand surface resulted from larger jet angle is greater than that caused by smaller jet angle. Similarly, in the stage of full fluidization, the fluidized region resulted from larger angle is also larger. After the jet velocity reached the critical sediment scouring velocity, the length of the scouring pit produced by smaller jet angle gradually get close to that resulted from the larger jet angle, whereas the width of the pit produced by the smaller jet angle tends to be larger than or equal to that of larger angle. The pit generated by smaller jet angle is deeper than that of larger angle.

3.4. The Ability of Scouring Deposited Sediment by Different Arrangement of Multi-port Diffuser

Typically, the ports of the multi-port diffuser are arranged in one row or two rows. Different port sizes and arrangement are selected from feasible plans to test their abilities of scouring deposited sediment. In the experiments, the deposited sediment thickness is 0.5m over the centerlines of the lower row of ports. The minimum design jet velocity is 1.4m/s and the average design velocity is 2.4m/s. For each feasible scheme deposited by sediment, the ability of scouring deposited sediment is stronger with larger port diameter for the same jet velocity, and the ability of riser with two rows of ports is stronger than that with one row. When deposited sediment thickness over the centerline of the ports (lower row of ports for two rows arrangement) is 0.5m, stable and clear scouring pits are formed when the jet velocity is 2.4m/s for all scenarios; plum blossom shaped pit is formed for riser with one row of ports, and circular pit is formed for riser with two rows of ports. When the jet velocity is 1.4m/s, stable and clear scouring pit is formed for all configurations except the one with one row of smaller ports (260mm) that only reaches the stage of full fluidization.

4. REGULATION OF THE JET SCOURING AMBIENT SEDIMENT IN SEDIMENT-LADEN WATER

The sediment-laden water tests are conducted to examine the ability of different design schemes to scour depositing sediment when the ambient current has the maximum sediment content but the average minimum flow rate. The ability of riser with different nozzle lengths to scour depositing sediment is also compared. The sediment content of ambient current is $S = 1.10$ kg/m^3, and the flow rate is Ua = 0.35m/s and the simulated period is 3 days in the test.

4.1 The Ability of Scouring Depositing Sediment with Different Types of Multiport Diffuser

Under such experimental condition, the thickness of sediment deposited outside the region influenced by the jets is about 10cm in 3 days. When the port height of lower row from centerline to bed is 0.5m, and the jet velocities are equal to or larger than 1.4m/s, a fairly wide range of irregular circular, deposit free region is formed around the risers (see figure 4). If the ports are 1.0m above the bed, the jets have no effect upon the sediment deposition. This shows that the lower row port height from centerline to bed being 0.5m is feasible and there will be no significant sediment deposition over the risers.

FIG.4. Non-deposited Regions Scoured Around the Riser (n=2×10, d=280mm, v$_j$=1.4m/s)

4.2 The Ability of Scouring Depositing Sediment of Multi-port Diffuser with Different Nozzle Lengths

The relevant studies discovered that the nozzle length of ports had influence on the near-field dilution and headloss (Tongji University, 1998). To determine the appropriate nozzle length (b) comprehensively, the ability to scour depositing sediment of risers with different nozzle length is examined through sediment-laden water test. In the range of the experiments undertaken (0.57 ≤ b/d ≤ 2.86, where d is the port diameter), the ability of scouring depositing sediment increases with the nozzle length. The non-deposited area is small when the relative nozzle length (b/d) is less than 1 (see figure 5(a)), and the non-deposited area increases obviously when the relative nozzle length is 1.14 (see figure 5(b)), whereas the non-deposited area further increases but not as significant when the relative nozzle length is increased beyond 1.14.

| (a) b/d=0.57 | (b) b/d=1.14 |

FIG.5. Non-deposited Regions Scoured by Jets with Different Nozzle Lengths (n=10, d=175mm, v$_j$=2.4m/s)

5. Analysis of the Ability of Souring Ambient Sediment for Feasible Schemes

Seven feasible outfall designs are proposed for the Stage 2 of Shanghai Sewerage Project (Table 1). All the schemes have three types of port arrangements except the first one that has just one. Based on the experimental result described above, the following conclusions can be obtained for these seven schemes.

When the ambient current is sediment-laden, the port height of the centerline of lower row over river floor being more than 0.5m is feasible for all the designs of every scheme, the diffuser will not be covered by deposited sediment if the jet velocity is greater than 1.4m/s, the minimum design jet velocity.

Considering 1m thickness of sediment deposition has occurred on the river floor, every feasible scheme is studied with the port height of centerline of lower row over river floor at 0.5m level (Table 2). After the diffusers are covered by the deposited sediment, larger areas of clear scouring pits can be formed for every design if the jet velocity is 2.4m/s, the average design velocity. The larger the port diameter, the bigger is the scouring pit. If the jet velocity is 1.4m/s, clear scouring pit is formed for all outlets except the two near-shore risers of the 5th scheme that can only reach the stage of full fluidization. Under these conditions, the jet velocities of two risers of the 5th scheme are less than the critical sediment scouring velocity. However, if proper operational procedures are followed to ensure the jet velocities of those two risers can reach the level of 2.4m/s, the 5th scheme will still be a feasible option.

Table 1 Summary of feasible schemes

Feasible Scheme	No. Of Risers	Type 1			Type 2			Type 3		
		No. of Ports	Port Dia.(mm)	Rows of ports	No. Of Ports	Port Dia.(mm)	Rows of ports	No. of Ports	Port Dia.(mm)	Rows of ports
1st	6	12	380	two	-	-	-	-	-	-
2nd	6	8	390	one	16	310	two	16	340	two
3rd	7	6	410	one	16	310	two	16	340	two
4th	7	8	360	one	16	310	two	16	340	two
5th	11	8	250	one	16	250	two	16	290	two
6th	11	8	290	one	16	250	two	16	270	two
7th	7	8	360	one	12	360	two	14	360	two

Table 2 Analysis of the Abilities of Scouring Ambient Sediment for Feasible Schemes

Schemes	1st	2ed	3d	4th	5th	6th	7th
No. of risers	6	6	7	7	11	11	7
Shape and length of min. scouring pit* when vj=1.4m/s	Circular Clear Pit 12.3m	Circular Clear Pit 10.3m	Circular Clear Pit 10.3m	Circular Clear Pit 10.3m	Fluidized Blossom Shaped Pit 6.3m	Circular Clear Pit 8.6m	Clear Blossom Shaped Pit 11.5m
Scouring abilities	Strong	Strong	Strong	Strong	General	Rather Strong	Strong

*Ports heights of centerline of lower row over river floor are 0.5m, and the deposited sediment over floor is 1m.

6. CONCLUSIONS

Two series of experiments on jet scouring of ambient sediment have been conducted. The first group is to test with sediment deposited over the diffuser and clean ambient current. The second group is to test with the sediment laden ambient current. The results show that if the port height of the centerline of the lower row over river floor being more than 0.5m, all proposed multi-port diffuser design are feasible. The risers will not be covered by deposited sediment if the sewage flow reaches the minimum design flow rate of 1.0 million m^3/day or 11.6 m^3/s (with the corresponding jet velocity = 1.4m/s).

7. REFERENCES

Lee, J.H.W., and P. Neville-Jones, (1987). "Initial dilution of horizontal jet in crossflow", *J. Hydr. Engrg.*, ASCE, 113(HY5), 615-629.

Tongji University, (1998). *Report on optimizing the diffuser design parameters for the Second Stage of Shanghai Sewerage Project.*

Wuhan Institute of Hydraulics and Power, (1981). *River sediment engineering*, Hydraulics Publisher House, Beijing, China, 37-63.

Xu Liang, (1991). "The study on the dilution of pure jet in a current", in: *Environmental hydraulics*, J H W Lee & Y K Cheung(Eds). A A Balkema Publishers, Rotterdam, 223-228.

Zuo Dongqi, (1981). *Theories and methods of model test*, Hydraulics Publisher House, Beijing, China, 2-10, 109-113.

Environmental Hydraulics, Lee, Jayawardena & Wang (eds) © 1999 Balkema, Rotterdam, ISBN 90 5809 035 3

Near field dilution study of bailong gang wastewater outfall of stage 2 Shanghai sewerage project

Heping Wei & Gaotian Xu
Institute of Environmental Engineering, Tongji University, Shanghai, China

J.H.W. Lee & K.W. Choi
Department of Civil Engineering, The University of Hong Kong, China

ABSTRACT: This paper describes a hydraulic model study on near field mixing of the Bailong Gang wastewater outfall in cross-flow. The effects of diffuser parameters including horizontal and vertical port discharge angle, number of nozzles, port arrangement, leading tube on the port and riser spacing are examined. The experimental results show that diffuser design parameters have significant influence on near field dilution. It is found that the vertical jet angle should be controlled from $0°$ to $5°$, nozzle number should not exceed 14, and ports should be arranged in two rows in deep water and single row in shallow water.

1. INTRODUCTION

Shanghai has China's largest port and is one of its most important financial and industrial centres. Currently, about 5.5 millions tonnes of wastewater are generated per day, out of which 54% is discharged directly into Huangpu River and its tributaries without treatment. This causes serious pollution problems and hinder further developments of the city. The Stage 2 of the Shanghai Sewerage Project is the largest wastewater treatment project in China; it is intended to improve the water quality in the upper and middle sections of Huangpu River. It will serve the regions including Xuhui, Luwan, two districts in Puxi and Minhang, Wujing industrial zones and Lujiazui, Yangjin, Jinqiao, Zhangjiang, Beicai, Zhoujiadu, Yunlian, Yangsi, Liuli in Pudong. The total area serviced is 271.7 km^2, with a population of 3.56 million; the wastewater flow is about 5.6 million tonnes/day (64.8 m^3/s). The design dry weather flow is 1.7 million tonnes/day (19.7 m^3/s).

The main interception pipes are newly built at both Puxi and Pudong by which the wastewater is transported to Bailong Gang and discharged into the Yangtze River Estuary after pre-treatment to make good use of the dilution, dispersion and transportation capacity of the large water body. The water depth in the Bailong Gang region is rather shallow; it is only 4.7m to 8.2m deep at low level along the diffuser. The density difference is rather small with a salinity differential only about 1.5 ppt. There is strong ambient current with the averaged tidal velocity in the range of 0.6 – 1.0 m/s. Furthermore, restricted by the presence of the main navigation channel, the maximum length of the diffuser is limited to about 200m.

In order to come up with the most appropriate diffuser design, an extensive physical and numerical modelling exercise have been undertaken. The numerical models employed include the Lagrangian model JETLAG and an integral model developed by Tongji University. The physical and numerical model results agreed reasonably with each other [1]. The following is a summary of the 73 physical model tests undertaken to study the relationships between near field dilution and the major outfall design parameters.

2. MODEL TEST

2.1 Experimental layout

The hydraulic physical model tests are carried out in a rectangular glass flume to determine the reasonable engineering parameters of the diffuser. Flow conditions of Bailong Gang zone are simplified. Each factor influencing the dilution and dispersion of wastewater is analyzed individually, and provide the basis for the diffuser design. The experiment layout is shown as figure 1.

2.2 Model design

Due to the large velocity difference between the jet and ambient current near the outfall, and the three-dimensional entraining interaction between the jet and the ambient current, an undistorted physical model must be used. To preserve densimetric Froude similarity and maintain turbulent jet flow conditions, the model length scale is taken as $L_r = 50$, with the following scale ratios for velocity, discharge, time and roughness.

$$V_r = L_r^{1/2} = 7.07$$
$$Q_r = L_r^{5/2} = 17664.33$$
$$T_r = L_r = 7.07$$
$$n_r = L_r^{1/6} = 1.92$$
$$(Re_j)_{min} = 2566$$

The roughness scale, n_r, is required to be 1.92, so the model should be smoother than the prototype. The outfall will be built by reinforced concrete with a roughness of 0.014, thus, the roughness of the model should be 0.0073. It is found that glass is the suitable material for constructing the model.

3. MODEL TEST RESULTS AND ANALYSIS

3.1 Vertical Jet Angle

The longitudinal movement of wastewater is related to the vertical jet angle (α). The larger is the jet angle, the path of the discharge from the port will be bent downstream more quickly due to the forced entrainment of the ambient crossflow. At the same time the jet and current are intermixing, the jet moves along the current with its width becomes larger and larger. When water depth is shallow, and with a large jet angle, the effluent impinges onto the surface with a greater momentum; it will cause the instability of the flow and form a whirlpool as shown in figure 2. The mixed wastewater will be re-entrained into the plume due to the vertical circulation, which hinder the initial dilution, and will also bring the polluted water to the bottom.

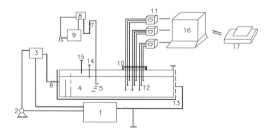

1-Storage basin 2-Pump
3-High elevated water tank
4-Glass flume 5-Port 6-Inflow valve
7-Rotameter 8-High elevated solution tank
9-Solution Storage box 10-Measuring frame
11-Turbidimeter 12-Measuring electrode
13-Tailgate 14-Velocity meter
15-Water level recorder 16-Microcomputer
17-Printer

Figure 1. Experimental layout

262

Figure 3 shows the model test results for different vertical jet angles (α). From the figure, it can be seen that the greater the jet angle, the shorter is the horizontal distance traveled by wastewater when reaching the surface. Dilution of larger jet angle is greater than that of smaller jet angle at the same distance away from the ports before reaching the surface, but discharge from the larger jet angle will reach the surface earlier. Especially in shallow water, such as h = 4.7m to 8.2m, with larger jet angle such as 10° and 15°, wastewater can form the unstable state on the surface shown as figure 2, where the resulted dilution is worse than that of smaller angle. For various water depths, the highest dilution is achieved with α = 0°. According to the test results, the initial dilution is 10 percent greater with α = 0° instead of 5°, for a current velocity of 0.2 m/s.

3.2 Horizontal Discharge Angle

The jet path and dilution are also closely related to horizontal discharge angle (β). The jet trajectory is almost a straight line when discharged directly downstream (coflow, β = 0°), while a curved line resulted when

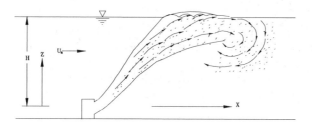

Figure 2. Observed flow for large jet angle discharge in shallow water

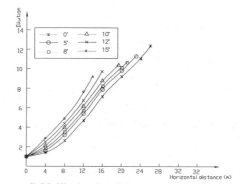

Figure 3. Measured near field dilution for different vertical jet angle (u_a = 0.2 m/s, β = 0°)

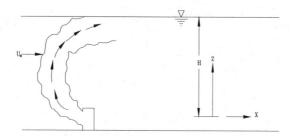

Figure 4. Observed jet trajectory in counterflow (β = 180°)

discharged against the current (counterflow, β = 180°). For the latter case, there is a turning point on the streamline after which the wastewater begins to flow downstream. The spreading width in a counterflow is greater than that in a coflow, which may be more beneficial to dilution. Due to the shallow water depth at Bailong Gang, some turning points have approached the surface, which is not beneficial to dilution.

It can be seen from figure 5 that when β = 90°, i.e. perpendicular to ambient current, the dilution is greatest when wastewater reaches the surface. This is because the discharge from the port spreads out quickly on the current section by the strong ambient water, which quickly mixes with the wastewater and achieve the dilution efficiently. In the outfall design, it is preferred to have the jets become perpendicular to the ambient current as soon as possible. The worst horizontal angle for the dilution is β = 0°, i.e. along the current direction. Under this condition, the horizontal distance is the longest when reaching the surface, but wastewater cloud cannot spread out as quickly by the current, the contact area with the ambient current is small, full dilution and dispersion is not achieved. β = 90° and 0° are the best and the worst angle for dilution, and other angles ranging from good to bad are β = 120°, 150°, 45° and 30°.

3.3 Port Number

Maintaining the discharge and port area unchanged, i.e., jet velocity is constant, configurations with different port number are tested. The model results are shown in figure 6. It shows that the initial dilution increase as

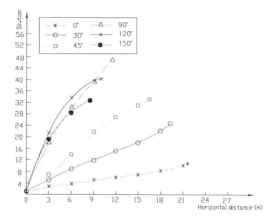

Figure 5. Measured near field dilution for different jet horizontal angle (u$_a$ = 0.5 m/s, α = 5°)

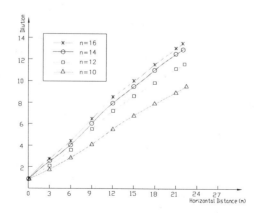

Figure 6. Measured near field dilution for different port numbers (u$_a$ = 0.5 m/s, α = 5°)

the port diameter becomes smaller. Providing more ports on each riser can reduce the number of costly risers and more discharge ports will be beneficial to the dilution of wastewater. However, when the port number increases from 14 to 16, the gain in dilution is not as significant as that obtained by increased from 10 to 12. This is because when more ports are installed on the each riser, the discharge from adjacent ports can interfere more quickly, and hinder the mixing with the ambient water. So, it is not recommended to have more than 14 ports in each riser.

3.4 Port Arrangement

For the nearshore risers, both a single row and double row port arrangement are studied. The model results are shown in figure 7. Owing to shallow water, changing from two rows to one row will increase the effective submerged depth of the ports, which is beneficial to the dilution of wastewater. The results show that for the two risers near the shore, when port arrangement is one row, the effective port depth increases by 0.5m. The dilution result of one row is evidently better than that of two rows, it increases from 5.4 to 6.5.

3.5 Leading Tube to the Port Opening

As the riser wall will not be very thick, there is a question on whether the ports should be perforated directly

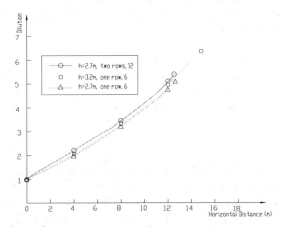

Figure 7. Comparison of port arrangement of one row and two rows (u$_a$ = 0.5 m/s, α = 5°)

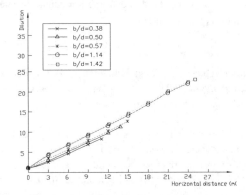

**Figure 8. Observed dilution for different leading tube length to port diameter ratio
(u$_a$ = 0.5 m/s, α = 5°)**

265

on the riser or a leading tube should be added. Thus, the performance of outfall with different leading tube length (b) to port diameter (d) ratio is studied. The model result (figure 8) shows that the dilution increases by 76% from 12.5 to 22, when the ratio of leading tube length and port diameter (b/d) increases to 1.14 from 0.57 for h = 5.5m. It also shows that the dilution does not change significantly, when b/d is less than 0.57 or greater than 1.14. So the optimal design ratio of leading tube length to port diameter (b/d) on the diffuser should be around 1.14 for Bailong Gang outfall.

3.6 Riser Spacing

For a multi-port diffuser, depending on the riser spacing, water depth and ambient velocity, the adjacent jets may merge before reaching the end of the near field region. The merging may hinder the mixing of the wastewater with the ambient water, and hence lower the dilution level that can be achieved. For fixed diffuser length, the optimal riser spacing would be the one with the adjacent jets merged just towards the end of the initial mixing zone, as it will maximize the number of ports but avoid the unfavourable re-entrainment of the wastewater. Configurations with different discharges, port arrangements, depths and ambient current velocities are examined.

For the 6-riser scheme, the design riser spacing is 40m. The jet spreading width is biggest, when the ambient current velocity is 0.2m/s. The adjacent jets do not interfere before the wastewater rises to the surface. The adjacent jets begin to merge long after the near field ends. So, the riser spacing can be decreased to enable the jets to overlap just before the near field ends.

For the 7-riser scheme, when velocity is 0.2m/s, the discharge from the risers basically overlap together except the riser in the shallower water. Compared with the 6-riser scheme, the diffuser is more efficiently utilized. For the 11-riser scheme, when velocity is 0.2m/s, all the adjacent jets merge together before reaching the surface. Even when the velocity increases to 0.5m/s, the jets still overlap together before they reach the surface.

4.CONCLUSION

Based on the results of the physical model tests, it is found that, for Bailong Gang outfall in Yangtze River, the vertical jet angle should be in the range from 0° to 5°; the number of ports should not exceed 14, which are arranged in two rows in deep water, and one row in shallow water. Leading tube should be installed on the port; the ratio of leading tube length and port diameter is approximately 1.14. Two rows of port arrangement are used in place of one row to increase the effective depth of the risers near the shore. It is suggested that discharge ports are arranged laterally on both sides of the main current direction in order to gain better dilution and dispersion, to avoid the two extreme horizontal discharge angles, $\beta = 0°$ and 180°, which are not favourable to dilution of wastewater.

5.REFERENCES

[1] Institute of Environmental Engineering, Tongji University. (1998). *Report on optimizing the diffuser design parameters for Shanghai Sewerage Treatment Secondary Phase Project.*
[2] Heping Wei. (1993). *Study on dilution and dispersion of pure jet in cross flow*, Science Publishing House.
[3] Changzhao Yu.(1992) *Guideline on environmental hydraulics,* Tsinghua University Publishing House.
[4] Shunong Zhang.(1988). *Environmental hydraulics,* Hohai University Publishing House.
[5] Heping Wei.(1992)."Study on marine treatment technology", *China environmental science*, Vol.12, No.2
[6] Lee J H W and Jirka G H. (1981). "Vertical round buoyant jet in shallow water", *J. Hydr. Div.*, ASCE, 107(12), 1651-1675.
[7] Lee J H W and Neville-Jones P. (1987). "Initial dilution of horizontal jet in cross flow", *J. Hydr. Div.*, ASCE, 113(5), 615-629.

Environmental Hydraulics, Lee, Jayawardena & Wang (eds) © 1999 Balkema, Rotterdam, ISBN 90 5809 035 3

Application of thermal plume modelling and establishment of initial mixing zone

Yiping Chen
Department of Civil and Environmental Engineering, University of Bradford, UK

Graham Thompson
Binnie Black and Veatch, Redhill, UK

ABSTRACT: The well known Brooks' simple surface plume model (Brooks, 1960) assumed a steady flow of constant speed. Such an assumption may not be true for nearshore coastal waters. An improved and more general surface plume model, without such a restriction, has been developed by the authors (Chen and Thompson, 1995). Further improvement and refinement to the model have been made to include diffusion effects in the vertical direction. A surface thermal discharge study has been carried out using the refined surface plume model for a proposed coastal thermal discharge scheme. It has been shown that an initial mixing zone near the outfall discharge can be established for a given temperature rise threshold value by varying the cross-current speed and direction. A number of sensitivity tests were also carried out, including for example diffusion coefficients, surface heat exchange coefficient and the variations in ambient cross-current.

1. INTRODUCTION

Thermal discharges into the coastal water will cause the local receiving water to experience some degree of temperature rise above the ambient value. Such temperature excess persists until the discarded heat is advected, dispersed and dissipated through the movement of the plume and its associated mixing processes, as well as further heat loses through the water surface due to wind and evaporation effects. The rising temperature of the receiving water has some adverse impact on the local marine environment, as it affects both water quality and aquatic life. A thermal impact study is therefore required in order to assess the likely changes in water temperature close to the thermal outfall discharge and to establish the initial mixing zone, within which the temperature rise will exceed a threshold value. Further management decisions, such as increase or decrease of the thermal discharge capacity, can then be made based upon the results.

The well known Brooks' surface plume model (Brooks, 1960) has been widely used for evaluation of surface plume mixing effects in coastal and oceanic waters (Tchobanoglous and Burton, 1991). However, the Brooks surface plume model has to be restricted to a constant plume speed flowing in one direction only, assuming three predefined lateral diffusion coefficient variations with the plume width, including (a) constant, (b) linear, and (c) four-third power law relationships. An improved and more general surface plume model, without these restrictions, has been developed by the authors (Chen and Thompson, 1995). All the Brooks' solutions for the plume characteristics can be obtained as the special simplified cases from the newly developed model. Further improvement and modification have been made recently by the authors to include diffusion effects in the vertical direction. Details of a surface thermal plume study, using the improved plume model, for a proposed coastal thermal discharge through an open channel is provided in the paper. The pattern of the temperature rise in the receiving water has been obtained, with the corresponding initial mixing zone also being established based upon the specified threshold value of temperature rise. A number of sensitivity tests were also carried out, including for example diffusion coefficient, surface heat exchange coefficient and the variations in ambient cross-current.

2. SURFACE PLUME MODEL DEVELOPMENT AND REFINEMENT

The surface thermal discharge may be represented by a uniform source element of strength C_o, length $2b$ and depth h at the origin when t=0 in a 3-D domain, with the x, y, and z being defined as longitudinal, lateral and vertical directions respectively. Considering the total travel time of the plume to be consisted of many small time steps Δt_i where i=1,2,3, ..., etc., with the plume speed being a constant value of U_i within each small time step and being variable between different time steps, then the total travel distance of the plume at any time t

(within time steps t_{i-1} and t_i) may be expressed as $x = U_i\,(t - t_{i-1}) + x_{i-1}$, where $x_{i-1} = \sum_{k=1}^{i-1} (U_k \Delta t_k)$,

$t_0 = 0$ and $t_i = \sum_{k=1}^{i} \Delta t_k$. Assuming that the longitudinal mixing is insignificant in comparison with the

longitudinal advection, the basic governing equation for the surface plume mixing processes may be written, for the ith time step Δt_i (within the sub-region Δx_i where $\Delta x_i = x_i - x_{i-1} = U_i \Delta t_i$) as:

$$\frac{\partial C_i}{\partial t} + \frac{\partial U_i C_i}{\partial x} = D_{yi} \frac{\partial^2 C_i}{\partial y^2} + D_{zi} \frac{\partial^2 C_i}{\partial z^2} - k C_i; \qquad (t_{i-1} \le t \le t_i) \qquad (1)$$

where k = surface heat exchange rate, C_i, U_i, D_{yi} and D_{zi} are respectively temperature, speed of the plume, lateral and vertical diffusion coefficients within the sub-region Δx_i. The temperature rise distribution obtained at $t = t_{i-1}$, i.e. $x = x_{i-1}$, will serve as the initial condition for equation (1). The solution to equation (1) for the first time step when i=1 depends upon the initial distribution of the source element when t=0.

The diffusion coefficient can be assumed to be dependent upon the effective width of the plume ($2W_i(t)$), which varies with time, such that:

$$D_{yi} = D_i [W_i(t)/b]^m \qquad (2)$$

$$D_{zi} = \beta D_{yi} = \beta D_i [W_i(t)/b]^m \qquad (3)$$

where $D_i = \alpha U_{*i} H_i$ = lateral mixing coefficient, in which α = eddy viscosity coefficient ($\alpha \approx 0.6$), U_{*i} = local shear velocity and H_i = local water depth, $\beta = D_{zi}/D_{yi}$, m = general exponent factor, with $m = 0, 1$, and 4/3 corresponding to the constant, linear and four-third power law diffusivity variations respectively. The vertical mixing coefficient has been estimated as $0.067 U_{*i} H_i$ for non-stratified currents (Fischer et. al. 1979), which is about 11 percent of the lateral mixing coefficient. However, the vertical mixing effect will be suppressed for stratified flows, as observed for the surface thermal plumes. Using the vertical mixing coefficient relationship suggested by Pritchard (Pritchard 1960), the ratio between the vertical and lateral mixing coefficients for stratified flows may be obtained as $\beta = 0.01$ when H_i=15 m and $\beta = 0.011$ when H_i=45 m, indicating a much slower spreading in vertical direction.

Equation (1) may be transformed into a pure diffusion equation, by considering the advection process separately and representing the diffusions occurring in a semi-infinite y-z-t frame of reference moving at the advective speed U_i in the x-direction. Following the same analytical solution procedure as described by the authors for the 2-D surface plumes (Chen and Thompson, 1995), the solutions to equation (1) may be obtained, after the solution of the pure diffusion has been advected to the appropriate longitudinal location, as:

$$C_i = \frac{C_0}{4} \exp(-kt) \left[erf(\frac{b-y}{A}) + erf(\frac{b+y}{A}) \right] \left[erf(\frac{h-z}{A\sqrt{\beta}}) + erf(\frac{h+z}{A\sqrt{\beta}}) \right] \qquad (4)$$

268

in which

$$A=2\sqrt{D_i\tau+\sum_{k=1}^{i-1}D_k\tau_k} = \frac{2b}{\sqrt{6}}\sqrt{\left[\left(1+\frac{6}{b^2}\sum_{k=1}^{i-1}D_k\tau_k\right)^{\frac{2-m}{2}}+\frac{3(2-m)D_i}{b^2}(t-t_{i-1})\right]^{\frac{2}{2-m}}-1} \qquad (5)$$

and $erf(\eta)=\frac{2}{\sqrt{\pi}}\int_0^{\eta}\exp(-\xi^2)\,d\xi$ is well-known as the error function. The temperature at the plume centre line may be obtained from equation (4) by setting y=0 and z=0, giving

$$C_i\,(centre\,line,t) = C_0\exp(-kt)\,erf\left(\frac{b}{A}\right)erf\left(\frac{h}{A\sqrt{\beta}}\right) \qquad (6)$$

The effective width of the plume may be defined as $2W_i(t)=2\sqrt{3}\,\sigma_i(t)$, which satisfies the initial condition $W_1(0)=b$, with $\sigma_i(t)$ being the lateral standard deviation of the temperature distribution C_i. The effective width of the plume can be calculated as:

$$\frac{W_i(t)}{b}=\left[\left(1+\frac{6}{b^2}\sum_{k=1}^{i-1}D_k\,\tau_k\right)^{\frac{2-m}{2}}+\frac{3(2-m)D_i}{b^2}(t-t_{i-1})\right]^{\frac{1}{2-m}} \qquad (7)$$

in which

$$\frac{6}{b^2}\sum_{k=1}^{i}D_k\tau_k=\left[\left(1+\frac{6}{b^2}\sum_{k=1}^{i-1}D_k\tau_k\right)^{\frac{2-m}{2}}+\frac{3(2-m)D_i}{b^2}(t_i-t_{i-1})\right]^{\frac{2}{2-m}}-1 \qquad (8)$$

It can be shown that the Brooks' surface plume model for a constant plume speed with (a) constant, (b) linear and (c) four-third power law diffusivity variations can be obtained directly from equations (4) - (8), setting the value of m to be 0, 1, and 4/3 respectively. It can also be seen that solutions for W_i and C_i within $t_{i-1}\leq t\leq t_i$ depend upon the results obtained at all previous time steps.

3. MODEL APPLICATION

A thermal plume study has been carried out using the refined surface plume model. The purpose of the study is to establish the temperature rise pattern in the receiving water as a result of discharging heated cooling water from a cooling system into the sea via an open channel, and to establish the initial mixing zone within which the temperature rise will exceed the given threshold value. The initial thermal plume at the outlet of the outfall channel is 100 metre wide (b=50m) and 4 metre deep, with the initial plume speed and the initial temperature rise being 0.11 m/s and 10°C respectively. With a mean tidal range of approximately 0.3 metre, tidal influences on the initial depth of the plume are assumed to be insignificant. The ambient cross-current U_{i1} is assumed to be zero at the shore line and is increased linearly over a set distance and then remains as a constant value (U_c) further offshore. The variations in the strengths and directions of the cross-current have also been considered. The plume has its own momentum even without any ambient current and will progressively expand as it moves into the sea with a speed $U_{i2}=Q/(2W_iH_i)$, where Q= discharge rate of the

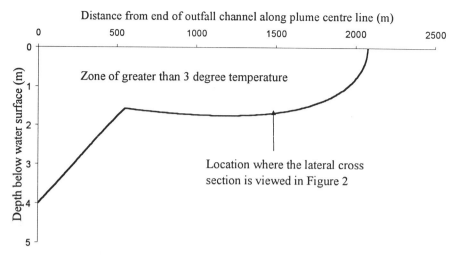

Figure 1 Typical longitudinal section along plume centre line

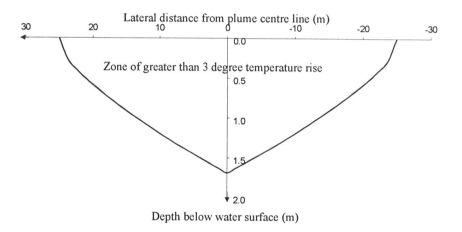

Figure 2 Cross section through plume at longitudinal distance of 1500 m

thermal water. The resultant plume speed is obtained by combining the speed of the ambient cross-current U_{i1} and the plume's own momentum speed U_{i2}. The plume speed and the plume path will therefore vary with time and need to be adjusted at regular intervals. The time step was chosen to be one minute. The surface heat exchange rate k may be determined by $k = E/(\rho\lambda H_i)$, where $E =$ surface heat exchange coefficient, $\rho =$ sea water density and $\lambda =$ specific heat of sea water.

Typical longitudinal and lateral cross-sections through the thermal plume can be seen in Figures 1 and 2 for a moderate cross-current of -0.08 m/s. Only the region which exhibits a temperature rise exceeding 3°C is shown in Figures 1 and 2, with the longitudinal section being taken through the plume centre line and the lateral cross-section being taken at 1500 metre from the end of the outfall channel where the plume starts. Various plume paths from the thermal outfall channel with the cross-current varying from −0.15 m/s to 0.25 m/s can be seen in Figure 3.

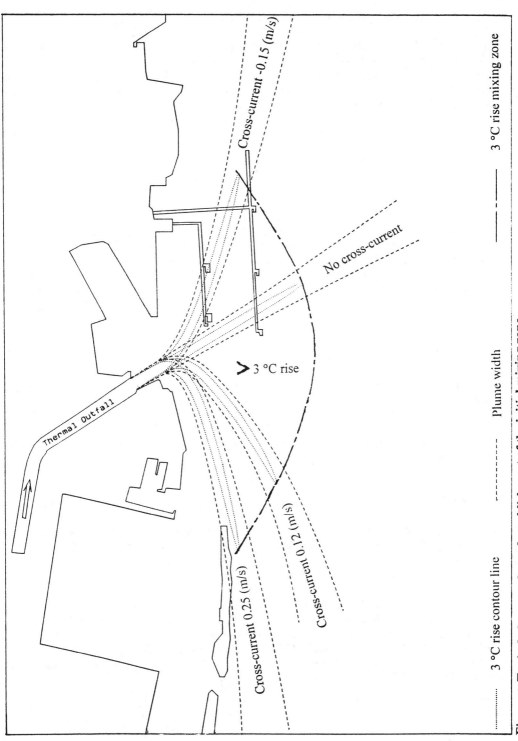

Figure 3 Typical plume paths and establishment of the initial mixing zone

Thermal Outfall

Cross-current -0.15 (m/s)

No cross-current

3 °C rise

Cross-current 0.12 (m/s)

Cross-current 0.25 (m/s)

.... 3 °C rise contour line

– – – – – Plume width

————— 3 °C rise mixing zone

271

4. ESTABLISHMENT OF THE INITAL MIXING ZONE

A computer program has been written to obtain the surface plume characteristics described above due to the complexity of the analytical solutions. The plume path was found to be very sensitive to the cross-current. This is because of the advection of the plume due to the current. For each plume path, the 3°C rise contour and the plume width lines are also shown in Figure 3. By joining together the far most point of the 3°C rise contour for each plume path, the initial mixing zone for the specified temperature rise can be established as shown in Figure 3. In interpretation of the results, it must be stressed that locations within the initial mixing zone will experience a temperature rise of 3°C temporarily as the plume of warm water passes through the location. Nevertheless such information on the initial mixing zone with regard to its area and location resulting from the thermal discharge is useful for the environmental impact assessment.

5. SENSITIVITY STUDY

A number of sensitivity tests have been carried out, including the eddy viscosity coefficient, the surface heat exchange coefficient, the cross-current pattern, the initial depth of the plume and the initial plume momentum effects. It has been found that the eddy viscosity coefficient is the most sensitive parameter to affect the area of the initial mixing zone, with the mixing zone boundary being about 400 metre further away from the outfall channel if the lower bound value of $\alpha = 0.4$ is used and about 250 metre closer to the outfall channel if the upper bound value of $\alpha = 0.8$ is used. The effects on the initial mixing zone area by other parameters tested were found to be insignificant.

6. CONCLUSIONS

The previously developed surface plume model, based upon variable plume speeds and a general diffusion coefficient relationship, has been further improved and refined to include the vertical mixing process. The refined plume model has been successfully applied to a thermal plume study, showing great potential in establishing the initial mixing zone area for a specified threshold value of temperature rise. Through the sensitivity tests, it has been found that the eddy viscosity coefficient is the most sensitive parameter to affect the area of the initial mixing zone.

7. REFERENCES

[1] Brooks, N.H. (1960), 'Diffusion of Sewage Effluent in an Ocean Current', Proc. 1st International Conference on Waste Disposal in Marine Environment, University of California, (ed. E.A. Pearson), Pergamon Press, pp246-267.

[2] Tchobanoglous, G. and Burton, F.L. (1991), Wastewater Engineering: Treatment, Disposal, and Reuse, 3rd edition, MacGraw-Hill.

[3] Chen, Y. and Thompson, G. (1995), "Mathematical Modelling of Surface Plumes in Oceanic Waters with Varying Velocities and Generalised Diffusion Coefficients", Proceedings of IAHR XXVI Congress HYDRO 2000, Vol. 2, *Industrial Hydraulics and Multi-phase Flows*, ed. M.A. Leschziner, pp52-57.

[4] Pritchard, D.W. (1960), "The Movement and Mixing of Contaminants in Tidal Estuaries", Proceedings of 1st International Conference on Waste Disposal in Marine Environment, University of California, Pergamon Press, pp512.

2   Transport and mixing processes
2.1 Turbulent shear flows

Environmental Hydraulics, Lee, Jayawardena & Wang (eds) © 1999 Balkema, Rotterdam, ISBN 90 5809 035 3

Keynote lecture: Turbulent transport and mixing in a stratified lake

J. Imberger
Centre for Water Research, University of Western Australia, Nedlands, W.A., Australia

ABSTRACT : Over the last 10 years we have been carrying out detailed measurements of the turbulent transport in stratified lakes. These include measurement in the surface layer, the subsurface layer, the epilimnion, the metalimnion, the hypolimnion and the benthic boundary layer. An overall picture has now emerged as to the energy transfer pathways from the wind to the turbulence in the water column. Further, not only has the energy flux path been clarified, but also the efficiency of mixing and the role of the irreversible and reversible fluxes is beginning to become clearer. These findings have major consequences for the biological and chemical processes operating in the various parts of the water column. The talk will aim to provide a quantitative overview of the flux path, the buoyancy flux partitioning and the implications of these for the biological and chemical processes. As an illustration results from a numerical simulation will be shown.

Environmental Hydraulics, Lee, Jayawardena & Wang (eds) © 1999 Balkema, Rotterdam, ISBN 90 5809 035 3

Invited lecture: Setting the stage: Large eddy simulation of laboratory scale flows*

Robert L. Street
Environmental Fluid Mechanics Laboratory, Stanford University, Calif., USA

ABSTRACT: Large eddy simulations are shown to produce quantitatively accurate results for complex flows at a laboratory scale. Preliminary results from an extension of these simulations to unsteady and three-dimensional sediment transport are given. The bases for further extensions to field scale are presented, and, thus, the stage is set for this next step.

1. MOTIVATION

Our approach is motivated by the concept that our simulations should contain as few approximations as possible and should be true to the assumptions underlying the simulation concepts. We have chosen large eddy simulation [LES] because, in the long run, it offers promise of being able to simulate field-scale flows accurately. At present, to simulate field scales, one must make significant approximations in addition to the assumptions inherent in the LES. For example, two underlying bases for LES are that the unresolved [SGS] fluid motions do not contain a significant portion of the flow energy and that the unresolved motions can be accurately modeled. At field scale this is not currently possible, and in addition crude approximations need to be made to the boundary conditions at solid surfaces [Cederwall and Street, 1997 & 1999]. Accordingly, we have undertaken a series of simulations of laboratory-scale flows in which we can meet all of the requirements of LES (Zang et al, 1994; Calhoun, 1998; Cui and Street, 1997&1998; Hodges, 1997; Yuan, 1997; Ding et al., 1998), including proper representation of the flow near boundaries. We have demonstrated that these simulations yield quantitatively accurate results. This paper reports on two sets of simulations and describes how these simulations, together with a fruitful means of describing the unresolved [SGS] motions proposed by Shah and Ferziger [1995] and Ferziger [1997], set the stage for accurate, unsteady, and three-dimensional simulation of sediment transport and the extension of the simulation scale to field values. In particular we examine the work of Calhoun (1998) on flow over complex bedforms and Hodges (1997) on turbulent channel flow beneath finite-amplitude water waves, these works providing the essential features needed for simulation of sediment transport beneath surface waves and over complex bedforms. A preview is given of the sediment transport scenario and the calculations currently underway (Zedler et al. ,1998).

2. LARGE EDDY SIMULATION

2.1 Basics and governing equations

The LES equations are derived by spatially filtering the instantaneous equations, and it is intended that all of the resolved motions will be fully and accurately simulated. Accordingly, the governing equations given here are the spatially-filtered, constant density, incompressible form for Cartesian [physical space] variables and [for simplicity] in Cartesian coordinates. Here, u_i represents the fluid velocity components, while Q is any scalar quantity, e.g. sediment concentration. The settling velocity w_s for the sediment is taken to be a constant. In addition [Hodges and Street, 1996; Hodges, 1997], for the free surface case a kinematic and dynamic boundary condition pair is required and a grid acceleration term appears in the momentum equations to account for the moving, wave-following and curvilinear grid. The details of mapping these equations to the curvilinear, boundary-following grid are presented in

---

* The simulations described herein are from the doctoral research efforts of Drs. R. J. Calhoun [currently at Stanford University] and B. R. Hodges [currently at the Centre for Water Research, U. West. Australia] and of E. A. Zedler [Research Assistant, Environmental Fluid Mechanics Laboratory, Stanford U.].

Hodges [1997] or Calhoun [1998]. The τ_{ij} and χ_j terms represent, respectively, the subgrid-scale stress and scalar flux. These terms cannot be calculated and must be modeled.

$$\frac{\partial \overline{u}_i}{\partial x_j} = 0 \tag{1}$$

$$\frac{\partial \overline{u}_i}{\partial t} + \frac{\partial}{\partial x_j}\left(\overline{u}_i\overline{u}_j\right) = -\frac{\partial \overline{p}}{\partial x_i} + \nu \frac{\partial^2 \overline{u}_i}{\partial x_j \partial x_j} - \frac{\partial \tau_{ij}}{\partial x_j} \qquad\qquad \tau_{ij} = \overline{u_i u_j} - \overline{u}_i\overline{u}_j \tag{2;3}$$

$$\frac{\partial \overline{Q}}{\partial t} + \frac{\partial}{\partial x_j}\left([\overline{u}_j - \delta_{i2}w_s]\overline{Q}\right) = D_Q \frac{\partial^2 \overline{Q}}{\partial x_j \partial x_j} - \frac{\partial \chi_j}{\partial x_j} \qquad\qquad \chi_j = \overline{Qu_j} - \overline{Q}\overline{u}_j \tag{4;5}$$

2.2 The LES code

Equations 1 through 5 are the governing LES equations. The equations are transformed into a boundary-conforming and general curvilinear coordinate system, using 3DGRAPE/AL (Sorenson and Alter, 1996) and SRAP (NASA Chimera Grid Tools). Both Coriolis terms and a scalar equation [e.g., for density, sediment concentration, or temperature] are implemented. Our full formulation has demonstrated its ability to describe (1) ocean upwelling experiments [Zang (1993) & Zang and Street (1995b)] and (2) dynamics of a jet in a crossflow (Yuan, 1997, and Yuan et al., 1998). This code has been extended to operate in massively-parallel mode on the IBM SP-2 or SGI Origin 2000, using the MPI [message passing interface] software (Cui and Street, 1997&1998).

Solutions of the incompressible Navier-Stokes equations are obtained by employing a co-located finite volume method [Zang et. al. (1994)]. A fractional step projection is used in which the velocities at the cell centers are advanced without the pressure term. A pressure-Poisson equation [PPE] is then obtained from the intermediate velocities and solved using a multigrid method. The resulting pseudo-pressure is used to correct the intermediate velocities to their mass conserving projection. The code employs second-order accurate differencing in space and time, but due to the projection step, is actually only first-order accurate in time.

The ability to subdivide a domain into overlapping component grids adds flexibility to the grid system and prevents grid cells from becoming either excessively skewed or stretched in convoluted geometries. In our approach, the momentum and transport equations are solved independently on each subdomain, and the pressure is computed simultaneously on the entire flow field. The multigrid technique is coupled with an intergrid iteration method to solve the pressure Poisson equation over composite domains [Zang and Street (1995a); Yuan and Street (1996 & 1997)].

2.3 SGS models

Dynamic subgrid-scale models use information from the smallest resolved scales to derive locally accurate parameters for modeling the turbulence. Our basic SGS model has been the dynamic mixed model (DMM) of Zang et al. (1993). Salvetti and Banerjee (1995) improved on that model. Salvetti et al. (1997) describes large-eddy simulations of decaying turbulence in an open channel, using different dynamic subgrid-scale models, viz. the dynamic model of Germano et al. (DSM), the DMM, and the dynamic two-parameter model of Salvetti and Banerjee (DTM). These models were incorporated into the Zang et al. (1994) finite-volume, Navier-Stokes-equation solver. The addition of the second model coefficient in the DTM improves the agreement with direct simulations. When the DSM was used, significant discrepancies are observed between the large-eddy and the direct simulations.

Through the use of variable grid spacing and the behavior of the SGS model itself, the representation of the flow transitions properly as a solid boundary is approached to yield direct numerical simulation accuracy in the direction normal to the boundary. Thus, single and two parameter dynamic models are significant improvements over models which must use *ad hoc* damping functions to achieve correct behavior near solid boundaries. This assures us of the kind of resolution that we need to properly treat the sediment transport there.

2.4 Implementation of a free surface and a variable bed geometry

In these implementations we have worked with flows that are periodic in the streamwise and spanwise directions and that have the appropriate boundary conditions applied on the bed and on the upper surface. In both cases, an appropriate mean pressure gradient was applied to drive the flows. In the free surface case the coordinate system was put in motion to follow the waves; in the variable bed case the flow was a channel flow with no-slip conditions on the bed and channel top.

A free surface version

Hodges (1997){See also: Hodges, et al. (1996) and Hodges and Street (1996)} focuses on the kinematics of the interaction between surface waves and a turbulent current. This research was directed at increasing our understanding of the processes that occur where surface waves and turbulent undercurrents interact. The primary motivation was two-fold: (1) to investigate the physical processes of wave turbulence interaction in the near-surface region that affect both the "signature" of the turbulence at the free surface and the mixing beneath the surface, and (2) to develop a numerical simulation method that can be used for future investigations of free-surface phenomena where finite-amplitude waves, turbulence, and structures interact with viscous and nonlinear effects. In this work, we showed that the turbulence in the near-surface region is enhanced by the interactions between non-breaking waves and a turbulent shear current.

The simulations are large-eddy simulations [LES] and employ the dynamic two-parameter model described above. Hodges (1997) includes a detailed study of the numerical method developed for a time-accurate simulation of an unsteady turbulent free-surface flow in three space-dimensions with a progressive, finite-amplitude, free-surface wave. The simulation uses the time-dependent Navier-Stokes equations with the nonlinear kinematic and dynamic free-surface boundary conditions. A new treatment of the kinematic boundary condition in mapped space is presented and a new grid is generated at each time step with a Poisson equation method adapted from the 3DGRAPE/AL code. Comparisons with experimental studies [Cowen (1996) & Cowen and Monismith (1995 & 1997)] validate the method for turbulent free-surface flows and provide insight into the flow behavior seen in laboratory flumes. Visualization of instantaneous and phase-averaged flow variables provides insight into the dynamics of the wave-turbulence interactions.

Figure 1 shows the nondimensional flow domain and the grid for case of a single finite-amplitude wave and a wave-following coordinate system. This simulation was performed at a shear Reynolds number $Re_\tau = 171$ with a resolution of 32 x 32 x 64 grid points. The surface and bottom boundary layers [Fig. 2] are well resolved, while the wall-streak spacing [Fig. 3] is on the order of 170 wall units. This is greater than that found in DNS or experiments, but typical of LES and dependent on the spanwise grid spacing. Figure 2 is a plot of the instantaneous velocity in a streamwise/vertical plane on the channel centerline; the shading indicates the spanwise velocity component variation. Figure 3 shows a spanwise/vertical plane of instantaneous data beneath the crest in the wave/current simulation; the shading indicates the streamwise velocity component variation. Both Figs. 2 and 3 reflect clear three-dimensional motions that are significantly wave-influenced as comparison with the current-only case shows (Hodges, 1997).

From this work several conclusions can be drawn: (1) simulation of three-dimensional, unsteady

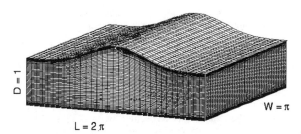

FIG. 1. Initial computational domain for turbulent channel flow with a wave

279

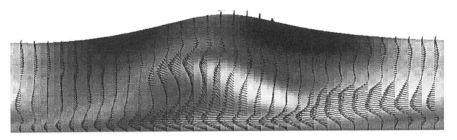

FIG. 2. Instantaneous velocity fluctuations in a streamwise/vertical plane for wave/current flow

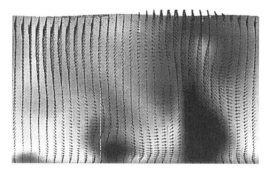

FIG. 3. Instantaneous velocity fluctuations on a spanwise/vertical plane for wave/current flow

turbulent free-surface flows is practical using our LES method; (2) a finite-amplitude surface wave can cause vertical stirring of the fluid where a surface wave propagates over a current with a strong shear; (3) turbulence in the near-free surface region is enhanced by interaction of the wave straining field acting on turbulent structures; and (4) rapid distortion of the turbulence by the wave straining field plays an important role in the wave/turbulence interactions.

Variable Bedforms

Calhoun and Street (1997) and Calhoun (1998) reported on the LES of a laboratory-scale channel flow. Both a wavy and a three-dimensional bed (Fig. 4) were studied. In some simulations, the solver and SGS model discussed above were used with two non-aligned, overlapping component grids. This domain decomposition allows fine-scale resolution near the wavy bottom and coarse grid scales in the interior of the flow. The quantitative agreement with available experimental data for flow over the wavy bed was excellent. The simulation results shown in Figs. 4 and 5 were obtained at a bulk Reynolds number $Re = 7400$ [based on bulk velocity and mean channel depth] and with a single grid with a resolution of 82 x 98 x 82 grid points. The wave length is 51 cm and the wave height is 5 cm. The ability to give detailed estimates of the friction and pressure drag on the bed surfaces was clearly demonstrated (e.g., see Fig. 5). The difference between a bed composed of plane waves (i.e., straight ripples) and a bed composed of a three-dimensional surface (somewhat like sinuous ripples) was demonstrated. The drag on the straight ripples was dominated by the pressure drag component. The viscous drag was dominant on the sinuous ripples, and high speed streaks were observed between ripple crests.

3. PRELIMINARY RESULTS FOR SEDIMENT TRANSPORT OVER A WAVY BEDFORM

The scalar equation in the code of Calhoun (1998) has been modified by Zedler (Zedler and Street, 1998) to treat heavy sediment, with particle to fluid density ratios of order unity [actually 2.5 in this case]. The sediment is treated here as a passive scalar, with a dilute concentration on the order of 0.001. The sediment is assumed not to affect the fluid motion. However, gravitational effects on the sediment are significant, so the settling velocity is subtracted from the vertical fluid velocity in the

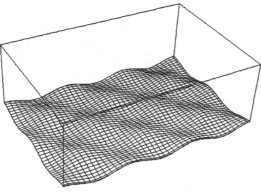

Figure 4. Channel flow over bedform composed of two sinusoids at right angles. Wave height/wave length = 0.1

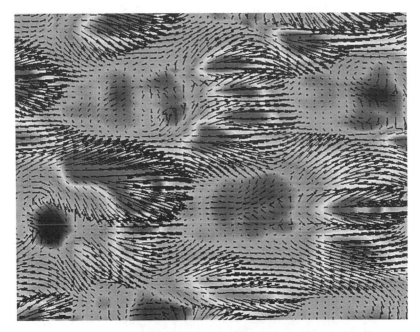

Figure 5. Aerial [plan] view of velocity vectors very close to surface of Fig. 4 with shear stress patterns superposed in shading. Notice close correlation of positive stresses [white; near crests] and negative stresses [black; near troughs] to velocity directions and intensities

advection term as shown in Eq. 4. Van Rijn's (1993) expression for the settling velocity for non-spherical sand grains with diameters in the 100-1000 micrometer range is used. An unsteady form of Van Rijn's (1984) pickup function is employed on the bottom boundary to specify the sediment concentration gradient at the bed. The required Shields' parameter is calculated from the local bottom shear stress, evaluated at each time step and at each bedform grid point; Van Rijn's (1993) expression is used for the critical Shields' parameter which is expressed [following Yalin (1977)] in terms of the nondimensional particle diameter D^*.

The simulation has been carried out on a laboratory-scale flow and is three-dimensional and unsteady. In Eq. 4, Q becomes the sediment concentration by volume fraction and for this illustration, the flow parameters for sediment transport are the same as those used in the wavy bed simulations reported above, except that a rigid, but free-slip, upper boundary condition and a single grid with 130 x 98 x 50 grid points are used on the channel. The flow domain is shown in Fig. 6.

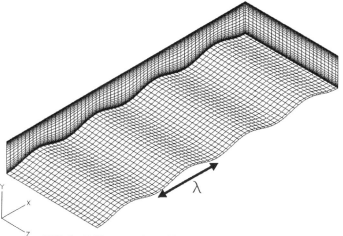

FIG. 6. Grid system for sediment transport case

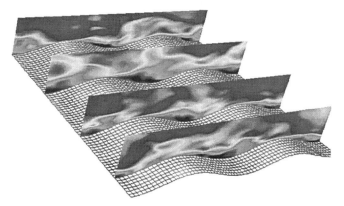

FIG. 7. Contours of the streamwise component of velocity

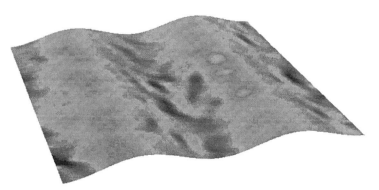

FIG. 8. Instantaneous contours of shear stress at the bed

To assist in assessing the sediment transport results, two plots from an equivalent fluid flow simulation by Calhoun (1998) are shown in Figs. 7 and 8. Figure 7 shows vertical planes of instantaneous streamwise velocity contours to illustrate the three-dimensional nature of the flow, recirculation areas in the lee of the wave crests, etc.; Fig. 8 illustrates the shear-induced (viscous) drag at the same instant.

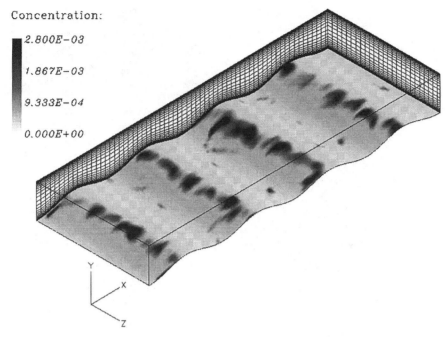

FIG. 9. Contours of instantaneous sediment concentration at the bed

Figures 9 and 10 show two views of the sediment concentration. Figure 9 shows the instantaneous sediment concentration at the bed; note the clear correlation with the bed stresses, particularly the concentration peaks near the crests and the so-called "spotted carpet effect" [cited by Nielsen (1992) and caused by the horizontally non-uniform sediment concentrations]. Figure 10 shows iso-contours of the sediment in the channel flow at an instant. The inclined "plumes" of sediment are consistent with streamwise vortices found by Calhoun and illustrate clearly the ability of this method to capture the four-dimensional details of the flow.

4. THOUGHTS ON THE FUTURE

The ultimate future will see a field-scale simulation of sediment transport; the project is underway! Three key elements will allow us to carry out the effort, namely, synthesis of the components described above, computer capacity and speed, and an improved and entirely rational model for the unresolved scales.

4.1 Synthesis of components

To simulate coastal sediment transport realistically it will be necessary to join the code components that treat the water waves and free surface to those that handle the complex bottom topography. There are no new principles to be developed to accomplish this, but the coding and code optimization effort requirement is large.

4.2 Computer performance

In Cui and Street (1998), we demonstrate that the computer capacity and speed exist to do large-scale simulations. There we report on the simulation of buoyant flow on a "continental" shelf in a rotating tank. The calculations have been carried out on an SGI Origin 2000 system using 40 processors and 11 million grid points; the code speed is about 1.6 GFLOPS.

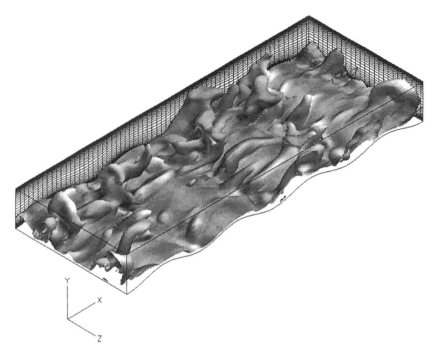

FIG. 10. Iso-contours of the sediment concentration in the flow: $Q = \sim 1e\text{-}11$

4.3 Modeling the unresolved scales

There are, of course, many models for the unresolved scales. However, for non-spectral codes and anisotropic flows, the most widely-used and validated models for the unresolved scales in LES have three common characteristics. First, they are dynamic mixed models (e.g., Zang et al., 1993 & Salvetti and Banerjee, 1995), employing a modified Leonard term [also known as the Bardina scale-similarity term] to capture back-scatter, and a term [usually a variant of the Smagorinsky term] to capture dissipation. Second, the dissipation term is proportional to the square of the spatial filter size, which is almost invariably equal to the grid spacing [hence, the term "subgrid-scale" model]. Third, the numerical computation scheme in curvilinear coordinates is likely to be second-order accurate in the grid spacing. As a consequence, there is some question as to whether one should retain the SGS term, but neglect the truncation error terms in the numerical scheme; note that the modified Leonard term is of order unity.

Shah and Ferziger (1995) have provided a new model for the unresolved scales, which requires no adjustable constants, adds little to the computational cost of the base code, and produces results for shear flows in a channel which are better than those from the mixed models and Smagorinsky model. In addition, our implementation of this Shah and Ferziger model shows that it successfully reproduces the results in more complex flow situations that we have studied with mixed models, but at a lower cost. In the process of providing insight to the bases for this new Shah-Ferziger non-eddy viscosity model for the subgrid-scale stresses, Ferziger (1997) has produced a simple means for generating the stress tensor for the unresolved motions and for separating the size of the grid from the size of the spatial filter. In Shah and Ferziger [1995] and Ferziger [1997], we find the essence of a rational approach to LES and the creation of unresolved-motion models. In addition, they demonstrate that one of the most effective terms [the Leonard or Bardina term] in current SGS models is a natural consequence of a formal mathematical approach.

Following Ferziger (1997), the basic idea is to expand the actual fluid velocities in a Taylor series about a point x and introduce this expansion into Eq. 3. For this derivation, the filter is assumed to be a top hat symmetric filter of the form [here $x = \{x_i\}$]

$$\overline{q}(\mathbf{x}) = \frac{1}{\Delta^3} \int_{x_3 - \Delta/2}^{x_3 + \Delta/2} \int_{x_2 - \Delta/2}^{x_2 + \Delta/2} \int_{x_1 - \Delta/2}^{x_1 + \Delta/2} q(\mathbf{x}')d\mathbf{x}' \tag{6}$$

284

where q is any variable and Δ is the filter size. The Taylor series for the velocity is

$$u_i(\mathbf{x}') \approx u_i(\mathbf{x}) + (x_j' - x_j)\frac{\partial u_i}{\partial x_j} + \frac{(x_j' - x_j)(x_k' - x_k)}{2}\frac{\partial^2 u_i}{\partial x_j \partial x_k} + \dots \tag{7}$$

Substituting Eq. 7 into Eq. 6 yields

$$\bar{u}_i(\mathbf{x}) \approx u_i(\mathbf{x}) + \frac{\Delta^2}{24}\nabla^2 u_i(\mathbf{x}) + O(\Delta^4) \tag{8}$$

Inversion by successive substitution yields

$$u_i(\mathbf{x}) \approx \bar{u}_i(\mathbf{x}) - \frac{\Delta^2}{24}\nabla^2\bar{u}_i(\mathbf{x}) + O(\Delta^4) \tag{9}$$

Substituting Eq. 9 into Eq. 3 produces

$$\tau_{ij} = \overline{u_i u_j} - \bar{u}_i \bar{u}_j \approx \left[\overline{\bar{u}_i \bar{u}_j} - \bar{\bar{u}}_i \bar{\bar{u}}_j\right] - \frac{\Delta^2}{24}\left[\overline{\bar{u}_i \nabla^2 \bar{u}_j} + \overline{\bar{u}_j \nabla^2 \bar{u}_i} - \left(\bar{\bar{u}}_i \nabla^2 \bar{\bar{u}}_j + \bar{\bar{u}}_j \nabla^2 \bar{\bar{u}}_i\right)\right] + O(\Delta^4) \quad (10)$$

Similarly, expanding Q in a Taylor series and using the same procedure gives

$$\chi_j = \overline{Q u_j} - \bar{Q}\bar{u}_j \approx \left[\overline{\bar{Q}\bar{u}_j} - \bar{\bar{Q}}\bar{\bar{u}}_j\right] - \frac{\Delta^2}{24}\left[\overline{\bar{Q}\nabla^2 \bar{u}_j} + \overline{\bar{u}_j \nabla^2 \bar{Q}} - \left(\bar{\bar{Q}}\nabla^2 \bar{\bar{u}}_j + \bar{\bar{u}}_j \nabla^2 \bar{\bar{Q}}\right)\right] + O(\Delta^4) \quad (11)$$

All of the terms on the RHS of the $\approx$ in Eqs. 10 and 11 may be calculated in terms of the resolved quantities; the version of Eq. 10 implemented by Shah and Ferziger (1995) provides both the correct backscatter and the correct dissipation. Because the lead term in Eq. 10 is the modified Leonard or the Bardina scale-similarity term, which is known to properly represent the backscatter, it is not surprising that this model is successful. Its most important characteristic is that it requires no *ad hoc* terms or expressions (e.g., the Smagorinsky term) as are normally required in models of the unresolved scales and no dynamic procedure is needed. The representations expressed by Eqs. 10 and 11 appear to be an ideal basis for simulations of larger-scale flows since there are also no *a priori* assumptions about the character of the flow (cf., the usual isotropic and inertial subrange assumptions for SGS models). Finally, since this approach is computationally cheap, it is logical to implement it with the filter scale equal to twice the grid scale, thus rendering the second terms in Eqs. 9 and 10 approximately 4 times larger than the truncation error in a second-order accurate finite-volume discretization of the Navier-Stokes equations.

5. SUMMARY

Two large eddy simulations illustrated the quantitatively accurate results produced for complex flows at a laboratory scale and related to sediment transport. Preliminary results from an extension of these simulations to unsteady and three-dimensional sediment transport were given. Synthesis of the water-wave and wavy-bottom code components, computer performance in a parallel mode, and a new modeling technique for the unresolved scales are the bases for further extensions to field scale sediment transport. Thus, the stage appears to be set for the next step in the four-dimensional simulation of sediment transport, namely, accurate simulation of realistic small- and larger-scale field conditions.

6. REFERENCES

Calhoun, R.J. and Street, R.L. (1997). "A comparison of turbulent flow over two- and three-dimensional sinusoidal topographies," *12th Symp. on Boundary Layers and Turb.*, AMS, pp. 528-529.

Calhoun, R.J. (1998). "Numerical investigations of turbulent flow over complex terrain," Ph. D. Diss., Dept. Civil&Env. Engr., Stanford University, 274 pages. [http://www-leland.stanford.edu /~calhoun/Wavy.html].

Cederwall, R.T. and Street, R.L. (1997). "Use of a Dynamic Subgrid-Scale Model for Large-Eddy Simulation of the Planetary Boundary Layer," *12th Symp. B. Layers & Turb.*, AMS, pp. 215-216.

Cederwall, R.T., and Street, R.L. (1999). "Turbulence modification in the evolving stable boundary layer: a large-eddy simulation.," *13th Symp. B. Layers & Turb.*, AMS, in press.

Cowen, E.A. (1996). "An experimental investigation of the near-surface effects of waves traveling on a turbulent current," Ph. D. Diss., Dept. Civil Engineering, Stanford University, 175 pages.

Cowen, E.A., and Monismith, S.G. (1995). "Digital particle tracking velocimetry measurements very near a free surface," *Air Water Gas Transfer*, pp. 135-144.

Cowen, E.A., and Monismith, S.G. (1997). "A hybrid digital particle tracking velocimetry technique," *Experiments in Fluids*, 22, 3, pp. 199-211.

Cui, A. and Street, R.L. (1997). "Parallel Computing of Upwelling in a Rotating Stratified Flow ," *1997 Fluid Dynamics Division Meeting*, APS, San Francisco.

Cui, A., and Street, R.L. (1998). "Initial instabilities in a rotating convective flow - a large-eddy simulation," *Eos, Trans.*, AGU, 79, Fall Meeting Abstr., in press.

Ding, L., Calhoun, R.J., and Street, R.L. (1998). "Numerical simulation of stratified flow over a three-dimensional hill," *Eos, Trans.*, AGU, 79, Fall Meeting Abstr., in press.

Ferziger, J.H. (1997). "Extended scale similarity model," *Personal Note*, 1 page.

Hodges, B.R. (1997). "Numerical simulation of nonlinear free-surface waves on a turbulent open-channel flow," Ph. D. Diss., Dept. Civil Engineering, Stanford University, 235 pages.

Hodges, B.R., Street, R.L., and Zang, Y. (1996). "A method for simulation of viscous, nonlinear, free-surface flows," *20th Symp. Naval Hydrodynamics*, pp. 791-809.

Hodges, B.R., and Street, R.L. (1996). "Three-dimensional, nonlinear, viscous wave interactions in a sloshing tank," *Proc. Fluids Engineering Summer Meeting*, 3, FED-Vol. 238, ASME, pp. 361-367.

Nielsen, P. (1992). *Coastal bottom boundary layers and sediment transport*, World Scientific Pub. Co.

Salvetti, M.V., and Banerjee, S. (1995). "*A priori* tests of a new dynamic subgrid-scale model for finite-difference large-eddy simulations," *Physics of Fluids*, 7, pp. 2831-2847.

Salvetti, M.V., Zang, Y., Street , R.L., and Banerjee, S. (1997). "Large-eddy simulation of free-surface decaying turbulence with dynamic subgrid-scale models," *Physics of Fluids*, 9, 8, pp. 2405-2419.

Shah, K.B., and Ferziger, J.H. (1995). "A new non-eddy viscosity subgrid-scale model and its application to channel flow," *Cen. Turb. Res. Ann. Res. Briefs*, NASA-Ames & Stanford U. Jt. Cen., pp. 73-90.

Sorenson, R.L., and Alter, S.J. (1996). "3DGrape/AL: the Ames/Langley technology upgrade," *5th Intl. Conf. on Numerical Grid Gen. in Comp. Field Simulations*, pp. 343-352.

Van Rijn, L.C. (1993). *Principles of sediment transport in rivers, estuaries and coastal seas*, Aqua Pubs.

Van Rijn, L.C. (1994). "Sediment pickup functions," *J. Hydr. Engr.*, ASCE, 110 (2), pp. 1494-1502.

Yalin, M.S. (1977). *Mechanics of sediment transport*, 2nd. Ed., Pergamon Press.

Yuan, L., and Street, R. L. (1996). "Large eddy simulation of a jet in crossflow," *Proceedings of the ASME Fluids Engineering Division, ASME*, FED-Vol. 242, pp. 253-260.

Yuan, L., and Street, R. L. (1997). "Using domain decomposition with co-located finite volume methods: two problems and two solutions," *Proc. ASME Fluids Engr. Div.*, FEDSM97-3645, pp. 1-8.

Yuan, L., Street, R.L., and Ferziger, J.H. (1998). "Large eddy simulations of a round jet in a crossflow," *J. Fluid Mech.*, in press.

Zang, Y. (1993). "On the development of tools for the simulation of geophysical flows," Ph.D. Diss., Dept. Civil Engineering, Stanford University, 224 pages.

Zang, Y., Street, R.L., and Koseff, J.R. (1993). "A dynamic mixed subgrid- scale model and its application to turbulent recirculating flows," *Physics of Fluids A*, Vol. 5, No. 12, 3186-3196.

Zang, Y., Street, R.L., and Koseff, J.R. (1994). "A non-staggered grid, fractional step method for the incompressible Navier-Stokes equations in curvilinear coordinates," *J. Comp. Physics*, 114, 18-33.

Zang, Y. and Street, R.L. (1995a). "A composite multigrid method for calculating unsteady incompressible flows in geometrically complex domains," *Int'l. J. Num. Meth. Flds.*, 20, 5, 341-361.

Zang, Y., and Street, R.L. (1995b). "Numerical simulation of coastal upwelling and interfacial instability of a rotating and stratified fluid," *Journal of Fluid Mechanics*, 305, pp. 47-75.

Zedler, E..A., Calhoun, R.J., and Street, R.L. (1998). "Large-eddy simulation of sediment transport over a wavy bedform," *Eos, Trans.*, AGU, 79, Fall Meeting Abstr., in press.

Environmental Hydraulics, Lee, Jayawardena & Wang (eds) © 1999 Balkema, Rotterdam, ISBN 90 5809 035 3

Invited lecture: Large scale flow structures and mixing processes in shallow flows

Gerhard H. Jirka

Institute for Hydromechanics, University of Karlsruhe, Germany

ABSTRACT: Shallow turbulent flows occurring in wide rivers, estuaries, lakes or coastal regions, as well as the atmosphere, are readily susceptible to transverse disturbances that lead to two-dimensional coherent structures. The shallow jet, the shallow wake and the shallow mixing layers are examples of such flow patterns. Three types of generation mechanisms are proposed for these flows. The large-scale coherent structures greatly influence the mixing and transport of pollutants that are released into such flows. They may also play a crucial role in transverse momentum exchange that controls friction in wide channel flows.

1. INTRODUCTION AND DEFINITION

Shallow flows are ubiquitous in nature. We define shallow flows as predominantly horizontal flows in a fluid domain for which the two horizontal dimensions greatly exceed the vertical dimension. Flows in wide rivers, in estuaries and coastal waters, in shallow lakes or in the upper mixed layer of deep stratified lakes or reservoirs are important examples of hydraulic or environmental engineering concern. Yet larger scale geophysical flows in the shallow atmosphere surrounding the earth or in ocean basins - in both cases with or without the influence of stratification that further enhances layer formation - are also in that category.

Attention is limited herein to predominantly one-dimensional shallow flows (without significant curvature effects as in river bends) that are at scales below the Rossby radius (thus, free of Coriolis force influences) and that are fully turbulent. The turbulence condition is measured by a depth Reynolds number $Re_h = UH/\nu$, in which U is the characteristic velocity, H the flow depth and ν the kinematic viscosity, that is sufficiently greater than 10^3. This is readily satisfied by the real-world flow, but may pose some difficulties in laboratory experimentation. The base flow is governed by wall turbulence, produced by the shear effect at the solid bottom. The structure of this turbulence is 3-D produced by ejection and sweep events near the solid boundary and characterized in the mean by a logarithmic-law velocity profile. Some types of coherent turbulent structures appear to be present in this flow (for a review see Nezu and Nakagawa, 1993) but the length scale of these vortical elements is of the order of or less than the water depth and their axes are aligned in the mean flow direction. Thus, these structures still represent 3-D turbulence quite distinct from the features discussed in the following.

Whenever such shallow flows - that may be uniform and wide in both horizontal directions - are subjected to localized or distributed disturbances they tend to undergo internal oscillations that grow into large-scale instabilities characterized by coherent structures. This is exemplified by Fig. 1 in the field or laboratory. Large-scale vortical structures are generated downstream of relatively small island obstacles. These structures are essentially two-dimensional, i.e. their horizontal extent ℓ is much larger than the depth H, $\ell/H \gg 1$, they have vertically aligned vorticity vectors and the distribution of turbulent kinetic energy among smaller scales (but still larger than H) is governed by the laws of "2-D turbulence".

In the following, a classification is developed for the generation mechanisms that produce large-scale 2-D coherent structures in shallow flows. Furthermore, recent results for two types of transversely sheared flows, namely the shallow wake and the shallow jet, are presented elucidating the role of the coherent structures in affecting momentum and mass transfer (mixing).

(a)

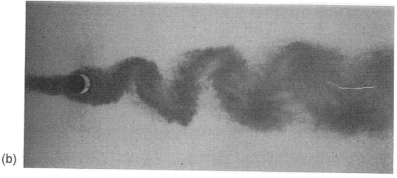

(b)

Fig. 1:   Shallow flow containing large-scale coherent structures consisting of 2-D vortical elements. a) Flow downstream of small island on Great Barrier Reef, water depth 10 to 20 m, horizontal extent several tens of km (photograph courtesy of E. Wolanski), b) Laboratory simulation, water depth 3 cm, cylinder diameter 17 cm, horizontal extent 3.5m (from Chen and Jirka, 1995). Flow is from left to right.

2. CLASSIFICATION OF 2-D COHERENT STRUCTURES IN SHALLOW FLOWS

2-D coherent structures (2DCS) are defined herein as connected, large-scale turbulent fluid masses that extend uniformly over the full water depth and contain a phase-correlated vorticity (with the exception of a thin near-bottom boundary layer). This definition is an adaptation of Hussain's (1983) definition for general (3-D) coherent structures.

The vorticity contained in 2DCS emanates from the initial transverse shear that has been imparted on these flows during their initial generation. Accordingly one can define three types of **generation** mechanisms for 2DCS, listed in order of their strength:

Type A: Topographical forcing: This is the most severe generation mechanism in which topographic features (islands, headlands, jetties, groynes etc.) lead to local flow separation, formation of an intense transverse shear layer and return velocities in the lee of the feature. An example of this flow type is given in Fig. 2, in which the growth of 2DCS within the shear layer that starts at the separation point can be clearly seen.

Type B: Internal transverse shear instabilities: Here velocity variations in the transverse directions that exist in the shallow flow domain give rise to the gradual growth of 2DCS. Such lateral velocity variations can be caused by a number of causes: due to source flows representing fluxes of momentum excess or deficit (shallow jets, shallow mixing layers, shallow wakes) or due to gradual topography changes or roughness distributions (e.g. flow in compound channels). Fig. 3 shows a shallow jet entering a laboratory basin in which the flow takes on a meandering character.

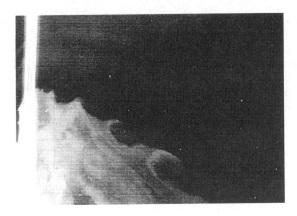

Fig. 2: Tidal inflow into shallow lagoon shown by thermal imagery. Cooler (dark) ocean water enters under bridge on top left, warm (light) water is blocked by causeway on bottom left. A series of large 2-D coherent structures develops in shear layer.

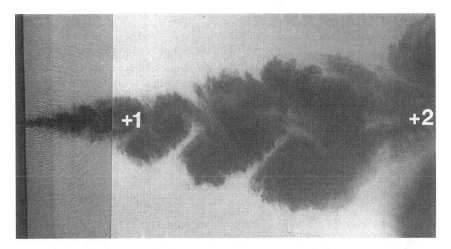

Fig. 3: Shallow jet (exit width B = 1 cm, H = 2.5 cm, Re = $U_oB/\nu \approx 10^4$) entering large flow basin (width of view about 3.5 m) (after Chen and Jirka, 1998)

<u>Type C: Secondary instabilities of base flow:</u> This is the weakest type of generating mechanism and experimental evidence is still limited. As remarked earlier, the nominal base flow is a uniform wide channel flow that is vertically sheared and contains a 3-D turbulence structure controlled by the bottom boundary layer. Slight imbalances in this flow process can lead to a wholesale redistribution of the momentum exchange processes at the bottom boundary, including as an extreme case separation of the bottom boundary layer. The distortion of the vortex lines caused by these flow imbalances lead ultimately to 2DCS. Contributing factors may be localized roughness zones or geometrical elements (underwater obstacles). The studies of Lloyd and Stansby (1997b) on submerged islands point in that direction (though there is also some connection to Type A mechanisms in their case). Gradual decelerations in the base flow (spatial or temporal, vis. tidal oscillations) can also lead to a breakdown of the base flow into 2DCS. The simulations by van Hijst et al. (1996) on cell formation in a shallow water tank seem to be examples of that. In either instance, the transverse momentum exchange induced by these flow patterns may explain the perplexingly high friction factors (Darcy-Weisbach coefficients) that have been found necessary when hindcasting numerical model results for flows in very wide open channels.

Whenever dealing with these generating mechanisms (especially Types A and B) it must be recognized that the generation of 2DCS always necessitated some travel time or convective distance from the origin of generation. A detailed analysis by Dracos et al. (1992) for the shallow jet and by Uijttewal and Tukker (1997) for the mixing layer actually shows three regions of development. In the "near-field" ($x/H \leq 1$) the transverse shear flow is 2-D in the mean, but contains highly 3-D small scale turbulence. The "middle-field" has a significant interaction of the mean and turbulent flow with bottom and surface, producing a strongly 3-D mean flow and 3-D turbulence. In the "far-field" ($x/H \geq 10$) the turbulence structures have grown to sizes greater than the water depth and for purely kinematic reasons these large eddies have vertical axes and hence 2-D character. Also, the mean flow, whose transverse scale is of the same order as the large turbulence eddies, becomes 2-D.

Following their generation the **growth** of 2DCS is governed by various processes. As they constitute turbulent vortical elements they grow by entrainment or engulfment of outside non - or less - turbulent fluid. Also pairing of separate structures leads to larger structures as observed in shallow jets (Dracos et al., 1992). In this fashion the 2DCS grow larger over time. The distribution of turbulent kinetic energy over different scales of the 2-D eddies is governed by the tenets of "2-D turbulence" theory (e.g. Kraichnan, 1967). Following the concept of the "enstrophy" cascade the flux of rotational momentum is constant among the different eddy scales leading to a k^{-3} distribution of turbulent kinetic energy in which k is the wavenumber. In 2-D turbulence there also exists the possibility of an inverse energy transfer in which energy is transferred from smaller to large scales. The vortex stretching mechanism that is the key attribute to usual 3-D turbulence by which energy is transferred to smaller and smaller scales until it is dissipated by viscosity does not exist in 2-D turbulence conditions.

The major mechanism that leads to the final **decay** of 2 DCS in shallow flows is the bottom friction at the base of the vortical elements. This friction is described by a shear stress $\tau_b = \rho c_f U^2/2$ in which ρ is the fluid density and c_f a quadratic law friction coefficient. Typically, $c_f \approx 0.005$ (field) to 0.01 (laboratory). One can readily show that for an eddy size $\ell_{max} \approx 2H/c_f$ the eddy looses its rotational energy during one turnover. Thus, ℓ_{max} can be considered as the maximum eddy size that 2DCS can obtain in shallow flows. With the above values of c_f, this corresponds to a relative size $\ell_{max}/H \approx 0(100)$.

3. METHODS OF INVESTIGATION

While the occurrence and the gross features of shallow flow instabilities are well established in hydraulic and environmental flows - mostly from aerial photographs - it is difficult to observe them in the field in a systematic fashion in order to ascertain their dynamic properties. This is best accomplished by laboratory experiments supported by analytical methods and numerical simulations.

Experiments on shallow flows require wide laboratory basins or "water tables" of sufficient size and with good flow control. A number of laboratories world-wide have used such installations to successfully simulate different flow features. This includes ETH Zurich (Giger et al., 1991), Cornell University (Chen and Jirka, 1995), University of Manchester (Lloyd and Stansby, 1997a,b) and Delft Technological University (Uijttewal and Tukker, 1997). Another shallow water basin has recently been installed at the University of Karlsruhe. Ideally, the basin dimensions should be at least 4 m wide and 10 m long in order to be free of boundary effects on the 2DCS. Also, Reynolds numbers Re_h exceeding 2000 can be readily maintained in such installations providing a fully turbulent base flow. Experiments with smaller size facilities (e.g. Chu and Babarutsi, 1988, on the mixing layer) may provide some qualitative insight into these processes. Their results, however, do deviate in quantitative terms from the large size installations and, supposedly, from the field behavior. Modern observational techniques are dye concentration measurements by means of planar laser-induced fluorescence (LIF) and velocity measurements by LDA or by PIV-field methods.

Stability analyses of the shallow water flows provide useful insight into the mechanisms for onset and growth of the 2DCS. For that purpose the depth-integrated momentum equations are linearized, leading to a form of Orr-Sommerfeld equations containing bottom shear stress terms. This approach was first suggested by the work of Chu and co-workers (e.g. Chu et al., 1983) and later extended by Chen and Jirka (1997, 1998) for wakes and jets, respectively. Despite the limitations of this approach, such as linearization, assumption of purely parallel (rather than expanding) flow and simple eddy viscosity formulation for the lateral shear stresses, it yields important diagnostic information on the conditions for growth and/or suppression of the instabilities that lead to 2DCS. As an example, Chen and Jirka (1997) have shown that the two forms of growth mechanisms that occur in a shallow wake, namely absolute and convective instabilities, correspond to different wake structures (Type A and Type B, as discussed in the following section) that can be observed experimentally. It appears that many additional questions, such on the effects of lateral boundaries or the role of slight transverse shear, can be usefully explored with this technique.

Numerical simulation models are, of course, another promising method for the understanding of these flows. Since the Reynolds number domain is outside the realm of current direct numerical simulation (DNS) techniques, some form of turbulence closure method needs to be employed. Furthermore, though the flows are 2-D in their gross feature, in their detail they are, of course, 3-D. While a depth-integrated model offers much computational simplicity, it may suppress important details on the vortex generation (i.e. the aforementioned progression from the "near-" to the "middle-" to the "far-field" of the initial shear layer). These questions need to be explored in the development of appropriate models. Ultimately, the large-eddy-simulation (LES) technique appears most obviously suited for a realistic simulation of these flows, but has not yet been implemented. Strongly forced (Type A) flows have been modeled with a k-ε technique by Lloyd and Stansby (1997a) showing reasonably good, although not fully satisfactory, agreement with observations on vortex-street wake flows. Whether a k-ε model would be successful for predicting more moderate Type B flows is quite doubtful. Only LES models appear appropriate for that.

4. SHALLOW WAKES

A few salient results from recent work are presented here to show the different types of 2DCS that occur in the shallow wake and their role in the mass exchange of advected material.

The behavior of shallow wake flow in the lee of an obstacle of diameter or width D is controlled by the wake parameter

$$S = c_f \frac{D}{H} \tag{1}$$

(Ingram and Chu, 1987; Chen and Jirka, 1995). As shown in Fig. 4, Chen and Jirka have classified three forms of wake patterns: i) the vortex street wake with unsteady separation at the cylinder, $S < 0.2$, ii) the unsteady bubble wake with a recirculating bubble attached to the cylinder that becomes unstable with sinuous oscillations further downstream, $0.2 \leq S \geq 0.5$, and iii) the steady bubble wake that shows no oscillations, $S \geq 0.5$. In essence, the vortex street pattern represents a Type A generation of 2DCS due to the flow separation at the cylinder periphery, while the unsteady bubble wake is a Type B generation due to the transverse shear imposed by the wake momentum deficit.

The wake behavior has been found to be independent of both depth Reynolds number Re_h - provided that $Re_h \geq 1500$ - as well as cylinder Reynolds number $Re_d = UD/\nu$ that are noted in the figure caption. The behavior seems Reynolds-invariant. The vortex street pattern (see also Fig. 1) looks remarkably similar to that for a von Karman vortex street in the laminar/turbulent transition of unbounded cylinder wakes, in the range $50 < Re_d < 300$. Secondary, spanwise instabilities rapidly destroy that vortex street pattern for higher Re_d in the unbounded case. In the shallow, bounded case, however, these primary instabilities remain the dominant ones because of the kinematic constraint and the flow becomes governed by these.

Whenever the shallow wake parameter S becomes large, bottom friction suppresses the vortex shedding process and any subsequent transverse instabilities leading to full stabilization.

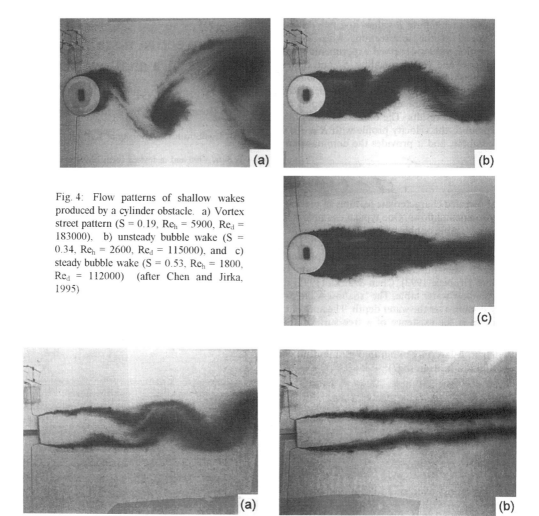

Fig. 4: Flow patterns of shallow wakes produced by a cylinder obstacle. a) Vortex street pattern (S = 0.19, Re_h = 5900, Re_d = 183000), b) unsteady bubble wake (S = 0.34, Re_h = 2600, Re_d = 115000), and c) steady bubble wake (S = 0.53, Re_h = 1800, Re_d = 112000) (after Chen and Jirka, 1995)

Fig. 5: Shallow wake patterns produced by porous plate device that allows for some ambient flow through obstacle. a) Unsteady bubble wake (S = 0.16, Re_h = 4850, Re_d = 121000), and b) steady bubble wake (S = 0.26, Re_h = 1780, Re_d = 55000) (after Chen and Jirka, 1995)

Qualitatively similar observations can be made for wakes produced by solid plate-like obstacles. With the purpose of eliminating the initial flow separation (and thus, Type A generation) Chen and Jirka also employed a porous plate device that allows for some through-flow. In that case the vortex street pattern never occurred, but as shown in Fig. 5 only the unsteady bubble wake (Type B generation) or the steady wake are possible. The limiting value between two patterns was about S ≈ 0.19 for the particular porosity (about 50%) of the porous plate. These results support the selectivity of the flow patterns as a consequence of type of generation as well as the shallow wake parameter S.

The different types of turbulent shallow wakes obviously have greatly differing mixing characteristics for any material, such as instantaneous or continuous discharge of pollutants that enter the wake region. As an example Fig. 6 shows the concentration pattern obtained by a planar LIF measurement in a cylinder wake that is near the transition between the vortex street and unsteady bubble flow pattern (S ≈ 0.2). The dye that has been continuously injected at the two cylinder shoulders has been concentrated into vortical blobs while the concentration in-between is much lower. In fact, systematic measurements (Chen and Jirka, 1991)

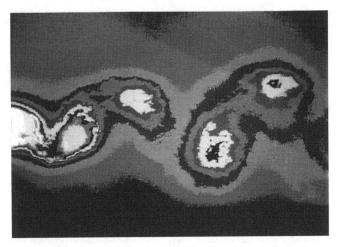

Fig. 6:   Instantaneous view of concentrated dye pattern in a shallow cylinder wake,
S ≈ 0.2, as measured by planar LIF technique. Cylinder is located at left edge of picture.

have shown that these instantaneous concentration peaks are about 4 to 6 times larger than the time-averaged maxima that occur at a particular downstream distance. This behavior has substantial bearing on the interpretation of pollutant data monitoring as may be employed by environmental enforcement agencies. Signals at a fixed monitoring point exhibit considerable intermittency. On the other hand, there may be very high exposure levels for organisms that are advected along in these concentrated blobs and that may have a sustained lifetime until they disintegrate.

The existence of attached bubble flows for wakes with high S values has further influences on pollutant trapping in the lee of islands. The flushing time for such trapped pollutants (stemming for example from a treatment plant located on the island) can be a factor of 10 larger than for shallow wake flows with the more active vortex street pattern (Chen and Jirka, 1991; Lloyd and Stansby, 1997a).

Considerably weaker types of instabilities have been found for headland wakes in which because of the symmetry imposed by the shoreline the more vigorous sinuous oscillations are suppressed and only varicose instabilities are possible (MacDonald and Jirka, 1997).

5. SHALLOW JETS

The shallow jet (see Fig. 3) has been investigated through detailed velocity measurements (Dracos et al., 1992) and LIF concentration measurements (Chen and Jirka, 1998). These results have been summarized by Jirka (1994).

As mentioned above a shallow jet enters its "far-field" at a distance, $x/H \approx 10$, beyond which it takes on its distinct meandering character with alternating vortical elements. Fig. 7 shows examples for one-dimensional energy spectra of the transverse velocity component v' measured on the jet axis. The spectra are scaled with the local centerline velocity U_m and half-width b; the non-dimensional frequency f abscissa can therefore be seen as a Strouhal number defined as $St = fb/U_m$. Fig. 7a shows near/middle field spectra. Their overlapping is indicative of similarity. There is a distinct energy peak around $St = 0.10$. This value agrees with some observations of weakly energetic large-scale structures in these types of flows (e.g. Thomas and Goldschmidt, 1986). These three-dimensional structures, however, break down through transverse instabilities and undergo a vortex stretching mechanism thereby, transferring their energy to smaller scales. A universal equilibrium subrange with a -5/3 wavenumber dependence, typical for three-dimensional cascading turbulent flow, is therefore observed at higher wavenumbers.

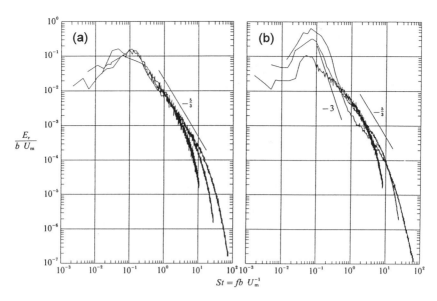

Fig. 7: One-dimensional normalized spectra of turbulent fluctuations of the transverse velocity. a) Spectra for three locations in the near/middle field, and b) spectra for three locations in the far field with increasing distance x/H

Power spectra in the far field (Fig. 7b) behave quite differently. Here dynamic self-similarity no longer exists. Rather the maxima in the non-dimensional energy density become more pronounced with increasing jet distance. At the same time, a range develops in which the energy transfer follows a -3 wavenumber dependence. Such dependence is consistent with a quasi-two-dimensional turbulence characterized by an enstrophy cascade (Batchelor, 1969). The increase of energy at the low wavenumbers is associated with a depletion of the energy content at higher wavenumbers. At even larger wavenumbers the energy transfer gradually relaxes back to that for three-dimensional turbulence. The shapes of the spectra suggest that some energy is extracted from the inertial subrange of the spectrum at the location where the enstrophy cascade begins and is transferred in an inverse cascade back towards the peak which increases in magnitude.

The Strouhal number of the peak, $St = 0.08$, as obtained from the spectra, can also be observed by a variety of other means, such as autocorrelation functions and counting of the passage of the visible vortical structures, all of which agree closely (Dracos et al.). As the dimensional frequency is decreasing with increasing distance there must be a loss in the number of vortical elements. Indeed, this occurs through a pairing mechanism of similarly rotating elements on a given jet side.

The power density spectral distributions for the concentration fluctuations are shown in Fig. 8 for the two jet locations, labeled "1" and "2" in Fig 3. Again a transition from a -5/3 frequency dependence to a -3 dependence at large distances in the shallow jet is apparent. Thus, there is a consistent indication that the meandering shallow jet with its 2DCS is governed by the energy transfer characteristics of 2-D turbulence.

This also has repercussion on the mean and rms concentration fields as shown in Fig. 9. The lateral distance y is normalized as y/x which not only shows the self-similarity of the mean profiles (Fig. 9a), but also indicates a linear concentration half-width, $b_c/x = 0.17$, that is considerably larger than the velocity half-width spreading coefficient, $b/x = 0.10$ (Giger et al., 1991). In fact, the dispersion ratio $\lambda = b_c/b = 1.7$ greatly exceeds that for the usual unbounded plane jet, $\lambda \cong 1.35$. This aspect seems to be an indication of the occasional passage of larger vortical elements that contain high concentrations at or beyond the mean jet periphery. The rms-profile (Fig. 9b) shows maximum activity at a location equal to the concentration width b_c. Again compared to the unbounded case (Davies et al., 1975) there seems to be an increasing fluctuation activity caused by these intermittent 2DCS with increasing distance along the shallow jet.

294

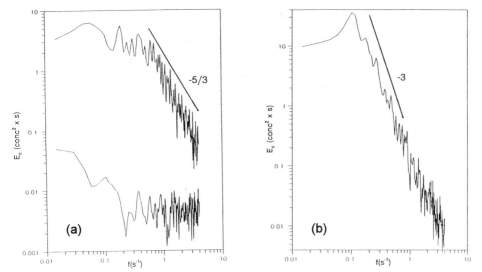

Fig. 8: Concentration fluctuation spectra for a shallow jet (B = 1 cm, H = 2.5 cm) obtained from LIF images for the two locations indicated in Fig. 3. a) Point 1 in the "middle-field". The lower trace measures the background noise. b) Point 2 in the "far-field"

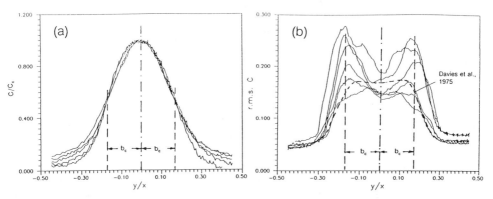

Fig. 9: Lateral profiles of a) mean concentration and b) rms concentration, both normalized by the centerline concentration, for different locations in the shallow jet. b_c indicates the position of the concentration half-width.

6. CONCLUSIONS

The study of shallow turbulent flows represents a fruitful line of inquiry in environmental fluid mechanics. These flows are readily susceptible to different types of disturbances that trigger transverse instabilities leading to 2-D coherent structures (2DCS). Three types of generation mechanisms have been proposed herein: topographic forcing (Type A), internal transverse shear (Type B) and secondary instabilities of the base flow (Type C).

The growth of the 2DCS and the internal turbulent kinetic distribution among different scales is governed by principles of "2-D turbulence" theory. Their ultimate decay is controlled by turbulent bottom friction at the base of the vortical elements.

The behavior of these 2DCS has a profound influence on the pollutant transport characteristics in such shallow flow. The vortical elements may contain strongly concentrated pollutant levels and may advect them over considerable times. Also, pollutant trapping behind topographic features (such as islands or peninsulas) appears to be influenced by these flow features.

Finally, these 2DCS seem to play an important role in the lateral momentum exchange in wide channel flow such as rivers, estuaries or coastal zones. Here, moderate topographic disturbance, uneven roughness distribution or deceleration effects in the flow may trigger momentum exchanges. The complete understanding of these processes appears to be an important goal for the future.

Acknowledgments: Research on shallow flow instabilities at the University of Karlsruhe is supported by the Deutsche Forschungsgemeinschaft (DFG Grant Ji 18/4-1).

REFERENCES

Batchelor, G.K. (1969). "Computation of the energy spectra in homogenous two-dimensional turbulence", *Phys. Fluids*, 233-238.

Chen, D. and Jirka, G.H. (1991). "Pollutant mixing in wake flows behind islands in shallow water", Proc. *Int. Symp. on Environmental Hydraulics* (J.H.W. Lee and Y.K. Cheung, Ed.s), Balkema, 371-377.

Chen, D. and Jirka, G.H. (1995). "Experimental study of plane turbulent wake in a shallow water layer", *Fluid Dynamics Research*, 16, 11.

Chen, D. and Jirka, G.H. (1997). "Absolute and Convective Instabilities of Plane Turbulent Wakes in a Shallow Water Layer", *J. Fluid Mechanics*, 338, 157-172.

Chen, D. and Jirka, G.H. (1998). "Linear Instability Analyses of Turbulent Mixing Layers and Jets in Shallow Water Layers", *J. Hydraulic Research*, 36, No.5.

Chen, D. and Jirka, G.H. (1998). "A Laser-Induced Fluorescence Study of a Plane Shallow Jet", *J. Hydraulic Eng.* (in press).

Chu, V.H. and Babarutsi, S. (1988). "Confinement and bed-friction effects in shallow turbulent mixing layers", *J. Hydraulic Eng.*, 114, 1257-1274.

Chu, V.H., Wu, J.H. and Khayat, R.E. (1983). "Stability of turbulent shear flows in shallow channel", *Proc. XX Congress IAHR*, Moscow, 3, 128-133.

Davies, A.E., Keffer, J.F. and Baines, W.D. (1975). "Spread of a heated plane turbulent jet", *Phys. Fluids*, 18, 770.

Dracos, T., Giger, M. and Jirka, G.H. (1992). "Plane Turbulent Jets in a Bounded Fluid Layer", *J. Fluid Mechanics*, 214, 587-614.

Giger, M., Dracos, T. and Jirka, G.H. (1991). "Entrainment and Mixing in Plane Turbulent Jets in Shallow Water", *J. Hydraulics Research*, 29, No.4, 615-643.

Hussain, A.K.M.F. (1983), "Coherent structures - reality and myth", *Phys. Fluids*, 26, 2816-2850.

Ingram, R.G., and V.H. Chu (1987). "Flow around islands in Rupert Bay: An investigation of the bottom friction effect", *J. Geophys. Res.*, 92(C13), 14521-14533.

Jirka, G.H. (1994). "Shallow Jets", in: *Recent Advances in the Fluid Mechanics of Turbulent Jets and Plumes*, P.A. Davies and M.J. Valente Neves (Ed.s), Kluwer Academic Publishers, Dordrecht.

Kraichnan, R. (1967). "Inertial ranges in two-dimensional turbulence", *Phys. Fluids*, 10, 1417-1428.

Lloyd, P.M. and Stansby, P.K. (1997a). "Shallow-water flow around model conical islands of small side slope. I: Surface piercing", *J. Hydraulic Eng.*, 123, No. 12, 1057-1067.

Lloyd, P.M. and Stansby, P.K. (1997b). "Shallow-water flow around model conical islands of small side slope. II: Submerged", *J. Hydraulic Eng.*, 123, No. 12, 1068-1077.

MacDonald, D.G. and Jirka, G.H. (1997). "Characteristics of Headland Wakes in Shallow Flow", *Proc. XXVII Congress IAHR*, San Francisco, Vol.1, 88-93.

Nezu, I. und Nakagawa, H. (1993). *"Turbulence in Open-Channel Flows"*, A.A. Bakema, Rotterdam.

Thomas, F. O. and Goldschmidt, V.W. (1986). "Structural characteristics of developing turbulent planar jet", *J. Fluid Mech.*, 63, 227-256.

Uijttewaal, W.S.J. and Tukker, J. (1998). "Development of quasi two-dimensional structures in a shallow free-surface mixing layer", *Experiments in Fluids*, 24, 192-200.

Van Heijst, G.J., Clerx, H. and Maassen, S. (1996). "Stably stratified flow in a rectangular container: cell pattern formation and anomalous diffusion", *5th IMA Conference on Stratified Flows*, Dundee, Scotland.

Environmental Hydraulics, Lee, Jayawardena & Wang (eds) © 1999 Balkema, Rotterdam, ISBN 90 5809 035 3

Mutual-interaction between bursts and boils very near the free-surface of open-channel flows

Iehisa Nezu & Tadanobu Nakayama
Department of Civil and Global Environment Engineering, Kyoto University, Japan

ABSTRACT: In an open-channel flow, a free surface has a large effect on the turbulent energy redistribution near the free surface. When the Froude number increases and the surface-wave fluctuations occur, the damping effect is gradually lost out and the turbulent energy redistribution near the free surface changes largely. It has been pointed out that these characteristics are closely related to the coherent structures. In this study, the damping characteristics are compared between the theoretical model and the experimental data. The effect of the surface-wave fluctuations is then considered by using a function of the Froude number. Furthermore, Reynolds Stress Model (RSM) is used to calculate the turbulent statistics in comparison with the experimental data obtained with a laser Doppler anemometer (LDA). Finally, a relationship between the "*bursting phenomenon*" generated near the wall and the "*surface renewal eddies*" near the free surface was evaluated in open-channel flows.

1. INTRODUCTION

In an open-channel flow, a free surface has a large effect on the turbulent energy redistribution near the free surface. Davies (1972) and Hunt and Graham (1978) have evaluated theoretically the damping characteristics of the vertical component of turbelence intensity. These effects of the free surface are very important in an open-channel flow and therefore, these characteristics are quite different from those in a duct flow near the symmetric surface. The turbulent structure becomes quite different from that in a quiet flow and a damping effect on the vertical motions is gradually lost out when the Froude number increases and surface-wave fluctuations occur. Consequently, "*Hunt's theory*" which was proposed in the flow of a smaller Froude number does not hold good near the free surface. At a higher Froude-number flow, a coherent structure, so-called the "*surface renewal eddies*", can be seen clearly at the free surface.

A closer interrelation between the wall region and the free-surface region, i.e. the bursting and boil phenomena in open-channel flows, has been suggested by some researchers. Jackson (1976) described the features of boil vortices from field observations of boils in rivers. He speculated that the bursting motions generated in the wall region would move toward the free surface and form a boil there because the bursting period for boundary layers was roughly equal to the boil period. Komori *et al.* (1989) have related the surface renewal and the bursting motions with each other in view of their frequencies. It is very interesting that Banerjee (1992) have suggested two modes of the interaction between bursts and the free surface motions, i.e., the "*splat pattern*" (pancake-shaped structure) and the "*attached vortex pattern*" (spinning structure), as shown in Fig.1. They have implied that both structures seem to occur and persist for some time at the free surface. Komori *et al.* (1993) have used the DNS adopting a boundary fitting condition of the free surface in open-channel flows and clarified the relationship between the generation of surface-renewal eddies and mass-transfer mechanism. Furthermore, Perot and Moin (1995) have estimated that there exist the surface renewal eddies due to the interaction between the splatting structure and the free surface, and that their structures promote the scalar transfer across the air-water interface.

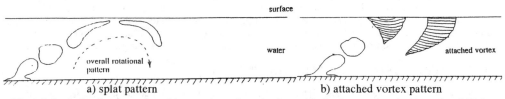

Fig.1 Schematic of two-modes of interaction of an ejection with the free surface by Banerjee (1992).

Table 1 Hydraulic Conditions.

case	S_b	h (cm)	B/h	A_m/h $(\times 10^{-2})$	U_m (cm/s)	$U*$ (cm/s)	Fr	Re $(\times 10^3)$
L-FR02	1/6000	5.0	8.0	0.0	16.0	0.99	0.23	8.0
M-FR06	1/1200	5.0	8.0	0.6	40.0	2.10	0.57	20.0
M-FR08	1/600	4.0	10.0	1.5	50.0	2.50	0.80	20.0
M-FR10	1/316	3.5	11.4	2.0	57.1	2.90	0.98	20.0
H-FR15	1/167	4.0	10.0	2.5	93.8	4.75	1.50	37.5
H-FR23	1/77	3.0	13.3	3.3	125.0	6.09	2.30	37.5
S-FR31	1/41	3.0	13.3	6.7	166.7	8.47	3.07	50.0

S_b=channel slope, h=flow depth, B=channel width, A_m=wave height, $Fr=U_m/(gh)^{0.5}$, $Re=U_m h/\nu$

In this study, the 3-D turbulence measurements have been conducted accurately with a laser Doppler anemometer (LDA) in order to reproduce the characteristics of turbulence redistribution near the free surface in open-channel flows by using Reynolds Stress Model (RSM). We clalify then the mutual interaction between bursts and boils near the free surface by estimating the event contribution to the Reynolds shear stress and the period of bursts in the whole depth. Furthermore, we evaluate the spatial scaling of bursts and boils by making use of PIV(Particle-Image Velocimetry) methods.

2. THEORETICAL CONSIDERATION

Davies (1972) derived a linear expression (1) for the vertical intensity v' from continuity equation supposing that the streamwise intensity u' is independent of $y'=h-y$ near the free surface. In this expression, λ represents the thickness of the surface layer of damped turbulence and $v'=v_0$ at $y' \geqq \lambda$ near the free surface.

$$v' / v_0 = y' / \lambda \quad ; \quad y' < \lambda \tag{1}$$

Hunt and Graham (1978) have analyzed a grid turbulence above the moving bed by using a boundary layer theory and spectral methods and deduced the expression (2), where ε is the turbulent energy dissipation near the free surface.

$$v' / U_* = \gamma \left(\varepsilon \cdot h / U_*^3 \right)^{1/3} \left(y' / h \right)^{1/3} \quad ; \quad \gamma = 1.34 \tag{2}$$

Furthermore, Brumley and Jirka (1987) supposed the expression (3) by using Hunt's theory (where p=1.54, λ_2=1.4, g_2=0.558).

$$v' / v_0 = \left[\left(\lambda_2 \left(y' / h \right)^{2/3} \right)^{-p} \exp \left(-g_2^{1/2} y' / h \right) + 1 \right]^{-1/2p} \tag{3}$$

All these expressions are derived in a quiet flow and it has been pointed out that in particular the expression (2) and (3) reproduce well the damping characteristics of v' near the free surface, as shown later.

3. EXPERIMENTAL AND COMPUTATIONAL PROCEDURES

The experiments were conducted in a 10m long, 40cm wide and 40cm deep tilting flume. The side and bed walls of the test section 6m downstream of the channel entrance were made of optical glass for LDA (DANTEC-made) and PIV (KANOMAX-made) measurements. The 3-D measurements were made in the center of the channel by using a fiber-optic LDA. The sampling time was 60sec and the sampling frequency was about 200Hz. On the other hand, PIV was conducted by using Nylon-12 particles (50μm diameter and 1.01 specific gravity) uniformly scattered in the circulating water of the flume. In this study, one pixel of the image corresponds to the area of 0.4×0.4mm. More detailed information on the present PIV system is available in Nezu and Nakayama (1998). Hydraulic conditions for the experiments are shown in Table 1.

Calculation was conducted in a 2-D fully-developed open-channel flow (aspect ratio was greater than 5 in the experiment as pointed out by Nezu and Nakagawa (1993)) by using Reynolds Stress Model (RSM). At that time, Daly-Harlow model (GGDH) and Speziale-Sarkar-Gastki model (SSG) were adopted in the diffusion term and the pressure-strain term, respectively. In particular, the free-surface damping function was added in the pressure-strain term in the same manner as Celik and Rodi (1984). The calculations use 50 grid points in the vertical direction and sufficient lengths (over 200 flow depth) in the longitudinal direction to ensure fully-developed flow. Boundary conditions for the mean velocity and the turbulent energy dissipation rate were symmetrical at the free surface. The Reynolds stress terms at the free surface were defined by the experimental data with LDA, as mentioned later.

4. CHARACTERISTICS OF TURBULENCE INTENSITY v'

It has been pointed out that the turbulence intensity changes in a complicated manner near the free surface, and in particular, the vertical component v' is greatly affected by the free surface. Fig.2 shows the distribution of the vertical

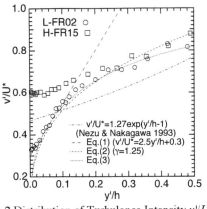

Fig.2 Distribution of Turbulence Intensity v'/U_*.

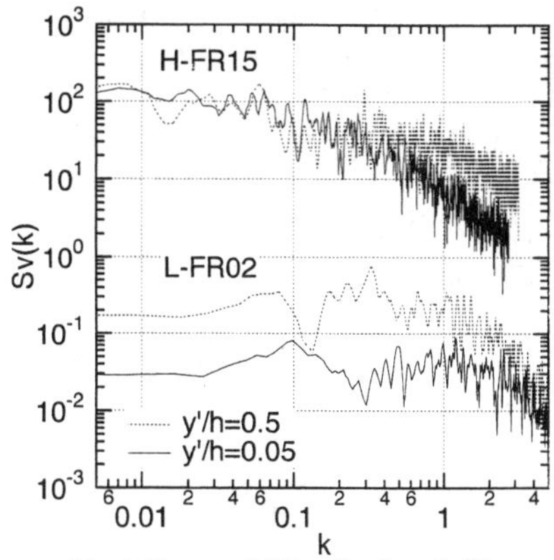

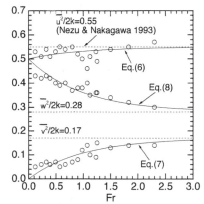

Fig.3 Spectral Distribution $S_v(k)$.

Fig.4 Turbulent Energy Redistributions.

component of turbulence intensity v' near the free surface normalized by the friction velocity U_*. In this figure, the expressions ((1), (2) and (3)) and the semi-empirical formula in open-channel flow derived by Nezu and Nakagawa (1993) are plotted together, where ε in (2) was the constant value near the free surface ($\varepsilon h/U_*^3 = 0.8$). It can be seen that (2) and (3) coincide well with the experimental value in a tranquil flow (L-FR02). However, v' increases rapidly and Hunt's theory does not hold good when the Froude number increases. Fig.3 shows some examples of the spectral distributions $S_v(k)$ for the fluctuations $v(t)$. The value of low wave-number near the free surface ($y'/h=0.05$) decreases in L-FR02, as has been pointed out by Hunt and Graham (1978). However, in a super-critical flow (H-FR15:$Fr=1.50$), the region of low wave-number does not decrease due to the turbulence eddies near the free surface, which is closely related to the increases of v' near the free surface.

5. VARIATION OF TURBULENT ENERGY REDISTRIBUTIONS NEAR THE FREE SURFACE

For calculating the RSM, it is necessary to define the boundary conditions of turbulent energy redistributions at the free surface. The boundary conditions at the free surface in a tranquil flow are given explicitly by Eq. (4) and therefore, the turbulent redistributions at the free surface can be expressed by Eq. (5).

$$\partial U/\partial y = \partial W/\partial y = V = 0 \;\; ; \;\; y'/h = 0 \tag{4}$$

$$\overline{u^2}/2k = a, \overline{v^2}/2k = 0, \overline{w^2}/2k = b, a+b=1 \;\; ; \;\; y'/h = 0 \tag{5}$$

Furthermore, Davies (1972) has pointed out that the thickness where the damping characteristics are predominant becomes thinner as the vertical fluctuations increase. So, it is predictable that the turbulent redistributions near the free surface approach the constant ratio in an intermediate region derived by Nezu and Nakagawa (1993) with an increase of the Froude number due to the decrease of surface-influenced layer. Therefore, it is better to express the turbulent redistributions by the exponential function. Fig.4 shows the turbulent energy redistributions near the free

299

surface (y'/h=0.02) versus Froude number, together with the constant values (dotted line) derived by Nezu and Nakagawa (1993). Furthermore, solid line is an approximate curve produced by the above-mentioned method ((6), (7), (8)).

$$\overline{u^2}/2k = 0.55 - 0.05\exp(-Fr) \tag{6}$$

$$\overline{v^2}/2k = 0.17(1 - \exp(-Fr)) \tag{7}$$

$$\overline{w^2}/2k = 0.28 + 0.22\exp(-Fr) \tag{8}$$

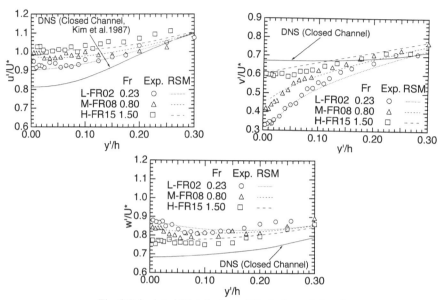

Fig.5 Calculated Distribution of Turbulence Intensity.

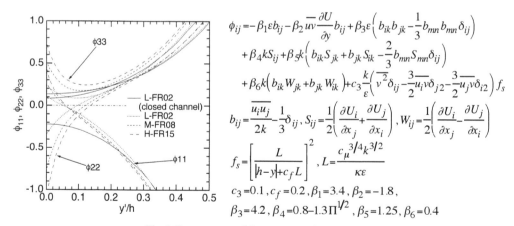

$$\phi_{ij} = -\beta_1 \varepsilon b_{ij} - \beta_2 \overline{uv}\frac{\partial U}{\partial y}b_{ij} + \beta_3 \varepsilon\left(b_{ik}b_{jk} - \frac{1}{3}b_{mn}b_{mn}\delta_{ij}\right)$$

$$+ \beta_4 k S_{ij} + \beta_5 k\left(b_{ik}S_{jk} + b_{jk}S_{ik} - \frac{2}{3}b_{mn}S_{mn}\delta_{ij}\right)$$

$$+ \beta_6 k\left(b_{ik}W_{jk} + b_{jk}W_{ik}\right) + c_3\frac{k}{\varepsilon}\left(\overline{v^2}\delta_{ij} - \frac{3}{2}\overline{u_i v}\delta_{j2} - \frac{3}{2}\overline{u_j v}\delta_{i2}\right)f_s$$

$$b_{ij} = \frac{\overline{u_i u_j}}{2k} - \frac{1}{3}\delta_{ij}, \; S_{ij} = \frac{1}{2}\left(\frac{\partial U_i}{\partial x_j} + \frac{\partial U_j}{\partial x_i}\right), \; W_{ij} = \frac{1}{2}\left(\frac{\partial U_i}{\partial x_j} - \frac{\partial U_j}{\partial x_i}\right)$$

$$f_s = \left[\frac{L}{|h-y| + c_f L}\right]^2, \; L = \frac{c_\mu^{3/4}k^{3/2}}{\kappa\varepsilon}$$

$$c_3 = 0.1, c_f = 0.2, \beta_1 = 3.4, \beta_2 = -1.8,$$

$$\beta_3 = 4.2, \beta_4 = 0.8 - 1.3\Pi^{1/2}, \beta_5 = 1.25, \beta_6 = 0.4$$

Fig.6 Components of Pressure-Strain Term.

It can be seen that the streamwise redistribution almost equals the spanwise redistribution when the Froude number becomes close to zero, so called the "*hydrostatic*". Furthermore, the turbulent redistributions approach the constant ratio proposed by Nezu and Nakagawa (1993), with an increase of the Froude number. In this way, it is because the thickness of the damping predominance becomes thinner than y'/h=0.05 that the region of low wave-number does not decrease in a super-critical flow in Fig.3

6. VARIATION OF PRESSURE-STRAIN TERMS IN TRANSPORT EQUATIONS

Fig.5 shows the distribution of all three components of turbulence intensity u'/U_*, v'/U_*, w'/U_* near the free surface when the above boundary conditions were taken in the computation. The boundary condition of ε at the free surface is symmetric. In this figure, DNS data in duct flow by Kim *et al.* (1987) are also plotted. Of particular significance is that v' increases greatly near the free surface with an increase of the Froude number and the calculation reproduces these characterictics. It can be seen that w' decreases and u' increases slightly. These facts indicate that the present calculation model is fairly correct on the basis that the effect of the Froude number is considered reasonably.

Perot and Moin (1995) have studied the structure of "*splats*" and "*antisplats*" near the free surface and evaluated the relation between these structures and the pressure-strain term. Fig.6 is the components of pressure-strain term when the Froude number changes. In this figure, the value in closed-channel flows is also shown. It can be seen that ϕ_{22} in open-channel flows is quite different from ϕ_{22} in closed-channel flows near the free surface, as Komori *et al.* (1993) pointed out. When the Froude number increases, ϕ_{22} near the free surface decreases greatly and simultaneously ϕ_{33} increases considerably. On the other hand, ϕ_{11} does not change so more greatly than ϕ_{22} and ϕ_{33}. This indicates that the turbulent energy redistribution between the vertical velocity fluctuation and the spanwise velocity fluctuation becomes more active with an increase of Froude number.

7. MUTUAL INTERACTION BETWEEN BURSTS AND BOILS

In this way, the turbulent structure in a steep open-channel flow is closely related with the mutual interaction between the surface-renewal eddies and bursting motions. It is possible that these relations may change when the Froude number increases.

It has been pointed out that the larger-scale coherent structures can be seen in the streamwise direction when the Froude number is larger than one in a super-critical flow and the surface-wave fluctuations occur. Rashidi and Banerjee (1988) have found that the coherent structures bound near the free surface and go back towards the wall in such a super-critical flow. Fig.7 shows instantaneous velocity vector fields for S-FR31 ($Fr=3.07$) in a steep open-channel flow that are viewed in movable coordinates of the bulk mean velocities in the center of channel. It can be seen that the two coherent vortices (C and D) are generated at the interval of $TU_{max}/h=3.3$ (U_{max} is a maximum velocity), which coincides well with the mean bursting period $T_B U_{max}/h=1.5\sim3.0$ (Nezu and Nakagawa, 1993). It was also true for the case of L-FR02 in a tranquil flow. Of particular significance is the agglemeration (G) of the coherent vortices at $t=0.0$s. The period of this vortex is quite large and represents the equivalent value of "*boils*

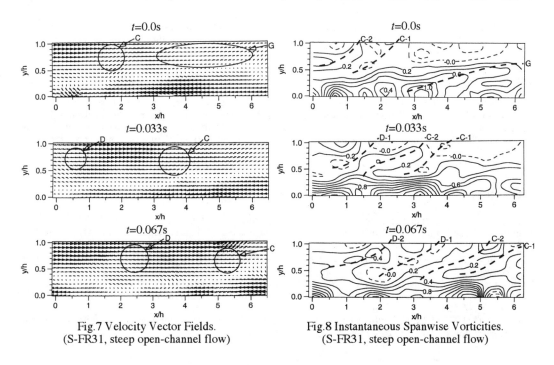

Fig.7 Velocity Vector Fields.
(S-FR31, steep open-channel flow)

Fig.8 Instantaneous Spanwise Vorticities.
(S-FR31, steep open-channel flow)

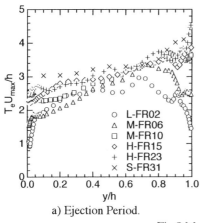

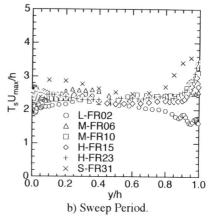

| a) Ejection Period. | b) Sweep Period. |

Fig.9 Mean Bursting Period.

(*surface renewal eddies*)" in rivers (TU_{max}/h=7.6), that is proposed by Jackson (1976). The contours of spanwise vorticity are shown in Fig.8, which are normalized by the maximum velocity and the flow depth. The front parts of upward region (C-1 and D-1) attain negative values and back parts (C-2 and D-2) attain positive values in the same way as L-FR02. Nevertheless, the other vortices (G) are very different and attain a negative value near the free surface. The region of positive value near the wall is blocked by the generation of negative vorticity near the free surface. In this way, the agglemeration is predominant when the surface-wave fluctuations occur, and therefore boils can be seen clearly on the free surface. Furthermore, it has a possibility that this agglemeration is closely related to the decrease of the surface-influenced layer (the thickness of the damping predominance).

The mean bursting period normalized by the maximum velocity U_{max} and the flow depth h is shown in Fig.9, where T_e and T_s are the ejection and sweep periods in the Quadrant techniques, so-called the "*half-value threshold levels*" by Nezu and Nakagawa (1993). It can be seen that T_e increases with y/h at about y/h<0.7, regardless of the Froude number, but T_s is almost constant in the same region. This indicates that the upward eddies that are generated by bursting motions near the wall become greater to the extent, but the outer (high-speed) flow goes downward with little deformation. On the other hand, the distributions of T_e and T_s change largely at about y/h>0.7. The value decreases near the free surface when the Froude number is small in a tranquil flow. However, the value increases with an increase of the Froude number. This is closely related with the effect of free surface and the agglemeration occurs in a super-critical flow.

8. CONCLUSIONS

In this study, Reynolds Stress Model was used to calculate the turbulent statistics near the free surface in open-channel flows. At that time, the boundary conditions at the free surface were defined as the function of Froude number on the basis of LDA data. Furthermore, it became clear that the bursting phenomena are comparatively related with the surface renewal eddies near the free surface and also that the agglemeration is predominant in a super-critical flow. These phenomena have a close relationship with the complexity of the turbulence redistribution near the free surface and it is necessary to evaluate these relations.

9. REFERENCES

Banerjee, S.(1992): "Turbulence structures", *Chemical Engineering Science*, Vol.47, No.8, pp.1793-1817.
Brumley, B. H. and Jirka, G. H.(1987): "Near-surface turbulence in a grid-stirred tank", *J. Fluid Mech.*, Vol.183, pp.235-263.
Celik, I. and Rodi, W.(1984): "Simulation of free surface effects in turbulent channel flows", *Physico Chemical Hydrodynamics*, Vol.5, pp.217-227.
Davies, J. T.(1972): "Turbulence Phenomena", Academic Press, San Francisco .
Hunt, J. C. R. and Graham, J. M. R.(1978): "Free-stream turbulence near plane boundaries", *J. Fluid Mech.*, Vol.84, pp.209-235.
Jackson, R. G.(1976): "Sedimentological and fluid-dynamic implications of the turbulent bursting phenomenon in geophysical flows", *J. Fluid Mech.*, Vol.77, pp.531-560.

Kim, J., Moin, P. and Moser, R.(1987): "Turbulence statistics in fully developed channel flow at low Reynolds number", *J. Fluid Mech.*, Vol.177, pp.133-166.

Komori, S., Murakami, Y. and Ueda, H.(1989): "The relationship between surface-renewal and bursting motions in an open-channel flow", *J. Fluid Mech.*, Vol.203, pp.103-123.

Komori, S., Nagaosa, R., Murakami, Y., Chiba, S., Ishii, K. and Kuwahara, K.(1993): "Direct numerical simulation of three-dimensional open-channel flow with zero-shear gas-liquid interface", *Phys. Fluids A*, Vol.5, No.1, pp.115-125.

Nezu, I. and Nakagawa, H.(1993): "Turbulence in Open-Channel Flows", IAHR-Monograph, Balkema.

Nezu, I. and Nakayama, T.(1998): "Separation between turbulence and water-wave effects in an open-channel flow", *the 7th Int. Symposium on Flow Modeling and Turbulence Measurement*, Taiwan. (submitted)

Perot, B. and Moin, P.(1995): "Shear-free turbulent boundary layers. Part 1", *J. Fluid Mech.*, Vol.295, pp.199-227.

Rashidi, M. and Banerjee, S.(1988): "Turbulence structure in free surface channel flows", *Phys.Fluids*, Vol.31, No.9, pp.2491-2503.

303

Environmental Hydraulics, Lee, Jayawardena & Wang (eds) © 1999 Balkema, Rotterdam, ISBN 90 5809 035 3

3-D turbulent structures in partly vegetated open-channel flows

Iehisa Nezu & Kouki Onitsuka
Department of Civil and Global Environment Engineering, Kyoto University, Japan

ABSTRACT: The effects of the vegetation growing in the water and at the edge of water on the turbulent structures are not so clear as yet, in spite of their importance. In this study, the turbulence measurements of open-channel flows with vegetated zone at a half channel width were conducted by making use of a laser Doppler anemometer(LDA). It was found that the intensity of the secondary currents and the turbulence energy increase with an increase of the Froude numbers. The turbulence is advected spanwisely near the free surface by the secondary currents, which are generated by the surface vortices. The horizontal vortices near the free surface are generated by the shear instability which increases with an increase of the Froude numbers. The wall shear stress in the vegetated zone is calculated by the equation of motion with the aid of the empirical drag coefficient.

1. INTRODUCTION

The vegetation growing in the water and at the edge of water has effects on the turbulent structure such as the mean velocity profiles, the Reynolds stress distributions, the secondary currents and also on the water resistance. The importance of vegetated zone in rivers has recently been recognized not only for river management but also for water flow environment. From this point of view, some researchers have investigated vegetated open-channel flows to clarify resistance law and turbulent structures.

As the turbulent structure in partly vegetated open-channel flows may somewhat be similar to that in two-stage compound open-channel flows, it is useful to compare the former with the latter which is recently investigated by many researchers; LDA measurements and numerical simulations of algebraic stress model (ASM), Reynolds stress model (RSM) and large eddy simulation(LES). Naot, Nezu and Nakagawa (1996) have extended an ASM for compound open-channel flows which calculates successfully secondary currents and turbulence (Nezu, 1996) to a partly vegetated open-channel flow. The ASM data showed a qualitative agreement with the existing low-accurate experimental data measured by a hot-film anemometer and an electromagnetic flow meter. However, accuracy of these data are not so high. Therefore, it is quite necessary to measure vegetated open-channel flows accurately with a laser Doppler anemometer(LDA) and to offer LDA database of vegetated flows in order to develop refined flow modeling.

In this study, the turbulence measurements of open-channel flows with vegetated zone at a half channel width were conducted by making use of a laser Doppler anemometer(LDA) to investigate 3-D turbulent structure such as the mean velocity profiles, the Reynolds stress distributions, the secondary currents and the water resistance.

2. THEORETICAL CONSIDERATIONS

The continuity equation and the equation of motion in the partly vegetated open-channel flows are shown as follows:

$$\frac{\partial V}{\partial y} + \frac{\partial W}{\partial z} = 0 \tag{1}$$

$$V\frac{\partial U}{\partial y} + W\frac{\partial U}{\partial z} = gI_e - F_x + \frac{\partial(-\overline{uv})}{\partial y} + \frac{\partial(-\overline{uw})}{\partial z} + v\left(\frac{\partial^2 U}{\partial y^2} + \frac{\partial^2 U}{\partial z^2}\right) \tag{2}$$

in which, U is the mean-velocity component in the streamwise direction, x, V is the mean-velocity component in the vertical direction, y, and W is the mean-velocity component in the spanwise direction, z. g is the gravitational acceleration, I_e is the energy gradient, v is the kinematic viscosity and F_x is the resistance force by the vegetation. Eq.(2) is transformed by Eq.(1) as follows:

$$gI_e = -\frac{\partial}{\partial y}\left(-UV - \overline{uv} + v\frac{\partial U}{\partial y}\right)$$
$$-\frac{\partial}{\partial z}\left(-UW - \overline{uw} + v\frac{\partial U}{\partial z}\right) + F_x \tag{3}$$

Upon integration of Eq.(3) with respect to y from the bed ($y=0$) to the free surface ($y=H$), Eq.(3) reads:

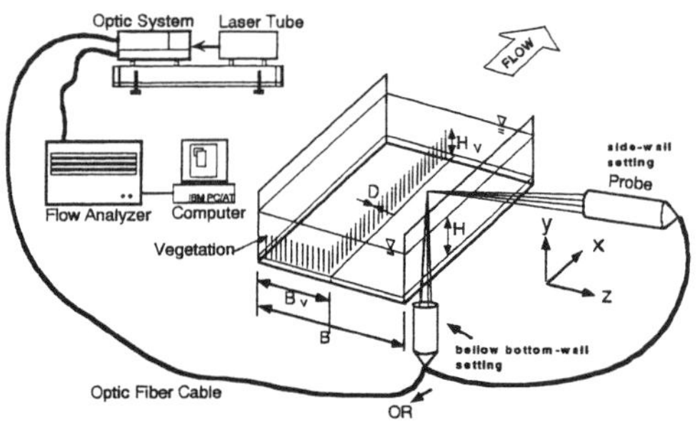

$$\frac{\tau_w}{\rho} = gHI_e + \int_0^H\left\{\frac{\partial}{\partial z}\left(-UW - \overline{uw} + v\frac{\partial U}{\partial z}\right)\right\}dy - \int_0^{H_v}F_x dy \tag{4}$$

in which, τ_w is the bed wall shear stress. The integration range of the third term in the right hand side in Eq.(4) is from the bed ($y=0$) to the top of the vegetation ($y=H_v$), because the vegetation affects on the water as the water resistance only in these regions. With the aid of Leibnitz rule, Eq.(4) is reduced as follows:

$$\frac{\tau_w}{\rho} = gHI_e + H\frac{d}{dz}(T - J) - \int_0^{H_v}F_x dy \tag{5}$$

$$T = \frac{1}{H}\int_0^H\left(-\overline{uw}\right)dy \tag{6.a}$$

$$J = \frac{1}{H}\int_0^H(UW)dy \tag{6.b}$$

Fig.1 Experimental Setup

Table 1 Hydraulic Conditions

Case	S	S_v (cm)	H (cm)	Q (ℓ/s)	Fr	Re ($\times 10^3$)	λH_v
FR1	1/3100			2.23	0.10	5.8	
FR2	1/2700	1.0	7.0	5.50	0.24	13.3	1.00
FR4	1/2600			8.93	0.40	23.5	

S: Bed Slope, S_v: Spacing of Vegetation, H: Depth, Q: Discharge, Fr: Froude Number, Re: Reynolds Number, λH_v: Density of Vegetation

3. EXPERIMENTAL SETUP AND HYDRAULIC CONDITIONS

The experiments were conducted in a 10m long, 40cm wide, and 30cm deep tilting flume, as shown in Fig.1. Three kinds of experiments were conducted as indicated in Table 1; the Froude numbers were changed. In Table 1, S is the bed slope, S_v is the spacing of vegetation, Q is the flow discharge, $Fr = U_m/\sqrt{gH}$ is the Froude number, $Re = U_m H/v$ is the Reynolds number and U_m is the bulk mean velocity. $\lambda = D/S_v^2$ is the density of vegetation (simulated by bronze cylinder rod with $D = 2$mm diameter and $H_v = 50$mm length).

Two components of instantaneous velocities, the streamwise velocity $\tilde{u} = U + u$ and the vertical velocity $\tilde{v} = V + v$, were measured with a four-beam LDA system. The LDA was located at 8m downstream of the channel entrance so that the flow was fully developed. All output signals of the LDA were recorded in a

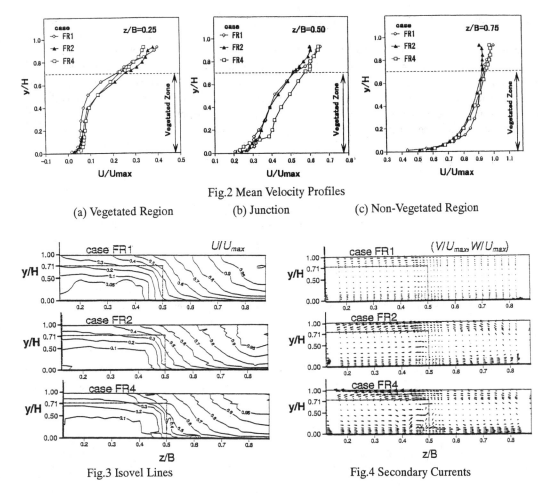

Fig.2 Mean Velocity Profiles

(a) Vegetated Region   (b) Junction   (c) Non-Vegetated Region

Fig.3 Isovel Lines   Fig.4 Secondary Currents

digital form with a sampling frequency more than 100Hz into the HDD of a personal computer. After experiments, all of the experimental data were transferred to the work station through the LAN network.

4. EXPERIMENTAL RESULTS

4.1 Mean Velocity Profiles

Fig.2 shows the mean velocity profiles of (a) in the vegetation region(z/B =0.25), (b) at the junction of vegetated and non-vegetated regions(z/B =0.50), and (c) in the non-vegetated region(z/B =0.75). In the non-vegetated region, the mean velocity profiles resemble those of normal flow, i.e., non-vegetation flow. This fact implies that the velocity distributions in non-vegetated region are expressed by the log-law. In contrast, the shape of the velocity profiles in the vegetated region is complicated. The mean velocity is almost constant in and around the inner region(y/H <0.1 to 0.4), and increases suddenly near the junction of vegetated and non-vegetated regions and the upper region of the vegetated region(y/H >0.71). These profiles agree well with those of canopy flow(Thompson (1979) and Raupach & Thom(1981)). At the junction (z/B =0.50), the velocity profiles are almost linear, because this region is affected both the vegetated and non-vegetated regions.

4.2 Isovel Lines

Fig.3 shows the isovel lines normalized by the maximum velocity U_{max} of the mean velocity for all cases. The dashed line shows the edge between the vegetation region and the non-vegetation region. In the vegetation region, the isovel lines are almost parallel to the bed. This shows that the flow in the vegetation region is almost two dimensional. However, near the free surface of the junction of the vegetated region and the non-vegetated region, the isovel lines bulge considerably toward the upper region of the vegetation. This characteristic is discussed in detail in chapter 4.3.

4.3 Secondary Currents

Fig.4 shows the velocity vectors of the secondary currents V and W normalized by the maximum mean velocity U_{max}. The pattern of secondary currents is quite different from that of the normal flow (Nezu & Rodi(1985)). The strength of the secondary currents near the free surface around the junction is considerably large. In the case of the normal flow, the secondary currents are generated by the vorticity generation term, $(\overline{w^2} - \overline{v^2})$ as shown by Nezu & Nakagawa(1984). However, in the vegetated flow, the secondary currents are generated not by the vorticity generation term, but by the surface vortex (Ikeda $et\ al.$(1992)). The strength of the secondary currents increase with an increase of the Froude numbers, because the surface vortices are generated more frequently. Therefore, the isovel lines of the mean velocity are deformed.

4.4 Reynolds Stress

Fig.5 shows the contour lines of the vertical Reynolds stress ($-\overline{uv}$). The shadow area means the negative region. It can be seen that the Reynolds stress ($-\overline{uv}$) increases extremely at the boundary of the vegetated and non-vegetated region (y/H =0.71). This is because the shear between the vegetated and the non-vegetated region is produced by the surface vortices.

Fig.6 shows the contour lines of the transverse Reynolds stress ($-\overline{uw}$). It was found that the values of the Reynolds stress ($-\overline{uw}$) are almost negative in the vegetated region, and that the maximum values occur near the boundary of vegetated and non-vegetated regions (z/B =0.50). The maximum value increases with an increase of the Froude numbers.

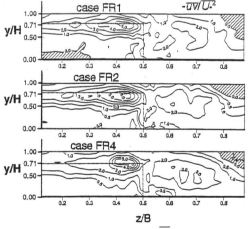
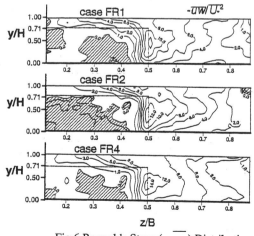

Fig.5 Reynolds Stress($-\overline{uv}$) Distribution   Fig.6 Reynolds Stress($-\overline{uw}$) Distribution

4.5 Generation of Turbulence

Fig.7 shows the turbulent generation rate normalized by the water depth H and the friction velocity $\overline{U_*}$. which is averaged over the smooth wetted bed in the non-vegetated region. The turbulent generation rate increases along the edge of the vegetated and non-vegetated regions, namely at the level of $y/H = 0.71$ and of $z/B = 0.5$. This is because, nevertheless, the velocity gradient is almost constant irrespective of the Froude numbers (see Fig.3), and the strength of the Reynolds stress increases with an increase of the Froude numbers(see Fig.6). In contrast, the turbulence is not generated in the vegetated region so much.

Fig.8 shows the distributions of the turbulent energy normalized by the mean friction velocity $\overline{U_*}$. . The turbulent energy increases where the turbulent generation rate increases (see Fig.7). In spite of the turbulent generation rate is quite small in the vegetated region (smaller than 4% comparing with the maximum value), the turbulent energy has a value to some extent (about 10 to 20% of the maximum value). This may be caused by the secondary currents. The secondary currents transport the turbulence from the upper region of the vegetation($y/H > 0.71$) in to the vegetated region($y/H < 0.71$) (see Fig.4) through the boundary.

4.6 Bed Shear Stress

Estimation of the bed shear stress is quite important because it is necessary not only for the normalizing flow characteristics but also for sediment problems. In the normal flow, the bed shear stress can be estimate easily from the log-law, the Reynolds stress distribution, and so on. However, these methods are not valid in vegetated flow, because these laws do not hold depending on the conditions.

In the vegetated flow, the bed shear stress can calculate from Eq.(5). The resistance force F_x is estimated as follows:

$$F_x = \frac{1}{2}\lambda C_d U \sqrt{U^2 + V^2 + W^2} \tag{7}$$

The drag coefficient C_d is substituted for that of the cylinder in an uniform flow. Fig.9(a) shows the bed shear stress distributions estimated by the log-law shown as follows:

$$\frac{U}{U_*} = \frac{1}{\kappa}\ln\frac{yU_*}{\nu} + A \tag{8}$$

in which κ (=0.41) is the Karman constant and A is the integration constant. In vegetated region, the bed

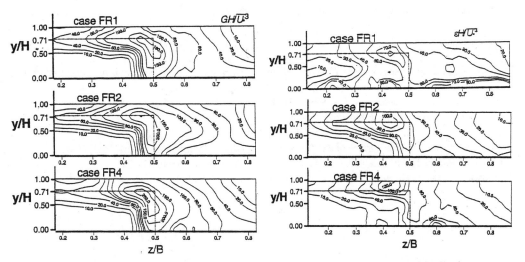

Fig.7 Turbulent Generation Rate   Fig.8 Turbulent Energy Distribution

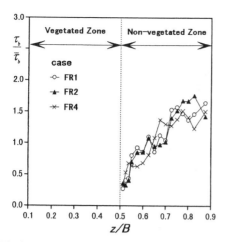 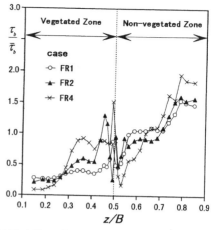

Fig.9(a) Bed Shear Stress Calculated by Log-Law   (b) Bed Shear Stress Calculated by Momentum Eq.

shear stress is not calculated, because the log-law(8) does not hold in this region. Fig.9(b) shows the bed shear stress distributions estimated by the momentum method(Eq.(5)). These two distributions agree well with each other in non-vegetated region. In vegetated region, the bed shear stress decreases in comparison with non-vegetated region, because the vegetation takes charge of the flow resistance. The behavior near the junction(z/B =0.5) is complicated, but it is quite important. These aspects will be investigated in the near future.

5. CONCLUSIONS

The turbulence measurements of open-channel flows with vegetated zone at a half channel width were conducted by making use of a laser Doppler anemometer(LDA). It was found that the intensity of secondary currents and the turbulence energy increase with an increase of the Froude numbers. The turbulence is advected spanwisely near the free surface by the secondary currents, which are generated by the surface vortices and after that it is transferred into the vegetated region. Therefore, the pattern of secondary currents is quite different from that of the normal flow. The wall shear stress in vegetated zone is calculated by the equation of motion with the aid of the empirical drag coefficient.

REFERENCES

Ikeda, S., Ohta, K. and Hasegawa, H. (1992): *Journal of Hydraulic, Coastal and Environmental Engineering*, Japan Society of Civil Engineering, No.443/II-18, pp.47-54(in Japanese).
Naot, D., Nezu, I. and Nakagawa, H.(1996): "Hydrodynamic behavior of partly vegetated open channels", *Journal of Hydraulic Engineering* , ASCE, vol.122, pp.671-673.
Nezu, I. and Nakagawa, H. (1984), *J. Hydraulic Eng.* Vol.110, pp.173-193.
Nezu, I. and Rodi, W. (1985): *Proc. of 21st Congress of IAHR*, Melbourne, Vol.2 pp.115-119.
Raupach, M.R. and Thom, A.S. (1981), *Annual Rev. Fluid Mech.* Vol.13, pp. 97-129.
Thompson, N. (1979) , *Boundary-Layer Meteorology* Vol.16 pp.293-310.

Environmental Hydraulics, Lee, Jayawardena & Wang (eds) © 1999 Balkema, Rotterdam, ISBN 90 5809 035 3

Coherent flow structures in unsteady open-channel flows over dune bed

Akihiro Kadota
Department of Civil and Environmental Engineering, Ehime University, Matsuyama, Japan

I. Nezu
Department of Civil and Global Environment Engineering, Kyoto University, Japan

K. Suzuki
Ehime University, Matsuyama, Japan

ABSTRACT : In the present study, the specific three-dimensional behaviors of coherent vortices in unsteady open-channel flows over dune bed were accurately measured by means of simultaneous use of two sets of laser Doppler anemometer. The conditional sampling analysis, which is applicable for the unsteady open-channel flows, is examined with a definition of some fixed-times of flood period. By using the analysis, conditional space-time correlation was evaluated, so that differences of convection properties on coherent vortices between the rising and falling stages of flood period, and interactions of coherent vortices, were recognized.

1. INTRODUCTION

There exists a strong upward-tilting streamwise vortex motion, called the *kolk*, near the bed, in a analogy to a tornado motion in the atmosphere, and the kolk vortex develops up to the free surface and then becomes a *boil*. A typical boil starts life with high sediment concentration, as a raised circular or oval patch on the water surface; in the course of time, it widens and gradually settles back down toward the river level until it subsides completely and merges with the surrounding, less-agitated parts. A kind of the boils is generated downstream of the crest of sand dunes and is related to the separated vortex from the crest. These vortices often occur especially in flooded fluvial rivers (Coleman 1969, Kinoshita 1984). The hydrodynamic behavior of flood structures is quite different between the rising and falling stages. In the rising stage, the concentration of suspended sediment increases due to the kolk-boil vortex because the rising stage corresponds to the growing period of dunes and because separated vortex from the dune collides intermittently at the next dune and washes out bed materials around there. Therefore, it is very important to investigate unsteadiness effects on the coherent vortices so that the growing process of the dunes and the behavior of the suspended sediment can be explained dynamically.

In the present study, the three dimensional behaviors of coherent vortices in unsteady open-channel flows over dunes were accurately measured by means of simultaneous use of two sets of laser Doppler anemometer. The differences of convection properties on coherent vortices between the rising and falling stages were discussed by evaluating the conditional space-time correlation.

2. EXPERIMENT & CONDITIONAL SAMPLING ANALYSIS

Experiments were conducted in a 10m long, 40cm wide and 50cm deep tilting flume. Fig. 1 illustrates the measuring system for space-time correlation analysis. Two sets of LDA, water-wave gauge and an automatic traversing system were used with connected to two sets of IBM personal computers which enable to control every parameter such as bias, Bragg cell, position of the movable probe. One is the Argon-Ion fiber-optic LDA with a 2W high-power water-

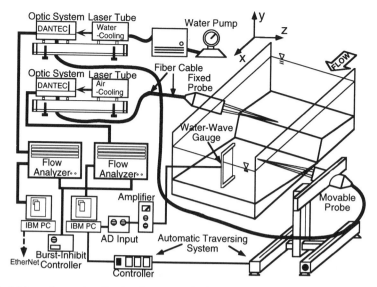

Fig.1 Measuring system for simultaneous velocity measurements by two sets of LDA.

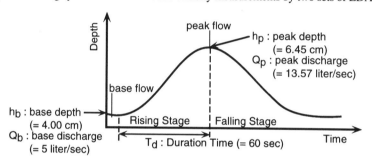

Fig.2 Hydraulic condition.

cooling laser light and 4 beams backscattering mode. This LDA was adopted for the movable probe. The other LDA for the fixed probe has the same specification as the water-cooling LDA except for 100mW power and the air-cooling laser light. The automatic traversing system has only an error of 0.1mm. The fixed-probe measuring points adopted here are both the separation point (crest of the dune) and the reattachment point. The latter is defined by the point at which the streamwise mean velocity becomes just zero on the channel bed. The experimental condition are shown in Fig.2. In the present study, the time t is normalized by the flood period T_d, i.e., $T=t/T_d$. Therefore, the period from $T=0$ to 1 corresponds to the rising stage, whereas the period from $T=1$ to 2 corresponds to the falling stage.

The conditional averaging of an arbitrary sampling signal $q(x+\Delta x, y+\Delta y, z+\Delta z)$ is defined as:

$$\left\langle q\left(x_0, y_0, z_0, \Delta x; \Delta y; \Delta z; t_{fix}, \tau\right)\right\rangle = \frac{\int_T q\left(x_1, y_1, z_1, t_{fix} + t + \tau\right) \cdot I\left(x_0, y_0, z_0, t_{fix} + t\right) dt}{\int_T I\left(x_0, y_0, z_0, t_{fix} + t\right) dt} \qquad (1)$$

where, the parenthesis < > indicates the averaged value after conditional sampling. The subscripts 0 and 1 are fixed and movable points, respectively. τ is the lag-time and (Δx, Δy, Δz) are the distances from the fixed point (0) to the

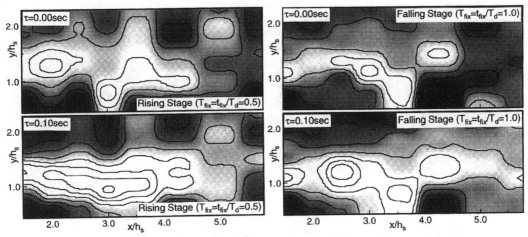

Fig.3 Conditional space-time correlation of streamwise velocity $<u>$ for separated vortex.

movable one (1). $I(x, y, z, t)$ is the detection function of coherent vortices. For the purpose of the strict conditional sampling analysis, the detection function and threshold value should be reasonably defined. At first, fixed-time points t_{fix} were defined in arbitrary phase of hydrograph shown in Fig.1. Then, the Reynolds numbers Re at each fixed time were evaluated in order to determine the generation period T_g of the separated and kolk-boil vortices by using the approximated relations between Re and T_g (Nezu et al., 1996). These relations consist of exponential functions on the basis of the assumption that the value of T_g may tend to the constant value as Re becomes larger. Next, instantaneous Reynolds stress $-uv(t)$ was adopted as detection functions $I(t)$. The expressions of detection functions for separated vortex I_s and kolk-boil vortex I_k are expressed in the followings :

$$I_s(t) \equiv \begin{cases} 1 : u > 0, v < 0 \ \& \ |uv/u'v'| \geq H_s \\ 0 : otherwise \end{cases}, \quad I_k(t) \equiv \begin{cases} 1 : u < 0, v > 0 \ \& \ |uv/u'v'| \geq H_k \\ 0 : otherwise \end{cases} \quad (2)$$

in which, H is the threshold value. Subscripts s and k denotes for the separated and kolk-boil vortices, respectively. The I_s means a condition that the high-speed separated vortex generated from the dune crest descends to the reattachment point, whereas the I_k indicates that the low-speed fluid ascends suddenly from the reattachment point. The threshold values H_s and H_k for these detection functions are determined so that the sample number which exceeds the threshold value coincides with the generation frequency.

4. RESULTS AND DISCUSSION

4.1 Conditional Space-time Correlation Coefficient in Vertical Section

Fig.3 shows the distributions of conditionally sampled streamwise velocity $<u>$ for separated vortex. The sequence of figures indicates the evolution of separated vortex generated at lag-time $\tau=0$sec from the location of fixed probe. The high positive parcel of vortex is convected in the streamwise direction. Especially in the falling stage, the high parcel drops down to the reattachment point ($x/h_s \approx 4$, h_s : height of dune crest) and streamwise convection velocity becomes larger as compared with the rising stage. On the contrary, the high parcel in the rising stage is enlarged to vertical direction from dune bed to free-surface and it is restrained in streamwise direction. The conditional space-time correlation of vertical velocity $<v>$ for the evolution of kolk-boil vortex is shown in Fig.4. The high-correlation

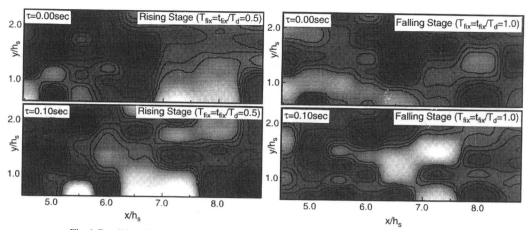

Fig.4 Conditional space-time correlation of vertical velocity $<v>$ for kolk-boil vortex.

region of $<v>$ indicates the strong upward flow from the reattachment point in the falling stage. The high correlation existing near the free surface in the rising stage is not convected in the streamwise direction as compared with the falling stage and the same tendency as field observations can be recognized. The kolk-boil vortex with weak energy is reasonably sampled in the outer region.

4.2 Conditional Space-time Correlation Coefficient in Horizontal and Spanwise Section

Fig.5 shows horizontal $(x-z)$ and spanwise $(y-z)$ contours of conditionally-sampled space-time correlation of streamwise velocity $<u>$ for separated vortex generated from the dune crest at $\tau=0$ and 0.2sec. The left figure indicates the convection process in the period from the rising stage to the peak flow ($T=0.5$), whereas the right figure indicates the process in the period from peak flow to the falling stage ($T=1.0$). By using the above-mentioned conditional sampling, the structure of low and high-speed streaks is clearly seen. From the horizontal view, the region of high positive correlation exists along the centerline $(z/h_s=0)$ near the dune crest $(y/h_s=2)$. In the neighborhood of the high-positive region, a high-negative region is seen and positive region exists next to the negative region. In the rising stage ($T=0.5$), high speed parcel which is presented by the high positive figure is convected not only in the vertical direction but also in the horizontal plane from the dune crest. The significant difference between the rising stage ($T=0.5$) and the falling stage ($T=1.0$) is that the convection of the high-positive parcel in the falling stage is larger than in the rising stage. From the horizontal view in the falling stage, the high parcel which existed around $x/h_s=4$ at $\tau=0$sec has already disappeared at $\tau=0.2$sec. On the other hand, in the rising stage, the high-positive parcel around $x/h_s=4$ at $\tau=0$sec remains even at $\tau=0.2$sec. Therefore, it can be understood that the separated vortex in the rising stage spreads in all three directions, whereas the vortex in the falling stage is dominant for the streamwise convection and does not spread in other directions, because the rising stage corresponds to the spatial deceleration and the falling stage to the spatial acceleration (Kadota, 1993).

In the same manner as Fig.4, Fig.5 shows the conditionally-sampled space-time correlation of vertical velocity $<v>$ for kolk-boil vortex generated from the reattachment point. At the lag-time $\tau=0$sec and near the reattachment point from the view of the horizontal and cross-sectional plane, the streak structure filled with the positive and negative correlations, is relatively narrow compared to the separated vortex in the rising stage. Further downstream, the high positive region which indicates high speed parcel is spread in horizontal plane, whereas the negative low-speed region becomes narrower. Therefore, it can be seen that a horseshoe-like vortex is generated behind the dune.

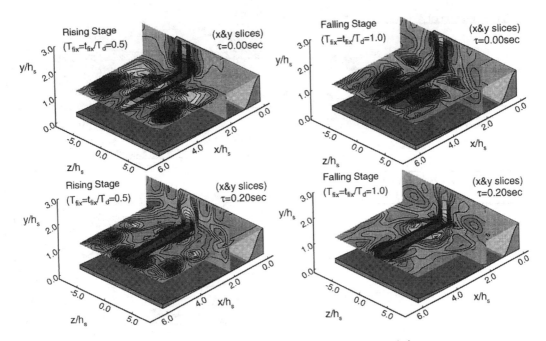

Fig.5 Horizontal and spanwise conditional space-time correlation
of streamwise velocity <u> for separated vortex.

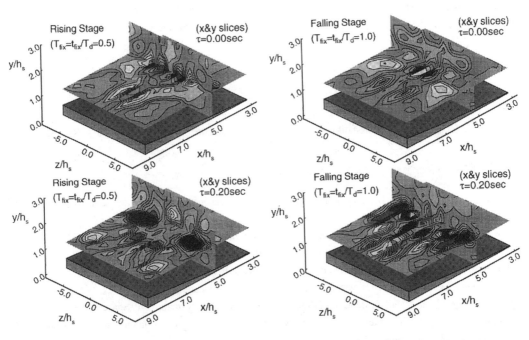

Fig.6 Horizontal and spanwise conditional space-time correlation
of vertical velocity <v> for kolk-boil vortex.

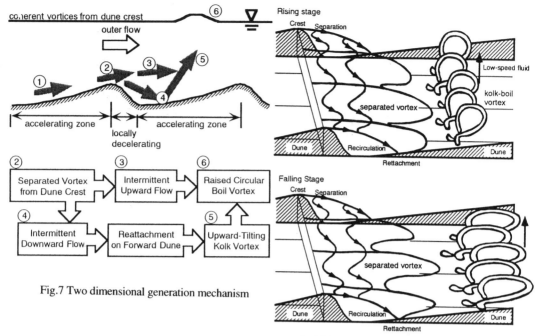

Fig.7 Two dimensional generation mechanism

Fig.8 Three dimensional generation mechanism

In the falling stage, the high speed parcel is not raised so much to the free surface and it shows strong streamwise convection as compared with the rising stage due to the spatial acceleration. The above mentioned coherent vortices are schematized as shown in Fig.7 and Fig. 8 to show their convection structure.

5. CONCLUSIONS

In the present study, the turbulence measurements in unsteady open-channel flows over dunes were conducted by means of 2 sets of LDA and water-wave gauge. The three dimensional structures of conditional space-time correlations for separated and kolk-boil vortices were examined and their convection structure was considered.

REFERENCES

1) Coleman, J. M. (1969), Brahmaputra River, Channel Process and Sedimentation. Sediment. Geol., vol.3, pp.129-239.
2) Kadota, A. (1993), Significant Difference between Turbulence Characteristics of Unsteady Flows in Open-Channels and Pipes, Proc. XXV Congress of IAHR, Student Paper Session, pp.49-56.
3) Kinoshita, R. (1984), Present Status and Future Prospects of River Flow Analysis by Aerial Photograph, J. Hydraulic, Coastal and Environmental Eng., JSCE, pp.1-19, (in Japanese)
4) Nezu, I., Kadota, A. and Kurata, M. (1996), Free-surface Flow Structures of Space-Time Correlation of Coherent Vortices Generated behind Dune Bed, Flow Modeling and Turbulence Measurements, IAHR, Balkema, Rotterdam, pp.695-702.

Environmental Hydraulics, Lee, Jayawardena & Wang (eds) © 1999 Balkema, Rotterdam, ISBN 90 5809 035 3

Coherent structures in unsteady open-channel flows

Kouki Onitsuka & Iehisa Nezu
Department of Civil and Global Environment Engineering, Kyoto University, Japan

ABSTRACT: Turbulence measurements of unsteady open-channel flows over a flat rough bed were conducted by making use of a laser Doppler anemometer (LDA). The contribution rates of each event to the Reynolds stress and the bursting period were investigated. The value of RS_4/RS_2, which means the contribution ratio of the sweep event against the ejection event, decreases with an increase of the duration time in rising stage. The bursting phenomena increase in the rising stage of flood, while they decrease in the falling stage. This shows that the bursting phenomena occur more frequently in the rising stage. From this study, the characteristics, that the concentration of suspended sediment increases in the rising stage of natural river, are substantiated.

1. INTRODUCTION

The open-channel flows consist of both the inner layer and the outer layer. In the inner layer, the bursting phenomena, which consist of four events; the outward interaction RS_1, the ejection RS_2, the inward interaction RS_3 and the sweep RS_4, occur quasi-periodically. In contrast, the kolk-boil vortices occur in the outer layer. In the case of 2-D steady open-channel flow, Nakagawa & Nezu(1977) have revealed that the bursting period is almost constant irrespective of the Reynolds numbers and Froude numbers. The bursting phenomena in the unsteady flows may be affected by the unsteadiness. These characteristics have not been investigated as yet, in spite of their importance. It is necessary to make clear the bursting phenomena in unsteady open-channel flows such as flooded rivers.

Until recently, the flood analysis had been dealt with under the assumption that the flow is quasi steady. However, Hayashi *et al.* (1988) have showed that the turbulence becomes stronger in the rising stage than in the falling stage by making use of a hot-film anemometer. Tu & Graf(1992) have measured unsteady open channel flow over gravel beds by making use of micropropeller flow meters and examined the unsteadiness effects on turbulent structures by using Clauser's equilibrium pressure gradient parameter. Song & Graf(1996) have measured by the use of an acoustic Doppler velocity profiler(ADVP) and proposed the prediction method of the Reynolds stress distributions by making use of the power law. Nezu & Nakagawa (1991, 1997) have measured unsteady smooth and rough open-channel flows by making use of a laser Doppler anemometer(LDA) with high accuracy, and have revealed that a peak discharge appears before a time of peak depth in proportion to unsteadiness and also that the mean velocity distribution up to the free surface can be expressed well by the log-wake law.

The mean flow structures of unsteady open-channel flows were made clear to some extent by many researchers as mentioned above. However, the bursting phenomena in unsteady open-channel flows are not investigated at all. In this study, the turbulence measurements of unsteady open-channel flows over a rough bed were conducted by making use of a two-component fiber-optic LDA system.

2. THEORETICAL CONSIDERATIONS

The instantaneous Reynolds stress $w(t) \equiv u(t)v(t)$ is analyzed in the quadrant technique; the outward interaction ($u > 0$, $v > 0$), the ejection ($u < 0$, $v > 0$), the inward interaction ($u < 0$, $v < 0$) and the sweep ($u > 0$, $v < 0$). u and v are the velocity fluctuations in the streamwise(x) and the vertical(y) directions, respectively. By an introduction of a threshold level H in the u-v plane, the contribution of each event to the Reynolds stress is classified into five events. The time fraction $T_i(H)$ and the contribution rate $RS_i(H)$ to the Reynolds stress are defined by Nakagawa & Nezu(1977) as follows:

$$T_i(H) = \begin{cases} \int_H^\infty p_i(w)dw, & (i = 2,4) \\ \int_{-\infty}^{-H} p_i(w)dw, & (i = 1,3) \end{cases} \quad (1)$$

$$RS_i(H) = \begin{cases} \int_H^\infty w \cdot p_i(w)dw \geq 0, & (i = 2,4) \\ \int_{-\infty}^{-H} w \cdot p_i(w)dw \leq 0, & (i = 1,3) \end{cases} \quad (3)$$

$$T_5(H) = \sum_{i=1}^4 \int_{-H}^H p_i(w)dw = 1 - \sum_{i=1}^4 T_i(H) \quad (2)$$

$$RS_5(H) = \sum_{i=1}^4 \int_{-H}^H w \cdot p_i(w)dw = 1 - \sum_{i=1}^4 RS_i(H) \quad (4)$$

in which, $p_i(w)$ is the probability density fraction of w.

The bursting period T_b can be evaluated by several method such as the VITA method and the u-v quadrant threshold method. In this study, the latter method was used. The threshold level H_e was defined in which $RS_2(H_e)$ has a value of $RS_2(0)/2$. This is truce of the thresholds level H_e of sweeps. This threshold level is called the 'half-value threshold level'. The ejection period T_e is determined at the level H_e and the sweep period T_s is determined at the level H_s.

3. EXPERIMENTAL SETUP AND HYDRAULIC CONDITIONS

3.1 Experimental Setup

The experiments were conducted in a 10-m-long, 40-cm-wide, and 50-cm-deep tilting flume. In this water flume, the discharge Q can be automatically controlled by a personal computer in which the rotation speed of a water-pomp motor involving an inverter transistor is controlled by the feedback from the signals of an electromagnetic flow-meter. The bed has roughness elements which have 12.5mm-diameter glass beds.

Two components of instantaneous velocities, the streamwise velocity $\tilde{u} = U + u$ and the vertical velocity $\tilde{v} = V + v$, were measured with a four-beam LDA system. The LDA was located at 8m downstream of the channel entrance so that the flow was fully developed. The water-wave gauge was located 20cm downstream of the velocity measuring point. All output signals of the LDA and the water-wave gauge were recorded in a digital form with a sampling frequency more than 100Hz into the HDD of a personal computer. After experiments, all of the experimental data were transferred to the work station through the LAN network.

3.2 Hydraulic Conditions

Experimental conditions are shown in Table 1. T_d is the duration from the base discharge Q_b to the peak discharge Q_p in flood. h_b is the water depth before the flood; base flow depth, h_p is the maximum water depth in the flood; peak flow depth. Q_b is the base discharge, while Q_p is the peak discharge. α is the unsteadiness parameter defined by Nakagawa et al. (1997) as follows:

Table-1 Hydraulic Conditions

Case	T_d (s)	h_b (cm)	h_p (cm)	Q_b (ℓ/s)	Q_p (ℓ/s)	α $\times 10^{-4}$
3C	90		8.0			4.81
3D	180	5.7	8.0	10.0	20.0	2.40
3E	270		8.1			1.68
3F	360		8.1			1.26

318

$$\alpha = \frac{2}{U_{mb} + U_{mp}} \frac{h_p - h_b}{T_d} \tag{5}$$

in which, U_{mb} and U_{mp} are the bulk mean velocity at the base and peak flows, respectvely.

3.3 Definition of Mean Velocity Component

In general, the methods of separating the mean component from the turbulent fluctuations are as follows: (a) the ensemble average method, (b) the moving time average method and (c) the Fourier component method. Nezu & Nakagawa(1991) pointed out that the method (c) is the most reasonable for unsteady open-channel flows. In this study, the method(c), therefore, was adopted. More detailed information is available in Nezu et al.(1997).

4. RESULTS AND DISCUSSION

4.1 Fractional Contributions to Reynolds Stress

Fig.1 shows the fractional contributions to the Reynolds stress $-\overline{uv}$ against the threshold level H as a function of time t/T_d and position y/h. It can be seen that the value RS_4 is greater than RS_2 in all ranges of H near the bed (y/h =0.05). In contrast, the RS_2 is greater than RS_4 in all ranges of H near the free surface (y/h =0.5). The theoretical curves of Nakagawa & Nezu(1977) also shown in Fig.1 The characteristics near the bed may be dominated by not unsteadiness but the roughness effect, because the unsteadiness does not have a significant influence upon the turbulence intensity and the Reynolds stress distributions normalized by the friction velocity so much, as pointed out by Song & Graf(1996) and Nezu *et al.*(1997). In steady rough open-channel flows, Nakagawa & Nezu(1977) have revealed that the sweeps are more dominated event than the ejections near the bed. The results of this study agree well with those of steady rough open-channel flow.

We discuss the unsteadiness effect on the fractional contributions. At the position of y/h =0.2, the RS_4 is greater than RS_2 in the time t/T_d =0.0 to 0.5, while the RS_4 is smaller than RS_2 in the time t/T_d =0.5 to 1.5. After that, the RS_4 becomes greater than RS_2 in t/T_d >2.0. Of course, the values of RS_2 and RS_4 after the flood approach to the values of before the flood. It can be generally said that the ejections activate turbulence in the outer layer of flood. The theoretical curves are in a good agreement with experimental data.

4.2 Contribution Rates

It is quite important to investigate the contribution rates of the ejections and sweeps, because the turbulence is almost generated in these events. Fig.2 shows the value of contribution rate of the sweeps against the ejections at the threshold level H =0. The ratio of RS_4/RS_2 slightly decreases near the bed and distinctly decreases in the outer region with an increase of the normalized time t/T_d. This fact means that the ejection event is the most dominant one as compared with the sweep event in flood flows. Although the profiles of RS_4/RS_2 against y/h are much complicated, here, we approximate these profiles as linear distributions. Such approximate linear distributions are described in Fig.2 by straight lines. It can be seen that the cross point between the straight line and RS_4/RS_2 =1.0 changes with the normalized time t/T_d.

Fig.3 shows the behavior the elevation of RS_4/RS_2 =1.0 against the normalized time t/T_d. The elevation decreases with an increase of t/T_d in the rising stage ($0<t/T_d<1.0$) and increases in the falling stage ($1.0<t/T_d<2.0$). Although there is some scatter in data, it can be said that the contribution rate of the ejections has a maximum value when the water depth has a peak depth.

$$3F\left(T_d = 360\text{s}, \quad \alpha = 1.26 \times 10^{-4}\right)$$

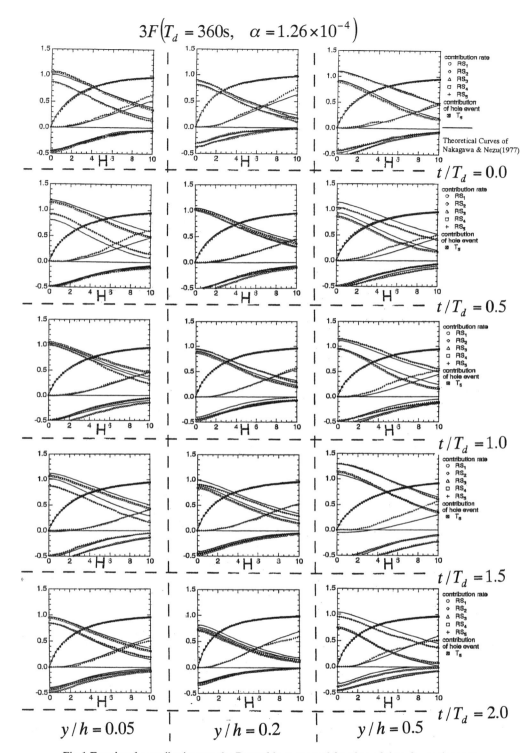

Fig.1 Fractional contributions to the Reynolds stress and fraction of time for each event.

4.3 Fraction Time

Fig.4 shows the ratio of the fraction time T_4 to the fraction time T_2 against the flow depth y/h, which were calculated by Eq.(2). This ratio gradually increases in the rising state $(0 < t/T_d < 1.0)$ and decreases in the falling stage $(1.0 < t/T_d < 2.0)$ with respect to the normalized time t/T_d. This indicates the fraction time of ejections decreases as compared with the sweeps in the rising stage. It was found that in the rising stage the fractional contribution of ejections increases but their fraction time decreases, as judged from Figs.3 and 4. This relationship feature of the fractional contribution and the fractional time agrees with that of 2-D steady open-channel flows, as shown by Nakagwa & Nezu(1977). This shows the strength of the ejection event in an unit time increases. These characteristics may substantiate the fact that the concentration of the suspended sediment increases in the rising stage in flood. This is because, when the ejections increases near the bed, sediments are rolled up from the bed by ejections and transported by the water.

4.4 Bursting Period

In the case of 2-D open-channel flow, the bursting periods, which are normalized by the maximum streamewise velocity U_{max} and the water depth h, are almost

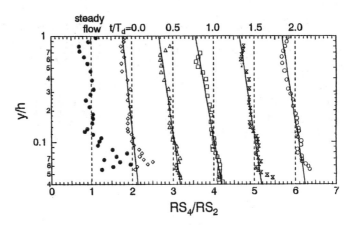

Fig.2 Variations of ratio RS_4/RS_2 in the case of 3F.

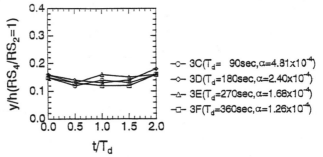

Fig.3 Positions normalized by h when $RS_4/RS_2 = 1$.

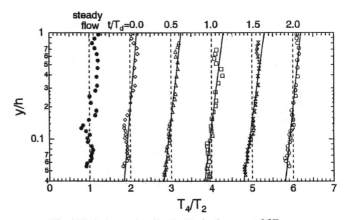

Fig.4 Variations of ratio T_4/T_2 in the case of 3F.

constant independent of the Reynolds numbers and the Froude numbers. Fig.5 shows the bursting period of the ejections and sweeps. These two period agree with each other, i.e., $\overline{T_e} \approx \overline{T_s} \approx \overline{T_{burst}}$. The period increases significantly near the bed. However, the period is almost constant in the whole depth expect near the bed and

321

the free surface.

Fig.6 shows the depth mean averaged bursting period T_b against of the normalized time. The value of T_b decreases in the rising stage, while it increases slightly in the falling stage. This means that the bursting occurs more frequently in the middle of rising stage.

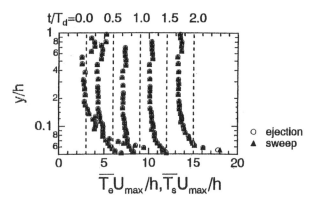

Fig.5 Mean bursting period normalized by outer variables.

5. CONCLUSIONS

Turbulence measurements of unsteady open-channel flows over a flat rough bed were conducted by making use of a laser Doppler anemometer(LDA). It was found that the fractional contribution rates of ejection event increases and the fraction time decreases in the rising stage as compared with the sweep event. The bursting period decreases in the rising stage, while it increases slightly in the falling stage. This shows that the bursting phenomena occur more frequently in the rising stage. These characteristics substantiate field observation results in which the

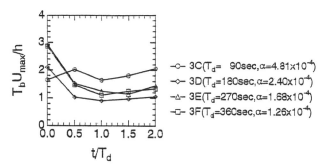

Fig.6 Mean bursting period against normalized time.

concentration of suspended sediment more increases in the rising stage of flooded natural rivers than in the falling stage.

REFERENCES

Hayashi, T., Ohashi, M., and Oshima, M.(1988), *Proc., 20th Symp. on Turbulence*, pp.154-159(in Japanese).

Nakagawa, H. and Nezu, I.(1977): *J. of Fluid Mech.*, Vol.80, pp.99-128.

Nezu, I. and Nakagawa, H.(1991), *Proc. of Int. Symp. On Transport of Suspended Sediments and its Math. Modeling*, IAHR, Firenze, pp.165-189.

Nezu, I. and Nakagawa, H. (1993): *"Turbulence in Open-Channel flows"*, IAHR Monograph, Balkema.

Nezu, I., Kadota, A. and Nakagawa, H.(1997), *J. of Hydraulic Eng.*, ASCE, Vol.123, No.9, Sep., pp.752-763.

Song, T. and Graf, W.H.(1996), *J. of Hydraulic Eng.*, ASCE, Vol.122, No.3, March, pp.141-154.

Tu, H. and Graf, W.H.(1993), *J. of Hydraulic Research*, Vol.31, No.1, pp.99-110.

Environmental Hydraulics, Lee, Jayawardena & Wang (eds) © 1999 Balkema, Rotterdam, ISBN 90 5809 035 3

Body-force effect of friction on quasi-two-dimensional turbulent flows in shallow waters

Wihel Altai & Vincent H.Chu
Department of Civil Engineering and Applied Mechanics, McGill University, Montreal, Que., Canada

ABSTRACT: Friction is considered as a body force with a stabilizing effect on the large-scale two-dimensional turbulent motions in shallow open-channel flows. To maintain a turbulent motion, the turbulent energy production by the transverse shear must be greater than the negative work done against the body force of the friction. Numerical calculations were conducted of the shallow turbulent flows, in the form of the starting jets, for several depths using a two-length-scale turbulence model developed to study the body-force effect. The results of the calculations are compared with a recent experimental observations of the friction effect on the starting jets of small depths.

1. INTRODUCTION

When a three-dimensional (3D) open channel flow is averaged over the depth, the friction force on the channel bed becomes a 'body force' of the two-dimensional (2D) depth-averaged motion. The body force of the friction

$$f_i = -\frac{c_f}{2h}\tilde{u}_i\sqrt{\tilde{u}_k\tilde{u}_k}, \quad i = 1, 2 \tag{1}$$

acts in the opposite direction of the depth-averaged velocity $\tilde{u}_i$. It slows down the mean flow U_i and exerts a stablizing influence to the two-dimensional turbulent motion u_i'. The part of the friction force associated with the mean flow is $\overline{f}_i = (c_f/2h)U_i\sqrt{U_kU_k}$. The part associated with the turbulent motion is obtained by Taylor's expansion of the force function $f_i(\tilde{u}_i)$ about the mean velocity U_i. The first term of this expansion is

$$f_i' = u_\ell'\frac{\partial f_i}{\partial \tilde{u}_\ell} = -\frac{c_f}{2h}[u_i'\sqrt{U_kU_k} + u_\ell'\frac{U_iU_\ell}{\sqrt{U_kU_k}}]. \tag{2}$$

The 2D turbulent motion must do work against the friction force. The rate of this frictional work is

$$F' = f_i'u_i' = -\frac{c_f}{2h}[\overline{u_k'u_k'}\sqrt{U_kU_k} + \overline{u_i'u_\ell'}\frac{U_iU_\ell}{\sqrt{U_kU_k}}]. \tag{3}$$

To maintain the 2D turbulent motion, the turbulent energy production

$$P' = \tau_{ij}\frac{\partial U_i}{\partial x_j} \simeq \nu_T(\frac{\partial U_i}{\partial x_j} + \frac{\partial U_j}{\partial x_i})\frac{\partial U_i}{\partial x_j} \tag{4}$$

must be greater than the negative work done by friction.

This concept of the depth averaging and the body-force effect is fundamental to understanding the 2D turbulence in open channel flows of small water depth. Chu, Wu and Khayat (1991) introduced the concept to study the stability of transverse shear flows of small depth. A two-length-scale turbulence

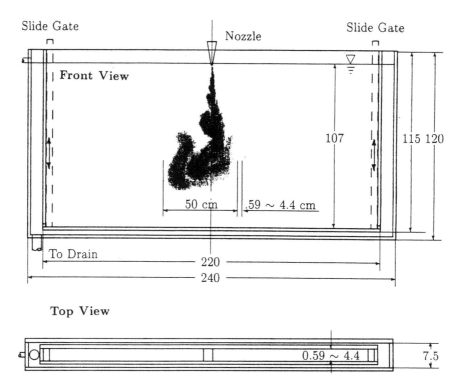

Figure 1: The double-tank apparatus for shallow turbulent flow simulation. Starting jets are produced in the inner tank by injection of dye into the small space between the tank walls. The lateral dimension of the turbulent motion (sim 50 cm) is very large compared with the small space (0.59 cm ~ 4.4 cm) between the parallel walls in the tank.

model was developed by Chu and Babarutsi (1989) to calculated the friction effect on the turbulent flows. Only a limited set of experimental data were available at the time to evaluate the performance of the turbulence model. For the friction to be effective, the length scale of the turbulent motion must be much greater than the depth of the flow, but that is a condition difficult to achieve in the laboratory. Recently Zhang (1997) was able to produce an unsteady turbulence flow in a large tank of small depth covering the entire range of friction effect. The tank as shown in Figure 1 is constructed of two large parallel walls (108 cm x 220 cm) with a small space (0.59 cm to 4.4 cm) between the walls. The shallow turbulent flow in the form of a starting jet was produced by injecting dye into the small space between walls. Concentration profiles of the turbulent flow was determined using a video imaging technique. In this paper, the data obtained from the starting-jet experiments are used to evaluate the performance of the two-length-scale turbulence model.

2. TURBULENCE MODEL

The depth-averaged velocity and dye concentration distributions in the shallow turbulent flows are obtained by the numerical solution of the momentum equations

$$\frac{\partial U}{\partial t} + U\frac{\partial U}{\partial x} + V\frac{\partial U}{\partial y} = -g\frac{\partial \zeta}{\partial x} + \frac{\partial}{\partial x}(\nu_T\frac{\partial U}{\partial x}) + \frac{\partial}{\partial y}(\nu_T\frac{\partial U}{\partial y}) \tag{5}$$

$$\frac{\partial V}{\partial t} + U\frac{\partial V}{\partial x} + V\frac{\partial V}{\partial y} = -g\frac{\partial \zeta}{\partial y} + \frac{\partial}{\partial x}(\nu_T\frac{\partial V}{\partial x}) + \frac{\partial}{\partial y}(\nu_T\frac{\partial V}{\partial y}) \tag{6}$$

and the mass conservation equation of the dye

$$\frac{\partial C}{\partial t} + U\frac{\partial C}{\partial x} + V\frac{\partial C}{\partial x} = \frac{\partial}{\partial x}(\nu_T\frac{\partial C}{\partial x}) + \frac{\partial}{\partial y}(\nu_T\frac{\partial C}{\partial y}) \tag{7}$$

where (U, V) = mean velocity, C = mean cocentration of the dye, ζ = water surface elevation, ν_T = turbulence viscosity, g = gravity constant, and c_f = bed-friction coefficient. The turbulence viscosity $\nu_T = \nu_{Ts} + \nu_{TL}$ of the two-length-scale turbulence model has two parts:

$$\nu_{TL} = c_\mu\frac{k'^2}{\epsilon'} \tag{8}$$

and

$$\nu_{Ts} = c_\nu U_* h \tag{9}$$

where $c_\nu \simeq 0.08$ and $c_\mu = 0.09$. The part due to the small-scale bed-generated turbulence, ν_{Ts}, is related to the shear velocity, $U_* = \sqrt{c_f/2}\sqrt{U^2 + V^2}$, and water depth, h, through a local equilibrium assumption. The part due to the large-scale transverse-shear-generated turbulence, ν_{TL}, is determined by the numerical solution of k'- and ϵ'-equations:

$$\frac{\partial k'}{\partial t} + U\frac{\partial k'}{\partial x} + V\frac{\partial k'}{\partial y} = \frac{\partial}{\partial x}(\frac{\nu_T}{\sigma_k}\frac{\partial k'}{\partial x}) + \frac{\partial}{\partial y}(\frac{\nu_T}{\sigma_k}\frac{\partial k'}{\partial y}) + (P' - F') - \epsilon', \tag{10}$$

$$\frac{\partial \epsilon'}{\partial t} + U\frac{\partial \epsilon'}{\partial x} + V\frac{\partial \epsilon'}{\partial y} = \frac{\partial}{\partial x}(\frac{\nu_T}{\sigma_\epsilon}\frac{\partial \epsilon'}{\partial x}) + \frac{\partial}{\partial y}(\frac{\nu_T}{\sigma_\epsilon}\frac{\partial \epsilon'}{\partial y}) + \frac{\epsilon'}{k'}c_{1\epsilon}[P' - (1 - c_{3\epsilon})F'] - c_{2\epsilon'}\frac{\epsilon'^2}{k'}. \tag{11}$$

where k' is the turbulent energy and ϵ' the dissipation rate associated with the large-scale turulence. The source and sink terms in these equation are the production term, P', and dissipation terms, F' and ϵ'. The production of the horizontal turbulence by the transverse shear is

$$P' = \nu_{TL}[4S_{xy}^2 + (S_{xx} - S_{yy})^2] - k'(S_{xx} + S_{yy}) \tag{12}$$

where

$$S_{xy} = \frac{1}{2}(\frac{\partial V}{\partial x} + \frac{\partial U}{\partial y}), \quad S_{xx} = \frac{\partial U}{\partial x}, \quad S_{yy} = \frac{\partial V}{\partial y}. \tag{13}$$

are the components of the strain-rate tensor. The rate of the negative work done by the large-scale motion against the force of friction is

$$F' = \frac{cf[\overline{u'^2}(2U^2 + V^2) + 2\overline{u'v'}UV + \overline{v'^2}(U^2 + 2V^2)]}{2h\sqrt{U^2 + V^2}}. \tag{14}$$

where

$$\overline{u'^2} = \nu_{TL}(S_{yy} - S_{xx}) + k', \quad \overline{u'v'} = -2\nu_{TL}S_{xy}, \quad \overline{v'^2} = \nu_{TL}(S_{xx} - S_{yy}) + k'. \tag{15}$$

The dissipation of the large-scale turbulence, ϵ', through the nonlinear process of energy cascade, is determined by the model equation, Equation 11.

The model coefficients $c_\mu = 0.09$, $c_{1\epsilon} = 1.44$, $c_{2\epsilon} = 1.92$, $\sigma_k = 1.0$, $\sigma_\epsilon = 1.3$, and $c_{3\epsilon} = 0.8$ are recommended for shallow turbulent flow simulation. These are identical to the set used by Hossain (1980) and Rodi (1985) to model the gravity-stratified flows. The selection of the identical set is based on an assumed analogy between friction and gravity forces. The same set of model coefficients

325

Table 1: Summary of the conditions of the laboratory experiments.

Test	$2h$	d_o	q_o	m_o/ρ
Numbers	cm	cm	cm^2/s	cm^3/s^2
W1	4.40	0.173	14.3	1189.9
M1	1.27	0.708	74.3	8676.6
N1	0.59	0.273	106.9	41890

were used successfully in the simulation of the turbulent mixing layers by Barbarutsi and Chu (1998).

3. NUMERICAL COMPUTATION

The numerical computations of the starting jets were conducted using a staggered grid. The velocity (U, V) is obtained by numerical solution of the momentum equation using a standard predictor and corrector procedure. The dye concentration distribution, C, the values of k' and ϵ' are calculated using the generalized second moment method (GSMM). The GSMM is unconditionally stable and positive definite. The oscillation and the negative value, known to be the problem of the conventional advection schemes, are completely eliminated. The details concerning implementation, stability and convergence of the GSMM are given in Chu and Altai (1998).

The 216 cm x 108 cm tank is divided into 384x192 computation cells. This corresponding to a cell size $\Delta x = \Delta y = 0.5625$ cm. The nozzle width is $d_o = 2\Delta x = 1.125$ cm that is greater than the nozzle sizes $d_o = 0.173$ cm, 0.273 cm and 0.708 cm in the laboratory experiments. Further reduction of the computational cell size is not necessary as the jet behaviour is primarily dependent of the momentum flux at the source. So the computation were conducted for open-channel flows of three depths: $h = 0.295$ cm, 0.635 cm and 2.2 cm and with the momentum flux at the source, $m_o = d_o V_o^2 = 1189$ cm^3/s^2, 8677 cm^3/s^2, and 41890 cm^3/s^2. These numerical simulations are corresponding to the laboratory experiments referred to by Zhang as tests N1, M1, and W1 for narrow, moderate and wide, respectively (see Table 1). The distances between the walls in the laboratory experiments ($2h = 0.59$ cm, 1.27 cm and 4.4 cm) are twice the depth of the open-channel-flow simulation.

Figure 2 shows the dye concentration profile of the starting jet obtained from the laboratory experiment W1 and the numerical simulation W. The starting jet is characterized by the formation of a 'head' which is twice as wide as the 'jet' feeding behind the head. To normalize the data, a friction length scale $= \ell_s = 2h/c_f$, a time scale $= t_s = \ell_s^{\frac{3}{2}}(m_o/\rho)^{-\frac{1}{2}}$ and a volume volume flux scale,

$$q_s = \frac{\ell_s^2}{t_s} = \ell_s^{\frac{1}{2}}(\frac{m_o}{\rho})^{\frac{1}{2}} = (\frac{2h}{c_f})^{\frac{1}{2}}(\frac{m_o}{\rho})^{\frac{1}{2}}, \tag{16}$$

are introduced. Figure 3 shows the simulation of the total entrainment flow rate into the starting jet, q, normalized by q_s. The initially development follows the relation, $(q/q_s)^2 \simeq 0.23(x_f/\ell_s)$, that after elimination of ℓ_s is

$$q \simeq 0.48\sqrt{m_o x_f/\rho} \tag{17}$$

independent of the friction effect. Friction become the a dominant effect in the later the stage of the development. The entrainment into the jet is limited by the asymptote $(q/q_s)^2 \simeq 0.03$, that is

$$q_{max} \simeq 0.17(\frac{2h}{c_f}\frac{m_o}{\rho})^{\frac{1}{2}}. \tag{18}$$

Calculations were conducted to find the dye concentration in other regions of the starting jet. The results were all in close agreement with the experimental observations. The friction effect becomes

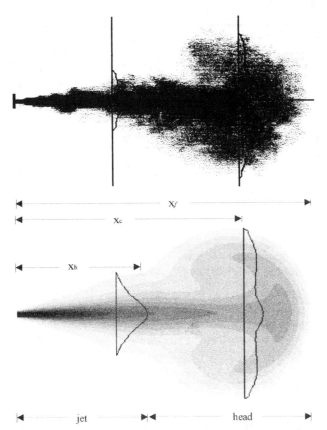

Figure 2: (a) Video image and (b) numerical simulation of the starting jet showing the concentration profiles in the 'head' and the 'jet' regions of the unsteady turbulent flow.

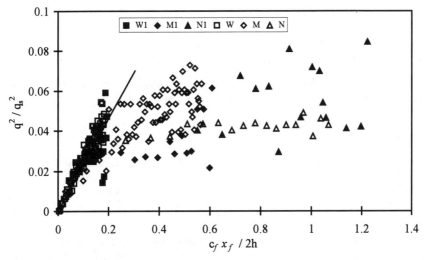

Figure 3: Volume rate of the fluid entrained into the starting jet. The data are normalized by the discharge scale, q_s.

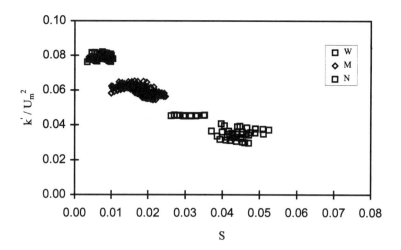

Figure 4: Relative level of the energy of the large-scale tubulence, k'/U_m^2, and its relation with the bed-friction number, **S**.

significant as the jet penetration distance is comparable to the friction length scale. Figure 4 shows the relatively level of turbulence energy, k'/U_m^2, obtained from the three simulations, W, M and N, with different ranges of the bed-friction number. The turbulence energy diminishes at a critical bed-friction number

$$\mathbf{S}_{\text{critical}} = \frac{c_f}{2h} \frac{\overline{U}}{(d\overline{U}/dy)_{\text{max}}} \simeq 0.08. \tag{19}$$

This critical value is identical to the value obtained from the experimental observation of the shallow turbulent mixing layer by Chu and Babarutsi (1988).

5. REFERENCES

Chu, V. H. and Altai, W. (1998). "Generalized second moment method for advection and diffusion processes," *Proc. 2nd Int. Symp. Envir. Hydraulics,* December 1998, Hong Kong.

Chu, V. H. and Babarutsi, S. (1989). "Modeling of the turbulent mixing layers in shallow open channel flows," *Proc. 23rd Congress of IAHR, Ottawa,* **A,** 191-198.

Chu, V. H. and Babarutsi, S. (1988). "Confinement and bed-friction effects in shallow turbulent mixing layers," *J. of Hydraulic Engineering, ASCE,* **114,** 1257-1274.

Chu, V. H., Wu, J.-H., and Kahyat, R. E. (1991). "Stability of transverse shear flows in shallow open channels," *J. Hydraulic Engineering.* **117,** 1-19.

Babarutsi, S. and Chu, V. H. (1998). "Modeling transverse mixing layer in shallow open-channel flows," *J. Hydraulic Engineering,* **124**(7), 718-727.

Rodi, W. 1985. Calculation of stably stratified shear-layer flows with a buoyancy-extented k-ϵ turbulence model. In *Turbulence and Diffusion in Stable Environments,* Edited by J. C. R. Hunt, Clarendon Press, Oxford.

Environmental Hydraulics, Lee, Jayawardena & Wang (eds) © 1999 Balkema, Rotterdam, ISBN 90 5809 035 3

Effects of vegetation on flow structures and bed profiles in curved open channels

A. Tominaga & M. Nagao
Department of Civil Engineering, Nagoya Institute of Technology, Japan

I. Nezu
Department of Civil and Global Engineering, Kyoto University, Japan

ABSTRACT: Flow in curved open channels is characterized by a generation of the pressure-driven secondary flow and the scour along the outer bank. It is expected that the vegetation can control the local scour and accumulation in curved open channels. In this study, three-dimensional mean flow structures were measured in curved open channels with various vegetation arrangements. Furthermore, experiments were conducted in a movable bed and the effects of vegetation on the local scour were examined. The secondary flow is generated only in the region outside the vegetated zones and the lateral scale of the vortex structures in bends are reduced. The main flow indicates complicated behaviors affected by the strong lateral shear, a centrifugal force and the momentum transport by the secondary flow. The vegetated area reduced the scour along the outer bank, but another scour around the vegetated zone was generated.

1. INTRODUCTION

A curved reach of a river is an important part for both flood management and ecological aspects. Flow in a curved open channel is characterized by a generation of the pressure-driven secondary flow and the scour at the side of the outer bank and the accumulation at the side of the inner bank. It has been a substantial subject in river engineering to clarify and control flow structures and bed configurations in bends (e.g., Rozovskii, 1967). The basic characteristics of flow in curved open channels are fairly understood, theoretically (e.g., Ikeda, Yamasaka and Kennedy, 1990) and numerically (e.g., Leshziner and Rodi, 1979). However, a response of curved flows to various boundary conditions has not been grasped well. In order to lessen the scour at the outer-bank bed, it should be effective to reduce the secondary flow by adding counter-rotating vortex. For example, Iowa vane method (Odgaard and Kennedy, 1983) and strip roughness attached on the outer bank (Sekine and Kikkawa, 1997) were designed and investigated. In a stream of a river with vegetation, it is conceivable that they make resistance to flow and they influence the 3-D flow structures including pressure-driven secondary flows. Also, it is expected that the vegetation can control the local scour and accumulation in curved open channels. These effects of vegetation on the flow in curved open channels have not been examined yet. In this study, the 3-D structures of the main flow and the secondary flows were investigated systematically in curved channel with vegetated zones. Furthermore, experiments were conducted in a movable bed and the effects of vegetation on the bed configuration were examined.

2. EXPERIMENTAL SETUP

The experiments were conducted in a 17.2m long, 0.9m wide and 0.3m deep flume. The curved reach was set between an upstream 10.8m long straight flume and a downstream 3.6m long straight flume as shown in Fig.1. The radius of curvature is 2.7m along the centerline of the channel and the angle of the circumference θ is

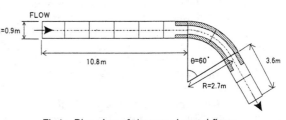

Fig.1 Plan view of the experimental flume

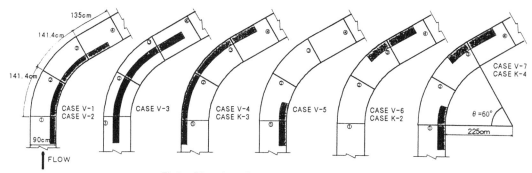

Fig.2 Plan view of vegetated zone arrangements

60 degree. As a model of tree, wooden sticks with a diameter of 5mm was used. The sticks were set to cover 225mm width on 50mm interval grid arrangement. Only the case of V-1 was densely arranged by adding one stick at the center of 50mm grid. The vegetated zones were placed in various arrangements; for instance, inside, center and outside of the channel at the curved section. The arrangements of vegetated zones are shown in Table.1 and Fig.2. The cases R-1 and K-1 are the reference cases with no vegetation for fixed bed and movable bed, respectively. The discharge Q was set to 0.04m³/s or 0.034m³. The test sections were located at x=-90cm upstream from the bend entrance, x=0 as the bend entrance, ϕ=15°, ϕ=30°, ϕ= 45° as a central angle and 60° at the bend outlet, and x'=45cm, 90cm, 135cm downstream from the bend outlet. The symbols, x and x', are the streamwise coordinates measured from the bend inlet (ϕ=0°) and from the bend outlet (ϕ=60°), respectively. The flow depth was set approximately 150mm at x=-90cm in all cases.

Three components of the velocity were measured by using two-component or three-component electromagnetic velocimeter. As for the two-component electromagnetic velocimeter, I-type and L-type probes were used for measuring streamwise/ spanwise components (u and v) and streamwise/ vertical components (u and w), respectively.

Next, in order to investigate the effects of vegetation on movable bed, 0.5mm diameter sand grains were spread on the bed with 100mm thickness from x=-180cm to x'=200cm. Four types of vegetation arrangement were conducted. The cases of K-1 to 4 correspond to R-5 (no vegetation), V-6, V-4 and V-7, respectively. After 5 hours from a start of the flow, bed configurations were measured by using bed configuration visualizing equipment. The velocity was measured on this transformed bed which was solidified by using cement.

Table 1 Vegetated zone arrangements

CASE	arrangement		type of bed
R-5	no vegetation		fixed
V-1	inside (dense)	-90cm to +90cm	fixed
V-2	inside	-90cm to +90cm	fixed
V-3	center	-90cm to +90cm	fixed
V-4	outside	-90cm to +90cm	fixed
V-5	inside	-90cm to ϕ=15°	fixed
V-6	outside	ϕ=45° to +90cm	fixed
V-7	inside and outside	-90cm to ϕ=15° + ϕ=45° to +90cm	fixed
K-1	no vegetation		movable
K-2	outside	ϕ=45° to +90cm	movable
K-3	outside	-90cm to +90cm	movable
K-4	inside and outside	-90cm to ϕ=15° + ϕ=45° to +90cm	movable

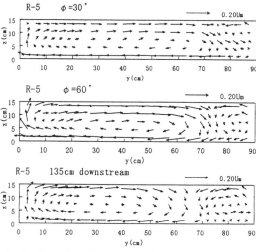

Fig.3 Secondary flow vectors for rectangular channel without trees (CASE R-5)

3. EXPERIMENTAL RESULTS FOR FIXED BED

3.1 Secondary Flow Structures

Firstly, the secondary flow structures in a rectangular curved channel with no vegetation are shown in Fig.3. The obtained structures of pressure-driven secondary flow sustained the common results, but an existence of a counter-rotating vortex near the outer bank is clearly shown. These multiple structure of the secondary flow in bends was also suggested by Bathrust et al.(1979) and Vriend(1983) and was predicted numerically by Sugiyama et al. (1997). The inner major vortex is called as the "main vortex" and the outer minor vortex is called as the "outer-bank vortex" in this paper.

Figs.4 (a)-(c) show the secondary flow vectors at representative sections for 3 types of vegetated zone arrangement. At the section of θ=30°, the secondary flows characteristic of bend are generated in all cases. The secondary flow becomes maximum at the section of θ=60° in the same manner as the case with no vegetation. The lateral flow across the vegetated zone is observed but no clear vortex structures are detected in the vegetated zone. As a result, the secondary vortices caused by the centrifugal force are confined within the areas outside the vegetated zones. In the case that vegetated zone is placed along the inner bank (CASE V-2), the same structures as the rectangular channels are generated in the outside part of the channel, and the outer-bank vortex is also clearly observed. Because of the inner-bank vegetated zone, the lateral scale of the main vortex becomes smaller. The results of densely arranged case (V-1) were almost similar to the case of V-2. In the case that vegetated zone is placed along the centerline (V-3), two independent secondary vortices are generated in both the inner and outer areas, which has the same rotation direction. The magnitude of secondary velocity of inner vortex is larger than that of outer vortex. The outer-bank vortex is observed in the outer area. In the case that vegetated zone is placed along the outer bank (V-4), the outer-bank vortex is diminished by the effect of the outer-bank vegetated zone and the main vortex becomes largest in both strength and the lateral scale. In the vegetated zone at the location of φ=60°, a strong inward flow is observed. At the location of x'=135cm, however, the lateral scale of the main vortex becomes smaller. In the case of short vegetation arrangements (V-5 and V-6), the secondary flow structures were almost similar

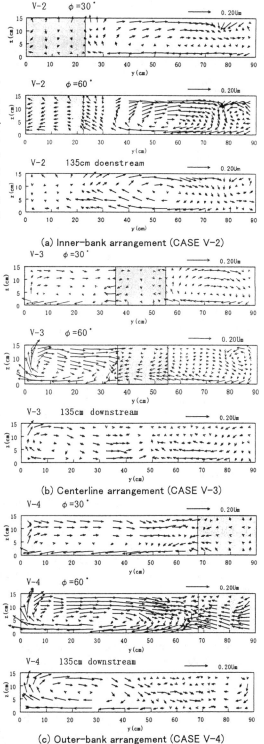

Fig.4   Secondary flow vectors for vegetated channel

331

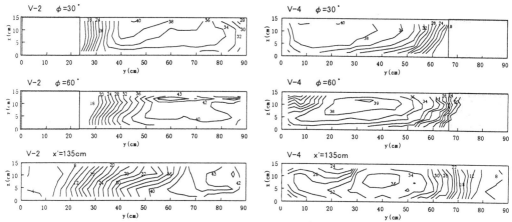

Fig.5 Contours of primary mean velocity

to the corresponding cases of long vegetation arrangements, i.e. the case of V-2 and V-4, respectively.

Fig. 5 shows the streamwise variation of the secondary flow strength, V_{max}, which is defined as the half value of the difference between the lateral mean velocities nearest to the free surface and to the bottom. The developing/decaying process of the secondary flow is almost similar in all cases including the case without vegetation. The peak value of the strength appears at the location of $\phi=60°(x=282.8cm)$. The maximum value of V_{max}/U_m is about 0.15 in most cases, but it is 0.24 in the cases of outer-bank arrangements (V-4 and V-6). The decrease of the strength from the bend outlet to the downstream is very small in the cases of V-1, V-2 and V-7.

3.2 Primary mean velocity

Fig.5 shows the contours of primary mean velocity in the case of V-2 and V-4. In order to extract the lateral structure of the mean flow, lateral distributions of the depth-averaged primary mean velocity are shown in Fig.6. The primary mean velocity is much decelerated in the vegetated zone and a strong lateral shear layer is generated in the interfacial region by a dynamic momentum exchange. In the lateral range of no vegetation, the higher velocity region moves from the inner side toward the outer side, as the flow advances downstream through the bend reach. The lateral shear layer develops toward the downstream direction in the case of inner-bank arrangement (V-2). As a result, the primary velocity is much accelerated near the outer bank. In the case of outer-bank arrangement (V-4), the lateral shear layer moves into the vegetated zone with maintaining the same lateral velocity gradient. In the case of centerline arrangement (V-3), a variation of the primary velocity in the streamwise direction is relatively smaller. The similar streamwise developing processes are observed in each half of the channel, i.e., the inner distributions are like those of the outer-bank arrangement and .the outer ones

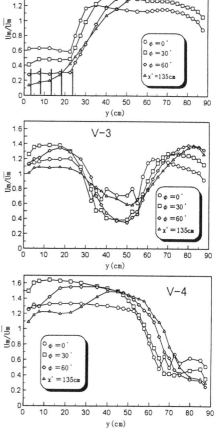

Fig.6 Depth-averaged primary velocity

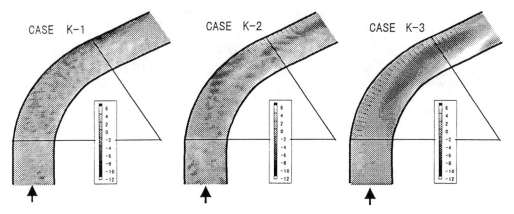

Fig.7 Bed configurations after 5 hours

are like those of the inner-bank arrangement, respectively. As the effect of the momentum transport by the secondary flow, the primary velocity near the free surface of the inner boundary is decelerated and the vertical velocity profile of this area indicates an inverse distribution. Consequently, the lateral shear layer due to the drag of the vegetation, a centrifugal force and the momentum transport by the secondary flows govern the complex structure of the main flow velocity and the bed shear stress. It was also observed that large-scale horizontal vortices were produced periodically in the shear region close to the vegetated zones.

4. EXPERIMENTAL RESULTS FOR MOVABLE BED

Fig 7 shows the measured bed configurations after 5 hours passed from the start of the flow. In the case of K-1 (no vegetation), common result is obtained and this is a basic bed configuration for evaluating the effects of vegetation. The scour starts near the outer bank just downstream from the bend outlet and develops toward upstream gradually. The maximum depth of scour is about 10cm at the outer bank in the range of of x'=0 to 50cm. The scoured region appears from about $\phi=35°$ to x'=120cm. The accumulated region

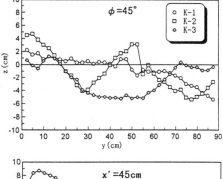

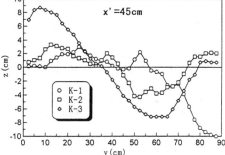

Fig.8 Lateral distribution of bed elevation

occurs along the inner bank from $\phi=35°$ to $\phi=60°$, and moves toward the centerline from $\phi=60°$ to the downstream. In the case of short outer-bank arrangement (K-2), the outer-bank scour near the bend outlet disappeared due to the vegetation, but another scour occurs in front of and beside the vegetated zone. At the section of $\phi=35°$, the magnitude of the scour becomes considerably large. The scoured region extends into the vegetated zone and a stability of trees becomes a problem. In the case of long outer-bank arrangement (K-3), the scoured region appears along the centerline from about $\phi=30°$ to $\phi=45°$, and beside the vegetated region from about $\phi=50°$ to x'=80cm. The maximum depth of scour is about 8cm at the bend outlet. In this case, the magnitude of accumulation becomes higher from the bend outlet to the downstream. Fig. 8 shows comparisons of the lateral distribution of the bed elevation at the section of $\phi=45°$ and x'=45cm. At the section of $\phi=45°$, the scour depth is almost the same value among three cases, but the width of scour is larger in the case of K-3. A high frequency fluctuation is observed in the case of K-2. At the section of x'=45cm where the maximum depth of scour occurs, the scour depth is reduced by the vegetation. However, the magnitude of the accumulation

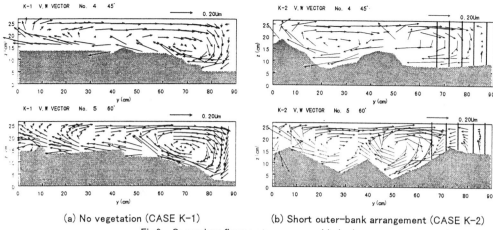

(a) No vegetation (CASE K-1)　　　(b) Short outer-bank arrangement (CASE K-2)

Fig.9　Secondary flow vectors on movable bed

becomes very large and the lateral variation of the bed elevation increases in the case of K-3.

Fig.9 shows the secondary flow vectors in the case of movable bed with and without the vegetated zone. The strong secondary flows are clearly observed. The structure of the secondary flow is adapted to the bed configuration and the outer-bank vortex disappears. The magnitude of the downflow at the scoured region becomes larger as compared with the cases of the fixed bed. The strength of the secondary flow becomes about 1.5 times larger than that of the fixed bed cases. Then, the development of the secondary flow becomes faster.

5. CONCLUSIONS

The effects of vegetation arrangements on 3-D mean flow structure and the bed configuration in bends were clarified experimentally. The boundary between the vegetated zone and the main channel acts as a kind of wall for the secondary flow formation, and reduces the lateral scale of the vortex structure. The main flow structures in bend with vegetation are governed by the strong lateral shear, a centrifugal force and the momentum transport by the secondary flow. The vegetation arranged along the outer bank reduced the scour just inside the vegetated zone, but another scour around the vegetated zones becomes remarkable. The vegetation in the channel changes the flow structures in bend and can decrease the scour, but is likely to change the position of scour. In order to control the bed evolution in bend, it should be examined about the most favorable arrangement of the vegetated zones.

6. REFERENCES

Bathurust, J.C., Throne, C.R. and Hey, R.D. (1979). "Secondary flow and shear stress at river bends", J. Hydr. Div., ASCE, 105(10), 1277-1295.

Ikeda, S., Yamasaka, M. and Kennedy, J. F. (1990). "Three-dimensional fully developed shallow-water flow in mildly curved bends", Fluid Dynamics Research, JSFM, 6, 155-173.

Leshziner, M. A. and Rodi, W. (1979). "Calculation of Strongly Curved Open Channel Flow", J. Hydr. Div., ASCE, 105(10), 1297-1314.

Odgaard, A. J. and Kennedy, J.F. (1983). "River-bend bank protection by submerged vanes", J. Hydr. Eng., ASCE, 109(8), 1161-1173.

Rozovskii I., L.(1957). "Flow of water in bends of open channels", Academy of Science of U.S.S.R., Kiev, U.S.S.R.

Sekine, M. and Kikkawa, H. (1997). "Secondary current control in river bend by the action of obliquely arranged square bar elements", J. Hydraulic, Coastal and Environmental Engineering, JSCE, 558, 61-70.

Sugiyama, H., Akiyama, M. and Kamezawa, M. (1997). "The numerical study of turbulent structure in curved open-channel flow", J. Hydraulic, Coastal and Environmental Engineering, JSCE, 572, 11-21.

Vriend, H.J. and Geldof, H.J. (1983). "Main flow velocity in short river bends", J. Hydr. Eng., ASCE, 109(7), 991-1011.

Environmental Hydraulics, Lee, Jayawardena & Wang (eds) © 1999 Balkema, Rotterdam, ISBN 90 5809 035 3

Experimental study of wakes behind submerged solid plate and porous plate

Daoyi Chen & Cheng-Chung Cheng
School of Engineering, University of Manchester, UK

ABSTRACT: Wake flows of a submerged solid plate and a porous plate have been studied experimentally at sufficiently large ambient Reynolds numbers. Behind porous plates of relatively small porosity, $\beta = 19\%$ to 30%, an unsteady bubble (UB) wake was formed. As porosity increased to a critical value, $\beta = 51\%$, flow meandering stopped occurring, a stable recirculating bubble (SB) was formed. The wakes of a 7.5 cm high submerged solid plate were investigated with different water depth H and water layer above the plate, H_1. For a thin layer of overflow, $H_1/H = 0.12$, vortex shedding was observed in the wake. The vigour of the shedding reduced becaming a less vigorous unsteady bubble (UB) as H_1/H was increased to 0.21. When the overflow depth is larger than the height of the submerged plate, for instance $H_1/H = 0.56$, a stable bubble (SB) occurred. A strong three-dimensional effect was observed in the near wake and it appears to be a quasi-two dimensional flow at a downstream distance of 1.5 to 2.0 times of diameter. It is also shown that flow stability theory can be applied to the flow pattern formation for both cases.

1. INTRODUCTION

Air flow over mountains is of great concern in atmospheric studies due to the need for the estimation of drag force in GCM modelling and possible adverse environmental effect in pollutant dispersion. Some efforts have contributed to the investigation of the flow over 2D and 3D hills (Hunt and Snyder, 1980). Other examples of wakes behind submerged objects are flow over islands, shoals and sandbanks. The effect of the submerged object on downstream flows and sediment motions is still ambiguous. Lloyd and Stansby (1997) have performed an investigation experimentally, providing information on wake formation and structure behind submerged three-dimensional models. It is interesting that the wake became quasi 2-D when flow reached a distance of 1.5 - 2.0 times diameter from the 3-D model.

Recently, a wake stability parameter S=$c_f D/H$ has been introduced by Ingram and Chu (1987) for shallow wake flows, where c_f is the friction coefficient, D is the cross stream dimension of the object and H is the water depth. This S parameter has been successfully applied to classify flow patterns, including vortex shedding, unsteady bubble and steady bubble (Chen and Jirka, 1995). From the absolute and convective instability analysis of Chen and Jirka (1997), critical values of S were determined theoretically as a threshold to classify flow patterns observed in Chen and Jirka (1995) and good agreement was reached, where the maximum return velocity U_m/U_0 has been used to formulate the velocity profiles. A strong returning flow is always associated with a vortex shedding flow pattern and a weaker one with a stable wake and flow over an object will reduce the velocity of returning flow but in a complicated 3-D way. From the point of reducing returning flow velocity, this is similar to the wake flow behind a porous plate where the flow is more close to 2-D. Flow-visualization of wakes behind porous plates by Inoue (1985) has shown that a sudden transition took place from vortex shedding to no vortex shedding. This is entirely consistent with the concept of global instability when the porosity increases over a critical value. It is hypothesesed that the 2-D flow instability analysis and the depth averaged return velocity may be useful in catching the main flow features (or its depth averaged flow features) for a more complicated 3-D wake flow behind submerged objects. A series of experiments has been conducted to make a comparison of wakes behind porous plates and submerged models and determine the wake turbulence of the flows when the water depth is greater than the model height.

2. EXPERIMENTAL APPARATUS AND METHODS

A total of 62 experimental runs, 20 for the porous plate and 42 for the submerged solid plate were conducted in a $3.4m \times 13m$ shallow water table in the Hydraulics Laboratory at the University of Manchester. The surface of the table was smooth, and was painted with white epoxy paint. The flow velocity was measured by using an ultrasonic velocimeter Sensordata Minilab; both still photograph and video images were recorded. A solid plate, 50cm in length, 7.5cm wide and 0.7cm thick, was used as the submerged object. A porous plate with 0.6cm diameter holes, producing about 51% porosity,

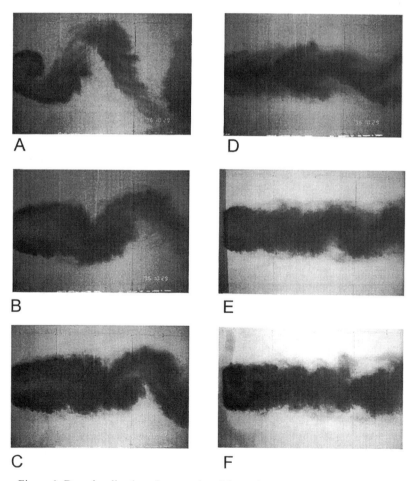

Figure 1. Dye visualization photographs of the wake produced by a submerged solid plate (D=50cm). Depth ratio: 0.12, 0.21, 0.29, 0.38, 0.5, 0.56

was chosen for the test. By overlapping two identical porous plates, 20cm wide and 50cm long, a range of porosity β between 0.19 - 0.51 was obtained. A depth Reynolds number was defined as $Re_h = U_0 H/\nu$ where U_0 is the free stream velocity, H the water depth and ν the kinematic viscosity. Efforts were taken to ensure that the values of the depth Reynolds number were greater than 1500 (from 2420 to 14580), and the Froude number of the flow was less than 0.5, ensuring that the ambient flow was both sufficiently turbulent and subcritical. Values of the S parameter for porous plates and submerged solid plates are smaller relative to other shallow water studies, covered a range from 0.023 to 0.069 which is far below the critical values in Chen and Jirka (1995) and the bed friction will be ignored in this study.

3. WAKE BEHIND SUBMERGED SOLID PLATES

Figure 1 shows six photographs of the dye distribution in the wakes from submerged solid plates, all with transverse width D = 50cm, and height $H_0 = 7.5cm$. This sequence of photographs shows the wakes generated with different flow depth H from 8.5cm to 17cm, making the height H_1 ($=H - H_0$) of free surface flow over the plate from 1cm to 9.5cm, which produce different types of flow pattern. For all tests, the ambient Reynolds number is Re_h=8650. In Figure 1a, little water flowed over the solid plate into the near-wake region with $H_1/H = 0.12$. A vigorous well organized vortex shedding wake was formed with flow recirculating right behind the plate, leading to the pronounced mechanism of a vortex street (VS) suggested by Chen and Jirka (1995).

Increasing the flow depth to $H_1/H = 0.21$ (Figure 1b), this represents a transition from vortex shedding to an unsteady bubble wake (UB). An attached near-wake bubble with recirculating flow was observed, but without a vortex street. The recirculating bubble length was about 1.3 plate transverse width (L=1.3D), and the periodic shedding began to grow at the end of the bubble, resulting from the flow instability. The transverse disturbances downstream were still vigorous in this case. Vortex shedding occurred, although the rolled-up vortex contained relatively low velocities compared to the previous case. Figure 1c represents a wake flow with $H_1/H = 0.29$. Here, the recirculating bubble zone extends downstream to about 2 times the plate transverse width (L=2D). The wake had narrowed, but clear periodic vortex shedding still appeared downstream from the end of the bubble, which displayed an increased frequency from the previous observation. With the flow depth increased to $H_1/H = 0.38$ (Figure 1d), the recirculating wake bubble moved further downstream and an even narrower wake formed, but the frequency of the vortex shedding increased. The recirculating bubble length increased to about L=3D and the position from which shedding occurs has moved further downstream. The interaction between the horizontal shear layers is weak. For the flow with $H_1/H = 0.5$ (Figure 1e), a steady recirculating bubble extended further downstream, with shedding generating from the downstream end, at x/D=5. The bubble length increased further with increasing depth. These transverse oscillations result from the instabilities in the horizontal shear layers at either side of the plate. However, the frequency of the shedding was relatively difficult to determine. When the flow depth was increased to $H_1/H = 0.56$, as shown in Figure 1f, any form of well organized vortex shedding has ceased to exist, displaying no vortex shedding characteristics. The wake now is characterized as a steady bubble (SB). The recirculating bubble zone persisted throughout the whole downstream water table, appearing as a stable turbulent wake, without any growing instabilities. In general, the vortex shedding decreases in strength from a vortex street (VS) to an unsteady bubble (UB) and then ceases as it becomes a stable bubble (SB), as the depth of the water above the plate increases from $H_1/H = 0.12$ to $H_1/H = 0.56$. The general trend observed for the relationship between flow depth and flow pattern is similar to the trend found by Lloyd and Stansby (1997) for the case of the wake behind a submerged island model. Due to the significant differences in geometry between the present plate and the model of Lloyd and Stansby, direct comparison of results may not be prudent.

337

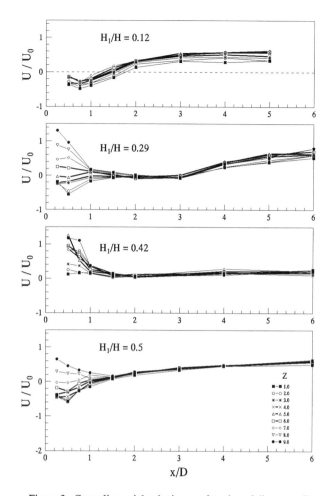

Figure 2. Centerline axial velocity as a function of distance x/D.

Mean axial velocity profiles obtained along the wake centerline (y/D=0), for various measuring points using the Minilab for different flow depths are shown in Figure 2a-d. For the case with $H_1/H = 0.12$ (see Figure 2a), significant return velocities dominate the near-wake field. It can be seen that the mean wake closure occurs around at x/D=1.5 behind the plate centerline. This extent of mean flow reversal is similar to the wake behind a circular cylinder, which has a value of 1.3 diameters. The maximum mean return velocity U_m/U_0= -0.31 at x/D= 0.75 compares very favorably with those results for the plane wake behind a blunt body (U_m/U_0= -0.35 for cylinder wake in Chen and Jirka 1995), however, this value is much greater than for the porous plate in the following section. Increasing the depth of the mean flow to a higher range from $H_1/H = 0.29$ to $H_1/H = 0.5$ (Figure 2b-d), we see that a significant change in the velocity profile and the flow pattern occurs as the volume of the water flowing over the top of the plate increases. In the near wakes, the flow in the upper layer move downstream and only return in the bottom 5 cm layer. But immediately after a distance of about 1.5 times the plate width, the flows, in the downstream direction, becomes uniform across the depth. This corresponds to an unsteady bubble (UB) with an attached recirculating wake bubble formed in these cases. Although the return velocity very close to the bed (about 1 cm above the bed) has the same magnitude as a blunt wake, the depth averaged velocity became -0.05 to 0.129. For all four cases with a range of H_1/H from 0.12 to 0.5, the turbulent intensities are generally high

338

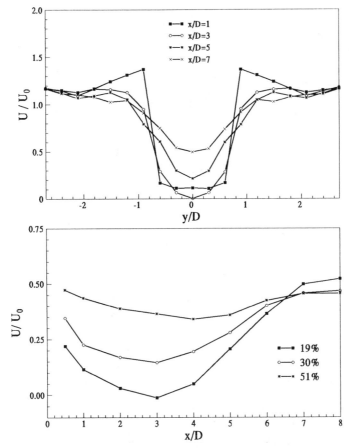

Figure 3. a). Mean velocity profiles for the wake flow of 19% porosity.
b). Centerline mean velocity variation at various porosity.

in the near wake before x/D = 1.5. Towards the downstream direction, after x/D = 2.0, the turbulent intensities dramatically decrease to a low level of $rms(U)/U_0$ less than 0.2. This observation is
consistent with the early discussion of uniform vertical velocity distribution after x/D = 1.5 to 2.0
in Figure 2.

The results discussed above indicate that the wake turbulent intensity of a submerged flat plate is
influenced mainly by the mass flux over the plate. This confirmed that for the cases with low depth
of overflow as H_1/H less than 0.12, the vortex shedding is still the dominant factor for turbulence
generation. As the flow depth increases further, the vortex shedding fades away and mainly three
dimensional flow structures generated by the overflows contribute to the velocity fluctuation on the
centerline.

4. WAKES BEHIND POROUS PLATE AND DISCUSSIONS

Four values of porosity $\beta = 19\%, 22\%, 30\%$ and 51% were created by overlapping two identical porous
flat plates. The characteristics of the near-wake from the porous plate have been found to fall into
one of two classes: unsteady bubble when $\beta = 19\%, 22\%$ and 30%; steady bubble when $\beta = 51\%$.
As a whole, with the increasing porosity, the wake bubble are located further downstream and the

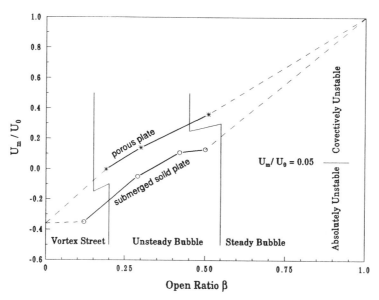

Figure 4. Maximum return velocity at various open-area ratio and depth ratio.
Flow patterns and instabilities are also illustrated accordingly.

mixing layers develop toward downstream. The features of the flow patterns characterized above are quite similar to those introduced by Inoue (1985) for flows past a screen. In order to obtain a simple picture of the flow field behind each plate, the velocities were measured along a line across the flume which is perpendicular to the free stream direction. The mean velocity profiles obtained in various open-area ratio of porous plates are quite similar and an example is shown in Figures 3a. The decay of the wake defect velocity is shown in Figure 3b. The maximum return velocity is almost equal to zero behind the lowest porosity plate (19%), which is lower than the strong flow reversal measured in the wake of blunt bodies ($U_m/U_0 = -0.35$ see Chen and Jirka, 1995). This significant reduction in the return velocity within the attached wake bubble has caused an obvious suppression of absolute instability which normally would lead to a vortex shedding in the wake. This result appears similar to the suppression of the von Karman vortex street in a wake flow once the plate porosity exceeds a critical value (see Inoue, 1985; Castro, 1971). It is clear that the reversed flow bubble which exists behind an ordinary flat plate has detached and moved downstream, so that there are two stagnant points, one at either end of the bubble. It coincides with a dip in the centerline mean velocity shown in Figure 3b. However, a convective, sinuous far-wake instability, which may be mistaken for vortex shedding, appears in general further downstream from the object.

The flow classification depends on an open-area ratio for the porous plate and a flow depth ratio (H_1/H) for the submerged solid plate. Actually, the flow depth ratio H_1/H is equivalent to the open-area ratio β so that the depth averaged return velocities are plotted against the open-area ratio β in Figure 4. The blunt body is represented at the bottom-left extreme where $\beta = 0$ and U_m/U_0 = -0.36. On the other hand, the open channel flow would be approached at the upper-right corner where $\beta = 1$ and $U_m/U_0 = 1.0$. For the same open ratio, the return velocity in the wake of the porous plate is smaller than in the wake of submerged solid plate. The reason may be the 3-D near wake flows behind the submerged solid plate is not so efficient in terms of "base-bleed". The dividing lines between VS, UB and SB are illustrated for the submerged solid plate and the porous plates respectively and the values for the former are slightly higher: vortex shedding when $\beta < 0.1$ to 0.2; unsteady bubble when $0.2 < \beta < 0.5$; and steady bubble when $\beta > 0.5 - 0.55$. Based on the instability analysis of Chen and Jirka (1997), if the bed friction is neglected, R = $(U_m - U_0)/(U_m + U_0)$ =

340

- 0.907, which is corresponding to $U_m/U_0 = 0.05$, would be the dividing line between absolutely unstable and convectively unstable and this falls into the range of unsteady bubble flow patterns. This means that the concept of "base-bleed" can be used to describe a submerged wake where the "base-bleed" is from the overflow layer on the top of the plate.

5. ACKNOWLEDGMENTS

We are grateful to Dr. P.M. Lloyd for his kind help in this work.

6. REFERENCES

Castro, I. P. (1971) Wake characteristics of two-dimensional perforated plates normal to an air-stream, *Journal of Fluid Mechanics*, Vol. 46, 599-609.

Chen, D. and G. H. Jirka (1997) Absolute and convective instabilities of plane turbulent wakes in a shallow water layer, *Journal of Fluid Mechanics*, Vol. 338, 157-172.

Chen, D. and G. H. Jirka (1995) Experimental study of plane turbulent wakes in a shallow water layer, *Fluid Dynamics Research*, Vol. 16, 11-41.

Hunt, J. C. R. and W. H. Snyder (1980) Experiments on stably and neutrally stratified flow over a model three-dimensional hill, *Journal of Fluid Mechanics*, Vol. 96, Part 4, 671-704.

Ingram, R. G., and V. H. Chu (1987) Flow around islands in Rupert Bay: An investigation of the bottom friction effect, *Journal of Geophysical Research*, Vol. 92, No. C13, 14521-14533.

Inoue, O. (1985) A new approach to flow problems past a porous plate, *AIAA Journal*, Vol. 23, 1916-1921.

Lloyd, P. M. and P.K. Stansby (1997), Shallow-water flow around model conical islands of small side slope. II. Submerged. *ASCE Journal of Hydraulic Engineering*, V.123, 1068-1077.

2.2 Turbulent mixing in open channel flow

Environmental Hydraulics, Lee, Jayawardena & Wang (eds) © 1999 Balkema, Rotterdam, ISBN 90 5809 035 3

Mixing processes at river confluences: Field informed numerical modelling

Stuart N. Lane, Kate F. Bradbrook, Steve W. B. Caudwell & Keith S. Richards
Department of Geography, University of Cambridge, UK

ABSTRACT: This paper reports combined results from field monitoring and numerical modelling of mixing processes at river channel confluences. The field results use an Acoustic Doppler Velocimeter to measure simultaneously three-dimensional flow velocity and suspended sediment concentration in a small measuring volume 5cm below the sensor head. The velocimeter was deployed in a confluence of three channels each with different suspended sediment concentrations. Data were collected from along a mixing layer interface between two of the confluent channels. The results illustrate that mixing occurs on a number of temporal scales, but that the nature of the link between periodic flow field fluctuations and suspended sediment transport varies with distance through the confluence. There is a strong correlation between cross-stream velocity variation and suspended sediment transfer at the entry to the confluence, associated with the presence of a stagnation zone. After rapid development of these instabilities, their further evolution is associated with eddy stretching, such that whilst there remain strong fluctuations in suspended sediment concentration further downstream, these are no longer correlated with velocity variation. Preliminary attempts to model these processes numerically are presented based upon Large Eddy Simulation.

1. INTRODUCTION

This paper is concerned with the way in which two river channels with different suspended sediment characteristics mix. Research has shown that river channel confluences are associated with a number of temporal and spatial scales of variability. These include: Kelvin-Helmholtz instabilities in the mixing layer (Best and Roy, 1991; Biron *et al.*, 1993); longer term mixing layer migration (Biron *et al.*, 1993) and periodic upwelling of fluid from one tributary within fluid from the other (Biron *et al.*, 1993, 1996). The nature of these periodicities is related to tributary momentum ratios (e.g. Best, 1986) as well as confluence morphology, notably junction angle (e.g. Mosley, 1976) and discordance between the elevations of the two tributary channels (e.g. Biron *et al.*, 1996). It follows that when tributaries with different suspended sediment or solute characteristics meet, the nature of the flow structures present could exert a major control upon the mixing process. This paper reports preliminary results from two different approaches to understanding this problem. The first is based upon acoustic Doppler velocimetry, specifically designed for use in shallow water, to obtain simultaneous, instantaneous estimates of relative suspended sediment concentration and three-dimensional velocity from within a measuring volume located below the sensor head, away from any instrument-associated flow disturbance (Lane *et al.*, 1998). The second set of results is based upon preliminary attempts to model this process numerically using Large Eddy Simulation (LES), a methodology that has seen widespread application in some areas of the environmental sciences, but not in terms of river channel hydraulics. It allows unsteady solution at the scale of the computational grid used for the numerical solution. As a first attempt, this paper applied LES to a simplified representation of the junction between two of the tributaries, to allow further interpretation of the field results.

2. FIELD MONITORING

2.1 Methods

Acoustic Doppler velocimetry is based upon the principle of backscatter from small measuring particles within the flow. Previous research has illustrated the potential for acoustic backscatter in generating reliable estimates of suspended sediment concentration (e.g. Thorne and Hardcastle, 1997). In theory, the intensity of individual scatterers within an instrument's measuring volume (i.e. their concentration) is a positive exponential function of the backscatter amplitude (Lohrmann, pers. comm.). The shape of this function depends upon the sizes of particles in suspension and there is a straightforward relationship between backscatter amplitude and suspended sediment concentration for any one particle size. However, in most rivers, there will be a wide variety of particle sizes in transport, each with different velocities. The instrument used in this study is designed to be most sensitive to a narrow range of very small particle sizes, whose velocities are likely to be a reflection of the flow field, with a peak sensitivity to 48μm diameter particles. The amplitude signal should decline as a function of the cube of particle size for sizes less than this, and as a function of the inverse of particle size for sizes greater than this. Thus, whilst back scatter amplitude will be more sensitive to certain particle sizes than others, which may prove a problem for calibration with concentration records where rivers of different particle sizes are mixing, or where there are strong turbulence-related gradients in grain-size, the backscatter amplitude record should at least provide a suspended sediment mixing signature. As this signature is obtained simultaneously with three-dimensional velocity information, it provides an important possibility for investigating the nature of mixing processes in shallow streams.

Field data were collected from the mixing layer that formed between the confluence two tributaries each with different suspended sediment characteristics: a low turbidity true left channel; and a high turbidity true left channel. A downwards-looking acoustic Doppler velocimeter (ADV) was mounted on a specially-designed wading rod and data were recorded at 25Hz. The geometry of the ADV used in this study (Figure 1) was such that data on three-dimensional velocities and signal amplitude were obtained for a point that was 5cm below

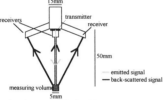

Figure 1. The downwards-looking ADV.

the sensor head, so minimising flow field interference. Field data collection involved sampling at 10 cm above the bed at a series of points, 50cm apart along the mixing layer, downstream from the confluence apex. Sampling was undertaken for 3 minutes at 10Hz., at 10cm above the bed, during a period of constant stage. In all cases, the instrument was aligned such that the x-direction of the instrument was parallel to the dominant downstream direction. Following the recommendations of Roy *et al.* (1996), all velocity measurements obtained along the mixing layer were rotated such that the mean cross-stream velocity was zero.

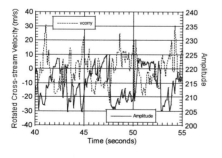
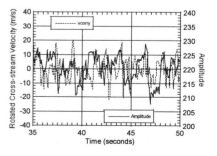

2a   2b

Figure 2. Plots of signal amplitude and cross-stream velocity through time at two points, one close to the apex (2a) and one further downstream (2b).

2.2 Results

Figure 2 shows two plots of amplitude obtained from within the mixing layer: in both plots, it is clear that there are periodicities present. At point 8 (Figure 2a), the periodicities seem to have two dominant frequencies: a shorter periodicity of about 0.5 to 1s, and a longer periodicity of between 4.5 and 5 seconds, with corresponding length scales according to the average downstream velocity of 0.4 to 0.8m and 3.6 to 4m.

Further downstream at point 11 (Figure 2b) the larger-scale periodicities seem to be less coherent. There are clearly longer duration concentration fluctuations, but these seem to be associated with a more gentle rise and a very rapid fall. Again, high frequency fluctuations seem to be imposed upon this, but identification of a dominant high frequency periodicity is much more difficult.

Figure 2 also shows cross-stream velocities. There is a striking inverse correlation between amplitude and cross-stream velocity at point 8, close to the apex (Figure 2a). In this situation, positive velocities are towards the true right, and the true left contains more turbid water than the true right. These patterns are associated with both the low and the high frequency amplitude fluctuations: peaks in the amplitude signal seem to be associated with the cross-stream advection of turbid water from the true left at two different scales. The first is a high frequency scale, associated with small length-scale features which, from field observations, seem to be Kelvin-Helmholtz instabilities. The larger scale features seem to be associated with longer-term migration of the mixing layer. By point 11, the velocity-concentration association is less clear. There is little association between the low frequency amplitude fluctuations, and cross-stream velocity. There is some association between the high frequency amplitude fluctuations and cross-stream velocity, but there seems to be some phase difference between the two signals. Figure 3 shows changes in the correlation between the three velocity components and amplitude with position along the mixing layer. There is a strong coherence between both cross-stream and downstream velocity records and amplitude close to the confluence apex, but this decays with distance downstream. Correlation with cross-stream velocity drops quite suddenly between positions 9 and 10, whereas that with downstream velocity decays more gradually. These results begin to provide insight into the possible manner in which tributary flows mix in confluences. First, there seems to be a high frequency mixing process associated with Kelvin-Helmholtz instabilities, which form at the bend apex. There is a very strong inverse correlation between these fluctuations and cross-stream velocity, suggesting that cross-stream advection of fluid is critical for this process. Further downstream, there is still some correlation with these periodicities, but this seems to involve some form of phase difference. There was also a very strong

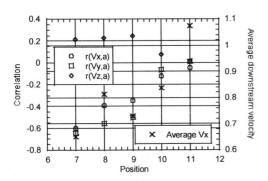

Figure 3. Correlations between velocity records and the amplitude signal with position along the mixing layer.

correlation between cross-stream velocity and amplitude at a longer time-scale, close to the confluence apex. This seemed to be associated with longer-term migration of the mixing layer, in response to longer-term changes in cross-stream and downstream velocity. Correlations between the latter and amplitude were negative, suggesting that when the downstream flow is reduced, cross-stream migration of the mixing layer takes place. There is virtually no association between this larger-scale amplitude fluctuation and any velocity fluctuations further downstream. The next stage of this paper investigates the extent to which these mixing periodicities can be modelled using large eddy simulation.

3. NUMERICAL MODELLING

3.1 Methods

Numerical modelling was based upon a finite volume solution of the three-dimensional elliptic form of the

Navier-Stokes equations, in which a turbulence model is used only to solve for the effects of turbulence which occurs at scales smaller than those used to discretise the spatial domain of the area of interest. A Smagorinsky-type turbulence model is used for this (Samgorinsky, 1963). In addition, numerical simulation was undertaken for a numerical tracer subject to advection and turbulent diffusion, to represent concentration. Numerical solution is based upon coupling of the pressure and momentum equations using the SIMPLEST algorithm of Patankar and Spalding (1972) and an interpolation scheme which is hybrid-upwind, where upwind-differences are used in high convection areas (Peclet number>2) and central differences where diffusion dominates (Peclet number<2). Although this scheme can suffer from numerical diffusion, it is very stable, and stability is important when periodic flow fields are being investigate so as to avoid the introduction of spurious oscillations in the solution which can occur with some higher-order numerical schemes. As a first attempt at modelling the field data observed above, a 2m(downstream)x 1m(cross-stream)x 0.4m(vertical) block of fluid

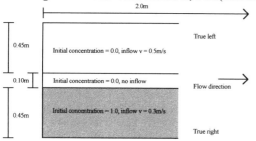

was used (Figure 4). The corresponding number of grid cells used was 150x100x20. No boundaries were specified, and the slower flow on the true right was labelled with the numerical tracer of concentration 1.0. Inflow velocities were chosen as typical of the two tributaries, although they were parallel rather than angled with respect to the downstream flow direction. They were not perturbed as a function of time in any way. A stagnation zone, as is commonly observed at confluence apices was also included. The model was run with a 0.1s time step for 50 seconds. The model results were sampled at two points in the middle of the channel, 0.20m

Figure 4. The simplified geometry used to model Kelvin-Helmholtz instabilities.

and 0.53m downstream from the bend apex respectively.

3.2 Results

Figure 5 shows time-series of model output for two points in the mixing zone. Close to the apex, concentration fluctuations are of a higher magnitude and longer duration, with a much more rapid concentration change. Further downstream, the magnitude and duration of the fluctuations is reduced, and the concentration changes are much more gradual. There is clear evidence that close to the apex these correlate with velocity fluctuations, as per the field data, and there is a very strong positive correlation between concentration and both cross-stream (r=0.427) and downstream (r=-0.820) velocity. This can be illustrated in terms of an eddy trajectory (Figure 6a), in which time is substituted by relative position in space. This shows how, as the flow moves towards the

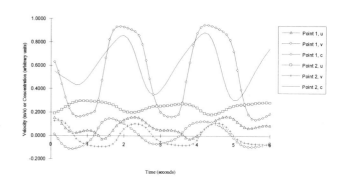

Figure 5. Time-series predicted using the LES model for two points, 0.20m (Point 1) and 0.53m (Point 2) downstream from the confluence apex.

true right (negative y position), the concentration falls, and as the flow moves towards the true left, the concentration rises. The eddies are asymmetric with flow towards the true left of greater magnitude and duration than towards the true right, such that the concentration changes are also asymmetric (Figure 6a). Between each eddy, there is a short-lived period of negligible downstream flow, during which period the cross-stream flow reverses (Figure 5). As the cross-stream flow reverses, the downstream flow does likewise. This

348

reversal separates the end of the previous eddy and the start of the next one (Figure 6a), and as the next eddy starts, concentrations rapidly rise. Once concentrations are high, they are maintained as high for some time, no doubt aided by upstream transport of high concentration fluid during the temporary reversal of downstream flow (Figure 5). Concentration falls rapidly towards the end of the eddy, as the cross-stream velocity reverses once more towards the true right. In summary, the concentration signal is associated with: (i) eddy creation, with strong cross-stream and weak down-stream flow reversal; (ii) sustained concentration due to the advection of high concentration material associated with this eddy in the downstream direction; until (iii), the cross-stream flow reverses, and low concentration fluid is introduced. (iii) lasts until another eddy is created. Both the eddy and the concentration signal are in phase, and asymmetric in nature.

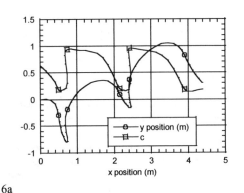 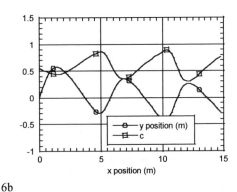

6a   6b

Figure 6. Plot of eddy trajectory with concentration superimposed for 0.20m downstream (6a) and 0.53m downstream (6b).

Further downstream, the patterns are different. Most notably, both the concentration and the eddy trajectory patterns are somewhat more symmetrical (Figure 6b), but now almost exactly 180° out of phase. There is now a much lower correlation between cross-stream velocity and concentration changes (r=0.035), and a negative correlation between downstream velocity and concentration change (r=-0.345). This is reflected in Figure 6b where as water within the eddy moves *towards* the true right (negative *y*), the concentration rises, and *vice versa*, even though concentration is lower on the true left. The 180° phase change with respect to the point close to the apex reflects the fact that eddies are no longer being generated at this point, but advected downstream by the main flow. As a wave of increasing concentration begins to pass the sampling point, the flow must necessarily be towards the true right, along the front of the wave, in order to maintain the wave form. The wave form, generated upstream as an eddy, is a reflection of the stretching of this eddy as it moves downstream, and reflected in the longer length scale associated with the concentration fluctuations in Figure 6b as compared with Figure 6a. In summary, these results suggest that the mixing zone dynamics comprise the initial creation of a small-scale eddy, associated with strong cross-stream mixing of fluid, followed by stretching of this eddy as it is transported downstream, with the nature of the correlation between concentration and velocity characteristics changing to reflect this.

4. CONCLUSIONS

Combining the results obtained from Sections 2 and 3 allows a number of conclusions to be made. First, these results provide the first numerical simulation of Kelvin-Helmholtz instabilities at river channel tributary confluences, according to the authors' knowledge. They allow basic understanding of how these instabilities form. This suggests that they are initiated close to the confluence apex, possibly associated with a local stagnation point. The eddies initiate mixing between the two tributaries. Concentration fluctuations were also seen further downstream. These were of a larger-scale, associated with stretching of the eddy in the downstream direction and explained by the advection of eddies created upstream through the sampling point. Thus, initial coherence between the cross-stream velocity field and the amplitude of concentration changes was lost further downstream. Second, there is a surprising level of correspondence between the evidence that is provided from the LES study and the high frequency amplitude fluctuations reported in section 2, given that the numerical modelling experiment is based upon a highly-idealised representation of the field situation. Strong correlations between cross-stream velocity fluctuations and the ADV amplitude signal were also observed at the apex. These were not observed further downstream, where there seemed to have been some sort of phase shift between cross-stream velocities and amplitude signals, as was evident in the LES results. Third, there was no longer-term concentration fluctuation in the LES predictions, unlike in the field data. This may reflect the simplified numerical experiment presented here. However, it may also reflect the fact that this longer-time fluctuation is not generated by local instabilities associated with the stagnation point at the confluence apex, but processes occurring at the confluence scale, which result in mixing layer migration. This migration may be as a result of high frequency discharge pulsation in the tributaries, or as a result of larger-scale internally-generated instabilities. Increasing the complexity of confluence boundary condition specification to model this effect is the subject of current research. Finally, the LES results do not have the irregularity of the field data, presumably because of the scale-dependence of the eddies that are being resolved, and also because of the nature of boundary condition specification in this model. However, and partly as a result, they do allow much more detailed interpretation of the mixing process.

In summary, these are preliminary results. They show how the combination of field measurement and numerical modelling experiments can be used to improve understanding of the mixing patterns often observed when two tributaries join. The ADV is capable of improving our understanding of how tributaries with different suspended sediment concentrations mix, as it provides instantaneous and simultaneous velocity and suspended sediment information from the same measuring volume. LES, which is only now seeing more widespread application to fluvial hydraulic problems, is a complimentary technique which may be used to understand and explain field-observed temporal instabilities. Current research is using LES to explore the effects of different confluence configurations upon tributary mixing processes.

5. ACKNOWLEDGEMENTS

This research was supported by NERC Grant GR3/9715 awarded to SNL, KSR and Dr. J.H. Chandler. Discussions with Andre Roy and Pascale Biron have helped in generating the ideas presented in this paper.

6. REFERENCES

Best, J.L., 1986. The morphology of river channel confluences. *Progress in Physical Geography*, **10**, 157-74.

Best, J.L. and Roy, A.G., 1991. Mixing-layer distortion at the confluence of channels of different depth. *Nature*, **350**, 411-3.

Biron, P., De Serres, B., Roy, A.G. and Best, J.L., 1993. Shear layer turbulence at an unequal depth confluence. In Clifford, N.J., French, J.R. and Hardisty, J. (eds) *Turbulence: Perspectives on Flow and Sediment Transport*, Wiley, Chichester, 197-214.

Biron, P., Roy, A.G., and Best, J.L., 1996. Effects of bed discordance on flow dynamics at open channel confluences. *ASCE Journal of Hydraulic Engineering*, **122**, 676-82.

Lane, S.N., Biron, P.M., Bradbrook, K.F.,Butler, J.B., Chandler, J.H., Crowell, M.D., McLelland, S.J., Richards, K.S. and Roy, A.G., 1998. Integrated three-dimensional measurement of river channel topography and flow processes using acoustic doppler velocimetry. Forthcoming in *Earth Surface Processes and Landforms*.

Mosley, M.P., 1976. An experimental study of channel confluences. *Journal of Geology*, **84**, 535-62.

Patankar, S.V. and Spalding, D.B., 1972. Heat and mass transfer in boundary layers. 3rd edition, London.

Roy, A.G., Biron, P. and De Serres, B., 1996. On the necessity of applying a rotation to instantaneous velocity measurements in river flows. *Earth Surface Processes and Landforms*, **21**, 817-27.

Smagorinsky, J., 1963. General circulation experiments with the primitive equations. *Monthly Weather Review*, **91**, 99-164.

Thorne, P.D. and Hardcastle, P.J., 1997. Acoustic measurement of suspended sediments in turbulent currents and comparison with *in-situ* samples. *Journal of the Acoustical Society of America*, **101**, 2603-2614.

Environmental Hydraulics, Lee, Jayawardena & Wang (eds) © 1999 Balkema, Rotterdam, ISBN 90 5809 035 3

Numerical modelling of two-dimensional turbulent air-water flows with evaporation

H.Q. Ni & Z.J. Zou
China Institute of Water Conservancy and Hydroelectric Power Research, Beijing, China

L.X. Zhou
Department of Engineering Mechanics, Tsinghua University, Beijing, China

ABSTRACT: A theoretical model of two-dimensional turbulent air-water flows with evaporation is proposed based on the concept of Stefan flux, and the numerical simulation has been carried out by using the turbulence model and the SIMPLE algorithm. The results show effect of gas velocity, air-water temperature difference and relative humidity on evaporation. Most of the predicted results are in qualitative agreement with experiments, and very useful in oil and water environment engineering.

1. INTRODUCTION

The evaporation on a natural water surface has been experimentally studied (Yue, 1979; Chen, 1989),   and a series of empirical formulas are obtained. These results give the effect of wind velocity, air temperature and humidity on water-surface evaporation. However, up to now, most of them are limited to experimental studies, and it is still lack of theoretical studies. In this paper a theoretical model of two-dimensional turbulent air-water flows with evaporation based on the concept of Stefan flux and the turbulence model is proposed. Numerical predictions have been done by using the SIMPLE algorithm. Predicted results are compared with experiments and the effect of various physical factors on evaporation is discussed.

2. MATHEMATICAL MODEL

The air-water flow with evaporation to be studied is shown in Fig.1. There is a turbulent air flow in the channel which causes a recirculating water flow. Evaporation takes place on the water surface. The heat is transferred from the surface of hot water to the air by convection, and the water vapor diffuses from the surface to the air. The proposed theoretical model is based on: (1) quasi-steady two-dimensional turbulent air flow and laminar water flow with evaporation on the surface; (2) k-ε gas turbulence model (Zhou, 1993); (3) boundary-layer type air flow; (4) neglected buoyancy force;

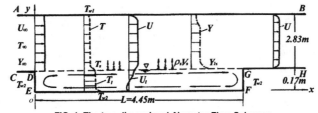

FIG. 1. The two-dimensional Air-water Flow Schemes

Thus, the basic conservation equations of gas and water flows can be expressed in the following generalized form:

$$\frac{\partial}{\partial x}(\rho U\varphi)+\frac{\partial}{\partial y}(\rho V\varphi)=\frac{\partial}{\partial x}(\Gamma_\varphi\frac{\partial\varphi}{\partial x})+\frac{\partial}{\partial y}(\Gamma_\varphi\frac{\partial\varphi}{\partial y})+S_\varphi \qquad (1)$$

where φ is the generalized dependent variable, Γ_φ and S_φ denote the transport coefficient and source term of respectively. The physical meaning of φ, Γ_φ and S_φ for each equation is given in Tab.1.

TAB.1.

Equation	φ	Γ_φ	S_φ
Gas continuity	1	0	0
Gas momentum	U, V	μ_e	$-\dfrac{\partial P}{\partial x}, -\dfrac{\partial P}{\partial y}$
Gas energy	h	μ_e / σ_h	0
Vapor diffusion	Y_1	μ_e / σ_Y	0
k - equation	k	μ_e / σ_k	$\mu_T\left(\dfrac{\partial U}{\partial y}\right)^2 - \rho\varepsilon$
ε - equation	ε	$\mu_e / \sigma_\varepsilon$	$\dfrac{\varepsilon}{k}\left[c_1\mu_T\left(\dfrac{\partial U}{\partial y}\right)^2 - c_2\rho\varepsilon\right]$
Water continuity	1	0	0
Water momentum	U_l, V_l	μ_l	$-\dfrac{\partial P_l}{\partial x}, -\dfrac{\partial P_l}{\partial y}$
Water energy	T_l	λ_l / c_{pl}	0

where $\mu_e = \mu+\mu_T, \mu_T = c_\mu\rho\, k^2/\varepsilon$, $\sigma_h, \sigma_k, \sigma_\varepsilon, \sigma_Y, c_\mu, c_1, c_2$ are universal constants that values are given in Tab.2.

TAB.2

σ_h	σ_k	σ_ε	σ_Y	c_μ	c_1	c_2
0.90	1.00	1.33	1.00	0.09	1.44	1.92

3. BOUNDARY CONDITIONS

The boundary conditions are taken in a similar way as that used by Ni and Wang (1987), except the conditions at the air-water interface.
(1)Gas-phase boundary conditions: uniform distribution of all variables at the inlet; fully developed flow condition at the exit; No-slip condition and zero diffusion flux at the walls; the wall function approximation for near-wall grid nodes.
(2)Liquid-phase boundary conditions: no-slip and zero-flux conditions at the walls.
(3)Air-water interface conditions: The interface conditions are of vital importance to this problem. Based on the concept of Stefan flux (Zhou, 1993) the condition for diffusion equation at the interface can be taken as

$$\rho_s V_s = -\frac{1}{1-Y_{ls}}\left(D\rho\frac{\partial Y_1}{\partial y}\right)_s \qquad (2)$$

The mass conservation at the interface gives

$$\rho_s V_s = \rho_{ls} V_{ls} \qquad (3)$$

The condition for momentum at the interface is actually coupling of gas and liquid phase velocity and stress

352

$$U_s = U_{ls} \tag{4}$$

$$\mu_s \left(\frac{\partial U}{\partial y} \right)_s = \mu_s \left(\frac{\partial U_1}{\partial y} \right)_s \tag{5}$$

The condition for energy equation is thermal equilibrium at the interface

$$\left(\lambda \frac{\partial T}{\partial y} \right)_s + \rho_s V_s q_e = \lambda_s \left(\frac{\partial T_1}{\partial y} \right)_s = q_s \tag{6}$$

$$T_s = T_{ls} \tag{7}$$

and Eqs. (2) and (6), D_s, and λ_s are determined by

$$D_s = \frac{V_{es}}{\sigma_Y}, \quad \lambda_s = c_p \frac{\mu_e}{\sigma_h}$$

Besides, the relation between the concentration and the temperature at the interface is

$$Y_{ls} = B \exp \left(-\frac{E}{RT_s} \right)$$

where $\quad Y_{ls} = 0.622 x_{ls} / (1 - 0.378 x_{ls}); \quad x_{ls} = e_0/p$

4. SOLUTION PROCEDURE

The whole computed domain is divided into 16×16 grid nodes (Fig.2). The differential equations are integrated in the control volume to obtain finite difference equations by using an upwind difference scheme for gas phase and a hybrid difference scheme for liquid phase. The FDE's were solved by using the SIMPLE algorithm (Patankar, 1980), i. e., p-v correction with TDMA line-by-line iterations and under-relaxation. The computer code consists of 2097 statements in FORTRAN 77 language, and the CPU time in a VAX-2020 computer is about 4 hours.

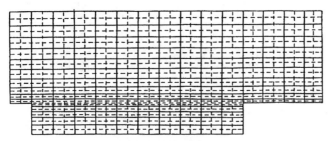

FIG.2. Grid System

5. RESULTS AND DISCUSSION

The predicted velocity, temperature and vapor concentration distributions in gas and liquid phases are shown in Fig.3 to Fig.8. Due to lack of detailed measurements these results have not yet been verified by experiments, so the comparison between predictions and experiments was made only for the effect of various parameters on the global evaporation characteristics: the mean evaporation rate and the evaporation coefficient , which are defined as

$$E = \frac{\partial \rho_s V_s dS}{\int_s \rho_i dS} \quad (m/s)$$

$$\alpha = \frac{E}{e_o - e_a} \quad \left(mm \cdot d^{-1} \cdot mb^{-1}\right)$$

where e_o is the saturated vapor pressure at the surface temperature T_s and e_a is the vapor partial pressure in air at the level of 1.5 meter high above the surface. Fig. 9 to 12 give the evaporation coefficient α and evporation rate E vs. as the air velocity for different temperature differences and relative humidity. The predicted linear relationship of E with U_∞ is in general agreement with that observed in experiment. Predictions show decreasing E, α with increasing relative humidity RH (Fig.13, Fig.14) which is plausible. Figures 15 to 18 give the effect of air-water temperature difference ΔT on the evaporation rate E and evaporation coefficient α. E increases with increasing ΔT (Fig.15,Fig.16). This is in agreement with the observed tendency in experiments. However, the tendency of decreasing α with increasing ΔT (Fig.17, Fig.18) is not in agreement with experiments. Perhaps, there is some uncertainty in selecting e_o and e_a in experiments.

It is worthwhile to note that the temperature T_s at the water surface, may be either greater (Fig.5) or smaller (Fig.4) than the air inlet temperature T_∞. This result implies that at lower wind velocity, the evaporation rate is rather small, the heat for evaporation is supported only by heat conduction from the water surface, while

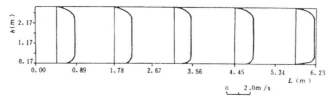

FIG.3. The predicted Velocity Distributions in Gas Phase

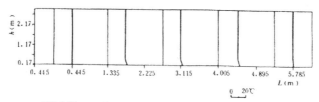

FIG.4. The predicted Temperature Distribution in Gas Phase
(U_o = 0.58m/s, ΔT = 15°C, RH = 90%)

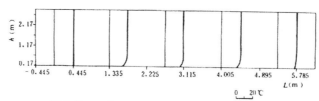

FIG.5. The predicted Velocity Distributions in Gas Phase
(U_o = 1.4m/s, ΔT = 15°C, RH = 90%)

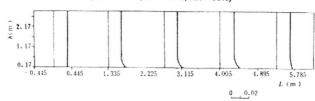

FIG.6. The predicted Mass Fraction Distribution in Gas Phase

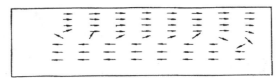

FIG.7. The predicted Velocity Vector Diagram in Water Phase

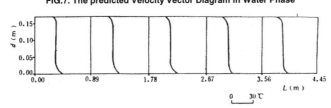

FIG.8. The predicted Temperature Distribution in Water Phase

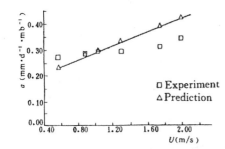

FIG.9. The Evaporation Coefficient α vs. As the Air Velocity U (ΔT=10°C, RH=90%)

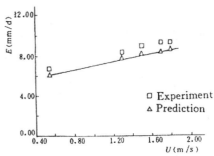

FIG.10. The Mean Evaporation Rate E vs. As the Air Velocity U (ΔT=10°C, RH=90%)

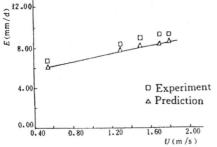

FIG.11. The Evaporation Coefficient α vs. As the Air Velocity U (ΔT=15°C, RH=90%)

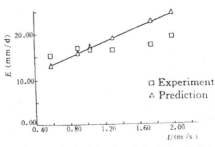

FIG.12. The Mean Evaporation Rate E vs. As the Air Velocity U (ΔT=15°C, RH=55%)

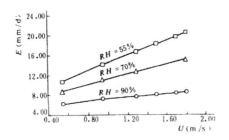

FIG.13. The Mean Evaporation Rate E vs. as Relative Humidity RH (ΔT=10°C)

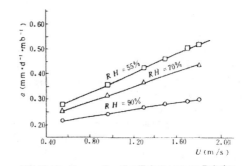

FIG.14. The Evaporation Coefficient α vs. as Relative Humidity RH (ΔT=10°C)

355

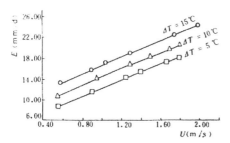

FIG.15. The Mean Evaporation Rate *E* vs. as Air water Temperature Difference *ΔT (RH=55%)*

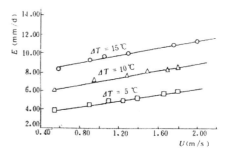

FIG.16. The Mean Evaporation Rate *E* vs. as Air water Temperature Difference *ΔT (RH=90%)*

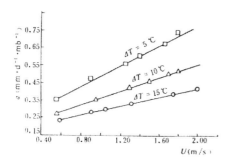

FIG.17. The Evaporation Coefficient *α* vs. as Air water Temperature Difference *ΔT (RH=55%)*

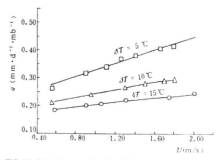

FIG.18. The Evaporation Coefficient *α* vs. as Air water Temperature Difference *ΔT (RH=90%)*

at higher wind velocity, the evaporation rate is rather large, the heat for evaporation is supported both by heat conduction from the water surface and heat convection from air to the water surface.

6. CONCLUSIONS

(1)The two-dimensional turbulent air-water flows with evaporation at the interface can be successfully simulated by the turbulence model combined with the concept of Stefan flux at the interface.
(2)Many predicted results, such as the effect of wind velocity, air-water temperature difference on evaporation rate, are in general agreement with experiments, but there is discrepancy between predictions and experiments in the effect of temperature difference on the evaporation coefficient.
(3)For fixed temperature difference and relative humidity there is a critical wind velocity above which the temperature at the water surface will be lower than the air inlet temperature.

7. REFERENCES

Chen,H.Q.(1989). "J. Hydraulics(China)", No.10, 27-36.
Ni,H.Q. and Wang.N.G. (1987) "J. Hydraulics (China)", No.6, 11-16.
Patankar, S.V. (1980) "Numerical Heat Transfer and Fluid Flow", Hemisphere, N.Y.
Yue, J.T. (1979) "Determination of the heat transfer coefficient of the water surface", Res.Rep. IWHR.
Zhou,L.X. (1993) "Theory and Numerical Modeling of Turbulent Gas-particle Flows and Combustion", Science Press (Beijing) and CRC Press (Florida).

Environmental Hydraulics, Lee, Jayawardena & Wang (eds) © 1999 Balkema, Rotterdam, ISBN 90 5809 035 3

Generalized second moment method for advection and diffusion processes

Vincent H.Chu & Wihel Altai
Department of Civil Engineering and Applied Mechanics, McGill University, Montreal, Que., Canada

ABSTRACT: A semi-Lagrangian numerical scheme known as the second moment method is generalized to simulate the advection and diffusion processes. By making a correction to the procedure for the diffusion process, the stability and convergence of the method are improved. The generalized method is positive definite and unconditionally stable. Numerical experiments were conducted and compared with the exact solution of the advection-diffusion equation. The condition for convergence of the numerical solutions to the exact solution is determined for the entire range of Courant and Péclet numbers.

1. INTRODUCTION

The second moment method (Egan and Mahoney, 1972) is an accurate and stable numerical method for the simulation of the advective process that have completely eliminated numerical diffusion. The mass, in the form of a 'block', is advected by a Lagrangian procedure and then re-distributed onto the Eulerian grid. New blocks are formed at the end of each time step by conserving the zero, first and second moments of the masses in the Eulerian grid. The problems of oscillation and negative mass, known to be associated with most of the advection schemes, are completely eliminated. The method is perfect for advective process but is often perceived to be not suitable for diffusion process (see, for example, Kowalik and Murty, 1993).

In the original implementation by Egan and Mahoney (1972), the numerical solution of the advection-diffusion equation was obtained in two steps using a hybrid scheme. The advective process was simulated by the second moment method (SMM) while the diffusion process by a separate step using a conventional finite difference scheme. The employment of a hybrid scheme have led to false diffusion and numerical instability in situation when the process is dominated by diffusion (Marchand, 1997).

The hybrid scheme and the difficulty with the diffusion process has hindered the acceptance of the SMM as a general computational procedure. Although an attempt had been made in the appendix of the paper by Egan and Mahoney (1972) using the SMM to calculate the diffusion process, most of the subsequent analyses and extensions of the method were concerned with simulation of the advective process (Pedersen and Prahm, 1974; Pepper and Long, 1978; Kerr and Blumberg, 1979; Pepper and Baker, 1980). The purpose of this paper is to show that the SMM can be generalized so that both the advection and the diffusion processes can be simulated with equal precision by a single step.

2. SECOND MOMENT METHOD (SMM)

The computational elements of the SMM by Egan and Mahoney (1972) are the cells and the blocks. The cells are fixed in space. The mass in the cell are assumed to be contained in a block. The block moves with the velocity $u(x_c, y_c, t)$ and $v(x_c, y_c, t)$[1]. Over the period of one time step, from $n\Delta t$ to

---

[1]Extension of the method for three dimensional problem is straight forward but will not be given in this paper

$(n+1)\Delta t$, the mass center of the block moves from (x_c^n, y_c^n) to (x_c^{n+1}, y_c^{n+1}). The displacements in x- and y-directions over the period are given by the integrals

$$x_c^{n+1} - x_c^n = \int_{n\Delta t}^{(n+1)\Delta t} u(x_c, y_c, t)dt, \tag{1}$$

$$y_c^{n+1} - y_c^n = \int_{n\Delta t}^{(n+1)\Delta t} v(x_c, y_c, t)dt. \tag{2}$$

These integrals are determined numerically by iteration method (see, e.g., Staniforth and Côté, 1991). As the block moves, masses are transported to the neighbouring cells. The fraction of the mass from a block into a cell is determined by the 'portioning parameters', (P_x, P_y), which are related to the widths of the block, (w_x^{n+1}, w_y^{n+1}), and the location of the mass center, (x_c^{n+1}, y_c^{n+1}), at the end of the time step as follows:

$$P_x = \frac{1}{w_x^{n+1}}[s_x x_c^{n+1} + \frac{1}{2}w_x^{n+1} - \frac{1}{2}\Delta x] \tag{3}$$

$$P_y = \frac{1}{w_y^{n+1}}[s_y y_c^{n+1} + \frac{1}{2}w_y^{n+1} - \frac{1}{2}\Delta y] \tag{4}$$

where the s_x and s_y are equal to either $+1$ or -1 depending on the direction of the flow. In the original formulation of the SMM, the block is divided into four portions and then re-distributed to the neighbouring cells at the end of the time step. The size of each portions are calculated using the portioning parameters. Implicit in this division scheme is the assumption that the blocks are equal to or smaller in size than the cells. This however is not generally true. During the calculations, the block may become greater than the cell, and may extend beyond the upstream border of the cell if the velocity is small. Under this condition, the masses that would have stay behind in the upstream cells is artificially carried forward. To circumvent this difficulty of 'false advection', a procedure was introduced by Pedersen and Prahm (1971) to correct for the size of the block. Subsequent analysis of the correction procedure by Pepper and Long (1978) and Marchand (1977) however has shown that the correction do not alway give better results. The error associated with false advection worsen when the diffusion process is included in a hybrid scheme. False advection leads to false diffusion and ultimately to instability and breakdown of the hybrid scheme.

3. GENERALIZED SECOND MOMENT METHOD (GSMM)

With diffusion, the blocks will almost alway greater than the size of the cells. Figure 1 shows the general advection and diffusion problem. Initially the block is contained within the cell as shown in Figure 1a. At the end of the time step, the block increases in size due to diffusion and extent beyond cell boundary as shown in Figure 1b. In the generalized method, the mass in the block is divided into the 9 portions and re-distributed to all its neighbouring cells at the end of the time step. This division scheme of the generalized procedure is in contrast to the original implementation of Egan and Mahoney (1972) in which the block is divided into 4 portions by the portioning parameters.

The rates that the size of the block increases with time, are proportional to the diffusion coefficients, D_x and D_y as follows:

$$D_x = \frac{1}{2}\frac{d\sigma_x^2}{dt} \tag{5}$$

$$D_y = \frac{1}{2}\frac{d\sigma_y^2}{dt} \tag{6}$$

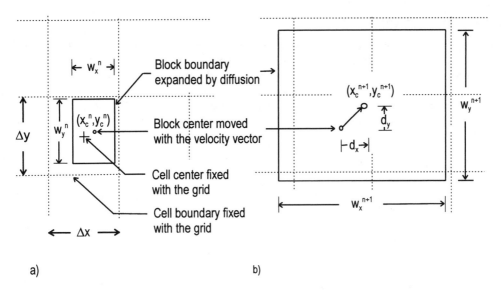

Figure 1: The generalized second moment method. (a) The block in the cell at time $t = n\Delta t$. (b) The block at time $t = (n+1)\Delta t$ after enlargement and relocation of the block by diffusion and advection. At the end of the time step, the masses in the block is divided by the grid into 9 portions which are then distributed to all the neighbouring cells.

where (σ_x, σ_y) are the second moments that are related to the widths of the block, (w_x, w_y), by

$$\sigma_x^2 = \frac{1}{12}w_x^2, \tag{7}$$

$$\sigma_y^2 = \frac{1}{12}w_y^2. \tag{8}$$

Integrating Equations 5 and 6 with time from time $= n\Delta t$ to time $= (n+1)\Delta t$, gives

$$w_x^{n+1} = \sqrt{(w_x^n)^2 + 24\int_t^{t+\Delta t} D_x(x, y, t)dt} \tag{9}$$

$$w_y^{n+1} = \sqrt{(w_y^n)^2 + 24\int_t^{t+\Delta t} D_y(x, y, t)dt} \tag{10}$$

By moving the block according to Equations 1 and 2 and calculating the width according to Equations 9 and 10, both the advection and diffusion processes are taken care in one single advection-diffusion step.

At the end of the time step, masses in the blocks are re-distributed onto the Eulerian gird. New blocks are formed in each cell by assembling the mass staying in the cell and the masses entering from all its neighbouring blocks. In general, each new block is formed from portions of 9 different blocks. These are the portion of the block in the (i, j) cell and the portions of the neighbouring $(i-1, j-1)$ block, $(i-1, j)$ block, $(i-1, j+1)$ block, $(i, j-1)$ block, $(i, j+1)$ block, $(i+1, j-1)$ block, $(i+1, j)$ block, and $(i+1, j-1)$ block.

4. CONVERGENCE

The computation algorithm of the GSMM is based on the assumption that the boundary of the block at the end of the computational step stay within the immediate neighbouring cells. This assumption leads directly to a neccessary condition for convergence of the numerical solution. The derivation of the condition follows. Assuming that the initial widths of the block were $w_x^n = \Delta x$ and $w_y^n = \Delta y$, due to diffusion, the widths at the end of the time step Δt would be

$$w_x^{(n+1)} = \sqrt{(dx)^2 + 24 D_x \Delta t} \tag{11}$$

$$w_y^{(n+1)} = \sqrt{(dy)^2 + 24 D_y \Delta t} \tag{12}$$

(see Equations 9, and 10). The displacements of the block are $U\Delta t$ and $V\Delta t$. The combined displacement by advection and diffusion must not exceed $(\frac{3}{2}\Delta x, \frac{3}{2}\Delta y)$. So the neccessary conditions for a convergent solution become

$$\frac{1}{2}\sqrt{(\Delta x)^2 + 24 D_x \Delta t} + |U|\Delta t \leq \frac{3}{2}\Delta x \tag{13}$$

$$\frac{1}{2}\sqrt{(\Delta y)^2 + 24 D_y \Delta t} + |V|\Delta t \leq \frac{3}{2}\Delta y \tag{14}$$

which can be re-written in terms of the Courant numbers, Co_x and Co_y and the Péclet numbers, Pe_x, and Pe_y, as follows:

$$\sqrt{\frac{1}{4} + 6\frac{\mathrm{Co}_x}{\mathrm{Pe}_x}} + \mathrm{Co}_x \leq 1.5 \tag{15}$$

$$\sqrt{\frac{1}{4} + 6\frac{\mathrm{Co}_y}{\mathrm{Pe}_y}} + \mathrm{Co}_y \leq 1.5 \tag{16}$$

where

$$\mathrm{Co}_x = \frac{|U|\Delta t}{\Delta x}, \quad \mathrm{Co}_y = \frac{|V|\Delta t}{\Delta y}, \tag{17}$$

$$\mathrm{Pe}_x = \frac{|U|\Delta x}{D_x}, \quad \text{and} \quad \mathrm{Pe}_y = \frac{|V|\Delta y}{D_y}. \tag{18}$$

5. NUMERICAL EXPERIMENT

The numerical experiments were conducted for flows with a uniform velocity and constant diffusion coefficient. The governing equation

$$\frac{\partial c}{\partial t} + U\frac{\partial c}{\partial x} = \frac{\partial}{\partial x}[D_x \frac{\partial c}{\partial x}] + \frac{\partial}{\partial y}[D_y \frac{\partial c}{\partial y}]. \tag{19}$$

is linear for the constant advection and diffusion process and has an exact solution which, due to a line source of mass, dM, is

$$dc(x, y, t) = \frac{dM}{4\pi t \sqrt{D_x D_y}} \exp[-\frac{(x - Ut)^2}{4 D_x t} - \frac{y^2}{4 D_y t}]. \tag{20}$$

The concentration distribution due to a finite initial source is obtained by superposition of the line sources. The calculations were conducted for a finite source $c(x, y, 0) = c_o(x, y)$ defined in the region

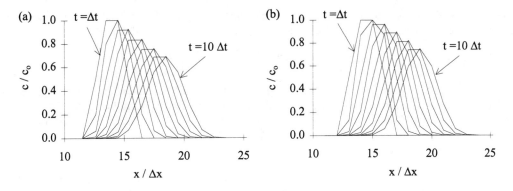

Figure 2: (a) The concentration profiles on the plane of symmetry obtained by GSMM for a period of 10 Δt at interval of Δt. (b) The exact solution. Courant and Péclet numbers are Co = 0.5 and Pe = 10.

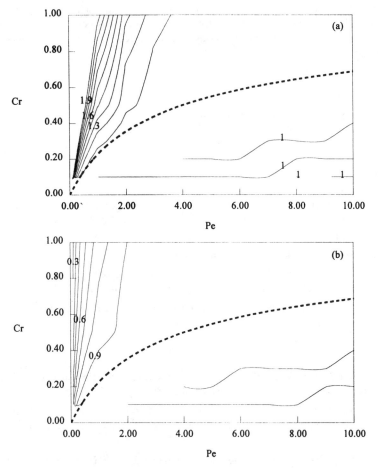

Figure 3: (a) Contours of the maximum concentration ratio, $(c_{\max})_{\text{GSMM}}/(c_{\max})_{\text{Exact}}$ at time $t = \Delta t$. (b) Contours of the half-width ratio, $\delta_{\text{GSMM}}/\delta_{\text{Exact}}$ at time $t = \Delta t$. Dash line is the necessary condition for convergence.

$-a < x < a$ and $-b < y < b$. The exact solution for this finite source is obtained by superposition of the line sources of mass $dM = c_o(\xi, \eta)d\xi d\eta$ located at $x = \xi$ and $y = \eta$ as follows:

$$c(x, y, t) = \int_{-a}^{+a} \int_{-b}^{+b} \frac{c_o d\xi d\eta}{4\pi t \sqrt{D_x D_y}} \exp[-\frac{(x - \xi - Ut)^2}{4D_x t} - \frac{(y - \eta)^2}{4D_y t}]. \tag{21}$$

Figure 2a shows the concentration profiles obtained by the GSMM on the plane of symmetry for a period of ten time steps. These are compared with the exact solution shown in Figure 2b. The simulation were made for an initial rectangular block of mass centered around the origin of the coordinates. The width and length of the block are initially $3\Delta x$ and $3\Delta y$ corresponding to a condition of $c_o = 1$ and $a = b = 1.5$ in Equation 21.

The maximum concentration c_{max} and the half-width δ defined at $c = \frac{1}{2}c_{\text{max}}$ are determined as parameters characterizing concentration profiles for a range of Courant and Péclet numbers

$$\text{Co} = \frac{|U|dt}{dx} \quad \text{and} \quad \text{Pe} = \frac{|U|dx}{D}. \tag{22}$$

The ratios of $(c_{\text{max}})_{\text{GSMM}}$ obtained by the GSMM and $(c_{\text{max}})_{\text{Exact}}$ of the exact solution given by Equation 21 are evaluated. Also evaluated are the ratios of the half-width, $\delta_{\text{GSMM}}/\delta_{\text{Exact}}$. Figure 3 shows the contour lines of these ratio on the Co abd Pe plane. Perfect simulations, as these ratios approach the value of unity, are shown in the figures to be defined in a region below the dash line defined by the equation

$$\sqrt{\frac{1}{4} + 6\frac{\text{Co}}{\text{Pe}}} + \text{Co} \leq 1.5 \tag{23}$$

(i.e., Equation 15). Simulations conducted with a Courant and Péclet number below the dash line should expect an accuracy of 2% after the first time step and an accuracy of 0.2% after 8 time steps.

6.REFERENCES

1. Egan, B. A. and Mahoney, J. R. (1972) Numerical modeling of advection and diffusion of urban area source pollutants. *J. Appl. Meteorology*, Vol. 11, pp. 312-322.

2. Kowalik, Z. and Murty, T. S. (1993) *Numerical Modeling of Ocean Dynamics*. Advanced Series on Ocean Engineering, World Scientific, Singapore, 481 pp.

3. Kerr, C. and Blumberg, A. (1979) An analysis of a local second-moment conserving quasi-Lagrangian scheme for solving the advection equation. *J. Comp. Phys.*, Vol. 32, pp. 1-9.

4. Marchand, P. (1997). Hydrodynamic modeling of shallow basins. M. Eng. thesis, Dept. of Agricultural and Biosystem Engineering, McGill University, Montreal, Canada, 106 pp.

    Pedersen, L. B. and Prahm, L. P. (1974) A method for numerical solution of the advection equation. *Tellus*, Vol. 26, pp. 594-602.

5. Pepper, D. W. and Baker, A. J. (1980) High-order accurate numerical algorithm for three dimensional transport equations. *Computers and Phys.*, Vol. 8, pp. 371-390.

6. Pepper, D. W. and Long, P. E. (1978) A comparison of results using second-order moments with and without width correction to solve the advection equation. *J. Appl. Meteorology*, Vol. 17, pp. 228-233.

7. Marchand, P. (1997) Hydrodynamic modeling of shallow basins. M. Sc. thesis, Dept. of Agricultural and Biosystems Engineering, McGill University, Montreal, Canada.

8. Staniforth, A. and Côté, J. (1991) Simi-Lagrangian integration schemes for atmospheric models - a review. *Monthly Weather Rev.*, Vol. 119, pp. 2206-2223

Environmental Hydraulics, Lee, Jayawardena & Wang (eds) © 1999 Balkema, Rotterdam, ISBN 90 5809 035 3

Transverse mixing of solute and suspended sediment from a river outfall during over bank flow

K.J.Spence & I.Guymer
Department of Civil and Structural Engineering, University of Sheffield, UK

J.R.West
School of Civil Engineering, University of Birmingham, UK

B.Sander
Lehrstuhl und Institut für Wasserbau und Wasserwirtschaft, Rheinisch-Westfälische Technische Hochschule, Aachen, Germany

ABSTRACT:

Laboratory experiments have been undertaken on a large scale compound channel to investigate the transverse mixing of solute and suspended sediment in effluent plumes under river flood conditions. Attention has been focused on the effect of a cross-stream depth variation from main channel to flood plain on the lateral variation of local transverse mixing coefficient in the near and mid field zones. Preliminary experimental results are presented for the transverse variation of solute and suspended sediment concentrations. Both tracers were injected simultaneously above the mid point of a side slope of the main channel. These show in this particular case that there is little difference between the transverse mixing of solute and suspended sediment. An analytical solution provides a potentially powerful technique to obtain the width averaged transverse mixing coefficient that may also determine the lateral variation of transverse mixing coefficient.

1. INTRODUCTION

Pollutants, both soluble and suspended particulate, are routinely discharged into water courses as effluent from storm water overflows during flood events. In addition, contaminated sediment lying dormant in the river bed and banks may also be mobilised and incorporated into the flow. Storm water overflows are often positioned at the river bank to minimise construction costs. During over bank flow, this location may have an adverse effect on water quality in the mid-field zone before the effluent is transversely well mixed. Here, the rapid change in depth that occurs between the main channel and flood plain regions in a two stage channel leads to the formation of large horizontal eddies as momentum is exchanged. As a result, lateral transport of highly concentrated pollutants onto the flood plain can occur. The lower depth and longitudinal velocity present on the flood plain can cause these high concentrations to persist for a considerable distance downstream of the outfall site. An understanding of the lateral transport of both solute and suspended sediment is therefore basic to the prediction of pollutant transport.

2. PREVIOUS WORK

Laboratory or field data may be employed in standard solutions (Rutherford, 1994) to obtain estimates of depth averaged transverse mixing coefficients. The transverse mixing coefficient, ε_y can be obtained from transverse solute concentration profiles assuming an infinitely wide channel, that is the depth mean longitudinal velocity u, depth h and transverse mixing coefficient are constant with change in transverse distance, y. This information can be used in the change in moment method, which is described by

$$\varepsilon_y = \frac{u}{2}\frac{d\sigma^2}{dx} \tag{1}$$

where σ^2 = transverse variance of the concentration distribution and x = longitudinal distance.

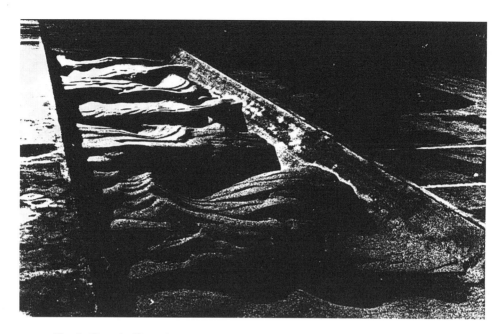

Fig. 1   Dune bedforms in main channel formed during an over bank flow of 450 l/s

A generalised change in moment method, first developed by Holley *et al.* (1972), may be adopted where there are lateral variations in depth mean longitudinal velocity, depth and local transverse mixing coefficient. It can account for the effect of the compound channel shape on these variables and for the presence of transverse velocities. The solution is given by

$$\frac{d}{dx}\left[\frac{\int_A^B huc(y-\overline{y})^2\,dy}{\int_A^B hucdy}\right] - \frac{2\int_A^B hwc(y-\overline{y})dy}{\int_A^B hucdy} = -2K\frac{\int_A^B h\varphi\frac{\partial c}{\partial y}(y-\overline{y})dy}{\int_A^B hucdy} \qquad (2)$$

where c = tracer concentration, w = depth averaged transverse velocity, K = a constant (the integral of the local transverse mixing coefficient over the limits of integration), φ = a function of the local transverse mixing coefficient, A and B = limits of integration in the transverse direction, for example the channel width. The value φ is an assumed local mixing coefficient divided by the integral of these mixing coefficients over the limits of integration in the transverse direction, such that

$$\int_A^B \varphi\,dy = 1 \qquad (3)$$

The first term in equation 2 is the longitudinal rate of change of mass flux variance. The second term accounts for the effects of transverse velocities and the third term considers the lateral variation of local transverse mixing coefficients. If channel bathymetry, the transverse variation of longitudinal velocities and transverse solute concentration profiles are available, then the only unknown in equation 2 is K. Hence the width averaged transverse mixing coefficient can be obtained by dividing K by B-A. We now have two values of width averaged transverse mixing coefficients, one value from the assumed transverse variation of local transverse mixing coefficient and a calculated value from equation 2. The former can therefore be adjusted to minimise the difference between these width averaged mixing coefficients.

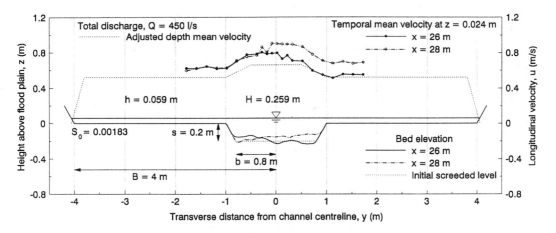

Fig. 2 Cross section of Flood Channel Facility

3. LABORATORY STUDY

An investigation of this issue has been conducted at the Flood Channel Facility (FCF), which is funded by the UK Engineering and Physical Science Research Council (EPSRC) and located at HR Wallingford Ltd. For the results presented, a large scale trapezoidal compound channel was used. This had smooth mortared flood plains that were flat in the transverse direction. The main channel side walls were composed of the same material and set at 45 degrees to the horizontal. Sand (D_{50} = 0.835 mm) was initially screeded to a depth of 0.2 m below the flood plain. An over bank discharge of 450 l/s allowed large three dimensional dunes to form naturally over a 24 hr period (Fig. 1). The facility was then drained and the dune bedforms immobilised using sprays of sodium silicate and sodium bicarbonate solutions. Fluorescent Rhodamine WT dye was used as the solute tracer and a fine sand (D_{50} = 0.109 mm) was used for the suspended sediment. Due to the potential of suspended sediment modifying the turbulence characteristics of any flow, the sand and solute were introduced together in well-mixed suspension as a 70 mm wide line source above the mid point of a channel bank. The technique for immobilising the large dune bedforms only allowed test periods of a limited duration before disintegration of the dunes and transport of the sand bed occurred. A three dimensional study of concentrations and velocities was not feasible. Development of the plume was determined in the transverse direction at one vertical elevation (0.4 x the depth of water on the flood plain) and at 5 longitudinal positions. This was performed by discrete sampling. The concentrations of solute and suspended sediment were subsequently determined using a Series 10 Turner Designs fluorometer and gravimetric analysis respectively. Longitudinal velocities were measured using a miniature propeller meter at the same elevation at which the discrete samples were taken and are shown in Fig. 2. No sand settled on the flood plains, but a small amount of sand settled in the lee of some of the dunes.

4. PRELIMINARY EXPERIMENTAL RESULTS

Longitudinal measurement of the water surface indicated that its average slope was parallel to that of the flood plain. Uniform flow was therefore assumed to a first approximation, as the dune bed forms will result in a complex flow structure within the main channel that will cause some longitudinal variation in depth. The longitudinal change in the transverse variation of bed elevation and longitudinal velocity (Fig. 2) illustrates interesting features. The longitudinal velocity is similar on the left hand flood plain but a change occurs in the main channel with higher velocities being observed on the right hand side further downstream. This is caused by the change in main channel bed elevation. The average velocity for both flood plains is 0.58 m/s from 60 measurements at various transverse and longitudinal positions, whilst in the central region of the main channel the width averaged velocity is 0.74 m/s near the flow surface (Spence *et al.*, 1997).

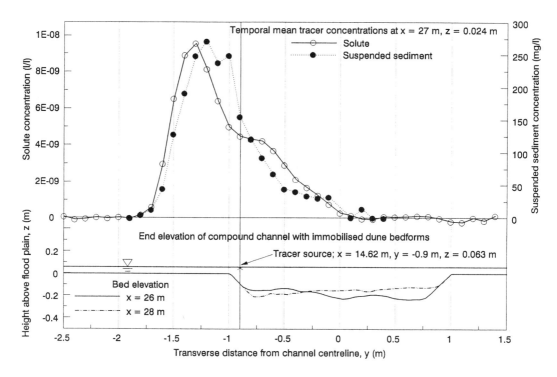

Fig. 3 Experimental data from a solute and suspended sediment injection

Flow continuity was assessed to establish the relative magnitudes of flood plain and main channel velocities. The average bed elevation in the main channel should be the same as the initial screeded level. Transverse measurement of the bed at 8 longitudinal positions shows that this average elevation is 4% higher than the initial screeded level of 0.2 m. This is considered acceptable based on the complexity of the bed topography and shows that no bed material was lost from the recirculating apparatus for the bed material as the dunes were being formed. The general shape of the transverse variation of point longitudinal velocity is judged to be representative of depth mean values (Fig. 2). It is therefore assumed that the longitudinal velocity on both flood plains is equal and constant with transverse distance, there is a linear variation of velocity from the edge of each flood plain to a distance of 0.5 m into the main channel and that the velocity is constant in the central region of the main channel. Flow continuity was satisfied in this complex three dimensional flow by multiplying the observed point velocities by 0.89 to give estimated depth mean values (Fig. 2). The same factor was applied to the average depth within main channel in recognition of the blocking effect of the dunes.

Results are presented of the transverse variation of both solute and suspended sediment at one longitudinal distance downstream, with the background concentrations removed (Fig. 3). The observed skewed shape of the concentration distributions is caused by the plume extending from the low mixing zone on the flood plain to the high mixing zone in the main channel. Despite the source being at the midpoint of a main channel side slope, the data indicates an apparent injection point on the flood plain. This results in much higher than expected concentrations occurring on the flood plain with an associated higher potential for an adverse effect from any harmful pollutant. This may be caused by local flow conditions at the injection site or general flow behaviour in the interaction zone between the flood plain and main channel. This requires further investigation. Inspection of the data clearly shows that shape of both solute and suspended sediment is similar, but that in both the main channel and flood plain regions, the majority of the suspended sediment has not travelled as far laterally as the solute.

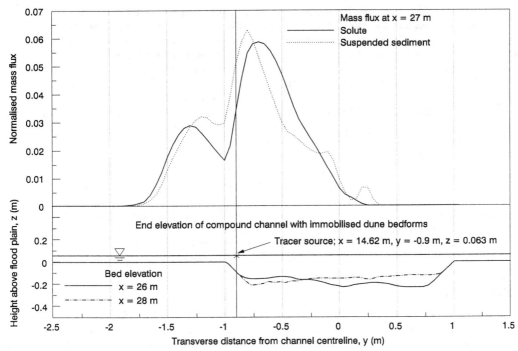

Fig. 4 Transverse variation of mass flux for solute and suspended sediment

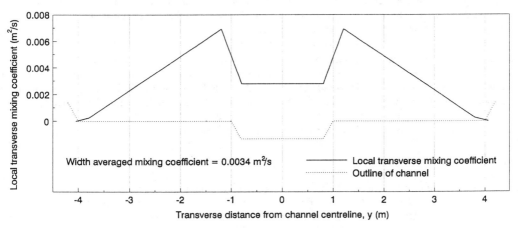

Fig. 5 Lateral variation of local transverse mixing coefficient

When concentrations extend over regions of different depth and velocity, the variance should be calculated from the transverse variation of mass flux (Equation 2). Transverse variation of mass flux is illustrated for both tracers, using the adjusted average values of depth and velocity (Fig. 4). This gives a clearer indication of the amount of tracer transported in both flood plain and main channel than is shown by the transverse variation of measured concentration. The mass flux variance for the solute is 0.156. This is 2.5% smaller than the value of 0.160 obtained from the suspended sediment concentrations. This higher variance for the suspended sediment was not expected from the qualitative inspection of the concentration distributions. It is caused by the small sediment concentrations at $y = 0.2$ m. Thus, the variances of solute and suspended sediment can be taken as equal here, within the limitation of experimental measurement.

To obtain a channel width averaged transverse mixing coefficient using equation 2, the longitudinal rate of variance is required. An estimate can be obtained by assuming the variance at the injection position is zero, which gives $d\sigma^2/dx = 0.0126$ m. This assumption has been shown to be justified from analysis of other data. The potential influence of transverse velocities due the presence of dunes in the main channel is set aside here, so the second term in equation 2 is zero. The form of the transverse variation of local transverse mixing coefficient has to be decided. It has been shown from previous work that the longitudinal rate of variance in the main channel region from a centreline injection of solute is about 0.00849 m (Spence et al., 1997). Using equation 1 and the adjusted value of longitudinal velocity in the main channel, this gives a mixing coefficient of 0.00276 m^2/s in this region. Studies of the transverse variation of eddy viscosity in the longitudinal and transverse planes (Knight & Shiono, 1990) show a peak value on the flood plain near the channel step, due to the formation of large horizontal eddies caused by the interaction between main channel and flood regions of flow. The transverse mixing coefficient at the far edges of the flood plains, away from the main channel, is likely to be described by $\varepsilon_y = 0.13du_*$ (Rutherford, 1994), where d is the depth and the shear velocity, u_* is given by $\sqrt{(gdS_o)}$ with g = gravitational acceleration and S_o = longitudinal channel bed slope. Based on this information, a lateral variation of local transverse mixing coefficient that satisfies equation 3 is illustrated. This is not a unique solution. However, with further analysis of the concentration distributions at other longitudinal positions, a more accurate description of the longitudinal rate of change of variance and the lateral variation of local transverse mixing coefficient should be obtained.

5. CONCLUSIONS

Simultaneous measurement of the transverse mixing of both solute and suspended sediment ($D_{50} = 0.109$ mm) has been performed within a trapezoidal compound channel during an over bank flow of 450 l/s. Large dune bed forms were present in the main channel region. The peak suspended sediment concentration near the surface of the flow was 280 mg/l. The transverse mixing of the solute and suspended sediment from an injection above a side slope of the main channel were found to be similar. This shows that the transport of suspended sediment may be approximated by using solute transport studies. The transverse mixing coefficient averaged over the channel width has been obtained using an analytical solution and is equal to 0.0034 m^2/s. This has been achieved by a first approximation of the lateral variation of local transverse mixing coefficient. Estimates of the non-dimensionalised transverse mixing coefficient ε_y/du_* for the main channel region and for the peak value observed on the flood plain are 0.19 and 3.6 respectively. The former value is within generally accepted limits for the non-dimensionalised transverse mixing coefficient (Rutherford, 1994), whilst the larger peak value observed on the flood plain reflects the influence of the large horizontal eddies generated by the interaction between the main channel and flood plains waters. Further analysis should allow a more accurate description of the lateral variation of local transverse mixing coefficient.

6. ACKNOWLEDGEMENTS

The authors gratefully acknowledge the support of the Engineering and Physical Science Research Council under Research Grant Ref. GR/K/04651 and the technical assistance provided at HR Wallingford Ltd by Mary Johnstone and Dave Wilmer. The final author was generously supported by European Community Human Capital Mobility funds.

7. REFERENCES

Holley, E.R., Siemons, J. & Abraham, G. (1972). "Some aspects of analysing transverse diffusion in rivers", J. Hyd. Res., 10, 27-57.

Knight, D.W. & Shiono, K. (1990). "Turbulence measurements in a shear layer region of a compound channel", J. Hyd. Res., 28, 175-196.

Rutherford, J.C. (1994). *River Mixing*, John Wiley & Sons Ltd, Chichester.

Spence, K.J., Potter, R. & Guymer, I. (1997). "Transverse solute mixing from river outfalls during over bank flows", 3rd Int. Conf. River Flood Hyd., 5-7 Nov 1997, 485-494.

Environmental Hydraulics, Lee, Jayawardena & Wang (eds) © 1999 Balkema, Rotterdam, ISBN 90 5809 035 3

Lateral turbulent dispersion in compound channels: Some experimental and field results

Xingnian Liu & Ban Jiu
State Key Hydraulics Laboratory, Sichuan Union University, China

Donald W. Knight
School of Civil Engineering, University of Birmingham, UK

Tongliang Gong
Hydrology Agency of Xizhang (Tibet), Lasha, China

ABSTRACT: This paper presents and discusses some results of experiments in simple channels, then investigates the lateral variation of the mixing coefficients λ in compound channels. Average λ values in both simple channels and in the main channels of compound channels are obtained. For compound channels, the relative depth (H-h)/H and the relative width, are used to analyze the influence of the compound channel on the λ value. The experiments on the lateral distribution of turbulent mixing coefficients were conducted at the University of Birmingham. In this paper, some new parameters are introduced to describe lateral and mixing dispersion in both simple and compound channels. Some of field data on turbulent dispersion in the Lasha River (Tibet, China) is also listed and analyzed in this paper.

1. INTRODUCTION

As a river changes from inbank to overbank flow, which frequently occurs in flood flow conditions, not only does the cross section shape of the channel change significantly, but also the streamwise pathways for flow may also alter considerably. Usually river engineers are only concerned with the parameters at the channel boundaries and with lateral distributions, and therefore various methods have been proposed for determining the boundary shear stress and eddy viscosity, and solving the one-dimensional or two dimensional hydrodynamic equations (e.g. Knight & Shiono, 1990; Shiono & Knight, 1989; Shiono & Knight, 1990; Knight & Demetriou, 1983; Knight & Hamed, 1984). For the simple channel there has been a lot of experimental work and analysis (Lau & Krishanappan, 1977; Elder, 1959; Holley, 1971; Knight & Shiono, 1990), These form the basis of analyzing the experimental data in compound channels.

Both the theoretical and numerical modeling of sediment transport rely on having correct values for the turbulent mixing coefficients. For lateral dispersion, it is usual to use a lateral mixing coefficient which is proportional to the product of shear velocity and average depth, called the dimensionless eddy viscosity, λ. In compound channels, the depth is chosen as the depth of main channel. Unfortunately, the value, λ, has shown considerable variations, even for simple straight channels. Experiments for simple channel show that the λ value varies with the width to depth ratio, W/H, and roughness. For compound channels, there are considerately little researches about the lateral dispersion coefficients. So this paper presents and discusses some results of experiments in simple channels, then investigates the lateral variation of the mixing coefficients in compound channels. A comparison is made between λ values in both simple channels and compound channels. For compound channels, some parameters, such as relative depth (H-h)/H and relative width, are added into analyze the influences of the compound channel on the λ value.

The experiments of the lateral distribution of turbulent mixing coefficients were conducted in the University of Birmingham. The experimental data from the SERC-FCF are also used to determine dimensionless eddy viscosity ratios.

2. ANALYSIS OF EDDY VISCOSITY DATA IN SIMPLE CHANNEL

The calculation of the concentration in simple open channels is usually based on solutions of the mass conversation equation, and the accuracy of these solutions depends on having correct values for the turbulent mixing coefficients. It is customary to assume that the lateral mixing coefficient is proportional to the product of shear velocity and water depth, that is, $\varepsilon = \lambda u_* H$, in which λ, the proportionality constant is called the dimensionless eddy viscosity coefficient. This has been measured in the laboratory as well as in the field. In straight rectangular open channels, the parameters affecting the lateral dispersion coefficient are mean velocity u, depth h, width w (or B, b), shear velocity u_*, density ρ, and viscosity μ, that is

$$\varepsilon = \phi(u, h, w, u_*, \rho, \mu) \tag{1}$$

If different parameters or dimensions of the cross section are chosen as the basic parameters, then equation (1) can be nondimensionlised as follows:

$$\frac{\varepsilon}{u_* h} = \phi(f, \frac{w}{h}) \tag{2}$$

$$\frac{\varepsilon}{u_* w} = \phi(f, \frac{w}{h}) \tag{3}$$

$$\frac{\varepsilon}{u_* \sqrt{wh}} = \phi(f, \frac{w}{h}) \tag{4}$$

In which $f = 8(u_* / u)^2$ is the Darcy-Weisbach friction factor. It can be seen that the dimensionless eddy viscosity coefficient is a function of both the friction factor and the width-to- depth ratio, since the friction factor indicates the bottom shear that generates the turbulence in the flow and the width-to-depth ratio affects the secondary circulation in the channel.

The previously published data on rectangular flumes, varying in width from 36 cm to 238 cm, with either smooth or artificially roughened beds, are from Elder, Engmann, Holly, Kalinske & Pien, Okoye, and Sayre & Chang, etc. Fig.1 is a plot of $\varepsilon / (u_* h)$ versus w/h using all the data points. The values of $\varepsilon / (u_* h)$ vary from 0.08 to 0.24, while w/h varied from 4.37 to 65.1, consequently, many engineers have used a constant value, regardless of the flow configuration.

As shown in Fig.2, there appears to be no clear relation between $\varepsilon / (u_* h)$ and friction factor. Because $\varepsilon / (u_* h)$ is not a constant or because of the scatter in the values of $\varepsilon / (u_* h)$ versus w/h, Lau & Krishnappan suggested that the secondary circulation appears to be a more important factor governing the transverse spreading, and think that it is more appropriate to use the channel width as the characteristic length scale, that is, Eq.(3). Fig.2 shows the plots of $\varepsilon / (u_* w)$ versus w/h and $\varepsilon / (u_* w)$ versus f respectively. It can be seen that the relationship between $\varepsilon / (u_* w)$ and w/b is fine, but that the relationship between $\varepsilon / (u_* w)$ and f is poor. It is apparent that the characteristic length scale should include both width and water depth, and so the product of width and depth was chosen as the length scale, that is, $\sqrt{wh}$.

Fig.3 shows the relationship of $\varepsilon / (u_* \sqrt{wh})$ versus w/h, and $\varepsilon / (u_* \sqrt{wh})$ versus f respectively. From Fig.3, it can be seen that both relationships are fine with little scatter. Comparison of Fig.1 to Fig.3 indicates that all $u_* h$, $u_* w$ and $u_* \sqrt{wh}$ maybe the suitable representation of ε, which depends on the usage. This means that if you want to use a constant value for λ, $u_* h$ should be chosen, but if you want to study the influence of friction factor upon the eddy viscosity, then $\varepsilon / (u_* \sqrt{wh})$ should be chosen instead.

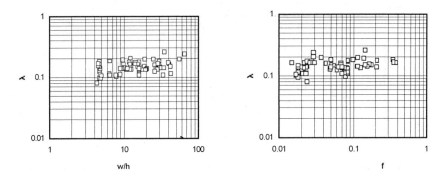

Fig.1 Relation of dimensionless eddy viscosity $\varepsilon / (u_* h)$ with relative width w/h, and friction factor f

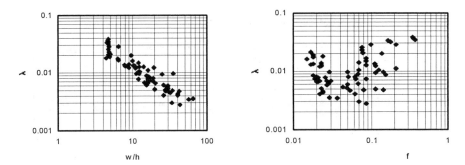

Fig.2 Relation of dimensionless eddy viscosity $\varepsilon / (u_* w)$ with relative width w/h, and friction factor f

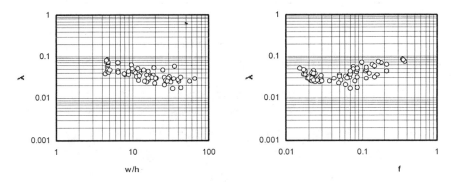

Fig.3 Relation of dimensionless eddy viscosity $\varepsilon / (u_* \sqrt{wh})$ with relative width w/h, and f

3. EXPERIMENTAL DATA IN COMPOUND CHANNEL

The lateral distributions of boundary shear stress and velocity have been measured in compound channels with a trapezoidal main channel and varied bank side slopes at HR Wallingford, England. All the data used in this paper are from the Science and Engineering Research Council Flood Channel Facility (SERC-FCF). The flume has a main channel width of 1.5 m (b=0.75 m, B varies from 1.2b to 6.67b), and the values for bank side slopes is 0, 1 and 2. For each experiment, the shear stress and velocity are measured at the distance intervals of 0.05 m or 0.1 m away from centerline.

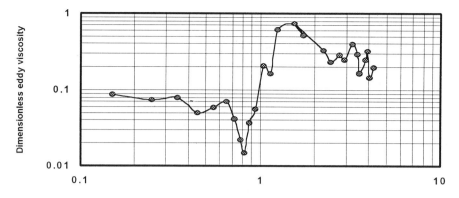

(a) SERC-FCF 010201 with Dr=0.10 and B/b=6.67

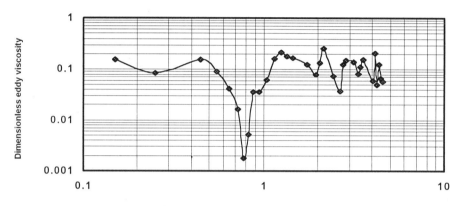

(b) SERC-FCF 010401 with Dr=0.20 and B/b=6.67

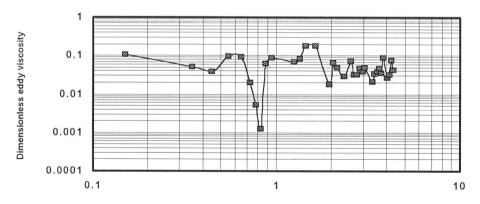

(c) SERC-FCF 010701 with Dr=0.40 and B/b=6.67

Fig.4 Typical lateral distribution of dimensionless eddy viscosity (SERC-FCF 0102)

4. LATERAL DISTRIBUTION OF DIMENSIONLESS EDDY VISCOSITY

If it is assumed that the depth averaged eddy viscosity is non-dimensionlised by the local friction velocity, then this gives:

$$\varepsilon = \lambda u_* H \tag{5}$$

Introducing a depth averaged eddy viscosity $\rho\varepsilon \dfrac{\partial U_d}{\partial y} = \tau_V$ gives:

$$\lambda = \frac{\tau_V}{\rho u_* H \dfrac{\partial U_d}{\partial y}} \tag{6}$$

Where τ_v is the apparent shear stress and is determined by $\rho g H S_0 + \dfrac{\partial}{\partial y}\{H\tau_V\} - \tau_b = 0$, U_d is depth-averaged velocity in flow direction. See Knight and Shiono for more details.

The dimensionless eddy viscosity can then be computed from the experimental data by the above method. Typical lateral distributions of dimensionless eddy viscosity λ are shown in Fig.4, which shows that the dimensionless eddy viscosity on the floodplains is much larger than that in main channel when the relative depth Dr=(H-h)/H is small. With an increase in relative depth, the values of dimensionless eddy viscosity in main channel and floodplains get close together as shown in Fig.4 (c). The average dimensionless eddy viscosity ratios in the main channel versus relative depth are shown in Fig.5, which shows a average value of around 0.07.

5. THE INFLUENCE OF RELATIVE WIDTH AND DEPTH ON THE LATERAL EDDY VISCOSITY

The relationship between the ratio of dimensionless eddy viscosity in the main channel and on the floodplains, and relative depth Dr is shown in Fig.6, from which it can be seen that the ratio decreases with an increase in relative depth. This may reflect the influence of secondary flows after overbank flow occurs.

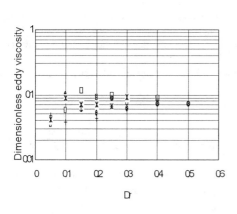

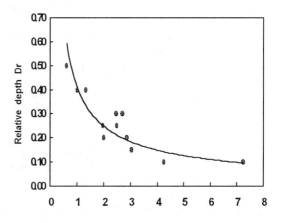

Fig.5 $\varepsilon / (u_* h)$ in main channel versus relative depth for compound channel

Fig.6 Influence of relative depth on the ratio of eddy viscosity in main channel and floodplains

373

6. CONCLUSIONS

For simple channels, the eddy viscosity may be expressed as the form s of $\varepsilon/(u_*h)$, $\varepsilon/(u_*w)$ and $\varepsilon/(u_*\sqrt{wh})$, which depend on the usage or research purposes. If $\varepsilon/(u_*h)$ is chosen as the expression, then the dimensionless eddy viscosity value λ is roughly a constant although there is some scatter. If $\varepsilon/(u_*\sqrt{wh})$ is chosen, then, the influence of friction factor is clearly seen. For compound channels, the dimensionless eddy viscosity λ value is generally greater than that on floodplains. The ratio of dimensionless eddy viscosity in the main channel and on floodplains decreases with an increase in relative depth. The average value of dimensionless eddy viscosity in main channel of compound channel is around 0.07, which is smaller than the average value of dimensionless eddy viscosity in simple channel.

7 ACKNOWLEDGMENTS

This study was completed while a Visiting Honorary Research Fellow at the University of Birmingham, sponsored by the British Council, whose support is gratefully acknowledged.

8 REFERENCES

Elder, J. W., (1959), The dispersion of market fluid in turbulent shear flow, Journal of Fluid Mechanics, Vol.5, 544-560.

Holley, E. R., (1971), Transverse mixing in rivers, Report No. S132,Delft Hydraulics Laboratory, Delft, The Netherlands.

Knight, D. W. and Yuen, W. H., (1994), Boundary shear stress distributions in open channel flow (Ed. K. J. Beven, etc.), Wiley & Sons Ltd., 51-87.

Knight, D. W. (1981) Boundary shear stress in smooth and rough channels, J. Hydr. Div. ASCE, Vol.107, No.HY7, 839-851)

Knight, D. W. and Hamed, M. E., (1984), Boundary shear in symmetrical compound channels, J. Hydra. Engin. ASCE, Vol. 110, 1412-1430.

Knight, D. W. and Patel, H. S., (1985), Boundary shear in smooth rectangular ducts, J. Hydr. Engin. ASCE, Vol. 111, 29-47.

Knight, D. W. and Shiono, K., (1990), Turbulence measurements in a shear layer region of a compound channel. J. Hydr. Res. 28, 175-196.

Knight, D. W., Demetriou, J. D. and Hamed, M. E., (1984), Boundary shear in smooth rectangular channels, J. Hydr. Engin. ASCE, Vol. 110, 405-422.

Lau, Y. L. & Krishnappan, B. G., (1977), Transverse dispersion in rectangular channels, J. of Hydrau. Div. ASCE, vol.103, No.HY10, pp.1173-1189.

Liu, X. & Chen, Y., (1987), Nonuniform bedload transport, J. of Chengdu University of Science and Technology, No.02, 21-30.

Miller, A. C. and Richarde, E. V., (1974), Diffusion and dispersion in open channel flow, J. of Hydrau. Div. ASCE, Vol. 100, No. hy1, 159-171.

Shiono, K. and Knight, D. W., (1989), Transverse and vertical measurements of Reynolds stress in a shear layer region of a compound channel. In: 7th Inter. Symposium on Turbulent Shear Flows, Stanford, USA, August 1989, 28.1.1-28.1.6, Stanford University Press.

Shiono, K. and Knight, D. W., (1990), Mathematical models of flow in two or multi-stage straight channels, in: Proceedings of the Inter. Conf. on River Flood Hydraulics (Ed. W. R. White). Wiley,, 229-238.

Environmental Hydraulics, Lee, Jayawardena & Wang (eds) © 1999 Balkema, Rotterdam, ISBN 90 5809 035 3

Numerical investigation of zero-mean flow turbulence near a free surface

Yuji Sugihara & Nobuhiro Matsunaga
Department of Earth System Science and Technology, Kyushu University, Kasuga, Japan

ABSTRACT: A numerical analysis has been made to investigate the behaviors of zero-mean flow turbulence near a free surface. A Reynolds stress model proposed by Launder et al. (1975) has been applied to oscillating-grid turbulence in a homogeneous fluid with finite depth. The model predicts a strong influence of free surface on turbulent structure within the range of the order of integral length scale. The numerical solutions normalized by using the analytical solutions for the infinite depth have been compared with experimental data. A good agreement is seen between the both. This agreement shows that the Reynolds stress model is useful for the turbulence near the free surface.

1. INTRODUCTION

In open channel flows, the distribution of axial velocity is influenced considerably by the anisotropy of turbulence near a free surface. Turbulent eddies close to a free surface also play an important role in momentum, heat and gas transfer processes across an air-water interface. The improvement of accuracy in numerical simulation for near-surface turbulence is significant to predict these hydraulic phenomena.

Reynolds stress models have been developed to simulate reasonably anisotropic turbulent flows. Applicability of the models to open channel turbulent flows has been already investigated in previous studies (e.g., Gibson and Rodi, 1989; Kawahara and Tsuneyama, 1994). However, since the structure of open channel flow is very complicated, how accurately the model can analyze the anisotropic turbulent distortion has not been revealed in detail.

Zero-mean flow turbulence forms when a square grid is oscillated vertically in a fluid at rest. The oscillating-grid turbulence, which is referred to as OGT hereafter, has horizontal homogeneity and isotropy. Komatsu et al. (1995) investigated the behaviors of OGT near a free surface based on LDV measurements, and gave the vertical profiles of the turbulent intensity and the integral time scale. Their data seem to be useful for a closure modeling of turbulence near a free surface.

The purposes of the present study are to numerically investigate the behaviors of OGT near a free surface by using a turbulence closure model, and to compare the model solutions with the experimental data by Komatsu et al.. A Reynolds stress model proposed by Launder et al. (LRR model) has been used in the numerical analysis. It should be noted that a weak anisotropy exists between the horizontal and vertical fluctuating velocities for the infinite depth. In order to simulate accurately the turbulent distortion due to free surface, the anisotropy should be related to model constants. The values have been determined on the basis of experimental data and analytical solutions of the model for the infinite depth.

2. ANALYTICAL SOLUTIONS OF REYNOLDS STRESS MODEL FOR INFINITE DEPTH

Fig. 1 shows schematically an experimental apparatus of OGT. The turbulence field is in equilibrium between the diffusion and the dissipation of turbulent energy, and it is homogeneous and isotropic in the horizontal plane. For OGT, the governing equations of LRR model are given by

$$C_S \frac{d}{dz}\left\{\frac{k}{\varepsilon}\left(w^2\frac{dk}{dz}+w^2\frac{dw^2}{dz}\right)\right\} = \varepsilon, \quad C_S\frac{d}{dz}\left\{\frac{k}{\varepsilon}3w^2\frac{dw^2}{dz}\right\} = C_{\phi 1}\frac{\varepsilon}{k}\left(w^2 - \frac{2}{3}k\right) + \frac{2}{3}\varepsilon$$

$$\text{and } C_\varepsilon \frac{d}{dz}\left\{\frac{k}{\varepsilon}w^2\frac{d\varepsilon}{dz}\right\} = C_{\varepsilon 2}\frac{\varepsilon^2}{k}, \tag{1}$$

where k is the turbulent energy, w the intensity of vertical fluctuating velocity, ε the dissipation rate of the energy, z the coordinate taken vertically upward from the center of the grid oscillation. $C_S, C_{\phi 1}, C_\varepsilon$ and $C_{\varepsilon 2}$ are model constants. When the turbulent energy k_0 and the dissipation rate ε_0 are generated steadily at z = 0 and w^2 is given there by rk_0, boundary conditions become as follows.

$$k = k_0, \ w^2 = rk_0, \ \varepsilon = \varepsilon_0 \quad at \ z = 0$$

$$\text{and } k \to 0, \ w^2 \to 0, \ \varepsilon \to 0 \quad as \ z \to \infty, \tag{2}$$

where r is assumed to be constant. The governing equations and boundary conditions are non-dimensionalized with k_0 and ε_0. Dimensionless quantities are defined by

$$k_* = \frac{k}{k_0}, \ w_*^2 = \frac{w^2}{k_0}, \ \varepsilon_* = \frac{\varepsilon}{\varepsilon_0} \quad \text{and} \quad z_* = \frac{z}{\sqrt{k_0^3\,\varepsilon_0^{-2}}}. \tag{3}$$

Non-dimensionalized governing equations are the same as Eq. (1) except for the addition of the asterisks. The boundary conditions at z = 0 become $k_* = \varepsilon_* = 1$ and $w_*^2 = r$ at $z_* = 0$. Here, let us introduce a new independent variable ζ_* defined by

$$\frac{d\zeta_*}{dz_*} = \frac{\varepsilon_*}{C_S k_* w_*^2} > 0 \quad for \ 0 < z_* < \infty,$$

$$\zeta_* = \zeta_0 \quad at \ z_* = 0 \text{ and } \zeta_* \to \infty \quad as \ z_* \to \infty. \tag{4}$$

Transformation of the basic equations and boundary conditions into the ζ_*-coordinate system gives

$$\frac{d^2 k_*}{d\zeta_*^2} + \frac{d^2 w_*^2}{d\zeta_*^2} = C_S k_* w_*^2, \quad \frac{d^2 w_*^2}{d\zeta_*^2} = \frac{C_S C_{\phi 1}}{3}w_*^4 - \frac{2}{9}C_S\left(C_{\phi 1}-1\right)k_* w_*^2,$$

$$\frac{d^2\varepsilon_*}{d\zeta_*^2} = \frac{C_S^2 C_{\varepsilon 2}}{C_\varepsilon}\varepsilon_* w_*^2, \tag{5}$$

$$k_* = 1, \ w_*^2 = r, \ \varepsilon_* = 1 \quad at \ \zeta_* = \zeta_0$$

$$\text{and } k_* \to 0, \ w_*^2 \to 0, \ \varepsilon_* \to 0 \quad as \ \zeta_* \to \infty. \tag{6}$$

The analytical solutions are assumed to be a power-law form, i.e.,

$$k_* = \left(\frac{\zeta_0}{\zeta_*}\right)^\alpha, \quad w_*^2 = r\left(\frac{\zeta_0}{\zeta_*}\right)^\beta \quad \text{and} \quad \varepsilon_* = \left(\frac{\zeta_0}{\zeta_*}\right)^{3+\gamma} \tag{7}$$

in which $\alpha, \beta > 0$ and $\gamma > -3$. Substitution of Eq. (7) into the first and second relations in Eq. (5) gives

376

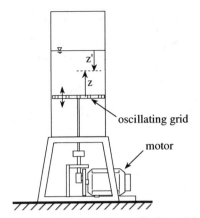

oscillating grid

motor

Fig. 1 Schematic diagram of an oscillating-grid tank

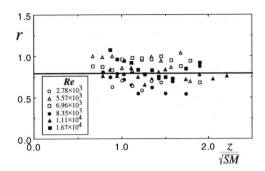

Fig. 2 Vertical profile of r

$$\alpha = \beta = 2 \; , \quad \frac{6}{C_S \zeta_0^2} = \frac{r}{1+r} \quad \text{and} \quad C_{\phi1} = \frac{7r-2}{(1+r)(3r-2)} \, . \tag{8}$$

This implies that the ratio of w^2 to k becomes spatially constant and the value takes r, and that $C_{\phi1}$ must be related to the anisotropy of turbulence.   It should be also noted that Eq. (7) is valid under the condition of the third relation in Eq. (8).   Substitution of Eq. (7) into the third relation in Eq. (5) also gives

$$\gamma = -\frac{7}{2} + \sqrt{\frac{1}{4} + \frac{6 \, C_S \, C_{\varepsilon2} \, (1+r)}{C_\varepsilon}} \, . \tag{9}$$

Using the definition of Eq. (4) and the derived relations, we find a relationship between ζ_* and z_*, i.e.,

$$\left(\frac{\zeta_*}{\zeta_0} \right)^\gamma = \frac{z_*}{z_0} + 1, \tag{10}$$

where z_0 is defined by $C_s \zeta_0 r / \gamma$.   Therefore, by substitution of Eq. (10) into Eq. (7) and the relation $\alpha = \beta = 2$, the analytical solutions for the infinite depth are expressed by

$$k_* = \left(\frac{\zeta_0}{\zeta_*} \right)^2 = \left(\frac{z_*}{z_0} + 1 \right)^{-\frac{2}{\gamma}} , \quad w_*^2 = r \left(\frac{\zeta_0}{\zeta_*} \right)^2 = r \left(\frac{z_*}{z_0} + 1 \right)^{-\frac{2}{\gamma}}$$

$$\text{and } \varepsilon_* = \left(\frac{\zeta_0}{\zeta_*} \right)^{3+\gamma} = \left(\frac{z_*}{z_0} + 1 \right)^{-\frac{3+\gamma}{\gamma}} . \tag{11}$$

In the case when OGT is not influenced by a free surface, r should be confirmed to be spatially constant. Fig. 2 shows a vertical profile of r obtained from experimental data.   Here, S and M are the stroke and the mesh size of the grid, respectively and Re a Reynolds number defined by fS^2/v, where f is the frequency of the grid and v the kinematic viscosity. From the figure, we can read $r \doteqdot 0.8$ though some scattering is seen.   The model analysis gives some relationships existing among the model constants. The values of γ and z_0 have been obtained from turbulence measurements, and they are 0.4 and 1.82, respectively (Matsunaga et al., 1998). The efficiency of $C_{\varepsilon 2} = 1.90$ has been also verified experimentally for OGT (Sugihara et al., 1994). Therefore, using Eqs. (8), (9) and $z_0 = C_s \zeta_0 r / \gamma$, we can determine the model constants as follows.

$$C_s = 0.0613, \ C_{\phi 1} = 5.0, \ C_{\varepsilon} = 0.0841,$$
$$C_{\varepsilon 2} = 1.90 \ \text{and} \ r = 0.8. \tag{12}$$

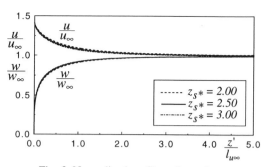

Fig. 3 Normalized profiles of u and w near the free surface

These values are considerably different from those by Launder et al. and may be peculiar to OGT.

3. NUMERICAL ANALYSIS WITH REYNOLDS STRESS MODEL FOR FINITE DEPTH

In order to introduce the effect of free surface on turbulent redistribution, a closure model of the pressure-strain correlation is added into the right hand of the second relation in Eq. (1). The correlation model is expressed by

$$\Phi_S = 2 \, C_d \, \frac{\varepsilon}{k} \, w^2 f_S, \tag{13}$$

where C_d is a model constant. The surface-proximity function f_s is given by following a model of Gibson and Rodi, i.e.,

$$f_S = \frac{L}{z_S} \frac{Z^2}{1-Z}, \quad L \equiv \frac{k^{3/2}}{\varepsilon} \quad \text{and} \quad Z \equiv \frac{z}{z_S}, \tag{14}$$

where z_s is the distance between the center of the grid oscillation and the free surface. In the numerical analysis, the equations and boundary conditions have been non-dimensionalized by using k_0 and ε_0. The dimensionless boundary conditions become as follows.

$$\left. \begin{array}{l} k_* = 1, \quad w_*^2 = r, \quad \varepsilon_* = 1 \qquad at \ z_* = 0 \\[2mm] \text{and} \ \dfrac{dk_*}{dz_*} = 0, \ w_*^2 = r_S k_*, \ \dfrac{d\varepsilon_*}{dz_*} = 0 \qquad at \ z_* = z_{S*}, \end{array} \right\} \tag{15}$$

where r_s is an empirical constant. It should be noted that boundary conditions at free surface have not been sufficiently established at present. An unsteady numerical calculation has been made by using the governing equations including the unsteady terms. The numerical solution was assumed to be steady when the ratio of the unsteady term to the dissipation one had been smaller than 10^{-3}.

4. RESULTS AND DISCUSSIONS

Fig. 3 shows normalized numerical solutions of the intensities of horizontal and vertical fluctuating velocities u and w, where u_∞ and w_∞ are the intensities for the infinite depth obtained from Eq. (11), z'the

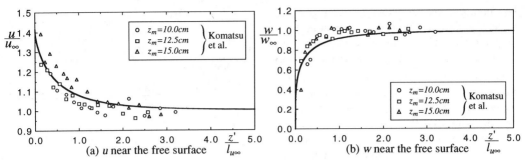

(a) u near the free surface   (b) w near the free surface

Fig. 4 Comparisons between experimental data and numerical solutions for
the intensity of turbulent velocity : ———, numerical solutions

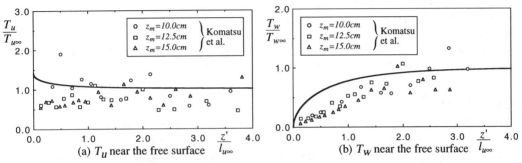

(a) T_u near the free surface   (b) T_w near the free surface

Fig. 5 Comparisons between experimental data and numerical solutions for the
integral time scale of turbulent velocity : ———, numerical solutions

distance taken downward from the free surface and $l_{u\infty}$ ($= 0.14z$) the integral length scale. Here, u has
been obtained from k and w on the basis of the horizontal isotropy. The values of C_d and r_s which give a
good agreement between numerical solutions and experimental data are 0.05 and 0, respectively. It is seen
from the figure that the normalized profiles are independent of z_{s*} and the influence of the free surface on
turbulent structure is confined within the order of the integral length scale.

In Figs. 4 (a) and (b), normalized experimental data of u and w obtained by Komatsu et al. are compared
with the numerical solutions. Their data were obtained by varying the water level above a fixed
measuring point z_m and correspond to those in the range of $z_s = 10 \sim 15$ cm. Dimensionless range based
on empirical relations for k_0 and ε_0 by Matsunaga et al. becomes $z_{s*} = 1.8 \sim 2.7$. Since the normalized
profiles are independent of z_{s*}, the experimental data are compared with the numerical solutions for $z_{s*} =$
2.5. Approaching the free surface, a rapid-damping of w/w_∞ is seen, and u/u_∞ increases due to the
effect of the pressure-strain correlation. The behaviors of the solutions agree well with those of the
experimental data. This agreement shows that the Reynolds stress model is effective for the turbulence
near the free surface.

Figs. 5 (a) and (b) show comparisons for the integral time scales of horizontal and vertical fluctuating
velocities T_u and T_w, where $T_{u\infty}$ and $T_{w\infty}$ denote the time scales for the infinite depth. Though $T_u/T_{u\infty}$
is approximately constant, $T_w/T_{w\infty}$ decreases rapidly in a surface-influenced layer. These trends can be
predicted well by the model.

The k-ε turbulence model seems to be most reasonable among various numerical ones. We should know how the k-ε model can predict the turbulence close to the free surface. Nezu and Nakagawa (1987) proposed a modified boundary condition at free surface $k = D_w k_a$, where k_a is the turbulent energy obtained under the symmetry condition and D_w a damping coefficient. Their condition has been applied to a numerical simulation with the k-ε model. However, the symmetry condition has been used for the dissipation rate. The optimum value of D_w for OGT has been determined on the basis of experimental data, and it becomes 0.625. In Fig. 6, the numerical solutions of the turbulent energy by both the LRR and k-ε models are compared with experimental data. Though the

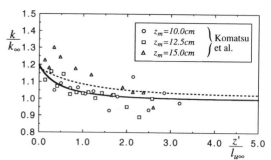

Fig.6 Comparison between experimental data and numerical solutions for the turbulent energy: ———, LRR model solution ; ...., k-ε model solution

behavior of the experimental data agrees well with that of the solutions, the LRR model gives better results than the k-ε model. The data also supports that the turbulent energy becomes larger than that for the infinite depth by the presence of the free surface.

5. CONCLUSIONS

In order to investigate the behaviors of zero-mean flow turbulence near a free surface, the LRR model has been applied to oscillating-grid turbulence. The analytical solutions for the infinite depth have been derived. The values of model constants have been determined on the basis of the analytical solutions and the experimental data. The numerical solutions of the turbulent intensity, the turbulent energy and the integral time scale for finite depth have been obtained. The behaviors of their solutions normalized by using the analytical solutions agree well with those of the experimental data. This agreement shows that the Reynolds stress model is applicable to predict the turbulence near the free surface.

6. REFERENCES

Gibson, M.M. and Rodi, W. (1989) " Simulation of free surface effects on turbulence with a Reynolds stress model" , *J. Hydraul. Res.*, 27, 233-244.

Kawahara, Y. and Tsuneyama, S. (1994) " Modeling of free surface effects on anisotropy of turbulence" , *Proc. Hydraul. Eng., JSCE*, 38, 821-824. (in Japanese)

Komatsu, T., Shibata, T., Asai, K. and Takahara, K. (1995) " Turbulent characteristics near free surface of an oscillating-grid turbulence field" , *Annu. J. Hydraul. Eng., JSCE*, 39, 819-826. (in Japanese)

Launder, B.E., Reece, G.J. and Rodi, W. (1975). " Progress in the development of a Reynolds-stress turbulence closure" , *J. Fluid Mech.*, 68, 537-566.

Matsunaga, N., Sugihara, Y., Komatsu, T. and Masuda, A. (1998) " Quantitative properties of oscillating-grid turbulence in a homogeneous fluid" , *Fluid Dyn. Res.* (accepted)

Nezu, I. and Nakagawa, H. (1987) " Numerical calculation of turbulent open-channel flows by using a modified k-ε turbulence model" , *Proc. JSCE*, 387/II-8, 125-134. (in Japanese)

Sugihara, Y., Matsunaga, N. and Komatsu, T. (1994) " A temporal decay of zero-mean-flow turbulence" , *J. Hydraul., Coast. Environ. Eng., JSCE*, 503/II-29, 215-218. (in Japanese)

Environmental Hydraulics, Lee, Jayawardena & Wang (eds) © 1999 Balkema, Rotterdam, ISBN 90 5809 035 3

Improvements of depth-averaged κ-ε turbulence model

S. Yano
Department of Civil Engineering, Nagasaki University, Japan

T. Komatsu & Y. Nakamura
Department of Civil Engineering, Kyushu University, Fukuoka, Japan

ABSTRACT: The modification of depth-averaged k-ε turbulence model by introduction of "correction factors" was attempted. It was assumed that each physical quantity has a semi-empirical vertical profile of a fully-developed turbulent flow in an open-channel. Formulae for several correction factors, that express the effects of correlations of physical quantities in each term, were obtained by substitution of those profiles into 3-dimensional full k- and ε- equations. Simultaneously two modified formulae for model constants of bottom shear production terms of k and ε were also obtained.

1. INTRODUCTION

In order to predict water pollution and water purification in a nearly-horizontal coastal flow field, which has a very large horizontal scale compared with a vertical one, such as a bay, a harbour, and so on, the depth-averaged model (2-dimensional horizontal model) has been usually used for numerical simulations of tidal current and contaminant diffusion. Then, it is necessary to estimate the eddy viscosity and the eddy diffusivity, but it is very difficult to do it simply and accurately.

Many researchers have assumed that the horizontal eddy viscosity has a constant value spatially and temporally in a tidal current simulation and they have often used an empirical value. However, some researchers has reported the importance of accurate estimation of the eddy viscosity for a tidal current simulation. Oonishi (1977) reported that patterns of the tidal residual current, which are related to the tidal exchange ratio between an inner bay and an outer sea area, depend on a magnitude of the eddy viscosity. Abe *et al.* (1995) also systematically investigated the difference in results of tidal current simulations in a bay with various constant eddy viscosity and reported that the calculated tidal exchange ratio is closely related to those values. Thus, it is necessary to estimate the eddy viscosity accurately for reliable simulations of tidal current and pollutant diffusion.

Rastogi and Rodi (1978) firstly adopted the depth-averaged k-ε turbulence model for use in depth-averaged flow simulations. McGuirk and Rodi (1978) had a more detailed discussion. The researches in which the eddy viscosity was estimated by the depth-averaged k-ε turbulence model have so far been proliferating (e.g., Booji (1989, 1991), Chu and Babarutsi (1989), Younus and Chaudhry (1994)). They more or less modified Rastogi and Rodi model (or McGuirk and Rodi model, which is different from Rastogi and Rodi model as to modeling of the Reynolds stress and the production term due to the horizontal gradient of velocity). Especially, Booji and Chu and Babarutsi modified it to solve an essential problem on the standard k-ε model which assumes isotropic eddy viscosity.

As is generally well known, because of the vertical profile of each physical quantity such as U_i: horizontal velocity components, k: turbulent kinetic energy, ε: dissipation rate of k and v_t: eddy viscosity, depth-integrations of terms composed of two or more physical quantities in k- and ε- transport equations generate "correlations" (or "moments") of those deviations from the depth-averaged values. However, the effects of those correlations are not considered in the standard depth-averaged k-ε turbulence model which was proposed by Rastogi and Rodi (1978). In addition, the model constants used in those researches were the same as ones in 3-dimensional k-ε model.

In this study, the consideration of the effects of those correlations were attempted. Firstly, we assumed that each physical quantity in the k-ε model has a semi-empirical vertical profile of a fully-developed turbulent flow in an open-channel. Next, we attempted to obtain formulae for several correction factors that considered the effect of the correlation of each term shown up by depth-integration of 3-dimensional full k- and ε- equations with those profiles. Simultaneously two model constants of bottom shear production terms of k and ε were also modified. Those factors and constants are expressed as functions of the non-dimensional Chezy number φ, which is the ratio of a depth-averaged velocity to a bottom shear velocity. Finally, the necessity of corrections of the depth-averaged k-ε model will be discussed.

2. DERIVATION OF DEPTH-AVERAGED k-ε TURBULENCE MODEL

2. 1 3-Dimensional Standard k-ε Turbulence Model

The form of the 3-dimensional standard k-ε turbulence model is as follows (using the Einstein's summation convention) :

$$\frac{\partial k}{\partial t} + U_i \frac{\partial k}{\partial x_i} = \frac{\partial}{\partial x_i}\left(\frac{\nu_t}{\sigma_k}\frac{\partial k}{\partial x_i}\right) + P - \varepsilon , \tag{1}$$

$$\frac{\partial \varepsilon}{\partial t} + U_i \frac{\partial \varepsilon}{\partial x_i} = \frac{\partial}{\partial x_i}\left(\frac{\nu_t}{\sigma_\varepsilon}\frac{\partial \varepsilon}{\partial x_i}\right) + C_{1\varepsilon}\frac{\varepsilon}{k}P - C_{2\varepsilon}\frac{\varepsilon^2}{k} , \tag{2}$$

with $P = \nu_t \left(\frac{\partial U_i}{\partial x_j} + \frac{\partial U_j}{\partial x_i}\right)\frac{\partial U_i}{\partial x_j} ,$ (3) $\nu_t = C_\mu \frac{k^2}{\varepsilon} ,$ (4)

where i, j: dummy indices (=1, 2, 3), x_i: coordinates (x_1 and x_2 are the horizontal directions and x_3 is the vertical one), U_i: velocity component in the x_i-direction, ν_t: eddy viscosity, C_μ, $C_{1\varepsilon}$, $C_{2\varepsilon}$, σ_k, and σ_ε are empirical model constants and have the values 0.09, 1.44, 1.92, 1.0 and 1.3, respectively, for non-buoyant situations.

2. 2 Vertical Profiles of U, k, and ε

In order to estimate the effects of correlation in the depth-averaged k-ε model, it is assumed that the physical quantities have semi-theoretical profiles of fully-developed turbulent flow in an open-channel; the horizontal velocity components U_i has the logarithmic profile, the turbulent kinetic energy k and the k's dissipation rate ε have the semi-theoretical profiles proposed by Nezu and Nakagawa (1993) as the following non-dimensional forms :

$$\frac{U_i}{U_{*i}} = \frac{1}{\kappa}\ln\left(\frac{\xi}{\xi_s}\right) + A , \qquad \frac{k}{U_*^2} = D\exp\left(-2\,\xi\right), \quad \frac{\varepsilon h}{U_*^3} = \frac{E\exp\left(-3\,\xi\right)}{\sqrt{\xi}} , \qquad (5, 6, 7)$$

where $\xi = y/h$, y: vertical coordinate positive upward from the bed, h: water depth, $\xi_s = k_s/h$, k_s: equivalent roughness, κ: Karman constant (= 0.4), U_{i*}: bottom friction velocity in x_i-direction, $U_* = (U_{1*}^2 + U_{2*}^2)^{1/2}$, $A = f(U_* k_s/\nu)$: empirical function clarified from the pipe-flow experiments by Nikuradse (1933), $D = 4.78$ and $E = 9.76$.

Although Nezu and Rodi (1986) pointed out a necessity of introduction of Coles' wake function to the velocity profile of an open-channel flow, it is not done here for simplicity. It has been clarified that Eqs. (6) and (7) are valid in the whole region of open-channel flow except a wall region $(0 < \xi < 0.15)$, but we apply these profiles even to the wall region in this study. Non-dimensional depth-averaged quantities are obtained by integrating Eqs. (5)-(7) over the whole region from $\xi = 0$ to 1 as follows:

$$\frac{\overline{U}}{U_*} = \frac{1}{\kappa} \ln\left(\frac{1}{\xi_s}\right) - \frac{1}{\kappa} + A = \varphi \ , \qquad \frac{\overline{k}}{U_*^{\,2}} = a_k = 2.07 \ , \qquad \frac{\overline{\varepsilon}\,h}{U_*^{\,3}} = a_\varepsilon = 9.84 \ , \qquad (8, 9, 10)$$

where φ : dimensionless Chezy coefficient.

2. 3 Vertical Profile of v_t

As is generally well known, the eddy viscosity v_t has a parabolic profile in case of an uniform turbulent flow under the assumption that the logarithmic velocity profile holds in a whole region from the bottom to the free surface, as the following non-dimensional form :

$$\frac{v_t}{U_* h} = \kappa \, \xi (1 - \xi). \qquad (11)$$

Comparison of this profile with the experimental results by Nezu and Rodi (1986) reveals that Eq. (11) holds well in the whole region. Thus, Eq. (11) is adopted as the vertical profile of the eddy viscosity. The depth-averaged non-dimensional eddy viscosity obtained by integrating Eq. (11) from $\xi = 0$ to 1 is expressed as follows :

$$\frac{\overline{v_t}}{U_* h} = a_{v_t} = \frac{\kappa}{6} = 0.0667 \ . \qquad (12)$$

By analogy with the 3-dimensional k-ε model, Rastogi and Rodi (1978) assumed that the local depth-averaged eddy viscosity can be estimated by the same relation as Eq. (4), but they didn't correct the model constant C_μ. Firstly, we attempt to modify the estimation of $\overline{v_t}$ by introduction of correction factor β_{v_t} as follows :

$$\overline{v_t} = \beta_{v_t} \, C_\mu \frac{\overline{k}^2}{\overline{\varepsilon}} \ . \qquad (13)$$

Eqs. (9), (10) and (12) give $\beta_{v_t} = a_{v_t} a_\varepsilon / C_\mu a_k^2 = 1.71$. Although there was not a large difference between the modified estimation Eq. (13) and Rastogi and Rodi model's one, it is expected that Eq. (13) and β_{v_t} can estimate more accurate depth-averaged eddy viscosity $\overline{v_t}$ which has the parabolic profile expressed by Eq. (11).

2. 4 Modified Depth-Averaged k- and ε- Transport Equations

Depth-averaged k- and ε- transport equations are obtained by integrating 3-dimensional transport equations (1) and (2) from the bottom to the free surface. In order to consider the effect of correlations, several correction factors are introduced to the depth-averaged k-ε turbulence model as shown in the following depth-averaged transport equations of k and ε :

$$\frac{\partial \overline{k}}{\partial t} + \beta_{kHA} \, \overline{U_i} \frac{\partial \overline{k}}{\partial x_i} = \frac{\partial}{\partial x_i}\left(\beta_{kHD} \frac{\overline{v_t}}{\sigma_k} \frac{\partial \overline{k}}{\partial x_i} \right) + \beta_{kHP} \, P_H + C_k \frac{U_*^3}{h} - \overline{\varepsilon} \ , \qquad (14)$$

$$\frac{\partial \overline{\varepsilon}}{\partial t} + \beta_{\varepsilon HA} \, \overline{U_i} \frac{\partial \overline{\varepsilon}}{\partial x_i} = \frac{\partial}{\partial x_i}\left(\beta_{\varepsilon HD} \frac{\overline{v_t}}{\sigma_\varepsilon} \frac{\partial \overline{\varepsilon}}{\partial x_i} \right) + \beta_{\varepsilon HP} \, C_{1\varepsilon} \left(\frac{\overline{\varepsilon}}{\overline{k}}\right) P_H + C_\varepsilon \frac{U_*^4}{h^2} - \beta_{2\varepsilon} \, C_{2\varepsilon} \frac{\overline{\varepsilon}^2}{\overline{k}} \ , \qquad (15)$$

with $\ P_H = \overline{v_t}\left(\frac{\partial \overline{U_i}}{\partial x_j} + \frac{\partial \overline{U_j}}{\partial x_i} \right) \frac{\partial \overline{U_i}}{\partial x_j} \ , \qquad (16)$

383

where β_{kHA}, $\beta_{\varepsilon HA}$, β_{kHD}, $\beta_{\varepsilon HD}$, β_{kHP}, $\beta_{\varepsilon HP}$, $\beta_{2\varepsilon}$: correction factors, C_k, C_ε: empirical constants related to the productions of k and ε due to the bottom shear stress and $i, j = 1, 2$.

2. 5 Estimations of Correction Factors

The vertical profiles of physical quantities in the k-ε model, such as U_i, k, ε and v_t, can be written in the form of $\Phi = F_\phi \overline{\Phi}$ by combining assumed profiles [Eqs. (5), (6), (7), and (11)] with the depth-averaged values [Eqs. (8), (9), (10), and (12)]. All correction factors are given by integrating the 3-D transport equations (1) and (2) with those profiles as follows :

$$\beta_{kHA} = \frac{D}{a_k\,\phi}\left(-1.65 + 0.432\,A'\right), \qquad (17) \quad \beta_{\varepsilon HA} = \frac{E}{a_\varepsilon\,\phi}\left(-7.84 + 1.01\,A'\right), \qquad (18)$$

$$\beta_{kHD} = \frac{0.0677\,\kappa D}{a_{v_t}\,a_k} = 0.939, \qquad (19) \quad \beta_{\varepsilon HD} = \frac{0.0924\,\kappa E}{a_{v_t}\,a_\varepsilon} = 0.549, \qquad (20)$$

$$\beta_{kHP} = \frac{1}{\phi^2\,a_{v_t}}\left(0.44 - 0.278A' + 0.167\kappa A'^2\right), \qquad (21)$$

$$\beta_{\varepsilon HP} = \frac{a_k\,E}{a_\varepsilon\,a_{v_t}\,D\,\phi^2}\left(0.985 - 0.44\,A' + 0.178\,\kappa A'^2\right), \qquad (22)$$

$$\beta_{2\varepsilon} = \frac{a_k\,E^2}{a_\varepsilon^2 D}\left[\int_{\xi_s}^{1}\frac{exp\left(-4\,\xi\right)}{\xi}d\xi + exp\left(-6\,\xi_s\right)\right], \qquad (23)$$

where $A' = -(1/\kappa)\ln\xi_s + A = \phi + 1/\kappa$. It was assumed that $k = DU_*^2 = \text{const.}$ and that $\varepsilon = \varepsilon(\xi_s) = \text{const.}$ in the region from $\xi = 0$ to ξ_s for the calculation of $\beta_{2\varepsilon}$, otherwise the integral would diverge. As shown above, those correction factors except β_{kHD} and $\beta_{\varepsilon HD}$ are expressed as functions of ϕ. ξ_s is related to ϕ by Eq. (8).

2. 6 Estimations of C_k and C_ε

The production term $P_{kv} = C_k\,U_*^3/h$ due to the bottom shear stress in Eq. (14) is calculated by integration of only the part related to vertical shear in a production term in Eq. (1). By calculating this integration with Eqs. (5) and (11) under the assumptions of constant stress ; $v_t\,(\,dU_i/dx_3) = U_*^2 = \text{const.}$ and the linear velocity profile ; $U_i/U_{*i} = A_r\,\xi/\xi_s$ over the region $\xi = 0 - \xi_s$, the model constant C_k is obtained :

$$C_k = \phi + \xi_s/\kappa \approx \phi \ . \qquad (24)$$

In the same manner, the model constant C_ε is obtained under the same assumptions as those for calculations of $\beta_{2\varepsilon}$ and C_k as follows :

$$C_\varepsilon = \frac{C_{1\varepsilon}E}{D}\left[\frac{1}{\kappa}\int_{\xi_s}^{1}\frac{exp\left(-\xi\right)}{\xi\sqrt{\xi}}\left(1 - \xi\right)d\xi + A\,\frac{exp\left(-3\,\xi_s\right)}{\sqrt{\xi_s}}\right]. \qquad (25)$$

3. INVESTIGATIONS OF THE EFFECTS OF MODIFICATION

3.1 Effects of the correction factors

Usually, φ has a value within the range of 8 to 25. Thus, the values of correction factors were calculated as shown in **Fig. 1** in case of rough bed, that is, $A = A_r = 8.5$.

The effects of correlations of advection terms in $\bar{k}$ - and $\bar{\varepsilon}$ - transport equations, namely the dispersive effects, are evaluated from the values of β_{kHA} and β_{eHA}, respectively. As shown in **Fig. 1a)**, β_{kHA} is nearly equal to unity. Therefore, the dispersive effect is not so significant for the transport of $\bar{k}$ in the depth-averaged model. On the contrary, as β_{eHA} is small apart from 1.0, the effect on $\bar{\varepsilon}$ is very significant but is not so dispersive. It is also clear from **Figs. 1b), 1c)**, Eqs. (19) and (20) that the corrections for diffusion, production and dissipation terms are significant for $\bar{\varepsilon}$ but not for $\bar{k}$.

3.2 Model Constants C_k and C_ε

The productions of k and ε due to the bottom shear stress are related to the local bottom friction velocity U_* and the local water depth h by the empirical constants C_k and C_ε described as Eqs. (24) and (25) in the modified depth-averaged k-ε model. The widely used, standard depth-averaged k-ε model [Rastogi and Rodi (1978)] has the following constants :

$$C_k = \varphi \ , \qquad\qquad\qquad C_\varepsilon = 3.6 \ \varphi^{3/2} C_{2\varepsilon} \ C_\mu^{1/2} \ . \qquad\qquad (26, 27)$$

Rastogi and Rodi estimated C_k from the bulk energy balance in uniform flows. They took as empirical input the depth-averaged eddy viscosity determined from the experimental results by Laufer (1951) and estimated C_ε.

Eq. (26) is almost the same form as Eq. (24), while Eq. (27) is quite different from Eq. (25). Coincidence of C_k between Eq. (26) and Eq. (24) indicates that the assumptions for convenient calculations are justifiable. On the other hand, the difference of C_ε between Eq. (27) and Eq. (25) is derived from the difference in experimental results used for the estimations of C_ε. Comparisons between the present modified model and Rastogi and Rodi's one show that we can use the uncorrected $\bar{k}$ -equation, while it is necessary to modify the $\bar{\varepsilon}$ - equation for accurate estimations of depth-averaged $\bar{k}$ and $\bar{\varepsilon}$.

4. CONCLUSIONS

The modified depth-averaged k-ε turbulence model was established under the several empirical and simplified assumptions. As a result of this research, it is expected that the depth-averaged $\bar{k}$, $\bar{\varepsilon}$ and $\overline{v_t}$ will be obtained more accurately by introduction of the correction factors and two modified model constants.
In this paper, we didn't carry out a numerical feasibility study of the present new model. In a further research, we will try it and tune up the correlation factors and model constants by modifying assumptions for the derivation of them. However, several further problems remain for the feasibility study as follows :

(1). highly accurate measurements of spatial distributions of $\bar{k}$ and $\bar{\varepsilon}$ in nearly horizontal flow fields,

(2). evaluation of effects of spatial grid size in numerical simulations.

5. ACKNOWLEDGMENTS

The authors express sincere appreciation to Emeritus Professor of Kyushu University, T. Tsubaki, for his helpful comments on the present research.

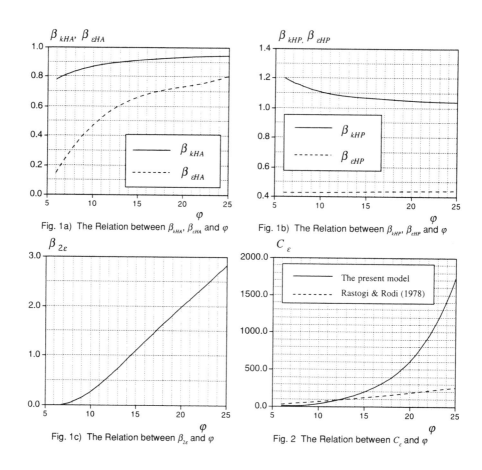

Fig. 1a) The Relation between β_{kHA}, $\beta_{\varepsilon HA}$ and φ

Fig. 1b) The Relation between β_{kHP}, $\beta_{\varepsilon HP}$ and φ

Fig. 1c) The Relation between $\beta_{2\varepsilon}$ and φ

Fig. 2 The Relation between C_{ε} and φ

6. REFERENCES

Abe, T., Hukuda, H. and Yoon, J. H. (1995) . "Tidal Current Simulation of Ariake Bay", *Bull. Res. Inst. Appl. Mech.*, *Kyushu Univ.*, 78, 63-82.

Booij, R. (1989) : "Depth-averaged K-ε-modelling", *Proc. XXIIIrd IAHR Cong.*, *Ottawa*, A, 199-206.

Booij, R. (1991) : "Eddy Viscosity in a Harbour", *Proc. XXIVth IAHR Cong.* , *Madrid*, C,83-90.

Chu, V. H. and Babarutsi, S. (1989) : "Modelling the Turbulent Mixing Layers in a Shallow Open-Channel", *Proc. XXIIIrd IAHR Cong.*, *Ottawa*, A, 191-198.

Laufer, J. (1951) : "Investigation of Turbulent Flow in a Two-Dimensional Channel", *NACA Report*, 1053, 1247-1266.

McGuirk, J. J. and Rodi, W. (1978): "A Depth-averaged Mathematical Model for the Near Field of side Discharges into Open-channel Flow", *J. Fluid Mech.*, 86 (4), 761-781.

Nezu, I. and Nakagawa, H. (1993) : *Turbulence in Open-Channel Flows*, IAHR Monograph, BALKEMA.

Nezu, I. and Rodi, W. (1986): "Open-Channel Flow Measurements with a Laser Doppler Anemometer", *J. Hydr. Engrg.*, ASCE, 112, 335-355.

Nikuradse, J. (1933): "Stromungsgesetze in rauhen Rohren", Forschg. Arb. Ing.-Wes., 361.

Oonishi, Y. (1977) : "A Numerical Study on the Tidal Residual Flow", *J. Oceanogr. Soc. Japan*, 33, 207-218.

Rastogi, A. K. and Rodi, W. (1978) : "Predictions of Heat and Mass Transfer in Open Channels", *J. Hydr. Div.*, ASCE, 104 (HY3), 397-420.

Younus, M. and Chaudhry, M. H. (1994) : "A depth-averaged k-ε turbulence model for the computation of free-surface flow", *J. Hydr. Res.*, 32 (3), 415-444.

386

Environmental Hydraulics, Lee, Jayawardena & Wang (eds) © 1999 Balkema, Rotterdam, ISBN 90 5809 035 3

Vertical mixing processes due to waves and currents

J.M. Pearson & I. Guymer
Department of Civil and Structural Engineering, The University of Sheffield, UK

ABSTRACT: Investigations reported relate to a series of hydrodynamic and fluorometric tracer experiments undertaken on a large scale wave-current facility. Detailed measurements were made under controlled laboratory conditions using non-intrusive measurement techniques, to elucidate the magnitude of vertical mixing under a wave-current environment. Results are presented for a current only and a wave-current condition. The spatial vertical concentration distributions, measured for the current only condition indicate that the vertical mixing can be adequately described by conventional mixing theories. The addition of waves onto the current only condition results in additional mixing mechanisms which increase the vertical mixing.

1. INTRODUCTION

Industrial and sewage treatment works often discharge their effluent into estuarine or coastal waters. With increased awareness by the general public and control by regulatory bodies, it has become vital to assess the importance of pollutants in these regions. Frequently, the tools employed for the management of the coastal environment are numerical simulations. However, although considerable advances have been made in the understanding of the hydrodynamic processes occurring within this zone, few experimental data are available on the spatial distribution of solutes.

In numerical models of the coastal region, the dispersal of plumes are often simulated using predicted velocity distributions and an estimated mixing coefficient. Even fewer data are available, particularly with respect to direct tracer measurements from which to estimate the mixing coefficient. A further complication is that the vertical structure of the water column is associated with different levels of turbulence and hence mixing. This results in an inability to accurately predict the mixing occurring within the coastal zone. This paper describes a series of fluorescent tracer experiments for two wave-current conditions undertaken on the 'Scheldegoot' wave-current facility, at Delft Hydraulics, to elucidate the vertical mixing processes.

2. PREVIOUS WORK

For solute transport studies, the analysis of Taylor (1954), remains the most frequently employed description to quantify turbulent diffusion. The analogy of turbulent mixing processes to Fickian diffusion has been made by many researchers. Reviews of this field are provided by Fischer et al. (1979) and more recently Rutherford (1994).

The theoretical solution for the vertical spreading of a solute injected at the bed, from a continuous transverse line source in uniform plane open channel flow, can be given by (Fischer et al., 1979);

$$c(x,z) = \frac{2Q}{\bar{u}\sqrt{4\pi e_z(x/\bar{u})}} \exp\left(-\frac{z^2 \bar{u}}{4 e_z x}\right) \tag{1}$$

where c = solute concentration; x & z = longitudinal and vertical directions (x-axis is orientated along the principle axis of the flow); $\bar{u}$ = longitudinal depth averaged velocity; e_z = diffusion coefficient in the z direction; Q = mass inflow rate of solute per unit width.

The concentration profile predicted by Eq. (1) is a Gaussian distribution mirrored at the bed. This has the properties such that the measure of spread about the bed can be determined by calculating the spatial variance, σ_z^2 of the distribution, given by;

$$\sigma_z^2 = \frac{\int_0^\infty z^2 c(z)\,dz}{\int_0^\infty c(z)\,dz} \tag{2}$$

A feature of a Gaussian distribution is that the spatial variance, σ_z^2 increases linearly with distance from the source. This allows the mixing coefficient, e_z to be defined as the rate of change of variance according to

$$e_z = \frac{\overline{u}}{2}\frac{d\sigma_z^2}{dx} \tag{3}$$

In open channel flow, in the vicinity of boundaries, the longitudinal velocity profile varies in the vertical direction. Conventionally, in infinitely wide channels, the vertical velocity distribution in plane turbulent shear flows can be given by the logarithmic law of the wall, expressed by (Kironoto & Graf, 1994);

$$u = \frac{u_*}{\kappa}\ln\left|\frac{z}{z_o}\right| \tag{4}$$

where κ = von Karmans constant, which is usually quoted 0.4 (Kironoto & Graf, 1994); u_* = bed shear velocity; z_o = bottom roughness parameter, which over a rough bed can be given by $z_o = d_{50}/30$, where d_{50} is the median grain diameter on the bed; and z = vertical co-ordinate above the bed, positive upwards.

Elder (1959) utilised the logarithmic law of the wall [Eq. (4)] to investigate the effects of vertical variations in the longitudinal velocity on the vertical mixing processes. It was postulated that the vertical diffusivity could be given by (Rutherford, 1994);

$$e_z = \kappa u_* z\left(1 - \frac{z}{d}\right) \tag{5}$$

Rutherford (1994) referred to the experimental laboratory investigations of Jobson and Sayre (1970), who concluded that for the majority of practical mixing studies it was sufficiently accurate to use the depth averaged value, as it was found that predicted concentration profiles were not particularly sensitive to the variation of vertical diffusivity with depth. Thus, by assuming that $\kappa = 0.4$ and integrating Eq. (5) with depth, yields

$$\overline{e_z} = \frac{\kappa}{6}du_* = 0.067du_* \tag{6}$$

The oscillatory nature of the particle velocities under waves may influence the mixing of solutes by inducing a near bed shear. There have been comparatively few experimental studies to investigate aspects of mixing under waves and currents. Some site specific field studies using tracers, have been undertaken, however the lack of detailed hydrodynamic data has made it difficult to interpret the mixing processes. Bowen & Inman (1974) summarised the important aspects of mixing within the nearshore zone. They suggest that outside the surf zone (where the waves are not breaking), the contribution by waves to the turbulence generated is small compared to the turbulence generated by the currents.

De Vriend & Stive (1987) developed a theoretical solution to near-shore hydrodynamic simulations, and suggested that the eddy viscosity caused by the oscillatory wave motion over the bed, could be given by;

$$\nu_t = 0.208\kappa du_* \tag{7}$$

where u_* = bed friction velocity in the direction of wave propagation.

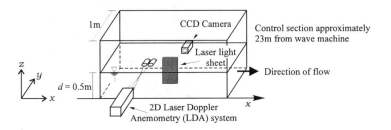

FIG. 1. The control section of the 'Scheldegoot' wave-current facility at Delft Hydraulics, The Netherlands.

This brief review suggests that although some advancement has been made in the understanding of the near-shore environment, little experimental data are available. This paper therefore describes a series of experimental tests which to utilise fluorometric and hydrodynamic measurements with the aim of parameterising the diffusive and dispersive vertical solute transport processes.

3. LABORATORY STUDY

The control section of the 'Scheldegoot' wave-current facility is shown in Fig. (1), a detailed description of the facility set-up is given by Klopman (1994). The walls of the facility were glass and the facility had an operating depth of 0.5m, length of 46m and was 1m wide. The flow within the facility was generated by an electronic pump, controlled by a butterfly value, in a re-circulating flow system of constant discharge 80 l/s. The facility consisted of a plane concrete bed which was roughened with coarse sand (d_{50} = 2.0 mm) embedded into paint. An absorption wave system was utilised to reduce the reflections from the incident waves.

The two test conditions presented within this present study were investigated; a current only condition (no waves), and a wave-current condition. The wave-current case had a measured wave height (H) of 120mm, and constant period (T) of 1.44s. The generated wave was monochromatic, where the crest travelled in the same direction as the current.

The wave-current facility was equipped with a LDA velocity system. The LDA system was a non-intrusive measurement device, which enabled detailed turbulent velocity measurements to be undertaken in close proximity to the facility bed. The LDA had a measuring volume of less than 1mm and enabled the simultaneous measurement of the longitudinal and vertical velocity components. The output from the laser was logged onto a PC equipped with a data acquisition system. In order to determine the hydrodynamic conditions, one detailed vertical velocity profile for each test condition was obtained using the LDA system from the central region of the facility (approximately 23m from the upstream wave machine and 0.5m from the glass wall). Approximately 25 points were measured in the vertical for each test condition, logging at each point for 10 minutes at a data rate of 200Hz.

The advective processes under waves are relatively small, thus when performing laboratory tracer studies any discrete sampling technique will have a significant effect on the mixing characteristics. Thus, a non-intrusive measurement technique was used. Laser induced fluorescence (LIF) is the technique by which a fluorescent substance is made to emit light of a higher wavelength to that of the laser beam which strikes it. A laser is used to provide a coherent beam of light with little divergence containing a known number of discrete wavelengths. In this application, the laser beam was converted into a light-sheet. Using an arrangement of mirrors, the light-sheet was directed through the bottom of the facility to produce a vertical sheet approximately 100mm wide and 1mm deep, running parallel to the main flow.

The emitted light from the light-sheet was detected by a CCD camera and manipulated via an image processing system, enabling the full 2D concentration field to be recorded. The resulting grey scale value from the image provided the relative concentration of the dye. The excitation wavelengths of light from the laser were separated from the emitted wavelengths from the dye using a bandpass filter fitted to the lens of the camera. To ensure that the dye was not excited by natural or artificial light, the whole facility had to be

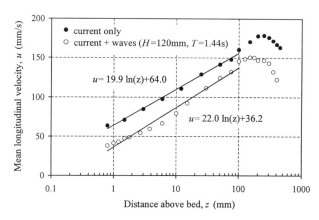

FIG. 2. Mean vertical longitudinal velocity profiles for two wave conditions

blackened during the testing period. This was achieved using black polythene sheeting. A more detailed description of the LIF system developed by the University of Sheffield is given by Guymer & Harry (1996).

A constant injection of Rhodamine 6G fluorescent dye was introduced into the facility from a small tapping point flush with the bed and the subsequent vertical concentration distribution was measured. For each injection location, three minutes of image data received from the CCD camera were recorded onto sVHS video tape for later analysis. For each vertical injection location, approximately 10 detailed fluorometric measurements were performed at different longitudinal locations from the injection point. The problem with a re-circulating flow system is the continual build-up of background dye concentrations, thus during testing, a background concentration image was monitored and recorded every 15 minuets. An in-situ calibration was performed during the testing period by adding approximately 20 known concentrations of dye into the facility and recording the resultant images.

4. RESULTS

Figure (2) illustrates the mean longitudinal vertical velocity profiles for the two test conditions. For the current only condition, the depth averaged velocity, $\bar{u} = 165$ mm/s, while for the combined wave and current condition, $\bar{u} = 133$ mm/s, a reduction presumably as a result of the wave-induced undertow. By adopting the theoretical logarithmic law of the wall [Eq. (4)], the bed shear velocity (u_*) has been calculated by solving a least squares linear regression on the velocities in the lower 20% region of the flow depth. For the current only condition, the bed shear velocity, $u_* = 8.0$ mm/s and bottom roughness parameter, $z_0 = 0.04$ mm, by recalling that $z_0 = d_{50} / 30$ a, gives $d_{50} = 1.2$ mm, a value which compares well to the bed material used taking into account that it was embedded into paint. By adopting a similar analytical procedure, for the combined wave and current condition, $u_* = 8.8$ mm/s, and $z_0 = 0.19$ mm. Thus, the mean bed shear velocity (u_*) remains approximately constant for the two conditions, while the roughness parameter (z_0), increases by an approximate factor of 5 when the waves are superimposed on the current.

For the fluorometric tests, post processing of consecutive images from the videotape was performed using a high speed data transfer image processing system. This allowed a region of interest measuring 700 by 30 pixels wide (550x25mm) to be digitally recorded onto a PC at a rate of 25Hz. A two minute data series was recorded for each consecutive set of images which was then temporally averaged. Each grey scale pixel value within the region of interest was converted into a concentration value by the use of the in-situ calibration, which showed that within set limits, the relationship between the grey scale value detected by the camera and dye concentration was approximately linear. The average vertical concentration profiles (with background dye concentration subtracted) measured at different locations from the injection point, for the two test conditions are illustrated in Fig. (3).

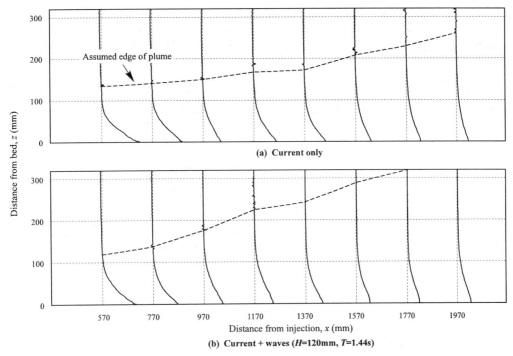

(a) **Current only**

(b) **Current + waves (H=120mm, T=1.44s)**

FIG. 3. Vertical concentration profiles for (a) current only; (b) current + waves

The spatial variance (σ_z^2) of each concentration distribution [Eq. (2)], with longitudinal distance from the source is illustrated in Fig. (4). The results demonstrate that the increase in vertical variance, although showing some scatter is approximately linear.

5. CONCLUSIONS

The hydrodynamic measurements for the current only condition, indicates that the vertical velocity gradient of the longitudinal flow can be adequately described by Eq. (4). Thus by application of Elder's (1959) analysis [Eq. (6)], the depth averaged vertical mixing coefficient ($\overline{e_z}$) has a calculated value of 270 mm²/s. For the fluorometric data, by the application of Eq. (3), $\overline{e_z} = (165/2) \times 3.0 = 250 \mathrm{mm}^2/\mathrm{s}$. Thus it can be

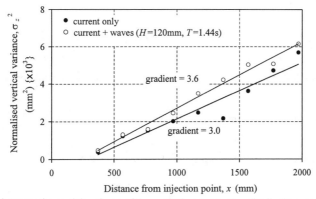

FIG. 4. Relationship between the spatial variance of the vertical concentration distributions and longitudinal distance

concluded that, within experimental error, for the current only condition, the vertical mixing can be adequately described by existing semi-empirical knowledge. Additionally, laser induced fluorescence is a viable measurement technique to obtain high spatial and temporal resolution of tracer concentrations with no disturbance to the flow regime.

For the case of waves superimposed on the current, the mixing processes become more complex, which makes it difficult to isolate the effects of any one mixing influence. The discharge of the re-circulating flow system remained constant during both test conditions, thus to enable a realistic comparison of the mixing generated between the two test conditions, the depth averaged velocity measured for the current only condition has been adopted. Thus, for the fluorometric dye concentration measurements, a mixing coefficient of $\overline{e_z} = (165/2) \times 3.6 = 295$ mm^2/s is obtained, an increase of approximately 20%, when compared to the current only condition. The mixing coefficient obtained by the measurement technique adopted may include the effects of two mechanisms. This first mechanism, adopted De Vriend & Stive (1987) [Eq. (7)] is the increased mixing generated by the oscillatory wave motion over the bed, generating a velocity shear. The second mechanism it that adopted by Koole & Swan (1994), who investigated the dispersive properties of a jet discharged into oscillatory wave motion. They found that the temporal averaging of the jet's sinusoidal displacement under the motion of the waves produced an apparent additional mixing mechanism. For the solute transport studies adopted in this present study, the Eulerian vertical concentration measurement at given sections from the injection point, will incorporate this increase. Analysis of the turbulent Reynolds stresses are currently underway which may elucidate some of the suggested mixing mechanisms so far identified.

ACKNOWLEDGEMENTS

The authors gratefully acknowledge the European Communities TMR Programme for their financial support for the research undertaken at Delft Hydraulics.

REFERENCES

Bowen A. J. & Inman D. L., 1974. Near-shore mixing due to waves and wave-induced currents. *Rapp. P.-v. Reun. Cons. int. Explor. Mer*, 167: 6.

De Vriend H. J. & Stive M. J. F., 1987. Quasi-3D modelling of near-shore current. *Coastal Engineering*, 11, 565.

Elder J. W., 1959. The dispersion of marked fluid in turbulent shear flow. *J. Fluid Mech.* 5, 546.

Fischer H. B., List J. E., Koh R. C. Y., Imberger J., Brooks N. H., 1979. Mixing in inland and coastal waters. New York: *Academic Press*.

Guymer I & Harry A., 1996. Use of laser induced fluorescence and video imaging techniques in an investigation of mixing across the dead zone/flow boundary. *Proc. IMechE Int. Seminar on Optical Methods and Data Processing in Heat and Fluid Flow. London*. 419.

Kironoto B. A. & Graf W. H., 1994. Turbulence characteristics in rough uniform open-channel flow. *Proc. Inst Civ. Engrs Wat., Marit. & Energy, ICE*, 106, 333.

Klopman G., 1994. Vertical structure of the flow due to waves and currents. Delft Hydraulics, *Progress Report H 840.30, Part II*.

Koole R. & Swan C., 1994. Dispersion of pollution in a wave environment. *Proc. 24th Int. Conf. on Coastal Engineering*, 3071.

Rutherford J. C., 1994. River Mixing. *J. Wiley & Sons, Chichester, England*.

Taylor G. I., 1954. The dispersion of matter in turbulent flow through a pipe. *Proc. R. Soc. London Ser. A*, 233, 446.

Environmental Hydraulics, Lee, Jayawardena & Wang (eds) © 1999 Balkema, Rotterdam, ISBN 90 5809 035 3

A theoretical study on complete mixing distance in open-channel uniform flow

Ke-Zhong Huang
Department of Geography, Zhongshan (Sun Yatsen) University, Guangzhou, China

Tao Jiang
Department of Geography, The Chinese University of Hong Kong, China

ABSTRACT: A theoretical formula of the complete mixing distance in open-channel uniform flow for an instantaneous point source is derived by the principle of maximum entropy in this paper. The formula can be well used to the cases of continuous point source and intermittent point source. An indirect experiment verification method about the formula and some comparisons with the previous studies are also given.

1. INTRODUCTION

The longitudinal dispersion distance from the source to the initial complete mixing section in open-channel may be simply referred to as the complete mixing distance (CMD). The establishment of more reliable criteria of CMD is useful for measuring river discharge by the dilution method because the measurement of flow by the method may only be made if sampling is carried out at distance greater than the complete mixing distance. Otherwise, in order to determine the division between the stage of lateral mixing and the stage of longitudinal dispersion in the environmental engineering design of river, we also need the CMD.

Some results about the CMD derived by the traditional advective diffusion equation have been given in the literature, e.g. Ward (1973), and Fischer et al.(1979). However, there was doubt about them since the values of CMD calculated by their formulas are too large as well as lack of direct experiment verification. In order to obtain a more reliable criteria of the CMD, a theoretical formula is presented through the principle of maximum entropy (POME) in this paper.

2. THEORY

Consider the advective diffusion process of the conservative substance in the open-channel uniform flow. We are still to use the traditional assumptions as follows: Assume that the instantaneous velocity of an effluent particle is the same as the instantaneous velocity of a liquid particle located at the same point; the longitudinal time-mean velocities are equal to the cross-sectional average velocity U of the uniform flow; the vertical instantaneous velocities, the lateral time-mean velocities, and the longitudinal fluctuating velocities may be neglected; the concentrations along the depth h are the same. In a word, it is a two- dimensional problem of dispersion. If the source is an instantaneous point source, we may average the concentrations along the x-axis over a distance L, in which L is greater than the length of the dye cloud (see Fig.1),and c denotes the average concentration along L. Thus, the problem is now reduced to an one-dimensional (i.e. lateral) problem and the concentrations c is the function of lateral coordinate y and time t.

The instantaneous point source is located at the point which is $m_1 B$ and $m_2 B$ apart from the right and left channel-side, respectively, where B denotes the width of channel, $m_1 + m_2 = 1$, and m_1, $m_2 \geq 0$. Let the origin 0 of the coordinate system (x, y) overlapping with the point source 0', the x-direction and the y-direction shown in Fig.1.

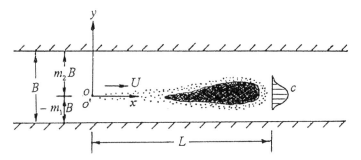

Fig.1   Two-dimensional dye cloud in open-channel uniform flow

On the basis of uncertainty view and the preceding assumptions, the position of effluent particle in certain time is a random event which results only from the lateral fluctuation velocity.   The lateral position of an effluent particle thus is random variable Y described by the conditional probability density function $f(y|t)$, in which the time is also assumed temporarily as a random variable..   Information entropy is the measure of uncertainty of random event. The problem of CMD can be dealt with the POME. According to the definition of information entropy, the joint information entropy $H(y,t)$ of the particle position can be expressed as

$$H(y,t) = -\int_{-m_1B}^{m_2B}\int_0^\infty f(y,t)\ln f(y,t)dtdy \tag{1}$$

where $f(y,t)$ is the joint probability density function.   It selects the function $f(y,t)$ which maximizes the joint information entropy $H(y,t)$ subject to the normalized condition

$$\int_{-m_1B}^{m_2B}\int_0^\infty f(y,t)dtdy = 1 \tag{2}$$

and the additional mean value condition

$$\int_{-m_1B}^{m_2B}\int_0^\infty \frac{y^2}{t} f(y,t)dtdy = E(\frac{y^2}{t}) \tag{3}$$

which is chosen according to the characteristics of the problem.

By using the method of the Lagrange multipliers, we can obtain the joint probability density function $f(y,t)$, the marginal probability density function $f(t)$, and the conditional probability density function $f(y|t)$ as follows:

$$f(y,t) = \exp(-a_0 - a\frac{y^2}{t}) \tag{4}$$

$$f(t) = \int_{-m_1B}^{m_2B} f(y,t)dy = \sqrt{\frac{\pi t}{a}} \frac{erf(m_1 B\sqrt{a/t}) + erf(m_2 B\sqrt{a/t})}{2\exp(a_0)} \tag{5}$$

and

$$f(y|t) = \frac{f(y,t)}{f(t)} = 2\sqrt{\frac{a}{\pi t}} \frac{\exp(-a y^2/t)}{erf(m_1 B\sqrt{a/t}) + erf(m_2 B\sqrt{a/t})} \tag{6}$$

where a_0 and a are multipliers, and the dimension of a is $[T/L^2]$. Note that the meaning of $f(y|t)dy$ is the probability of an effluent particle being between y and $y+dy$ on a given time, thus, the concentration is

$$c(y,t) = \frac{M}{hdy} f(y|t)dy = 2Bc_m \sqrt{\frac{a}{\pi t}} \frac{\exp(-ay^2/t)}{erf(m_1 B\sqrt{a/t}) + erf(m_2 B\sqrt{a/t})} \qquad (7)$$

where M is the mass of instantaneous point source; $c_m = M/(Bh)$ is the concentration on the complete mixing section.

The variance of the conditional distribution of position of an effluent particle is

$$\sigma_y^2 = \int_{-m_1 B}^{m_2 B} y^2 f(y|t)dy = \frac{t}{2a} - B\sqrt{\frac{t}{\pi a}} \frac{m_1 \exp(-am_1^2 B^2/t) + m_2 \exp(-am_2^2 B^2/t)}{erf(m_1 B\sqrt{a/t}) + erf(m_2 B\sqrt{a/t})} \qquad (8)$$

Otherwise, the variance that relates to the lateral mixing coefficient E_y can be expressed as (Taylor,1921)

$$\sigma_y^2 = 2E_y t \qquad (9)$$

It follows that

$$2E_y t = \frac{t}{2a} - B\sqrt{\frac{t}{\pi a}} \frac{m_1 \exp(-am_1^2 B^2/t) + m_2 \exp(-am_2^2 B^2/t)}{erf(m_1 B\sqrt{a/t}) + erf(m_2 B\sqrt{a/t})} \qquad (10)$$

Let dimensionless variables

$$\xi = \frac{2E_y t}{B^2} \qquad (11)$$

and

$$\eta = \frac{aB^2}{t} \qquad (12)$$

Eq.(10) becomes

$$\xi = \frac{1}{2\eta} - \sqrt{\frac{1}{\pi\eta}} \frac{m_1 \exp(-m_1^2 \eta) + m_2 \exp(-m_2^2 \eta)}{erf(\sqrt{m_1^2 \eta}) + erf(\sqrt{m_2^2 \eta})} \qquad (13)$$

From Eq.(13), we know that η is the decreasing with ξ increasing. The limit of η decreasing is zero because η is positive, and at the same time ξ is approaching a maximum ξ_{max}. Thus, taking the limit of η approaching to zero for Eq.(13), We have

$$\xi_{max} = \lim_{\eta \to +0} \left[\frac{1}{2\eta} - \sqrt{\frac{1}{\pi\eta}} \frac{m_1 \exp(-m_1^2 \eta) + m_2 \exp(-m_2^2 \eta)}{erf(\sqrt{m_1^2 \eta}) + erf(\sqrt{m_2^2 \eta})} \right]$$

$$= \frac{1}{3}(m_1^3 + m_2^3) = \frac{1}{3} - m_1 + m_1^2 \qquad (14)$$

Taking note of Eq.(11), Eq.(14) becomes

$$t_{max} = \frac{1}{2}(\frac{1}{3} - m_1 + m_1^2)\frac{B^2}{E_y} = K\frac{B^2}{E_y}$$

(15)

where the coefficient of CMD

$$K = \frac{1}{2}(\frac{1}{3} - m_1 + m_1^2)$$

(16)

Substituting $t_{max} = x_{max} / U$ into Eq.(15), yields

$$x_{max} = K\frac{B^2 U}{E_y}$$

(17)

Eq.(7) can be written as

$$c(y,t) = 2c_m\sqrt{\frac{\eta}{\pi}}\frac{\exp\left[-(y/B)^2\eta\right]}{erf(\sqrt{m_1^2\eta}) + erf(\sqrt{m_2^2\eta})}$$

(18)

Taking the limit of η approaching to zero for Eq.(18), we have

$$c(y,t)_{x=x_{max}} = \lim_{\eta\to+0}\left\{2c_m\sqrt{\frac{\eta}{\pi}}\frac{\exp\left[-(y/B)^2\eta\right]}{erf(\sqrt{m_1^2\eta}) + erf(\sqrt{m_2^2\eta})}\right\} = c_m$$

(19)

It proves that the mixing on the x_{max} section is complete mixing. Therefore, x_{max} is the CMD and Eq.(17) is its evaluated formula.

A continuous point source in the flow is equivalent to a succession of equal instantaneous point source. Eq.(17) thus can be applied to the continuous point source and intermittent point source. The formula of CMD about the continuous point source given by Huang and Jiang (1996) is the same with Eq.(17).

3. VERIFICATION

The previous experiments about the concentration of plume in open-channel to our knowledge were lack of the information near the x_{max} section, hence we can't verify Eq.(17) by the previous experiments. Still now it is difficult to decide the CMD with accuracy by experiment since the longitudinal change of lateral concentration distribution near the x_{max} section is very little. Sayre et al.(1968) gave three straight lines of the relationship between lateral displacement variance σ_y^2 and longitudinal dispersion distance x in their lateral dispersion experiments (see Fig.2). The hydraulic conditions of the experiments are shown in Tab.1. Because the effluent particles becomes distributed uniformly across the cannel on the downstream of x_{max}, the probability distribution of effluent particles in y-direction on these sections is uniform, and their lateral variance is

$$\sigma_y^2 = \frac{B^2}{12}$$

(20)

In terms of the value of B in Sayre's experiments (see Tab.1), we have $\sigma_y^2 = 5.11\ ft^2$. In other words, the values of σ_y^2 in the experiments should have not greater than $5.11\ ft^2$, by the way, we thus throw over an experiment point that has $\sigma_y^2 = 6.37\ ft^2$ in the run LA-D-3 of the experiments.

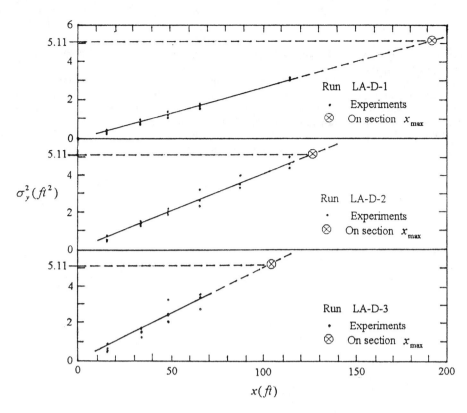

Fig.2　Relationship between lateral displacement variance and longitudinal dispersion distance

Tab.1 Hydraulic conditions of dispersion experiments and CMD

Run	B (ft)	h (ft)	U (ft/s)	E_y (ft²/s)	Source	m_1	x_{max} (ft) Eq.(17)	x_{max} (ft) Ward	x_{max} (ft) Fischer
LA-D-1	7.83	0.486	0.77	0.0103	Cont. point	0.5	191	389	458
LA-D-2	7.83	0.814	1.12	0.0230	Cont. point	0.5	126	257	302
LA-D-3	7.83	1.217	1.56	0.0385	Cont. point	0.5	104	213	250

The values of CMD calculated by Eq.(17) , Fischer's formula, and Ward's method are also shown in Tab.1.　By using the values of CMD calculated by Eq.(17) and the value of σ_y^2 , we obtain three points which are on the extrapolated lines of the experiment straight lines, respectively. Therefore, Eq.(16) and (17) can be regarded as right.

4. CONCLUSIONS

Eq.(17) derived by the POME is the theoretical formula of CMD in open-channel uniform flow.　It is the same in form as Ward's and Fischer's formulas. However, the coefficient K calculated by Eq.(16) is different from their formulas and methods, and the values of K calculated by former are generally much shorter than the values of K estimated by latter.　Eq.(16) is better than the previous formulas and methods by its strict theoretical background and generality of fitting arbitrary position of point source.

5. REFERENCES

Fischer, H.B., List, E.J., Koh, R.C.Y., Imberger, J., and Brooks, N.H. (1979). Mixing in inland and coastal waters, Academic Press, New York.

Huang, K. Z. and Jiang T. (1996). Maximum communication entropy theory of transverse spreading of the plume in open-channel uniform flow,   J. of   Hydraulic Engineering, No.5, pp.61-68. (in Chinese).

Sayre, W.W. and Chang, F.M. (1968).   A laboratory investigation   of   open-channel   dispersion processes for dissolved, suspended, and floating dispersant, Geological survey professional paper 443-E.

Taylor, G.I. (1921). Diffusion by continuous movements, London Math. Soc. Proc. Ser. A.20. 196-211.

Ward, P.R.B. (1973). Prediction of mixings for river flow gaging, J. Hyd. Div., ASCE, 99(7), 1069-1081.

Environmental Hydraulics, Lee, Jayawardena & Wang (eds) © 1999 Balkema, Rotterdam, ISBN 90 5809 035 3

Measuring velocities in a suspension

T. Dreier, J. Bühler, W. Kinzelbach & M. Virant
Institute of Hydromechanics and Water Resources Management, ETH, Zürich, Switzerland

ABSTRACT: An optical method for determining the velocity fields of both phases in dilute particle - laden flows is presented. The method was developed to study the sedimentation of particles through a rapidly evolving mixing layer. A laser light sheet in a vertical plane through the flow was used to illuminate both the sediment, and small, neutrally buoyant fluorescent spheres which served as markers for the fluid motion. The fluorescent light was separated from the reflected laser light by an optical low-pass filter. Both images were viewed through a glass wall of the tank and recorded on separate CCD cameras. To obtain a high correlation between subsequent images of the rapidly evolving flow field, a second pair of cameras was used, and triggered a sufficiently short time after the first pair. The velocity fields for both phases were then determined by least square matching of corresponding particle groups on subsequent frames.

1. INTRODUCTION

Particle imaging velocimetry (PIV) is a popular method for measuring velocity fields in fluid flows. The fluid is seeded with small marker particles which are illuminated by a laser light sheet, and velocity fields are obtained by correlating subsequent images. Conversely, a tracer like a fluorescent dye is used to map velocity and other fields by the Laser Induced Fluorescence (LIF) method. The two methods were reviewed by Adrian (1991) and Dracos (1996). In many industrial and environmental processes the flow contains another phase, and it is desirable to map the velocity fields of both phases simultaneously, which is possible by using optical methods when the flows are sufficiently dilute. A simple solution would be to use the PIV technique for the discontinuous phase and the LIF technique for the continuous one, provided that the fluorescent light can be separated sufficiently well from the reflected laser light.

The only effort we are aware of to map velocity fields of both the continuous and the discontinuous phase in two-phase flow by optical methods is the digital mask technique proposed by Gui and Merzkirch (1996), by which small flow markers and sediment particles are distinguished on the basis of their size. Another option is to use flow markers with special optical properties, such that the light they emit can be separated from that of the discontinuous phase by optical discrimination methods. This approach was pursued in the present study, and we used an advanced PIV technique to determine the velocity fields in both phases of slowly as well as rapidly evolving flows.

2. VISUALISATION AND IMAGE CORRELATION

The motion of both phases can be made visible by introducing small markers into the fluid which closely follow the flow, and by making use of a laser light sheet for illumination. A main feature of the proposed technique is that the markers are fluorescent and re-emit the incident laser light at a considerably larger wavelength, such that the two types of particle images can be be recorded on separate cameras by making use of filters. This technique allows a separation of the two velocity fields and has the advantage that it can be used for a wide range of particle types and sizes. Other methods, like the digital mask technique proposed by Gui and Merzkirch (1996), require that the flow markers and particles have a distinguishable size distribution. Moreover, a reliable distinction between the two phases is guaranteed with the present approach even when flow markers and particles would be overlapping if viewed without filters by a single camera. This advantage of optical filtering is particularly important for high particle concentrations. An optical separation of the two types of particles requires the use of at least two cameras, which have to be carefully aligned to assure that the same observation area is imaged. For rapidly evolving flows the temporal resolution of the motion can be improved by triggering two light pulses with a short time interval within the exposure time of the cameras. Another option is to make use of an additional pair of cameras, which eliminates the directional ambiguity problem inherent in doubly-exposed images, and allows for much higher particle or marker concentrations. The velocity fields for both phases were obtained by applying a technique introduced by Maas et al. (1994), which uses least-squares matching to correlate consecutive images of a group of particles (a 'pattern'). The corresponding algorithm is based on a two-dimensional affine transformation, i.e. a translation, a linear deformation, and rotation of the pattern. Only a small number of particles (typically five) is required for determining the velocity vectors. According to Gui and Merzkirch (1996) this procedure results in an improved spatial resolution compared to the classical cross-correlation methods. In addition, the least square matching technique immediately provides information on the vorticity and strain fields of the flow.

To test the proposed method, it was used in a particle-laden mixing layer which is formed when two fluid streams of different velocity are initially separated by a horizontal splitter plate, and start interacting downstream from it. The sediment particles were introduced into the upper one of the two emerging layers. Numerical simulations of particle dispersion in a mixing layer were recently performed by Ling et al. (1998), as well as Raju and Meiburg (1995). For such simulations it is generally assumed that the motion of each particle is solely affected by the fluid (one-way coupling), i.e. particle-fluid and particle-particle interactions are neglected. The experiments help to gain a better understanding of the interactions and allow to check the reliability of the numerical models.

3. EXPERIMENTAL ARRANGEMENT

The experiments were conducted in a rectangular tank 3000 mm in length, 1400 mm in height and 250 mm in width. Water was pumped into two overflow vessels, which evened out pump-induced variations of the flow. The water from each vessel entered a separate diffuser and was conveyed to the test section through a flow homogenizing chamber which contained a honeycomb to reduce the swirl of the fluid, two grids for turbulence reduction, and a flow contraction unit (Fig. 1). The velocity of each layer could be selected in the range between 0 and 30 cm/s. The test section was illuminated by an Argon ion laser with a total light

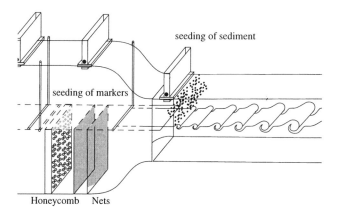

Fig. 1. Test section. Flow homogenizing devices shown for
lower layer only.

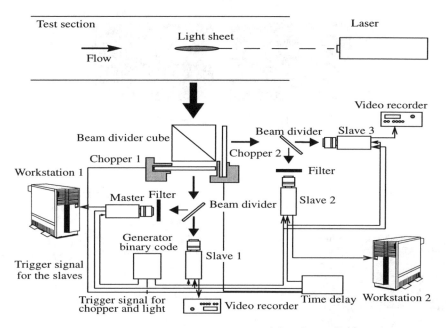

Fig.2 PIV - system with four cameras and three beam dividers

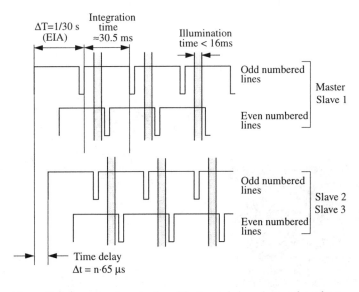

Figure 3 Camera synchronization. Master and slave 1 record markers,
slaves 2 and 3 sediment particles

power of 5 W (of which 2.0 W in the 514.5 nm line, and 1.5 W at 488 nm). The optical components for the generation of a light sheet were located beneath the test section. The system consisted of five lenses and a mirror, and allowed us to vary the beam-waist of the light sheet over a wide range (0.01 mm - 5 mm), and to locate the focus at any position in the test section. A continuous wave laser was used but the light was pulsed mechanically by a rotating disk with a slit (chopper), which limited the illumination time to less than 2 ms.

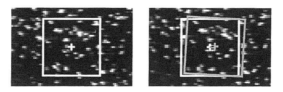

Fig. 4 Translation, rotation and deformation of patch

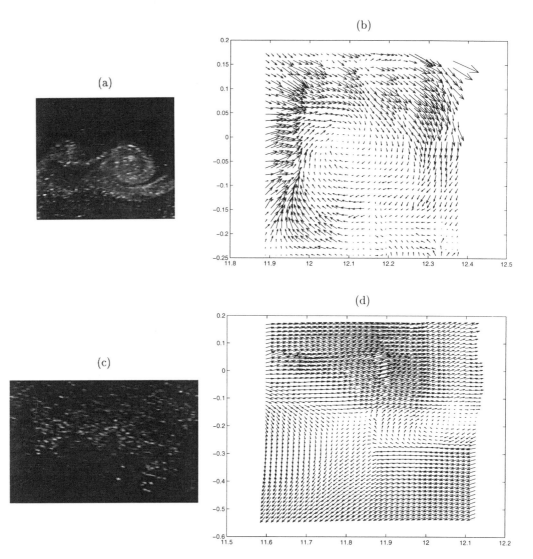

Fig. 5 (a), and b): Image and velocities of markers; c) and d): Image and corrected velocities of sediment particles, scales are in cm

Small and neutrally buoyant beads of diameter 63 - 70 μm filled with fluoresceine and rhodamine were used as flow markers. Fluoresceine has an absorption maximum near 514.5 nm and an emission maximum at 530 nm, the corresponding values for rhodamine are 530 and 560 nm. The beads were purchased from Prof. J Katz at John Hopkins Engineering, Department of Mechanical Engineering, G.W.C. Whiting School of Engineering, 3400 N. Charles Street, Baltimore, MD 21218-2686, USA. The marker particles absorb laser light at 514.5 nm and emit at wavelengths near 560 nm. This allows the insertion of a suitable filter into the beam, which blocks any scattered light from the sediment particles and lets pass only the frequency-shifted light from the markers (see Fig.2). As the markers were small, their fluorescent emission was weak. Therefore, the sensitivity of the cameras imaging the sediment could be reduced to the extent that they detected only the sediment particles but not the markers. This discrimination method eliminates the need for high - pass filters for the reflected 514 and/or 488nm light.

The motion of the sediment particles and flow markers was recorded by four commercially available CCD cameras (IMAC XC77, EIA norm) with a frame rate of $1/\Delta T = 30Hz$ and a spatial resolution of 480 x 640 pixels in a field of view of about 4cm x 5 cm. The interline transfer CCD sensors were run in the interlaced full frame mode (Fig. 3). This required that the light pulse fell within the overlapping part of the integration times of the two fields. An electronic device was built to delay the recording time of slave cameras 2 and 3 relative to the master camera and slave 1 by a multiple of 65 μs (see Fig.3). A total delay of about 4 to 6 ms was generally used. To allow an identification of pairs of simultaneous pictures during processing, a sequential binary code was superimposed on all images. The phase of the rotating choppers was also synchronized by means of the master camera.

The images were digitized on two DEC α workstations 3000 AXP equipped with a Sound & Motion J3000 Turbo Channel Interface. Each frame grabber digitized the output of one CCD camera by applying a dynamic JPEG-compression. The data were stored temporarily on an extended RAM (200 MB) of the workstation and then written on hard disk. With this configuration image sequences of up to 60 s could be digitized in real-time. The images of the markers were digitized directly, whereas the output of the CCD cameras 1 and 3 were first stored on video tape and then digitized after the experiments.

4. EXPERIMENTAL PROCEDURE

In order to obtain reliable measurements an *in-situ* calibration is required. For this purpose a metal coated plane glass plate with photochemically etched reference points on a regular grid was submerged in the test section, and the orientation of the cameras was adjusted until the images of the grid were identical.

The accuracy of this mechanical procedure is limited to within a few pixels, and improvements could only be achieved by image processing. The reference points were recorded by each camera and the four images were made congruent by an affine transformation, which resulted in corrected video frames. This calibration procedure also has the advantage that other errors, like those originating from lens distortion and misalignment of the beam dividers, can be reduced.

Commercially available quartz sand with a density of ρ_p=2700 kg/m^3 was chosen for the sediment particles. The sand was sieved in order to obtain less polydisperse size distributions. Two fractions were used: one with a mean diameter of d = 205 μm and one with d = 320 μm. The upper, fast flowing stream was seeded by releasing sand from its upper boundary. The fluorescent particles were added to both streams near the splitter plate shown in Fig. 1.

5. IMAGE PROCESSING

The digital images were analysed based on the cross-correlation technique described by Maas et al. (1994) for 3-dimensional velocity fields. For this purpose, the images were subdivided into small patches (typically 21 x 21 pixels) each containing a pattern of a few particles. Each of these patches was matched with the corresponding patch on the subsequent image taken at the next time step. The corresponding affine transformation for plane fields can be derived along the following lines.

$$\mathbf{x_2 = A\ x_1 + T} \qquad\qquad (1)$$

$\mathbf{x}_1$ and $\mathbf{x}_2$ are the position vectors of the pixels of the patch at times t and t + Δt, respectively. The vector $\mathbf{T}$ represents the translation, whereas the 2 x 2 matrix $\mathbf{A}$ comprises the parameters for scaling, rotation and shear (Fig. 4). By following Maas et al. (1994) the six unknowns $\mathbf{A}$ and $\mathbf{T}$ are determined by an iterative procedure until the sum of the squares of the grey-value differences reaches a minimum. From the shift parameter $\mathbf{T}$ we immediately obtain the velocity for the origin of the patch, $\mathbf{x}_1 = 0$

$$\mathbf{u} = \frac{\mathbf{x}_2 - \mathbf{x}_1}{\Delta t} = \frac{\mathbf{T}}{\Delta t} \tag{2}$$

Further flow parameters can be derived from the spatial derivative of the velocity field, which is given by

$$\nabla \mathbf{u} = \frac{\mathbf{A} - \mathbf{I}}{\Delta t} \tag{3}$$

where I is the unit matrix.

6. RESULTS

Images of both the fluorescent and the settling particles at the same instant of time are shown in Figs. 5 (a), and (c), respectively. The velocity of the upper layer was 15.1 cm/s, and that of the lower layer 7.8 cm/s, both layers were 7 cm thick. The flow markers were introduced near the splitter plate, such that the Kelvin-Helmholtz rollers are made visible. Fig. 5 (b) shows the velocity field of the flow markers obtained by least-squares matching. Scales are in cm, with x being the downstream distance from the edge of the splitter plate and y the transverse coordinate. The mean velocity in the flow direction was subtracted, such that the velocity of the center of the eddy at x=12.1 cm vanishes.

The lower pictures focus on the braid region to the left of the eddy in the upper ones, and cover a wider field of view. As the sediment was released in the upper layer, Fig 5 (c) shows a front below which no grains are present. In Fig. 5 (d) the velocity field of the sediment particles is shown. Again the mean flow velocity was subtracted. In addition, the vertical velocity component was reduced by the individual settling velocity of the sediment particles of 2.1 cm/s. This has the effect that velocity vectors can be directed upwards as well. A vortical motion of the sediment particles can be seen near x = 12.05 cm, y = - 0.2 cm, whereas a stagnation point is observable at x = 11.85 cm, y = -0.2 cm.This indicates that sediment particles move to the braid region, where they settle.

7. REFERENCES

Adrian R. J. (1991). Particle-imaging techniques for experimental fluid mechanics. *Ann. Rev. Fluid Mech.*, 23, 261-304.

Gui, L.C.; Merzkirch, W. (1996). A method of tracking ensembles of particle images. *Exp. Fluids,* 21, 465-468.

Ling, W.; Chung, J. N.; Troutt, T. R., and Crowe, C.T. (1998). Direct numerical simulation of a three-dimensional temporal mixing layer with particle dispersion. *J. Fluid Mech.,* 358, 61—85.

Maas, H. G.; Stefanidis, A; and Grün, A. (1994). From pixels to voxels: tracking volume elements in sequences of 3-D digital images. *ISPRS Com. III Intercongress Symposium*, München, IAPRS, 30, Part 3/2, 539-546.

Dracos, Th. (ed.) (1996) *Three-dimensional velocity and vorticity measuring and image analysis techniques.* ERCOFTAC Series, Kluwer Academic Publishers, Dordrecht.

Raju, N.; Meiburg, E. (1995). The accumulation and dispersion of heavy particles in forced two-dimensional mixing layers. Part 2: The effect of gravity. *Phys. Fluids* 7, 1241—1264.

2.3 Stratified flows

Environmental Hydraulics, Lee, Jayawardena & Wang (eds) © 1999 Balkema, Rotterdam, ISBN 90 5809 035 3

On the 'stratification drag'

A. N. Srdić-Mitrović & H. J. S. Fernando
Department of Mechanical and Aerospace Engineering, Arizona State University, Tempe, Ariz., USA

ABSTRACT: It is well known that particles introduced into density stratified regions of atmosphere and ocean settle slower than they would in neutrally stratified environments. It is shown that this retarded descent can be due to the presence of an additional drag force, known as the "stratification drag", as a result of the existence of a fluid tail attached to the particle at low and moderate Reynolds numbers. The fluid tail contains low density fluid incorporated from upper layers, and hence is subjected to an additional buoyancy force that is responsible for the stratification drag. Laboratory experiments carried out with three–layer fluids, containing a thick inter-facial layer sandwiched between two homogeneous layers, indicate that the stratification drag is pronounced in the Reynolds number range $1.5 < Re_1 < 15$, where $Re_1 = U d_p / \nu$, U and d_p are the velocity and diameter of particles, respectively, and ν is kinematic viscosity of the fluid. For a given Re_1, the stratification drag on a particle is an order of magnitude larger than the total drag on the particle in a homogeneous fluid.

1. INTRODUCTION

Turbulence and stratification are two factors that have significant influence on aerosol distribution in the atmosphere and fine–particle distribution in oceans. Turbulence, in general, tends to keep particles in suspension by reentraining the settled particles and by laterally dispersing the suspended particles, thus increasing the time where particles are in suspension (however, in the presence of organized structures in turbulence, a different scenario may occur; see Nielsen 1994). It has been long known that the particle concentration in stably strati-fied layers tends to be higher than that in neutrally stratified environments, eventhough the turbulence is usually suppressed in such layers due to stabilizing buoyancy forces. An example is the volcanic ashes ejected into the stratosphere, which are known to persist for weeks, travel around the world several times before settling and cause spectacular sunsets. One might surmise that this retarded settling is related to the nature of particle–stratification interaction, but details of such interactions have not been studied hitherto. In this paper, we will discuss an interesting phenomenon that appears during the descent of dense particles in stratified fluids, specif-ically through density interfaces, that leads to a sharp increase in the drag on particles.

An additional drag on a particle in a stratified fluid can occur due to several mechanisms. For example, internal waves generated by the disturbances introduced by the particle may exert a wave drag. As pointed out by McIn-tyre *et al.* (1995), porosity of particles often found in the upper layers of ocean can lead to absorption of lighter fluid into the pores when particles descend through upper low density layers; this reduces the effective density of particles, which appears in the settling velocity measurements as if there is an additional drag. Another possibility is the generation of a tail (caudal fluid column behind the particles) at sufficiently low Reynolds numbers, which may lead to an additional buoyancy force as described below.

2. A GENERATION MECHANISM OF STRATIFICATION DRAG

Consider a solid spherical particle of diameter d_p and density ρ_p descending into a stably stratified fluid (strati-fication specified by the buoyancy frequency N) of kinematic viscosity ν (see Figure 1). The Reynolds number

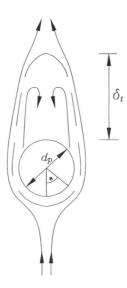

FIG. 1. Schematic sketch showing problem considered

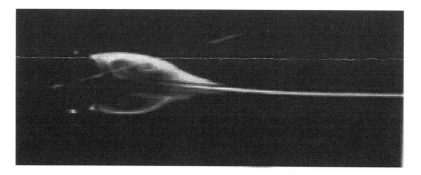

FIG. 2. A picture of the caudal fluid lump taken during the horizontal travel of a sphere in a linearly stratified fluid (from Lin. *et al.* 1992)

$Re_1 = U d_p / \nu$, where U is the velocity, is such that both the inertial and viscous effects are important in the flow surrounding the particle (i.e., above the creeping flow limit). During its descent, the particle experiences a continuous change of ambient (undisturbed) density, but the fluid in the immediate vicinity of the particle cannot renew so rapidly because of the attachment of neighborhood fluid to the particle via the no slip boundary condition. The change of the environment or density is communicated to the particle through viscous stresses and ensuing boundary layer. Downstream of the frontal stagnation point, the flow is determined by the polar angle θ, d_p, ν and N. The viscous boundary layer around the particle is expected to grow with θ as $\delta_n \sim (\nu d_p \theta / U)^{1/2}$, with the boundary layer thickness in the lee of the particle being on the order $\delta_b \sim (\nu d_p / U)^{1/2}$.

Due to the influence of viscous effects, the flow cannot simply separate behind the particles, because viscous effects transmit the influence of no slip boundary condition to a distance δ_t determined by the balance $\nu U / d_p^2 \sim U^2 / \delta_t$ or $\delta_t \sim Re d_p$ (Figure 1). The fluid contained in this region travels with the particle as a caudal fluid lump, until it is separated by either the rupture of the fluid column at the surface of the particle (Hartland 1968; Shah et al. 1969), instabilities of the caudal column (Maru *et. al.*1978) or by buoyancy forces. The last mechanism becomes dominant when the particle descends in a stratified fluid. In this case, the caudal fluid is developed by incorporation of fluid from upper layers where the particle originates its descent. As the particle enters layers with increasing density, the caudal fluid encounters an ever increasing buoyancy force due to the growing density difference between the interior and exterior of the caudal fluid lump. This appears as an extra drag force on the particle, which we refer to as the "Stratification Drag". Although the caudal fluid lump is held attached to the particle by viscous forces, it can be detached from the particle if the buoyancy force on the caudal fluid exceeds viscous stresses, which is expected to occur when $g(\rho_2 - \rho_1)d_p^2\delta_t \sim \rho_1\nu U(d_p\delta_t)/d_p$ or $\Delta b \sim (U^2/d_p)Re^{-1}$, where Δb is the difference between buoyancy of the caudal fluid lump and buoyancy of its surroundings. This corresponds to a vertical particle traverse of $\Delta z_p \sim \Delta b/N^2 \sim d_p Fr^2 Re^{-1}$, where $Fr = U/Nd_p$ is the Froude number. Thus, it is expected that particles descending in stratified fluids at low and moderate Reynolds numbers shed caudal fluid periodically, at depth intervals of order Δz_p. The maximum of the drag coefficient is expected to be on the order of $\Delta C_{DS} \sim \Delta b d_p^2 \delta_t/S_p U^2$, where S_p is the crossectional area of the particle, tantamounting to $\Delta C_{DS} \sim$ constant, independent of Re and Fr.

3. EXPERIMENTS

To demonstrate the existence of stratification drag, a series of experiments were conducted in a transparent tank of dimensions $30 \times 60 \times 60$ cm. The working fluid was prepared by first introducing a layer of lighter fluid into the tank and then slowly feeding a layer of dense fluid underneath the lighter layer as a sheet via a capped hole. The lighter fluid was a layer of 200–proof ethyl alcohol and water mixture of thickness 25 cm. The bottom layer was a solution of salt water of approximately the same thickness. The concentrations of alcohol and salt were selected so that the refractive indices of the two layers were the same. The density interface between the layers was thick due to local mixing at the interface, of the order of 2–4 cm, and the fluid within it had a continuous density distribution. The ratio of densities of the lower (ρ_2) to upper layer (ρ_1) was varied in the range 1.025 – 1.065. The density measurements were taken by traversing a conductivity probe, properly calibrated to account for the conductivity variation in alcohol–salt mixtures.

Visualization of descending particles was performed by illumination of the vertical center plane, along the long axis of the tank, using a 0.5 cm thick sheet of light. The descent of particles were video recorded from a direction normal to the light sheet. A particle tracking system was used to obtain velocity measurements and video recordings were processed using the DigImage software package. Details of this particle tracking system can be found in Dalziel (1992,1993) and Drayton (1993). Some experiments were also carried out to probe the behaviour of the fluid layer surrounding the particle. A small amount of dye (sodium fluorescein) was injected at the level where the density begins to increase at the base of the top layer. The plane of descent was illuminated by a sheet of Argon–ion laser light, and a precision video microscope (Infinity/VAR) was used to magnify the region of interest. In addition to velocity measurements, the volume of the upper–layer fluid drifted into the interfacial region was also measured using images obtained by the video microscope. The above measurements, together with the vertical density distribution measured before the experiment by the traversing conductivity probe, were used to calculate the buoyancy force acting on the caudal fluid column drifted from the upper layer.

The drag calculation was performed by modifying the conventional equation of motion of particle to include the 'total stratification drag' which includes the drag due to fluid drift and internal waves, viz.,

$$\rho_p V_p \frac{dU}{dt} = (\rho_p - \rho_f)V_p g \tag{1}$$
$$-(C_D^H + C_S)\frac{1}{2}\rho_f U^2 S_p$$
$$+(F_A + F_H),$$

where C_S is the drag coefficient introduced to account for the density stratification. Here V_p is the volume of the particle, ρ_f is undisturbed density of the fluid at the position of the particle, C_D^H is the drag coefficient of the particle in the absence of stratification, F_A is acceleration force due to added mass and F_H is the Basset drag force. Using the distance–time trajectories obtained in the experiments, the terms in (1) could be calculated and thus C_S could be evaluated. Detailed measurements showed that the internal wave drag is negligible and hence C_S includes only the contribution from the caudal fluid. The measurements of Basset history term showed that its contribution is also negligible.

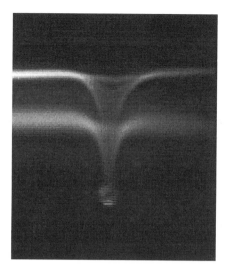

FIG. 3. A particle settling through the density interface at $Re_1 = 5$. Two horizontal dye lines were used to visualize the caudal fluid

4. OBSERVATIONS

Visualization using the video microscope indicated that indeed a caudal fluid column exists in the Reynolds number range $1.5 < Re_1 < 15$, where $Re_1 = d_p U_1 / \nu$ is the Reynolds number of the flow around the particle approaching the stratified layer. The caudal fluid column penetrates into the density interface (Figure 3), thus increasing the drag on the particle and reducing its velocity of descent (Figure 4). As expected from the discussion in Section 2, after descending some distance into the stratified layer, the particle separates from the caudal fluid thus reducing the drag and increasing the velocity.

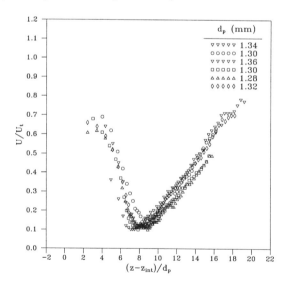

FIG. 4. The non dimensional velocity of a particle versus the dimensionless distance traveled from the interface. U_t is the instantaneous steady settling velocity of the particle if the stratification were to be removed

410

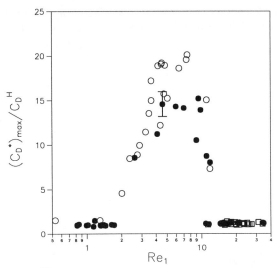

FIG. 5. The ratios of $(C_D^*)_{max}/C_D^H$ as a function of Re_1, for different Fr_1 indicated. $\circ$, $Fr_1 = 3$; $\bullet$, $Fr_1 = 5$; $\square$, $Fr_1 = 10$

The development of a second caudal fluid column predicted in Section 2 was not observed in these experiments because of the entry of the particle to the lower homogeneous layer before the development of such a fluid lump.

Figure 5 shows a plot of the maximum drag coefficient C_{Dmax}^* observed during individual stratified experiments normalized by its homogeneous–fluid counterpart as a function of Re_1 and $Fr_1 = U_1/Nd_p$. The stratification drag is well evident in $1.5 < Re_1 < 15$, with narrow $C_{Dmax} \approx const$ regions for both runs. However, the database was not comprehensive enough to check whether $(C_D^*)_{max}$ is independent of Fr_1. Future work should consider such issues in detail.

5. REFERENCES

Dalziel, S.B. (1992). "Decay of rotating turbulence: some particle tracking experiments", *Appl. Sci. Res.*, 49, 217–244.

Dalziel, S.B. (1993). "Rayleigh–Taylor instability: experiments with image analysis", *Dynamics of Atmospheres and Ocean*, 20, 127–153.

Drayton, M.J. (1993). "Eulerian and Lagrangian studies of inhomogeneous turbulence generated by an oscillating grid", *PhD thesis, Cambridge University, DAMTP*.

Hartland, S., (1968). "The profile of draining film between a rigid sphere and a deformable fluid–liquid interface", *Chem. Engng. Sci.*, 24, 987–995.

Huppert, H.E., Turner, J.S. and Hallworth, M.A. (1995)."Sedimentation and entrainment in dense layers of suspended particles stirred by an oscillating grid", *J. Fluid Mech.*, 289, 263–292.

Kellog,(1990). "Aerosols and Climate", *In: Interaction of Energy and Climate, eds. Bach, W., Pankrath, J. and Williams, J.*, Dordrecht: Reidel.

Lande, R. and Wood, A.M. (1987). "Suspension times of particles in the upper ocean", *Deep Sea Res.*, 34, 61–72.

Lin, Q., Boyer, D.L. and Fernando, H.J.S. (1992). "Stratified flow past a sphere", *J. Fluid Mech.*, 240, 315–355.

MacIntyre, S., Alldredge, A.L. and Gotschalk, C.C. (1995). "Accumulation of marine snow at density discontinuities in the water column", *Limnol. Oceanogr.*, 40, 449–468.

Maru, H.C., Wasan, T.D. and Kintner, R.C. (1971). "Behaviour of a rigid sphere at a liquid interface", *Chem. Engng. Sci.*, 26, 1615–1628.

Nielsen, P., (1992). "Effects of turbulence on the settling velocity of isolated suspended particles", *11th Austral–*

asian Fluid Mechanics Conference, University of Tasmania, Hobart, Australia, 179–182.

Shah, S.T., Wasan, D.T. and Kintner, R.C. (1967). "Paper No. 26a", *66th National Meeting of the* A.I.Ch.E..

Environmental Hydraulics, Lee, Jayawardena & Wang (eds) © 1999 Balkema, Rotterdam, ISBN 90 5809 035 3

Interfacial instabilities in exchange flows

David Z. Zhu
Department of Civil and Environmental Engineering, University of Alberta, Edmonton, Alb., Canada

Gregory A. Lawrence
Department of Civil Engineering, University of British Columbia, Vancouver, B.C., Canada

ABSTRACT: This paper reports findings from an experimental study of interfacial instabilities in exchange flow through a channel. Kelvin-Helmholtz instabilities were observed at both ends of the channel while Holmboe instabilities in the middle region. The development of the Holmboe instabilities was controlled by the vertical shift of the density interface from the shear center. The amount of the shift changed during the experiments. Measurements of the lengths and speeds of the Holmboe waves compared well with the linear stability theory. Holmboe waves were stabilized when the bulk Richardson number exceeded about 0.8.

1. INTRODUCTION

The exchange of fluids between two basins containing fluids of different densities is a common natural phenomenon. Many of these exchange flows have attracted considerable attention because of their impact on water quality and circulation. Studies of exchange flows have shown that the exchange flow rate is controlled by the interfacial shear stress (Zhu 1996). The magnitude of the interfacial shear stress is primarily determined by turbulent fluctuations which are primarily the result of interfacial instabilities. In the present study we seek a better understanding of these interfacial instabilities in the hope that this will ultimately lead to an improvement in our ability to predict exchange flow rates and vertical mixing rates.

The interfacial region between the two layers in an exchange flow can be characterized using the shear layer thickness, δ, and the density layer thickness, η, see Fig. 1. The strength of the shear is characterized by the bulk Richardson number $J = g'\delta/(\Delta U)^2$, where $g' = g(\rho_2 - \rho_1)/\rho_2$ is the reduced gravity with ρ_1 and ρ_2 being the densities of the upper and lower layer respectively, and ΔU is the velocity difference between the two layers. Many studies of stably stratified shear flows have assumed that $\eta \approx \delta$, in which case the primary mode of instability is the Kelvin-Helmholtz instability, see Turner (1973). Holmboe (1962), however, showed that for inviscid flows with $R = \delta/\eta >> 1$, another type of instability, later known as the Holmboe instability, can be generated no matter how weak the shear is. The Holmboe instability is characterized by two sets of waves having the same growth rate, and propagating in opposite directions at the same speed with respect to the mean flow velocity. Hazel (1972) found that Holmboe instabilities can be generated when $R > 2$. Recent studies by Smyth & Peltier (1989) and Haigh (1995) show that R needs to be larger than 2.4. Instead of symmetric Holmboe waves, non-symmetric instabilities are usually observed (Koop & Browand 1979; Lawrence *et al.* 1991, Pouliquen *et al.* 1994; Yonemitsu *et al.* 1996). Lawrence *et al.* (1991) explained the non-symmetric Holmboe instabilities using non-symmetric flow fields where the density interface is shifted from the center of the shear layer. The effect of viscosity on Holmboe instabilities was studied by Nishida & Yoshida (1987) and Yonemitsu *et al.* (1996). Viscosity is found to stabilize the Holmboe instabilities when the shear Reynolds number $Re = \Delta U \cdot \delta/\nu$ is sufficiently small. Holmboe instabilities are also stabilized when the Richardson number exceeds a critical value.

Despite recent advances in theoretical and numerical studies, good experimental realizations of symmetric Holmboe instabilities have been limited. Thorough experimental studies are still needed in order to understand the generation, development and evolution of Holmboe instabilities, as well as the stability conditions for Holmboe instabilities. In the present paper, we will mainly focus on the Holmboe instabilities observed in a laboratory model of exchange flow over an obstacle.

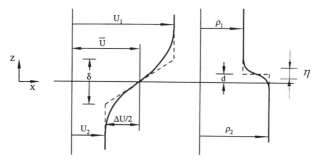

Fig. 1. Definition diagrams for the velocity and density profiles. δ and η are the thickness of the shear and density layer, respectively. d is the vertical shift of the density interface from the shear center.

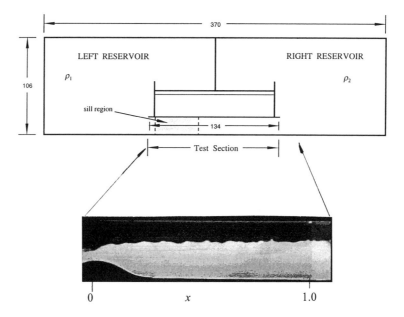

Fig. 2. (a) Plan view of the experimental setup. All dimensions are in centimeters. (b) Photo showing an experiment of exchange flow over a sill. The flow is from right to left in the lower layer, and from left to right in the upper layer.

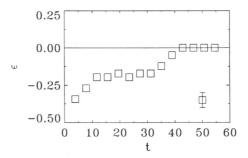

Fig. 3. Change of the shift ε, ($\varepsilon = 2d/\delta$), with time in Exp. A with $g' = 1.6$ cm/sec^2.

414

2. EXPERIMENTS

Laboratory experiments of exchange flows were conducted in a rectangular channel of 10 cm wide connecting two large reservoirs containing water of slightly different density, see Fig. 2. A sill of the form $h(x) = h_m \cos^2(\pi x/L_s)$ (for $|x/L_s| \leq \frac{1}{2}$), with $h_m = 8$ cm and $L_s = 50$ cm, was centered 31 cm from the left hand end of the channel. The driving buoyancy force was obtained by dissolving salt in the right reservoir. An experiment (Exp. A) was conducted with the reduced gravity $g' = 1.6$ cm/sec^2 in a channel with a total depth $H = 28.5$ cm and a length $L = 175$ cm. (L is measured from the sill crest to the right hand end of the channel.) Additional series of experiments (Exp. B) were conducted using a short channel of $L = 103$ cm with $H = 28$ cm and g' varying from 1.1 to 12.5 cm/sec^2. These experiments lasted for a sufficient long period and were quasi-steady, thus ideal for the study of Holmboe instabilities.

Flow visualization and image processing techniques were used to make direct measurements of the flow. The position of the interface and its deformation by flow instabilities were visualized by dissolving a fluorescent dye into the lower layer and illuminating it with a thin sheet of laser light. The velocity field of the flow was obtained by tracking the movements of neutrally buoyant particles in the light sheet. Unlike previous studies, we obtained continuous and simultaneous measurements of the density field and velocity fields. Additional density profiles were taken using a conductivity probe. Further details on experimental techniques can be found in Zhu (1996).

3. RESULTS

3.1 Wave Development

Experiments were started when a gate separating the left reservoir from the denser right reservoir was pulled out from the middle of the channel. During the initial unsteady start-up period significant mixing occurred. Gradually the flow became steady with two hydraulic controls. (A detailed discussion on the hydraulics of exchanges can be found in Zhu & Lawrence, 1998). As the experiments proceeded, one control became submerged and the flow became quasi-steady. Significant interfacial wave activities were observed during the period of the steady and quasi-steady stages. In the following discussion, parameters are non-dimensionalized with respect to the horizontal length scale L, the velocity scale $(g'H)^{1/2}$, and the time scale $L/(g'H)^{1/2}$.

At the interface of exchange flows, both Kelvin-Helmholtz (K-H) instabilities and Holmboe instabilities were observed. K-H instabilities were observed to the left of the sill crest, since the shear was strong due to a large lower layer velocity. These K-H waves had zero propagating speed with respect to the mean flow, and thus were washed down the sill due to the left-moving mean velocity. In the middle region, the shear was weaker with J much larger than that required for K-H instabilities and only Holmboe instabilities were observed. This study focuses on the Holmboe instabilities observed in the middle region where the flow conditions varied gradually with location and the flow was relatively parallel, given a small interface slope.

The thickness of the interfacial density layer and shear layer was obtained from the time-averaged density and velocity profiles. The density layer thickness, η, was initially large due to the significant interfacial mixing occurred at the start of the experiment, with $R = \delta/\eta$ being about 2. The mixed fluid was gradually swept downstream in both the upper and lower layers, and the density layer was sharpened. Thus R increased to about 15 after $t \approx 10$, and remained high for the rest of the experiment. The density interface (the point with the maximum density gradient) was displaced vertically from the center of the shear layer. The variation of the non-dimensional shift ε, $\varepsilon = 2d/\delta$, (d is the vertical shift), with time was plotted in Fig.3. The shift was large initially, being about - 0.25, with the negative sign corresponding to the shear center lower than the density interface. It gradually decreased to zero at $t \approx 40$.

The development of interfacial waves was studied using wave characteristic plots. A time sequence of the interface positions was first obtained. The development of interfacial waves was then shown by displaying these interface positions in an x - t characteristic plot, similar to those used in open channel flows. A typical wave characteristics plot is shown in Fig. 4, which was constructed by stacking the obtained interface positions along the channel as rows. Each strip contained 120 such rows with each row representing the interface positions at every second, thus covering a period of 2 minutes. The light intensity within each row represents the relative height of the interfacial position, with the bright and dark points representing high and low elevations, respectively. Thus, we see the characteristics of the positive waves (upward cusps) and the negative waves (downward cusps) as oblique bands of light and dark.

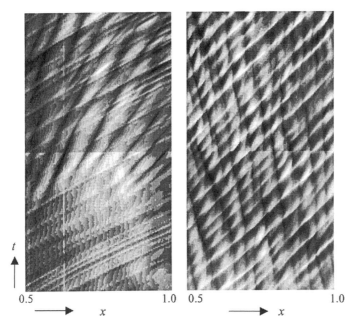

0.5   1.0   0.5   1.0

x x

Fig. 4. Characteristics of Holmboe waves in Exp. A with $g' = 1.6$ cm/sec^2 in the region $x = 0.5 \sim 1.0$. The left and right strip evolves from $t = 8 - 16$ and $36 - 44$, respectively. Each strip contains 120 rows, with each row representing the interface position along the channel. The intensity is a measure of the height of the interface: brighter shading stands for higher interface. Oblique bands of dark and light show the propagation of positive and negative waves, respectively.

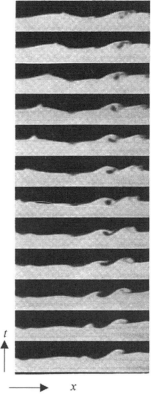

t

x

Fig. 5. Series of images showing the deformation of interfacial waves in Exp. B with $g' = 2.3$ cm/sec^2 starting at $t = 19$ (5 minutes). Images were captured at $\Delta t = 0.5$ seconds. The horizontal length is 18 cm.

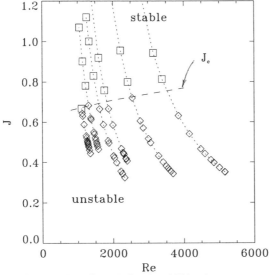

Fig. 6. Stability conditions for Holmboe instabilities from Exp. B. Flow evolves along the dotted line during each experiment. $\Diamond$, flow conditions with Holmboe instabilities observed; $\square$, without Holmboe instabilities. $----$, stability boundary from Nishida & Yoshida (1987). From left to right, g' is 1.1, 1.6, 2.3, 3.1, 6.2 and 12.5 cm/sec^2, respectively.

416

While some positive waves were observed soon after the flow became steady, the negative waves were generated much later (Fig. 4). This is because the density layer thickness in the middle region was large initially given R being about 2, thus flow was stable to the Holmboe waves. The positive waves, on the other hand, were generated in the sill region where the flow conditions were different, and propagated through the middle region. Once the density layer was sharpened and R increased to about 15, negative waves were quickly generated, indicating that the negative waves had large growth rates. The difference in the growth rate between the positive and negative waves was due to the large negative shift (Fig. 3), which gave the negative waves larger growth rates.

At a later time, the shift gradually decreased to zero, thus the growth rate of positive waves increased. Both the positive and negative waves had the similar growth rate and grew to approximately equal amplitude. The symmetric Holmboe waves were then observed in the middle region, with waves becoming regularly spaced. This observation of symmetric Holmboe waves is important since it has been reported difficulty in realizing symmetric Holmboe waves (Lawrence et al. 1991). Later, the shear decreased rapidly and the Richardson number increased when the exchange flow slowed down. Also the viscous force increased with the decrease of the shear Reynolds number in the quasi-steady stage. The growth rate of interfacial instabilities gradually decreased. Finally, the waves died down and the interface was stabilized when $J \approx 0.8$ and $Re \approx 1300$, unlike the inviscid theories which predict Holmboe waves for any large J.

The lengths and speeds of the Holmboe waves were measured and compared with the linear theory of Haigh (1995). In general, the experimental measurements compare well with the linear theory, with the measurements within the unstable region predicted by the theory. The measurements of the wave numbers centered in the region with the maximal growth rate predicted by the linear theory. The wave speeds, however, was found to be under-predicted by the linear theory by about 0.1 - 0.2. For further details see Zhu (1996).

A sequence of photos showing the non-linear development of Holmboe waves is presented in Fig. 5. It is clearly shown that some of the waves rolled up while they were propagating, even though J was about 0.3 to 0.5, much larger than that required for K-H waves ($J < 0.07$). Eventually, these waves broke down. The observations of this wave deformation and breaking are important since they are not expected for the flow with such strong stratification. Previous studies such as that by Browand & Winant (1973) indicate that Holmboe waves do not concentrate vorticity into lumps but rather break at sharply peaked wave crests, different from K-H waves which break by concentrating the available vorticity into discrete lumps along the interface. Thus it is believed that the mixing due to Holmboe waves is much less vigorous than that due to K-H waves. From Fig. 5 we see that Holmboe waves can in fact concentrate vorticity into lumps, and thus cause significant mixing.

3.2 Stability Conditions

The stability conditions for Holmboe instabilities were examined using the experimental data and plotted on a (Re, J) plot in Fig. 5. There is a minimum value of experimental Richardson number being about 0.3 to 0.4. This minimum value is related to the experimental setup since the forcing in our experiments was generated by the density difference. For each experiment, the Reynolds number decreased while J increased during the experimental process. The interface finally became stable after J exceeded a critical value. An upper boundary for the unstable region can be established from these measurements. This boundary is a function of both J and Re: the flow is stabilized when J is larger than a critical value J_c or Re is smaller than a critical value. From Fig. 5, J_c is found to increase slightly from 0.65 - 0.75 (with an error of ± 0.1) when Re increased from about 1000 to 5000. This can be explained by the fact that viscous effects are more important when Re is small, thus stronger shear or smaller J_c is required to generate Holmboe instabilities.

The above results compare very well with the numerical study of Nishida & Yoshida (1987) where they found that for the flow with a shear Reynolds number ranging from 200 to 4000, the flow is stable when J exceeds a critical value of about 0.6 to 0.75. Viscosity is found to have little effect on Holmboe instabilities when Re is larger than 1000 (Nishida & Yoshida 1987; Smyth et al. 1988; Haigh 1995). This is also confirmed as J_c is only weakly dependent on Re when Re is larger than 1000. Previous experimental studies also reported the existence of a critical Richardson number J_c: Browand & Wang (1972) observed Holmboe instabilities only when $J < 0.7$ in their experiment with $Re \approx 100$; Koop & Browand (1979) found J_c of about 1.0 in their flow with $Re \approx 300$ and $\varepsilon \approx 1.0$; Yonemitsu et al. (1996) gave J_c a value of about 1.0 for his flow with $Re \approx 400$ and $\varepsilon \approx 0.25$. From these studies and our experiments, it can be concluded that there exists a critical Richardson number J_c for Holmboe instabilities. The value of J_c changes from about 0.6 to 0.8 when Re changes from 100 to 5000, and J_c increases to about 1.0 when the density interface is shifted from the shear center.

417

Some previous studies (Keulegan 1949; Grubert 1989) used the Keulegan number, $K = (\Delta U)^3/g'\nu$ to study the stability conditions. The onset of instabilities for a laminar flow, as judged by the appearance of waves on the interface, was shown to depend mainly on K, and occurred when K exceeds a critical value of about 500 (Keulegan 1949). Here the usefulness of the Keulegan number for the Holmboe instabilities can be re-examined using the above results. Note that K can be rewritten as $K = Re/J$. Given that J_c changes from about 0.6 to 0.8 when Re increased from about 100 to 5000, the critical K then increased from about 170 to 6200. Therefore there is no single critical K value for this instability study.

4. CONCLUSIONS

We conducted an experimental study on the interfacial instabilities in a natural exchange flow situation. Both Kelvin-Helmholtz and Holmboe instabilities were observed. The center of the shear layer was shifted vertically from the density interface. This shift resulted in non-symmetric Holmboe waves. As the experiments progressed the shift gradually decreased to zero, and symmetric Holmboe waves were observed. The measurements of wave speeds and wave lengths compared well with the linear theory of Haigh (1995). The laboratory observation of the non-linear evolution of the Holmboe waves showed that these waves can concentrate the available vorticity into lumps and cause significant mixing during their break. The Holmboe waves were stabilized when J exceeded a critical value J_c. J_c is weakly dependent on the shear Reynolds number with J_c increasing slightly from about 0.6 to 0.8 when Re increases from about 100 to 5000. The use of Keulegan number as a stability criterion was found to be unsuitable.

5. REFERENCES

Browand, F.K. and Wang, Y.H. 1972 An experiment on the growth of small disturbances at the interface between two streams of different densities and velocities. *Int. Sym. Stratified Flows*, Novosibirsk, USSR, 491-498.

Browand, F.K. and Winant, C.D. 1973 Laboratory observations of shear-layer instability in a stratified fluid. *Boundary-Layer Meteorology, 5*, 67-77.

Grubert, J.P. 1989 Interfacial stability in stratified channel flows. *J. Hydraulic Eng., 115*(9), 1185-1203.

Haigh, S.P. 1995 Non-symmetric Holmboe waves. Ph.D. thesis, Dept. of Mathematics, Univ. of British Columbia.

Hazel, P. 1972 Numerical studies of the stability of inviscid stratified shear flows. *J. Fluid Mech., 51*, 39-61.

Holmboe, J. 1962 On the behavior of symmetric waves in stratified shear layers. *Geofysiske Publikasjoner, 24*, 67-113.

Keulegan, G.H. 1949 Interfacial instability and mixing in stratified flows. Res. Paper RP2040, U.S. Nat. Bureau of Standards, 43, 487-500.

Koop, C.G. and Browand, F.K. 1979 Instability and turbulence in a stratified fluid with shear. *J. Fluid Mech., 93*, 135-159.

Lawrence, G.A., Browand, F.K. and Redekopp, L.G. 1991 The stability of a sheared density interface. *Physics of Fluids A, 3*(10), 2360-2370.

Nishida, S. and Yoshida, S. 1987 Stability and eigen functions of disturbances in stratified two-layer shear flows. *Proc. of Third Intl. Symposium on Stratified Flows*, Pasadena, California, 3-5 February 1987, 28-34.

Pouliquen, O., Chomaz, J.M., and Huerre, P 1994 Propagating Holmboe waves at the interface between two immiscible fluid. *J. Fluid Mech., 266*, 277-302.

Smyth, W.D., Klaassen, G.P. and Peltier, W.R. 1988 Finite amplitude Holmboe waves. *Geophys. Astrophys. Fluid Dynamics, 43*, 181-222.

Smyth, W.D., Peltier, W.R. 1989 The transition between Kelvin-Helmholtz and Holmboe instability; An investigation of the over reflection hypothesis. *J. Atmospheric Sciences, 46*(24), 3698-3720.

Turner, J.S. 1973 *Buoyancy effects in fluids*. Cambridge University Press.

Yonemitsu, N., Swaters, G.E., Rajaratnam, N., and Lawrence, G.A. 1996 Shear instabilities in arrested salt-wedge flows. *Dynamics of Atmospheres and Oceans, 24*, 173-182.

Zhu, Z. 1996 *Exchange Flow Through a Channel With an Underwater Sill,* Ph.D. thesis, Dept. of Civil Engineering, Univ. of British Columbia.

Zhu, Z. and Lawrence, G.A. 1998 Flow regimes of exchange flows. *The 2nd International Symposium on Environmental Hydraulics*, Hong Kong.

Environmental Hydraulics, Lee, Jayawardena & Wang (eds) © 1999 Balkema, Rotterdam, ISBN 90 5809 035 3

Laboratory investigation of a three dimensional wall attached density current down an inclined slope in a rotating tank

N.E. Kotsovinos & P. Kralis
School of Engineering, Democritos University of Thrace, Xanthi, Greece

ABSTRACT : The time dependent three dimensional ,wall attached ,density current down an inclined plane surface in a large tank , rotating anticlockwise , is studied experimentally. As expected, the density current is deflected to the right .There is a initial regime (regime A) where the angle of the deflection is almost constant and obtains large values (up to 70 degrees) .The regime A is followed by another regime (regime B) , where the angle of deflection is smaller . The transition time from regime A to regime B is equal to about one rotation period .The angle of the deflection depends from the Rossby radius of deformation. For the configuration of our experiments the predominant driving force is the gravity .

1.INTRODUCTION

The objective of this paper is the laboratory study of the spreading of three dimensional dense gravity current down an inclined plane surface in a large tank rotating anticlockwise with angular velocity Ω. This flow simulates the bottom outflow of dense deep ocean water from an elevated strait .The dense outflow sinks and spreads over the sloping ocean bottom. The dynamics of this flow is influenced by the earth rotation due to the scale of the problem . The rotation of the earth alters the path and structure of the gravity current. This flow is important from the environmental point of view because the density current may transport sediment or pollutant deposited at the sea floor.The geometry of the problem is shown in Figures 1a, 1b and 1c . The outflow volume flux that feeds the gravity current is Q .The entrainment in the gravity current is assumed small so that we may assume that to the first approximation the conservation of mass gives:

$$HSB \sim Qt \qquad (1)$$

where H is the thickness , B the width and S(t) the distance of the foremost point of the front , measured along the deflected centerline longitudinal trajectory . The trajectory of the spreading gravity current dawn an inclined plane surface of slope with the horizontal plane equal to θ (see Figure 1a) results from the balance of five forces: the gravity force Fg , the pressure force Fp, the inertia force Fi , the Coriolis force Fc and the friction (mainly bottom friction) force Fd. The forces which drive the flow are two: the gravity force F_g , and the pressure (or buoyancy) force F_p .The forces which retard (or resist) the flow are three : the inertia of the gravity current F_i , the Coriolis force Fc and the friction force F_d . The methodology that we follow to find the asymptotic growth rate of the length S(t) with time is based on the balance of the forces which drive and retard the flow. Similar methodology has been used previously by Chen and List (1976) , and Lemkert and Imberger (1993). Subsequently we find the scaling of the above mentioned forces, where the continuity equation (1) has been considered and where the typical horizontal velocity U of the front is given by S/t , where t is the time .

The pressure inside the gravity current is clearly greater than the pressure outside .To the first approximation the excess pressure force F_p is given by

$$F_p = \text{pressure (or buoyancy) force} = O(\rho'gH^2 B \cos\theta) = O(\rho'gB^{-1}Q^2 S^{-2} t^2 \cos\theta) \qquad (2)$$

where ρ' is the excess density . The scaling of the other forces is given subsequently.

Fg= gravity force =$O(\rho'g\ H\ BS\ \sin\theta)$=$O(\rho'g\ Qt\ \sin\theta)$ (3)

F_i = inertia force=$O(\rho\ H\ B\ U^2)=O(\rho\ Q\ S\ t^{-1})$ (4)

Fc=Coriolis force=$O(\rho H\ B\ S\ f\ U)=O(\rho\ Q\ f\ S)$ where f= the Coriolis parameter=2Ω (5)

We assume that the drag force is mainly due to bottom shear

$$F_d = O(\mu\ U\ H^{-1}\ BS\ )=O(\mu B^2 S^3 Q^{-1}\ t^{-2})\quad \text{where}\quad \mu\ \text{is the dynamic viscosity}$$
                                                                         (6)

It is apparent that the magnitude of the above forces vary with time and with the length S . The ratio of the Coriolis force becomes larger than the inertia force when

$$\frac{Fc}{Fi} = ft >1 \quad \text{or when}\quad t >0.08 \ \text{rotation periods}$$
         (7)

The driving gravity force becomes larger than the pressure force when

$$\frac{Fg}{Fp} = \frac{Qt\sin\theta}{Q^2 B^{-1} S^{-2} t^2 \cos\theta} = Q^{-1} B S^2 t^{-1} \tan\theta = SH^{-1}\tan\theta >1 \quad \text{or}\quad \tan\theta > H/S .$$
     (8)

The experiments of this study satisfy the above relationship (e.g. tan θ= 0.06 and 0.12 ; visual observations indicate that in these experiments the thickness H is less than 0.06 S).
We examine subsequently the following balance of the driving and resisting forces:
i)Regime A:
In this regime the driving gravity force is larger than the pressure driving force ,and the resisting Coriolis force is larger than the inertia force and larger than the drag force . We have therefore a balance of the driving gravity force F_g and the resisting Coriolis force F_c : i.e. $F_g = F_c$, so that we obtain:

$$S_A = C_A\ (\rho'g/\rho_s)\ f^{-1} \sin\theta\ t \qquad \text{where}\quad C_A\ \text{is an experimental constant}$$
  (9)

ii) Regime B:
In this regime we have a balance of the gravity driving force F_g and the resisting friction force F_d , and we obtain:

$$B^2 S^3 \approx (\rho'g/\rho)\ v^{-1} Q^2 t^3$$
             (10)

In these experiments is not a good approximation to assume self similarity, and it is not easy to assume a relationship between B and H . We prefer to estimate from the contours the growth rate of the area of spreading , which is assumed equal to BS . We found in these experiments that

$$B\ S\ \approx t^m \quad \text{where}\quad m\sim 1\ \text{to}\ 1.15\quad \text{Therefore}\quad \text{equation 10 gives}$$
    (11)

$$S \approx(\rho'g/\rho)\ v^{-1} Q^2 t^n \quad \text{where}\quad n=0.85\pm 0.15$$
    (12)

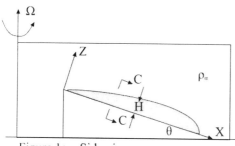

Figure 1a : Side view

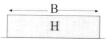

Figure 1b : section cc

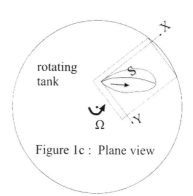

Figure 1c : Plane view

3. EXPERIMENTAL PROCEDURE - EXPERIMENTAL RESULTS

The experiments were made in a large rotating tank (diameter 5.2 m) containing a inclined glass wall mounted on the tank floor . The tank was filled with tap water and was rotating for about three hours to achieve ambient fluid motionless and rotating as a solid body. The bottom gravity current was produced by continuously injecting , at the top of the inclined plane ,colored salt water of density ρ_o at constant volume flux through a pipe of diameter D=3.5 mm . A substantial number of experiments were performed (80 experiments) , where we vary the slope of the inclined plane (inclination angle with the horizontal θ=3.52° and =6.9 °) , the initial volume flux of the injected salt water Q from 2.73 cm³/s to 5.16 cm³/s and the Coriolis parameter f from 0.105 sec⁻¹ to 0.465sec⁻¹). For comparison we run a few experiments without rotation of the tank.The initial Richardson number of the gravity current Ro is defined as $Ro = \dfrac{\mu\beta^{1/2}}{m^{5/4}}$ where μ , β and m the kinematic fluxes of mass ,

buoyancy and momentum at the origin of the gravity current and varied from 0.048 to 0.13 .The Rossby number =U/fS , calculated using the mean velocity of the front ,varied from 0.082 to 0.51 . The characteristic baroclinic Rossby deformation radius ro
$ro \approx \left(\dfrac{\rho'g}{\rho}H \right)^{1/2}/f$ calculated for a typical thickness H of the gravity current equal to 1 cm , varied from 9 to 62 cm. The characteristic inertial radius ri=U/f varied from 1.96 to 12.27 . The ratio ri/ro (~densimetric Froude number) varied from 0.09 to 0.39 .

The main emphasis of these experiments is to determine the trajectory foll. current as a function of time . The gravity current was recorded continuously using a color video camera . Typical contours of the spreading of the gravity current every 10 sec is shown in Figure 2 .

The Cartesian Coordinates X(t) and Y(t) of the position of the front of the spreading are plotted in Figure 3 (for non rotating tank the coordinate Y(t) of the foremost point of spreading front is zero) . It is observed that there are at least two basic regimes . The initial regime A is characterized by a substantial deflection of the trajectory to the right. The angle of the deflection ω_A in this regime from various experiments is plotted in Figure 4 as a function of the Rossby deformation radius ro .We observe that the deflection angle ω_A decreases with increasing ro.

The regime A is followed by the regime B where the deflection angle is ω_B . We observe that $\omega_A > \omega_B$ (see Figures 2 , 3 and 4) . The best fit line between the deflection angles ω_A and ω_B and the deformation radius gives

ω_A=360/ro^0.6 and ω_B=160/ro^0.64

The transition time Tc which characterizes the transition from regime A to regime B , divided by the

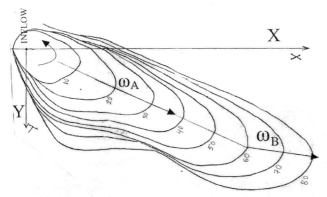

Figure 2 Contours of the gravity current as a function of time; Q=4.33cm³/s, θ=6.9° ,rotation period 45.5 sec , ρ'=0.029 gr/cm³

421

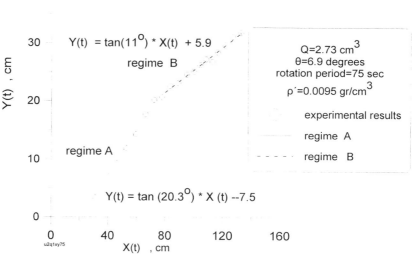

Figure 3 The Cartesian Coordinates Y(t) and X(t) of the front of the gravity current as a function of time t . The transition time Tc from regime A to regime B is about one rotation period ,i.e. about 75 sec. In regime A the deflection angle is ω_A =20.3° ,and in regime B is ω_B =11°

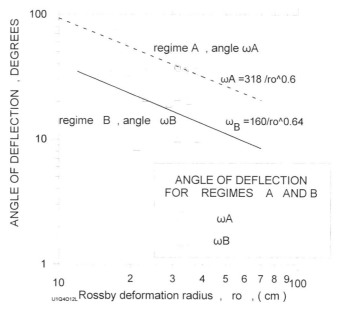

Figure 4 The angle of deflection of the trajectory of the gravity current as a function of the Rossby deformation radius .

rotation period T , is plotted in Figure 4 as a function of the Rossby deformation radius ro . We observe that there is a weak dependence of this transition from ro , but that to the first approximation the transition from regime A to regime B occurs at a time equal to about one rotation period .

The position of the foremost point of the front of the gravity current as a function of time is found for each experiment from the corresponding contour . These foremost points determine a trajectory which is used to measure the length S(t) of the foremost point of the front . The length of the trajectory of the front spreading S(t) is plotted as a function of time t in Figure 5 using the experimental data which were used to plot Figure 3 . We observe the appearance of two regimes in the growth rate of S(t) . The transition occurs at the same location where we observed from Figure 3 the transition from regime A to regime B , i.e . at the location where we have a change in the magnitude of the deflection angle. Two straight lines are fitted to the experimental data in Figure 4 , one for the regime A and the other for the regime B .The corresponding equations of the fits are printed in the Figure 4 and are given subsequently:

regime A (gravity-Coriolis) $S(t)=0.74\,t+17.2$ (13)

regime B (gravity- viscous) $S(t)=0.5t+37.1$ (14)

We observe that the fitting equation for the regime A is of the form $S(t)\sim t$, and is compatible with the theoretical equation (9) which for a balance of the driving gravity force and resisting Coriolis force give that S(t) increases linearly with time . The growth rate of S(t) in regime B (gravity - viscous) is in this example linear function of time ; however in this regime the longitudinal growth S(t) depends directly from the growth of the width B (see equation 11) .

Finally, in some experiments we observed the lateral growth (to the right of the flow) of large instabilities which used to grow with time to produce minor secondary trajectories , with small volume flux , but large deflections angles to the right , almost perpendicular to the main stream of flow.

6. CONCLUSIONS

The macroscopic experimental observations of the trajectory and the contours of the gravity current down the inclined plane surface in a rotating tank reveal the appearance of various regimes and lateral instabilities (or secondary flows). This preliminary study indicate a correlation of the deflection angle with the Rossby deformation radius ro. The inclination of the plane bottom in these experiments was

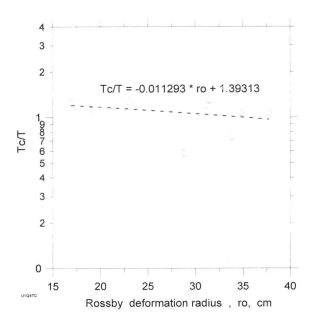

Figure 5. The transition time Tc from regime A to regime B (transition of deflection angle from ωA to ωB) , divided by the rotation period of the tank T , is plotted as a function of Rossby deformation radius.

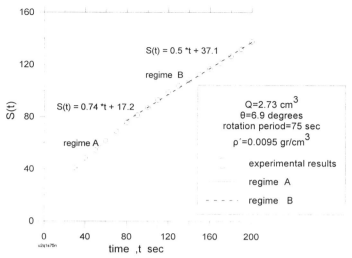

Figure 6 The growth rate S(t) of the axial distance of the front of the gravity current as a function of time t . The rotation period is 75 sec and therefore for this plot 0.26<t/T<2.6 . The transition from regime A to regime B occurs at time equal to one rotation period.

relatively high , so that the driving gravity force was balanced with the resisting Coriolis force , giving a regime where the distance along its trajectory of the foremost point of the front grows linearly with time. It appears also that at time t equal about to one rotation period there is a substantial change in the deflection angle of the trajectory. It seems that at that time the gravity -Coriolis regime changes to gravity -viscous regime.

REFERENCES

Chen,J.C.,and List,E.J. (1976), "Spreading of buoyant discharges", ICHMT Conference, Dubrovnik , Yugoslavia ,171-182 .
Lemkertt, C.J. and Imberger ,J. (1993) , "Axisymmetric Intrusive Gravity Currents in linearly Stratified Fluids " , J. Hydraulic Engrg., ASCE, vol 119 (6) ,662-679.

Environmental Hydraulics, Lee, Jayawardena & Wang (eds) © 1999 Balkema, Rotterdam, ISBN 90 5809 035 3

Vertical velocity profiles of turbulent two-layer flow

Y. Wang, A. Dittrich & F. Nestmann
Institut für Wasserwirtschaft und Kulturtechnik, Universität Karlsruhe, Germany

ABSTRACT: Experiments were conducted in a rectangular flume with two layer steady, unidirectional flow of different density fluid. A relationship based on the assumptions of *McCutcheon (1981)* and *Lawrence (1985)* was developed and tested with experimental data to describe the vertical velocity and density profile of turbulent two-layer flow. The results show a good agreement between measured and calculated profiles.

1. INTRODUCTION/THEORY

The vertical, time-averaged velocity profile of two-layer flow deviates significantly from the homogeneous flow case. It is characterized by a distinct point of inflection at the interface, as a result of the damped turbulence in this region. In the literature, many information are given to describe the vertical distribution of the time-averaged velocity and the turbulence intensity of homogeneous, fully developed open channel flow(see e.g. *Nezu and Rodi, 1986*). In contrast, almost no information are available in the case of two-layer flow with an interface between both layers.

Thus, in this study a relationship was developed and tested with experimental data to describe the velocity profile of turbulent two-layer flow. It bases on the assumptions and relationships of *McCutcheon (1981)* and *Lawrence (1985)*. According to *McCutcheon (1981)* the following expression was obtained for the velocity gradient $\partial u/\partial y$ in stratified flow

$$\frac{\partial u}{\partial y} = \frac{u^*}{\kappa y}\left(\frac{1}{1+\dfrac{\alpha g \kappa^2 y^2}{\overline{\rho}u^{*2}}\dfrac{\partial \rho}{\partial y}}\right) \tag{1}$$

where u is the time-averaged velocity in horizontal direction and u* the shear velocity, y is the coordinate orientated in vertical direction, α is a constant and κ the *von Kármán* constant, g is the acceleration due to gravity, $\overline{\rho}$ is the depth-averaged density and $\partial \rho/\partial y$ is the density gradient.

For unstratified flow, in which $\partial \rho/\partial y = 0$, Eq. 1 reduces to the well known *Prandtl-von Kármán* velocity law. The integration of Eq. (1) over rough surfaces results in

$$\frac{u(y)}{u^*} = \frac{1}{\kappa}\int_{k}^{y}\frac{dy}{y\left(1+\dfrac{\alpha g}{\overline{\rho}u^{*2}}\kappa^2 y^2 \dfrac{\partial \rho}{\partial y}\right)} \tag{2}$$

where k is the roughness height. In the case of smooth surfaces, k equals zero. The application of Eq. 2 is restricted to the knowledge of the density gradient $\partial \rho/\partial y$. According to *Lawrence (1985)*, the vertical density distribution in two-layer flow can be described by the following relationship

$$\frac{\rho(y) - \rho_1}{\rho_2 - \rho_1} = \frac{1}{2}\left[1 - \tanh\left(\frac{y - \dfrac{Q_2}{Q}H}{\delta_\rho}\right)\right] \qquad (3)$$

Where ρ_1 and ρ_2 are the densities of the upper and lower layer, Q and Q_2 are the total discharge and the discharge of the lower layer, $\delta\rho$ is the thickness of the interlayer and H is the water depth. Based on shear layer theory, the thickness of the interlayer $\delta\rho$ can be expressed as *(Lawrence,. 1985)*

$$\delta_\rho = 0.3 \cdot \Delta U^2 / g' \qquad (4)$$

where ΔU is the velocity difference between both layers and g' = $(\rho_1 - \rho_2)/\rho_2 \cdot g$ is the density reduced acceleration of gravity. Furthermore, two additional requirements are to fulfill for the unknown α and κ ($\kappa \neq 0,4 \neq$ constant in stratified flow) to solve Eq.2. The two requirements are

$$q_2 = \int_0^{H_2} u(y)dy \qquad \text{and} \qquad q_1 = \int_{H_2}^{H} u(y)dy \,. \qquad (5)$$

Where q_1 and q_2 are the known discharges of unit width of the upper and lower layer, and H_2 is the Water depth of lower layer.

2.   EXPERIMENTS

To verify Eqs. 2, 3 and 4, a series of laboratory experiments were conducted in a specially designed flume with installations generating homogeneous as well as density stratified flow. The length of the flume is 9.50m, and the aspect ratio of width to height is 0.2m × 0.4m (**Fig 1**). Additional information of the flume are given in *Plate et al (1987)*. Vertical velocity profiles were measured at three locations 3.65m, 3.95m and 4.65 m downstream of the flume entrance by Laser-Doppler-anemometry (LDA). A conductivity probe was applied to determine salinity and thus, in connection with temperature measurements, to calculate density. Details are given in *Loy (1990)* and *Kertzscher (1994)*.

Three different flow rates (total discharge) of 1.0 l/s, 1.5 l/s and 2.0 l/s were adjusted to cause two-layer flow over smooth and rough surfaces. The roughness elements consisted of fine gravels with a mean grain

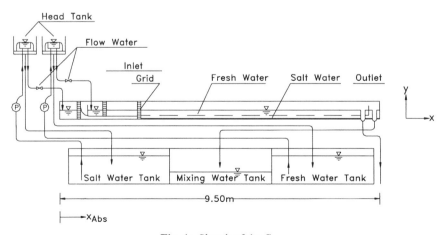

Fig. 1: Sketch of the flume

426

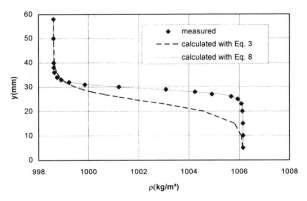

Fig. 2 Comparison of measured and calculated density profiles for $q_1/q_2 = 6/4$, $H = 60$mm and $(\rho_2 - \rho_1)/\rho_2 = 0.7$ % (= density difference between both layers)

diameter of $d_{50} = 2.8$ mm. They were spread uniformly over the entire flume bottom. Furthermore, three different salt concentrations of 0.6 %, 1 % and 2 % were used in the experiments. The external Froude number varied between 0.07 and 0.17 and the internal between 1.2 and 5.4. The shear velocity u* was calculated from the slope of the vertical velocity profile close to the boundary.

3. RESULTS AND DISCUSSION

The density gradient $\partial\rho/\partial y$ (it is a function of the thickness $\delta\rho$) has to be determined to calculate the vertical velocity distribution with the aforementioned procedure. Therefore, the relationship of Eq.3 that had been derived by *Lawrence (1985)* was tested with own data in the first step. **Fig. 2** shows that no good agreement exists between measured and calculated profiles. The differences are twofold: First, the calculated interlayer is located remarkably below the measured curve. Second, the calculated density profile yields in a smaller gradient in the region of the interlayer than the measured one, i.e. the calculated interlayer is thicker. One reason for the deviations result from the assumption $Q_2/Q\cdot H = H_2$ that is not in agreement with the observations in the flume and therefore, the thickness of the lower layer H_2 does not correspond with the relationship $Q_2/Q\cdot H$. Another reason may be explained by the not existing validity of the shear layer theory that from the base of the derivations of Eqs.3 and 4. Obviously, other mechanisms like boundary induced turbulence cause the vertical distributions of density and velocity.

To improve *Lawrence*'s relationship, the expression $Q_2/Q\cdot H$ was replaced by H_2. The effect that the calculated interlayer is thicker than the measured one can be corrected by the insertion of $\delta\rho/N$ instead of $\delta\rho$, where N is a variable. This variable was used to define the standard deviation of the measured and calculated data as

$$\sigma(N) = \left[\frac{\sum\limits_{i=1}^{n}(\rho_{me\beta} - \rho_{rech})^2}{n}\right]^{0.5} . \qquad (6)$$

Where n is the whole number of measuring points, $\rho_{me\beta}$ is the measured density and ρ_{rech} is the density calculated with the formula

$$\rho_{rech} = \frac{\rho_2 - \rho_1}{2}\left[1 - \tanh\left(\frac{y - H_2}{\delta_\rho / N}\right)\right] + \rho_1 \qquad (7)$$

A value of N = 2 results from Eg.7 by theoretical considerations. However, a value of 2.6 resulted from the analysis of the data for $\sigma(N)_{min}$, i.e. for the minimum of the standard deviation. Thus, by replacing $Q_2/Q \cdot H$ and $\delta\rho$ in Eq.3 with H_2 and $\delta\rho/2.6$ the following relationship

$$\rho = \frac{\rho_2 - \rho_1}{2}\left[1 - \tanh\left(\frac{y - H_2}{\delta_\rho \, / \, 2.6}\right)\right] + \rho_1 \tag{8}$$

is recommended to determine ρ.

Fig. 2 shows that a good agreement exists between the measured density profile and the one calculated with Eq.8 .The derivation of ρ with y in Eq.8 yields in the vertical density gradient

$$\frac{\partial\rho}{\partial y} = -\frac{1}{2}(\rho_2 - \rho_1)\frac{\left[1 - \tanh^2\left(\frac{y - H_2}{\delta_\rho \, / \, 2.6}\right)\right]}{\delta_\rho \, / \, 2.6} \tag{9}$$

To verify Eq.4 ,the experimentally determined interlayer thicknesses are plotted versus the calculated values of $\delta\rho$ (with Eq.4) in **Fig. 3** . All points should fall on the bisector of the diagram in the case of an optimal

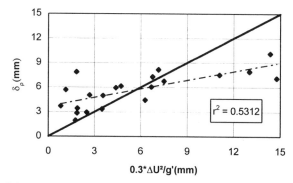

Fig. 3 Comparison of the measured and calculated (Eq.4) thicknesses $\delta\rho$ of the interlayer (determined at the location 3.65 m downstream of the flume entrance)

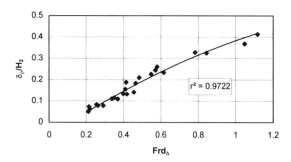

Fig. 4 The relative interlayer thickness $\delta\rho/H_2$ as function of the densimetric differential Froude number Frd_Δ

428

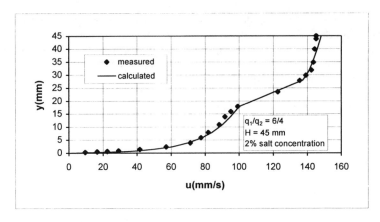

a)

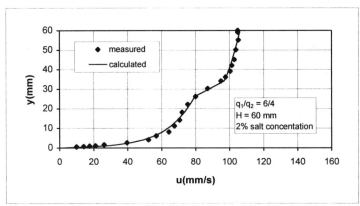

b)

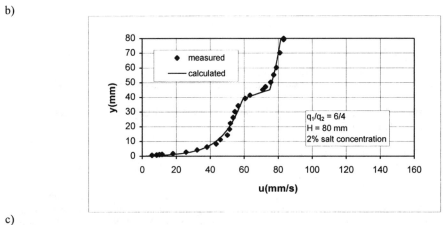

c)

Fig. 5. Measured and calculated velocity profiles over smooth surfaces for different water depths H and a discharge ratio of $q_1/q_2 = 6/4$ between the upper and lower layer and a salt concentration of 2% of the lower layer

agreement between measured and calculated values. **Fig. 3** shows that the calculated values are remarkable higher then the measured ones in the case of small relative velocities and are remarkable smaller in the case of high relative velocities. The regression coefficient is very low ($r^2 = 0.5312$) and the straight regression line does not fit the zero point. The plot illustrates that the relation between $\delta\rho$ and $\Delta U^2/g'$ (ΔU = difference of the depth averaged velocities in both layers) is not constant. Eq.4 includes the velocity and density differences between the upper and lower layer only. However, the experiments showed that a dependency of $\delta\rho$ on the water depth H and the thicknesses of the upper and lower layer exists as well. Thus, the thickness of the interlayer was described by the following parameters

$$\delta_\rho = f(\Delta U, \ g', \ H, \ H_2) \qquad .$$ (10)

The transformation of Eq.11 in dimensionless parameters yields in

$$\frac{\delta_\rho}{H_2} = f\left(\frac{\Delta U}{\sqrt{g'H}}\right) = f(\mathrm{Frd}_\Delta) \qquad .$$ (11)

In **Fig. 4**, the dimensionless interlayer thicknesses $\delta\rho/H_2$ are plotted as a function of the densimetric differential Froude number Frd_Δ (regression coefficient: $r^2 = 0.9722$). As expected, the relative interlayer thickness $\delta\rho/H_2$ increases with increasing Frd_Δ. According to the functional relationship in **Fig. 4**, the thickness $\delta\rho$ can be determined with a high accuracy by the knowledge of Frd_Δ and the thickness of the lower layer H_2. With Eq.9 and the curve in **Fig. 4** the density gradient can be calculated as well.

Finally, the vertical velocity distributions are obtained by inserting Eq.9 into Eq.2 . As already mentioned, the validity of these relationships was tested for different flow cases with density differences of 0.4 %, 0.7 % and 1.4 % between both layers over smooth and rough surfaces. **Fig. 5** shows that a good agreement exists between measured and calculated profiles.

4.   ACKNOWLEDGEMENT

The study was funded by the Deutsche Forschungsgemeinschaft (DfG). We wish to express our thanks to the DfG for their financial support.

5.   REFERENCES

Lawrence, G. A., 1985: "The Hydraulics and Mixing of Two-Layer Flow over an Obstacle", Report No. UCB/HEL-85/02, Berkeley, California, USA.

Lawrence, G. A., 1993: "The Hydraulics of Steady Two-Layer Flow over a fixed Obstacle", Journal of Fluid Mechanics, 254, 605-33.

McCutcheon, s. C., 1981: "Vertical velocity profiles in stratified flows", Journal of the Hydraulics Division, Vol. 107, No. HY8, pp. 973-988

Monin, A.S., and Yaglom, A.M., 1971: "Statistical Fluid Mechanics: Mechanics of Turbulence", J. L. Lumley, ed., Vol. 1, The M.I.T. Press, Cambridge, Mass., pp. 425-486.

Nezu, I., Rodi, W., 1986: "Open-Channel Flow Measurements with a Laser Doppler Anemometer", Journal of Hydraulic Engineering, Vol. 112, No. 5.

Plate, E., Friedrich, R., Loy, T., Wacker, J. 1987: "Homogeneous and Two-Layered Open Channel Flow Over Sills", Symposium of New Technology in Model Testing in Hydraulic Research, Poona, India.

Prandl, L., 1924: "The Mechanics of Viscous Fluids", Aerodynamic Theory, Vol. 3, Springer Verlag, Berlin, Durand, W:F (Ed.).

Environmental Hydraulics, Lee, Jayawardena & Wang (eds) © 1999 Balkema, Rotterdam, ISBN 90 5809 035 3

Regimes of exchange flows

David Z. Zhu
Department of Civil and Environmental Engineering, University of Alberta, Edmonton, Alb., Canada

Gregory A. Lawrence
Department of Civil Engineering, University of British Columbia, Vancouver, B.C., Canada

ABSTRACT: Regimes of two-layer exchange flow are studied theoretically and experimentally. Internal hydraulic theory is extended to include the effects of friction and non-hydrostatic forces, and regimes of exchange flow are classified. Laboratory experiments were conducted to study the evolution of the flow through different regimes. Measurements of the flow rate and interface positions during the maximal and submaximal exchange flows compared well with the theory.

1. INTRODUCTION

When two water bodies of different densities are connected by a strait or a channel, a two-layer densimetric exchange flow occurs. Many of these exchange flows have attracted considerable attention because of their impact on water quality and circulation; *e.g.*, the exchange flow through the Strait of Gibraltar (Armi & Farmer 1988), and the exchange flow through the Burlington ship canal (Hamblin & Lawrence 1990). Two-layer exchange flows are commonly modeled as homogeneous layers of inviscid fluid with negligibly small vertical velocities. Consequently, the pressure distribution can be considered hydrostatic, and the hydraulic (or shallow water) equations can be applied to each layer, see for example Armi (1986). This extension of the hydraulic equations to two-layer flows is called internal hydraulic theory, and has been used to study exchange flow problems (Farmer & Armi 1986; Dalziel 1991). However, the effect of friction can be important in exchange flows (Bormans & Garrett 1989; Hamblin & Lawrence 1990), and non-hydrostatic forces cannot always be neglected (Lawrence 1993; Zhu & Lawrence 1998).

As in one-layer open channel flows, two-layer exchange flows can have up to two internal hydraulic controls where the internal Froude number is one (Farmer & Armi 1986; Dalziel 1991). The regime of exchange flows is determined by the number of controls. Maximal, submaximal or uncontrolled flow occurs with two, one, or no controls, respectively. The regime of exchange flows needs to be identified before exchange flows can be predicted. In this study, the internal hydraulic theory will be extended to include the effects of non-hydrostatic forces and friction. The flow regimes observed in the laboratory experiments are compared with the theory.

2. THEORY

2.1 Extended Internal Hydraulic Theory

We study steady, irrotational, two-layer flow through a rectangular channel with a two-dimensional sill, see Fig. 1. The flow is shallow, with $\sigma = (H/L)^2 \ll 1$, where H is the total depth and L is the channel length. The sill is hydraulically smooth with its height much less than its length. The density difference between the two layers is small and hence $\varepsilon = (\rho_2 - \rho_1)/\rho_2 \ll 1$, where ρ_i is the density and subscripts 1 and 2 refer to the upper and lower layer, respectively. With $\varepsilon \ll 1$, the slope of the free surface is negligible, and is commonly referred as the "rigid lid" or Boussinesq approximation. In the following study, the flow variables are normalized with respect to the vertical scale H, horizontal scale L, pressure scale $\rho_2 gH$, and

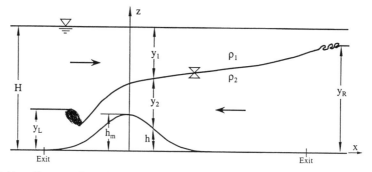

Fig. 1. Definition diagram of two-layer exchange flow through a channel with an underwater sill.

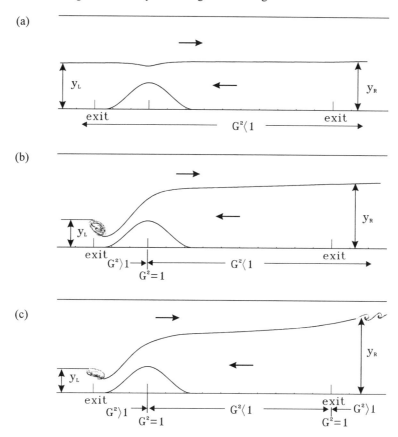

Fig. 2. Regimes of exchange flow through a channel of a constant width with a sill: (a) uncontrolled flows, (b) submaximal exchange flow, (c) maximal exchange flows.

velocity scale $(g'H)^{1/2}$, where $g' = \varepsilon g$ is the reduced gravitational acceleration.

We define the internal energy E as $E = (E_2 - E_1)/\varepsilon$, where E_i is the mechanical energy (or Bernoulli constant) for layer i. Zhu & Lawrence (1998) show that the internal energy can be written as the sum of the hydrostatic component E_H and the non-hydrostatic correction due to the streamline curvature E_{NH}, i.e.,

$$E = E_H + E_{NH} \tag{1}$$

with
$$E_H = y_2 + h + \tfrac{1}{2}(U_2^2 - U_1^2) \tag{2}$$

$$E_{NH} = \tfrac{1}{6}\sigma\{U_2^2(2y_2 y_{2xx} + 3y_2 h_{xx} + 3h_x^2 - y_{2x}^2) + U_1^2(2y_1(y_{2xx} + h_{xx}) + (y_{2x} + h_x)^2)\} \tag{3}$$

where y_i is the thickness of layer i, h is the sill height, $U_i = Q_i / by_i$ with Q_i being the volumetric flow rate and b the channel width. Note, x is the horizontal coordinate, and the subscript x denotes the differentiation with respect to x.

When friction is also considered, the change of the internal energy can be expressed as

$$dE / dx = S_f \tag{4}$$

where S_f is the friction slope. For a rectangular channel of a width b,

$$S_f = \sigma^{-1/2}\{[\tfrac{1}{2} f_w U_2^2(2y_2 + b) + \tfrac{1}{2} f_I(\Delta U)^2 b]/y_2 b + [\tfrac{1}{2} f_w U_1^2(2y_1) + \tfrac{1}{2} f_I(\Delta U)^2 b]/y_1 b\} \tag{5}$$

where f_w and f_I are the friction factors for the wall and interface, respectively, and $\Delta U = |U_1| + |U_2|$.
Substituting (1) and (2) into (4), the slope of the interface position, $y = y_2 + h$, can be written as:

$$dy / dx = (S_f - S_0 - S_c)/(1 - G^2) \tag{6}$$

where S_0 is the topographic slope, $S_0 = \sigma^{\frac{1}{2}}[F_2^2 h_x + (y_1 F_1^2 - y_2 F_2^2) b_x / b] \tag{7}$

and S_C is the curvature slope, $S_C = d(E_{NH})/dx$. $\tag{8}$

$G^2 = F_1^2 + F_2^2$ is the composite (internal) Froude number, where $F_i^2 = U_i^2/y_i$ is the densimetric Froude number for layer i. In two-layer flows, G^2 serves the same role as the classical Froude number in single-layer flows, namely, the locations where $G^2 = 1$ are internal control points, and the flow is supercritical (or subcritical) when $G^2 > 1$ (or $G^2 < 1$).

The location of the controls can be determined from Eq. (6). When the effects of friction and streamline curvature are negligibly small, the controls (where $G^2 = 1$) are located at the points where $S_0 = 0$. From Eq. (7), controls can only be located at the sill crest (where $h_x = 0$) or the narrowest point in the channel (where $b_x = 0$) when the sill and the contraction are displaced. For a channel of constant width, the control can also occur at the channel entrance. However, when friction and curvature are important, the control points are shifted from the point where $S_0 = 0$ to a new point where $S_f - S_0 - S_c = 0$.

2.2 Flow Regimes

In the following study, we will concentrate on the flow without barotropic forcing, *i.e.*, $Q_1 = Q_2$. Flows with barotropic forces are not fundamentally different (Farmer &Armi 1986). We will illustrate exchange flows using a channel of constant width with an underwater sill. Results will be applicable to flow through a channel of more general topography. In the following, we will examine regimes of exchange flows by varying the difference in the interface heights between the two reservoirs.

Exchange flow with no controls (uncontrolled flow): When there is no difference in the interface heights between the two reservoirs, there is no driving force and no flow is being exchanged. However, when the interface in the right reservoir (y_R) increases while that in the left (y_L) decreases, exchange begins with the upper layer moving from left to right and the lower layer from right to left, Fig 2(a). The flow rate increases when the difference between y_R and y_L increases. Nevertheless, there will be a range of interface height difference over which the flow remains subcritical throughout the channel, *i.e.*, $G^2 < 1$ everywhere. This flow is called uncontrolled flow, and can be predicted using Eq. (4) together with the interface level at both reservoirs (y_R and y_L).

Exchange flow with one control (submaximal exchange flow): When the difference in y_R and y_L further increases, the flow rate continues to increase, while the slope of the interface becomes steeper. At a certain point, an internal hydraulic control will be established at either the sill or the channel exit. Exchange flow with only one control is called submaximal. For the moment, we will assume that it is a sill control, see Fig.

(a)

(b)

(c)

(d)

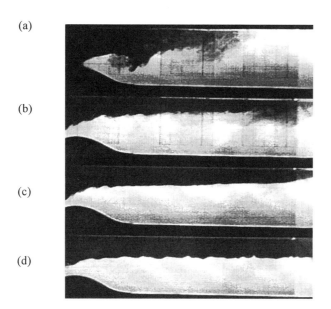

Fig. 3. Images showing the flow evolution in Exp. 2 with $g' = 1.6$ cm/sec^2. (a) Gravity current, (b) unsteady exchange, (c) maximal exchange, (d) submaximal exchange. The flow in the upper layer is from left to right, while the lower layer from right to left. The grids in (a) and (b) are 5 cm apart.

(a)

(b)

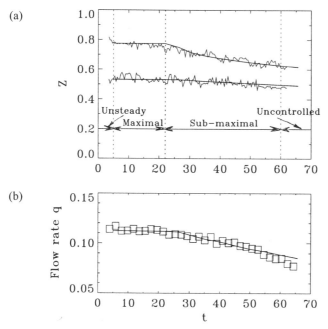

Fig. 4. Comparison of the measurements of Exp. 2 with the predictions of the extended internal hydraulic theory. (a) Variations of the interface position at the channel exit and at the sill crest (shown as the upper and lower line respectively). The smooth lines are the theoretical predictions. (b) Flow rate: the measurements are shown in symbols and the predictions shown in line.

434

2(b). The argument remains the same if it is an exit control. Given control at the sill, the flow is supercritical ($G^2 > 1$) to the left of the control as the lower layer accelerates down the sill. The interface then matches that of the left-hand reservoir through an internal hydraulic jump or turbulent mixing. The supercritical flow down the sill prevents the variation of y_L from affecting the exchange. Meanwhile exchange flow is still affected by the variation of y_R given the flow is subcritical there. The amount of the exchange is determined by the control condition at the sill as well as the value of y_R.

Exchange flow with two controls (maximal exchange flow): When the flow rate is further increased, an exit control will also be established, see Fig. 2(c). The flow is then subcritical between the two controls. Upon moving out of the right hand exit, the flow is supercritical, and the interface eventually matches that of the right hand reservoir through internal hydraulic jump or turbulent mixing. The existence of the two controls prevents the interface levels in both reservoirs from affecting the exchange flow. The exchange flow with two controls is maximal. A maximal exchange flow has the maximum flow rates for the given channel geometry. The flow rate, as well as the interface level between the controls, is independent of the flow conditions in the reservoirs.

3. COMPARISON WITH EXPERIMENT

Experiments were conducted in a channel of a constant width of 10 cm connecting two reservoirs of different densities. An underwater sill of a length of 50 cm and a maximum height of 8 cm was placed in the left portion of the channel. Experiments were conducted with a total water depth H of 28 cm and a channel length L of 103 cm. The driving buoyancy force was obtained by dissolving salt in the right reservoir. A total of nine experiments were conducted with the reduced gravity g' varying from 1.1 to 12.5 cm/sec^2. Detailed on the experimental setup and techniques can be found in a companion paper by Zhu & Lawrence (1998b).

3.1 Evolution of the Experiments

Experiments were started when a gate separating the left and right reservoirs was pulled out from the middle of the channel. Fig. 3 shows the flow evolution for Exp. 2. Upon removal of the gate, two gravity currents developed, (Figure 3a). Once these gravity currents reached the reservoirs, an exchange flow was established (Figure 3b). Gradually, the sill control was established, and the exit control was established at a later time. The period from the start of the experiments to the establish of two controls is called the unsteady start-up regime. Significant mixing occurred during this period, and the mixed fluid was gradually advected into the supply reservoirs.

When both the sill control and the exit control (Figure 3c) were established, the flow was then in the maximal exchange. The maximal exchange flow remained steady even though the interface level at the right hand reservoir (y_R) decreased steadily, while that at the left reservoir (y_L) increased steadily.

As experiment progressed, the exit control was submerged due to the falling of the interface level in the right reservoir, and the flow became submaximal (Figure 3d). The submaximal flow depended on the remaining sill control as well as the interface level in the right reservoir, y_R. Due to the decrease of y_R, the flow rate, as well as the interface level along the channel, decreased steadily during this regime. This submaximal flow lasted until the sill control was also submerged due to the rise of the interface level in the left reservoir.

When the sill control was also submerged, the flow became uncontrolled, with the flow rate and interface position depending solely on the interface levels in both reservoirs. The actual flow at this stage was quite complicated: significant mixing occurred downstream of the sill. Some of the mixed fluid was brought to the middle region of the channel by advection of the upper layer, and a three-layer density structure was formed.

One of the important features of the flow is the formation of interfacial instabilities during the maximal and sub-maximal exchange. Kelvin-Helmholtz instabilities were observed to the left of the sill crest, where the shear is strong due to the high velocity of the lower layer. Holmboe instabilities develop in the middle region of the channel, where the shear is not as strong. These instabilities are important since they determine the interfacial friction and the turbulent mixing in exchange flows. These instabilities are examined in detail in Zhu & Lawrence (1998b).

3.2 Flow Predictions

Detailed measurements of Exp.2 are shown in Fig. 4, where time is non-dimensionalized with respect to a time scale, $T = L/(g'H)^{1/2}$. After a short period of the unsteady start-up regime, which lasted till t about 5, the flow was in the maximal exchange regime with constant flow rate and interface positions. Quasi-steady submaximal exchange follows at t about 22, and lasted until t about 60. The fluctuation of the interface position was due to the interfacial instabilities generated during the experiment.

The maximal exchange was predicted using Eq. (6) together with two control conditions. The wall friction factor f_w was predicted from the boundary layer theory, while the interfacial friction factor f_i was obtained experimentally (see Zhu 1996). By including friction, the predicted flow rate is reduced by about 20 %, while the effects of the non-hydrostatic forces increase the prediction by about 6 %. With both friction and curvature effects included, the predicted flow rate and interface positions are in very good agreement with the laboratory measurements.

The submaximal flow was predicted using Eq. (6) together with the conditions of the sill control and the interface position at the right hand reservoir y_R. The flow rate decreased as y_R dropped during the experiment. The prediction of the flow rate and the interface positions compared well with the experimental measurements until $t \approx 50$. After that, the flow rate was over-predicted. This is probably the result of using the same friction factors obtained from the maximal exchange. At a later stage, the flow slows down significant and the effects of friction became more important. Nevertheless, the difference between the predicted and measured flow rate is still within 8 % at $t \approx 60$ (see Fig. 4), indicating that the flow predictions were reasonable.

4. CONCLUSIONS

We conducted a theoretical and experimental study of the regimes of two-layer exchange flows. Internal hydraulic theory is extended to include the effects of friction and non-hydrostatic forces. The measurements of the maximal and submaximal exchange flows were obtained and compared very well with the predictions of the extended theory.

5. REFERENCES

Armi, L. 1986 The hydraulics of two flowing layers with different densities. *J. Fluid Mech.*, **163**, 27-58.

Armi, L. and Farmer, D.M. 1988 The flow of Mediterranean water through the Strait of Gibraltar. *Progress in Oceanography*, **21**, 1-105.

Bormans, M. and Garrett C. 1989 The effects of non-rectangular cross section, friction, and barotropic fluctuations. *J. Physical Oceanography*, **19**(10), 1543-1557.

Dalziel, S.B. 1991 Two-layer hydraulics: a functional approach. *J. Fluid Mech.* **233**, 135.

Farmer, D.M. and Armi, L. 1986 Maximal two-layer exchange over a sill and through the combination of a sill and contraction with barotropic flow. *J. Fluid Mech.*, **164**, 53-76.

Hamblin, P.F. and Lawrence, G.A. 1990 Exchange flows between Hamilton Harbour and Lake Ontario. *Proc. of 1990 Annual Conf. of Canadian Society for Civil Eng.*, **V**: 140-148.

Lawrence, G.A. 1993 The hydraulics of steady two-layer flow over a fixed obstacle. *J. Fluid Mech.*, **254**, 605-633.

Zhu, D.Z. 1996 *Exchange Flow Through a Channel With an Underwater Sill,* Ph.D. thesis, Dept. of Civil Engineering, Univ. of British Columbia.

Zhu, D.Z. and Lawrence, G.A. 1998a. Non-hydrostatic effects in layered shallow water flows, *J. Fluid Mech.* **355**, 1-16.

Zhu, D.Z. and Lawrence, G.A. 1998b. Interfacial instabilities in exchange flows, *The 2nd International Symposium on Environmental Hydraulics*, Hong Kong.

Environmental Hydraulics, Lee, Jayawardena & Wang (eds) © 1999 Balkema, Rotterdam, ISBN 90 5809 035 3

Laboratory model studies of flushing of trapped salt water from a blocked tidal estuary

P.A. Davies, Y. Guo & J. Cremers
Department of Civil Engineering, The University of Dundee, UK

M.J. Coates
School of Ecology and Environment, Deakin University, Warnambool, Vic., Australia

ABSTRACT: Results are presented from a series of laboratory model studies of flushing of saline water from a partially- or fully-closed estuary mouth. Experiments have been carried out to determine the response of the trapped saline volume as a function of the fresh water river discharge Q, the density difference $\Delta\rho$ between the saline and fresh water and the estuary bed slope α. Flow visualisation and density probe data confirm that the transient behaviour of the plume can be represented well by the use of length and time scales $Q^{2/5}/(g')^{1/5}$ and $Q^{1/5}/(g')^{3/5}$ respectively, where $g' = g(\Delta\rho)/\rho_0$ is the modified gravitational acceleration and ρ_0 is the density of the freshwater flow.

1. INTRODUCTION

Conditions in many shallow, microtidal estuaries are such that variations in water salinity are determined primarily by the strength of the river inflow. For certain well-documented cases in South Africa and Australia (e.g the Palmiet River near Cape Town (Largier *et al*, 1992), the Murray River and the Glenelg River in Western Victoria), low river flows in summer conditions can cause the formation of a sediment bar at the estuary mouth and the consequent temporary closure of the estuary. Under such circumstances, saline water in the estuary becomes decoupled from its source of supply, with significant implications for the estuary ecology and aquaculture. Flushing of this saline volume can then only take place naturally by the restoration of a purging river flow, either steady or unsteady (such as, for example, from flooding or storm events and/or upstream releases of water from reservoirs or dams).

The flushing or purging process for estuaries of this type is the subject of the present paper. A laboratory model has been built to simulate the essential features of the estuarine dynamics and the motion of the trapped salt wedge within the estuary has been measured for various external forcing conditions. In the present paper only steady purging flows have been included, though further data are also available for cases in which the purging flow is intermittent. The study is related to other investigations by Armfield and Debler (1993), Debler and Armfield (1997) and Debler and Imberger (1996) on the flushing of saline pools from cavities by steady and "one shot" purging flows respectively and recent work by Grigg and Ivey (1997) on shear-induced mixing in a salt wedge estuary.

2. PHYSICAL SYSTEM

Fig 1 shows a schematic view of the experimental arrangement, consisting of a long rectangular section channel of plan dimensions 8.2 m x 8.2 cms and height 25 cms. Initially, the tank is filled with salt water of prescribed initial density ($\rho_0 + \Delta\rho$) to the level H_s of the top of the false base and the exit weir at the far downstream end of the channel. In the experiments, false bottom sections having slope angles α of 2% and 4% have been studied and the quantity H_s has been varied between 10 cms and 15 cms. With the system at rest, fresh water of density ρ_0 is pumped from a reservoir at a prescribed constant flow rate Q and the

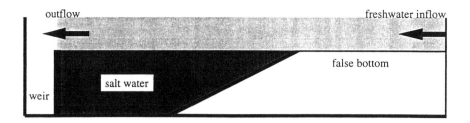

Fig 1: Schematic sketch of initial arrangement.

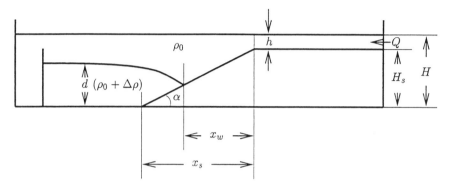

Fig 2: Schematic representation of salt wedge formation, with defining notation.

response of the system to the imposition of such a flow is monitored by means of video-based flow visualisation records and sequential vertical density profiles taken at the toe of the bed slope with a microconductivity probe mounted on an automated traverse. Internal waves initiated by the onset of the freshwater flow are damped by the placement of a sloping beach at the weir. Dye is added to the salt water to aid direct flow visualisation and shadowgraph techniques are also applied to look more closely at the shear interface between the salt water and the purging freshwater flow. At some time t after the initiation of flow, the salt water initially filling the whole of the lower part of the channel has been shifted downstream, to form the wedge-like structure typified by Fig 2.

3. EXPERIMENTAL RESULTS

In the experiments to be described below, quantitative measurements were made of the distance x_w moved by the saltwater/freshwater interface in an elapsed time t, for a range of values of Q, $\Delta\rho$, α and H_s. Examples of such measurements are shown on the plots on Fig 3, for slope angles of 4% and 2% respectively. In each case there is a clear tendency shown for the front of the wedge to migrate downstream in response to the freshwater shear flow at a rate that is systematically dependent (for a given slope angle α) upon both Q and $\Delta\rho$. In order to compare such data with prototype behaviour, it is convenient to derive from a dimensional analysis of the problem the length and time scales of the flow as $Q^{2/5}/(g')^{1/5}$ and $Q^{1/5}/(g')^{3/5}$ respectively, where $g' = g(\Delta\rho)/\rho_0$ is the modified gravitational acceleration and ρ_0 is the density of the freshwater flow. Following Grigg and Ivey (1997), the position of the front of the wedge at any elapsed time t is conveniently parameterised by the ratio $(x_s - x_w)/x_s$. With the above scalings, the complete data set can be plotted in composite form for each of the bottom slopes α, as illustrated in Fig 4. For ease of comparison, the data are plotted at two dimensionless reference times $t' = t(g')^{3/5}/Q^{1/5} = 500$ and 1500.

In order to diagnose the temporal changes to the density structure within the interfacial shear layer between

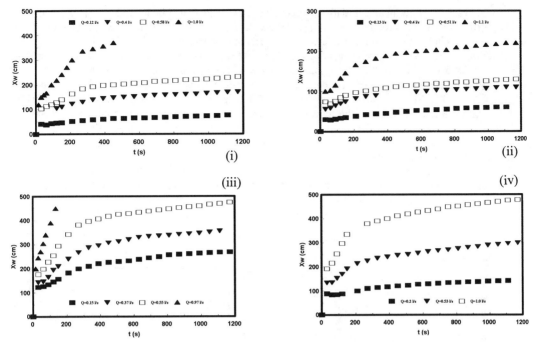

Fig 3: Plots of x_w versus t (see text) for slope angles of 4% (top row) and 2% (bottom), Q ($\ell.s^{-1}$) and $\Delta\rho$ (kg.m^{-3}) values [$\blacksquare$,$\blacktriangledown$,$\square$,$\blacktriangle$] = (i) [0.12,0.40,0.58,0.99];10, (ii)[0.13,0.40,0.51,1.06];30, (iii) [0.15,0.37,0.55,0.97];10 and (iv) [0.20,0.53,1.02];30.

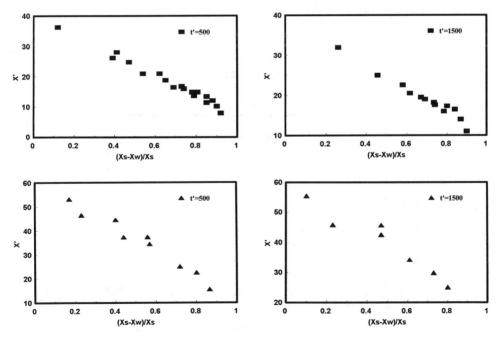

Fig 4 Plots of x' (= $x(g')^{1/5}/Q^{2/5}$) versus $(x_s - x_w)/x_s$ (see text) for α = ($\blacksquare$) 4%, ($\blacktriangle$) 2% and t' = $t(g')^{3/5}/Q^{1/5}$ = 500 (left) and 1500 (right)

439

the imposed freshwater flow and the initially-quiescent lower saline layer within the wedge, sequential vertical density profiles were taken throughout a given experiment at a fixed location in the channel. The toe of the bottom slope was chosen for the reference location. Typical microconductivity probe data showing temporal changes of the vertical density profile at the reference station are shown in Fig 5i. For each density profile $\rho(z)$, the derived buoyancy profile $N(z)$ ($= (g/\rho_0)(\partial\rho/\partial z)$) was also computed, in order to determine the location of the interface mid-level ($z = d$), defined as the elevation above the base ($z = 0$) of the channel for which $N = N_{max}$. Fig 5(ii) shows the time variation of d corresponding to the composite density profiles in Fig 5(i). At this reference location, the form of the decrease in d, namely an initial rather sharp decrease from the fully homogeneous profile at $t = 15$ s followed by a more gradual change at large times, is in accordance with the behaviour of the wedge front on the slope itself, as deduced from the dye flow visualisation data presented above.

Finally, comparisons are made between the erosion of the interface (as deduced by the reduction in d with elapsed time t at the reference location) for different bottom slope values α but otherwise fixed conditions. Such a typical comparison is shown on Fig 6, from where it can be seen that there is no significant difference in the behaviour of the d versus t data for the different slopes.

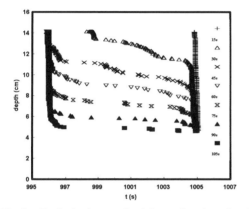 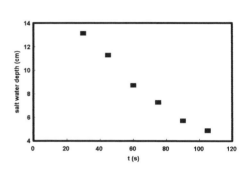

Fig 5: Typical microconductivity probe plots showing temporal variation of (i) density profiles $\rho(z)$ and (ii) interface elevation d (see text) for $Q = 1.02$ $\ell.s^{-1}$, $\Delta\rho = 5$ kg.m^{-3} and $\alpha = 4\%$. In (i) profile data are taken every 15 s from $t = 15$ s (+) to $t = 105$ s ($\blacksquare$).

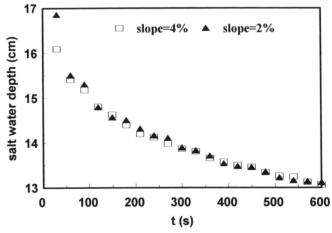

Fig 6 Comparison of plots of d versus t for different slopes $\alpha = (\square)$ 4% and ($\blacktriangle$) 2% and $Q = 0.225$ $\ell.s^{-1}$, $\Delta\rho = 50$ kg.m^{-3}

440

4. SUMMARY AND CONCLUSIONS

The results have shown that for steady purging flows, the downstream advection of the nose of a trapped salt wedge depends systematically upon the quantities Q, $\Delta\rho$ and α and that for any dimensionless reference time $t(g')^{3/5}/Q^{1/5}$ the dimensionless position $x(g')^{1/5}/Q^{2/5})$ of the nose position scales well with the parameter $(x_s - x_w)/x_s$. The dependence of the above relationship upon slope angle α is relatively weak, as is the dependence for the decrease in the depth of the interface as measured at the toe of the bottom slope.

The sequence of events described above is similar in many respects to those recorded by Debler and Armfield (1977) and Debler and Imberger (1996), who derived a scaling to describe successfully their data on the purging of saline cavities. However, the experiments described in these papers were conducted in channels that had quite high angles (8.5° - 90°) on the cavity walls; in the experiments reported in the present communication, the geometry employed could be regarded as an extreme case of a cavity, though slope angles were substantially less (1.4° - 2.8°). Thus, the slope effects recognised as being important by Debler and Imberger and Debler and Armfield (*ibid*) were less significant here.

5 ACKNOWLEDGEMENTS

The initial experiments reported in this study were carried out during a summer visit to the Dundee laboratory by one of the authors (MJC), with the support of the Exchange Programme of the Australian Academy of Science and Australian Academy of Technological Sciences and Engineering with The Royal Society of London. The authors are grateful for this support. JC participated in the study as part of her undergraduate project for the Department of Physics, Eindhoven University of Technology, Netherlands. Technical support from Mr D Ritchie is gratefully acknowledged.

REFERENCES

Armfield, S.W., and Debler, W. (1993). "Purging of density stabilized basins", *Int. J. Heat Mass Transfer*, 36(2), 519-530.

Debler, W., and Armfield, S.W. (1977). "The purging of saline water from rectangular and trapezoidal cavities by an overflow of turbulent sweet water", *J. Hydraul. Res.*, 35(1), 43-62.

Debler, W., and Imberger, J., (1996). "Flushing criteria in estuarine and laboratory experiments", *J. Hydr. Engng, ASCE*, 122(12), 728-734.

Grigg, N.J., and Ivey, G.N. (1997). "A laboratory investigation into shear-generated mixing in a salt wedge estuary", *Geophys. Astrophys. Fluid Dynamics*, 85(1,2), 65-96.

Largier, J.L., Slinger, J.H., and Taljaard, S. (1992). "The stratified hydrodynamics of the Palmiet - a prototype bar built estuary", in ***Dynamics and Exchanges in Estuaries and the Coastal Zone*** (ed D. Prandle), American Geophysical Union, Washington DC, USA

Environmental Hydraulics, Lee, Jayawardena & Wang (eds) © 1999 Balkema, Rotterdam, ISBN 90 5809 035 3

Simulation of steady three-dimensional boundary-attached density currents

G.M.Horsch
Department of Civil Engineering, University of Patras, Greece

G.C.Christodoulou
Department of Civil Engineering, National Technical University of Athens, Greece

M.Varvayanni
National Centre for Scientific Research 'Demokritos', Aghia Paraskevi, Greece

ABSTRACT: Density currents emanating from a slot and flowing downslope on an inclined bottom while spreading also laterally have been analyzed, using an integral model as well as three-dimensional simulations. Width integration of the depth-averaged equations, performed to formulate the integral model, creates the need for an additional equation to achieve closure. An equation already used by previous researchers was chosen, and the resulting integral model produced results in qualitative agreement with laboratory experiments. The quantitative agreement was, however, limited. The closure equation was evaluated by solving an inverse problem, which consisted of using the experimentally measured width for closure and back calculating parameters of the original closure equation. This method produced no strong support for the closure equation. Preliminary results of fully three-dimensional simulations using the parabolic equations governing fluid flow and mass transport produced detailed features of the currents which complement the experimental observations.

1. INTRODUCTION

Density currents in the aquatic environment may originate from differences in salinity, temperature, dissolved or suspended matter concentration. Heavier than ambient density currents may form as a result of the inflow of rivers into lakes and reservoirs, or of the disposal of certain industrial wastes - such as from geothermal plants, mining operations etc - and dredging material. Such currents tend to flow along the bottom and may be either confined laterally within a submarine canyon, or unconfined on an open slope. In the latter case their evolution is three-dimensional and their most interesting feature is the rapid lateral buoyant spreading, as observed experimentally.

Two-dimensional density currents on sloping boundaries have been extensively studied since the pioneering work of Ellison and Turner (1959). Three-dimensional density currents, i.e. those emanating from a source of finite length and proceeding downslope with no lateral restriction, have, however, received considerably less attention. An experimental study of three-dimensional density currents was presented by Fietz and Wood (1967); their study, performed in a small laboratory tank, was limited to the zone of establishment of the currents. Hauenstein and Dracos (1984) presented results of experiments of three-dimensional plunging currents in a large laboratory tank. They analyzed the experimental results using an integral model, based on depth and width integration, and, in order to produce a closed system of equations, they supplemented their model with an equation previously used by Stolzenbach and Harleman (1971) in a study of heated surface discharges. Their results are limited by the assumption that bottom friction can be neglected. Alavian (1986) presented a limited number of experiments in a relatively small laboratory tank, while Tsihrintzis (1988) performed a large number of experiments in the same tank and generalized the integral model of Hauenstein and Dracos (1984) to include bottom friction. The currents were classified as laminar, transitional and turbulent. Tsihrintzis and Alavian (1986) presented results of the integral model for different values of the governing parameters, without including, however, comparison with the experimental results. Finally, a large number of experiments have been performed by Christodoulou and Tzachou (1994) (see also Christodoulou, 1994) for a wide range of the independent parameters in a large laboratory tank.

In the review article by Alavian et al. (1992) it is concluded that three-dimensional density currents present rapid lateral spread, the rate of which reduces until the current reaches a normal state at which the Richardson number becomes nearly constant and the flow behaves as two dimensional. The behavior of the density current during the stage of lateral spreading and up to the point where equilibrium is reached determines its geometry as well as the dilution of the substances it carries, so that the study of this region is of considerable importance.

In the present work, two models of laterally unconfined density currents are presented: the first is based on depth- and width-integration of the flow equations, while the second uses a finite volume technique to extract solutions of the fully three-dimensional parabolic flow and mass transport equations.

2. INTEGRAL MODEL

After postulating the validity of the Boussinesq approximation, the depth-averaged equations for a steady density current on the inclined x, y plane, are:
Continuity equation:

$$\frac{\partial(u\,h)}{\partial x} + \frac{\partial(v\,h)}{\partial y} = E\,u \tag{1}$$

x, y momentum equations respectively:

$$\frac{\partial(u^2\,h)}{\partial x} + \frac{\partial(u\,v\,h)}{\partial y} = g'\,h\,\sin\theta - \frac{1}{2}\frac{\partial(g'\,h^2)}{\partial x} - c_d\,u^2 \tag{2}$$

$$\frac{\partial(u\,v\,h)}{\partial x} + \frac{\partial(v^2\,h)}{\partial y} = -\frac{1}{2}\frac{\partial(g'\,h^2)}{\partial y} - c_{dv}\,v^2 \tag{3}$$

conservation of buoyancy equation:

$$\frac{\partial(g'\,u\,h)}{\partial x} + \frac{\partial(g'\,v\,h)}{\partial y} = 0 \tag{4}$$

where h is the depth of the current, u, v the x, y components of the velocity respectively, E the entrainment coefficient, which is a function of the local Richardson number, $g' = g\Delta\rho\,/\,\rho$ the effective acceleration of gravity, and c_d, c_{dv} are bottom drag coefficients.

Eq. 4 is essentially an equation for g' (equivalently $\Delta\rho$), so that the problem consists of a system of four differential equations (1), (2), (3) and (4) which includes as unknowns u, v, h, g'.

In order to prepare for width integration, we assume self-similarity for u, v, h in cross-sections of the current:

$$u(x,y) = U(x)\,f_u(y/b), \quad v(x,y) = V(x)\,f_v(y/b), \quad h(x,y) = H(x)\,f_h(y/b) \tag{5}$$

where b is the width of the current. Then width integration from -b/2 to b/2 of equations (1)-(4) produces:

$$\frac{d}{dx}\left(c_8\,U\,H\,b\right) = c_9\,b\,U\,E \tag{6}$$

$$\frac{d}{dx}\left(c_1\,U^2\,H\,b\right) = c_2\,\overline{g}'\,\sin\theta\,H\,b - \frac{1}{2}\frac{d}{dx}\left(c_3\,\overline{g}'\,b\,H^2\right) - c_d\,c_4\,b\,U^2 \tag{7}$$

444

$$\frac{d}{dx}\left(c_5\, U\, V\, b\, H\right) = \frac{1}{2}\, c_6\, \overline{g}'\, H^2 - c_{dv}\, c_7\, b\, V^2 \qquad (8)$$

$$B = \overline{g}'\, U\, b\, H = \text{const.} = B_0 \qquad (9)$$

We note that the system consisting of the four ordinary differential equations (6)-(9) contains now five unknowns, namely U, V, H, b, $\overline{g}'$;thus, one more equation is needed for closure. According to previous researchers (Hauenstein and Dracos, 1984, Tsihrintzis, 1988) the following empirical equation can be used:

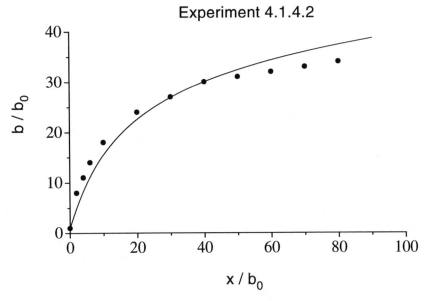

FIG 1. Comparison of width "b" predicted by the integral model (solid line) with experimental measurements (circles)

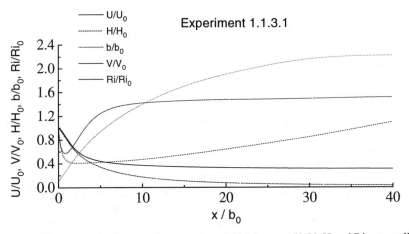

FIG. 2 Inverse problem: using experimental b(x) curve, U, V, H and Ri are predicted

$$\frac{db}{dx} = a_1 \frac{V}{U} + a \tag{10}$$

This equation was originally introduced by Stolzenbach and Harleman (1971) for the analysis of heated surface discharges. The constant coefficients c_1 to c_9 were evaluated by making reasonable assumptions about the shape of the functions f_h, f_u, f_v. The value of the coefficient "a" (Eq. 10) was set equal to 0, while that of "a_1" was originally set equal to 2, in accordance to Stolzenbach and Harleman (1972). Equations (6) through (10) were solved using a Runge-Kutta fourth order method. Simulation of selected experiments reported in Christodoulou (1994) was performed using the corresponding initial conditions U_0, b_0, h_0, $\Delta\rho_0$. Rather than providing V_0, however, the value of $(db/dx)_0$ was extracted from photos of experiments.

The success of the integral model in reproducing quantitatively such major features of the density currents as the experimentally observed downstream decline of the rate of growth of the width b, to the point where b becomes almost constant, was deemed as limited. From all the coefficients that enter the equations, solutions proved most sensitive to the value of "a_1". No optimum, universal value for "a_1" could be determined. For example, simulation of the width "b" of the current in experiment 4.1.4.2 (Christodoulou, 1994), shown in Fig. 1, required $a_1=0.2$. In that experiment the Reynolds (Re_0) and Richardson (Ri_0) numbers of the current at the entrance were 2002 and 0.783 respectively, and the bottom slope $\theta=5°$.

The behavior of the model is similar to that of the model of Tsichrintzis (1988), which is based on equivalent assumptions. In order to evaluate the role played by Eq. 10 (which was used to close the system of Eqs 6-9) in the discrepancy between observed and simulated features, the following inverse problem was devised. For a given experiment, the observed curve b(x) was reproduced (through a combination of photographically determined experimental values and interpolation by cubic splines), and this curve was then used to close the system of Eqs 6-9. An example of the outcome of this procedure is shown in Fig. 2, for experiment 1.1.3.1, for which $Re_0=1086$, $Ri_0=0.233$ and $\theta=15°$ (Christodoulou, 1994). Solution of this system produced then U, V, H, $\overline{g}'$, as well as the coefficient "a_1" not as a constant but as a function $a_1=a_1(x)$. Application of this method to selected experiments produced no convincing evidence that "a1" has a universal value, or that it is a constant or has a universal shape at all.

In conclusion, it can be said that the above described integral model, while reproducing qualitatively features of boundary attached density currents, leaves much to be desired in terms of quantitative prediction.

3. THREE-DIMENSIONAL SIMULATION

Further study of the modeling of density currents led to full three-dimensional simulation, which, while much more demanding in terms of computing, requires, in return, no a priori adoption of restrictive hypotheses. As a byproduct, simulations of this sort provide solid ground for the evaluation of hypotheses entering the formulation of any integral model.

More specifically, the on-going study aims at producing simulations of laminar density currents that behave parabolically throughout their development. The restriction to parabolic behavior is needed to make possible three-dimensional simulations performed at each cross-section and marching downstream - a computational task possible on workstations or high power PCs. In contrast, complete three-dimensional elliptic calculations, which require solution throughout the whole domain simultaneously, would tax even strong Supercomputers. Besides, according to experimental observations (Tsichrintzis, 1988, Christodoulou, 1994) the behavior of the density currents considered herein is indeed parabolic, with the possible exception of the entrance region, where plunging and/or internal jump render the phenomenon locally elliptic. Elliptic simulations for plunging regions have been reported by Farrel and Stefan (1985), in two dimensions, however. Simulations of currents which experience plunging and/or internal jump at the entrance using the parabolic equations could be accomplished by starting the simulation downstream of the locally elliptic region; the difficulty then would be shifted to estimating the initial conditions.

The three-dimensional parabolic system of equations consists of the equations of continuity, momentum and salt transport (salt is considered as agent which renders the entering jet negatively buoyant), where the streamwise diffusion terms have been ignored in the momentum and mass transport equations. This system is being solved using the finite volume method of Patankar and Spalding (1972), based on the improved

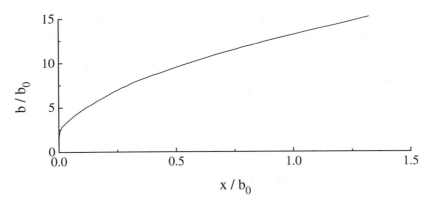

FIG. 3 The width b of the current as predicted by the three-dimensional simulation

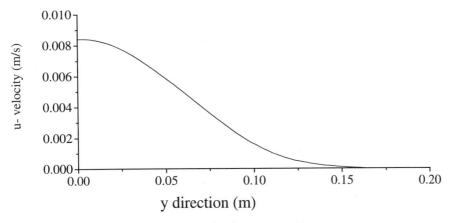

FIG.4 A transverse profile of the axial velocity of the three-dimensional simulation

SIMPLER algorithm of Patankar (1980). Iterations within a space step were, however, imposed. The code was tested and found to reproduce perfectly the analytical solution which represents spreading of the two-dimensional laminar free jet.

In order to avoid problems with initial plunging, internal jump, as well as boundary-attachment problems, such as those encountered by Farrel and Stefan (1986), the preliminary simulations reported herein are of low Reynolds number and high Richardson number. Since the current is considered to spread symmetrically around a vertical plane tracing the centerline of the sloping bottom, simulation within only half of the domain was considered. Concerning the domain of the solution, the original intent to mach the dimensions of the tank where the experiments described in Christodoulou (1994) were performed (5 m width x 0.6 m maximum depth), proved unduly wasteful. This is because, by restricting the domain while letting the computational boundaries open to entrainment, the computational time was reduced without harming the accuracy of the solution appreciably. The latter was established by numerical experiments with the two-dimensional free jet. The grid consisted of 100 (horizontal) x 50 (vertical) points, and was uniform in the horizontal direction, but in the vertical direction it was considerably finer near the bottom. The symmetry computational boundary was considered a slip boundary for the streamwise velocity (u) and the vertical velocity (w) and a zero velocity boundary for the horizontal velocity (v). At the side vertical boundary, u was set to zero and no gradient was set for v and w. At the top horizontal boundary u was set to zero, and no gradient was set for v, w, while the bottom was an impervious no slip boundary. The convergence criterion of the iteration algorithm was based on the global (i.e. concerning the entire domain) satisfaction of continuity as well as local (i.e. concerning a grid cell) satisfaction of continuity. At the same time the global satisfaction of buoyancy flux conservation (the equivalent of Eq. 9) was monitored. The marching step, Δx, was let to adjust in accordance with the number of iterations required to reach convergence; a typical step was between 5 and 10 mm.

Various features of a simulation with Re=10, and Ri=2940 are presented in Figs. 3, 4. In Fig. 3 the spreading of the current is depicted by defining the outline of the current as the locus of points for which $u \leq 0.1$ mm, while in Fig 4 a transverse profile of the streamwise component of the velocity (u) is shown at a distance from the bottom such that the value of u was the largest.

4. CONCLUSIONS

- The integral model presented reproduces qualitatively the main features of boundary attached density currents; the quantitative agreement with laboratory experimental data is, however, limited.
- The inverse problem, devised to test the equation used to close the integral model, gave no strong support for that equation.
- Preliminary results of three-dimensional parabolic simulations of laminar density currents complement the observations of laboratory experiments.
- More research is needed to produce a reliable integral model which would produce quantitatively acceptable solutions, without adjusting the shape factors of the model. Considerable support to this effort will be provided when further results of the three-dimensional parabolic simulations become available.

5. REFERENCES

Alavian, V. (1986). "Behavior of density currents on an incline", J. Hydraulic Eng., ASCE, 112(1), 1473-1496.

Alavian, V., Jirka, G.H., Denton, R.A., Johnson, M.C. and Stefan H.G. (1992). "Density currents entering lakes and reservoirs", J. Hydraulic Eng., ASCE, 118(11), 1464-1489.

Christodoulou, G.C. and Tzachou, F.F. (1994). "Experiments on 3-D turbulent density currents", Preprints 4th Intern. Symp. on Stratified Flows, Vol3, Grenoble, France.

Christodoulou, G.C (1994) *Three-dimensional density currents resulting from the disposal of heavier- than - ambient fluids in receiving waters*, Final report of research project, NTUA, Athens (in Greek).

Ellison, T.H. and Turner, J.S. (1959). "Turbulent entrainment in stratified fluids", J. Fluid Mech., 6, 423-448.

Farrel, J. and Stefan, H.G. (1985) "Numerical simulation of plunging reservoir inflow", Proceedings, International Symposium on Refined Flow Modeling and Turbulence Measurements, Iowa City, 16-18 Sept., p E 24.

Farrel, J. and Stefan, H.G. (1986). *Buoyancy induced plunging flow into reservoirs and coastal regions*, St. Anthony Falls Hydraulic Laboratory, Report 241, pp251.

Fietz, T.R. and Wood, I.R. (1967). "Three-dimensional density current", J. Hydr. Div., ASCE, 93(HY 6), 1-23.

Hauenstein, W., and Dracos, Th. (1984). "Investigation of plunging density currents generated by inflows in lakes", J. Hydraulic Res., 22(3), 157-179.

Patankar, S.V. and Spalding, D.B. (1972) "A calculation procedure for heat, mass and momentum transfer in three-dimensional parabolic flows", Int. J. Heat Mass Transfer, 15, 1787-1792.

Patankar, S.V. (1980) *Numerical heat transfer and fluid flow,* McGraw-Hill, pp 197.

Tsihrintzis, V.A. and Alavian, V. (1986). "Mathematical modeling of boundary attached gravity plumes", Proc. Intern. Symp. on Buoyant Flows, 1-5 Sept, Athens, Greece, Frame Publ. Co., 289-300.

Tsihrintzis, V.A. (1988), *Theoretical and experimental investigation of three-dimensional boundary attached gravity plumes*, Ph.D. Thesis, Univ. of Illinois at Urbana-Champagn, USA, pp 351.

Environmental Hydraulics, Lee, Jayawardena & Wang (eds) © 1999 Balkema, Rotterdam, ISBN 90 5809 035 3

Turbulent gravity current of lock release type: A numerical study

G.Q.Chen
Centre for Environmental Science, Peking University, Beijing, China

J.H.W.Lee
Department of Civil Engineering, University of Hong Kong, China

ABSTRACT: The time evolution of a turbulent gravity current of lock release type, formed by a finite volume of homogenous fluid released instantaneously into another fluid of slightly lower density, is studied numerically via the renormalization group (RNG) $k - \epsilon$ model for Reynolds-stress closure to characterize the flow with transitional and highly localized turbulence. Consistent with previous experimental observations, the numerical results show that the gravity current passes through two distinct phases, an initial slumping phase in which the current head advances steadily, and a second self-similar phase in which the front velocity decreases like the negative third power of the time after release. An overall entrainment ratio proportional to the distance from the release point is found and compares well with available experimental data for the slumping phase.

1. INTRODUCTION

Gravity currents, sometimes called density currents or buoyancy currents, are often generated by a density difference of less than a few percent and are primarily horizontal flows. In the laboratory, a common procedure for generating gravity currents is the sudden removal of a vertical partition separating two bodies of fresh water with temperature difference or of salt and fresh water, as illustrated in Fig.1. A typical case which has been studied frequently is the lock release gravity current, for which the initial depth h_0 of salt water is the same as H_0, that of fresh water, in a rectangular channel. If viscosity of the dense fluid is very small, the lock release gravity current is highly turbulent in its head, for a substantially long duration before eventually becoming laminar. The lock release gravity current within its turbulent regime is now referred to as turbulent gravity current of lock release type (TLRGC).

Many experimental observations based on shadowgraph were consistently successful in determining the general picture and the front location as a function of time after release for the TLRGC (Simpson 1997). Intensive mixing and turbulence are seen in the upwind side in the current head. It has been consistently confirmed that when the partition is removed the dense fluid quickly accelerates and then forms a gravity current passing through two distinctive phases, an initial slumping adjustment phase during which the front advances with constant velocity, and a second self similar phase during which the front velocity decreases with the negative third power of the time after release. But due to time-dependent localization of the TLRGC, there were only a very few experimental investigations on the mixing characteristics, an essential aspect commonly concerned. To tentatively quantify the bulk mixing in the slumping phase in the experiment by Simpson & Britter (1979) (Simpson 1997), the current head was maintained in a steady state by means of a special apparatus with a moving floor, then mean velocity and salinity measurements were made for some sections within the current head, using a hot-film probe combined with a constant temperature anemometer and using a conventional conductivity meter, respectively. Relied on the neutralization of the head by entrainment of acidic

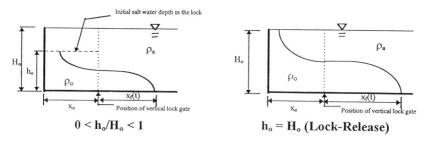

Fig. 1   Formation of gravity current and related definitions

Fig. 2   Contour plots for computed mean concentration of $0.05C_0$, $0.1C_0$, $0.2C_0$, $0.3C_0$, $0.4C_0$, $05C_0$, $0.6C_0$, $0.7C_0$ and $0.8C_0$ for case C7

ambient fluid and visualization with a pH indicator, Hallworth et al. (1993) attempted to quantify the entrainment of ambient fluid into the current head, rather than into the current as a whole. With modern image processing technique and sophisticated calibration, Hacker et al. (1994) found that substantial mixing occurs even in the early stage of the current, and Lam (1995) obtained detailed data for the internal salinity structure for the current as a whole in the slumping phase. At the same time, a lot of theoretical efforts were made to explain the characteristics of gravity currents, though the essential mixing effect has not been accounted for. Numerical efforts to qualitatively simulate some characteristics of gravity currents were made.

In the present study, we perform a CFD (Computational Fluid Dynamics) simulation to reproduce the TLRGC as a whole. The numerical solutions are then compared with experimental data to yield self-similarity relations, and related flow and mixing mechanism is discussed.

2. FORMULATION AND COMPUTATION

Consider a laboratory flume of length L with a vertical partition placed at a distance of x_0 from the left end wall. The channel is filled with dyed salt water with density ρ_0 on the left side of the partition, and fresh water with density $\rho_a = \rho_0 - \Delta\rho$ on the other side. Initially, the water depth is h_0 throughout the channel. Note that $\rho = \rho_a(1 + \beta C)$, where β = constant, and C stands for concentration (salinity) with an initial value of C_0. At $t = 0$, the partition is removed, and the density difference gives rise to a turbulent gravity current which is primarily advected along the channel floor while mixing with the ambient fluid. The flow and mixing of this current are investigated. For convenience, current front location x_f and time t after release are nondimensionlized, by the lock length x_0 and a time scale $t_c = x_0/(g'h_0)^{1/2}$ respectively, as $x_f^* = x_f/x_0$ and $t^* = t/t_c$.

By virtue of the fact of very small density difference, i.e., $\Delta\rho/\rho_a \ll 1$, we take the following Reynolds-averaged equations for incompressible flows:

$$\frac{\partial U_i}{\partial x_i} = 0 \tag{1}$$

$$\frac{\partial U_i}{\partial t} + U_j \frac{\partial U_i}{\partial x_j} = -\frac{1}{\rho}\frac{\partial p}{\partial x_i} + \frac{\partial \tau_{ij}}{\partial x_j} + \beta C g_i \tag{2}$$

where U_i = fluid velocity in x and y direction (denoted as u and v respectively), ρ= density, p = dynamic pressure, τ_{ij} = stress tensor, and g_i is the gravitation vector. τ_{ij} is equal to the sum of the viscous and Reynolds-stress, given by the eddy-viscosity model as $\tau_{ij} = \nu_{eff}(\partial U_j/\partial x_i + \partial U_i/\partial x_j) - 2/3 \cdot k\delta_{ij}$, with the effective viscosity ν_{eff} equal to the sum of the molecular viscosity ν and the turbulent viscosity ν_t. To reasonably account for the turbulent wake, separation and stagnation characteristic of the gravity current with horizontal mainstream, we adopt the renormalization group (RNG) $k - \epsilon$ model (Yakhot 1986, Fluent Inc. 1995), with modification for buoyancy effect, for turbulence closure:

$$\nu_{eff} = \nu \left[1 + \sqrt{\frac{c_\mu}{\nu}\frac{k}{\sqrt{\epsilon}}}\right]^2 \tag{3}$$

$$\frac{\partial k}{\partial t} + U_i\frac{\partial k}{\partial x_i} = \nu_t S^2 - \epsilon + \frac{\partial}{\partial x_i}\alpha_p\nu_t\frac{\partial k}{\partial x_i} + G_b \tag{4}$$

$$\frac{\partial \epsilon}{\partial t} + U_i\frac{\partial \epsilon}{\partial x_i} = c_{1\epsilon}\frac{\epsilon}{k}\nu_t S^2 - c_{2\epsilon}\frac{\epsilon^2}{k} - R + \frac{\partial}{\partial x_i}\alpha_p\nu_t\frac{\partial \epsilon}{\partial x_i} - R \tag{5}$$

where $G_b = -g_i\beta(\nu_t/Sc_t)\partial C/\partial x_i$ and $C_\mu = 0.0845$, k is the turbulence kinetic energy, ϵ is referred to as dissipation rate of k, and α_p = the inverse Prandtl number for turbulent transport as computed via the relation $|(\alpha_p - 1.3929)/0.3929|^{0.6321}|(\alpha_p + 2.3929)/3.3929|^{0.3679} = \nu/\nu_{eff}$, the rate-of-strain term $R = c_\mu\eta^3(1 - \eta/\eta_0)/(1 + \beta\eta^3) \cdot \epsilon^2/k$ with $\eta = Sk/\epsilon$, $\eta_o = 4.38$, and $S^2 = 2S_{ij}S_{ij}$ is the modulus of the rate of strain tensor expressed as $S_{ij} = (\partial U_i/\partial x_j + \partial U_j/\partial x_i)/2$. The RNG theory gives values of the constants $C_{1\epsilon} = 1.42$, $C_{2\epsilon}=1.68$. In addition to the flow equations, the study of the current necessitates the calculation of a passive scalar field from the conservation equation:

$$\frac{\partial C}{\partial t} + U_j\frac{\partial C}{\partial x_j} = \frac{\partial}{\partial x_j}(\frac{\nu_t}{Sc_t}\frac{\partial C}{\partial x_j}) \tag{6}$$

where Sc_t stands for the turbulent Schmidt number, equal to $1/\alpha_p$ in the RNG theory.

The free open surface is approximated by a free-slip boundary at $y = h_0$ with $\partial u/\partial y = 0$ and $v = 0$ imposed. At the solid boundaries, the non-slip condition is imposed, as $u = v = 0$. A zero value is set for k, ϵ and normal gradient of concentration over all the boundaries. Such a simple treatment of solid boundaries normally requires very fine near-wall grid cells, for an accurate numerical simulation

451

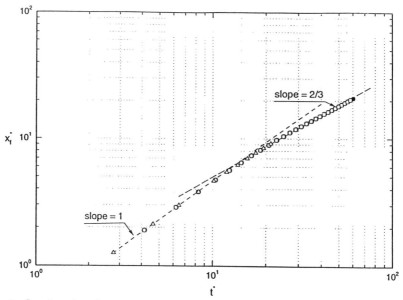

Fig. 3   Log-log plot of computed current front location as a function of time after release
△ for case C2,   □ for case C3, and   ● for case C7

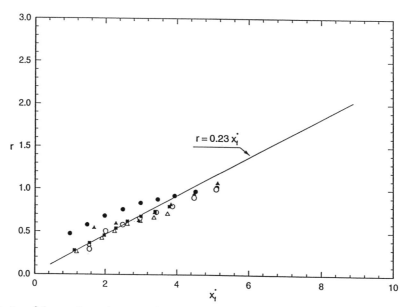

Fig. 4   Measured entrainment ratio as a function of front location, by Lam (1995)
■ for case C1,   △ for case C2,   ● for case C3,   O   for case C4
and   ▲ for case C5

based on RNG $k - \epsilon$ model. But in the present first approximation, we have no intention to resolve the near-wall details, for which laboratory measurement is still in short for comparison purpose. At $t = 0$, u and v are set to zero, and negligibly small initial values of (k, ϵ) are given, for numerical reasons, such that $\nu_t < 0.001\nu$, for the overall fluid domain; Values of C_0 corresponding to original salt water and of zero to fresh water are set for concentration over the zones of $x \leq x_0$ and $x \geq x_0$, respectively. Fluid properties are taken as those of fresh water. As will be seen from the numerical results below, the solution shortly after the release is rather insensitive to the initial value of k and ϵ assumed over a reasonable range.

The governing equations are solved numerically using the finite difference method (Patankar 1980) as embodied in the code FLUENT (Fluent Inc. 1995). A 200×20 (x-y) orthogonal grid, with sizes of x ranging from 0.004 to 0.008 and $\Delta y = 0.005$, is used for this problem of simple geometry. Convergence is declared when the normalized residuals are less than 5×10^{-4}. In all the results reported herein, the following parameters are adopted: L = 1.316 m, H_0 = 0.1 m, exactly corresponding to those in the laboratory experiment performed by Lam (1995). Three typical cases, with different lock aspect ratio $R \equiv h_0/x_0$ and density difference $\Delta\rho$, of C2 (R=0.75, $\Delta\rho = 15.5kg/m^3$), C3 (R=1.00, $\Delta\rho = 6.5kg/m^3$) and C7 (R=1.68, $\Delta\rho = 15.5kg/m^3$) among those experimentally observed by Lam (1995) have been numerically simulated in this work. Extensive tests have confirmed that the general features and characteristics reported are unrelated to the numerical procedure adopted.

3. RESULTS AND DISCUSSION

The concentration evolution of the current is typically shown in Fig.2 for case C7. It is shown that, after the partition is removed the dense fluid begin to collapse, followed by a counter flow of the less dense fluid beginning in opposite direction, and forms a gravity current moving away from the left end wall, with a current head developed. The initial acceleration from rest happens very rapidly, gives an essentially constant velocity of the current front when $t^* \geq 0.2$. The current head, once well formed after about $t^* \geq 5.0$, is observed to occupy about half of the water depth and has an inclination of roughly $\pi/3$ for the steepest slope of the front, consistent with the theoretical results in Benjamin (1968) (Simpson 1997). The head front has a foremost point, generally referred to as nose, slightly raised above the floor. The height of the nose is observed to be about 1/8 of the overall head depth, in agreement with the experimental result of Keulegan (1957) (Simpson 1997) for cases with Reynolds-number (based on current front velocity and head height) greater than 3000 as the present cases are. Behind the head, a tail of dense fluid developed. After the current has propagated about twelve lock lengths, when the front head is overtaken by a disturbance generated at the left end wall as 'reflection' of the counter flow of less dense fluid, the self-similar phase begins in which the front velocity decreases. An intensive mixing zone is seen in the upwind side of the head, as remarked by the great concentration gradient there. It is remarkable that in the intensive mixing zone the vertical distribution of concentration can be non-monotonic, that is, away from the floor, concentration decreases at first, but then can increases to some degree before finally decreasing to zero beyond the interface. This feature could be reasonably contributed to a combined effect of intensive turbulence, primarily corresponding to wave like billows observed experimentally, and the overtaking of the disturbance generated at the left end wall. All the computed features can be well compared with corresponding experimental observations (Lam 1995), though instantaneous as they are.

Fig.3 shows the computed current front position as a function of the time after release, in a logarithmic plot. It is clear that the front travels at a constant velocity initially and then at a decreasing velocity after traveling about 12 lock lengths for all cases. For the initial slumping phase, the constant velocity of the head is equal to $0.45(g'h)^{1/2}$. This value of 0.45 compares very well with typical experimental data of 0.43 given by Lam (1995), 0.45 by Rottman & Simpson (1983) (Simpson 1997) and 0.46 by Barr (1967) (Simpson 1997). Data points during the second phase are on the straight line with slope of 2/3, showing front location varying with $t^{*2/3}$, corresponding to velocity decreasing with $t^{*-1/3}$. No matter which contour among those defined by $0.05C_0$, $0.1C_0$ and $0.2C_0$ is the front defined by, the well known front propagation relationship is perfectly reproduced, considering the moderate scattering of experimental data.

To quantify the mixing between the current and the ambient, we define a dilution or entrainment ratio $r = V_a/V_b$ as V_a, the volume (per unit width for the two-dimensional case) of ambient fluid entrained into the current, over V_b, that of original dense fluid within the current. The value of r must be initially zero at the release point, then increases downstream. Because r is dimensionless, it must be independent of the initial reduced gravity with dimension LT^{-2}, as already indicated by Hallworth et al. (1993). Then, we satisfactorily fitted the data, for all the computed cases with different density difference and lock aspect ratio, with a linear relationship between r and the dimensionless front location x_f^* as $r = 0.23x_f^*$ for the overall evolution process with two distinct phases. This result is based on current shape defined by $0.05C_0$ contour. Redefining the shape by $0.1C_0$ and $0.15C_0$ contours still results in a linear relationship, though with somewhat different proportional coefficients. It is promising to compare this result with the data of Lam (1995), the unique quantitative measurement for the current as a whole. A good agreement is shown in Fig. 4 for the reliably measured range of $x_f^* \leq 5.13$. A complete validation is open for further measurement covering the whole process.

4. CONCLUDING REMARKS

The flow and mixing of turbulent gravity current of lock release type have been studied using the renormalization group (RNG) model. The salient features of the predicted flow and scalar mixing are well-supported by experimental observations. The current is shown to pass through two phases, an initial slumping adjustment phase in which the head front velocity is constant, and a second self-similar phase in which the front velocity decreases with the negative third power of the time after release. The transition between the two phases occurs when a disturbance generated at the end wall overtakes the current head at a front location of about 12 lock lengths. A dilution or entrainment ratio for the current is found proportional to the front location, as partially supported by available experimental data for the slumping phase.

5. REFERENCES

FLUENT INC. (1995). *FLUENT (version 4.3) User's Guide*, Lebanon, New Hampshire.
HACKER, J. et al. (1994). "Mixing in lock-release gravity currents", *Proc. 4th. Int. Sym. on Stratified Flows*, Grenoble, July 1994.
HALLWORTH, M.A. et al. (1993). "Entrainment in turbulent gravity currents", *Nature* 362, 829-831.
LAM, S.T. (1995). *Experimental investigation of lock-release gravity current*. M.Sc. dissertation, Dept. of Civil & Structural Engineering, University of Hong Kong.
PATANKAR, S.V. (1980). *Numerical Heat Transfer and Fluid Flow*, Hemisphere Publishing Co., New York.
SIMPSON, J.E. (1997). *Gravity currents in the environment and the laboratory*, 2nd edition. Cambridge University Press.
YAKHOT et al. (1986). "Renormalization group analysis of turbulence, I. Basic theory", *J. of Sci. Comput.* 1,1-51.

Environmental Hydraulics, Lee, Jayawardena & Wang (eds)© 1999 Balkema, Rotterdam, ISBN 90 5809 035 3

A computational fluid dynamics (CFD) investigation of stratified flows

B. Donnelly, G. A. Hamill, D. J. Robinson, P. Mackinnon & H. T. Johnston
Department of Civil Engineering, The Queen's University of Belfast, UK

ABSTRACT: A computational fluid dynamics (CFD) code, FLUENT v4.4, using Volume of Fluid (VoF), SPECIES and k-ε turbulence models was used to investigate the development of an entrapped saline wedge. Initially a stratified flow situation exists of fresh water overlying a denser saline layer. The development of the wedge and an interfacial mixing layer between the stratified fresh and saline water was monitored and compared with experimental results.

1. INTRODUCTION

Situations exist, both naturally occurring and man made whereby fluids of differing densities are brought into contact. Most commonly, river estuaries where fresh flowing water meets a denser salt water ocean or the discharge from industries such as chemical works, power generation or waste water treatment plants are examples. The end product of this contact is often not an homogenous fluid of well mixed fresh / saline solution but the stratification and inefficient mixing of the two.

In the case of a river estuary the fresh water flow tends to override the denser salt water, with oxygen uptake at the river surface due to natural wind / wave action prevented from penetrating to the lower levels due to stratification. As a result an oxygen starved saline layer develops. This layer of low dissolved oxygen content then leads to damage of the marine and aquatic environment, Millington (1997).

The use of weirs and tidal barrages to control water levels within such estuaries for recreational and industrial usage has also caused the isolation or entrapment of the saline layer from its source, thus hampering the natural mixing and removal process by underlying currents and tidal action.

2. REVIEW

The investigation of arrested saline wedges (Fig.1) has dominated this field of mixing and stratification . However the problems of entrapped wedges (Fig.2) are becoming increasingly common and an understanding of their behaviour is vital as more and more cities return to using their rivers, old dockland areas etc, for industrial and recreational usage. Such regeneration schemes often involve constructing weirs and tidal barrages, examples being the Thames Barrage (London) and Belfast's Lagan Weir, creating the conditions for entrapped wedges to occur.

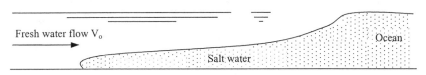

Fresh water flow V_o

Salt water

Ocean

FIG. 1. Arrested Saline Wedge

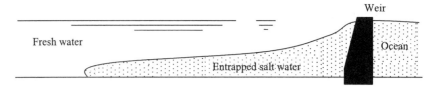

FIG. 2. Entrapped Saline Wedge

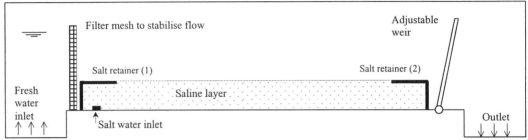

FIG. 3. Experimental Test Facility

Table 1 - Test parameters

Fresh water velocities	9 – 96 mm/s
Fresh water depths	200 – 400 mm
Salt water depth	250 mm

The use of computational fluid dynamics (CFD) packages and major advances in computer facilities is enabling engineers to construct computational models of such flow situations without the need for the construction of large scale prototype or expensive test rigs.

2.1 Analytical Studies

Previous investigators of such density stratified flows have shown that due to frictional forces being developed along the density interface, the lower saline mass will develop into a wedge shape pointing upstream, Keuleugan (1966). If the velocity of the fresh water exceeds a certain critical value internal waves begin to break intermittently and eject plumes of water across both sides of the sharp interface. As mixing of the solutions begins a third or interfacial layer is created by the mixing process, with the properties of this layer dependent upon such parameters as the relative density and velocity between layers, Grubert (1980).

Physical model tests performed by Walker (1996) using a 20m long by 0.75m wide by 0.75m deep rectangular channel (Fig.3) on an isolated saline impoundment 17 m long, with fresh water overflow at varying velocities and depths, suggested two possible mechanisms of erosion for the lower saline mass. Both mechanisms concerned the mixing processes within the interfacial layer and in particular how thickness and density of the mixing layer varied as fresh water depth, velocity and relative density between layers changed. The test facility used by Walker is modelled by the current CFD study with physical testing also occurring for comparison and calibration with CFD work.

2.2 Equipment & Procedure

Initially the tank is filled to a depth of 250mm with a salt water solution of density 1025 kg/m^3. Fresh water (density 1000kg/m^3) is then allowed to slowly enter the tank and lies above the salt water layer. As Laser

456

Doppler Anemometry is used in the recording of velocity, seeding of the different solutions occurs to aid the laser in recognition and to enable visual monitoring of the process. A salinity probe is then programmed to scan in 5 mm intervals over a distance of 120 mm around the expected interfacial layer. The fresh water is then forced to flow down the tank and over the weir. A range of fresh water velocities and depths are used, allowing flows within both the laminar and turbulent regimes.

2.3 Numerical Studies

Computational fluid dynamics (CFD) is a combination between the fields of fluid mechanics and computer simulation. This rapidly growing area of engineering has developed substantially in the last decade with the availability and development of advanced computing facilities. Numerical simulation of the flow problem is based on solving the basic flow equations of conservation of mass, momentum and energy. This requires the volume being investigated to be divided up into a number of small cells forming a grid or mesh. Thus it is the representation of any continuous field in a computer by a set of points (Fig. 4).

These points have values for the physical characteristics of the simulated flow. The solution to the problem being found by iteration over the grid with values of the variables such as velocity and pressure being guessed initially. The values found after each iteration are used as the next starting value until the difference in successive values is negligible.

The finer the grid the more accurate the approximation to the actual flow field. However costs increase with grid density, with an increase in computing time required to produce a solution to the problem. Thus a balance between speed and accuracy must be achieved. The current study uses a commercially available CFD package FLUENT v4.4. This code uses the well validated finite volume method for solution of the unknown variables.

2.4 Modelling Parameters

The importance of mesh density on accuracy and running time has already been discussed. The current study makes use of Body Fitted Co-ordinates (BFC) to 'mould' the mesh to the geometry of the tank whilst also concentrating on increasing grid density in areas of interest and change of flow variables. In particular

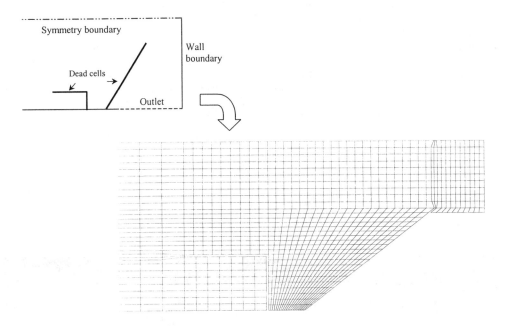

FIG. 4. Body Fitted Grid Around Salt Retainer and Weir Gate

457

the region behind salt baffle two and at the initial interface between fluids (Fig. 4). The problem also lends itself to 2D simulation and not the more computationally intensive 3D model.

Within the model set-up two approaches are used to model the three different phases, (fresh water, salt water and air). To enable modelling of the free surface boundary between water and air a VoF or Volume of Fluid model is employed with the primary phase determined as water, and a symmetry boundary for the top boundary giving the condition of a frictionless plane. A thorough investigation and advantages of the VoF model over similar models is given by Hirt et al (1979).

As the salt and fresh water are two chemically different solutions. A SPECIES transport model (non-reacting) is used to monitor the diffusion and entrainment of fresh / salt water across the interface. It is here that one of the benefits of using a CFD approach to the problem occurs as the monitoring of velocity and density can occur at any points of interest compared to the limited approach with the physical apparatus.

As some of the flows involved are within the turbulent region a standard k–ε turbulence model is used to assist the Navier-Stokes equation in calculating characteristics of the flow.

3. RESULTS AND DISCUSSION

The test cases within this study investigate the practicality of VoF/SPECIES modelling along with an investigation into the density profiles within the model over time. The flow patterns around the salt retainers, weir gate and within the saline layer are also investigated.

Figure 5 shows a density contour plot produced from the CFD model. The developed wedge shape of the lower saline layer is clearly visible along with an interfacial zone between the stratified fresh and saline water. The interfacial layer is thickest at the start of the wedge and then follows the profile of the lower saline layer. This initial mixing is due to the incoming fresh water falling over salt retainer one, which is acting as a backward facing step. This can lead to the creation of a forward facing wedge, trapped beneath the salt baffle, as is also shown in Fig. 5.

As velocity of the fresh water is increased, a turbulent shear layer is developed along the fresh / saline interface and visual waves are formed between layers. This leads to an increase in the rate of removal of salt water from the system. It is clear therefore that in times of low flow within estuaries, that a stable wedge shall develop with hampered erosion due to the lack of turbulent flow. This is indicative of summer months when rain fall is less and river levels generally lower. Such situations shall call for mechanical means of destratification and removal of the saline layer should dissolved oxygen content fall below the required values for a sustainable river environment.

Figure 6(a) indicates a scanned density profile, (obtained from physical experimentation) along a vertical plane at the centre of the saline impoundment. On this scale '0' indicates fresh water (density 1000 kg/m^3) with '1' indicating fully saline conditions. The presence of the interfacial layer is clearly present and comparison with Fig. 6(b), produced by monitoring density values within the CFD package shows a similar trend. Although both profiles mirror the same trend, from experience the computer model has tended to predict a faster rate of diffusion / erosion than physical experiments.

Previous investigators have also commented upon a negative flow (in opposing direction to the fresh water flow) within the saline layer. Actual investigation of such velocities and their presence has been limited due to physical constraints within the experimental test rig. The velocity vectors shown in Figure 7 however indicate the presence of a recirculation cell within the lower saline layer and behind salt baffle two. The creation of this circulation is linked to the shear stress developed along the wedge interface and is dependent upon fresh water velocity and density difference between layers.

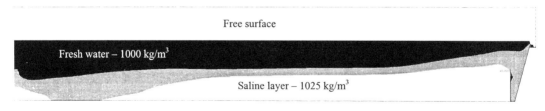

Free surface

Fresh water – 1000 kg/m^3

Saline layer – 1025 kg/m^3

FIG. 5. Density Contour Plot From Within CFD Package

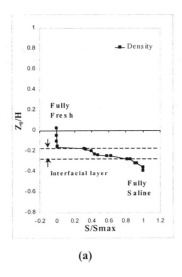

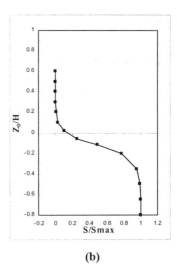

(a)   (b)

FIG. 6. Density Profiles (a) Physical Experiments   (b) CFD Test

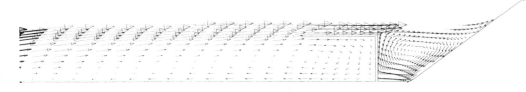

FIG. 7. Velocity Vectors Within Saline Wedge

The circulation zone behind the saline impoundment is formed when the heavier saline water is initially lifted from the impoundment by the fresh water overflow. On encountering the region of fresh water behind the baffle, the heavier solution falls, displacing the fresh water from within this region. The mixed layer flowing over this entrapped region then maintains the circulation.

The undercurrents within the wedge itself extend backwards in the order of a few metres and their affect on saline wedge removal requires further investigation. However results have shown that the position of reattachment to the main flow and the strength of such currents varies with time, overflow velocity and density between layers.

4. CONCLUSION

Although development of the computer model is ongoing, preliminary results show trends corresponding strongly to experimental work and the feasibility of modelling stratified flows with a CFD approach. The development of a distinct interfacial layer generated by mixing of the fresh and saline water has also been shown along with the corresponding wedge profile of the saline layer due to interfacial friction. The presence of a circulation cell within the physical apparatus along with opposing flow within the entrapped wedge has also been highlighted.

5. REFERENCES

Grubert, J.P. (1980). "Experiments on arrested saline wedge", *J. Hyd. Div.,* ASCE, 106, 15484, 945-960

Hirt, C.W., and Nichols, B.D. (1981). "Volume of Fluid (VoF) Method for the Dynamics of Free Boundaries", *Journal of Computational Physics*, 39, 201-225.

Keulegan G.H., (1966). "The mechanism of an arrested saline wedge", *Estuary and coastline hydrodynamics,* (11), Ippen, A.T., ed., McGraw Hill Book Co., New York., 546-574.

Millington, G.S. (1997). "Development of the River Lagan in Belfast"*, Proc Inst. Civ. Eng.*, 120(4), 165-176.

Walker, S.A., Hamill, G.A., and Johnston, H.T. (1996). "Development and erosion of a mixing layer in stratified flows", 5th IMA Conference on Mixing and Dispersion in Stably Stratified Flows, September, Dundee, Scotland.

Environmental Hydraulics, Lee, Jayawardena & Wang (eds) © 1999 Balkema, Rotterdam, ISBN 90 5809 035 3

Application of computational fluid dynamics (CFD) to model entrainment across a density stratified interface

Y. Oh
Korean Ocean Research and Development Institute, Korea

R. Burrows & K. H. M. Ali
University of Liverpool, UK

ABSTRACT: A flexible floating bottomless containment has been considered as a potential means of keeping a freshwater separated from seawater in a river mouth or estuary. Viability assessment for such a scheme calls for study of entrainment across the density-stratified interface between the freshwater and the saltwater flowing underneath. Earlier research has considered this phenomenon by laboratory experiment. In the work reported here the commercial computational fluid dynamics modelling package FLUENT is used to simulate the entrainment process and the results are compared with earlier experimental findings. For this multi-phase flow process, the entrainment rate so synthesised numerically has been found to be capable of yielding good agreement with the test data only following a process of model parameter calibration and empirical adjustment. For this situation and at this point, FLUENT is, therefore, found to be capable of qualitative synthesis of the entrainment between the two fluids and discussion is made of problematic aspects of the current implementation of the numerical procedure.

1. THEORETICAL BACKGROUND

Turbulent entrainment by shear-flow across a stratified interface has been studied by many researchers following the pioneering work by Richardson (1920) and Keulegan (1949). Where the high interfacial shear leads to internal hydraulic jump formation, however, the results reported by different investigators, including Ellison and Turner (1959, 1960), Chu and Baddour (1984), Burrows, Ali and Crapper (1994) and Rajaratnam and Subramanyan (1987), have not always been consistent. This has arisen partly because of the variety of flow conditions examined and the specific experimental set-ups used, but also as a consequence of the complexity of the phenomenon.

The stability of the interface is commonly measured in terms of a Richardson number, R_i, with a higher Richardson number indicating a more stable interface, and consequently less mixing. The Richardson number is defined as $R_i = g\Delta\rho h/\rho_0 U^2$ where g is the gravitational acceleration, ρ_0 is the density of the flowing fluid. $\Delta\rho$ is the density difference of the two fluids, h is the depth of the moving layer and U is its mean velocity. Four basic entrainment mechanisms can generally be distinguished over different ranges of the Richardson number, as described by Christodoulou (1986). R_i is the reciprocal of the square of the densimetric Froude number so the flow condition for Richardson number less than 1 is supercritical and when greater than 1 is subcritical. At very low R_i, mixing takes place as in homogenous fluid and influence of stratification is not significant. At slightly higher R_i where flow conditions are supercritical "vortex" entrainment occurs, the intense mixing then depending on internal turbulence generated within the sheared interface. As R_i increases further, subcritical flow conditions prevail, and the "cusp" entrainment mechanism becomes dominant, characterised by formation of sporadic undulation at the interface. At very high R_i, turbulence cannot be sustained within the interface, and molecular diffusion is the dominant process causing buoyancy flux.

2. NUMERICAL MODELLING BY FLUENT

The FLUENT, computational fluid dynamics program suite has been designed to synthesise incompressible and turbulent flow. FLUENT relates Reynolds stresses to mean flow quantities via one of three turbulence models, a κ-ε model, a ReNormalization Group (RNG) κ-ε model or a Reynolds Stress Model (RSM). These three turbulence models have been designed to be widely applicable, however, there are practical considerations that dictate the appropriateness of each to the problem under consideration. The RSM involves a seven-equation model of turbulence. This can be compared to the two-equation model framework utilised in the κ-ε model and RNG models. The RSM relaxes the assumption made by the two-equation models that turbulence is isotropic (or quasi-isotropic in the case of RNG) but also imposes a greater computational burden on the solver. The standard κ-ε model has been the workhorse of engineering turbulence models for more than two decades and has the desirable properties of robustness, economy and a fairly wide domain of applicability. The RNG model, as implemented in FLUENT, has many of the characteristics of a good engineering turbulence model, including universality, economy, and robustness and has been adopted here.

FLUENT can model not only single phase fluid motions but also multiple phases, and is able, therefore, to represent the interaction of two or more phases. The multiple-phase model has two alternatives, each applicable to mixing problems with differing characteristics of the density interface. The Volume of Fluid (VOF) model is designed for two or more immiscible fluids, where the position of the interface between the fluids is of interest. The Eulerian model allows for the modelling of multiple separate, yet interacting, phases. Whereas the Eulerian model makes use of multiple momentum equations to describe the individual fluids, the VOF model does not. Suitable applications of the VOF model include stratified immiscible fluid flows, free surface flows, sloshing and dam break. On the other hand, the Eulerian multiple-phase model is best suited for general applications involving liquid-liquid and liquid-gas interfaces. The interaction between the freshwater and seawater is well suited to the Eulerian multiple-phase model. To change from a single-phase model, where a single set of conservation equations for momentum and continuity is solved, to a multiple-phase model, additional sets of conservation equations must be introduced. In the process of introducing additional sets of conservation equations, the original set must also be modified. The modifications involve the introduction of the volume fraction ε_1, ε_2, ... ε_n for the multiple phases, as well as a mechanism for the exchange of momentum between the phases. Volume fractions represent the space occupied by each phase. The volume of phase q, V_q, is defined by

$$V_q = \int_V \varepsilon_q dV \qquad \text{where} \qquad \sum_{q=1}^n \varepsilon_q = 1 \qquad\qquad (1)$$

The basic equations solved by FLUENT for fluid-fluid flows are presented in general form as follows,

$$\frac{\partial}{\partial t}\varepsilon_q \rho_q + \nabla.\left(\varepsilon_q \rho_q u_q\right) = \dot{m}_q \qquad\qquad \sum_{q=1}^n \dot{m}_q = 0 \qquad\qquad (2)$$

for conservation of mass, where ρ_q is the physical density of phase q, ∇ is the gradient function, u_q is the velocity of phase q and $\dot{m}_q$ is the rate of creation or destruction of phase q, and

$$\frac{\partial}{\partial t}\left(\varepsilon_q \rho_q u_q\right) + \left(\nabla.\left(\varepsilon_q \rho_q u_q\right)u_q\right) = \nabla.\overline{S}_q + \varepsilon_q \rho_q F_q + f_q \quad ; \quad \sum_{q=1}^n f_q = 0 \qquad\qquad (3)$$

for conservation of momentum, where $\overline{S}_q$ is the qth phase stress tensor, F_q is an external body force, and f_q is an interaction force. The interaction force f_q depends on the drag between the phases, the added mass force and some other effects. FLUENT uses a simple interaction term of the form

$$f_q = \sum_{r=1}^{n} K_{qr}\left(u_r - u_q\right) \quad ; \quad K_{qr} = \frac{3}{4} C_D \varepsilon_r \frac{\rho_q \left|u_q - u_r\right|}{d_r} \tag{4}$$

where K_{qr} is the interphase momentum exchange coefficient, u_r is the velocity of phase r, C_D is the Drag coefficient described as a function of Reynolds number and d_r is the particle diameter of the secondary phase r. For fluid-fluid flow, each secondary phase is assumed to form droplets or bubbles as it passes through the primary fluid. For example, in flows where there are unequal amounts of two fluids, the predominant fluid should be modelled as the primary fluid, since the sparser fluid is more likely to form droplets.

3. ENTRAINMENT OF SALTWATER IN A FRESHWATER CONTAINMENT

Storage of freshwater inside a bottomless tank in the ocean has been considered by Foo et al. (1995). They completed experimental tests in a flume for different lengths of the tank and under different current speeds and found good agreement with the results of Christodoulou (1986) and Atkinson (1988). For the severe prototype conditions considered in their final discussion they raised question over the practicability of such systems and follow up field studies on an open cylindrical tank by Shuy et al (1998) further this doubt. Whilst, at this point, the situation may not be conclusive for other potential applications, it remains of interest to further the investigation by numerical modelling.

The present study is concerned with the entrainment characteristics across a density-stratified interface inside a bottomless tank (Figure 1). In a practical application the left hand side can be assumed as the downstream of a river influenced by the tide and the saltwater flow is allowed underneath the tank from the tidal current running upstream during the flood and downstream during the ebb. To check whether FLUENT can simulate this phenomenon, the freshwater fills the entire domain at the beginning and the saltwater is introduced from

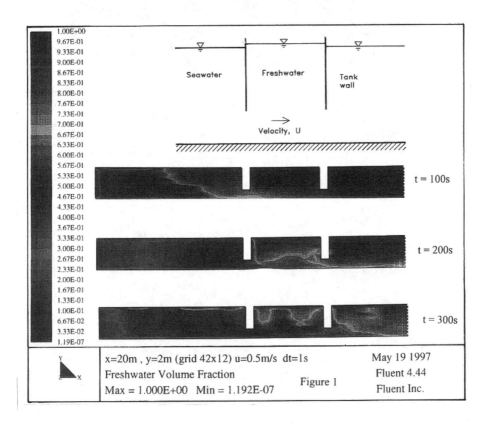

Figure 1

x=20m , y=2m (grid 42x12) u=0.5m/s dt=1s
Freshwater Volume Fraction
Max = 1.000E+00 Min = 1.192E-07

May 19 1997
Fluent 4.44
Fluent Inc.

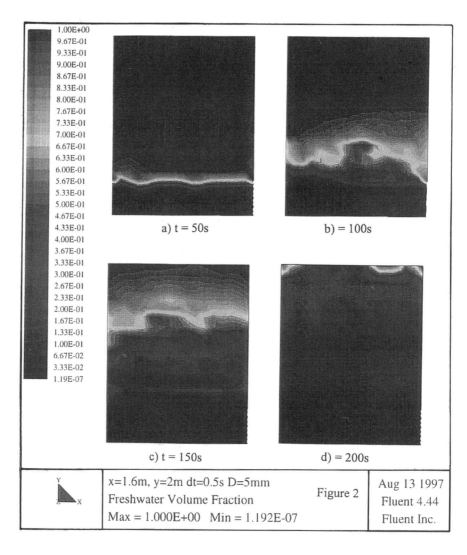

a) t = 50s   b) = 100s

c) t = 150s   d) = 200s

⊿ Y Z X	x=1.6m, y=2m dt=0.5s D=5mm Freshwater Volume Fraction Max = 1.000E+00 Min = 1.192E-07	Figure 2	Aug 13 1997 Fluent 4.44 Fluent Inc.

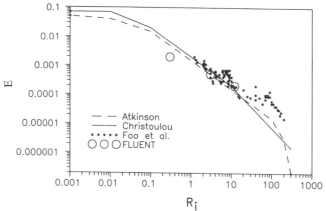

Figure 3

the left hand boundary during the simulation period. Figure 1 shows the results of entrainment in a 20m long, 2m high storage domain with an upstream flow velocity of 0.5m/s. It illustrates the evolution of the freshwater volume fraction as the external freshwater (of density $1000Kg/m^3$) is displaced from the left by the incoming saltwater (of density $1030Kg/m^3$). This result shows a first attempt at use of FLUENT to simulate the mixing phenomenon. However, it was not considered to be realistic due to the apparently high speed of the mixing process. From equation (4) the mixing rate is closely related to the interphase momentum exchange coefficient. This is a function of the particle diameter of the secondary phase, so it is possible to adjust mixing rate by changing the particle diameter in an attempt to fit to experimental observations.

In the next phase of the investigation the entrainment rate was modified in an attempt to match with the experimental observations by Foo et al (1995) by changing the particle diameter of the saltwater. In this modelling the computational domain was confined in horizontal extent to the tank only for computational efficiency because, otherwise, in the early part of the simulation the underflow is a mixed fluid as opposed to the pure saltwater in the real situation. The flow domain selected was of a bottomless tank 1.6m long and penetrating 1.6m down through a 2.0m water depth whilst subjected to a saltwater ($1030 Kg/m^3$) underflow of chosen velocity, U. The entrainment process can be characterised through the entrainment velocity, $W_e = (z/t)$, and entrainment coefficient, $E = W_e/U$, where z is the vertical rise of the interface in the elapsed time, t, and U is the saltwater underflow velocity. For the numerical synthesis, the entrainment rate was calculated for a range of runs selected by considering different Richardson numbers, $R_i = 0.3$, 3 and 12 [defined in terms of the driving flow velocity (U) and under baffle gap (h)], to represent supercritical flow, transitional flow and subcritical flow, respectively. This was achieved by variation in the undershear velocity in successive simulations. After many early trial runs, the saltwater particle diameter was finally chosen as 0.5cm. This value intuitively seems too large, however, smaller values were found to make the mixing too fast. Figure 2 shows snapshots of the entrainment process through a simulation for $R_i = 3$. The entrainment rate here is much faster than that arising from the experimental observations. Consequently, the diameter of saltwater 'particles' was increased above 0.5cm even though this becomes more unrealistic physically.

Another critical problem experienced in application of FLUENT comes from the occurrence of an inconsistent entrainment rate if the particle diameter is increased above a certain size. For example, when the particle diameter was taken as 0.9cm the entrainment was found to progress very slowly initially then abruptly change to very rapid mixing at a certain time. As a result it was eventually concluded that the diameter of 0.5cm is close to the upper limit necessary to keep the entrainment rate consistent with time and so represent, at least qualitatively, the observed physical behaviour. Test runs for three different flow conditions (hence R_i) were, therefore, completed even though the mixing rate simulated was known to be too rapid. Methods were then sought to match the numerical model results (based on tracking interface rise over the central 100 seconds of the 200 second simulation) with the experimental results simply by factoring the elapsed time. The entrainment coefficients were found to be well matched, against Foo's (1995) experimental findings only when numerical time scales were factored by 100, as shown in Figure 3. This adjustment seems optimal for the sub-critical conditions ($R_i = 3$ and 12) but for supercritical conditions ($R_i = 0.3$) an adjustment factor nearer to 30 would have most closely fitted the experimental data. Scale effects are known, however, to have substantial influence on entrainment, as amply illustrated by Shuy et al (1998), who quote field estimates of entrainment velocity more than 4 orders of magnitude larger than their comparable laboratory results. Whilst the numerical results (unadjusted) from FLUENT lie midway between these widely differing estimates, it is clear that further investigations are required to properly resolve the issue.

4. DISCUSSION AND CONCLUSIONS

As a preliminary to serious consideration of the viability of freshwater storage by open bottomed containment systems in rivers affected by seawater intrusion, it is indispensable to study the interfacial entrainment phenomenon resulting from the turbulent shear flows. FLUENT which is designed for incompressible and turbulent flow has been used to assess its ability to synthesise the entrainment. In the Eulerian model applied, the mixing rate is greatly dependent on the particle size selected for the secondary fluid phase. The bigger the size the slower the mixing rate, however, there seems to be a limit to the increase of size. If this is exceeded the mixing process synthesised does not evolve with time consistently and there seems to be no choice,

465

therefore, but to impose a limiting size to simulate the mixing phenomenon. In this situation, to match the mixing rate with that of available experimental data addition factoring (reduction) of the numerical time frame has been found necessary here.

In the model as implemented herein a single horizontal frictionless 'lid' has been imposed to define the surface boundary. With differing density distributions either side of the containment's walls, the hydrostatic conditions do not precisely match reality, as the water level in the containment will vary as the salinity of the water column changes. These hydrostatic imbalances will translate into velocity field errors, which might be expected to influence the entrainment to some degree. To overcome this problem a third fluid (air) would need to be added to the simulation so as to force adherence to the surface boundary condition. With the computational power available (networked SUN Sparc-station IPX operating under UNIX) this far more complex problem was judged to be impracticable, given the severe limitations experienced in completing the simpler studies described herein.

The main findings of the present study can be summarised as follows: i) The entrainment generated by the numerical model matches well with earlier experimental findings only when its timescale is factored by 100; ii) The particle size of the secondary phase is critical to the entrainment rate; iii) Beyond a certain particle size limit the entrainment process evolves sporadically and contrary to expectation (this appears to be between 0.5 and 0.9cm from the trials completed here and the particle size at the lower limit is suggested for applications); iv) The switch from the single-phase model to multiple-phase modelling greatly increases computation time and the Eulerian model, which allows interaction across an interface, needs much more time than the VOF model; v) For reasons of practicability, careful consideration must be given to the determination of the grid size and computational domain size when using the Eulerian model; vi) At the present time the computational requirements are a severe restriction to the progressing of studies of this nature; vii) Introduction of a third 'phase' (air) to satisfy the surface boundary according to instantaneous hydrostatic conditions was judged to be impracticable with the computational resources available for the study.

5. ACKNOWLEDGEMENT

The authors wish to express thanks to the Korea Science and Engineering Foundation and Korean Ocean Research & Development Institute which supported this study by their release of Dr Oh for a 12 month secondment as a Research Fellow in the Department of Civil Engineering at the University of Liverpool.

6. REFERENCES

Atkinson, J.F., Note on interfacial mixing in stratified flows, J.Hyd. Res., Vol. 206., No. 1, 1988, pp227-31.

Burrows, R., Ali, K.H.M. and Crapper, M., Entrainment from a buoyant surface layer created by an under baffle wall-jet, Recent Res. Adv. in the Fluid Mech. of Turbulent Jets and Plumes, 1994, pp489-501.

Christodoulou, G.C., Turbulent mixing in stratified flows, J. Hyd. Res. Vol. 24, No. 2, 1986, pp77-02.

Chu, V.H. and Baddour, R.E., Turbulent gravity stratified shear flows, J.F.M. Vol.138, 1984, pp353-78.

Ellison, T.H. and Turner, J.S., Turbulent entrainment in stratified flows, J.F.M. Vol 6, Parts 3, 1959, pp423-48.

Ellison, T.H. and Turner, J.S., Mixing of dense fluid in a turbulent pipe flow, Parts 1 and 2, J.F.M. Vol. 8, 1960, pp514-45.

Foo, M.H., Shuy, E.B. and Chen, C.N. 'Entrainment across a density interface inside a flume compartment'. J Hydraulics Research, Vol 33, 1995, No 2, 181-196.

Keulegan, G.H., Interfacial instability and mixing in stratified flows, RP2040, US Nat. Bureau of Standards, Vol. 43, 1969, pp487-500.

Richardson, L.F., The supply of energy from and to atmosphere, Proc. R. Soc., London, 1920, pp354-73.

Rajaratnam, N. and Subramanyan, S. 'Plane turbulent denser wall jets and jumps'. J. Hydraulics Research, Vol. 24, 1986, No 4, 281-295.

Shuy, E.B., Chui, P.C., Chua, H.C. and Chen, C.N., 'Entrainment across a density interface inside a cylindrical tank with a concentric base opening'. J. Hydraulic Research, Vol 36, 1998, No 2, 253-267

Environmental Hydraulics, Lee, Jayawardena & Wang (eds) © 1999 Balkema, Rotterdam, ISBN 90 5809 035 3

Contaminant transport in a temperature-stratified reservoir

Ruochuan Gu & Se-Woong Chung
Department of Civil and Construction Engineering, Iowa State University, Ames, Iowa, USA

ABSTRACT: An unsteady 2D reservoir hydrodynamics and transport model is used to simulate mixing and transport of a spilled conservative chemical plume in a stratified reservoir. The occurrence of three flow regimes (plunging flow, underflow and interflow) are identified and their behavior are captured by analyzing the model results. The 2D simulations gain insight into the flow field such as the spread of the spill and transport processes of the plume. Simulation results are compared with field data for water temperature and contaminant concentration collected in the reservoir during the emergency response to the spill. This study also quantifies the effects of inflow boundary condition, ambient stratification, and geometry on the behavior of a density-induced current in a stratified reservoir through numerical experiments with the validated 2D simulation model. The results can assist in contamination control and remediation after a toxic chemical spill, guide field sampling during the spill, and provide information useful for water quality management.

1. INTRODUCTION

A reservoir's water quality conditions can be attributed to physical processes as well as biochemical reactions. Reservoir transport mechanism impact water quality and reservoir limnology, especially when the flow contains point and nonpoint source pollutants (Thorton et al. 1990). After a contaminated flow enters the reservoir, the ultimate fate depends not only on the nature of the chemicals but also on characteristics of ambient and boundary conditions of the reservoir (Johnson et al 1989). The density differences between incoming river water and ambient reservoir water and between reservoir layers may alter behavior of the contaminant plume in the reservoir. The flow may plunge and form an underflow (Hebbert et al. 1979; Alavian et al. 1992), depending on the buoyant force resulting from density differences (Fig. 1). In a density-stratified reservoir, an interflow occurs after the density current separates from the riverbed and intrudes into a layer where the flow is neutrally buoyant (Fischer et al. 1979; Imberger and Hamblin 1982; Ford and Jonhson 1983). Different flow patterns result in different processes of transport and mixing and dilution of toxic chemicals. An integral (1D) simulation model was developed by Gu, et al (1996) for describing gross behavior of the contaminant plume and the general mixing and transport pattern of spilled chemicals in the stratified reservoir. The model was employed during a chemical spill into the Sacramento River and the Shasta Reservoir, California, USA in 1991 to provide timely information on the location and dilution of the contaminant plume, assist emergency response, confirm the remediation measure, and guide data collection. The simplicity of the model makes it suitable as a screening tool for assessing contamination levels at different longitudinal locations as a first approximation during a spill emergency.

In this study, an unsteady 2D reservoir hydrodynamics and transport model is used to simulate mixing and transport of a spilled conservative chemical plume in the stratified Shasta Reservoir. The reservoir is simulated as a whole flow domain by means of using the laterally averaged model. The simulated velocity fields and concentration distributions are used to determine plunging distance, separation depth of underflow, intruding thickness of interflow, dilution of contaminants, travel time and propagation speed of the contaminant plume. Predicted reservoir stratification and chemical dilution are compared with Field measurements. The 2D simulations and analyses improve understanding and predictions of the movement of a conservative contaminant plume in a stratified reservoir.

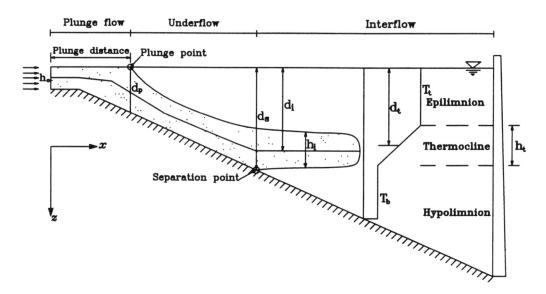

Fig. 1 Schematic of density flow regimes in a stratified reservoir

This paper also quantifies the effects of inflow boundary condition, ambient stratification, and geometry on the behavior of a density-induced current in a stratified reservoir through numerical experiments with the validated 2D simulation model. The inflow boundary conditions, ambient stratification, and reservoir geometry are characterized by five independent parameters. Reservoir flow behavior is described and represented by five dependent variables. The flow variables are derived from the simulated 2D distributions of velocities, concentration, temperature and density. Model results are analyzed to evaluate reservoir flow sensitivity to variations of the parameters, using an influence coefficient method. The results can assist in contamination monitoring and reservoir management for improvement of water quality.

2. NUMERICAL SIMULATIONS

Used in this study is a laterally averaged (longitudinal-vertical) reservoir hydrodynamics and transport model (Cole and Buchak 1994). In the model, 2D advection-diffusion equations describing laterally averaged fluid motion and mass transport are solved by using the finite difference method. The dependent variables are water surface elevation (n), pressure (P), density (ρ), horizontal and vertical velocities (U, W), and constituent concentrations (C). The independent variables are longitudinal distance (x), flow depth (z), and time (t). The governing equations in a Cartesian coordinate system are

$$\frac{\partial(UB)}{\partial t} + \frac{\partial(UUB)}{\partial x} + \frac{\partial(WUB)}{\partial z} = -\frac{1}{\rho}\frac{\partial(BP)}{\partial x} + \frac{\partial}{\partial x}(B A_x \frac{\partial U}{\partial x}) + \frac{\partial(B \tau_x)}{\partial z} \qquad \text{for horizontal momentum} \quad (1)$$

$$\frac{\partial(BC)}{\partial t} + \frac{\partial(UBC)}{\partial x} + \frac{\partial(WBC)}{\partial z} - \frac{\partial}{\partial x}(B D_x \frac{\partial C}{\partial x}) - \frac{\partial}{\partial z}(B D_z \frac{\partial C}{\partial z}) = q_c B + S_c B \qquad \text{for constituent transport} \quad (2)$$

$$\frac{\partial(B_n \cdot n)}{\partial t} = \frac{\partial}{\partial x}(\int_n^h UBdz) - \int_n^h qBdz \qquad \text{for free water surface elevation} \quad (3)$$

$$\frac{\partial P}{\partial z} = \rho g \qquad \text{for hydrostatic pressure} \quad (4)$$

$$\frac{\partial(UB)}{\partial x} + \frac{\partial(WB)}{\partial z} = qB \qquad\qquad \text{for continuity} \quad (5)$$

$$\rho = f(T_w, C_{TDS}, C_{SS}) \qquad\qquad \text{for density} \quad (6)$$

A finite difference grid system was generated, consisting of 32 segments with lengths of 200-1000 m in the longitudinal direction and 52 vertical layers with thickness of 0.5-4 m. The nonuniform grid system has finer grids in the plunging region and coarser grids in other regions. Reservoir widths are obtained from a map. Variable time-steps were used in the simulations, which are a fraction of the maximum time step calculated from the numerical stability criterion with an autostepping algorithm. Unsteady MITC concentrations were specified at the upper end of the computational domain as the upstream boundary condition. The flow boundary condition at the upstream end of the reservoir was a constant discharge of 7.5 m^3/s during the simulation period.

Simulations were conducted for water temperature, density, flow velocity, and toxic chemical methyl isothiocyanate (MITC) concentration throughout the reservoir over time. MITC was simulated as a conservative tracer. It is assumed that the dilution of MITC in different reservoir flow regimes was caused mainly by transport processes, i.e. diffusion and advection. The tracer was used to evaluate the effects of mixing on material distributions and to predict flow behavior of the contaminant plume. The locations of plunging point and separation point, intruding depth and thickness, and chemical dilution were determined from the simulated velocities and MITC concentrations.

The mixing and transport of the plume are displayed by the simulated contours of MITC concentrations during the period of July 17-20 (Fig. 2). Plunging and underflow can be seen in Fig. 2a. Separation and intrusion of the density plume on July 18 and 19 is seen in Figs. 2b and 2c. On July 20, 1991, the contaminated density plume was isolated in the reservoir by the successive re-intrusion of fresh river water and the core of the plume was located 7 km downstream from the reservoir head (Fig. 2d). The plume separated from the bottom, formed an interflow, and propagated horizontally at a depth of 7-8 m in the reservoir on July 17,1991. The contour plots of MITC in the reservoir clearly show that dilution of the plume by mixing with ambient water occurred while it spread vertically and propagated in the longitudinal direction.

Observed and simulated vertical profiles of MITC concentrations at selected stations are presented in Fig. 3. It is shown that the chemical plume plunged near the head of reservoir and formed an underflow and an interflow in the

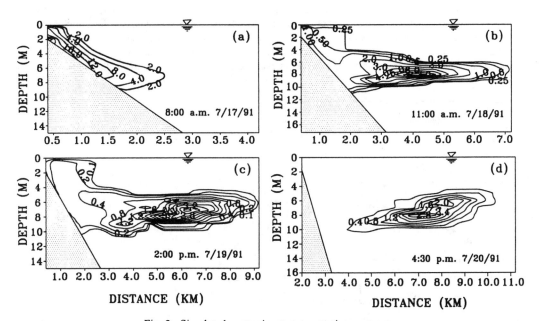

Fig. 2 Simulated contaminant concentration contours

469

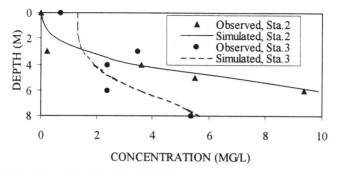

Fig. 3 Observed and simulated MITC concentration profiles at sampling stations 2 and 3

layers of 6.0 to 9.0 m below the water surface. The development of underflow can be seen at stations 2 and station 3. The interflow propagate downstream.

3. SENSITIVITY ANALYSIS

Using the validated 2D simulation model, the effects of inflow and ambient parameters on the behavior of a density-induced contaminant current in various flow regimes in a stratified reservoir are quantified through numerical experiments and sensitivity analysis. The inflow boundary conditions, ambient stratification, and reservoir geometry are characterized by five independent parameters. They are: densimetric Froude number (Fr_o), Reynolds number (Re), stratification number (St), aspect ratio (r_a) and half reservoir expansion angle (S_w). The governing equations (1)-(6) are non-dimensionalized and expressed in terms of the independent (characterizing) parameters. The upstream inflow boundary conditions are characterized by inflow densimetric Froude number for buoyancy and momentum and Reynolds number for turbulence. The aspect ratio characterizes the geometry of river entering the reservoir at the upstream boundary. The half-angle is adopted to describe reservoir geometry (side expansion). A linear expansion is assumed. The stratification number (St) is used to describe the density or thermal structure of the ambient reservoir.

Reservoir flow behavior is described and represented by five dependent variables, including plunge depth, underflow separation depth, intruding thickness of interflow, dilution along the plunging flow, underflow and interflow, and travel time of the contaminant plume. The flow variables are derived from the simulated 2D distributions of velocities, concentration, temperature and density. Model results are analyzed to evaluate reservoir flow sensitivity to variations of the parameters using an influence coefficient method. The parameters to

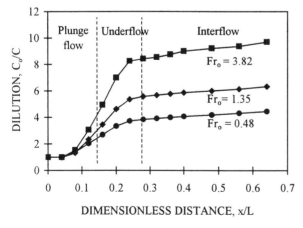

Fig. 4 Variations of dilution factor (C_o/C) with longitudinal distance and Froude number

470

which flow is most sensitive are identified. As an example, a density flow with a higher Fr_o separates at a shallower depth than that with lower Froude number because of higher entrainment rate of ambient water into the plume. More dilution in the plunge flow and underflow account for the shallower separation. The dilution along the horizontal distance (three flow regimes) for Fr_o = 0.48, 1.35 and 3.82 are shown in Fig. 4. Approximately 70-90 % of contaminant concentration reduction $(1-C/C_o)$ occurs before underflow separation. Dilution increases dramatically while the plume propagates from plunging point to separation point because of high entrainment across the interface between the ambient water and the underflow.

Flow sensitivity to each of the aforementioned parameters is determined by computing sensitivity coefficients. A sensitivity coefficient represents the change of a dependent variable that is caused by a unit change of a independent parameter while holding the rest of the parameters constant. To meaningfully compare different sensitivities, the sensitivity coefficient is normalized by reference values, i.e. the average of lowest and highest values of the selected range, which is called sensitivity index. A total sensitivity index is a sensitivity index corresponding to the total change for a selected range of independent parameter. Summarized in Table 1 is the total sensitivity index (s_i), equivalent to the change in each flow variable in response to the selected range of change in each independent parameter. The total sensitivity index varies from -1.06 to 1.12. The role of Fr_o is mainly in plunge depth, entrainment or dilution, and travel time. Re and r_a primarily influence entrainment and dilution. St affects mainly separation depth and intruding thickness only. Travel time (t/t_r) is most sensitive to Fr_o and not sensitive to Re. Plunge depth is sensitive to Fr_o only. In summary, if the number of absolute values of s_i greater than 0.40 in Table 1 is compared, it is found that reservoir flow is sensitive, from most to least, to Fr_o, Re, r_a, S_w, and St for the selected variation ranges.

4. CONCLUSIONS

The occurrence and features of three distinct flow regimes of the contaminated density currents (plunging flow, underflow, and interflow) were successfully captured through 2D numerical simulations of velocity fields and concentration distributions. It was demonstrated that negative buoyancy and ambient stratification played a dominant role during the processes of plunging, separation, and intrusion. The effects of inflow boundary, reservoir water and geometry on mixing and transport processes of a reservoir density flow were investigated and quantified through 2D numerical experiments. The simulation results were analyzed to determine the sensitivity of flow variables to the characterizing parameters. Plunge depth or distance is controlled by Fr_o only. Separation depth and intruding thickness are mainly affected by St. Entrainment and dilution are sensitive to all parameters except St. Travel time (t/t_r) is most sensitive to Fr_o and not sensitive to Re. In general, reservoir flows are sensitive, from most to least, to Fr_o, Re, r_a, S_w, and St.

Table 1. Summary of numerical experiment results: total sensitivity index*

Reservoir flow variable	Boundary or ambient parameter				
	Fr_o	Re	St	r_a	S_w
Plunge depth, d_p/h_o	0.79				
Plunge distance, x_p/L	0.43				
Concentration at plunge, C_p/C_o	-0.25	0.71		0.20	-0.24
Dimensionless entrainment $(Q_p-Q_o)/Q_o$	1.05	-1.06		-0.77	1.12
Dilution at plunge C_o/C_p	0.25	-0.71		-0.20	0.24
Separation depth, d_s/d_t	-0.21	0.37	-0.56	0.13	-0.28
Concentration at separation, C_s/C_o	-0.50	0.98	0.15	0.68	-0.49
Dimensionless entrainment $(Q_s-Q_o)/Q_o$	0.58	-1.04	-0.18	-0.60	0.66
Dilution at separation, C_o/C_s	0.50	-0.97	-0.15	-0.68	0.49
Intruding thickness, h_i/h_t	-0.19	0.45	-0.52	0.22	-0.31
Travel time, t/t_r	0.97	-0.08	-0.28		0.43

*Total sensitivity index, $s_i = (\Delta F/F_m)/(\Delta p/p_m)$, where subscript F_m and p_m denotes reference value (average) and ΔF and Δp denotes variation range of flow variable and boundary or ambient parameter, respectively.

5. REFERENCES

Alavian, V., Jirka, G. H., Denton, R. A., Johnson, M. C., and Stefan, H. G. (1992). "Density currents entering lakes and reservoirs", *J. Hydr. Engrg.*, ASCE, 118(11), 1464-1489.

Cole, T. M., and Buchak, E. M. (1994). *CE-QUAL-W2: A two-dimensional, laterally averaged, hydrodynamic and water quality model, version 2.0 user's manual*, U.S. Army Corps of Eng. Waterways Experiment Station, Vicksburg, Miss.

Fischer, H. B., List, E. J., Koh, R. C. Y., Imberger, J., and Brooks, N. H. (1979). *Mixing in inland and coastal waters*, Academic Press, New York.

Ford, D. E., and Johnson, M. C. (1983). "An assessment of reservoir density currents and inflow processes", Tech. Rep. E-83-7, U.S. Army Corps of Eng. Wat. Exp. Sta., Vicksburg, Miss.

Gu, R., McCutcheon, S. C., and Wang, P. (1996). "Modeling reservoir density underflow and interflow from a chemical spill", *Water Resources Research*, 32(3), 695-705.

Hebbert, B., Imberger, J., Loh, I., and Patterson, J. (1979). "Collie river underflow into the Wellington reservoir", *J. Hydr. Div.*, ASCE, 105(5), 533-545.

Imberger, J., and Hamblin, P. F. (1982). "Dynamics of lakes, reservoirs and cooling ponds", *Ann. Rev. Fluid Mech.*, 14, 153-187.

Johnson, T. R., Ellis, C. R., and Stefan, H. G. (1989). "Negatively buoyant flow in a diverging channel. IV: Entrainment and dilution", *J. Hydr. Engrg.*, ASCE, 115(4), 437-456.

Thornton, K. W., Kimmel, B. L., and Payne, F. E. (1990). *Reservoir limnology: Ecological perspectives*, A Wiley-Interscience Publication, John Wiley & Sons, Inc., New York.

Environmental Hydraulics, Lee, Jayawardena & Wang (eds) © 1999 Balkema, Rotterdam, ISBN 90 5809 035 3

Numerical calculation on generation process and diffusion phenomenon of muddy water caused by sand dumping into water

T.Shigematsu & K.Oda
Department of Civil Engineering, Osaka City University, Japan

S.Horii
Construction Technique Research Company Limited, Japan

ABSTRACT: A calculation method to predict the generation and diffusion process of turbidity by sand dumping in real sea is developed. Some calculated results are presented and the relationship between the transportation of turbidity and the soil dumping condition are discussed.

1. INTRODUCTION

In order to make effective use of ocean space, a large number of developments have arisen in coastal zones such as reclaimed land for airports and/or disposal waste sites. A large volume of sand would be sometimes deposited from a hopper barge in a short period of time for economic reasons. When the construction site is in an enclosed sea, it is often pointed out that muddy water has a bad influence upon the ecosystem around the site because fine sand particles would adhere to gills of the fish and cover the sea bottom which has an important role for benthic organisms. Under construction, however, generating of concentrated muddy water can not be avoided.

To inhibit the effect of construction on the water environment, some measures should be taken. The installation of flexible membranes, called silt protectors, around the construction site is considered to be one of them. The silt protector could be installed in order to prevent muddy water from spreading. Nevertheless, effective deployment and arrangement of these silt protectors have not been detected yet. This is because the process of generating muddy water and ambient fluid motion driven by soil dumping which have a large effect on the initial spreading of muddy water.

Some studies on the behaviour of the particle motion have been shown. Oda et al.(1992) developed the method by combining the Discrete Element Method (DEM) for the calculation of particle motion and the Marker and Cell (MAC) method for the calculation of the ambient fluid flow. Nadaoka et al.(1996) developed the Grid-Averaged Lagragian (GAL) model for new formulation of solid-liquid motion in a manner suitable for the LES modeling. These studies are only comparisons to the laboratory experiments (Murota et al., 1988; Tamai et al., 1993).

The only study of the turbidity diffusion generated by the soil dumping in the actual sea was approached by Tamai et al. (1998). Tamai et al. carried out two-dimensional numerical simulation by using the two-fluid model with the $k - \varepsilon$ turbulence model based on the two fluid model by Elghobashi et al.(1983).

The aims of this study are to develop a calculation method to predict the turbidity concentration generated immediately after soil dumping sufficiently and accurately and to investigate the scale of the diffusion phenomena including the fluid motion driven by sand dumping at the initial stage in the actual sea.

2. CALCULATION METHOD

The sand particles are assumed to be composed of two kinds of sand. One of them is coarse sand, which contributes to the generation of the ambient fluid motion. The other is fine sand, which forms muddy water.

Moreover the assumptions in this study are
(1) The collision of particles with the sea bottom and the process of deposition are not considered.
(2) The water surface displacement can be negligible.
(3) The resuspension of turbidity from the bottom is not taken into account.

The developed calculation method for the prediction of the generation process and diffusion phenomena has three steps. At the first stage, the motion of the coarse sand will be calculated. The ambient fluid motion driven by the settling particles will be calculated at the next stage. Finally, the diffusion of turbidity will be calculated based on the calculated result of the fluid flow.

2.1 Particle Motion

The authors have already developed a liquid-solid model which can predict the motion of the settling particles and the ambient fluid flow by means of the Discrete Element Method (Cundall, 1974). It has been shown that the model is in good agreement with the laboratory experimental data of the settling velocity and dispersion width in settling of the cluster of particles. The DEM can obtain the velocity and location of each particle by solving the equation of motion of each particle. In other words, when it is required to predict the phenomena in an actual construction site such as a land filling, the equations of motion beyond number have to be solved. It is impossible. In this study, the DEM is extended to solve a semi-infinite number of particle motions.

(1) The number of particle is calculated from the void fraction of particles to fluid and the position of the center of gravity of particles in a cell is calculated.
(2) The particles, which are called "representative particles" and the number of particle is n, are arranged uniformly as the center of gravity of those particles would be at the position of the center of gravity of the all particles in the cell. The distance between the adjacent particles is decided by the void fraction ratio.
(3) The motions of the representative particles are solved by using the DEM taking the particle-particle and particle-fluid interaction into consideration. When one of the representative particles moves the distance Δl, N/n particles move the same distance with the representative particles. After all of the representative particles has been moved, the new gravity center of the particles in each cell is calculated. By the same calculation iteratively, the location of all the particles in the cell at each time step is decided.
(4) For every cell the calculation above is carried out.

2.2 Ambient Fluid Flow

The motions of an incompressible fluid can be described by the conservation equation and the Navier-Stokes equations:

$$\frac{\partial \varepsilon u}{\partial x} + \frac{\partial \varepsilon w}{\partial z} = 0 \tag{1}$$

$$\frac{\partial \varepsilon u}{\partial t} + \frac{\partial \varepsilon uu}{\partial x} + \frac{\partial \varepsilon uw}{\partial z} = -\frac{1}{\rho}\frac{\partial \varepsilon p}{\partial x} + \frac{\partial}{\partial x}\left\{2(v+v_t)\frac{\partial \varepsilon u}{\partial x}\right\} + \frac{\partial}{\partial z}\left\{(v+v_t)\left(\frac{\partial \varepsilon w}{\partial x}+\frac{\partial \varepsilon u}{\partial z}\right)\right\} + \frac{1}{\rho}F_{px} \tag{2}$$

474

$$\frac{\partial \varepsilon w}{\partial t} + \frac{\partial \varepsilon u w}{\partial x} + \frac{\partial \varepsilon w w}{\partial z} = -\frac{1}{\rho}\frac{\partial p}{\partial z} + \varepsilon g + \frac{\partial}{\partial x}\left\{(v+v_t)\left(\frac{\partial \varepsilon w}{\partial x} + \frac{\partial \varepsilon w}{\partial z}\right)\right\} + \frac{\partial}{\partial z}\left\{2(v+v_t)\frac{\partial \varepsilon w}{\partial z}\right\} + \frac{1}{\rho}F_{pz} \quad (3)$$

where ε is void ratio, ρ is the fluid density, u and w are velocity components of fluid, g is the gravitational acceleration, v is the coefficient of kinematic viscosity of the fluid and v_t is the coefficient of kinematic eddy viscosity F_p is the drag force and the force in the i direction is calculated

$$F_{pi} = \frac{\beta}{\rho}\left(u_{pi} - u_i\right) \quad (4)$$

$$\beta = \begin{cases} \dfrac{1-\varepsilon}{d_p \varepsilon^3}\left[150\dfrac{(1-\varepsilon)\mu\varepsilon}{d_p} + 1.75\rho\varepsilon^2\left|u_{pi}-u_i\right|\right] & \varepsilon \leq 0.8 \\[4mm] \dfrac{3}{4}C_D\dfrac{\rho(1-\varepsilon)\left|u_{pi}-u_i\right|}{d_p}\varepsilon^{-2.7} & \varepsilon > 0.8 \end{cases} \quad (5)$$

where μ is the coefficient of viscosity of the fluid and d_p is a particle diameter. C_D is the coefficient of the drag force and is given as

$$C_D = \begin{cases} \dfrac{24(1+0.15\,\mathrm{Re}^{0.687})}{\mathrm{Re}} & \mathrm{Re} \leq 1000 \\[4mm] 0.43 & \mathrm{Re} > 1000 \end{cases} \quad (6)$$

where $\mathrm{Re} = d_p\left|\vec{u}_p - \vec{u}\right|/v$ is the Reynolds number.

The turbulence energy k is given by solving the transport equation

$$\frac{\partial k}{\partial t} + \frac{\partial \overline{u}_j k}{\partial x_j} = R_{ij}\frac{\partial \overline{u}_i}{\partial x_j} + \frac{\partial}{\partial x_i}\left\{(v+v_t)\frac{\partial k}{\partial x_i}\right\} - C_e\frac{k^{3/2}}{\Delta} + P_S + P_G \quad (7)$$

$$\Delta = (\Delta x \Delta z)^{1/2}, \qquad v_t = C_s\Delta k^{1/2} \quad (8)$$

where $\overline{u}$ is the average velocity in the grid scale, Δx and Δz are the grid distances, R_{ij} is the Reynolds stress, P_S and P_G represent the energy production by the particles in the sub-grid scale and the energy dissipation in the grid scale respectively and are obtained by

$$P_G = -\alpha\beta\left|\vec{u}_p - \vec{u}\right|^3, \qquad P_S = \beta\left(\overline{u'_{pi}u'_i} - \overline{u'^2_i}\right)\left|\vec{u}_p - \vec{u}\right| \quad (9)$$

The details of the calculation should be referred to Nadaoka et al.(1996).

2.3 Diffusion of Turbidity

On the basis of the calculated results of the ambient fluid flow, The distribution of turbidity concentration caused by the fine sand included sands dumped into water is calculated by solving the diffusion equation

$$\frac{\partial C}{\partial t} + \frac{\partial uC}{\partial x} + \frac{\partial (w+w_0)C}{\partial z} = \frac{\partial}{\partial x}\left(D_x\frac{\partial C}{\partial x}\right) + \frac{\partial}{\partial z}\left(D_z\frac{\partial C}{\partial z}\right) + q \quad (10)$$

where C is turbidity concentration, w_0 is settling velocity of fine particle, q is the turbidity quantity per unit time and unit volume. D_x and D_z are the turbulence diffusion coefficient and decided to be equal to the eddy viscosity coefficient. In this study, the settling velocity of the fine sand is set to zero. As the boundary condition, the gradients of the concentration in the vertical direction at the water surface and the bottom were equal to zero and those in the horizontal direction at the both side of the calculation region were equal to zero.

3.CALCULATION CONDITIONS AND RESULTS

The condition of sand dumping was decided as shown in Table. 1 with reference to the data that Furudoi(1989) investigated the sands in a field experiment. The hopper barge is assumed 50m long, 4m width. The time for the soil dumping in a real operation is almost 10 seconds. In this study, the calculations with three different dumping times were carried out in order to investigate the effect of

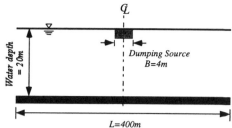

FIG. 1. Calculation Domain

the dumping time on the turbidity transportation. Moreover, the effect of the total volume of dumped soil was also examined. The calculation domain is 400m wide and 20m deep. The grid size is $\Delta x = 2\,\mathrm{m}$ and $\Delta z = 1\,\mathrm{m}$. The dumping source is at the water surface of the center of the calculation domain as shown in FIG.1.

FIG.2 and FIG.3 show on the calculated results of the ambient fluid motion and the turbidity distribution respectively for Case-2. These figures show the result of only the right half of the calculation domain. It is found that the turbidity would generate but would slightly spread until 5 seconds after dumping. After coarse sand reaches the bottom, the large circulation would be generated near the bottom. By this large circulation, the turbidity would begin to spread rapidly to horizontal and vertical directions. The circulation would move away from the center axis of the sand dumping while enlarging and slowing with the elapse of time. The turbidity would spread with the circulation in horizontal direction. The height of the muddy water mass would become almost constant after 40 seconds.

FIG.4 shows the distance between the tip of muddy water mass and the center of the dumping and the height of tip of the muddy water mass. The horizontal distances of 10ppm contour from the center of dumping for all case would increase with time. The horizontal distance for the Case-1 would be larger than that for the other cases. It implies that the diffusion of the turbidity would be affected by the dumping condition. The height of the muddy water mass would be almost constant after 50 seconds. It is estimated that the height of the muddy water mass would be restrained by the water depth.

Fig.5 shows the vertical distribution of the load ratio of the fine sand weight, obtained by integration of the concentration of turbidity at each level, to the total weight of the dumped fine sand after 200 seconds. FIG. 5

TABLE 1. Calculation Conditions of Sand dumping

		Case-1	Case-2	Case-3	Case-4
Total volume of dumped sand	V [m³]	3000	3000	3000	1000
Total volume of fine sand	q [ton]	0.954	0.954	0.954	0.318
Void fraction	ε	0.4	0.4	0.4	0.4
Time for sand dumping	t_q [s]	5	10	20	3.3
Settling velocity of dumping sand	W_0 [m/s]	3.0	1.5	0.75	1.5

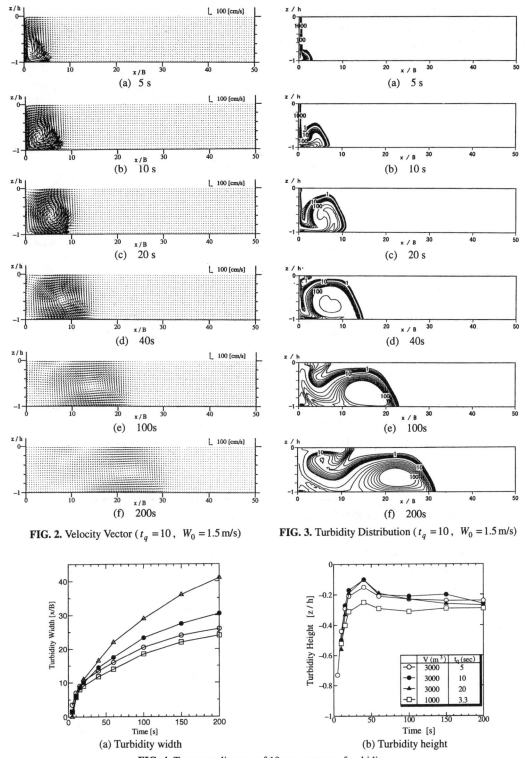

FIG. 2. Velocity Vector ($t_q = 10$, $W_0 = 1.5$ m/s)

FIG. 3. Turbidity Distribution ($t_q = 10$, $W_0 = 1.5$ m/s)

(a) Turbidity width

(b) Turbidity height

FIG. 4. Transport distance of 10ppm contour of turbidity

implies that 65~70 % of the total load suspends in $z/h \leq -0.6$, 20~25% suspends in $-0.6 \leq z/h \leq -0.4$, and that less than 10% of is distributes in $z/h > -0.2$. It is found from FIG. 5 that the distribution curve of the load ratio in the case that the dumping velocity is smaller would be less variable.

4.CONCLUSION

The results obtained in this study by the presented calculation method would offer very useful knowledge in
calculating the diffusion in the larger scale as an initial condition. Moreover the calculated results imply that the condition of sand dumping effects on the fluid motion driven by the sand settling and the diffusion phenomena. It must be noted, however, that the data of the turbidity and the ambient fluid motion in an actual sea should be investigated.

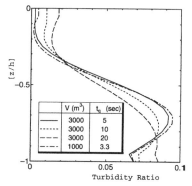

FIG. 5. Vertical distribution of load ratio

5.REFERENCES

Cundall, P.A.(1974), *Rational design of Tunnel Supports—Interactive Graphics for the Input and Out put of Geometrical Data*, Technical Report MRD-2-74, Missouri River Division, US. Army Corps of Engineers.

Elghobashi, W.E. and W.T. Abou-Arab(1983), *"A Two-equation Turbulence Model for Two-phase Flow"*, Phys. Fluids, No.26(4), 931-938.

Murota A., K. Nakatsuji, M. Tamai, and H. Machida (1988), *"Formation of Solid-Fluid Buoyant Cloud in Reclamation Works"*, Proc. of the 35th Japanese Conference on Coastal Engineering, JSCE, 777-781.

Nadaoka K., Y. Nihei et al. (1996), *"An LES Modeling for Solid-Fluid Phase Turbulent Flow Based on New Formulation of Solid-Particle Motion"*, J. Hydraulic, Coastal and Environmental Engineering, JSCE, No.533, II-34, 61-74.

Oda K., T. Shigematsu, N. Onishi, and M. Inoue (1992), *"Descent and Dispersion Analysis of Particles Dumped into Waters by Means of an Improved DEMAC Method"*, Proc. of Coastal Engineering, JSCE, Vol.39(2), 971-975.

Tamai M., Y. Shimoya, and K. Muraoka (1993), *"Experimental Study on Turbulence properties of Particle Plume"*, Proc. Hydraulic Engineering, JSCE, Vol.37, 433-438.

Tamai M. and K. Muraoka (1998), *"Numerical Simulation on Characteristics of Turbidity Transport Generated in Direct Dumping of Soil"*, Annu. J. Hydraulic Engrg., JSCE, Vol.42, 541-546.

Furudoi M(1989), *"A study on the behaviour of turbidity in the port and harbour construction site"*, Ph. D. thesis, Osaka University, 37-65.

2.4 Air-water interaction

Environmental Hydraulics, Lee, Jayawardena & Wang (eds) © 1999 Balkema, Rotterdam, ISBN 90 5809 035 3

Accuracy of tracer measurement of gas-desorption rates

Charles S. Melching
Water Resources Division, US Geological Survey, Urbana, Ill., USA

ABSTRACT: Methods for measuring the gas-desorption rate for flowing water in streams involve the use of different gases (krypton-85, propane, and ethylene), injection procedures (slug and continuous), and computational approaches. For slug injection, the gas-desorption rate may be computed by determining the change in the peak concentration at upstream and downstream measurement locations (peak method) or by determining the change in the area under the gas concentration versus time curve at upstream and downstream measurement locations (area method). This paper summarizes the results of comparisons of paired measurements of the reaeration-rate coefficient (K_2) utilizing the different tracer-gas methods as follows: 108 K_2 pairs from the propane- and ethylene-peak methods, 32 K_2 pairs from the propane- and ethylene-area methods, 43 K_2 pairs from the propane-peak and -area methods, 52 K_2 pairs from the ethylene-peak and -area methods, 8 K_2 pairs from the propane-peak and krypton-85 methods, and 9 K_2 pairs each from of the krypton-85 and the propane-area and ethylene-peak and –area methods. The methods utilizing propane and/or ethylene yield unbiased estimates of K_2. The mean absolute relative percent difference in the estimated K_2 values obtained from tracer-gas methods ranged between 11.3 and 20 percent. Sixty-seven percent of the paired K_2 measurements had relative percentage differences (RPD) with absolute values less than 20 percent and errors relative to the mean of the measurements less than 10 percent. On the basis of the data analyzed, the standard error of tracer-gas measurements of gas desorption is approximately 10 percent.

1. INTRODUCTION

Tsivoglou et al. (1965, 1968) developed a direct method to estimate the gas-desorption or -absorption rates for flowing water in streams. They found that the desorption rate for gases that are not present in substantial quantities in natural streams are directly proportional to the desorption or absorption rates of other gases that may be important to the environmental health and water quality of the stream, particularly dissolved oxygen (DO) concentrations. The method developed by Tsivoglou et al. was later modified by Rathbun et al. (1975) and Yotsukura et al. (1983). These methods generally may be described as tracer-gas (or simply tracer) measurement methods of gas-desorption rates. These methods involve injection of a soluble, biologically inert gas that is not present in substantial concentrations in natural streams and injection of a conservative tracer that is used to account for the effects of dispersion and dilution (dispersion/dilution tracer) and to indicate gas-sampling times. By determining the change in the concentration of the injected gas as it travels downstream, adjusted for dispersion and dilution as necessary, the desorption rate of the gas in the stream to the atmosphere can then be computed. These methods were developed so that the reaeration-rate coefficient (K_2) of flowing water could be determined through the proportionality between gas desorption and absorption of oxygen from the atmosphere. These methods also have been applied to estimate desorption rates of gases important to other biochemical reactions. For example, Longsworth (1991) used these methods to determine the desorption of carbon dioxide and adsorption of oxygen; processes that appear to control precipitation of manganese oxides and which affect the movement and mobility of other metal compounds.

The accuracy of the tracer-gas methods is important to environmental studies and water-quality management, particularly with respect to DO. The concentration of DO in natural waters is the primary indicator of the overall water quality and viability of the aquatic habitat. Reaeration is the physical absorption of oxygen from the atmosphere by water, and it is the fundamental process by which streams affected by waste inputs replenish the DO concentration and purify themselves (Tsivoglou et al., 1968). The reaeration-rate coefficient is typically the dominant parameter affecting the reliability of the simulation of DO concentrations in streams (Brown and Barnwell, 1987, p. 175; Melching and Yoon, 1996), and the tracer-gas methods have been the primary means to measure the reaeration-rate coefficient since their development.

Tsivoglou et al. (1968) studied the accuracy and reproducibility of K_2 measurements made with krypton-85 as the tracer gas in 14 reaches on the Jackson River in Virginia, United States (U.S.). They made replicate measurements 2 days apart on 11 of these reaches, and they made 3 measurements at 2-day intervals on the other 3 of these reaches. The error relative to the mean of the measurements for each reach ranged from 0.7 to 18.2 percent with an average for all of the reaches of 8.3 percent. These results indicate that the tracer-gas methods may be sufficiently accurate for reliable water-quality management. However, results for other gases and computational methods for a wider range of streamflow conditions is needed to evaluate the general accuracy of tracer-gas methods. The U.S. Geological Survey (USGS) has made paired measurements of K_2 with different gases or computational methods on many streams throughout the U.S. This paper summarizes the comparisons among the results of the USGS measurements utilizing different gases (propane, ethylene, or krypton-85) or computational methods (peak or area).

2. DATA AVAILABLE

The USGS has made paired measurements of K_2 on stream reaches utilizing propane, ethylene, and/or krypton-85 as the desorbing gas and using the peak and area methods of gas-desorption-rate computation. These data are available in 17 USGS reports and in project files available at USGS District Offices. Because of the size limitations on this paper, it is not possible to list all the reports in the references. However, the author has recently completed a Cooperative Synthesis project for the USGS wherein all the measurements of K_2 made by the USGS, and the associated stream-hydraulic and water-quality characteristics (where available), were compiled. This information may be obtained from the author as a Microsoft Excel spreadsheet. Data from K_2 measurements on 493 stream reaches in the U.S. are included in the spreadsheet.

The number of stream-reach measurements, number of streams, and states for which data are available are listed by the methods compared in Table 1. The ranges of selected hydraulic characteristics of streamflow measured or estimated during the K_2 measurements for the paired tracer-gas methods are listed in Table 2.

Table 1. Tracer-gas methods compared and the number of streams and reaches and the location by State of the data available for comparison.

Methods compared	Number of reaches	Number of streams	Stream location by State
Propane peak and ethylene peak	108	44	Colorado, Florida, North Dakota, Ohio, Texas, Utah, and Wisconsin
Propane area and ethylene area	32	10	Colorado and Wisconsin
Propane peak and area	43	16	Colorado, Kentucky, New Mexico, Wisconsin, and Wyoming
Ethylene peak and area	52	15	Colorado, Illinois, South Carolina, and Wisconsin
Krypton-85 and propane peak	8	4	Wisconsin
Krypton-85 and all other methods	9	4	Wisconsin

Table 2. Ranges of hydraulic characteristics measured or estimated for streamflow during the measurement of the reaeration-rate coefficient (K_2) with the paired application of the various tracer-gas methods.

Methods compared	Width (m)	Depth (m)	Discharge (m³/s)	Slope (m/m)	Velocity (m/s)	Length (km)
Propane peak and ethylene peak	1.55-58.2	0.062-1.64	0.0028-16.7	0.000072-0.018	0.0147-1.21	0.365-12.6
Propane area and ethylene area	3.26-51.8	0.10-0.87	0.048-5.86	0.00068-0.012	0.049-0.48	0.45-5.79
Propane peak and area	1.98-108	0.062-0.99	0.0076-22.6	0.00013-0.014	0.028-0.55	0.284-24.3
Ethylene peak and area	2.45-51.8	0.055-1.14	0.048-15.2	0.000085-0.017	0.049-0.78	0.40-16.9
Krypton-85 and all propane and ethylene methods	3.26-27.0	0.10-0.31	0.048-1.53	0.00072-0.0029	0.049-0.36	0.73-1.90

The comparison data are representative of a wide range of streamflow characteristics for predominately low flows (< 25 m³/s) in small to medium sized (1.5- to 110-m wide) streams with low to mild slopes (< 0.02 m/m). Data for comparison of tracer-gas methods are not available for high flows (> 25 m³/s) or for large streams (> 110-m wide) or streams with steep slopes (> 0.02 m/m).

3. STATISICAL ANALYSIS OF THE PAIRED MEASUREMENTS

The relative percentage difference (RPD) was computed for each paired measurement of K_2 made with the methods compared as

$$RPD = 100[2(K_{2,1} - K_{2,2})/(K_{2,1} + K_{2,2})]$$

where $K_{2,i}$ is the value of K_2 determined with method i. The value of the RPD is twice the value of the error relative to the mean of the measurements reported by Tsivoglou et al. (1968). The RPD is applied to cases where the mean of the measurements may not be the best estimate of the measured variable. For each comparison of tracer-gas methods, the real and absolute values of the mean and median of the RPD values are listed in Table 3 along with the percentage of absolute RPD values that are less than 20 percent.
propane-area and ethylene-area methods indicates much closer agreement between these values than the values from the peak methods for these gases. The mean and median absolute RPD values are 14.0 percent and 11.6 percent, respectively, with 78.1 percent of the RPD values less than 20 percent. The K_2 values from the propane-area and ethylene-area methods result in better agreement with the K_2 values from the krypton-85 method than do the K_2 values from the propane-peak and ethylene-peak methods. Thus, because there is less variability in the values of K_2 determination of K_2 using the area method may yield more reliable estimates of K_2 than using only the peak-gas concentrations.

The relatively large differences between the results of the propane-peak and ethylene-peak methods also may have resulted because ethylene gas is more likely to biodegrade in the natural environment and is more chemically reactive than propane in the presence of halogens, such as fluorine and chlorine, that may be found downstream of wastewater-treatment plants (Kilpatrick et al.,1989). For these reasons, Kilpatrick et al. note that propane has become the more commonly used gas for measurement of K_2. The comparison of the propane-peak and ethylene-peak methods includes 76 more stream reaches than the comparison of the propane-area and ethylene-area methods, and, thus, may be more likely to include reaches in which biochemical processes affect ethylene measurements. The value of K_2 determined using ethylene measurement was higher for propane for 34 of the 51 paired measurements with RPD values greater than 20 percent; 12 of these 34 measurements have RPD values greater than 50 percent. However, the propane

Table 3. Statistics of the relative percentage difference (RPD) between the measurements made with the methods compared (negative values indicate that the values obtained from the second method are higher than that obtained with the first method).

Methods compared	Mean	Median	Absolute mean	Absolute median	Absolute RPD < 20 percent (percent)
Propane peak and ethylene peak	-8.37	-4.90	26.4	18.0	52.8
Propane area and ethylene area	-0.36	2.70	14.0	11.6	78.1
Propane peak and area	-7.20	-6.20	14.6	11.4	79.1
Ethylene peak and area	-0.08	-1.30	12.4	10.0	80.8
Propane peak and krypton-85	1.11	-4.60	16.5	17.8	62.5
Propane area and krypton-85	7.70	8.60	11.3	8.6	77.8
Ethylene peak and krypton-85	10.2	2.70	20.0	7.6	55.5
Ethylene area and krypton-85	17.1	12.2	19.9	12.5	66.7

The comparison of the K_2 values from the propane-peak and ethylene-peak methods yields the largest absolute mean and median RPD values (26.4 and 18.0 percent, respectively). Also, only 52.8 percent of the RPD values were less than 20 percent in this comparison. The comparison of the K_2 values from the measurement was higher than the ethylene measurement with RPD values greater than 50 percent for five steam reaches. Thus, the high mean and median absolute RPD values between the results of the propane-peak and ethylene-peak methods cannot entirely be attributed to known problems with application of ethylene as the tracer gas.

The K2 values determined using the peak and area methods for both propane and ethylene generally are in good agreement. For approximately 80 percent of the paired measurements the absolute RPD is less than 20 percent for each gas. The mean and median absolute RPD values for propane are 14.6 and 11.4 percent, respectively, and for ethylene these are 12.4 and 10.0 percent, respectively.

If the error relative to the mean of the measurements was used for comparison, as was done by Tsivoglou et al. (1968), the mean and median of the error would be less than 10 percent for all comparisons except the mean absolute error in the comparison of the results of the propane-peak and ethylene-peak measurements. Further, 181 of the 270 paired measurements (67 percent) would have errors less than 10 percent. Thus, two-thirds of the measurements made with tracer-gas measurements would be expected to be within 10 percent of the true value of K_2 for the specific reach. For normally distributed data, one-standard deviation error bounds include approximately 68 percent of the data. Therefore, a standard error of 10 percent appears reasonable for the tracer-gas methods on the basis of the data analyzed in this study. These statistics indicate the high reliability and accuracy of the tracer-gas methods relative to each other. This high reliability and accuracy is further illustrated in the scattergram of the K_2 measurements made with the propane-area and ethylene-area methods shown in Figure 1.

The Wilcoxon signed ranks test (Conover, 1980, p. 278-292) was used to determine whether differences between the K_2 values measured by the different tracer-gas methods were statistically significant. In this test, the hypothesis examined was that the central values of the data series did not differ. For a two-sided Wilcoxon signed ranks test, the critical value of the test statistic above which (in absolute value) the hypothesis can be rejected is 1.96 at the 5-percent level of significance and 2.58 at the 1-percent level of significance. These critical values are based on a normal distribution approximation that is valid for data comparisons involving more than 30 sample pairs (Helsel and Hirsch, 1992, p. 104). The values of the Wilcoxon signed ranks test statistic were –0.81 for propane peak versus ethylene peak, 0.42 for propane area versus ethylene area, -2.32 for propane peak versus propane area, and –1.11 for ethylene peak versus ethylene area. Therefore, the hypothesis that the paired data are from the same distribution and the methods

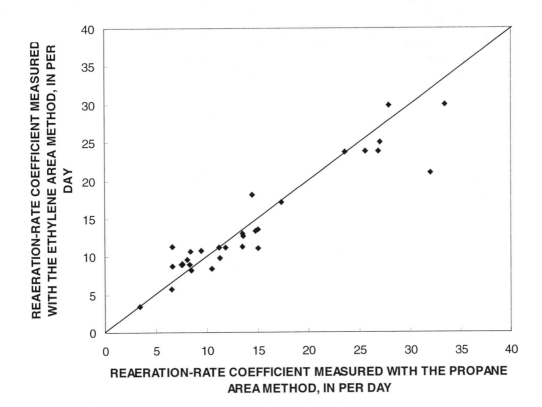

Figure 1. Reaeration-rate coefficients measured with the propane-area and ethylene-area methods

are unbiased cannot be rejected at the 1-percent level for propane peak versus propane area, and at the 5-percent level for all other comparisons.

The Wilcoxon signed ranks test was not applied to comparisons of either propane or ethylene with krypton-85 because of the limited number of available paired data. The ethylene-area method resulted in a higher K_2 value than the krypton-85 method for 8 of the 9 reaches sampled. For the other propane or ethylene methods, the distribution of high and low values among propane or ethylene and krypton-85 was more even. In general, these results indicate that the K_2 values estimated using the various tracer gases and computational methods are unbiased.

4. PRACTICAL CONSIDERATIONS

Since the development and promotion of the continuous-rate injection (CRI) method by Yotsukura et al. (1983), nearly all stream measurements of K_2 made by USGS personnel have utilized the CRI method. The CRI method is now the USGS standard method for measuring K_2 because less gas sampling is required and the dye injection is simpler than the other methods (Kilpatrick et al., 1989). No paired measurements are available comparing the measurements made with the CRI method with measurements made with any other tracer-gas method. The accuracy of the methods compared is a function of the accuracy of sampling the dye and gas, and the accuracy of the first-order kinetic model of gas desorption. For the peak methods, sampling errors result in the identification of the peak gas concentration and in the determination of the dye recovery. For the area methods, sampling errors result in the measurement of the entire range of gas concentrations. In particular, measurement of very low gas concentrations is difficult and errors in measuring low concentrations can affect the area under the curve of gas concentration versus time. The CRI method generally is not affected by these key sampling errors in the peak and area methods. The CRI method results could be affected by the errors associated with measuring low gas concentrations, but probably to a lesser degree and less frequently than the area method. Therefore, the accuracy of the CRI method would be expected to be at least as good as that for the peak or area methods.

5. CONCLUSIONS

The methods utilizing propane and/or ethylene yield unbiased estimates of reaeration-rate coefficient (K_2), as indicated by the Wilcoxon signed ranks test. Consideration of 270 paired measurements of K_2 using tracer-gas methods on small, mild sloped streams during low-flow conditions indicated that 67 percent had relative percentage differences (RPD) with absolute values less than 20 percent and errors relative to the mean of the measurements less than 10 percent. The mean absolute RPD in the estimated K_2 values obtained from tracer-gas methods ranged between 11.3 and 26.4 percent, and the median absolute RPD ranged between 7.6 and 18.0 percent. The mean and median errors relative to the mean of the measurements are one half the corresponding values for the RPD. On the basis of the data analyzed, the standard error of tracer-gas measurements of gas desorption is approximately 10 percent. Thus, tracer-gas methods should provide acceptable accuracy for environmental analysis and water-quality management.

6. REFERENCES

Brown, L.C., and Barnwell, T.O., Jr. (1987). "The enhanced stream water quality models QUAL2E and QUAL2E-UNCAS: Documentation and user manual," *Rep. EPA/600/3-87/007*, U.S. Environmental Protection Agency, Athens, Ga.

Conover, W.J. (1980). *Practical nonparametric statistics*, John Wiley and Sons, New York.

Helsel, D.R., and Hirsch, R.M. (1992). *Statistical methods in water resources*, Elsevier, Amsterdam.

Kilpatrick, F.A., Rathbun, R.E., Yotsukura, N., Parker, G.W., and DeLong, L.L. (1989). "Determination of stream reaeration coefficients by use of tracers." *U.S. Geological Survey Techniques of Water-Resources Investigations, Book 3, Chapter A18*, Reston, Va.

Longsworth, S.A. (1991). "Measurement of stream reaeration at Pinal Creek, Arizona," U.S. Geological Survey Toxic Substances Hydrology Program, Proceedings of the technical meeting, Monterey, Calif., March 11-15, 1991, G.E. Mallard and D.E. Aronson, eds., *U.S. Geological Survey Water-Resour. Investigations Rep. 91-4034*, Reston, Va.

Melching, C.S., and Yoon, C.G. (1996). "Key sources of uncertainty in QUAL2E model of Passaic River," *J. Water Resour. Plng. And Mgmt.*, ASCE, 112(2), 105-113.

Rathbun, R.E., Shultz, D.J., and Stephens, D.W. (1975). "Preliminary experiments with a modified tracer technique for measuring stream reaeration coefficients," *U.S. Geological Survey Open-File Rep. 75-226*, Bay St. Louis, Miss.

Tsivoglou, E.C., Cohen, J.B., Shearer, S.D., and Godsil, P.J. (1968). "Tracer measurement of stream reaeration. II. Field studies," *J. Water Pollution Control Federation*, 40(2), 285-305.

Tsivoglou, E.C., O'Connell, R.L., Walter, C.M., Godsil, P.J., and Logsdon, G.S. (1965). "Tracer measurement of stream reaeration. I. Laboratory studies," *J. Water Pollution Control Federation*, 37(10), 1343-1362.

Yotsukura, N., Stedfast, D.A., Draper, R.E., and Brutsaert, W.H. (1983). "Assessment of a steady-state propane-gas tracer method for reaeration coefficients—Chenango River, New York," *U.S. Geological Survey Water-Resour. Investigations Rep. 83-4183*, Albany, N.Y.

Environmental Hydraulics, Lee, Jayawardena & Wang (eds) © 1999 Balkema, Rotterdam, ISBN 90 5809 035 3

A relationship between spray quantity above white caps and wind shear stress

Nobuhiro Matsunaga & Kenichi Uzaki
Department of Earth System Science and Technology, Kyushu University, Kasuga, Japan

Misao Hashida
Department of Civil Engineering, Nippon Bunri University, Oita, Japan

ABSTRACT: Spray quantity generated when a strong wind blows over white caps has been investigated experimentally by using a wind-wave tank. It depends strongly on the friction velocity on the water surface and the settling velocity of spray. An empirical curve is obtained by examining the dependence of the drag coefficient on the wind-wave factor. The graphic data processing of spray diameters shows that the settling velocity decreases linearly with height. The vertical profiles of spray concentration are expressed universally by using these results.

1. INTRODUCTION

When a large amount of sea water spray is transported to a coastal zone, salt damage often occurs and causes over the wide area disasters such as the corrosion of steel structures, the decrease of agricultural products and the stoppage of electric power. Making it clear how much sea water spray is transported from sea to shore is indispensable to predict the scale of salt damage. Some studies by Toba（1958）, Toba & Tanaka (1963) and Hama & Takagi (1970) have been already made about the generation of sea-salt particles and their transport process. However, there is still much left to be studied hereafter about the spray generation from breaking waves or coastal structures under a strong wind. Matsunaga et al. (1996) have carried out an experimental study on the landward transport of spray generated from a wave absorbing sea wall. As a result, it has been obtained that the spray concentration decays exponentially both in the leeward direction and in the vertical direction. They also start on developing a new type of revetment method to reduce the spray quantity.

In this study, the quantity of spray generated from white caps under a strong wind has been investigated ‣experimentally, and a trial to obtain a universal expression for vertical profiles of spray concentration has been made.

2. EXPERIMENTAL METHODS

Experiments were carried out by using a wave tank equipped with an inharation-typed wind tunnel. It is drawn schematically in Fig. 1. The tank was 32 m long, 0.6 m wide and 0.94 m high. The mean water depth was 0.3 m at the horizontal bed section. Two-dimensional regular waves with high steepness were made by a wavemaker. White caps were formed by a strong wind blowing over the waves. Wave parameters, wind velocity and spray quantity were measured at positions 1 to 9. Their intervals were 1.80 m. A sloping bed was placed at the leeward end of the tank to prevent the wave reflection.

Nine experiments were made by varing wind velocity and wave parameters. The parameters H, L, T and c shown in Table 1 are respectively the wave height, wave length, wave period and phase velocity of

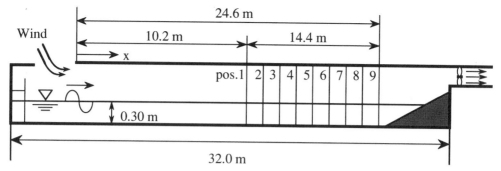

Fig. 1   Schematic diagram of experimental set-up.

Table 1   Experimental parameters.

Run	$\overline{H}$ (cm)	$\overline{L}$ (cm)	$\overline{T}$ (s)	$\overline{c}$ (m/s)	$\overline{H}/\overline{L}$	$\overline{U}_m$ (m/s)	$\overline{u}_*$ (m/s)	$\overline{u}_*/\overline{c}$
1	12.5				0.0912	16.5	1.39	1.01
2	12.5	137	1.00	1.37	0.0912	17.5	1.59	1.16
3	12.4				0.0905	18.4	2.13	1.55
4	13.1				0.0740	16.5	1.12	0.758
5	13.1	177	1.20	1.47	0.0740	17.5	1.32	0.893
6	12.7				0.0718	18.4	1.77	1.20
7	14.9				0.0693	16.5	1.14	0.738
8	14.5	215	1.40	1.53	0.0674	17.5	1.30	0.846
9	14.2				0.0660	18.4	1.67	1.09

predominant waves in the field where wind waves are superposed on the regular waves. Wind and wave fields were quasi-uniform in the test section. The overbars denote averaged values of data obtained at positions 1 to 9. A propeller current meter was used for the measurements of mean wind velocity. The cross-sectionally averaged wind velocity $\tilde{U}_m$ was varied from 16.5 m/s to 18.4 m/s. The friction velocity on the water surface u_* was obtained by fitting the logarithmic law to the wind velocity profile at each position. Settling velocity of spray was estimated by analyzing the marks of spray on droplet sampling paper.

3. EXPERIMENTAL RESULTS AND DISCUSSION

3.1 Drag Coefficient of White Caps

The friction velocity u_* on the water surface is one of the most important factors which determine the spray quantity. Vertical profiles of mean wind velocity U was logarithmic near the water surface. Therefore, the values of u_* were estimated by fitting the logarithmic law

$$\frac{u}{u_*} = \frac{1}{\kappa} \ln\left(\frac{z}{z_0}\right)$$
(1)

to the vertical profiles, where κ (=0.4) is von Karman's constant, z the height from the mean water level and z_0 the roughness parameter. The drag coefficient C_D is defined by

$$C_D = (u_* / U_{10})^2, \tag{2}$$

where U_{10} is the mean wind velocity at 10 m height. Though many empirical relationships between C_D and U_{10} have been proposed, some disagreement is seen among them. Such a data arrangement has been made for the utility. However, relationships between C_D and dimensionless parameters should be examined to obtain a universal expression for C_D. From such a viewpoint, the values of C_D are plotted against the wind-wave factor u_*/c in Fig. 2. These data include others obtained by Matsunaga et al. (1996). The data in this study are plotted by open circles. All the data collapse well on the data-based curve and C_D seems to increase monotonically with u_*/c.

3.2 Vertical Profiles of Spray Concentration

Figure 3 shows a leeward variation of vertical profiles of spray concentration C. The values of C were calculated by dividing the spray quantity transported per unit time and area into the water density and the mean wind velocity. The leeward variation is very small in comparison with the vertical one because the wind and wave fields are almost uniform in the leeward direction. Therefore, we can guess that the governing equation of spray concentration is given by

$$\frac{d}{dz}(w_0 C) + \frac{d}{dz}(K \frac{dC}{dz}) = 0, \tag{3}$$

where w_0 is the settling velocity of spray and K is the turbulent diffusivity. The solution of Eq. (3) is given by

$$C = C_* \exp \int_1^{\tilde{z}} -\frac{w_0 H}{K} d\tilde{z}, \tag{4}$$

where $\tilde{z} = z / H$ and C_* is the spray concentration at $\tilde{z} = 1.0$. To obtain analytically the vertical profile of C,

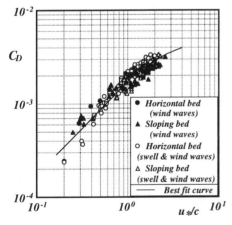

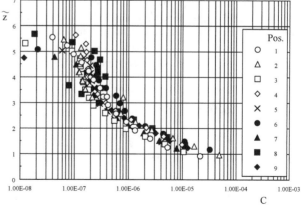

Fig. 2   Relationship between C_D and u_*/c.   Fig. 3   Vertical profiles of spray concentration in Run 1.

489

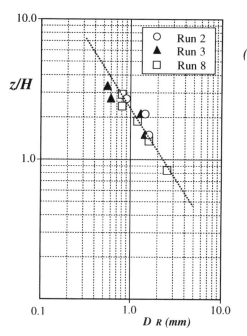

Fig. 4   Vertical profile of representative spray diameter D_R.

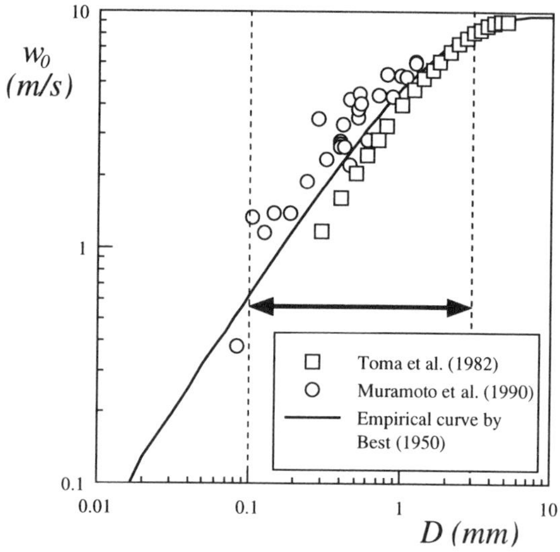

Fig. 5   Relationship between w_0 and D. Double arrow indicates the range of spray diameter in this study.

z-dependences of w_0 and K should be known. The authors tried to read the spray diameter and the number by using droplet sampling paper. As the spray leaves blue marks when it collides with the sampling paper, the diameters of marks Ds and the number of spray colliding per unit time and area N can be analyzed by means of graphic data processing. Spray diameters in the air D were estimated by using the empirical relationship D $= 0.42Ds^{2/3}$ proposed by Sasho et al. (1989). From this analysis, it was revealed that the values of N decrease exponentially with the increase of D. The quantity of spray with the diameter D transported per unit time and area is given by M= ρ (π /6)D^3N. By defining the spray diameter at which M takes the maximum value as a representative diameter D_R, z-dependence of D_R is shown in Fig. 4. From this figure, it is seen that D_R decreases linearly with z / H.

Figure 5 shows empirical relationships between the settling velocity of raindrop w_0 and its diameter D. The data obtained by Toma et al. (1982) and Muramoto et al. (1990) are approximated well by the empirical curve by Best (1950). If the settling velocity in the range indicated by the double arrow is expressed approximately by $w_0 = 5.50D$, w_0 is given by

$$w_0 = 5.50 D_* (z/H)^{-1} \tag{5}$$

where D_* is the spray diameter at z = H, and the units of w_0 and D_* are m/s and mm, respectively.   If K is given by κ u$_*$z, Eq. (4) becomes

$$C = \frac{C_*}{\exp(\alpha/\kappa)} \exp \frac{\alpha}{\kappa \tilde{z}} \tag{6}$$

490

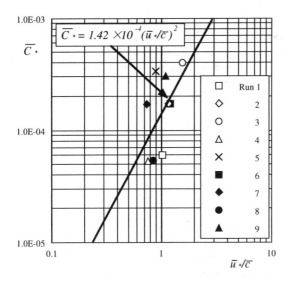

Fig. 6   Relationship between C_* and u_*/c.

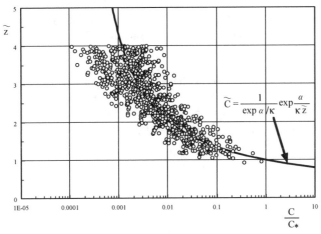

Fig. 7   Comparison of semi-empirical solution and experimental data.

($\alpha = 3.6$, $\kappa = 0.4$)

where $\alpha = 5.50\, D_*/u_*$. The values of C_* and α were obtained by fitting Eq. (6) to all the vertical profiles of C. As a result, it was seen that α, being the rate of the settling velocity at $\tilde{z} = 1.0$ to the friction velocity, takes 3.6 approximately in all the experiments. The dependence of $\bar{C}_*$ on $\bar{u}_*/c$ is shown in Fig. 6 and a relationship of $C_* = 1.42 \times 10^{-4}(u_*/c)^2$ can be read. Figure 7 shows the comparison between the experimental data and the semi-empirical relation with $\alpha = 3.6$. A good agreement is seen between the both.

4. CONCLUSIONS

The experiments were carried out to estimate the spray quantity generated from white caps under a strong wind. The drag coefficient on the water surface is given by a monotone increasing function of the wind-wave factor. From that the representative diameter of spray decreases linearly with height, it is deduced that the settling velocity also decreases linearly. The ratio of the settling velocity at the elevation of wave height to the friction velocity takes about 3.6 and the spray concentration at the elevation of wave height is proportion to the second power of wind-wave factor. If mean wind velocity at 10 m height and predominant wave velocity are known, the spray concentration at an arbitrary height can be estimated by using the above mentioned results.

5. REFERENCES

Best, A. C. (1950). "Empirical formulae for the terminal velocity of water drops falling through the atmosphere", Quart. J. Roy. Meteor. Soc, Vol.76, pp.302-311.

Hama, K. and Takagi, N. (1970). "Measurement of sea-salt particles on the coast under moderate winds", Papers in Meteorology and Geophysics, Vol.21, No.4, pp.449-458.

Matsunaga, N., Hashida, M. and Irie, I. (1996). "Landward transport of spray generated from a wave absorbing sea wall", Proc. of 25th ICCE, ASCE, pp.1022-1033.

Matsunaga, N., Hashida, M. and Kawakami, Y. (1996). "Wind-induced waves and currents in a nearshore zone", Proc. of 25th ICCE, ASCE, ASCE, pp.3363-3377.

Muramoto, K., Shiina, T., Nakata, K. and Doai, M. (1990). "Settling velocity of a raindrop and its diameter by graphic data processing", Proc. of IEICE Transactions on Information and Systems, pp.7-186, (in Japanese).

Sasho, Y., Mori, T., Onozaki, O., Saito, T. and Tsutsui, K. (1989). "On the accuracy of the method for measuring the mass of precipitation particles by using filter paper", Tenki, 37(1), pp.61-66, (in Japanese).

Toba, Y. (1958). "Observation of sea water droplets by filter paper", Jour. Oceanogr. Soc. Japan 15, pp.121-130.

Toba, Y. and Tanaka, M. (1963). "Study on dry fallout and its distribution of giant sea-salt nuclei in Japan", Jour. Met. Soc. Japan, Ser. II. 41, pp.135-144.

Toma, K., Ihara, T., Yamamoto, H. and Manabe, T. (1982). "A performance test of a microphonic raindrop-sizemeter by artificial waterdrops and photographing of the drop shapes", Rept. of CRL, Vol.28, No.147, pp.503-519, (in Japanese).

Environmental Hydraulics, Lee, Jayawardena & Wang (eds) © 1999 Balkema, Rotterdam, ISBN 90 5809 035 3

Numerical modeling of wind-induced currents in shallow lake

Li Jinxiu, Liu Shukun, Li Shujun & Yu Xuezhong
Department of Hydraulics, China Institution of Water Resources and Hydropower Research, Beijing, China

ABSTRACT: Wind field is the key factor for simulation of the water flow in a shallow lake. In this paper, a three-dimensional airflow model and a two-dimensional water flow model are developed and applied to simulate the wind-field over lake, wind-induced currents and wind-induced water level oscillation in the Dianchi lake.

1. BACKGROUND

It has become increasingly evident that physical mixing processes in lakes and reservoirs have significant effects on water quality, in particular, wind-induced currents are responsible for the transport of nutrients and the dispersion of effluents in shallow lake, so it is very necessary to study the wind induced currents. In the simulation of wind-induced currents in past 2-3 decades, a spatially uniform and timely constant wind field over the simulated area were normally assumed. However these simplifications are not valid in most cases, and can lead to serious error in the computed flow field(J.Józsa, et al, 1990). In fact, when wind flows across a lake, there must be a non-uniform wind field over lake because of the effects of complex topography around lake and the surface roughness difference between water and land. In order to precisely quantify the wind-induced currents, it is signification to develop an air-water coupled model.

In this paper, we present a three-dimensional airflow model suitable for simulating wind field over lake and a two-dimensional water model suitable for simulating flow field in shallow lakes. The two models were coupled and applied for simulating the wind-induced currents in the Dianchi lake.

2. INTRODUCTION OF MODELS

2.1 A 3D Airflow Model (3DAM)

To meet complex topography surrounding lake, we develop a terrain-following (x, y, $\bar{z}$)coordinate three dimensional micro-meteorology model. Define $\bar{z}$ as :

$$\bar{Z} = H(Z - Zg) / (H - Zg) \tag{1}$$

where H is the top height of initial model, Zg is the terrain height.

Based on assumption of dry and incompressible in the modeled atmo-sphere, the motion of atmosphere in (x, y, $\bar{z}$) coordinate system can be described as follow:

$$\frac{du}{dt} = -\theta\frac{\partial\pi}{\partial x} + g\frac{\overline{z}-H}{H}\cdot\frac{\partial Zg}{\partial x} + f\cdot v + F_u \tag{2}$$

$$\frac{dv}{dt} = -\theta\frac{\partial\pi}{\partial y} + g\frac{\overline{z}-H}{H}\frac{\partial Zg}{\partial y} - f\cdot u + F_v \tag{3}$$

$$\frac{\partial u}{\partial x} + \frac{\partial v}{\partial y} + \frac{\partial \overline{w}}{\partial z} - \frac{u}{H-Zg}\cdot\frac{\partial Zg}{\partial x} - \frac{v}{H-Zg}\cdot\frac{\partial Zg}{\partial y} = 0 \tag{4}$$

$$\frac{dq}{dt} = F_q \tag{5}$$

$$\frac{\partial p}{\partial z} = -\frac{H-Zg}{H}\cdot\frac{g}{q} \tag{6}$$

where,

$$\frac{d}{dt} = \frac{\partial}{\partial t} + u\frac{\partial}{\partial x} + v\frac{\partial}{\partial y} + \overline{\omega}\frac{\partial}{\partial z} \tag{7}$$

$$\overline{\omega} = \omega\frac{H}{H-Zg} + \frac{\overline{z}-H}{H-Zg}\cdot u\cdot\frac{\partial Zg}{\partial x} + \frac{\overline{z}-H}{H-Zg}\cdot v\cdot\frac{\partial Zg}{\partial y} \tag{8}$$

and u, v, ω are the wind velocities in x, y, z directions respectively, ϖ is wind velocity in $\overline{z}$ direction, f is the coriolis parameter, θ is the potential temperature, $\pi = C_p\left(P/P_0\right)^{R/C_p}$ is Exner function representing atmosphere pressure, P_0=1000hPa, F_u, F_v, F_θ are turbulent eddy term representing the u, v, θ respectively. Where:

$$F_\phi = K_H\cdot(\frac{\partial^2\phi}{\partial x^2} + \frac{\partial^2\phi}{\partial y^2}) + \left(\frac{H}{H-Zg}\right)^2\cdot\frac{\partial}{\partial\overline{Z}}\left(K_V\cdot\frac{\partial\phi}{\partial\overline{Z}}\right) \tag{9}$$

and K_H, K_V are turbulent eddy coefficients in horizontal and vertical directions respectively.

The 3DAM was compared with analytical solutions for three idealized airflow before being calibrated and verified with field data from Dianchi lake. The turbulent eddy coefficients in horizontal and vertical directions respectively were given,

$$K_H = 50m^2/s$$

$$K_V(z) = K_V(z_1)\frac{z}{z_1}exp\left[-\rho(z-z_1)/h\right] \tag{10}$$

K_v was taken by Shir and Shieh(1974) empirical relationship, where z_1=10m which represents the height of wind station, ρ is the stability coefficient of atmosphere, and was given $\rho = 2.0$, h is the height of mixing layer.

2.2 A 2D Hydrodynamic Model (2DHM)

In the majority of shallow lakes, the horizontal spatial flow patterns is more important than the vertical spatial patterns, the water body can be treated as a two-components system which is of practicality. Therefore, the depth-integrated 2D plane non-steady flow equations are applied for describing the water movement of lake.

$$\frac{\partial h}{\partial t} + \frac{\partial M}{\partial x} + \frac{\partial N}{\partial y} = 0 \tag{11}$$

$$\frac{\partial M}{\partial t} + \frac{\partial}{\partial x}(u \cdot M) + \frac{\partial}{\partial y}(v \cdot M) = -g \cdot H \frac{\partial H}{\partial x} + f \cdot N + \frac{1}{\rho}(\tau_{x(s)} - \tau_{x(b)}) \tag{12}$$

$$\frac{\partial N}{\partial t} + \frac{\partial}{\partial x}(u \cdot N) + \frac{\partial}{\partial y}(v \cdot N) = -g \cdot H \frac{\partial H}{\partial y} - f \cdot M + \frac{1}{\rho}(\tau_{y(s)} - \tau_{y(b)}) \tag{13}$$

where, h is the water depth, h=Z-Z$_b$, Z$_b$ is the bed elevation, Z is the water surface elevation, u, v are the velocity components in x, y directions respectively, M, N are the discharges per unit width in x, y directions respectively, f is the Coriolis parameter, $\tau_{x(b)}, \tau_{y(b)}$ are the bottom friction stresses in x, y directions, $\tau_{x(s)}$ $\tau_{y(s)}$ are the wind stresses at surface in x, y directions.

$$\tau_{x(s)} = \rho_a \cdot c_d \cdot w \cdot w_x ; \quad \tau_{y(s)} = \rho_a \cdot c_d \cdot w \cdot w_y \tag{14}$$

where, ρ_a is air density, w_x, w_y are the wind velocity components in x, y direction respectively, and w is wind speed at an elevation of 10m above the lake surface, c_d is a surface drag coefficient. According to Bendtsson(1973), the drag coefficient was about 1.0×10^{-3} in most fresh water, and the most convenient wind drag coefficient was set at about 1.3×10^{-3} in a shallow lake, $\tau_{x(b)}, \tau_{y(b)}$ are the bottom friction stresses in x, y directions respectively.

$$\tau_{x(b)} = c_b \cdot \rho \cdot u \cdot \sqrt{u^2 + v^2} ; \quad \tau_{y(b)} = c_b \cdot \rho \cdot v \cdot \sqrt{u^2 + v^2} \tag{15}$$

c_b is the bottom friction coefficient, and is given by Chezy formula as $c_b = \frac{1}{n} \cdot h^{\frac{1}{6}}$, n is the bed roughness of lake.

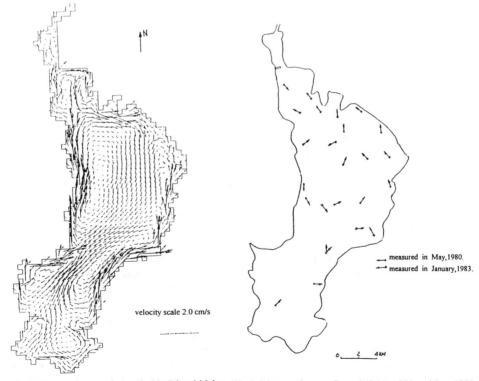

Fig.1: Simulated flow field under uniform wind in Dianchi lake   Fig.2: Measured water flows in May 1980 and Jan. 1983
(Swwind,4.2m/s)

495

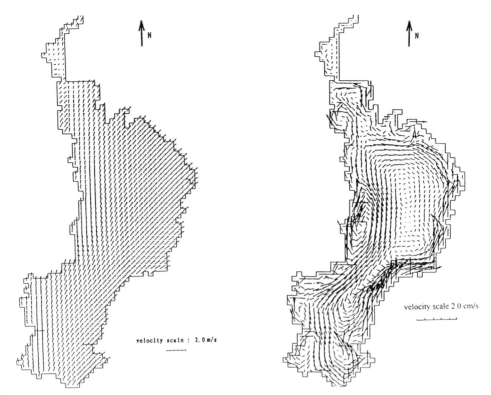

Fig.3:(a)Simulated horizontal wind field (SW,4.2m/s)   (b)Simulated flow field under simulated wind field (a)
at 10m height from the surface of Dianchi lake

3. NUMERICAL SIMULATING ON THE WIND-INDUCED CURRENTS IN THE DIANCHI LAKE

3.1 The Description of the Dianchi Lake

The Dianchi lake is a famous fresh lake in the Yungui plateau in China, it is located in the southern suburbs from Kunming city in Yunnan province, with a surface of 298km^2, and a average depth of 3m-5m. The length of the lake from south to north is about 40km and the maximum width from east to west is 12.5km, the lake is in the shape of a bow-from north to south. In the western side of the lake there are hills which are 100m-500m high from the water surface and form a wind sheltered zone close to the west side.

3.2 The Simulated Wind-Induced Currents

The river inflows and outflow entering the Dianchi lake with low quantity have relatively low effect on the generally circulation pattern. The water movements of lake are therefore greatly induced by wind over lake. The Dianchi lake is located in the south-west monsoon climate zone, the south-west wind is the annual prevailing wind direction. Numerical tests with 2DHM indicated that the flow patterns in the Dianchi lake varied obviously with the variance wind of direction and velocity. In this paper we only introduce the research results of wind-induced currents under the case SW wind.

At first, we simulated the wind-induced current by the approximation of a uniform SW wind field, the simulated flow field is shown in Fig.1, the flow directions in the shallower water zones not far from shores are agreement with the wind direction, and then in the deeper water zones in the middle part of lake the flow

496

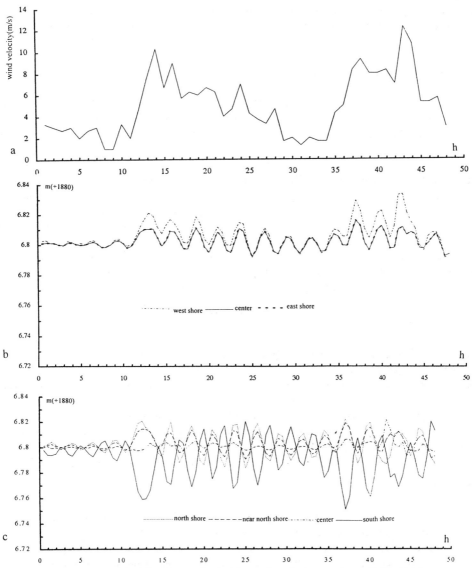

Fig.4: (a)Measured the wind directions and velocities per hour in the east shore of Dianchi lake
from 26 Apr. to 28 Apr.

(b)Simulated water level variation at three sites in lateral section with max. width

(c)Simulated water level variation at four sites in longitudinal section, the initial level: 1886.8m

497

directions are opposed to the wind direction. This flow field is alike with the theoretical analysis solution of circulation in ideal shallow lake. But the simulated flow field in Fig.1 is deviated from the measured flow field show in Fig.2.

If we use the coupled model of 3DAM and 2DHM, Computed wind field over the lake and flow field (see Fig.3(a) and Fig.3(b) respectively) show that the wind field is really non-uniform because of the effects of west hills sheltering, and the computed flow field under non-uniform is fitting better the measured lake flows than the computed flow field with uniform wind field.

By comparison in Fig.2 with Fig.3(b), it is clear that the hills sheltering affect largely the flow field in the the Dianchi lake.

3.3 Simulated the water level oscillation wave

Because of the complex topography around the Dianchi lake, the wind field over lake usually varies in a day and even in hour. In order to understand the complex hydrodynamic characters, the water level oscillation in Dianchi lake were simulated by air-water coupled model under unsteady and non-uniform wind field.

Fig.4(a) shows the measured wind directions and wind velocities per hour in the east shore of Dianchi lake from 26 Apr. to 28 Apr. in 1995. During the two days, the wind directions fluctuated around the SW direction, but the wind velocities fluctuated largely from min. of 1m/s to max. of 12.4m/s. Under the unsteady wind-field, the simulated water level variations at seven representative sites are shown in Fig.4(b) and Fig.4(c). There are obvious periodic oscillation wave on the water surface in the Dianchi lake. The period of oscillation is about 2.5 hours that has nothing to do with the wind speed, and the amplitude of water level increases with the increases of wind speed. The max. amplitude is 6cm appeared at the site of south shore. These results are correspondence with the natural phenomenon in the Dianchi lake.

4. CONCLUSION

(1) The wind field over lake are obviously non-uniform because of the effect of hills sheltering in the west shore in the Dianchi lake;

(2) The simulated flow field with air-water model is fitting better the measured lake flows than the computed flow field with uniform wind field;

(3) The topography surrounding the Dianchi lake affect largely on the flow field ;

(4) The wind induces the periodical oscillation of water level in the Dianchi lake, the period of oscillation is about 2.5 hours that has nothing to do with the wind speed.

5. REFERENCE

Bengtsson, L.(1973). "Models of wind generated circulation in lakes-comparison with measurements", University of Lulea, Sweden,page,6.

J.Józsa, J., Sarkkula, R. and Tamsalu (1990). "Calibration of modeled shallow lake flow using wind field modification", Proceedings of the 8th international conference on computational methods in water resources, p165-170.

Shir, C.C., Shieh, L.J.(1974). "A generalized urban air pollution model and its application to the study of SO^2 distribution in St.Louis metropolitan area", *J. Appl. Meteor.*, 13:185-204.

Environmental Hydraulics, Lee, Jayawardena & Wang (eds) © 1999 Balkema, Rotterdam, ISBN 90 5809 035 3

Heat balance analysis in the Bay of Tokyo

Pranab J. Baruah & So Kazama
WEM/SCE, Asian Institute of Technology, Bangkok, Thailand

ABSTRACT : The energy budget model with bulk aerodynamic formula can describe air-sea exchange of heat in large spatial scale without much complication. Sea surface temperature(SST) distribution pattern of Tokyo Bay is studied and mean yearly net heat content of Tokyo-Bay is estimated using NOAA/AVHRR remote-sensing data from 1990 to 1994. Distribution of SST over Tokyo bay is not uniform in all months of a typical year. In the winter season, the inlet of the bay has higher temperature than the farthest point and in the summer season this pattern reverses. For net heat content, the inlet has lower value than the farthest point in the summer.

1. INTRODUCTION

Japanese Tokyo Bay is the inlet of the Pacific Ocean. It lies on the east central coast of Honshu island, Japan. Along the northwestern shore of the bay are the major cities of Tokyo, Yokohama, and Kawasaki. The bay is a closed type bay with about 1000 km² area. The bay is linked to the Pacific ocean by the Uraga channel, which lies between the Miura(west) and Boso(east) peninsulas. The climate is hot all over the year except December to January with a temperature between 0°C to 4°C in January. In April it is between 12°C to 20°C and in October it ranges from 16°C to 26°C. The August is the hottest month with temperature ranging from 26°C and up.

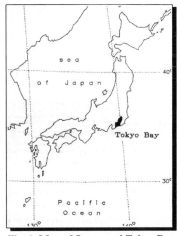

Fig. 1 Map of Japan and Tokyo Bay

The purpose of this paper is to study the temperature distribution pattern in the Tokyo Bay in a typical time period and to study the net heat content of the bay using climatological and NOAA/AVHRR data. The utilization of climatological data is not well justified in calculating heat fluxes because of the variability of atmospheric conditions due to synoptic disturbances, and also that transfer formulas are actually nonlinear because transfer coefficients are determined generally from synoptic data rather than climatic data (Haney, 1971). A simple model based on climatological data gives the approximate result for a stationary atmosphere which is more than enough for assessing different required quantities for practical purposes.

2. DATA USED

For the study of the heat environment of the Tokyo bay, the remote sensing Sea Surface Temperature (SST) data is used. This data is taken from the AVHRR/NOAA image database(JAIDAS). Center for Atmospheric and Ocean studies and the Computer center of the Tohoku University are producing JAIDAS (Japan International DAtabaSe). This dataset has 2 channels which are near infrared of channel 2 can express similar visible ray and for infra-red of channel 4 can express SST. Channel 4 data is used in this study which has albedo of 0.1. Each pixel represents an area of 1.1 km square on ground. The size of one pixel is 1 byte. The SST data from JAIDAS is in binary format. A total number of 814 pixels represent the Tokyo-Bay. SST is available on daily basis for the period 1990 to 94.

As the distributed meteorological data are not available, uniform values are assumed over the whole bay. These meteorological data are which are directly observed at a ground station are wind speed at reference height, Air Temperature at the reference height, vapor pressure, sunshine hours, SST measured directly at a coastal station. The reference height is 1.5 m.

3. BASIC DATA PROCESSING

a.   JAIDAS data collection :

For this study, at first fine satellite images over the bay of Tokyo are taken into consideration. The area considered is about 60 km from east to west and 70 km from north to west. After that full resolution thermal images of channel 4 were download and finally part scenes of 100 line . 100-pixel were collected. Altogether, 244 dates were selected from July, 1990 to December, 1994.

b.   Correction of Sea Surface Temperature

The channel 4 image of the JAIDAS data is a thermal one and its pixel value can be converted to the brightness temperature. Due to imperfect transfer function and contamination of the observed data by the fluctuation of atmospheric aerosol, the pixel values have some errors. Correction is applied by shifting the values of brightness temperature by amount of difference between the satellite data and the real sea surface temperature observed off the Kannonzaki Cape by the Marine Laboratory of Kanagawa Prefecture.

4. STUDY OF THE SST DISTRIBUTION PATTERN OF THE BAY OF TOKYO

Color SST distribution diagrams of Tokyo Bay have been developed for each of 244 days of remote sensing data. Also, mean monthly SST distribution diagrams for 1990-94 are generated for study changing pattern in a typical year. The diagrams show that, in the summer season Tokyo Bay experiences higher temperature at the farthest point followed by middle region and the inlet of the bay. This pattern reverses as the winter season comes. The Tokyo bay is shallower in the inner part and deeper towards the inlet. This is the cause for changing pattern of temperature distribution over the bay Because of lower depth, the inner parts of the bay is easily cooled down or become hot which is reflected in the diagrams. Moreover, the inlet is under influence of oceanic currents in the pacific ocean, such as warm Kuroshio current. This connection to the pacific ocean keeps the inlet-temperature almost constant almost all over the year which is obvious from the distribution pattern diagrams

Upwelling of cold and anoxic water, which is termed as "blue tide" is also visible in the satellite images. A blue tide was reported on September 9, 1992. The image of September 9, 1992 shows high temperature in the inner bay and lower one at the inlet, whereas, image of September 7 shows low temperature at the inner most part of the bay. This cold water mass corresponds to upwelled water by blue tide.

5. HEAT BALANCE OF THE TOKYO BAY

The energy budget model with bulk aerodynamic formula can describe air-sea exchange of heat in a large spatial scale without much complications. The simple one dimensional energy balance model is considered here is as follows :

$$Q = SW - Q_{sens} - Q_{lat} - LW \qquad (1)$$

where, Q: net surface heat flux into the ocean, Q_{sens} : sensible heat flux, Q_{lat} : evaporative heat flux, SW: net downward short wave heat flux, LW: net upward long wave radiative heat flux. The above equation is valid if processes such as heating from chemical and biological processes, heat induction through the bottom, advected energy, and transformation of kinetic energy to heat are neglected. Furthermore, the heat flux due to direct rainfall and from the rivers are not considered in this paper.
The sensible heat flux is estimated by the following equation :

$$Q_{sens} = \rho_a\, C_p\, WC_H\, (T_s - T) \qquad (2)$$

where, ρ_a : density of air, C_p : specific heat capacity of the atmosphere, W : wind speed, C_H : turbulent exchange coefficient for sensible heat, T_s : sea surface temperature, T : atmospheric temperature.
The following conventional formula compute upward latent heat which takes into account the effect of wind :

$$Q_{lat} = \rho_a\, W\, C_E\, L_v\, (q_{sat}(T_s) - RH.q_{sat}(T)) \qquad (3)$$

where, Q_{lat} : Potential latent heat flux from a fully saturated, C_E : turbulent exchange coefficient for latent heat, L_v : Latent heat of vaporization, RH: Relative humidity, $q_{sat}(T)$: Saturation specific humidity at air temperature T.
The net long wave radiation can be computed from the empirical formula which takes into account the temperature difference between surface layer and lower layer of air (Rosati and Miyakoda, 1988) :

$$LW = \varepsilon\, \sigma\, T_s^4(0.39 - 0.05\,\sqrt{e_a}\,)\,(1 - \chi.n_c^2) + 4\varepsilon\, \sigma\, T_s^3\, (T_s - T) \qquad (4)$$

There are numerous empirical relations to predict the attenuation of solar radiation by the clouds. The one given below is after Reed (1977) and taken by Rosati and Miyakoda (1988) :

$$SW = Q_{TOT}(1 - 0.62.n_c + 0.0019\beta)(1 - \alpha_g) \qquad (5)$$

The state of the atmosphere is assumed to be neutral and the transfer coefficients C_E and C_H are calculated out using the parameterization proposed by Kondo (1975).
To analyze the results and to draw an effective conclusion, three pixels are selected along the bay which represent the inlet, end-point and the middle of the bay (fig. 2).

It is noted in the winter season both latent heat and sensible heat fluxes are positive whereas in the summer season both of them are negative. This shows condensation at the air-water interface in summer. Time series of net heat content, Q and SST at end-point and inlet is shown in the fig. 3 and fig. 5 respectively. From the difference of SST between the inlet and the end-point (fig. 4) the winter and summer season can easily be determined. In the winter season this difference (end point - Inlet) is negative as SST at the end point is lesser. The difference becomes positive as the summer season comes. Difference of net heat content, Q between the end point and the inlet is also shown in the same fig. 4. It is observed that the sign of difference (end point - Inlet) of Q is always opposite to that of SST (fig. 4). Fig. 6 shows these time variations of average

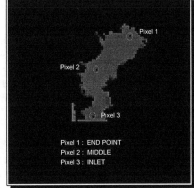

Fig. 2 Three positions of Tokyo Bay

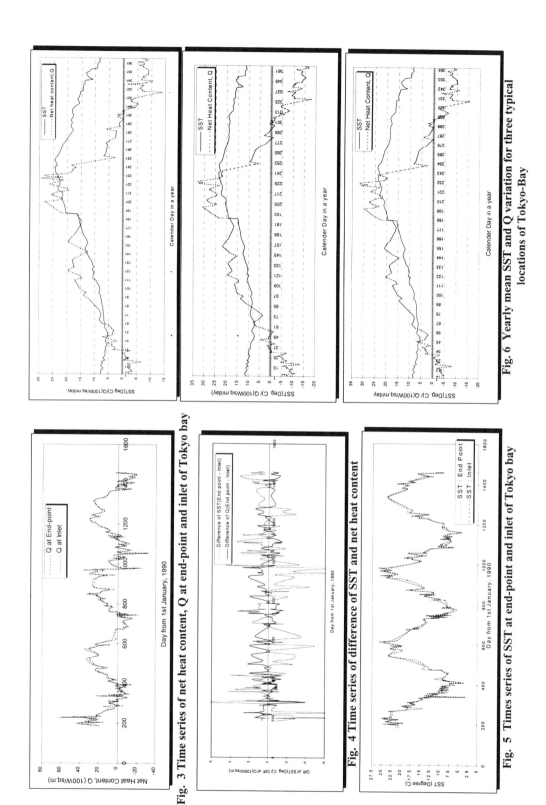

Fig. 6 Yearly mean SST and Q variation for three typical locations of Tokyo-Bay

Fig. 3 Time series of net heat content, Q at end-point and inlet of Tokyo bay

Fig. 4 Time series of difference of SST and net heat content

Fig. 5 Times series of SST at end-point and inlet of Tokyo bay

502

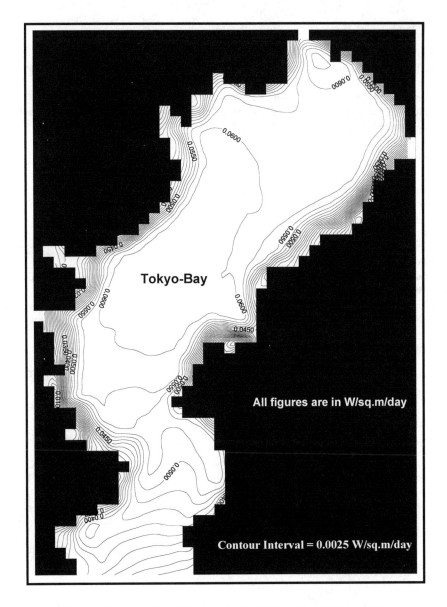

Fig. 7   Mean yearly net heat content of Tokyo Bay (W/sq.m/day)

yearly SST and net heat content at the mentioned positions. The three positions have almost same pattern of changing SST and Q over the year with different magnitude. In the summer season net heat content is highest at the middle point followed by inlet and end point. When it comes to SST, this pattern is reversed. Fig. 7 shows the contour diagram of the mean annual net heat content of the Tokyo-Bay. The diagram clearly shows higher values in the middle of the bay and having a sharp gradient of net heat content towards middle of the bay alongside the coastal line and towards the inlet. The mean yearly net heat content for the whole Tokyo-Bay is estimated to be around 40 W/m^2/day which is the summation of mean yearly net heat content values for all the 814 pixels constituting the bay.

6. CONCLUSION

SST distribution pattern of Tokyo-Bay is affected by the depth gradient to notable extent. Consideration of parameters and terms which takes into consideration the depth of water will give quite accurate estimation of heat content of the bay. The coastal-line and the inlet need more complex models to estimate the net heat content more accurately than one used in this Remote sensing data can be effectively used in the study of heat environment in the coastal area. However, to effectively study short time scale phenomena like upwelling data are needed for different times of a day.

7. REFERENCE

Aase, J. K., and Idso, S. B. (1978). "A comparison of two formula types for Calculating Long-wave radiation from the Atmosphere", Water Resources Research, Vol.14, No.4, pp.623-625.

Atlas of Surface Marine Data (1994). Disc.1 : Directly Observed Quantities, Disc.2: Heat, water and fresh water fluxes, U.S. Department of Commerce, NOAA, NODC.

Bruetsaert, W. (1984), *Evaporation into the atmosphere ,Theory, History & Applications*, 1st Edition, D. Reidel Publishing Company, pp.299.

Burman R., and Pochop, L. O. (1994). *Evaporation, Evapotranspiration and Climatic Data*, 1st Edition, Development in Atmospheric Science, 22, ELSEVIER, pp.278.

Friehe, Carl A., and Schmidth, Kurt F. (1976). "Parameterizations of AIR-Sea Interface Fluxes of Sensible Heat and Moisture by the Bulk Aerodynamic Formulas", Journal of Physical Oceanography, Vol.6 , pp. 801-809.

Holton R. (1992). *An Introduction to the Dynamic Meteorology*, Third Edition, Academic Press Inc., pp.507.

Haney, Robert L. (1971). "Surface Thermal Boundary Conduction for Ocean Circulation Models", Journal of Physical Oceanography, Vol.1, No. 4, pp.241 - 248.

Kondo, J., (1975). "Air-Sea Bulk Transfer Coefficients In Diabetic Conditions", Boundary Layer Meteorology, 9, pp.91-112.

Large, W., and Pond, S. (1982). "Sensible and latent heat flux measurements in moderate to strong winds", J. Physical Oceanography, 11, 324-336

Overhuber, J.M. (1988). *An Atlas based on the COADS Data set : The Budget of heat, Bouyancy, and Turbulent Kinetic energy at the surface of the Global Ocean. Report No. 15*, Max-Planck Institute fur Meteorology.

Reed, R.K. (1977). "On estimating insolation over the oceans", Journal of Physical Oceanography, 7, 482-485.

Rosati, A. and Miyakoda, K. (1988), *Marine Atmosphere*, Intl. Geophys. Ser., Vol.7, 426 pp.

Simpson, J.J. and Paulson, C.A. (1979). "Mid-Ocean observations of atmospheric radiation", Quarterly Journal R. Met. Society, 105, 487-502.

2.5 Environmental fluid mechanics

Environmental Hydraulics, Lee, Jayawardena & Wang (eds) © 1999 Balkema, Rotterdam, ISBN 90 5809 035 3

Large eddy simulation of oscillating flows over a flat plate

Chin-Tsau Hsu
Department of Mechanical Engineering, The Hong Kong University of Sciences and Technology, China

Xiyun Lu
Department of Mechanical Engineering, The Hong Kong University of Sciences and Technology and Department of Modern Mechanics, University of Science and Technology of China, Hefei, Anhui, China

ABSTRACT: Oscillatory turbulent flow over a flat plate is studied by using large eddy simulation (LES). A dynamic subgrid scale (SGS) model is employed. The profiles of the mean velocity and the wall shear stress are computed for the accelerating and decelerating phases during the oscillating cycle. The present results show that the surface shear oscillation in the laminar regime has an amplitude 30% less than and a phase angle $35°$ lag from those of surface shear oscillation in the turbulence regime.

1. INTRODUCTION

An oscillatory turbulent flow over a flat plate has its theoretical and practical significance. This problem is relevant to the interaction between surface gravity waves and sea bottom for the understanding of wave damping and sediment transport in shallow waters. As the waves propagate from the generating area towards the coast, usually the flow near the bottom develops from laminar to turbulent. In addition, the characteristics of turbulence in such an oscillatory flow is quite different from that of wall turbulence in a steady mean flow. Therefore, an investigation of the oscillatory turbulent flow is needed to understand the aspects of unsteady transition from laminar to turbulence, coherent turbulent structures, profiles of the mean velocity, turbulence intensities and the shear stress on the wall during the accelerating and decelerating phases.

Although there exist a few experimental investigations performed by Hino et al. (1983), Sato et al. (1987), and others, little numerical simulations were done for oscillatory turbulent flow over a wall. Blondeaux (1987) studied numerically the turbulent Stokes layer generated by an oscillating flat plate of infinite extent in a fluid at rest by using the Reynolds-average Navier-Stokes (RANS) method based on Saffman's (1970) turbulence model. During recent years large eddy simulation (LES) has been developed to become one of the most powerful computational tools available for the calculation of turbulent flows. Most of the LES works were done for studying steady mean flows with simple geometry (Galperin and Orszag 1993). Lu *et al.* (1997), however, employed LES method to calculate oscillating flows past a circular cylinder.

In this study, an oscillatory turbulent flow over a flat plate is studied by using large eddy simulation (LES) method, to obtain the quantitative information on the transition from laminar to turbulent. In LES, the filtered time-dependent three-dimensional incompressible Navier-Stokes equations are solved using the non-staggered-grid, fractional step method (Zang *et al.* 1994), and a dynamic subgrid-scale model of Germano et al. (1991) was employed.

2. MATHEMATICAL FORMULATION

The filtered time-dependent three-dimensional incompressible Navier-Stokes equations used for LES are given by

$$\frac{\partial \bar{u}_j}{\partial x_j} = 0 \qquad (1)$$

$$\frac{\partial \bar{u}_j}{\partial t} + \frac{\partial}{\partial x_j}(\bar{u}_j \bar{u}_i) = -\frac{\partial \bar{p}}{\partial x_i} + \frac{\partial}{\partial x_j}(v\frac{\partial \bar{u}_i}{\partial x_j} - \tau_{ij}) \qquad (2)$$

where $i, j = 1,3$, $\bar{u}_i$ represents the filtered Cartesian velocities, $\bar{p}$ is the filtered pressure divided by fluid density, v is the kinematic viscosity and τ_{ij} represents the unresolved subgrid scale (SGS) stress term defined as

$$\tau_{ij} = \overline{u_i u_j} - \bar{u}_i \bar{u}_j \qquad (3)$$

This SGS quantities are modeled by using the dynamic subgrid-scale eddy viscosity model (Germano *et al.* 1991; Zang *et al.* 1993). This is different from a classical eddy viscosity model (Smagorinsky 1963) for the subgrid-scale stress which has a prescribed model coefficient. Instead, the dynamic SGS eddy viscosity model calculates the model coefficient using the resolved variables by filtering the governing equations at two different scales. This allows for the determination of the eddy viscosity dynamically to cover the range from laminar to turbulent. Therefore, the dynamic subgrid-scale eddy viscosity model is most suitable for the investigation of flow transition (Piomelli and Zang 1991, Kleiser and Zang 1991).

For simulating the oscillating flows, we use the maximum velocity U and amplitude A of flow oscillation as velocity and length scales, respectively. Thus, $U=2\pi fA$ and the time scale is $A/U=1/2\pi f$, where f is the frequency of the flow oscillation. Eqs.(1) and (2) are then transformed through curvilinear coordinate system into the following nondimensional conservation form,

$$\frac{\partial U_m}{\partial \xi_m} = 0 \qquad (4)$$

$$\frac{\partial}{\partial t}(J^{-1}\bar{u}_i) + \frac{\partial}{\partial \xi_m}(U_m \bar{u}_i) = -\frac{\partial}{\partial \xi_m}(J^{-1}\frac{\partial \xi_m}{\partial x_i}\bar{p}) + \frac{\partial}{\partial \xi_m}[(\frac{1}{Re} + \frac{1}{Re_T})(G^{mn}\frac{\partial \bar{u}_i}{\partial \xi_n})] \qquad (5)$$

where Re is the Reynolds number, defined as $Re = 2\pi fA^2/v$ and Re_T is turbulent viscosity calculated from the dynamic subgrid-scale eddy viscosity model (Germano *et al.* 1991; Zang *et al.* 1993); J^{-1} is the inverse of Jacobian, U_m is the contra-variant velocity along ξ_m multiplied by J^{-1}, and G^{mn} is the "mesh skewness tensor"; they are defined, respectively, by

$$J^{-1} = \det(\frac{\partial x_i}{\partial \xi_j}), \quad U_m = J^{-1}\frac{\partial \xi_m}{\partial x_j}\bar{u}_j, \quad G^{mn} = J^{-1}\frac{\partial \xi_m}{\partial x_j}\frac{\partial \xi_n}{\partial x_j} \qquad (6)$$

In the present calculation, we use periodic boundary conditions in streamwise and lateral directions and no-slip boundary condition on the plate surface. At upper boundary far from the plate, nondimensional velocity along streamwise direction is set as $\overline{u}_\infty = \sin(t)$.

3. NUMERICAL PROCEDURE

Numerical solutions to the equation system (4-5) are obtained by the non-staggered-grid, fractional step method. Following Zang et al. (1994), a semi-implicit time-advancement scheme is adopted with the Adams-Bashforth method for the explicit terms and the Crank-Nicolson method for the implicit terms.

Except for the convective terms, all the spatial derivatives are approximated with second-order center differences. The convective terms are discretized by a variation of QUICK shown in Perng and Street (1989). Using the fractional step method, the time advancement in Eq.(8) is solved with a predictor-corrector procedure. The pressure is then obtained by solving the pressure Poisson equation using the multigrid method (Perng and Street 1989).

In the present computation, the number of mesh points for LES calculations was 65×65×65 in streamwise, spanwise and wall-normal directions, respectively, with a time step of 0.001. In order to increase the resolution of mesh near the wall, stretching transformation is used along the wall-normal direction. It has also been verified that the computed results are independent of the time steps and the grid sizes.

To validate the LES code, we first calculated a fully developed turbulent channel flow at Re_C=3300, based on the mean centerline velocity and channel half-width. The profile of the steady mean velocity is calculated by time-average and space-average over horizontal planes (homogeneous directions). The profile in terms of wall coordinate (u^+, y^+) as non-dimensionalized by the friction velocity is shown in Fig.1. Also shown in the figure is the mean-velocity profile from the experimetal results of Eckelmann (1974). The dashed line represents the law of the wall and the semi-log law. Within the viscous sublayer, y^+<5, both the experimental and the computational results follow the linear law of the wall. For y^+ >40, the computed mean-velocity profile shows the semi-logarithmic relation and is in good agreement with the experimental results. From the mean

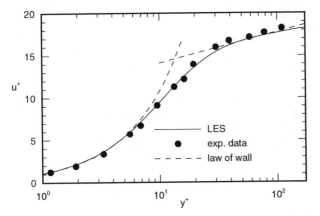

FIG.1. Mean-velocity profile in the vicinity of the lower wall for fully developed turbulence channel flow computed by LES at Re_C=3300, and comparison with experimental data of Eckelmann(1974).

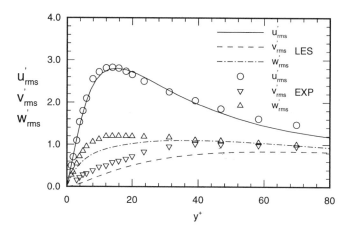

FIG.2. Resolvable turbulence intensities for channel flow computed by LES at Re_C=3300 in wall coordinates and comparison with experimental data of Kreplin and Eckelmann (1979).

velocity profile, the Reynolds number based on the friction velocity and channel half-width is Re_τ=180, which is in excellent agreement with the DNS result by Kim et al.(1987). The turbulence intensities normalized by the friction velocity are shown in Fig.2. The comparison of the present calculated results with the experimental data by Kreplin and Eckelmann (1979) for Re_τ=194 shows that they are in good agreement.

4. RESULTS AND DISCUSSION

With the LES code validated, we then proceed to compute the oscillating flow over flat plate with $Re=4\times10^5$ and the results are presented below. The space-averaged velocity profiles over the x-z plane parallel to the wall are plotted on semi-log diagrams at various phases as shown in Fig.3. During the half cycle, a layer that obeys the semi-logarithmic law exists above a sublayer, as marked by the tangential lines in Fig.3. This feature is similar to the viscous sublayer of steady wall turbulent flows. However, the semi-loglaw regimes for the accelerating phase are very narrower, while those for the decelerating phase are broader. As indicated by the slope of the tangential lines, we also observed that the turbulent shear stress is much higher during the accelerating phase than the decelerating phase. The behaviors mentioned above indicate the completely different turbulence features between the accelerating and decelerating phases. These features are qualitatively in good agreement with the experimental results of Hino et al. (1983).

Fig.4 shows the variations of the friction coefficient C_F with time, as calculated by LES and laminar prediction, respectively. We see that the amplitude of surface shear stress oscillation for turbulent results is higher than the laminar prediction by about 30%. The phase angle of the turbulent result however leads about 35° to that of laminar result. In reference to the phase angle of the ambient oscillation, the phase angle of the surface shear oscillation is 45° in laminar regime and is about 10° in turbulent regime.

5. CONCLUDING REMARKS

An oscillatory flow over a flat plate is studied by LES. The filtered time-dependent three-dimensional incompressible Navier-Stokes equations are solved using the non-staggered-grid, frictional step method in LES coupled with a dynamic subgrid-scale model. Based on our calculated results, the different features of turbulence structures between the accelerating and decelerating phases can be identified. Excellent agreement among theoretical prediction, experimental data and the LES computation are found. The features in the friction coefficient and phase shift of the oscillatory surface shear stress are investigated. The transition from a laminar oscillating flow to a turbulent oscillating flow leads not only to the change in the amplitude of shear stress oscillation, but also the phase angle. The LES result at $Re=4\times10^5$ shows that the amplitude of the friction coefficient for turbulent oscillating flow is about 30% higher than that of the laminar flow. On the other hand, the phase of turbulent flow leads that of laminar flow by about 35°.

ACKNOWLEDGEMENT

The authors sincerely thank Prof. R. L. Street of Stanford University for providing the initial version of the LES code. This work is supported by the Hong Kong Government under the RGC Grant Nos. HKUST708/95E and HKUST815/96E , and under the RIG Grant No. RI95/96.EG15.

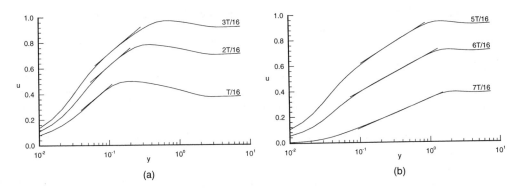

FIG.3. The x-z plane space-averaged velocity profiles calculated by LES for $Re=4\times10^5$ plotted on semi-log diagrams: (a) accelerating phase; (b) decelerating phase.

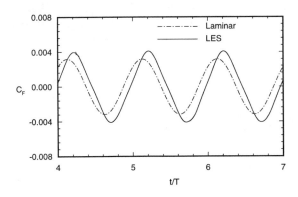

FIG.4. Variations of the friction coefficient C_F with time at $Re=4\times10^5$.

6. REFERENCES

Blondeaux, P. (1987). "Turbulent boundary layer at the bottom of gravity waves", *J. Hydro. Res.*, 25, 447-461.

Eckelmann, H. (1974). "The structure of the viscous sublayer and the adjacent wall region in a turbulent channel flow", *J. Fluid Mech.*, 65, 439.

Galperin B., and Orszag, S. A. (1993). "Large eddy simulation of complex engineering and geophysical flows", *Cambridge Univ. Press*.

Germano, M., Piomelli, U., Moin, P., and Cabot, W.H. (1991). "A dynamic subgrid-scale eddy viscosity model", *Phys. Fluids*, 3, 1760-1765.

Hino, M., Kashiwayanagi, M., Nakayama, A., and Hara, T. (1983). "Experiments on the turbulence statistics and the structure of a reciprocating oscillatory flow", *J. Fluid Mech.*, 131, 363-400.

Kamphuis, J.W. (1975). "Friction factor under oscillatory waves", *J. Waterways Harb. Div.*, ASCE, 101, 135.

Kim, J., Moin, P., and Moser, R. (1987). "Turbulence statistics in fully developed channel flow at low Reynolds number", *J. Fluid Mech.*, 177, 133-166.

Kleiser, L., and Zang, T. A. (1991). "Numerical simulation of transition in wall-bounded shear flows", *Ann. Rev. Fluid Mech.*, 23, 495-537.

Kreplin, H., and Eckelmann, H. (1979). "Behavior of the three fluctuating velocity components in the wall region of a turbulent channel flow", *Phys. Fluids*, 22, 1233.

Lu,X.Y., Dalton,C., and Zhang,J. (1997). "Application of large eddy simulation to an oscillating flow past a circular cylinder", *J. Fluids Engng.*, ASME, 119, 519-525.

Perng, C.-Y., and Street, R.L. (1989). "Three-dimensional unsteady flow simulations: alternative strategies for a volume-averaged calculation", *Int. J. Numer. Methods Fluids*, 9, 341-362.

Piomelli, U., and Zang, T. A. (1991). "Large-eddy simulation of transitional channel flow", *Computer Phys. Comm.*, 65, 224-230.

Saffman, P.G. (1970). "A model for inhomogeneous turbulent flow", *Proc. Roy. Soc. London A*, 317, 417-433.

Sato, S., Shimosako, K., and Watanabe, A. (1987). "Measurements of oscillatory turbulent boundary layer flow above ripples with a laser-Doppler velocimeter", *Coast. Engng Japan*, 30, 89-98.

Smagorinsky, J. (1963). "General circulation experiments with the primitive equations, I. The basic experiment", *Mon. Weath. Rev.*, 91, 99-164.

Zang, Y., Street, R.L., and Koseff, J. R. (1993). "A dynamic mixed subgrid-scale model and its application to turbulent recirculating flows", *Phys. Fluids*, 5, 3186-3196.

Zang, Y., Street, R.L., and Koseff, J. R. (1994). "A non-staggered grid, fractional step method for time-dependent incompressible Navier-Stokes equations in curvilinear coordinates", *J. Comput. Phys.*, 114, 18-33.

Environmental Hydraulics, Lee, Jayawardena & Wang (eds) © 1999 Balkema, Rotterdam, ISBN 90 5809 035 3

Steady streaming in an oscillatory flow over a periodic wavy surface

Xiyun Lu
Department of Mechanical Engineering, The Hong Kong University of Sciences and Technology, and Department of Modern Mechanics, University of Science and Technology of China, Hefei, Anhui, China

Chin-Tsau Hsu
Department of Mechanical Engineering, The Hong Kong University of Sciences and Technology, China

ABSTRACT: Oscillating flow over a periodic wavy surfaces is numerically investigated by solving the 2-D time-dependent incompressible Navier-Stokes equations in curvilinear coordinates. When the amplitude and the frequency of ambient oscillations are sufficiently small, solutions which are periodic in time and space exist and steady circulation cells due to the streaming effect form. The structure of the streaming pattern becomes great important since the steady circulation can help to redistribute the suspended sediments. Based on the present calculated results, the steady streaming patterns with either two cells or four cells over one period of the wavy surface are observed. The number of circulation cells is found to depend only on α and Re, but not on ka.

1. INTRODUCTION

Sand ripples are frequently found on beaches under the influence of ocean tides, currents and waves. When the velocity of the tides, currents and waves is sufficiently large, vortices are generated in the lee of every crest. These vortices are effective in dislodging sand particles and maintaining them in suspension; hence, they contribute to the evolution of the ripples as well as the transport of sediments. At high velocity, the flow above the natural ripples is usually turbulent. Theoretical analysis of turbulent flows is extremely difficult especially when coupled with sand motions. A laminar approach was adopted traditionally as a helpful first step towards a better understanding of the turbulent, two-phase flows. To delineate the effect of time scale on the surface forces and to better understand the process of sediment transport due to ocean waves, the problems were further simplified by considering oscillatory flows over periodic wavy ripples. The important physical and geometrical properties in the laminar problem are the frequency ω and amplitude A of the flow oscillations, the ripple amplitude a and wavelength l, and the fluid viscosity v. The combined results of ω and v leads to a dynamic length scale called the Stokes layer thickness, $\delta_s = (v/\omega)^{1/2}$. From these length scales, one can find three non-dimensional governing parameters: the Keulegan-Carpenter number $\alpha = 2\pi A/l$ (which is the non-dimensional oscillation period), the ripple slope $ka = 2\pi a/l$, and the viscous diffusion parameter $\sigma = \delta_s/l$. Alternatively, one can define the Reynolds number by Re=$U_o l/v$, where U_o is the amplitude of velocity oscillation in free stream and consider Re as an independent parameter to replace σ with Re=$\alpha/2\pi\sigma^2$. Because of the nonlinear nature in the problem, theoretical treatments were basically performed under the conditions when one of these non-dimensional parameters is small. Therefore a linearization perturbation technique becomes admissible.

The problem of oscillatory flows over wavy surfaces had been studied by Lyne (1971), Sleath (1974), Kaneko and Honji (1979), Vittori (1989), Hara and Mei (1990a,b), Blondeaux (1990),

Vittori (1989), and Vittori and Blondeaux (1990, 1992). Most of these theoretical studies used perturbation methods based on small parameters. The first theoretical treatment was given by Lyne (1971) who gave a perturbation analysis for small ka/Re. He calculated the steady streaming and found stationary cells inside the Stokes layer for both small and large α. These cells can keep the sediments in suspension. Sleath (1974) assumed both small α and samll ka, and obtained the cellular pattern of Lagrangian mass transport. Kaneko and Honji (1979) introduced a perturbation analysis for small α but finite ka/Re. Vittori (1989) has extended Lyne's analysis, and investigated only the steady streaming. Recently, Hara and Mei (1990a,b) also studied the same problem for samll α but finite ka and $Re \geq O(1)$ based on the perturbation analysis. For higher values of ka, α and Re, discrete numerical methods have been applied to initial-value problems of the full Navier-Stokes equations by Shum (1988).

In this study, we present a numerically accurate solution of the Navier-Stokes equations for an oscillatory flow over a sinusoidal wavy surface in a wider range of the parameters of ka, α and Re. The two-dimensional time-dependent incompressible Navier-Stokes equations in curvilinear coordinates are solved using the non-staggered-grid, fractional step method. The formation of a steady streaming, which has relevance to the ripples formation at the bottom of ocean waves, is investigated.

2. MATHEMATICAL FORMULATION

Considerations were given to sinusoidally oscillating viscous incompressible flows past periodic wavy surfaces (ripples) by assuming that both the oscillatory flows and the wavy surfaces are two-dimensional in the plane normal to the crest direction of the wavy surfaces. The governing equations are then given by the well-known two-dimensional Navier-Stokes equations. After normalization by the velocity U_0 and the wavelength l, the Navier-Stokes equations in Cartesian coordinate system x_j are transformed through a curvilinear coordinate system ξ_m into non-dimensional forms given by

$$\frac{\partial U_m}{\partial \xi_m} = 0 \tag{1}$$

$$\frac{\partial}{\partial t}(J^{-1}u_i) + \frac{\partial}{\partial \xi_m}(U_m u_i) = -\frac{\partial}{\partial \xi_m}(J^{-1}\frac{\partial \xi_m}{\partial x_i}p) + \frac{1}{Re}\frac{\partial}{\partial \xi_m}(G^{mn}\frac{\partial u_i}{\partial \xi_n}) \tag{2}$$

where u_i is the Cartesian velocity components and p is the pressure divided by fluid density. In Eqs.(1) and (2), J^{-1} is the inverse of Jacobian, U_m is the contra-variant velocity along ξ_m multiplied by J^{-1}, and G^{mn} is the "mesh skewness tensor"; they are defined, respectively, by

$$J^{-1} = \det(\frac{\partial x_i}{\partial \xi_j}), \quad U_m = J^{-1}\frac{\partial \xi_m}{\partial x_j}u_j, \quad G^{mn} = J^{-1}\frac{\partial \xi_m}{\partial x_j}\frac{\partial \xi_n}{\partial x_j} \tag{3}$$

In this study, the wavy surface is described by $y = a\cos kx$; therefore, the curvilinear coordinates are adopted from the "terrain-following" coordinate system of Clark (1971), which generally is non-orthogonal.

3. NUMERICAL PROCEDURE

Numerical solutions to the equation system (1-3) are obtained by the non-staggered-grid, fractional step method. Following Zang et al. (1994), a semi-implicit time-advancement scheme with the Adams-Bashforth method is adopted for the explicit terms and the Crank-Nicolson method for the implicit terms. Except for the convective terms, all the spatial derivatives are approximated with second-order center differences. The convective terms are discretized using a variation of QUICK given in Perng and Street (1989). Using a fractional step method, the time advancement is solved with a predictor-corrector solution procedure. The pressure is obtained by solving the pressure Poisson equation using the multi-grid method as described in Perng and Street (1989). To resolve the highly concentrated vorticity near the wavy surface, the curvilinear coordinate normal to the wavy surface is further stretched during discretization based on the scheme described in Moin and Kim (1982). The initial conditions were set when the sinusoidal varying negative pressure gradient is at maximum and the velocity field is zero. The number of mesh points in the present calculation is 130×130 in streamwise and wall-normal directions. While the computational domain in the streamwise direction is one wave length using a periodic boundary condition, the domain in the wall-normal direction is $2.5l$ with a periodically oscillating velocity at the free stream and a solution to Eqs.(1)-(2) are first computed over several oscillating periods to eliminate the transient behavior and then non-slip condition at the wavy wall. The time step is 0.0005. In order to increase the resolution of mesh near the ripple boundary, a stretching transformation is used. The code validation has been done by reconstructing the results obtained by earlier investigators using the present new scheme. It has also been verified that the present computed results are independent of the time steps and the grid sizes.

4. RESULTS AND DISCUSSION

Steady streaming over wavy surface occurs when the flow varies periodically at small values of α, ka and Re. To calculate the steady streaming flow, solutions to Eqs.(1)-(2) are first obtained over several oscillation periods to eliminate the transient behavior and then time-average is taken over one period of the periodically varying flow. We first discuss the steady streaming of viscous oscillatory flow over a sinusoidal ripple. Fig.1 shows the streaming patterns at $\alpha=1$ and $ka=0.5$ with different Reynolds numbers of Re=50, 100, 300, 500. The streaming patterns consist either of two or four recirculating cells within one wavelength depending on Re. At smaller Re, e.g., Re=50 and 100, two recirculating cells are found in Figs.1a,b. When Re increases, four cells are formed at Re=300 and 500, as shown in Figs.1c,d, and the lower two cells become thinner. We have also calculated the flow at higher Reynolds numbers with the same values of α and ka; but only to find that the flow oscillation (not shown here) becomes non-periodic. Based on these calculated results, there exist two threshold values Re_1 and Re_2. When Re is smaller than Re_1, two recirculating cells are formed. As Re increases, the recirculating cells becomes thinner, with the center of the cell moving toward the wavy surface. Further increase of Re to $Re>Re_1$ leads to the generation of counter rotating recirculating cells near the free stream so that the total number of cells becomes four. If Re is increased further, the four recirculating cells move further toward the wavy surface and eventually become unstable when $Re>Re_2$. The flow oscillation then becomes non-periodic. This evolution of the recirculating cells with Re is also clearly demonstrated by Fig.2 which shows the y-component velocity profiles along the symmetry line of the wavy surface. The counter rotation of the recirculating cell is also evident from Fig.2.

Fig.3 shows the streaming patterns for Re=500 and $ka=0.5$ at $\alpha=0.1$, 0.5, 2 and 3. For samller α, such as $\alpha=0.1$ and 0.5 (Fig.3a,b), four cells are found. As α increases, the strength and the thickness of the lower cells grows and the cells tend to spread when the stagnation point which separates the

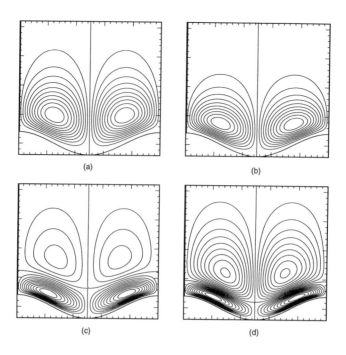

FIG.1. Steady streaming patterns for $\alpha=1$, $ka=0.5$, and Re=(a)50, (b)100, (c)300, (d)500.

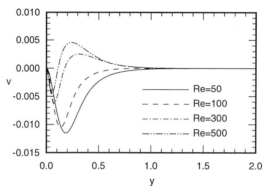

FIG.2. The y-component velocity profiles along the symmetry line of the wavy surface.

cells moves toward free stream as shown in Fig.3c for $\alpha=2$. The strength of the upper cells are weakened and disappearing till only two lower are left, as indicated in Fig.3d for $\alpha=3$. Based on the present results, when α is smaller than a threshold value α_1, there are four cells. When α is higher than α_1, two cells are formed. This evolution of the recirculating cell structure with α is also demonstrated by the y-component velocity along the symmetric line shown in Fig.4. The magnitude of y-component velocity is a measure of the strength of the recirculating cell. The comparison of Fig.4 to Fig.2 indicates that α has the opposite effect on the recirculating cell pattern to Re.

The streaming patterns for Re=500 and $\alpha=1$ with $ka=0.05$ and 1.0 are shown in Fig.5, where four recirculating cells are found. Although the ripple slope changes from 0.05 to 1.0, the steady streaming patterns are similar except the geometric deformation caused by the ripple slope difference. We also calculated the case for Re=100 and $\alpha=1$ with different ka. Always two cells in the streaming patterns (not shown here) are formed. Based on those results shown above, it is clear

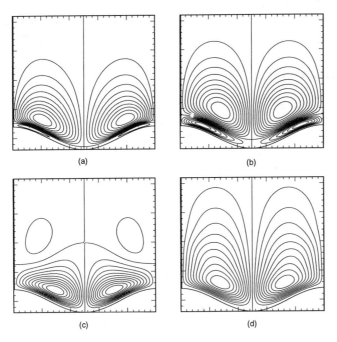

FIG.3. Steady streaming patterns for Re=500, ka=0.5 at α=(a)0.1, (b)0.5, (c)2, (d)3.

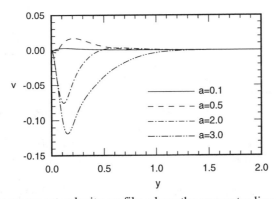

FIG.4. The y-component velocity profiles along the symmetry line of the wavy surface.

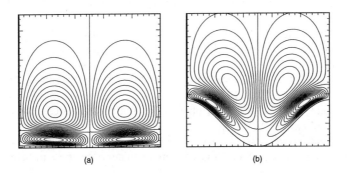

FIG.5. Steady streaming patterns for Re=500, α=1 and ka=(a)0.05, (b)1.0.

517

that the steady streaming pattern with either two cells or four cells depends only on α and Re, but not on ka.

5. CONCLUDING REMARKS

A numerical study of an oscillatory flow over periodic ripples has been performed by solving the two-dimensional time-dependent incompressible Navier-Stokes equations in curvilinear coordinates with the non-staggered-grid, frictional step method. In this study, we can accurately take numerical calculations over a wide range of the parameters (i.e. Re, α, and ka) to delineate the evolution of the steady streaming patterns, which are unattainable through the theoretical analysis. Two and four recirculating cells are found in the steady streaming flow patterns. The steady streaming pattern with either two cells or four cells depends only on α and Re, but not on ka.

ACKNOWLEDGEMENT

This work is supported by the Hong Kong Government under the RGC Grant Nos. HKUST708/95E and HKUST815/96E , and under the RIG Grant No. RI95/96.EG15.

REFERENCES

Blondeaux, P. (1990). " Sand ripples under sea waves. Part 1. Ripple formation", *J. Fluid Mech.*, 218, 1-17.

Clark, T. (1977). "A small-scale dynamic model using a terrain-following coordinate transformation", *J. Comput. Phys.*, 24, 186-215.

Hara, T. , and Mei, C. C. (1990a). "Oscillating flows over periodic ripples", *J. Fluid Mech.*, 211, 183-209.

Hara, T., and Mei, C. C. (1990b). "Centrifugal instability of an oscillatory flow over periodic ripples", *J. Fluid Mech.*, 217, 1-32.

Kaneko, A., and Honji, H. (1979). "Double structures of steady streaming in the oscillatory viscous flow over a wavy wall", *J. Fluid Mech.*, 93, 727-736.

Lyne, W. H. (1971). "Unsteady viscous flow over a wavy wall", *J. Fluid Mech.*, 50, 33-48.

Moin, P., and Kim, J. (1982). "Numerical investigation of turbulent channel flow", *J. Fluid Mech.*, 118, 341-377.

Perng, C.Y., and Street, R.L. (1989). "Three-dimensional unsteady flow simulations: alternative strategies for a volume-averaged calculation", *Int. J. Numer. Mthods Fluids*, 9, 341-362.

Shum, K. T. (1988). "A numerical study of vortex dynamics over rigid ripples", *Ph.D. Dissertation*, MIT, USA.

Sleath, J. F. A. (1974). "Mass transport over a rough bed", *J. Mar. Res.*, 32, 13-24.

Vittori, G. (1989). "Non-linear viscous oscillatory flow over a small amplitude wavy wall", *J. Hydraul. Res.*, 27, 267-280.

Vittori, G., and Blondeaux, P. (1990). "Sand ripples under sea waves. Part 2. Finite-amplitude development", *J. Fluid Mech.*, 218, 19-39.

Vittori, G., and Blondeaux, P. (1992). "Sand ripples under sea waves. Part 3. Brick-pattern ripple formation", *J. Fluid Mech.*, 239, 23-45.

Zang, Y., Street, R.L., and Koseff, J.R. (1994). "A non-staggered grid, fractional step method for time-dependent incompressible Navier-Stokes equations in curvilinear coordinates", *J. Comput. Phys.,* 114, 18-33.

Environmental Hydraulics, Lee, Jayawardena & Wang (eds) © 1999 Balkema, Rotterdam, ISBN 90 5809 035 3

Dispersion of vapour in a layered unsaturated zone

Chiu-On Ng
Department of Mechanical Engineering, The University of Hong Kong, China

ABSTRACT: We present a derivation of the macroscopic equations by the homogenization method for chemical vapour transport in an unsaturated zone with a periodic structure of heterogeneity. The effective specific discharge and hydrodynamic dispersion coefficient are expressible in terms of some cell functions, whose analytical solutions are sought for the simple case of a bi-layered composite.

1. INTRODUCTION

The spread of vapour in an unsaturated zone often plays an important role in the contamination or decontamination of an aquifer, if the contaminants are volatile enough. Very often, the contaminated zone is stratified; layers of soil with different properties are embedded in one-another. For modelling purposes, it is desirable if effective equations can be developed for the vapour transport in such a multi-layered heterogeneous system, where advantage is taken of the fact that the layer thickness is usually much smaller than the global length scale for the transport.

In this work, we adopt the homogenization method to develop macroscopic equations for the transport of chemical vapour in a multi-layered unsaturated zone. We emphasize the direct upscaling from the mesoscopic or formation scale. To enable the analysis, we shall assume an ideal soil structure in which the heterogeneity is periodic. Macroscopic description of the heterogeneous system is then obtained upon averaging over a unit cell of periodicity. On further assuming each cell being composed of two horizontal layers each with distinct properties, we may determine the effective coefficients which appear in the macroscopic equations analytically in terms of the basic flow variables and soil properties. No calibration or statistical parameters need to be introduced. Some applications of the homogenization method have recently been reviewed by Mei et al. (1996).

We consider an unsaturated zone where the moisture is at the residual level. Water with dissolved and sorbed chemicals is held immobilized in the micropores, while air with chemical vapour can flow readily in the macropore space. It is also assumed that residual non-aqueous phase liquid (NAPL) is not present in the soil matrix, or has already been volatilized into vapour phase. Equilibrium chemical phase partitioning on the pore scale is assumed. That is, the local phase exchange among the aqueous, vapour and sorbed phases occurs much faster than the transport on the formation scale. It also means that rate-limiting effects such as aqueous diffusion in aggregates are minimal at the microscopic scale.

In the subsequent analysis, only reference to the formation (mesoscopic) scale (ℓ) and the regional (macroscopic) scale (ℓ') will be required. The small ratio $\varepsilon = \ell/\ell' \ll 1$ will be used as the ordering parameter in the analysis. The coordinates at the formation and regional scales are denoted by $\boldsymbol{x} \equiv (x_1, x_2, x_3) \equiv (x, y, z)$ and $\boldsymbol{x}' \equiv (x'_1, x'_2, x'_3) \equiv (x', y', z')$ respectively. The medium heterogeneity is periodic so that it can be divided into unit cells Ω. We further assume that the mesoscale Peclet number is of order unity: $Pe \equiv U\ell/D = O(1)$ where U and D are respectively the characteristic air velocity and bulk molecular diffusion coefficient of chemical vapour in pore air. Physically $Pe = O(1)$ means that over the formation scale vapour diffusion is as important as advection. It also implies that the regional Peclet number is much greater than unity: $Pe' \equiv U\ell'/D = O(1/\varepsilon) \gg 1$, or the vapour advection dominates at this upper scale. Two time scales are therefore required. While the global advection time scale $T_1 = \ell'/U$ is the primary time scale for the transport, the longer time scale $T_2 = \ell'^2/D$ is for the global diffusion/dispersion: $T_2 = \ell'^2/D = (U\ell'/D)(\ell'/U) = Pe'T_1 = O(\varepsilon^{-1}T_1) \gg T_1$.

2. EFFECTIVE TRANSPORT EQUATIONS

We start from the vapour transport equation on the layer scale:

$$\beta\frac{\partial C}{\partial t} + \nabla \cdot (\boldsymbol{u}C) - \nabla \cdot (\boldsymbol{D} \cdot \nabla C) = 0, \tag{1}$$

where C is the vapour concentration (i.e, mass of chemical in vapour phase per unit volume of pore air), $\boldsymbol{u}$ is the specific discharge, $\boldsymbol{D}$ is the second-order tensor of bulk molecular diffusion coefficient in the pore air, and β is the retardation factor. Note that by virtue of the high vapour diffusivity in air, the mechanical dispersion is relatively unimportant over the mesoscale when $Pe = O(1)$ is true, as shown by Ng and Mei (1996). Therefore $\boldsymbol{D}$ depends only on the soil micro-structure (typically given by the product of the tortuosity tensor and the molecular diffusion coefficient in pure air), but not the flow kinematics. Also, the retardation effect arises from the local equilibrium partitioning among the vapour, aqueous and sorbed phases. Hence, the retardation factor, a function of soil and chemical parameters, may be expressed as follows: $\beta = \theta_g + (\theta_w + K_d\theta_s\rho_s)/K_H$ where θ_g, θ_w and θ_s are respectively the air-filled porosity, water and solid volume fractions, ρ_s is the solid density, K_d is the sorption partition coefficient and K_H is the Henry's law constant.

Upon substituting the standard multiple-scale expansions into (1) and recognizing that the first term is of order ε, we may obtain the following perturbation equations:

$O(1)$:

$$\nabla \cdot (\boldsymbol{u}^{(0)}C^{(0)}) - \nabla \cdot (\boldsymbol{D} \cdot \nabla C^{(0)}) = 0, \tag{2}$$

$O(\varepsilon)$:

$$\beta\frac{\partial C^{(0)}}{\partial t_1} + \nabla \cdot (\boldsymbol{u}^{(1)}C^{(0)} + \boldsymbol{u}^{(0)}C^{(1)}) + \nabla' \cdot (\boldsymbol{u}^{(0)}C^{(0)})$$
$$- \nabla \cdot [\boldsymbol{D} \cdot (\nabla C^{(1)} + \nabla'C^{(0)})] - \nabla' \cdot (\boldsymbol{D} \cdot \nabla C^{(0)}) = 0, \tag{3}$$

$O(\varepsilon^2)$:

$$\beta\left(\frac{\partial C^{(0)}}{\partial t_2} + \frac{\partial C^{(1)}}{\partial t_1}\right) + \nabla \cdot (\boldsymbol{u}^{(2)}C^{(0)} + \boldsymbol{u}^{(1)}C^{(1)} + \boldsymbol{u}^{(0)}C^{(2)}) + \nabla' \cdot (\boldsymbol{u}^{(1)}C^{(0)} + \boldsymbol{u}^{(0)}C^{(1)})$$
$$- \nabla \cdot [\boldsymbol{D} \cdot (\nabla C^{(2)} + \nabla'C^{(1)})] - \nabla' \cdot [\boldsymbol{D} \cdot (\nabla C^{(1)} + \nabla'C^{(0)})] = 0. \tag{4}$$

Let us first show that $C^{(0)}$ is independent of the mesoscale. On multiplying (2) with $C^{(0)}$ and using the product rule of differentiation, we obtain

$$(1/2)\nabla \cdot (\boldsymbol{u}^{(0)}C^{(0)^2}) + (1/2)C^{(0)^2}\nabla \cdot \boldsymbol{u}^{(0)} = \nabla \cdot [(\boldsymbol{D} \cdot \nabla C^{(0)})C^{(0)}] - \boldsymbol{D} : \nabla C^{(0)}\nabla C^{(0)}. \tag{5}$$

The second term on the left-hand side vanishes by virtue of air flow continuity. Further taking Ω–average of (5) and invoking the Ω–periodicity, we finally obtain

$$\int_\Omega \boldsymbol{D} : \nabla C^{(0)}\nabla C^{(0)}\mathrm{d}\Omega = 0. \tag{6}$$

Since $\boldsymbol{D}$, a diffusion coefficient, is positive definite, (6) must imply that $\nabla C^{(0)}$ is zero and $C^{(0)}$ is constant with respect to $\boldsymbol{x}$. Hence we can write that

$$C^{(0)} = C^{(0)}(\boldsymbol{x}', t_1, t_2). \tag{7}$$

We next derive the leading order effective transport equation. With (7), (3) reduces to

$$\beta \frac{\partial C^{(0)}}{\partial t_1} + \nabla \cdot (\boldsymbol{u}^{(1)} C^{(0)} + \boldsymbol{u}^{(0)} C^{(1)}) + \nabla' \cdot (\boldsymbol{u}^{(0)} C^{(0)})$$
$$- \nabla \cdot [\boldsymbol{D} \cdot (\nabla C^{(1)} + \nabla' C^{(0)})] = 0. \tag{8}$$

On taking Ω–average of (8) and using Gauss theorem, we obtain the macroscopic transport equation at the leading order:

$$\langle \beta \rangle \frac{\partial C^{(0)}}{\partial t_1} + \nabla' \cdot (\langle \boldsymbol{u}^{(0)} \rangle C^{(0)}) = 0, \tag{9}$$

where $\langle \beta \rangle$ denotes the Ω–average of β, and so on. We further find a representation for $C^{(1)}$. By eliminating the unsteady term $\partial C^{(0)} / \partial t_1$ from (8) and (9), we obtain

$$\nabla \cdot [\boldsymbol{u}^{(0)} C^{(1)} - \boldsymbol{D} \cdot \nabla C^{(1)}] = \left[\nabla \cdot \boldsymbol{D} - \tilde{\boldsymbol{u}}^{(0)} \right] \cdot \nabla' C^{(0)}$$
$$+ \left[-\nabla \cdot \boldsymbol{u}^{(1)} - \nabla' \cdot \boldsymbol{u}^{(0)} + (\beta / \langle \beta \rangle) \nabla' \cdot \langle \boldsymbol{u}^{(0)} \rangle \right] C^{(0)} \tag{10}$$

where

$$\tilde{\boldsymbol{u}}^{(0)} = \boldsymbol{u}^{(0)} - (\beta / \langle \beta \rangle) \langle \boldsymbol{u}^{(0)} \rangle \tag{11}$$

is the velocity fluctuation on the mesoscale. By linearity, we may put

$$C^{(1)} = \boldsymbol{M} \cdot \nabla' C^{(0)} + E C^{(0)} \tag{12}$$

where $\boldsymbol{M}(\boldsymbol{x}, \boldsymbol{x}')$ is a vector and $E(\boldsymbol{x}, \boldsymbol{x}')$ is a scalar. The mesocell problems for these functions follow from (10):

$$\nabla \cdot [\boldsymbol{u}^{(0)} \boldsymbol{M} - \boldsymbol{D} \cdot (\boldsymbol{I} + \nabla \boldsymbol{M})] = -\tilde{\boldsymbol{u}}^{(0)}, \tag{13}$$

$$\nabla \cdot [\boldsymbol{u}^{(0)} E - \boldsymbol{D} \cdot \nabla E] = -\nabla \cdot \boldsymbol{u}^{(1)} - \nabla' \cdot \boldsymbol{u}^{(0)} + (\beta / \langle \beta \rangle) \nabla' \cdot \langle \boldsymbol{u}^{(0)} \rangle, \tag{14}$$

and $\boldsymbol{M}$ and E are Ω–periodic. For uniqueness, we further require these functions to satisfy

$$\langle \beta \boldsymbol{M} \rangle = \langle \beta E \rangle = 0. \tag{15}$$

Note that (15) implies $\langle \beta C^{(1)} \rangle = 0$; the rationale is to ultimately eliminate $\partial C^{(1)} / \partial t_1$ in the $O(\varepsilon)$ effective transport equation. Physically (12) gives corrections to $C^{(0)}$ owing to mesoscale spatial variation in advection velocity as in the first term, and in air compressibility as in the second term. Obviously for incompressible flows E will be identically zero.

Finally we may derive the $O(\varepsilon)$ effective transport equation as follows. Using Gauss theorem and the uniqueness condition (15), we obtain the Ω–average of (4):

$$\langle \beta \rangle \frac{\partial C^{(0)}}{\partial t_2} + \nabla' \cdot [\langle \boldsymbol{u}^{(1)} \rangle C^{(0)} + \langle \boldsymbol{u}^{(0)} C^{(1)} \rangle]$$
$$- \nabla' \cdot [\langle \boldsymbol{D} \cdot \nabla C^{(1)} \rangle + \langle \boldsymbol{D} \rangle \cdot \nabla' C^{(0)}] = 0. \tag{16}$$

On further substituting (12), the above equation becomes

$$\langle \beta \rangle \frac{\partial C^{(0)}}{\partial t_2} + \nabla' \cdot \left(\bar{\boldsymbol{u}}^{(1)} C^{(0)} \right) - \nabla' \cdot \left(\boldsymbol{D}' \cdot \nabla' C^{(0)} \right) = 0, \tag{17}$$

where

$$\bar{\boldsymbol{u}}^{(1)} \equiv \langle \boldsymbol{u}^{(1)} - \boldsymbol{D} \cdot \nabla E + \boldsymbol{u}^{(0)} E \rangle \tag{18}$$

is the $O(\varepsilon)$ correction to the specific discharge, and

$$\boldsymbol{D}' \equiv \langle \boldsymbol{D} \cdot (\boldsymbol{I} + \nabla \boldsymbol{M}) - \boldsymbol{u}^{(0)} \boldsymbol{M} \rangle \tag{19}$$

is the effective hydrodynamic dispersion coefficient. Finally on combining (9) with (17) we obtain the macroscale effective transport equation which is valid up to $O(\varepsilon)$:

$$\langle \beta \rangle \frac{\partial C^{(0)}}{\partial t} + \nabla' \cdot \left(\boldsymbol{u}' C^{(0)} \right) - \nabla' \cdot \left(\boldsymbol{D}' \cdot \nabla' C^{(0)} \right) = 0, \tag{20}$$

521

where

$$\boldsymbol{u}' \equiv \langle \boldsymbol{u}^{(0)} \rangle + \bar{\boldsymbol{u}}^{(1)} \tag{21}$$

is the combined effective specific discharge. Note that while advection is dominant at the leading order, dispersion is significant only over a long time scale comparable to T_2. Also note that mechanical dispersion now emerges on the macroscale owing to the mesoscale heterogeneity.

3. LAYERED FORMATION

To obtain analytical solutions for the cell functions, we further consider the simple yet practically important composite meso-structure; the medium is composed of two types of horizontal layers, A and B, stacking alternately in the vertical direction x_3 or z. A periodic unit cell Ω consists of one Layer A $(0 < z < d_a)$ and one Layer B $(d_a < z < d_a + d_b)$ where d_a and d_b are the respective layer thicknesses. Within each layer the medium is isotropic and homogeneous, and the material properties are known. Specifically, the conductivity and diffusion tensors are

$$k_{ij} = k\delta_{ij}; \qquad D_{ij} = D\delta_{ij} \tag{22}$$

where δ_{ij} is the Kronecker delta, and

$$(k, D, \beta) = \begin{cases} (k_a, D_a, \beta_a) & \text{in Layer A} \\ (k_b, D_b, \beta_b) & \text{in Layer B} \end{cases}, \tag{23}$$

where k_a, k_b, D_a, D_b, β_a and β_b are all constants. Since strictly horizontal layers are considered, the dependence on the horizontal coordinates (x, y) on the mesoscale can be omitted. The macroscale permeability tensor is given by:

$$k'_{11} = k'_{22} = \langle k \rangle = \frac{k_a d_a + k_b d_b}{d_a + d_b}, \qquad k'_{33} = \langle k^{-1} \rangle^{-1} = \frac{k_a k_b (d_a + d_b)}{k_a d_b + k_b d_a}. \tag{24}$$

From the air flow equations, the following relations can also be obtained:

$$(u^{(0)}, v^{(0)}) = (k/\langle k \rangle)(\langle u^{(0)} \rangle, \langle v^{(0)} \rangle), \qquad w^{(0)} = \langle w^{(0)} \rangle \tag{25}$$

where we have denoted (u_1, u_2, u_3) by (u, v, w). Therefore while the horizontal components of the specific discharge are discontinuous across the layers, the vertical component is continuous and equal to that of the macroscale. Unless stated otherwise, we shall from here on omit the leading order superscript, and use subscripts a and b to distinguish medium properties in Layers A and B respectively.

The problems for $\boldsymbol{M}(z)$ and E are now reduced to one-dimensional boundary value problems with second-order differential equations, whose solutions can be found analytically as a function of z. After some manipulation, we may finally obtain the hydrodynamic dispersion tensor and the effective specific discharge as follows:

$$D'_{ij} = D^*_{ij} + \gamma_i \eta_j \langle u_i \rangle \langle u_j \rangle, \qquad u'_i = (1 - \gamma_i f_E) \langle u_i \rangle. \tag{26}$$

In above expressions, D^*_{ij} is the effective diffusion coefficient given by

$$D^*_{11} = D^*_{22} = \langle D \rangle = \frac{D_a d_a + D_b d_b}{d_a + d_b}, \quad D^*_{33} = \langle D^{-1} \rangle^{-1} = \frac{D_a D_b (d_a + d_b)}{D_a d_b + D_b d_a}, \qquad D^*_{ij} = 0 \ (\text{for } i \neq j), \tag{27}$$

while

$$\eta_1 = \eta_2 = \left(\frac{\beta_a - \beta_b}{\beta_a d_a + \beta_b d_b} - \frac{k_a - k_b}{k_a d_a + k_b d_b} \right), \qquad \eta_3 = \left(\frac{\beta_a - \beta_b}{\beta_a d_a + \beta_b d_b} + \frac{D_a - D_b}{D_a d_b + D_b d_a} \right), \tag{28}$$

$$\gamma_1 = \gamma_2 = \frac{(D_a d_b + D_b d_a)(k_b \beta_a - k_a \beta_b)}{(\beta_a d_a + \beta_b d_b)(k_a d_a + k_b d_b)} \zeta_1, \qquad \gamma_3 = \left(\frac{\beta_a D_a - \beta_b D_b}{\beta_a d_a + \beta_b d_b} \right) \zeta_1, \tag{29}$$

522

where

$$\zeta_1 = \frac{d_a d_b}{\langle w \rangle^2} - \frac{(e^{\frac{\langle w \rangle d_a}{D_a}} - 1)(e^{\frac{\langle w \rangle d_b}{D_b}} - 1)(D_a d_b + D_b d_a)}{(e^{\frac{\langle w \rangle d_a}{D_a} + \frac{\langle w \rangle d_b}{D_b}} - 1)\langle w \rangle^3}, \tag{30}$$

and

$$f_E = -\left[\eta_1 \langle u \rangle \frac{\partial p}{\partial x'} + \eta_2 \langle v \rangle \frac{\partial p}{\partial y'} + \eta_3' \langle w \rangle \frac{\partial p}{\partial z'}\right] p^{-1} \tag{31}$$

in which p is the air pressure and

$$\eta_3' = \left(\frac{\beta_a - \beta_b}{\beta_a d_a + \beta_b d_b} + \frac{k_a - k_b}{k_a d_b + k_b d_a}\right). \tag{32}$$

Note also the limit

$$\zeta_1 \to \frac{d_a^2 d_b^2}{12 D_a D_b} \quad \text{as} \quad \langle w \rangle \to 0. \tag{33}$$

Obviously the hydrodynamic dispersion coefficient $\boldsymbol{D}'$ is the sum of the effective molecular diffusion coefficient $\boldsymbol{D}^*$ and the mechanical dispersion coefficient which is proportional to the square of the velocity. While it is trivial to see that $\gamma_1 \eta_2 = \gamma_2 \eta_1$, it can also be shown after some algebra that $\gamma_1 \eta_3 = \gamma_3 \eta_1$ and $\gamma_2 \eta_3 = \gamma_3 \eta_2$. Therefore the dispersion coefficient tensor is symmetric: $D_{ij}' = D_{ji}'$.

4. THE DISPERSION COEFFICIENT

We stress that the effective dispersion coefficient is a function of both the convection velocity and the layer properties. The vapour dispersion is enhanced at the macroscopic scale because of the layered heterogeneity. Let us now further examine the factors of the hydrodynamic dispersion. For this purpose, we concentrate on the longitudinal and the transverse components of the dispersion coefficient, whose dimensionless forms $\hat{D}_{11}'$ and $\hat{D}_{33}'$ are as below:

$$
\begin{aligned}
\hat{D}_{11}' =\ & \frac{1 + \delta_D \delta_d}{1 + \delta_d} + \frac{(\delta_k - \delta_\beta)^2 (1 + \delta_d)(\delta_d + \delta_D)\delta_d}{(1 + \delta_\beta \delta_d)^2 (1 + \delta_k \delta_d)^2} \\
& \times \begin{cases} \left[1 - \frac{(e^{\langle \hat{w} \rangle Pe_a} - 1)(e^{\langle \hat{w} \rangle Pe_b} - 1)(Pe_a + Pe_b)}{(e^{\langle \hat{w} \rangle (Pe_a + Pe_b)} - 1)\langle \hat{w} \rangle Pe_a Pe_b}\right]\left(\frac{\langle \hat{u} \rangle}{\langle \hat{w} \rangle}\right)^2 & \text{for } \langle \hat{w} \rangle \neq 0 \\ \left(\frac{Pe_a Pe_b}{12}\right)\langle \hat{u} \rangle^2 & \text{for } \langle \hat{w} \rangle = 0 \end{cases}
\end{aligned} \tag{34}
$$

$$
\begin{aligned}
\hat{D}_{33}' =\ & \frac{\delta_D (1 + \delta_d)}{\delta_d + \delta_D} + \frac{(1 - \delta_\beta \delta_D)^2 (1 + \delta_d)\delta_d}{(1 + \delta_\beta \delta_d)^2 (\delta_d + \delta_D)} \\
& \times \begin{cases} \left[1 - \frac{(e^{\langle \hat{w} \rangle Pe_a} - 1)(e^{\langle \hat{w} \rangle Pe_b} - 1)(Pe_a + Pe_b)}{(e^{\langle \hat{w} \rangle (Pe_a + Pe_b)} - 1)\langle \hat{w} \rangle Pe_a Pe_b}\right] & \text{for } \langle \hat{w} \rangle \neq 0 \\ 0 & \text{for } \langle \hat{w} \rangle = 0 \end{cases}
\end{aligned} \tag{35}
$$

The normalization is $\hat{u}' = u'/U$, $\hat{\boldsymbol{D}}' = \boldsymbol{D}'/D_a$ where U is a characteristic flow velocity scale, and the ratios of layer properties are

$$\delta_d = d_b/d_a, \quad \delta_k = k_b/k_a, \quad \delta_D = D_b/D_a, \quad \delta_\beta = \beta_b/\beta_a, \tag{36}$$

and the Peclet numbers are

$$Pe_a = U d_a/D_a, \qquad Pe_b = U d_b/D_b = Pe_a \delta_d/\delta_D. \tag{37}$$

Note that the second (mechanical dispersion) terms in (34) and (35) are always non-negative; they are zero only when the corresponding velocity component vanishes or when $\delta_k = \delta_\beta$ for $\hat{D}_{11}'$, $\delta_\beta \delta_D = 1$ for $\hat{D}_{33}'$. Therefore, as expected, in the absence of heterogeneity the mechanical dispersion of vapour in a porous medium becomes subdominant. Note also that these two dispersion components are even functions of $\langle \hat{w} \rangle$, i.e., $\hat{D}_{11}'(\langle \hat{w} \rangle) = \hat{D}_{11}'(-\langle \hat{w} \rangle)$ and $\hat{D}_{33}'(\langle \hat{w} \rangle) = \hat{D}_{33}'(-\langle \hat{w} \rangle)$.

It can readily be seen from $\hat{D}'_{11}$ that the longitudinal mechanical dispersion is of the magnitude $U^2 d_b^2/D_b$, which is the form of Taylor dispersion in a capillary tube (Taylor, 1953). This is reasonable since in our model the dispersion is essentially caused by the effect of velocity variation across the layers interacting with molecular diffusion that promotes lateral transfer between layers with different retardation factor and flow velocity. Similar results were also obtained for the special case of structured media composed of mobile and immobile regions (e.g., Passioura, 1971). The diffusive exchange between these two regions can effectively increase the dispersion coefficient by the amount proportional to $U_m^2 d_{im}^2/D_{im}$, where U_m is the velocity in the mobile region, d_{im} is a characteristic dimension of the immobile region, and D_{im} is the molecular diffusion coefficient in the immobile region. It suggests that the longitudinal dispersion has a strong dependence on the velocity along the more permeable layer, and the diffusion time scale across the less permeable layer.

We also remark that the transverse mechanical dispersion vanishes when the flow is strictly horizontal. This is consistent with the analysis by statistical means which yields a zero transverse asymptotic macrodispersivity when the heterogeneous structure of the medium does not create transverse dispersion (Gelhar and Axness, 1983). In the present case, transverse mechanical dispersion occurs only in the presence of transverse flow, but with a rather weak dependence on it.

5. CONCLUDING REMARKS

In this paper we have presented a mathematical derivation of the macroscopic equations for the transport of chemical vapour in a multi-layered unsaturated zone. Based on the assumptions of cell periodicity and a sharp contrast in length scales, the method of homogenization is applied in order to identify the range of validity of the theory. The effective transport equation (20), in which the specific discharge and dispersion coefficient are expressed in terms of some cell functions, is good for all forms of periodic heterogeneity with a characteristic dimension much smaller than the global length scale. Analytical solutions to the cell functions have been obtained for the simple case of bi-layered unit cell. In terms of flow velocity and medium parameters, the specific discharge and the hydrodynamic dispersion coefficient are given in (26). The longitudinal dispersion is found to be caused by Taylor mechanism. That is, the coefficient varies with the square of the longitudinal velocity, and linearly with the diffusion time across the less permeable layer. A larger contrast in the retardation factor and conductivity can only modestly increase the dependence of the dispersion on the flow velocity. Also, an increase in the transverse flow velocity will lower the value of longitudinal dispersion coefficient. On the other hand, owing to the ideal layering structure, the transverse mechanical dispersion does not depend on the longitudinal flow, and only weakly depends on the transverse flow velocity.

6. ACKNOWLEDGMENTS

This research has been supported by the CRCG research grant 335/064/0082 awarded by the University of Hong Kong.

7. REFERENCES

Gelhar, L. W., and Axness, C. L. (1983). "Three-dimensional stochastic analysis of macrodispersion in aquifers", *Water Resour. Res.*, 19(1), 161–180.

Mei, C. C., Auriault, J. L., and Ng, C. O. (1996). "Some Applications of the homogenization theory", in *Adv. in Applied Mech.*, Vol. 32, ed. J. W. Hutchinson and T. Y. Wu. Academic Press, California, 277–348.

Ng, C. O., and Mei, C. C. (1996). "Homogenization theory applied to soil vapor extraction in aggregated soils", *Phys. Fluids*, 8(9), 2298–2306.

Passioura, J. B. (1971). "Hydrodynamic dispersion in aggregated media 1. Theory", *Soil Sci.*, 111(6), 339–344.

Taylor, G. I. (1953). "Dispersion of soluble matter in solvent flowing slowly through a tube", *Proc. R. Soc. Lond.*, **A** 219, 186–203.

Environmental Hydraulics, Lee, Jayawardena & Wang (eds) © 1999 Balkema, Rotterdam, ISBN 90 5809 035 3

Shear flows in shallow water: Foundation and accuracy of the rigid-lid assumption

M.S.Ghidaoui & J.Q.Deng
The Hong Kong University of Science and Technology, China

A.A.Kolyshkin
Riga Technical University, Latvia (Presently: Hong Kong University of Science and Technology, China)

ABSTRACT: Linear stability analysis of lateral motions in open channel flows is frequently performed under the rigid-lid assumption. The objective of this paper is to study the foundation of the rigid-lid assumption and determine its accuracy. It is shown that implicit in the rigid-lid approximation is the fact that Froude number tends to zero. Comparisons of the stability domains obtained without the rigid-lid assumption (i.e., Froude number $\neq 0$) with those obtained with the rigid-lid assumption (i.e., Froude number $= 0$) show that the rigid-lid assumption works well for weak shear flows and/or small Froude numbers.

1. INTRODUCTION

Large-scale lateral turbulent motions occur frequently in shallow free surface water flows in compound open channels with floodplains. The fact that the extent and rate of spreading of this lateral mixing is mainly governed by the ratio of bed friction dissipation and turbulent energy production has been determined experimentally by Chu and Babarutsi (1988) and theoretically by Chu et al. (1991). Similar conclusions were arrived at by Chen and Jirka (1997). The work of Chu et al. (1991) and Chen and Jirka (1997) is based on the rigid-lid equations. The objective of this paper is to study the foundation of the rigid-lid assumption and determine its accuracy. It is shown that implicit in the rigid-lid approximation is the fact that Froude number tends to zero. Comparisons of the stability domains obtained without the rigid-lid assumption (i.e., Froude number $\neq 0$) with those obtained with the rigid-lid assumption (i.e., Froude number $= 0$) show that the rigid-lid assumption works well for weak shear flows and/or small Froude numbers.

2. DIMENSIONLESS NAVIER-STOKES EQUATIONS IN OPEN CHANNELS

The Navier-Stokes equations along with their boundary conditions for free surface water flows are as follows (Chaudhry 1993):

$$\frac{\partial u}{\partial x} + \frac{\partial v}{\partial y} + \frac{\partial w}{\partial z} = 0 \tag{1}$$

$$\frac{\partial u}{\partial t} + u\frac{\partial u}{\partial x} + v\frac{\partial u}{\partial y} + w\frac{\partial u}{\partial z} = -\frac{1}{\rho}\frac{\partial p}{\partial x} + \frac{\mu}{\rho}\nabla_z^2 u + \frac{\mu}{\rho}\frac{\partial^2 u}{\partial z^2} \tag{2}$$

$$\frac{\partial v}{\partial t} + u\frac{\partial v}{\partial x} + v\frac{\partial v}{\partial y} + w\frac{\partial v}{\partial z} = \frac{1}{\rho}\frac{\partial p}{\partial y} + \frac{\mu}{\rho}\nabla_z^2 v + \frac{\mu}{\rho}\frac{\partial^2 v}{\partial z^2} \tag{3}$$

$$\frac{\partial w}{\partial t} + u\frac{\partial w}{\partial x} + v\frac{\partial w}{\partial y} + w\frac{\partial w}{\partial z} = -g - \frac{1}{\rho}\frac{\partial p}{\partial z} + \frac{\mu}{\rho}\nabla_z^2 w + \frac{\mu}{\rho}\frac{\partial^2 w}{\partial z^2} \tag{4}$$

$$w = \frac{\partial h}{\partial t} + u\frac{\partial h}{\partial x} + v\frac{\partial h}{\partial y} \quad \text{at} \quad z = h \tag{5}$$

$$w = 0 \quad \text{at} \quad z = -b \tag{6}$$

$$p = 0 \quad \text{at} \quad z = h \tag{7}$$

where (x, y, z) are the spatial coordinates with z being the vertical axis; t is the time; u is the x velocity component; v is the y velocity component; w is the z velocity component; μ is the dynamic viscosity of the fluid; ρ is the density of the fluid; p is the normal component of pressure; g is the gravitational acceleration; h is the distance from $z = 0$ to the water surface; b is the distance from $z = 0$ to the bottom of the channel and $\nabla_z^2 = \frac{\partial^2}{\partial x^2} + \frac{\partial^2}{\partial y^2}$. Let L be the horizontal length scale, H be the vertical length scale, T be the time scale, A be the scale of wave amplitude and U be the horizontal velocity scale. According to Stoker (1957) the relationships between the dimensionless and dimensional quantities are $u = u^*U$, $v = v^*U$, $w = w^*UL/H$, $x = x^*L$, $y = y^*L$, $z = z^*H$, $t = t^*T$, $p = p^*\rho U^2$, where the superscript $*$ is used to denote dimensionless variables. Note that $u/w = u^*/w^*H/L$. At first, this expression for u/w appears paradoxical as it seems to imply that the horizontal velocity u is smaller than w by a factor of H/L! However, in this paper it is shown that u^*/w^* is order L^2/H^2 implying that u/w is of order L/H and thus the paradox is resolved.

Substituting the above relationships between the dimensionless and dimensional quantities into (1) to (5) gives

$$\frac{U}{L}\frac{\partial u^*}{\partial x^*} + \frac{U}{L}\frac{\partial v^*}{\partial y^*} + \frac{UL}{H^2}\frac{\partial w^*}{\partial z^*} = 0 \tag{8}$$

$$\frac{U}{T}\frac{\partial u^*}{\partial t^*} + \frac{U^2}{L}u^*\frac{\partial u^*}{\partial x^*} + \frac{U^2}{L}v^*\frac{\partial u^*}{\partial y^*} + \frac{U^2 L}{H^2}w^*\frac{\partial u^*}{\partial z^*} = -\frac{U^2}{L}\frac{\partial p^*}{\partial x^*} + \frac{\mu}{\rho}\frac{U}{L^2}\nabla_{z^*}^2 u^* + \frac{\mu}{\rho}\frac{U}{H^2}\frac{\partial^2 u^*}{\partial z^{*2}} \tag{9}$$

$$\frac{U}{T}\frac{\partial v^*}{\partial t^*} + \frac{U^2}{L}u^*\frac{\partial v^*}{\partial x^*} + \frac{U^2}{L}v^*\frac{\partial v^*}{\partial y^*} + \frac{U^2 L}{H^2}w^*\frac{\partial v^*}{\partial z^*} = -\frac{U^2}{L}\frac{\partial p^*}{\partial y^*} + \frac{\mu}{\rho}\frac{U}{L^2}\nabla_{z^*}^2 v^* + \frac{\mu}{\rho}\frac{U}{H^2}\frac{\partial^2 v^*}{\partial z^{*2}} \tag{10}$$

$$\frac{UL}{TH}\frac{\partial w^*}{\partial t^*} + \frac{U^2}{H}u^*\frac{\partial w^*}{\partial x^*} + \frac{U^2}{H}v^*\frac{\partial w^*}{\partial y^*} + \frac{U^2 L^2}{H^3}w^*\frac{\partial w^*}{\partial z^*} = -g - \frac{U^2}{H}\frac{\partial p^*}{\partial z^*} + \frac{\mu}{\rho}\frac{U}{LH}\nabla_{z^*}^2 w^* + \frac{\mu}{\rho}\frac{UL}{H^3}\frac{\partial^2 w^*}{\partial z^{*2}} \tag{11}$$

$$U\frac{L}{H}w^* = \frac{A}{T}\frac{\partial h^*}{\partial t^*} + \frac{UA}{L}u^*\frac{\partial h^*}{\partial x^*} + \frac{UA}{L}v^*\frac{\partial h^*}{\partial y^*} \quad \text{at} \quad z^* = A\frac{h^*}{H} \tag{12}$$

where $\nabla_{z^*}^2 = \frac{\partial^2}{\partial x^{*2}} + \frac{\partial^2}{\partial y^{*2}}$. Note that $L = UT$, $F_r = \frac{U}{\sqrt{gH}}$, $R_e = \frac{\rho U H}{\mu}$ where F_r is the Froude number and R_e the Reynolds number. In addition, the horizontal velocity is driven by a height difference which is of the same order as the wave amplitude. That is, U scales as $\sqrt{gA}$. Therefore, $F_r^2 = A/H$. Letting $\sigma = H^2/L^2$. Rearranging (8) to (12) and multiplying the results by σ yields

$$\sigma\left(\frac{\partial u^*}{\partial x^*} + \frac{\partial v^*}{\partial y^*}\right) + \frac{\partial w^*}{\partial z^*} = 0 \tag{13}$$

$$\sigma\left(\frac{\partial u^*}{\partial t^*} + u^*\frac{\partial u^*}{\partial x^*} + v^*\frac{\partial u^*}{\partial y^*} + \frac{\partial p}{\partial x^*} - \frac{1}{R_e}\nabla_{z^*}^2 u^*\right) = -w^*\frac{\partial u^*}{\partial z^*} + \frac{1}{R_e}\frac{\partial^2 u^*}{\partial z^{*2}} \tag{14}$$

$$\sigma\left(\frac{\partial v^*}{\partial t^*} + u^*\frac{\partial v^*}{\partial x^*} + v^*\frac{\partial v^*}{\partial y^*} + \frac{\partial p}{\partial y^*} - \frac{1}{R_e}\nabla_{z^*}^2 v^*\right) = -w^*\frac{\partial v^*}{\partial z^*} + \frac{1}{R_e}\frac{\partial^2 v^*}{\partial z^{*2}} \tag{15}$$

$$\sigma\left(\frac{\partial w^*}{\partial t^*} + u^*\frac{\partial w^*}{\partial x^*} + v^*\frac{\partial w^*}{\partial y^*} + \frac{\partial p}{\partial z^*} - \frac{1}{R_e}\nabla_{z^*}^2 w^*\right) = -w^*\frac{\partial w^*}{\partial z^*} + \frac{1}{R_e}\frac{\partial^2 w^*}{\partial z^{*2}} \tag{16}$$

$$w^* = \sigma F_r^2\left(\frac{\partial h^*}{\partial t^*} + u^*\frac{\partial h^*}{\partial x^*} + v^*\frac{\partial h^*}{\partial y^*}\right) \quad \text{and} \quad p^* = 0 \quad \text{at} \quad z^* = F_r^2 h^* \tag{17}$$

$$w^* = 0 \quad \text{at} \quad z^* = -\frac{b}{H} \tag{18}$$

3. MASS BALANCE, FROUDE NUMBER AND THE RIGID-LID ASSUMPTION

Camassa et al. (1997) obtained the rigid-lid equations from the Euler equations by taking the limit as Froude number tends to zero. The objective of this section is to briefly summarize and illustrate the relationship between the rigid-lid assumption and the Froude number.

Integrating equation (13) with respect to z^* gives:

$$\sigma \int_{-b^*}^{F_r{}^2 h^*} \left(\frac{\partial u^*}{\partial x^*} + \frac{\partial v^*}{\partial y^*} \right) dz^* + w^*|_{F_r{}^2 h^*} - w^*|_{-b^*} = 0 \tag{19}$$

where the notation $f|_x$ means the value of f at x. Using the fact that $w^*|_{-b^*} = 0$ and applying Leiblitz's rule reduces (19) to the following expression:

$$\sigma \frac{\partial}{\partial x^*} \int_{-b^*}^{F_r{}^2 h^*} u^* dz^* + \sigma \frac{\partial}{\partial y^*} \int_{-b^*}^{F_r{}^2 h^*} v^* dz^* - F_r{}^2 \sigma \left[u^* \frac{\partial h^*}{\partial x^*} + v^* \frac{\partial h^*}{\partial y^*} \right] + w^*|_{F_r{}^2 h^*} = 0 \tag{20}$$

or

$$\sigma \left[\frac{\partial}{\partial x^*} \left(\bar{u}^* \frac{F_r{}^2 h + b}{H} \right) + \frac{\partial}{\partial y^*} \left(\bar{v}^* \frac{F_r{}^2 h + b}{H} \right) \right] - F_r{}^2 \sigma \left[u^* \frac{\partial h^*}{\partial x^*} + v^* \frac{\partial h^*}{\partial y^*} \right] + w^*|_{F_r{}^2 h^*} = 0 \tag{21}$$

where the overbar is used to denote vertically averaged velocities. Invoking boundary condition (17) simplifies (21) as follows:

$$\frac{\partial}{\partial x^*} \left(\bar{u}^* \frac{F_r{}^2 h + b}{H} \right) + \frac{\partial}{\partial y^*} \left(\bar{v}^* \frac{F_r{}^2 h + b}{H} \right) - F_r{}^2 \sigma \frac{\partial h^*}{\partial t^*} = 0 \tag{22}$$

Taking the limit of (22) as $F_r \Rightarrow 0$ produces the following mass balance equation:

$$\frac{\partial}{\partial x^*} \left(\bar{u}^* \frac{b}{H} \right) + \frac{\partial}{\partial y^*} \left(\bar{v}^* \frac{b}{H} \right) = 0 \tag{23}$$

This is the rigid-lid continuity equation. That is, the rigid-lid assumption implicitly assumes that Froude number is zero. More discussion is available in Camassa et al. (1997). However, it must be noted that the above derivation alone cannot determine or justify the accuracy of the rigid-lid assumption nor guarantee uniform convergence of the solution of the shallow water equations to the solution of the rigid-lid equation as F_r is reduced. That is, the above derivation alone cannot provide an estimate of the accuracy of using of the rigid-lid assumption in studying the stability of shear flows in open channels. The section below numerically determines the accuracy of the rigid-lid assumption in shear flows and shows that the convergence is uniform in this particular problem.

4. ACCURACY OF THE RIGID-LID ASSUMPTION IN SHEAR FLOWS

Stoker (1957) showed that the shallow water equations can be derived from (13), (14), (15), (16) and (17) by expanding all dependent variables as a power series in σ and ignoring terms of order σ^2 and higher. That is, as $\sigma \Rightarrow 0$ for finite F_r, (13), (14), (15), (16) and (17) gives the following dimensionless form of the shallow water (Stoker 1957):

$$\frac{\partial h}{\partial t} + \frac{\partial}{\partial x}(uh) + \frac{\partial}{\partial y}(vh) = 0, \tag{24}$$

$$\frac{\partial u}{\partial t} + u\frac{\partial u}{\partial x} + v\frac{\partial u}{\partial y} + \frac{\partial h}{\partial x} - (S_{0x} - S_{fx}) - \frac{1}{Re}\nabla^2 u = 0, \tag{25}$$

$$\frac{\partial v}{\partial t} + u\frac{\partial v}{\partial x} + v\frac{\partial v}{\partial y} + \frac{\partial h}{\partial y} - (S_{0y} - S_{fy}) - \frac{1}{Re}\nabla^2 v = 0, \tag{26}$$

where $*$ are dropped and S_{0x} and S_{0y} are the bed slopes in the x and y directions, S_{fx} and S_{fy} are

527

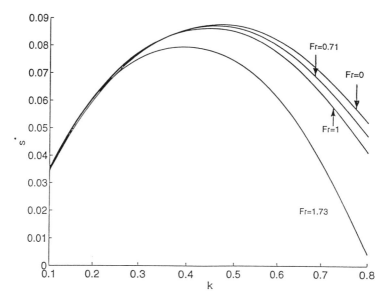

FIG. 1 s^* versus k for different values of Froude number.
The case $F_r = 0$ corresponds to the rigid-lid assumption.

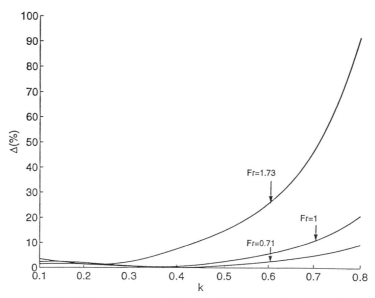

FIG. 2 The percentage difference Δ between the values of s^*
with and without the rigid-lid assumption.

the components of the bed friction forces in the x and y directions. Linearizing this set of equations around a baseflow of $(U(y), 0, H(y))$ gives:

$$\frac{F_r^2}{U^2}\frac{\partial h'}{\partial t} + \frac{\partial u'}{\partial x} + \frac{F_r^2}{U}\frac{\partial h'}{\partial x} + 2\frac{U_x}{U}v' + \frac{\partial v'}{\partial y} = 0 \tag{27}$$

528

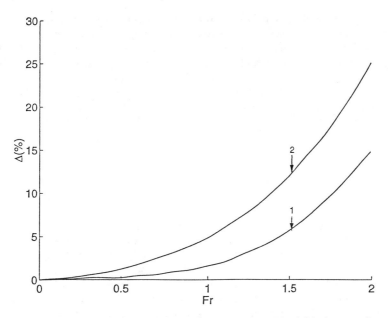

FIG. 3 The percentage difference Δ of the critical bed-friction number s^* from the rigid-lid assumption as a function of Froude number for two base velocity profiles: $1 - U = 1 + 0.5(1 + tanhy), 2 - U = 1 + 1/cosh^2y$.

$$\frac{\partial v'}{\partial t} + U\frac{\partial v'}{\partial x} + v'\frac{fF_r^2}{U} + \frac{\partial h'}{\partial y} = 0 \tag{28}$$

$$\frac{\partial u'}{\partial t} + \frac{\partial h'}{\partial x} - h'\frac{fF_r^4}{U^2} + U\frac{\partial u'}{\partial x} + v'U_y + \frac{2fF_r^2}{U}u' = 0 \tag{29}$$

where $'$ is used to denote the perturbations from the baseflow. Multiplying Equation (27) by h', (28) by v' and (29) by u', adding these relationships together and rearranging produces

$$\frac{\partial k}{\partial t} + U\frac{\partial k}{\partial x} + \{\frac{\partial p_e}{\partial t} + U\frac{\partial p_e}{\partial x}\}\frac{F_r^2}{U} = -\frac{\partial h'u'}{\partial x} - \frac{\partial h'v'}{\partial y} - \frac{2U_y}{U}v'h' - u'v'U_y - [2u'^2 + v'^2]\frac{C_fU}{2H} \tag{30}$$

where k is the turbulent kinetic energy and p_e is the turbulent potential energy. The terms in the right hand side of (30) are: first two terms represent the transport of turbulent energy by velocity and pressure fluctuations, the third and fourth terms represent production of turbulent energy and the last term represents the dissipation of turbulent energy by the bed friction. Note that as F_r tends to zero, one recovers the rigid-lid turbulent energy equation (see Chu et al. 1991).

In order to estimate quantitatively the accuracy of the rigid-lid assumption we perform computations for TANH and SECH velocity profiles with $U_2 = 1$, $U_1 = 1$, respectively, for different values of Froude numbers and compare the results with the similar computations with the rigid-lid assumption. The stability curves for a TANH velocity profile are plotted against the wavenumber (k) in FIG. 1 for different values of F_r. The curve corresponding to the rigid-lid assumption is indicated as $F_r = 0$. To compare our results with the results of Chu et al. (1991) the bed-friction number $s^* = sF_r^2$ is used. It is seen from FIG. 1 that the error is quite small for small F_r, however, the error increases as the Froude number grows, especially in the region of larger k. The percentage difference Δ between the values of s^* with and without the rigid-lid assumption is given in FIG. 2. It follows from FIGs. 1 and 2 that the rigid-lid assumption works well for small Froude numbers since the error in the stability domain due to the rigid-lid assumption is limited to 20% for Froude numbers smaller than 1.0. However, for Froude numbers larger than one, the error in the stability domain due to the rigid-lid assumption becomes large. For example, when the Froude number is 1.7 the rigid-lid assumption error is 80% to 90% for wavenumbers close to one. The percentage departure Δ of the critical bed-friction number s^* from the rigid-lid approximation as a function of Froude number is shown in FIG. 3 for two velocity profiles. One can conclude from FIG. 3 that the rigid-lid assumption works very well for small F_r. However, for larger values of F_r the difference increases as F_r grows. Another conclusion is that stability domain determined without the rigid-lid assumption converges uniformly to the stability domain determined from the rigid-lid assumption as F_r is continuously reduced.

5. CONCLUSIONS

Linear stability analysis of lateral motions in open channel flows is frequently performed under the rigid-lid assumption. The objective of this paper is to study the foundation of the rigid-lid assumption and determine its accuracy. It is shown that implicit in the rigid-lid approximation is the fact that Froude number tends to zero. Comparisons of the stability domains obtained without the rigid-lid assumption (i.e., Froude number $\neq$ 0) with those obtained with the rigid-lid assumption (i.e., Froude number = 0) show that the rigid-lid assumption works well for weak shear flows and/or small Froude numbers. In addition, the stability domain determined without the rigid-lid assumption converges uniformly to the stability domain determined from the rigid-lid assumption as F_r is continuously reduced. This is confirmed by the monotonicity of the error Δ.

6. ACKNOWLEDGMENT

The financial support of this work by the Research Grant Council (RGC) of Hong Kong under project number HKUST718/96E is gratefully acknowledged.

7. REFERENCES

Camassa, R., Holm, D. D., and Levermore, C. D. (1997). "Long-time shallow-water equations with a varying bottom". J. Fluid Mech., 349, 173–189.

Chaudhry, M. H. (1993). Open-channel flows, Prentice Hall, Englewood Cliffs, N.J.

Chen, D., and Jirka, G. H. (1997). "Absolute and convective instabilities of plane turbulent wakes in a shallow water layer". J. Fluid Mech., 338, 157–172.

Chu, V. H. and Babarutsi, S. (1988). "Confinement and bed-friction effects in shallow turbulent mixing layers". J. Hydr. Engrg., ASCE, 114(10), 1257-1270.

Chu, V. H., Wu, J. H., and Khayat, R. E. (1991). "Stability of transverse shear flows in shallow open channels". J. Hydr. Engrg., ASCE, 117(10), 1370–1388.

Stoker J. J. (1957). "Water Waves: The mathematical theory with application". Interscience Publishers, Inc., N. Y.

Environmental Hydraulics, Lee, Jayawardena & Wang (eds) © 1999 Balkema, Rotterdam, ISBN 90 5809 035 3

Dispersion of a small solid spherical particle in turbulent flows

Dahe Jiang
School of Environmental Engineering, Tongji University, Shanghai, China

Ziping Huang & Weimin Qian
Department of Mathematics, Tongji University, Shanghai, China

Weiguo Wang
Department of Atmospheric Sciences, Nanjing University, China

ABSTRACT: The dispersion of both heavy and non-heavy particles is analyzed and numerically studied. The effect of crossing trajectories is properly illustrated by distinguishing the dispersion properties of the particle, the fluid element seen by the particle, and a marked fluid element.

1. INTRODUCTION

The dispersion of a small solid spherical particle is a fundamental problem in fluid mechanics (Tchen 1947). However, since the "effect of crossing trajectories" was emphasized (Yudine 1959), successful studies on this problem have been restricted to heavy particle cases. The key difficulty is, in the force term representing the stress of the undisturbed flow, the flow velocity acceleration is now considered in the D/Dt form, but the equation of particle motion is in the sense of d/dt. d/dt refers to particle trajectory, but D/Dt refers to the trajectories of fluid elements which incidentally cross the particle trajectory. The analysis can only be carried out when this term can be neglected, such as in heavy particle cases (Stock 1996).

The present study reconsiders the situation. It is proposed that the d/dt versus D/Dt problem is still open for the following reasons. (1) Since the controversial term has been neglected in the studies for heavy particle cases, the success in discussing the effect of crossing trajectories in these studies in fact revealed that such an effect is not necessarily related to the use of the D/Dt form. (2) In the well known study by Csanady (1963), the asymptotic eddy diffusivities of a heavy particle in the horizontal and vertical direction were shown significantly different from that of fluid elements. However, an important assumption was used: the velocity fluctuations of the particle would instantly follow that of the neighboring fluid, while the particle trajectory was considered to cut the trajectories of fluid elements. However, for a heavy particle, it must possess strong inertia. How could its velocity fluctuations follow that of the fluid instantly ? This dilemma may be solved by introducing an alternative velocity correlation function. In our study, it is shown, with the new velocity correlation function, there is no contradiction between the results by Tchen and Csanady. (3) In the rational derivation for the equation of particle motion by Maxey and Riley (1983), the actual flow was split into an undisturbed flow $w^{(0)}$ and a disturbed flow $w^{(1)}$. For the undisturbed flow, the particle is not considered as present. Therefore, the stress term was naturally found in the form of D/Dt. On the other hand, with the nonlinear terms neglected, the equation of the disturbed flow was shown to lead to the force terms: Stokes drag, added mass, Basset memory, even the Faxen terms (Gatignol 1983). A simple combination of these force terms lead to the motion equation of the particle, and the stress term was thus in the form of D/Dt. However, all the force terms from the disturbed flow can be found from earlier studies that a particle is moving relative to a still uniform fluid. The extension to unsteady non-uniform flow is solely on the force term from the undisturbed flow, but the latter is derived without the presence of the particle. This is somewhat contradictory with the one-way coupling approximation: the particle is small enough so that the effect of its movement on the flow can be neglected, but the motion of the fluid has a significant effect on particle motion. Then how is the effect by the movement of the non-uniform undisturbed flow on the motion of the particle ? In other words, the force term in the D/Dt form is only for the fluid element with the same size as that of the particle and at the same location, but not for the particle. We consider this problem might be solved by reconsidering the equation for the disturbed flow. In fact, it might not be appropriate to neglect all the nonlinear terms, especially when both $w^{(0)}$ and $w^{(1)}$ appear in the same term. Although we have not succeeded in derivation at present, such a modification might lead to the d/dt form of the respective force term.

With above considerations, we propose the d/dt versus D/Dt problem is still open. In the present study, the stress term is presumably assumed in the d/dt form, and the turbulence is assumed stationary and homogeneous. It is to demonstrate that the effect of crossing trajectories can be properly described by the use of a new velocity correlation function. This might be considered as indirect evidence to support the d/dt form.

2. FORMULATION

Suppose a particle is released at $(\vec{x}_0, t_0)$. After $\tau = t - t_0$, it makes a displacement $\vec{y} = \hat{\vec{Y}}(\tau)$ to a new position $(\vec{x}, t) = (\vec{y} + \vec{x}_0, \tau + t_0)$. The hat '∧' in $\hat{\vec{Y}}$ indicates its random behavior. The symbols will be used: '$<>$' for ensemble average, '—' and '~' for the mean and fluctuating part of a random variable. Then particle velocity is written as:

$$\hat{V}_i(\tau) = d\hat{Y}_i(\tau)/d\tau.$$ (1)

For mathematical simplicity, the Basset and Faxen terms are dropped, so that,

$$\frac{4}{3}\pi a^3 \rho_P \frac{d\hat{V}_i}{d\tau} = \frac{4}{3}\pi a^3 \rho_f \frac{d\hat{u}_i^P}{d\tau}$$

$$- 6\pi a v \rho_f (\hat{V}_i - \hat{u}_i^P) - \frac{2}{3}\pi a^3 \rho_f \frac{d}{d\tau}(\hat{V}_i - \hat{u}_i^P) - \frac{4}{3}\pi a^3 (\rho_P - \rho_f)g\delta_{i3},$$ (2)

where $\delta_{i3} = 1$ when $i = 3$, otherwise zero. On the right hand side, the force terms are in turn: stress (or pressure gradient), Stokes drag, added mass, and buoyancy. Note that, $\hat{u}_i^P$ is the flow velocity evaluated along the particle trajectory. In this sense, $\hat{u}_i^P$ is also a Lagrangian quantity. We shall use the symbol $\hat{u}_i^E$ for the Eulerian velocity such that $\hat{u}_i^P(\tau) = \hat{u}_i^E[\vec{Y}(\tau) + x_0, \tau + t_0]$. Let $\Gamma \equiv \rho_P / \rho_f$, $\alpha \equiv 3/(2\Gamma + 1)$, and $\beta \equiv 3av/a^2$. This equation is easily integrated to find

$$\bar{V}_i(\tau) = \alpha \bar{u}_i^P(\tau) + (1-\alpha)\beta \int_0^\tau d\tau' \bar{u}_i^P(\tau') \exp[-\beta(\tau - \tau')] - (1-\alpha)(g/\beta)\delta_{i3}.$$ (3)

$$\tilde{V}_i(\tau) = \alpha \tilde{u}_i^P(\tau) + (1-\alpha)\beta \int_0^\tau d\tau'' \tilde{u}_i^P(\tau'') \exp[-\beta(\tau - \tau'')].$$ (4)

Since only the stationary properties of the particle motion that are to be studied, which correspond to $\tau \to \infty$, all the terms with the factor $\exp(-\beta\tau)$ have been neglected. After a lengthy derivation, the velocity correlation function of the particle is found as:

$$Q_{ij}^P(\Delta\tau) = \lim_{\tau' \to \infty} <\tilde{V}_i(\tau')\tilde{V}_j(\tau' + \Delta\tau)>$$

$$= \alpha^2 Q_{ij}^{fP}(\Delta\tau) - (1-\alpha^2)\beta \int_0^{\Delta\tau} d\tau'' \sinh[\beta(\Delta\tau - \tau'')]Q_{ij}^{fP}(\tau'') + (1-\alpha^2)\cosh(\beta\Delta\tau)A_{ij}^{fP}$$ (5)

where

$$A_{ij}^{fP} \equiv \beta \int_0^\infty d\tau' \exp(-\beta\tau')Q_{ij}^{fP}(\tau')$$ (6)

is explicitly a function of α and β, but implicitly a function of the whole history of the particle motion. On the other hand, a new velocity correlation function,

$$Q_{ij}^{fP}(\tau'') \equiv <\widetilde{u}_i^P(0)\widetilde{u}_j^P(\tau'')>, \tag{7}$$

is introduced for the fluid velocity fluctuations seen by the particle. Now it is obvious that there exist three kinds of Lagrangian velocity correlation functions: Q_{ij}^P is for the particle. Q_{ij}^{fP} is also Lagrangian but for the fluid element seen by the particle, which may change from time to time. Based on (5) and (7), Q_{ij}^P is generally different from Q_{ij}^{fP}. The third may be noted as Q_{ij}^f, i.e., the Lagrangian velocity correlation function of a 'marked' fluid element. Note that Q_{ij}^f is not at all related to the particle trajectory so that it is generally different from either Q_{ij}^P or Q_{ij}^{fP}. With the stationary and symmetry properties of Q_{ij}^P, the mean-squared particle displacement is written as

$$R_{ij}^P(\tau) \equiv <\widetilde{Y}_i(\tau)\widetilde{Y}_j(\tau)> = \int_0^\tau d\tau' \int_0^\tau d\tau'' <\widetilde{V}_i(\tau')\widetilde{V}_j(\tau'')> = 2\int_0^\tau (\tau - \tau')Q_{ij}^P(\tau')d\tau'. \tag{8}$$

The particle diffusivity is conventionally defined as:

$$K_{ij}^P(\tau) \equiv \frac{1}{2}\frac{d}{d\tau}R_{ij}^P(\tau) = \int_0^\tau Q_{ij}^P(\tau')d\tau', \tag{9}$$

so that

$$K_{ij}^P(\tau) = \int_0^\tau d\tau' Q_{ij}^{fP}(\tau') - (1-\alpha^2)\int_0^\tau d\tau' \cosh[\beta(\tau-\tau')]Q_{ij}^{fP}(\tau') + \frac{(1-\alpha^2)}{\beta}\sinh(\beta\tau)A_{ij}^{fP}. \tag{10}$$

This expression is closely comparable to that found by Reeks (1977). However, ours includes the α parameter, which measures the density ratio of the particle and fluid. And the second term on the right hand side is in a more correct form. Note that we may introduce an 'eddy diffusivity' for the 'fluid element' seen by the particle:

$$K_{ij}^{fP}(\tau) \equiv \int_0^\tau Q_{ij}^{fP}(\tau')d\tau'. \tag{11}$$

The short term results are similar to other studies. Only the long term results are provided here for discussion. For $\tau \to \infty$, it is easy to show $\lim_{\tau\to\infty} Q_{ij}^P(\tau) \to 0$. For diffusivity,

$$\lim_{\tau\to\infty} K_{ij}^P(\tau) \to K_{ij}^P(\infty) = \int_0^\infty d\tau' Q_{ij}^{fP}(\tau'), \tag{12}$$

which is formally comparable to that of a 'marked' fluid element (Roberts 1961, Jiang 1985):

$$\lim_{\tau\to\infty} K_{ij}^f(\tau) \to K_{ij}^f(\infty) = \int_0^\infty d\tau' Q_{ij}^f(\tau'). \tag{13}$$

According to the definition, $\lim_{\tau\to\infty} K_{ij}^{fP}(\tau) \to K_{ij}^{fP}(\infty) = \int_0^\infty d\tau' Q_{ij}^{fP}(\tau')$. Therefore,

$$K_{ij}^P(\infty) / K_{ij}^{fP}(\infty) \to 1. \tag{14}$$

This is in fact the result obtained by Tchen (1947). However, as in the expressions (12) and (13), Q_{ij}^{fP} and Q_{ij}^f contain the knowledge of different trajectories. The ratio $K_{ij}^P(\infty) / K_{ij}^f(\infty)$ and $K_{ij}^{fP}(\infty) / K_{ij}^f(\infty)$ generally do not approach unity, especially when the particle is heavy. In other words, the 'fluid element' seen by the particle is so specific that its properties generally do not represent that of the fluid. Therefore, although D/Dt is replaced by d/dt in the motion equation, the result in (14) does not lead to any inconsistency with the concept of the effect of crossing trajectories. In the next section, numerical results will be shown in agreement with the well known relationships by Csanady. Seen from the definition of K_{ij}^{fP}, it is not necessarily to require a fluid element always accompanying the solid particle. The particle is allowed to change its neighborhood during dispersion.

3. NUMERICAL TEST WITH A MODAL TURBULENCE SPECTRUM

Similar to the treatment for a marked fluid particle (Roberts 1961), a probability density function (PDF), $\overline{P}_P(\bar{y}, \tau)$, is introduced as the probability density for the particle to make a displacement within $\bar{y}$ and $\bar{y} + d\bar{y}$, in the time interval τ. With the independence conjecture (Corrsin 1959),

$$Q_{ij}^{fP}(\tau') \equiv <\tilde{u}_i^{P}(0)\tilde{u}_j^{P}(\tau')> \approx \int d\bar{y}' Q_{ij}^{E}(\bar{y}', \tau')\overline{P}_P(\bar{y}', \tau'), \qquad (15)$$

where the integration $\int d\bar{y}'$ is over the whole space, and

$$Q_{ij}^{E}(\bar{y}', \tau') \equiv <\tilde{u}_i^{E}(\bar{x}_0, t_0)\tilde{u}_j^{E}(\bar{x}_0 + \bar{y}', t_0 + \tau')> \qquad (16)$$

is the Eulerian velocity correlation function of the undisturbed turbulent fluid flow. Then

$$K_{ij}^{P}(\tau) = \int_0^{\tau} d\tau' \int d\bar{y}' \overline{P}_P(\bar{y}', \tau') Q_{ij}^{E}(\bar{y}', \tau')$$

$$- (1-\alpha^2)\int_0^{\tau} d\tau' \cosh[\beta(\tau - \tau')] \int d\bar{y}' \overline{P}_P(\bar{y}', \tau') Q_{ij}^{E}(\bar{y}', \tau') + \frac{(1-\alpha^2)}{\beta} \sinh(\beta\tau) A_{ij}^{fP}, \qquad (17)$$

where $A_{ij}^{fP} = \beta\int_0^{\infty} d\tau' \exp(-\beta\tau') \int d\bar{y}' \overline{P}_P(\bar{y}', \tau') Q_{ij}^{E}(\bar{y}', \tau')$. The corresponding equations for a 'marked' fluid element have been shown (Roberts 1961, Jiang 1985) as:

$$Q_{ij}^{f}(\tau') \equiv <\tilde{u}_i^{f}(0)\tilde{u}_j^{f}(\tau')> \approx \int d\bar{y}' Q_{ij}^{E}(\bar{y}', \tau')\overline{P}_f(\bar{y}', \tau'), \qquad (18)$$

$$K_{ij}^{f}(\tau) = \int_0^{\tau} d\tau' \int d\bar{y}' \overline{P}_f(\bar{y}', \tau') Q_{ij}^{E}(\bar{y}', \tau'). \qquad (19)$$

Assume $\bar{\bar{u}} = 0$. With the spatial Fourier transformations for the PDF and Q_{ij}^{E}, and the PDF is assumed Gaussian. The numerical tests are performed with the modal spectrum (Kraichnan 1970):

$$E(k, t) = 16\sqrt{2/\pi}\, u_1^2 k_0^{-5} k^4 \exp(-2k^2/k_0^2)\exp[-(u_1 k_0 t)^2]. \qquad (20)$$

The detailed formulation is referred to Jiang et al. (1998). In the computation, the carrying fluid is assumed as air with kinetic viscosity $v = 1.569 \times 10^{-5}\,\mathrm{m}^2\,\mathrm{s}^{-1}$. For the turbulence, it is assumed that $u_1 = 0.2\,\mathrm{m\,s}^{-1}$ and $k_0 = 20\,\mathrm{m}^{-1}$.

Figure 1 illustrates the results when $a = 100\,\mu$ m and $\Gamma = 1000$ ($\alpha \approx 0.0015$, $\beta' \approx 1.764$), a heavy particle. Without gravity, shown in 1(a) and 1(b), the effect of crossing trajectories is solely due to the particle inertia. K^P and K^{fP} are shown to approach the same value, but both larger than K^f. In figures 1(c) and 1(d), gravity is included. The effect of crossing trajectories is shown even stronger. K^P, K^{fP}, Q^P and Q^{fP} are significantly reduced. The reduction is stronger in the horizontal, K_H^P, K_H^{fP}, Q_H^P and Q_H^{fP}. Note that negative values are found in the Q_H^{fP} curve. This is in agreement with to the results by Nir and Pismen (1979) and the discussion by Csanady. However, their curves were supposed for Q_H^P. The existence of Q_H^{fP}, and the difference between Q_H^{fP} and Q_H^P were not mentioned in their papers.

In figure 2, $a = 2000\,\mu$ m and $\Gamma = 2$ ($\alpha = 0.6$, $\beta' \approx 1.765$), a non-heavy particle. The deviation of particle dispersion from that of the flow is not as strong as that in figure 1. In this case, the particle inertia is mainly due to its size. In figure 1(a), the effect of unsteady forces is negligible, because the particle is heavy with a large Γ or a small α. Figure 2(a) shows that, since α is now comparable to unity, the inclusion of the unsteady force terms reduces the deviation of particle dispersion from that of a fluid element. On the other hand, since the particle is not heavy, it is shown in figure 2(b), the effect of gravitational settling is accordingly reduced.

Restricted by the size of the paper, other figures can not be included. Here we only note that, the increase either in Γ or in a strengthens the effect of crossing trajectories, and the effect is even

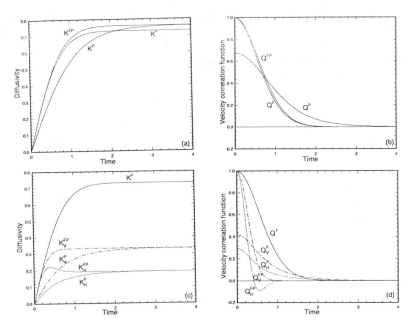

Figure 1. Calculated eddy diffusivities and velocity correlation functions of the case
$\Gamma = 1000$, $a = 100\mu$ m . (a) and (b) gravity not included; (c) and (d) gravity included,
dashed lines - horizontal direction, dash-dot lines - vertical direction

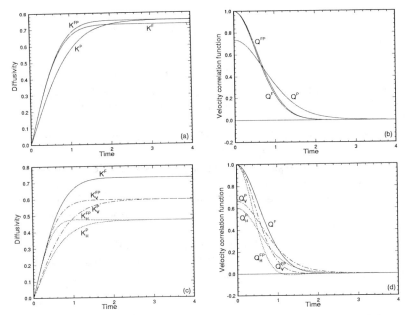

Figure 2. Calculated eddy diffusivities and velocity correlation functions of the case
$\Gamma = 2$, $a = 2000\mu$ m . (a) and (b) gravity not included; (c) and (d) gravity included,
dashed lines - horizontal direction, dash-dot lines - vertical direction

535

stronger when gravity is included. Secondly, if $\Gamma > 1$, i.e., $\rho_p > \rho_f$. $K^P(\infty)$ is found generally larger than $K^f(\infty)$. However, if $\Gamma < 1$, i.e., $\rho_p < \rho_f$, it is expected that $K^P(\infty) < K^f(\infty)$, but the overall effect of the unsteady force terms and the crossing trajectories is so weak that it can hardly be seen in the figures. Thirdly, it is found, with a fixed particle size, as its density increases, $K_H^P(\infty) / K_V^P(\infty)$ approaches 0.5. This is the result predicted by Csanady.

4. CONCLUSTION

The problem of non-heavy particle dispersion is practically as important as that of heavy particles. By considering the D/Dt versus d/dt problem still open, the present study extends the analysis to both heavy and non-heavy particles. It is shown analytically and numerically, the effect of crossing trajectories can be illustrated with the three kinds of the velocity correlation functions and eddy diffusivities. The results are in agreement with previous well known studies. Comprehensive and rigorous analysis based on first principles is needed for more reliable conclusion.

This study was supported by the National Foundation of Natural Sciences of China, but completed when the first author was teaching at the Dept. of Mech. Engn., the University of Hong Kong .

5. REFERENCES

Corrsin, S. (1959). "Progress report on some turbulent diffusion research", *Adv. in Geophysics* **6**, 161-184

Csanady, G.T. (1963). "Turbulent diffusion of heavy particles in atmosphere", *J. Atmos. Sci.* **20**, 201-208

Gatignol, R. (1983). "The Faxen formulae for a rigid particle in an unsteady non-uniform Stokes flow", *Journal de Mechanique theoretique et appliquee* **1**, 143-160

Jiang, D., Wang, W., Huang, Z., and Qian, W. (1998). "On the effect of crossing trajectories of the dispersion of a small spherical particle", submitted to *J. Fluid Mech.*

Kraichnan, R. H. (1970). "Diffusion by a random velocity field", *Phys. Fluids* **13**, 22-31

Maxey, M. R. and Riley, J. J. (1983). "Equation of motion for a small rigid sphere in a nonuniform flow", *Phys. Fluids* **26**, 883-889

Nir, A. and Pismen, L. M. (1979). "The effect of a steady drift on the dispersion of a particle in turbulent fluid", *J. Fluid Mech.* **94**, 369-381

Reeks, M. W. (1977). "On the dispersion of small particles suspended in an isotropic turbulent fluid", *J. Fluid Mech.* **83**, 529-546

Roberts, P. H. (1961). "Analytical theory of turbulent diffusion", *J. Fluid Mech.* **11**, 257-283

Stock, D. E. (1996). "Particle dispersion in flowing gases - 1994 Freeman Scholar Lecture", *Trans. of the ASME* **118**, 4-17

Tchen, C. M. (1947). *Mean values and correlation problems connected with the motion of small particles suspended in a turbulent fluid*, Ph.D. Dissertation, University of Delft

Yudine, M.I. (1959). "Physical considerations on heavy-particle diffusion", *Advances in Geophysics* **6**, 185-191

3 Hydraulics and water quality

3.1 Water quality model studies

Environmental Hydraulics, Lee, Jayawardena & Wang (eds) © 1999 Balkema, Rotterdam, ISBN 90 5809 035 3

Invited lecture: Modelling salt, sediment and heavy metal fluxes in estuarine waters

R.A. Falconer & B. Lin
Cardiff School of Engineering, Cardiff University, UK

ABSTRACT: Details are given of the development and application of refined two-dimensional and three-dimensional depth and layer integrated numerical models for predicting water elevations, velocity components, water quality constituents, cohesive and non-cohesive sediment fluxes and the fate of trace metals in estuarine waters. The modelling framework allows for: (i) advection and diffusion of contaminants in the dissolved phase, (ii) partitioning of contaminants between the dissolved and adsorbed phases, (iii) input of adsorbed contaminants to the water column due to re-suspension of bed sediments, (iv) accumulation of contaminants in the bed sediments due to deposition of suspended sediment, and (v) losses due to volatisation and biodegredation for organic contaminants. The models use the finite difference technique to solve the equations of mass and momentum conservation and the transport equation for water quality constituents, sediments and trace metals, with the ULTIMATE QUICKEST scheme being used to represent the advective transport processes more accurately and without the occurrence of grid scale oscillations. The models outlined have been applied to the Humber Estuary in the UK and verified against extensive data.

1. INTRODUCTION

The transport of suspended sediments - both cohesive and non-cohesive - has historically been of considerable interest to scientists and engineers involved in estuarine and coastal water management, primarily from the viewpoint of erosion, deposition, navigation and flood defence. However, in recent years there has been a growing interest in other aspects relating to estuarine and coastal sediment dynamics, including long term morphological processes, estuarine and coastal inlet stability and, in particular, in the transport of heavy metals and toxic waste via adsorption onto sediment particles and re-suspension into the water column. Estuaries are favourable sites for industrial and urban development but are physically and chemically highly complex aquatic environments. Hence, more stringent legislation relating to industrial discharges has been increasingly introduced (e.g. Commission of the European Communities, 1993) and estuarine managers have therefore sought accurate models to predict the transport and fate of trace metals and organic contaminants.

In addition to estuarine suspended sediment dynamics being a complex phenomenon, with sediment transport including such processes as erosion, deposition, flocculation, advection and diffusion etc., the transport of heavy metal contaminants - either in solution or via suspended particles - is even more complex. The partitioning of trace metals between these two phases is dependent on estuarine water column variables, including: salinity, pH, availability of complex species and the physical and chemical characteristics of suspended particles (Ng et al, 1996). Thus, heavy metal fluxes are difficult to predict from theoretical considerations alone.

Details are given herein of the refinement of 2-D and 3-D numerical models for predicting suspended sediment and heavy metal fluxes in estuarine waters. For the 2-D model an orthogonal boundary fitting curvilinear coordinate grid has been deployed to replicate complicated and irregular estuarine boundaries. Likewise, for the 3-D model, an operator splitting algorithm has been used to split the 3-D sediment transport equation into two parts; one for solving the vertical differential terms and the other for solving the horizontal terms, with the ratio of the vertical to the horizontal length scale being small for most estuarine studies.

The parameterisation of estuarine metal-water interactions of pollution may be achieved empirically by examining the time-dependent behaviour of a contaminant with its radioactive counterpart, added to natural samples maintained under carefully controlled laboratory conditions (Turner and Millward, 1994). This approach yields particle-water interaction functions relevant to reaction conditions encountered in estuaries, together with the quantification of adsorption kinetics through the derivation of reaction time constants. In this study, the application of site specific and empirically derived partition coefficients to the modelling of estuarine contaminants has been investigated, incorporating experimentally derived partition coefficients for Cadmium (Cd) and Zinc (Zn). The site studied most was the Humber Estuary, along the North East coast of England, with this being one of Britain's largest and industrially active estuaries.

2. MATHEMATICAL MODEL FORMULATION

In order to predict the water quality constituent distributions and suspended sediment and heavy metal fluxes in estuarine waters, the equations to be solved included:- the fluid continuity and momentum conservation equations, including appropriate turbulence closure equations, and the solute transport equation for salinity, suspended sediments - both cohesive and non-cohesive - and heavy metal concentration distributions.

2.1 Hydrodynamic Model

The governing equations for the hydrodynamic models consisted of the 2-D and 3-D horizontal momentum equations and the continuity equation, with these equations being obtained by integrating the Navier-Stokes equations over the depth and layer of the water column respectively. The equations were written in their full conservative form and included the effects of the earth's rotation, wind action, bed friction and turbulent diffusion. For turbulence equations two models were considered, namely a zero-equation two-layer mixing length model and a two-equation $k - \varepsilon$ turbulence model (Rodi, 1984). The bed resistance stress was represented in the form of Darcy's equation, with the friction factor being evaluated using the Colebrook-White equation, and with Reynolds number effects being included and found to be particularly significant for shallow water flood plain flows (Falconer and Chen, 1996). The surface wind stress was represented using a quadratic friction law, with the resistance coefficient being represented in a piecewise manner (Wu, 1969).

The governing 3-D hydrodynamic equations were solved using a combined explicit and implicit finite difference scheme. An alternating direction implicit scheme was used to solve the depth integrated hydrodynamic equations to give the water elevation field. The layer-integrated equations were then solved to obtain the layer-averaged velocities, using the water elevation field predicted from the depth-integrated equations. In solving the depth-integrated hydrodynamic equations the layer-averaged velocities were integrated to obtain the depth averaged velocity. The Crank-Nicolson scheme was used to solve the layer-integrated hydrodynamic equations, with the vertical diffusion terms being treated implicitly and the remaining terms treated explicitly. Two iterations were performed to solve the coupled depth - and layer - integrated equations. The flooding and drying processes were modelled using a robust scheme developed by Falconer and Chen (1991) and with full details of this hydrodynamic model being given by Lin and Falconer (1997a).

2.2 Solute Transport Model

For the 3-D model predictions of water quality constituents, suspended sediment and heavy metal fluxes in estuarine waters, the generalised 3-D solute transport equation was expressed as:-

$$\frac{\partial \phi}{\partial t} + \frac{\partial (u\phi)}{\partial x} + \frac{\partial (v\phi)}{\partial y} + \frac{\partial}{\partial z}\left[\left(w - W_s\right)\phi\right] - \frac{\partial}{\partial x}\left(\varepsilon_x \frac{\partial \phi}{\partial x}\right) - \frac{\partial}{\partial y}\left(\varepsilon_y \frac{\partial \phi}{\partial y}\right) - \frac{\partial}{\partial z}\left(\varepsilon_z \frac{\partial \phi}{\partial z}\right) = 0 \qquad (1)$$

where ϕ = solute concentration (for salinity, suspended sediments and heavy metals), u,v,w = layer averaged velocity components in x, y, z, directions, W_s = apparent sediment settling velocity (=0 for salinity and heavy metal fluxes), and $\varepsilon_x, \varepsilon_y, \varepsilon_z$ = turbulent diffusion coefficients in x, y, z, directions.

In solving equation (1) four different boundary condition types needed to be prescribed. These included:

(i) Open boundary conditions: For an inflow condition the concentration profiles were prescribed using either available field data or assumed conditions (e.g. for salinity) or equilibrium concentration profiles for suspended sediment fluxes, related to the local bed shear stress. For an outflow boundary condition, the concentration profiles were obtained by extrapolation using a first order upwind difference scheme.

(ii) Bank boundary conditions: At bank boundaries the solute fluxes were set to zero, i.e. the normal derivatives of the concentration were set to zero.

(iii) Surface boundary condition: At the free surface the net vertical solute flux was assumed to be zero, giving a zero gradient with respect to z for salinity and for sediment and trace metal fluxes, we get:-

$$\left[W_s\phi + \varepsilon_z \frac{\partial \phi}{\partial z} \right]_{\zeta} = 0 \tag{2}$$

where ζ = water surface elevation relative to datum.

(iv) Bed boundary condition: At the bed the flux of both salinity and heavy metals was zero, whereas for sediment fluxes the bed boundary was specified at a small height 'a' (i.e. the reference level) above the bed, and the concentration or its gradient was prescribed by its equilibrium value at this elevation, that is:-

$$\text{(i)} \quad S_a = S_{ae} \tag{3a}$$

$$\text{(ii)} \quad E_a = \left(-\varepsilon_z \frac{\partial S}{\partial z} \right)_a = E_{ae} \tag{3b}$$

in which the subscript 'ae' denotes the equilibrium value at the reference level 'a'.

A wide variety of relationships exist in the literature for predicting the near bed reference concentration of suspended sediment, from which the entrainment rate of bed sediment flux into suspension can be obtained. Garcia and Parker (1991) carried out a detailed comparison of seven relationships against a common set of experimental data, for which direct measurements of the bed concentrations were available. They found that the relationship given by van Rijn (1984) gave one of the best fits and is given as:-

$$S_{ae} = 0.015 \frac{D_{50} T^{1.5}}{a D_*^{0.3}} \tag{4}$$

where D_{50} = sediment diameter, of which 50 % of bed material is finer, T = transport stage parameter and D_* = particle parameter. The reference concentration τ_{ae} is a function of the bed shear stress τ_b and the critical bed shear stress $\tau_{b,cr}$ for the initiation of motion. If the bed shear stress exceeds the critical bed shear stress, then particles settled on the bed will re-suspend into the flow and τ_{ae} is positive. Likewise, if the bed shear stress is less than the critical shear stress, then no erosion occurs and τ_{ae} becomes zero. The upward diffusive sediment flux is therefore given by:-

$$E_{ae} = -W_s S_{ae} \tag{5}$$

For cohesive sediment transport, the bed boundary fluxes were represented in the model as follows:-

$$-W_s\phi - \varepsilon_z \frac{\partial \phi}{\partial z} = q_{dep} \qquad \text{when} \qquad \tau_b \leq \tau_d \qquad \text{(deposition)}$$

(a) analytical solution

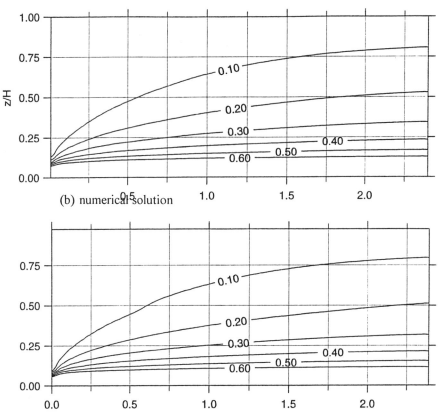

Figure 1 Adjustment of concentration profiles on a sand bed in
a horizontally uniform flow

(b) numerical solution

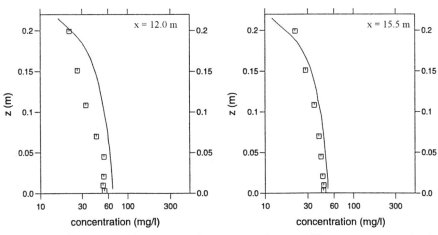

Figure 2. Adjustment of concentration profiles on a porous bed
in a horizontally uniform flow

$$-W_s\phi - \varepsilon_z \frac{\partial\phi}{\partial z} = q_{eor} \qquad \text{when} \qquad \tau_b \geq \tau_e \qquad \text{(erosion)}$$

$$-W_s\phi - \varepsilon_z \frac{\partial\phi}{\partial z} = 0 \qquad \text{when} \qquad \tau_d < \tau_b < \tau_e \qquad \text{(equilibrium)}$$

where τ_b = bed shear stress, τ_d = critical shear stress for no further deposition, τ_e = critical shear stress for erosion and q_{dep} and q_{ero} = deposition and erosion rates respectively at the bed. In this study the deposition rate proposed by Krone (1962) was used, as given by:-

$$q_{dep} = -W_s\phi_b\left(1 - \frac{\tau_b}{\tau_d}\right) \qquad (6)$$

where ϕ_b = near bed concentration and τ_b was in the range 0.04 - 0.15 N/m^2. Likewise, the erosion rate for soft natural mud was represented by the following empirical expression (Raudkivi, 1990):-

$$q_{ero} = q_f \exp\left[\alpha(\tau_b - \tau_e)^{1/2}\right] \qquad (7)$$

where q_f = 4.2 x 10^{-6}, α = 8.3 and τ_e was in the range 0.07 - 0.17 N/m^2.

Finally for heavy metals the distribution of contaminants between the dissolved and adsorbed particulate phases was defined by an empirically-derived equilibrium partition coefficient, K_D, given as:-

$$K_D = \frac{\phi_P}{\phi_s} \qquad (8)$$

where ϕ_P and ϕ_s = concentrations of contaminants adsorbed on suspended particles and in solution respectively. The partition coefficient K_D was assumed to vary in one of the two following ways:
(i) Using the relationship of Turner and Millward (1994), that is

$$\ln K_D = b \ln(s + 1) + \ln K_D^o \qquad (9)$$

where b is constant, s = salinity, and K_D^o = partition coefficient in fresh water (i.e. s = 0).
(ii) Using explicit tabulations of measured values of K_D as a function of salinity

In estuarine and coastal waters, the horizontal length scale is generally much larger than the vertical scale and an operator splitting algorithm was used to split the 3-D advective diffusion equation into a horizontal 2-D equation and a vertical 1-D equation. The 2-D advective diffusion equation was first solved horizontally, followed by the 1-D vertical advective-diffusion equation. To be consistent with the 3-D layer-integrated hydrodynamic model, the horizontal 2-D equation was also integrated through the layers, to give the layer integrated 2-D equation. The 2-D QUICKEST scheme extended from Leonard's 1-D QUICKEST scheme (1979) was used to solve the horizontal 2-D layer-integrated equation. The modified 1-D ULTIMATE algorithm was also used to prevent non-physical numerical oscillations, which frequently arise due to numerical dispersion when large concentration gradients exist, see Lin and Falconer (1997b).

3. MODEL APPLICATIONS

The numerical models have been tested against analytical solutions and laboratory measurements for different types of boundary conditions, and have also been applied to predict tidal currents and salinity, suspended sediment and heavy metal concentration distributions in the Humber Estuary.

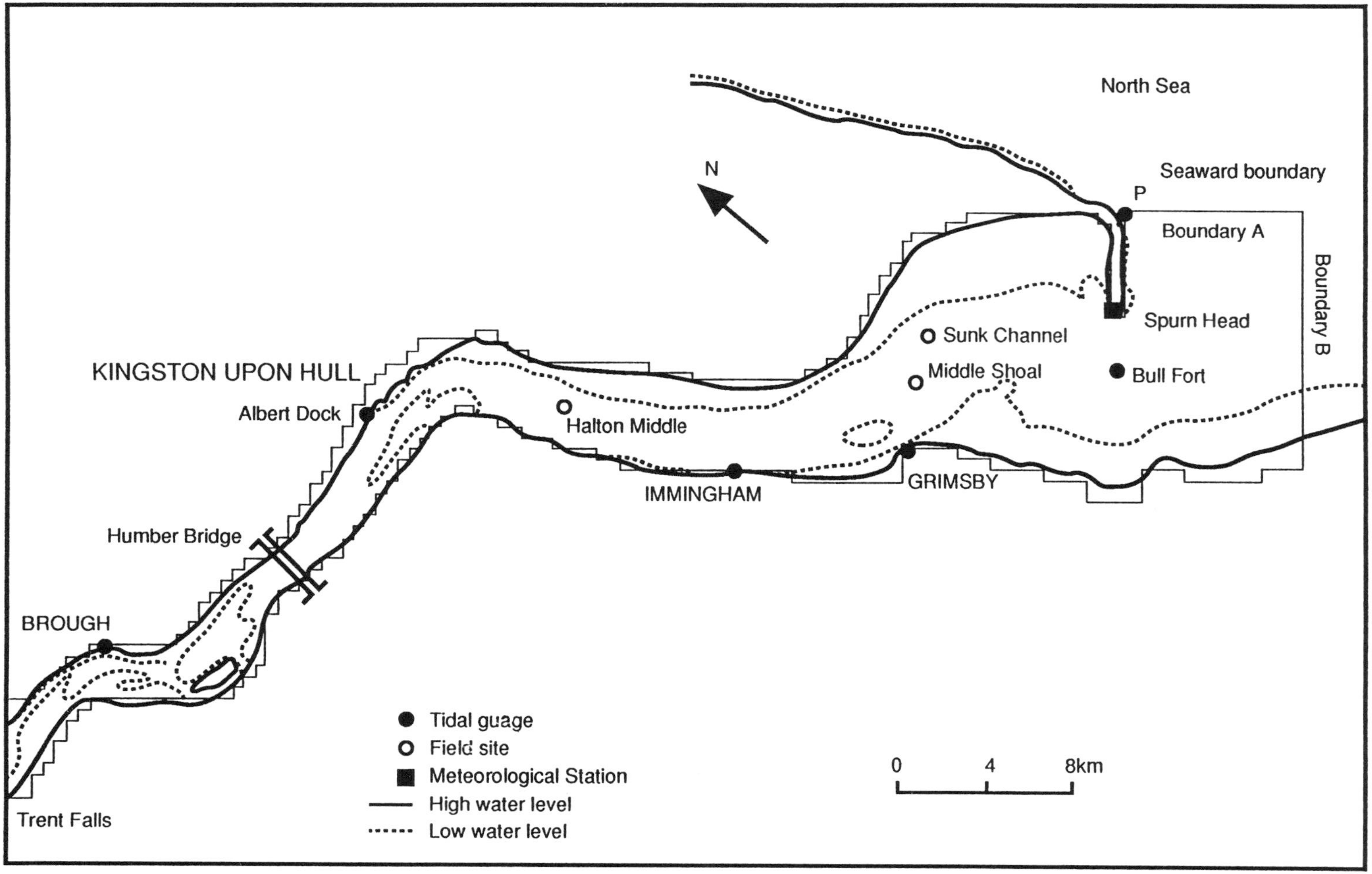

Figure 3   Plan of the Humber Estuary

An example test used for comparisons with analytical solutions was the development of the vertical sediment concentration profiles on a sand bed in a horizontally uniform flow and with the inlet sediment concentration being zero. The 3-D model was used to predict the concentration profiles. Figs 1(a) and 1(b) show the results of the analytical solution and the numerical model predictions respectively. Another test used for comparisons with the laboratory measurements was the adjustment of the vertical sediment concentration profiles on a porous bed in a horizontally uniform flow and with a constant supply at the inlet. Figs 2(a) and 2(b) show the resulting comparisons for the 3-D model predictions and measured data at two downstream locations. The sediment mass balance was also checked for this test and the error was found to be less than 0.15 %.

The numerical models were then extensively tested for the Humber Estuary to study the spatial and temporal distributions of both suspended sediments and dissolved and particulate trace metals along the estuary. The Humber Estuary is a large well mixed estuary, situated along the north-east coast of England and providing an outlet to the North Sea for the rivers Trent and Ouse and shipping access to a number of ports. It has the largest catchment in the UK, draining over 20 % of England. The main estuary stretches from Spurn Head in the east to Trent Falls in the west, a distance of approximately 62 km (see Fig 3). There are numerous sewerage and industrial effluent inputs along the estuary, with some being metal bearing. It has a relatively large tidal range, of up to 7 km, giving well mixed waters and extensive flooding and drying. The intertidal mudflats are environmentally sensitive areas, supporting naturally rare bird and plant populations. There are also five sites of special scientific interest along the estuary.

In applying the 3-D numerical model to the Humber Estuary, a regular finite difference grid of 118 x 56 cells, equally spaced at 500 m intervals, was set up over the horizontal plane, covering the estuary from Trent Falls at the head to about 6 km seawards of Spurn Head. Fifteen layers were included in the vertical. During flood tide the flow was southwards, with a section of the flow separating from the main current and flowing into the estuary. In contrast, on ebb tide this flow pattern was reversed, with the seaward boundary parallel to Spurn Head effectively acting as a streamline. Hydrodynamic data were also taken from a courser grid model of the North Sea. Full details of the estuary and boundary locations are given in Fig 3.

In order to drive the hydrodynamic model, water elevation data recorded by ABP Research and Consultancy Ltd. were used at both the seaward and landward boundaries, i.e. just beyond Spurn Head and Trent Falls. Field measurements of water elevations, velocities, water quality constituents and suspended sediment fluxes were also available at several sites along the estuary for model calibration and verification. For the various effluent and industrial discharges along the estuary, mean daily rates were input at the outfall cells and linear interpolation assumed to obtain intermediate values. For the water quality constituent suspended sediment and heavy metal fluxes from the rivers Wharfe, Aire, Don, Trent and Ouse, the boundary values used were obtained by summing the individual water quality components and from field data by Edwards et al (1987) for heavy metals. The predictions of the hydrodynamic parameters, i.e. water elevations and velocities, and the salinity and suspended sediment fluxes agreed well with field data along the estuary. Full details are given in Lin and Falconer (1997a) and with a comparison of the suspended sediment fluxes being given in Fig 4.

The performance of the contaminant transport module for heavy metals was assessed by running the model for the following partitioning scenarios, under a typical spring tide and a high river discharge: (i) no partitioning of contaminants between the dissolved and adsorbed phases, i.e. $K_D = 0$, (ii) no salinity dependence of K_D (i.e. by setting b in Eq. (9) to zero), (iii) salinity dependence of K_D based on empirically derived partitioning results, and (iv) increasing the salinity dependence of K_D (i.e. by increasing b to -1.5). The boundary values of dissolved contaminant concentrations were based on measured values of dissolved cadmium and were 0.5 µg/l at the freshwater gauging station on the River Ouse and 0.06 µg/l at the seaward boundary. The heavy metal module was calibrated against dissolved and particulate trace metal data measured at Bull Fort, i.e. at a site as remote as possible from the input sites. The model results and observed data were compared rigorously, with observed distributions of dissolved cadmium and zinc being reasonably well produced by the model.

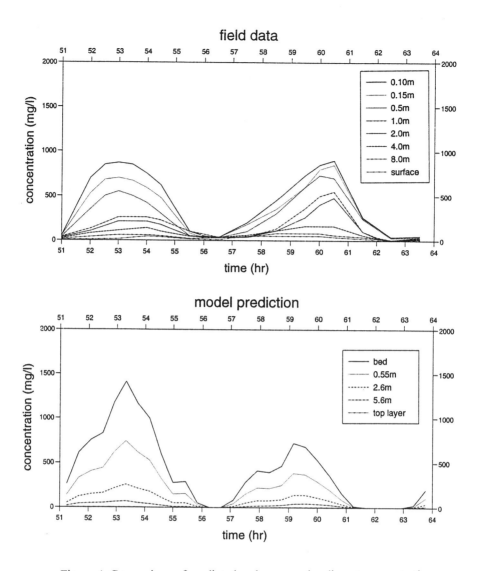

Figure 4 Comparison of predicted and measured sediment concentrations

4. CONCLUSIONS

Details are given of the increasing application of numerical models for predicting hydrodynamic, water quality, sediment and contaminant transport parameter distributions in estuarine waters. Particular emphasis is focused herein on the refinement of a 3-D model to include a higher order accurate numerical scheme for accurately modelling the advective processes and further consideration has been given to the representation of closed boundary conditions. In particular, the model has been extended to include trace metal flux predictions, with emphasis being focused on including a dynamic salinity based partition coefficient, defining the proportion of trace metal in the dissolved and particulate phases.

The models have been applied to analytical and laboratory test cases, for which idealised data exist, and to the Humber Estuary, in northern England, for which extensive hydrodynamic, water quality, suspended sediment and trace metal data exist. The hydrodynamic model predictions agree well with independently acquired field data for both regular (3-D) and curvilinear (2-D) distributions of a range of water quality indicators and suspended sediment fluxes along the estuary, with the models now being used as management tools to predict the hydro-environmental conditions of the estuary. For the heavy metal predictions along the Humber Estuary, with a novel salinity dependent partition coefficient, the model has shown good qualitative agreement with the field data, with the model predicting higher levels of trace metals in the dissolved state seawards of the estuary. This model is now being refined further at present in applying the model to the Mersey Estuary in the UK.

5. ACKNOWLEDGMENTS

This paper outlines briefly a number of research projects funded by the Natural Environment Research Council LOIS programme, the Engineering and Physical Sciences Research Council, BP Chemicals Ltd., Yorkshire Water plc and BMT Ports and Coastal Ltd. The authors are also grateful to the Environment Agency (Yorkshire Region) and ABP Research and Consultancy Ltd. for the provision of field data.

6. REFERENCES

Commission of the European Communities (1993). "Proposal for a council directive on integrated pollution prevention and control", COM (93), 423 final, Brussels, Belgium

Edwards, A., Freestone, R. and Urquhart, C. (1987). "The water quality of the Humber Estuary", Report of the Humber Estuary Committee, Yorkshire Water plc.

Falconer, R.A. and Chen, Y. (1991). "An improved representation of flooding and drying and wind stress effects in a 2-D numerical model", Proc. Institution of Civil Engineers, Part 2, 91, 659-687

Falconer, R.A. and Chen Y. (1996). "Modelling sediment transport and water quality processes on tidal floodplains", in Floodplain Processes (Eds MG Anderson et al), John Wiley & Sons Ltd.

Garcia, M. and Parker, G. (1991). "Entrainment of bed sediment in suspension", J. Hyd. Eng., ASCE, 117(4), 414-435

Krone, R.B. (1962). "Flume studies of the transport of sediment in estuarial processes", Final Report, Hydraulic an Sanitary Engineering Res Lab, University of California, Berkely

Lin, B. and Falconer, R.A. (1997a). "Three-dimensional layer integrated modelling of estuarine flows with flooding and drying", Est. Coast. and Shelf Science, 44, 737-751

Lin, B. and Falconer, R.A. (1997b). "Tidal flow and transport modelling using the ULTIMATE QUICKEST scheme", J. of Hyd. Eng., ASCE, 123(4), 303-314

Ng, B., Turner, A., Tyler, A.O., Falconer, R.A. and Millward, G.E. (1996). "Modelling contaminant geochemistry in estuaries", Water Research, 30 (1), 63-74

Raudkivi, A.J. (1990). "Loose boundary hydraulics", Third Edition, Pergammon Press plc, Oxford

Rodi, W. (1984). "Turbulence models and their application in hydraulics", IAHR Publication (Second Edition), Delft.

Turner, A. and Millward, G.E. (1994). "The partitioning of trace metals in a macrotidal estuary: implications for contaminant transport models", Est. Coast. and Shelf Science, 39, 45-58

van Rijn, L.C. (1984). "Sediment transport part 2: suspended load transport", J. of Hyd. Eng., ASCE, 110(10), 1613-1641

Wu, J. (1969). "Wind stress and surface roughness at air-surface interface", J. Geophysical Res, 74, 444-455

Environmental Hydraulics, Lee, Jayawardena & Wang (eds) © 1999 Balkema, Rotterdam, ISBN 90 5809 035 3

A post audit of nutrient reductions in two estuaries

Wu-Seng Lung
Department of Civil Engineering, University of Virginia, Charlottesville, Va., USA

ABSTRACT: Significant reductions of nutrient loads from point sources have taken place in the Chesapeake Bay region during the past two decades. Both nutrient loading rates from the point sources and the response in the receiving water have been monitored closely, accumulating a significant amount of data. Two estuaries tributary to the Bay which have the most nutrient reduction among the major tributaries are the focus of this study. Water quality data are first reviewed and analyzed. Water quality models are then used to quantify the response in a post audit effort.

1. INTRODUCTION

Point source nutrient reductions have been progressively implemented in the Chesapeake Bay region (Figure 1) for the last two decades. The control measures include phosphate detergent bans and nutrient removal at wastewater treatment plants. Most of the wastewater treatment plants are located along major estuaries tributary to the Bay. For example, there are eight municipal wastewater treatment plants located above the fall line of the Patuxent Estuary in Maryland. Phosphorus removal process was first installed at these plants in 1980's. Later a phosphate detergent ban was in effect in Maryland. Starting in 1990, nitrogen removal has been installed at several of these plants. In 1988, the Commonwealth of Virginia established a phosphate detergent ban. Municipal wastewater treatment plants in the James River basin, Virginia started phosphorus removal in 1990's to meet an effluent total phosphorus limit of 2 mg/L.

While all these control initiatives were being implemented, receiving water monitoring has continued through the last decade, collecting a significant amount data for the Chesapeake Bay and its tributaries. An analysis of the data indicates that while phosphorus and nitrogen concentrations in these two estuaries have been decreasing steadily over the last decade, clearly due to reductions of nutrient input, the algal biomass levels in the water column have not shown any appreciable reduction. Further, the bottom water of the Patuxent Estuary continues to display an anoxic condition during the summer months. Modeling analyses also substantiate this observation by reproducing the receiving water data collected before and after the nutrient controls.

2. NUTRIENT LOAD REDUCTIONS

Figure 2 shows the total phosphorus and nitrogen loads from point sources to the Patuxent Estuary from 1984 to 1995. The impact of the Maryland phosphate detergent ban is clear, reducing the phosphorus loads by 50% in 1985-1986. Continuing improvements in wastewater treatment have further reduced the phosphorus loads by another 50%, to about 100 kg/day. Nitrogen load reductions took place in 1991 when biological nutrient removal was installed at the wastewater treatment plants. Note the seasonal trend in the nitrogen loads due to the seasonal nature of the nitrogen removal process. A factor of four in load reduction is obtained by nitrogen removal (Figure 2).

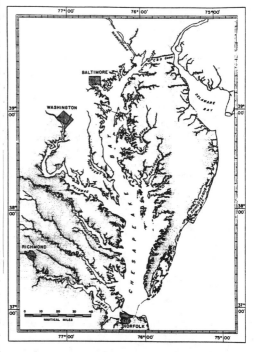

Figure 1. Chesapeake Bay and Tributaries

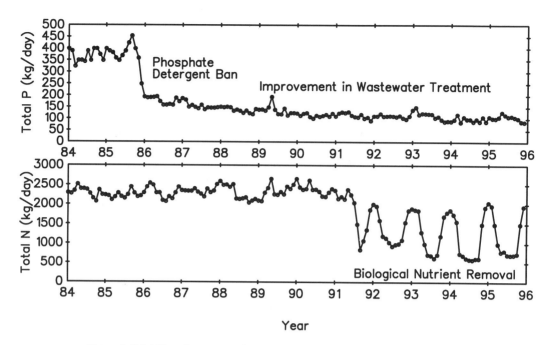

Figure 2. Total Phosphorus and Nitrogen Loads to the Patuxent River Estuary, 1984-1995

In the James Estuary basin, only phosphorus removal has been implemented at the point sources. Its impact on phosphorus load can be shown by one of the major point sources, the Chesapeake Elizabeth wastewater treatment plant. Figure 3 shows that the trend in total phosphorus concentrations of the effluent from 1980 to 1995. Prior to the phosphate detergent ban in Virginia, the effluent concentration is about 6 mg/L. It is reduced to below 2 mg/L at the present time, meeting the limit set by the regulatory agency. Other plants in the James River basin have also achieved similar reductions in phosphorus concentrations (i.e., by a factor of 3) as reported by Riverson (1997).

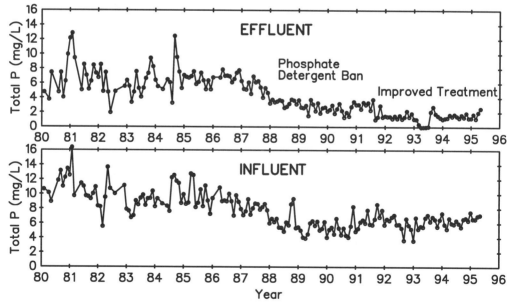

Figure 3. Effluent and Influent Total P Concentrations at Chesapeake Elizabetn Plant

3. RECEIVING WATER RESPONSE

The receiving water data have been reviewed to detect any significant trend in water quality. As indicated, the size of the data base for nutrients, algal biomass, and dissolved oxygen for the last decade is considerable. Due to the limited space in the manuscript, only the salient features of the data are presented. The algal biomass (chlorophyll a) peak has been consistently observed at Nottingham prior to the nutrient reduction and the bottom dissolved oxygen concentrations are near anoxic near Broomes Island in the Patuxent Estuary. Figure 4 shows these two water quality constituents from 1985 to 1996, covering the period of the phosphate detergent ban, treatment improvement, and biological nitrogen removal. Figure 4 shows that there is no significant improvement in these two parameters. High chlorophyll a levels are being measured from year to year following the nutrient controls. In the meantime, the anoxic condition near Broomes Island has not improved. A similar observation is found in the James River Estuary, where consistent peak chlorophyll a levels are measured near Hopewell during the summer months. Figure 5 shows that there is a very small reduction in chlorophyll a levels in the James River Estuary following the implementation of a phosphate detergent ban and phosphorus removals at treatment plants.

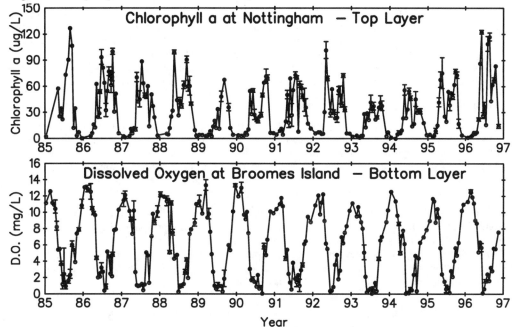

Figure 4. Chlorophyll a and Dissolved Oxygen Concentrations in Patuxent Estuary, 1985-1996

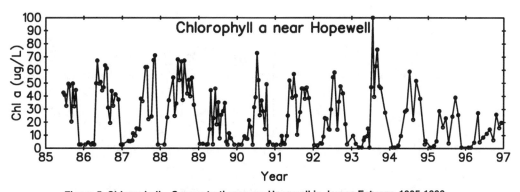

Figure 5. Chlorophyll a Concentrations near Hopewell in James Estuary, 1985-1996

4. MODEL POST-AUDIT ANALYSIS

4.1 The Patuxent Estuary

The Patuxent Estuary water quality model developed by Lung (1992) is used in this study to analyze the water quality data in recent years. The model was originally calibrated with three years of data from 1983 to 1985. The water quality column in the model is divided into 38 segments in two layers (i.e., 19 segments in each layer). There are two sediment layers (aerobic and anaerobic) below the water column. In each of the segments, 14 water quality variables are simulated, including two phytoplankton groups, zooplankton biomass, organic nitrogen, ammonium, and nitrite/nitrate nitrogen, particulate and dissolved organic and inorganic phosphorus, CBOD, dissolved oxygen, salinity, and suspended solids. The model is reconfigured with the 1995 environmental and boundary conditions to drive the calculation in this model post-audit analysis. Note that nutrient loads in 1995 for the Patuxent Estuary are much lower than the 1983-85 levels and are incorporated into the model. Water column and sediment kinetic coefficient values used in the original model calibration analysis are kept in the modeling analysis of the 1995 data. Figure 6 shows the model results from the 1995 data compared with the model results from 1985, i.e., comparing post-control condition with the condition prior to nutrient controls. The comparison is presented in a succinct fashion, showing the algal biomass (in chlorophyll *a*) levels at Nottingham and dissolved oxygen concentrations at Broomes Island. The comparison indicates that the model reproduces the data in 1985 and 1995 closely. Perhaps the most important result is that algal biomass levels in the the Patuxent remain high and the dissolved oxygen levels in the bottom water of Broomes Island continue to be low during the summer months. There is a very insignificant change in water quality during the last decade following considerable nutrient reductions.

4.2 The James Estuary

The James River Estuary model (Lung and Testerman, 1989) are used to analyze the seasonal steady-state conditions in the James River Estuary. It should be pointed out that unlike the Patuxent Estuary, the James Estuary does not have a dissolved oxygen problem. The low dissolved oxygen condition in the James Estuary has been eliminated over two decades ago following the wastewater treatment improvement to remove organic materials. The primary water quality problem in the James is high algal biomass, consistently observed near the Hopewell area (see Figure 5).

The James Estuary is divided into 61 segments along the main stem from Richmond to the mouth of the river in a one-dimensional fashion. Additional segments are included in the Appomattox River which a major tributary to the James. Together, there are 103 segments in the model. Water quality variables simulated are CBOD, DO, organic nitrogen, ammonium, nitrite/nitrate, organic phosphorus, orthophosphate, and chlorophyll *a*. The water quality model is configured to analyze the August 20, 1996 data in the James Estuary when chlorophyll *a* levels in the estuary reach a peak for the year.

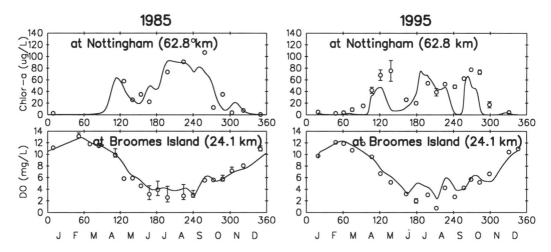

Figure 6. Comparing Water Quality Conditions with Model Runs (Before and After Nutrient Controls)

552

Model results from this post-audit analysis are compared with the results of September 20, 1983, prior to any nutrient controls in the James River basin. Again, water column kinetic coefficient values are not adjusted between these two water quality conditions. The major difference in driving the model is the significant phosphorus load reductions from the point sources to the estuary as noted in the orthophosphate concentrations, immediately below the Richmond treatment plant (Figure 7). That is, peak orthophosphate concentrations in the James Estuary have been reduced by a factor of 8, from 0.4 mg/L to 0.05 mg/L. Yet, the impact on algal biomass is small with the peak chlorophyll *a* level reduced from 50 µg/L to 40 µg/L.

5. DISCUSSIONS

Nutrient controls in the Patuxent and James Estuaries have resulted in very small, if any water quality improvement in the last decade. While both inorganic nitrogen and phosphorus concentrations have significantly reduced in the receiving water, chlorophyll *a* levels in the estuaries have declined only slightly. A brief review of the nutrient limitation factors in the model reveals that the relatively low nutrient levels now in the estuaries do not reach any alarming level of nutrient limitation. For example, the inorganic nutrient levels are still well above the Michaelis-Menton half-saturation levels for nitrogen and phosphorus.

On the other hand, light attenuation in the estuaries is a major factor controlling eutrophication. That is, the phytoplankton in the water column is light-limited. Figure 8 presents a historical data of Secchi depth in the James Estuary near Hopewell, where significant algal biomass has been observed consistently. The data shows no temporal trend in Secchi depth over the last ten years in the water column of the James Estuary. It is not surprising that the algal biomass is not much affected by nutrient controls to date.

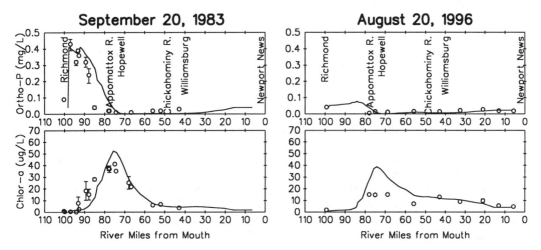

Figure 7. Model Results vs. Data for Chlorophyll *a* and Ortho-P in James Estuary, 1983 and 1996

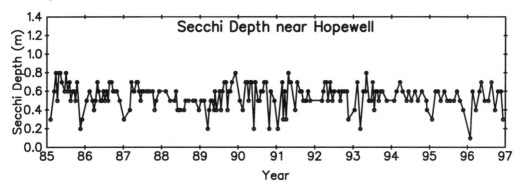

Figure 8. Secchi Disk Depths in the James Estuary near Hopewell, 1985-1996

In an earlier study of the James Estuary eutrophication and nutrient control, Lung (1986) applied the same James Estuary water quality model to evaluate a number of nutrient control alternatives for the James River basin. Lung's calculation indicates that a phosphate detergent ban would only reduce a peak chlorophyll a level in the James Estuary from 70 μg/L to 60 μg/L under a 7-day 10-year low flow condition. Phosphorus removal meeting an effluent limit of 2 mg/L would further reduce the peak chlorophyll a levels to about 40 μg/L, consistent with this model post-audit result (see Figure 7). To achieve a 20 μg/L chlorophyll a level, a phosphorus limit of 0.2 mg/L would be needed (Lung, 1986).

6. SUMMARY AND CONCLUSIONS

The review of the nutrient loading data to the Patuxent and James Estuaries indicates that substantial reductions of nutrient loads have achieved in the past decade through phosphate detergent bans, phosphorus removal, and biological nutrient removal at wastewater treatment plants in these two river basins. However, an analysis of the receiving water data suggests that very insignificant improvement in water quality of the receiving water has been measured while nutrient concentrations in the water column have reduced significantly. For example, algal biomass levels in chlorophyll a at Nottingham, Maryland have remain high during the summer months in last ten years despite the nutrient reductions. At Broomes Island, low dissolved oxygen levels (near anoxic) in the bottom water of the Patuxent Estuary still exist. In the James Estuary, peak chlorophyll a levels near Hopewell have declined slightly in recent years following the point source phosphorus controls and subsequent reductions in orthophosphate concentrations in the receiving water.

Water quality modeling analyses of the Patuxent and James Estuaries confirm field measurements. More importantly, the models calibrated with data prior to the nutrient controls are able to reproduce the field data observed in 1995 and 1996 (after nutrient controls) for the Patuxent and James Estuaries, respectively. Additional insights from the models clearly indicate that eutrophication in these estuaries are controlled by light attenuation in the water column instead of nutrients, before and after nutrient reductions.

A recent modeling study of the Chesapeake Bay tributaries has reached the similar conclusion, suggesting controlling suspended solids in the water column as one of the options to restore the living resources such as submerged aquatic vegetation in these tributaries.

7. REFERENCES

Lung, W.S. (1986). "Assessing phosphorus control in the James River basin", *J. Envir. Engrg.*, ASCE, 112 (1), 44-60.

Lung, W.S. and Testerman, N. (1989). "Modeling fate and transport of nutrients in the James Estuary", *J. Envir. Engrg.*, ASCE, 115(5), 978-991,

Lung, W.S. (1992). A Water Quality Model for the Patuxent Estuary. Environmental Engineering Research Report No. 8, Department of Civil Engineering, University of Virginia, Charlottesville, VA.

Riverson, J.D.N. (1997). Evaluation of Phosphorus Control Initiative at the Hampton Roads Sanitation District. Senior Thesis submitted to University of Virginia, Charlottesville, VA.

Environmental Hydraulics, Lee, Jayawardena & Wang (eds) © 1999 Balkema, Rotterdam, ISBN 90 5809 035 3

Transmission of nitrogen and phosphorus in the Maizuru Bay and proposal of improvement method for water quality

Atsuyuki Daido
Department of Civil Engineering, Ritsumeikan University, Shiga, Japan

Hiroshi Miwa
Maizuru National College of Technology, Kyoto, Japan

Hidetoshi Ikeno
Himeji Institute of Technology, Hyogo, Japan

ABSTRACT: The Maizuru Bay is the typical closed estuary, which is located at center of Japan, is faced the Japan Sea. Nitrogen (N) and phosphorus (P) have been increasing according as the population increases as other estuaries. In order to keep the better natural environment, an actual condition of water quality has been measured and the transmission of N and P has been researched. Moreover, in order to decrease the level of N and P, the improvement methods which harness the action of marine products have presented according to the idea of zero emission for the water pollution.

1. INTRODUCTION

The suitable treatment for the excesses of nitrogen, phosphorus and other materials in closed estuary is one of the important problems, which confront the many closed estuaries. The treatment methods differ in inflow materials, their volumes and geological conditions of bays. The treatments are hopeful treated by the natural method in each area according to the idea of zero emission. This paper explains the actual condition of the water quality in the Maizuru Bay first. The pollution degree in the bay is not so high compared with other closed estuaries and the inflow materials are organic ones only. However, the pollution levels of N and P at the East bay in the Maizuru Bay are near the upper limit of the environmental standards in Japan. The balances of N and P in the area are calculated by the box model, and the targets of decrease for N and P are determined to keep the environmental standard. In order to achieve the targets, the following plans are introduced, that is, an increase in production of the shellfish and/or the seaweed and introduction of porous revetment and/or a reformation of wharf. The estimations of these effects are shown. Moreover, the residual flow vector and the distributions of N, P, COD (Chemical oxygen demand) and DO (Dissolved oxygen) are calculated by using three-dimensional flow model with considering reaction of materials in the bay.

2. ACTUAL CONDITIONS IN THE MAIZURU BAY

2.1 Natural Environment

The catchment area of the Maizuru Bay (shown in Fig.1) is 178 km^2 and the sea area of it is 23.35 km^2 (East bay: 10.88, West bay: 8.25, bay mouth: 4.22). The depths of the East and the West bay are approximately 10 m and depth of the bay mouth is approximately 27 m. The water volume is 387×10^6 m^3 (East bay: 163×10^6, West bay: 123×10^6, bay mouth: 101×10^6). The tidal range is 0.35m. The wave height is not so large.

2.2 Seawater Environment

The sewage from the drainage basin is treated by the two sewage disposal plants as shown in Fig.1. Distributions of DO at 1 m above the seabed from the bay mouth to the East bay are presented in Fig.2 (Kansai Electric Power Co., 1988). The environmental standards are added for reference. Fig.2 shows DO in the inner bay reaches the level 2 in summer and fall, and that it should not decrease DO any more. Distributions of N and P at 0.5 m blow the sea level are shown in Fig.3. The figure

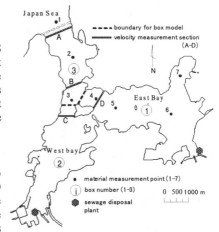

FIG. 1. Map of Maizuru Bay Showing Locations of Velocity Measurement sections, Material Measurement points, etc.

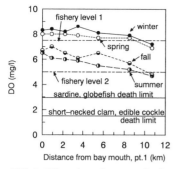

FIG.2. DO at 1 m above seabed.

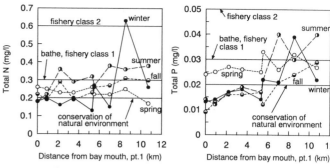

FIG. 3. Total N and Total P at 0.5 m below Sea Level

shows P and N in the inner bay have to be decreased below the present level. The poisonous materials in the environmental standard are not detected from the present measurement.

3. ESTIMATION OF FLOW IN THE BAY BY BOX MODEL

3.1 Law of Conservation for Flow and Salinity

The schema of flow in bay is presented in Fig.4. The low salinity flow flows to the outer bay at the surface layer of the cross section and the salinity flow flows to the inner bay at the sub-layer of the cross section. The discharges at each section are calculated according to the conservation law of the water and the salinity.

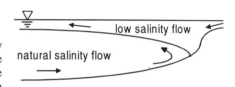

FIG.4. Circulation of Flow in Bay

For each bay area in the Maizuru Bay as shown in Fig.1, the continuous equations for the water are expressed as follows:

For East bay:① $\quad q_1 + Q_{31} = Q_{13}$ (1)

For West bay:② $\quad q_2 + Q_{32} = Q_{23}$ (2)

For bay mouth:③ $\quad q_3 + Q_{13} + Q_{23} + Q_i = Q_{31} + Q_{32} + Q_o$ (3)

in which q_1, q_2 and q_3 are the inflow discharge from land, Q is the discharge, suffix 1, 2 and 3 express the East bay, the West bay and the bay mouse, for example, Q_{13} means the discharge from the East bay to the bay mouse, Q_i and Q_o are the inflow discharge form the outer bay and outflow to it, respectively.

The continuous equations for the salinity are expressed as follows:

For East bay:① $\quad d(V_1 C_1)/dt = q_1 C_i - Q_{13} C_1 + Q_{31} C_3$ (4)

For West bay:② $\quad d(V_2 C_2)/dt = q_2 C_i - Q_{23} C_2 + Q_{32} C_3$ (5)

For bay mouth:③ $\quad d(V_3 C_3)/dt = q_3 C_i + Q_{13} C_1 + Q_{23} C_2 + Q_i C_o - Q_{31} C_3 - Q_{32} C_3 - Q_o C_3$ (6)

in which V_1, V_2 and V_3 are the volume of each bay, C is the salinity concentration, suffix i and o mean the land and the outer bay, respectively.

Assuming the steady state, the discharges are yielded as follows:

$$Q_o = C_o(q_1 + q_2 + q_3)/(C_o - C_3) \quad (7) \qquad Q_i = Q_o - (q_1 + q_2 + q_3) \quad (8)$$

$$Q_{23} = \{Q_{13}(C_3 - C_1) - C_3(q_1 + q_2)\}/(C_2 - C_3) \quad (9) \qquad Q_{32} = Q_{23} - q_2 \quad (10)$$

$$Q_{13} = \{Q_{23}(C_3 - c_2) - C_3(q_1 + q_2)\}/(C_1 - C_3) \quad (11) \qquad Q_{31} = Q_{13} + q_1 \quad (12)$$

Equation 7 means the outflow discharge to the outer bay is proportional to the salinity concentration at the outer bay and the inflow discharge from the land, and is inverse proportion to the difference of salinity concentration between the outer bay and the bay mouth.

3.2 Verification of Proposed Equations by Measured Velocity

Table 1 shows the measured salinity concentration for the seasons in the each bay. The total inflow discharge from the lands is the 5.02 m³/sec for the mean discharge and 5.74 m³/sec for the 95day discharge. The velocity was measured at four sections as shown in Fig.1 at autumn, 1973 (Maritime Safety Agency, 1974). The examples of velocity distribution at A and B section are shown in Fig.5. The results show the outflow in upper section and the inflow in lower section. These figures do not satisfy the conservation of mass between A and B section, since the measurement lasted for few days and the tidal condition changed in the duration. Table 2 shows the comparison between the measured discharges and the calculated ones by Eqs. (7)~(12) using the salinity concentration at autumn and the mean inflow discharge. The calculated results do not coincide with the measured ones, as above explained reason and the inflow discharge at 1973 is not always clear. However, it may be concluded Eqs. (7) ~(12) are suitable for the calculation of the discharge in those conditions. From the calculation results, it can be found that the out flow discharge from the East bay is a quarter of it from the West bay. This is a cause of bad water quality in the East bay compared with the West bay.

Table 1 Salinity Concentration in each Bay in 1973 (mg/cm³)

Bay salinity	East bay C_1	West bay C_2	Bay mouth C_3	Out of bay C_0
Spring	32.07	33.00	33.20	34.10
Summer	32.40	33.00	33.23	33.50
Fall	32.50	32.40	33.00	33.30
Winter	32.33	33.50	34.00	34.20

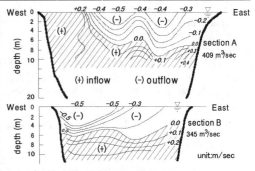

FIG.5. Measured velocity at A and B section

Table 2 Comparison between Measured and Calculated Discharge

Bay discharge	Bay mouth		West bay		East bay	
	Q_o	Q_i	Q_{23}	Q_{32}	Q_{13}	Q_{31}
Line	A	B	C		D	
Meas. (m³/s)	409	345	254		64.6	
Cal. (m³/s)	359.7	356.5	282	280.3	77.7	76.4

3.3 Balance of N and P in Basin

Figure 6 shows the balance of N and P in the basin. The amounts of N and P from the lands are calculated by the published figures from the sewerage agency and the measured value in rivers. The difference between the inflow and the outflow volume for N and P at the boundary is the consumable amount by the phytoplankton, following series of action by the biological circulation and deposition on the bay-bed after those actions. The consumption is advanced by

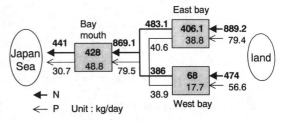

FIG.6. Balance of N and P in Basin

the actions due to the benthos and the seaweed's in the whole sea area and the shoreline. The consumable amounts due to these factors are estimated by the experience in Japan as follows. That is, 0.008 kg/m/day for the concrete seawall for 0.35 m tidal range, 0.015 kg/m/day for the natural shoreline for 0.35 m tidal range and 0.256 g/m²/day for the seaweed. Applying these indexes to the East Bay, 33.6 kg/day for the concrete seawall (4.2km), 278kg/day for the natural seawall (18.7km) and 19 kg/day for the seaweed (0.075 km²) are estimated. Then, 380.6 kg/day can be consumed in the East bay. This estimation is comparatively equal to the consumption of N in the East bay. This figure may be effective for the design of improvement method.

4. TREATMENT FOR EXCESS OF N AND P

4.1 Target of N- and P-Cut

N is mainly treated in this paper, since P is proportional to N by the ratio of 1 to 8. For the target of N-cut in the East bay, we adopt that the concentration of N in the East bay keeps 0.19 mg/l (Environmental standard for the bathe or the fishery class2 is 0.2mg/l for N). The present concentration of N in the East bay is 0.2 mg/l at middle layer, 0.28 mg/l at surface layer, its mean concentration is 0.24 mg/l. In order to decrease to 0.19 mg/l, the decrease of 0.05 mg/l is demanded. The decrease of 0.05 mg/l is equivalent to the decrease of 60 kg/day for N, as the water volume of the East bay, 1.63×10^8 m^3, is removed for 27 days.

4.2 Means of N-Cut

The means of N-cut are planed by the idea of mitigation. The East bay is not fully development for the marine product. The seaweed products are low levels compared with the West bay. Therefore, the physical and chemical reasons that the seaweed stay in low production must be researched, since the breeding of marine products is hopeful of decreasing N and P. The means of N-cut are the introduction of the shellfish farm of 0.5 km^2 (72 kg/day cut), the culture of seaweed of 0.25km^2 (8.0 kg/day), the concrete seawall reform to the rip rap type, 1.0km (6.9 kg/day) and the liner surface of concrete wharf reform to uneven type 1.0 km (8 kg/day). By the combination of these methods, the decrease of 60 kg/day of the target may be achieved. The shellfish are cultivated by the local government office at the man-made shallows experimentally.

5. ESTIMATIONS OF FLOW AND MATERIAL CONCENTRATION BY SIMULATION

5.1 Governing Equations

The numerical model for the estimations of flow water and materials is introduced, since the box model is not effective for the change of the circumferential condition. On the Navier-Stokes equation for three-dimensional, the acceleration term, the non-linear term and the viscosity term for z component are negligible to the gravity acceleration term and the pressure term. Then,

$$\frac{\partial u}{\partial t} + u\frac{\partial u}{\partial x} + v\frac{\partial u}{\partial y} + w\frac{\partial u}{\partial z} = fv - \frac{1}{\rho}\frac{\partial p}{\partial x} + v_h\left(\frac{\partial^2 u}{\partial x^2} + \frac{\partial^2 u}{\partial y^2}\right) + v_v\frac{\partial^2 u}{\partial z^2} \tag{13}$$

$$\frac{\partial v}{\partial t} + u\frac{\partial v}{\partial x} + v\frac{\partial v}{\partial y} + w\frac{\partial v}{\partial z} = -fu - \frac{1}{\rho}\frac{\partial p}{\partial y} + v_h\left(\frac{\partial^2 v}{\partial x^2} + \frac{\partial^2 v}{\partial y^2}\right) + v_v\frac{\partial^2 y}{\partial z^2} \tag{14}$$

$$0 = -g - \frac{1}{\rho}\frac{\partial p}{\partial z} \tag{15}$$

in which, f is the Coriolis coefficient. p is the pressure, g is the gravity acceleration, ρ is the density of seawater, u, v and w are the each velocity component for x, y and z direction, respectively, v_h and v_v are the component of eddy viscosity for horizontal component and vertical component, respectively.

The continuous equation is as follows:

$$\frac{\partial u}{\partial x} + \frac{\partial v}{\partial y} + \frac{\partial w}{\partial z} = 0 \tag{16}$$

The diffusion equation is introduced for each polluted substance concentration c, which presents in the seawater, in the form:

$$\frac{\partial c}{\partial t} + \frac{\partial(uc)}{\partial x} + \frac{\partial(vc)}{\partial y} + \frac{\partial(wc)}{\partial z} = \frac{\partial}{\partial x}\left(D_x\frac{\partial c}{\partial x}\right) + \frac{\partial}{\partial y}\left(D_y\frac{\partial c}{\partial y}\right) + \frac{\partial}{\partial z}\left(D_z\frac{\partial c}{\partial z}\right) \tag{17}$$

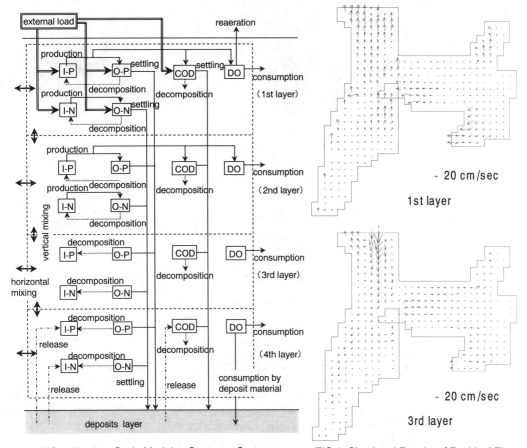

FIG.7. Nutrient Cycle Model to Seawater System   FIG.8. Simulated Results of Residual Flow

in which, D_x D_y and D_z are the each component of turbulent diffusion coefficient. The first term in the left side of Eq.17 means reactions of materials in seawater. The reaction equation of Organic N, for example, is expressed as following form (Horie, 1987):

$$\frac{\partial c_{ON}}{\partial t} = \{\text{Production by phytoplankton } (k =1, 2)\} - \{\text{Consumption by organic decomposition}$$
$$(k =1 \sim K\,)\} - \{\text{Settling } (k = K\,)\} + \{\text{External Load (k=1)}\} \qquad (18)$$

in which, k is the layer-order from sea surface and K is the bottom layer as shown in Fig.7 (Kuramoto, T. and Nakata, K., 1992). The reaction equation for other materials, DO, COD, Inorganic N, Organic P and Inorganic P, is also composed by the terms of production, decomposition and settling and/or release and external load (except DO) or consumption by deposit material and reaeration (DO only).

5.2 Simulated Results

Figure 8 shows the simulated results for the vector of the horizontal residual flow in the first and the third layer. The residual flow is the outflow at the first and the second layer, the inflow at the third layer and the stagnation flow at the bottom layer as same as measured velocity. The flow at the bottom layer is very small except around the boundary between the bay mouth and the outer bay. This is a one of cause for the bed water quality. The vertical residual flow is very small except the northeast side of the bay mouth.

559

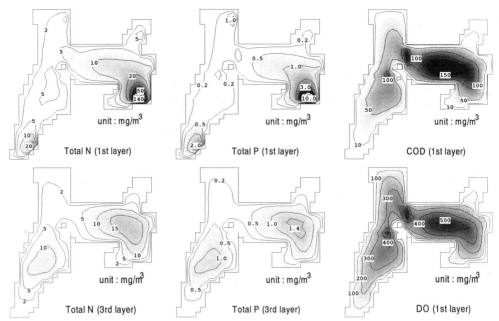

FIG.9. Simulated Results for T-N, T-P, COD and DO

The simulated results for the horizontal distribution of total N, total P, COD and DO are presented in Fig.9. The concentration of N and P of first layer at the south parts of the East bay and the West bay is very high because there is the sewage disposal plant at the two points. The states of the horizontal distribution for total N and total P show the measured tendency. These concentration decrease gradually as toward the bay mouth from the inner bay. The concentration of total P at seabed is high. This means the release of P from the seabed is active. N, P, COD and DO show the high value at the center in the East and West bay. These areas correspond to the stagnation area of flow.

6. CONCLUSIONS

1. Measurements of water quality show the necessity for decreasing N and P for better environment.
2. Outflow and inflow discharges at the boundary between bays were calculated with the box model that is considered the salinity concentration of the each box.
3. The balances of N and P in the basin were shown by the discharge and the measured N and P.
4. The target to N-cut was determined as 60 kg/day in the East bay for keeping the environmental standard.
5. According to the idea of zero emission, the means of consumption of N was planed by the increase of the marine products.
6. The numerical model was introduced for the estimation of state due to change of circumferential conditions, and the simulated results showed the actual tendency for distributions of substances in the bay.

7. REFERENCES

Horie, T. (1987) "Modeling for the Prediction of the effects of sea bed sediment treatment on the improvements of ecological conditions and seawater quality", *Report of the Port and Harbor Research Institute*, Vol.26, No.5, 175-214.

Kansai Electric Power Co. (1988) "Report of environment assessment for Maizuru Power Station", 2.39-2.56 (in Japanese).

Kuramoto, T. and Nakata, K. (1992) "Physical and chemical model", *Environmental Capacity in Fishing Grounds*, Hirano, T. ed., Koseishakoseikaku (in Japanese).

Maritime Safety Agency (1974) "Report on measurement of tidal current in the Maizuru Bay", 1-20 (in Japanese).

Environmental Hydraulics, Lee, Jayawardena & Wang (eds) © 1999 Balkema, Rotterdam, ISBN 90 5809 035 3

Modelling the effects of macrophyte decomposition on the nutrient budget in shallow lakes

Takashi Asaeda, Truong Van Bon, Takeshi Fujino & Vu Kien Trung
Department of Environmental Science and Human Technology, Saitama University, Urawa, Japan

ABSTRACT: A model of the annual life cycle of submerged macrophytes, *Potamogeton pectinatus L.*, and their decomposition process was developed. Their growth and decay were considered for five fractions simultaneously: main and secondary shoots, roots, tubers and new tubers of *P. pectinatus* in the growth submodel. In the decomposition submodel, different decomposition rates were incorporated for the easily and hardly degradable fractions of the above ground biomass. The model was applied to the field experiment in Swartvlei Lake (South Africa) rich in *Potamogeton pectinatus* L.. The biomass of each fraction, standing stock of phosphorus and the phosphorus content in shoots, secondary shoots and roots were reproduced well. The model was also able to simulate successfully the decomposition process, the remaining dry mass, and the phosphorus and nitrogen content of the remaining biomass. A series of numerical experiments were carried out for an imaginary lake to investigate the long term effects of macrophyte decomposition on the lake ecosystem.

1. INTRODUCTION

Field studies showed that the submerged macrophytes have an important role in the restoration of the shallow eutrophic lake. During the growing season macrophytes accumulate a large amount of nutrients from both sediments and the overlying water. When the macrophytes die they fall to the sediment on the bottom. In the succeeding decomposition process, the accumulated organic matter and nutrients are released, which have a significant influence on the lake ecosystem for a long time.

This study is, thus, aimed at developing a model of macrophytes growth, which is combined with the decomposition process to evaluate the effects on the nutrient budget in the lake. *Potamogeton pectinatus L.*, one of the most common and widespread species of submerged macrophytes in the world, has been intensively studied recently, and a quantitative evaluation of the growing process is now possible. As a first step, therefore, the species of *P. pectinatus* was analyzed in this study.

2. THE PROCESS OF MACROPHYTE GROWTH

Based on the detailed characteristics of *P. pectinatus*, the growth of *P. pectinatus.* was modeled, using five fractions of the entire biomass, shoots, secondary shoots, roots, tubers and new tubers, quantified in grams by ash-free dry weight. The net growth of each fraction is given as a result of photosynthesis, respiration, mortality, reallocation from dead biomass, reserves from the main tubers to the secondary shoots and new tubers, respectively (Hootsmans, 1994). These are given in Table 1, where a set of 22 parameters were selected from other reported studies and seven others were calibrated (Asaeda & Bon, 1997). The growth was assumed to start when the biological time, the sum of the daily average temperature from the 0th Julian day in the northern hemisphere or from July 1st in the southern hemisphere, exceeded 1471 day °C (Williams , 1978) for Swartvlei Lake. The number of original shoots from the tuber bank, the main shoots, NO is assumed as constant through the growing season, given by $NO = B_{tb}(j)/B_o$, where, $B_{tb}(j)$ is the tuber biomass on the j th day of the growing season, and B_o is the initial biomass of the plant. The main shoots receive the tuber flow of $NO = B_{tb}(j)/B_o$ until $B_{tb}(j) = 0$.

The increase in the above-ground biomass is allocated to the elongation and the growth of the main shoots and the existing secondary shoots in each layer and the formation of new shoots. The elongation rates of the main and secondary shoots, $\Delta h_{st,sc}$, are given by $\Delta h_{st,sc} = ET\,\Delta B_{st,sc}(top)\big/B_r$, where, B_r (=0.1) is the biomass per unit length (g/m^3); $\Delta B_{st,sc}$(top) are the increases in the main and the secondary shoot biomasses; ET is the coefficient according to the light intensity reaching the top of the vegetation. The roots start to grow from biological day 38126 (day $^\circ$C). (Hootsman, 1994) . Because little information is available, the root biomass was distributed evenly with the biomass density of 25 g/m^{3}over the corresponding thickness to satisfy the computed total root biomass (Asaeda & Bon, 1997). After subtracting the reflection component at the water surface, the insolation at the depth z , $I(z)$, is given as $I(z) = I_0 e^{-(\eta_w + \eta_c Chl + \eta_m MAC)z}$, where, I_o is the transmitted light intensity through the water surface; η_w, η_c and η_m are the extinction coefficients due to the water turbidity, the total phytoplankton concentration and the total macrophyte biomass, MAC is the macrophyte biomass above depth z of the lake. The total water column was divided into 20 sublayers. The photosynthetic characteristics of the plant tissue are dependent on the daily average insolation intensity received by the biomass in each layer. The growth of shoots and secondary shoots was considered only as the growth in the same layer, except for the elongation at the top of the shoots and secondary shoots. Because little information is available for the concentration of sediment nutrients, analyses were made assuming a sufficient amount of nutrients is stocked in the sediments. Thus, the nutrient dynamics in the sediment were excluded from the model. Also excluded were the effects of other organisms to simplify the system for easier understanding.

3. THE DECOMPOSITION PROCESS

In this study, dead macrophyte biomass was divided into two components of easily and hardly decomposable materials, $V1(j)$ and $V2(j)$, respectively, as evaluated by oxygen demand, 2.67 g O$_2$ g^{-1} C (Pereira et al., 1994). Also, suspended organic matter, SOM, resulting from the decomposition of $V1$ and $V2$, was considered.

The easily decomposable fraction of biomass is composed mainly of leaf biomass and a small fraction of shoot biomass, whereas the shoot biomass is made up mainly from the hardly decomposable materials. The ratio of leaf biomass to the above-ground biomass is dependent on the age of the plants, light intensity and the macrophyte species. The governing equations for the decomposition processes are, thus, given, on the j th day by Table 2.

The ratios of nitrogen nc_i and phosphorus pc_i contents to carbon in the decomposed material increase in the decomposition process (Williams and Davies, 1979). Thus, the form of the decreasing function with respect to time was applied, such that $nc_i = nc_{oi}(Hn_i/(Hn_i + t))$ and $pc_i = pc_{oi}(Hp_i/(Hp_i + t))$, where, nc_{oi} and pc_{oi} ($i = 1,2$) are the ratios of nitrogen to carbon and phosphorus to carbon for organic matter type 1 and type 2 (-), respectively, at the beginning stage of decomposition; Hn_i and Hp_i are the half saturation constants of time for nitrogen and phosphorus, respectively (Williams and Davies , 1979), in the decomposed material (d).

4. THE ANNUAL VARIATION IN MACROPHYTE DEVELOPMENT AND STANDING STOCK OF PHOSPHORUS

To obtain the insight into the annual cycle of the standing stock of phosphorus, *P. pectinatus* and phosphorus circulation in Swartvlei Lake was simulated for July 1975 to June 1976. The input data such as temperature and the solar radiation were obtained from Williams (1978). Figure 1 compares the simulated total biomass and the standing stock of phosphorus with the observed values. Both total biomass and standing stock of phosphorus had a satisfactory agreement. The maximum phosphorus storage in *P.pectinatus* in the lake reached 2 gm^2 at the end of February. Since then, however, the phosphorus stock in the standing macrophytes decreased immediately as the total biomass reduced. Figure 2 gives the observed and calculated phosphorus content in the shoots, secondary shoots and roots from July 1975 to June 1976, as well as the total phosphorus content in each part. The large difference in the root content in April is due to

Table 1. The macrophyte Growth Model

$$\frac{dB_{st}}{dt} = Ph_{st} - R_{st} - De_{st} + C_d.De^P{}_{st} + T_f.f_t(i)f_s(i) - L_{sc}.max[0, F_{st}] - f_{st}.\varepsilon_{st}.B_{st} \tag{1}$$
$$- f_r.G_{rt}.B_{rt} - f_{nt}.G_{nt}.B_{nt}$$

$$\frac{dB_{sc}}{dt} = Ph_{sc} - R_{sc} - De_{sc} + C_d.De^P{}_{sc} + L_{sc}.max[0, F_{st}] - f_{st}.\varepsilon_{st}.B_{st} \tag{2}$$
$$- (1 - fr).G_{rt}.B_{rt} - (1 - fnt).G_{nt}.B_{nt}$$

$$\frac{dB_{rt}}{dt} = G_{rt}B_{rt} - R_{rt} - De_{rt} + C_d.De^P{}_{rt} - N_{tf}.f_{tf}(i) + T_f.f_t(i)(1 - f_s(i)) \tag{3}$$

$$\frac{dB_{tb}}{dt} = -R_{tb} - De_{tb} - T_f.f_t \tag{4}$$

$$\frac{dB_{nt}}{dt} = G_{nt}B_{nt} - R_{nt} - De_{nt} + C_d.De^P{}_{nt} + N_{tf}.f_{tf}(i) + f_{sc}.\varepsilon_{sc}.B_{sc} + f_{st}.\varepsilon_{st}.B_{st} \tag{5}$$

$$F_{st} = Ph_{st} - R_{st} - De_{st} + C_d.De^P{}_{st} + T_f \qquad \text{for the insolation} > 200\mu Em^2 d^{-1} \tag{6}$$

$$fst = 0 \text{ if } \quad Ph_{st} - R_{st} - De_{st} + C_d.De^P{}_{st} + T_f - L_{sc}.max[0, F_{st}] \geq 0 \tag{7}$$

$$fst = 1 \text{ if } \quad Ph_{st} - R_{st} - De_{st} + C_d.De^P{}_{st} + T_f - L_{sc}.max[0, F_{st}] \prec 0 \tag{8}$$

$$fsc = 0 \text{ if } \quad Ph_{sc} - R_{sc} - De_{sc} + C_d.De^P{}_{sc} + L_{sc}.max[0, F_{st}] \geq 0 \tag{9}$$

$$fsc = 1 \text{ if } \quad Ph_{sc} - R_{sc} - De_{sc} + C_d.De^P{}_{sc} + L_{sc}.max[0, F_{st}] \prec 0 \tag{10}$$

$$Ph_{st,sc} = k_{co}.P_m \cdot \frac{PO_{av}}{K_p + PO_{av}} \cdot \frac{NH_{av}}{K_N + NH_{av}} \cdot \frac{PAR}{K_{par} + PAR} \cdot \frac{K_{age}}{K_{age} + Age} \cdot B_{st,sc} \tag{11}$$

$$R_{st} = \beta_{st}.B_{st} \quad ; \quad R_{sc} = \beta_{sc}.B_{sc} \quad ; R_{rt} = \beta_{rt}.B_{rt} \text{ and } \quad R_{nt} = \beta_{nt}.B_{nt} \tag{12a,b,c,d}$$

$$De_{st} = \gamma_{st}.B_{st} ; \quad De_{sc} = \gamma_{sc}.B_{sc} ; De_{rt} = \gamma_{rt}.B_{rt} \text{ and } \quad R_{nt} = \beta_{nt}.B_{nt} \tag{13a,b,c,d}$$

$$\gamma_{st,sc} = r_{st,sc} \cdot \frac{k_{st,sc}}{k_{st,sc} + Ph_{st,sc} - R_{st,sc} + C_d.De^P{}_{st,sc}} \tag{14}$$

where, B is the biomass (g.m^{-2}) with st, sc, rt, tb and nt the main shoots, secondary shoots, roots, tubers and new tubers, respectively (g.m^{-2}); p is the quantity of the previous day; Ph is the gross photosynthesis rate of the main shoots (g.m^{-2}d^{-1}); R($=\beta_{st,sc,rt,tb,nt}B_{st,sc,rt,tb,nt}$, $\beta_{st,sc,rt,tb,nt}$ =0.033T+0.108) are the respiration rates (g.m^{-2}d^{-1}); C_d is the fraction of the dead biomass used for growth; De are the mortality rates per day (g.m^{-2}d^{-1}); T_f is the tuber flow for the growth of main shoots per day (g.m^{-2}d^{-1}); L_{sc} is the effect of light on the development of new secondary shoots; ε_{st}, ε_{sc} and N_{tf} (=0.01) are the fractions of main shoot, secondary shoot, and root biomasses transferred to new tubers, respectively; G_{rt} (=0.2d^{-1} after 38126dayC) and G_{nt}(=0.2d^{-1} after 30 days) are the growth rate of roots and new tubers, respectively; f_{tf}(j) is the function indicating 0 for Julian day j<jst+30 and 1 for j≥jst+30, where jst is the Julian day at the start of the growing season and 30 is the number of days (empirical value) to form new tuber; f_s(j)=1 if Julian day j≤j_{st}+7 and f_s(j)=0.75 if Julian day j>j_{st}+7, i.e. after 7 days 25% of tuber flows is transferred to the root system f_t(j)=0 if j<j_{st} and f_t(j)=1 if j≥j_{st}; P_m is the maximum rate of gross photosynthesis (g.O$_2$g^{-1} h^{-1})(=0.408+10.12T); Age is the age of the main or secondary shoots (d); K_N and K_P are the half-saturation ammonium and phosphate concentrations, respectively, in the water of the sediment pores (g N. m^{-3}); K_{age} and K_{par} (=349-4.99T) are the half-saturation constants of the age and PAR, respectively; k_{co} is the conversion constant from oxygen to the ash-free dry weight (g.g^{-1}O$_2$); PAR is the insolation of photosynthetically active radiation averaged for a day (mE m^{-2} d^{-1}); PO$_{av}$ and NH$_{av}$ are average phosphate phosphorus and ammonium concentrations in the sediments,respectively. are the respiration rates (d^{-1}) , and $k_{st,sc}$ are the half saturation constants for the mortality (g m^{-2}d^{-1}) of the main and secondary shoots.

the assumption that the phosphorus content is dependent on the concentration in the overlying water, which decreased from March, 1976 (Williams, 1978).

Table 2. The decomposition model

$$\frac{dSOM}{dt} = ir.\sum_{i=1}^{j-1} r_{(j-i)}.\left[rm_{12}.V_1(i) + rm_{22}.V_2(i)\right]\frac{1}{h_{av}} + ir.\sum_{i=1}^{j-1} r_{(j-i)}.rm_{12}.F_{in} \tag{15}$$

$$- \sum_{i=1}^{j-1} r_{(j-i)}.(rm_{11} + rm_{12}).F_{out} - \frac{V_s}{h_{av}}.SOM$$

$$\frac{dV_1(j)}{dt} = -\sum_{i=1}^{j-1} r_{(j-i)}.(rm_{11} + rm_{12}).V_1(i) + \sum_{i=1}^{j-1} r_{(j-i)}.rd.V_2(i) \tag{16}$$

$$\frac{dV_2(j)}{dt} = -\sum_{i=1}^{j-1} r_{(j-i)}.(rm_{21} + rm_{22}).V_2(i) - \sum_{i=1}^{j-1} r_{(j-i)}.rd.V_2(i) \tag{17}$$

$$\frac{dNH_w}{dt} = \sum_{i=1}^{j-1} r_{(j-i)}.\left[nc_{1(j-i)}.(rm_{11} + rm_{12}).V_1(i) + nc_{2(j-i)}.(rm_{21} + rm_{22}).V_2(i)\right]\frac{1}{h_{av}} \tag{18}$$

$$+ \sum_{i=1}^{j-1} r_{(j-i)}.nc_{1(j-i)}.(rm_{11} + rm_{12}).SOM$$

$$\frac{dPO_w}{dt} = \sum_{i=1}^{j-1} r_{(j-i)}.\left[pc_{1(j-i)}.(rm_{11} + rm_{12}).V_1(i) + pc_{2(j-i)}.(rm_{21} + rm_{22}).V_2(i)\right]\frac{1}{h_{av}} \tag{19}$$

$$+ \sum_{i=1}^{j-1} r_{(j-i)}.pc_{1(j-i)}.(rm_{11} + rm_{12}).SOM$$

$$\frac{dDO_w}{dt} = -\sum_{i=1}^{j-1} r_{(j-i)}.rm_{11}.SOM - \sum_{i=1}^{j-1} r_{(j-i)}.\left[rm_{11}.V_1(i) + rm_{21}.V_2(i) + rd.V_2(i)\right]\frac{1}{h_{av}} \tag{20}$$

in which, SOM is the suspended organic matter (g O_2 m^{-3}); ir : the anaerobic intermediate organic matter release (-); $r_{(j-i)}$, $(= e^{-bj}\theta^{T-10})$ is the coefficients of decay rates dependent on the ambient temperature $T(°C)$(Godshalk and Wetzel, 1978) with θ being the Arrhenius or van't Hoff constant($=1.09$) (-) and b the exponential constant; $V_1(i)$ and $V_2(i)$: the easily and the hardly degradable organic matter of the dead macrophyte biomass on the i-th day named type 1 and type 2 (g O_2 m^{-2}); V_s : the settling velocity of suspended organic matter; rd : the decomposition rate of organic matter from type 2 to type 1 (s^{-1}); F_{in} , F_{out} : the inflow and the outflow of suspended organic matters (g O_2m^{-3}.s^{-1}); rm$_{i1}$ (= λ_{i1}max(0,(DO$_w$-aer)/(DO$_w$+aer)), rm$_{i2}$ (= λ_{i2}max(0,(ana-DO$_w$)/(ana+DO$_w$)) : the transition between aerobic and anaerobic processes depending on the dissolved oxygen concentration with aer being 0.1 gm^{-3}, ana 1 g.m^{-3}; DO_w : the dissolved oxygen concentration in the overlying water; i (=1 or 2) referring to easily or hardly decomposable materials, λ_{i1}, $\lambda_{i2}(i=1,2)$: the maximum degradation rate of organic matter of type 1 and type 2 in aerobic $(aer;i=1)$ and anaerobic $(ana;i=2)$ processes (s^{-1}) (Pereira et al., 1994); NH_w : the ammonium concentrations in the overlying water; PO_w : the phosphate concentrations in the overlying water; nc_i and pc_i $(i=1,2)$: the ratios of nitrogen to carbon and phosphorus to carbon for organic matter type 1 and type 2 (-), respectively; and h_{av} : the average depth of the lake(m).

The fact is notable that the secondary shoot biomass is largest, resulting in the secondary shoots having the largest total phosphorus contents, not the primary shoots. The phosphorus content is highest in October for roots, two months after the start of growing, but is already decreasing in the hottest month, December. However, for shoots and secondary shoots it is highest in February, two months after the hottest month.

5. THE HARVESTING EFFECTS OF MACROPHYTES ON THE LONG-TERM NUTRIENT BUDGET

A series of computational experiments for an imaginary lake in a temperate zone was made to obtain more insight into the nutrient cycle in the lake ecosystem rich in the submerged macrophyte *P. pectinatus* L. as a cosmopolitan submerged plant (Williams and Davies, 1979; Hootsmans, 1994). The assumed lake is on latitude 52° N with an area of 30 km^2 and the maximum depth of 1 m. The solar radiation was computed according to Asaeda and Ca (1993) and the water temperature on the jth day by the empirical formulus proposed by van Vierssen et al.(1994), $T_j = (T_{ma} + T_{mi})/2 - (T_{ma} - T_{mi})/2 \cos[2\pi(J - J_p)/360]$, where, T_{ma}(=25 °C) and T_{mi}(=13 °C) are the maximum and the minimum water temperature and J_p=10. The bottom is completely covered with *P. pectinatus* having a standing stock of 300 g m^{-2} (dry weight) for the above-ground biomass (AGB).

Figure 3 compares the easily decomposable $V1$ and the hardly decomposable $V2$ macrophyte biomasses, as well as the total quantity of $V1 + V2$ for 10 years with or without harvesting the above-ground biomass at the end of the growing season. In with harvesting case, the above ground biomass was harvested at the time of highest biomass. Clearly, the accumulation of the organic matter resulted mainly from the hardly decomposable fraction of macrophytes, which is more than 30 times that of the easily decomposable fraction. If the above-ground biomass is not removed, the accumulated detritus is 50g dw/m^2 within 10 years, which is more than three times that with harvesting. Figure 4 gives the daily phosphorus fluxes from the detritus. Without harvesting the above-ground biomass, the release of phosphorus from the detritus continues to increase except for the annual periodicity. I takes more than ten year to achieve the saturated value. Whereas, if the above-ground biomass is removed, they will be in an almost steady condition within three years. Therefore, within 10 years, for example, the annual maximum phosphorus flux increases only 2 % with harvesting, whereas it increases 280% without havesting. Most of the difference between these two cases is due to the release from the hardly decomposable material.

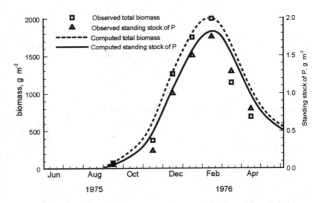

Figure 1 Observed and computed total biomass and the standing stock of phosphorus

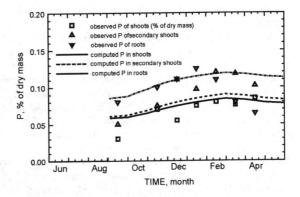

Figure 2 Observed and computed phosphorus content in the shoots, secondary shoots and roots

565

6. CONCLUSION

A model of an annual cycle of submerged macrophytes was developed for *P. pectinatus L.*. The model successfully reproduced the biomass during the growing season, and the remaining dry mass, phosphorus and nitrogen during the process of the decomposition. Total macrophyte biomass, standing stock of phosphorus, and phosphorus content in shoots, secondary shoots and roots were computed successfully for the Swartvlei Lake experiments. The model was applied for a successive 10 years to the reproduction of a 10 years transition of an imaginary lake. A comparison was made in the results between the cases of with and without harvesting the above-ground biomass at the end of the growing season. We found that the remaining biomass significantly increases if the above-ground biomass is not removed, mainly because of the accumulation of hardly decomposable material, whereas the 10 year change was only 20 % if the above-ground biomass is removed. Phosphorus release also increases dramatically if the above-ground biomass is not removed, which is also from hardly decomposable materials. This model will provide valuable information about the optimum time to cut macrophytes, as well as an additional insight into the contribution of macrophyte decomposition on nutrient budget in lakes.

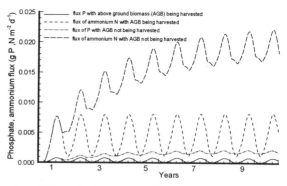

Figure 3 The easily decomposable $V1$, the hardly decomposable $V2$ and the total quantity of $V1+V2$ macrophyte biomasses for 10 years with or without harvesting the above-ground biomass

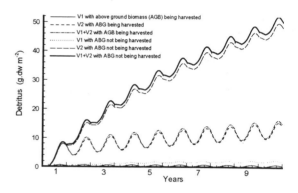

Figure 4 The daily phosphorus fluxes from the detritus

7. REFERENCES

Asaeda, T. and Bon, T.V., 1997. Modelling the effects of macrophytes on algal blooming in eutrophic shallow lakes. *Ecol. Model.*104, 261-287.

Asaeda, T. and Ca, V.T., 1993. The Subsurface Transport of Heat and Moisture and its Effect on the Environment: A Numerical model. *Bound. Layer Meteorol.*, **65**, 159-179.

Hootsmans, M.J.M., 1994. A growth analysis model for *Potamogeton pectinatus L.*. In: W.V. Vierssen, M.J.M. Hootsmans, J. Vermaat (Editors), Lake Veluwe, A Macrophyte-dominated System under Eutrophication Stress. Kluwer Academic Publishers, pp. 250-286.

Pereira, A., Tassin, B.and Jφrgensen, S.E., 1994. A model for decomposition of the drown vegetation in an Amazonian Reservoir. *Ecol. Model.*, **75/76**, 447-458.

Williams, C.H., 1978. Growth and production of aquatic macrophytes in a south temperate saline lake. *Verh. Internat. Verein. Limnol.*, **20**, 1153-1158.

Williams, C.H., 1981. Studies on the ability of a Potamogeton pectinatus community to remove dissolved nitrogen and phosphorus compound from lake water, *J. of Appl. Ecol.*, **18**, 619-637.

Williams, C.H. and B.R. Davies, 1979. The rates of dry matter and nutrient loss from decomposing *Potamogeton pectinatus* in a brackish south-temperate costal lake. *Freshwat. Biol.* **9**, 13-21.

Vierssen, W.V., 1994. A growth analysis model for *Potamogeton pectinatus L.*. In: W.V. Vierssen, M.J.M. Hootsmans, J. Vermaat (Editors), Lake Veluwe, A Macrophyte-dominated System under Eutrophication Stress. Kluwer Academic Publishers, pp. 250-286.

Environmental Hydraulics, Lee, Jayawardena & Wang (eds) © 1999 Balkema, Rotterdam, ISBN 90 5809 035 3

Water quality simulation in reaches of Hanjiang River by 2D model

D.H.Zhao & C.Qi
Nanjing Institute of Hydrology and Water Resources, China

H.W.Shen
University of California at Berkeley, Calif., USA

ABSTRACT: This paper presents an application of a two-dimensional depth-averaged flow-pollutant coupled model to the simulation of changes of water quality in some reaches of middle-downstream of the Hanjing River due to water transferring from the river. The 2D flow-pollutant coupled model incorporates the pollutant advection-dispersion equation into shallow water equations, which are solved simultaneously on an unstructured grid using finite volume method (FVM). The numerical flux in the normal direction to and across the interface of neighboring elements is estimated by using Osher type scheme, an approximate Riemann solver. Simulated water quality results have shown the effects of transferring water from the Danjiangkou Reservoir on the flow velocity field, pollution concentration and polluted area in the middle-downstream reaches of the Hanjiang river. The figures on flow velocity field and contours of pollutant COD in the river reaches are given in the paper.

1. INTRODUCTION

The middle-downstream of Hanjiang River receives water from the Danjiangkou Reservoir and ends at the confluence with the Yangtze River, the largest river in China. There are several industrial hubs along this river. The waste discharge and industrial effluents from the cities are known to adversely impact the water quality and environment of the river. The pollution concentration and its polluted area along the river near the major cities are paid attention. In addition, it is planned to transfer water from the Danjiangkou Reservoir to the north China through a large canal, which will change the flow regimes in middle-downstream of Hanjiang River. The associated pollution concentration, polluted area and water environment in the river will be changed accordingly. Those changes in the vicinity of a city can hardly be simulated by a one-dimensional water quality model. Therefore, a depth-averaged two-dimensional flow-pollutant coupled model was developed and used to predict the changes of the water quality variables, such as COD, BOD_5 and NH_3-N, due to transferring water from Dannjiangkou Reservoir.

The 2D flow-pollutant coupled model adopts shallow water equations and advection-dispersion equation to describe the flow and pollutants in the river, respectively. The model incorporates the advection-dispersion equation into shallow water equations, which are solved simultaneously on an unstructured grid using finite volume method (FVM). For each element, the numerical flux in the normal direction to and across the interface of neighboring elements is estimated by using Osher type scheme, an approximate Riemann solver. First, The parameters of the model were calibrated against observed data. Then the model was verified by the monitored data in 1994. The calculated results show a good agreement with the measured data. After that, the model was used to predict the changes of the water quality variables, COD, BOD_5 and NH_3-N, due to water-transferring project. The figures on flow velocity field and contours of pollutant COD in the river reaches are given in the paper.

2. TWO-DIMENSIONAL FLOW-POLLUTANT MODEL

2.1 Governing Equations

The conservative (divergence) form of the two-dimensional shallow water equations and advection-dispersion equation is given by:

$$\frac{\partial q}{\partial t} + \frac{\partial f(q)}{\partial x} + \frac{\partial g(q)}{\partial y} = b(q) \tag{1}$$

where $q = [h, hu, hv, hC]^T$ is the conserved physical vector, $f(q) = [hu, hu^2 + gh^2/2, huv, huC]^T$ is the flux vector in the x-direction and $g(q) = [hv, huv, hv^2 + gh^2/2, hvC]^T$ is the flux in the y-direction. Here h is the water depth; u and v denote the depth-averaged velocity components in the x- and y-directions, respectively; C is the depth-averaged constituent concentration in water column; and, g is the gravitational acceleration. The source/sink term b(q) is written as:

$$b(q) = [b_1, b_2, b_3, b_4]^T \tag{2}$$

where

$$b_1 = 0, \quad b_2 = gh(s_{ox} - s_{fx}), \quad b_3 = gh(s_{oy} - s_{fy}), b_4 = \nabla \cdot (D_i \nabla(hC)) + khC \tag{3}$$

where s_{ox} and s_{fx} are the bed slope and friction slope in the x-direction, respectively; s_{oy} and s_{fy} are the bed slope and friction slope in the y-direction, respectively. Other external forces such as baroclinic effect of density gradients, wind-induced and the Coriolis forces can be accounted for in the source/sink term but are omitted here. D_i are dispersion coefficients and k is a synthetic pollutant decay rate.

2.2 Formulation Of The Finite Volume Method

Upon integrating Eq. (1) over an arbitrary element Ω, the basic equation of the FVM obtained using the divergence theorem is given by:

$$\iint_\Omega q_t d\omega = -\int_{\partial\Omega} F(q) \cdot n dL + \iint_\Omega b(q) d\omega \tag{4}$$

in which n = a unit outward vector normal to the boundary $\partial\Omega$ of an element; and, $d\omega$ and dL = the area and arc elements, respectively; the vector quantity q is an average value and is assumed to be constant within an element in the first-order scheme. By using a rotational invariance property of the flux vectors f(q) and g(q), the discretized Eq. (4) is expressed by (Spekreijse, 1988)

$$A \frac{\Delta q}{\Delta t} = -\sum_{j=1}^{m} T(\Phi)^{-1} f(\bar{q}) L^j + b_*(q) \tag{5}$$

where $T(\Phi)^{-1}$ = the inverse transformation matrices by rotation of coordinate axes; Φ = the angle between vector n and the x-axis (measured counterclockwise from the x-axis). $\bar{q}$ = the quantity transformed from q, with components in the normal and tangential directions.

An explicit time discretization method is used in the model. Eq.(5) can be solved as a series of local one-dimensional Riemann problem. The estimate of the normal flux $f(\bar{q})$ will be addressed in the ensuing sections.

2.3 Estimation of Normal Flux

The local Riemann problem is an initial-value problem written as:

$$\bar{q}_t + [f(\bar{q})]_{\bar{x}} = 0 \tag{6}$$

The $\bar{x}$-axis is directed to the outward normal of a side in a cell.
Similar to 2D shallow water equations, the approximate solution of the Riemann problem is:

$$
\begin{aligned}
f_{LR}(q_L, q_R) &= f^+(q_L) + f^-(q_R) \\
&= f(q_L) + \int_{q_L}^{q_R} J^-(q)dq \\
&= f(q_R) - \int_{q_L}^{q_R} J^+(q)dq
\end{aligned} \tag{7}
$$

where $f^+(\bar{q})$ and $f^-(\bar{q})$ = forward and backward fluxes respectively, $J^+(\bar{q})$ and $J^-(\bar{q})$ = Jacobian matrices associated to the positive and negative eigenvalues of J. It is assumed that two given states q_L and q_R are connected by a piecewise arc consisting of four segments of characteristic curve.

According to Eq. (7) and the sign of eigenvalues λ_k with upwindness, the approximate Riemann solver by Osher scheme is obtained (Spekreijse, 1988):

$$
f_{LR}(\bar{q}_L, \bar{q}_R) = \begin{cases}
f(q[\varsigma]) - f(q[s]) + f(q[0]) & \text{if} \lambda_k(q[0]) > 0, \lambda_k(q[\varsigma]) < 0 \\
f(q[s]) & \text{if} \lambda_k(q[0]) < 0, \lambda_k(q[\varsigma]) > 0 \\
f(q[0]) & \text{if} \lambda_k(q[0]) \geq 0, \lambda_k(q[\varsigma]) \geq 0 \\
f(q[\varsigma]) & \text{if} \lambda_k(q[0]) \leq 0, \lambda_k(q[\varsigma]) \geq 0
\end{cases} \tag{8}
$$

The eigenvalues $\lambda(\bar{q}[p])$ for k = 1, 2, 3 and 4, and the variables $\bar{q}[0]$ and $\bar{q}[\zeta]$ are the quantities at the two endpoints of each segment. The quantity q[s] is the state at critical point s, where λ changes its sign, satisfying $\lambda(\bar{q}[s]) = 0$. For given hydraulic properties (i.e. u and c), there are 16 possible solutions of Eq.(8) to estimate the normal flux $f_{LR}(\bar{q}_L, \bar{q}_R)$ (Zhao et al, 1994).

2.4 Second-Order Oscillation Free Scheme

In the proposed schemes, Roe's superbee flux limiter was used as a second-order space-accuracy interface numerical normal flux $f_{LR}^{(2)}$. It is defined as:

$$
f_{LR}^{(2)} = f_{LR} + \frac{1}{2}\left[\phi(r_L^+)\cdot \alpha_{LR}^+ \cdot \delta f_{LR}^+ - \phi(r_L^-)\cdot \alpha_{LR}^- \cdot \delta f_{LR}^-\right] \tag{9}
$$

where

$$
r_L^+ = \frac{\alpha_{LR-1}^+ \delta f_{LR-1}^+}{\alpha_{LR}^+ \delta f_{LR}^+}; \alpha_{LR}^+ = 1 - \frac{\delta f_{LR}^+}{\delta q_{LR}}\sigma; r_L^- = \frac{\alpha_{LR+1}^- \delta f_{LR+1}^-}{\alpha_{LR}^- \delta f_{LR}^-}; \alpha_{LR}^- = 1 + \frac{\delta f_{LR}^-}{\delta q_{LR}}\sigma; \tag{10}
$$

$$
\delta f_{LR} = f_R - f_L = \delta f_{LR}^+ + \delta f_{LR}^-; \delta q_{LR} = q_R - q_L
$$

In the above equation f_{LR} is the first-order numerical flux solved by Osher approximate Riemann solver; ϕ is called limiter which physically corrects the second-order terms to ensure the TVD property; σ is equal to $\Delta t / \Delta x$. The time step Δt for first-order explicit upwind scheme is subject to Courant condition, that is Cr = local max($\sigma|f|$) ≤ 1, f = $\Delta f(q)/\Delta q$. For the second-order upwind scheme with limiter, Sweby (1984) showed that the Courant condition is

$$Cr^* = \left(\sigma|f'|\right)_{max} \le \left(\frac{2}{2+\beta}\right)$$ (11)

2.5 Boundary Conditions

The boundary conditions are given by solving a boundary Riemann problem. Basically, the unknown state can be determined by selecting the outgoing characteristic relations based on the local flow regime (either sub- or super-critical), or by specifying the physical boundary condition. The details of determination of flow boundary conditions refer to papers by Zhao et al (1994 and 1996). With regard to concentration, the possible boundary condition can be given by (1) a concentration time series or (2) $C_R = C_L$.

3. APPLICATIONS

Two sample applications were performed for a preliminary investigation of the applicability of the model.

3.1 Gaussian Concentration Distribution

The model was tested in a steady uniform flow field with water depth and flow velocity of 0.5 m and 0.5 m/s, respectively. The computed area is consisted of 50 x 50 square grids ($\Delta x = \Delta y = 100m$). A hypothetical Gaussian concentration distribution was used for this test. The isotropic dispersion coefficients are given $D_{xx} = D_{yy} = 5\,m^2/s$ and $D_{xy} = D_{yx} = 0$. The maximum concentration of initial distribution was 1 mg/l. The boundary condition of concentration was set to $C_R = C_L$. The computational time steps for flow and solute were set to 10 sec and 100 sec, respectively.

The simulated results by 2nd-order scheme with Roe's superbee limiter at 1.0 and 2.0 hours are shown in figure 1. It shows a good agreement between simulated and exact solutions. The simulated error is within 3.5 %.

3.2 Simulation of COD Concentration in Reaches of Hanjing River

The city locates in the middle stream of Hanjiang River, whose preliminary treated wastewater and industry effluent are discharged into the river directly.

For simulation of COD distribution in the vicinity of Xiantao, the computational domain covers about 1.532 km². A total of 1056 elements with 1139 nodes were set in the domain. Based on an 1D model's calibration, the corresponding Manning roughness were set to 0.025 and 0.015 for river main channel and flood plan, respectively. The dispersion coefficients E_x and E_y as well as the synthetic pollutant decay rate k were calibrated with observed data. Water discharge of 900 m³/s and water stage of 23.02 m were set as upstream

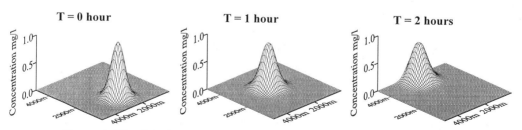

Figure 1. Simulated concentration distributions by second-order scheme with Roe's superbee limiter

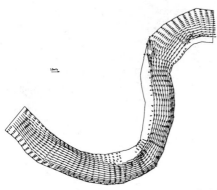

Figure 2. The Simulated Velocity Field

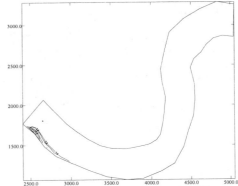

**Figure 3.   The Calculated COD
Concentration Contours**

and downstream boundary conditions, respectively. A point pollutant source with COD load of 193 g/s was given on right side of the river. In that case, the time step was set to five seconds. The simulated flow velocity field and calculated COD concentration contours in this area are shown on figures 2 and 3. The results agree well with observed data. When water is transferred from the Dangjiankou Reservoir, the inflow reduces about 28 % under the same COD load conditions. It is found that the length of pollution area is extended about 8.8 % and the width will be reduced by 4.5 %.

4. CONCLUSIONS

(1)   The characteristic-based coupled model solves both the flow and constituent equations.
(2)   In the framework of the FVM with the Osher scheme, the model follows full conservation, upwindness and high resolution in discontinuity.
(3)   The model uses an explicit scheme. The time step is constrained by the Courant condition.
(4)   In sample applications, the numerical solutions give a good match with analytical and observed data.

5. ACKNOWLEDGMENT

We wish to thank the Research Institute of Water Resources Protection of Yangtze River Water Resources Commission to sponsor this study and provide the essential data.

6. REFERENCES

Spekreijse, S.P., (1988), Multigrid Solution of Steady Euler Equations, CWI Tract 46, Amsterdam.
Sweby, P. K. (1988), "High resolution Schemes Using Flux Limiters For Hyperbolic Conservation Laws", SIAM J. Numer. Anal. Vol. 21 (5), 995-1011.
Zhao D. H., Shen, H. W., Tabios III, G.Q., Lai, J. S., and Tan, W. Y., (1994), "Finite-Volume Two-Dimensional Unsteady-Flow Model For River Basins." J. Hydr. Engrg., ASCE, 120 (7) 863-883.
Zhao D. H., Shen, H. W., Lai, J. S., and Tabios III, G.Q., (1996), "Approximate Riemann Solvers in FVM for 2D Hydraulic Shock Waves Modeling" J. Hydr. Engrg., ASCE, 122 (12).

Environmental Hydraulics, Lee, Jayawardena & Wang (eds) © 1999 Balkema, Rotterdam, ISBN 90 5809 035 3

An improved phosphorus budget model and its application to Lake Yanaka

Guangwei Huang & Nobuyuki Tamai
Department of Civil Engineering, University of Tokyo, Japan

ABSTRACT: For a lake with water supply and flood regulation functions, a dynamical trophic state model which takes the storage change into consideration is explored. Application of the model to lake Yanaka shows a good agreement between observation and calculation of the total phosphorus variation with time.

1. INTRODUCTION

Lake Yanaka is part of the Watarase retarding basin, north of Tokyo, $36^0 13'$ N in latitude and $139^0 40'$ in longitude. Its main purpose is three-fold: flood protection, drinking water supply to Tokyo area and maintenance water supply for downstream rivers. It has a surface area of 4.5 km^2 and an average depth of six meters with seasonal changes of about three meters for flood control. The lake is divided into three blocks by levees and connected by gaps as depicted in Fig. 1. Inflow and outflow are regulated at the pumping station which is located in South Block. The water is taken from Watarase river and Yata river.

In recent years, the eutrophication problem has surfaced up in Lake Yanaka. High inputs of nutrients led to excessive growth of phytoplankton. For example, the concentration of chlorophyll a ranges between 50 μg/l and 150 μg/l with peaks of up to more than 250 μg/l as shown in Fig.2. To improve water quality in lake Yanaka, some initiatives have been taken. For instance, a project of using a reed-wetland as filter to remove nutrients is now progressing. To understand how the lake may respond to various improvement work, a simple but effective model is needed for quick estimation of trophic level of the lake under different conditions. In the present study, the characteristics of water quality of lake Yanaka is reviewed, then a dynamic mass balance model is constructed and tested for lake Yanaka.

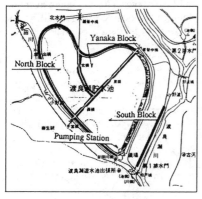

Fig. 1 Lake Yanaka

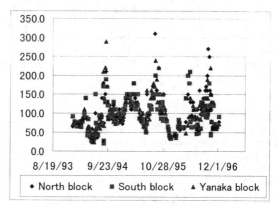

Fig. 2 Chlorophyll a concentration

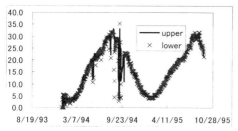

Fig. 3 Water temperature variation

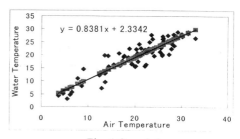

Fig. 4 Correlation

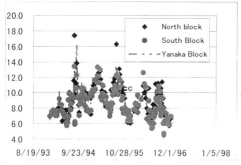

Fig. 5 Variation of COD in three blocks

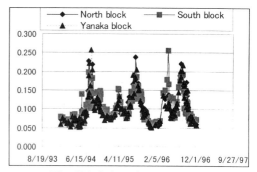

Fig. 6 Variation of TP in three blocks

2. CHARACTERISTICS OF WATER QUALITY IN LAKE YANAKA

Characterization of water quality in lake Yanaka is based on continuous monitoring data obtained from the Tone River Upstream Work Office. The field measurement was done at one point in each block at two different depths, namely 50cm below the surface and 1m above the bottom. Measured Quantities are: water level, air temperature, water temperature, transparency, odor, color of the water, wind direction and speed, pH, DO, BOD, COD, suspended solids, Colon bacillus, turbidity, EC, total phosphorus, PO_4-P, dissolved total phosphorus, dissolved total PO_4-P, total nitrogen, NH_4-N, NO_2-N, NO_3-N, organic nitrogen, dissolved total nitrogen, dissolved organic nitrogen, total organic carbon, total chlorophyll, chlorophyll a, chlorophyll b, chlorophyll c, organic phosphorus and various toxic substances.

Figure 3 shows the variation of water temperatures for the upper and lower layer. The maximum average water temperature is about 33^0C, and the minimum average is about 4^0C. As can be seen, the difference in water temperature between upper and lower layer is quite small. And the variation of water temperature follows closely the air temperature as indicated in Fig. 4.

Figure 5 shows the seasonal cycle of COD in 1996 and 1997. COD usually ranged between 5 and 17 mg/l. Peaks occurred in spring and summer when the water level are kept low for flood control. The observed pattern of the seasonal cycle of COD is quite similar to that of chlorophyll a. Figure 6 shows the variation of TP in three blocks, high concentration occurred in spring and autumn. The seasonal behavior of chlorophyll a can be observed in Figure 2. It is clear that there were two peaks, a spring bloom and an autumn bloom. It is also obvious that the difference between blocks is negligible on average. In order to investigate the spatial variation of water qualities in lake Yanaka, field measurements were carried out on August 6th to 9th , 1997 and December 3rd to 5th, 1997. Data were collected at five points in each block every three hours. The data suggests that the water quality indexes do not change significantly through space during the period of field measurement. However, it should be mentioned that there are some occasions when the temperature difference between blocks may be appreciable.

3. TROPHIC STATE MODEL

In the last two decades, the complexity of physical, chemical and biological processes in lakes has prompted the development of modeling as an instrument for understanding the workings of the lake system. Models such as 2-D hydrodynamic-water-quality model, 3-D layer integrated model, and ecology-oriented model have been proposed and validated. Those models are able to make predictions in the face of varying physical and biological conditions. However all of them contain quite a number of parameters, and require calibration to suit site-specific conditions. On the other hand, the simple input-output model of Vollenweider has been widely used to predict trophic conditions of lakes and reservoirs because of its simplicity. The model can be derived based on a mass balance analysis around a fully mixed lake on annual scale, which implies that the storage of lake does not vary significantly. In lake Yanaka, 10 million cubic meters of the total planned storage capacity of 26 million cubic meters are dedicated to flood control. During flood season from July 1st to September 30th the water depth is lowered to be 3 or 4m. Fig.7 shows the water level changes greatly from season to season in lake Yanaka. Therefore, a dynamic model which takes into account the storage change should be used for lake Yanaka. The dynamic mass balance model can be set up as following:

$$\frac{dVP}{dt} = Q_{in}P_{in} - Q_{out}P_{out} - SL \tag{1}$$

in which V is the storage of the lake, Q_{in}, P_{in} are inflow rate and its TP concentration. Q_{out}, P_{out} are outflow rate and its TP concentration. SL is the rate of in-lake loss. One way to account for this loss is to represent it as a monomolecular decay:

$$SL = \sigma VP \tag{2}$$

where σ the first-order decay rate. Although Eq.2 has been shown to be an adequate approximation of the long-term phosphorus dynamics in lake, it seems not well suited for a lake with variable volume. The alternative is to rewrite Eq.2 as below:

$$SL = sAP \tag{3}$$

where A is the surface area of the sediments(which is assumed equal to the lake's surface area).

The remaining difficulty is the determination of the apparent settling velocity s. It may very over a wide range. Normally, it can be estimated with filed data on an annual basis. For example, s is estimated to be 25.9m/y(or 0.07m/d) in lake Kasumigaura. However, we take the view that the annual average settling velocity does not fit the dynamic mass balance model. In this study, the following formula for the settling velocity is proposed for use in the dynamic mass balance model.

$$s = \begin{cases} \sqrt{q_s h} & \text{for Auguest and September} \\ 11.6 + 0.2q_s & \text{for other period of a year} \end{cases} \tag{4}$$

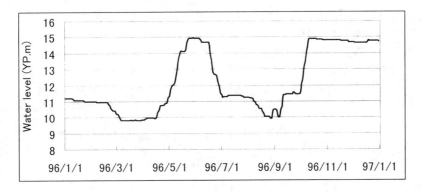

Fig. 7 Variation of water level in lake Yanaka

In which $q_s = Q_{in}/A$ is the hydraulic loading rate, and h water depth.

According to this formulation, the settling velocity is proportional to the water depth. The rationale is that when the water level is low, wind-induced re-suspension may be enhanced, so that the apparent settling velocity decreases.

By considering water balance, the Eq.1 can be rearranged as

$$\frac{dP}{dt} = fP_{in} - (f + s/h)P \qquad (5)$$

where f is the flushing rate.

4. MODEL VALIDATION

The validity of this unsteady model is examined by using this model to simulate the variation of TP concentration in lake Yanaka for 1995, 1996 and 1997. Figure 8, 9, 10 show the comparisons of simulated results with observed data for 1995, 1996 and 1997. The time interval used in this case is 1 month. The dotted lines in the figures indicate the model output obtained with an annual average settling velocity. It can be seen clearly that the model output based on an annual average settling velocity is far from satisfaction, the use of formula (4) in the dynamic mass balance model improved the simulation results considerably.

For 1994 and 1995, the highest value of TP concentration was well predicted by the model, however, the highest value of TP concentration for 1996 was overestimated by the model. For Watarase retarding basin, 1996 was a dry year, and in dry year, the ratio of water quantity taken from Yata river to that from Watarase river is unclear (although the total was recorded). Since the water quality of Yata river is quite different from that of Watarase river, this might be the reason of the overestimation. If the attention was given to the variation of two-month average of the total phosphorus in lake Yanaka, in other words, when the time interval was taken to be two months, better results were obtained for 1995 as shown in Figure 11.

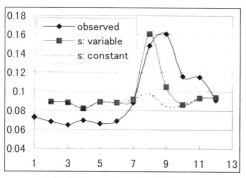

Fig. 8 Comparison for 1994

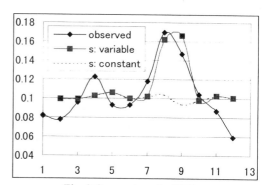

Fig. 9 Comparison for 1995

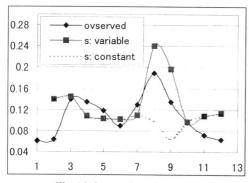

Fig. 10 Comparison for 1996

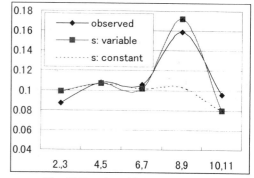

Fig. 11 Results with 2-month interval

5. CONCLUSION

Based on the scrutinizing of filed data, it is realized that the assumption of well mixing is applicable to lake Yanaka. Nevertheless, its volume changes greatly from season to season. Therefore, it is necessary to consider the volume change when applying the mass balance approach to lake Yanaka. By comparing the calculated results with observed data, it is found that the combination of the dynamic model with a constant apparent settling velocity fails to work properly, and considerable improvement can be achieved by use of a variable settling velocity as given in Eq.4.

With the Eq.4, the apparent settling velocity takes a lower value during summer than that in other seasons. This may be explained as the consequence of the increasing role of the likely internal loading during summer in lake Yanaka. Since the internal loading is not explicitly included in the present model, therefore the apparent settling velocity should be decreased in order to account for the contribution from the internal source. It might be postulated that if the internal loading were correctly estimated and included in the model, a constant apparent settling velocity derived on a yearly base would work well.

The approach outlined in the present study is shown to be promising in predicting the seasonal change of total phosphorus concentration. Due to its simplicity and the ability to trace the short time variation of phosphorus level, it is a useful tool for quick prediction of the lake response to particular management decisions. Finally, it should be mentioned that further model refinement work is needed to minimize model error, and test in other lakes should be pursued.

6. ACKNOLOWGEMENT

Thanks should be given to the Upper Tone River Construction Office for their kind assistance in field observation and data collection. Valuable assistance was also provided during the field study by Mr. Morita and many other members of the River and Environmental Engineering Lab (REEL), University of Tokyo.

7. REFERENCES:

Chapra S. C. (1977), "Total phosphorus model for the great lakes", J. Envir. Div., ASCE, 103, 147-161.

Dillon P.J., F.H. Rigler (1974), "A test of a simple nutrient budget model predicting the phosphorus concentration in lake water", J. Fish Res. Board Can., 31, 1771.

Vollenweider R.A. (1975), "Input-output models with special reference to the phosphorus loading concept in limnology", Hydrologie, 37, 53.

Osborne, P.L. (1980), "Prediction of phosphorus and nitrogen concentration in lakes from internal and external loading rates", Hydrobiologia, 69, 229-233.

Carrick, H.K., F.J., Aldridge, (1993), "Wind influences phytoplankton biomass and composition in a shallow, productive lake", Limnol. Oceanogr., 38, 1179-1192.

Simons T.J., D.C.L. Lam (1980), "Some limitations of water quality models for large lakes: a case study of Lake Ontario, Water Resour., 16, 105.

Weiyan, Tan, (1992), "Shallow Water Hydrodynamics", Amsterdam, Elsevier.

Uesaka, T., H. Sinmyou, and Horibe M., (1995), "A measure for water quality conservation using reed-wetland", 6th International Conference on the Conservation and management of Lakes-Kasumigaura'95, 418-421.

Huang G.W., N.Tamai, and Matuzaki H., (1998), "Water quality in Lake Yanaka", Environmental Conservation Engineering, 27, No.8, 548-552.

Environmental Hydraulics, Lee, Jayawardena & Wang (eds) © 1999 Balkema, Rotterdam, ISBN 90 5809 035 3

Water quality model and management information system for a shallow water sea bay

Jianhua Tao & Zeliang Wang
Department of Mechanics, Tianjin University, China

ABSTRACT: To deal with the character of semi-enclosed, shallow water sea bay, the water quality model has been investigated. In order to meet the requirement of ocean management, an Environment Management, Informational System (EMIS) included water quality model has been set up.

1. INTRODUCTION

In most countries, the rapid economy developing areas are located around coast. In order to maintain the rate of economic development, the protection of aquatic environment must be paid much attention. In this paper, based on the background of Bohai Bay where the third largest city Tianjin of China is near by, the 2-D water quality model was investigated. Because the wide mild slope beach, the treatment of moving boundary condition is a key problem to simulate hydrodynamic and transport process. In this model the Slot Method [Tao, 1983] has been used to solve this problem. The bio-chemical decay of the pollutant parameter has also been introduced into the transport model. The decay formula for COD has been carried out by laboratory experiments. In order to meet the requirement of aquatic environmental management and give a full play of the water quality model, EMIS(Environmental Management Information System)has been set up for the bay. This EMIS consists of four parts: Maps, Data Base, Numerical Models and Assess Management. The numerical model plays an important role in this system, which can be used to predict the impact of economic development on aquatic environment and the water quality in the near field and far field of the bay.

2. WATER QUALITY MODEL

Water quality model plays an important role in the EMIS, with this model, the distribution of pollutants in the bay can be predicated in case of sudden increasing a mount of pollutants drained into the bay, the impact assessment can be done quickly with the computation results and using of the assessment model. Necessary measure can be taken in time to protect the environment of the bay. Because there is a mild slope beach in Bohai Bay, moving boundary must be use in the model. In order to treat the moving boundary, the Slot method has been used in both the hydrodynamic and the addiction-diffusion model.

2.1 Hydrodynamic Model

For a shallow water bay the horizontal scale is much larger than the vertical scale ,there for a 2-D depth integrated hydrodynamic models can be used. The basic equations considered for the slots are as follows:

Continuity equation:

$$\frac{\partial P}{\partial x} + \frac{\partial Q}{\partial y} + F(z)\frac{\partial z}{\partial t} = 0$$

$$F(z) = \begin{cases} 1 & z_s > z_b \\ 2f(z) - f^2(z) & z_s \le z_b \end{cases}, \quad f(z) = \begin{cases} \varepsilon + (1-\varepsilon)e^{\alpha(z_s - z_b)} & z_s \le z_b \\ 1.0 & z_s > z_b \end{cases}$$

(1)

where z_s is the water surface elevation; z_b is the elevation of the beach at the slot location; ε is a parameter defining the minimal width of the slot; α is a smoothing parameter which determines the shape of the cross-section of the slot (in this study $\alpha = 2.3$); and z_0 is the slot bottom elevation, which is set to be lower than the lowest water elevation.

Momentum equations:

$$\frac{\partial P}{\partial t} + \frac{\partial}{\partial x}\left(\frac{P^2}{H}\right) + \frac{\partial}{\partial y}\left(\frac{PQ}{H}\right) + gh\frac{\partial z}{\partial x} = -g\frac{P\sqrt{P^2+Q^2}}{C_z^2 H^2} + fQ + E\left(\frac{\partial^2 P}{\partial x^2} + \frac{\partial^2 P}{\partial y^2}\right)$$

(2)

$$\frac{\partial Q}{\partial t} + \frac{\partial}{\partial x}\left(\frac{PQ}{H}\right) + \frac{\partial}{\partial y}\left(\frac{Q^2}{H}\right) + gh\frac{\partial z}{\partial y} = -g\frac{Q\sqrt{P^2+Q^2}}{C_z^2 H^2} - fP + E\left(\frac{\partial^2 Q}{\partial x^2} + \frac{\partial^2 Q}{\partial y^2}\right)$$

(3)

where $H = \begin{cases} \varepsilon(z_s - z_0) - \dfrac{1-\varepsilon}{\alpha}e^{\alpha(z_s-z_b)}(e^{\alpha(z_s-z_b)} - 1) & z_s \le z_b \\ \varepsilon(z_b - z_0) - \dfrac{1-\varepsilon}{\alpha}(1 - e^{\alpha(z_0-z_b)}) + (z_s - z_b) & z_s > z_b \end{cases}$,

P, Q are the unit width discharges in the x and y directions respectively; z is the surface elevation; H is the equivalent water depth including the slots; C_z is the Chezy coefficient; E is the dispersion coefficient; and f is the Coriolis coefficient, where $f = 2\omega\sin\varphi$, and ω is the angle speed of the earth's rotation and φ is the latitude.

2.2 Transport Model

For the addiction-diffusion equation[Wang, 1997]:

$$\frac{\partial(PC)}{\partial x} + \frac{\partial(QC)}{\partial y} + \frac{\partial A}{\partial t} = \frac{\partial}{\partial x}\left(E_x H\frac{\partial C}{\partial x}\right) + \frac{\partial}{\partial y}\left(E_y H\frac{\partial C}{\partial y}\right) + HS_m - KHC,$$

(4)

$$A = \begin{cases} 2HC - \int_{z_0}^{z_s} Cf^2(z)dz \\ HC + H'C - \int_{z_0}^{z_b} Cf^2(z)dz \end{cases}; \quad H' = \varepsilon(z_b - z_0) + \frac{1-\varepsilon}{\alpha}(1 - e^{\alpha(z_0-z_b)})$$

where C is the depth averaged concentration; E_x, E_y are the dispersion coefficients in the x and y directions respectively; S_m is the source term and K is the bio-chemical degradation coefficient.

2.3 Experimental Decay of COD

In order to determine the bio-chemical degradation coefficient, laboratory studies of the degradation of COD have been undertaken by using the bay water. The experiments were undertaken for different temperatures, since the coefficient k varies with temperature. After a detailed statistical analysis, the formula for the degradation of COD has been established , it is $C = C_0 e^{(-Kt)}$. where K is the degradation coefficient for COD; C_0 is the initial concentration of COD; t is the time; and $\ln K = \dfrac{-5374}{T} + 15.35$, where T is the absolute temperature.

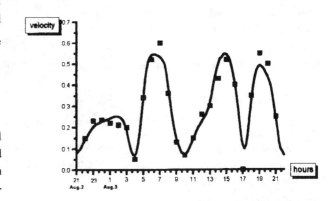

Fig. 1 Comparison of observation data and numerical model results for velocities at station F1
(— Computation; ■ Observation)

2.4 Hydrodynamic Model Tests

In order to calibrate the numerical models for Bohai Bay, a large scale field measurement program was organized in August 1993. There are nine Current-Water Quality station and three Tracking station in this program. The measurements were undertaken for 50 hours and more than 20,000 data records were obtained. Fig. 1, Fig. 2 shows a comparison of the predicted history of the velocity and measured data at station F1, F3, in August 1993. The numerical results are in good agreement with the field data.

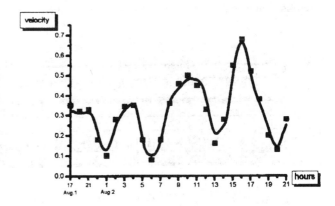

Fig. 2 Comparison of observation data and numerical model results for velocities at station F3
(— Computation; ■ Observation)

2.5 Water Quality Tests

The Tianjin Environmental Monitoring Center has a regular monitoring program for Bohai Bay. There are 22 measuring station within the Bay. Three sets of field observations were undertaken in May, August and October each year. Fig 3 shows the numerical results of average COD in Bahai Bay 1995.

In the comparison of the 22 measuring station data with the numerical results ,the average error is 10%. It is good agreement between the numerical simulation and field measurement.

3. ESTABLISHMENT OF THE EMIS FOR BOHAI BAY

In general ,the main parts of a Ocean EMIS Consist of data base and assessment module. By using histories and present field monitoring data, the aquatic environmental quality can be analyzed and assessed .According to this results, the policy and measure of environmental protection will be made .

Recent years, China's economy developed rapidly and there has been a corresponding increase in the industrial and domestic development around Bohai Bay. In order to control the water quality of Bohai Bay, to establish a water quality model for prediction the pollution of the bay in the EMIS is much important .

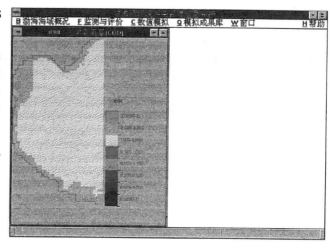

Fig. 3 Numerical results of concentration distribution of COD in Bohai Bay 1995

In the EMIS of Bohai Bay the numerical water quality model combined with data base and assessment module. The input and output of the numerical model is taken from the data base and used by assessment module .In this way, beside the normal function of EMIS, it has strong prediction ability. This EMIS will support the decision of economical maintainable development in the coastal area.

There are four modules in the EMIS, the structure is showed in Fig. 4.

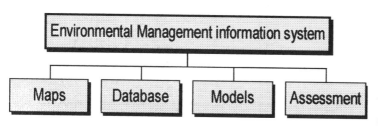

Fig. 4 Structure of EMIS

The maps module included geographic data of Bohai Bay, rivers, wastewater channels and the position of monitoring station. The data base contains basic data, monitoring data and simulated data. The model module has all the numerical models. The assessment module can be used to analyze whether the present or prediction water quality satisfied the requirement of the water quality standard of the ocean environmental management. The assessment models in the EMIS of Bohai Bay consist of three parts. 1) assessment pollutants sources. 2)assessment sea water quality. 3)assessment of bottom materials

Fig. 5 shows the topography of the Bohai Bay ,Fig. 6、7、8 plotted some information from data base and assessment.

582

Fig. 5 Map of Bohai Bay

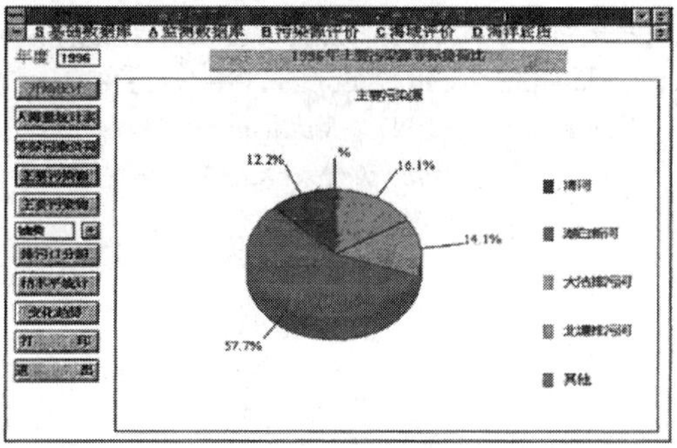

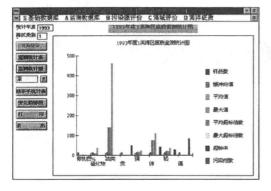

Fig. 7 Heavy metals in Bohai Bay

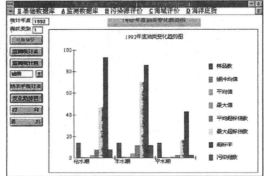

Fig. 6 Main pollutants of Bohai Bay

Fig. 8. Main pollutant outlets of Bohai Bay

4. CONCLUSIONS

(1) EMIS is and will be a powerful tools for the ocean environment management and protection. With the help of this scientific tools, prompt and valid measurement can be taken for the sudden pollution case.

(2) Bohai Bay is a typical semi-enclosed shallow water bay to the west of the Bohai Sea and has a mild bed slope beach with a heavy pollutant loads, the EMIS based on the Bohai Bay has been used successfully in the environment management of Bohai bay in Environmental Protection Bureau of Tianjin, China.

(3) The tidal exchange of the basin to the Bo sea and east sea is very weak, therefore the effects of the bio-chemical processes on the water quality are very important. In this study, the formula for the degradation of COD has been set up from laboratory data, and the bio-chemical processes of other pollutant parameters will be investigated further in the future.

5. REFERENCES

Tao Jianhua (1983). "Report of Danish Hydraulic Institute". Denmark

M.B. Abbott (1991). Hydroinformatics, the academic publishing group.

Toshimitsu (1994). "Estimation of 1-D and 2-D Dispersion Coefficient in a Bay", *Proc. of China-Japan*

Bilateral Symposium on Fluid Mechanics and   Management Tools for Environment, Beijing.

Falconer, R.A. and Chen, Y. (1996). "Modeling Sediment Transport and Water Quality Processes on Tidal Floodplains", in Floodplain Processes, Eds. M.G. Anderson et al., J. Wiley & Sons Ltd., London.

Wang Zeliang (1997). *Study on the Theory, Method and Informationalization of Numerical Simulation of Water Quality for the Sea-bay*, Doctoral thesis, Tianjin University, China.

Environmental Hydraulics, Lee, Jayawardena & Wang (eds) © 1999 Balkema, Rotterdam, ISBN 90 5809 035 3

La Laguna Las Peonias – A case study of an inverse estuary

A. Koenig, J.H.W. Lee & F. Arega
Department of Civil Engineering, The University of Hong Kong, China

ABSTRACT: La Laguna "Las Peonias", situated along the Strait of Maracaibo in Venezuela, is a shallow coastal lagoon exhibiting characteristics of an inverse estuary and advanced hypertrophication. Salinity increases landward due to evaporation of water, while other water quality parameters show different concentration gradients, indicating sources or sinks within the lagoon such as waste discharges or algal uptake. A numerical hydraulic model is presented to simulate the salinity profile taking into account evaporation. The model could be extended to incorporate nutrient uptake due to algal growth as well as the effect of waste discharges at the landward end of the lagoon, thus explaining the different inorganic nitrogen and phosphorus profiles in the lagoon. The main factors responsible for the unique hydrodynamic and water quality characteristics of the lagoon are identified as (i) high evaporation, (ii) negligible lateral mixing effects, (iii) long, narrow basin and shallow depth, (iv) low mass exchange with the main estuary because of a long connection channel, and (v) uncontrolled discharge of animal waste.

1. INTRODUCTION

Coastal lagoons are generally defined as semi-enclosed bodies of water connected with the open sea by inlets through a system of barriers (physical, chemical, hydrodynamical) (SCOR, 1980). These transitional zones between land and sea are usually zones of high biological productivity, but also sensitive to disturbance by human impact. Coastal lagoons have attracted much interest because of their ecosystems, considered to be a part of the national heritage in many countries (Kerambrum, 1986). One early such example is the Laguna "Las Peonias", a shallow coastal lagoon situated in the dry tropical region to the Northeast of the City of Maracaibo, Venezuela. Due to its unique ecological characteristics and surrounding mangrove vegetation, the lagoon and its adjacent area have been declared a National Park. In this connection, a comprehensive study was carried out to investigate the water quality of the lagoon and the major processes controlling it. It was hoped, that the obtained data would serve as important reference to evaluate the impact of human activities and of future changes in the use of the lagoon. This paper reports some interesting results on its water quality and attempts to provide a theoretical analysis of the observed hydrodynamic phenomenon known as "inverse estuary" (Hedgpeth, 1967).

2. CHARACTERISTICS OF THE LAGUNA "LAS PEONIAS"

The Laguna "Las Peonias" is situated to the Northeast of the City of Maracaibo, Venezuela, at approximately $10^0 45'$ N latitude and $71^0 40'$ W longitude. The main characteristics of the lagoon are: irregular shape with an area of 649 ha, maximum length 5900 m, maximum width 2200 m, mean depth 0.65 m, maximum depth 1.20 m, approximate volume 4,150,000 m^3. The lagoon is connected to the Strait of Maracaibo through a natural channel of about 1.3 km length, with a mean depth of 1.2 m and a maximum depth of 3 m in the centre. The Strait of Maracaibo itself is a large estuary of 450 km^2 which connects Lake Maracaibo (12,013 km^2) with the Gulf of Venezuela. The Strait therefore exhibits seasonal variations of salinity and is also subject to tidal influences (Parra-Pardi, 1980). Strictly speaking, the Laguna "Las Peonias" can thus be considered a shallow estuarine lagoon. The mean yearly temperatures in the area of the lagoon fluctuate between 27^0 C and 30^0 C, with an average of about 28.4^0 C. The highest and lowest mean monthly temperatures vary from 23.4^0 C to 33.4^0 C. Mean annual precipitation amounts to approximately 400 mm, but may vary largely from year to year in a range between 200 mm and 800 mm. Precipitation shows an irregular pattern; generally 100 mm rain are recorded during the dry season (November to April) and 300 mm during the wet season (May to October). Mean annual evaporation was estimated as 2626 mm. Runoff from the catchment basin into the lagoon is quite limited and occurs only sporadically. Wind speeds are low and remain between 5 and 15 km/h; during long periods of the year almost no wind occurs (Rodriguez, 1973).

3. MATERIALS AND METHODS

Based on preliminary work in 1978, a series of field surveys were carried out in the Laguna "Las Peonias" between 1981 and 1982, mainly during the dry season, to (i) establish the general physico-chemical water quality (temperature, pH, alkalinity, chloride, dissolved oxygen and BOD_5), (ii) determine the nutrient conditions (NH_4-N, NO_3-N and PO_4-P concentration), and (iii) assess the productivity conditions (chlorophyll-a concentration, algal productivity) for classification of the trophic state. Depending on the scope and purpose of the different surveys, between 6 and 16 marked sampling stations were employed to obtain a representative pattern of the spatial water quality distribution. In addition, hourly, daily or weekly sampling programmes were conducted to establish important temporal variations of various parameters due to dynamic physical, chemical or biological processes. Figure 1 shows a general map of the lagoon and the location of the 10 sampling stations used for the determination of most parameters. Temperature and pH were measured in situ, all other physico-chemical parameters, nutrient and chlorophyll a concentrations were determined in the laboratory according to the standard methods of APHA (1980). Algal productivity was measured in the field using the classical oxygen light and dark bottle technique as described by Russell-Hunter (1970).

4. RESULTS

Table 1 summarises selective results of the daily sampling programme carried out during the first week of March 1981. Similar results were found during other sampling periods. A pronounced spatial variation was observed, with concentrations apparently exhibiting a longitudinal distribution pattern dependent on distance from the mouth of the lagoon. In Figure 2, the ratios of the concentrations in the lagoon to the concentrations at the mouth (taken as equal to the concentration in the Strait of Maracaibo) are plotted versus relative distance of the different sampling stations from the mouth. Considering chloride as a conservative substance not undergoing any chemical or biological transformation, deviations of the plot for other substances from the plot for chloride indicate addition (for higher ratios) or losses (for lower ratios) of these substances due to chemical or biological transformations in the lagoon,

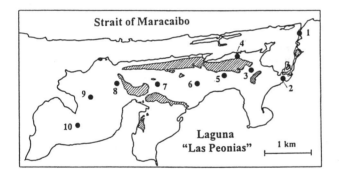

Figure 1. General map of the Laguna "Las Peonias" showing location of sampling stations.

Table 1. Mean concentration of chloride (mg/L) and inorganic nutrients (microgram/L) during dry season of 1981

Station	Cl⁻	NH_4–N	NO_3–N	Inorganic N (NH_4–N+NO_3–N)	PO_4–P	N/P
1 entrance	5291	598	121	719	37	19.4
2	5410	564	160	724	46	15.7
3	5474	537	21	558	54	10.3
4	7026	618	198	816	82	10.0
5	7560	506	150	656	98	6.7
6	7819	442	139	581	135	4.3
7	7873	536	159	695	174	4.0
8	7783	572	161	733	173	4.2
9	8661	598	126	724	168	4.3
10	8705	524	103	627	288	2.2
Mean (2–10)	7368	544	135	679	135	5.0

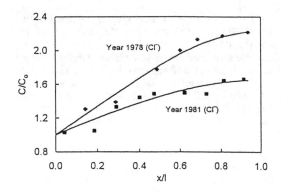 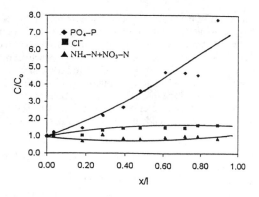

Figure 2. Longitudinal distribution of concentration ratios for different water quality parameters

Table 2. Primary productivity of phytoplankton

Station	Date	Gross primary productivity			Net primary productivity			Respiration	
		max. hourly rate	mean rate	daily rate	max. hourly rate	mean rate	daily rate	max. rate	mean rate
		g C/m^3–h	g C/m^2–h	g C/m^2–d	g C/m^3–h	g C/m^2–h	g C/m^2–d	g O$_2$/m^3–h	g O$_2$/m^3–h
1 entrance	08.10.80	0.23	0.10	1.20	0.15	0.06	0.72	0.32	0.25
2	25.09.80	0.31	0.17	2.04	0.20	0.08	0.96	0.55	0.37
3	24.09.80	0.58	0.31	3.72	0.54	0.27	3.24	0.20	0.15
4	23.09.80	0.54	0.23	2.76	0.41	0.16	1.92	0.35	0.22
5	26.09.80	0.46	0.20	3.00	0.32	0.19	2.28	0.50	0.41
6	30.09.80	0.53	0.29	3.48	0.50	0.25	3.00	0.27	0.16
7	27.09.80	0.67	0.28	3.36	0.47	0.16	1.92	0.63	0.52
8	28.09.80	1.15	0.41	4.92	1.00	0.34	4.08	0.38	0.32
9	01.10.80	1.08	0.40	4.80	0.96	0.32	3.84	0.42	0.33
10	02.10.80	0.45	0.32	3.84	0.44	0.16	1.92	0.15	0.10
Mean		0.60	0.28	3.31	0.50	0.20	2.39	0.38	0.28

internal sources or sinks, or external sources. The longitudinal increase in chloride concentration itself can be attributed to the predominance of evaporative effects (dry tropical climate) and the unique hydraulic characteristics of the lagoon (long connection channel, long extension, shallow depth and no landward influent).

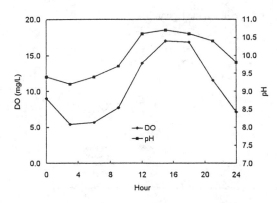 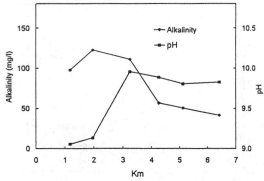

Figure 3. Diurnal variation of dissolved oxygen and pH at sampling station 5 on 1.4.1982

Figure 4. Longitudinal distribution of alkalinity and mean pH

The effect of the high algal photosynthetic activity on physico-chemical water quality parameters is manifested by the large diurnal variations of dissolved oxygen concentration and pH (Figure 3) as well as by the elevated mean pH values and decreased alkalinity in the inner sections of the lagoon (Figure 4).

Table 2 summarises the results of primary productivity measurements carried out at 10 different sampling stations during fall 1980. Chlorophyll a concentrations which were determined during a different sampling programme in 1982 varied from approximately 50 mg/m^3 to 150 mg/m^3 from the seaward to the landward end of the lagoon, with an average of 103.8 mg/m^3. BOD$_5$ concentrations averaged consistently about 20 mg/L. Based on the extremely high net primary productivity rates of up to 4 g C/m^2-d as well as the chlorophyll a and BOD$_5$ concentrations, the waters of the lagoon clearly exhibit advanced hypertrophic conditions according to common classification criteria (Wetzel, 1975).

5. HYDRODYNAMIC AND WATER QUALITY MODELLING

The tides in the Strait of Maracaibo are mixed and mainly semi-diurnal with a maximum tidal range of 0.35 m (Parra-Pardi, 1981); this is in agreement with published tidal harmonic constants for M2, S2, O1, K1 for the Maracaibo Approaches. The tidal elevation and velocity are first computed by a pseudo-two dimensional link-node hydrodynamic model embodied in the USEPA water quality model WASP5 (Ambrose et al., 1993). Based on topographic data, the lagoon is schematized into six segments, which includes the 1300 m long and narrow channel which connects the lagoon with the Strait of Maracaibo. Assuming a M2 tidal forcing of 0.3 m range at the Strait, the continuity and momentum equations of the fluid motion are solved numerically until a dynamic steady state is obtained. A Courant number of 0.1 (corresponding to a time step of 20 s) and a Manning bottom roughness coefficient of 0.04 are used. Figure 5 shows the computed variation of tidal level and velocity. The model predicts that the tidal amplitude is significantly dampened to 1-2 cm at the landward end. The velocity also varies from a maximum of about 0.3 m/s at the connecting channel to a few mm/s at the most inner section of the lagoon. The model predictions are in general agreement with previously reported observations (Parra-Pardi, 1981). The hydrodynamic model provides the advective mass transport coefficients across any segment boundary for any water quality variable.

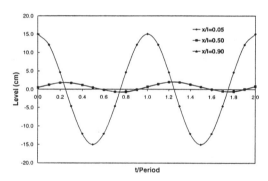 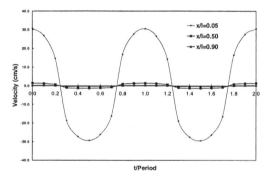

Figure 5. Predicted tidal level and velocity in the lagoon

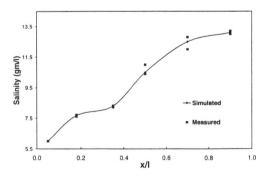 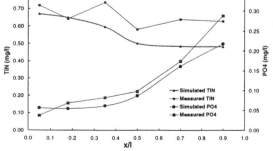

Figure 6. Predicted longitudinal salinity concentration in the lagoon

Figure 7. Predicted inorganic nitrogen and phosphorus concentrations in the lagoon

588

At the heart of the water quality model is a conservation of mass equation which embodies: mass transport by advection and dispersion, sources and sinks, and mass transformations. A constant dispersion coefficient of 5 m^2/s is used. The evaporation is treated in the mass balance equation as an internal outflow; the average advective evaporation rate of 8.32E-8 m/s (or 2.6 m/a) is adopted. It is assumed that no mass other than water is transported with evaporation and that only volumes are adjusted to maintain continuity. Figure 6 shows the computed longitudinal distribution of salinity (as Cl$^-$). The predicted salinities (which are insensitive to the assumed dispersion coefficients in the range of 5-25 m^2/s) are well supported by the field data. Model simulations (not shown) also demonstrate conclusively that the increase of salinity towards the landward end is due to the high rate of evaporation. Attempts have also been made to simulate the phytoplankton and nutrient dynamics based on the well established WASP framework; details of the formulation can be found elsewhere (Thomann and Mueller, 1987; Lee et al., 1991). In view of the lack of data for model calibration and verification, model simulation was carried out with the estimated pollution loads and environmental forcing functions; observed concentrations in the Strait are used as boundary conditions. While there is no synoptic data for for comparison, Figure 7 shows the predicted spatial trends in total inorganic nitrogen (TIN) and orthophosphate PO$_4$-P are well supported by the limited field data. There is an increasing trend of total inorganic phosphorus and decreasing trend in inorganic nitrogen towards the landward end.

6. DISCUSSION

The hydrodynamic predictions show that the long and narrow connection channel would is limiting any tidal effect on the hydraulic regime of the lagoon, dissipating the tidal influence a few hundred meters from the Strait. Actual measurements strongly supported this conclusion. The main body of the lagoon behaved therefore like a unilaterally open-ended static water body with limited communication with the open Strait for exchange of water and dissolved materials. The low value of the estimated coefficient of dispersion further supports the argument of negligible tidal influence.

The nutrient input to the lagoon results from two sources: (i) water from the Strait for evaporative makeup, and (ii) effluent discharges from piggeries at the landward end of the lagoon. Preliminary estimates show that the input from the Strait amounted to 10,168 kg/a N and 581 kg/a P (assuming an annual volumetric turnover of the lagoon by evaporation of 3.5 times and N and P concentrations in the Strait water of 0.7 g/m3 and 0.04 g/m3, respectively), whereas the input from pig wastes amounted to 9000 kg/a N and 3000 kg/a P, respectively (assuming 400 pigs @ 50 kg with daily waste generation of 450 g N and 150 g P per 1000 kg liveweight, respectively [Loehr, 1984]). The net primary productivity of 2.4 g C/m^2-d would require an algal uptake of 1,001,174 kg/a N and 138,555 kg/a P (assuming an algal composition of $C_{106}H_{180}O_{46}N_{16}P$ [Schwoerbel, 1987]). Since the potential algal uptake exceeds the estimated external N and P supply by 40 to 50 times, internal nutrient recycling forms the main nutrient source for the algae. Strait water supplies approximately 50% of external N and 16% of external P. The high proportion of P from the piggeries explains the skewered orthophosphate concentration in the lagoon towards the landward end.

Compared with other lagoons and estuaries, the Laguna "Las Peonias" represents a unique combination of the factors of longitudinal salinity increase and hypertrophic water quality in a shallow estuarine lagoon in a dry tropical climate. It differs from hypersaline lagoons which exhibit saline excess over oceanic salinity, because it exhibits salinity increase only above the relatively lower estuarine levels and hence cannot be considered a hypersaline system. Hypersaline lagoons have been described for several arid climatic zones, especially along the shores of the Gulf of Mexico (Hedgpeth, 1967; Britton and Morton, 1989), Australia (Arakel, 1981; Atkinson, 1990) and the Mediterranean Sea (Kerambrum, 1986; Castel et al., 1996); however, the main research interest generally focused on ecological aspects of hypersalinity. On the other hand, the Laguna "Las Peonias" also differs from conventional estuaries because of the limited freshwater input which leads to the phenomenon of an "inverse estuary". Seasonal occurrences of salinity increase in estuary water due to evaporative effects have been reported in Australia (McComb and Lukatelich, 1995) and Africa (Savenije, 1986). Savenije (1988) also presented an estuarine salt intrusion model taking into account the effects of evaporation and local rainfall which simulated well the tropical, seasonally hypersaline estuary of the Casamance River in Gambia, Africa. A different, but interesting example of evaporative salinity increases has been reported by Anwar et al. (1996) for Indonesian fishponds, which may be considered as modified shallow tropical lagoons. The above cited literature demonstrates that, despite much recent research on the phenomena of hypersalinity of lagoons and estuaries, the early work on the Laguna "Las Peonias" still retains its validity.

7. CONCLUSION

Based on the results of field observations and mathematical modelling, the following can be concluded:

- The lagoon exhibits unique hydrodynamic characteristics, which are manifested by a distinct increase in salinity with distance from the mouth to the Strait of Maracaibo. The salinity distribution as well as spatial trends in nutrient concentration are well predicted by a numerical hydrodynamic and water quality model.

- The causes for the strong longitudinal increase in salinity are (i) high net evaporation, (ii) long extension and shallow depth,(iii) negligible lateral dispersion due to very low average water and wind speed, and (iv) low mass exchange with the Strait because of the dampening effect of the long connection channel. Since these conditions persist throughout the year, the lagoon functions hydrodynamically as an "inverse estuary".
- The water quality in the lagoon could be considered in an advanced hypertrophic state, as shown by high concentrations of BOD_5, inorganic nitrogen, orthophosphate and chlorophyll-a as well as extremely high primary productivity of phytoplankton, associated with large diurnal fluctuations of dissolved oxygen and pH. The hypertrophication was caused by uncontrolled discharge of nutrients from piggeries at the landward end of the lagoon as well as by internal recycling of nutrients.

8. ACKNOWLEDGEMENTS

The water quality studies of this research were supported in part by the then Council for the Development of Arts and Sciences of the University of Zulia (CONDES). Hidroconsult Ltd provided a rare fan driven boat for free, making sample collection possible. The first author also wishes to thank his former colleague T. Perruolo and the following former students for their enthusiasm in carrying out field work under the most torrent and dangerous conditions: L.A. Acevedo, H. Canas, H.R. Castro, R. Fehervary, J. Falcon, O. Guanipa, M. Molero, N. Salcedo, E. Sanchez and R. Scrivante. Part of the modelling work is also supported by a grant from the Hong Kong Research Council.

REFERENCES

Ambrose, R.B. et al. (1993). *The Water Quality Analysis Simulation Program, WASP5*, Environmental Research Laboratory, USEPA, Athens, Georgia.

APHA (1980). *Standard methods for the examination of water and wastewater*, 15th Edition, American Public Health Association, Washington, DC.

Anwar, N., Anggrahini, Ogihara, K., Fukui, Y. and Tanaka, S. (1996). "Response of water level and salt density in fish ponds by tidal motion", In: *Hydrodynamics: Theory and applications*, Proceedings of the Second International Conference on Hydrodynamics, Hong Kong, 16-19 December 1996, (Chwang A.T., Lee, J.H.W. and Leung, D.Y.C., eds.), A.A. Balkema, Rotterdam, Netherlands, 1099-1104.

Arakel, A.V. (1981). "Geochemistry and hydrodynamics in the Hutt and Leeman evaporitic lagoons, Western Australia: a comparative study", *Marine Geology*, 41, 1-35.

Atkinson, M.J. (1990). "Dynamics of phosphate in Shark Bay, Western Australia", In: *Research in Shark Bay: Report of the France-Australe Bicentenary Expedition Committee* (Berry, P.F., Bradshaw, S.D., and Wilson, B.R., eds.), Western Australian Museum, Perth, Australia, 49-60.

Castel, J., Caumette, P. and Herbert, R. (1996). "Eutrophication gradients in coastal lagoons as exemplified by the Bassin d'Arachon and the Etang du Prevost", In: *Coastal lagoon eutrophication and Anaerobic processes (C.L.E.A.N.)*, (Caumette, P., Castel, J., and Herbert, R., eds.), Kluwer, Dordrecht, Netherlands, ix-xxviii.

Hedgpeth, J.W. (1967). "Ecological aspects of the Laguna Madre, a hypersaline estuary", In: *Estuaries:ecology and populations*, (Lauff, G.H., ed.), American Association for the Advancement of Science Publications, No. 83, Washington, DC, 407-419.

Kerambrum, P. (1986). *Coastal lagoons along the Southern Mediterranean coast (Algeria, Egypt, Libya, Morocco, Tunisia): Description and bibliography*, Unesco reports in marine science, Unesco, Paris.

Lee, J.H.W., Wu, R.S.S., and Cheung, Y.K. (1991). "Forecasting of dissolved oxygen in marine fish culture zone", *Journal of Environmental Engineering*, 117(6), 816-833.

Loehr, R. (1984). *Pollution control for agriculture*, Second edition, Academic Press, Orlando, USA.

McComb, A.J. and Lukatelich, R.J. (1995). "The Peel-Harvey estuarine system, Western Australia", In: *Eutrophic shallow estuaries and lagoons* (ed. McComb, A.), CRC Press, Boca Raton, USA, 5-17.

Parra-Pardi, G. (1980). "Modeling pollution in Strait of Maracaibo", *Journal of Environmental Engineering Division*, 106(5), 959-976.

Rodriguez, G. (1973). *The system of Lake Maracaibo: Biology and environment* (in Spanish). Instituto Venezolano de Investigaciones Cientificas (IVIC), Caracas, Venezuela.

Russell-Hunter, W.D. (1970). *Aquatic productivity: an introduction to some basic aspects of biological oceanography and limnology*. Macmillan, New York.

Savenije, H.H.G. (1986). "A one-dimensional model for salinity intrusion in alluvial estuaries", *Journal of Hydrology*, 85, 87-109.

Savenije, H.H.G. (1988). "Influence of rain and evaporation on salt intrusion in estuaries", *Journal of Hydraulic Engineering*, 114(12), 1509-1524.

Schwoerbel, J. (1987). *Handbook of limnology*, Ellis Horwood, Chichester, Great Britain.

SCOR (Scientific Committee on Oceanic Research)/Unesco ad hoc Advisory Panel on coastal Lagoons, 1976-1978 (1980). *Coastal lagoon survey*, Unesco technical papers in marine science 31, Unesco, Paris.

Thomann, R.V. and Mueller, J.A. (1987). *Principles of surface water quality modelling and control*, Harper and Row, New York.

Environmental Hydraulics, Lee, Jayawardena & Wang (eds) © 1999 Balkema, Rotterdam, ISBN 90 5809 035 3

Non-dimensional steady-state model for volatile toxics in a lake

Carlo Gualtieri & Guelfo Pulci Doria
Hydraulic and Environmental Engineering Department 'Girolamo Ippolito', University of Naples 'Federico II', Italy

ABSTRACT: A non-dimensional steady-state model for volatile toxics in a lake was proposed in order to gain a broader insight into contaminant transport and transformation phenomena; furthermore, simplified solutions were obtained for a quick estimation of system response to toxics input.

1. FOREWORD

During last years water quality models have been applied increasingly to represent the interaction between pollutants and aquatic environment due to transport and transformations; these models can be used to better understand pollution phenomena and to choose among different, alternative management strategies (USEPA, 1991). The different complex physical, chemical and biological phenomena simulated by the water quality models are related to the characteristics of the pollutants taken into consideration; particularly, toxic substances differ from conventional pollutants in that they are partitioned into two different forms, dissolved and particulate, i.e. associated with solid matter in the water column and in the bed sediments (Thomann and Mueller, 1987; O'Connor, 1988a; Chapra, 1997; Gualtieri 1998). This distinction has an impact on toxicant transport and fate in the sense that certain removal mechanisms act selectively on one or the other of the two forms; in fact, particulate fraction may be exchanged between water column and sediments through settling, resuspension and burial processes, while dissolved fraction could be subjected to volatilization and diffusion.

2. DIMENSIONAL MODEL FOR VOLATILE TOXICS IN A LAKE

In the modeling effort, a lake with its underlying sediments can be characterized as two well-mixed reactors, which are connected and each other interact.

Consider now a water column-sediments system of a lake, where there are an inflow rate Q_{in} (m³/year) with a contaminant concentration c_{in} (g/m³) and an outflow rate Q_{out} (m³/year); the contaminant is partitioned into particulate and dissolved fraction. The former one is subjected only to settling, resuspension and burial with velocities, respectively, of v_s, v_r and v_b (m/year), while the latter could volatilize across the air-water interface, with net transfer velocity v_{vol} (m/year), and diffuse between water column and sediments layer, with diffusive mixing velocity v_{diff} (m/year), (Fig.1). If V_{lake} and V_{sed} are, respectively, water column and sediments volumes (m³) and A is the water column and active sediments area (m²), mass balances for contaminant in the water column and in the active sediment layer can be written as (Chapra, 1997; Gualtieri, 1997):

$$V_{lake} \frac{dc_{lake}}{dt} = Q_{in} c_{in} - Q_{out} c_{lake} - v_s A F_{p\text{-}lake} c_{lake} - v_{vol} A F_{d\text{-}lake} c_{lake} +$$
$$+ v_r A F_{p\text{-}sed} c_{sed} + v_{diff} A \left(F_{d\text{-}sed} c_{sed} - F_{d\text{-}lake} c_{lake} \right) \tag{1a}$$

$$V_{sed} \frac{dc_{sed}}{dt} = v_s \, A \, F_{p\text{-lake}} \, c_{lake} - v_r \, A \, F_{p\text{-sed}} \, c_{sed} - v_b \, A \, F_{p\text{-sed}} \, c_{sed} +$$
$$+ v_{diff} \, A \left(F_{d\text{-lake}} \, c_{lake} - F_{d\text{-sed}} \, c_{sed} \right) \tag{1b}$$

where $F_{d\text{-lake}}$, $F_{p\text{-lagke}}$ $F_{d\text{-sed}}$ and $F_{p\text{-sed}}$ are, respectively, dissolved and particulate fractions in the lake and in the sediments and c_{lake} and c_{sed} are, respectively, water column and sediments contaminant concentrations (g/m^3).

It should be pointed out that the active sediments layer represents the bed volume which is involved in transport exchange phenomena with water column, i.e. settling and resuspension; thus, this layer could be considered having a constant volume. Furthermore, the water column volume is assumed to be constant too. If diffusion mechanism could be considered quantitatively negligible and, thus, could be skipped, (1a) and (1b) yield:

$$V_{lake} \frac{dc_{lake}}{dt} = Q_{in} \, c_{in} - Q_{out} \, c_{lake} - v_s \, A \, F_{p\text{-lake}} \, c_{lake} - v_{vol} \, A \, F_{d\text{-lake}} \, c_{lake} + v_r \, A \, F_{p\text{-sed}} \, c_{sed} \tag{2a}$$

$$V_{sed} \frac{dc_{sed}}{dt} = v_s \, A \, F_{p\text{-lake}} \, c_{lake} - v_r \, A \, F_{p\text{-sed}} \, c_{sed} - v_b \, A \, F_{p\text{-sed}} \, c_{sed} \tag{2b}$$

If the system is subjected to a constant loading for a sufficient time, it will attain a dynamic equilibrium condition called steady-state, where the concentrations remain the same over the time; thus, at steady-state, dividing for A, (2a) and (2b) yield:

$$v_f \, c_{in} - v_f \, c_{lake} - v_{vol} \, F_{d\text{-lake}} \, c_{lake} - v_s \, F_{p\text{-lake}} \, c_{lake} + v_r \, F_{p\text{-sed}} \, c_{sed} = 0 \tag{3a}$$

$$v_s \, F_{p\text{-lake}} \, c_{lake} - v_r \, F_{p\text{-sed}} \, c_{sed} - v_b \, F_{p\text{-sed}} \, c_{sed} = 0 \tag{3b}$$

where, for steady-state condition, we put $Q_{in} = Q_{out} = Q$ and $v_f = Q/A$; it should be pointed out that the ratio Q/A is, from a physical point of view, the filling rate of the lake, at first empty, for the inflow rate Q_{in}. Therefore, the ratio Q/A is called filling rate v_f.

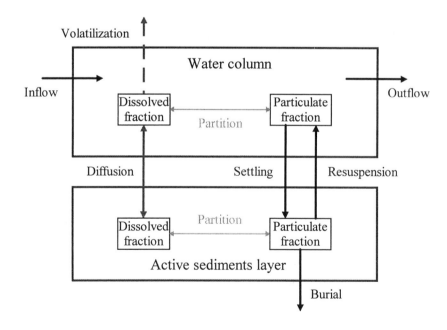

Fig.1 - Water column-sediments system in a lake

3. NON-DIMENSIONAL STEADY-STATE MODEL FOR VOLATILE TOXICS IN A LAKE

In this paper, the Authors propose a simplified non-dimensional solution for (3a) and (3b) in order to obtain a quick estimate for the contaminant concentrations in the water column and in the sediments layer, which does not require the full knowledge of the input parameters involved in the model; thus, this solution appears to be very useful in the management context.

Particularly, non-dimensional solution provides, instead of water column and sediments concentrations c_{lake} and c_{sed} values, their ratios to inflow concentration c_{in}; these ratios are the so-called transfer functions $\beta_{lake}=c_{lake}/c_{in}$ and $\beta_{sed}=c_{sed}/c_{in}$. The transfer functions specify how the system input c_{in} is transformed or transferred to an output, as represented by c_{lake} or c_{sed}.
Generally speaking, if $\beta_{lake} \ll 1$, it means that the lake's removal mechanism is acting to greatly reduce the level of pollutant in the lake/sediments; conversely, if $\beta_{lake} \to 1$, the water column can remove very weakly the inflow loading and the pollutant level will approach that of the inflow.

The transfer functions β_{lake} and β_{sed} could be attained through a dimensional analysis (Gualtieri and Pulci Doria, 1998). If one assume as fundamental quantities the filling rate v_f [L T^{-1}] and the inflow contaminant concentration c_{in}.[M L^{-3}], then, (3a) and (3b) can be rewritten, after some simplifications, as:

$$\left(1+\frac{v_{vol}}{v_f} F_{d-lake} + \frac{v_s}{v_f} F_{p-lake}\right) \beta_{lake} - \left(\frac{v_r}{v_f} F_{p-sed}\right) \beta_{sed} = 1 \tag{4a}$$

$$\left(\frac{v_s}{v_f} F_{p-lake}\right) \beta_{lake} - \left(\frac{v_r}{v_f}+\frac{v_b}{v_f}\right) F_{p-sed} \beta_{sed} = 0 \tag{4b}$$

where β_{lake} and β_{sed} are the non-dimensional unknowns.

Now, assuming that $\eta=v_b/v_r$, (4a) and (4b) can be solved firstly in order to obtain β_{lake}:

$$\beta_{lake} = \frac{1}{1+\dfrac{\eta}{\eta+1}\dfrac{v_s}{v_f}+\eta_{pv}} \tag{5}$$

where η_{pv} is:

$$\eta_{pv} = F_{d-lake}\left(\frac{v_{vol}}{v_f} - \frac{\eta}{\eta+1}\frac{v_s}{v_f}\right) \tag{6}$$

It should be pointed out that η_{pv} is formed by two terms; the former, $F_{d-lake}\times v_{vol}/v_f$, is related to the volatilization effect, while the latter $F_{d-lake}\times(\eta/\eta+1)\times v_s/v_f$, could be considered as the influence of the presence of the dissolved fraction on contaminant removal mechanism by solids settling.
Fig.2 shows the distribution of β_{lake} vs the ratio v_s/v_f for different values of η and η_{pv}; particularly, for each value of η_{pv}, we have two plots, for $\eta = 1$ and $\eta = \infty$.
In fact, we suppose that $v_b > v_r$, so that η belongs to the range $[1, \infty]$; so that, obviously, for each value of η_{pv} we have a band. This assumption appears to be representative of the majority of lakes and reservoirs, where the shear at water-sediment interface does not achieve the critical value to produce resuspension (O'Connor, 1988b). Periodically, during intense storms and winds, resuspension could be present, but, on the annual average or possibly seasonal basis, it would be likely that in most lakes and reservoirs $v_r \approx 0$ and, thus, $\eta \to \infty$.

It should be pointed out that for $\eta_{pv} \geq 0$, the plots corresponding to different values of η give always $\beta_{lake} \leq 1$; furthermore, they define a narrow band, which has a maximum width of 0.1716 for $\eta_{pv}=0$; therefore, β_{lake} estimate appears to be acceptable also in lack of exact knowledge of η, i.e. of v_b and v_r. The plots for $\eta_{pv}<0$

could have, on the left-side of the x-axis, i.e. for low v_s/v_f values, $\beta_{lake}>1$, while it could be proved that, as obvious, β_{lake} cannot exceed 1; therefore, the plots should be cut at $\beta_{lake}=1$ ordinate, which identifies the threshold abscissa $(v_s/v_f)_{lim}$. The condition, $\beta_{lake}\leq 1$ could be expressed as:

$$1+\frac{\eta}{\eta+1}\frac{v_s}{v_f}+\eta_{pv}\geq 1 \tag{7}$$

which yields:

$$\frac{v_s}{v_f}\geq-\frac{\eta+1}{\eta}\eta_{pv}=\left[\frac{v_s}{v_f}\right]_{min} \tag{8}$$

which represents a threshold condition for v_s/v_f.

Generally speaking, Fig.2 shows that β_{lake} and, thus, c_{lake} reduce as η_{pv} increases; furthermore, they decrease with the increasing ratio v_s/v_f; in other words, β_{lake} becomes smaller for heavier solids.

Through a similar way, we solved (3a) and (3b) for:

$$\beta_{sed}=\frac{\dfrac{v_s}{v_f}F_{p\text{-}lake}}{\dfrac{v_r}{v_f}\left[1+\dfrac{v_{vol}}{v_f}F_{d\text{-}lake}+\eta\left(1+\dfrac{v_{vol}}{v_f}F_{d\text{-}lake}+\dfrac{v_s}{v_f}F_{p\text{-}lake}\right)\right]} \tag{9}$$

If $\eta\rightarrow\infty$, i.e. if $v_b>>v_r$, or, anyway, if $\eta>1$, it is possible to skip the two first terms in the squared brackett at the denominator of (9), which becomes:

$$\beta_{sed}=\frac{\dfrac{v_s}{v_b}F_{p\text{-}lake}}{1+\dfrac{v_{vol}}{v_f}F_{d\text{-}lake}+\dfrac{v_s}{v_f}F_{p\text{-}lake}} \tag{10}$$

where the resuspension rate v_r, which is often extremely difficult to measure, is absent.

In order to gain insight into (10), consider two opposite situations, $v_s\approx\infty$ and $v_s\approx0$. For the first case, (10) yields:

$$\beta_{sed}\approx\frac{v_f}{v_b} \tag{11a}$$

where high values for settling rate v_s imply a quick settling of the inflow contaminant; thus, β_{sed} depends mainly on the filling rate v_f and the lake acts as contaminant-trap.

In the second case, if $F_{d\text{-}lake}\times v_{vol}/v_f >> 1$, (10) can be solved for:

$$\beta_{sed}\approx\frac{\dfrac{v_s}{v_b}F_{p\text{-}lake}}{1+\dfrac{v_{vol}}{v_f}F_{d\text{-}lake}} \tag{11b}$$

where the contaminant concentration in the sediments layer assumes the maximum value for $F_{p\text{-}lake}=1$ and $F_{d\text{-}lake}=0$; in this condition, $\beta_{sed}\approx v_s/v_b$ and is very little.

Anyway, Fig.3 shows the distribution, according to (10), of the normalized transfer function $\beta_{sed}/(F_{p\text{-}lake}\times v_s/v_b)$ vs the sum $(F_{p\text{-}lake}\times v_s/v_f + F_{d\text{-}lake}\times v_{vol}/v_f)$; it should be pointed out that, if this sum is known, one plot expresses all the situations. The plot has an asymptotic fashion.

The sum $(F_{p\text{-}lake}\times v_s/v_f + F_{d\text{-}lake}\times v_{vol}/v_f)$ represents the overall contaminant removal mechanism from the water column; therefore, the normalized β_{sed} is as lower as that removal mechanism is greater.

Therefore, we can conclude that, through (5) and (10), it is possible to attain reliable values for β_{lake} and β_{sed} also in lack of data about resuspension velocity v_r, and, for β_{lake} only, about burial velocity v_b.

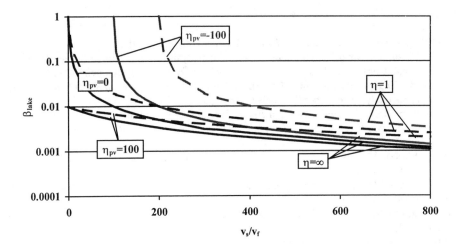

Fig.2 - Transfer function β_{lake}

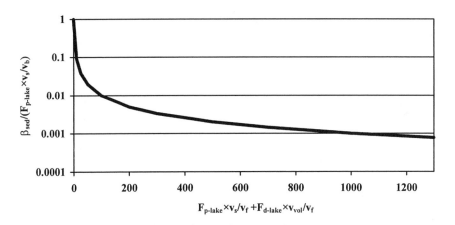

Fig.3 - Normalized transfer function $\beta_{sed}/(F_{p-lake} \times v_s/v_b)$

4. CONCLUSIONS

In last years, public concern for toxics pollution has been increased; potentially toxic substances have been largely produced since about 50 years and may be transferred to humans with subsequent short-term or long-term impact on public health. Toxics effects are also related to their concentrations, which could be predicted by mathematical models.

The paper proposes a non dimensional steady-state model for volatile toxics in a lake; first at all, non-dimensional approach offers a broader insight into toxics transport and transformation phenomena because provides, instead of water column and sediments concentrations c_{lake} and c_{sed} values, their ratios to inflow concentration c_{in}, i.e. the transfer functions β_{lake} and β_{sed}, which point out system response to contaminant input. Then, non-dimensional solutions point out the relevance of dissolved/particulate fractions and of removal mechanisms considered.

Finally, the simplified solutions obtained provide reliable values of transfer functions also if it is not available a full knowledge of the input parameters involved in the model; this appears to be a good technical result in lake management.

REFERENCES

Chapra S.C. (1997). *Surface water quality modeling.* McGraw-Hill, New-York

Gualtieri C. (1997). Surface water quality models: simplified analysis of water column-sediments interaction in a lake (in italian). Conference *Hydraulics and Environment*, Bari (Italy), October 13/14, 1997

Gualtieri C. (1998). Toxics modeling in a lake (in italian). *XXVI Convegno di Idraulica e Costruzioni Idrauliche*, Catania (Italy), September 9/12, 1998

Gualtieri C. and Pulci Doria G. (1998). Non-dimensional steady-state solutions for solids balance in a lake. IAWQ Conference *Application of models in water management*, Amsterdam (The Netherlands), September 24/25, 1998

O'Connor D.J. (1988a). Models of sorptive toxic substances in freshwater systems. I. Basic equations. *J.Env.Eng.Div. ASCE*, **114**, 3, pp.507-532

O'Connor D.J. (1988b). Models of sorptive toxic substances in freshwater systems. II. Lakes and reservoirs. *J.Env.Eng.Div. ASCE,* **114**, 3, pp.533-551

Thomann R.V. and Mueller J.A. (1987*). Principles of surface water quality modeling and control.* Harper Collins, New York

USEPA (1991). *Technical Support Document for Water Quality-based Toxics Control.* U.S. EPA, Office of Water, Washington, DC

Environmental Hydraulics, Lee, Jayawardena & Wang (eds) © 1999 Balkema, Rotterdam, ISBN 90 5809 035 3

Field observations of heat budget on a tidal flat

Nobuhiro Matsunaga, Masashi Kodama, Yuji Sugihara & Kazuyo Fukuda
Department of Earth System Science and Technology, Kyushu University, Kasuga, Japan

ABSTRACT: In order to investigate the heat budget on a tidal flat, systematic measurements of wind velocity, air, water and soil temperatures, air humidity and radiation fluxes were made on Wajiro tidal flat at the east side of Hakata Bay in Kyushu Island, Japan. The observations were carried out in both summer and winter of 1997. The relationship between the albedo and water depth, the properties of radiation fluxes and the budget among sensible, latent and subsurface heat fluxes have been revealed. In the daytime, the net radiation to the tidal flat is nearly equal to the net solar radiation and is distributed to the latent heat flux and the subsurface one. As the net radiation in the nighttime becomes almost zero, the subsurface heat flux is balanced against the latent one.

1. INTRODUCTION

In tidal flats, sea water can be exchanged vigorously by the tidal movement and enough solar radiation can reach to the sea bed. Therefore, many kinds of benthos inhabit there. They play an important role in the water purification of a nearshore zone. The tidal flats are also a favorable place for migratory birds to overwinter or take a rest. Recently, their value begins to be recognized from the standpoint of natural conservation as well as that of the water purification. In Japan, however, some tidal flats are disappearing with the coastal zone utilization. Therefore, a few of trials (e.g., Lee et al., 1997) have been made to conserve natural tidal flats or to resuscitate vanished ones.

The activity of benthos depends strongly on the temperature near the ground surface. Understanding the heat budget on a tidal flat and predicting the water and soil temperatures are very important to evaluate the natural function of water purification. Lim et al. (1996) conducted a field observation of heat budget when the sea ebbed away at low tide. Comparing the numerically analyzed soil temperature with the observed one, they obtained a good agreement between the both.

In this study, systematic measurements have been carried out at a point where the state that the sea bed is under water or appears is repeated, and the thermal properties have been discussed.

2. OUTLINE OF OBSERVATIONS

The observations were made at one point of Wajiro tidal flat. It is located at the deepest part of Hakata Bay and is developing at the mouth of the Tounoharu River. Figure 1 shows the location of the observation point. The bed area appearing at the spring tide is about 0.8 km². The bed materials are almost sandy soil. The observation periods were from 18:00 Aug. 20 to 12:00 Aug. 22 and from 18:00 Dec. 14 to 12:00 Dec. 16 in 1997. In the summer observation, the sea bed appeared for 5 hours and submerged for 7 hours. In the winter observation, 4-hour appearance and 8-hour submergence were repeated. The water depths at high tide were about 0.9 m and 0.5 m, respectively. The vertical profiles of wind velocity, air, water and soil temperatures and air humidity have been obtained by arranging vertically measuring instruments. Downward and upward solar radiations and downward and upward long-wave radiations were measured respectively by using two pyranometers and two pyrgeometers.

3. THERMAL PROPERTIES

3.1 General weather condition

Figures 3 (a) and (b) show the time series of wind speed U, wind direction WD, air temperature T_a, specific humidity q_a, surface temperature T_s and water level which were obtained in the summer observation. The data of U, WD, T_a and q_a are ones at 2.5 m above the sea bed. T_s denotes water surface temperature when the tidal flat is under water and bed surface temperature when the sea has ebbed away. During the observation period, Kyushu Island was covered with a high atmospheric pressure developing above the Pacific Ocean, and stable fine weather continued for a week. The south-easterly wind, i.e., the land breeze was observed from the midnight to the early morning. On the other hand, the north-westerly sea breeze occurred in the daytime. T_s was higher than T_a except a short time in the midnight. This means the atmosphere above the tidal flat is unstable in the summer season. The time series of q_a shows a large variation under the influence of the water level variation and the strong solar radiation.

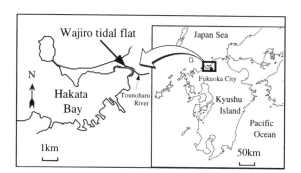

Fig. 1   Location of Wajiro tidal flat.

Fig. 2   View of measuring system.

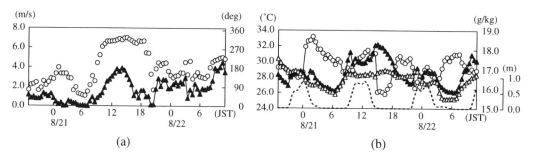

(a)   (b)

Fig. 3   Time series of wind speed U, wind direction WD, air temperature T_a, specific   humidity q_a and surface
temperature T_s in summer observation.   The dashed line indicates water level variation.
(a) ▲ ;U , ○ ; WD   , (b) △ ; T_a , ▲ ; T_s , ○ ; q_a

Figures 4 (a) and (b) show the time series of U, WD, T_a, T_s and q_a in the winter observation. On the whole, they are smoother than the summer ones. The southerly land breeze was formed in the nighttime and north-westerly sea breeze in the daytime. The land breeze was almost steady. The land and sea breeze system is more obvious than the summer one as seen from the comparison between Figs. 3 (a) and 4 (a). The reason seems to be because the local atmospheric instability was relatively small in winter. We can guess that the atmosphere above the tidal flat was stable because T_a was larger than T_s. The stability seems to suppress the variation of q_a.

Figures 5 (a) and (b) show the time series of soil temperature. The soil temperature at 40 cm depth from the bed surface is not influenced by solar heating and water covering. In summer, the soil temperature decreases with the depth and the reverse is seen in winter. When the sea ebbs away in the nighttime, soil temperature increases gradually with depth in both the seasons. As described later, this is due to the latent heat fluxes from the bed surface into the atmosphere.

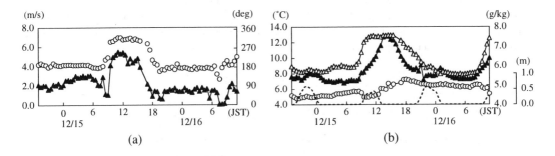

Fig. 4   Time series of wind speed U, wind direction WD, air temperature T_a, specific   humidity q_a and surface

temperature T_s in winter observation.   The dashed line indicates water level variation.

(a) ▲ ;U , ○ ; WD   , (b) △ ; T_a , ▲ ; T_s , ○ ; q_a

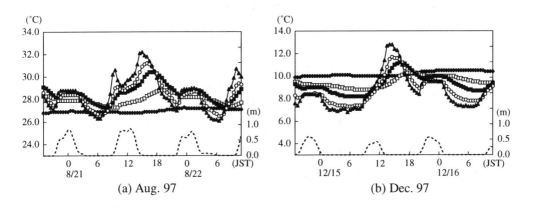

(a) Aug. 97   (b) Dec. 97

Fig. 5   Time series of soil temperatures.   The dashed lines indicate water level variation.

▲ ; 0 cm , ○ ; -4 cm , ■ ; -10 cm , □ ; -20 cm , ◆ ; -40 cm

599

3.2 Radiation fluxes

The net radiation R_{net} at the surface of a tidal flat is given by

$$R_{net} = S_d + S_u + L_d + L_u ,$$ (1)

where S_d is the downward solar radiation, S_u the upward reflected solar radiation, L_d the downward long-wave radiation and L_u the upward long-wave radiation. The fluxes toward the tidal flat surface are assumed to be positive. Figures 6 (a) and (b) show the daily variations of R_{net}, S_d, S_u, L_d and L_u. The value of S_d reached to 700 W/m^2 in the summer observation and 400 W/m^2 in the winter observation. The long-wave radiations L_d and L_u are approximately in equilibrium in both the seasons. Therefore, R_{net} is given by $S_d + S_u$ and becomes almost zero in the nighttime.

The relationship between the value of albedo α_w at water surface and the water depth is shown in Fig. 7. The data by other observation made in Dec. 1996 are also included. When the sea ebbs away, α_w takes about 0.18. The value is nearly equal to that of grassland or naked land. As the water depth increases, α_w decreases monotonically and approaches to 0.07. Therefore, it is seen that almost solar radiation is absorbed in water when the depth becomes more than 0.8 m.

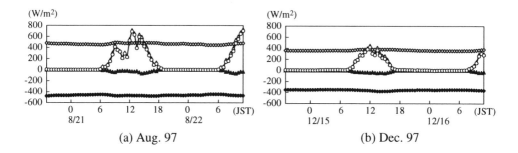

(a) Aug. 97   (b) Dec. 97

Fig. 6   Daily variation of radiation fluxes. $\bigcirc$; R_{net} , $\triangle$; S_d , $\blacktriangle$; S_u , $\diamondsuit$; L_d $\blacklozenge$; L_u

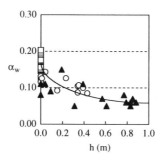

Fig. 7   Relationship between albedo and water depth.

$\square$; Dec. 96 , $\blacktriangle$; Aug. 97 , $\bigcirc$; Dec. 97

3.3 Heat fluxes

The net radiation R_{net} is distributed into sensible heat flux H, latent heat flux lE and subsurface heat flux G, i.e. ,

$$R_{net} = H + lE + G .$$
(2)

Signs of the fluxes are taken to be positive when their directions are outward from the surface. Figures 8 (a) and (b) show daily variations of the fluxes in the summer and winter observations, respectively. The sensible and latent heat fluxes are calculated by using the bulk transfer method. The equations are given by

$$H = - C_H C_p \rho (T_{2.5} - T_0) U_{2.5}$$
(3)

and

$$lE = - C_E l \beta (q_{2.5} - q_0) U_{2.5},$$
(4)

where $C_p \rho$ is the volume heat capacity $(1.25 \times 10^3 \, JK^{-1} m^{-3})$, l the latent heat of evaporation $(2.50 \times 10^6 \, Jkg^{-1})$, β the efficiency of evaporation and C_H and C_E the bulk transfer coefficients for heat and water vapor, respectively. The subscript 2.5 denotes physical quantities at 2.5 m above the bed surface and 0 those on the water surface or bed surface. The value of G is calculated by subtracting H and lE from R_{net}. H is almost zero except the daytime since T_s is nearly equal to T_a in the nighttime as shown in Figs. 3 (b) and 4 (b). In the nighttime, therefore, the relationship of $G \sim - lE$ is derived. As T_s is larger than T_a in the daytime of summer, H takes a positive value. In winter, H becomes negative because of the smaller value of T_s than T_a. However, the value of H is relatively small and R_{net} in the daytime is distributed into lE and G.

4. CONCLUSIONS

The systematic field observations were made to understand the thermal properties and the heat budget on a tidal flat. The atmosphere above tidal flats becomes unstable in the daytime of summer and the specific humidity varies considerably in response to the water level variation and solar radiation. In winter, the variation of the specific humidity seems to be suppressed because of the stability of

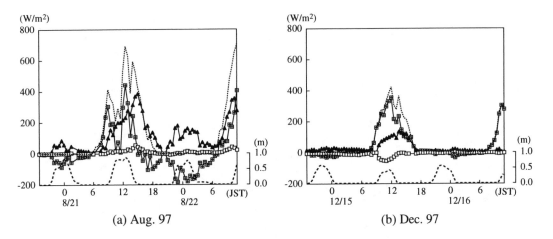

(a) Aug. 97   (b) Dec. 97

Fig. 8   Daily variations of heat fluxes.   Dotted lines indicate net radiation and the dashed lines indicate water level variations.   □ ; H , ▲ ; lE, ⊞ ; G

atmosphere. The soil temperature decreases with the depth in the daytime of summer, and the reverse is seen in the nighttime of winter. The soil temperature at the place deeper than 40 cm is not influenced by the solar heating and the water covering. In summer and winter seasons, the upward long-wave radiation from the surface is in equilibrium with the downward one from the atmosphere. Therefore, the net radiation to the tidal flat is equal to the net solar radiation. The value of albedo decreases monotonically with the increase of water depth and approaches to 0.07. In the daytime, the net radiation is distributed mainly into both the latent heat flux and the subsurface heat flux. In the nighttime, the subsurface heat flux becomes nearly equal to the minus latent heat flux.

5. REFERENCES

Lee, J.G., Nishijima, W., Mukai, T., Takimoto, K., Seiki, T., Hiraoka, K. and Okada, M. (1997), " Comparison for structure and function of organic matter degradation at natural and constructed tidal flats" , Jour. Japan Society on Water Environment, Vol. 20, No. 3, pp. 175-184, (in Japanese).
Lim, B.K., Akiyama, S., Sakurai, N. and Tanaka, M. (1996), " Field observation on the thermal environment of tidal flat estuary" , Proc. Meeting of Japan Society of Fluid Mechanics, pp. 443-444, (in Japanese).

Environmental Hydraulics, Lee, Jayawardena & Wang (eds) © 1999 Balkema, Rotterdam, ISBN 90 5809 035 3

Analysis of water quality variation along a regulated stream in Sri Lanka

K.D.W.Nandalal
Department of Civil Engineering, University of Peradeniya, Sri Lanka

ABSTRACT: Man modifies water quality not only directly through discharges but also indirectly through such activities as adjusting flow regimes, impounding and abstracting. The Mahaweli Development Scheme in Sri Lanka comprises of a complex network of reservoirs, diversion structures, canals, tunnels, hydropower plants etc. The paper presents an analysis of water quality in one of its subsystems, the system from the Polgolla barrage to the Kalawewa reservoir. Water passes through tunnels, hydropower plants, reservoirs, canals and natural streams in this stretch. The investigation is limited to physical and chemical parameters of water quality. The results indicate that some water quality parameters depend on the quantity of water, which could be controlled by the various structures in the subsystem. However, the quality of water is observed to be within the acceptable limits with respected to the investigated quality parameters.

1. INTRODUCTION

The hydraulic structures have a significant influence upon the natural flow regime driving to the negative or positive effect upon the environment. It is very vital to monitor water quality in such systems continuously so that measures to negate adverse effects on the environment due to water quality changes could be taken immediately. The Mahaweli Development Scheme is a multipurpose water resources scheme that harnesses the

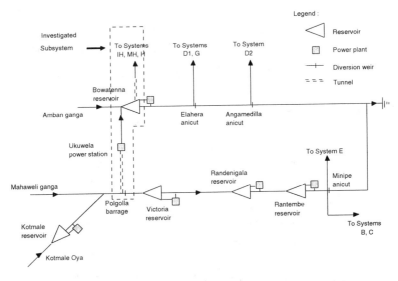

Figure 1 Schematic Diagram of the Mahaweli Development Scheme

hydroelectric and irrigation potential of the Mahaweli Ganga (river) in Sri Lanka. It comprises of a complex network of regulating reservoirs and diversion structures built on the main stem of the Mahaweli river as well as on its tributaries and diversion routes as shown in Figure 1.

A research project was carried out to monitor water quality in a subsystem of the Mahaweli Development Scheme in Sri Lanka to identify problems related to the quality of water and investigate the impact of the modifications to the operation patterns of the structures on the quality of water. This investigation was limited to examine the physical and chemical qualities of water.

There are three reservoirs in the main stem of the Mahaweli river: Victoria, Randenigala and Rantembe, each with a hydro-electric power plant. These reservoirs regulate the flow for satisfying irrigation demands of the Systems B, C and E at the Minipe anicut. The uppermost Kotmale reservoir is situated on the Kotmale Oya (creek), a tributary of the Mahaweli river. Downstream of the Kotmale reservoir is the Polgolla diversion structure, which plays a vital role in the Mahaweli Water Resources System. It is used to divert water from the Mahaweli river to the Amban Ganga (river) basin through a diversion tunnel. The diverted water is used to generate hydro-power at the Ukuwela power station before it reaches the Bowatenna reservoir. The water diverted from the Bowatenna reservoir for irrigation reaches the Kalawewa reservoir, which regulates flow for satisfying irrigation demands of the System H.

2. WATER QUALITY MONITORING PROGRAMME

The subsystem from the Polgolla barrage to the Kalawewa reservoir (i.e., trans basin diversion of the Mahaweli river via Polgolla and Bowatenna schemes) was considered in this study. The Kalawewa reservoir receives Mahaweli water since 1976. Figure 2 shows the selected subsystem in detail. It includes the locations of the sampling sites. Water quality monitoring in this area was started in July 1996 and has been continuing

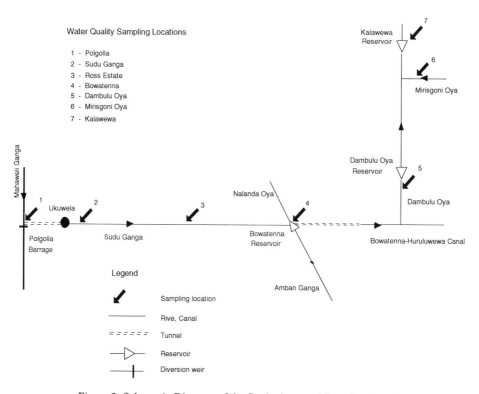

Figure 2 Schematic Diagram of the Study Area and Sampling Locations

on a monthly basis. The parameters chosen to characterize water quality conditions are temperature, pH, conductivity, turbidity, total dissolved solids, dissolved oxygen and salinity.

There have been very few water quality studies conducted for surface waters in the Mahaweli Development Scheme area. TAMS (1980) presents some water quality data for the streams, rivers and reservoirs in this area, which have been gathered from several sources. However, in that report water quality parameters at Polgolla for three months in 1979 and at the Ukuwela hydro-power station (Sudu Ganga) for one month in 1979 are only available within the study area shown in Figure 1. The physio-chemical properties of the Kalawewa reservoir waters on monthly basis are available in Silva (1996) for the year 1983.

3. ANALYSIS

The analysis of water quality conditions in the subsystem is based on a record of data collected over a period of two years on monthly basis from July 1996 to June 1998. Data on the diversion at the Polgolla barrage, irrigation release at the Bowatenna reservoir, release from the Dambulu Oya reservoir and rainfall in the area under investigation were collected from the Headworks Administration, Operation and Maintenance Division of the Mahaweli Authority of Sri Lanka.

3.1 Temperature

Temperature is basically important for its effect on other properties, e.g., speeding up of chemical reactions, reduction in solubility of gases, amplification of tastes and odours, etc. Most of the biochemical and biological processes are temperature dependent. Table 1 presents the variation of temperature in the regulated stream from the Polgolla barrage to the Kalawewa reservoir.

The temperature slightly increases towards the downstream direction. However, the average values indicate that the variation along the stream is not large. It is further noted that the temperature at all the stations in November, December and January to be low and in March, April and May to be high.

3.2 Dissolved Oxygen

Dissolved oxygen (DO) is a major stream water quality component that affects the aquatic environment and ecosystem. Table 2 presents the DO variation along the stretch from the Polgolla barrage to the Kalawewa reservoir.

The average temperature in this stretch varies from 25°C to 29°C and therefore, the DO solubility is about 8.4 mg/l to 7.6 mg/l. Clean surface waters are normally saturated with DO. Though the observed DO concentrations are less than the solubility levels, the values are quite satisfactory. The DO availability is

Table 1 Temperature variation from the Polgolla barrage to the Kalawewa reservoir in °C

Station	Polgolla	Sudu Ganga	Ross Estate	Bowatenna	Dambulu Oya	Mirisgoni Oya	Kalawewa
Average	25.2	24.6	25.2	27.3	27.9	27.7	29.0
Standard Deviation	1.4	1.4	1.4	1.5	1.6	0.8	1.5
Maximum	28.3	26.9	27.9	30.6	31.3	29.7	31.8
Minimum	23.1	22.5	22.8	25.0	25.4	25.8	26.9

Table 2 Dissolved oxygen variation from the Polgolla barrage to the Kalawewa reservoir in mg/l

Station	Polgolla	Sudu Ganga	Ross Estate	Bowatenna	Dambulu Oya	Mirisgoni Oya	Kalawewa
Average	5.71	5.72	5.91	6.00	6.00	5.39	5.81
Standard Deviation	0.86	0.90	0.69	0.86	0.99	0.58	0.82
Maximum	7.34	7.67	7.12	7.44	7.80	6.42	6.92
Minimum	4.10	4.30	4.50	4.50	4.60	4.16	4.28

desirable to maintain conditions favourable for the growth and reproduction of a normal population of fish and other organisms. A detailed analysis of DO availability with the hydrological regime revealed that the DO concentrations are high during "Yala" (South-west monsoon) and "Maha" (North-east monsoon) cultivation seasons. During these periods, the releases made towards the Kalawewa reservoir from the Polgolla barrage are high. The "Hydrilla" plants are reported to be growing rapidly during water release periods in the Bowatenna-Huruluwewa canal. The relatively high concentration of DO, which promotes photosynthetic activity may be one of the factors that affects this high growth of the "Hydrilla" plant. The thickly grown "Hydrilla" hinders the water flow in the canals.

3.3 Conductivity

A rapid estimation of the dissolved solids content of water can be obtained by specific-conductance measurements. The analysis of the collected data indicates that the conductivity increases towards the downstream direction starting from the Polgolla barrage. The conductivity of water in the Polgolla reservoir and the Sudu Ganga monitoring station do not show a significant correlation. However, water at the Sudu Ganga, Ross Estate and Bowatenna stations are highly correlated. This suggests that the water in this stretch of about 42 km is not polluted by external sources. However, downstream of the Bowatenna reservoir, starting from its outlet upto the Dambulu Oya reservoir, the conductivity of water increases at a faster rate. The specific conductivity of the Mirisgoni Oya water, which joins the canal/stream from the Dambulu Oya to the Kalawewa reservoir is very high throughout the whole period, the average value being about 415 μ°S.

The observations indicate that the specific conductivity values of water are very high in March and April. During these periods the diversions from the Polgolla barrage towards the Bowatenna reservoir are minimum. It was observed that the conductivity at the Bowatenna reservoir rises to about 650 μ°S during this low release periods.

The specific conductivity of the water from the Kalawewa reservoir varied between 144 μ°S and 357 μ°S during the present observation period (1996-1998). Silva (1996) reports that the specific conductivity in the waters of the Kalawewa reservoir varied between 175 μ°S to 300 μ°S during the period from 1980 to 1981. These results indicate that the specific conductivity has not changed during this period of nearly 15 years.

3.4 Total Dissolved Solids

The amount of total dissolved solids (TDS) present in water is a consideration in its suitability for specially domestic use. In general, waters with a total dissolved solids content of less than 500 mg/l are most desirable for such purposes. Table 3 presents the averages and standard deviations of TDS measured over the whole period at the seven stations shown in Figure 2.

From the Polgolla barrage to the Ross Estate, the TDS seems to be low. From the Bowatenna reservoir onwards, TDS increases rapidly. The TDS in the Mirisgoni Oya is relatively high throughout the whole period. During the months March and April, in which only very small diversions are made from the Polgolla barrage, the TDS at the Bowatenna and Dambulu Oya stations were high. However, as TDS was always less than 500 mg/l, the water in this stretch could be considered as of acceptable quality.

Table 3 Total Dissolved Solids variation from the Polgolla barrage to the Kalawewa reservoir in mg/l

Station	Polgolla	Sudu Ganga	Ross Estate	Bowatenna	Dambulu Oya	Mirisgoni Oya	Kalawewa
Average	38.8	44.8	53.6	79.3	88.8	200.6	112.2
Standard Deviation	18.2	25.8	26.5	77.2	65.7	48.8	24.9
Maximum	96.0	126.0	132.0	316.0	335.0	308.0	169.0
Minimum	20.3	24.5	21.5	31.2	37.5	139.0	68.0

3.5 pH

The averages of pH at all the stations showed that the waters along this stretch of the system are alkaline as presented in Table 4.

The waters in the Polgolla and Sudu Ganga stations showed slightly acidic in a very few occasions. The system downstream of the Sudu Ganga station was always alkaline, pH being greater than 7. Silva (1996) reports that the pH of water in the Kalawewa reservoir was between 7.5 and 8.5 in 1980-1981. Table 4 shows that the pH of the water from the Kalawewa reservoir ranges from 7.46 to 8.70 throughout the present study period. This implies that the alkalinity of the reservoir has not changed during the past 15 years period.

3.6 Turbidity

The turbidity causes by a wide variety of suspended materials, which range in size from colloidal to coarse dispersions, depending upon the degree of turbulence. The measurements showed that the turbidity in this regulated stream stretch sometimes increases to very high values. Table 5 presents the averages, minimum and maximum values of turbidity observed at the seven stations.

The variation of turbidity at each station is very high during the observed period Further, it is observed that the turbidity increases to very high levels during the periods of high streamflows, the turbidity being mainly due to relatively coarse dispersions.

4. CONCLUSIONS

The analysis of the water quality in the system from the Polgolla reservoir to the Kalawewa reservoir was a long felt need to assess its suitability for the various purposes it is presently being used. The present study was limited to analyze physical and chemical water quality parameters only. The DO concentration seems to be satisfactory from the Polgolla barrage to the Kalawewa reservoir. The specific conductivity is observed to be dependent on the flow quantity. The flows could be regulated at several locations in this stream stretch. This capability can be used to control the specific conductivity, which is a measure of the dissolved solids content in water. The water in this sybsystem is alkaline. The data collected over a period of two years indicate that the water quality in this system has not changed significantly. Also, the physical and chemical water quality parameters are observed to be within acceptable limits.

The Mirisgoni Oya is observed to be a highly polluted confluence. It joins the outflow from the Dambulu Oya reservoir before reaching the Kalawewa reservoir. However, this has not affected much the quality of water in the Kalawewa reservoir due to its small flow contribution.

Table 4 pH variation from the Polgolla barrage to the Kalawewa reservoir

Station	Polgolla	Sudu Ganga	Ross Estate	Bowatenna	Dambulu Oya	Mirisgoni Oya	Kalawewa
Average	7.33	7.28	7.49	7.77	7.95	7.79	8.09
Standard Deviation	0.35	0.25	0.14	0.22	0.14	0.10	0.37
Maximum	8.20	7.84	7.80	8.10	8.17	8.00	8.70
Minimum	6.92	6.88	7.30	7.38	7.75	7.64	7.46

Table 5 Turbidity variation from the Polgolla barrage to the Kalawewa reservoir in NTU

Station	Polgolla	Sudu Ganga	Ross Estate	Bowatenna	Dambulu Oya	Mirisgoni Oya	Kalawewa
Average	19.7	16.6	18.7	18.6	18.4	13.1	13.5
Maximum	134.1	64.9	91.7	108.0	66.6	39.0	28.3
Minimum	2.4	2.6	3.0	2.1	3.1	4.0	2.2

A complete spectrum of water quality data (physical, chemical and biological) is needed to describe the status of water quality completely. Therefore, it is recommended to carry out a monitoring programme in this stretch with reference to biological parameters also.

5. REFERENCES

Silva, E.I.L. (1996). "Water Quality of Sri Lanka: a review on twelve water bodies", Institute of Fundamental Studies, Sri Lanka.

TAMS (1980). "Environmental Assessment, Accelerated Mahaweli Development Programme", Volume III, Aquatic Environment, Tippetts-Abbette-McCarthy-Stratton, New York.

6. ACKNOWLEDGMENTS

This work was funded through a research grant (Research Grant RG/96/E/04) from the Natural Resources, Energy & Science Authority of Sri Lanka (NARESA).

Environmental Hydraulics, Lee, Jayawardena & Wang (eds) © 1999 Balkema, Rotterdam, ISBN 90 5809 035 3

Contaminant transport model of confined dike facilities

T. Kuppusamy
Virginia Tech., Blacksburg, Va., USA

ABSTRACT: The contaminant transport through Confined Dike Facilities (CDF) are analyzed by finite element method. The unsaturated fluid flow coupled with convective dispersive mass transport of chemicals are solved. Various practical cases of the CDF are analyzed to arrive at the effectiveness of the protective measures.

1. INTRODUCTION

The U.S. Army Corps of Engineers has been building confined dike facilities to dispose contaminated soils near Great Lakes area. The soils, contaminated by industrial waste such as heavy metals and other toxic chemicals, were dredged and disposed off close to the bay in a landfill form. The only protection against pollution was a dike built between the bay and the contaminated materials with a layer of cover material on the waste dump. A typical CDF is shown in figure 1 drawn not to scale. The CDF is 300 m long and 20 m deep. Most of the old existing facilities do not have either the clay liner or the cut-off wall. The water level of the lake varies and fluctuates due to tidal effect and the water table level is very high on the side of the landfill. It creates a back flow and leads to the pollutant transport back to the bay.

2. THEORY

A Computer Program called POLUT2D, based on the theories of two dimensional unsaturated/saturated fluid flows and mass transport of chemicals through porous media, is used here to analyze the problem. First, the flow equation is

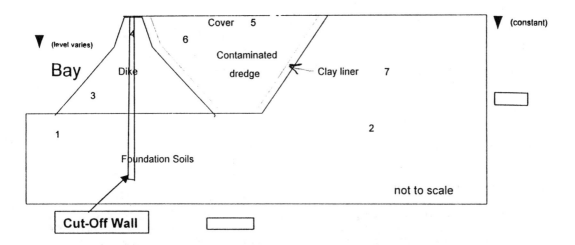

Figure 1. A Typical CDF (not to scale)

solved to obtain the seepage velocities within each element of the finite element mesh. Then the mass transport equation is solved to obtain the contaminant concentration at each node. Nonlinear iterations by modified Chord method and efficient time integration schemes are used (Kuppusamy et al.,1991; Bear,1979). The governing equation for each of these processes is given below.

The Flow equation can be written as (Kuppusamy et al.,1991; Kuppusamy,1991; Bear,1979):

$$\frac{\partial}{\partial x}(k_x \frac{\partial \phi}{\partial x}) + \frac{\partial}{\partial y}(k_y \frac{\partial \phi}{\partial y}) = C_h \frac{\partial \phi}{\partial t}$$ (1)

Where Φ = the total head, the hydraulic conductivities k_x, k_y, and C_h, the moisture capacity in the above equation are functions of the soil type (Van Genuchten, 1980). But more importantly, each is a non-linear function of pressure head. When the pressure head is positive (thus the soil is completely saturated), the saturated hydraulic conductivities rnutinely determined in the field or laboratory may be used for k, and k,, and C, is zero. However, when the pressure head is negative, the unsaturated permeability and the coefficient of moisture capacity of the soil must be evaluated. Various models have been developed to define the unsaturated values of k, and k,, and the value of C,. A very popular model that has been used extensively is the one developed by Van Genuchten(Van Genuchten,1980; Klute,1986; Dane,1980; Stephens et al.,1985; Parker et al.,1985; Milks et al.,1989).

The mass transport process is governed hy the following equation (Bear,1979):

$$D_x \frac{\partial^2 C}{\partial x^2} + D_y \frac{\partial^2 C}{\partial y^2} - V_x \frac{\partial C}{\partial x} - V_y \frac{\partial C}{\partial y} + F = R \frac{\partial C}{\partial t}$$ (2)

Where C = concentration, V_x, V_y = velocity in x and y direction, F= source or sink term,

R = retardation factor due to adsorption , D_x, D_y = hydrodynamic dispersion Coefficient in x and y directions.

Problems Studied

The following cases of the problem are studied:
Case I: The CDF without a clay liner or cut-off trench
Case 2: The CDF with a cut-off trench, 19 m deep
Case 3: The CDF with a clay liner
Case 4: The CDF with clay liner as well as cut-off trench

The mesh

The Finite Element mesh has 391 elements and 431 nodes. We have a finer mesh, with small elements in the dike, under the dike, and in the left part of the contaminated dredge materials. We have a coarser mesh, with big size elements, in the right part of the domain. This mesh was chosen after a parametric study.

Materials

The material properties used are shown in the table 1.

3. ANALYSIS OF THE DIFFERENT CASES

The maximum concentration levels reaching the bay as a percentage of the initial concentration of the heavy metal concentration of the dredge material are shown in Table 2. The clay liner is thicker than the cut-off trench (1m for the clay liner and 0.7 m for the cut-off trench) and the cut-off trench is only 19 m deep in the foundation soil 01: thus the flow easily gets under the cut-off trench and hence concentrations increase into the bay quickly. And the clay liner is all around the contaminated dredge material and so it is more effective.

Table 1. Material Properties of the CDF

Soil#	Permeability (cm/s)	Dispersivity (m)	Distribution Coefficient (1/kg)	Van Genuchten Parameter 1 (1/m)	Van Genuchten Parameter 2
1	0.01	16	0.5	2.4	0.6
2	0.02	16	0.5	2.4	0.6
3	0.03	16	0.5	2.4	0.6
4	0.003	16	0.5	2.4	0.6
5	0.02	16	0.5	2.4	0.3
6	0.003	16	0.5	0.15	0.3
7	0.003	16	0.5	2.4	0.6
8	0.0000001	3	0.5	0.15	0.3
9	0.0000001	3	0.5	0.15	0.3

Table 2. Migration of the Maximum Concentrations on to the Bay in Percentage of Initial Concentration.

Time	Case 1	Case2	Case 3	Case 4
3 months	10	0	0.2	0
1 1/2 year	65	10	6.6	0.4
2 years	70	17	10	0.9
5 years	85	54	26	10

When there is a cut-off trench, the concentrations reaching the bay side is nearly the same (about 50%) at all the nodes on the slope of the dike compared to the case with clay liner where the concentration is about 25% at the top and 0% at the bottom node of the dike. It is very clear that Case 3(clay liner) is more effective than case 2(cut-off) while case 4(liner + cut-off) is the most effective and should be preferred for the future CDF constructions.

4. EFFECTS DUE TO THE VARIATION OF THE HORIZONTAL PERMEABILITY

To study the influence of the coefficients of permeability of the foundation soils, the horizontal permeability values of the materials 1,2 and 3 are changed by 10 and 100 fold. For the Case I.when K increases, concentrations reaching the bay increases as expected. At 90 days, the maximum concentration on the bay for K multiplied by 100 is 9 times higher than the initial values. At 1950 days, with K multiplied by 10 or by 100, the bay side is completely contaminated with nearly all the nodes have a concentration over 95%

For the Case 2. Half of the nodes on the bay side reach a concentration level of 90% for the similar study. For the Case 3. Only few nodes reached to a concentration level of 40%. For the Case 4. Again only few nodes reached to a concentration level of 10 %.

5. CONCLUSIONS

It is found that the 10% COncentration contours reach the bay after three months when no protection structure is present, while it takes one and half years when there is a cut-off trench and two years when there is a clay liner; and five years and four months when a cut-off trench and a clay liner are both installed. So, we can notice that a cut-off trench is less effective than a clay liner. Two reasons can be given to explain that: (1) The concentration of pollutant at the front between the dike and the lake increases more quickly than for the clay liner because the pollutant spreads under the cut-off trench, whereas the clay liner constitutes an impermeable enclosure for the pollutant and (2) the reaching concentrations on the slope of the dike is more uniform, thus the pollution spreads more quickly and more easily into the lake. On the other hand, to be on safer side, it would be better to put both the clay liner and the cut-off trench.

Moreover, if there is already a clay liner and then a cut-off trench is installed, the 10% concentration contours reach the bay in three years and four months later. It is the same when there is first a cut-off trench built; installing a clay liner prevents the pollutant from being at a concentration of 10% for three years and ten months instead of two years. The effectiveness due to both the clay liner and the cut-off trench present together, is much more than the sum of the effectiveness due to the clay liner and the cut-off trench present individually. It is difficult to install clay liners in existing confined dike facilities but the cut-off trench is much more easier to construct. Of course, if it is a new facility to be constructed then a clay liner must be recommended.

6. REFERENCES

Kuppusamy, T., and Ahmad, F., (1991), POLUT2D – "A Finite Element Program for Vertical 2-D Advective Dispersive Pollutant Transport Problems: User's Manual", Department of Civil Engineering, Virginia Polytechnic Institute and State University.

Van Genuchten, M.Th., (1980), "A C."losed-form Equation for Predicting the Hydraulic Conductivity of Unsaturated Soils", Soil Science Society of America Journal, Vol 44.

Kuppusamy, T., (1991), "Multi Phase Flow And Transport Through Soils", A Short Course for the U.S. Army Corps of Engineers, Vicksburg, Mississippi.

Klute, A., (1986), "Water Retention: Laboratory Models", *Methods of Soil Analysis Part 1:Plrysieaf aml Mineralogical Methods,* A. Klute (Ed.), 2nd Edition, Number 9, AmericanSociety of Agronomy, Inc., Madison Wisconsin, pp. 635-662.

Bear, J., (1979), *Hydraulics of Groundwatar,* McGraw-Hill Inc.

Dane, J., (1980), "Comparison of Field and Laboratory Determined Hydraulic Conductivity Values", Soil Science Society of American Journal, Vol. 44, No. 2, pp. 228-231.

Stephens, D., and Rehfeldt, K., (1985), "Evaluation of Closed-Form analytical Models to Calculate Conductivity in Fine Sand", Soil Science Society of America Journal, Vol. 49, No. 1, pp. 12-19.

Parker, Kool and van Genuchten, (1985), "Determining Soil Hydraulic Properties from One-Step Outflow Experiments by Parameter Estimation: II, Experimental Studies", *Soil* Science Society of America Journal, Vol. 49, No. 6, pp. 1354-1359.

Milks, R., Fonteno, W., and Larson, R., (1989), "Hydrology of Horticulture Substrates: I. Mathematical Models for Moisture Characteristics of Horticultural Container Media", Journal of the American Society for Horticultural Science, Vol.114, No. 1, pp. 48-52.

Environmental Hydraulics, Lee, Jayawardena & Wang (eds) © 1999 Balkema, Rotterdam, ISBN 90 5809 035 3

A method of estimating the effective hydraulic conductivity for moisture transport in porous media

A.W.Jayawardena & P.B.G.Dissanayake
Department of Civil Engineering, The University of Hong Kong, China

ABSTRACT: A method of estimating the effective hydraulic conductivity for flow through partially saturated porous media is proposed. It is done by modifying the governing moisture transport equation by introducing a velocity term to make it analogous to the longitudinal dispersion equation. The approach then follows that for deriving the longitudinal dispersion coefficient.

KEY WORDS: Effective and local hydraulic conductivity, large-scale moisture transport, longitudinal dispersion coefficient, moving co-ordinate system, partially saturated porous media

1. INTRODUCTION

The complex natural heterogeneities in soils pose a major problem when the small scale classical moisture transport behaviour in homogeneous soils has to be extrapolated to large scale field problems. The usual approach in attempting to overcome this problem is by interpreting the field as an equivalent vertical soil column with a set of effective hydraulic parameters which are found by some averaging procedure. Several studies have been carried out in the past to formulate expressions for the effective hydraulic conductivity of porous media using approaches based on equivalent homogeneity assumption and stochastic behaviour (For example, Yeh et al. (1985a); Mantoglou and Gelhar (1987); Yeh (1989); Yeh and Harvey (1990); Polmann et al. (1991); Russo (1992); amongst others).

In this paper an expression for the effective hydraulic conductivity (spatially averaged in the horizontal dimension) is derived using a different approach. The method is partly based on the approach taken in the derivation of the longitudinal dispersion coefficient [Taylor, (1953); Fischer et al. (1979)]. In the study, the moisture transport equation is modified by introducing a quantity which can be thought of as the downward rate of advance of soil moisture. This and the soil moisture content are expressed respectively as the addition of a horizontally averaged mean and a fluctuating component. The equation is solved using a transformed co-ordinate system where the origin is moving with a 'velocity' equal to the mean of the downward rate of advance of soil moisture. This approach of defining an effective hydraulic conductivity also implies a reduction of the dimensionality of the moisture transport problem.

2. DERIVATION OF EFFECTIVE HYDRAULIC CONDUCTIVITY

The θ–based two-dimensional moisture transport equation is of the form [Richards (1931)]

$$\frac{\partial \theta}{\partial t} = \frac{\partial}{\partial z}\left[D(\theta)\frac{\partial \theta}{\partial z}\right] + \frac{\partial}{\partial x}\left[D(\theta)\frac{\partial \theta}{\partial x}\right] - \frac{\partial K(\theta)}{\partial z} \tag{1}$$

in which θ is the volumetric moisture content, z is the vertical distance measured positive downwards, x is the horizontal distance, t is the time, $K(\theta)$ is the hydraulic conductivity and $D(\theta)$ is the soil moisture diffusivity (= $K(\theta)\partial\psi/\partial\theta$ where ψ is the soil water pressure).

By substituting $u = \dfrac{\partial K}{\partial \theta}$, Eq. 1 can be written as

$$\frac{\partial \theta}{\partial t} + u\left(\frac{\partial \theta}{\partial z}\right) = \frac{\partial}{\partial z}\left[D(\theta)\frac{\partial \theta}{\partial z}\right] + \frac{\partial}{\partial x}\left[D(\theta)\frac{\partial \theta}{\partial x}\right] \tag{2}$$

where u can be thought of as a rate at which moisture is transported. Eq. 2 has the same mathematical form as the two-dimensional dispersion equation. When describing the large scale moisture transport problem, the quantities of interest are the values of the moisture content averaged over the horizontal or the vertical direction.

If averaging over the horizontal direction is carried out, the mean moisture content $\bar{\theta}$ and the mean gradient of the K-θ characteristic $\bar{u}$ in the horizontal direction can respectively be defined as

$$\bar{\theta} = \frac{1}{w}\int_0^w \theta dx \tag{3}$$

and

$$\bar{u} = \frac{1}{w}\int_0^w u dx \tag{4}$$

where w is the horizontal distance up to which averaging is done. With respect to the mean values, θ and u in Eq. 3 and 4 can be written as

$$\theta(x) = \bar{\theta} + \theta'(x), \text{ and, } u(x) = \bar{u} + u'(x) \tag{5 \& 6}$$

where $\theta'(x)$ and $u'(x)$ denote the deviations of θ and u from their respective means. Substituting Eq. 5 and 6 in Eq. 2, the moisture transport equation becomes

$$\begin{aligned}
\frac{\partial}{\partial t}\left(\bar{\theta} + \theta'\right) + \left(\bar{u} + u'\right)\frac{\partial}{\partial z}\left(\bar{\theta} + \theta'\right) &= \frac{\partial}{\partial z}\left[D(\theta)\left(\frac{\partial}{\partial z}\left(\bar{\theta} + \theta'\right)\right)\right] \\
&+ \frac{\partial}{\partial x}\left[D(\theta)\left(\frac{\partial}{\partial x}\left(\bar{\theta} + \theta'\right)\right)\right]
\end{aligned} \tag{7}$$

which can be simplified by a transformation to a co-ordinate system (ξ,x,τ) whose origin moves at the 'mean velocity' $\bar{u}$ in the z direction as follows

$$\xi = z - \bar{u}t \tag{8}$$
$$x = x \tag{9}$$
$$\tau = t \tag{10}$$

The derivatives $\dfrac{\partial \theta}{\partial z}$ and $\dfrac{\partial \theta}{\partial t}$ in the old co-ordinate system thus can be transformed to equivalent derivatives in the new co-ordinate system using the chain rule of differentiation as follows:

$$\frac{\partial \theta}{\partial z} = \left(\frac{\partial \theta}{\partial \xi}\right)\left(\frac{\partial \xi}{\partial z}\right) + \left(\frac{\partial \theta}{\partial x}\right)\left(\frac{\partial x}{\partial z}\right) + \left(\frac{\partial \theta}{\partial \tau}\right)\left(\frac{\partial \tau}{\partial z}\right) = \frac{\partial \theta}{\partial \xi} \tag{11}$$

$$\frac{\partial \theta}{\partial t} = \left(\frac{\partial \theta}{\partial \xi}\right)\left(\frac{\partial \xi}{\partial t}\right) + \left(\frac{\partial \theta}{\partial x}\right)\left(\frac{\partial x}{\partial t}\right) + \left(\frac{\partial \theta}{\partial \tau}\right)\left(\frac{\partial \tau}{\partial t}\right) = -\bar{u}\frac{\partial \theta}{\partial \xi} + \frac{\partial \theta}{\partial \tau} \tag{12}$$

Substituting Eq. 11 and 12 and using the property that $\dfrac{\partial \bar{\theta}}{\partial x} = 0$ (from Eq. 3), Eq. 7 becomes,

$$\frac{\partial \bar{\theta}}{\partial \tau} + \frac{\partial \theta'}{\partial \tau} + u' \frac{\partial \bar{\theta}}{\partial \xi} + u' \frac{\partial \theta'}{\partial \xi} = \frac{\partial}{\partial \xi}\left[D(\theta)\left(\frac{\partial \bar{\theta}}{\partial \xi} + \frac{\partial \theta'}{\partial \xi} \right) \right] + \frac{\partial}{\partial x}\left[D(\theta)\frac{\partial \theta'}{\partial x} \right] \tag{13}$$

which is independent of the term $\bar{u}$.

In advective dominated dispersion problems, the concentration gradient in the flow direction due to velocity variation is very much greater than that due to molecular diffusion. If the same simplification is made for the moisture transport equation by assuming that the moisture transport in the vertical direction by u [u = u(K)] is very much greater than that by diffusion, i.e. $u\dfrac{\partial \theta}{\partial \xi} >>> \dfrac{\partial}{\partial \xi}\left(D(\theta)\dfrac{\partial \theta}{\partial \xi} \right)$, then Eq. 13 simplifies to

$$\frac{\partial \bar{\theta}}{\partial \tau} + \frac{\partial \theta'}{\partial \tau} + u' \frac{\partial \bar{\theta}}{\partial \xi} + u' \frac{\partial \theta'}{\partial \xi} = \frac{\partial}{\partial x}\left[D(\theta)\frac{\partial \theta'}{\partial x} \right] \tag{14}$$

To simplify Eq. 14 further, as it is still intractable, the operator $\dfrac{1}{w}\int\limits_0^w (\ )dx$ is applied to each term of Eq. 14.

Then,

$$\frac{1}{w}\int\limits_0^w \left(\frac{\partial \bar{\theta}}{\partial \tau} \right)dx + \frac{1}{w}\int\limits_0^w \left(\frac{\partial \theta'}{\partial \tau} \right)dx + \frac{1}{w}\int\limits_0^w \left(u'\frac{\partial \bar{\theta}}{\partial \xi} \right)dx + \frac{1}{w}\int\limits_0^w \left(u'\frac{\partial \theta'}{\partial \xi} \right)dx$$

$$= \frac{1}{w}\int\limits_0^w \left(\frac{\partial}{\partial x}\left[D(\theta)\frac{\partial \theta'}{\partial x} \right] \right)dx \tag{15}$$

It can be shown that in Eq. 15,

$$\frac{1}{w}\int\limits_0^w \frac{\partial \bar{\theta}}{\partial \tau}dx = \frac{\partial \bar{\theta}}{\partial \tau}; \quad \frac{1}{w}\int\limits_0^w \frac{\partial \theta'}{\partial \tau}dx = 0; \quad \frac{1}{w}\int\limits_0^w \left(u'\frac{\partial \bar{\theta}}{\partial \xi} \right)dx = 0 \text{ (if w is sufficiently large } \int\limits_0^w u'dx = 0) \text{ and that}$$

$$\frac{1}{w}\int\limits_0^w \left(u'\frac{\partial \theta'}{\partial \xi} \right)dx = \overline{u'\frac{\partial \theta'}{\partial \xi}} \text{ (the over bar indicates the horizontally averaged value)}$$

If it is further assumed that $\dfrac{\partial \theta'}{\partial x} = 0$ at x = 0 and x = w (This implies zero flux boundary at x=0 and x = w ; also used in Polmann et al. (1991)), then the RHS of Eq. 15 is zero. Therefore, the simplified form of Eq. 15 is

$$\frac{\partial \bar{\theta}}{\partial \tau} + \overline{u'\frac{\partial \theta'}{\partial \xi}} = 0 \tag{16}$$

Subtracting Eq. 16 from Eq. 14

$$\frac{\partial \theta'}{\partial \tau} + u' \frac{\partial \bar{\theta}}{\partial \xi} + u' \frac{\partial \theta'}{\partial \xi} - \overline{u'\frac{\partial \theta'}{\partial \xi}} = \frac{\partial}{\partial x}\left[D(\theta)\frac{\partial \theta'}{\partial x} \right] \tag{17}$$

It can be assumed that $u'\dfrac{\partial \theta'}{\partial \xi} - \overline{u'\dfrac{\partial \theta'}{\partial \xi}} = 0$ (Fischer et al., 1979) if u' and θ' are small, well behaved and their variation in the vertical direction is small. Further assuming quasi-steady state condition, $\dfrac{\partial \theta'}{\partial \tau}$ term can also be neglected (In Fischer et al. (1979), the variation of the horizontally averaged mean concentration with respect to flow direction was assumed constant after a certain initial period. This results in the variation of the fluctuating component with time to be zero). Eq. 17 then simplifies to

$$u' \frac{\partial \overline{\theta}}{\partial \xi} = \frac{\partial}{\partial x}\left[D(\theta) \frac{\partial \theta'}{\partial x} \right] \tag{18}$$

which has the solution

$$\theta'(x,\xi) = \left(\frac{\partial \overline{\theta}}{\partial \xi} \right) \int_0^x \frac{1}{D(\theta)} \int_0^x u'dxdx + \theta'(0,\xi) \tag{19}$$

Considering a soil column of unit width extending from 0 to w in the x direction, the rate of moisture flow (m^3/m/s) in the vertical direction at a depth $\xi = \xi_0$ is given by

$$Q = \int_0^w u'\theta'(x,\xi_0)dx + \overline{u}\,\overline{\theta}w \times 1 \tag{20}$$

in which the first term represents the moisture flux in the moving co-ordinate system and the second term is added to transform back into the original co-ordinate system.

Substituting for θ' from Eq. 19 in Eq. 20 and simplifying

$$Q = \left(\frac{\partial \overline{\theta}}{\partial z} \right) \int_0^w u' \int_0^x \frac{1}{D(\theta)} \int_0^x u'dxdxdx + \overline{u}\,\overline{\theta}w \tag{21}$$

Using Darcy's law for the vertical moisture flux through the same soil column

$$Q = -K_{eff}\left[\frac{\partial \overline{\phi}}{\partial z} \right]w \tag{22}$$

where K_{eff} is the effective hydraulic conductivity for the soil column in the vertical direction and $\overline{\phi}$ is the total potential averaged in the x direction. Substituting $\overline{\phi} = \overline{\psi} + z$ ($\overline{\psi}$ is the soil suction averaged in the x direction) and considering the vertical downward movement of moisture (substituting Q=-Q (Q is an inward flux) and z = -z), Eq. 22 can be written as

$$Q = K_{eff}\left[-\frac{\partial \overline{\psi}}{\partial z} + 1 \right]w \tag{23}$$

By comparing Eq. 21 and Eq. 23, K_{eff} can be derived as

$$K_{eff} = \frac{\left(\dfrac{\partial \overline{\theta}}{\partial z} \right) \int_0^w u' \int_0^x \dfrac{1}{D(\theta)} \int_0^x u'dxdxdx + \overline{u}\,\overline{\theta}w}{\left\{ -\dfrac{\partial \overline{\psi}}{\partial z} + 1 \right\}w} \tag{24}$$

In Eq. 24, K_{eff} is computed for a particular value of z. It is valid only when the denominator is positive. This situation occurs when $\dfrac{\partial \overline{\psi}}{\partial z} < 1$. i.e. when the moisture flow is in the downward vertical direction.

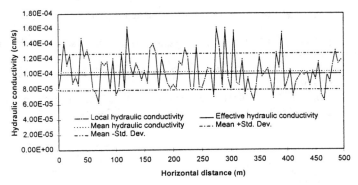

Figure 1:Variation of hydraulic conductivity with distance at Z= 0 cm

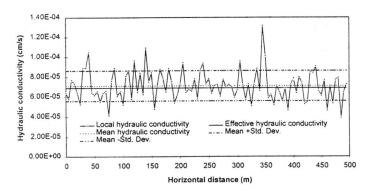

Figure 2:Variation of hydraulic conductivity with distance at Z=30 cm

3. APPLICATION

The proposed method has no direct verification as there is no reference value to compare with. It can however be indirectly verified if sufficient field data are available. A crude comparison can however be made with reference to the arithmetic or geometric means of the local hydraulic conductivity values. Such a comparison was made for random data and the results are close to the assumed reference values at shallow depths where the assumptions made in the derivation are valid (Figs 1 & 2). In the computation of K_{eff} it is necessary to know the initial, residual and saturated soil moisture contents, soil moisture retention characteristics, and saturated hydraulic conductivity values on a suitably spaced vertical grid.

4. CONCLUSION

A method of estimating the effective hydraulic conductivity for moisture transport through partially saturated porous media based on the longitudinal dispersion coefficient derivation is proposed. It has its limitations arising from the assumptions made. The method has no direct verification but a crude comparison is made using the arithmetic and geometric means as the reference hydraulic conductivity values.

5. REFERENCES

Fischer, H. B., List, E. J., Koh, R. C. Y., Imberger, J. and Brooks, N. H. (1979). "Mixing in inland and coastal waters." Academic Press Inc. (London) Ltd., 82-86.

Mantoglou, A and Gelhar, L. W.(1987). "Capillary tension head variance, mean soil moisture content and effective specific soil moisture capacity of transient unsaturated flow in stratified soils." Water Resour. Res., 23(1) 47-56.

Mantoglou, A and Gelhar, L. W. (1987). "Effective hydraulic conductivities of transient unsaturated flow in stratified soils." Water Resour. Res., 23(1) 57-67.

Mantoglou, A and Gelhar, L. W. (1987). "Stochastic modeling of large-scale transient unsaturated flow systems." Water Resour. Res., 23(1) 37-46.

Polmann, D. J., McLaughlin, D., Gelhar, L. W. and Ababou R. (1991). "Stochastic modeling of large-scale flow in heterogeneous unsaturated soils." Water Resour. Res., 27(7) 1447-1458.

Richards, L. A. (1931). "Capillary conduction of liquids through porous media". Physics, 1, 318-333.

Russo, D. (1992). "Upscaling of hydraulic conductivity in partially saturated heterogeneous porous formation." Water Resour. Res., 28(2) 397-409.

Taylor, G. I. (1953). "Dispersion of soluble matter in solvent flowing slowly through a tube." Proc. R. Soc. London Ser. A, 219, 186-203.

Yeh, T. C. J. and Harvey, D. J. (1990). "Effective unsaturated hydraulic conductivity of layered sands.' Water Resour. Res., 26(6), 1271-1279.

Yeh, T. C. J., Gelhar, L. W. and Gutjahr, A. L. (1985). "Stochastic analysis of unsaturated flow in heterogeneous soils 1. Statistically isotropic media." Water Resour. Res., 21(4), 447-456.

Yeh, T. C. J., Gelhar, L. W. and Gutjahr, A. L. (1985). "Stochastic analysis of unsaturated flow in heterogeneous soils 2. Statistically isotropic media with variable α." Water Resour. Res., 21(4), 457-464.

Yeh, T. C. J. (1989). "One-dimensional steady state infiltration in heterogeneous soils." Water Resour. Res., 25(10), 2149-2158.

3.2 Cooling water studies

Environmental Hydraulics, Lee, Jayawardena & Wang (eds) © 1999 Balkema, Rotterdam, ISBN 90 5809 035 3

The generating method of river-regime-fitted or river-mainstream-fitted orthogonal grid and its applications

Y. H. Dong
River Research Department, Yangtze River Scientific Research Institute (YRSRI), China

ABSTRACT: Based on Hermite ter-polynomial interpolation, a generating method of river-regime-fitted or river-mainstream-fitted orthogonal quadrangular grid is presented in this paper. Applying this grid-generating method, adapting simplified depth-average flow governing equations under boundary layer coordinates system and using SIMPLER algorithm, 2-D flow verified and applied calculations for 2 typical river reaches of the Changzhou power plant and the Qinglongzhou embankment displacement are carried out. Properties of the grid-generating method are discussed finally.

1. INTRODUCTION

Two key problems about grid-generating methods for 2-D numerical simulation of river channels are fitting between grid and river channels, and matching among grid, governing equations and numerical methods. Based on Hermite ter-polynomial interpolation, a generating method of river-regime-fitted or river-mainstream-fitted orthogonal quadrangular grid is presented in this paper. Its essential ideas can be described that, firstly, by using river-regime-fitted or river-mainstream-fitted grid, complicated boundaries of river banks, bars and shoals can be avoided, difficult problems such that grid is fitted with river embankments but not with river regime or mainstream, caused by current-used-frequently Thompson grid-generated method, can also be solved; secondly, using orthogonal grid and 2-D depth-average governing equations of fluid mechanics under boundary layer coordinates system, associated with proper numerical discrete and iterate method such as SIMPLER algorithm, various 2-D planar numerical calculations for simulation of flow, sediment, temperature, concentration etc. in river channels may be implemented.

Applying this grid-generating method, adapting simplified depth-average flow governing equations under boundary layer coordinates system and using SIMPLER algorithm, 2-D flow verified and applied calculations for river reaches of the Changzhou power plant, which is located in lower reaches of the Yangtze River, and for river reaches of the Qinglongzhou embankment displacement, which is located in lower part of the Zishui River, a tubutary of the Yangtze River, are carried out. The feasibility and the applicability of above-mentioned grid-generating method is verified and proved. Finally, properties of the proposed grid-generating method are discussed.

2. GRID-GENERATED METHOD

2.1 Hermite Ter-polynomial Interpolation

Hermite interpolating polynomials are fitting curves that are satisfied with equal conditions of not only functional values but also 1-order or higher order derivatives at interpolating nodes. Hermite ter-polynomial, which is used in this paper, is one of these fitted curves with equal functional values and equal 1-order derivatives at interpolating nodes. The mathematical expressions can be described in following (only for 2 interpolating nodes as example)(Li, 1982):

Conditions:

$$H_3(x_1) = y_1 \qquad H_3(x_2) = y_2 \qquad H_3{'}(x_1) = m_1 \qquad H_3{'}(x_2) = m_2 \tag{1}$$

Ter-polynomial:

$$H_3(x) = y_1 \alpha_1(x) + y_2 \alpha_2(x) + m_1 \beta_1(x) + m_2 \beta_2(x) \tag{2}$$

in which:

$$\alpha_1(x) = \left(1 + 2\frac{x - x_1}{x_2 - x_1}\right)\left(\frac{x - x_2}{x_1 - x_2}\right)^2 \tag{3-1}$$

$$\alpha_2(x) = \left(1 + 2\frac{x - x_2}{x_1 - x_2}\right)\left(\frac{x - x_1}{x_2 - x_1}\right)^2 \tag{3-2}$$

$$\beta_1(x) = (x - x_1)\left(\frac{x - x_2}{x_1 - x_2}\right)^2 \tag{3-3}$$

$$\beta_2(x) = (x - x_2)\left(\frac{x - x_1}{x_2 - x_1}\right)^2 \tag{3-4}$$

2.2 Grid Generation

The grid-generating method is specifically including 3 steps. First step: determination of grid-control sections and points. Inlet/outlet cross-sections of river reaches, and hydrographic or key cross-sections which can reflect and control river regime or river mainstream, are chosen as grid-control sections. Grid-control points, which may be left, right, middle or arbitrary points on these grid-control sections are selected subsequently.

Second step: generation of grid-control curve. By means of Hermite ter-polynomial interpolation given above, a grid-control curve, which passes through all grid-control points and is perpendicular to the grid-control sections, is generated. The grid-control sections and points are chosen and adjusted repeatedly until the grid-control curve is fitted with the river regime or the river mainstream. This is the key step of this grid-generating method.

Final step: fabrication of grid. The orthogonal quadrangular grid is fabricated by curves parallel to the grid-control curve and sections perpendicular to the grid-control curve(including the grid-control sections).

2.3 Governing Equations and Numerical Method

Obviously, the orthogonal grid, generated by above-mentioned method, is perfectly matched with governing equations of fluid mechanics under boundary layer coordinates system(Wu, 1982), and SIMPLER algorithm (Patankar, 1980).

Simplified depth-average flow governing equations under boundary layer coordinates for 2-D numerical simulation of river channels, which are used in following examples of application, are given here:

$$\frac{\partial Z}{\partial t} + \frac{R}{R + y}\frac{\partial}{\partial x}(Uh) + \frac{\partial}{\partial y}(Vh) + \frac{Vh}{R + y} = 0 \tag{4}$$

$$\frac{\partial}{\partial t}(Uh) + \left[\frac{R}{R + y}U\frac{\partial}{\partial x}(Uh) - \left(\frac{R}{R + y}\right)^2 v_t \frac{\partial^2}{\partial x^2}(Uh)\right] + \left[V\frac{\partial}{\partial y}(Uh) - v_t \frac{\partial^2}{\partial y^2}(Uh)\right] +$$

$$\frac{R}{R + y}gh\frac{\partial Z}{\partial x} + \frac{n^2 gU\sqrt{U^2 + V^2}}{h^{1/3}} + \frac{hUV}{R + y} - fhV = 0 \tag{5}$$

$$\frac{\partial}{\partial t}(Vh) + \left[\frac{R}{R + y}U\frac{\partial}{\partial x}(Vh) - \left(\frac{R}{R + y}\right)^2 v_t \frac{\partial^2}{\partial x^2}(Vh)\right] + \left[V\frac{\partial}{\partial y}(Vh) - v_t \frac{\partial^2}{\partial y^2}(Vh)\right] +$$

$$gh\frac{\partial Z}{\partial y} + \frac{n^2 gV\sqrt{U^2 + V^2}}{h^{1/3}} - \frac{hU^2}{R + y} + fhU = 0 \tag{6}$$

622

in which, x,y,t : planar(2-D) and time coordinate; u,v: velocities of flow ; R: curvature radius of x-axis; h,Z: depth and water level; g,n,v$_t$,f: gravitational constant, river roughness, turbulent diffusion coefficient and Coriolis force coefficent.

Boundary conditions for numerical solution of these equations are discharge profile given at inlet cross-section for the simulated river reahces, and water stage profile given at outlet cross-section. Other boundary reatments, used in the mathematical model, include that:

$$\frac{\partial Z}{\partial n}\Big|_{bank} = 0 \qquad U\Big|_{bank} = V\Big|_{bank} = 0 \qquad \frac{\partial U}{\partial n}\Big|_{outlet} = \frac{\partial V}{\partial n}\Big|_{outlet} = 0 \qquad (7)$$

in which, n: normal coordinates of corresponding boundaries.

Other dealings neccessary to 2-D numerical simulation for natural river channels, such as technique of movable river-bank treatment, modelling of piers and intake/drainage structures etc., are used the model (Dong, 1995 ;Dong, 1996)

3. EXAMPLES OF APPLICATION

3.1 2-D Flow Simulation for River Reaches of the Changzhou Power Plant

The river reaches of the Changzhou power plant is located at the lower reaches of the Yangtze River, approximate 230Km away from the estuary. The river course for simulation belongs to gentle-bend bifurcated tidal reaches and is about 12km in length(from cross-section Cs1 to cross-section Cs4). The project of the Changzhou power plant consists of power plant embankment, 6 ash embankments, piers and intake/drainage structures(Fig.1).

According to above-mentioned grid-generating method , 4 cross-sections, which are Paozizhou(inlet section, Cs1), Liuweigang(Cs2),Tianshengang(Cs3) and Ligang(outlet section, Cs4), are selected as grid-control sections, and the left points of the 4 sections are chosen as grid-control points. The fabricated nodes for 2-D numerical simulation are 120 × 53, grid interval distance along the river regime is 100 ~ 200m and grid interval distance along sections is 100m(Fig.2).

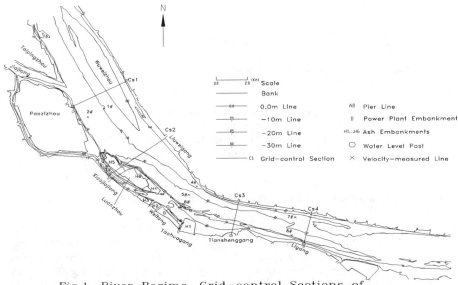

Fig.1. River Regime, Grid-control Sections of the River Reaches and Layout of the Changzhou Power Plant

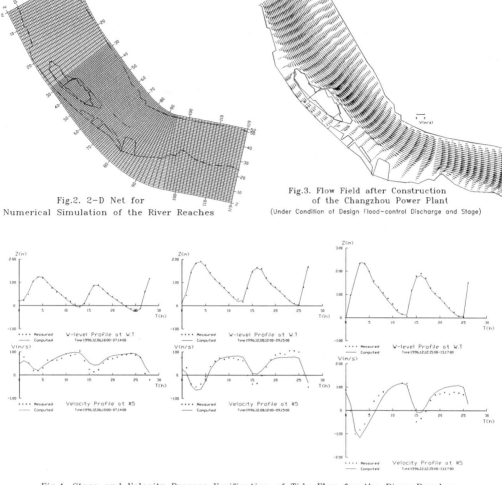

Fig.2. 2-D Net for
Numerical Simulation of the River Reaches

Fig.3. Flow Field after Construction
of the Changzhou Power Plant
(Under Condition of Design Flood-control Discharge and Stage)

Fig.4. Stage and Velocity Process Verification of Tide Flow for the River Reaches

Using the 2-D generating grid, depth-average flow governing equations under boundary layer coordinates system and SIMPLER algorothm, 2-D unsteady tidal verified and engineering effecting calculations are carried out(Dong, 1997, "2-D flow Numerical Simulation and Analysis of Effects of the Changzhou Power Plant on the River Channels", research report of YRSRI). Partial results of the calculation are shown as Fig.3 and Fig.4.

3.2 2-D Flow Simulation for River Reaches of the Qinglongzhou Embankment Displacement

The river reaches of the Qinglongzhou embankment displacement is located at the lower reaches of the Zishui River which is one of the four main branches of the Dongtinghu lake in the middle reaches of the Yangtze River. The river stream for calculation is about 13Km in length(from Licanggang cross-section to Yiyang water stage post), belongs to alternative widening/converging bifurcated reaches(Fig.5). The aim of the embankment displacement is to reduce the length of flood-control dykes so as to improve the flow capacity of the river channel in flood seasons.

624

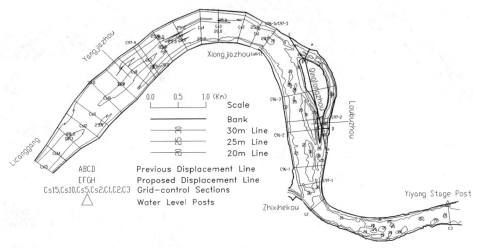

Fig.5. River Regime, Grid-control Sections for the River Reaches
and Sketch of the Embankment Displacement

7 grid-control sections(i.e. cross-sections of Cs15, Cs10, Cs5, Cs2, C1, C2 and C3) are selected, and the middle points of the 7 sections are regarded as grid-control points. The fabricated net for 2-D calculation consists of 140 × 56 nodes, grid interval distance along river regime is 50 ~ 100m, grid interval distance along sections is 20m(Fig.6). Obviously, the 2-D net is satisfactorily fitted and agreed with the river regime of the river reaches, although not rigorously fitted with the river banks.

2-D steady flow verified and engineering effecting calculations are implemented, the proposed design line of the embankment displacement is developed by the numerical simulation through optimising local flow velocity condition and avoiding recirculation(Dong,1997, "2-D flow Numerical Simulation and Analysis of Effects of the Qinglongzhou Embankment Displacement on the River Channels", research report of YRSRI). Partial results of the calculation are shown as Fig.7 and Fig.8.

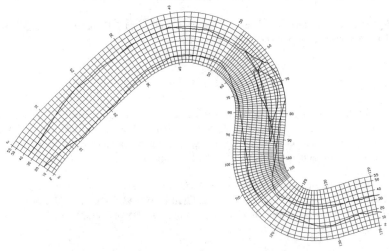

Fig.6. 2-D Grid for Calculation of the River Reaches

625

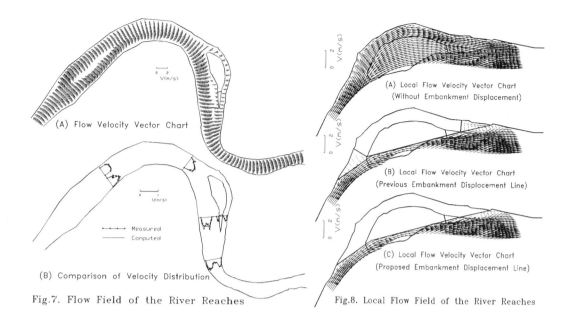

(A) Flow Velocity Vector Chart

(B) Comparison of Velocity Distribution

Measured
Computed

Fig.7. Flow Field of the River Reaches

(A) Local Flow Velocity Vector Chart
(Without Embankment Displacement)

(B) Local Flow Velocity Vector Chart
(Previous Embankment Displacement Line)

(C) Local Flow Velocity Vector Chart
(Proposed Embankment Displacement Line)

Fig.8. Local Flow Field of the River Reaches

4. PROPERTIES OF THE METHOD

Description of the grid-generating method and two examples of application show that, compared with other orthogonally river-fitted grid-generating methods(e.g. Thompson grid-generating method), this net generating method has following main advantages.

(1) fitting between the net and the river regime or the river mainstream;
(2) matching among the grid-generating method, the governing equations and the numerical method;
(3) harmony between grid-control sections and river hydrographic or key cross-sections;
(4) relatively uniform grid interval and convenient manipulation.

Thus, this grid-generating method is expected to be widely applied to 2-D planar numerical calculations for simulation of flow, sediment, environment etc. in river channels.

5. REFERENCES

Dong, Y.H. etc.(1995). "The Research and Its Application of a 2-D Unsteady Flow Mathematical Model in Estuarine Tidal Reaches"(in chinese), *J. Yangtze River Scientific Research Institute*, 12(1), 31-39.
Dong, Y.H.(1996). "The Analysis on Effects of the Liujiagang Embankment in Taicang City on the River Channel by Using 2-D Mathematical Model of Tidal Flow"(in chinese), *Journal of Hydraulic Engineering*, No.12, 62-69.
Li, Q.Y. etc.(1982). "Numerical Analysis"(in chinese), *Huazhong Technology University Press,* Wuhan.
Patankar,S.V.(1980) . "Numerical Heat Transfer and Fluid Flow", *McGraw-Hill, U.S.A.*.
Wu, W.Y.(1982). "Mechanics of Fluid"(in chinese), *Beijing University Press,* Beijing.

Environmental Hydraulics, Lee, Jayawardena & Wang (eds) © 1999 Balkema, Rotterdam, ISBN 90 5809 035 3

Numerical simulation and experimental study of cooling water project and thermal pollution of HL Thermal Power Plant

Jiang Wei, Lin Youjin & Wang Zhangjun
Guangdong Provincial Research Institute of Water Conservancy and Hydropower, China

ABSTRACT: The cooling water project and thermal pollution of HL Power Plant are studied by both the mathematical and physical models. The best outlet-intake scheme is given and the temperature of cooling water of the plant and the areas of thermal pollution on studied seas are predicted. Furthermore, the methods of numerical simulation and model experiment are discussed in detail, and the differences between the calculated and experimental results are enumerated and the reasons affecting the differences are analyzed in detail. These conclusions and calculative and experimental methods in the paper are of universal significance for cooling water projects of tidal currents, and are also valuable for the study of diffusion of other waste water in bays or estuaries

1.INTRODUCTION

A vast amount of waste heat energy will exhaust with cooling water when a thermal power plant is working. Generally, plants constructed in coastal areas always make full use of seawater as the cooling water source for economizing investment. Because the characteristics of tidal current and heat flow are very complex due to unsteady features, the arrangement of cooling water projects in bay and estuary is more difficult than in river or reservoir. Therefor, for large-scale power plants the cooling water project must be determined together through physical and mathematical models in our country. In recent years, a great deal of studies on cooling water projects of tidal current are carried out, and some results have been obtained. But many respects are still not clear, and have much room for improvement. This paper is an applied achievement in the field.

The HL Thermal Power Plant, located near the JH Bay and with 8 ×600MW installed capacity, is a large-scale project. Its discharge of cooling water is 200 m³/s, the temperature difference between the outlet and inlet is 8 ℃ and the seawater of JH Bay is used as the cooling water source. The power plant sets up a 20,000-tonnage coal wharf, a dock basin about 20 m in depth, a deepwater channel 200 m in width and a breakwave 1515 m in length (seeing Fig.1). The inlet of cooling water of the plant, which altitudes of top and bottom edges are separately -5 m and -10 m, is put up in the dock basin. The outlet is considered to arrange in two places: one is at the southeast corner of the site of the plant, called as the near drainage scheme, and another is at the northeast corner of the ashery, called as the far drainage scheme. The outlets of two schemes are the square pipes 3×2 m² in height and width.

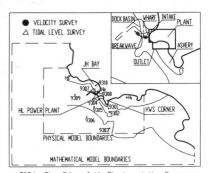

FIG.1 The Site of HL Plant and the Regions of the Mathematical and Physical Models

In the paper, the tidal current fields in JH Bay and its surroundings are first simulated by the physical and mathematical models. And the tide and velocity curves of both numerical calculation and physical model test are verified through comparing the prototype measurement data. Then, under satisfying similarities the calculation and experiment of the heated water is carried out. And the temperature curves at the intake and thermal pollution areas on studied seas are predicted. Two outlet-intake schemes are carefully compared and the better is applied to the practical engineering. Finally, some important conclusions are obtained through comparing the results of numerical calculation and model test. These conclusions and methods of numerical and experimental studies in the paper are of universal significance for cooling water projects of tidal currents, and are also valuable for the study of diffusion of other waste water in bays or estuaries.

2.METHOD OF NUMERICAL SIMULATION

The sea area around the HL Power Plant is a typical shallow bay. Judging from observed field data, there is not difference on the whole between the surface and bottom velocities. Therefore, the two-dimensional mathematical model averaged along the direction of water depth can be used to present the movement of the tidal current and the heated discharge in the shallow bay.

2.1 Governing Equations

The full shallow water equations describing the movement of the tidal current and the heated discharge are expressed as follows:

Continuity equation:

$$\frac{\partial H}{\partial t} + \frac{\partial Hu}{\partial x} + \frac{\partial Hv}{\partial y} = 0 \tag{1}$$

Momentum equations:

$$\frac{\partial u}{\partial t} + u\frac{\partial u}{\partial x} + v\frac{\partial u}{\partial y} = fv - g\frac{\partial \eta}{\partial x} + \frac{\tau_{wx}}{\rho H} - \frac{\tau_{bx}}{\rho H} + \varepsilon(\frac{\partial^2 u}{\partial x^2} + \frac{\partial^2 u}{\partial y^2}) \tag{2}$$

$$\frac{\partial v}{\partial t} + u\frac{\partial v}{\partial x} + v\frac{\partial v}{\partial y} = -fv - g\frac{\partial \eta}{\partial y} + \frac{\tau_{wy}}{\rho H} - \frac{\tau_{by}}{\rho H} + \varepsilon(\frac{\partial^2 v}{\partial x^2} + \frac{\partial^2 v}{\partial y^2}) \tag{3}$$

Energy equation:

$$\frac{\partial T}{\partial t} + u\frac{\partial T}{\partial x} + v\frac{\partial T}{\partial y} = -\frac{KT}{\rho C_p H} + \frac{1}{H}[\frac{\partial}{\partial x}(HE_x\frac{\partial T}{\partial x}) + \frac{\partial}{\partial y}(HE_y\frac{\partial T}{\partial y})] \tag{4}$$

where u and v are the depth averaged velocity in the horizontal axis x and y; H=h+ η is water depth, h is the depth under the calculated base level, η is the height above the same base level; ρ is water density; f is Coriolis coefficient; g is gravitational acceleration; ε is coefficient of eddy diffusion; T is super-temperature (temperature difference between heat-affected and no heat-affected water bodies); K is composite coefficient of heat emission of water surface, which includes evaporation, convection and heat radiation; C is specific heat; Ex and Ey are coefficients of heat diffusion in axis x and y; τ_{wx} and τ_{wy} are wind stresses in axis x and y; τ_{bx} and τ_{by} are bottom stresses.

Initial condition: assuming that water body in whole calculating field Ω is in static when t=0,and super-temperature is 0,e.g.

$\qquad$ u(x,y,0)=v(x,y,0)=0, η(x,y,0)=0, T(x,y,0)=0 $\qquad\qquad$ (x,y) $\in \Omega$.

Boundaries condition: dividing the boundaries over the calculating field into solid boundary $\Gamma_1$and fluid boundary Γ_2, and giving over Γ_1

$\qquad$ **v** • **n**=0, $\qquad$ gradT • **n**=0 $\qquad\qquad\qquad\qquad$ (x,y) $\in \Gamma_1$,

where **v** is the velocity vector, **n** is the unit vector of normal direction of boundaries, and allowing over Γ 2

$\qquad$ η = η$_0$(t), $\qquad$ gradT • **n**=0 $\qquad\qquad\qquad\qquad$ (x,y) $\in \Gamma_2$,

and providing discharges, velocities and heat fluxes for the outlet and intake of the plant.

2.2 Numerical Method

An operator-splitted finite element method is applied to solve the governing equations. The main features of the method are that the convection and diffusion terms of the equations are solved stage by stages over nets of irregular triangular elements. In the fore half of time step width, the convection term is computed by a retrofitted characteristic curve technique and in the later half the diffusion term is solved by a finite element technique of lumped mass. And the continuity equation is calculated by a finite volume method (Wu,1988). The advantages of the method are of the better computative stability and the faster convergence speed and the wider adaptability to various complex topographies in bays and estuaries. Thus it is fit for calculations of long-time unsteady flows (Jiang and Zhao,1992), and changing and increasing elements for the major studied region is easy to handle (Jiang and Chen,1994).

3. EXPERIMENTAL METHOD OF PHYSICAL MODEL

Strictly speaking, cooling water movement is a three-dimensional thermal density flow. Heated water drifts on the surface of cold seawater. There is temperature difference between the upper and lower layer water. Especially in the region near the outlet of cooling water, such temperature difference is extremely obvious. Exactly to simulate such three-dimensional density flow, physical model is a proper choice.

3.1 Design of Model

Cooling water models in bays or estuaries should satisfy the following hydro-thermal similarity criterions: tidal current abides by gravity similarity criterion, i.e. $(F_r)_r = 1$, where $F_r = V / \sqrt{gZ}$; heated water flow satisfies $(F_\Delta)_r = 1$ where $F_\Delta = V / \sqrt{\Delta \rho g Z / \rho}$; heated balance similarity on water surface requires $Q_r = K_r L_r^2$; and Renold number in models must exceed the critical Renold number.

For ensuring the similarity of the temperature field, the model must contain a larger sea region so as to simulate the accumulation of quantity of heat when the heated water drifts back and forth with the tidal current. Otherwise, the model geometric scale should be as small as possible for decreasing the effects of the model reduced scale and distortion. According to the aforementioned model similarity criterions, the conditions of the testing hall and the range of the sea region studied, the model scales are determined: horizontal scale Lr=800, vertical scale Zr=150, distortion ratio e=5.33, relevantly velocity scale Vr=12.25, time scale tr=65.3, discharge scale Qr=1470000, temperature difference scale T=1. Therefore, a tidal period 24 hours in prototype corresponds to 22.15 minutes in the model. The range of the sea region contained in model sees Fig.1.

3.2 Testing Method

In order to reappear the prototypal tidal current in the model, a tide generator needs to be installed on the model water boundaries. Seeing that there are more than three open boundaries, a new multi-boundaries tide generator has been designed. It consists of hundreds of miniature pumps that are divided into two groups. One group controls the tide flood and another controls the tidal fall. All pumps run on the working procedure that is designed in advance and automatically given by a computer. The preliminary conditions of open boundaries are provided by the mathematical model. Then the boundary conditions are continuously readjusted until the model has been similar to the prototype in the tidal level and the velocity field. The advantage of the multi-boundaries tide generator is that it can better run when water contains the sediment. Therefore this tide generator can be also used in sediment models affected by tide.

The tidal level, period and velocity distribution in the model are automatically measured with tidal gauges and velocity meters controlled by the computer. The collection of temperature field can be rapidly finished in about three seconds by a multipoint temperature meter which has 200 sensors. All velocity and temperature sensors are fixed on a set of lift-fall equipment, which can always keep these sensors in some regular water depth. The heated discharge is provided by a set of heating equipment of circulation which can simulate circulating water of power plants.

4. RATING OF MATHEMATICAL AND PHYSICAL MODELS

For the study of cooling water project, two large-scale synchronous hydrologic surveys in the studied seas were carried out separately in the summer and the winter of 1993.There were three tidal level stations and ten flow gauging stations in the surveys (seeing Fig.1). The observed contents contained meteorological element, tide, current, water temperature, salinity and sediment. According to the observed data, the tide in the studied

629

seas is an irregular daily type tide, but the tidal current has an irregular half-daily tide feature and presents flowing back and forth. In tidal flood, seawater flows towards northeast and flows towards southwest in tidal fall. The flowing direction is essentially parallel to the isobath of the studied seas.

Generally, it is necessary to choose some typical tidal stencils for rating mathematical and physical models and simulating the cooling water movements. The tidal stencils must be able to represent the tide and current features in the studied seas, and should be most disadvantageous to the determination of the outlet-intake schemes of the power plant. In consideration that the water temperature is higher in the summer, the tidal stencil from 14:00 on September 20 to 14:00 on September 21 is selected as an experimental and calculative condition.

The rating procedures of mathematical and physical models are almost the same, that is first checking up the tidal levels, then checking up the tidal currents in various gauging stations, and finally rating whole velocity field. The key to the examination of the similarity of the models mainly lies in the correct determination of the opening boundary conditions. For the mathematical model, we can suppose some tidal curves in the open boundaries referring to prototype observation data near the boundaries and according to the characteristics of propagation speed and deformation of shallow water wave, and finally determine them through repeatedly revising the tide and velocity fields. The open boundary conditions of the physical model are provided by the mathematical model, but the proper regulation is still necessary. Through many calculations and experiments, we do not think the models have been similar to the prototype until all three tidal curves and ten current curves in the models and prototype have essentially no

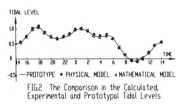

FIG.2 The Comparison in the Calculated, Experimental and Prototypal Tidal Levels

difference. One tidal curve and one current course among all gauging stations are given in Fig.2 and Fig.3. Comparing the results of the calculation and experiment with prototype data shows that the models are better similar to the prototype. Therefor they can be used for the simulations of cooling water project and thermal pollution.

5. RESULTS OF CALCULATIONS AND EXPERIMENTS

5.1 The Tidal Current Field

On the whole, water in the studied seas presents a typical reciprocating flow. The flow in far regions from the coastline is smoother, but in near coastline the flow is changeful under the influence of topography. In nearby regions where the coastline bulges obviously to sea, the velocities are largest. And there are the backflows at both sides of the bulging coast. The velocities are generally smaller in concave shore. Because the breakwave, wharf, ashery and a part of regions of the plant will be built in sea, the coastline will be changed and thus the state of flow in the regions by the plant will also be changed. Judging by the results of the mathematical and physical models after building the plant, the velocity field in the places far from the plant is hardly affected because the breakwave and ashery are essentially parallel to the direction of the original tidal current. But there are two obvious variations in waters near the coastline: one is that due to the hindrance of the breakwave, the velocity in the JH Bay decreases more obviously than before so as to be hardly seen; another is that the

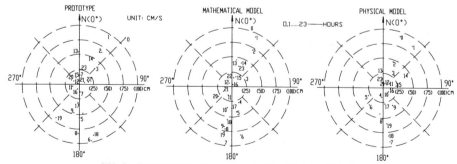

FIG.3 The Comparison in the Calculated, Experimental and Prototypal Velocities

backflows at both sides of the bulging coast become weaker. These variations will affect the flow state of heated water.

5.2 The Temperature Field

Fig.4 shows the temperature distributions at a moment in the tidal flood and the temperature curve at intake obtained by the calculations and experiments. As can be seen from the figures, the heated water discharged from the outlet of the plant is easier to diffuse with the strong tidal current, thus the regions of the higher temperature are not too large and the temperature fields of both near and far drainage schemes are essentially distributed at both sides of the outlets. Comparatively speaking, the heated water for the far drainage scheme is easy to accumulate in the bay on the north of the plant because there is the backflow and the outlet is too near the backflow; for the near drainage scheme, the heated water is easy to be transported by the tidal current because the outlet is near the mainflow. Therefore areas of the higher temperature of the far drainage scheme are a little larger than that of the near drainage scheme.

By comparison, it is not difficult to see that the temperature fields from the physical modal are clearly wider than that from the mathematical model. The main reason is that the outflow momentum of the outlet in the mathematical simulations is smaller than that in physical experiments. In addition, the areas of temperature distributions obtained by the physical model are larger than that obtained by mathematical model because the heated water in the physical model floats on the upper region of the sea, but the temperatures in the upper and lower regions of the body of water are always assumed to be the same in the mathematical model (seeing Table 1). All in all, the regions of the higher temperature are smaller according to the results of the calculations and experiments, therefore the heated water from the plant will not seriously pollute the sea environment around the plant.

5.3 The Water Temperature at the intake

For the cooling water projects, the lower the temperature of water flowing into the intake is, the better. Judging from the data in Fig.4 and Table 1,the water temperatures obtained by the calculations and experiments at the intake are all lower, so it can meet the plant's demands on the circulation water. Comparatively speaking, the average water temperature of the mathematical model in a tidal cycle is higher than

Table 1 The comparison between the calculated and experimental results

Models	Schemes	Temp.at Intake(℃)			Area of Higher Temp.(km²)			
		Max.	Min.	Ave.	>3℃	>2℃	>1℃	>.5℃
Math.Mod	Near Drainage Sch.	1.	.9	.97	1.5	4.5	19	45
	Far Drainage Sch.	.65	.57	.6	2.5	6.8	17	43
Phys.Mod.	Near Drainage Sch.	.5	.4	.45	1.9	8	24	67
	Far Drainage Sch.	.4	.3	.35	2.4	7.6	25	

that of the physical model, and the further the distance between the outlet and intake is, the bigger the difference of temperature between the calculations and experiments. The reason is that the intake installed at the lower altitude can draw the lower temperature body of water since the water temperature in the physical model is by layer in the vertical and the nearer to the outlet, the more obvious the gradient of temperature is. Therefore, the temperature curves obtained by the experiments are closer to the real water temperature at the intake in the most conditions than that from the calculations. Usually, the temperature curves obtained by calculations may be higher than real water temperature, specially when the intake is very close to outlet the difference will be quite remarkable.

The temperature curves given in Fig.4 are quite smooth and do not fluctuate with the tidal current. That present the heated water will not be directly carried by the tidal current into the inlet. In fact, the body of water in the JH Bay flows quite slowly so that it needs longer time to finish the interchange of the body of water inside and outside the Bay because the breakwave plays the better part in obstructing the tidal current and heated current into the JH Bay. Thus the heated water for both far and near drainage schemes will not diffuse into the inlet until it has be transferred back and forth by the tidal current.

6.CONCLUSION

Through discussion on the above results of the mathematical and physical models, we can gain the conclusions
a). It is necessary to carry out both mathematical and physical models for the cooling water projects of large-scale thermal power plants constructed in the coastal areas. The physical model can more accurately simulate

631

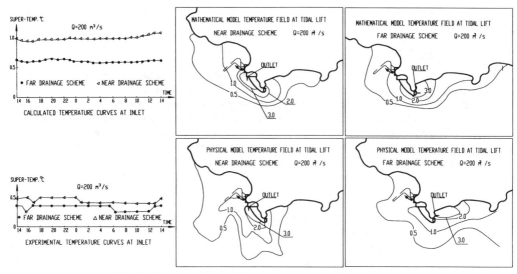

FIG.4 The Calculated and Experimental Temperature at Intake and Temperature Field

the three-dimensional flows of heated water near the outlet-intake, and the vertical distributions of temperature of water body can be gained through it. But a part of heated water from the outlet may escape from the model with the tidal current in the most situations because the regions of the model, which are limited by the area of testing hall, may not be enough large. The mathematical model can include larger waters and meteorological elements in the calculations are easier to be controlled than in the experiments, but it cannot better simulate the heated currents near the outlet-intake. Therefore, the best cooling water project should be determined by both calculations and experiments.

b). The biggest differences between the mathematical and physical models are for them to having the different initial momentum of outflow and vertical distributions of temperature. It is more difficult for the mathematical model accurately to simulate these features of heated flow. Generally, the outflow momentum of calculations is smaller than one of experiments, and it may remarkably affect the temperature fields near the outlet-intake. Thus the numerical simulations are suitable to consider the temperature distributions of the regions far from the outlet.

c). In the most situations, the calculated temperature curves of water flowing into the inlet are slightly higher than the experimental those, but the calculated areas of waters covered by higher temperature are usually smaller than the experimental those. That obviously tallies with the practical conditions of the movement of cooling water. Generally speaking, the further from the outlet the intake is, the closer the calculated and experimental temperature curves are. In practically predicting the temperature curves at the inlet and the areas of thermal pollution of outflow of plants, It is necessary to take a comprehensive view of the results of the mathematical and physical models.

d). According to the calculated and experimental data, we consider that to install $8 \times 600MW$ capacity in the HL Plant will be feasible for the cooling water project. The temperature of the cooling water of the plant and the areas of the thermal pollution on the studied regions for both the near and far drainage schemes will be able to fulfil the requirements. Comparatively speaking, the near drainage scheme is the better one since its effluent pipes are shorter and more economical.

7.REFERENCES

Jiang Wei and Zhao Wenqian (1992). "Two-dimensional numerical modeling of oil pollution in coastal zone", *Proc. of Sec. Inter. Offshore and Polar Eng. Conf.*, San Francisco, USA.

Jiang Wei and Chen zhuoying (1994). "A study of the numerical simulation of the cooling water project of DG Power Plant" (Chinese), *Proc. of Third National Environmental Hydraulics Conf.*, Xian, China.

Wu Jianghang (1988). *The theory, method and application of computational fluid mechanics* (Chinese), Science Press, Peking.

Environmental Hydraulics, Lee, Jayawardena & Wang (eds) © 1999 Balkema, Rotterdam, ISBN 90 5809 035 3

Numerical simulation for eutrophication in lakes with cooling water circulation

Chen Kaiqi & Li Pingheng
Center of Environment Impact Assessment and Research, IWHR, Beijing, China

ABSTRACT: The paper presents a mathematical approach in assessing the influence of eutrophication of lakes or shallow reservoirs with cooling water circulation. The algorithm of modified hybrid method of fractional steps is used. Based on the correct simulation of flow and temperature fields, the water quality of the relevant water region is predicted with single step one level ecological dynamic model and the influence of alga growth processes resulted from the temperature increment is analyzed. The method has been well tested and used in the environmental assessment and engineering design of a large power project.

1. INTRODUCTION

Temperature has a profound effect on all forms of matter and governs the rate and mode of chemical reactions. It is therefor an important environmental factor governing biological process. The release of artificial heat to the water environment, especially to the water environment with closed boundary such as lake and shallow reservoir will induce in various degrees the negative response to the water quality, including its ramification on economic utilization of the water bodies, other water uses and esthetic values.

Cooling water, functioning as the cold source in the electric generation system, is an absolute requisite for the operation of thermal power plants. The cooling water required with the terminal temperature rise of 8-12°C is characterized by the fact that the heat energy to be dissipated is larger than the total energy generated by the plant. Such large amount of heat injected into the lake will change the thermal load of the receiving water body. Besides that ,one of the features of the cooling water system is the coexistence of the intake and the outlet with the same discharge. The effect of the circulating water flow would redistribute the velocity and temperature pattern of the water region and induce further complicity to the change of biological system. The main purpose of the paper is to demonstrate a method to predict the eutrophication in lake or shallow reservoir with the addition of heated effluence from large thermal plant. The proposed method has been used with its calculated values well fit with the data observed on site.

2. MATHEMATICAL MODEL

2.1 Shallow Water Circulation Model in Lakes and Reservoirs

For lakes and shallow reservoirs, the depth of water is not very deep compared with the other length dimensions of the lake. It is reasonable to describe the distribution of flow and temperature fields using depth-averaged viscous dynamic equations as follows:

$$\frac{\partial \xi}{\partial t} + \frac{\partial (H u)}{\partial x} + \frac{\partial (H v)}{\partial y} = 0 \tag{1}$$

$$\frac{\partial u}{\partial t} + u \frac{\partial u}{\partial x} + v \frac{\partial u}{\partial y} = -g \frac{\partial \xi}{\partial x} + f v - \frac{g}{c^2} \frac{\sqrt{u^2 + v^2}}{H} u + \frac{\tau_{sx}}{\rho H} + \varepsilon_x (\frac{\partial^2 u}{\partial x^2} + \frac{\partial^2 u}{\partial y^2}) \tag{2}$$

$$\frac{\partial v}{\partial t} + u\frac{\partial v}{\partial x} + v\frac{\partial v}{\partial y} = -g\frac{\partial \xi}{\partial x} - f u - \frac{g}{c^2}\frac{\sqrt{u^2+v^2}}{H}v + \frac{\tau_{sy}}{\rho H} + \varepsilon_y\left(\frac{\partial^2 v}{\partial x^2} + \frac{\partial^2 v}{\partial y^2}\right) \tag{3}$$

$$\frac{\partial \theta}{\partial t} + u\frac{\partial \theta}{\partial x} + v\frac{\partial \theta}{\partial y} = \frac{\partial}{\partial x}\left(K_x\frac{\partial \theta}{\partial x}\right) + \frac{\partial}{\partial y}\left(K_y\frac{\partial \theta}{\partial y}\right) - \frac{K_s\theta}{\rho C_p H} \tag{4}$$

with boundary conditions: $\vec{v}\cdot\vec{n} = 0$, $\partial\theta / \partial\vec{n} = 0$ ($\vec{n}$ is the direction perpendicular to the boundary) Among (1)-(4), t = time; u ,v = fluid velocity components for the x ,y direction respectively; ξ= height of disturbed water surface; H= $\xi+H_0$ = depth of water; H_0 = initial depth of water; $f = 2\omega\sin\varphi$ = Coriolis force; g =gravitational acceleration; $C=n^{-1}(H)^{1/6}$=Chezy coefficient; ρ = water density; C_p = water specific heat; $\theta = T - T_\infty$ = extra-temperature of water; T = water temperature; T_∞= natural water temperature; symbol $\varepsilon_x, \varepsilon_y$; τ_{sx}, τ_{sy}; K_x, K_y= turbulent diffusion coefficient; inner stress of water surface; heat diffusion coefficient in the x ,y direction respectively.

2.2 Ecological dynamic model

The key of ecological dynamic model is the accurate simulation of the process of ecological dynamics. This should be a system model with fractional steps, multiple levels. Considering observation data on site are very limited when applied into practical engineering, the model should be simplified to be single step-one level, relaxing the demand of choosing and rating parameters. Not only the affect of nutrients migration, but also the influences of growth, mortality, respiration, grab have been taken into account in this model and quantitative analysis for some environmental factors controlling algae and nutrients could be made .(Chen,1995)

The transfer and change of algae (Chl-a), total nitrogen (TN), total phosphorus (TP) can be described as following on the basis of mass conservation principle:(Chen,1994;Yu,1989;Sha and Liao, el al. 1994)

$$\frac{\partial \phi}{\partial t} + u\frac{\partial \phi}{\partial x} + v\frac{\partial \phi}{\partial y} = \frac{\partial}{\partial x}\left(\Gamma_x\frac{\partial \phi}{\partial x}\right) + \frac{\partial}{\partial y}\left(\Gamma_y\frac{\partial \phi}{\partial y}\right) + S_\phi \tag{5}$$

in which:

Equation	ϕ	Γ_x , Γ_y	S_Φ
Chl-a	C_{chl-a}	$\Gamma_x= \alpha\Delta u\delta$ $\Gamma_y= \alpha\Delta v\delta$	$\mu C_{chl-a}-(r+m+G+S)C_{chl-a}$
TN	C_{TN}	$\Gamma_x= \alpha\Delta u\delta$ $\Gamma_y= \alpha\Delta v\delta$	$(D_p- \mu)PA_{np}-0.35JC_{TN}$
TP	C_{TP}	$\Gamma_x= \alpha\Delta u\delta$ $\Gamma_y= \alpha\Delta vs\delta$	$(D_p- \mu)PA_{pp}-I_1C_{TP}+S_p$

in which:

C_{chl-a}: concentration of Chl-a;

μ: growth rate of algae;

r: respiration rate;

m: mortality decomposition rate;

G: grab rate of phytoplankton capturing algae;

S: settling rate;

C_{TN}: concentration of TN

D_p : mortality rate of algae;

P: concentration of phytoplankton

A_{np}: ratio of nitrogen to phosphorus in algae;

J: denitrification rate constant;

C_{TP}: concentration of TP;

A_{pp}: ratio of phosphorus to carbon in phytoplankton;

I_1: settling rate of phosphorus;

S_p : release rate of sediment;

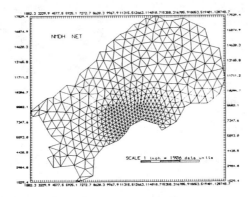

Fig.1   The computation meshes

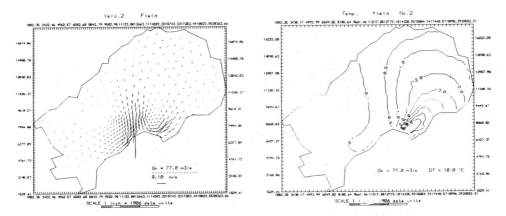

Fig.2-(a)(b)   The distribution graph of flow field and temperature field

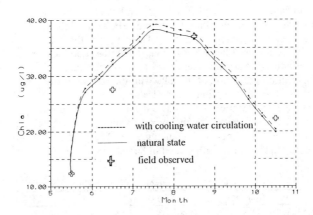

Fig.3   The effect of cooling water on Chl-a

635

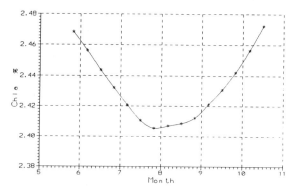

Fig..4   The development of eutrophication in spring and autumn is more

sensitive to cooling water

3.   NUMERICAL SIMULATION METHOD

Noticing that lakes affected by cooling water circulation will arrive at dynamic equilibrium state after a period of time, in order to save calculation time and convenient in engineering application, the flow and temperature fields of the water region are solved first with no consideration of its concentration distribution ,the latter is then solved through the substance transport equation and the variant biochemical reaction functions. As the shape of lakes and reservoirs is usually irregular, finite element analysis method with mesh automatic generation technique is used .The finite element algorithm of modification hybrid method of fractional steps is applied .Before the calculation of the concentration of algae (Chl-a), total nitrogen (TN) and total phosphorus (TP), some environmental factors in biochemical reaction functions need to be rated. In case of no such data provided , such parameters could be adjusted with the application of Runge-Kutte method referencing the spot observation data.(Chen, et al. 1994)

4. PRACTICAL APPLICATION

The proposed method has been used in the eutrophication predication for Daihai lake with cooling water of 77.0 m^3/s discharged from Daihai power plant. The length and width of the lake are about 20 and 10km with average water depth about 8m. The computation meshes are shown in figure 1. Figure 2 is the distribution graph of flow field and temperature (extra-temperature),Table 1 gives the averaged values of Chl-a, TN, TP in lake in natural state with no artificial heated effluence. The rather good agreement of the predicted value with that observed in site indicates the parameters used are basically reasonable. The same parameters are then applied in the eutrophication prediction of the lake with the cooling water circulation. Fig3,4 reveal the comparison of Chl-a, with and without cooling water discharges. It is obvious that the impoundment of the cooling water do enhance the eutrophication process also the difference seems to be not so significant. This is due to the fact that the comparison is made for the average value of the whole lake . For the water region nearer to the heat entrance, the difference will be much greater. The predicted patterns of Chl-a distribution for different month from May to October are illustrated in Fig.5.

Table 1   Averaged values of Chl-a, TN, TP

Month	Chl-a (μg / l)		TN (mg / l)		TP (mg / l)	
	Predicted	Observed	Predicted	Observed	Predicted	Observed
June	32.88	27.61	1.650	1.652	0.058	0.059
July	39.19	None	1.972	None	0.159	None
August	37.60	37.84	2.197	2.276	0.183	0.182
September	29.85	None	1.744	None	0.126	None
October	20.40	22.32	1.212	1.066	0.076	0.048

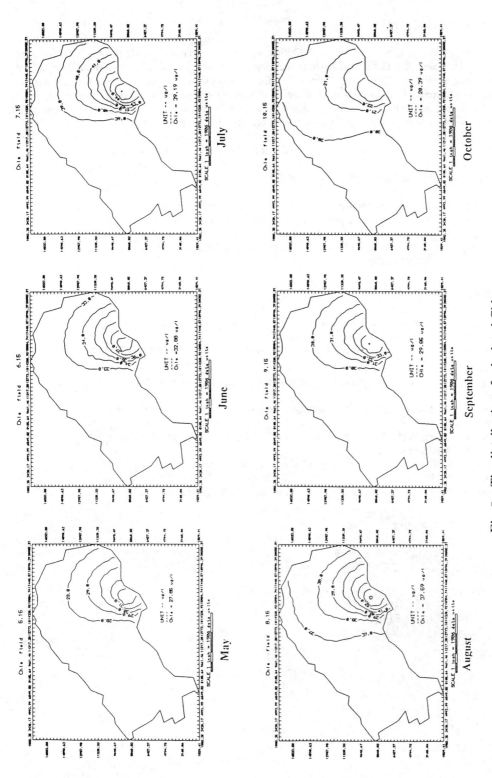

Fig.5   The distribution of calculated Chl-a
         in the whole lake from May to October.

637

5. CONCLUSIONS

As natural ecosystem is highly dynamic, exhibiting significant variations on daily or cycles with other disturbance superposed. It is difficult to separate accurately the effect of artificial heat addition from naturally occurring changes. However ,for cooling lakes where the amount of heat effluencted to the surrounding water shares not a small part of the thermal load of the lake, it is reasonable and more appropriate to assess the eutrophication state in a better way . The presented numerical simulation of the eutrophication conducted by solving the substance transport equation together with verdant biochemical reaction equations on the basis of temperature distribution of the lake obtained from modeling of the cooling water circulation could better reflect the eutrophication dynamics. It is not only good for getting a better understanding of the lakes environmental behavior, but also useful for planning the position of the cooling water structures which in turn directly govern the flow and temperature patters of the lake.

The numerical simulation of eutrophcation in Daihai lake caused by cooling water shows that the cooling water from power plants speeds up the eutrophication development at a certain level, but its affect is localized and limited on the whole. Only when the temperature-rising affected area is large enough and the time of affect is long enough may the cooling water make an obvious influence to the accelerated development of eutrophication in lakes and reservoirs.

6. REFERENCES

Chen Kaiqi,(1995), The Numerical Simulation of Flow Field and Thermal Pollution for Water Environmental Impact Assessment, Zejiang University.

Chen Kaiqi,(1994),Water Environmental Impact Assessment Report of Daihai Coal-fired Power Plant in Inner Mongolia , IWHR.

Ye Changming,(1989), The Theory and Control of Water Pollution , Academic Book Press.

Kinsberg.N,(1987), Numerical Simulation of Water Environment, Chinese Architectural Industry Press.

Peng Jinxin,Chen Huijun etc.,(1984),Water Quality Eutrophication and Control, Chinese Environmental Sciences Press.

Sha Huiwen, Liao Wengen etc.,(1994), Bajiao Lake Water Environmental Impact Assessment, IWHR.

Torgenson, (1983), Application of Ecology in Environmental Management.

3.3 Environmental impact studies/Three Gorges project

Environmental Hydraulics, Lee, Jayawardena & Wang (eds) © 1999 Balkema, Rotterdam, ISBN 90 5809 035 3

Hong Kong's worst red tide

M. D. Dickman
Department of Ecology and Biodiversity, The University of Hong Kong, China

Abstract
Red tide dinoflagellates often concentrate along vertical convergences produced by wind generated or tide generated currents. These vertical fronts may contain 100 times more dinoflagellates than are found in nearby open waters. At times these fronts are further concentrated at the leward end of small bays. On 15 April, 1998, Hong Kong's Chief executive, Tung Chee-hwa, visited some of the fish farms on outlying Lamma Island to see, first hand, the damage to fish held in cages in the Sok Kwu Wan mariculture area. Phytoplankton samples taken near these enclosures on the same day by the Dept. of Agriculture and Fisheries contained over one million cells per litre of *Gymndinium mikimotoi*. Samples taken on the same day by University of Hong Kong researchers were from the nearby Mo Tat Wan area located near the opening of the same bay. These samples contained *Gymndinium mikimotoi* at densities of $107,000\pm18,600$ cells/L. As a result of these lower dinoflagellate densities at Mo Tat Wan, many Lamma Island fish farmers moved their enclosures from Sok Kwu Wan to the deeper waters of Mo Tat Wan in order to avoid the dense aggregations of *G. mikimotoi* concentrated at the leeward end of the bay. At high densities, the *G. mikimotoi* formed dense clumps of individuals which produced a sticky mucus. When the mucus touched the gill filaments of fish or shellfish it coated them resulting in asphyxiation. On 21 April, strong winds disbursed the red tide. The Hong Kong press reported that the worst fish kill in Hong Kong's history had wiped out 1,500 tonnes of fish stocks worth over HK$200 million ($25.8 million U.S.).

1. Introduction
A dinoflagellate referred to as *Gymndinium mikimotoi* (formerly referred to as *Gyrodinium aureolum)* produced copious amounts of a sticky mucus during its peak abundance in March and April of 1998. The mucus coated the gills of fish and shellfish at many locations in Hong Kong. The present report is the first to document temporal changes in the population density of this dinoflagellate during bloom development in Hong Kong at a site near the Lamma Island fish farms. Phytoplankton samples were taken at Mo Tat Wan (Fig. 1) near Lamma Island each week before, during and after the *G. mikimotoi* red tide. Changes in the abundance of the dominant species as well as some of the subdominant species as the red tide developed were examined and the results plotted. The study also attempts to exploain why mucilage production by *G. mikimotoi* might reduce its chances of being consumed by predators.

2. Methods
Phytoplankton samples were taken by passing 10-100 litres of water through a 10μm mesh plankton net and concentrating it to a volume of 50-100 ml. Lugol's IKI solution was added as a preservative. An Olympus research microscope equipped with Nomarski interference optics was used to count the sample. All organisms under the coverslip were counted.

A minimum of 500 individuals were counted for each sample. The density of each algal species was calculated using the formula:

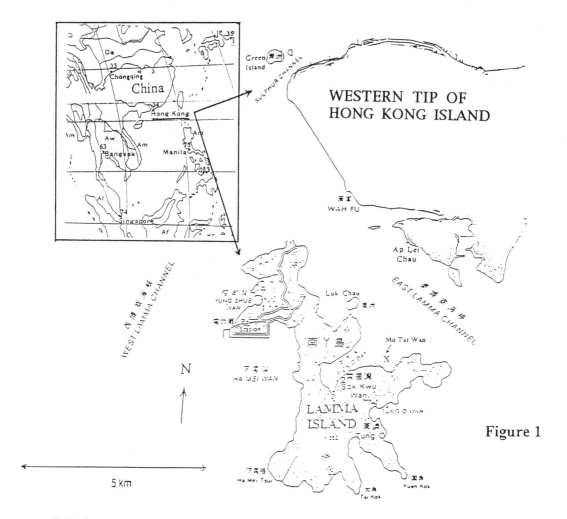

Figure 1

$$n = \frac{(a)(c)}{L}$$

n = the number of plankters per litre of original water
a = the estimated number of plankters on each slide
c = the volume of the concentrate
L = volume of original water

Only diatoms and dinoflagellates are reported here as these were the most abundant phyla.
Species from these two phyla were identified using the references of Yamaji (1984); Fukuyo *et al.* (1990); Jin *et al.* (1992); Taylor *et al.* (1995) and Tomas (1995).
Samples were also analysed using a scanning electron microscope (SEM) in order to determine the presence of harmful species of diatoms and dinoflagellates (*e.g. Pseudonitzschia* species).

3. Results and Discussion

Phytoplankton was sampled weekly at Mo Tat Wan and the results of these samples are plotted for the period 18 March to 21 April, 1998. Changes in the total abundance of dinoflagellates (Fig. 2a) total diatom abundance (Fig. 2b), *Gymndinium mikimotoi* (fig. 2c) and *Prorocentrum sigmoides* abundance (Fig. 2d) are plotted.

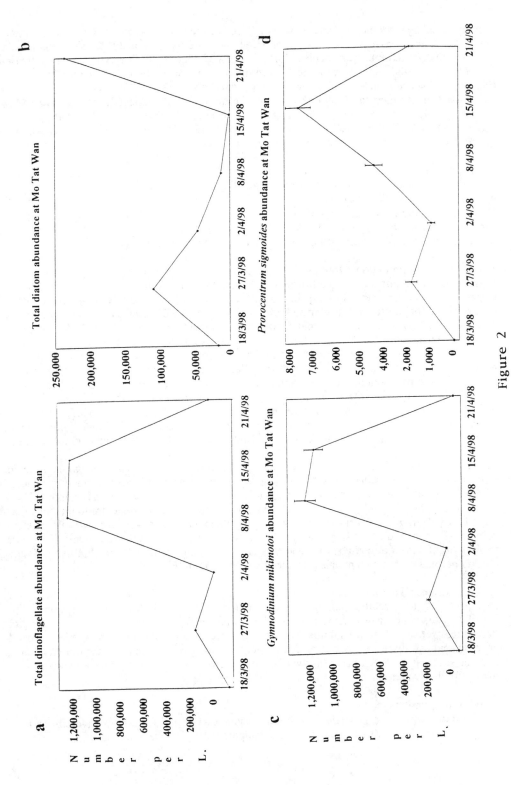

Figure 2

643

In mid March, *Gymndinium mikimotoi* was observed at slightly less than 2,600 cells per litre and the phytoplankton assemblage was dominated by diatoms, primarily *Bacillaria paradoxa* (2,300/L) and *Pseudonitzschia* species (1,450/L). By 27 March, the *G. mikimotoi* had increased ten fold to 23,000 cells/L and the dominant diatoms were now *Chaetoceros curvisetum* (90,000/L), *C. costatum* (15,000/L) and *Pseudonitzschia* species (1,380/L). By 2 April, the *G. mikimotoi* had declined to 10,000 cells/L and the dominant diatoms, *Chaetoceros curvisetum* (10,300 per litre), *C. costatum* (3,500 per litre) and *Pseudonitzschia* species (14,000/L) had also declined. On 8 April, the *G. mikimotoi* had increased to 107,000 cells/L, a ten fold increase from the previous week and the dominant diatoms, *Chaetoceros curvisetum* (1,800 per litre), *C. costatum* (rare) and *Pseudonitzschia* species (6,000/L) continued their decline. A second dinoflagellate, *Prorocentrum sigmoides*, reached 4,400 cells/L (Fig. 2d). Large colonies of *Gymndinium mikimotoi* enveloped in mucilage (Fig 3 c,d&e) were common. The following week, *Gymndinium mikimotoi* remained relatively unchanged at 102,400 ($\pm$ 6,500 cells/L, n = 3). News stories about the devastation of the fish in the Lamma Island area fish farms continued with some fish farms claiming up to 90% of their fish had been asphyxiated by the dense slime (mucilage) that had coated the gills of their fish turning the gills white.

4. Mucus Production and its Impacts

Most copepod nauplii pass through 6 instars before their transformation into copepodites. The Copepodites look and behave much like adult copepods but are sexually immature. After six instars the copepodite is transformed into an adult (McConnaughey 1974). Diatoms and dinoflagellates constitute the principal food of most planktonic copepods (*ibid*). The copepod's second antennae vibrate at 600 to 2,600 times per minute producing a vortex along both sides of the body. Food particles are concentrated in the center of these two vortices and come in contact with the myriad fine setae on the second maxillae. Once trapped in these setal hairs the food particles are scraped from the second maxillae by the endites of the first maxillae and passed to the mandibles and the mouth (McConnaughey 1974). If algal mucus coats the setal hairs it becomes difficult for the endites to scrape them clean. I have observed copepods which stop filter feeding when they are near mucus covered objects. Presumably, this is done to avoid fouling their maxillae with mucus. Thus mucus production may enhance the survival of mucus secreting dinoflagellates.

Copepods were common in the water column at the time of the *Gymndinium mikimotoi* bloom. Copepodites and adult copepods were rare relative to the copepod nauplii. The mucus produced by *Gymndinium mikimotoi* was often observed attached to the sides of the copepod nauplii (Figs. 3 a & b). This gave rise to the hypothesis that mucilage production by some dinoflagellates makes them less vulnerable to crustacean predation. Mucilage associated with *Gymndinium mikimotoi* colonies (Fig. 3c) and fibres (Figs. 3 d & e) indicate its sticky characteristics. It appears to attach to anything it comes in contact with and it is easy to imagine how fish and shellfish gills might quickly become coated with mucus produced by *G. mikimotoi* (Fig. 3 f).

5. Wind Induced Turbulence

In late April, the weather changed and strong winds mixed the waters dispersing the dinoflagellates which were previously abundant at the surface and rare below 3 m where light intensities were too low for photosynthesis. The role of the wind in dinoflagellate red tide development and dispersal is often critical. After the strong winds dispersed the dinoflagellates, diatoms sucha as *Skeletonema costatum* and *Chaetoceros* species suddenly regained dominance in the turbulent water.

6. Acknowledgments

This research was made possible with funding from CRCG and from the University of Hong Kong. I am grateful to Ms Manna Wan for her support and assistance in this study.

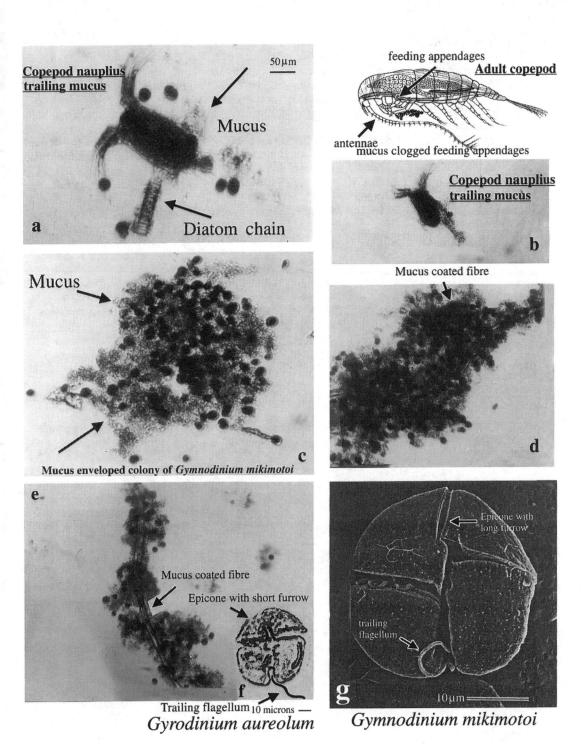

a Copepod nauplius trailing mucus — 50μm — Mucus — Diatom chain

b feeding appendages — Adult copepod — antennae — mucus clogged feeding appendages — Copepod nauplius trailing mucus

c Mucus — Mucus enveloped colony of *Gymnodinium mikimotoi*

d Mucus coated fibre

e, f Mucus coated fibre — Epicone with short furrow — Trailing flagellum 10 microns — *Gyrodinium aureolum*

g Epicone with long furrow — trailing flagellum — 10μm — *Gymnodinium mikimotoi*

Figure 3

7. References Cited

Fukuyo, Y., Takano, H., Chihara, M. & Matsuoka, K. 1990. *Red Tide Organisms in Japan - An illustrated taxonomic guide*. Uchida rokakuho, Tokyo, Japan, 430pp.

Jin, D. X., Cheng, Z. D., Lin, J. M. & Liu, S. C. 1992. *Marine Benthic Diatoms in China (II)*. China Ocean Press, Beijing, China, 437pp (in Chinese).

McConnaughey, B. H. 1974. *Introduction to Marine Biology*. C. V. Mosby Co. Saint Louis, MO. USA 544pp.

Taylor, F, J. R., Fukuyo, Y. & Larsen, J. 1995. Taxonomy of harmful dinoflagellates. In: *Manual on Harmful Marine Microalgae* (G. M. Hallegraeff, D. M. Anderson & A. D. Cemberlla, eds). IOC Manual and Guides No. 33, pp. 283-316. UNESCO.

Tomas, C. R. (ed.) 1995. *Identifying Marine Diatoms and Dinoflagellates*, Academic Press, San Diego, Calif. U.S.A. 598pp.

Yamaji, I. 1984. *Illustrations of the Marine Plankton of Japan*. Hoikusha, Osaka, Japan. 540pp

Environmental Hydraulics, Lee, Jayawardena & Wang (eds) © 1999 Balkema, Rotterdam, ISBN 90 5809 035 3

Fisheries at Shimonoseki area and waste management

T. Kano
Department of Marine Science and Technology, Tokyo University of Fisheries, Japan

K. Torii
Kyoto University, Japan

H. Yamakawa
Department of Bio-Science, Tokyo University of Fisheries, Japan

ABSTRACT: Koya river estuary at Shimonoseki city had brought good income to seaweed farmers by seaweed and short-necked clam. But successive withdrawal from seaweed aqua-culture continues in recent 20 years, and harvest of short-necked clam decreases quickly. The decline is caused by change of aqua-environment, especially quantity and quality of river water. Moreover, waste management had been in operation from 1995 at reclaimed land beside the mouth of Kanda river. The outfall discharge from the waste management is faced to sea and it directly flow into the beach at ebb tide. Discharge decrease of Koya river between 1985 and 1997 is illustrated by the comparison of two satellite images taken by Landsat TM. The sea water exchange of this estuary is very low and river water contains lot of fine suspended sediment by the actual site observation and hearing from farmers.

1. INTRODUCTION

Location of area to be studied is shown in Fig.1. Shimonoseki City situates at the western end of Honsyu island(main island of Japan) and faces to Kyusyu island across Kanmon strait which being the dangerous path of the navigation due to over 7 knots of tidal current. Photo. 1 is a satellite image of Shimonoseki area

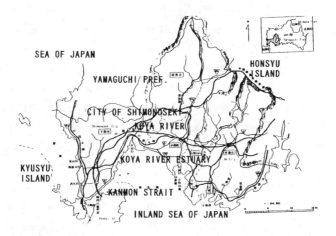

Fig. 1 Yamaguchi Prefecture and Shimonoseki City

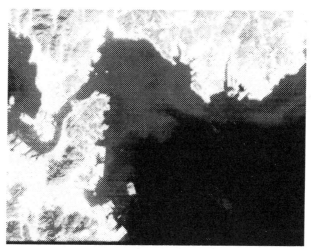

Photo. 1. Satellite image from JERS_1 on 15 Nov. 1995

on 15 Nov. 1995. The photograph expresses large eddy circulation in front of Koya river estuary. This implies the tidal current from and into Kanmon strait formulates complex patterns of flow at the both ends of the strait. And its implies that the flow in the strait acts very important rolls to water quality of Koya river estuary. Authors are going to reproduce this eddy circulation in computer, but this is not a main subject of them at present.

2. HARVEST OF SEAWEED AND SHORT-NECKED CLAM AT KOYA RIVER ESTUARY

Seaweed and Short-necked clam grows in similar environments. That is, fresh water inflow and sandy tidal land are necessary. And 4 hours drying up of culture bed when it being ebb especially for seaweed aquaculture. The farmer of seaweed culture decreasing continuously during 1970 to 1998. Over 2000 farmers worked at the beach at ebb tide at Koya river estuary, only some tens of farmers engaged in seaweed aquculture there at present. And harvest of short-necked clam also shows the same tendency as the numbers of seaweed culture farmer. Especially, there is no harvest at all in recent 3 years. It should be considered that the environments of the basin might be changed drastically. One should be fine suspension sediment due to the earth works of reservoir construction and relative works of river training at Koya river, and the other should be the outfall of waste management operating at the shore close to the mouth of Koya river. At first, authors considered that the affection of waste management could be dominate, and they got other satellite images than Photo. 1 before and after the operation of waste management starts. The images are taken by the Landsat TM on 1985 before waste management begin to work and on 1997 after it have been worked. Surface width of Koya river is evidently became narrower in the photograph taken in 1997 than that of in 1985. It expresses the decrease of river discharge. Authors got flood routing data of Koya river after 1990 when the reservoir upstream of Koya river completed, but they do not do the discharge observation before 1990 because Koya river is not the main river to execute river training by national government. So that they

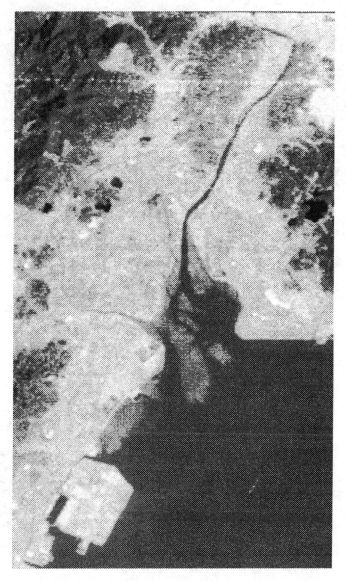

Photo. 2. Landsat TM image of Koya river estuary on 2 May 1985

can not compare the discharge decrease with actual discharge data. Existence and the shape of shore at ebb tide at Koya river estuary observable by the satellite image. The large eddy current recognized by Photo. 1 does not appear in Photos. 2 and 3. It might be the difference of time and used band range to take the images. The waste management had been situated at the left corner of shore line. Designed outfall dichare is $51000\,m^3\,/\,day$ [City of Shimonoseki, 1983], but sewer system connecting to the waste management is working 6.7% at present. Then outfall discharge is $1500\,m^3\,/\,day$. So that, affection of outfall discharge at present is not so serious at present. Farmers said that the rivers which flow into the estuary transport lot of fine suspended sediment and it

Photo. 3. Landsat TM image of Koya river estuary on 1 Apr. 1997

suffocates short-necked clam andreduces the quality of seaweed. As to seaweed, they improved the way of culture and resulting amount of seaweed harvest increased, but the quality of seaweed still reducing. It should accelerates to abandon farming. Photo. 4 indicates the boundary of sea water and river water when it is flood tide at shore of waste management. It is observable that almost horizontal thick white line. It is the boundary of sea water(upper part) and river water(lower part). Comparing water color of upper and lower part, upper sea water is blue, and lower river water colored light braown. River water colored by the fine suspension sediment. Upstream of the river they executing river training and intake works for irrigation. This sediment might be originated from these earth works. Though it is necessary to be supplied sediment from the upstream region of

Photo. 4 sea and river water boundary at Koya river estuary when it is flood tide

rivers, sandy sediment should be necessary for the culture of seaweed and short-necked clam. It is undesirable shortage of river water as if it being to be drought, because it should bring shortage of nutrients to bio-resources at the shore in ebb tide and fine suspension sediment to suffocate also bio-resources there.

3. WATER QUALITY OF OUTFALL FROM THE WASTE MANAGEMENT

The way of sewage treatment at the waste management is activated mud method. Quality of outfall designed

Photo. 5 Water color in front of the outfall

less than $20\,mg/l$ of BOD and less than $70\,mg/l$ of SS. Quantity of outfall at present is $17.4\,l/s$, that is, 6.7% of designed quantity after completion, it gives not so big impact to the environments. Discharge from the waste management is much more clean in water color as shown in Photo. 5. Outfall is equipped in lower side of the Photo. 5. Then, it is observable that lower water being transparent but upper being turbid. As mentioned already above, amount of outfall being small, it might not affect much to the environments at present, precise monitor of the water quality of outfall and environments should become necessary. They executed some laboratory tests to investigate the affection of treated sewage to seaweed and short-necked clam. According to tests, seaweed in sea water mixed with treated sewage is better and worth in its growth and color depending on chlority concentration. And affection to short-necked clam is generally worth existence of treated sewage in the environments.

CONCLUSIONS

River discharge of Koya river in Shimonoseki city decreased obviously by the satellite images, comparing that of 1985 and 1997. Though authors collected satellite images in order to clear the affection of outfall from waste management, discharge decrease is recognized. It should be resulted shortage of sediment transport capacity and the river cease to sand supply to shore in ebb tide. And this might brought desolation of aqua-culture at the shore in ebb tide at Koya river estuary. And authors found large scale eddy tidal current in Koya river estuary unexpectedly by satellite image. Farmers at the basin did not recognized the eddy in the estuary in their farming works, it could bring some influence to the aqua-environments of this area. The reproduction of this eddy in the computer is going to now. The silt protector sheets or any other considerations should be necessary in order to cut off artificial sediment from earth works in the river, such as the constructions of Kansai International Airport in Japan and New Hong Kong International Airport in Hong Kong to preserve fisheries.

REFERENCE

Sewage system construction bureau ed. (1983) "Report of environmental assessment for new waste manegement at Shimonoseki City" City of Shimonoseki, Japan(in Japanese)

Environmental Hydraulics, Lee, Jayawardena & Wang (eds) © 1999 Balkema, Rotterdam, ISBN 90 5809 035 3

Impact of quarrying coarse gravel from river channel

Peng Junshan
Gezhouba Hydroelectric Station, Hubei, China

ABSTRACT: The construction of the Gezhouba Project consumed a total concrete volume of 11.04million m^3, in which the aggregates are sand and gravel from the river bed . To meet the needs of the concrete production and other aspects, a large amount of sand and gravel was quarried in downstream river stretch from Zhenchuanmen to Huyatan during the construction of the Project, with a total quarrying amount of 33million m^3. As a consequence, the wet season water level has been lowered evidently in the following years. For the discharge of 4 000 m^3/s, the water level at the Yichang station has dropped. from 39.69 m to 38.66 m. The minimum design navigable water depth of the shiplocks and the downstream approach channels can only meet the current navigation conditions, but the design water depth requirement of 4.5 m.

KEY WORDS: River bed   Excavation   Coarse gravel   Drop of water level

1. INTRODUCTION

The Yangtze river is the largest river in China and ranks third in the world. It is a key water way from the east to the west of China. The Gezhouba project, the first one built on the mainstream of the Yangtze river, is mainly composed of the spillways, power stations and navigation structures. The navigation structures include the No. 1 shiplock located at the main river, the No. 2 and No. 3 shiplocks located at the third river channel, and corresponding the upstream and downstream approach channels as well. The minimum design navigable discharge is 3 200 m^3/s, correspondingly, with a downstream water level of 39.00m at the Yichang station. The minimum navigable water depth along the third river channel and the main river, respectively, is 4.5 m with an elevation of 34.5 m for the bottom of No. 2 shiplock and the downstream approach channel, and 5.5 m with an elevation of 33.5 m for the bottom of No. 1 shiplock and the downstream approach channel.

The construction of the Gezhouba project was started on Dec. 30, 1970. The river closure was accomplished on Jan. 4, 1981, the impounding of the reservoir started on May 23 and the first-phase works were put into operation and the construction of the second-phase works was commenced. The second-phase cofferdams were removed in Dec.1985 and the turbine-generator No. 8 was put into service. The discharging and flushing started on July 16, 1988 through the approach channels along the main river. The trial navigation started in November. From then on , the Gezhouba was put fully into operation.

The Gezhouba has on enormous size, with a total concrete volume of 11.04million m^3. The aggregates in concrete came from the course sand and gravel on the downstream river bed. According to an investigation, the sand and gravel quarryed from the Gezhouba site to the Jiangkou town since 1970's with a length of 109.2 km. The small-size quarrying was performed, by the local people, along the flood plains or shallow-water areas between Huyatan and the Jiangkou town, with a less impact on the river channel.

2. DESCRIPTION OF DOWNSTREAM REACH OF GEZHOUBA

With a length of 22.71 km, the downstream reach between the Gezhouba dam site and Huyatan is a transition stretch from the mountain river to the plain river, the section of the Yichang station is a bend with a width of 800 m and an apex of the concave shore near Zhuanqiaohe. The river channel from Yichang to Huyatan is slightly straight in wet seasons, with a width of 800 m for the upstream   of Yanzhiba and of 900 m ~ 1 400 m for the downstream of it. (See Fig.1). Due to the beach near the Yichang city the Yanzhiba bar and the Linjiangxi beach of jigsaw pattern, there is still a bend in dry season. There exist the deep pit alternating with the high bed along the main thalweg. The low bed is located against the beach and the high bed is the transition zone between deep pits. The transition zones located both the ends of the Yanzhiba bar are the main parts of this stretch.

The banks downstream of the dam are formed mainly by the steep and solid rock mountain at the right side, and the protection slope for the Yichang city or hills at the left side, with a less variability. However, the river bed has a complicated composition, including rock zones, sand and gravel zones, moderate and fine sand zones, and so on. Through investigation within the area between the dam site and the downstream end of the Yanzhiba bar, the sand and gravel zones account for 64% of the total area, the moderate and fine sand zones 27% and the rock zones 9%. Correspondingly, within the downstream area of the Yanzhiba bar, they are 38.2%, 60.4% and 1.4%, respectively. The thickness of the sand and gravel layer, in general, varies between 5 m and 15 m, with a maximum thickness of 36 m ~ 55 m. They have various grain sizes, d_{50}= 0.2 - 0.6 mm for the sandy river bed, d_{50}= 35 mm for the sand and gravel on the Yanzhiba bar with a maximum size of 412 mm (34% for $d_{50} > 60$ mm cobble, 44% for d_{50} =2 -60 mm for gravel and 22% for coarse sand), It is suitable for the mix of the concrete.

3. QUARRY

By investigation, the quantity of quarrying within the stretch between Yichang and the Jiangkou town was smaller before the commencement of the Gezhouba project. But since the beginning of construction of the Gezhouba project, a large-size quarrying by heavy-duty quarry boats was launched by the construction contractor within the area between the dam site and Huyatan, to meet the need for the construction. Additionally, the sand and gravel were quarried by the light equipment or hand tools within the bars or shallow water areas downstream of Huyatan, for the need of municipal works along the river. From the statistical data, the quarrying quantity of coarse sand and cobble from the river bed between Yichang and the Jiangkou town was 37.20million m^3 during the period from 1972 to 1987, equivalent to a compact volume (just the same to the following) of 27.90million m^3 at a compact coefficient of 0.75, indicating a numerous quarry (See the attached table).

The quarrying distribution along the river includes 25.2million m^3, accounting for 90.3% of the total, from the area upstream of Huyatan, with an average of 1.1million m^3/km; 2.703million m^3 from the area of 86.49 km, downstream of Huyatan, with an average of 21 600 ~ 38 800 m^3/km, concentrating mainly near the Yinzhiba bar of 6.25 km long with a total quarrying quantity of 11.25million m^3 and an average of 1.8million m^3/km. In wet season, the quarrying was performed on the beach near the boundary of main current; and in dry season, on the main river

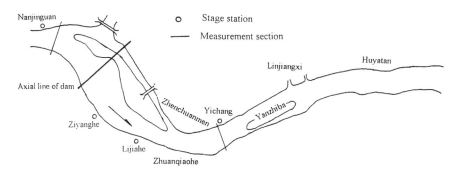

Fig.1   The stretch between Yichang and Huyatan

bed. With the progress of the Gezhou- ba project, the yearly quarrying quantity was increased following the requirements for the construction with a maximum of 4.277million m^3 in 1980, and then decreased towards the completion of the project. By the end of 1980's the yearly quarrying amount for the whole stretch was up to 1.0 million m^3.

4. EFFECTS OF QUARRYING ON THE RIVER CHANNEL

With the sediment transport, a river is outcome of the interaction between water flow and river bed. The water flow may mould the river bed, and *vice versa*, the alternating river bed may change the structure of water flow. After the water control works is completed the souring and deformation of the river bed before the dam occur, resulting from the sediment deposition in the reservoir and less incoming sediment. The change of river bed downstream of the dam was influenced by the quarrying activities.

4.1 River Bed Downcut and Water Level Lowered due to Quarrying

As shown in the underwater topographic maps measured since the completion of the Gezhouba project, the cumulative scouring volume including the quarry within the area between Zhenchuanmen and Huyatan between 1971 and 1987 was 26.768million m^3 (See table.1), with an average scouring depth of 1.9 m, correspondingly a river width of 784 m in dry season. During the same period, the compact quarrying volume was 24.76 million m^3 with an average quarrying depth of 1.76 m, equivalent to 92.5% of the scouring volume and scouring depth, respectively. Especially, at the stretch near the Yanzhiba bar, having a significant impact on the change of the river bed and water level near the downstream of the dam, the average quarrying depth was 2.30 m, for the local sections up to 5.3 m.

Table.1   Effect of Quarry and Scouring at the Area between Zhenchuanmen and Huyatan on Water Level

Year	Bulk quarrying volume /M m³	Compact quarrying volume /M m³	Scouring volume / Mm³	Water level at Yichang / m
1972	0.34	0.255	0.47	
73	0.47	0.353	7.287	39.69
74				
75	0.468	0.351	-6.904	39.64
76	0.693	0.520	1.411	39.60
77	1.169	0.877	1.083	39.52
78	2.465	1.849	5.303	39.40
79	2.684	2.013	-1.385	39.46
80	5.703	4.277		39.44
81	4.287	3.215	16.188	39.27
82	2.955	2.216	-5.916	39.18
83	4.102	3.080	0.324	39.12
84	4.216	3.162	4.072	39.09
85	1.678	1.259		39.05
86	1.772	1.329	4.835	39.04
87				38.77
88				38.73
89				38.76
90				38.66
92				38.65
95				38.67
Total	33.002	24.756	26.768	

Notes: 1. Compact quarrying volume = Bulk quarrying volume × 0.75

2. "+" for the scouring volume; "-" for the deposition volume.

655

The change of the river bed during the first- and second- phase construction also demonstrates the effect of the quarrying on the downcut of river bed. Under the natural conditions, the main river course was the main way of the transport of the coarse sand and cobble. During the first-phase construction, although the quarrying amount within the area between Zhenchuanmen and Huyatan was up to 10.49million m^3, the coarse sand and cobble could still pass down through the main river course that was still in the natural state, compensating for the quarrying and even depositing. According to the measurement, the cumulative scouring volume including the quarrying was 7.26million m^3 and the deposition volume was 3.23million m^3, with an average downcut of 0.52 m and a drop of water level of 0.25 m, mostly resulting from the quarrying activities. During the second- phase construction of the Gezhouba project, the quarrying amount from the section between Zhenchuanmen and Huyatan was 14.26million m^3, with an average quarrying depth of 1.01 m. Because of impounding of the reservoir the coarse sand and cobble were deposited mostly in the reservoir. In addition, the quarrying damaged the natural compact sand and gravel layer, aggravating the scouring. So, the total scouring amount including the quarrying was up to 19.50million m^3, with a scouring depth of 1.39 m and a drop of water level of 0.67 m. Following 1987 the quarry at the stretch was basically stopped. The downcut of the river bed suffered mainly from the flow scouring. Since operation for several years the river bed suffered from the scouring, but the water level was lowered only by 0.11 m and the river bed were kept stable. It is concluded that since the impounding of the reservoir the sediment has deposited in the reservoir and the source of coarse sand and cobble have been decreased dramatically, finally, a new scouring- depositing balance will occur when the river bed is downcut and the water level is decreased stably. As mentioned above, the excessively quarrying activities accelerated its development process and extended its impact level. By calculation, if the mean quarrying depth, at the section of about 40 km downstream of Yichang, is up to 2 m, the water level at Yichang station may be reduced by 0.78 m for the river discharge of 5 010 m^3/s.

If the drop of water level is calculated according to the quarry amount and scouring volume, separately, when drop of water level is 1.03 m, 0.74 m resulted from the quarrying activities and 0.29 m came to the scouring after the impounding of the reservoir. Thus it can be seen that the downcut of river bed and dropping of water level downstream of the Gezhouba project mainly result from the excessively quarrying on the river bed.

4.2 Effects from Downcut of River Bed

The minimum design navigable water depth is 5.5 m and 4.5 m, respectively, in the downstream approach channels for the main course and the third river channel at the Gezhouba. Equivalent to a drop of 1 m for the discharge of 4000 m^3/s, when the design navigable discharge is 3200 m^3/s. the minimum navigable water depth is 4.5 m for the main course and is 3.5 m for the third channel, which could not meet the design requirements. However, no obstacle to the navigation occur, in fact, because the maximum draught for the ships is currently only less than 2.8 m. For power generation, the dropping of water level has not only effect on the power generation, but also may increase the productive head and also the generation benefits. But as the design head for the lock gate is 27 m, the pool level should be lowered when the water head is more than 27 m. Therefore, the dropping of water level has no significance for power generation, but no losses. The dropping water level will improve the standard of flood control, will be favorable for the flood control.

It should be noted that the river bed of 100 km , downstream of Gezhouba, is composed of sand and cobble, the sediment concentration is basically in a secondary saturation state and the cumulative suspended load deposit may not occur. The coarse sand and cobble are deposited in the reservoir after impounding and the incoming amount is decreased dramatically. Therefore, the process will be very long and may be impossible when the completion of the Three Gorges project that the natural conditions are resumed when the scouring-depositing balance is reached of the bed load deposited in the reservoir. The Gezhouba project has operated for 17 years and withstood all trials under sorts of sediment transport conditions. As shown from the measurements, during the period from 1990 to 1996 the water level at the Yichang station had been stable, the small change in scouring or depositing took place and a new scouring-depositing balance occurred. In order to control the dropping of water level and keep the current navigation conditions for the downstream approach channels, the quarrying activities on the long stretch downstream of the dam are not allowed, especially, forbidden within the section upstream of Huyatan, because the river bed variation of the transition sections upstream and downstream of the yanzhiba has dominantly influence on the dropping of water level downstream of the dam. In fact, the coarse sand and cobble quarrying at the section upstream of Huyatan has been stopped.

5. CONCLUSION

A large number of quarrying has been performed on downstream of the dam site. since the construction of the Gezhouba project. Only at the section from Zhenchuanmen to Huyatan the quarrying amount was up to 24.76million m^3, with a mean quarrying depth of 1.76 m; after the impounding of reservoir the downcut of the river bed was up to 1.9 m including flow scouring, which has a direct effect on the minimum design navigable water depth in the downstream approach channels. Form now on, therefore, any quarrying activities should be stopped within the long stretch downstream of the dam, to avoid the further downcut of river bed and the further lowering of the water level due to artificially imposed activities.

6. REFERENCES

Chen Shiruo, (Oct., 1989). "Analysis on water surface profile between Yichang and Shashi".

Long Yinhua and Huang Lihua, (Oct., 1989). "Quarrying at the section between Gezhouba dam to Jiangkou and its effects".

Sun Changwan, (Oct., 1989). "Basic features of composition of river bed between Yichang and Jiangkou".

Environmental Hydraulics, Lee, Jayawardena & Wang (eds) © 1999 Balkema, Rotterdam, ISBN 90 5809 035 3

Impact of the Yangtze Three Gorges Project on hydraulics and water quality

Fang Ziyun
Yangtze Water Resources Protection Bureau, China

ABSTRACT: The paper studies the impacts of TGP on hydraulics and water quality and introduces the peculiar pollution patterns. The pollution tendency after impoundment and the water quality management strategies are discussed.

1. INTRODUCTION

The Three Gorges Project (TGP) is a key project in the development and harnessing of the Yangtze River. The dam site is situated in Sandouping of Yichang city, Hubei Province, with a distance of 40 km upstream from the Gezhouba dam. The TGP is a multipurpose hydro-project mainly for flood control, power generation, navigation and water supply improvement.

The main characteristic of the Three Gorges reservoir is:
- The relative volume of the reservoir is small and its regulation capacity for runoff is low. It is a seasonal regulation reservoir.
- The reservoir is of gorge type, with a length of 660 km and an average width of 1.1 km.
- The operation program for reservoir regulation considers various requirements including the environmental aspect.

Now the total wastewater and sewage discharged into the Reservoir reach amounts to more than 1 billion ton annually. The water quality, however, remains good in general due to huge quantity of stream flow, except for pollution belts along the banks near cities. Impoundment of TGP will aggravate shoreline pollution.

2. IMPACT OF TGP ON HYDRAULICS AND WATER QUALITY

After regulation of the reservoir, the annual quantity of discharge downstream of the dam and at the estuary will remain unchanged. Only the distribution pattern of discharge among seasons will be changed. Without the project the highest stage within the reservoir occurs in the summer season, the lowest stage occurs in the winter season. However, with the project in place, the reverse applies, it would make oncomalania to breed difficulty in the reservoir area.

In the reservoir area the reduced velocity of flow will decrease the ability of reoxygenation and dispersion of pollutants, thereby threatening to worsen existing local near shore pollution belts. Therefore, it is necessary to control strictly the pollution of the reservoir by industrial effluent, agricultural runoff and domestic sewage, and to mitigate the pollution by waste water treatment and pollution prevention measures.

The thermal stratification of the reservoir water body begin around April and end in May. It would take 20 more days for the temperature of downstream water to rise to the fish spawning temperature of 18 ℃.

The reservoir will regulate the flood flow and reduce the catastrophic flood damage. It has the significant effect of reducing the flood peak. At the beginning of operation period of the reservoir , around 70% of the sediment of the previous 0.5 bill tons would be settled to the bottom of the reservoir and the fluctuating area of backwater, the discharge of clear water from the reservoir would erode the downstream bank as well as the bottom of the river, but the live span of Dongting Lake would be extended.

Due to the fluctuation of daily output of the powerplant, the flow discharge passing the turbines would have a big fluctuation within a day during the dry season, especially in the dry period of dry year. It would cause the daily max. fluctuation of the water stage in the Gezhouba reservoir of 3 m and 1.4 m at Yichang, but the water surface of unsteady flow would be smoothed down very rapidly and have no significant effect below Zhicheng station.

3. 2-D MODEL PREDICTION OF THE RESERVOIR POLLUTION TENDENCY

The theoretical concentration of a pollutant increases in the reservoir downstream from a point source because of reduction in flow velocity. The basic equation is :

$$C(X,Z) = \frac{m}{U\sqrt{4\pi E_Z X}} \exp[\frac{-UZ^2}{4E_Z X}] \tag{1}$$

where $m = q_o \cdot c_o / H$, q_o is initial effluent flow (m³/sec), c_o is initial pollutant concentration (mg/l), H is mean depth of receiving water body (m), U is river mean velocity (m/sec), X is longitudinal distance (from effluent) (m), Z is lateral distance (m), E_z is lateral diffusion co-efficient (m²/sec).

The following table shows the result obtained by considering a natural velocity of 2 m/sec and by setting X at 1 m.

TABLE 1   Relationship of Pollutant Concentration with Reduction of Flow Vel. in Reservoir

	Natural conditions (U=2m/sec)	20% reduction of flow vel. (U=1.6m/sec)	40% reduction of flow vel. (U=1.2m/sec)	70% reduction of flow vel. (U=0.6m/sec)	85% reduction of flow vel. (U=0.3m/sec)
Theoretical concentration of a pollutant immediately downstream of a source	C	1.11C	1.28C	1.82C	2.58C

The effect of TGP on dispersion of Chongqing and Wanxian point sources of pollution would be as table 2.

TABLE 2   Effect of TPG on Dispersion of Chongqing and Wanxian pt. Sources

Incoming river flow	River Section	Conditions mean vel. (m/sec)		Distance from effluent at which Pb water quality standard is respected (m)	Effluent concentration of Pb to respect water quality standard 1 km downstream (mg/l)
5000	Chongqing	Natural	1.7	1	8.9
		NPL170	0.5	12	2.6
		NPL180	0.2	50	1.3
	Wanxian	Natural	1.0	1	5.2
		NPL170	0.1	100	0.5
		NPL180	0.1	123	0.5

Table 2 shows the results of calculations for post impoundment conditions. First it gives the distance downstream necessary to dilute an effluent sufficient so that water quality standards are respected, assuming unchanged effluent concentrations, and predicted post impoundment flow velocities. Second, it gives the effluent concentrations necessary to maintain water quality standards at 1000 m downstream. In this table Pb has been chosen as a pollutant example for which the water quality standard is 0.1 mg/l and the current concentrations in the river are usually below the detection limit of 0.001 mg/l. The results show that downstream distance required to obtain third class surface water quality increases considerably with flow velocity reductions caused by the project. For example, under the NPL 180 scenario at chongqing there standards would be met 50 km downstream instead of 1 km under natural conditions. This effect can nonetheless be mitigated by better water treatment. For the same example, water quality standards could be met 1 km downstream if effluent concentration was reduced from 8.9 mg/l to 1.3 mg/l.

4. INTEGRATED STRATEGIES TO WATER QUALITY MANAGEMENT

4.1 Peculiar Pollution Pattern

The Yangtze River is about 6,300 km long with drainage area(D.A) of 1.8 million square km, equal to about one fifth of the total of the country. The mean annual runoff is 450 billion m^3 at Yichang gaging station, and 914 billion m^3 at Datong station. In the D.A. there is a population of 410 million. Its water quality, on the whole, is good or rather good, but the reaches along the industrialized city banks, the tributaries and the lakes of the basin have now been polluted heavily. The sources of pollution are from urban, industry, agriculture and moving sources of navigation, but now are mainly from industries and municipalities. The pattern of pollution in Yangtze is the occurrence of polluted belts along the banks of cities with 60 ~ 100 m in width. The polluted belts are all located downstream from the outlets of sources of pollution. Now the polluted belts are discontinuous in the longitudinal direction of the river. If further pollution should not be checked, the discontinuous polluted belts near the cities would connect into a continuous one. Should the polluted belts of the river extend to become continuous, it would mean the pollution of the whole river. It is expected that the abundant water resources and hydroelectric resources as well as the navigation superiority of the Yangtze will attract more industries to locate along the river, so that the tendency for pollution will be great.

4.2 Use 2-D Model to Investigate the Pollution of Main Stem of Natural River

The model used is as above.
For E_z in Wuhan Reach; it was found that E_z is portional to the discharge of the river, and also E_z is proportional to the side slope of the bank within the pollution belt as shown in Table 3.

TABLE 3   Relationship of E_z with River Discharge and River Bank Side Slope

Name of source of effluent	Discharge of the River (m^3/sec)	Side slope of river bank	E_z (m^2/sec)
(1)	11500	0.017	0.153
(2)	9430	0.047	0.15
	18800	0.05	0.589
(3)	8730	0.14	0.528
	16700	0.24	0.950

4.3 Protection Strategies

The strategies for protection of Yangtze should focus on checking the further elongation of polluted belts and reducing gradually the existing length and the pollutant concentration of the belts. The principles and measures for fulfilling the task are as follows:

(1)Applying comprehensive pollution control planning and the complex remedial biological and engineering measures.
The main principle for river pollution control should be to restrict waste water effluent to reduce the loading of pollutants and to exploit clear water sources to increase the self-purifying(assimilative) capacity of the aquatic environment. In planning to rationally deal with the above two aspects and to coordinate regulation of the river reaches, the whole main stem of the river and the whole basin are very important.

(2)Control strictly the pollution sources coming from factories, mines, enterprises and village industries by:
a.   economic use of water,
b.   diminishing the waste water discharge,
c.   controlling the emission of pollutants by quantity,
d.   good planning and siting for the development of village industries.

(3)Take care of the main sources of pollution, and control polluted water from cities by the following measures:
a.   improving the drainage and sewer system and associated facilities,
b.   construction of concentrated sewage treatment plants by municipalities to treat mainly for domestic sewage or construction of a plant by a big factory to treat industrial polluted water and sewage, or construction of a plant by several factories jointly,

c.   choose suitable control techniques for each special locality,
d.   turn the waste in the polluted water into resources to be reused.

(4) Rationally site new industries and urban developments, preventing further pollution from new sources and regulating existing sources.
(5) Extend both industrial discharges and sewage outfalls into the middle of the river in certain reaches.
(6)Prevent soil erosion to reduce the turbidity and the organic content of the river.
(7)Control oil pollution and other pollutants from navigation.

According to Chinese Regulations, the outlets of industrial polluted effluent and domestic sewage discharges should not be located at a distance nearer than 1000 m upstream and 1000 m downstream from the site of the inlet to a water supply. Ordinarily many alternatives for pollution control of the Yangtze along the shore are formulated among different distances(less than 1000 m) downstream from polluted water outlets in order to restore the quality of the river, such as the 300 m recovery plan, the 500 m recovery plan, the 800 m plan and the 1000 m recovery plan, etc. The shorter the recovery distance, the more expensive is the cost of the plan. The plans selected are different at different times. In general, at present the longer distance recovery plan is adopted, whereas in the further a smaller distance recovery plan will be selected. Of course, the fundamental principle would be to minimizing harmful fluxes from land to water in Yangtze finally.

4.4 Difficulties of Minimizing Harmful Fluxes and Their Overcoming Ways

The difficulties of minimizing harmful fluxes might be as follows:
(1)The popular environmental consciousness is not so high that human life-styles and activities can not link closely with minimizing harmful fluxes from land to water.
(2)" The Water Law " and " The Water Pollution and Control Law of China "have not been worked strictly. The law systems have not been established perfectly.
(3)Funds limited to take measures.
(4)The production technology and equipments of some old industries are outdated.
(5)The advanced and new technology has not been popularly applied.
(6)Floods often break and overtop the dykes to make the alluvial plains inundated.
(7)Sudden occurrence of pollution accidents, etc.

As to the way for overcoming the above barriers it is considered that full communication of information between scientists and decision makers, education on all levels, coordinating the economic development and water quality management, increasing funds for regulation and successful integration of land and water use might be efficient.

5. REFERENCES

Fang, Z.Y. (1993), "Integrated Approach To Water Quality Management Of Yangtze", *Proceedings of Stockholm Water Symposium*, Stockholm Water Company, p379-383
Fang, Z.Y. (1996), "Environmental Perspective: Beneficial and Adverse Effects of Three Gorges Project", *International J. Sediment Research*, IRTCES, p1 ~ 21

Environmental Hydraulics, Lee, Jayawardena & Wang (eds) © 1999 Balkema, Rotterdam, ISBN 90 5809 035 3

The contents, methods and progresses of water pollution control of the Three Gorges Reservoir on the Yangtze River (WPC-TGR)

Z.L.Huang
Executive Office of State Council Three Gorges Project Construction Committee, Beijing, China

Y.C.Chen, Y.L.Li & C.B.Jiang
Department of Hydraulic Engineering, Tsinghua University, Beijing, China

ABSTRACT: This paper summarizes the research work of Water Pollution Control of Three Gorges Reservoir(WPC-TGR), which includes the research contents, the methods, the present progresses and the main results obtained. Related papers are reported as well in this proceedings.

1. INTRODUCTION

The Three Gorges Project(TGP) on the Yangtze River is located at Sandouping Town of Yichang county, Hubei Province, which is 40km away from the downstream Gezhouba project, and its control drainage area is about $1,000,000\ Km^2$. It is a vital important and backbone project in harnessing and developing of the Yangtze River. The Construction was started on December, 1984, the main river course was closed on November 8, 1997, and the whole project will be completed by 2009.

Based on the priliminary design report of TGP, the Project consists of three major parts: the large dam across the Yangtze River, the hydroelectric power houses and the navigation structures. The dam will be of a concrete gravity type, with the crest elevation of 185m above the sea level(Wusong Elevation), the normal water level is about 175m, the flooding control water level is 145m. The lower (dead) water level is 155m. For the normal water elevation and in case of 20-year flood frequency, the back water reaches Mudong Town of Ba county, Chongqing City, which is about 565.7Km from the dam site, forming a typical river-like reservoir as shown in Fig. 1. The surface area of the reservoir will reaches $1084\ Km^2$ with average width of 1100m which will be about twice of the width of natural river channel. The total capacity of TGR amounts to 39.3 billion m^3, among which flood control storage capacity is 22.15 billion m^3, and effective capacity is 16.5 billion m^3 accounting for 3.7% of the annual total discharge rate through the dam site. The reservoir is of seasonal regulation with low regulation capacity. Through the analysis and computation of selected typical annual wet, regular and dry season inflows, it is found that the annual runoff downstream of the dam will not change after operation of the reservoir. The project will not affect the annual inflow entering into the sea.

The major benefits of the TGP include flood control, electric power generating and the improvement of navigation etc.. However, during the process of argument and decision-making, the impacts of TGP on environment and ecological system attract world wide attention. The Chinese government has also given highly concern about this problem. While in the feasibility study stage, the comprehensive evaluation has been carried out about the environmental problems of the TGP , and the corresponding policies has been made.

The water quality is one of the keynote problems which attracted much attention. At present, the serious threaten to the future TGR comes from that the great amount of pollutant materials in the reservoir which may be discharged into the river. The management has been strengthened now(NEPA of China,1998). Based on the investigation on the 94 industrial pollutant sources and 80 urban waste water drainages in 1997, it is shown that: the annual amount of industrial waste water is 1.12 billion tons, and the urban waste water is 0.38 billion tons, totally 1.5 billion tons. The main water quality indexes are: BOD, COD, volatile phenol, total phosphorus, total nitrogen, oils and ammonia-nitrogen etc., most of them are directly discharged into Yangtze River. Due to the large runoff rate and its strong capabilities of dilute and self-clean, the monitoring data collected from the observed sections of the reservoir show good water quality in the reservoir, it rank in the national class 2 water quality standards on most sections. The main problems are the local pollution regions near the large and middle cities, forming pollution belt along the river bank, especially in Chongqing, Wanxian , Fuling cities. In addition, during the agriculture activities along the river bank, huge amount of nitrogenous fertilizer are applied, among which only 30 to 35% are used by the crop, about 11.9 thousand yearly ton were run off , which is discharged into Yangtze river along with the ground water. The problem of oil pollution and the trash from the ships can also not be ignored.

The impacts of TGP to water quality need to further study. According to the evaluation of environmental affect, it is shown that: the TGP will be harmful to the water quality of the reservoir region, and will beneficial to the water quality of middle and downstream region during the dry season. Thus, most concerns and studies are focused on the reservoir region, which is the harmful affective region. Generally speaking, after the dam being built up, the flow velocity and turbulent diffusion capacity will decrease, this will reduce the capacity of water environment, and aggravate the pollution belt along the river bank and the local pollution in the static region of reservoir.

In order to further understand the effects of TGP to water quality in the reservoir region, predict the changes of water quality in the reservoir region under the changed pollution loads in the upstream and reservoir regions during the initial operation stage of the reservoir, and propose a practical countermeasure controlling the reservoir pollution, so that the water quality will not change after the formation of the reservoir, Since 1995 under the organization of Executive Offilce of State Council TGP Construction Committee and China Yangtze TGP Development Corporation, several universities/institutes,such as Tsinghua Univ., Chinese Inst. of Hydr. and Hydropower Res. have carried out the study of WPC-TGR. This research project will be completed by Dec., 2000.

2. RESEARCH CONTENTS

The main contents of WPC-TGR include: Investigation and observation of water quality in the TGR at present situation; Numerical modeling for analysrs and computation of water quality from the late period of construction to the initial stage of operation; Comparison of the water environmental capacity before and after the construction of the reservoir; Zone Classification according to the water body function of TGR, and identification of the river bank and total water environmental capacity, proposition and optimization of the total amount of pollution loads; proposition of countermeasure for the control of water pollution; Upbuilding the information management system about TGR water pollution which can serve as a technical support to the TGR water environmental management. According to the above demands, this study will closely combine the practical situation of TGR, the systematical research will be carried out based on field investigation/observation, theoretical analysis, numerical computation, quantity and quality methods. The details of the study are shown as the following three aspects.

2.1 Field Investigation of Pollutant Sources,Hydrology and Water Quality in the Reservoir

2.1.1 1-D synchronous investigations of hydrology and water quality in the reservoir region

In order to further understand the global status of water quality in the reservoir reaches during the flood season and non-flood season, and to identify the parameters and verify the numerical model of one dimensional hydraulic and water quality formulations, to mﾃke the mathematical model suitable for the practical application, the synchronous inwestigations of both hydrology and water quality(including the pollution sources) should be carried out two times at the reservoir reaches. One is for the small runoff rate at dry season, which reflects the pollution situation of the point sources. others are carried out for large flood coming mainly from the upstream, which reflects the pollution of plane sources.

2.1.2 The investigations of typical pollution belt

The pollution belt along river bank of the Yangtze are mainly at the river reaches nearby large and middle cities, which can be classified as two types: One kind of belt pollution comes from the waste water of industry and daily life(point source); Another kind of belt pollution is due to the polluted sub-branches. Therefore, the two kinds of typical pollution belt should be observed separately. The measured data are used to verify the two dimensional and three dimensional numerical models.

2.1.3 The investigation and prediction of pollutant sources

With the development of economics and progress of pollution treatment in the reservoir reaches, especially the environment protection during the resettlement, the pollution sources in the reservoir will be changed greatly. The boundary conditions for transport quantities are also needed for the numerical modeling. Therefore, it is necessary to carry out the site investigations and observations for the main inflow sections and the pollution sources along the river bank, which include the point sources, line sources(due to the ships) and the plane sources. Based on the points above, the numerical model should be established to predict the changes of pollution sources in the future, including the main pollutant materials, concentration, loads and its distribution. This will provide the basic information for both the numerical model of water quality and the water pollution control.

2.2. The Calculation of Hydraulics, Water Quality, Water Environment Capacity and the Study of Water Pollution Control Countermeasures in the Reservoir

Mathematical model: As the TGR is river-like and the pollutant outlets are well dispersed, the following mathematical models are developed. '1-dimensional hydraulics and water quality model of TGR', 'The computation of mixing zones nearby the main urban outlets and the important converges of sub-branches ' and 'The numerical analysis of 3-D velocity field, 3-D temperature and concentration field in the reservoir'. The 1-D model is mainly applied to study the global transport and diffusion properties of all kinds of pollutant materials in the reservoir, which includes 1-D unsteady hydraulic model of river net and 1-D dynamic model of water quality; The 2-D model is employed for the regions near local outlets and the converges of sub-branches; The 3-D model is applied for the regions near dam where strongly three dimensional flow characters are presented, special attention should be made about the short time thermal stratification during May and June, in order to prevent or reduce the thermal stratification. As the TGP is under construction, what we have done for the field investigations can only reflect the present natural situations, and then use these data to detemine the parameters of the numerical models and verify them. Owing to the very complicated situations of Yangtze river, the model and its parameters may probably not be suitable for the future TGR. Therefore, the physical model experiments or the analogy study should be made. The real verification will be made by field observation after the TGP is completed.

The countermeasure of pollution control: According to the demand for the water resources of TGR, the reservoir should be zoned according to the function of water body, the environmental capacity of the whole reservoir and along the river bank should be computed using the above numerical models. Finally, the programs for the pollution load distribution, and countermeasures for pollution control should be proposed and optimized.

2.3. Information Management System for Water Pollution

The information management system for the water pollution is a comprehensive result of present study, which includes data base, model library, dynamic display, strategic decision support and management system et al.

3. RESEARCH PROGRESSES

Up to now, the achievements gained based on the plan include the following several aspects:

(1). 1-D synchronous investigation of hydrology and water quality in the reservoir(the segment from Qingxi to Fengjie).

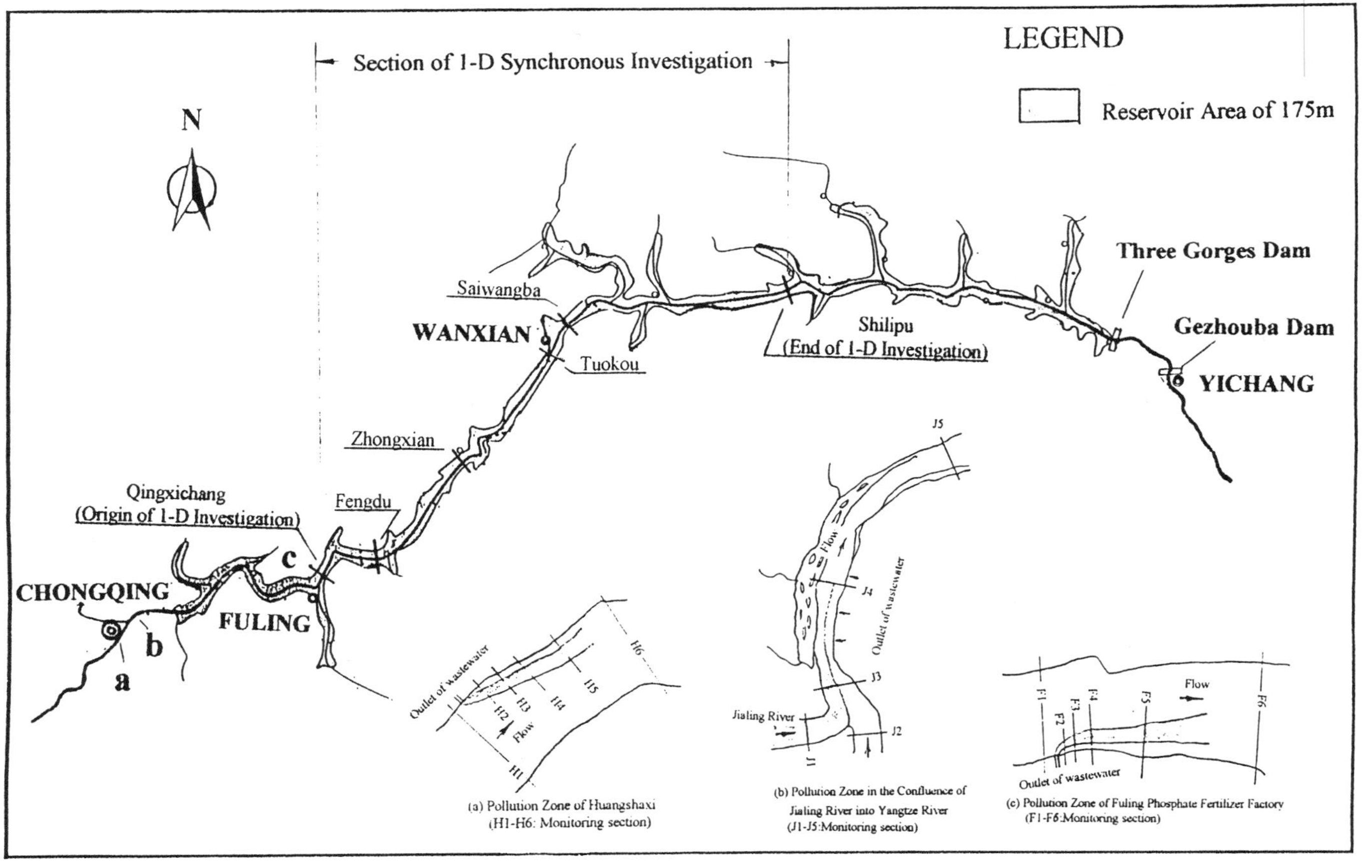

FIG.1 Schematic of The Sections and Locations of 1-D Synchronous Investigation and Pollution Zones

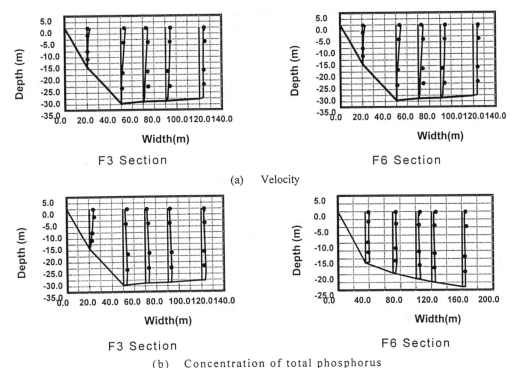

(a) Velocity

(b)   Concentration of total phosphorus

FIG.2   The Comparison for the Velocity and Concentration for Side Discharge of
Wastewater of Fuling Phosphate Fertilizer Factory

Based on the expense and require of project, The 1-D synchronous investigation was carried out in the segment which is 300 km long from Qingxi to Fengjie, as shown in Fig. 1. The synchronous investigation includes not only Qingxi, Fengdu, Zhong county, Tuokou, and Saiwangba, the six investigation sections located on the main reach, but also the important converges of sub-branch (such as the zhuxi River) and the important pollutant sources(such as the Tuokou Electric Power Factory). The main hydrological parameters include water elevation, velocity, flow and sediment etc.. The parameters of water quality include water temperature, pH, SS, DO, COD_{Mn}, BOD_5, NH_3-N, NO_2-N, NO_3-N, TN, mercury, Cr^{6+}, oils and Volatile Phenol etc.. The investigation during flood season was carried out from Aug. 1 to 8, which include the start point, the mid rise point, peak value, mid drop point and the terminal point. The non flood season observation was carried out at 9 am. and 4 pm., two times everyday, from March 15 to 18, 1998. The investigation data has set up a solid foundation for verifying the numerical model and analyzing the present status of the water quality in reservoir region, which will be reported in the neat future.

(2). The investigation of typical pollution belt.
The investigation of pollutant belt near river bank formulated from the pollutant sources of industry was carried out near Fuling Phosphate Fertilizer Factory, which situated in the mid of reservoir region. Six sections (5 to 9 vertical lines per section and 3 points per line) and the discharge outlet are measured synchronously. The hydrological parameters include water elevation, velocity, flow. The index of water quality include COD_{Mn}, NH_3-N, total phosphate, total iron etc.. The investigation during ordinary season was held on Nov. 24, 1997, and the investigation during dry season was carried out on Jan. 17, 1998. The distribution of pollutant on every section and the trend of variation measured through the observation can be applied in study of current situation of water quality of typical pollutant   belt in reservoir region and verification of numerical model, which will be reported in the near future.
The investigation of pollutant belt near river bank due to the waste water of daily life and polluted sub-branches was carried out at Huangshaxi outlet which situated at the end of reservoir region and the converge

of Jialing River, as shown in Fig. 1 (a) and (b). Six section and Five section are arranged at the Huangshaxi outlet and the converge of Jialing River respectively. The pollutant source was measured synchronously. The main parameters of water quality include water elevation, velocity and flow. The index of water quality include COD_{Mn} and NH_3-N. The measurement was carried out on Oct. 31, Nov. 7, Dec. 15, and Dec. 22, and the corresponding observation data will be reported in the near future. . The distribution of COD_{Mn} and NH_3-N measured on every section which reflect the transport and diffusion properties of pollutant near converges and outlet, can serve as a solid foundation for verification of numerical model.

(3). The tsaks of numerical models include three parts,namely 1-dimensional hydraulic and water quality model of TGR', 'The computation of mixing zones nearby the main urban outlets and the important converges of sub-branch 'and 'The numerical analysis of 3-D velocity field, 3-D temperature and concentration field in the reservoir'.

The 1-D numerical model was applied to study the global transport and diffusion properties of pollutant in the reservoir region. Up to now the parameter of numerical model has been verified tentatively based on the investigation data(Huang J.C.& Li J.X,1998).

The mixing zone near outlet and converges of sub-branches was simulated using 2-D numerical model. A 3-D multi-layers numerical model using upwind FEM is employed to simulate the flow field and concentration field of pollutant near the outlet of Fuling Phosphate Fertilizer Factory. Thus the complicated 3-D flow problem is simplified as several vertically coupled 2-D problems. The computational result agrees well with the field observation as shown in Fig. 2. (Li C.,1998)

The 3-D model is applied for the regions nearby the dam where strongly three dimensional flow characters are presented. At present, the numerical model has been chosen, and the related data of topography, hydrology and water quality are being prepared for verifying the model.

4. CONCLUSION

The WPC-TGR is a practical application research project, which shows that the environmental hydraulics has not only abundant content but also wide application field. It is the first time to carry out such large scale 1-D synchronous investigation of hydrology and water quality, and the investigation of typical pollution belt in the TGR region, the transport and diffusion characters of pollutant materials in the reservoir reaches and the characters of belt pollution along the river bank are obtained, which is very important for the effective water pollution control . Progresses have also been achieved in the choice and comparison of the mathematical models, and determination of the parameters, these have all provided a reliable guarantee for the prediction of water quality in the future.

5. REFERENCES

The EIA Dept.of CAS and the Res. Inst. for Prot. of Yangtze Water Resour.(1995), *Environmental impact statement for Yangtze Three Gorges Project(A brief Edition)*,Science Press, Beijing.

NEPA of China(1998). *Bulletin on ecological and environmental monitoring results of the Three Gorges Project(1998)*, May 1998, Beijing .

Huang J.C. and Li J.X., Impact of sediment movement on the water quality variation of the Three Gorges Reservoir, *Proc. of Seventh Symp. on River Sedimentation & Second Inter. Symp. on Envir. Hydraulics*, Dec. 16-18, 1998, HONGKONG.

Huang Z.L., On several problems of environmental hydraulics for the Three Gorges Project, In *Ecological and Environmental Protection on large-size dam of Yangtze river in 21st century*, Ed. by Huang Z.L., Chinese Environmental Science Press, 1998, Beijing(in Chinese).

Huang Z.L. and Li. J.X., Preliminary study on the longitudinal dispersion coefficient for the Three Gorges Reservoir, *Proc. of Seventh Symp. on River Sedimentation & Second Inter. Symp. on Envir. Hydraulics*, Dec. 16-18, 1998, HONGKONG.

Li C.(1998). Computation and analysis of mixing zone of pollutant for sid discharge into natural river using FEM, Thesis for Master Degree, Dept. of Hydraulic Engineering, Tsinghua Univ.

Environmental Hydraulics, Lee, Jayawardena & Wang (eds) © 1999 Balkema, Rotterdam, ISBN 90 5809 035 3

Preliminary study on longitudinal dispersion coefficient for the Three Gorges Reservoir

Huang Zhenli
Department of Technology and International Cooperation, Executive Office of State Council Three Gorges Project Construction Committee, Beijing, China

Li Jinxiu & Huang Jinchi
Department of Hydraulics, China Institution of Water Resources and Hydropower Research, Beijing, China

ABSTRACT: Based on theoretical analysis on the longitudinal dispersion coefficient K_x and field measurements of cross-section velocity distributions from Three Gorges Reservoir, the formula of K_x for the Three Gorge Reservoir is preliminarily proposed in this paper, .

1. INTRODUCTION

Longitudinal dispersion coefficient K_x, one of the important coefficient representing the basic mixing characteristics in natural river and stream, has become increasing necessary for engineers and scientists to be able to predict the impact when pollutants will released in river. A method of rapidly predicting the dispersion is needed for pollution control or warning system on river where data are limited.

With the widespread research on the longitudinal dispersion characteristics, many empirical and semi-empirical formulas are available to make the estimating of longitudinal dispersion, but none can be used with confidence before calibration and verification to the particular river reach. The longitudinal dispersion coefficient is very complex and with a large variation ranges in natural water environment, which mainly depends on the conditions of hydrology, form and regime of river etc.. Fischer(1975, 1979) has deduced longitudinal dispersion coefficients from data taken in a range of environments, from laboratory flumes to major rivers, he found values for $\alpha(=K_x/hu_*$, h is the depth, u_* is the friction velocity) usually in the range 100-500 with an extreme value of 7500 in the Missouri. So it is very difficult to get the suitable value for the Yangtze river.

The Three Gorge Reservoir will be of a river channel type with a length more than 600km. It is located on the upper reach of Yangtze river and in the regions of hilly land and mountain valley. The form and regime of river is very complex. The water level will varies in a range in year. In particular, when the reservoir will store water in Oct. and discharge water in the end of May, the water level will change between the $\nabla 145m$ and $\nabla 175m$. All of the river hydrological and topographical features will greatly change, so the longitudinal dispersion coefficient for the Three Gorge Reservoir varies with time and space. In this paper, based on theoretical analysis and field investigations of hydrology, the formula computing the longitudinal dispersion coefficient for the Three Gorge Reservoir is preliminarily proposed.

2. ANALYSIS ON THE CALCULATION FORMULAS OF K_x

2.1 Integral Formula

If the detailed hydraulic data are available in the specific river reach, it is suitable to calculate the K_x by the integral formula:

$$K_x = -\frac{1}{A} \int_0^B q'(y)dy \int_0^y \frac{1}{K_y h(y)} dy \int_0^y q'(y)dy \qquad (1)$$

where A is the area of cross-section, B is the width of section, h(y) is the depth in lateral coordinate y, K_y is the transverse mixing rate given by

$$K_y = \alpha_z h(y)u_* \qquad (2)$$

Where α_z is the cross mixing coefficient, and q'(y) is the unit discharge deviation from mean given by

$$q'(y) = h(y)(u(y) - \bar{u}) \qquad (3)$$

where u(y) is vertical average velocity in lateral coordinate y, $\bar{u}$ is the average velocity of cross-section.

Although the formula (1) is usually capable of calculating the K_x with a fair degree of accuracy, this hydraulic data is seldom available, so it is signification to develop a simple formula to be used to predict K_x.

2.2 Empirical Formulas

The first calculation of the longitudinal dispersion coefficient K_x in turbulent flow was that of Taylor(1954) for flow in a cylindrical pipe, he concluded that

$$K_x = 10.1au_* \qquad (4)$$

where a is the pipe radius.

Elder(1959) extended this result to a wide shallow channeland concluded that vertical shear and diffusion resulted in a longitudinal dispersion given by

$$K_x = 5.93hu_* \qquad (5)$$

Fischer(1975) investigated the dispersion arising from transverse shear and concluded that in natural river, it would be the dominate process and that in many cases dispersion due to vertical shear could be neglected. He derived a formula predicting K_x by combing laboratory measurements, theoretical derivation for ideal cases and from a few real streams:

$$K_x = 0.011B^2 u^2 / hu_* \qquad (6)$$

The Fischer's formula is of more applicability for the natural river than Elder's, because in the natural river usually with B/h>10, the transverse shear is much more important in determining K_x than vertical shear, which can be ignored.

The followings are other famous semi-empirical formulas of K_x:

Mcquivey and Keefer(1974): $K_x = 0.058 \dfrac{Q}{JB}$ \qquad (7)

where J is the hydraulic slop, Q is the river discharge.

Liu and Chen(1977): $K_x = \gamma \dfrac{Au_*}{h^3}, \gamma = 0.5 - 0.6$ (8)

Liu(1980) $K_x = \beta \dfrac{u^2 B^2}{Au_*}, \beta = 0.18\left(\dfrac{u_*}{u}\right)^{1.5}$ (9)

Iwasa and Aya(1991) $K_x = 2.0\left(\dfrac{B}{h}\right)^{1.5} hu_*$ (10)

Where 2<B/h<20 for open channel flow in laboratory, 10<B/h<100 for natural river.

Analyzing those empirical formulas described above, it is shown that: although there are clear different among these formulas, but they have one property in common: the coefficient $\alpha(K_x/(hu_*))$ will be increased with increasing of u/u_* or B/h, and the α for these formulas can be presented respectively again by following forms:

Mcquivey and Keefer: $\alpha = \dfrac{0.058}{J}\dfrac{u}{u_*}$

Fischer: $\alpha = 0.11\left(\dfrac{u}{u_*}\right)^2\left(\dfrac{B}{h}\right)^2$

Liu and Chen: $\alpha = (0.5 - 0.6)\left(\dfrac{B}{h}\right)^2$

Liu: $\alpha = 0.18\left(\dfrac{B}{h}\right)^2\left(\dfrac{u}{u_*}\right)^2$

Iwasa and Aya: $\alpha = 2.0\left(\dfrac{B}{h}\right)^{1.5}$, when 6<$\alpha$<2000

Herein, the normal function representation for α can be described as formula(11) in this paper:

$$\alpha = a\left(\dfrac{B}{h}\right)^b\left(\dfrac{u}{u_*}\right)^c \qquad (11)$$

where three coefficients of a, b, c may be derived from the measurement data of cross-sectional velocity distributions.

3. VELOCITY PROFILE DATA FROM THREE GORGES RESERVOIR AND DEVELOPMENT OF PREDICTION EQUATION

Recently, detailed measurements of cross-sectional velocity distribution have become available on two reachs, one of the measured reach is 300km length on the Three Gorges Reservoir, in where twice measurements have made respectively in Aug. 1997 (the high-water period) and Mar. 1998(the low-water period) at three cross-sections, i.e.Qingxichang(QXC), Tuokou(TK) and Fengjie(FJ); the other measured reach is 2km length on the Jialing river near the junction with Yangtze river which is the largest branch river on the Three Gorges Reservoir, in where twice measurements have made respectively in Apr.1989 (at river discharge level of 1720m³/s) and May. 1989(at river discharge level of 2760m³/s) at five cross-sections. In the high-water period of Three Gorge Reservoir, four times detailed measurements have made at the site of QXC respectively at river discharge levels of 22300m³/s, 26900m³/s,20200m³/s and 15300m³/s; in the low-water period, the measured river discharges are of 4100m³/s at QXC, 4120m³/s at TK and 4580m³/s at FJ. The transverse depth profiles and the transverse unit discharge profiles at two typical measured section are

671

ploted in Fig.1 to Fig.4.

Based on these cross-sectional velocity distribution data, the longitudinal dispersion coefficients K_x in each measured section are computed by integral equation(1), where only one coefficient, i.e. transverse mixing coefficients α_y on each river reach are needed given by artificially, and according to the research results from Huang(1998), the values of α_y are taken 0.63 on the Jialing river reach and 0.75 on the Three Gorges Reservoir reach respectively. The computed results see Tab.1.

It is shown that the values of K_x varies largely at different measured sections, and even at the same measured section but at the different measured periods. By mean of analysis of multiple linear regression on the formula(11) with the samples in Tab.1, the simple calculation formula for α is obtained:

$$\alpha = 0.2(\frac{B}{h})^{1.3}(\frac{u}{u_*})^{1.2} \tag{12}$$

The root-mean-square error of the prediction formula(12) is 2.2, with variance σ^2 of 0.37.

Herein, the calculation formula of K_x for Three Gorges Reservoir can be given:

$$K_x = 0.2(\frac{B}{h})^{1.3}(\frac{u}{u_*})^{1.2} hu_* \tag{13}$$

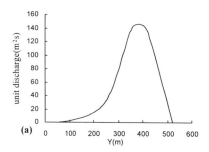

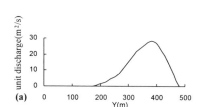

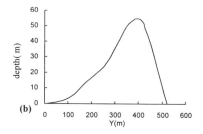

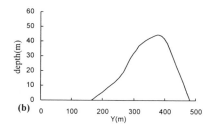

FIG.1 QXC: measured on 6 Aug.1997,
   at flood peak, Q=26900m³/s.
   (a)Transverse unit discharge profile;
   (b)Depth profile

FIG.2.QXC:measured on 15.Mar.1998,
   at low water, Q=4100m³/s.
   (a)Transverse unit discharge profile;
   (b)Depth profile

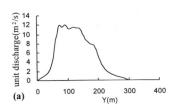

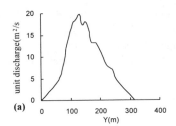

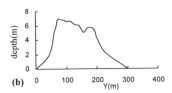

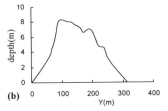

FIG.3 Jialing river:measured in May1989, Q=1720m³/s. (a)Transverse unit discharge profile;(b)Depth profile

FIG.4 Jialing river:measured in May 1989, Q=2760m³/s. (a) Transverse unit discharge profile; (b) Depth profile

Tab.1: The computed results of longitudinal dispersion coefficient for Three Gorges Reservoir

Measured reach	Measured period	Cross-section	Depth (m)	Width (m)	Average velocity (m/s)	Friction velocity (m/s)	K_x (m²/s)	α
Jialing river	Apr.1989	1	11.0	194	1.00	0.13	61.2	47.8
		2	7.4	374	0.90	0.12	163.4	22.1
		4	6.0	330	1.09	0.10	415.6	692.7
		7	6.8	301	1.56	0.10	671.3	987.2
		12	5.5	340	1.03	0.10	154.9	281.6
	May.1989	2	9.8	380	1.22	0.12	737.7	627.3
		3	6.6	363	1.23	0.10	640.9	971.1
		4	7.0	386	1.25	0.10	821.4	1173.4
		7	7.7	315	1.89	0.10	279.3	362.7
		12	5.8	352	1.17	0.10	126.1	217.4
Three Gorges Reservoir	Mar.1998	QXC	23.3	319	0.55	0.21	434.3	88.7
		TK	10.9	398	0.95	0.14	419.3	343.4
		FJ	6.3	349	2.09	0.11	2883.5	4160.9
	Aug.1997	QXC	22.3	510	1.96	0.20	1150.1	257.9
		QXC	23.2	520	2.23	0.20	839.1	180.8
		QXC	22.2	499	1.82	0.20	427.7	98.72
		QXC	25.1	403	1.5	0.21	333.6	63.3

4. CONCLUSION

A formula predicting the longitudinal dispersion coefficient for Three Gorges Reservoir is proposed relied on measurements of cross-sectional velocity distributions. Because the river hydrological and topographical features on Three Gorges Reservoir reach are very complex and variable, it is believed that existing data from measured samples is too little to represent the whole river reach, the prediction formula of K_x is needed further calibration and verification from more field data. On the others hand, the Three Gorges Reservoir is now under construction, the real Three Gorges Reservoir doesn't appear. After operation of the Project, whether the formula of K_x proposed in this paper may be used or not, that is also needed to be verified.

5. REFERENCES

Elder, J.W. (1959). "The dispersion of marked fluid in turbulent shear flow", *J. Fluid Mech.* 5, 544-560.

Fischer, H.B. (1975). "Discussion of 'simple method for predicting dispersion in stream' by R.S.Mcquivey and T.N. Keefer ", *J. Environ. Eng.. Div.* Proc. Amer. Soc. Civ. Eng. 101, 453-455.

Fischer, H.B. et al. (1979). *Mixing in inland and coastal waters,* Academic Press, New York, N.Y.

Huang Z.L(1998). " On several problems of environmental hydraulics for the Three Gorges Project", In: *Ecological & environmental projection on the large-size dam of Yangtze river in 21st century,* Ed. By Huang Z.L. et al , Chinese Environmental Science Press, 1998, Beijing (in Chinese).

Iwasa Y. and Aya S.(1991). " Predicting longitudinal dispersion coefficients in open-channel flow " , In: *Environmental Hydraulics,* Ed. By J.H.W. Lee et al., A. Balkema, Rotterdam, NL.

McQuivey, R.S. ,and Keefer, T.(1974). " Simple method for predicting dispersion in stream", *J. Environ. Eng. Div.* Proc. Amer. Soc. Civ. Eng., 100, 997-1011.

Taylor, G.I.(1954). "The dispersion of mater in turbulent flow through a pipe", Proc., Royal Soc., London, England, Ser. A, 223, 445-468.

Environmental Hydraulics, Lee, Jayawardena & Wang (eds) © 1999 Balkema, Rotterdam, ISBN 90 5809 035 3

Impact of sediment movement on the water quality of Three Gorge Reservoir

Huang Jinchi & Li Jinxiu
Department of Hydraulics, China Institute of Water Resources and Hydropower Research, Beijing, China

ABSTRACT: Field observed data shows that the impact of sediment movement on water quality can not be ignored. This paper gives a special consideration, with the special case of Three Gorge Reservoir as a background. Based on the analysis of mass equilibrium of pollutants, a basic equation sets reflecting the function of sediment transport and bed erosion is deduced. Observation in flood season shows that the sediment transport plays an important role on the pollutant movement. The data analysis shows that water quality indexes variation can not be explained if the impact of sediment transport haven't be reasonably considered. Some data has been collected to build a calculating mode on the exchange between the sediment particles and pollutants. Adaptation of the mode on a one dimensional numerical model has been discussed.

1. BACKGROUND

In recent years many numerical models have been developed with the requirements in water pollution assessment. But many of them don't take account of the impact of sediment transport on the quantitative statement, In most cases with little sediment movement. In China many rivers carry heavy sediment load, and the impact of sediment transport on the water quality cannot be ignored. Water quality of the Three Gorge Project is concerned by scientists all over the world. It is necessary to built a valid one-dimensional numerical model for estimation on the water pollution development. But as well known the sediment concentration is relatively high in Yangtze River, compared to those rivers in developed countries, especially in the flood season. From the observed data analysis it can be seen that the sediment concentration usually is higher than 1 kg/m^3 in flood season in Yangtze River. Just because of the high sediment laden flow, the concentration of pollutants can changes in a large extent and this gives a special difficult for water quality estimation. For example, a flood season observed COD concentration reaches as high as 10 mg/L but only 0.5 mg/L in dry season. For developing a forecasting water quality model which is still effective in flood season, it is necessary to give a careful consideration on the impact of sediment laden flow on the water quality variation.

2. BASIC EQUATION

From the consideration on the material conservation, most of the pollutants can exist in river system as following forms:

1. absorbed to the particles
2. dissolved into the water body
3. absorbed to the bed material

For the sediment laden flow rivers, the pollutants cycle in a river can be shown as Fig.1, except the chemical

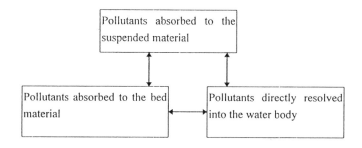

FIG.1 Pollution material exchange flow

purification process. It can be seen that any pollutant in natural waters can exists in various forms. In order to estimate the flux of any pollutant, It is essential to have a quantitative understanding of the pathways of all forms of the pollutants, including particulate movement, dissolved material transport, or sediment deposition. Up to now some analysis on the water quality variation have been made but most of them is on the principles of exchanges and emigration of heavy metal. Little studies can be found on more extensive pollution terms. This paper gives a preliminary consideration on the problem, especially concerned with Three Gorge Project.

According to the observation, some toxic substances, such as phenol, arsenic, chromium, will be adsorbed and transported with sediment, especially with the particle size less than 0.005mm. Zhang et,al [Zhang,1986] had found that a fact as shown in Fig.2. For the case of a natural river, such a fine material will be directly suspended in the water and will be purified during the transport processes. But in a reservoir the sluggishness of the water body will made such material settling in a short distance and these toxic substance will be collected together, and carried up to a harmful level . For the Three Gorge Reservoir, that a fact can not be ignored when the water quality problems need to be considered.

Usual equation of one-dimensional water quality model can be written as:

$$\frac{\partial Ac}{\partial t} + \frac{\partial Qc}{\partial x} = \frac{\partial}{\partial x}((D_x + E_x)A\frac{\partial c}{\partial x}) + AS \tag{1}$$

where A is the cross section area; Q is the water discharge; c is the concentration of the pollutants;

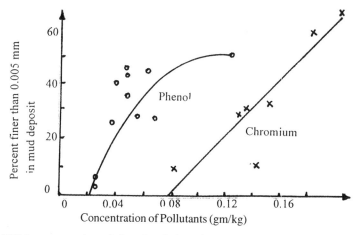

FIG.2 concentration of phenol and chromium vs percentage of fine particles

D_x and E_x are longitudinal diffusive coefficients respectively, and S represents a total sources which is the most variable term. For a numerical model the solution of equation (1) is not so difficult in solving method except the determination of those coefficients, D_x, E_x, and the source term S. Here we are mainly interested in the determination of source term S. Usually this term includes three parts as: C_s, a contribution by sediment deposition or erosion; C_c, degradation by biological-chemical function; and C_q, addition of outer pollution sources, including natural release from bed material and some factories pollutants drainage. Actually in the water flow with a certain of sediment movement, some pollutants will varied with the concentration of the sediment content. So the equation (1) can be rewritten as:

$$\frac{\partial Ac}{\partial t} + \frac{\partial Qc}{\partial x} = \frac{\partial}{\partial x}((D_x + E_x)A\frac{\partial c}{\partial x}) + C_s + C_c + C_q \tag{2}$$

2.1 Cs, Pollutants Concentration Variation in Unite Time and River Length by Sediment Erosion and Deposition

For the river system with enough sediment transport and extensive alluvial process, the concentration of sediment is a varied variable which induces a obvious change of pollutant concentration. Such variation is based on the erosion of bed material erosion or deposition of the suspended material. Owing to the absorption or desorption, the contaminated material in water body mainly comes from: a. the bed material absorbed with some toxic material eroded by the water flow; b. sediment deposition of the suspended sediment witch will reduce the concentration of the pollutants in the water body; and c. Sediment flux variation. Under the assumption that a linear relation exists between the concentrations of pollutants and sediment particulate, the function of sediment movement can be expressed as:

$$C_s = k_c S_c + k_d S_e \tag{3}$$

Here, Sc and Se are sediment concentration variation caused by bed erosion and deposition respectively, kc and kd are coefficients corresponding to erosion and deposition process respectively, obviously the unsteady variation of sediment movement is ignored.

2.2 Cc, Degradation by Biological-Chemical Function

This term is a usual consideration on most numerical model and here we simply use those common results. The typical function can be expressed as:

$$C_c = C_{co}e^{-k_1 t} \tag{4}$$

where, C_{co} is a enter section concentration of some pollutants; k_1 is a coefficient reflecting the biological-chemical process. All these parameters can be previously determined by an empirical test which is not the focal point of this paper.

2.3 Cq, Addition of Outer Pollution Sources

Outer pollution sources include direct addition of factory pollution drainage and natural release from bed material. For the initial running stage of Three Gorge Reservoir, the sediment deposition is not so serious and the bed material is still mainly consisted of coarse sediment particles. As expressed above that the toxic material is usually absorbed to fine particles, so the natural release of pollutants from the bed material can be ignored here. The term C_q, then, can be replaced by observed outer pollution sources directly.

3. RELATION BETWEEN THE SUSPENDED SEDIMENT AND POLLUTANT

3.1 Determination on the Exchange Coefficients

For a given river reach, the sediment flux variation is caused by river bed adjustment, bed erosion or deposition. The scoured material is mostly an accumulative result of the deposition process and the concentration of the pollutants is relatively higher than suspended sediment. Thus the coefficients Kc and Ks in Eq.(3) is different in most of the flood process. But for a given river reach, in a limited time period, the sediment properties keep a relative-consistency, and we have:

$$k_c = k_d = k_t \quad \text{and}$$
$$C_s = k_t s_t \tag{5}$$

where, k_t is a comprehensive exchange coefficient between the suspended sediment concentration s_t and pollutantl content C_s. For practical running of a numerical model, coefficient k_t need to be determined firstly. Determination on the variation of k_t is a difficult task for all the pollutants, even under the   linear variation assumption. Here based on the data observed in Yangtze river, an empirical relation on the values of k_{COD} of COD with the concentration of suspended sediment is found as shown in Fig.3 . The figure can be simply described by linear relation as:

$$C_{cod} = k_{cod} S_s \tag{6}$$

Through a regression analysis, also referring to other research[Chu Junda, Xu Huici, 1994], the coefficient k_{COD} can be obtained as constant 0.0053 (mg/L/mg/m^3).
Fig.3 is the most simple situation in the relation of sediment concentration and pollutants content. Actually, most of other contaminated materials such as TN, TP do not always show such a simple relation. Fig.4 is a data set of TN and SS(suspended sediment) observed in a reach of Three Gorge Reservoir from which a

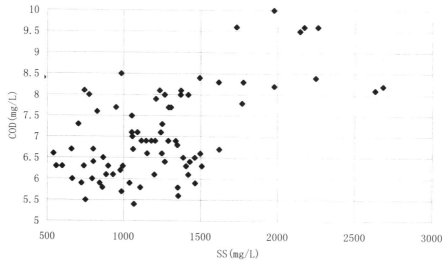

FIG.3 observed relation of SS and COD

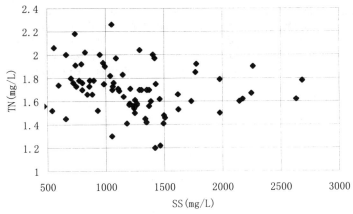

FIG. 4 observed relation of SS and TN

linear relation is not so obvious. But it should be noted that the function of sediment content on the variation of TN concentration still exists. How to reflect such relation in numerical model is a further research target.

3.2 Determination on the Sediment Concentration

For a natural river, the sediment transport is usually a non-equilibrium process and the erosion or deposition depended on the water flow conditions and composition of sediment particulate. From previous research, for a given river reach in a one-dimensional flow, the sediment concentration of the lower section of the reach can be calculated by:

$$S_l = S_* + (s_0 - s_{*0})\exp(-\frac{\alpha wL}{q}) + (s_{*0} - s_*)\frac{q}{\alpha wL}[1 - \exp(-\frac{\alpha wL}{q})] \tag{7}$$

where α is a coefficient which varies depending on the erosion or deposition and takes a value 0.25 in bed erosion or 1.0 in sediment deposition. s_0, s_{0*} are the sediment concentration and transport capacity at the enter section of the studied river reach respectively; s_t, s_* are the sediment concentration and transport capacity at the lower section of the studied river reach respectively; L is the river reach length; w is the settling velocity.

For numerically resolving equation (7), a transport capacity calculating method need to be developed. Owing to the complexity of the problem, most of the previous investigations are concentrated on the empirical relations. Before a more successful theoretical relation is found an empirical relation contributed by other sediment model[Chen, Z.C., Fu, R.S., and Zhang, R.1994] is employed in present model:

$$S_* = 0.147 \cdot (\frac{V^3}{gHw})^{0.93} \tag{8}$$

where, V is the flow velocity; H is water depth and g is the specific gravity . Eq.(8) has been verified by an extensive data sets observed in Yangtze River and here the reasonability used in a water quality model is believed to be naturally confirmed.

4. CONCLUSION

The concentration of pollutants is obviously affected by the sediment movement, including bed material erosion, exchange between the sediment particulate and water body, and sediment deposition. From the data observed in field it is found that the function of sediment movement on the water quality is obvious. With the data of COD variation depending on sediment concentration in Yangtze River as an example, a one-dimensional numerical model is developed. For estimating the function of sediment movement on pollutants transport, a non-equilibrium suspended sediment transport mode is introduced into present water quality model. Owing to present research is still in its early stages, a further verification on the model need to be done in following research work.

5. ACKNOWLEDGMENT

Some data used in this paper was provided by Bureau of Hydrology, Changjiang Water Resources Committee, China. Permission was granted by Dept. of Tech. & Inter. Coop., Executive Office of State Council Three Gorges Project Construction Committee to publish this information.

6. REFERENCES

Chen, Z.C., Fu, R.S., and Zhang, R.(1994), The impacts of 172m scheme on Sedimentation and Navigation Problems in Three Gorges Project, Journal of Sediment Research, No.1 (in Chinese)

Zhang, Q.S., Jiang, L.S., and Lin, P.N.(1986), Environmental problems Associated with sediment deposition in Guanting Reservoir, International Journal of Sediment Research, Vol.1 No.1

Chu, J.D., and Xu H.C.(1994), Study for the effect of scouring and settling of river sediment on water quality, Journal of Hydraulic Engineering, No.11 (in Chinese)

Environmental Hydraulics, Lee, Jayawardena & Wang (eds) © 1999 Balkema, Rotterdam, ISBN 90 5809 035 3

Numerical simulation of flow in scour of downstream of Three Gorge spillway

Yongcan Chen & Junrui Dong
Department of Hydraulic Engineering, Tsinghua University, Beijing, China

ABSTRACT: The flow in the scour of downstream of Three Gorge spillway induced by the impinging jet is simulated by an implicit numerical model, which is developed by using the K-ε two equation turbulence model and irregular boundary-fitted coordinate system. The computed maximum velocity near the bottom and the pressure of every measured point in the scoured region are compared with experimental data at five special sections. It is shown that the computed values   agree well with the measured data. The flow velocity and pressure obtained by numerical simulation can describe the changes of velocity in the whole scour,   the whirling state on two sides of the jet, and the energy exchange directly. The strong correlation revealed in the comparisons between the computed and experimental results   indicates that the present method is one of an effective tools for the study of safety in high dam construction.

1. INTRODUCTION

Many high dams, such as the Three Gorge Project are being constructed or will be constructed in   China. It is very important to simulate the scour by overfall jet from the spillway because it is dangerous to the dam if the scour is deep and large. The numerical method to simulate the complicated flow in the scour is very useful and effective because it is difficult to measure the velocities in the scour but the flow characteristics in the scour is the main reason to cause the riverbed scouring. One of the most important steps required to accurately solve the problem involves the proper location of the nodal points in the flow region. A variety of techniques are developed for the task of numerical grid generation(Thompson, 1982; Chu, et al 1991). The method that generates the grid by solving   a system of elliptic equations is most popular. The flow characteristics in a special scour was   simulated by a upwind finite element method by Chen and Xu(Chen, et al 1993) but the numerical results had not been verified because there is no measured data for the given scour. A boundary-fitted numerical method was employed to simulated the velocities and pressure field in scour by Wang and Chen(Wang, et al 1995). Based on these the flow characteristics in the scour downstream of Three Gorge spillway was simulated in this paper.   The numerical results of   pressure and velocities are verified by the measured data. It also provides a dependable quantitative basis for analyzing the flow characteristics in the scour and the reason of scouring in downstream of the spillway.

2. MATHEMATICAL FORMULATION

The governing equations in both Cartesian coordinates and general curvilinear coordinates can be written in conservation law form. The governing equations can be written in Cartesian coordinates :

$$\frac{\partial}{\partial x}\left(\rho U\phi - \Gamma_\phi \frac{\partial \phi}{\partial x}\right) + \frac{\partial}{\partial y}\left(\rho V\phi - \Gamma_\phi \frac{\partial \phi}{\partial y}\right) = R_\phi(x,y) \tag{1}$$

where ϕ is a general variable, ρ, Γ_ϕ and R_ϕ are the density, diffusion coefficient and source term

Table 1 The governing equations

Equations	ϕ	Γ_ϕ	R_ϕ
Continuity	1	0	0
X- momentum	U	$\mu + \mu_t$	$-\dfrac{\partial P}{\partial x} + \left(\dfrac{\partial}{\partial x}[(\mu + \mu_t)\dfrac{\partial u}{\partial x}] + \dfrac{\partial}{\partial y}[(\mu + \mu_t)\dfrac{\partial v}{\partial x}] \right)$
Y- momentum	V	$\mu + \mu_t$	$-\dfrac{\partial P}{\partial y} + \left(\dfrac{\partial}{\partial x}[(\mu + \mu_t)\dfrac{\partial u}{\partial y}] + \dfrac{\partial}{\partial y}[(\mu + \mu_t)\dfrac{\partial v}{\partial y}] \right)$
Turbulent energy	K	$\mu + \dfrac{\mu_t}{\sigma_k}$	$\rho\left\{ \mu_t \left(2[(\dfrac{\partial u}{\partial x})^2 + (\dfrac{\partial v}{\partial y})^2] + (\dfrac{\partial u}{\partial y} + \dfrac{\partial v}{\partial x})^2 \right) - \varepsilon \right\}$
Turbulent energy dissipation	ε	$\mu + \dfrac{\mu_t}{\sigma_\varepsilon}$	$\dfrac{\varepsilon}{k}\left\{ C_1\mu_t \left(2[(\dfrac{\partial u}{\partial x})^2 + (\dfrac{\partial v}{\partial y})^2] + (\dfrac{\partial u}{\partial y} + \dfrac{\partial v}{\partial x})^2 \right) - C_2\rho\varepsilon \right\}$

respectively, when ϕ stands by 1, U, V, k or ε, then the corresponding Γ_ϕ, R_ϕ(x,y) can be written as in table 1.

Where μ is the viscosity, μ_t is eddy viscosity, $\mu_t = c_\mu k^2/\varepsilon$.

When new independent variables ξ and η are introduced by employing an orthogonal grid system in transformed plane, equation (1) changes according to the general transformation $\xi=\xi(x,y)$, $\eta=\eta(x,y)$. A schematic illustration of the relations between the physical domain for the flow in the scour and the transformed domain is shown in Fig.1. Equation (1) can be rewritten in (ξ,η) coordinates as follows:

$$\frac{\partial}{\partial x}\left(\rho G_1 \phi - \Gamma_\phi q_1 \frac{\partial \phi}{\partial x} \right) + \frac{\partial}{\partial y}\left(\rho G_2 \phi - \Gamma_\phi q_2 \frac{\partial \phi}{\partial y} \right) = S_\phi(\xi,\eta) \tag{2}$$

where $G_1 = u\xi_x + v\xi_y$; $G_2 = u\eta_x + v\eta_y$; $q_1 = \xi_x^2 + \xi_y^2$; $q_2 = \eta_x^2 + \eta_y^2$, G_1, G_2 are the contravarinat velocities written without matrix normalization, $S(\xi, \eta)$ is the source term of the governing equation in (ξ,η) coordinates.

The continuity equation can be transformed to the (ξ,η) plane as

$$\frac{\partial}{\partial x}(\rho G_1) + \frac{\partial}{\partial y}(\rho G_2) = 0 \tag{3}$$

The solution algorithm used in the calculation is extension of the SIMPLE algorithm which was developed by Patankar(Patankar, 1980). After some modification, the SIMPLE solution procedure is extend to the present curvilinear coordinate system. It is noted that, after suitable interpolations for variables whose values are unknown on the control surface, a relation between ϕ_p and the neighboring variables can be written as follows:

$$a_P\phi_P = a_E\phi_E + a_W\phi_W + a_N\phi_N + a_S\phi_S + b \tag{4}$$

Since the velocity distribution is obtained by momentum equation with a given pressure field, and usually does not satisfy the continuity equation. the discretized momentum equation for U and V velocities can be expressed as

$$a_p^u U_p = \Sigma a_{nb}^u U_{nb} + b^u + \left(P_\xi \xi_x - P_\eta \eta_x \right)A^u \tag{5}$$

$$a_p^v V_p = \Sigma a_{nb}^v U_{nb} + b^v + \left(P_\xi \xi_y - P_\eta \eta_y \right)A^v \tag{6}$$

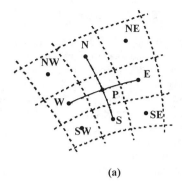

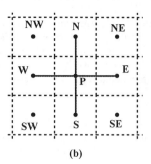

(a)   (b)

(a) Grid on physical domain   (b) Grid on computational domain
Fig.1 Types of grid system in the physical and computational domains

where b^U, b^V are the source term without the pressure gradients, and A^U, A^V are the areas of control-volume faces.

In the above two equations, the velocity components and pressure will not satisfy the continuity equation, the estimated pressure must be corrected, and the velocity components are also corrected correspondingly. The corrected forms for U and V can be obtained by

$$U'_p = -(P'_\xi \xi_x + P'_\eta \eta_x)\frac{A^u}{a^u_p} \tag{7}$$

$$V'_p = -(P'_\xi \xi_y + P'_\eta \eta_y)\frac{A^v}{a^v_p} \tag{8}$$

The corresponding correction formulas for the contravariant velocities are

$$G'_1 = -(\frac{A^u \xi_x^2}{a^u_p} + \frac{A^v \xi_y^2}{a^v_p})P'_\xi \tag{9}$$

$$G'_2 = -(\frac{A^u \eta_x^2}{a^u_p} + \frac{A^v \eta_y^2}{a^v_p})P'_\eta \tag{10}$$

By substituting Eqs.(9) and (10) into Eqs(3), the pressure corrected equation is obtained

$$a_p P'_p = a_E P'_E + a_W P'_W + a_N P'_N + a_S P'_S + b \tag{11}$$

where b represents a mass imbalance and is

$$b = [(\rho G'_1)_W - (\rho G'_1)_E]\Delta\eta + [(\rho G'_2)_S - (\rho G'_1)_N]\Delta\xi \tag{12}$$

It can be seen that the solution of the problem is obtained when   b is less than a given minimizing value (10^{-4} is given in this paper).

3. THE VERIFICATION AND ANALYSIS OF THE COMPUTATIONAL RESULTS

Under the experimental simulation condition of the three-Gorge spillway when the reservoir water level is 175.0m and the downstream water level is z=77.05m, and there is no hummock, the time-averaged flow field and pressure on the bottom of the scour of downstream of Three Gorge spillway has ever been measured (Chai, 1991). A numerical simulation method is applied to calculate the flow field in scour in this paper, and the computational results has been verified according to the experimental data. The parameter of model in the experiment and calculation is, the model scale λ=1:100, then the volume of flow per width q=0.171m^{2}s^{-1}, the velocity of coming flow u_0=3.94ms^{-1}, the angle of dip β=38°, the depth of downstream water h_t=0.37m. The time-averaged velocity field in numerical simulation is shown in figure 2, and the computational results of the time-averaged pressure contour in the scour are shown in figure 3. The computational maximum velocity and measured maximum velocity of every section are compared in table 2. Correspondingly it can be seen that the computed values agree well with the measured data. The flow velocity and pressure obtained by numerical simulation can describe particularly the changes of velocity in the whole scour, the whirling state on two sides of the jet, the advection-diffusion process and the energy exchange directly. One of the most important problems in the scouring process to riverbed induced by jet, is the pressure on the bottom as well as the velocity near the bottom in the scouring region. It can be seen the time-averaged pressure increases gradually with the decrease of the velocity of the jet in the downstream scour. The pressure contour of the deflecting flow in the scour comes closer and closer to the scour bottom, that is to say the pressure changes very dramatically. At the same time, due to the effect of the entertainment between the jet and the surrounding water in the scour, the kinetic energy of the jet and pressure energy exchanges to each other, the flow lost part of its energy. The pressure contour is parallel to the water surface when the flow is out of the scour, that is to say, the distribution of pressure is approximately the same as static fluid. The comparison between the computational and experimental pressure of every measured point is also shown in table 3. The values of the computational pressure is a little bigger than measured, but this is acceptable and reasonable. One reason is that the effect of gas isn't considered in calculation, but in fact, there is a lot of gas involved in flow during the process of jet into the scour and the gas bulb went up from the water surface, a part of energy is also lost. Another reason is the hypothesis of fixed surface can't simulate the suitable position of the free surface.

4. CONCLUSIONS

The flow field and pressure field in the scour of downstream of Three Gorge spillway are successfully simulated by using a numerical method and the results is verified by the experimental data. This computational results are very important for analyzing the effect of scour in downstream of the spillway. The strong correlation revealed in the comparisons between the computed and experimental results for such a kind of complicated flow in the scour indicates that the present method is an effective tool and is a compensatory method to the experiment for the study of the safety of high dam construction. It also provides a dependable quantitative foundation for analyzing the flow characteristic in the scour and the effect on scour due to the jet into the flow in downstream of the spillway. At the same time the numerical method developed in this paper is easy to be used in the process of selecting and optimizing design schemes,

Table 2   Comparison between the computed and measured value of the velocity near the bottom of scour

Measured sections	1-1	2-2	3-3	4-4	5-5
Measured value u_m (ms^{-1})	1.40	1.50	1.60	1.45	1.0
Computed value u_c (ms^{-1})	1.21	1.42	1.62	1.37	1.10

Table 3   Comparison between the computed and measured pressure on the bottom of the scour

measured points	A	B	C	D
measured value p_m (10^4Nm^{-2})	0.521	0.611	0.591	0.526
computed value p_c (10^4Nm^{-2})	0.608	0.666	0.627	0.539

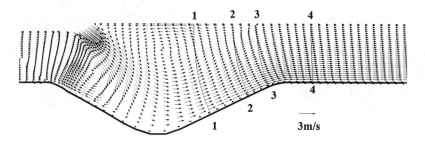

Fig.2 The velocity in the scour of downstream of three gorge spillway

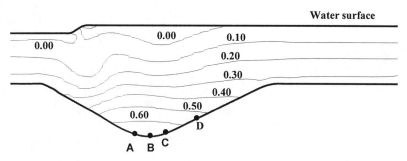

Fig.3 The pressure in the scour of downstream of three gorge spillway

and much more time can be saved in numerical comparisons of the schemes because it is easier to change the parameters in numerical simulation than in experiment.

5. REFERENCES

Cai, Hua (1991). *Study on pressure and velocity fluctuations of scour pool flow downstream the Three Gorge spillway*, Master Thesis, Dept. of Hydraulic Eng., Tsinghua Univ., Beijing, China.

Chen, Yongcan and XU, Xieqing, (1993). "Numerical simulation of the flow characteristics of a free overfall jet into the downstream scour", Journal of Hydraulics, No.4, 48-54.

Chu, Cheng, Ping & Wu, Tser Son, (1991). "Study on the flow fields of irregular-shaped domain by algebraic grid-generation technique", JSME, 34(1), 69-77.

Wang, Yanmin, Chen, Yongcan and Dong, Junrui, (1995). "Numerical simulation of the flow in the scour pool downstream a high spillway", Journal of Hydrodynamics, 10(2), 125-134.

Patankar, S.V., (1980). *Numerical heat transfer and fluid flow*. McGraw Hill Book Co., New York.

Thompson, J.F.(ed.), (1982). *Numerical grid generation*, North-Holland, New York.

Environmental Hydraulics, Lee, Jayawardena & Wang (eds) © 1999 Balkema, Rotterdam, ISBN 90 5809 035 3

Effects on the channel by diversion from the Yellow River

Hou Suzhen, Chen Xiaotian, Li Yong & Yue Dejun
Institute of Hydraulic Research of YRCC, Zhengzhou, China

ABSTRACT: The paper studies development status and characteristics of distribution of diverting water and sediment from the Yellow River basin. The effects on water and sediment of the lower reaches and evolution of erosion and deposition of the channel by the diversion are analyzed with a great amount of data of hydrology and sediment .

1. INTRODUCTION

The Yellow River is the important water source of the Northwest and North China. The irrigation history was as early as in West Han Dynasty. After 1949, the amount of diversion had been increased tremendously along with the development of national economy. Viewing from the whole river, the average water consumption before 50's was about 5 billion m^3 and about 30 billion m^3 in 90's , the utilization ratio of water resources has reached to about 50%.

The most part of the Yellow River course belongs to an alluvial channel. The characteristic of evolution mainly depends on conditions of coming water and sediment and boundary condition of the riverbed. In the last few years, the development and utilization of the water resources have brought huge benefits to industrial and agricultural productions along the river. But it has also caused changes of water and sediment conditions and effects on riverbed evolution. Therefore, the paper makes a special study on this issue and tries to provide a scientific basis for further development and utilization of water resources of the Yellow River.

2. BASIC CHARACTERISTICS OF WATER AND SEDIMENT DIVERSION

2.1 Continuous Increase of the Amount of Diversion in the Upper and Lower Reaches

After 1949, the Yellow River basin had conducted irrigation facilities construction in a large scale, making the amount of diversion increased continuously. Comparing with 40's , the amount of diversion had been increased 4 times in 80's , among which, agricultural consumption makes 90 - 95% of the total. Before 50's, the development of the region in the upper reaches (above Hekouzhen) was slow, while the water consumption was increased from 7.9 billion m^3 in 50's to 14 billion m^3 in 90's ; the average diversion in the middle reaches (the region between Hekouzhen and Sanmenxia) was 2.1 billion m^3 in 50's to 4 billion m^3 in 90's, almost 2 times of that in 50's and in the area of the lower reaches downstream of Sanmenxia had been developed from no diversion to diversion and the amount of diversion has reached to 12 billion m^3 in 90's (see Fig.1).

The amount of sediment diversion, which is closely related to the sediment concentration of coming water does not increase simultaneously with water diversion. Under the conditions of same amount of water diversion, if it is a year of more sediment contained, and more sediment will be diverted, then if it is a year of less sediment contained and less sediment will be diverted. Take the conditions in 50's and 60's for an example, the mean annual diversion is 13.5 billion m^3 and 17 billion m^3 respectively, while the mean annual sediment diversion is 170 million tons and 140 million tons respectively; another example is in 70's and 80's that the mean annual water diversion is 23.5 billion m³ and 27.7 billion m³, while the sediment diversion is 270 million tons and 194 million tons respectively.

The diversion proportion has been changed along with the development of irrigation in each region. In 50's, the diversion of the upper reaches makes up 58% of the total, middle reaches area 15% and lower reaches area 26%. In 80's , the diversion of the upper, middle and lower reaches was 44%, 14% and 42% of the total respectively and the sediment diversion at the upper, middle and lower reaches was 16%, 22% and 61% of the total respectively. Thus, it can be seen that the water diversion of the Yellow River concentrates in the upper and lower reaches and sediment diversion in the lower reaches.

2.2 Uneven Distribution within a Year

Taking 1980 - 1992 period for an example, diversion of the whole river concentrates from May to August, making up 52.2% of a whole year and the mean annual diversion is 3.2~4 billion m³ each month. The industrial consumption and domestic use are close for each month. The seasonal irrigation has made water consumption relatively concentrated within a year. Due to the difference of geographical position and regional crops that makes the allocation of diversion in different areas along the Yellow River difference. See Fig.2. The upper stream diversion concentrates in the period of May - July, making up 53.2% of the total. The amount of diversion in May and June is 3.4 times and 2 times of that of the observed volume. In 90's, the diversion has reached more than 900 - 1000 m³/s. Due to freezing of Ningxia - Inner Mongolia section during Jan. - Feb. and Dec., there is no diversion. 45% of diversion of the lower reaches is concentrated in March, April and May and the amount of diversion is 4.74 billion m³, making up about 60% of the observed coming water at the same period. There are many tributaries in the middle reaches and usually more heavy storms happened in flood season. During flood season and rainy season, the amount of diversion is less. While during dry season and droughts, the amount of diversion is relative greater. All these characteristics have caused dispersed diversion of that region and the difference of water consumption of each month of a year is not great.

3. EFFECTS OF DIVERSION ON WATER AND SEDIMENT

3.1 The Process of Runoff of the Main Channel Changed

The majority water of the Yellow River comes from the region above Huayuankou. In the natural conditions, water comes from the region above Hekouzhen occupies 57% of that of Huayuankou, while water diversion is concentrated in the regions above Hekouzhen and downstream of Huayuankou. The difference of

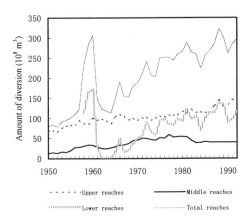

Fig.1. The Process of Development of Diversion at Different Sections in Each Year

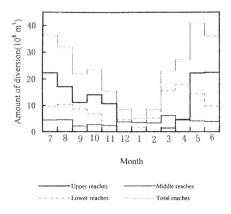

Fig. 2 Allocation of Diversion Within a Year During the Period of 1980 - 1992

688

concentrated regions of coming water and diversion makes the contradiction of supply and demand of water resources of the Yellow River increasingly acute. For an example, the observed water volume at Lanzhou Station in 80's is a little greater than that of in 50's. It has been reduced somewhat at Hekouzhen, Sanmenxia makes up 84% of that in 50's and Lijin 60%. The water volume along the river has become less and less, while the natural runoff along the river is close to that of in 1950s. In 1992, the water volume at Lanzhou Station is close to that of in 50's, 65% of that of 50's at Hekouzhen, 62% at Huayuankou and only 32% at Lijin. This has been shown fully that water volume has become less and less because of diversion along the river.

Due to concentrated and large amount of diversion, uneven distribution within a year and different ratio of monthly diversion over relevant natural runoff, the process of annual runoff flowed into the lower reaches has been changed. For instance, during the period of 1980~1992, diversion at the upper stream is 36% of natural runoff and concentrated in May, June and July, making up 71% of relevant natural runoff in May and June. Reservoir regulation has made water volume reduced in flood season, while is replenished in non-flood season. But, water consumption is great in April-June and relatively to say that proportion of water in Jan.-feb. and Dec. has been increased. Comparing with the situations in 50's or with long series of much less diversion of 1919 - 1949, the monthly hydrographic curve within a year is tending even.

3.2 Increased Opportunity of High Sediment Contained Floods

Different source of water and sediment is one of the main traits of coming water and sediment of the Yellow River. The overwhelming majority of sediment of the Yellow River comes from the reach between Hekouzhen and Tongguan where the duration of flood is short, small flood volume and high sediment content. The flow with low sediment content and comes from the region above Hekouzhen can dilute the floods come from the middle reaches and greatly reduced the sediment concentration of flow. According to the statistical data of water and sediment volume of the middle reaches in 1960 - 1989 that 90% of sediment of He-Long reach concentrates in flood season, the volume of coming sediment is 616 million tons and coming water only 2750 million m^3. The average sediment concentration in flood season is 224 kg/ m^{3,} which usually concentrates in a few events of heavy storms in July and August. The average diversion during that period in the region above Hekouzhen reaches to 600 - 900 m^3/s and the capability of sediment delivery of the channel is direct proportion with the high power of discharge. Therefore, the diversion of the upper reaches has greatly effected the dilution of hyper sediment concentrated floods of the middle reaches, creating the opportunities of happening of hyper sediment concentrated floods.

3.3 Effects on No-flow by Diversion

The reduction of water along the river and severe shortage of water in the lower channel have caused no-flow year by year. In 23 years from 1972 - 1994, there are 17-year no-flow, accumulating 400 days no water at Lijin station. It has become even worse in 1990s. According to the statistical data that the most part of no-flow are occurred in the period of March - July. It is shown by the 23-year data that the no-flow days from May to July makes up 76.8% of the total days, in which, 43% in June, almost no water through out a whole year since 1992 and once in a while no water during autumn sowing.

The coming water from the upper stream of the Yellow River forms the base flow of the middle and lower reaches which has been reduced because of heavy diversion in those regions. Especially the rapid increase of diversion in May - July has made the water volume entered into the middle and lower reaches greatly reduced. For an instance, the total diversion occupies 53% of a whole year during 1980s and diversion in May and June occupies 90% and 60% of natural runoff, averaging 2 - 4 times of the observed volume.

The coming water of the lower reaches region downstream of Sanmenxia only makes up 11% of the total. Thus, the water demands of the lower reaches region are met mainly depended on the coming water from the upper and middle reaches. Since irrigation has been put on the right track in 1965, the diversion in the lower reaches has being developed at full speed, in March - May is the maximum which is shown by Fig.2.

The result of diversion at the upper stream has reduced the base flow of the middle and lower reaches. The water volume of the main river while flowing through the middle reaches can be replenished at different degrees, while in the lower reaches where it is heavily diverted it has no source to be replenished, making the volume further less until the water is totally consumed and no-flow happened. Therefore, it can be seen that the accumulated repetition of consumption at the upper and lower reaches and concentrated diversion in May - July are the direct reason of cut off flow in the lower Yellow River.

4. EFFECTS ON EROSION AND DEPOSITION OF THE CHANNEL

4.1 Mechanism Analysis

According to the study conducted by Mai Qiaowei, Zhao Yean and Pan Xiandi that the sediment delivery of the Yellow River can be basically shown by the following formula when sediment content is high:

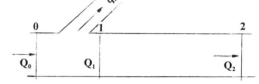

$$G_s = KQ^\alpha S_0^\beta \qquad (1)$$

Fig. 3. A Sketch Map of Water and Sediment Diversion

in which, Gs - coefficient of sediment delivery; t/s;Q - discharge, m³/s; S_0 - sediment content of an upper station, kg/m³; K - coefficient; α, β - indexes which are related to the morphology of riverbed. Based on the study conducted by many Chinese experts that when the sediment content of flow is very low, $\alpha = 2$, $\beta = 0$; when the sediment content of flow is higher, $\alpha = 1.13 - 1.33$, $\beta = 0.7 - 0.9$. In Fig.3, suppose the diverted sediment content is

$$S_i = \tfrac{\lambda}{2}(S_o + S_1) \qquad (2)$$

The diversion discharge is $q_i = \eta Q_0$.

It is derived according to the balance of sediment that after diverting water and sediment, the specific value (called increased deposition ratio) of increased volume of deposition and the coming sediment at intake cross section of the river section 0 - 1 is:

$$\Delta_{0-1} = \eta \frac{S_1}{S_0}[\alpha_1 - \frac{\alpha_1(\alpha_1 - 1)}{2}\eta + \frac{\alpha_1(\alpha_1 - 1)(\alpha_1 - 2)}{6}\eta^2 - \frac{\lambda}{2}(1 + \frac{S_0}{S_1})] \qquad (3)$$

It can be seen that the increased deposition ratio after diverting is related to the erosion and deposition (S_1 / S_0), water diversion ratio (η), sediment diversion ratio (λ) and the characteristic of channel.

After diverting water and sediment, the increased deposition ratio of the lower river sections 1 - 2 is :

$$\Delta_{1-2} = \eta\{-\alpha_1 + [\alpha_2 + \beta_2(\alpha_1 - 1)]\frac{S_2}{S_1}\} \qquad (4)$$

It can be seen that the effects is mainly related to water diversion ratio (η), erosion and deposition of the original section and the characteristics of channel.

In order to analyze and compare the effected degree of diversion on the section and the section below, we suppose the erosion and deposition of the original section is basically balanced (under the conditions of without water and sediment diversion), then, $S_0 / S_1 = 1$ and $S_2 / S_1 = 1$, in the meanwhile it diverts the water which sediment content is 70% of that of the main river, when sediment content is higher that $\alpha_1 = \alpha_2 = 1.2$, $\beta_2 = 0.8$, from (3) and (4) it obtains that .

When sediment content is less, $\beta_2 \approx 0$, it is even more smaller. It has been shown that the effects of water and

$$\Delta_{1-2} / \Delta_{0-1} = 0.32 - 0.46 \tag{5}$$

sediment diversion is greater on this section than that of the lower ones.

4.2 Analysis and Calculation of Effects of Erosion and Deposition of the Channel by Water and Sediment Diversion

4.2.1 Effect on the Channels of the Middle and Lower by Diversion at the Upper Yellow River

The section from Hekouzhen to Longmen is a gorge type channel and the variations of erosion and deposition of the riverbed is not great. The erosion and deposition of the middle reaches mainly refers to the shifting channel from Longmen to Tongguan. It is shown by the calculation results of 1980 - 1992 that the average annual increased deposition in the middle reaches is 55 million tons, only 7 million tons in flood season and 48 million tons in non-flood season because of diversion at the upper stream.

Due to diversion and changes of the capability of sediment delivery of the upper stream, the annual mean reduction of sediment of Hekouzhen is 88 million tons and the sediment discharged to the lower reaches by Sanmenxia Project is reduced by 147 million tons. According to the calculation conducted by a mathematical model of the lower Yellow River that the average increased deposition of the lower reaches is 81 million tons in 13 years, but the increased deposition in each section is greatly different which is shown in Table 1. Viewing from a whole year that the increased deposition of the section above Gaocun is great which is mainly increased in non-flood season or the reduction of erosion and; the deposition of the section below Gaocun is increased in flood season and reduced in non-flood season. The diversion in non-flood season has kept away from the range of discharge which will cause severe deposition in the channel downstream Aishan, reducing the channel deposition.

The effect on erosion and deposition of channel by water and sediment diversion is a very complex problem. The results as studied above are related with the conditions of coming water and sediment, diversion ratio and status of erosion and deposition of the channel in early days. By the calculation and analysis of different combinations of coming water from the Upper stream and coming sediment from the Middle , it is shown that under the conditions of the same amount of diversion and when the amount of coming water of Hekouzhen is not in big difference that the more sediment comes from the Middle , the greater effect is on increased deposition of diversion ; when the amount of coming sediment is basically the same, the more water comes from the upper stream, the smaller effect is on the channel by diversion. Therefore, it can be seen that the effects on erosion and deposition of the channel by diversion are also closely to the combination of coming water and sediment.

4.2.2 Effects on erosion and deposition of the lower river course by diversion in the lower reaches

Take the water and sediment data of 1974 - 1987 as the basis of calculation and analysis that the average coming water is 40.59 billion m^3 and coming sediment 1.01 billion tons in 14 years. The annual average diversion of Sanmenxia - Lijin reach is 9.52 billion m^3 , among which, Sanmenxia -Huayuankou makes up 10.7%, Huayuankou - Gaocun 25%, Gaocun - Aishan 25.6% and Aishan - Lijin 38.7%. The annual average sediment diversion is 150 million tons. The results of analysis and calculation are shown in Table 1. The annual average increased deposition of the lower Yellow River is 16.4 million tons, making up 1.63% of the coming sediment and 25% of the actual deposition at the same period.

Table 1.   Increased Channel Deposition Caused by Diversion   Unit: 10^8 tons

Item	Effects of upper stream diversion			Effects of downstream diversion		
	Flood season	Non-flood season	Whole year	Flood season	Non-flood season	Whole year
Sanmenxia – Huayuankou	- 0.16	0.56	0.40	0.012	0.029	0.041
Huayuankou – Gaocun	0.130	0.24	0.36	0.027	0.081	0.108
Gaocun –Aishan	0.16	- 0.04	0.12	0.023	0.009	0.032
Aishan - Lijin	0.17	- 0.25	- 0.07	0.118	- 0.134	- 0.016
Sanmenxia – Lijin	0.30	0.51	0.81	0.18	- 0.016	0.164

5. CONCLUDING REMARKS

The water resources of the Yellow River are not inexhaustible. To heavily divert and consume the Yellow River water have caused water shortage in the lower channel and frequent cut off flow, affected industrial and agricultural productions and the improvement of natural ecological environment, narrowed the channel, reduced the capability of flooding and increased channel deposition. But, the facts have been shown that the development and utilization of water resources have brought huge economic benefits to the society, speeded up the development of socialist economic construction, which are impossible to be stopped or confined to be developed. For solving this present contradictions, to be suited to the requirements of economic development at a high speed and   advantageous to the safety of the channel, we must tap the latent power of the Yellow River water, greatly popularize water conservation, comprehensively utilize the resources of water and sediment and obtain overall benefits through the combination of characteristic of less water and more sediment of the Yellow River.

3.4 Water quality management

Environmental Hydraulics, Lee, Jayawardena & Wang (eds) © 1999 Balkema, Rotterdam, ISBN 90 5809 035 3

An application of the adjoint technique to locating wastewater discharges into a surface water system

J.Q. Deng & M.S. Ghidaoui
Department of Civil Engineering, The Hong Kong University of Science and Technology, Kowloon, China

D.A. McInnis
Komex International Limited, Calgary, Alb., Canada

ABSTRACT: Water resource quantity and quality issues are often the most crucial factors to consider when governments plan for medium and long term economic development. Frequently, planners designate specific regions for various types and levels of development according to each region's intrinsic attributes and relationship to those infrastructure systems required for development. The planning process is inherently difficult since the impact of development on water quality, and hence on the intended use of each area, is not known *a priori*. As a result, planning for development tends to follow a prescriptive rather than a proactive pattern. Inverse models for surface water quality in the context of planning appear to be advantageous since, for a given development strategy, the water quality standards needed to support and maintain a proposed development plan can be determined in advance. An explicit adjoint model for uniform flows is proposed in this paper that can be easily applied to 1- or 2-dimensional surface water systems to identify feasible regions for siting wastewater discharges while maintaining desired water quality standards at specified target regions.

1. INTRODUCTION

With the population growth, urban migration, and industrial development, the quantities of wastewater produced has increased dramatically. Since the water resources are essential for health, biodiversity preservation, recreation, tourism, fisheries and navigation, it has become a priority for many nations to ensure adequate levels of environmental quality for our aquatic environments. The problem of locating a pollutant source so as to minimize its impact on a receiving water body, such as lakes, rivers and coastal waters, can be solved in a "trial-and-adjustment" manner using hydraulic or hydrodynamic simulation techniques together with pollutant transport modeling. In other words, successive runs of the computer model(s) using a number of alternative locations of the pollutant source are specified, and the resulting pollutant concentrations at a designated "target" area(s) are evaluated with respect to some maximum allowable concentration of water quality parameter(s). In this approach, the physical system properties, initial and boundary conditions are given. The forward model is run to solve for the concentration field arising from the polluted discharge at the candidate location. Feasible locations for siting waste discharges are those which do not violate any of the presumed or established water quality standards at the target region. Problems with this method are that it is time-consuming, prone to subjective interpretation, and only gives solutions for the (usually small number of) alternative locations investigated. Far more useful in terms of providing information relevant to siting decisions would be to solve the inverse problem. Simply stated, given the water quality standard or maximum allowable concentration for a particular pollutant in some target region, and knowing the physical properties of the system, determine the set of all feasible discharge (source) locations which will not exceed the limiting concentration at the target location.

Inverse modeling has been successfully applied to many problems involving parameter estimation in other areas of science and engineering. Many researchers have shown that, subject to certain mathematical requirements, the adjoint formulation technique is often a more efficient method for solving inverse problems. Sykes (1985) and Sun and Yeh (1990) have used the adjoint method to identify parameters in a two-dimensional groundwater flow. Jarny et al. (1991) applied the adjoint technique to heat conduction problems.

Cacuci et al. (1984) and Hall (1986) applied the adjoint formulation method to meteorology and climate modeling. Marchuk (1995) applied the adjoint formulation technique to air pollution problems. The particular virtue of the adjoint method for the surface water quality problem is that the solution can be analytic for a number of practical problems. This offers a quick and efficient way for decision makers to evaluate various wastewater discharge sites. An explicit adjoint model for uniform flows is proposed in this paper that can be easily applied to 1- or 2-dimensional surface water systems to identify feasible regions for siting wastewater discharges while maintaining desired water quality standards at specified target regions.

2. SOURCE LOCATION IN A 1-DIMENSIONAL FLOW

2.1 Adjoint Model in a 1-Dimensional Flow

In a 1-D flow, the objective is to find the feasible region for locating the pollutant source(s) with known strength S subject to the constraint that the pollutant concentration C at the target point ξ at observation time τ is less than or equal to some specified maximum allowable pollutant concentration, i.e., $C(\xi, \tau) \leq C_{all}$. The 1-D transport equation, together with given boundary, initial, and constraint conditions, can be stated as follows:

$$\frac{\partial C}{\partial t} + u \frac{\partial C}{\partial x} = D \frac{\partial^2 C}{\partial x^2} + S\delta(x - x_0)$$

$$C(-\infty, t) = C(+\infty, t) = 0, \qquad C(x, 0) = 0$$

(1)

where C and u are cross-sectional averaged concentration and velocity, respectively; D is diffusion coefficient; S is the pollutant source term; x_0 is the position at which the pollutant source is located; and $\delta(x - x_0) = \begin{cases} 1 & for & x = x_0 \\ 0 & for & x \neq x_0 \end{cases}$ which is used to indicate the source position.

The inverse problem is to find a range of x_0 such that $C(\xi, \tau) \leq C_{all}$. To achieve this the adjoint technique provides an efficient method. To obtain the adjoint form of Eqn.(1), we multiply (1) by $\lambda(x, t)$ called adjoint function, integrate the result over the entire spatial and relevant temporal domain (i.e., x from $-\infty$ to $+\infty$; t from 0 to τ), and set $\lambda(x, t)$ to satisfy the following initial and boundary conditions:

$$\lambda(x, \tau) = 0, \text{ and } \lambda(-\infty, t) = \lambda(+\infty, t) = 0 \quad .$$

(2)

The final result can be written as follows:

$$\int_0^\tau \int_{-\infty}^{+\infty} C \left[-\frac{\partial \lambda}{\partial t} - u \frac{\partial \lambda}{\partial x} - D \frac{\partial^2 \lambda}{\partial x^2} \right] dx dt = \int_0^\tau S\lambda(x_0, t) dt$$

(3)

Let

$$-\frac{\partial \lambda}{\partial t} - u \frac{\partial \lambda}{\partial x} - D \frac{\partial^2 \lambda}{\partial x^2} = \delta(x - \xi)\delta(t - \tau)$$

(4)

then Eqn.(3) becomes

$$C(\xi, \tau) = \int_0^\tau S\lambda(x_0, t) dt .$$

(5)

Eqn.(5) now expresses the concentration at the target point ξ as a function of source location x_0, source strength S, observation time τ, velocity u and diffusion coefficient D. Letting $J(x_0, S, \tau, D, u)$ denote this function. For given constant D and u, this function can be simply expressed as $J(x_0, S, \tau)$. Thus, Eqn.(5) can be written in the following form:

$$C(\xi, \tau) = J(x_0, S, \tau) = \int_0^\tau S\lambda(x_0, t) dt \quad .$$

(6)

Once the solution of λ is known, the concentrations at the target point corresponding to all potential points for locating the source can be obtained by Eqn.(6). To determine λ, solving (4) subject to the conditions (2) gives:

$$\lambda(x_0, t) = \frac{\theta(t)}{\sqrt{4\pi D(\tau - t)}} e^{-\frac{[x_0 - \xi + u(\tau - t)]^2}{4D(\tau - t)}} \qquad \text{where} \quad \theta(t) = \begin{cases} 1 & \text{for} \quad t \leq \tau \\ 0 & \text{for} \quad t > \tau \end{cases} \tag{7}$$

Thus, the final solution for the problem of locating the pollutant source in a 1-D flow becomes:

$$C(\xi, \tau) = J(x_0, S, \tau) = S \int_0^\tau \lambda(x_0, t) dt = S \int_0^\tau \frac{1}{\sqrt{4\pi D(\tau - t)}} e^{-\frac{[x_0 - \xi + u(\tau - t)]^2}{4D(\tau - t)}} dt \tag{8}$$

The integral of Eqn.(8) may be evaluated numerically.

Since the concentration at the target point has been expressed as a function of the source location x_0, we can easily obtain feasible source locations by evaluating Eqn.(8) for a range of x_0 values and comparing the target concentration to the allowable concentration. In this manner, the entire feasible region of source locations can be identified. The adjoint model is efficient for this application because, once the flow velocity is provided, the solution is analytic and easy to evaluate.

The above procedure can be extended to the case of siting multiple waste sources. Let S be the set of sources and let x be the set of source locations. If n is the total number of pollutant sources, S_j is the strength of the jth source, x_j is the location of the jth source and λ_j is given by Eqn.(7), then $C(\xi, \tau)$ can be determined by:

$$C(\xi, \tau) = J(\mathbf{x}, \mathbf{S}, \tau) = \sum_{j=1}^n S_j \int_0^\tau \lambda_j(x_j, t) dt. \tag{9}$$

Once the profile of J along the spatial domain is obtained using model Eqn.(8) for a single pollution source or model Eqn.(9) for multiple sources, it is straightforward to delineate the feasible domain for siting the pollutant source(s) by imposing the condition $J \leq C_{all}$.

2.2 Demonstration of 1-D Adjoint Model

Locating a single pollutant source: Adjoint model Eqn.(8) is employed to solve the problem of locating one pollutant source. The following conditions are used: the 1-D channel under consideration is assumed to be infinite; advection velocity $u = 1.0$ *m/s*, the diffusion coefficient $D = 200$ *m²/s* and the pollutant source strength $S = 10$ *g/l/s*. The protective target point is located at 7,200 *m* from the origin. Fig.1 shows the computed result. Note that if the maximum allowable pollutant concentration at the target point is 7.0 *g/l*, then it is easy to see that the feasible region for the discharge location is $x < 4,500$ *m* and $x > 7,400$ *m*. From Fig.1, general confirmation of the model behavior is evident since proximity of the source to the target region results in increased concentrations at the target.

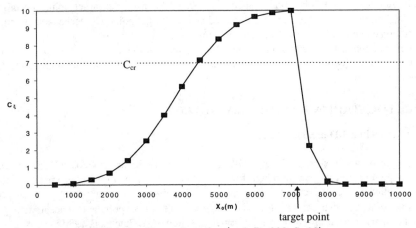

FIG.1 One Source Location (u=1, D=200, S=10)

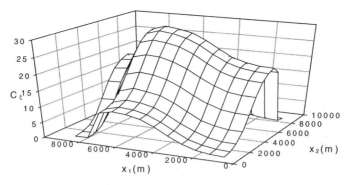

FIG.2 Two Source Location (a) (u=1, D=200, S1=20, S2=10)

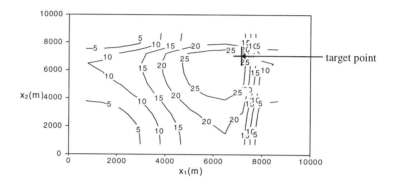

Fig.3 Two Source Location (b) (u=1, D=200, S1=20, S2=10)

Locating two pollutant sources: Adjoint model (9) is employed to solve the problem of locating two pollutant sources. The computational conditions are as follows: the advection velocity $u = 1.0$ *m/s*, the diffusion coefficient $D = 200$ *m²/s*, and the strength of the first source $S_1 = 20$ *g/l/s* while the strength of the second source $S_2 = 10$ *g/l/s*. The target region is located at 7,200 *m* from the origin. In Fig.2, axes x_1 and x_2 represent the possible locations of S_1 and S_2 respectively, and the coordinate C_ξ represents the concentration at the target point. This figure shows that when both sources are located far from the target point, the influence of the sources are at a minimum. If both sources are located near the target point, C_ξ reaches maximum. Plotting the iso-concentration contours of C_ξ, as has been done in Fig.3, permits one to locate the domain of feasible source locations. For example, if the maximum allowable concentration at the target point is 20 *g/l*, then the feasible domain for locating the two sources are the *x*-coordinate pairs represented by the region outside the 20 *g/l* contour line.

3. SOURCE LOCATION IN A 2-DIMENSIONAL FLOW

3.1 Adjoint Model in a 2-Dimensional Flow

For open bodies of water such as lakes, wide rivers, estuarine areas and coastal waters we need a 2-dimensional representation to give meaningful results. Hence, the problem now becomes to find the feasible region for a pollutant source location (x_0, y_0) such that the pollutant concentration at the target point at a specified time τ is less than or equal to the maximum allowable pollutant concentration C_{all}. The 2-dimensional primary transport problem can be written as follows (McCutcheon, 1989):

$$\frac{\partial C}{\partial t} + u\frac{\partial C}{\partial x} + v\frac{\partial C}{\partial y} = D_L\frac{\partial^2 C}{\partial x^2} + D_T\frac{\partial^2 C}{\partial y^2} + S\delta(x - x_0)\delta(y - y_0)$$

$$C(-\infty, y, t) = C(+\infty, y, t) = 0$$

$$C(x, -\infty, t) = C(x, +\infty, t) = 0, \qquad C(x, y, 0) = 0$$
(10)

where D_L and D_T are longitudinal (x-direction) and transverse (y-direction) diffusion coefficients, respectively, and (x_0, y_0) is a potential pollutant source location.

Following the same procedure employed in the 1-D case, i.e., multiplying Eqn.(10) by $\lambda(x, y, t)$, integrating the result over the appropriate spatial and temporal domain, and setting

$$\lambda(-\infty, y, t) = \lambda(\infty, y, t) = 0$$

$$\lambda(x, -\infty, t) = \lambda(x, \infty, t) = 0$$

$$\lambda(x, y, \tau) = 0$$
(11)

yields the following equation:

$$\int_0^\tau \int_{-\infty}^{+\infty} \int_{-\infty}^{+\infty} C\left[-\frac{\partial \lambda}{\partial t} - u\frac{\partial \lambda}{\partial x} - v\frac{\partial \lambda}{\partial y} - D_L\frac{\partial^2 \lambda}{\partial x^2} - D_T\frac{\partial^2 \lambda}{\partial y^2} \right] dx dy dt = \int_0^\tau S\lambda(x_0, y_0, t) dt$$
(12)

Let

$$-\frac{\partial \lambda}{\partial t} - u\frac{\partial \lambda}{\partial x} - v\frac{\partial \lambda}{\partial y} - D_L\frac{\partial^2 \lambda}{\partial x^2} - D_T\frac{\partial^2 \lambda}{\partial y^2} = \delta(x - x_\xi)\delta(y - y_\xi)\delta(t - \tau)$$
(13)

then, Eqn.(12) becomes:

$$C(x_\xi, y_\xi, \tau) = \int_0^\tau S\lambda(x_0, y_0, \xi) dt = J(x_0, y_0, S, \tau)$$
(14)

As long as the adjoint problem Eqn.(13) is solved, the feasible spatial domain for locating the source can be easily obtained by Eqn.(14). Solving Eqn.(13) subject to the conditions (11) produces the following result:

$$\lambda(x_0, y_0, t) = \frac{\theta(t)}{4\pi(\tau - t)\sqrt{D_L D_T}} e^{-\frac{[x_0 - x_\xi + u(\tau - t)]^2}{4D_L(\tau - t)} - \frac{[y_0 - y_\xi + v(\tau - t)]^2}{4D_T(\tau - t)}}$$
(15)

The final solution for pollutant concentration at the target point becomes:

$$C(\xi, \tau) = J(x_0, y_0, S, \tau) = \int_0^\tau S\frac{\theta(t)}{4\pi(\tau - t)\sqrt{D_L D_T}} e^{-\frac{[x_0 - x_\xi + u(\tau - t)]^2}{4D_L(\tau - t)} - \frac{[y_0 - y_\xi + v(\tau - t)]^2}{4D_T(\tau - t)}} dt$$
(16)

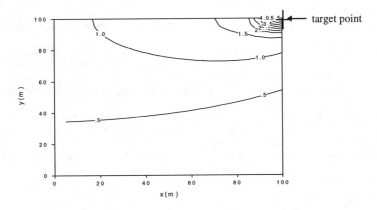

FIG.4 Locate One Source in 2-D Domain (Contour Plot)

The feasible domain for siting the pollutant source can be found by imposing the condition $J \le C_{all}$ once the contour profile of J in the spatial domain is obtained from the solution of the adjoint model Eqn.(16).

3.2 Demonstration of 2-D Adjoint Model

The 2-D adjoint model Eqn.(16) is used to find the domain of feasible locations for a single pollutant source in a two-dimensional spatial domain. Fig.4 shows the computed result for the following computational conditions: $D_L = 20 \; m^2/s$, $D_T = 10 \; m^2/s$, $u = 0.5 \; m/s$, $v = 0 \; m/s$, $x_t = 100 \; m$, $y_t = 100 \; m$, $S = 80 \; kg/m^3/s$ and τ =1001 s. This figure shows that a source located furthest from the target point, has the least influence on the target point pollutant concentration. Conversely, when a source is located near the target point, the maximum concentration at the target point is seen. For a maximum allowable pollutant concentration of 1.0 g/l at the target site, the feasible domain for locating the source can be seen in the contour plot as the region outside the closed contour line labeled 1.0.

4. CONCLUSIONS

Inverse models for surface water quality in the context of planning appear to be advantageous since, for a given development strategy, the water quality standards needed to support and maintain the potential development can be determined in advance. An analytic adjoint model for uniform flows is proposed in this paper that can be easily applied to 1- or 2-dimensional surface water systems to identify the feasible domain for siting wastewater discharges while maintaining desired water quality standards at specified target regions.

Extension of the model proposed in this paper to more complex situations is relatively straightforward. 2- and 3-dimensional flow fields simply add additional terms to the 1-dimensional adjoint formulation. Multiple sources are easily handled by superposition of results for each potential waste discharge location. Multiple target regions can also be accommodated as the intersection of sets of feasible domains for each target region. Extension of the current model to include nonuniform, steady flows would significantly extend the applicability of the model for real-world problems.

5. ACKNOWLEDGMENT

The authors acknowledge the financial support by the Hong Kong University of Science and Technology under the UPGC Research Infrastructure Grant Program, Project number RIG 94/95. EG14.

6. REFERENCES

Cacuci, D. G. and Hall, M.C.G. (1984). Efficient estimation of feedback effects with application to climate models. J. of the Atm. Scie. 13(2) 2063-2068.

Hall, M.C.G. (1986). Application of adjoint sensitivity theory to an atmospheric general circulation model. J. of Atm. Scie. 43(22), 2644-2651.

Jarny, Y., Ozisik, M. N. and Bardon, J. P. (1991). A general optimization method using adjoint equation for solving multidimensional inverse heat conduction. Int. J. of Heat and Mass Transfer, 34(11), 2911-2919.

Marchuk, G. I. (1995). Adjoint Equations and Analysis of Complex Systems. Kluwer Academic Publishers. London.

McCutcheon, S.C. (1989). Water Quality Modeling Vol.I: Transport and Surface Exchange in Rivers. CRC Press, Inc. Florida.

Sun N-Z. (1994). Inverse Problems in Groundwater Modeling. Kluwer Academic Publishers.

Sun, N-Z. and Yeh, W-G. (1990). Coupled Inverse Problems in Groundwater modeling 1. Sensitivity analysis and parameter identification. Water Resour. Res., 26(10), 2507-2525.

Sykes, J. F. (1985). Sensitivity analysis for steady state ground water flow using adjoint operators. Water Resour. Res., 21(3), 359-371.

Environmental Hydraulics, Lee, Jayawardena & Wang (eds) © 1999 Balkema, Rotterdam, ISBN 90 5809 035 3

Simulation of coastal water pollution with fuzzy parameters

H. Mpimpas, P. Anagnostopoulos & J. Ganoulis
Department of Civil Engineering, Aristotle University of Thessaloniki, Greece

ABSTRACT: Randomly changing values in time and space of different parameters such as pollutant load sources and physico-chemical coefficients, is the main difficulty in water pollution modeling. Deterministic modeling is not adequate because introduces crisp values of these parameters. To tackle the problem, fuzzy set theory is applied to define imprecise parameters which are used in a water pollution model. Physico-chemicals coefficients and pollutant load sources are expressed in the form of fuzzy numbers. A two-dimensional finite element algorithm combined with fuzzy arithmetic is used for the solution of the advection-dispersion equation for the BOD parameter. A characteristic-Galerkin scheme is also used in order to reduce the numerical diffusion due to the convective term. This model is applied for the study of pollutant transport in the Gulf of Thermaikos located in Northern Greece.

1. INTRODUCTION

The increasing discharge of wastewater in coastal areas in the last decades causes serious damages to the coastal environment. The knowledge of concentration distribution of the pollutant loads in the coastal area under different conditions, is necessary to face the pollution problem. Numerical modeling constitutes a powerful tool for the study of convection-dispersion of pollutants for different scenarios. The main difficulties in the numerical approach are due to the imprecise data related to the values of pollutant load sources and physicochemical coefficients like diffusion settling and deoxygenation coefficients because depend on factors which are changing randomly in time like turbulence, temperature and composition of organic mass (Ducstein and Plate, 1987; Ganoulis, 1994).

Fuzzy set theory and its derivative fuzzy arithmetic (Zadeh, 1965; Zimmerman, 1991), may be used to introduce imprecise data into a mathematical model in a direct way with minimal input data requirements (Silvert, 1996). While probabilistic methods like Monte Carlo simulation or Bayesian approach requires information about entire probabilistic distribution like variances, dispersion and correlation for all input variables, in fuzzy modeling only the range and the mean of the input variables are required, so it can be used successfully when the available data are too sparse for a probabilistic method (Ferson et al., 1994). A fuzzy set is a function μ to a set X. The function μ is called membership function and is considered to measure the degree to which an element belongs to the set X, and expressed by a number between 0 and 1. Number 0 represents the minimum possibility, while number 1 represents the maximum possibility (certainty), for the element to belong to the set X. A fuzzy number $\tilde{A}$ is a fuzzy set of real numbers that achieves unity and is convex. The simplest type of fuzzy number is the triangular fuzzy number (T.F.N). A T.F.N can be defined by a triplet (x_1, x_2, x_3) (Kaufmann and Gupta, 1985).

In this paper the advection-diffusion equation is used for the study of coastal pollution in the Thermaikos Gulf. To overcome the difficulty of numerical dispersion a characteristic-Galerkin finite element scheme was employed, combined with fuzzy arithmetic to represent imprecise parameters in data and functional relationships.

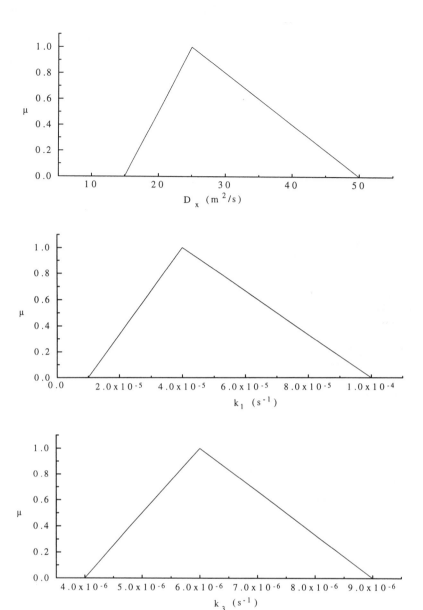

FIG. 1. Membership function of fuzzy coefficients

2. MATHEMATICAL MODELING

Pollutant loads contain many different organic carbon compounds like proteins and hydrocarbonates. Since it is not easy to determine the exact chemical composition of all these compounds, they are treated through the BOD (Biochemical Oxygen Demand) parameter. The equation which describes the variation of concentration for the BOD parameter with respect to the time is the well-known advection-dispersion equation in a non-conservative form:

$$\frac{\partial c}{\partial t} + u\frac{\partial c}{\partial x} + v\frac{\partial c}{\partial y} = D_x\frac{\partial^2 c}{\partial x^2} + D_y\frac{\partial^2 c}{\partial y^2} - k_1 c - k_2 c \tag{1}$$

where:

     c = concentration of BOD (mg/l).

     u = water velocity in the x direction (m/s).

     D_x = diffusion coefficient for the x direction (m/s^2).

     D_y = diffusion coefficient for the y direction (m/s^2).

     k_1 = deoxygenation rate coefficient (s^{-1}).

     k_2 = rate of loss coefficient of BOD, due to settling (s^{-1}).

All these coefficients are expressed in the form of T.F.N as shown in figure 1. Each function μ is the grade of membership function of x in each T.F.N. The values of $\mu(x)$ are in the closet interval [0, 1]. For example, the fuzzy diffusion coefficient may be expressed as 'The diffusion coefficient is approximately 25 m/s^2 and it is certainly above 15 m/s^2 but not greater than 50 m/s^2'. Similar interpretation can be given to the coefficients k_1 and k_2 presented as T.F.Ns in figure 1.

3. NUMERICAL SOLUTION

An explicit finite element scheme is used for the solution of equation (1). To overcome the problem of numerical dispersion which is caused by the advection term in equation (1), the characteristic-Galerkin method is used (Zienkiewicz and Taylor, 1991).

Since the coefficients and the load sources are expressed as T.F.Ns the output concentrations c_i are also fuzzy, therefore equation (1) becomes a fuzzy equation. The operations between the T.F.N $\tilde{A} = (x_1, x_2, x_3)$ and $\tilde{B} = (y_1, y_2, y_3)$ defined as: $\tilde{A} + \tilde{B} = (x_1+y_1, x_2+y_2, x_3+y_3)$, $\tilde{A} - \tilde{B} = (x_1-y_3, x_2-y_2, x_3-y_1)$. At a given α-level cut in the interval [0, 1] of the membership function $\mu(x)$ each T.F.N $\tilde{A}$ is represented by an interval number $\overline{A}_\alpha = [x_1, x_3]$, which is called interval of confidence (Kaufmann and Gupta, 1985). Since multiplication and division operations can be performed only in the intervals of confidence $\overline{A}_\alpha$, equation (1) becomes an interval equation at each α-level cut.

The operations on the intervals $\overline{A} = [\alpha_1, \alpha_2]$ and $\overline{B} = [\beta_1, \beta_2]$ may be calculated as: $\overline{A} + \overline{B} = [\alpha_1+\beta_1, \alpha_2+\beta_2]$, $\overline{A} - \overline{B} = [\alpha_1-\beta_2, \alpha_2-\beta_1]$, $\overline{A} * \overline{B} = [\min(\alpha_1\beta_1, \alpha_1\beta_2, \alpha_2\beta_1, \alpha_2\beta_2), \max(\alpha_1\beta_1, \alpha_1\beta_2, \alpha_2\beta_1, \alpha_2\beta_2)]$, $\overline{A} / \overline{B} = [\min(\alpha_1/\beta_1, \alpha_1/\beta_2, \alpha_2/\beta_1, \alpha_2/\beta_2), \max(\alpha_1/\beta_1, \alpha_1/\beta_2, \alpha_2/\beta_1, \alpha_2/\beta_2)]$, $-\overline{A} = -[\alpha_1, \alpha_2] = [-\alpha_2, -\alpha_1]$ (Moore, 1979; Alefeld, 1983; Nickel, 1986). For the solution of equation (1) interval calculations are carried out at five different α-level cuts (0, 0.25, 0.50, 0.75, 1) of T.F.Ns . At each α-level cut an interval value of concentration is computed. From these results the fuzzy concentrations are constructed (Dou et al., 1995; Ganoulis et al., 1995).

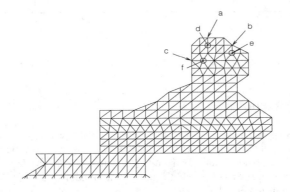

FIG. 2. The finite element mesh

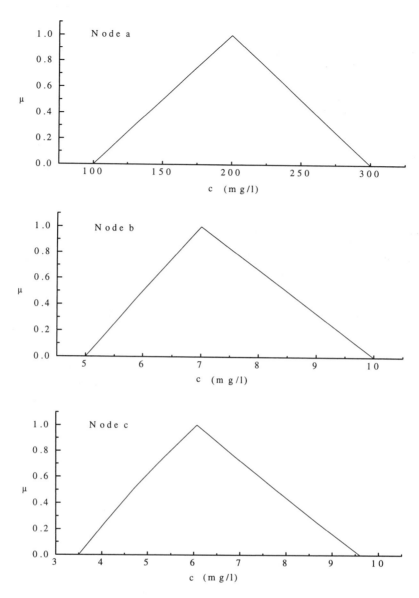

FIG. 3. The membership functions of pollutant load sources at nodes a, b and c

4. APPLICATION AND RESULTS

The pollutant loads expressed as T.F.N (fig. 3) were introduced at the nodes a, b, and c (fig. 2). The velocity field was obtained by solving the shallow water equations using triangular finite elements in the domain of Thermaikos Gulf for northern wind of velocity 10 m/s (Anagnostopoulos and Mpimpas, 1995). The time step was 500 s, the number of nodes was 989 and the number of iterations was 400. The advection-diffusion equation was solved at five different α-level cuts (0, 0.25, 0.5, 0.75, 1) of fuzzy input parameters. The concentration distribution for BOD, constructed from the interval output in the form of fuzzy numbers at the nodes d, e, and f are shown in figure 4.

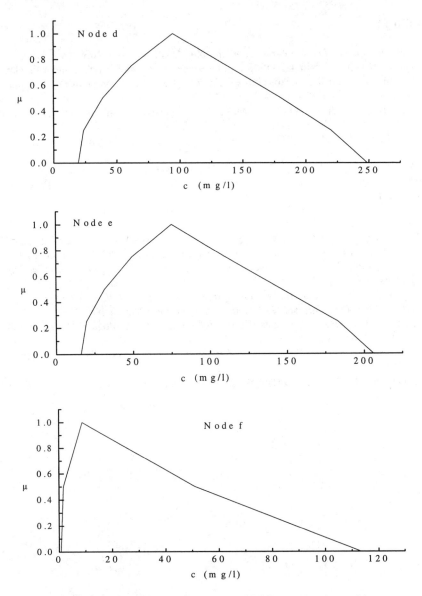

FIG. 4. Concentration distribution of BOD at nodes d, e and f

5. CONCLUSIONS

Fuzzy arithmetic approach combined with the finite element method was introduced to express imprecise parameters like biochemical coefficients and pollutant load sources for the study of pollution in the Thermaikos Gulf. With fuzzy modeling we can utilize imprecise data and produce imprecise output in the form of fuzzy numbers providing an alternative and versatile method to the probabilistic approach, with minimal input data requirements into the modeling process, and without performing a large number of computations.

6. REFERENCES

Alefeld, G. (1983). *Introduction to interval computations*, Academic Press, New York, USA.

Anagnostopoulos, P., and H. Mpimpas (1995). 'Finite element solution of wind-induced circulation and pollutant dispersion in the Thermaikos Gulf', *Proc. 2nd International Conference of Computer Modeling of Seas and Coastal Regions*, Cancun, Mexico, 133-140.

Dou, C., Woldt, W., Bogardi, I., and Dahab, M. (1995). 'Steady state groundwater flow simulation with imprecise parameters' *Water Resources Research*, Vol. 31, 2709-2719.

Ducstein, L., and Plate, E. (Eds.), (1987). *Engineering Reliability and Risk in Water Resources,* E.M. Nijhoff, Dordrecht, The Netherlands.

Ferson, S., Millstein, J., and Mauriello, D. (1994). '*Environmental risk assessment using interval analysis and fuzzy arithmetic',* In association with 1994 Annual Meeting of the Society for Risk Analysis, Baltimore, Maryland.

Ganoulis, J. (1994). *Engineering Risk Analysis of Water Pollution Probabilities and Fuzzy Sets,* VCH, Weinheim.

Ganoulis, J., Mpimpas, H., and Anagnostopoulos P. (1995). 'Coastal Water Assessment using Fuzzy arithmetic', *Proc. IEEE Workshop on Nonlinear Signal Image Processing,* Vol. II , 1015-1018.

Kaufmann, A., and Gupta, M.(1985). *Introduction to Fuzzy Arithmetic Theory and Applications*, Von Nostrand Reinhold, New York.

Moore, R.M. (1979). *Methods and Applications of Interval Analysis*, SIAM, Philadelphia.

Nickel, K. (ed.) (1986). Proc. *Interval Mathematics 1985,* Springer Verlag, New York.

Silvert, W. (1996). 'Ecological impact classification with fuzzy sets', *Ecological Modeling,* Vol. 96, 1-10.

Zadeh, L.A. (1965). 'Fuzzy sets', *Information and Control*, Vol. 8, 338-353.

Zimmerman, H.J. (1991). *Fuzzy set theory and its applications,* 2nd Ed., Kluwer Academic Publishers, Boston.

Zienkiewicz, O.C., and R.L. Taylor (1991). *The finite element method*, 4th Ed., McGraw-Hill, London.

Environmental Hydraulics, Lee, Jayawardena & Wang (eds) © 1999 Balkema, Rotterdam, ISBN 90 5809 035 3

Study in the method of predicting the water environmental capacity in a river basin

Liu Lanfen & Zhang Xiangwei
China Institute of Water Resources and Hydropower Research, Beijing, China

ABSTRACT: This paper emphasizes that predicting the water environmental capacity is the basis of the total amount control of pollutant in water environmental , and the foundation of water environmental management plan. A method of predicting water environmental capacity was obtained . The method suits in the relevant prediction for the basins of medium or small rivers.

1. INTRODUCTION

At present, in our country, total amount of sewage discharge is about 38.6 billions ton every year. A lot of sewage water do not be treatmented and are discharged into water environment. The water environment have be polluted by the sewage. Therefore, we have to control total amount of pollutant discharge for improving the water quality. Technical basis of water quality management is the combination of the ways of artificial treatment and rationale utilization of natural absorbability for a river. The key problem of total amount control is to know the absorbability of water for pollutants, or that the water environmental capacity. The water environmental capacity is the most capacity of absorbing pollutants of the river as it satisfies the given quality standards(Fang, 1988).

There have had a lot of studies in this field in developed countries such as USA, Japan and Germany, etc. They have carried out the control for total amount or the quota control for discharge in a river and achieved the profit in environment and economy(Kneese et al, 1979). It is shown in literature that there are many studies in this field in China, too. However, most of them study in some part or some section of a river or the estuary. A few of them study in the whole river and its basin(Zhang et al, 1991). This study shown in this paper focuses on the whole river and its basin. It obtains the method and models, which mainly suit in the medium or small river.

2. THE METHOD FOR PREDICTING THE WATER ENVIRONMENTAL CAPACITY

There are four steps in the method for predicting the water environmental capacity in a medium or small river as follows:

2.1 The Determination of The Premise Conditions and Relevant Data

They includes: (a). determining the year of present level and the year of prediction objective; (b). dividing the river into sections/elements; (c). defining water quality objective to be reached and analyzing the hydrological data of the river; (d). defining the upper water quality of the river and the background data of the water body in each river section; (e). analyzing the data of pollutants sources and selecting the pollutant factors; (f). collecting the data of development programs about the river basin.

2.2 The Prediction for the Changing Trends of Waste Amount and Pollutant Load

We employed the Grey System Analysis Method, GSAM, to develop the macro-systems information as many as possible in the case of lack of actual data, using the grey relations between systems. In this case employed the macro-model seems to be suitable (Deng, 1985). From this method, we obtained the dynamic state information of every sub-system. The one order model, GM(1,1), for predicting the dynamic state is:

$$\frac{dx_1^{(1)}}{dt} + a\kappa_1^{(1)} = \mu$$

(1)

where, $dx_1^{(1)}/dt$ is the grey derivative; $\kappa_1^{(1)}$ is the background value of grey derivative; α, μ, the grey parameters. The background value $\kappa_1^{(1)}$ can be written in dispersed series as

$$\kappa_1^{(1)}(k) = \alpha x_1^{(1)}(k) + (1-\alpha)x_1^{(1)}(k-1)$$

where, α is information weighing, $\alpha = 0.5$. The formula (1) can be written as

$$-a\kappa_1^{(1)}(k) + \mu = x^{(0)}(k), \qquad k = 1,2,\cdots,n$$

$\{x^{(0)}(k)\}$ is original sample series. The solution of formula (1) be written in dispersed response model as

$$x^{(0)}(t+1) = [x^{(0)}(1) - \hat{\mu}/\hat{a}]e^{-at} + \hat{\mu}/\hat{a}$$

(2)

To accumulate data and expanding predicte, the result was obtained

$$\hat{x}^{(0)}(t+1) = \hat{x}^{(1)}(t+1) - \hat{x}^{(1)}(t)$$

The main content of the Grey System Method as follows:
(1).To analyze the informations of pollution sources, amount of waste water discharge, socio economic factors and population, apply grey relation regress analysis and GM model to predict total amount of discharging waste water and main pollutants from industry, living in the area, and verify these models.
(2).To predict total amount of discharging waste water and main pollutants from industry, living in objective year by using GM(1,1) and GM(1,N) models.

(3).To determinate grey rate factors α_{ij}, as following:
$$\alpha_{ij} = w_{ij}/w_j$$

The waste water dischargs and pollution loading amount of every river sections for present year and objective year were predicted by

$$L_{ij}(k) = \alpha_{ij} L_j(k)$$

where, i - the river section; j - the project of pollution load; k shows the year. w_{ij}, L_{ij} - pollution load of i th river section j th project for present year and objective year, respectively; w_j, L_j - total amunt of pollution load of j th project for present year and objective year, respectively. Then have

$$w_j = \sum_{i=1}^{n} w_{ij}, \quad L_j = \sum_{i=1}^{n} L_{ij}$$

where, n is number of river sections.

2.3 The Prediction for Average Velocity and Water Quality of Each Section in the River

As the width and the depth are smaller for the medium or small river, the distribution of pollutant's concentration tends quickly even in a section of the river. Because there are the effect of convection, dispersion and dilution after the pollutants discharge in the water bady. Therefore, one can neglect the differences in the distribution of the concentration in a section. Neglecting the distribution of velocity in a section, the average velocity of a section represents the velocity in the section of the river. Thus, the one-dimensional model can be used to predict the water quality of the river.
The control equations of the one-dimensional model are as follows:

Continuity Equation: $\qquad B\dfrac{\partial z}{\partial t} + \dfrac{\partial(BHu)}{\partial x} = 0$

Momentum Equation: $\dfrac{\partial u}{\partial t} + u\dfrac{\partial u}{\partial x} + g\dfrac{\partial z}{\partial x} + g\dfrac{u^2}{C^2 R} = 0$

Concentration Equation: $\dfrac{\partial c}{\partial t} + u\dfrac{\partial c}{\partial x} = \dfrac{\partial}{\partial x}(D_L\dfrac{\partial c}{\partial x}) - Kc + s$

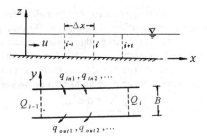

Fig.1 Finite difference symbol explanation

where, B is the width of the river; H, the average depth of a section; u, the average velocity of a section; z, water level; C, Chezy Coefficient; R, hydraulic radius; g, gravity acceleration; c, pollution concentration; K, degradation coefficient; D_L, concentration dispersion coefficient; s, source term of pollutants; t, time; x, direction along the river.

The equation can be writed in difference formula for some river section as

Continuity Equation: $Q_i - Q_{i-1} = \sum q_{in} - \sum q_{out}$

Momentum Equation: $z_{i-1} - z_i = \left[(\dfrac{Q_i + Q_{i-1}}{2})^2 \cdot n_{i-1}^2 / (\dfrac{A_i + A_{i-1}}{2})^2 / (\dfrac{R_i + R_{i-1}}{2})^{3/4}\right]\Delta x_{i-1} + \dfrac{1}{2g}(\dfrac{Q_i^2}{A_i^2} - \dfrac{Q_{i-1}^2}{A_{i-1}^2})$

Concentration Equation: $\dfrac{(Qc)_i - (Qc)_{i-1}}{(A_i + A_{i-1})\Delta x_{i-1}/2} = \dfrac{1}{\Delta x_{i-1}}\left[\dfrac{(D_L c)_{i+1} - (D_L c)_i}{\Delta x_i} - \dfrac{(D_L c)_i - (D_L c)_{i-1}}{\Delta x_{i-1}}\right] + S_i - K(c_i - c_0)$

where, c_0, background concentration; n, drag coefficient of river bottom; finite difference symbol were shown in Fig.1. The dispersion D_L wes obtained by Fisher Method.

2.4 The Prediction for Water Environmental Capacity in Each Section of the River and Its Basin

One can predict the water environmental capacity, basing on the pollutant load predicted, average velocity and quality of water. Moreover, in this procedure, it must consider the water level, water amount, designed water amount, and the water quality objective in each section and its basin. Specifically, the first is to predict the environmental capacity of the tributaries, next, to predict those of the mainstream. As it is always lack of data from the tributaries, one can think that the tributaries are the pollution sources of the mainstream.

(a). The formulation of water environmental capacity :
For i^{th} section of the mainstream or tributary , there is:

$$E_i = c^i{}_N(Q_{pi} + q_i) - c^i{}_{01}Q_{pi} - c^i{}_N(Q_{pi} + q_i)[1 - \exp(-K_i x_i / u_i)]$$

The total environmental absorbability of the river is the sum of those of its sections:

$$E = \sum_{i=1}^{n} E_i$$

where, E and E_i are the total environmental capacity of the river and that of its ith section, respectively; Q_{pi}, the upstream flow of ith section; q_i, the total waste effluent in i^{th} section; x_i, the length of ith section; u_i, the average velocity of ith section; K_i, the degradation coefficient; $c^i{}_{01}$ and $c^i{}_N$, the pollutant concentration and water quality standard in the upstream flow of ith section, respectively.
They are shown in Figure 2. The formula for computing the remained environmental capacity is:

$\Delta E_i = L_i - E_i$

where, L_i is the pollution load in i^{th} section; ΔE_i is the remained environmental capacity, when $\Delta E_i < 0$, and, ΔE_i is the decreasing amount of waste discharge, when $\Delta E_i \geq 0$.

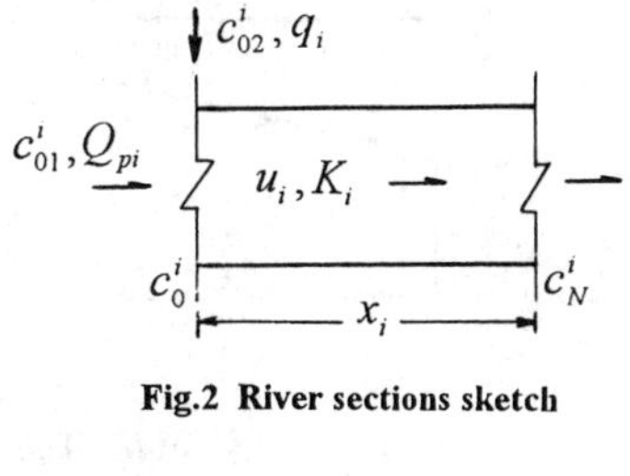

Fig.2 River sections sketch

(b).The main paramaters to be calculated for the formulation above include :

The concentration of upstream flow in each section: The concentration value in the end of each section is calculated using the one-dimensional model for predicting water quality, and it will become the initial concentration value of the upstream water in the next section. it is assumed that the pollution is homogeneously mixed in each cross-section. The mixed value is calculated using the S-P model.

The mixed concentration of effluent in each section:

$$c^i_0 = \frac{Q_{pi}c^i_{01} + q_i c^i_{02}}{Q_{pi} + q_i}$$

where, c^i_0 is the mixed concentration ; c^i_{02} , the effluent concentration in i^{th} section; c^i_{01} , the pollutant concentration of upstream for i^{th} section.

The degradation coefficient of the pollutant:

The degradation coefficient of the pollutants in the river can be calculated or estimated by expermental simulation in laboratory or by observation in the river sections.

3. EXAMPLE OF APPLICATION

This method has been applied to assess the environmental effect, in which the effect of the hydrological engineering on the water environment has been evaluated for the central line in the proposed South – North Water Transfer. These models were employed in predicting the environmental capacity of the basin of middle and lower reaches of the Hanjiang River. A lot of observation data were used in these models and the results were suitable.

3.1 General

Total length of Hangjiang River middle and lower reaches is about 640km. The basin area is $6.65 \times 10^4 km^2$. There are seventeen citys along the river. The industry and agriculture are flourishing and the density of population is heavy in this region.

South – North Water Transfer engineering plans to transfer $150 \times 10^8 m^3$ water from Danjiangkou Reservoir to north. The Transfering water engnreeing will effect on water amount and water quality of hanjiang middle and lower reaches. This method is applied to predict the water environmental capacity of Hanjiang River after the Transfering water engnreeing.

3.2 Conditions of Calculation

(1). The present level year: 1990 ; the predict objective year: 2000.
(2). The basin was divided into 13 sections: they are 1 Huangjiagang, 2 Laohekou, 3 Gucheng, 4 Xiangyang, 5

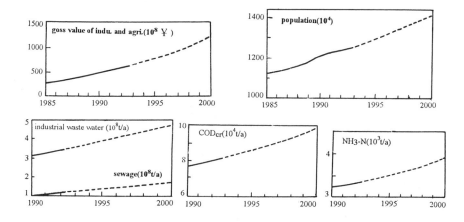

Fig.3 The predicative result using Gray System Method

Xiangfan, 6 Yicheng, 7 Zhongxiang, 8 Jingmen, 9 Qiangjing, 10 Tianmen, 11 Xiantao, 12 Hanchuan and 13 Wuhan.

(3).The water quality objective was difined as II level in the national standard GB3838-88.

(4).Some parameters were calculated, such as drag coefficient of river bottom, background consistency of every sections for objective year, degradation coefficient of pollutions, etc.

3.3 The Main Result Predicted

The result prediction of every year were shown in fig.3, they are gross value of industrial and agricultural output, population, industrial waste water, sewage from living, pollution load amount of COD_{cr} and NH_3-N.

The water environmental capacity of COD predicted of Hanjiang River middle and lower reches for dry season after the Transfering water engnreeing were shown in the table.

Table: Water Enviromental Absorbability of the river in 2000 , Transfering water $150 \times 10^8 m^3$ Unit: t/d

Section	COD Discharge	Enviromental Capacity	Remainder(-) Decrease(+)	Section	COD Discharge	Enviromental Capacity	Remainder(-) Decrease(+)
1	9.7	104.1	-94.4	8	11.4	58.5	-47.1
2	28.3	17.5	+10.8	9	24.5	27.9	-3.4
3	4.3	40.9	-36.6	10	4.0	28.4	-24.4
4	4.7	32.1	-27.4	11	14.8	16.2	-1.4
5	69.5	112.2	-42.7	12	2.4	20.5	-18.1
6	11.0	26.6	-15.6	13	138.4	7.8	+130.6
7	0.4	20.2	-19.8	Total	323.4	512.9	-189.5

4. SEVERAL CONCLUSIONS

From the stady and use of these models, several conclusions can be obtained:

(1).The prediction for water environmental capacity in a river can advance and promote the quality managemant in protecting the water environment. The the control for total amount of pollutants is important and helpful to the management.

(2).The prediction method involves several models, such as the grey system analysis model , river one-dimensional water quality model, and environmental capacity prediction model, etc. Through checking, these models are applicatable and their results are good in general. Naturally, it should be studied further to make complete and general .

(3).The key of predicting water environmental capacity of the river basin is the estimation of the parameters in these models. To keep the parameters to be suitable and to be actual, the experiments and checks are necessary. It had better to use the statistical data and observed data. The more complete are for these data, the more actual will be for prediction.

(4).We will make an effort in estimating the parameters, especially for some key parameters.

REFERENCES

Zhang, Y. L., Liu, P. Z.(ed.)(1991), A Comprehensive Handbook for the Environmental Absorbability of Water, Tsing Hua University Press, Beijing, China.

Fang, Z.Y.(eds.)(1988), A Handbook for Water Resources Protection, He Hai University Press, Wuhan, China.

Kneese, A., Bower, B.(1968), Managing Water Quality; Economics, Technology, Institutions, The Johns Hopkins University Press for RFF, 1616 P Street, N. W., USA

Kneese, A., Bower, B.(1979), Environmental Quality and Residuals Management, Resources of Future, U.S.A.

Liu, L. F. et al (1994), Study in Prediction to the Effect of the South-North Transfer Water engineering on Water Quality and the Water Environmental Capacity in the Central-downstream in the Hanjiang River, Research Report, China Institute of Water Resources and Hydropower Research, IWHR, Beijing, China.

Cai, Y.M., Guo, Z.Y. et al (ed.)(1987), Handbook of Evaluating for Environmental Effect, China Environment Science Press, Beijing.

Deng, J. L. (1985), Grey System Predicting Method, Hua Zhong Science Technology University Press, Wuhan, China.

4   Sediment-water-environment interactions/Other topics
4.1 Sediment/Contaminant transport

Environmental Hydraulics, Lee, Jayawardena & Wang (eds) © 1999 Balkema, Rotterdam, ISBN 90 5809 035 3

Invited lecture: Effect of sediment on pollutant transport-transformation

Wan Zhaohui
China Institute of Water Resources and Hydro-Power Research, Beijing, China

Suiliang Huang
State Key Sub-Laboratory of Water Environmental Simulation, Beijing Normal University, China

ABSTRACT: Transport-transformation of pollutants, such as heavy metals, toxic organic matter, nutrient, oil, etc., is closely related to sediment motion. The study on correlation among pollutants, water and sediment has particular significance. In this paper the adsorption-desorption of heavy metal pollutants, organic matter and nutrient by sediment is discussed and data obtained in China are analyzed. Transfer of sediment motion related pollutants in aquatic environment exhibits convective-diffusive law of common tracers and characteristics of transport-transformation induced by sediment motion. Integrating the research results in environmental chemistry and those in sediment motion will greatly promote the development and application of this science.

1. INTRODUCTION

A variety of pollutants, such as heavy metals, organic matter, nutrient, oil, etc. are prone to be absorbed onto sediment particles and move together with the latter. Therefore, correlation/interaction among pollutants, water and sediment is a topic worthwhile to study both in theory and in practice. Sediment particles in water play dual effects (Forstner and Wittmann, 1979). On the one hand, containing certain amount of clay mineral, organic and non-organic colloids, sediment particles can absorb a variety of pollutants entering rivers and make the water purified. It is also reported that sediment particles promote biological activity of microbe on their surface (Coughtrey et al., 1987, Hart, 1991). On the other hand, as pollutant itself and a carrier of pollutants, under certain conditions sediment particles release pollutants and cause secondary pollution of the received waters.

China is a vast country with numerous sediment-laden rivers. The Yellow River ranks first in the world both in annual sediment load and in average sediment concentration. Therefore the study on correlation among pollutants, water and sediment has particular significance in China. In Sanmenxia Reservoir, a large amount of samples taken at 11 sections were examined. COD_{cr} values of mud water all exceed 100 mg/l, with a maximum of 400 mg/l. After the samples were filtered and sediment in it was taken out, COD_{cr} values of clear water usually were less than 10 mg/l, and the BOD_5 values were less than 3.0 mg/l (Yang et al., 1995). Researchers at YRCC suggested that the qualities of mud water, clear water and sediment should be evaluated separately first and then an overall assessment can be made.

2. HEAVY METALS

2.1 Most of Heavy Metal ions Are Absorbed onto Sediment Particles

Heavy metal pollution has not been well controlled in China. Contamination monitor is widely carried out on main rivers in China. According to observation, in sediment-laden rivers heavy metal ions are mostly

713

Table 1 Partition factors [(mg/kg/(mg/l)] of some heavy metals in some Chinese rivers

River	Pb	Cu	Zn	Cr	Ni
Xiangjiang			$2*10^3-1.5*10^4$		
Jinshajiang	10^3-10^4	10^4-10^5	10_3-10^5		10^4-10^5
Yangtze River	$3*10^2-6*10^3$	10^3-9*10^3	10^3-4*10^4		
Yellow River	10^2-10^4	10^2-10^4	10^2-10^3	$5*10^2-5*10^3$	10^4-10^5

absorbed onto sediment particles. Partition factors, i.e., the ratio of the weight of heavy metal on specific suspended particles to the dissolved heavy metal concentration, of some heavy metals along several main rivers in China are listed in Table 1.

2.2 Adsorption of Heavy Metals by Sediment Particles in the Yellow River

The Yellow River is a heavily sediment-laden river. Large scale of monitoring and study on heavy metal pollutants have been carried out along the Yellow River, particularly, in nineties (YRCC, 1995). The main conclusions deduced from these work are as follows:

i) Entering the item or tributaries of the Yellow River, heavy metal pollutants either coprecipitate with or are adsorbed by sediment particles, mainly those finer than 25 μm, with extremely high speed. generally, adsorption equilibrium can be reached in 10 minutes. Over 99 percentage of heavy metal pollutants in the Yellow River are adsorbed or coprecipitate. Monitoring in Sanmenxia also shows: in bottom mud 60-70% of heavy metal pollutants are adsorbed onto sediment particles finer than 0.03 mm, and in suspended load 90% are adsorbed onto such fine particles (Yang et al., 1995).

ii) Most pollutants are concentrated on particulate phase, in which pollutants on suspended material are 15 times those on bottom mud. Partition factors of different heavy metal pollutants between suspended particulate and water phase are: 10^3-10^4 for Pb and Cu; 10^2-10^3 for Zn; $5*10^2-5*10^3$ for Cr and 10^4-10^5 for Ni. It clearly shows that sediment particles are the main carriers of heavy metal pollutants and play a dominant role in the transfer of the latter.

iii) The water in the Yellow River is generally weak alkaline with pH value of 7.9-8.3. Only if pH value is kept larger than 5, no heavy metals will be released from sediment. Therefore the water in the Yellow River possesses powerful buffer capacity of heavy metals in this sense.

iv) It is found that the pollutant contents in the surface layer of bottom mud is larger than that in the old deposit. It means pollution is recent years is getting worse.

2.3 Effect of Sediment Motion on Heavy Metal Pollutant Transport-Transformation

Huang et al. did a systematic study on the heavy metal pollutant adsorption-desorption by sediment both experimentally and theoretically (Huang, 1993). The main findings can be summarized as follows:

i) Experimental results in lab show that the heavy metal adsorption-desorption by sediment can be better described by Langmuir adsorption isotherm:

$$N_\infty = b\frac{c_\infty}{k + c_\infty} \qquad (1)$$

in which N_∞ and c_∞ are the equilibrium heavy metal adsorption content by specific weight of suspended sediment and the equilibrium dissolved concentration of heavy metal, respectively; b, saturated adsorption content, k, intensity of adsorption-desorption by suspended sediment. An example of such relationship is shown in Fig.1.

The adsorption-desorption dynamic equation is:

$$\frac{dN}{dt} = k_1 c(b - N) - k_2 N \tag{2}$$

Here N and c are the heavy metal adsorption content by specific weight of suspended sediment and the dissolved heavy metal concentration at time t, respectively; k_1, coefficient of adsorption rate and k_2, coefficient of desorption rate.

ii) By combining the adsorption-desorption dynamic equation and mass conservation equation and considering the initial conditions, formulas for the variations of N and c with time are established in both cases of adsorption and desorption (Huang and Wan, 1997a). Experimental results well coincide with the theoretical formulas. An example is shown in Fig.2.

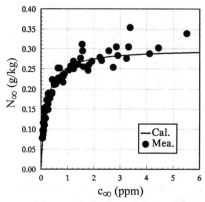

Fig.1 A comparison of Langmuir adsorption isotherm with experimental data

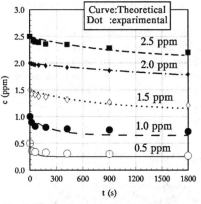

Fig.2 Variations of dissolved heavy metal concentrations c with time

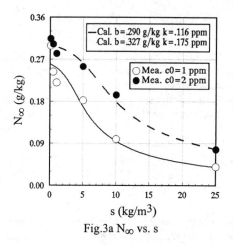

Fig.3a N_∞ vs. s

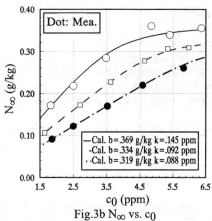

Fig.3b N_∞ vs. c_0

715

iii) Furthermore, formulas for calculating equilibrium adsorption content of specific weight of sediment N_∞ and equilibrium dissolved heavy metal concentration c_∞ in both cases of adsorption and desorption are also obtained (Huang and Wan, 1997a). The agreement between calculation and experimental data is satisfied, see Fig.3. It is deduced from both these formulas and experimental results that sediment concentration has no influence on characteristic parameters of sediment adsorption , b, k and k_1 (k_2).

iv) In natural rivers sediment is always non-uniform. The adsorption by non-uniform sediment is further studied, and a concurrent adsorption model for non-uniform sediment is developed (Huang and Wan, 1997b). The saturated adsorption content of non-uniform sediment is the summation of the saturated adsorption contents of each group of sediment (nearly uniform). That is, no competition adsorption or interference among different grain sizes of sediment exists.

v) A turbulence-simulation device, similar to that used by H. Rouse in developing the diffusion theory, is used (Huang and Wan, 1997c). In the experiments the influence of sediment motion patterns (suspended material, or bottom mud) is studied. Under low turbulence intensity and bottom mud keeping stationary, the adsorption speed of heavy metal ions by sediment is very low. Under strong turbulence and continuous interchange of suspended material and bottom mud the decrease of dissolved heavy metal concentration is much greater and faster, and the time for approaching equilibrium is much longer. Whether there is bottom mud on bed has great influence on the variation of the dissolved heavy metal concentration.

vi) By combining basic principles of environmental chemistry, hydraulics and mechanics of sediment motion and findings mentioned above, a mathematical model describing heavy metal pollutant transport-transformation in its entirety in fluvial rivers is developed (Huang and Wan, 1994). A preliminary application of the model to two simple cases shows its reasonableness and validity. The result of application reveals that the transfer of sediment motion related pollutants in aquatic environment possesses not only convective-diffusive generality of common tracers, but also characteristics of transport-transformation induced by sediment motion.

3 TOXIC ORGANIC MATTER

Nowadays there are more than seven million kinds of organic matter in the world. And it increases with a speed of dozens of thousands kinds each year. Pollution caused by toxic organic matter becomes more and more serious. Usually a toxic organic material possesses its threshold value, over which its toxicity appears, most toxic organic materials hardly degrade, and they often concentrate in biology fat. By this way it causes long-term accumulative harm to biology or environment.

3.1 Mechanism

Partition/adsorption and bioaccumulation are two main effects in the transformation of organic components from water phase into solid phase. By partition/adsorption non-ionic organic matters, which are hard dissolvable in water, are dissolved by organic matter in sediment. By bioaccumulation organic components in water are dissolved in fat of aquatic biology. It results in a decrease of the dissolved concentration of organic component, i.e., a transformation of the organic component in water phase into solid phase.

Partition factor of an organic component between solid phase and water phase is defined as:

$$K_d = \frac{C_0}{C_w} \tag{3}$$

in which C_0 and C_w are mole concentration of the organic component in organic material of solid phase and that in water phase, respectively. Usually partition factor is normalized by the organic material in sediment as follows:

$$K_{om} = \frac{K_d}{f_{om}} \qquad or \qquad K_{oc} = \frac{K_d}{f_{oc}} \tag{4}$$

in which f_{om} and f_{oc} are the organic material content in sediment and the organic carbon content in sediment, respectively.

3.2 Partition Factor of Several Representative Organic Components

Jin et al. (1992) selected seven representative non-ionic organic components and carried out systematic experiments on partition factors and related issues. The seven organic components are: hexacholorobenzene, bata-BHC, 1, 2, 4-tricholorobenzene, 1, 2-dichlorobenzene, 1, 2-dichloropropane and cholorobenzene. Experiments on their partitions show:

i) All the isotherms are linear, two examples are shown in Fig.4.

ii) Normalized partition factors K_{om} or K_{oc} almost keep constants, which are close to the theoretical values. An example is listed in Table 2. The consistency of the normalized partition factor log K_{oc} or log K_{om} and the agreement between the experimental data and theoretical value seems to verify the correctness of the partition theory.

3.3 Partition Factors of Organic Components between Sediment-Water in the Yellow River

Three toxic organic components, which have been detected in the Yellow River water, are selected as representatives and their partition factors between sediment and water are measured in YRCC. Results are shown in Table 3 (YRCC, 1995). Table 3 shows that for three different organic components log K_{oc} or log

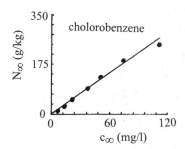 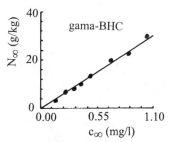

Fig.4 Isotherms of non-organic components

Table 2 Partition factor of gamma-BHC between water and sediment (or soil)

	K_d	K_{oc}	log K_{oc}	K_{om}	log K_{om}	Theoretical log K_{oc}
Soil A	10.07	1199.2	3.08	695.6	2.84	3.58
Surface soil	14.64	1190.2	3.08	690.4	2.84	3.58
Sediment	35.07	1498.8	3.18	869.4	2.94	3.58
Average		1296.1	3.11	751.8	2.88	3.58

Table 3 Experimental results on partition factors

	K_d	K_{oc}	log K_{oc}	K_{om}	log K_{om}
1, 2-dichlorobenzene	0.967	224.5	2.35	130.2	2.11
parathion methyl	3.073	713.7	2.85	414.0	2.62
p-nitrocholobenzene	2.83	243.1	2.39	141.2	2.15

K_{om} are close to each others. And these values are also close to the results obtained by Jin (Table 2). Besides, sediment samples taken from different reaches along the Yellow River were collected. Organic contents of these samples were measured firstly. And partition factors of 1, 2-dichlorobenzene between sediment-water for these samples were also measured, see Table 4. Table 4 clearly shows the tendency of K_d increasing with organic content. And the normalized K_{om} also increases with organic content increasing. All these verifies the important role played by organic content in dissolving toxic organic components.

Table 4 The effect of organic material on Kom of 1, 2-dichlorobenzene

	Longmen	Sample A	Hejin	Surface soil	Rock bed	Dry plant
Organic content (%)	0.743	1.448	2.51	2.12	4.49	73.99
Range of C (mg/l)	0-40	0-40	0-40	0-40	0-40	0-40
K_d	0.967	2.34	5.743	4.90	12.63	21.9
K_{om}	130.1	161.7	228.9	230.9	281.1	519.3

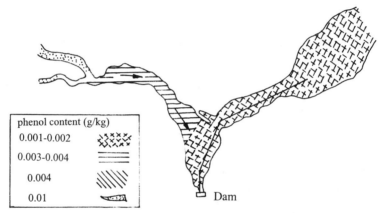

phenol content (g/kg)

0.001-0.002

0.003-0.004

0.004

0.01

Fig.5 Horizontal distribution of phenol in Guangting Reservoir

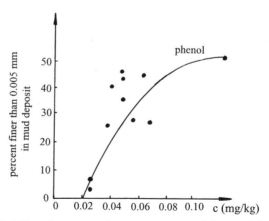

Fig.6 Phenol concentration vs. percentage of clay particles

3.4 Organic Pollutants in the Guangtin Reservoir

Guangtin Reservoir completed in 1953 is located on the upstream reach of the Yongding River. Located at northwest to Beijing, it plays dual role of flood-control and water supply. Since its completion, the reservoir is silted gradually and it was seriously polluted in seventies and eighties. According to observations, a horizontal distribution of phenol in the reservoir in 1974 is shown in Fig.5. In the figure the left branch is the item of the Yongding River and the right branch is just a small tributary with very little runoff. The figure shows that the phenol concentration decreases as sediment settles (Zhang et al., 1986). Furthermore, the phenol concentration in deposit increase with the percentage of clay particles (<5 µm) in deposit, as shown in Fig.6.

4 NUTRIENT (SALTS OF NITROGEN, PHOSPHORUS)

4.1 Nutrient in Bottom Mud of Some Lakes in China

Generally speaking, in China nutrient content of bottom mud in most lakes is rather high. Among them, nutrient contents of bottom mud in lakes near cities are higher than those in countryside. Some examples are listed in Table 5 (Jin, 1992). Large amount of nutrient in bottom mud deteriorates water quality and restricts the usage function of the water body.

Table 5 Nutrient in bottom mud of some Chinese lakes

Lake type	Name	TP (ppm)	TN (ppm)
Lakes by cities	Moshui lake	4504.7	25632.0
	Liuhua lake	1792.9	3479.0
	Xuanwu lake	2160.0	4825.0
	Dianchi	1715.9	4236.5
	West lake	1569.0	9008.6
	Liwan lake	1748.0	4594.3
	Dongshan lake	1285.4	2156.0
	Lu lake	1237.6	2629.0
Lakes in countryside	Cao lake	31.9	624.3
	Wuliangsuhai	382.7	799.8
	Bocitenhai	432.8	2140.0
	Dalai lake	405.9	2046.8
	Erhai	1002.0	2701.9

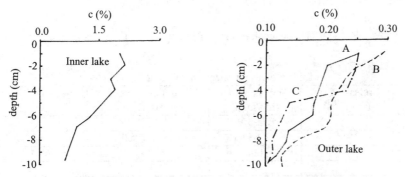

Fig.7 Profiles of nitrogen content in bottom mud of Dianchi

Some vertical profiles of nutrient contents show a common tendency, that is, nutrient content decreases with increasing depth. Two examples taken in Dianchi are shown in Fig.7. The higher content in the surface layer reflects the more serious pollution in recent years.

4.2 Study and Observation Conducted in the Yujiao Reservoir , Tianjin

Diverting water from the Ruan River to Tianjin is the first large-scale interbasin water diversion work in China. Functioning as a regulator, Yujiao Reservoir is located on the middle way of the water diversion line. Eutrophication is a significant problem for the reservoir and a series of observation and study have been done (Tianjin Environment Protection Institute, 1990).

4.2.1 Adsorption of Nutrients by Sediment

Natural colloid in sediment can absorb dissolved phosphate in water. Experiments conducted in the Yujiao Reservoir show that suspended sediment can adsorb phosphate strongly, and the adsorption can be well described by Freundlich isotherm, see Fig.8. Study on effect of environmental factors on adsorption has been carried out too. The following conclusion is deduced from experiments: adsorption content of phosphate increases with water temperature increasing and decreases with the increase of the negative ions in the system. Actually, it is a kind of competition adsorption.

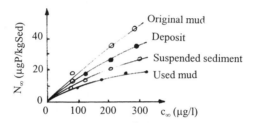

Fig.8 Adsorption of phosphate by suspended sediment

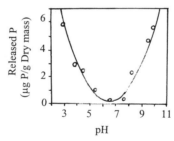

Fig.9 TDP concentration vs. pH values West Lake, Hangzhou

Table 6 Nutrient release in lakes in China

Lake	N-releasing rate (mg/m^{2}d)	P-releasing rate (mg/m^{2}d)	Amount of released N (t/a)	Amount of released P (t/a)
Guchenhu		7.74-8.10		
Erhai	55-90	2.2-5.6	485.8-795.0	194.3-494.6
West lake		1.02		1.346
Caohu			1705.2	220.38
Xuanwuhu				10.46

Table 7 Released TDP concentrations (mg/l) under different conditions

Time	June 27	June 29	July 3	July 6	July 14
anaerobic	0.335	0.415	0.619	0.739	0.778
aerobic	0.051	0.050	0.050	0.035	0.039

Remark: Observation was conducted in 1989.

4.2.2 Nutrient Release from Sediment

Nutrient will be released from deposited sediment once pollutant origins have been cut off. Usually released nutrient enters pore water first, then it is mixed and diffused with the cover water. Some examples of nutrient release in some lakes in China are shown in Table 6.

Main factors affecting the nutrient release are: dissolved oxygen, pH value, temperature, oxidation-reduction potential, biology and disturbance of the water, etc. Oxidation condition promotes the adsorption of phosphorus, and anaerobic condition promotes its release. Observation conducted in the Xuanwuhu shows that the releasing rate under different conditions differs greatly, see Table 7 (Nanjing Environment Monitoring Station, 1990).

Nutrient is released the least at neutral pH value, the release of nutrient increases at both higher and lower pH values, see Fig.9. Disturbance, temperature, chemical contents of the water, etc. also effect the release of nutrients. We will not discuss in detail due to limited space.

5. CONCLUSION

Being one of the necessary conditions for sustainable development, environment protection has attracted top concern of the society. Aquatic environment is one of the most important parts of the environment. Most natural rivers carry sediment. Pollutants, such as heavy metals, toxic organic matter, nutrient, etc. are sediment motion related ones. Studies and observation data obtained in China are presented in this paper. Following the historical development, before eighties water quality problem was generally studied in connection of oxygen-consuming pollutants, main water quality indexes are DO and BOD_5. Heavy metals, toxic organic matter, nutrient, etc. contribute little to such water quality indexes, but have a considerable negative effect on environment. Obviously, the concept of water quality has been enlarged and deepened. Existing results have shown that transport-transformation of such pollutants in sediment-laden rivers possesses not only the generality of common tracers, but also characteristics of transport-transformation induced by sediment motion. A combined effort by environment chemists, hydraulic engineers and biologists will certainly promotes the development of this discipline of science.

6. REFERENCES

Coughtrey, P. J., Martin, M. H. and Unsworth, M. H. (1987), Pollutant Transport and Fate in Ecosystems, Oxford, Boston: Blackwell Scientific Publications.

Forstner, Ulrich and Wittmann, G. T. W. (1979), Metal Pollution in the Aquatic Environment, Berlin, New York: Springer-Verlag.

Hart, B. T. (1989), Water Quality Management—Fate and Transfer of Pollutants in Aquatic Environment, Translated by Zhang Licheng et al., China Environment Press (1991).

Huang, S. L. (1993), Study on the Effect of Sediment Motion on Transport-Transformation of Heavy-Metal Pollutants (in Chinese), Ph D. Dissertation, China Institute of Water Resources and Hydro-Power Research.

Huang, S. L. and Wan, Z. H. (1997a), Study on Sorption of Heavy Metal Pollutants by Sediment, Journal of Hydrodynamics, Ser. B., No.3, pp.9~23.

Huang, S. L. and Wan, Z. H. (1997b), Concurrent Sorption of Heavy Metal Pollutants by Sediment with Different Grain Sizes, Journal of Hydrodynamics, Ser. B, No.2, pp.1~12.

Huang, S. L., Wan, Z. H. and Wai, Onyx W. H. Wai (1997c), Effect of Sediment Motion Patterns on Sorption of Cadmium Ions by Sediment Particles, International Journal of Sediment Research, Vol.12, No.4.

Huang, S. L. and Wan, Z. H. (1994), Study on the Mathematical Model of Heavy Metal Pollutant Transport-transformation in Fluvial Rivers, International Journal of Sediment Research, Vol.9, No.4, pp.36~45.

Jin, X. C. (1992), Chemistry of Contaminants and Sediments (In Chinese), Press of Chinese Environment Science.

Nanjing Environment Monitoring Station, 1990, Investigation and Study on Bottom Mud in Xuanwuhu (In Chinese), Report of Nanjing Environment Monitoring Station.

Tianjin Environment Protection Institute, 1990, Eutrophication in Yujiao Reservoir and its Prevention (In Chinese), Report of Tianjin Environment Protection Institute.

Yang, Q. A., Yu, Q. R. and Feng, J. M., 1995, Operational Study of the Sanmenxia Project on the Yellow River (In Chinese), Henan People' s Press.

YRCC, 1995, Study on the Effect of Sediment on Water Quality and Water Pollution Control In Main Reaches (In Chinese), Report of Yellow River Water Resources Protection Institute.

Zhang, Q. S., Jiang, L. S. and Lin, B. L. (1986), Pollution of Water by Sediment and The Ecological Problems, International Journal of Sediment Research, Vol. 1, No.1.

Environmental Hydraulics, Lee, Jayawardena & Wang (eds) © 1999 Balkema, Rotterdam, ISBN 90 5809 035 3

Invited lecture: The influence of fine sediments on water quality

N.V.M.Odd
HR Wallingford Limited, UK

ABSTRACT: The transport in suspension, settlement, consolidation and re-erosion of clay flocs (mud) and organic matter has an important influence on water quality, eutrophication and the dispersal of adsorbed micropollutants in a wide range of situations. The paper reviews the main processes by which fine sediments affect water quality by reference to practical case studies. The summarised conclusions are designed to aid those responsible for protecting or modelling the aquatic environment.

1. INTRODUCTION

Relatively few shallow (<30m) water bodies are entirely free of suspended sediment.

Natural mud consists of clay mineral platelets, including montmorillonite, illite and kaolinite with sizes of less than 1 micron, with a mixture of silt particles, including quartz and feldspars with sizes in the range 1-60 microns. It also contains organic matter, nutrients, a wide range of chemical substances including toxic pollutant, and living matter such as algae.

In freshwater almost devoid of dissolved salts (<20ppm), the clay platelets repel each other and form a colloidal suspension with a negligible settling velocity. With levels of dissolved salts exceeding about 100ppm, the clay platelets are attracted to each other when they collide to form mud flocs. The strength, size and settling velocity of the mud flocs depend on concentrations of mud in suspension and the rate of local shearing ($\partial u/\partial z$) in the flow. The settling velocity of weakly flocculated fluvial muds and fully flocculated marine muds are of the order of 10^{-6} and 10^{-3}m/s, respectively. The cohesive strength of settled muds varies with the percentage and type of clay mineral, their density and the degree of flocculation.

The three dimensional pattern of concentrations of mud in suspension in a water body and on the bed are governed by the sources of new sediment, the processes causing re-erosion of settled matter, the levels of turbulence in the water column and residual currents. Particles of raw sewage and organic detritus released into a water body are incorporated into the mud flocs and thereafter move with the mud in suspension. Suspended matter tends to settle and accumulate locally on the bed in sheltered (slack) zones where the velocities and bed stresses are below the threshold of erosion. Whereas the dissolved fractions are diluted and flushed away more easily.

Many toxic micropollutants, such as lead, are strongly attracted and adsorbed onto the surfaces of clay platelets, especially in aerobic marine conditions. Phosphates are also strongly bound to mud in freshwater. A deposit of weakly consolidated mud contaminated with raw sewage can exert a considerable oxygen demand on the overlying water including the release of hydrogen sulphide in a marine environment. A rapid re-suspension of large quantities of anoxic fluid mud can cause an immediate and large oxygen sag in the overlying water. Various reducing processes take place within the bed such as de-nitrification in the anaerobic pore water. Some pollutants desorb into the pore water. Low density organic sediments that remain undisturbed by waves, currents, shipping and dredgers may diffuse recycled nutrients back into the water column.

Quite small concentrations of suspended mud (5-10ppm) are effective at reducing the penetration of light into the water column thereby concentrating thermal energy in the surface layer and reducing algal growth in the water column. Detrital material may be incorporated into suspended mud flocs.

It is therefore important to consider the role of suspended solids every time one analyses a water quality problem or when setting up a water quality model of a given situation. This review uses a number of particular studies to highlight various impacts of fine sediments on water quality.

2.   TAI LAKE, CHINA

Tai Lake is part of a complex network of canals, which is managed by the Tai Lake Basin Authority (TBA). TBA commissioned the Nanjing Institute of Geography and Limnology (NIGL) to make a series of detailed synoptic surveys of the lake starting in 1990. An example of one such survey is shown in Figure 1. The following analysis is published with the permission of the director of TBA, Professor Huang Xuan Wue, but it does not necessarily reflect the views of TBA or NIGL.

Tai Lake is approximately circular with maximum width of 50km and an area of about 2300km^2. The lake is about 40km from the Yangtze River at its nearest point, and 70km from Shanghai. The depth varies from 1 to 1.5m near the shore to a maximum of 2.6m in the centre and is vertically well mixed most of the time. In the past, most of the inflow entered the lake from the western shore and flowed towards the Yangtze River with a flushing time was about 1.2 years. In the future, water from the Yangtze River will be introduced into the lake via Wang Yu (River) and up to 300m^3/s will be discharged via the Taipu (River) to provide water supplies for Shanghai.

The prevailing south easterly winds generate a downwind residual current in the shallower water near the shore and a return current in the deep water in the centre of the lake creating twin circulating cells, modified by the Coriolis effect and by the detailed geometry of the lake.   The wind driven circulation transports suspended mud, pollution, nutrients and algae around the lake. Moderate to strong winds (8m/s) cause heavy wave action on the lake. The prevailing wind tends to concentrate and trap buoyant algae in the Wuxi Bay region which is a major source of nutrients. Suspended sediment and algal detritus tends to move upwind in the bed layers in the deeper parts of the lake.

In 1990, the bed sediments in the centre of the lake were clean yellow clay becoming grey clay nearer the shore, both with a low organic content. Soft black mud up to 2m deep with a higher organic content, probably mainly algal detritus or cellulose from plants, was found in the areas protected from waves from the prevailing SW direction. Suspended solids in August 1990 varied from about 1-180ppm with an average Secchi disc depth of August 1990 was 0.3m compared to an average water depth of 2m. It appeared that the concentration of total phosphorus in the lake sediments had not increased in recent years but had remained constant at about an average of 0.05% by dry weight, equivalent to between 40,000 – 120,000 tonnes in the lake as a whole depending on the depth of bio-turbation and density of the surface sediments. This quantity of particulate phosphorus was large compared to the estimated inflow of phosphorus of about 1500t/yr, mainly from the western catchments which are difficult to regulate. Concentrations of dissolved total phosphorus appeared to have increased from about 0.03mg/l in 1980 to over 0.06mg/l in 1990. Thus, in 1990 it appeared that phosphorus was not in short supply in the lake.

Observations made since 1976, showed an increase in the average organic content of the lake mud from about 0.8 to 1.9% by dry weight. The area of lake muds with organic contents exceeding 1.2% had increased from 3% in 1976 to 12.5% in 1981 and 32% in 1986. Assuming an average thickness of sediment of 0.3m and a dry density of 400kg/m^3 there was about 4 x 10^6 dry tonnes of organic matter in the lake in 1990.

The total average influx of total nitrogen was estimated to be about 30,000 tonnes/y in 1991, most of which entered the lake from the NW catchments in the first flush of run-off in the early part of the wet season. The estimated average net outflow was about 25,000 tonnes/yr, giving a net retention of 5,000 tonnes/yr.

Total nitrogen concentrations had increased from about 1.0 to 2.3mg/l between 1980 and 1990. The average mass of nitrogen in the lake water in 1990 appeared to be 6000 tonnes. Much larger quantities of nitrogen are associated with the algae and macrophyte biomass and detrital matter in the lake. The annual average standing stock of algae which shows great variation (orders of magnitude) both temporally and spatially, appeared to be approximately 5 mg/l or 20,000 tonnes.

The pool of organic detrital matter in the lake sediments appeared to be rising at a rate of the order of 0.1 x 10^6 tonnes/annum, assuming a 0.1% annual increase in the organic content (1.1% in 15 years 1976-91). This would indicate an annual productivity of the order of 500 gC/m^2/yr, which is in the correct order for a eutrophic lake. The data implies an annual utilisation of 10,000 tonnes of nitrogen by algae each year.

The main cause of eutrophication in Tai Lake appeared to be an excessive influx of nitrogen early in the wet season from the rivers in the north western shores of the lake. Phosphorus does not appear to be a limiting factor except perhaps locally for short periods. It was estimated that only about 30% of the available nitrogen was utilised by algae in the lake. The maximum potential productivity of the lake including macrophytes could therefore been several times the 1990 level if it was not inhibited by the turbidity of the lake and by rapid flushing of the lake in the wet season..

The formation of floating algal mats in the Wuxi Bay region is caused by intense productivity in the deepened photic zone in calm periods lasting a few days, which allow the mud flocs to settle, and by gentle prevailing westerly winds accumulating algae in the surface waters of Wuxi Bay.

The lake reportedly did not suffer significant oxygen depletion or severe anaerobic conditions following an algal bloom. This is because frequent wave action increases turbidity, halting growth and mixing algae into the water column which are carried upwind at depth. Large quantities of fluorescent microcystic clusters were observed in the turbid waters of Wuxi Bay in rough conditions (0.5m, 3 second waves) following an algal bloom in September 1992.

It may be concluded that the level of eutrophication in Tai Lake is controlled in part by the unsteady pattern of transport of suspended solids in the lake.

Examples of similar lakes include:-

- Chao Lake also in China, which is eutrophic, turbid, 4m deep and has an area of 700km^2.
- Thale Luang at Songkhla in Thailand, which is a turbid oligotrophic lake with moderate concentrations of nutrients and no significant phytoplankton blooms. The lake is 2.5m deep and covers 500km^2.
- Lake Okeechobee in Florida in the USA which is 1-3m deep and has an area of 1600km^2.

In all the above nutrient rich lakes, the algal blooms are controlled in part by suspended sediments and effects of wind driven currents and wave action. To summarise:-

- Suspended sediments in shallow lakes are put into suspension and transported by wind and wave action.
- The suspended sediments control the level of eutrophication in nutrient rich lakes by limiting light penetration and storing phosphorus.
- Engineering works or actions which change the turbidity of a lake may also have an impact on water quality.

3. RUN-0F-THE-RIVER RESERVOIRS

The author has been unable to present data for any deep reservoirs for reasons of confidentiality. However, experience with run-of-the-river reservoirs has shown that quite small concentrations of dissolved salts with a conductivity of about 100μS/cm are sufficient to cause colloidal muds to form very small flocs. These flocs settle at rates of the order of 0.1m/day in the absence of significant vertical turbulent exchange. The critical velocity which allows the deposition of such river muds is of the order of 20mm/s compared to 0.3m/s in marine waters.

In the case of muddy rivers flowing into deep slow flowing reservoirs, even such a small settling velocity is sufficient to clear the surface layers and deepen the photic zone and trap mud and organic matter in the reservoir.

One can expect relatively cooler mud laden floodwaters to generate fairly strong density currents along the sloping bed of a reservoir, which could carry easily re-eroded low density mud deposits to the toe of the dam.

Low level sluices are an effective method of reducing the sediment and nutrient trapping efficiency of such reservoirs.

There are some extreme examples of massive floods flowing through reservoirs with huge sediment loads (400,000ppm) causing a non-Newtonian behaviour, which are outside the scope of this review. To summarise:-

- The settling velocity of weakly flocculated mud flocs in river water with a conductivity of about 100

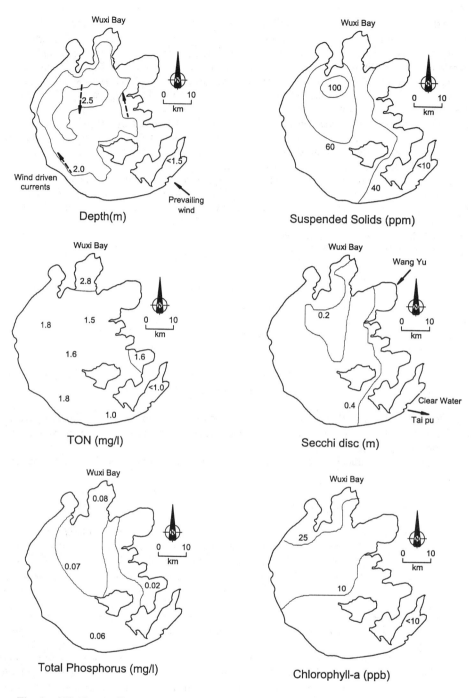

Fig.1. Water quality and suspended sediment in Tai Lake August 1990

μS/cm is of the order of 0.1m/day. This is sufficient to clear surface layers.

- The critical velocity which allows deposition of fine river muds is of the order of 20mm/s.
- Low level sluices can be an effective means of minimising the trapping of nutrients and fine sediments in run-of-the-river reservoirs, without excessive drawdown.

4. HONG KONG HARBOUR

The views expressed in this chapter are those of the author and do necessarily reflect the views of the Hong Kong Government. Raw and partially treated domestic sewage effluent from several millions of people is the main cause of the oxygen sag in Victoria and Western Harbours. The assimilative capacity of Hong Kong Harbour area depends on the rate at which the oxidising components of the sewage are diluted, flushed and dispersed from the sources of pollution. Every effort has been made to relocate outfalls to reduce concentrations of BOD and ammoniacal nitrogen in the Harbour area. Likewise, considerable efforts are made to streamline reclamations to maintain the rate of flushing of Victoria Harbour by tidal and gravitational flows.

However, unlike treated effluents, up to about 50% of the biological oxygen demand of screened sewage can be in the form of particulate matter (ie several hundred tonnes a day in Hong Kong). The particulate fraction of sewage effluent usually consists of different proportions of fast and slowly oxidising organic matter depending on the level of treatment. The sewage particles combine with natural suspended solids and behave in a similar fashion to mud flocs and have a settling velocity of the order of 1mm/s in seawater.

The sewage flocs are carried to and fro with the tide moving up and down the water column and in and out of suspension depending on the pattern of tidal currents and the strength of vertical turbulent exchange, oxidising all the while. Eventually the sewage flocs settle in a zone sheltered from velocities high enough to cause re-suspension, ie where the peak velocities are less than about 0.3m/s and the peak bed stress is less than about $0.1N/m^2$. The settled organic matter forms a soft low-density fluffy deposit and exerts a static sediment oxygen demand (SOD) on the overlying water. The distribution and magnitude of the sediment oxygen demand is governed by the same mechanisms that control mud siltation; namely, settlement, deposition and re-erosion by tidal currents, dredging and wave action. The sediment oxygen demand is often greatest in slack zones such as typhoon shelters and re-entrants along the shoreline. As a result, the overlying water may become anaerobic and sulphate in the seawater would be reduced to generate hydrogen sulphide as in the case of the Kai Tak Nullah. In contrast, the dissolved BOD is dispersed by tidal currents and flushed out of the Harbour area by the residual currents.

In 1987, HR Wallingford deployed a respirometer to measure SOD at various sites on the north shore of Hong Kong Harbour. The SOD values in still water varied from 2 – 20 gO_2/m^2/day. These values are significant. For example $10g/m^2/d$ is equivalent to 100tonnes/d over an area of $10km^2$, which is the order of the area of potential settlement in Hong Kong Harbour. As a result, HR Wallingford have always incorporated both dissolved and particulate SOD in all their water quality models of Hong Kong Harbour.

Ideally, to do this, one requires a knowledge of the composition of the sewage effluent in terms of fast and slowly decaying and dissolved and particulate fractions. Because of a lack of detailed information in the composition of the sewage effluents discharged to Hong Kong Harbour, these fractions have to be estimated or evaluated by trial and error. One also needs to prescribe the settling velocity and critical shear for deposition and erosion of the sewage particles. These are usually assumed to be the same as for marine mud flocs.

The application of HR's latest TELEMAC-3D water quality model to conditions in Hong Kong was an opportunity to compare predicted SOD distributions in the wet and dry seasons, which are shown in Figure 2. There was a significant difference between the distribution of settled BOD in the harbour area in the dry and wet seasons. The results are sensitive to the magnitude of the prescribed BOD particulate loads, which cannot be verified in retrospect. The simulations are shown only to illustrate the discussion.

In the dry season, the tidal velocity profiles are logarithmic with high velocities near the bed and the bulk of the particulate BOD discharged into Victoria Harbour was predicted to settle off the north shore of Hong Kong Island and on Kellet Bank, a zone of low bed stress. There is a secondary zone of deposition in Rambler Channel.

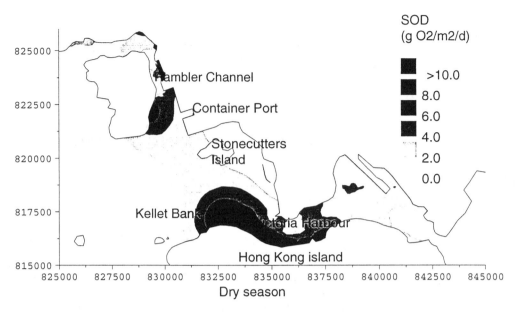

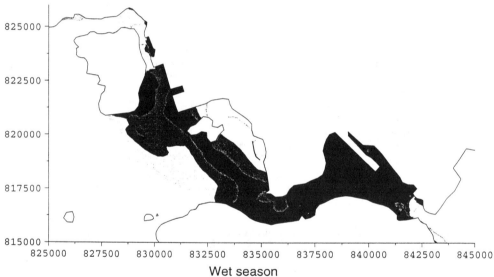

Fig. 2 Predicted SOD values in Hong Kong

In the wet season, the water column is stratified, vertical turbulence is heavily damped and the tidal velocities in the more saline bed layers are below the threshold of re-suspension of mud over most of Victoria Harbour much of the time. As a result, the particulate BOD discharged into Victoria Harbour tends to be trapped in the lower layer and to be more uniformly distributed across the bed of the harbour. It also follows that the sediment oxygen demand is exerted on a smaller and less well flushed volume of water compared to the dry season. In the Western Harbour in the wet season the model predicted less settlement of BOD solids in Kellet Bank, despite the increased western residual flow in the bed layer, and a zone of increased BOD settlement between Stonecutters Island and the container port area.

The distribution of settled particulate BOD is sensitive to the loading pattern, effluent composition, plume behaviour, settling and erosion characteristics of the particulate BOD fraction.

The sediment oxygen demand exerted by a given deposit on the overlying water depends on its composition, structure, density, porosity, temperature, bio-turbation and redox potential, as well as the oxygen content and speed of the overlying water. It can be measured using an in-situ respirometer, which simulates aerobic still-water conditions. Prescribed SOD values (ie not predicted) are of limited value in the case of Hong Kong where the coastline and sources of pollution are changing from year to year.

The most conservative and basic assumption is to assume that the settled sewage forms a thin low-density surface deposit in which oxygen in the overlying water is removed by oxidation of the full depth of the BOD deposit. This method proved to be successful in predicting the observed SOD's for Hong Kong Harbour within a factor of 2. Experience has shown that the SOD settles down to a state of dynamic equilibrium over a neap-spring cycle.

This simplified method is probably adequate in an area with large sources of partially treated sewage, relatively little natural sediment, low levels of primary productivity and with aerobic conditions in the lower layer of the water column-conditions that prevail in most of Hong Kong Harbour these days.

A more complex representation of the benthal processes may be needed in a closed body of water especially if recycling of nutrients is significant. To summarise:-

- Raw and partially treated sewage from large coastal cities contain a significant load of particulate BOD. Ideally, the particulate BOD load should be represented by at least two fractions with different oxidation rates.
- Particulate BOD behaves in a similar fashion to marine mud and quite different from dissolved BOD, which is usually dispersed more readily.
- The distribution of settled particulate BOD varies with the pattern of tidal and gravitational flow, saline/thermal stratification and also wave action in shallow water.
- The particulate BOD tends to settle and accumulate in zones, basins and re-entrants with a high mud trapping efficiency, which as a result may suffer a severe oxygen sag, even if they are reasonably well flushed.
- The static sediment oxygen demand from a thin low density deposit of settled particulate BOD in a slack zone may be predicted reasonably accurately from the total remaining BOD on the seabed at any one time by assuming that the whole of the potential demand is applied to the overlying water.
- The static oxygen demand in slack zones may be measured fairly accurately using a simple cylinder respirometer.

5. THE FATE OF HEAVY METALS IN THE NORTH SEA

In 1988, the Water Directorate of the UK Department of the Environment (DoE) commissioned HR Wallingford to develop and calibrate a three dimensional policy model of the Southern North Sea capable of simulating the main processes governing the transport, dispersal and deposition zones of heavy metals adsorbed onto particulate matter which have been discharged into the North Sea.. Cadmium and lead were selected to represent metals that are transported mainly in solution and adsorbed onto fine sediment, respectively. The model three dimensional seasonal flow, mud and metal transport models were calibrated by reference to the NERC North Sea Project data collected over a 5 month period in 1988/89.

The observed partition coefficient for lead was 10^7 which meant that about 98% of the metal was adsorbed onto the marine mud. The rates of adsorption and desorption which gave the best fit for lead were $0.2d^{-1}$ and $0.002d^{-1}$, respectively. In contrast, only about 20% of the cadmium was adsorbed.

At the 1990 North Sea Conference participating governments including the UK, agreed to adopt a set of common actions to further reduce the inputs of hazardous substances, ie those that are persistent, toxic and liable to bioaccumulate, which include cadmium and lead. In particular, the North Sea States agreed to a significant reduction (of 50% or more) of inputs of hazardous substances via river and estuaries between 1985 and 1995 and atmospheric emissions by 1995 or 1999 at the latest. However, inputs were not adequately monitored in 1985.

DoE required HR Wallingford to use the calibrated model to predict (Odd et al 1995) the beneficial effect of stopping cadmium and lead loadings from sewage sludge dumping and incineration at sea and reducing loadings from UK rivers, all North Sea rivers and atmospheric deposition of the same metals by 50% compared to 1990 loads as agreed at the Third International Conference on the protection of the North Sea in the Hague in March 1990. The 1990 loads for sewage sludge, dredging and incineration at sea were based on the Oslo and Paris Commission (1992) estimates for 1990. The UK agreed to stop the disposal of sewage sludge to the North Sea by 1998 at the very latest. Incineration at sea was phased out by 1992.

The three dimensional seasonal model predicted an anticlockwise and northerly transport of polluted mud from the English Channel and riverine and coastal sources of sediment. In the summer, mud settled out of suspension forming a transient deposit at the edge of the deeper water in the centre of the North Sea (Figure 3). The surface mud deposits in the shallower areas are resuspended by wave action in the winter months. The ultimate fate of the mud is to be accumulated the Norwegian Trench.

Particulate lead concentrations (per unit volume of water) were highest in the winter when the mud is in suspension. There were peaks in the concentrations in the Thames estuary, off the mouth of the Rhine and in the German Bight where concentrations exceeded about 2ppb in January. There was little particulate lead in suspension in the calm summer months when the mud settles on the seabed.

The predicted concentrations of adsorbed lead in the thin superficial deposits of fluffy mud on the bed of the North Sea at the end of the relatively calm summer period are also shown in Figure 3. The largest mass of lead is found in the muddy area offshore in the centre of the North Sea, a biologically rich zone, where concentrations exceeded 300ppm, which was the peak value that was observed in the suspended mud in the winter period during the 1988/89 NSP survey (Charnock et al, 1994). The prediction ignored the diluting effect of bioturbation. Observations of lead concentrations in the generally small clay fraction in the upper 100mm of the sea bed made in 1986 by Irion and Muller (1987) were generally in the range of 100-200ppm, with evidence of values exceeding 200ppm off the NE UK coast, mid sea and near the Norwegian Trench.

The mainly particulate lead settles into the Norwegian trench within about one year depending on the severity of winter storms. The mainly dissolved cadmium (~80%) from the UK is flushed into the North Atlantic within about one year.

The estimated atmospheric loads for lead totalled 1670t/yr in 1990. Nearly all the lead in the dredged spoil from UK ports in 1990 (790t/yr lead) was derived from coastal sediments containing historical deposits of metals from the last hundred years of industrial activity. This source will gradually reduce as the historical deposits of lead in the estuarine and coastal mud deposits are remobilised and flushed into the North Sea and as the contaminated mud is carried across the North Sea. New mud eroded from the cliffs contains much smaller concentrations of metals. The influx of lead from the English Channel is also large (1200t/yr) and fairly uncertain.

The sources of new mud are even more uncertain. This has a direct effect on the metal/mud concentrations but not on the speed or pattern of transport of adsorbed micro pollutants. In retrospect, the mud inputs from the UK coast (13 Mt/yr) were probably over estimated. The latter must be assumed to include dredged spoil remobilised by wave action in the winter months. The English Channel also provides a very large source of mud (~10Mt/yr). The model excluded areas of net deposition like the Wadden Sea on the Netherlands Coast, which is said to trap 2 Mt/yr of mud; or smaller areas of erosion such as the Flemish Banks (1 Mt/yr). To summarise:-

- The transport and ultimate fate of toxic micro-pollutants which are readily adsorbed by marine muds, such as lead, are controlled almost completely by the long term pattern of mud transport in shallow seas.
- An error in the magnitude of the sources of new mud or mud concentrations has a direct effect on metal/mud concentrations, but not necessarily on the speed or pattern of transport or the ultimate fate of adsorbed micro-pollutants.
- The rate and pattern of transport and ultimate fate of adsorbed micro-pollutants is sensitive to the seasonal variations in the strength of residual currents and of bed stresses exerted by waves and near bed tidal currents.

6. CONCLUSIONS

The author has used the selected case studies to demonstrate the main effects of fine sediments on water quality, which include limitation of light penetration, storing phosphorus, benthal oxygen demand and

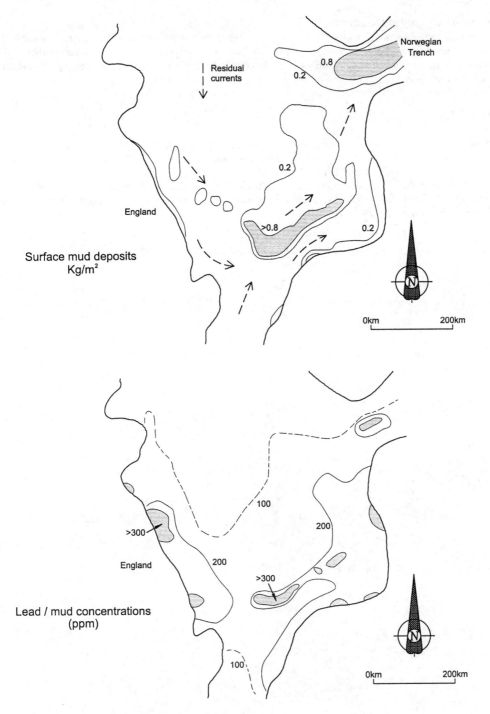

Fig.3. Predicted mud deposits and lead concentrations, North Sea, October 1990

transportation adsorbed micro-pollutants.

It is therefore important to consider the impact that engineering work and actions will have on the fine sediments, including particulate organic matter, if they are likely to impact on water quality. The summarised findings from each case study are designed to aid those responsible for protecting or modelling the aquatic environment.

7. REFERENCES

Charnock et al (1994). *Understanding the North Sea System*, Chapman and Hall for the Royal Society.

Irion and Muller G (1987). "Heavy metals in surface sediments in the North Sea". *Int. Conf. Heavy metals in the Environment*. New Orleans 1987.

Odd N.V.M. et al (1995). "Particulate Pollutants in the North Sea – Phase II, Final Report. Development and Application of NORPOLL (mk1.1) to predict the effect of load reductions on concentrations of Cadmium and Lead". HR Report SR 383, March 1995.

Environmental Hydraulics, Lee, Jayawardena & Wang (eds) © 1999 Balkema, Rotterdam, ISBN 90 5809 035 3

Modeling transport and fate of contaminants with sediment interaction and transient storage in streams

Yi Zhang & M.M.Aral
School of Civil and Environmental Engineering, Georgia Institute of Technology, Atlanta, Ga., USA

ABSTRACT: In this paper, a contaminant transport model was developed for modeling of transport and fate of radionuclides and various hazardous substances in stream and river systems. The model includes the sediment-contaminant interaction and transient storage in addition to various physical and chemical processes. Compared to other sediment-contaminant transport models, the main feature of the proposed model is to integrate the transient storage and related physical and chemical processes into the model in order to simulate contaminant fate and transport more accurately. In the model, all governing equations are written in the conservative form and an efficient algorithm was proposed for the solution of the model. A real case study is solved to show the validation and accuracy of the proposed model.

1. INTRODUCTION

Many contaminants, such as some pesticides, radionulides, heavy metals, and many toxic chemicals, have high distribution coefficients, and are therefore easily adsorbed by river sediments in surface water with high suspended sediment concentration. Contaminated sediments may be deposited on the river bed, and become a long-term source of pollution through desorption and resuspension processes [Onish et al., 1982]. To take this into account, some investigators [Onish, 1981, Onish et al., 1982, Tkalich, 1993] integrated sediment-contaminant interaction into the transport model. However, these models do not consider the transient storage mechanism, which is often an important factor for accurately simulating contaminant transport and fate in streams. A transient storage phenomenon has been often observed in many streams, and the storage zone is the portion of the stream that contributes to transient storage, that is, stagnant pockets of water and porous areas along the streambed or stream bank. By including this transient storage, solute transport models such as those by [Bencala, 1983; Bencala et al., 1984; Runkel et al.,1996] may simulate the solute transport process more accurately. However, the disadvantage of these models is that the sediment-contaminant interaction is not included.

In this paper, we developed a contaminant transport model, which includes the sediment-contaminant interaction and transient storage, in addition to various physical and chemical processes, for modeling of transport and fate of radionuclides and various hazardous substances in stream and river systems. Compared to other sediment-contaminant transport models, the main feature of proposed model is that the transient storage is included and all governing equations are written in the conservative form. An efficient algorithm was also proposed for the solution of the model. Finally, a real case study is solved to show the validation and accuracy of the proposed model.

2. MATHEMATICAL MODEL

The proposed model is based on some existing models [Onish, 1981; Tkalich et al.,1993], but it includes the transient storage mechanism, and was developed for transport and fate of radionuclides, trace metals and some toxic chemicals in streams or rivers. The model consists of several submodels which describe the different physical and chemical processes. The conservative forms of equations are used for all major transport processes since it is important for the solution of conservation law. The mathematical expressions of the model are described follows.

Sediment Transport Submodel:

The sediment transport submodel, which is similar to those [Onishi, 1981; Tkalich et al., 1993], can be expressed as following conservative form:

$$\frac{\partial(AS)}{\partial t} + \frac{\partial(QS)}{\partial x} = \frac{\partial}{\partial x}\left(AD\frac{\partial S}{\partial x}\right) + q_L(S_L - S) - A\left(\frac{q_D}{h} - \frac{q_R}{h}\right) \tag{1}$$

where, A is the cross-sectional area (L^2); Q is the flow discharge (L^3/T); h is water depth (L); S is the concentration of sediment (M L^{-3}); S_L is the concentration of sediment in lateral inflow (M L^{-3}); q_D and q_R are the vertical fluxes of sediments, i.e., sediment deposition rate per unit bed surface area and sediment erosion rate per unit bed surface area (M L^{-2} T^{-1}). For noncohesive sediment, sedimentation q_D and erosion q_R can be determined by some methods[Onishi, 1981, or Zheleznyak et al., 1992].

Dissolved Contaminant Transport Submodel in Main Channel:

Dissolved contaminant transport submodel in main channel, which includes the transient storage, may be expressed by following equation:

$$\frac{\partial(AC)}{\partial t} + \frac{\partial(QC)}{\partial x} = \frac{\partial}{\partial x}\left(AD\frac{\partial C}{\partial x}\right) + q_L(C_L - C) + \alpha A(C_S - C) - kAC + A\sum_{i=1}^{5}k_{ci}C$$

$$+ a_{13}A(C_b - K_d C)\rho(1-n_s)Z_b/h + a_{12}AS(C_p - K_d^p C) \tag{2}$$

where C_p is the particulate contaminant concentration (weight of contaminant per unit weight of sediment, M M^{-1}); C_S is the concentration in storage zone as defined before (M L^{-3}); C_b is particulate contaminant concentration in the streambed (weight of contaminant or radioactivity per unit weight of sediment, M M^{-1}); α is stream-storage exchange coefficient (T^{-1}). k is the first order radioactive decay or degradation coefficient (T^{-1}); k_{ci} (i=1,···,5) is the first order reaction rate of contaminant degradation due to hydrolysis, oxidation, photolysis, volatilization and biological activities (T^{-1}); K_d^P is the distribution (or partition) coefficient between dissolved contaminant and suspended particulate contaminant(L^3 M^{-1}); a_{12} is the exchange rate of contaminants between water (i=1) and suspended sediment (j=2) (T^{-1}); K_d is the distribution (or partition) coefficient between dissolved contaminant and streambed particulate contaminant (L^3 M^{-1}); a_{13} is the exchange rate of contaminants between water (i=1) and sediment in streambed (j=3) (T^{-1}); Z_b is the efficient thickness of the contaminated, upper, bottom deposition layer (m); n_s is the porosity of the streambed; ρ is the density of sediment (M L^{-3}).

Particulate(suspended sediment) Contaminant Transport Submodel in Main Channel:

Particulate contaminant submodel describes the transport and sorption/desorption processes of contaminants on the suspended sediment and may be described by following equation:

$$\frac{\partial(ASC_p)}{\partial t} + \frac{\partial(QSC_p)}{\partial x} = \frac{\partial}{\partial x}\left(AD\frac{\partial(SC_p)}{\partial x}\right) + q_L S_L(C_{Lp} - C_p) - kASC_p$$

$$+ a_{12}AS(K_d^P C - C_p) + A(q_R C_b - q_D C_p)/h - \alpha AS(C_p - C_{sp}) \tag{3}$$

where C_{sp} is the suspended particulate concentration in storage zone; C_{lp} is the suspended particulate contaminant concentration in lateral inflow.

Streambed Sorbate Contaminant Concentration Submodel:

Submodel for streambed sorbate of contaminants, which is similar to that[Tkalich et al., 1993], may be described by the equations:

$$\frac{\partial C_b}{\partial t} = a_{13}(K_d C - C_b) - (q_R C_b - q_D C_p)/(\rho(1-n_s)Z_b) - kC_b \tag{4}$$

734

$$\frac{\partial Z_b}{\partial t} = (q_D - q_R)/((1-n_s)\rho) \tag{5}$$

Concentration Exchange Submodel in Storage Zone:
By including transient storage, additional set of equations are required to describe similar processes in storage zone as in a main channel. These equations, which are not subjected to transport processes, may be described as follows:

$$\frac{\partial C_S}{\partial t} = \alpha \frac{A}{A_S}(C - C_S) + \rho_S^* \lambda_S(C_{Sb} - K_{dS}C_S) - k_S C_S + a_{s12}S_s(C_{sp} - K_{ds}^P C_S) \tag{6}$$

$$\frac{\partial C_{Sb}}{\partial t} = -\lambda_S(C_{Sb} - K_{dS}C_S) + k_D C_{Sp} \tag{7}$$

$$\frac{\partial C_{Sp}}{\partial t} = \alpha(C_p - C_{Sp}) - a_{s12}(C_{Sp} - K_{ds}^P C_S) - k_D C_{Sp} \tag{8}$$

where C_{Sb} is the streambed sediment contaminant concentration in storage zone (g g^{-1}); λ_S is the first-order rate coefficient for sorption in the storage zone (s^{-1}); ρ_S^* is the mass of "accessible" sediment per volume of water in the storage zone (g m^{-3}); K_{dS} is the distribution coefficient in the storage zone (m^3 g^{-1}). a_{s12} is the exchange rate of contaminants between water and suspended sediment in storage zone (s^{-1}); k_D is sedimentation rate in storage zone and can be calculated by $k_D = w_p / h_p$, in which w_p is settling velocity for the storage zone particulate phase and h_p is effective storage zone settling depth.

3. SOLUTION METHOD

The "global" model consists of the sediment transport model and the contaminant transport model. When flow is unsteady, the flow simulation model [Aral, Zhang and Jin, 1996] may be included to simulate flow regime. All submodels are related. In every time step of solution, hydrodynamic model and sediment transport model are solved independently. The outputs of hydrodynamic model are flow parameters, cross-sectional area A and discharge Q, which, in turn, are input of the sediment transport model. Then, sediment transport model is solved and output of the solution is sediment concentration S, sedimentation rate q_D and resuspended rate q_R. Based on the flow parameters and sediment concentration, the contaminant fate and transport models including main channel and storage zone can be solved for both dissolved and particulate concentration. We can see that above model mainly consists of transport equations. A fractional-step method was used for the solution of transport equation. In the first step, pure advection equation is solved and the intermediate values of variables are obtained, while the relaxation scheme [Aral, Zhang and Jin, 1996] was used as the solution method. In the second step, the pure diffusion equation is solved base on the intermediate value obtained in the first step and a central difference scheme or Crank-Nicolson method was used for the solution of the equation. This solution algorithm and procedure are efficient and flexible because we do not need to solve any algebraic equations and all equations can be solved explicitly so that it is easy to programing, and at the same time, has higher accuracy.

4. CASE STUDY

A real case, Rhine River point source release, was studied using the proposed model. The Rhine River originates in the Swiss Alps and flows through French, German, then reaches the Netherlands, and finally the North Sea. On November 1, 1986, a fire occurred at a chemical warehouse at Schweizerhalle near Basel, Switzerland, on the Rhine River. The fire lasted for about 12 hours and most of the stored chemicals were destroyed in the fire. Some of them went into the Rhine River, among which, disulfoton and thiometon accounted for approximately 75% of

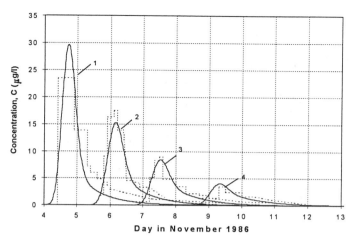

Figure 1. Simulated Concentration of Disulfotion in Rhine River (solid lines) and Observed data (dashed lines).
1 - 362 km, 2 - 496 km, 3 - 640 km, 4 - 865 km.

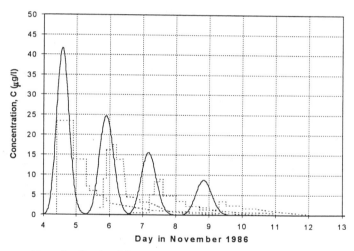

Figure 2. Simulated Concentration of Disulfotion in Rhine River (solid lines) and Observed data (dashed lines) without transient storage. 1 - 362 km, 2 - 496 km, 3 - 640 km, 4 - 865 km.

the total organo-P pesticides which were spilled to the Rhine River [Schnoor et al., 1992]. The chemicals entered the Rhine River and experienced an initial mixing until the plume reached the dam of hydroelectric power plant, approximately 4.7 km below the site of the spill. After the dam, transverse mixing of the chemicals was nearly complete [Capel et al., 1988]. Field measurements were taken at four locations as the plume passed. They are Maximiliansau, Germany (362 km), Mainz-Wiesbaden, Germany (496 km), Bad Honnef, Germany (640 km), and Lobith, the Netherlands (865 km). The total mass of organ-P ester pesticides, disulfoton and thiometon has been estimated to be approximately 7000 kg [Mossman et al.,1988], 4500 kg and 1800 kg respectively [Capel et al.,1988; Mossman et al.,1988]. Based on Mossman et al. [1988], a triangular input loading function for disulfoton and thiometon is used here. The input occurred over 12 hours, with the peak input occurring at 4 hours. The ascending part of the triangle has a slope of 1, and the descending slope is -0.5. The value of the function at any point is the mass loading rate. Flow field is steady state. The river width is assumed as constant, and the velocity field is assumed as a linear function increasing from 0.7 to 1.5 m/s [Mossman et al., 1988]. Sediment transport is

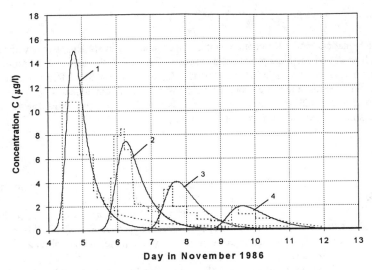

Figure 3. Simulated Concentration of Thiometon in Rhine River
(solid lines) and Observed data (dashed lines).
1 - 362 km, 2 - 496 km, 3 - 640 km, 4 - 865 km.

assumed as steady state and sediment concentration is taken as S = 10 mg/l [Schnoor et al., 1992]. Other
parameters for modeling are shown in Table 1.

The proposed model is used to simulate the transport of disulfoton and thiometon in the Rhine River after chemicals
spill into the river. From the measured data, we can see that concentration profiles have long tail. This was likely
due to a slow dissolution of an oily phase which remained on the bottom of the River after the accident, and dead
zones or relative stationary areas along the River banks which failed to mix well with the main channel [Schnoor
et al., 1992]. The simulations were run and Fig.1 and 2 show the results for disulfoton with and without transient
storage and sediment interaction considered in the model. The comparisons between simulated results and measured
data for disulfoton show that there are good agreements between measured data and the simulated results when
transient storage (dead zone) and sediment interaction were considered in the model and the long tail effect of a
concentration profile was simulated successfully in this case. However, the results without considering dead zone
and sediment interaction (Fig.2) did not generate satisfactory results, compared with field data. The simulated
results did not catch the long tail of the measured data. This fact indicates that the transient storage effect and
sediment interaction is very important and should be considered in the model. Figure 3 show simulated results and
measured data for thiometon with transient storage and sediment interaction. Similarly, the simulated results match
measured data very well. The model was calibrated for several parameters which are storage zone area A_s, stream-
storage zone exchange coefficient α, and first-order rate coefficient λ_s for sorption in the storage zone. These
parameters were taken as $A_s = 120$ m^2, $\alpha = 0.0001$ s^{-1} and $\lambda_s = 0.00003$ s^{-1}. Other data such as K_s and K_{ds} in
storage zone were taken as same as those in the stream, the porosity in the bed was taken as $n_s = 0.4$.

Table 1. Parameters in the Rhine River Chemical Spill [Schnoor, 1992]

k = 0.20 / day	S = 10 mg/l
w_s = 0.5 m/day	A = 1500 m^2
Z_b = 0.02 m	h = 5 m
D = 4 × 10^6 cm^2/s	ρ = 2.6 kg/l
K_d = 100 l/kg	a_{12} = 1 / day
K_{dp} = 200 l/kg	a_{13} = 1 / (100)day

737

5. CONCLUSION

In this paper, a mathematical model for contaminant transport and fate in streams was developed. The model includes the major physical and chemical processes in streams. Especially, the sediment-contaminant interaction and transient storage mechanism are included in the model. A case study for Rhine River release shows that the sediment interaction and transient storage are important factors which affect contaminant fate and transport process in surface water systems and should be included in the model. By including these processes, the model can simulate the contaminant transport process in a stream more accurately.

6. REFERENCES

Aral, M. M., Zhang, Y, and Jin, S. (1998). "A Relaxation Scheme for Wave Propagation Simulation in Open-Channel Networks." *J. of Hydraulic Engineering, ASCE, in publication.*

Bencala, K. (1984). "Interactions of solutes and streambed sediment, 2, A dynamic analysis of coulped hydrologic and chemical processes that determine solute transport", *Water Resources Res.*, 20, 1804-1814.

Bencala, K., and R. Walters (1983). Simulation of solute transport in a mountain pool-and-riffle stream: A transient storage model", *Water Resources Res.*, 19, 732-738.

Capel, P. D., W. Giger, P. Reichert, O. Wanner (1988). "Accidental Input of Pesticides into the Rhine River," *Environ. Sci. Technol.* 22, 992-996.

Mossman, D. J., L. Schnoor, and W. Stumm (1988). "Predicting the Effects of a Pesticide Release to the Rhine River," *J. Water Pollut. Control Fed.* 60, 1806-1812

Onishi Y. (1981). "Sediment-Contaminant transport model." J. of the Hydraulics Division, ASCE, Vol. 107(9), 1089-1107.

Onishi Y. and S.E. Wise (1982a). Mathematical model, SERATRA, for sediment-contaminant transport in rivers and its application to pesticides transport in Four Mile and Wolf creeks in Iowa. EPA-600/3-82-045, U.S. Environ. Prot. Agency, Environ. Research Lab., Athens, Georgia.

Runkel, R. L., D.M. McKnight, K.E. Bencala, and S. L. Chapra (1996). "Reactive solute transport in streams 2. Simulation of a pH modification experiment", *Water Resour. Res.*, 32, 419-430.

Schnoor J. L., Mossman D. J., Borzilov V.A., Novitsky M. A. and Gerasimenko A. K. (1992) "Mathematical Model for chemical spills and distributed source runoff to large rivers", in J. L. Schnoor (ed), Fate of Pesticides and Chemicals in the Environment, A Wiley-Interscience Publ., pp347-370.

Tkalich, P.V., M.J. Zheleznyak, G.B. Lyashenko and A.V. Marinets (1993). "RIVTOX-Computer code to simulate radionuclides transport in rivers." In A. Peters et al. (eds), *Computational Methods in Water Rersources X*, 1189-1196. *Kluwer Academic Publishers.*

Zheleznyak M. J., Demchenko R. I., Khursin S. L., Kuzmenko Y. I., Tkalich P. V. and Vitjuk N. Y. (1992) "Mathematical modeling of radionuclide dispersion in the Pripyat-Dnieper aquatic system after the Chernobyl accident", *The Science of the Total Environment,* 112, 89-114.

Environmental Hydraulics, Lee, Jayawardena & Wang (eds) © 1999 Balkema, Rotterdam, ISBN 90 5809 035 3

Artificial flood release, a tool for studying river channel behaviour

A. Krein & W. Symader
Department of Hydrology, University of Trier, Germany

ABSTRACT: The local water works enables us to create flood waves using their surplus water. According to our scientific hypothesis, we have the possibility to control the form and number of waves to gather information about flood water sediment interactions. An isolated examination of in-channel processes happens without superimposition by external influences. The activation and exhaustion of dissolved substances, as well as reason of mobilization and dynamic transport patterns for suspended sediment contaminents are focused upon. Another focal point is the influence of different types of floodwaves on changes in sediment quality. Areas of resuspension are identified. Adsorption processes are outlined as a reason for sinking heavy metal contents in the water body. The lag effect of various substances is exposed as well as a partial remobilization of heavy metals from the pore water system. During different stages of the hydrograph, particle size and density determine the content of some adsorbed contaminents which are also controlled by antecedent conditions in the whole basin.

1. INTRODUCTION AND OBJECTIVES

It is the entire basin that responds to a precipitation event. Therefore, the resulting flood wave is a complex pattern of many interrelated processes that take place within vegetation, soil, bedrock, human facilities, the channel and in different parts of the basin. Analysing the response pattern is a challenging task, and it is not always possible to prove the validity of the conclusions that have been drawn. The main advantage of field experiments is that some of the governing factors can be excluded and others can be controlled by the experimental design. As both amount and composition of the released water is known, it is possible to study the changes of water quality within its travel route. A simple mixing model using conservative ions can be used to assess storage processes within the channel. For this problem single peaked flood waves are used. Each natural flood response releases manganese and iron from at least three sources. They come from deep soil layers of the valley bottom, from interstitial water of remobilized river sediments, and from remobilized material that has deposited within the sewerage system. Artificial floods activate only the source of river bottom sediments. This phenomenon is investigated by using a succession of flood waves. The transport of suspended particles is supply controlled. Natural events mainly activate sources within the basin. Artificial floods can only activate sources within the channel. Both a succession of flood waves and floods of different magnitudes can give valuable information on how these sources are exhausted. Preliminary results of artificial flood waves can be presented. By comparing them with natural flood responses the basic structure of a flood wave can be understood. In addition, artificial flood waves are an important step in determining the location of sources of particle associated contaminants. Normally, the transport of particle associated contaminants is induced by flood events, but there are no indications that either discharge, particle concentration, particle size nor suspended organic carbon play a decisive role in this process during natural events. The pattern of single flood events varies more markedly than could be explained by the properties of the suspended sediments alone. It seems that there are some criteria of overriding importance that mask the expected relationships (Umlauf and Bierl, 1987). Contrary to those normal indications, these assumptions can be outlined in this investigation.

2. AREA UNDER INVESTIGATION

The size of the basin is about 35 km². It is located in the northern Hunsrück mountains near the city of Trier. Bedrock consists of schist and quartzite, the vegetation is a type of patchwork with arable land on the plateau, forests on the north and east facing slopes of the valley and vineyards on the south facing slopes. The valley bottom consists of pastures. Several villages and some minor industries as well as some roads with a high traffic density deliver a wide variety of suspended material into the river. Waste waters come from both diffuse and point sources.

3. THE SAMPLING PROGRAMMES

The municipal water works of Trier release their surplus water from their main reservoir into a small river that has been investigated since 1988. As only the amount of surplus water is a limiting factor, the number and design of the artificial floods can be chosen freely. Samples of river water and suspended sediments are taken during flood events. Additional samples were taken from the channel before and after the waves (4 points), as well as riverbank material along the brook axis (8 points). For a better comparison with suspended sediment, all samples were sieved for > and < 63 µm in diameter. A defined quantity of drinking water is let off from the water works turbine into the Olewiger Bach. Investigations are made by three measurement stations in the longitudinal river axis using gauging stations or subsonic discharge measurement tools (UNIDATA). The sampling and analyzing programme covers aspects of hydrologic conditions (water-level, conductivity, temperature), dissolved solids (nutrients, heavy metals), particle characteristics (grain size, organic carbon, nitrogen) and particle associated solids. It also encloses analyses of major ions, heavy metals and organic pollutants (PCBs, PAHs).

Flood water samples of 60 litres were taken midstream when visible changes of water-level, turbidity or colour were observed. The suspended particles were separated by centrifugation (Z 41, PADBERG) within 24 hours after sampling. Anions were determined by ion chromatography (690 IC, METHROM), cations and heavy metals by atomic absorption spectroscopy (SPECTRAA 640, VARIAN). For the analysis of phosphate, the molybdenum blue method was used. Suspended matter was digested with HNO_3 before analysis. Suspended sediment concentrations were determined by filtering the sample through a 0.45 µm filter. Total suspended carbon and nitrogen, as well as their inorganic and organic proportions were measured using a C-N analyser (CHN-1000, LECO). The particle size distribution was determined by a stream laser technique, using the GALAI CIS-1 system. The samples were spiked with internal standards, „enhanced solvent extracted" (AUTOPREP 44, SUPREX) and purified by column chromatography. For identification and quantification, gaschromatography / mass-spectrometry was used (MD 800, FISONS).

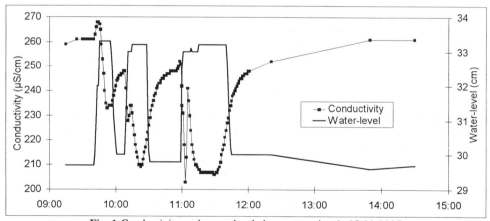

Fig. 1 Conductivity and water-level changes, station 1, 15.12.1997

4. THE LAG EFFECT OF CONDUCTIVITY AND PARTICLE BOUND SOLIDS

Figure 1 shows 3 small events with a base flow of 290 l/s (260 µS/cm) plus 3 releases of 170 l/s (86 µS/cm). It was one of the first screening events with changing wave durations and different interruptions inbetween. The results are impressive. Within the first kilometre a small mill-race with a base flow of about 20 l/s is deeply dug before the waves enter the natural river bed a few metres above station 1. All events show a great first flush in the rising peaks which is induced by sediment resuspension or uptake of soluble substances from nearby areas.

Travelling another 2 kilometres downstream, the events passed station 2. As a result of a higher base flow, the first flushes are covered by dilution. Figure 2 outlines variations of conductivity lagging behind associated changes in discharge which increases at points situated further downstream. After 3 kilometres, discharge and conductivity are divided. New water of the first wave arrives with the second discharge peak. In all experiments the reaction lag occurs after 30 to 40 minutes regardless of base flow amount or season.

This is induced by the morphology of the natural river bed and can be explained by kinematic wave theory. Therefore it is not only limited to larger basins as postulated by Glover and Johnson (1974). Figure 3 elucidates the same phenomenon related to suspended material. Particle-bound manganese is an appropriate tracer for resuspended sediment. It is to a large degree eroded in the mill-race above station 1 and travels at a different velocity than the water body. In the larger natural river channel below station 1, the first wave can only cause an unimportant new „peak" up to station 2 (10:15).

Those examples outline basic problems of natural wave interpretation. Only a multifingerprinting approach with hydrograph analysis and natural tracers can, for example, determine sources of pollutants.

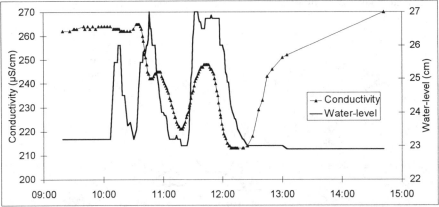

Fig. 2 Conductivity and water-level changes, station 2, 15.12.1997

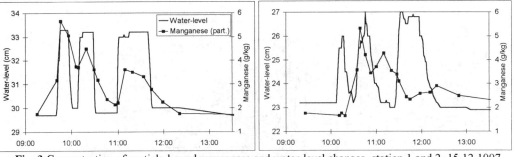

Fig. 3 Concentration of particle-bound manganese and water-level changes, station 1 and 2, 15.12.1997

5. SUSPENDED SEDIMENT CONCENTRATION

Figure 4 shows a double event with a 45 l/s base flow and two successive waves each with an additional 170 l/s discharge. The change in the concentration of suspended sediments is documented at 3 stations located respectively 1, 3 and 6 kilometres downstream from the Trier-Irsch water works. The concentrations of suspended sediments decrease along the brook's axis due to dwindling transport strength. Station 1 additionally shows that the second wave has less material at its disposal (b). The channel has been „cleaned" (a). This contrasts to station 2. At this point the channel has not been cleaned to a large degree, so that the second wave contains the same amount of material. At station 3 the sediment cloud has been so dispersed that no further separation can follow. At a certain point in the upper course only deposition takes place due to decreasing transport strength.

6. DISSOLVED HEAVY METALS

The remobilization of channel sediments and the impact of interstitial water on water quality is not well understood. Their release is controlled by hydraulic conditions and river bed morphology. Concentrations and types of heavy metals in interstitial water depend on biogeochemical activities, the degree of pollution and the amount of organic material in the sediment. Heavy metals are partly mobilized from pore water during resuspension of the sediments when anoxic sediments have contact with air oxygen or oxygenated water, and redox conditions change. Zinc, iron and manganese respond more distinctly to remobilization than other heavy metals. As each heavy metal responds individually to these processes, minor time lags between the concentration peaks of different heavy metals can occur. Although the proportions of interstitial water that contributes to total runoff and its influence on water quality cannot be calculated, the process of remobilization

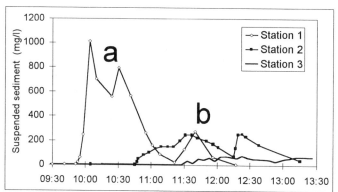

Fig. 4 Concentration of suspended sediment, 28.4.1997

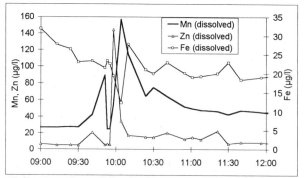

Fig. 5 Concentrations of dissolved heavy metals, station 1, 28.4.1997

itself can be identified. The disturbance of the sediment pore water system (Fig. 5) leads to a resuspension of zinc, manganese and on a minor level iron. A first flood wave (10:00) cleans the brook's interstitial so that a second flush (11:00) can not be measured. The occurance of various heavy metals is determined by the redox conditions in the sediment body and therefore by the antecedent hydrological conditions. Also meaningful is the opening of an impounding weir that led to a resuspension of deposited sediments (9:30). The event was conducted independently of the wave which followed shortly thereafter. It displayed however the same sequence of zinc, manganese and iron. This is most probably controlled by horizontal processes in the sediment, so iron goes with lower redox potentials, therefore in deeper sediment layers, into solution and logically is also later resuspended.

Figure 6 shows that adsorption processes are a reason for sinking heavy metal contents in the water body. The highest concentrations appear in the area of resuspension. At the second station the quantities relativize themselves again. This results in a depositing on the particle phase. Additionally, no renewed resuspension takes place.

7. PARTICLE BOUND SUBSTANCES

The transport of suspended sediments and their properties are supply controlled. Contents of particle associated contaminents in the waves are much higher than in sediment or river bank material (Fig. 7 a). This result is caused by a higher content of lightweight particles with more organic carbon and fine grained material

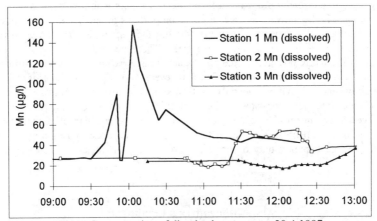

Fig. 6 Concentration of dissolved manganese, 28.4.1997

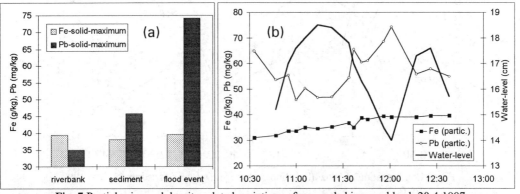

Fig. 7 Particle size and density related variations of suspended iron and lead, 28.4.1997

especially between and after the waves when transport strength is reduced. Sediment fractionization with several grain size classes showed, for example, a dependency with lead but not with iron. This result is confirmed by Figure 7 b. Iron is relatively constant and does not react to the wave. In contrast, particle bound lead reacts inversely to the water-level.

The origin of ubiquitous micropollutants like PAHs and PCBs is difficult to determine because they have both point and diffuse input sources. They are subjected to mixing processes, chemical transformations and selective erosion. Crucial to the understanding of catchment processes, and the ability to predict the fate of organic pollutants, is the identification of hydrologic pathways within the catchment, and the related transit times for particle bound solids. A multi-tracer technique is one way to answer some of these questions. There is also a clear tendency of a fast and distinct response of smaller PAHs with higher solubility which could be explained by a gradual enrichment of these partly mobile molecules at locations along the flow path. The patterns of the PAH chemographs are further determined by chemical alteration and the season. Figure 8 illustrates the behaviour of benzo(b+k)fluoranthene in relation to the water-level and the concentration of suspended sediment. After a first flush, a finer grain size with a higher specific surface causes an increase in the concentrations. The organic carbon content as well as the CN ratio fluctuate with this event only little, and are thereby ruled out as driving factors.

8. CONCLUSIONS

The results show a significant influence from sediment to the water body and vice-versa: Deep digging of the sediment body leads to a strongly pronounced flush effect in waters with low basic conductivities and longer sediment depositing. The lag effect can assume larger proportions in smaller flowing waters. The larger the concentration of suspended sediments, the smaller the transport strength. With larger reaches the sediment cloud dissipates and sedimentation processes take place. The concentration and type of resuspended heavy metals depends on the redox potential of the pore water. Longer sediment depositing times release even the difficultly soluble iron and copper in the water body. If several waves precede the artificial wave, then a more restricted release occurs. The redox potentials are then so high that no heavy metals in solution exist to any large degree. If pore water contacts with oxyginated water the redepositing on the particle phase constitutes a sink for dissolved heavy metal. The grain size dependency of certain particle bound toxic agents can be impressively proven with artificial flood waves because there is no superimposition by external influences. A cleaning up for sediment quality only occurs after larger and / or longer waves when the system is well flushed.

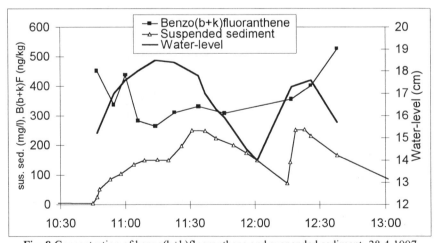

Fig. 8 Concentration of benzo(b+k)fluoranthene and suspended sediment, 28.4.1997

9. REFERENCES

Glover, B.J., and Johnson, P. (1974). "Variations in the natural chemical concentration of river water during flood flows, and the lag effect", *J. of Hydrology* 22, 303-316.

Umlauf, G., and Bierl, R. (1987). "Distribution of organic micropollutants in different size fractions of sediment and suspended solid particles of the river Rotmain", *Z. Wasser- Abwasser-Forsch.* 20, 203-209.

Environmental Hydraulics, Lee, Jayawardena & Wang (eds) © 1999 Balkema, Rotterdam, ISBN 90 5809 035 3

Turbulent effect on heavy metal release from river suspended sediments

X.D.Zhou & Y.Wang
School of Water Resource and Hydro-Electrical Engineering, Xian University of Technology, China

T.L.Huang
Department of Environmental Engineering, Xian University of Architecture and Technology, China

ABSTRACT: Heavy metal release is meainly determined by the capacity of water carrying sediments and the turbulent intensity of flow. A formula to calculate the capacity is obtained and the relationship between the concentration of heavy metal release and the turbulent intensity is established. Based on the kinetics of heavy metal release and in view of above relationship, a kinetic model considering water turbulence is developed. The kinetic model is satisfactorily verified by the kinetic process of heavy metal release.

1. INTRODUCTION

Many water bodies in densely populated countries contain large amounts of heavy metals. Two aspects are important of heavy metal pollution should be paid attention: firstly heavy metals are not usually eliminated from the aquatic ecosystems by natural processes, in contrast to most organic pollutants, and secondly, most metal pollutants concentrate in mineral and organic substances. Metal concentrations in water bodies varies in wide ranges (Tada, 1982; Lietz, 1989; Ye, 1990; Luan and Tang, 1990). The heavy metal load on sediments particles often reaches such high levels that a sudden desorption or release would results in a serious pollution of the water resource (Devies, 1970; Salomons, 1984). Heavily sediment-laden flow may results in increased heavy metal concentration in rivers. As also indirectly affects water quality of lakes or reservoirs.

2. CHARACTERISTICS AND KINETICS OF HEAVY METAL RELEASE

According to present studies (Huang and Zhou, 1992; Huang and Shen, 1992; Jin et al., 1987) a peak concentration of heavy metal released into water arises during heavy metal release, i.e. The concentration of heavy metal released initially increases with time up to the peak value, then it decreases down to release equilibrium. The first course is carried out much more rapidly than the second one. This result can not be explained with common adsorption-desorption theory. The characteristic indicates that there are other varieties of reactions of physical chemistry during the process of heavy metal release from sediments. With heavy metal releasing, other soluble substance in sediments, such as organic complexing agents, inorganic salts as well as some anions or organic particulates with which metal's coprecipitations may occur, dissolve into water. When the release and dissolve processed to a centain degree, released heavy metal concentration reaches the peak value, and then some reactions including complaxation adsorption, coagulation and flocculation, corprecipitation etc. So the metal concentration begins to decrease down to the equilibrium. On the basis of the characteristics, a kinetic model metal release is developed as follow (Huang, 1994).

$$\frac{dC_w}{dt} = k_d S(C_s - C_{se}) - (k_a + k_c + k_s) \cdot (C_w - C_e) \tag{1}$$

$$S\frac{dC_s}{dt} = -k_d S(C_s - C_{se}) + k_a \cdot (C_w - C_e) \tag{2}$$

And its solution is expressed as

$$C_w = \frac{Sk_d(C_{s0} - C_{se}) - (\beta - k_d)(C_{w0} - C_e)}{(\alpha - \beta)}\exp(-\beta t)$$
$$+ C_e\frac{Sk_d(C_{s0} - C_{se}) - (\alpha - k_d)(C_{w0} - C_e)}{(\alpha - \beta)}\exp(-\alpha t) \tag{3}$$

where

C_w - Concentration of soluble heavy metal in water, (μg/l);

C_{w0} -initial concentration of heavy metal in water; (μg/l);

C_e - equilibrium concentration of heavy metal in water; (μg/l);

C_s - concentration of heavy metal in sediments; (μg/g);

C_{s0} -initial concentration of heavy metal in sediments; (μg/g);

C_{se} - equilibrium concentration of heavy metal in sediment; (μg/g);

S - concentration of suspended sediments; (g/l);

k_d - rate coefficient of heavy metal desorbed from sediments; (l/h);

k_a - rate coefficient of heavy metal adsorbed from sediments; (l/h);

k_c, k_s - coefficients of complexation and coprecipitation of heavy metal respectively;

α, β - parameters related to above coefficients.

3. HEAVY METAL RELEASE AND RIVER SEDIMENT TRANSPORT

Heavy metal release from river sediments is in close relationship with the motion state of river sediments. According to the characteristics of sediment transport, it is considered that the increase of the suspended sediment concentration along the river course during the undersaturation transport is of greater significance for studying heavy metal release. The experimental results show that the time required to complete the transition of sediment concentration from undersaturation to saturation is very short, generally in several minutes (Qian, 1983). Therefore it is reasonable to consider the undersaturation transport as saturation transport in studying heavy metal release, since the period of heavy metal release generally required several or dozens of days. But it is also found that unequal exchange between suspended sediment and bed sediment leads to the fining of the suspended particle fraction to a certain extent along the river course at initial period of the experiment. The fining effect has a little effect on the kinetic process of heavy metal release, but it will greatly affect the transport of particulates bound metal.

4. TURBULENCE OF WATER FLOW AND SEDIMENT TRANSPORT

Since the sediment movement only depends on the hydraulic and hydrologic characteristics of rivers as well as sediment properties, it can be studies separately without regard to heavy metal release. Just as discussed above, the undersaturated sediment transport can be treated as saturated sediment transport in studying heavy metal release from river sediments.

Based on our experiment results, a formula to calculate the capacity of carrying sediment by river water is obtained as follows

$$S = 0.026\frac{v^3}{gR\omega} \tag{4}$$

where S is suspended sediment concentration (kg/m^3), v is velocity of river water (m/s), g is gravitational acceleration m/s^2, R is hydraulic radius (m). ω is mean settling velocity of sediment particulates (m/s).

5. TURBULENT INTENSITY AND HEAVY METAL RELEASE

On the basis of above analysis, heavy metal release is a process of physical chemistry while the water turbulence and saturated sediment transport only depend on the conditions river hydraulics and sediment transport. So the effect of turbulent intensity on the kinetic releasing process is solely embodied the effect on releasing intensity at some fixed time, and cannot fundamentally change the kinetic process. At time t, factors related to heavy metal release may be expressed as

$$C_t = f(C_s, v, R, \rho, \mu, S) \tag{5}$$

where S has been discussed above and C_s, which has been included in the kinetic equation, is independent of water flow conditions. So equation (4) becomes

$$C_t = f(v, R, \rho, \mu,) \tag{6}$$

where: C_t - concentration of heavy metal released into water at time t, (μg/L);

ρ - water density, (kg/m^3);

μ - water viscosity, (kg/m·s).

By means of the dimensional analysis method, the relationship between heavy metal release and water turbulent intensity can be derived as

$$C_t = k\rho \left(\frac{vR\rho}{\mu} \right)^{-d} = K \, Re^n \tag{7}$$

where d, n are powers (n=-d), K(=kρ) is parameter relative to the kinetics of heavy metal release, Re is Renold's number. From equation (6), if a log-log plot of C_t against Re give a straight line, the slope of line would equals to power n, It can be seen that Fig.1 shows a good linear relationship between $\log C_t$ and $\log Re$. The linear correlative coefficients, r are all over 0.92 at different time and the slopes take nearly the same value from 1.2 to 1.4, which indicates that the power n has little change under the experimental conditions. On the contrary, the intercepts of these lines are of great difference, since the intercepts are determined by the kinetic process of heavy metal release, This further supports above analysis and the presupposition. Let n=1.3 then equation (7) becomes

$$C_t = K \, Re^{1.3} \tag{8}$$

6. KINETIC MODEL IN CONSIDERING WATER TURBULENCE

According to the kinetic equation (3) and by considering the relational expression, equation (8), between heavy metal release and water turbulence, a modified kinetic model of heavy metal release under water turbulent conditions can be expressed as follows

$$
\begin{aligned}
C_{wt} &= (Re/Re_0)^{1.3} C_w \\
&= \left(\frac{Re}{Re_0} \right)^{1.3} \left\{ 0.026 \frac{v_0^3}{gR\omega} \frac{k_d (C_{s0} - C_{se})}{\alpha - \beta} [\exp(-\beta t) - \exp(-\alpha t)] \right. \\
&\quad \left. + \frac{C_{e0}}{\alpha - \beta} [(\beta - k_d) \exp(-\beta t) - (\alpha - k_d) \exp(-\alpha t)] + C_{e0} \right\}
\end{aligned}
\tag{9}
$$

where C_{wt} is the concentration of heavy metals released into water at time t, and Re_0 and v_0 represent a reference Renold's number and a reference velocity respectively. Equation (9) is the modified kinetic equation when the turbulent effects of water flow are taken into account.

The kinetic model, Equation (9) is verified with the experimental measured data, the results are shown in Fig.2. From Fig. 2 we can see the modified kinetic model developed above can satisfactorily describe the kinetic process of heavy metal release in rivers, and further perfects the kinetic model under static suspended conditions of the river sediment, Equation (9) can be used to describe or predict the kinetic process of heavy metal release in a dynamic test or natural rivers.

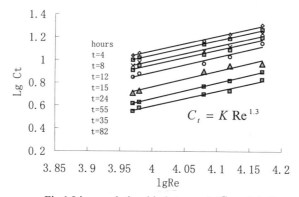

Fig.1 Linear relationship between $\lg C_t$ and. $\lg Re$

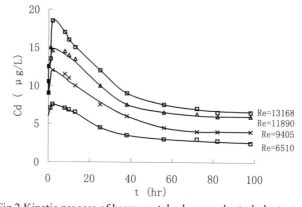

Fig.2 Kinetic process of heavy metal release under turbulent conditions

REFERENCE

Devies, A.G. (1970). J. Mar. Ass. U.K., 50: 65-86.
Tada, F. and Syzuki, S.(1982). J. Water Research, 16:1498-1494.
Qian, N. and Wan, Z. H. (1983). "Mechanics of Sediment Transport (Chinese)", Science Publication House.
Salomons, W. and Forstner, U. (1984). " Metals in Hydrological Cycle", Springer-Verlag, Heideberg.
Jin. X. C. Et al. (1987). China Environmental Science, 7(6): 11-26.
Lietz, W. and Galling, G, (1989). Water Research, 23: 247-252.
Ye, Y. Z. (1990). Environmental Chemistry (Chinese),9(5): 27-33.
Luan, Z. K. and Tang, H. X. (1990). Environmental Science (Chinese), 11(4): 18-2.
Huang, T. L. And Zhou, X. D. (1992). Youth Academic Proceeding of conservancy & Hydroelectric Engineering, Pub. House of Science and Technology of China.
Huang, T. L. And Shen, J.(1993). J. of Shaanxi Institute of Mechanical Engineering, 9 (4): 285-292.
Huang, T. L. (1994). Ph.D Dissertation, Xi'an University of Technology.

Environmental Hydraulics, Lee, Jayawardena & Wang (eds) © 1999 Balkema, Rotterdam, ISBN 90 5809 035 3

Desorption of cadmium from suspended particles in a turbulence tank

Huang Suiliang
State Key Sub-Laboratory of Water Environment Simulation, Beijing Normal University, China

Onyx W.H.Wai
Department of Civil and Structural Engineering, The Hong Kong Polytechnic University, China

ABSTRACT: Sediment is responsible for the fate and transport of many toxic metals in the aquatic environment. It is important to understand the reactivity of toxic metals in aquatic sediment for environmental planning and assessment. This paper presents the experimental results of Cadmium (Cd) desorption from artificially contaminated sediment particles in a turbulence tank. The turbulence is generated by harmonic grid stirred motion. The turbulence intensity is quantified in terms of eddy diffusivity which is equal to 9.84F, where F is the harmonic vibration frequency. In the present of turbulence, it is found that Cd contaminated sediment particles in suspension gradually release Cd into the water. The metal releasing behavior well conforms with a one-dimensional unsteady water-sediment-metal transport-transformation equation formulated for the turbulence tank. The determination of the equation's parameters for sediment adsorption-desorption of metals in the tank is discussed. With the help of the equation, the variations of the total and dissolved metal concentrations with time and the metal adsorption-desorption abilities in the present of suspended sediment particles in natural waters can be investigated.

1. INTRODUCTION

It is convenient and effective to use the indoor-static-experiment method (ISEM) to investigate the metal adsorption-desorption abilities of sediment particles (Mckinley and Jenne, 1991, Huang and Wan, 1995). However, this method pay little attention to the water and sediment motion interactions in natural rivers. Thus, to make use of the results obtained by this method to solve practical engineering problems in natural rivers needs further research. Generally, aquatic chemical conditions in reactors are artificially changed when studying heavy metal desorption from contaminated sediment particles. For example, acid is added to increase pH in reactors (Chuan and Liu, 1996) or other metal ions are added (Warren and Zimmerman, 1994) to change the total ionic concentration. In view of the present situation, we use a turbulence simulation tank (TST) with harmonic vibrations to simulate the suspension of Cadmium-contaminated sediment particles in which the turbulence is generated by the harmonic motion of the disturbed grids in the tank. This device has the advantage that water and sediment motions in the tank are similar to those in natural rivers and the design concept of the tank has been thoroughly investigated.

2. TURBULENCE SIMULATION TANK AND DESORPTION EXPERIMENT

TST, in which turbulence is generated by a harmonically vibrating set of grids, was first used by Rouse (1938) to study the mechanism of sediment particle suspension. In China, Yang and Chien (1986) used a similar device to investigate the flocculent structure of clay slurry. The device was improved by Huang (1993) by replacing all the submerged spares by plexiglass. The coefficient of the vertical turbulent diffusion E_y could be calculated by $E_y = 9.84F$, in which F is the vibrating frequency of the grids in 1/s, E_y has the units of cm^2/s. When doing experiments, a pre-determined volume of deionized water is added into the tank before the grids start to vibrate. A given amount of Cadmium-contaminated sediments is also added into the tank and this time is recorded as zero (t=0). During the experiment the muddy water samples are sampled intermittently and filtered with a 0.45 μm filter membrane. Determination of dissolved Cadmium concentrations in the filtrate are carried out by conventional flame atomic absorption spectrophotometer using standard techniques (Lacerda et al., 1984). Four plexiglass tubes are mounted to a wall of the tank at different depths, and they are being connected to a rubber tubes for sampling. Before each sampling, the remaining solution in the rubber tubes is discarded.

To determine the adsorption-desorption parameters of the experimental sediment samples, which are the saturation adsorption content in unit weight of sediment particles, b, and the constant for adsorption-desorption rate, k, in the Langmuir adsorption isotherm in equilibrium adsorption experiments at constant temperature (Langmuir, 1996). A detailed description of the procedures for the collection of sediment particles and preparation of Cadmium-polluted sediment samples used in the desorption experiment can be found in Wai and Huang (1998). Properties of sediment samples used in the desorption experiment are shown in Table 1.

The measured and calculated desorption results are shown in Table 2 and Figs. 1 and 2. It can be seen that the dissolved Cd concentrations are distributed uniformly along the depth and slowly increased with time due to considerable turbulent intensity, small water depth and slow desorption of Cd ions from sediment particles. Generally speaking, for common pollutants in trace amount, their concentrations uniformly distributed in the water column are easily found in shallow water (James, 1993). While pollutant concentrations increasing with time is a typical phenomenon of heavy metal pollutant transport-transformation, which is heavily related to the sediment motion. Thus, when flowing water is carrying metal contaminated sediment particles, attention needs to be paid to the water pollution due to the release of metals from the sediments.

3. MATHEMATICAL MODEL

The mathematical model of heavy metal pollutant transport-transformation consists of the equations of water motion, sediment motion, heavy metal transport-transformation (heavy metal continuity) and adsorption-desorption reaction kinetics of heavy metal (Huang, 1993; Huang and Wan, 1994).

3.1 Heavy metal transport-transformation equations in Polar Coordinates

Let u_r, u_θ and u_y be the velocities of the control volume $dr*rd\theta*dy$ in $\vec{r}$, $\vec{\theta}$, $\vec{y}$ directions, respectively. Here s is the concentration of suspended load, c denotes the dissolved Cd concentration, N is the suspended

Table 1 Properties of sediment samples used in the desorption experiment

particle sizes d (mm)	setting velocity ω (cm/s)	total weight w (g)	b mg/kg)	k_1 (1/mg.s) (10^{-6})	k_2 (1/s) (10^{-6})	$k=k_2/k_1$ (mg/l)	h_a (cm)	s_a (kg/m3)	water depth h (cm)
.0330~.0385	0.097	80	0.630	7.1	1.32	.186	45.5	3.5	50.5

Table 2 Measured and calculated dissolved Cd concentrations (mg/l)

time (hour)	water depth (cm)							
	10.1		20.2		35.2		45.5	
	Mea.	Cal.	Mea.	Cal.	Mea.	Cal.	Mea.	Cal.
0	0.00	0.000	0.00	0.000	0.00	0.000	0.00	0.000
1	0.00	0.006	0.01	0.006	0.01	0.006	0.01	0.006
2	0.01	0.012	0.01	0.012	0.02	0.012	0.02	0.012
3	0.02	0.017	0.02	0.017	0.02	0.017	0.02	0.017
4	0.02	0.022	0.02	0.022	0.02	0.022	0.02	0.022
5	0.03	0.026	0.02	0.026	0.02	0.026	0.02	0.026
6	0.03	0.030	0.03	0.030	0.03	0.030	0.03	0.030

particulate Cd concentration, ω is the mean settling velocity of the sediment particles. Based on the conservation of heavy metal in a control volume and neglecting the fluctuation of N, one can deduce that:

$$\frac{\partial}{\partial t}(\bar{c}+N\bar{s})+\frac{1}{r}\frac{\partial}{\partial r}(r\overline{cu_r}+r\overline{c'u_r'}+rN\overline{su_r}+rN\overline{s'u_r'})+\frac{1}{r}\frac{\partial}{\partial \theta}(\overline{cu_\theta}+\overline{c'u_\theta'}+N\overline{su_\theta}+N\overline{s'u_\theta'})$$

$$+\frac{\partial}{\partial y}\left[\overline{cu_y}+\overline{c'u_y'}+N\overline{s}(\bar{u}_y-\omega)+N\overline{s'u_y'}\right]=0 \qquad (1)$$

Introducing the conservation equations of water flow and sediment motion:

$$\frac{1}{r}\frac{\partial}{\partial r}(r\overline{u_r})+\frac{1}{r}\frac{\partial \overline{u_\theta}}{\partial \theta}+\frac{\partial \overline{u_y}}{\partial y}=0 \qquad (2)$$

$$\frac{\partial \overline{s}}{\partial t}+\frac{1}{r}\frac{\partial}{\partial r}(r\overline{u_r}\,\overline{s}+r\overline{u_r's'})+\frac{1}{r}\frac{\partial}{\partial \theta}(\overline{u_\theta}\,\overline{s}+\overline{u_\theta's'})+\frac{\partial}{\partial y}\left[\overline{s}(\overline{u_y}-\omega)+\overline{s'u'}\right]=0 \qquad (3)$$

The turbulent diffusion terms in Eqs. (1) and (3) are proportional to the gradients and the proportionality constants (turbulent diffusion coefficients) are denoted as E shown in Eq. (4).

$$\overline{c'u_r'}=-E_r\frac{\partial \overline{c}}{\partial r} \qquad \overline{c'u_\theta'}=-E_\theta\frac{\partial \overline{c}}{\partial \theta} \qquad \overline{c'u_y'}=-E_y\frac{\partial \overline{c}}{\partial y}$$

$$\overline{s'u_r'}=-E_r^s\frac{\partial \overline{s}}{\partial r} \qquad \overline{s'u_\theta'}=-E_\theta^s\frac{\partial \overline{s}}{\partial \theta} \qquad \overline{s'u_y'}=-E_y^s\frac{\partial \overline{s}}{\partial y} \qquad (4)$$

Substituting Eqs. (2), (3) and (4) into Eq. (1), one can get

$$\frac{\partial \overline{c}}{\partial t}+\overline{u}_r\frac{\partial \overline{c}}{\partial r}+\frac{1}{r}\overline{u}_\theta\frac{\partial \overline{c}}{\partial \theta}+\overline{u}_y\frac{\partial \overline{c}}{\partial y}-\frac{1}{r}\frac{\partial}{\partial r}(rE_r\frac{\partial \overline{c}}{\partial r})-\frac{1}{r}\frac{\partial}{\partial \theta}(E_\theta\frac{\partial \overline{c}}{\partial \theta})-\frac{\partial}{\partial y}(E_y\frac{\partial \overline{c}}{\partial y})$$

$$=-\overline{s}\frac{\partial N}{\partial t}-\overline{u}_r\overline{s}\frac{\partial N}{\partial r}-\frac{1}{r}\overline{u}_\theta\overline{s}\frac{\partial N}{\partial \theta}-(\overline{u}_y\overline{s}-\omega\overline{s})\frac{\partial N}{\partial y}+E_r^s\frac{\partial \overline{s}}{\partial r}\frac{\partial N}{\partial r}+\frac{1}{r}E_\theta^s\frac{\partial \overline{s}}{\partial \theta}\frac{\partial N}{\partial \theta}+E_y^s\frac{\partial \overline{s}}{\partial y}\frac{\partial N}{\partial y} \qquad (5)$$

Eq.(5) is for heavy metal pollutant transport-transformation in polar coordinates with the effect of suspended sediments. For common tracer pollutant convection-diffusion, the right hand side of Eq. (5) is equal to zero. The right hand side of Eq.(5) takes account of the effect of sediment motions on heavy metal transport-transformation for a simple case. Huang (1993) has applied the equation for a common situation.

3.2 Simplified Equation of Heavy Metal Transport-Transformation

For pure uniform turbulence caused by the up and down motions of the uniform grids, one can assume that

$$\overline{u}_r=0; \qquad \overline{u}_y=0; \qquad \overline{u}_\theta=0; \qquad \frac{\partial}{\partial \theta}(\ )=0 \qquad (6)$$

And Eq.(5) becomes

$$\frac{\partial \overline{c}}{\partial t}-\frac{1}{r}\frac{\partial}{\partial r}(rE_r\frac{\partial \overline{c}}{\partial r})-\frac{\partial}{\partial y}(E_y\frac{\partial \overline{c}}{\partial y})=-\overline{s}\frac{\partial N}{\partial t}+\omega\overline{s}\frac{\partial N}{\partial y}+E_r^s\frac{\partial \overline{s}}{\partial r}\frac{\partial N}{\partial r}+E_y^s\frac{\partial \overline{s}}{\partial y}\frac{\partial N}{\partial y} \qquad (7)$$

For suspended sediment motion in steady state, Eq. (3) can be simplified as follows.

$$\frac{1}{r}\frac{\partial}{\partial r}(rE_r^s\frac{\partial \overline{s}}{\partial r})+\frac{\partial}{\partial y}(\omega\overline{s}+E_r^s\frac{\partial \overline{s}}{\partial y})=0 \qquad (8)$$

753

It can be seen from Eq.(8) that the concentration of suspended sediment is varied in the vertical and radial directions due to the side wall effects, gravity and flow turbulence. Because the effect of the wall is generally in the vicinities of the wall, the radial distribution of the suspended sediment concentration is neglected, $\partial \bar{s}/\partial r = 0$. This further simplifies Eq.(8) to $\partial/\partial y\,(\omega \bar{s} + E_y^s\,\partial \bar{s}/\partial y) = 0$. Under the equilibrium condition, the upward sediment transport rate $E_y^s\,\partial \bar{s}/\partial y$ due to turbulent diffusion is equal to the rate of sediment settling through a unit area (Zhang and Xie, 1989), that is,

$$\omega \bar{s} + E_y^s\,\partial \bar{s}\big/\partial y = 0 \tag{9}$$

The effect of the wall on heavy metal transport-transformation in TST is also neglected, namely, $\partial \bar{c}/\partial r = 0$. Substituting these Eqs. into Eq.(7), one can get

$$\frac{\partial \bar{c}}{\partial t} - \frac{\partial}{\partial y}(E_y \frac{\partial \bar{c}}{\partial y}) = -\bar{s}\frac{\partial N}{\partial t} \tag{10}$$

E_y and E_y^s can be taken as constants due to relatively uniform turbulence caused by the grids in the tank and it is assumed that $E_y = E_y^s$ (Zhang and Xie, 1989). Therefore, Eqs.(9) and (10) can be simplified as follows.

$$s = s_a \exp\left[-\frac{\omega}{E_y}(y-a)\right] \qquad \frac{\partial \bar{c}}{\partial t} - E_y \frac{\partial^2 \bar{c}}{\partial y^2} = -\bar{s}\frac{\partial N}{\partial t} \tag{11, 12}$$

in which s_a is the concentration of suspended sediment at water depth y=a. E_y was equal to 44.6 cm^2/s during the experiments.

4. MATHEMATICAL MODEL AND APPLYING TO TST

The mathematical model for heavy metal transport-transformation applied to TST includes the following equations:
i) equation of water motion, namely, Eq.(6);
ii) equation of sediment motion, namely, Eq. (11);
iii) equation of heavy metal transport-transformation, Eq.(12);
iv) equation of adsorption reaction kinetics

$$\frac{\partial N}{\partial t} = k_1 \bar{c}(b-N) - k_2 N \tag{13}$$

in which k_1, k_2 and b are the coefficient of adsorption rate, the coefficient of desorption rate and saturation adsorption content for unit weight of suspended sediment, respectively.
v) The initial conditions
It is assumed that the water is initially "clear" and that the sediment has an initial Cd concentration No, namely,

$$\bar{c}\big|_{t=0} = 0 \qquad and \qquad N\big|_{t=0} = N_0$$

vi) Boundary conditions

At the water surface $E_y \frac{\partial \bar{c}}{\partial y}\big|_{y=h} = 0$ and at the bottom of the tank $E_y \frac{\partial \bar{c}}{\partial y}\big|_{y=0} = 0$

The calculated results from the above equations are shown in Table 2 and Figs. 1 and 2. Water surface in the tank is defined as y=0, the time step, dt, is 0.25 s and the space size is 5.05 cm.
It can be seen from the calculated results that the dissolved and particulate Cd concentrations (uniform size sediment particles are used) are uniformly distributed along the depth, and the distributions of adsorption

capacities (defined as the amount of particulate Cd in unit volume of water $Q = \bar{s} * N$) and the total Cd concentrations (defined as $c_T = s * N + \bar{c}$) along the depth are similar to that of the suspended sediment concentration. It also can be noted that at the same depth dissolved Cd concentrations are gradually increasing in time, while particulate Cd concentrations are gradually decreasing and come to stable after about 60 hours (desorption equilibrium). The variations of the adsorption capacities are similar to the particulate Cd concentrations as expected. Variations of the total Cd concentrations in unit volume of water are evidently related to the water depth. c_T close to water surface gradually increases, while that close to the bottom gradually decrease and they are approaching a stable value with time. It should be noted that the experiments were only conducted for 6 hours. This is not enough because it needs 60 hours to reach an equilibrium of Cd desorption condition according to the calculation.

5. RESULT COMPARISONS

It is shown in Table 2 and Figs. 1 and 2 that the calculated results agree fairly well with the measured one. It indicates that the deduced mathematical model for heavy metal transport-transformation is valid.

The parameters of sediment adsorption-desorption used in the calculation are determined as follows. It has been shown by Huang and Wan (1997) that the parameters, namely, the saturation adsorption content, b, coefficients of adsorption rate, k_1 and coefficients of desorption rate, k_2, do not change with sediment concentrations under same conditions (such as pH and temperature), thus b=0.630 (mg/kg), k_1=0.0000071 (1/ppm.s) and k_2=0.00000132 (1/s) are used in the calculation. It means that b and k ($=k_2/k_1$) applied in the calculation are the same as those determined in ISEM. Because considerable ISEM investigations have been carried out by chemists in the past, the above method for determining parameters of sediment adsorption-desorption is suitable to apply in natural waters.

It can be seen from the calculation and the experiment that uniform distributions of dissolved heavy metal concentrations under turbulent condition are achieved fairly fast, while much more time will be needed to arrive to an equilibrium condition for desorption of heavy metal from sediments. These agree well with those encountered in natural rivers, that is, little variations of dissolved heavy metal concentrations are usually found in natural rivers.

6. CONCLUSIONS

Using the turbulence simulation tank, Cd ion desorption from Cd-contaminated sediment particles was experimental and mathematically studied. Conclusions are summed up as follows.

1) It is indicated from the experiment that when metal contaminated sediment particles are disturbed and resuspended into the watercolumn, release of heavy metal ions from the sediments will take place and pollute the water.

2) Mathematical model of heavy metal transport-transformation in polar coordinates, in which only the effect of suspended sediments is considered, was deduced, simplified and numerically solved in connection with the experimental conditions. Experimental results agree well with those calculated by the model. This implies that the mathematical model is valid. The determination of the adsorption-desorption parameters of sediment particles was analyzed in the calculation. It is determined that the results obtained in ISEM investigations can be applied to solve practical engineering problems in natural waters.

3) Distributions of dissolved Cd concentrations, particulate Cd concentrations, adsorption capacities and the total Cd concentrations in unit volume of water along the depth were analyzed. The results be used as a base for analyzing vertical profiles of heavy metal desorption from contaminated sediment particles in natural rivers.

ACKNOWLEDGMENT

The work described in this paper was partially supported by two grants from the Research Grants Council of the Hong Kong Government (Proj. No. HKP62/95E) and from the Central Research Grant of the Hong Kong Polytechnic (Proj. No. G-V459).

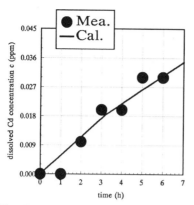

Fig.1 Comparisons of calculated and measured
dissolved Cd concentrations

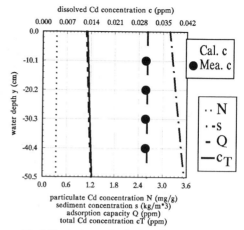

Fig.2 Variations of calculated and measured
results with water depth at Hour 6

REFERENCES

Chuan, M. C. and J. C. Liu (1996), Release behavior of chromium from tannery sludge, Wat. Res., Vol.30, No.4, pp.932-938.

Donald Langmuir (1996), Aqueous environmental geochemistry, Prentice-Hall International (UK) limited, London, pp.360-362.

Huang, S. L. (1993), The effect of sediment motion on heavy metal pollutant transport-transformation, Ph. D Dissertation at China Institute of Water Conservancy and Hydroelectric Power Research (in Chinese).

Huang, S. L. and Wan, Z. H. (1994), Study on the mathematical model of heavy metal pollutant transport-transformation in fluvial rivers, International Journal of Sediment Research, Vol.9, No.4, pp.36-45.

Huang, S. L. and Wan, Z. H. (1995), Present situation of heavy metal pollutant adsorption by sediment, International Journal of Sediment Research, Vol .10, No.3, pp.69-81.

Huang S. L. and Wan, Z. H. (1997), Study on sorption of heavy metal pollutants by sediment particles, Journal of Hydrodynamics, Ser. B, 3, pp.9-23.

James A. (1993), An introduction to water quality modelling, John Wiley and Sons, pp. 141-233.

Lacerda, L. D., C. M. M. Souza and M. H. D. Pestana (1984), Geochemical distribution of Cd, Cu, Cr, and Pb in sediments of Estuarine areas along the southeastern Brazilian coast, Marine Pollution Bulletin, No.4, pp.86-99.

Mckinley, P. James and Everett A. Jenne (1991), Eperimental investigation and review of the "solids concentration" effect in adsorption studies, Environ. Sci. Technol., Vol.25, No.12, pp.2082-2087.

Onyx, W. H. Wai and Huang, S. L. (1998), Determination of Cadmium ion desorption from non-uniform sediment particles, Proceedings of Environmental Strategies for the 21st Century, Singapore, pp.131-136.

Rouse, H. (1938), Experiments on the mechanics of sediment suspension, Proc., 5th. Intern. Cong. for applied Mech., pp.550-554.

Warren, A. Lesley and Ann, P. Zimmerman (1996), The influence of temperature and NaCl on cadmium, copper, and zinc partitioning among suspended particulate and dissolved phases in an urban river, Wat. Res., Vo.28, No.9, pp1921-1931.

Yang, M. Q. and Ning, Chien (1986), Effects of turbulence on flocculent structure of slurry composed of finer sediment, Journal of Hydraulic Engineering, No.8, pp.35-42 (in Chinese).

Zhang, R. J. and Xie, Q. H. (1989), Mechanics of sediment motion, China Water Conservancy and Hydroelectric Power Press, pp.181-220 (in Chinese).

4.2 Modelling of collection and treatment systems

Environmental Hydraulics, Lee, Jayawardena & Wang (eds) © 1999 Balkema, Rotterdam, ISBN 90 5809 035 3

Towards a model to understand foul flushes in combined sewers

R.M.Ashley, A.Fraser, T.McIlhatton & D.Phelan
Wastewater Technology Centre, University of Abertay Dundee, UK

S.Tait
Department of Civil and Structural Engineering, Sheffield University, UK

S.Arthur
Department of Building Engineering, Heriot-Watt University, Edinburgh, UK

ABSTRACT: Foul flushes are events in combined sewers which typically occur in the initial period of storm flows, and for which the concentration of suspended sediments and other pollutants are significantly higher than observed in the later stages of the storm. This paper summarises current knowledge regarding the mechanisms which create foul flushes and proposes modelling strategies which will be able to account for such large changes in suspended sediment concentration under time varying flow conditions.

1. INTRODUCTION

The problems caused by solids and associated pollutants in combined sewer systems are now well known. As yet however, the processes which cause 'pulses' of solids and pollutants in suspension, known as 'foul flushes' are not entirely understood (Ashley & Verbanck, 1996). These flushes can be observed when the flow is accelerating either due to storm inputs or during peak flows in dry weather (Fig.1). However, flushes are not always observed under what would appear to be identical hydraulic conditions even in the same sewer. Flushes are important as they cause highly concentrated 'waves' of solids and pollutants to be conveyed to the outlets of sewerage systems, potentially causing shock loadings.

In the UK, the past decade of research has seen integrated investigations of sewer sediments and transport as part of the Water Industry's drive to develop new tools to model systems. This paper draws on two current field based projects in Dundee, investigating solids transport and the control of the latter for sewer operation, and a laboratory study in Sheffield of the erosion and movement of sediment from bed mixtures in a large flume, includes synthetic cohesives.

2. UNDERSTANDING AND PREDICTION OF FOUL FLUSHES

Currently, there are two approaches to the estimation of foul flushes: (i) empirical catchment based relationships to predict the representative solids loads and peaks; (ii) detailed consideration of the physical processes causing flushes in order to develop deterministic relationships to describe these processes.

2.1 Catchment Based Studies

The occurrence or otherwise of a first flush is believed to be primarily dependent upon catchment size and patterns of flow history (Stotz and Krauth,1984), although there are a wide variety of definitions for flushes such that their occurrence has been questioned in recent French studies (Saget et al, 1996). Catchment based approaches to the prediction of flushes have been attempted using empirical models. Gupta and Saul (1996) for example, related flush characteristics to storm durations, rainfall intensity and antecedent dry weather period. The relationships derived from such studies appear to be catchment specific and so far generalised equations have been elusive.

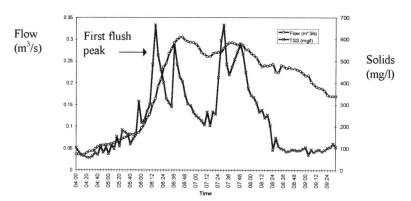

FIG.1. First and second foul flushes during rainfall in the main Dundee interceptor sewer

2.2 Detailed Process Based Studies

In an attempt to explain the physical processes for foul flushes in sewers, more detailed studies have investigated the effects of disturbance of the bed 'sediments' by increasing rates of shear stress along the invert. It is believed that erosion mobilises the surficial material, re-entraining particles and releasing the interstitial liquid trapped in the solids matrix. Low density wall slimes and biofilms attached to the solid particles may also be released increasing biochemical loads (Michelbach, 1995), but not being significant for the solids mass.

Observations in Belgium, Germany and the UK of the solids (and pollutants) in transport in combined sewers have revealed that there is a zone near the bed, or just above a clean invert, in which the moving material is particularly concentrated (Verbanck, 1995, Arthur & Ashley, 1997, Ristenpart, 1998). During dry weather this material is virtually entirely organic, comprising particles up to a few centimetres in size, and known as 'near bed' solids. It is believed that a significant part of the foul flush is from this near bed material being entrained. The material is a mix of solids moving along the bed (a form of bed load) together with what may be a 'dense suspension' in the lower part of the flow. During storm flows, the lighter material is believed to be taken into suspension over the full flow depth and in many cases the moving near bed material, becomes replaced by mostly granular bed load. Although in steeper sewers there is evidence that the near bed solids are always predominantly granular (Lin & LeGuennec, 1996). Unfortunately studies of the material moving near the bed in real sewers have been limited due to practical measurement difficulties. Results are based on observations of what is either collected in bed traps (UK and France), or has been obtained by vacuum extraction of the flow samples (usually with small bore tubes; Belgium, Germany). Indirect observations using sonar and thermal conductivity devices to monitor sediment bed erosion have shown the movement of a 'pulse' of dense material moving adjacent to the bed just prior to an eroding event (Ashley et al, 1993).

Current prediction methodologies applied to the movement of sediment in sewers are derived from sediment transport relationships developed in fluvial environments. However, there are several reasons which make the application of these methods inappropriate where transport of near bed solids in sewers is concerned (Ashley & Verbanck, 1996). Notwithstanding these limitations, a number of fluvial sediment transport methodologies have been applied to sewer sediment transport, both in suspension and near the bed (assumed as bed load), with varying degrees of success (e.g. Kleijwegt 1992; Verbanck, 1995a; Ackers et al., 1996, Arthur, 1996).

2.3 Analytical models

Detailed models for the determination of foul flushes at locations within sewer systems require estimates of the erosion of both the near bed solids and any erosion of the bulk underlying bed. Harrison and Holmes (1967) demonstrated that a storm flush wave overtakes a dry weather flow without significant dispersion of the storm water mass into the overtaken baseflow. This concept may be used to estimate the total suspended load within a flush by the summation of the individual sediment load components (Eqn 1). The total solids load in a flush (TSS_{FLUSH}) is made up of the sum of the total solids in transport during dry weather, which is the solid load that is present without the storm flow (TSS_{DWF}), plus the 'release' of any solids moving near the bed (NBS_{DWF}) together with any eroding solids from the bulk of the bed (BES_{WET}):

$$TSS_{FLUSH} = TSS_{DWF} + NBS_{DWF} + BES_{WET} \qquad (1)$$

Solids transported in the dry weather flow component (TSS_{DWF}): Determination of the first term in Eqn.1 is relatively straightforward. If a pattern of dry weather flow has been established, or assumed based on e.g. population, then the temporally varying flows and mean suspended solids concentrations throughout the flow column will be known. Typically this will be a small proportion of the total load in a flush.

Solids available in the near bed solids transport (NBS_{DWF}): Estimation of the near bed solids term in Eqn.1 is more difficult. The concept of the 'near bed solids' contribution to the first foul flush has been developed using field observations and data (Verbanck, 1995, Arthur & Ashley, 1998). At present it is not possible to precisely define, due to practical constraints, whether the material in transport near to the bed moves as a true suspension or as a bed load. Tests in Marseille with bed-load traps, collected mostly granular particles (up to 10mm in size) in the steeper (1.78%) sewer sections, whereas in the shallower sections (0.1%), the particles were more similar to those observed in suspension. Three invert traps in Dundee collected sediments in a steep sewer (4.6%) and at two locations on a shallower sewer (0.069%). In the former, the solids were largely granular (≈98.6%), and in the latter, mainly organic, and distinct from granular 'bed load'. The evolution of the bed deposits including both erosion and deposition in the Marseilles trunk sewer was shown to be predictable using a relationship derived from the Meyer-Peter and Muller formula, taking into account an assumed surface mixing layer and a number of grain size classes (Lin & Le Guennec, 1996).

Arthur (1996) attempted to apply a range of existing sediment transport relationships to both the organic and inorganic near bed solids fractions collected in invert traps in a number of sewers in dry weather flow conditions. The Ackers-White relationship (Ackers, 1991) produced disappointing results. Although an improved performance was obtained by adjusting the empirically derived coefficients. The approach that was found to produce the best results, was a modified form of the Perrusquia and Nalluri (1995) bed load formula for granular transport and a new empirical equation for the organic load (Arthur & Ashley, 1997). Subsequently, the latter was used to predict the likely contribution to the foul flush load increase (NBS_{DWF}) when the near bed material in transport during dry weather was re-entrained into the flow (Arthur & Ashley, 1998).

An alternative approach to predicting the near bed solids load has been utilised based on the work of (Verbanck, 1995a). A two layer model was suggested which assumes that the near bed solids are moving as a suspension. The concentration profile of the solids transported 'in suspension' in the layer nearest to the bed is given by (Coleman, 1969):

$$\frac{C_y}{C_{a*}} = \left(\frac{y}{a*}\right) \qquad (2)$$

Where the reference level ($a*$) from the bed correlates with a multiple of the displacement boundary layer thickness of the measured velocity profile, C_{a*} is the reference concentration, and C_y the concentration at height y. The solids moving in the upper region of the flow can be calculated using the exponential form of relationship proposed by Rouse (1937). A number of measured suspended solids concentration profiles recorded during dry weather in the main Dundee interceptor sewer by small bore tube/sampler extraction were used to determine the most appropriate value for the reference level ($a*$). The best fit was achieved using a value of 300mm in a flow depth of 450mm. This is rather higher than suggested by Coleman (1969). The reported values of the solids concentrations conveyed near the bed (Verbanck,1995a and Arthur & Ashley, 1998) were expected to be of limited accuracy because the solids were known to be large compared with the sample tube bore. This excluded many of the solids which should have been sampled, therefore these collection methods were assumed to underestimate the amount of near bed solids. Hence the amount of near bed solids transported were assumed to be equal to those collected in the invert traps. Estimates were then possible for the total mass of solids in transport during dry weather, both in suspension and near the bed (TSS_{DWF}, NBS_{DWF})

by calculating the concentration at various flow depths and integration across the entire depth. Table 1 shows the comparison between the measured near bed solids transport rate (invert trap), and the suspended sediment transport rate calculated from a concentration profile estimated using Eqn 2, and that from the measured suspended solids concentration profile; both profiles being integrated over the lower 25% of the flow. In general (with one exception) the estimated suspended sediment transport rate obtained using Eqn 2 is in closer agreement with the rate measured from material recovered from the invert traps than from the measurements of the suspended sediment concentrations at various depths in the flow. This indicated that by careful choice of a suitable reference height ($a*$) that the amount of near bed solids typically being transported in dry weather flow (NBS_{DWF}) could be estimated by the use of Eqn 2.

Solids eroded from the bed (BES_{WET}) - contribution to flush: Field studies undertaken in Dundee and Hildesheim, Germany in large sewers, have been used to develop and test a sewer sediment erosion model. The erosion of the bed H_e, from an initial depth of H_o can be related to the bulk density ρ_e (varying with depth), $\bar{\rho}_d$ average bed (dry) density, and the applied bed shear stress (τ_b) (Wotherspoon, 1994):

$$H_e = H_0\left[1 - \left(\frac{\rho_e}{\zeta\rho_d}\right)\left\{\frac{1 + \left(\frac{\tau_b}{9.659x10^7}\right)^{-\frac{1}{3.168}}}{1 + \frac{\rho_e}{\rho}\left(\frac{\tau_b}{9.659x10^7}\right)^{-\frac{1}{3.168}}}\right\}^{-\frac{1}{\xi}}\right] \tag{3}$$

Erosion during both dry and wet weather has been successfully predicted for data from Dundee and Hildesheim (Ristenpart, 1996). It is however, necessary to 'calibrate' the coefficients ξ and ζ for each event. Application of the erosion model to the deposits over an 82m length of main sewer in Dundee, showed a good correspondence between the measured mass eroded from the bed during a storm event with the difference between the cumulative suspended solids measured coming into the sewer inlet, and those exiting the length of sewer (Ashley et al, 1993).

Recent laboratory tests in a pipe channel have investigated the link between the prevailing hydraulic conditions and the erosion of fine grained (cohesive-like) organic material from in-pipe deposits into suspension. In the tests, the discharge was increased at a uniform rate and then held constant. As the discharge increased the suspended sediment transport rate and concentration was found to increase rapidly to a maximum. The peak values of suspended sediment concentration and transport were observed as the peak discharge was attained. Following this, the suspended sediment transport rate diminished significantly despite the peak discharge being maintained (Skipworth et al., 1996). Examination of the rate of change in suspended sediment transport in relation to the time varying flows indicated that there was a link between the rate of change of discharge and peak transport rate. The values of these peaks in suspended sediment appeared to be strongly dependent on the rate of change of the flow discharge and the characteristics of the weak surface layer. Experiments also indicated that the bed became more resistant to erosion with depth eroded until a point was reached at which the resistance of the bed to erosion was effectively constant. Modelling of the suspended sediment transport using an excess shear stress type relationship presumed a two layer bed structure; with a weaker surface layer overlying a stronger layer. Reasonable agreement was found with suspended sediment measurements made in a number of laboratory tests (Skipworth, 1996).

Table 1 Comparison of estimates of near bed solids transport rates during dry weather using predicted suspended solids profiles (Coleman, 1969) and measured profiles

Date/Time	Measured Rate (bed traps) kg/s	Calculated Rate (Coleman Profile)		Integrated Rate (Measured Profile)	
		kg/s	% diff.	kg/s	% diff.
28/6/95 07:30	4.892×10^{-4}	4.245×10^{-4}	-13.2	1.261×10^{-4}	-74.2
10/7/95 16:30	2.008×10^{-4}	4.139×10^{-4}	+106.1	1.466×10^{-4}	-27.0
1/8/95 09:30	4.564×10^{-4}	4.923×10^{-4}	+7.9	2.353×10^{-4}	-48.45
9/8/95 06:30	0.847×10^{-4}	0.612×10^{-4}	-27.8	0.362×10^{-4}	-57.19

In a study looking at the similarities between estuarine mud erosion and sewer sediments, a general erosion model has been developed for cohesive/non-cohesive sediments (Torfs, 1995). The onset of erosion can be predicted using alternative equations for muds or sewer sediments. The rate of erosion, depending upon whether this is surficial (weak), or mass erosion (consolidated bed) is given by an excess shear stress relationship, provided the mud content >3-15% (Parchure and Mehta, 1985; Michener and Torfs, 1996).

2.4 Combined First Flush model - An Example:

The first flush shown in Fig.1 has been analysed using the approaches described above. The flush has been taken as that defined by (Gupta & Saul, 1996). A start time was assumed of 05:00hrs and the end as 07:15hrs. This determined the flush as occurring over the first 78% of the storm event. The total suspended solids load measured (TSS_{FLUSH}) was 332kg. For this period of the day the total suspended solids during dry weather (TSS_{DWF}) were estimated from measurements as 29kg. Earlier calculation of the near bed solids (Arthur & Ashley, 1998) gave a total load (NBS_{DWF}) of 14kg. Thus the 'missing' solids presumed to be eroded from the bed: $BES_{WET} = TSS_{FLUSH} - (TSS_{DWF} + NBS_{DWF}) = 289kg$.

A major problem of confirming that any of the erosion models above can predict this mass, is that of the appropriate length of sewer over which the erosion occurs. In the laboratory experiments (Skipworth, 1996), was able to assume that the entire bed length contributed, in the sewer it is not clear how far upstream of the point under consideration should be included, as potentially a significant part of the network could be contributing to the increase in suspended solids. It is possible to show, using the laboratory based model, or that developed from the field studies, that typical erosion depths immediately upstream are 10mm, from the 100mm deposit. Hence, for the actual bed width, the erosive length necessary to contribute a mass of 289kg, for a bed density of $1500kg/m^3$ is 40m. The veracity or otherwise of this has not yet been tested, and is part of the current field experiments. The relative contribution of the components to the total flush load is: TSS_{DWF} 9%; NBS_{DWF} 4%; BES_{WET} 87%. The relative magnitude of the bed erosion component to the mass transported thus dominates, and explains why the current generation of commercial computational models, which ignore the near bed solids, provide reasonable estimates of the flush solids. However, it must be appreciated that the biochemical constituents of flushes are not so distributed, with the near bed solids (together with slimes) comprising a much higher proportion of this load.

3. CONCLUSIONS

The foul flush phenomenon, commonly found in combined sewers with time varying flows, can be described by a number of physical processes. The total load contained within a foul flush can be subdivided into components which correspond to different physical processes. It is proposed that the total load is comprised of contributions from the suspended and near bed solids already present during dry weather flow, and material eroded by the storm flow from existing bed deposits. Field observations indicate that the re-suspended material which is moving as near bed solids during dry weather flow provides the bulk of the chemical and biochemical fraction (with evidence that slimes also comprise an important element) and is therefore highly polluting. Comparison with field measurements of the total amount of material moving during dry weather flow, as both near bed solids and in suspension, with that measured during storm flows suggests that a significant amount of material is unaccounted for. The erosion of bed material is therefore believed to be the primary source of the solids load increase in the suspended flush. Both laboratory data and field data have indicated that significant amounts of a bed deposit can be eroded into suspension during periods of high flow. Laboratory testing has indicated a wide variation in erosion rates from organic cohesive-like beds even under similar flowrates. Modelling developed from the laboratory results indicate the importance of bed strength and character of the time varying flow in determining the amount and pattern of sediment release in response to time varying storm flows. As the erosion of bed deposits is believed to make a significant contribution to the total load within a foul flush this may explain the field observations that flushes are not always identical even under what would appear to be identical hydraulic conditions in the same sewer.

4. REFERENCES

Ackers P. (1991) Sediment aspects of drainage and outfall design. *Proc. Int. Symp. on environmental hydraulics, Hong Kong*, pub. A.A. Balkema (Rotterdam). (Ed. Lee J H W, Cheung Y K) pp19-29

Ackers J C., Butler D., May R W P. (1994). Design of sewers to control sediment problems. *CIRIA 141*

Arthur S. (1996) Near bed solids transport in combined sewers. PhD Thesis, *University of Abertay Dundee*

Arthur S., Ashley R M. (1997) Near Bed solids transport rate prediction in a combined sewer network *Wat.Sci.Tech.* Vol.36, No. 8-9 pp 129-134

Arthur S., Ashley R M. (1998) Near bed solids transport and first foul flush in combined sewers. *Wat.Sci.Tech. Vol.37 No. 1*

Ashley R M, Wotherspoon D J J, Coghlan B P, Ristenpart E. (1993). Cohesive sediment erosion in combined sewers. *Proc. 6th ICUSD, Niagara Falls,* Sept

Ashley R M, Verbanck M A (1996). Mechanics of Sewer Sediment erosion and transport. *J. Hydr. Res.* Vol.34, 1996, No.6 pp753-770

Ashley R M, Hvitved-Jacobsen T., Bertrand-Krajewski J-L (1998). Quo Vadis Sewer Process modelling? *Proc. Conf Urban Drainage Modelling.* London Sept.

Coleman N. (1969) A new examination of sediment suspension in open channels; *Jour of Hydraulic Research* 7(1) 67-82.

Gupta K., Saul A J. (1996) Specific relationships for the first flush load in combined sewer flows. *Water Res.* Vol.30, No.5

Harrison A J M., Holmes D W. (1967). The movement of storm water in combined sewers. *Report INT 73.* Hydraulics Research Station, Wallingford.

Kleijwegt R A. (1992). On sediment transport in circular sewers with non-cohesive deposits. PhD thesis, *Technical University of Delft,* Netherlands.

Lin H., Le Guennec B. (1996) Sediment transport modelling in combined sewer. *Water Science and technology*, 33, No.9 pp 61-68

Michelbach S. (1995) Origin, resuspension and settling characteristics of solids transported in combined sewage. *Wat.Sci.Tech.* Vol.31, No.7 pp 69-76

Mitchener H Torfs H (1996) Erosion of mud/sand mixtures; *Coastal Engineering*, 29 1-25

Perrusquia G, Nalluri C. (1995). Modelling of bed - load transport in pipe channels. *Proc. Int. Conf. On the transport and sedimentation of solid particles.* Prague.

Ristenpart E. (1998) Solids transport by flushing of combined sewers. *Wat.Sci.Tech.* Vol.37 No.1

Rouse H. (1937) Modern conceptions of the mechanics of sediment suspension. *transactions, ASCE*, Vol 102, 463-543.

Saget A, Chebbo G, Bertrand-Krajewski J L. (1996). The first flush in sewer systems, *Wat. Sci. Tech.* Vol 33, No 9.

Skipworth P.J. Tait S.J. Saul A.J. (1996) Laboratory Investigations into Cohesive Sediment Transport in Pipes, *Water Science and Technology* Vol. 33 No. 9

Skipworth P J (1996) The erosion and transport of cohesive-like beds in sewers; PhD Thesis *University of Sheffield*

Stotz G., Krauth K H., (1984) Factors affecting first flushes in combined sewers. *Proc. 3rd Int. Conf. on Urban storm drainage*, Goteborg, Sweden.

Torfs H.(1995) Erosion of mud/sand mixtures; PhD thesis, *University of Leuven*, Belgium, October.

Verbanck M. A. (1995) Capturing and releasing settleable solids - the significance of dense undercurrents in combined sewer flows. *Wat. Sci. Tech.* Vol 31, No.9

Verbanck M A. (1995a) Transferts de la charge particulaire dans L'egout principal de la ville de Bruxelles. DSc thesis. *Universite Libre Bruxelles.*

Wotherspoon D J J. (1994). The movement of cohesive sediment in a large combined sewer. *PhD thesis. University of Abertay Dundee.*

Environmental Hydraulics, Lee, Jayawardena & Wang (eds) © 1999 Balkema, Rotterdam, ISBN 90 5809 035 3

Modelling of critical flux in cross-flow microfiltration

D.Y. Kwon & S. Vigneswaran
Environmental Engineering Group, Faculty of Engineering, University of Technology Sydney, N.S.W., Australia

ABSTRACT: In this study, the concept of critical flux was introduced. Below this critical flux, there will be no membrane fouling. Laboratory - scale crossflow microfiltration (CFMF) experiments were performed to investigate the effects of operational factors such as suspended particle size, cross-flow velocity, membrane pore size on the critical flux in CFMF performance. The experimental results indicated that the critical flux increased with the increase in particle size and cross-flow velocity. A hydrodynamic force balance model developed in this study predicted successfully the critical flux value.

1. INTRODUCTION

One of the major drawbacks hindering the application of membrane processes in water and wastewater treatment is the reduction in the flux with time (below the theoretical capacity of the membrane). Under the conditions of constant transmembrane pressure (TMP) and cross-flow velocity, the flux in cross-flow microfiltration (CFMF) declines to a steady-state value which can be as much as two orders of magnitude lower than the initial or clean water value (Lokjin et al., 1992). In general, the flux declines rapidly at the initial stages. This is followed by a long but slow and gradual decline of flux till it reaches the steady-state flux.

It is well known that membrane fouling is one of the main phenomena responsible for this flux decline. The fouling mechanism is extremely complicated. The fouling affects the performance of the membrane either by deposition of a layer onto the membrane surface or, by complete or partial blockage of the membrane pores. This changes the effective membrane pore size distribution (Tarleton et al., 1993).

In this study, a critical flux was defined based on the TMP increase. Below this critical flux, the CFMF can be operated without membrane fouling (based on the CFMF experiments conducted under constant permeate flux operational mode). The factors affecting the critical flux values are discussed (such as particle size, membrane pore size, cross-flow velocity). A model was developed based on hydrodynamic force balance to calculate the critical flux.

2. EXPERIMENTS

Experiments were conducted at constant transmembrane pressure (TMP) mode. The schematic diagram of the microfiltration set-up used in this study is shown in Figure 1. Monodispersed suspension of spherical polystyrene latex particles (of pre-determined concentration) was delivered from a stock tank (equipped with a stirrer) to the CFMF cell by a variable speed tubing pump. Both the permeate and retentate lines were returned to the stock tank to maintain constant inlet concentration. The pressure of membrane was controlled by two valves and the transmembrane pressure (TMP) drop was monitored by using a pressure transducers (Model 19-626A from Devar Inc.). The membranes used are PVDF (polyvinyl flouride) membrane (MILLIPORE : Catalogue no. GVLP OMS 10) with nominal pore sizes of 0.1, 0.2 μm. In each experiment, new membrane was used to obtain reproducible results.

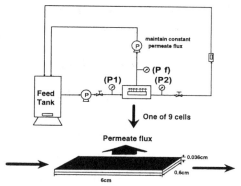

Figure 1. CFMF experimental set-up

3. THEORY

3.1 Definition of Critical Flux

The concept of critical flux has been recently introduced with a number of theoretical and experimental evidences. The concept of critical flux proposed by Field et al (1995) states that *"on start-up there exists a flux below which a decline of flux with time does not occur ; above it, fouling is observed. This is defined as critical flux and its value depends on the hydrodynamics and probably also on the other variables"*. This flux should be equivalent to the corresponding clean water flux at the same TMP. Theoretical calculations of critical flux for particles of different sizes indicated that different mechanisms govern the critical flux for different sizes of particles (Bacchin et al, 1995). For small particles of the order of 0.1 μm, Stoke-Einstein diffusion away from the membrane surface is important and the critical flux depends significantly on the surface charge effects. For particles over 1 μm, they are lifted from the surface by the shear-induced diffusion and the surface charge has insignificant effect.

The operation of CFMF below critical flux may provide a significant technical and economical advantage. If it is sustained, the cost of membrane cleaning can be removed and the life span of membrane can be prolonged significantly. Figure 2 compares the cumulative amount of water filtered when CFMF is operated (a) at a high initial filtrate flux and (b) at a flux below critical flux. It is clear from this figure that the mount filtered is higher in a longer run, when flux is below the critical flux. As can be seen in the figure, the accumulated volume of permeate from CFMF below critical flux exceeded that from normal CFMF after 600 minutes.

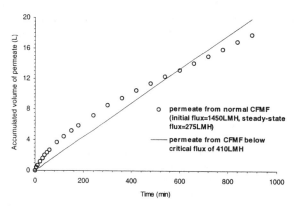

Figure 2. Comparison of the total permeate volume from CFMF experiments with high initial flux and with flux below critical flux (particles size of 0.46μm, membrane pore size of 0.2μm, membrane surface area of 0.00324m^2 and influent concentration of 40mg/l

766

3.2 Experimental Method to Define Critical Flux

The following equation can be used to describe the overall characteristics of flux decline (Aimar et al., 1989).

$$J = \frac{\Delta P}{\mu(R_m + R_f)}$$

(1)

where, J is the permeate flux (permeate flow per unit membrane area), ΔP is the TMP, μ is the viscosity and R_m is the clean membrane hydraulic resistance. The resistance, R_f accounts for the fouling effects on the flux. From this equation, it is evident that R_f is dependent only on the TMP if the CFMF is operated at a constant permeate flux. Thus, there will be no increase in the TMP with time if no or negligible membrane fouling occurs at a given constant flux. However, the TMP will increase with time if the flux is increased beyond a certain value. In other words, the membrane fouling can be detected in terms of the increase in the TMP.

3.3 Force Balance Model

To calculate the critical flux in CFMF, a theory based on a force balance (applied to particles deposited on the membrane surface) can be used. This model is an extension of the one proposed by Lu and Ju (1989). Figure 3 illustrates the forces acting upon a particle located in the deposited layer and the equations governing these forces are summarised in Table 1.
At a critical flux (u_{pcrit}), it is reasonable to assume that the forces on the particle are balanced or in equilibrium ;

$$F_n = F_t + F_l + F_{si} + F_B + k_l(F_e - F_v)$$

(2)

By substituting all the equations in Table 1 into (2), and rearranging the equation for u_p at critical flux, one obtains ;

$$u_{pcrit} = \frac{5.1\left(\frac{d_p}{L} - \frac{d_p^2}{2L^2}\right)u_s \tan\theta}{\phi} + \frac{2.6u_s^2 d_p^3 \rho}{16\mu L^2} + \frac{u_s d_p^2}{20 L^2} + \frac{KT}{3\pi\mu L d_p}$$

$$- \frac{0.0088 k_l H}{d_p^2}\left(\frac{2d}{h} - \frac{d_p^2}{h^2} - \frac{2d_p}{h+d_p} - \frac{d_p^2}{(h+d_p)^2}\right) + \frac{k_l \varepsilon_e \varepsilon_r \varphi^2 d_p k \exp(-kh)}{3\phi\mu(1-\exp(-kh))}$$

(3)

4. RESULT AND DISCUSSION

4.1 Effect of Particle Size

A series of experiments were conducted to study the effect of particle size on critical flux. Different sizes of latex particles (0.1, 0.3, 0.46, 0.816, 1.07 and 3.2 μm) were used in the experiments. The other experimental conditions were maintained constant. In each experiment, the permeate flux was increased step by step until an increase of TMP with time was observed. Figure 4 presents the critical flux values for different sizes of latex particles used. The maximum flux corresponding to no increase in TMP with time and the minimum flux which resulted in TMP increase in time were noted. The average of these two values is taken as critical flux.
It can be seen from the figure that the critical flux increased with the increase in the particle size. This might be due to the fact that the smaller particles have preferential tendency of deposition on the membrane at lower permeate flux. The TMP increase depends essentially on the fouling resistance at a constant permeate flux operational mode (Kwon and Vigneswaran, 1998). Therefore, the different critical flux value with different particle size is due to the variation in the resistance of the deposited particles to flux. The calculated values of the critical flux is also presented in Figure 4 together with the experimental values. It is clear from the figure that the flux can be predicted satisfactory using the model developed in this study.

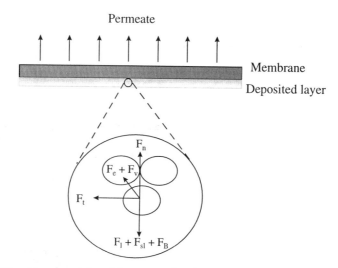

Permeate

Membrane
Deposited layer

Figure 3. Schematic representation of forces exerted on the particle in the deposited layer

Table 1. Summary of forces acting on a particle

Tangential force, $F_t = 1.7009(3\pi)\mu d_p\{6(d_p/2L)[1-(d_p/2L)]u_s\}$

Normal drag force, $F_n = 3\pi\mu d_p u_p[R_m d_p/3+(1.072)^2]^{0.5}$

Lateral migration force, $F_l = 3\pi\mu d_p u_l[R_m d_p/3+(1.072)^2]^{0.5}$, $u_l = 2.6u_s^2 d_p^3\rho/(16\mu L^2)$

Shear induced force, $F_{si} = 3\pi\mu d_p u_{si}[R_m d_p/3+(1.072)^2]^{0.5}$, $u_{si} = u_s d_p^2/(20L^2)$

Brownian diffusion force, $F_B = 3\pi\mu d_p u_B[R_m d_p/3+(1.072)^2]^{0.5}$, $u_B = KT/(3\pi\mu L d_p)$

Double layer repulsion force, $F_e = -\pi\varepsilon_e\varepsilon_o\varphi^2 k d_p exp(-kh)/[1-exp(-kh)]$

Van der Waals attraction force, $F_v = (H/12d_p)[2d_p/h-d_p^2/h^2-2d_p/(h+d_p)-d_p^2/(h+d_p)^2]$

where, d_p - particle diameter (m), h -distance between the particles surface (m)
        H - Hamaker constant (J), K - Boltzmann constant (J/K)
        L - clearance of crossflow channel (m), R_m - membrane resistance (1/m)
        T - temperature (K), u_B - Brownian diffusion velocity (m/s)
        u_l - lateral migration velocity (m/s), u_{si} - shear induced velocity (m/s)
        u_p - permeate flux (m/s), u_s - average crossflow velocity (m/s)
        ε_e - dielectric constant of water (-), ε_o - permittivity of the media (C²/Jm)
        φ - zeta potential of particle (V), μ - viscosity of fluid (kg/ms)
        ρ - density of particle (kg/m³)

4.2 Effect of Cross-Flow Velocity

The influence of cross-flow velocity on the critical flux is shown in Figure 5. As can be seen from the figure, the critical flux increased with the increase in cross-flow velocity. The critical flux value calculated by the hydrodynamic force balance model agrees very well with that obtained experimentally. A higher cross-flow velocity leads to a greater tangential and back transport force, thus resulting in a higher critical flux.

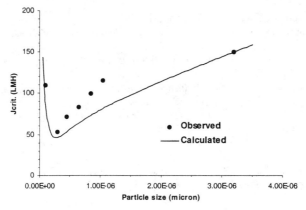

Figure 4. Effect of particle size on critical flux (membrane pore size of 0.1μm, influent concentration of 200mg/L and cross-flow velocity of 0.2m/s)

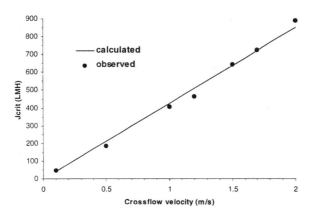

Figure 5. Effect of cross-flow velocity on critical flux (particles size of 0.46μm, membrane pore size of 0.2μm and influent concentration of 10mg/L)

5. CONCLUSIONS

The critical flux was defined as a flux below which no flux decline occurs with filter operation. This leads to a higher volume of filtrate in the long run. The membrane, in principle, will not be blocked and does not need elaborate cleaning. Their life span is also increased. From the experimental result, the followings could be concluded :

- The smaller the particle, the lower the critical flux (for a size range of 0.32 ~ 3.2 μm).
- The higher the cross-flow velocity, the greater the critical flux.

The hydrodynamic force balance model was developed. Its results agree very well with that obtained from the experiments.

Acknowledgements : This work is funded through the University of Technology, Sydney Key Research Strength on Water & Wastewater Management.

6. REFERENCES

Aimar, P. and Howell, J. A., (1989). *Effects of concentration boundary layer development on the flux limitations in ultrafiltration*, Chem. Eng. Res. Des., 67, 255-261

Bacchin, P., Aimar, P. and Sanchez, V., (1995). *Model for colloidal fouling of membranes*, AICHE J., Vol 41, No2, 368-376

Field, R. W. Wu, D., Howell, J. A. and Gupta, B. B., (1995). *Critical flux concept for microfiltration fouling*, J. Mem. Sci., 100, 259-272

Kwon, D. Y. and Vigneswaran, S., (1998). *Influence of particle size and surface charge on critical flux of crossflow microfiltraion*, J. Water Sci. and Tech., in press

Lokjine, M. H., Field, R. W. and Howell, J. A., (1992). *Crossflow filtration of cell suspensions : A review of models with emphasis on particle size effects*, Trans I Chem E, 70, Part C: 149

Lu, W. M. and Ju, S. C., (1989). *Selective particle deposition in crossflow filtration*, Sep. Sci, & Tech., 24, 517-540

Tarleton, E.S. and Wakeman, R.J., (1993). *Understanding flux decline in crossflow microfiltration - Part 3 : Effects of membrane morphology*, Trans I Chem E, 71, Part A : 521

Environmental Hydraulics, Lee, Jayawardena & Wang (eds) © 1999 Balkema, Rotterdam, ISBN 90 5809 035 3

Flocculation with the aid of hydraulic jets

N. Suresh Kumar
Department of Civil Engineering, College of Engineering, Osmania University, Hyderabad, India

S.G. Joshi & B.S. Pani
Department of Civil Engineering, IIT, Bombay, India

ABSTRACT: The use of free jets for flocculation is a simple and inexpensive technique. In the present study the effect of plan-wise location of the outlet for jet flocculators was examined in a square tank for floc growth and hydraulic behaviour of the flow. The results have indicated that the effect is insignificant.

1. INTRODUCTION:

Procurement of potable water from a source contaminated by naturally introduced or man made pollutants involve a series of unit operations. Coagulation and flocculation are important steps to achieve this objective. Flocculation is a gentle mixing phase that follows the rapid dispersion of coagulant by the flash mixing unit. Its purpose is to accelerate the rate of particle collisions, causing the agglomeration of electrolytically destabilized colloidal particles. The rate of collisions among the particles is dependent upon the number and size of the particles in suspension and the intensity of mixing in the flocculation chamber. The agglomerates called 'flocs', should be of settleable and filterable sizes.

Camp and Stein (1943) expressed the mean velocity gradient as the relative velocity of two flow lines divided by the perpendicular distance between them. Mathematically, it is expressed as

$$G = \left(\frac{P}{\mu V}\right)^{0.5} \tag{1}$$

where,
G = mean velocity gradient ;   P = total input power
V = volume of water to which power is applied ; μ = absolute viscosity of water

According to Argaman and Kaufman (1970), in a turbulent flow the mean square of fluctuating velocities and its spectrum are the energy parameters which affect the flocculation. Based on Argaman and Kaufman's data Cleasby (1984) has proved $(\bar{\varepsilon})^{2/3}$ as a superior power input function which reflects the above features of flocculation. Hence, power input per unit mass to the two-thirds exponent is rather, a more appropriate flocculation parameter than G for common water and waste-water flocculation practice. By definition,

$$(\bar{\varepsilon})^{2/3} = \left(\frac{P}{\rho V}\right)^{2/3} \tag{2}$$

where,
$\bar{\varepsilon}$ = average power dissipation per unit mass; ρ = mass density of water

In the present study the effect of plan-wise location of outlet on the floc growth and the consequences like volumetric utilization of a square tank for a fixed location of a jet inlet was examined. For this purpose, the inlet and outlet arrangements were made in the leading and trailing end-walls of the chamber which is referred to as 'normal arrangement'. In another arrangement, the inlet and outlet were placed in the same leading wall; i.e., the influent jet after reaching the down stream end of the chamber was allowed to flow backwards to reach the outlet which is referred to as 'reverse arrangement'.

2. DESCRIPTION OF EXPERIMENTAL SET-UP:

Details of the experimental set-up are depicted in Fig.1 and Fig.2. The flocculator was fabricated out of acrylic sheets for viewing purposes. Raw water stored in a tank was pumped to the flocculator with the help of a pump. The required flow rate was setup by adjusting the valves provided on the main and bypass pipe lines. The inlet to the flocculator was a jet emanating from a submerged pipe. The outflow from the flocculator takes place over a suppressed sharp crested weir whose crest level could be adjusted. The rate of flow entering and leaving the flocculator was measured volumetrically using a collecting tank. To ensure proper mixing of the coagulant with the raw water, the alum solution was injected as a co-current axial jet into the raw water pipe line 1.5 m ahead of the flocculator.

To determine out the total energy of the flow on the up-stream side of the flocculator, pressure tappings were provided on the influent pipe leading to the flocculator. In order to know the head over the outflow weir a pressure tapping was provided in one of the adjoining side walls of the flocculation chamber. All pressure head measurements were made with reference to the centre line of the influent pipe to the flocculator.

For proper mixing Holland and Chapman (1966) recommended that the liquid level in the tank should be less than 1.25 times the diameter of the tank so as to avoid multiple impellers or excessive power consumption. Van De Vusse (1955) observed that, the circulation pattern established by a jet system is similar to that established by a propeller stirrer. A jet submergence depth equal to 50% of liquid depth was adopted as a reference value. However, the optimum depth of jet submergence was obtained by observing the hydraulic behaviour of flocculator tanks with the jet located at 40% and 60% of liquid depths as well as by injecting the flow from the base of the tank.

The flow rates were preset to 0.1473 and 0.0982 lps for obtaining the detention times of 10 and 15 minutes respectively. The jet diameter was kept as 8 mm and a velocity range of 15 to 60 cm/s in the flow field was adopted. Fullers earth was used to prepare the required concentrations of 50 NTU and 100 NTU of turbidity. In order to prevent the settling of the soil particles, a stirrer with variable speed arrangement was provided in the raw water storage tank.

3. RESULTS OF EXPERIMENTAL OBSERVATIONS:

The amount of energy consumed is an indicator of the efficiency of mixing. Hence, the energy loss occurring in the chamber was measured and values are given in column 2 of Table 1. The variation in these values for normal and reverse outflow arrangements was insignificant. From these values the mean velocity gradient was computed and its values are given in column 3 of Table 1. Power inputs per unit mass to the two-thirds exponent are also computed and these values are presented in the column 5 of Table 1. For mechanical flocculators, Cleasby (1984) reported that these values can be upto 80. According to Skeat (1961) the energy loss in a hydraulic flocculator normally ranges from 23 to 183 cms. Even though the observed head loss at 15 minutes of detention time is slightly on the lower side, the performance of the flocculator, considering other parameters, is quite satisfactory. The detention times are low due to the fact that for an efficient agitation, hydraulic flocculators generally require one half of the time taken by the mechanical flocculators (Gregory and Thomas, 1990).

TABLE 1 HEAD LOSS AND OTHER FLOW PARAMETERS

Detention time (min)	Head loss (cm)	$G = (P/\mu V)^{0.5}$ (s^{-1})	GT	$(\bar{\varepsilon})^{0.67}$ $= (P/\rho V)^{0.67}$ $(cm^2/s^3)^{0.67}$
10	43.21	88.60	53160	17.33
15	19.20	48.22	43398	7.67

The intensity of agitation was indirectly evaluated from the GT values, T being the residence time. According to Camp (1955), the desirable values of G vary from 20 to 80 s^{-1} and GT from 2 to 6 x 10^4 for aluminium

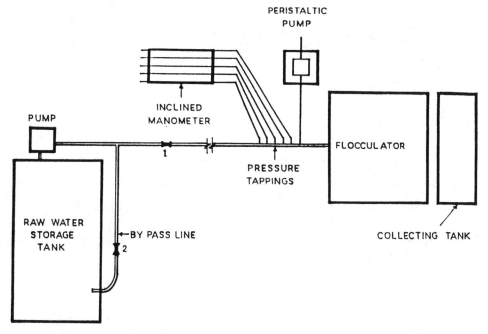

FIG.1. LAYOUT OF THE EXPERIMENTAL SET-UP

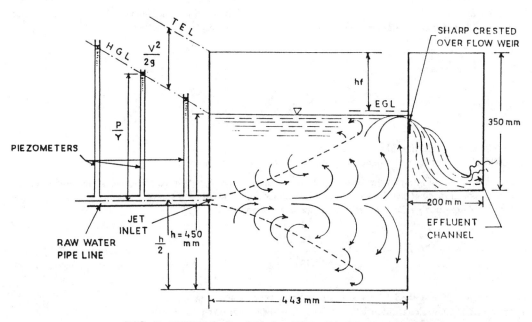

FIG.2. DETAILS OF A JET FLOCCULATOR

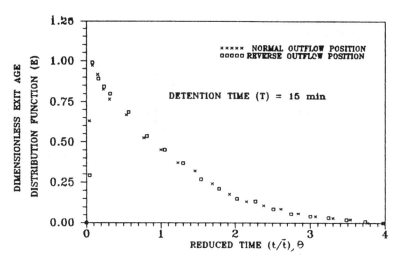

FIG.3 RESULTS OF TRACER STUDIES IN A SQUARE TANK

coagulants. Except the slightly higher G value at 10 minutes of detention time, the values obtained in the present study are in agreement with the above mentioned values.

4. TRACER TESTS:

Tracer studies were conducted in a square tank with sodium chloride as a tracer material. 500 ppm of salt was injected into the tank as a slug dose. The sodium concentration at the chamber outlet was measured with the help of a digital conductivity meter as a function of time. Material balance check was performed to ensure the complete recovery of the injected tracer material (Hudson, 1981). Results of tracer studies for the normal and reverse outflow arrangements are shown Fig.3. The dimensionless exit age distribution function (E) is defined in such a way that E dθ is the fraction of the tracer material in the fluid leaving the chamber with age between θ and θ + dθ. The mean residence time in the chamber was computed from

$$\bar{t} = \frac{\int_0^\infty tCdt}{\int_0^\infty Cdt} \qquad (3)$$

where,
t = time from the injection of tracer material
C = concentration of the tracer material at the chamber outlet at any time t

It can be seen from Fig.3 that the normalized values for both the layouts have collapsed on to a single curve. The important flow characteristics viz., plug flow, mixed flow, and dead space were determined making use of Rebhun and Argaman's (1965) expression which is given below.

$$1 - F(t) = \exp{-\left(\frac{1}{(1-p)(1-m)}\right)\left(\frac{t}{T} - p(1-m)\right)} \qquad (4)$$

F(t) = fraction of the applied dosage at basin outlet, for each value of time t
T = nominal or computed residence time = V/Q (Q is the rate of flow)
p = fraction of active flow volume acting as plug flow

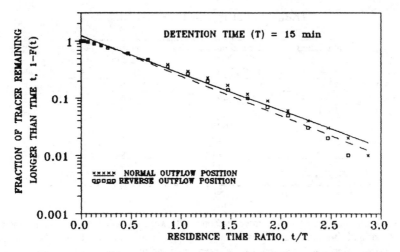

FIG.4 RESIDENCE TIME DISTRIBUTION IN A SQUARE TANK

TABLE 2 IMPORTANT RESIDENCE TIME CHARACTERISTICS

Outlet arrangement	Detention time (min)	Plug flow (%)	Mixed flow (%)	Dead space (%)	DI = t_{90}/t_{10}
Normal	10	16.90	83.10	16.92	13.15
	15	16.44	83.56	19.98	14.27
Reverse	10	18.59	81.41	12.00	11.51
	15	17.38	82.62	24.85	12.38

1-p = fraction of active flow volume acting as mixed flow
m = fraction of total basin volume that is dead space
1-m = fraction of total basin volume that is effective

Correspondingly, the fraction remaining longer than t is 1-F(t). The degree of mixing was evaluated from the percentage of mixed flow. The residence time distribution is shown in Fig.4. Important residence time characteristics for the normal and reverse outlet arrangements, derived from the above plot, are given in Table 2.

The percentage of mixed flow values for the normal outflow arrangement are slightly better than the reverse layout. The degree of mixing was also evaluated from dispersion index (DI) values. Morril (Hudson, 1981) defined DI as the ratio of the times taken for 90% and 10% of the tracer material to pass through the basin. Further, he observed that this index gives a qualitative picture regarding the degree of mixing. Rebhun and Argaman (1965) reported a DI value of about 12 in their study on the evaluation of the hydraulic efficiency of sedimentation basins without baffle walls.

5. FLOC STUDIES:

The optimum alum dosages of 30 mg/l for the turbidity of 50 NTU and 52 mg/l for 100 NTU were obtained from the jar tests. Flocculation studies were conducted at 3% strength of alum solution. After properly mixing the alum in water, alum dosing was started. Alum solution rate, as required for the given flow rate and turbidities, was controlled by a peristaltic pump. The alkalinity of the raw water was measured during each run and it was observed to be constant Alkalinity of raw water was about 50 mg /l of calcium carbonate. The floc laden samples were collected at the outlet at an interval equal to the detention time for all the runs and allowed

TABLE 3 RESULTS OF FLOC STUDIES

Description	Detention time (min) & Outlet position	Final turbidity (NTU)	Turbidity removal (%)
50 NTU	10	16.0	68.0
100 NTU	Normal	19.0	81.0
50 NTU	15	19.0	62.0
100 NTU	Normal	22.5	77.5
50 NTU	10	13.4	73.2
100 NTU	Reverse	17.6	82.4
50 NTU	15	18.4	63.8
100 NTU	Reverse	20.2	79.8

to settle for about 30 minutes in rectangular jars. Hudson (1981) recommended the jar dimensions as 11.5 cm x 11.5 cm x 21.0 cm and the outlet for the withdrawal of sample to be 10 cm below the liquid level. The turbidity of the settled water was measured and an average of the three samples is reported. The results of flocculation studies are given in Table 3. The turbidity removal in case of a reverse outflow arrangement was found to be marginally better than the normal outlet arrangement.

From these results it may be concluded that, the flocculation is not sensitive to the plan-wise location of the outlet. Hence, the outlet can be placed at any convenient location to suit the field constraints like the availability of space and the settling tank position.

6. REFERENCES:

1. Argaman, Y. and Kaufman, W. (1970), 'Turbulence and flocculation', J. San. Engg., ASCE, 96(2) : 223-241

2. Camp, T. R., and Stein, P. C. (1943), 'Velocity gradients and internal work in fluid motion', J. Boston Soc. Civil Engg., 30 : 219-237

3. Camp, T. R. (1955), 'Flocculation and flocculation basins', Trans. Am. Soc. Civil Engg. 120 : 1-16

4. Gregory, Ross and Thomas F. Zabel (1990), 'Sedimentation and flotation', in Water quality and treatment (A hand book of community water supplies) by American Water Works Association, McGraw-Hill Inc., New York : 367-453

5. Holland, F. A., and Chapman, F. S. (1966), 'Liquid mixing and processing in stirred tanks', Reinhold Publishing Corporation, Chapman and Hall Ltd., London : 1-65

6. Hudson, H. E. Jr. (1981), 'Water clarification processes, practical design and evaluation', Van Nostrand Reinhold Company, New York : 40-122

7. John L. Cleasby (1984), 'Is velocity gradient a valid turbulent flocculation parameter', J. Envr. Engg. 110(5) : 875-897

8. Rebhun, M. and Argaman, Y. (1965), 'Evaluation of hydraulic efficiency of sedimentation basins', J. San. Engg., ASCE, 91(5) : 37-45

9. Skeat, W. O. (1961), 'Manual of British water Engg practice', Heffer & Sons Ltd., Cambridge : 504-508

10. Van De Vusse, J. G. (1955), 'Mixing by agitation of miscible liquids', Chemical Engg. Sci., 4 : 178-220

Environmental Hydraulics, Lee, Jayawardena & Wang (eds) © 1999 Balkema, Rotterdam, ISBN 90 5809 035 3

Floating medium flocculator/filter: Is unsaturated flow regime superior to saturated flow regime?

S. Vigneswaran, V. Jegatheesan, S. Santhikumar & H. H. Ngo
Environmental Engineering Group, Faculty of Engineering, University of Technology, Sydney, N.S.W., Australia

R. Ben Aim & A. Shanoun
INSA, Toulouse, France

ABSTRACT: Requirement for a filter that can achieve superior effluent quality and longer filter run time has led to the use of floating media in filters, especially for direct filtration (under the entire solid/liquid separation occurs within the filter). This paper evaluates the performance of a floating medium flocculator/ filter that is operated under saturated and unsaturated flow conditions. From the results, it is found that the filter with floating media bed of shallow depth operated at a high rate of 30-40 m/h under unsaturated flow conditions can function as an excellent static flocculator.

1. INTRODUCTION

Deep bed filtration has a number of applications in water and wastewater treatment and in industrial water reuse. The type and the size of filter media are the most important parameters in affecting the retention of particle within the filter bed, to achieve superior effluent quality and longer filter run times (lengths of filtration cycle). In conventional deep bed filters, the filter media usually used are sand (single medium), anthracite and sand (dual media) or anthracite, sand and garnet (multi-media). To fluidize these highly dense media during backwashing, a large quantity of water at a high backwash rate is required. In a water treatment plant, the water used for backwahing could be upto 5 % of the total daily water production. Backwash requirement can become very high in the case of direct filtration (more specifically in contact flocculation/ filtration) where flocculants are added to achieve flocculation, and the entire solid- liquid separation is within the filter bed itself. With the objective of reducing the backwash requirement, the use of synthetic buoyant filter materials (less dense than water) has been proposed for flocculation and filtration in water and tertiary wastewater treatment (Ben Aim et al., 1993, Ngo and Vigneswaran, 1994, 1995). The backwashing of filters of this type can be achieved with a small quantity of water at a much smaller backwash velocity. Apart from this advantage, the floating filter media are also found to have high solid retention capacity and low head loss development. Hence the use of floating filter media in direct filters (where a contact flocculation/ filtration arrangement is adapted) is a promising solution. It overcomes the disadvantages of direct filtration, such as limited retention capacity, shorter filter runs and high energy requirement for backwashing.

A detail experimental study with a kaolin clay suspension, natural surface water and waste water in a saturated down flow polypropylene medium flocculator/ prefilter show that this system has a good pollutant removal capacity. Further, it led to very low head loss development and produced uniform, microflocs (26-40 μm) suitable for direct filtration (Ngo and Vigneswaran, 1994, 1995; Ngo et al. 1996). Since both flocculation and significant amount of solid-liquid separation take place within the filter bed itself, filter can be used as static flocculator and the prefilter in-place of conventional process of flocculation and sedimentation. In this system, flocculation occurs during the contact of raw water and flocculent within the pores of a medium (the mixing is created by the flow of raw water and flocculent through the filter pores). This is followed by the separation of particle sand flocs in the filter medium. A subsequent polishing filter (such as a sand filter) can then remove remaining solids.

Although the use of floating media in flucculation/ filtration has been proven to be a successful solid-liquid separation method for removing a wide range of pollutant (Ngo and Vigneswaran, 1995), the design has been conservative without any in-depth theoretical study.

Further, most of these studies adopted saturated flow conditions. An experimental study has shown that a floating medium flocculator/ filter operated under unsaturated flow conditions led to very high flow rates (30-40 m/h as compared to 10-15 m/h for saturated flow conditions) with comparable flocculation performance (Shanoun, 1995). This paper thus compare the merits and demerits of the floating medium flocculator with saturated and unsaturated flow conditions.

2. EXPERIMENTAL

A series of floating medium filter experiments carried out to study removal mechanism under saturated and unsaturated flow conditions with in-line flocculation arrangement (Figure 1). Polypropylene beads (diameter, d = 3.8 mm), is used as filter medium. Bentonite at known concentration is used to prepare synthetic water. The characteristics of bentonite clay is as follows: size (medium) = 5.23 μm; density = 2600 kg/m^3. A commercial product of poly aluminium chloride ($Al_n(OH)_m(SO_4)_kCl_{3n-m-k}$) is used as flocculent.

3. COMPARISON OF VELOCITY GRADIENT AND CAMP'S NUMBER FOR SATURATED AND UNSATURATED FLOW CONDITIONS

The velocity gradient (G) and Camp's number (Ca) (Velocity Gradient % flocculation time) can be given by the following equations

$$G = \sqrt{\frac{\Delta P}{\tau \mu}}$$

and

$$Ca = \sqrt{\frac{\tau \Delta P}{\mu}}$$

where τ = contact time, μ = viscosity.

And ΔP is the power input, which is given by the following equation

$\Delta P = \rho g H$

Where H = headloss.

The G and Ca values were calculated for saturated and unsaturated flow conditions and are presented in Figure 2. It can be seen from this Figure that, Ca is higher for unsaturated flow conditions and these values are close to the Ca value recommended by Camp for successful flocculation (10^4 to 10^5).

4. COMPARISON OF FLOCCULATION/ FILTRATION PERFORMANCE

4.1. At Low Velocity (8 m/h)

In the case of saturated flow conditions, there was a solids breakthrough (or leakage of solid mass) through the bed after 5, 8 and 16 hours of filter operation. During unsaturated flow case, the breakthrough was after 8 hours of filter run during the 14 hours of filter operation. In the later, the top portion of packed bed became saturated and this saturation layer depth increased with time. The solids retention was better for saturated flow condition during the first 5 hours of filter run. Beyond this period, the solid removal was similar for both the flow regimes.

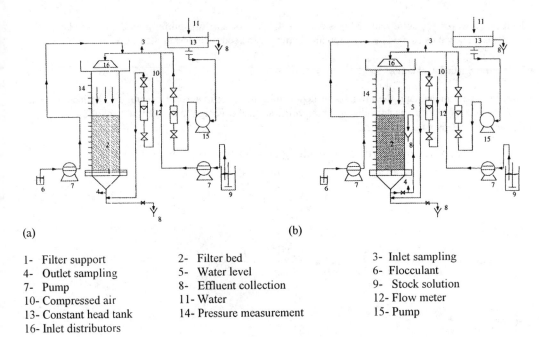

1- Filter support	2- Filter bed	3- Inlet sampling
4- Outlet sampling	5- Water level	6- Flocculant
7- Pump	8- Effluent collection	9- Stock solution
10- Compressed air	11- Water	12- Flow meter
13- Constant head tank	14- Pressure measurement	15- Pump
16- Inlet distributors		

Figure 1: Schematic of flocculator with (a) Saturated flow; (b) Unsaturated flow

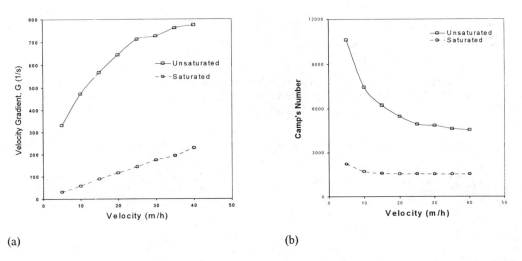

(a)   (b)

Figure 2: Variation of (a) Velocity gradient and (b) Camp's number; under saturated and unsaturated flow conditions (depth of floating medium, H= 36 cm)

4.2. At Higher Velocity (30 and 40 m/h)

At higher velocity range, the filter with saturated flow condition still led to solids leakage through the bed after 5 hours of filter operation. The flocs (or aggregates) were smaller and less uniform (25-40 microns) compared to the unsaturated flow conditions.

The filter bed with unsaturated flow conditions remained always unsaturated during the entire filter run of 28 hours. The solids retention, in general, was low for both the cases compared to the ones with low velocities. However, the solid removal was consistent. The floc size was uniform and relatively large (50 – 60 microns). The solids removal was marginally better for the bed with unsaturated flow.

4.3. Specific Surface Coverage

In order to better understand the progress of deposition, the specific surface coverage of collectors during the early stages of filtration is considered. The specific surface coverage can be defined as follows:
Specific Surface Coverage in time Δt

$$= \frac{\text{projected area of retained particles onto filter grains in a unit bed volume in } \Delta t}{\text{total surface area of collectors in unit bed volume}}$$

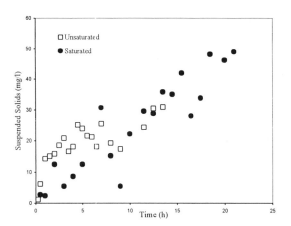

Figure 3: Comparison of flocculator performance for unsaturated and saturated flow conditions at low velocity (V = 8 m/h), H = 120 cm, Initial concentration = 52 mg/l, poly aluminum chloride (PAC) = 40 mg/l)

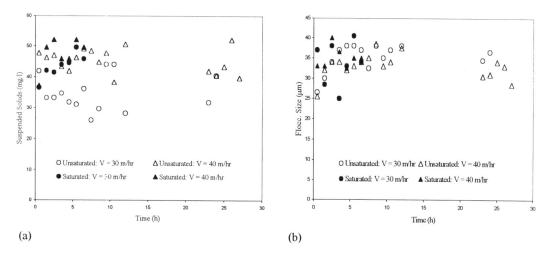

(a)   (b)

Figure 4: Comparison of (a) Flocculator performance and (b) Floc diameter; for unsaturated and saturated flow conditions at high velocities (H = 36 cm, Co = 52 mg/l, PAC = 40 mg/l)

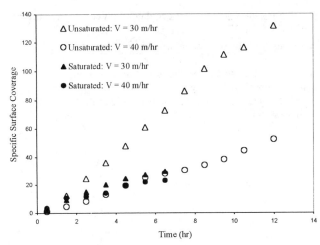

Figure 5: Variation of specific surface coverage under unsaturated and saturated flow conditions for high velocities (V = 30 & 40 m/h, H = 36 cm, Co = 52 mg/l, PAC = 40 mg/l)

This can be calculated in the following manner: if the filtration velocity is V,

Number of particles flowing towards the bed, in an incremental time $\Delta t = (\pi/4)D^2VN_0\Delta t$

Number of particles in the effluent in $\Delta t = (\pi/4)D^2VN\Delta t$

Surface area of particles retained on unit bed volume, in $\Delta t = (\pi a_p^2)VN_0(1-N/N_0)\Delta t/L$

Number of collectors in unit bed volume = $3(1-\varepsilon)/(4\pi a_c^3)$

Total surface area of collectors in a unit bed volume = $3(1-\varepsilon)/a_c$

Thus specific surface coverage in time $\Delta t = (\pi a_p^2)VN_0a_c(1-N/N_0)\Delta t/[3L(1-\varepsilon)]$

Then the Specific surface coverage from t = 0 to t = t:

$$= \left\{ \left(\pi a_p^2 \right) V N_0 a_c / [3L(1-\varepsilon)] \right\} \int_0^t (1 - N/N_0) dt$$

The surface coverage was higher for bed with unsaturated flow (Figure 5).

5. CONCLUSION

The floating polypropylene media bed of shallow depth of 36 cm is an excellent static flocculator in producing microflocs when it is operated at a high rate of 30 - 40 m/h under unsaturated flow conditions. These microflocs can successfully be removed by subsequent direct filtration.

Acknowledgements:- This work is funded by Australian Research Council on "Floating Filter", 1997-98.

6. REFERENCES

Ben Aim, R., Shanoun, A., Visvanathan, C. and Vigneswaran, S. (1993). "New Filtration Media and their use in Water Treatment", *Proceedings, World Filtration Congress*, Nagoya, Japan, 273-276.

Ngo, H. H, Vigneswaran, S. and Jegatheesan, V. (1996). "Mathematical modeling of down flow floating medium filter (DFF) with in-line flocculation arrangement", *Water Science and Technology*, 34(3-4), 355-362.

Ngo, H. H. and Vigneswaran, S. (1995). "Application of floating medium filter in water and wastewater treatment with contact-flocculation filtration arrangement", *Water Research*, 29(9),2211-2213.

Ngo, H. H. and Vigneswaran, S. (1995). "Floating medium down flow flocculator with coarse sand filter: A system for small community", *Water, AWWA*, 34-37.

Ngo, H. H. and Vigneswaran, S. (1994). "Application of floating medium filter in organic removal", *J. IAME*, 21(3), 55-62.

Shanoun, A. (1995). "Etude et conception d'un floculateur staique en ecoulment percolent", *PhD Dessertation*, Universite' de Technologie-Compiegne, France.

Vigneswaran, S. and Chang, J. S. (1986). "Mathematical modeling of the entire cycle of deep bed filtration", *Water, Air, Soil Pollut.*, 29, 155-164.

Environmental Hydraulics, Lee, Jayawardena & Wang (eds) © 1999 Balkema, Rotterdam, ISBN 90 5809 035 3

Mathematical modelling of declining rate filtration

U. K. Manandhar
Sinclair Knight Merz, Melbourne, Vic., Australia

S. Vigneswaran & H. H. Ngo
Faculty of Engineering, Environmental Engineering Group, University of Technology, Sydney, N.S.W., Australia

ABSTRACT: Modified O'Melia-Ali model based on particle collector concept and detachment mechanism has been extended for single filter unit operating at declining rate mode with constant inflow. The filter removal efficiency depends on variable velocity term which is unknown. The velocity is obtained by solving headloss and water balance equations. This model has been verified with a hypothetical case of declining rate filtration formulated based on past experiences. The two major advantages of declining rate filtration over constant rate filtration, namely prolonged working period and better filtrate quality for same terminal headloss development, are observed in simulations.

1. INTRODUCTION

Declining rate filtration has been widely used in practice for water treatment. There have been large number of experimental evidences that declining rate filtration is not only operationally convenient and economical, but also it produces equivalent or even better quality water as the constant rate filtration. The better water quality is achieved mainly during final stages of filter run for same terminal headloss development.

Optimisation of filter design and operation is required so that main objectives of filtration, namely water quality and quantity are met with minimum resources (capital and operational costs). Optimisation can be done either by performing extensive experimental study or by mathematical model with limited experimental data. Mathematical models are widely adopted approach since extensive experiments are expensive.

Most of past mathematical models of filtration are mainly for constant rate mode of operation. Therefore the main objective of this study is to develop mathematical model for declining rate filtration so that it can be used for filter optimisation.

2. MATHEMATICAL MODEL

The filter removal efficiency as obtained from modified O'Melia Ali model (O'Melia and Ali, 1978; Vigneswaran and Chang, 1986) based on particle collector concept and detachment mechanism is as below:

$$\eta_{ri} = \eta\alpha\left[1 + \eta_p\alpha_p\beta\frac{\pi}{4}d_p^2\sum_{i=1}^{t}n_{(i-1)0}V_{i-1}\Delta t.\exp\left[-\frac{3}{2}(1-f_0)\eta_{r(i-1)}\frac{\Delta L}{d_c}\right]\right]$$
$$-\left[\beta_2\frac{J_{i-1}}{n_0}\sum_{i=1}^{t}(\eta_r{}^n)_{i-1}\right] \tag{1}$$

The details on development of these equations and notations used is described elsewhere (Manandhar and Vigneswaran, 1991; Manandhar et.al., 1991).

783

It is to be noted that the filtration velocity is variable in case of declining rate filtration. The filtration velocity at the $(i-1)^{th}$ time step, V_{i-1} is known from previous time step and therefore η_{ri} can be calculated. The filtration velocity at i^{th} time can be calculated by solving headloss and water balance equations.

The headloss equation is based on Kozney's euation and is as follows:

$$h_{fi} = (VEL1).V_i \tag{2}$$

where,

h_{fi} = headloss at i^{th} time step;

$$VEL1 = 36K \frac{\mu}{g} \frac{V_i}{f} \frac{(1-f)^2}{f^3} \frac{1}{d_c^2} \left[\frac{S_2}{6}\right]^2 \left[\frac{1 + \beta' \left[\frac{N_p}{N_c}\right]\left[\frac{d_p}{d_c}\right]^2 \left[\frac{S_2}{S_1}\right]^2}{1 + \left[\frac{N_p}{N_c}\right]\left[\frac{d_p}{d_c}\right]^3 \left[\frac{S_2}{S_1}\right]^3}\right]^2 . L \tag{3}$$

Notations used are described elsewhere (Manandhar and Vigneswaran, 1991; Manandhar et.al., 1991).

The headloss in filter bed at a particular time step is thus function of the filtration velocity at that time step. The filtration velocity at a particular time step can be obtained by solving headloss and water balance equations simultaneously. The changes in water level, h in the filter box with respect to effluent level, which by definition is equivalent to the headloss is due to difference in inflow and variable filtration rate, $V = V(t)$. Thus the water balance equation is as follows:

$$\frac{\delta(hA')}{\delta t} = Q - V(t).A \tag{4}$$

where, $\quad Q \quad$ = inflow rate,
$\quad\quad\quad\quad h \quad$ = variable water level,
$\quad\quad\quad\quad A' \quad$ = water accumulation area, and
$\quad\quad\quad\quad A \quad$ = filter box cross-sectional area.

Therefore,

$$h(t) = h(0) + (Q/A')t - (A/A')\int_0^t V(t).dt \tag{5}$$

Representing $\int_0^t V(t).dt$ by P, Eq. 6 can be written as ,

$$h(t) = h(0) + (A/A').[(Q/A).t - P] \tag{6}$$

By definition, and using Eq. 2 for headloss, we get,

$$\frac{dP}{dt} = V(t) = \frac{h_{fi}}{VEL1} \tag{7}$$

From Eqs. 6 and 7, we get,

$$\frac{dP}{dt} = \frac{h(0) + A_r[(Q/A).t - P]}{VEL1} \tag{8}$$

where, $\quad A_r$ = Area ratio, A/A'.

Eq. 8 is a first-order differential equation which can be solved by fourth-order Runga-Kutta method of numerical integration with initial condition of P = 0 at t = 0. Once the solution of Eq. 8, P=P(t) is known, the velocity can be obtained by evaluating dP/dt in Eq. 8, and then the head variation from Eq. 6.

3. METHODOLOGY

3.1 Hypothetical Declining Rate Filter

This was formulated based on past experiences. The initial filtration rate is considered to be 9 m/h. The feed water to the filter is taken to be kaolinite clay suspension with 50 percentile average size on weight basis being 12 μm. The magnitude of influent concentration is taken as 80 mg/L. The filter media is single layer of sand and its characteristics are - depth = 60 cm; size = 1000 - 1190 micron; and porosity = 0.45.

3.2 Model Parameter Estimation

The model parameters are $\eta\alpha$, $\alpha_p\beta$, β_2, and β, and these are required to be evaluated before model could be used. The parameter β is the headloss coefficient. The initial removal efficiency, $\eta\alpha$ is calculated from "clean" filter performance, and other coefficients $\alpha_p\beta$ and β_2 are expressed as functions of particle size, d_p by relations - $\alpha_p\beta = (APB1).d_p^{(APB2)}$ and $\beta_2 = (BET1).d_p^{(BET2)}$ (Manandhar et.al., 1991). Thus the model coefficients to be calibrated are effectively APB1, APB2, BET1, BET2 and β.

Based on previous studies, the model coefficient values are chosen appropriately for hypothetical case of declining rate filter (Manandhar and Vigneswaran, 1991; Manandhar et.al., 1991). These are - $\eta_p =1$, $\eta\alpha = 0.0034$, APB1 = 0.02, APB2 = 1, BET1 = 0.00018, BET2 = 1, and $\beta = 0.1$.

4. RESULTS AND DISCUSSIONS

With chosen model parameters and filter characteristics, simulations have been carried out for different area ratios (A_r = 1.0, 0.5, 0.05, 0.01 and 0.0). The predicted filter performances (headloss, concentration and velocity profiles) are shown in Figs. 1, 2, and 3.

It is to be noted from velocity profiles (Fig. 1) that the filtration velocity remain at more or less constant values for higher values of area ratio, A_r, thus indicating the constant rate mode of operation. With increase in A_r values, the filter box area approaches to water accumulation area, and they are equal when $A_r = 1$. Since the

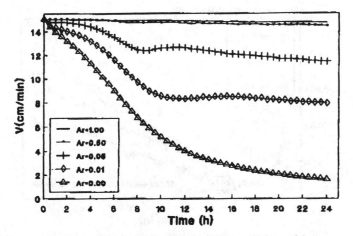

Figure 1 Velocity Profiles for Different Values of A_r

inflow to the filter is constant and permeability of filter media decreases as filtration proceeds, the water level in the accumulation area rises due to rapid increase in headloss across the filter. This results in rapid increase in driving force across the filter and hence tries to maintain the constant rate of filtration.

Smaller values of A_r indicate higher water accumulation area, A' as compared to filter box area, A. For such situations, the headloss profiles (Fig. 2) have mild slopes and there is drop in filtration rate (Fig. 1). The value of $A_r = 0$ represents true declining rate filter in which the filter has infinite storage and there is practically no change in total headloss as filtration proceeds.

In all the cases except $A_r = 0$, the velocity profiles showed some fluctuations. This is more significant for higher values of A_r. This could be explained in terms of breakthrough. For higher A_r values, the hydraulic gradient increases rapidly and there is tendency of breakthrough. This might lead to small increase in filtration rate and hence the fluctuation occurs. These results are in good agreement with previous study (Chaudhry, 1987).

For clarity, the concentration profiles are split into two - ripening period, and working and breakthrough period (Fig. 3). The ripening stage was same for all cases. As mode of operation approached that of true declining rate filtration, the working period prolonged and the better filtrate quality was obtained.

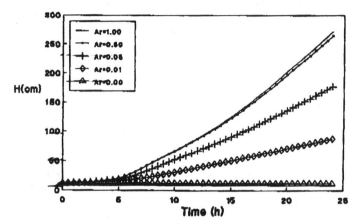

Figure 2 Headloss Profiles for Different Values of A_r

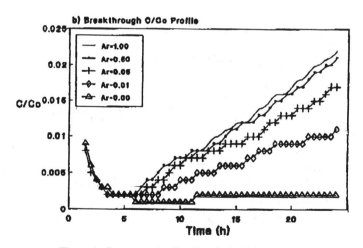

Figure 3 Concentration Profiles for Different Values of A_r

786

The model developed in this study is limited to declining rate filtration operating in a single filter unit with constant inflow. However, in actual practice, declining rate filters operate in a bank of two or more filters connected by a common influent channel and the common headloss in the filters is restricted to a certain maximum value.

5. SUMMARY AND RECOMMENDATIONS

Modified O' Melia and Ali model has been extended to declining rate filtration by incorporating water balance in the system. The new model successfully predicted behaviour of hypothetical filter operating in declining rate mode. Major advantages of declining filtration, namely prolonged filter run time, better effluent quality and smaller terminal headloss development are verified by model simulations.

Since this model is developed for declining rate filtration operating in single filter unit, it can not be used for case of declining rate filters operating in bank operation. It is therefore recommended that this model be further modified for case of declining rate filters in bank operation and tested with experimental data. Since backwashing is one of important operations in filtration, it should be mathematically modelled so that the model is more practical.

Acknowledgements - This work is funded through the University of Technology, Sydney, Key Research Strength on Water & Waste Management.

6. REFERENCES

Chaudhry (1987), "Theory of declining rate filtration I: continuous operation", *Journal of Environmental Engineering*, Vol. 113, No. 4.

Manandhar U. K. and Vigneswaran S. (1991), "Effect of media size gradation and varying influent concentration in deep-bed filtration: mathematical models and experiments", *Separations Technology*, Vol. 1, pp. 178-183, 1991.

Manandhar U. K., Vigneswaran S., Janssens J. G. and Ben Aim R. (1991), "Mathematical modelling of the effect of size distribution of suspended particles in deep-bed filtration - experimental testing", *Water Supply*, Vol. 9. Jönköping, pp. S85-S93, 1991.

O' Melia, C. R. and Ali, W. (1978), " The role of retained particles in deep bed filtration", *Prog. in Water Technology*, Vol. 10, No. 5/6, pp. 167-182.

Vigneswaran, S. and Chang J. S. (1986), "Mathematical modelling of entire cycle of filtration", *Water, Air and Soil Pollution*, 29, pp. 155-164.

Environmental Hydraulics, Lee, Jayawardena & Wang (eds) © 1999 Balkema, Rotterdam, ISBN 90 5809 035 3

Hydrodynamic and sediment problems of biological contact aeration process for water supply source pre-treatment

Huang Ben-sheng, Lai Guan-wen, Qiu Jing & Li Zi-hao
Guangdong Provincial Research Institute of Water Conservancy and Hydro-Power, Guangzhou, China

ABSTRACT: Biological contact aeration process technology consisting of two main systems: the filler and the aeration, as a developed technology for sewage treatment has been studied and applied to water supply source pre-treatment engineering in recent twenty years. Combining related tests of the East River-Shenzhen water supply source pre-treatment project, drag problem of biological contact aeration process technology and its test & research methods have been present in this paper. According to velocity distribution and the weight of silt deposited and gathered on fillers, hydrodynamic characteristics in biological treatment channel and possibility influence on the effect of water treatment have been analyzed.

1. INTRODUCTION

1.1 Brief Introduction of the Principle of Biological Contact Aeration Process Technology

The biological treatment methods are used in which many kind of preponderant microbes and animalcules were tamed and trained to eat nutriment in water including so-called pollutants, to reduce concentration of pollutants and purify water by their process of metabolism. There are two kinds of biological treatment methods: active sludge method and biological membrane method. Biological contact aeration process technology is one type of biological membrane method. The filler supplied for microbes and animalcules living and growing is a main part of this technology. The so-called biological membrane is poor-nutritious microbes and animalcules growing on the filler. The concentration of dissolved oxygen in polluted water is usually lower, in order to make microbes and animalcules can grow on the filler (i.e. so-called the filler with biological membrane) and purify water, aeration must be done to supply oxygen. Hence, the other main part of this technology is the system of aeration (see Fig.1).

1.2 Characteristics of the East River-Shenzhen Water Supply Source Biological Treatment Project

Applying biological contact aeration process technology to the East River-Shenzhen water supply source pre-treatment, its water capacity designed for treating amount to 4×10^6t/day, that will be a extraordinarily and the biggest water treatment project in the world, and of its specific hydrodynamic and sediment characteristics different from others. Among of them, drag and a great quantity of silt deposited and gathered on the filler are very importance problems. For push-flow type's treatment channel:

$$V = \frac{L}{T} \tag{1}$$

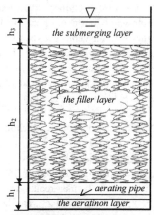

FIG.1. Cross Section of the Boilogcal Treatment Channel

$$Q = VBH \qquad (2)$$

$$h_f = \lambda \frac{L}{4H} \frac{V^2}{2g} \qquad (3)$$

Hence, the rising value of water level ΔZ at upstream due to the filler and the aeration can be expressed as:

$$\Delta Z \sim L^3 \qquad (4)$$

where V is sectional average velocity of the treatment channel, L is the length of the channel, T is the design retention period, B is the total width of the channel, H is design water depth in the channel, Q is treatment discharge, h_f is loss of water head, λ is drag coefficient, g is acceleration of gravity. The treating capacity of the biggest water source biological pre-treatment plants that have been built in China and Japan (the treatment channels lengths of them are about 100m) is about 1×10^5t/day, so it is about 40 times smaller than that of this project. If this project's length were of 360m, its rising value would be about 50 times bigger than that one. Hence, drag problem is very important for this project. First, laboratory experiment proved that the preliminary plan of 1(channel)×50m(width)×1080m(length) is not feasible and possible, then the plan was changed as 4×25m×360m, in order to reduce rising value of water level, the final plan was changed as 6×25m×240m (Huang, B.-S. et al, 1996,1997).

2. METHODS OF EXPERIMENT AND RESEARCH

It is obvious that drag of the biological contact aeration process technology consists of the form drag of the filler and the resistance caused by aerating. The filler is laid in the middle layer of flow, its form drag is difference from normal wall boundary resistance. Because of the interaction between aerating and water flow, flow in the channel is two-phase air and liquid flow, and even more air jetting out of perforating pipes in water forms two-phase air and liquid buoyant jets, all these are very complicated. Designed flow depth and average velocity in outlet of the channel are H=3.8m (h_1=0.4m, h_2=3.0m, h_3=0.4m) and V=0.0813~0.375m/s. These experiments had been done in two phases. First, drag experiment was done in laboratory. However, sectional model of the water depth 3.8m couldn't be built because laboratory facilities permitted, so we advance a superposition method of water level rising among the channel and an extending method by doing a series of depth experiments research into this problem. In the second phase, nature drag experiment of the filler with biological membrane & silt was done at Yantian nature flume.

2.1 The Extending Method by Doing A Series of Experiments

A series of difference depth experiments were made in a model, in which the filler is the same as the prototype's, but every parts of depth (including the height of the filler) and discharge were reduced by the ratio of water depth between prototype and model. Hence, sectional average velocities of all experiments in model and prototype are equal, and so are their Reynolds number ($R_{ed}=Vd/v$, where d is size of the filler, v is kinematic viscosity.) and form drag of the filler. Based on experimental results, the relationship between ΔZ and H(depth) can be plotted, the corresponding value of water level rising to prototype real depth can be extended by the relationship.

2.2 The Superposition Method of Water Level Rising among Treatment Channel

Because the length of experimental flume permitted, the rising value of water level at upstream in the channel of length 1080m or 360m can't be obtained by one time experiment. However, cross section of the channel is a regular rectangle, so we can use the superposition method to resolve this problem, at first, upstream water level Z_{up1} can be got within the effective length of the filler by controlling the downstream water level Z_{down1}(=3.8m), then controlling the downstream water level $Z_{down2}=Z_{up1}$, Z_{up2} can be got, and so on, the rising value of water level for any length of the filler in biological treatment channel can be obtained. However, it needs more number of times and longer time to do these experiments one by one to the length of 1080m or 360m (the effective length of the filler is 20m in the laboratory and 10m at the Yantian nature models). Hence, by doing proper number of times experiments, we can set up the corresponding relationship between rising values of water levels ΔZ_i and downstream flow depths H_i, it is found that they can be expressed as (see Fig.2):

$$\Delta Z_i = ae^{bH_i} \qquad (5)$$

where a and b are constants by relation analyzing, then we can calculate the total rising value of water level ΔZ for any length of treatment flume.

$$\Delta Z = \sum_{i=1}^{n} \Delta Z_i \qquad (6)$$

$$L = \sum_{i=1}^{n} \Delta L_i \qquad (7)$$

$$H_{i+1} = H_i + \Delta Z_i \qquad (8)$$

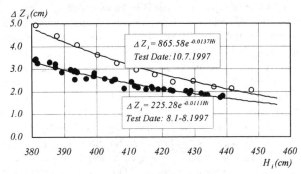

FIG.2. Relation between ΔZ_i and H_i of the Filler with Different Boilogical Membrane & Silt (V=0.0813m/s, Aerating)

2.3 Design & Construction of Models and System of Aeration & Controlling Devices

2.3.1 Laboratory Experimental Flume of GDRIWCH

A rectangle sectional flume of width 0.8m and height 2.0m and length 45m was built for doing this experiment at Guangdong Provincial Research Institute of Water Conservancy & Hydro-power, in which laying length of the filler is 30m and the effective length for superposition experiments is 20m. Aeration device is perforating pipes that have two rows micro-hopes of diameter 2mm and transverse spacing 14cm for the same row. Micro-hops of two rows are staggered off and down at 45° with vertical direction. The distance of each perforating pipe is 0.5m. An air compressor was used to supply air discharge, 8 difference types of rotameters and 20 pressure gauges were installed to control and measure air discharge. There were two branch valves at the end of A grade air supplying main pipe for controlling air discharge.

2.3.2 Nature Experimental Flume at Yantian

Two rectangle sectional flumes named as A and B of width 1.0m and height about 4.3m and length 30m were built at Yantian beside the East River-Shenzhen water supply channel. The original design flume only satisfied to the needs of doing biological contact aeration process effect's experiment, so water supplying discharge was smaller. In order to do this drag experiments, afterwards, the flume A was rebuilt up to height 5.0m. The laying length of filler is 20m and the effective length for superposition experiments is 10m. An air-blower was used to supply air discharge.

The superposition method is used to solve the problem that experimental flume is short, and the extending method is used to solve another problem that the experimental water depth in laboratory is not enough to prototype real water depth. Hence, superposition method had been only used for experiments at Yantian nature flume, but both methods had to be used for experiments at laboratory.

3. EXPERIMENTAL RESULTS

3.1 Comparison of Experimental Results between Laboratory and Nature

The great defect of the laboratory experiment is that the filler with biological membrane and silt can't be simulated. Moreover, water depth in the laboratory flume can't be reached to 3.8m, so the corresponding rising value of water level needs to be extended by doing a series of experiments. Comparing laboratory's experimental results with Yantian's nature experimental results in TABLE 1 (ratio of water and air is 1:1), it can be shown that the rising value of the nature is higher than that of the laboratory, in anther words, when the filler has biological membrane and silt, its drag increases. Experimental results are reasonable. Both rising values of water levels of per unit length are close, but the nature's is a little greater than the laboratory's. Because aeration device in the laboratory flume was a perforating pipe but in the Yantian nature flume was microporous and arranged as rectangle with a spacing of 1.0m longitudinally and 0.5m transversely, hence, intensity of per unit area aeration in the Yanyian flume is greater than that of the laboratory, and so is the aeration drag. Air bubbles flow out of microporous device are smaller than perforating pipe, velocity of small air bubble moves up in water is smaller than that of big bubble, so time of small bubble stay in water is longer, correspondingly, air concentration in water also greater. All of these prove that two of methods presented in this paper are reasonable and correct.

TABLE 1. Comparison of Rising Values of Water Levels Between Microbes Living on Fillers and Not

Date	L (m)	Velocity (m/s)	Place	State of Filler	No Aeration ΔZ_1	Aeration ΔZ	$(\Delta Z - \Delta Z_1)/L$ (cm/m)
1997.2	360	0.1220	Laboratory	Pure filler	68.0	86.3	0.051
1997.8	360	0.1220	Nature	with biological membrane & silt	82.3	109.4	0.075
1997.7	240	0.0813	Laboratory	pure filler	26.2	38.2	0.050
1997.8	240	0.0813	Nature	with biological membrane & silt	39.4	58.6	0.080
1997.10	240	0.0813	Nature	with *plumatella sp.*	52.8	70.1	0.072

3.2 Velocity distribution and Discharge Distribution

Due to the filler arranged in the middle layer of water, velocities along vertical direction in the treatment channel redistribute (see Fig.3). Velocities of above and under the filler layer (i.e. in the submerging water layer and the aeration layer) are (4~10) times bigger than velocities in the filler layer. This can prevent sediment depositing on the bottom of the biological treatment channel, however, it is probable disadvantageous to the effect of the biological contact aeration process. Because bottom slopes of experimental flumes was zero, among upstream, as laying length of the filler increase continuously, drag of flow and depths in the channel increase, hence, depths and discharges of the submerging water layer increase gradually, discharges of the filler and the aeration layers decrease gradually. By comparison with no aeration, when aerating, velocities in the submerging and the aeration layer decrease but velocities in the filler layer are increase correspondingly.

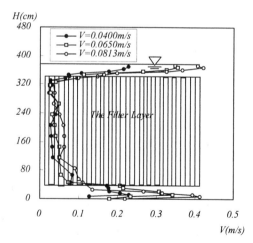

FIG.3. Velocity Distribution in Boilogical Treatment Channel (No Aeration)

3.3 The Problem of Fillers with Silt

Weight of a single pure filler of diameter 20cm and length 3.0m is about 0.25kg. From August 1, 1997, The flume at Yantian operated in accordance with sectional average velocity V=0.0813m/s. TABLE 2 is the comparison of weight of the filler with biological membrane & silt in different states. It can be seen that weight of filler with biological membrane & silt is greater than that of pure filler. Till middle of September, it had been found that a lot of *plumatella sp.* grown on the filler and weight of some single filler were one time greater than that of before. However, few *plumatella sp.* grew on the single filler at the end of flume (0+20m) and no air bubble current passed thought it, so weight of them were almost silt weight. Three times weights of this filler measured from August to October were almost the same, it is said that weight of the filler with silt were saturation when drag experiments were done in August. It is must be noticed that these single mobile fillers were often hung up to observe and weighed, that affected their weight, actually, weight of fillers fixed on supporter frame were much greater than these in TABLE 2, this is very importance for the filler supporter design.

The transport rate of sediment suspension can be expressed as (Chien, N. and Wan, Z-H., 1983):

$$S_{vm} = k \left(\frac{U^3}{gh\omega} \right)^m \tag{9}$$

where S_{vm} is depth averaged concentration of sediment in suspension, U is mean flow velocity, h is flow depth, ω is fall velocity of the sediment particles in water, constant k and m are a function of $\frac{U^3}{gh\omega}$.

Equation (9) shows that the transport rate of sediment suspension is in direct proportion to the cube of velocity. Because velocities in the filler layer are (4~10) times smaller than those of the submerging water layer or the aeration layer, so transport rate of sediment suspension in the filler layer is (64~1000) times smaller than the other layers correspondingly. When the flow with larger concentration of sediment suspension in the submerging water layer or the aeration layer flow into the filler layer, most silt of sediment suspension fell and gathered on fillers, others fell down to the aeration layer and became sediment suspension again. As time gone by, silt gathered on fillers were more and more and till saturation, but aerating can prevent silt from falling and gathering on the filler. Because velocity of the aeration layer is larger, seldom silt had been found to deposit on the bottom of the flume after the experiment had been done for ten days.

3.4 The Critical Slope of Uniform Flow of the Treatment Channel

Rising values of water levels in TABLE 1 were obtained under the condition of which bottom slopes of flumes were zero, so the water surface lines were rising curves, i.e. the flow depth at upstream was greater than that at downstream. If want flow depths in the treatment channel from downstream to upstream to be equal, we must make the bottom's slope of the channel to be the critical slope of uniform flow by reasonable designing. However, in the early periods of the biological treatment channel operating, drag of the filler varies with weight of biological membrane especially silt on the filler, and the water surface line also readjusts continuous. Under the condition of flow depth at downstream was equal to 3.8m, critical slopes and corresponding rising values at upstream of different lengths of channels and different states of the filler are listed in TABLE 3. It is clear that if the bottom slope of the channel were designed to be the critical slope, its rising value at upstream would be greater than that of horizontal slope.

TABLE 2. Comparison of Weight of Single Fillers with Different Biological Membrane & Silt

Date	Distance from upstream (m)	0	5	10	15	20
8/10/1997	Weight of filler with biological membrane & silt (kg)	4.70	5.65	6.0	3.10	7.10
9/21/1997	Weight of filler with *plumatella sp.* (kg)	11.20	10.35	8.75	3.25	6.85
10/12/1997	Weight of filler with *plumatella sp.* (kg)	6.25	12.25	8.25	3.90	6.00

TABLE 3. Critical Slopes and Rising Values of the Filler with Different Biological Membrane & Silt

L (m)	Velocity (m/s)	State of Filler	Critical slope (%)		Rising value (cm)	
			No Aeration	Aeration	No Aeration	Aeration
240	0.0813	pure filler	0.143	0.193	34.8	46.3
240	0.0813	with biologic membrane & silt	0.230	0.340	55.2	81.6
240	0.0813	with *plumatella sp.*	0.350	0.500	91.8	108.0
360	0.1220	pure filler	0.255	0.300	91.8	108.0
360	0.1220	With biologic membrane & silt	0.380	0.500	136.8	180.0

3.5 Comparison of Drags of the Filler with Difference Biological Membrane & Silt

We can obtain by Manning formula:

$$\frac{n_2}{n_1} = (\frac{V_1}{V_2})(\frac{R_2}{R_1})^{\frac{2}{3}}(\frac{i_2}{i_1})^{\frac{1}{2}} \tag{10}$$

where n is Manning roughness coefficient, R is hydraulic radius, i is slope of water surface, footnote "1" and "2" represent the filler with difference biological membrane & silt. If the designed bottom slope is the critical slope of uniform flow, then $V_1=V_2$ and $R_1=R_2$, according to TABLE 3, for $L=240$m and no aeration, we can obtain by equation (10): roughness of the filler with normal biologic membrane & silt was 1.27 times greater than that of the pure filler ($n_2/n_1=(i_2/i_1)^{1/2}=(2.3/0.143)^{1/2}=1.27$); furthermore, after many *plumatella sp.* had grown on the filler, its roughness was 1.23 times greater than that of before ($n_2/n_1=(i_2/i_1)^{1/2}=(3.5/2.3)^{1/2}=1.23$). When many *plumatella sp.* grew on the filler, critical slopes of uniform flow of the treatment channel had been measured and obtained as follows:

$$i(\%)=48.735V^2+0.3344V \qquad (\text{No aeration}) \tag{11}$$

$$i(\%)=25.607V^2+4.0549V \qquad (\text{Aeration}) \tag{12}$$

4. CONCLUSION

Drag experimental results of the pure filler at laboratory and the filler with biological membrane & silt at nature have the same lows, it is proved that the superposition method and the extending method presented in this paper are reasonable and correct. By comparison with the pure filler, drag of the filler with biological membrane & silt increase greatly, and it varies with difference biological membrane & silt. Although aerating causes drag of flow to increase, yet it plays a great role in exchanging and mixing of every layer flow each other. It must be noticed that a great quantity of silt deposited and gathered on the filler not only increase flow resistance, but also decrease effective area of the filler, it may influence the treating effect directly.

5. REFERENCES

Chien, N. and Wan, Z-H. (1983). *Mechanics of sediment transport*, Science Publication Co., Peiking, China. (in Chinese).

Huang, B-S. et al. (1996,1997). "Experimental research on hydrodynamic problems of the East River-Shenzhen water supply source biological pre-treatment project (Part 1, 2 and 5)", Technical Report, Guangdong Hydr. Res., Guangzhou, China. (in Chinese).

Huang, B-S. et al. (1997). "Experimental research on flow resistance of the East River-Shenzhen water supply source biological pre-treatment project", Proc., 11th Congr. of NCHD, Wuxi, China, 534-539. (in Chinese).

Environmental Hydraulics, Lee, Jayawardena & Wang (eds) © 1999 Balkema, Rotterdam, ISBN 90 5809 035 3

Water mixing and exchanging in biological contact oxidation process flumes of water supply works and wastewater engineering

Lai Yifeng, Lai Guanwen & Li Zihao
Guangdong Research Institute of Water Conservancy and Hydropower, Guangzhou, China

ABSTRACT: Characteristics of water mixing and exchanging in biological contact oxidation flumes are analyzed. The criteria of completely mixed regime and the length of the process flume are presented. This procedure is important in the context of predicting the conditions of water mixing and designing flumes. The initial results are also compared well with an experiment.

1. INTRODUCTION

The biological contact oxidation process has been developed and initially used in wastewater treatment engineering since the 19th century. During 1970s, it was improved in many aspects and then was widely used in practical projects. Recent years, it is taken on trial as preliminary treatment of light polluted water to reduce organic substances and nutrients (such as nitrogen) in some municipal water supply companies in China, and lots of good effects have been achieved. As adopted design treatment flowrates increase gradually, the extent of water mixing and exchanging exists some differences with the status of small flowrates in the process flume. Hereupon, some fundamental behaviors are studied.

2. BASIC STRUCTURE OF THE FLUME

The biological contact oxidation process flume mainly contains the inlet and outlet structures, the fixed contact bed (filter bed) and the aeration system. The purpose of inlet and outlet structures is to let the flow pass through the flume uniformly. Further more, some physical operations, such as screening and sedimentation, which could remove the floating and settleable solids in polluted water are also arranged in these structures. The contact beds provide the places where microorganisms could grow and live. The beds have many different ways, and some researchers in the world are still looking for the improving patterns of the beds. Basically, the good beds must have great specific surface areas. The aeration system supplies air to microorganisms, so the aerators are laid under the fixed contact beds. On the other hand, in order to avoid the effects of rain and sunshine upon the biological films, it usually keeps a depth water over the bed top. Fig. 1 shows the typical cross-section sketch of the flume.

The general principle of the biological contact oxidation process is to remove the biodegradable organic substances and some nutrients in polluted waters by biological activity. Basically, these substances are converted into gases that can escape to the atmosphere and into biological cell tissue that can be removed by settling. With proper environmental control, polluted waters can be treated biologically in most cases. Therefore, it is the responsibility of the engineer to ensure that the proper environment is produced and effectively controlled. As

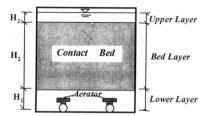

Fig. 1   Cross-section of the Biological Contact Oxidation Flume

shown in Fig.1, when polluted water flows through the flume, it would flow along three directions: over bed, bed and under bed. Here we define them as upper layer, bed layer and lower layer respectively. For the resistance of contact bed and the sediment on the bed, the flowrate in upper and lower bed layers would be greater than that in bed layer, the flowing-through time is also shorter than the time in bed layer flow. As the matter of fact, only the bed layer water touches with the biological films. Thus, the key to design an effective biological process flume is to make sure that all polluted water in the flume has chance to contact with the fixed biological films and also maintains a specified period of touching time.

3.   WATER MIXING AND EXCHANGING CHARACTERISTICS

3.1   POWER OF WATER MIXING AND EXCHANGING

Generally, there are two ways in water exchanging course, the first is through turbulence in flows; the second changes the position of water particles directly by exterior forces, of course, at the same time, the turbulence level is also enhanced.

When water passes through the flume, the flow in different layers has different turbulence level. The velocity and turbulence in upper and lower layers are greater and stronger than these in bed layer. When the aerators do not work, the quantities of exchanging water between the above three layers are very small during the water flowing forward. The majority of waters in upper and lower layers have no exchange with the bed layer water, they flow directly through the flume. Those parts of waters are defined as short-circuiting flow. This limited water exchange course belongs to the first way. The operation condition could been used to the disposal of sediment near the flume bed (Lai Yifeng et al. 1997). When the aerators work, the pumped air discharges into the crosscurrent. The gas bubble flow is buoyant jet flow at the beginning and then transfers to buoyant flow towards water surface. During the gas bubble flows, the diameter is changing and has momentum and position exchanges with the surrounding water. At the same time, the flow turbulence level is enhanced. In this case, the water exchanging quantities between the three layers increase greatly. All the above phenomena could be observed clearly in the experiment. Therefore, the aeration system not only supplies oxygen to microorganisms but also plays the key role in the water exchanging course between the three layers in the process flume.

3.2   QUANTITIES OF EXCHANGING AND SHORT-CIRCUIT WATERS

The water exchanging procedure between the three layers is very complicated. Here some simplifying assumptions are introduced, mainly the use of the average quantities of each variables. Because the exchanging procedure between lower layer and bed layer is the same as the procedure between upper layer and bed layer. Below considers the first status. Fig. 2 is the definition sketch of the exchanging procedure.

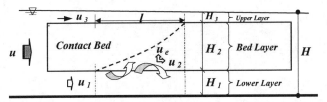

Fig. 2   Definition Sketch of the Exchanging Procedure

Suppose that through a length of l, the lower layer water begins to appear in the top of the fixed film bed, this procedure takes time t, then

$$t = \frac{l}{u_2} = \frac{H_2}{u_e} \tag{1}$$

During the time of t, the total water volume passed through lower layer in a specific width flume V_d is

$$V_d = u_1 t H_1 \tag{2}$$

The exchanged water volume in a specific width flume V_e may be approximately calculated as follows

$$V_e = \frac{1}{2} H_2 l \tag{3}$$

Thus the unexchanged water (the short-circuiting volume) in a specific width flume V_s is

$$V_s = V_d - V_e = u_1 t H_1 - \frac{1}{2} H_2 l \tag{4}$$

(4) is divided by t, then the short-circuiting flow Q_s is obtained in terms of flowrate:

$$Q_s = u_1 H_1 - \frac{1}{2t} H_2 l \tag{5}$$

Substituting the parameter t of (1) into (5), results in the following set equations:

$$\begin{cases} Q_s = u_1 H_1 - \frac{1}{2} u_2 H_2 & (6a) \\ Q_s = u_1 H_1 - \frac{1}{2} u_e l & (6b) \end{cases}$$

here $Q_1 = u_1 H_1$, is the apportion of flows in lower layer; $Q_2 = u_2 H_2$ is the flowrate in bed layer, and $Q_e = u_e l$ is the exchanged waters, so Eq. (6) is equals to

$$\begin{cases} Q_s = Q_1 - \frac{1}{2} Q_2 & (7a) \\ Q_s = Q_1 - \frac{1}{2} Q_e & (7b) \end{cases}$$

Therefore, the conditions that no short-circuiting flow occurs are:

$$\begin{cases} \dfrac{Q_2}{Q_1} \geq 2 & (8a) \\[2mm] \dfrac{Q_e}{Q_1} \geq 2 & (8b) \end{cases}$$

or

797

$$\begin{cases} \dfrac{u_2}{u_1} \geq \dfrac{2H_1}{H_2} & (9a) \\[3mm] \dfrac{u_e}{u_1} \geq \dfrac{2H_1}{l} & (9b) \end{cases}$$

The first item of Eq. (8) and Eq. (9) is the flow structure condition of the flume, the second term is related to aeration in the flume. Introducing the total flowrate Q, Eq. (8) then changed as:

$$\begin{cases} \dfrac{Q_2}{Q} \geq 2\dfrac{Q_1}{Q} & (10a) \\[3mm] \dfrac{Q_e}{Q} \geq 2\dfrac{Q_1}{Q} & (10b) \end{cases}$$

Suppose η is the ratio of air to water and $\eta = KQ_e/Q$, here K is the coefficient relating to the aeration rate and the flowrate in the flume. Thus the second term becomes:

$$\eta \geq 2\frac{1}{K}\frac{Q_1}{Q} \tag{11}$$

in which $k \geq 1$.

If the two terms in (8) - (9) are met at the same time, then flows would reach completely mixed regime. The other way round, only one term of the (8) - (9) is not met, it must exist short-circuiting flow in the flume, and the short-circuiting rate is expressed as:

$$Q_{sD} = Q_1 - \frac{Q_2}{2} \tag{12}$$

Similarly, obtained the conditions of water mixing completely between the upper layer and the bed layer waters:

$$\begin{cases} \dfrac{Q_2}{Q_3} \geq 2 & (13a) \\[3mm] \dfrac{Q_e}{Q_3} \geq 2 & (13b) \end{cases}$$

Here $Q_3 = u_3 H_3$. If the above conditions are not met at the same time, the short-circuiting rate in the upper layer flow is

$$Q_{sU} = Q_3 - \frac{Q_2}{2} \tag{14}$$

Usually, the flowrate in lower layer is greater than that in the up layer, so if the second term of Eq.(8) is met, the second term in Eq.(13) is met automatically. On the other hand, for the mixing length of the lower layer flow l is longer than that of the upper layer flow, if (8) - (13a) are not met at the same time, the total short-circuiting flowrate could be obtained in the following form:

$$Q_s = Q_{sD} + Q_{sU} = Q - 2Q_2 \tag{15}$$

3.3   TOUCHING TIME AND THE LENGTH OF THE FLUME

The touching time is the average time in statistical meanings, it is chosen according to the aims of the treatment and the practical experiences of the same kind process flumes. When the treatment flowrate and the touching time are determined, the cross section of the flume could be initially designed. Then Based on these data, the length of the flume could be calculated.

If water mixes completely in the flume, the length of the flume L is calculated by

$$L = u_2 T \frac{Q}{Q_2} \qquad (16a)$$

Here T is the necessary touching time. Eq.(16a) could be expressed in another way:

$$L = u \frac{u_2}{u} T \frac{Q}{Q_2} = uT \frac{H}{H_2} \qquad (16b)$$

If there exists the short-circuiting flow in the flume, then L is:

$$L = u_2 T \frac{(Q + Q_s)}{Q_2}$$

$$= u \frac{u_2}{u} T \frac{(Q + Q_s)}{Q_2} = 2uT \frac{H - \frac{u_2}{u} H_2}{H_2} \qquad (17)$$

4.   EXPERIMENT

The experiment was conducted in a 30.0 m long, 1.0 m wide and 4.3m deep flume in nature and raw water was used in the experiment (Lai Yifeng 1997). The contact bed is made of plastic thread, and was laid 20.0 m long in the flume. The main size of the cross-section is: H_1=0.55 m, H_2=3.0 m, H_3=0.45~0.25 m (at the front flume is 0.45 m, at the tail is 0.25 m). Need to say, the testing model was designed based on an actual water supply works, all the dimensions except the width and length are the same with the practical planning. The treatment flowrate of the prototype is 0.309 m^3/s.m, the necessary touching time is 36.3 minutes, and the flume length of is 240.0 m. The hydraulic behaviors in the flume were measured and calculated, the results were listed in Table 1.   They obviously revealed that the flowrate in the upper and lower layers are greater than that in the bed layer. Although the second terms of (8) - (13) may be met, but the first terms are not met in this case, thus there must exist the phenomenon of short-circuiting flow. In order to know water exchanging conditions, eight measuring sections were laid out along the flume, and each testing section have sixteen water samplers. By using Rodamine as tracer, the tracer concentration was detected. Fig.2 is the distribution of the concentration along the flume.

Using the Eq.(15) we could calculated the short-circulating flow rate Q_s is 0.1681 m^3/s.m, then we calculated it once more according to the following equation based on the detected concentration:

$$Q_{sTEST} = Q_1 \left(\frac{C_{2D}}{C_{1D}} + \frac{1}{2} \frac{u_2}{u_1} - \frac{1}{2} \frac{C_{2D}}{C_{1D}} \frac{u_2}{u_1} \right) + Q_3 \left(\frac{C_{2U}}{C_{1U}} + \frac{1}{2} \frac{u_2}{u_3} - \frac{1}{2} \frac{C_{2U}}{C_{1U}} \frac{u_2}{u_3} \right) \qquad (18)$$

where C_{1U}, C_{2U} is the concentration of the beginning and the end sections in a mixed length l as shown in Fig. 2 during putting tracer in upper layer; C_{1D}, C_{2D} is the values of the status during the lower layer.

Q_{sTEST}=0.1748 m^3/s.m, the error value with the above calculated Q_s is 3.99 %. Using Eq.(17) we calculate the touching time in 240 m long flume, T=24.59 minutes, the value is 32.26 % shorter than the designing touching

Run conditions	Table 1 Hydraulic Behaviors in the Flume (1)Q=0.309m³/s.m (2)u=0.0813m/s (3)Ratio of air to water: 1:1					
Water depth H (m)		383.79	384.22	401.55	403.23	421.39	437.70
Q_{layer}/Q_{total}	Upper layer	29.8	28.9	39.0	39.6	54.8	62.7
	Bed layer	22.8	25.4	17.4	16.7	8.9	5.2
(%)	Lower layer	47.4	45.7	43.6	43.7	36.3	32.1
Mean velocity	Upper layer	0.320	0.308	0.259	0.254	0.255	0.234
	Bed layer	0.023	0.026	0.018	0.017	0.009	0.005
(m/s)	Lower layer	0.361	0.348	0.332	0.333	0.277	0.245

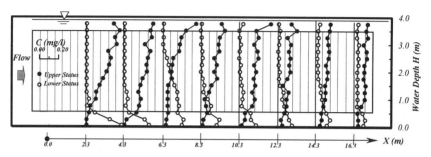

Fig. 3   Tracer Concentration Distributions in the Flume

time; If maintaining the necessary touching time, the length of the process flume needs 354.29 m, is 47.6% longer than the designing value.

5.   CONCLUSIONS

For some simplifying assumptions were made in deriving the criteria of completely mixed regime, mostly, these involved the use of average quantities in calculating various terms such as flowrates. The effect of these assumptions should be minor in practical engineering, of course, though it has compared well with the experiment described in this paper, its effect should be examined in great detail in future studies. On the other hand, the relations between the factors K and η, the resistance of contact bed and the flowrate distributing in the progressive flow flumes, should be studied more so that engineers could design the flumes efficiently.

6.   ACKNOWLEDGMENTS

The experiment was financially supported by the East River-Shenzhen Water Supply Bureau of Guangdong. The writers are grateful for every member of the research group and the State Research Center of Municipal Pollution Control of China in Tongji Universuty for their assistance in the experiment.

7.   REFERENCES

Lai Yifeng et al.(1997). *Hydraulic Model Study in Nature and Laboratory on Sediment Deposition and Dispo-sal Measures in the Biological Contact Oxidation Flume of the E.S. Water Supply Project,* Guangdong Research Institute of Water Conservancy & Hydropower.

Lai Yifeng (1997).   Water *Mixing and Exchanging Characteristics in the Biological Contact Oxidation Flume of the E.S. Water Supply Project,* Guangdong Research Institute of Water Conservancy & Hydropower.

4.3 Sediment management

Environmental Hydraulics, Lee, Jayawardena & Wang (eds) © 1999 Balkema, Rotterdam, ISBN 90 5809 035 3

River catchment sanitation by sediment management – A Flemish case study

M. Huygens & R. Verhoeven
Hydraulics Laboratory, University of Gent, Belgium

C. Janssens & M. Vangheluwe
Laboratory of Aquatic Research in Aquatic Pollution, University of Gent, Belgium

ABSTRACT: In the framework of an integral water management-policy, the sanitation of the water courses in the Flemish region is developped through small-scale projects to study local river catchments in a global environmental way. Indeed, the complex processes need a multidisciplinary approach to guarantee a successful implementation of water resources strategies. In the Zwalm-catchment, a small river catchment situated in the interfluvium between upper-Scheldt and Dender in the South-Western part of Flanders, this basic philosophy is explored in a pilot-project. Local morphological circumstances requires a study with focus on the sediment transport and its management.Indeed, a combination of sand-mud river beds, severe land erosion and discrete flood flows direct the sanitation program at a sediment management project. Sedimentation and transport of fine materials, as a favourite adhesive medium for pollutants, induce a general pollution of both water and river bed in the downstream reaches. In order to reduce the high treatment costs of the dredged (mud) material, some specific arrangements are necessary to realize an economical sanitation procedure.

1. INTRODUCTION

The general environmental policy in the Flemish region strives for an integral watermanagement to tackle the sanitation of the water courses. Small scale projects, based on this integrated conduct strategy, are developped to realize the global water sanitation of the Flemish river systems. Sand-mud river beds, severe run-off erosion and important flood flows direct the management study to a sediment transport evaluation. Indeed, mainly moveable sediments induce a general pollution of water and river bed in the downstream reaches of the water system. Therefore, the principal targets of the study can be formulated as :

- Sediment transport identification map of the river catchment; including the fractional distribution in fine (mud) and coarse (sand) materials with focus on the level of pollution for each sediment.
- Spatial and temperal evolution of the sediment transport over the region.
- Practical measures to avoid the mobility of sediments in the river environment.
- Engineering to obtain a separation of fine (mud) material and sand particles
- Formulation of sanitation strategies, applicable to other river catchments than the pilot project study area.

2. THE "ZWALM" RIVER CATCHMENT

The river Zwalm is a small tributary of the river Scheldt, situated in the South-Western part of Flanders in the interfluvium between the upper-Scheldt and Dender. The Zwalm-brook rises in the hill country around the city of Ronse (about 146 m above sea level) and debouches into the river Scheldt about 20 km upstream of Gent (8 m above sea level). The longitudinal trajectory of the main stream of the catchment is about 22 km long.

The Zwalm catchment (total surface = 116 km^2) is chosen as a pilot-project due to the intensive measuring activities in this area over the last years. Beside the extensive database on water quantity and quality, a detailed topographic and run-off characterisation is recorded for the river basin. Due to the relatively steep slopes in the upper reaches of the catchment and the local bottom structure (sand-silt-mud), the river environment is very sensitive to erosion and sediment transport. This local morphological conditions focus the project on a sediment transport management study, as a representative example for the Flemish river systems.

As a result the study is directed to four main objectives as a direct response to the formulated targets :
1. Detailed inventory of waterquantity, -quality and sediment transport characteristics.
2. Practical management measures to avoid erosive actions in the catchment (run-off, river bed degradation, land slide,...).
3. Engineering solutions for a proper sedimentation of respective sediment fractions.
4. Extrapolated practical applications in similar river catchments.

A tight collaboration between all partners of the project, driven by a stimulating governmental principal, ensures a succesfull multidisciplinar approach.

3. SEDIMENT TRANSPORT REGISTRATIONS

As already indicated before, due to local conditions on integral water resources engineering in the Zwalm-catchment shall be directed to a global sediment management approach. All aspects (from supply by erosion, over transport mechanisms to possible settlement of sediments) are covered in the monitoring and inventory program. Since flowing characteristics in the main brools of the catchment form the principal driving forces to transport and sedimentation of materials, a detailed inventory and on-line registration in discrete limnimetric measuring stations of local water levels and associated discharges are explored. As a

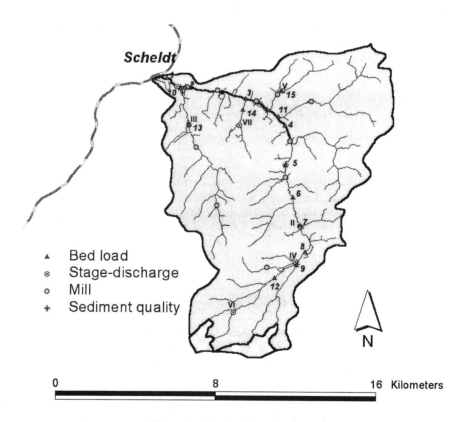

Figure 1: Zwalm river catchment

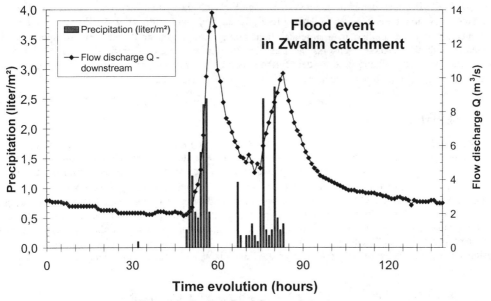

Figure 2: On-line flood flow registration.

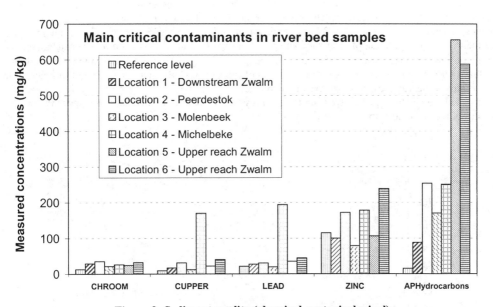

Figure 3: Sediment quality (chemical-ecotoxicological)

result of statistical analysis of the available limnimetric data (over period 1982-1997) the following flow parameters can be identified : $Q_{1\%} = 8.12$ m^3/s and $Q_{50\%} = 0.74$ m^3/s. An illustrative example of the on-line registration of a flood event on the downstream Zwalm-branch shows the sharp and quick response of both water level and discharge to the rain fall (figure 2).

Beside this traditional hydrodynamic records, a specific measuring program, to record sediment transport phenomena over the catchment, is developped. By register in 10 discrete locations on representative river branches of the catchment a complete view on the sediment mobility is formed. Sediment erosion, mobility and settlement is covered for the river system, together with the spatial dispersion of sediment material characteristics and their quality. A specially designed bed load sampler « BEDMAN » and specific turbidity sensors are applied in a practical field configuration to record sediment transport volumes. Together with the turbidity sensors, the traditional pump method to record the suspended sediment concentrations is refined and used to identify the general quality of the sensor registrations. The collected data lead to a general picture of the sediment balance over the river catchment.

4.   SEDIMENT QUALITY

Sediment and river bed quality is identified by both a chemical and eco-toxicologic analysis of the collected samples. Indeed, chemical analysis shows the actual state of contamination ; while the ecotoxico-logical study of the sediments reveals the potential availability of the present contaminants to quantify toxical rish potentials for the river environment. Next to a biological evaluation, both analysis results are integrated in a so-called TRIADE-judgement system. As indicated in figure 3 above, the main pollution in the river bed samples is due to the high level or organochloride pesticides and apolar hydrocarbon. The ammonium contamination (probably due to an intensive manuring) seems to be the highest threat for the river environment. As a result, the high level of pollution of the transported sediments in the Zwalm-river systems mortgages the engineering plans to create flood plains along the river branches.

5.   QUANTITATIVE EVALUATION OF SEDIMENT TRANSPORT

A detailed sieve analysis of respective sediment samples of both suspension fraction and bed material reveals a mainly silt-mud character. The sand fraction is very limited in all collected sediment samples. The mean grain size is given as $d_{50} = 30$ μm, while the organic content have a mean value of about 4 %. As a result, an explicit separation of clear sand and polluted fines is not really relevant for the catchment.

Bed load transport is recorded with the bed load sampler « BEDMAN ». Extensive laboratory calibration tests lead to a proper efficiency factor (about 20 %) of the sampler. By that, the collected samples during discrete measuring campaigns along the main river branch can be recalculated to an effective bed load transport volume in the record cross-section. Representative results of bed load transport in the main branches of the catchment are given in figure 5 below.

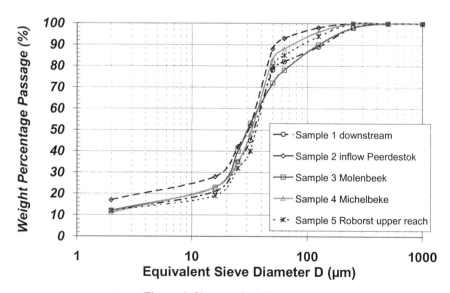

Figure 4: Sieve analysis Zwalm-sediments

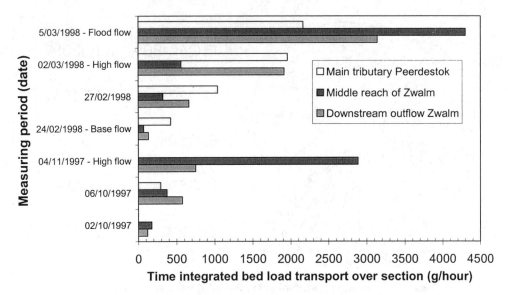

Figure 5: Bed load registrations with BEDMAN.

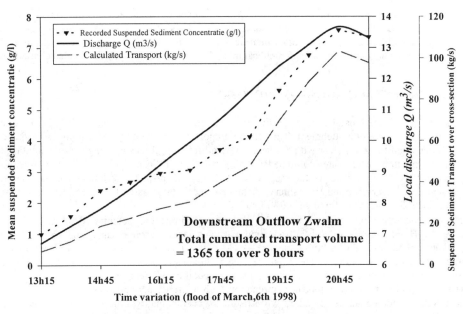

Figure 6: Suspended load under flood conditions.

Suspended sediment concentrations are measured by a calibrated turbidity sensor in discrete record spots over the cross-section to evaluate the concentration distribution. Due to the fine material, a quite uniform dispersion over the complete cross-section is found. From that, one single registration (in the middle of the cross-section, at 0.6 h under the water surface) as the mean value for the associated cross-section is valuable to quantify the total suspended load. A typical evolution of the suspension concentrations with the water flow conditions (flood) is shown in figure 6.

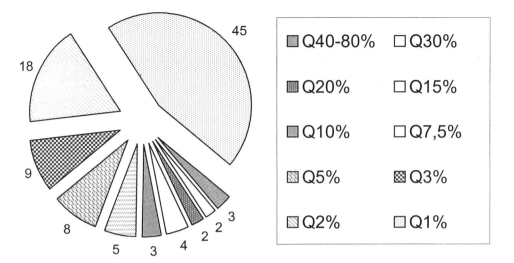

**Figure 7: Procentual contribution of suspended sediment transport volume
in relation to the flow regime (percentiel discharge Q%).**

Comparing the absolute transport volumes, it is clearly indicated that the suspension fraction is much more important than the bed load under flood conditions. At flood conditions, the transport of sediment in suspension is about 10 to 20 times higher than the bottom transport. Spatial distribution from upstream to downstream can't be identified : all sediment samples show similar randomly distributed characteristics.

6. SEDIMENT MANAGEMENT PROPOSAL

Based on all collected data and its preliminar analysis, some management strategy can already be formulated for the Zwalm-catchment. The relative importance of the absolute volume of suspended load during flood conditions, together with the very low sand fraction, forces any engineering solution to focus on the fine fraction in the water column. By that, separation of clean sand particles is not economically applicable. It is clearly indicated that mainly flood flows will move the sediment volume over the river catchment during very limited time intervals. Indeed, the biggest part of the sediment transport occurs during a small, concentrated flood period.

The high contamination level of this mud/silt material limits the potential deposition areas for the sediments. By that, separation of clean sand particles is not economically applicable. An effective extrampent of the polluted silt/mud fraction as close as possible to its origin should be worked out to sanitate the river environment. The input of fine sediments to the river by local land erosion in the upstream (rather steep) reaches should be avoided. A specific sediment-oriented management specifically towards the polluted mud fraction, in combination with a strictly applied source-oriented sanitation plan and a well-considered land-use policy will ensure a general quality improvement of the Zwalm river environment.

7. REFERENCES

ASTM (1993) Standard Guide for conducting sediment toxicity tests with fresh water.
Biswas A.K. (1997) « Water resources – environmental planning, management and development »,
  McGraw-Hill New York.
Misganaw D. (1996) « Sediment load during flood events for Illinois streams »,
  Water International Vol. 21, p. 131.
Raudkivi A.J. (1990) Loose Boundary Hydraulics 3rd edition, Pergamon Press New York

Van Rijn L.C. (1986) « Manual on sediment transport measurements », Delft Hydraulics Laboratory
Van Rijn L.C. (1993) « Principles of sediment transport in rivers, estuaries and coastal seas »,
 Aqua Publication Amsterdam

Environmental Hydraulics, Lee, Jayawardena & Wang (eds) © 1999 Balkema, Rotterdam, ISBN 90 5809 035 3

Disposal of contaminated mud in Hong Kong: A review of the environmental monitoring programme

A.L.C.Chan & A.Dawes
Water Policy and Planning Group, Environmental Protection Department, Hong Kong, China

ABSTRACT: Hong Kong has undertaken very extensive dredging works in recent years. It is estimated that in the 10-year period up to 2002, a total of some 40Mm³ of material classified as contaminated will have required dredging and disposal elsewhere. Contaminated dredged material has been disposed of into pits excavated in the bed of the Pearl River Estuary in an area of relatively shallow water depth and low current velocity since 1992. These pits are subsequently capped to ensure isolation of any potential pollutants from the wider marine environment. The disposal operations are subject to a rigorous monitoring and audit programme in order to confirm the effectiveness of the disposal strategy in containing contaminated material and to demonstrate its environmental acceptability. The monitoring work has included testing of the integrity of pit caps, sediment plume tracking, analysis for trace contaminants in sediments, marine water and biota, alongside study of benthic and epifaunal ecology. The key findings of the environmental monitoring programme are presented and the way forward for future work is discussed.

1. INTRODUCTION

Historically inadequate waste disposal arrangements have resulted in extensive contamination of Hong Kong's inland and marine sediments with pollutants originating from domestic, industrial and agricultural waste. Other non point sources such as road run off have also contributed to the pollutant load. Much of this mud must now be removed as part of the effort to sustain Hong Kong's economic development. In particular, many major reclamation projects have been undertaken in Hong Kong in recent years necessitating large scale dredging of sediments from waters adjacent to the territory's urban core. Dredging has also been undertaken for many other purposes, most notably to maintain marine navigation routes and port facilities and for improved flood protection of inland waterways.

Contaminated mud dredged in Hong Kong is disposed of into seabed pits at a site east of Sha Chau Island in the territory's North Western Waters. These pits are subsequently capped to ensure isolation from the wider marine environment. Notwithstanding Government policy that mud should be left in place whenever possible, it is estimated that some 40Mm³ of contaminated mud will have required disposal over the 10 year period 1992 - 2002. Environmental monitoring has been undertaken in and around the site to confirm that the disposal activities do not adversely affect the marine environment.

2. HISTORY OF CONTAMINATED MUD PITS IN HONG KONG

In response to a huge forecast volume of dredged material requiring disposal over the coming decade, the Environmental Protection Department (EPD) initiated the Contaminated Spoil Management Study (Mott MacDonald, 1991). The study recommended dredging, disposal and monitoring methods as part of an overall management system and provided the foundation for a sediment classification scheme that was adopted by the EPD in 1992 (Table 1). The study identified the use of seabed pits as the most suitable

Table 1   Criteria adopted by Hong Kong in 1992 to identify contaminated sediment

Metal	Cd	Cr	Cu	Hg	Ni	Pb	Zn
dry wt (ppm)	> 1.5	> 80	> 65	> 1.0	> 40	> 75	> 200

disposal option for contaminated sediment. Following evaluation of a number of possible sites Government decided that contaminated mud should be disposed of in dredged pits excavated specifically for the purpose at East Sha Chau (ESC). The principal attributes of this site were its shallow water depth and relatively low current velocities, which minimise the likelihood of sediment loss to the water column during disposal.
The first contaminated mud pit began receiving spoil in December 1992. Since that time a series of similar purpose dredged pits have now been filled and capped. Since November 1997 disposal of contaminated mud has been undertaken in an adjacent large exhausted borrow area known as CMP IV. This pit was formerly used to extract sand for the formation of the new Hong Kong International Airport and is expected to provide suitable disposal capacity for contaminated mud arisings until the year 2002.

Prior to operation of CMP IV, a detailed environmental impact assessment was undertaken to fully examine the nature and extent of possible environmental impacts (ERM, 1997). This work formed the basis for the facility operations plan that was subsequently implemented (ERM, 1998). In its conclusion, the EIA study laid down the following impact hypothesis:

> *Impacts associated with disposal of contaminated mud in the East Sha Chau CMP IV are not expected to result in exceedances of water quality objectives at sensitive receivers nor cause exceedances of applicable water quality standards. The operational design has been specified such that disposal of sediment quality shall not cause a detectable deterioration in sediment quality outside CMP IV. Physical impacts to fisheries and marine ecological sensitive receivers are not expected and no changes in contaminant levels in marine organism tissue are predicted to arise from this project.*

The hypothesis was tested predictively in the EIA study and will be verified through the currently ongoing environmental monitoring and audit programme (ERM, 1998).

3. ENVIRONMENTAL MONITORING PROGRAMME

The work has been undertaken by consultants working under the guidance of an interdepartmental working group, the membership of which includes technical experts from the Civil Engineering, Environmental Protection and Agriculture and Fisheries Departments and the Government Laboratory. The principal elements of the environmental monitoring programme are sediment and water quality, ecological community structure, biotic tissue contamination and ecotoxicology. The monitoring effort will continue for at least two years after completion of disposal and capping operations at East Sha Chau.

3.1 Water Quality

Salinity, temperature, DO and suspended solids concentrations of water have been routinely measured (Binnie and CES, 1994, 1995; CES and Binnie, 1994 and Mouchel, 1998). The results demonstrate the estuarine and seasonally variable nature of the site. The water temperature ranges between 15 °C and 30 °C over an annual cycle. The DO content of the water is generally between 4 and 10 mg/l (Table 2). The pH of the water typically varies between 8.1 and 8.4 (Lam, 1994). It has been reported elsewhere that biochemical oxygen demand (BOD) and nutrient levels in the North West Water Control Zone (including the ESC area) are only marginally influenced by local anthropogenic inputs (Lam, 1994). The dominant influence is the heavy nutrient load carried by the Pearl River. Suspended sediment concentrations measured at mid-water

Table 2 Ranges and means of some hydrological parameters, water metal levels and
sediment metal concentrations at the East Sha Chau contaminated mud disposal site.

		1994	1995	1996	1997
Water					
Temperature (°C)	Min-Max	16-30	15-28	16-30	15-29
	(Mean)	(22.8)	(22.7)	(23.7)	(22)
Salinity (ppt)	Min-Max	10-31	10-32	8-34	12-32
	(Mean)	(25.8)	(25.8)	(26.1)	(26.9)
Suspended solids	Min-Max	10-78	8-80	5-65	3-45
(mg/l)	(Mean)	(40)	(34.9)	(26.6)	(18.6)
D.O. (mg/l)	Min-Max	--	--	4-9.2	4-10
	(Mean)	--	--	(6.69)	(6.68)
Cadmium (Cd)	Min-Max	0.0001-0.0005	0.0015-0.006	0.0002-0.006	0.0003-0.006
(mg/l)	(Mean)	(0.0002)	(0.0029)	(0.0007)	(0.0015)
Chromium (Cr)	Min-Max	0.0012-0.002	0.002-0.008	0.0015-0.02	0.0025-0.06
(mg/l)	(Mean)	(0.0016)	(0.005)	(0.007)	(0.011)
Copper (Cu)	Min-Max	0.0018-0.007	0.0007-0.02	0.0006-0.005	0.0007-0.004
(mg/l)	(Mean)	(0.0035)	(0.0084)	(0.0024)	(0.0018)
Mercury (Hg)	Min-Max	0.0008-0.0018	0.0005-0.007	0.00003-0.0004	0.00006-0.0002
(mg/l)	(Mean)	(0.0013)	(0.0029)	(0.0002)	(0.0001)
Nickel (Ni)	Min-Max	0.0028-0.0045	0.005-0.8	0.005-0.1	0.005-0.1
(mg/l)	(Mean)	(0.0031)	(0.025)	(0.015)	(0.02)
Lead (Pb)	Min-Max	0.0001-0.011	0.005-0.04	0.002-0.006	0.003-0.006
(mg/l)	(Mean)	(0.005)	(0.017)	(0.003)	(0.004)
Zinc (Zn)	Min-Max	0.005-0.016	0.03-0.06	0.007-0.03	0.007-0.04
(mg/l)	(Mean)	(0.009)	(0.048)	(0.019)	(0.012)
Sediment					
Cadmium (Cd)	Min-Max	--	0.05-0.38	0.04-0.4	0.05-0.6
(mg/kg)	(Mean)		(0.19)	(0.1)	(0.12)
Chromium (Cr)	Min-Max	14-40	9-49	5-52	11-38
(mg/kg)	(Mean)	(27.3)	(24.9)	(25.8)	(28)
Copper (Cu)	Min-Max	10-45	12-75	5-55	8-40
(mg/kg)	(Mean)	(26.7)	(24.8)	(17)	(25.3)
Mercury (Hg)	Min-Max	--	--	0.001-0.6	0.006-0.11
(mg/kg)	(Mean)			(0.054)	(0.04)
Nickel (Ni)	Min-Max	12-33	13-36	2-20	5-25
(mg/kg)	(Mean)	(22.4)	(21.5)	(11.6)	(15.9)
Lead (Pb)	Min-Max	35-61	28-58	8-48	18-70
(mg/kg)	(Mean)	(47)	(41.9)	(26.6)	(32.7)
Zinc (Zn)	Min-Max	45-140	30-105	30-90	35-95
(mg/kg)	(Mean)	(92.2)	(85.6)	(53.3)	(75.3)
TOC (% dry wt)	Min-Max	0.3-1.6	0.26-1.3	0.3-2.5	1.3-4.8
	(Mean)	(0.90)	(0.95)	(1.02)	(3.14)

depth over the study period have varied from less than 10ppm to around 80ppm which is in line with
historical data reported by EPD. Suspended solids concentrations at the seabed are very variable, reaching
levels as high as 1000 mg/l, but exhibit no obvious trends.
Mid-depth water samples collected from sites adjacent to active pits have been analysed for suspended

sediment concentrations and heavy metals in solution and adsorbed onto sediment in suspension. The results are summarised in Table 2. The data do not display any correlation between detected metal concentrations and the capping and disposal operations.

3.2 Sediment

Seabed sediment has been sampled and chemically analysed at bimonthly intervals for heavy metals and TOC (Binnie and CES, 1994, 1995; CES and Binnie, 1994 and Mouchel 1998). The results are presented in Table 2. Copper, Chromium, Lead, Nickel and Zinc were detected in almost all samples whilst Cadmium and Mercury concentrations were below the limits of detection for a significant proportion of the samples. The limits of detection were set to detect concentrations at approximately 10% of the sediment quality criteria adopted by EPD to characterise contaminated sediment. No obvious increasing trend in heavy metal concentration has been detected over the study period.

3.3 Benthic Communities

Sampling of benthic invertebrates has been conducted at ESC since May 1993 (Binnie and CES, 1994, 1995; CES and Binnie, 1994 and Mouchel 1998). The fauna across the survey area is impoverished in both population density and species richness when compared with records from sedimentary habitats elsewhere in Hong Kong (Shin & Thompson, 1982). This is attributed to the estuarine nature of the area, which is influenced by the Pearl River outflows particularly in the wet season. Nevertheless, some 150 species have been identified in the trawl catches. Polychaetes are by far the dominant macrobenthic organisms in the survey area both in terms of number of species and individual numbers. 80% of the individuals recorded were polychaete specimens, 11% were crustaceans and other individual faunal groups made up only 1% or less of the total specimens collected.

High variability in benthic abundance near this site was recorded by Shin (1989) and the ability to differentiate the effect of disposal activities from natural variation is confounded by this inherent high variability. The mean numbers of species and individuals at each station has ranged from 1.6 to 4.5 and 1.7 to 10.6, respectively. The mean species diversity, H', was low (between 0.35 and 1.08) as was the mean species richness, SR (between 1.35 and 2.12), over the survey period. Species evenness, J, has remained in the range 0.86 to 0.98. There are no obvious spatial trends and no evidence that sites in the vicinity of the pits are impacted compared to more distant reference sites.

Samples from capped pits have been compared to the ambient seafloor since September 1994. The pit caps usually yield fewer benthic invertebrates and fewer taxa than the surrounding ambient seafloor but nevertheless recolonisation is evident.

3.4 Demersal (Trawl) Survey

Over 275 species of fish belonging to 55 families have been caught during the monitoring period. Five families of commercial value comprised 59% by number and 51% by weight of the total catch. Within these five families, members of the Leiognathidae (slipmouth) were the most abundant in terms of number, and members of the Platycephalidae (flathead) comprised the largest proportion by weight. Members of the Sciaenidae (croaker) were the second most abundant in terms of both numbers of individuals and weight, followed by the Cynoglosside (flatfish). The remainder of the catch was dominated by members of the Gobiidae and Centropomidae families.

Over 230 invertebrate species belonging to 41 families were obtained in the field survey. Families of commercial importance comprised 17% of the catch by number of individuals and 18% by weight. Overall, members of the shrimp family Penaeidae were predominant by numbers of individuals, and squillidae (mantis shrimps) by weight. Of the other invertebrate families caught, members of the Portunidae (swimming crabs) were predominant by both number and weight.

814

Generally, the number of species caught was low from January to June and increased dramatically in July reaching a peak around July to August. This was followed by a gradual decrease until December. This pattern was relatively consistent over the sampling period. In terms of abundance (number of individuals), a similar but less marked pattern was observed.

The community Diversity Index, H', has been calculated based on the catch data. The index values show little spatial variation and no temporal trend over the sampling period. Diversity is generally high for most of trawl stations. The calculated Evenness Index, J', data shows a medium to high evenness of the spread of individuals among species in most stations suggesting that no one species was numerically dominant in the communities sampled. There is little difference in evenness between the reference and pit areas. No clear pattern of disturbance or community damage was evident from the community analysis results. It is concluded that contaminated mud disposal is not affecting the demersal community.

3.5 Tissue Metal Concentrations

Relatively higher levels of copper have been found in crustaceans than in fish. Crustaceans naturally bio-regulate copper, which is an essential component of the respiration pigment haemocyanin. Chromium concentrations in some organisms have been found to be above the relevant Hong Kong Food Adulteration Regulations limit. However there is no evidence of any spatial trend and no indication that samples from the pits are more contaminated than samples obtained from reference stations distant from the site. Concentrations of other metals have been consistently low and not found to warrant concern.

4. CONCLUSIONS

As would be expected, suspended solids concentrations have been observed to increase during disposal. However, the increases are transient and usually not large relative to natural fluctuations. The monitoring work to date indicates that the disposal activities are not having a detectable environmental impact attributable to the loss of any contaminants of concern. Metal levels in the water column have not been elevated relative to reference locations as a consequence of disposal operations. The disposal of contaminated mud has not resulted in deterioration in sediment quality or contaminant levels in biota. There are no obvious spatial differences in ecological community structure attributable to the dumping. Recolonisation has already commenced at previously filled pits.

5. FUTURE DIRECTIONS

Early work in the programme essentially involved routine data collection intended to allow elucidation of any adverse temporal trends. This approach has been progressively refined with experience. More recent work is specifically designed to verify impact hypotheses arising from planning stage predictive studies. Scientifically testable hypotheses have been established and form the basis for statistically robust survey designs. The technical scope has been widened to encompass a broader range of contaminants of concern including organic micropollutants. Potential toxic impacts on sediments are assessed directly by ecotoxicological study. The field monitoring data is integrated with risk assessment and other evaluative processes to ensure that the programme delivers high quality information on the environmental acceptability of the disposal facility.

6. ACKNOWLEDGMENTS

The authors gratefully acknowledge the Director of Environmental Protection, Hong Kong SAR for permission to publish the results of the East Sha Chau Monitoring Programme. The views expressed in this paper are those of the authors and not necessarily those of the Government of the Hong Kong SAR.

7. REFERENCES

Binnie and CES (Consultants in Environmental Sciences [Asia] Limited). (1994 and 1995). *Environmental monitoring Programme East Sha Chau - Phase 2* (Agreement CE 35/93). Monthly progress reports from January 1994 to September 1995. Prepared for the Civil Engineering Department, Hong Kong Government.

CES and Binnie. (1994). *East Sha Chau Monitoring Programme*. Final Report (November 1992 – December 1993). Prepared for the Environmental Protection Department, Hong Kong Government.

ERM-Hong Kong, Ltd.. (1997). *Environmental Impact Assessment Study for Disposal of Contaminated Mud in the East Sha Chau Marine Borrow Pit*. Final report. Prepared for Civil Engineering Department, Hong Kong Government.

ERM-Hong Kong, Ltd.. (1998). *Environmental Monitoring and Audit for Contaminated Mud Pit IV at East of Sha Chau: Monitoring and Audit Manual*. Final report. Prepared for Civil Engineering Department, Hong Kong Government.

Lam, H.W. (1994). *Marine water quality in Hong Kong*: Results for 1993 from the marine monitoring programme of the Environmental Protection Department. Monitoring Section, Waste & Water Services Group, Environmental Protection Department, Hong Kong Government. Report no. EP/TR5/94.

Mott MacDonald (Mott MacDonald Hong Kong Ltd.). (1991). *Contaminated spoil management study*. Final report. Agreement CE 30/90. Volume 1. Prepared for the Environmental Protection Department, Hong Kong Government. Prepared by Mott MacDonald Hong Kong Ltd. in association with Dredging Research Ltd.

Mouchel-Hong Kong, Ltd.. (1998). *Environmental Monitoring and Audit for Contaminated Mud Pits II and III at East of Sha Chau*. Final Report (September 1995 – February 1997). Prepared for Civil Engineering Department, Hong Kong Government.

Shin, P.K.S. (1989). "Natural Disturbance of benthic infauna in the offshore waters of Hong Kong", *Asian Mar. Biol.* 6, 193-207.

Shin, P.K.S. and Thompson, G.B. (1982). "Spatial distribution of the infaunal benthos of Hong Kong", *Mar. Ecol. Prog.* 10, 37-47.

Environmental Hydraulics, Lee, Jayawardena & Wang (eds) © 1999 Balkema, Rotterdam, ISBN 90 5809 035 3

Development of Hong Kong's decision criteria for sediment disposal

Patrick C.K.Lei, Anthony W.K.Fok & Adrian Dawes
Environmental Protection Department of the Government of the Hong Kong Special Administrative Region, China

Peter G.D.Whiteside
Civil Engineering Department of the Government of the Hong Kong Special Administrative Region, China

ABSTRACT: In 1996, the Contracting Parties to the London Convention 1972 agreed to adopt a new protocol for assessing dredged material intended for marine disposal. The Hong Kong Government has conducted a comprehensive review of its existing sediment classification system to ensure its regulatory framework and practices are kept in line with international practices, as well as to optimise the allocation of its scarce disposal resources.

1. INTRODUCTION

For over a hundred years, Hong Kong has relied on marine reclamation to satisfy its demand for land to accommodate the needs of its increasing population. This has necessitated the dredging and disposal of vast quantities of marine sediment to improve the settlement characteristics of the reclaimed land, and to allow access to the underlying marine sand used to form land. The rate of land formation has increased dramatically in recent years. This has resulted in a corresponding increase in the rate of dredging which has generated an average volume of some 40 million cubic metres of marine sediment over the past four years (EPD, HK, 1997).

Traditionally, all dredged sediment was disposed of at open sea disposal sites. In 1992, Hong Kong introduced a sediment classification and disposal scheme in anticipation of the very large volume of dredged sediment arising from its major reclamation and airport projects (Mott MacDonald, 1992). This scheme requires all contaminated sediment be disposed of at a dedicated confined marine disposal site.

The 1992 scheme however has some drawbacks. Sediment is classified according to the level of a limited number of heavy metal contaminants, which is only a measurement of its potential toxicity and does not necessarily reflect the biological impact upon the marine ecosystem. The scheme also cannot identify the very toxic material which is not suitable for marine disposal. The scheme was reviewed in parallel with the adoption of the new protocol of the London Convention in 1996 (IMO, 1996).

2. EXISTING ARRANGEMENT FOR SEDIMENT CLASSIFICATION AND DISPOSAL

Hong Kong has a policy to minimise dredging activities by leaving marine sediment in place whenever possible (WB, HKG, 1992). This is necessary to preserve the limited disposal capacity and to reduce environmental impact. However, some dredging is still inevitable for port development and maintenance of navigational fairways and other facilities. As part of Hong Kong's programme of port, airport and urban development, an estimated 300Mm3 of marine sediment would have to be dredged between 1991 and 2000. Some 10% of the sediment might be too contaminated for open sea disposal (Whiteside et al., 1996). This necessitated the early development of a sediment classification scheme and additional marine disposal capacities for both contaminated and uncontaminated sediment.

In 1992, Hong Kong formally adopted a set of criteria to classify dredged sediment and identify sediment which is not suitable for open sea disposal (EPD, HK, 1992). For reasons of pressing necessity, the

Table 1   Hong Kong's existing criteria for open sea disposal of dredged sediment

Metal	Cd	Cr	Cu	Hg	Ni	Pb	Zn
ppm dry wt	<1.5	<80	<65	<1.0	<40	<75	<200

sediment criteria adopted were intended to be pragmatic and able to address immediate disposal problems. As such, the classification scheme was restricted to seven key metals (Cd, Cr, Cu, Pb, Hg, Ni, Zn) identified as particularly important contaminants in Hong Kong (Table 1). Although the effects of organic-micro pollutants were considered, these were not adopted at that time because of uncertainty regarding their likely toxicity and longer term impact.

Parallel to the development and implementation of the sediment classification criteria, additional marine disposal sites were identified to satisfy the projected demand. Suitable and exhausted marine sand borrow sites were converted into disposal sites. This also provided an opportunity to restore the seabed to its natural level. In 1992, a site in East Sha Chau area just north of the new airport was selected for the disposal of contaminated sediment (Mott MacDonald, 1992). This site was selected for its shallow water depth (5 to 6m) and the relatively low current velocities (CES and Binnie, 1993; Premchitt and Evans 1993). Contaminated sediment dumped in this site was covered by a thick layer of inert material, and the site itself was subjected to a very stringent environmental monitoring programme.

3. REVIEW OF THE EXISTING SEDIMENT CLASSIFICATION AND DISPOSAL ARRANGEMENT

The existing classification criteria have proved useful in identifying sediment containing contaminants arising from anthropogenic activity, but factors such as bioavailability and the likelihood of actual biological effects have not been considered. Therefore, sediment which exceeded any of these criteria has been classified as 'contaminated' irrespective of actual toxicity. This has provided a very environmentally secure but expensive approach to sediment management.

In 1996, the Contracting Parties to the London Convention 1972 agreed to adopt a new protocol for the assessment of dredged material (IMO, 1996). The protocol sets out generic guidelines for evaluating the need for dredging and disposal, dredged material characterisation, evaluation of disposal options, sea disposal site selection, impact assessment, permit issue and monitoring. To keep abreast of international developments and overseas practices, Hong Kong conducted a comprehensive review of the existing sediment management framework. The review comprised :

- An overview of the adequacy and appropriateness of the existing sediment classification criteria.
- An assessment of the environmental acceptability of using the existing sediment disposal site.
- An examination of recent sediment quality data to establish the local presence or absence of the full range of potential Contaminants of Concern.
- A literature review of the biological effects of chemical pollutants and sediment classification systems adopted in other countries.

The review (ERM, 1997; EVS, 1996a; EVS, 1996b) concluded that:

- The existing sediment disposal arrangement was environmentally acceptable.
- The existing sediment classification criteria should include a wider range of chemical pollutants, and should adopt biological tests to directly assess toxicity.

4. NEW SEDIMENT ASSESSMENT FRAMEWORK

In light of the review findings, Hong Kong has decided to replace its existing sediment classification criteria by a new Sediment Assessment Framework. The new Framework aims to more accurately determine the potential environmental impact of dumping sediment in the marine environment. It seeks to provide a stronger scientific basis to help decide on the most appropriate disposal option for dredged sediment by:

- broadening the existing chemical criteria to include all identified contaminants of local concern;
- using biological tests to assess the toxicity of contaminated dredged sediment; and
- identifying highly polluted sediment which requires special disposal.

The new Framework is based on a combination of chemical analysis to determine the presence of contaminants of concern and biological testing to directly assess sediment toxicity. The chemical data will provide an indication of whether contaminants are below levels at which biological effects have been reported or above levels at which biological effects would be expected. As biological testing is relatively expensive, it will mainly be undertaken for two purposes:

a)   To determine the toxicity of "moderately" contaminated sediment and its suitability for open sea disposal. In cases where the identified contaminants are not actually bioavailable then there is no environmental advantage in requiring this material to undergo the more expensive confined disposal.

b)   To determine the toxicity of the very small fraction of sediment which may be released in the process of confined disposal and settles in low concentrations outside the disposal site. This would facilitate decisions on the need for and scope of special treatment or alternative disposal arrangements.

For (a), Biological testing is relatively expensive and time consuming compared to chemical analysis. For pragmatic reasons, biological testing will not normally be required for material considered to have a high probability of either passing or failing the test. A Lower and an Upper Chemical Exceedance Levels (LCEL and UCEL) have been established to facilitate making such decisions. These levels were developed with reference to international co-ocurrence data (Long et al., 1995), Hong Kong's existing classification criteria and environmental monitoring data. Under this regime, sediment having chemical pollutant level below the LCEL is considered non-toxic and may be dumped into the open sea disposal sites without further analysis. Likewise, sediment which exceeds the UCEL is most likely toxic and would not be allowed for open sea disposal.

For (b), it is not possible to accurately determine the fate of the very small fraction of sediment released during the dumping operation. The new framework conservatively assumes that the dumped sediment would mix with the surrounding in-situ sediment and its pollutant concentration diluted to at least 10% of its original level. This assumption is based on a generic assessment of various factors including the hydrology of the confined marine disposal site with due consideration of the practicality of sample preparation. Under this regime, sediment having a pollutant concentration above 10 times that of LCEL is deemed potentially very toxic, and requires biological testing to help identify the most suitable disposal arrangement.

In line with the 1996 Protocol of the London Convention, the new Sediment Assessment Framework comprises a 3-tier screening procedure (Figure 1):

- Tier I screening of existing information.
- Tier II chemical screening of the level of eight specified heavy metals, Arsenic, PAHs, PCBs and TBT.
- Tier III biological screening of the toxicological responses of marine organisms .

Tier I   Screening of Existing Information

Tier I screening is conducted to establish whether available information is sufficient to make a decision on the disposal option without further testing. It comprises the examination of existing information on the characteristics of the sediment.

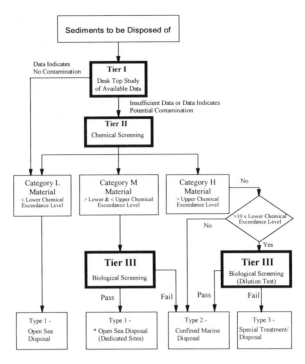

Figure 1.   Schematic Diagram of New Sediment Assessment Framework

If the available information explicitly indicates that the sediment to be dredged is not contaminated, the material can be classified as Uncontaminated Material. This material is acceptable for open sea disposal, and no more testing is necessary. Otherwise, the material must be subjected to Tier II Chemical Screening.

Tier II Chemical Screening

Tier II screening is conducted to determine whether the sediment is suitable for open sea disposal without further testing. It comprises the chemical analysis of the sediment for a suite of contaminants including eight specified heavy metals, Arsenic, PAHs , PCBs and, when necessary, TBT. Sediment will be categorised based on contaminant levels with reference to the Chemical Exceedance Levels (Table 2).
There are three categories of sediment :

Category L   Sediment with all contaminant levels below the Lower Chemical Exceedance Level is considered acceptable for open sea disposal, and no more testing is necessary.

Category M   Sediment with any one or more contaminant levels exceeding the Lower Chemical Exceedance Level and none exceeding the Upper Chemical Exceedance Level. The sediment must be subjected to Tier III Biological Screening before deciding on the marine disposal arrangement.

Category H   Sediment with any one or more contaminant levels exceeding the Upper Exceedance Level. The sediment is normally considered very contaminated and must be subject to confined marine disposal. However, if the sediment has exceeded the Lower Chemical Exceedance Level by more than a factor of l0, it must be subjected to Tier III Biological Screening to determine the need for special treatment or disposal arrangement.

Table 2 - Sediment Quality Criteria for the Classification of Sediment based on Chemical Contaminant Levels

Contaminants	Lower Chemical Exceedance Level (LCEL)	Upper Chemical Exceedance Level (UCEL)
Metals (mg/kg dry wt.)		
Cadmium (Cd)	1.5	4
Chromium (Cr)	80	160
Copper (Cu)	65	110
Mercury (Hg)	0.5	1
Nickel (Ni)	40	40
Lead (Pb)	75	110
Silver (Ag)	1	2
Zinc (Zn)	200	270
Metalloid (mg/kg dry wt.)		
Arsenic (As)	8	42
Organic-PAHs (µg/kg dry wt.)		
Low Molecular Weight PAHs	550	3160
High Molecular Weight PAHs	1700	9600
Total PAHs	4000	44800
Organic-non-PAHs (µg/kg dry wt.)		
Total PCBs	23	190
Organometallics (ug TBT/L in interstitial water)		
TBT	0.15	0.15

Table 3   Test endpoints and decision criteria for Tier III biological screening

Toxicity test	Endpoints measured	Failure criteria
10-days amphipod	Survival	Mean survival in test sediment < 90% mean survival in reference sediment and statistically significant (p≤ 0.05)[1] for test using *Leptocherius Plumulosus*, or Mean survival in test sediment < 80% mean survival in reference sediment and statistically significant (p ≤ 0.05)[1] for tests using *Ampelisca abdita* or *Eohaustorius estuarius*
20-days polychaete worm	Dry Weight	Mean dry weight in test sediment < 90% mean dry weight in reference sediment and statistically significant (p ≤ 0.05)[1]
48-hours larvae (bivalve or echinoderm)	Normality Survival[2]	Mean normality survival in test sediment < 80% mean normality survival in reference sediment and statistically significant (p ≤ 0.05)[1]

1. Statistically significant differences should be determined using appropriate two-sample comparisons (e.g., t-tests) at a probability of p ≤ 0.05.
2. Normality survival integrates the normality and survival end points, and measures survival of only the normal larvae relative to the starting number.

Tier III Biological Screening

Tier III biological screening is conducted to determine the most appropriate disposal arrangement based on the measured toxicity of the sediment. It comprises biological response tests (including survival rate, growth rate, and development deformity) for sediment using three specified marine organisms (amphipod, polychaete, bivalve or echinoderm). This screening is applicable to Categories M and H sediments, and is designed to check whether the sediment shows any adverse toxicological response.

The decision criteria for biological screening are set out in Table 3, and any sediment which fails the screening tests must be subjected to confined marine disposal to prevent any risk of adverse biological effects on the marine ecosystem. Special treatment or disposal arrangement may be required if the sediment is seriously contaminated and exhibits toxic effects even after dilution.

5. DISCUSSION

The new assessment framework will provide a more scientifically sound basis to classify the sediment and to determine the most appropriate disposal option. The framework is neither a relaxation nor a tightening of regulatory control but it will provide increased environmental security to protect the marine ecosystem from adverse impact, while at the same time ensuring the optimum use of the limited capacity of marine disposal sites.

Although the additional tests required by the new framework will result in higher analytical costs and lead time, these may be offset by savings elsewhere. A major benefit of the new framework is to identify sediment which although contaminated, for reasons of limited bioavailability, is not actually toxic. The non-toxic sediment may then be more cost effectively disposed at designated open sea disposal sites. Under the old framework, such sediment must be subjected to confined disposal at an unnecessarily high cost.

The success of the new framework depends very much on the reliability of the biological tests. At present, there is only limited local laboratory capability and no reliable tests based on indigenous local species. In the interim, it is necessary to rely on overseas protocols. In the longer term, several local tertiary institutions have plans to develop the techniques to conduct biological tests using local species. A database will also be maintained to enable future correlation analysis of chemical contaminant levels and toxicity.

ACKNOWLEDGEMENTS

The authors wish to express thanks to the Director of Environment Protection and the Director of Civil Engineering, the Government of the Hong Kong Special Administrative Region, for permission to publish this paper. The views expressed are those of the authors and do not necessarily reflect in any way official policy of the Government of the Hong Kong Special Administrative Region. Thanks are also expressed to many colleagues who are involved in the development of the new sediment assessment framework.

REFERENCES

Consultants in Environmental Science (Asia) Ltd (CES), and Binnie Consultants Ltd (1993). An Assessment of the Impact of Contaminated Dredged Material Disposal at Contained Disposal Facilities with Reference to East Sha Chau.

Environmental Protection Department, Hong Kong (EPD, HK) (1992). Classification of Dredged Sediment for Marine disposal, Technical Circular 1-1-92.

Environmental Protection Department, Hong Kong (EPD, HK) (1997). Environment Hong Kong 1997 (A Review of 1996).

Environmental Resources Management Hong Kong Ltd (ERM) (1997). Environmental Impact Assessment Study for Disposal of Contaminated Mud in the East Sha Chau Marine Borrow Pit.

EVS Environment Consultants (1996a). Classification of Dredged Material for Marine Disposal

EVS Environment Consultants (1996b). Review of Contaminated Mud Disposal Strategy and Status Report on Contaminated Mud Disposal Facility at East Sha Chau.

International Maritime Organisation (IMO) (1996). 1996 Protocol to the Convention on the Prevention of Marine Pollution by dumping of wastes and other matter, 1972.

Long, E.R., MacDonald D.D., Smith S.L.. and Calder F.D. (1995). Incidence of adverse biological effects within ranges of chemical concentrations in marine and estuarine sediments. Environ. Manage. 19: 81-97.

Mott Macdonald HK Ltd (1992). Contaminated Spoil Management Study.

Premchitt, J., and Evans, N.C. (1993). Stability of spoil and cap materials at East Sha Chau Contaminated Mud Disposal Area, Special Project Report 2/93 GEO, HK.

Whiteside, P.G.D., Ng. K. C. S. & Lee, W.P. (1996) Management of Contaminated Mud in Hong Kong, Terra et Aqua, Number 65: 10-17.

Works Branch, Hong Kong Government (WB, HKG) (1992). Marine Disposal of Dredged Mud, Technical Circular WBTC 22/92.

Environmental Hydraulics, Lee, Jayawardena & Wang (eds) © 1999 Balkema, Rotterdam, ISBN 90 5809 035 3

In situ sediment treatment in Kai Tak Nullah Approach Channel to control odours and methane gas production

J.M. Babin & J.T. Lynn
Golder Associates (HK) Limited, Hong Kong, China

T.P. Murphy
Aquatic Ecosystem Restoration Branch, Environment Canada, National Water Research Institute, Burlington, Ont., Canada

ABSTRACT: The sediments in Kai Tak Nullah Approach Channel (KTN) are responsible for the sulphide odour that is characteristic of the area around Kai Tak Airport. A method of controlling these sulphide odours by nitrate addition directly to the sediments was tested in the laboratory and an area of the KTN. Nitrate addition was found to increase sediment redox, oxidise the sulphides to sulphates, increase the rate of bacterial breakdown of organic materials, and reduce acute sediment toxicity. The laboratory and field tests of the treatments showed that nitrate addition could effectively control odours by oxidising the sulphides present and prevent future odour generation by providing an better and more efficient electron acceptor than is currently available for organic degradation. Nitrate addition will reduce or minimise future methane gas generation by increasing the degradation rate of the organic material accumulated at the bottom of KTN.

1. INTRODUCTION

Closure of Kai Tak Airport in July of this year has released a significant amount of land for reclamation and development. Part of the land scheduled for redevelopment is the Kai Tak Nullah Approach Channel (KTN) and Kwun Tong Typhoon Shelter (KTTS). Redevelopment of KTN and KTTS will involve infilling to the existing elevation of the area and development according to the Southeast Kowloon Development Masterplan.

The sediments in KTN and KTTS are highly organic and cause serious odour problems due to production of sulphides and release of hydrogen sulphide. Furthermore, the organic material present will degrade in the future via anaerobic processes, potentially creating methane gas problems.

The organic matter present in KTN and KTTS is primarily from current and historical sewage discharged to KTN. The restricted water circulation within KTN and high organic content of material discharged results in very limited dissolved oxygen in the water and sediments; consequently, little oxygen is available to bacteria in the sediments for the degradation of organic matter. Since there is little oxygen available for the bacteria, the most readily accessible oxidant is sulphate which is abundant in seawater. The preferred biodegradation pathway, with oxygen as the electron acceptor, is shown in Equation 1; the present anaerobic pathway, with sulphate as the electron acceptor is summarised in Equation 2:

$$C_7H_8 + O_2 \rightarrow CO_2 + H_2O \tag{1}$$

$$C_7H_8 + SO_4 \rightarrow CO_2 + H_2S \tag{2}$$

From Equation 2 it can be seen that sulphide is produced during the degradation pathway. It is this sulphide production that causes the odour problem characteristic of the area around Kai Tak Airport.

Odour control and increased bacterial degradation of organic matter can be achieved by supplying an alternate electron acceptor (*i.e.*, oxidant) to the bacteria. The preferred oxidant is oxygen; however, it has limited solubility in seawater at temperatures found in KTN. The next preferred oxidant is nitrate which is extremely soluble in seawater and is only slightly less energy efficient than oxygen.

Nitrate addition to sediments facilitates sulphide oxidation (equation 3) which removes bacterial toxicity and thereby increases the bacterial degradation rate of organic matter (equation 4).

$$S^- + Ca(NO_3)_2 \rightarrow CaSO_4 + N_2 \tag{3}$$

$$C_7H_8 + NO_3 \rightarrow CO_2 + N_2 \tag{4}$$

In the absence of most oxidants, methane formation (fermentation) occurs (equation 5):

$$4H_2 + CO_2 \rightarrow CH_4 + 2H_2O \tag{5}$$

During the planned reclamation of KTN, and to a lesser extent KTTS, the odour situation in the area will be exacerbated by sediment disruption during infilling. Furthermore, at the current biodegradation rate, the amount of organic material that will decompose between the present and when the reclamation begins will be small. Consequently there is potential that a methane problem in the future may exist if the amount of organic matter is not reduced.

Thus, the goal of the project is to:

- control the present sulphide odours by addition of an oxidant (Equation 3);
- prevent future odours from being produced (Equation 2);
- reduce future methane generation potential (Equation 5); and
- increase organic degradation rates (Equation 4).

2. BENCH-SCALE TESTING

A laboratory test was conducted in September 1997 on a composite sediment sample from KTN. The test was performed at the AQUAREF Laboratory of the National Water Research Institute of Environment Canada, located in Burlington, Ontario. The goal of this bench scale test was to determine if calcium nitrate addition to the sediments would:

- control present odours by oxidising sulphides to sulphate;
- control future odours by providing an alternate oxidant to sulphate for bacterial degradation of the organics;
- increase present bacterial activity rates thereby decreasing future methane gas generation potential;
- decrease sediment toxicity to bacteria;
- decrease sediment oxygen demand; and
- increase sediment redox potential.

The bench-scale test involved incubating a composite sediment sample from KTN in microcosms with two different concentrations of calcium nitrate added. A third incubation included lime $(Ca(OH)_2)$ with the objective to prevent potential metals released from sulphide oxidation. Gas produced in the microcosms was measured and analysed. Periodically, the sediments were sampled and analysed for the parameters of interest during the incubation.

Results of the bench-scale test showed that calcium nitrate addition alone:

- effectively oxidised 80 to 90% of the acid volatile sulphides (AVS) within 7 days of treatment, and 98% of the AVS within 14 days (Table 1);
- decreased sediment oxygen demand by 80%;
- decreased bacterial toxicity by 95%;
- increased bacterial activity as measured by gas production; and
- increased redox potential (corrected to standard hydrogen electrode) from -239 mV to + 60 mV.

Table 1: Acid Volatile Sulphides (mg S/g dry wt) in control and treated sediments. % reduction from initial concentration is given in parenthesis.

Treatment	Initial	Day 7	Day 9	Day 14
Control	5.08	5.80 (-14%)	5.30 (-4.4%)	4.76 (6.3%)
1000N	-	1.07 (79%)	0.88 (83%)	0.10 (98%)
1000N+L	-	1.93 (62%)	1.70 (66%)	0.87 (83%)
2000N	-	0.41 (92%)	0.15 (97%)	0.06 (99%)

Based on the bench-scale test results a pilot-scale test was performed in KTN to assess the feasibility of calcium nitrate addition to the sediments to control odours and methane generation.

3. PILOT-SCALE TEST

The field work for the pilot-scale test involved calcium nitrate injection to the sediments of a 50 x 50 m area of KTN. The treatments were undertaken during the periods of 20 to 22 November, and 13 to 14 December, 1997, when a total of 35 Tonnes of calcium nitrate was added to the sediments.

The field work involved injecting calcium nitrate solution directly into the sediments in the treatment area using a marine working boat equipped with a custom-designed/fabricated injection boom (8 m wide) with 40 cm injection tines. Dry calcium nitrate was stored on board the vessel and mixed with seawater as required. The calcium nitrate solution was injected into the sediments as the boat moved along a pre-determined injection path. Navigation was undertaken with a differential global positioning system accurate to ± 1 m.

A monitoring programme was implemented to determine the effectiveness of the treatment in reducing odour and future methane gas generation and to monitor potential deleterious environmental impacts of the treatment. Sediment core samples were collected from the treated and a control site with a modified K-B corer and subdivided at 0-5 cm, 5-10 cm and 10 cm intervals thereafter. The sediment slices were analysed for redox potential, AVS, sediment oxygen demand, total organic content, bacterial toxicity (*Microtox®*), and nitrate and sulphate concentrations.

4. PILOT SCALE RESULTS

The pilot scale sediment treatment encountered very few problems. The greatest problem involved the variable and shallow sediment depth. Sediment depth was anticipated to be greater than 1 m; however, during the treatments and subsequent monitoring programme it was found that sediment depth at the locations sampled was much less than 1 m. Many of the sediment cores collected had a clay plug at the bottom and the longest sediment core was 70 cm; the shallowest core collected was approximately 10 cm in length.

The variable and shallow nature of the sediments had implications on the achievable treatment depth and efficiency of the equipment used. The injection boom used in the pilot scale test was a rigid structure therefore if one section of the boom encounters shallow hard clay sediments, say at a depth of 15 cm, the penetration along the whole boom will only be to 15 cm.

In general, the treated sediments were markedly different than the untreated control sediments. Treated sediments were light brown in colour and had little to no detectable sulphide odour. The untreated control sediments were black and had the characteristic sulphide odour found in the vicinity of Kai Tak Nullah.

4.1 Redox

Generally, the redox values measured in the control/untreated sediments were lowest at the surface (0-5 cm) and increased with depth. The treated sediments had much higher redox values throughout although the numbers are variable. The overall mean and absolute redox values from 8 January, 1998, are presented in Table 2.

4.2 Acid Volatile Sulphides

Acid volatile sulphides (AVS) were significantly reduced in the treated sediments. In the surface 0-10 cm there was up to 99% less, and an overall average of 83% less, sulphide in the treated relative to control sediments. Notwithstanding the high variability, there is significantly less AVS in the top 40 cm of the treated sediments (an average of 60% less over the entire monitoring programme). For samples collect on 27 November and 22 January, the AVS concentrations and % reductions are given in Table 3.

4.3 Sediment Oxygen Demand

Sediment oxygen demand (SOD (mg DO/L/g sediment/min)) measurements showed a significant decrease (approximately 73 %) between untreated (control) and treated sediments. Mean ± standard deviation SOD of the control and treated samples were 0.165 ± 0.044 and 0.039 ± 0.018 respectively. The lower SOD values in the treated sediments were a result of lower AVS concentrations due to the nitrate addition.

4.4 Total Organic Carbon

Total organic carbon (TOC) decreased with depth from an average of 5.23 % (0-5 cm) to 1.58 % (30-40 cm). This decrease indicates that substantial degradation has occurred in the deeper sediments which

Table 2. Redox measurements (mV) in control and treated sediments from Kai Tak Nullah Approach Channel. Mean values are based on samples collected from 20 Nov. 1997 to 22 Jan. 1998. Absolute values are from 8 Jan. 1998.

Sediment	Mean		Jan. 8	
Depth (cm)	Control	Treated	Control	Treated
0-5	-191	34	-188	83
5-10	-188	19	-187	189
10-20	-183	-90	-171	131
20-30	-155	-159	-180	-70
30-40	-128	-127	-159	-148

Table 3. AVS (mg S/g dry wt.) in control (C) and Treated (T) sediment cores on two different sampling dates. % Reduction (Redn) for the treated cores on the individual days is also given.

Sediment	27 November			22 January		
Depth (cm)	C	T	Redn	C	T	Redn
0-5	4.02	0.05	99	1.03	0.01	99
5-10	6.08	0.83	86	3.05	0.01	99
10-20	5.10	0.77	85	5.13	3.12	39
20-30	6.11	1.85	70	4.67	2.49	47
30-40	4.29	2.81	34	3.40	2.43	28
40-50	-	-	-	2.94	1.59	46

Table 4. TOC concentration (%) in control (C) and treated (T) sediments from Kai Tak Nullah Approach Channel for both the entire monitoring programme and 22 Jan. %Reduction (Redn) is given for both cases.

Sediment	Programme Mean			22 Jan		
Depth (cm)	C	T	Redn	C	T	Redn
0-5	5.23	4.26	18	5.66	4.16	26
5-10	4.02	3.68	9	5.12	3.03	41
10-20	2.64	2.15	19	1.90	1.31	31
20-30	1.99	2.17	-9	1.59	0.92	42
30-40	2.27	1.18	48	-	-	-
40-50	1.58	1.09	31	-	-	-

Table 5. Microtox EC_{50} (dry weight basis) of control and treated sediment samples from Kai Tak Nullah Approach Channel

	Nov. 22	Nov. 27	Dec. 11	Dec. 18	Dec. 30	Jan. 8	Jan. 22
Control	0.023	0.020	0.025	0.021	0.008	0.016	0.022
Treated	0.107	0.252	0.405	0.343	0.270	1.770	1.3

implies that there is less potential for future methane gas generation from the deeper sediments since there is relatively less organic matter. Furthermore, the TOC remaining in the deeper sediments is more recalcitrant than the TOC that has already degraded, implying that future degradation by any method (aerobic or anaerobic) will be relatively slow in these deeper sediments.

Mean TOC from the entire monitoring programme show significant differences between the treated and untreated sediments (Table 4). The treated sediments have lower TOC as a result of increased bacterial degradation of the organics.

4.5 Bacterial Toxicity

Results of the Microtox® analyses (Table 5) show that sediment toxicity is significantly reduced in the nitrate-treated sediments as measured during this programme (wherein a higher value corresponds to a less toxic sample). Reduced sediment toxicity is expected since most of the AVS, which is extremely toxic to bacteria, has been removed from the sediments. This toxicity removal allows the naturally-occurring bacteria to flourish and increase the biodegradation rate of the organics.

4.6 Nitrate and Sulphate

Nitrate was routinely detected in the porewater of treated sediments, but not in the control untreated sediments, with values ranging from non-detectable to 12.4 g/L (18 December, 10-20 cm depth). In general, the results were extremely variable. The maximum depth where nitrate was detected was in the 20-30 cm interval. On 22 January, (the last sampling date) there was still a significant amount of nitrate detected in the treated sediments (up to 2.45 g/L) indicating the availability of a future long-term supply of oxidant.

Porewater sulphate concentrations (Table 6) show significant differences between the control and treated sites. Sulphate concentration consistently decreased with depth for both the control and the treated sediments indicating sulphate reduction is occurring. However, the mean sulphate concentration was higher in the treated sediments indicating that sulphides have been oxidised to sulphates from the nitrate addition.

Table 6. Porewater sulphate concentration (g/L) in control and treated sediments from Kai Tak Nullah Approach Channel. Sulphate concentration ratio between the two sites is also given.

Sediment Depth (cm)	Control Sediments	Treated Sediments	Sulphate Ratio
0-5	0.91	4.09	4.49
5-10	0.52	3.02	5.84
10-20	0.32	2.63	8.32
20-30	0.30	0.71	2.36
30-40	0.31	0.28	0.90

5. DISCUSSION

The sediments in KTN were less thick than anticipated, and variable with respect to depth and vertical concentration of the various parameters. Depth variability was due to the variable sediment deposition and hydraulic flow patterns in KTN. The AVS, sulphate, and TOC all decreased with increasing sediment depth. The implication of a thinner sediment layer, together with lower sulphide and organic concentrations in the deeper sediments, suggests that the surface (0-20 cm) sediments require a higher level of treatment than the deeper sediments (greater than 20 cm).

The sediment treatment monitoring results reflect those obtained during bench-scale testing. In the nitrate-treated sediments, redox increased, AVS decreased, sulphates increased and SOD, TOC and toxicity all decreased. The treated sediments were light brown in colour and there was no sulphide smell. By contrast, the untreated sediments were black in colour and had a noticeable sulphide odour.

6. CONCLUSION

Nitrate addition to sediments, as applied in KTN, controls odours as evidenced by the removal of sulphides (Equation 3) and the production of sulphates (Table 6). Evidence of enhanced organic content reduction (Equation 4) was observed by the decrease in TOC (Table 4) which will result in a decrease in future methane gas generation (equation 5). This enhanced organic degradation was facilitated by increased bacterial activity due to decreased bacterial toxicity (Table 5).

Full-scale treatment of KTN sediments by nitrate addition has been proposed as a method of mitigating the ongoing odour problem originating from KTN. Implementation of this recommendation will also result in decreasing future methane generation.

Environmental Hydraulics, Lee, Jayawardena & Wang (eds) © 1999 Balkema, Rotterdam, ISBN 90 5809 035 3

Managing contaminated sediment in a scenic river across a new town

D.S.W.Tang & L.M.C.Lau
Environmental Protection Department, Hong Kong, China

ABSTRACT: A substantial quantity of contaminated sediment has been accumulated in the Shing Mun River adversely affecting its water quality and suppressing the development of a balanced ecology. This paper provides an overview of a study for tackling the problems created by the contaminated sediment, with a view to improving the environmental conditions of the river in terms of water quality, ecology, odour and visual appearance for meeting its intended secondary contact recreation use. Various options, including aeration, chemical treatment, in-situ bioremediation, dredging, etc have been considered and a bioremediation pilot trial was conducted. The arriving of the selected improvement strategy involving phased bioremediation and dredging was discussed.

1. INTRODUCTION

The Shing Mun River (SMR) in Hong Kong is a 7.5 km long, straight artificial channel constructed in the 70's during the reclamation of the Sha Tin inlet. The river together with its tributaries are fed by a 37 km^2 catchment which covers rural areas and lower density developments at upstream; and high density developed Shatin New Town at downstream. Figure 1 shows the location plan of SMR and its main tributaries. By the late 80's, SMR had become excessively polluted due to indiscriminate discharges from various pollution sources. Collectively, the river received a high pollution load of organic materials (over 8,500 kg BOD_5/day) and heavy metals.

With the implementation of a number of pollution control measures over the past years such as the enactment of the Water Pollution Control Ordinance and the Waste Disposal (Livestock Waste) Regulations in 1987 and 1988 respectively, the organic pollution load in the river had been reduced to 1,300 kg BOD_5/day in 1996. The average water quality of SMR main channel was graded as "Good" (EPD, 1997). The remaining pollution loads are mainly originated from unsewered villages in the catchment, which will be curtailed upon completion of a sewerage connection scheme currently being implemented in the area (Balfours International Asia, 1990, Lei et al., 1996).

In addition to receiving pollution discharges, a substantial quantity of contaminated sediment has been accumulated on the riverbed. Contaminants are introduced into the SMR sediment from historic and present discharges. Due to their physio-chemical characteristics, contaminants would associate with either the particulate organic matter within the discharge, or with the fine particulate marine-derived material washed into the river. The SMR sediments are acting as a contaminant sink, a behaviour characteristic of all sediment (Yong, 1995).

The contaminated sediment not only adversely affects the river water quality due to its high oxygen demand and the possible release of contaminants, but also suppresses the development of a balanced nature ecology in the river. During low tide, the exposed sediment releases obnoxious odour which causes nuisance to nearby residents and is always the source of complaint from the local community. Figure 2 shows condition of the exposed sediment during low tide, which is obviously an eyesore to residents nearby.

In August 1996, a study was commissioned by the Environmental Protection Department (EPD), with the aims to determine the extent of sediment contamination and the environmental impacts created, and to formulate the most cost-effective and environmentally acceptable strategy to tackle the problems created by the sediment. The study has been substantially completed in early 1998.

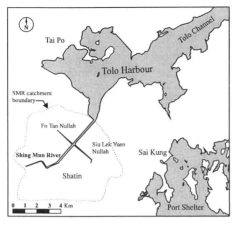

Figure 1. Location of the Shing Mun River

Figure 2. Conditions of the SMR Sediment

2. EXISTING ENVIRONMENTAL CONDITIONS OF THE SHING MUN RIVER

A comprehensive site investigation was carried out to determine the existing environmental conditions of SMR so that the associated environmental impacts could be accurately assessed. The key characteristics of the SMR environment are summarised in the following sections.

2.1 Sediment Chemical Characteristics

Sediment Oxygen Demand

Sediment oxygen demand (SOD) was tested *in-situ* using a respirometer (Wu, 1990) at 6 locations in the river. The testing locations and the respective SOD results are given in Figure 3.

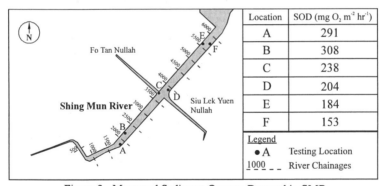

Location	SOD (mg O_2 m^{-2} hr^{-1})
A	291
B	308
C	238
D	204
E	184
F	153

Legend
● A   Testing Location
1000 _ _   River Chainages

Figure 3. Measured Sediment Oxygen Demand in SMR

All the SMR sediment samples tested had SOD values well in excess of 100 mg O_2 m^{-2} hr^{-1}, indicating that all sediments are organically polluted. Locations in the river can also be ranked in terms of SOD as follows: upstream > mid-stream > river mouth. The high sediment SOD has rendered non-compliance of the designated Water Quality Objectives (WQOs) of SMR in terms of dissolved oxygen (DO), which is set at 4 mg/l or 40% saturation, especially at upper reach of the river during low tide.

Biochemical Oxygen Demand and Chemical Oxygen Demand

Sediment concentration profiles of Biochemical Oxygen Demand (BOD) and Chemical Oxygen Demand (COD) within the SMR main channel are shown in Figure 4 and 5 respectively. Sediment BOD concentrations range from 500 mg/kg to 2,500 mg/kg, while sediment COD concentrations range from 10,000 mg/kg to 40,000 mg/kg. The maximum BOD and COD concentrations are observed in the main channel between Ch.3500 and Ch.4000, receiving Fo Tan Nullah that lined with industries on both sides.

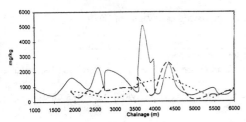

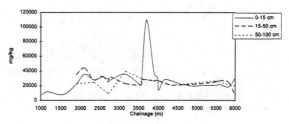

Figure 4. Sediment BOD Concentration Profile   Figure 5. Sediment COD Concentration Profile

Table 1. Maximum and Minimum Metal Concentrations recorded in SMR Sediment

Parameter	Maximum Recorded Concentration (mg/kg)	Minimum Recorded Concentration (mg/kg)
Copper	540	4.2
Lead	260	29
Zinc	1100	59

Metals

Sediments in SMR are particularly enriched with copper, lead and zinc as illustrated in Table 1. It is considered that these metals are present in the river as the result of historic effluent discharges from industrial premises in the area. The distribution of metal concentrations within SMR is fairly uniform, except at the junction between Fo Tan Nullah and the main channel where a peak of metal concentrations is recorded. In general, lower metal concentrations are recorded at the upper section where the sediment is more sandy, and at river mouth where SMR enters Tolo Harbour.

2.2 Shing Mun River Ecology

An ecological survey was conducted to determine the benthic and fish community characheristics in SMR. Although numerous fishes are present in SMR, the types of fish present (*Sarotherodon mossambicus, Mugil cephalus, Nematalosa nasus and tilapia*) and the low diversity as unveiled from the ecological survey results are indicative of a polluted environment. There was only one benthic organism (*Limnoperna* sp.) found in the mid-stream of the river. Overall, the benthic survey of SMR sediment illustrates that it does not have a benthic population, even of pollution tolerant species.

2.3 Sediment Impacts Upon Ambient Odour

An odour and air quality survey was conducted to assess the impact of the sediment on the ambient odour level. With a background odour concentration below 10 odour unit (ou) m^{-3}, the odour survey indicated that the highest odour concentration (35 ou m^{-3}) was detected at the upstream during low tidal conditions. The highest odour concentration (15 ou m^{-3}) during high tide was also detected at the intertidal zone near the upstream. Air quality analysis results indicated that the major air pollutants contributing to the odour nuisance were hydrogen sulphide (H_2S) and mercaptans (CH_3SH), with maximum concentrations detected at 12 ppb and 32.7 ppb respectively.

3. METHODS CONSIDERED FOR IMPROVING THE SHING MUN RIVER

Various sediment quality improvement methods have been considered in detail under the study. While different methods have their own merits and demerits, the methods as listed in Table 2 have been considered not suitable, with the principal disbenefits specified. With the exclusion of these unsuitable methods, bioremediation and dredging were evaluated and assessed as the most viable options to tackle the problems. It should be noted that regardless of the improvement techniques adopted, pollution source control is a critical factor affecting the quality of river water and the sediments accumulated. In addition to pollution source control by enforcing pollution control legislation, projects such as the Effluent Export Scheme

Method	Principal Disbenefits
Aeration	1. Efficiency limited due to the shallow water depth
	2. Presence of sludge pipes and gas pipes on riverbed obstructed installation of aerators
Chemical Stabilisation	1. Adverse impact on water quality and the river's ecology during works phase
Nutrient Addition	1. Further stimulate sediment oxygen consumption
	2. Would lead to eutrophication and uplifting of sediment
Engineering Solution	1. No immediate/short-term relief of existing adverse conditions

(Balfours International (Asia), 1988) which diverts treated effluent from Tolo Harbour to Victoria Harbour via the Kai Tak Nullah, and the sewerage connection scheme which intercepts domestic sewage from unsewered villages to treatment works, are vital for curbing pollution to the SMR.

4. BIOREMEDIATION VS DREDGING

4.1 Bioremediation

The bioremediation technique involves the *in-situ* inoculation of contaminated sediment with a specific microbial population, such that the contaminants therein are rapidly degraded into substances with no or minimum environmental impact. Although relatively new, the bioremediation technique has been successfully employed for treating contaminated sediments in other parts of the world especially in USA (USEPA, 1991, 1998). A schematic diagram illustrating bioremediation of contaminated sediment is shown as Figure 6 (NSTC, 1995). In order to successfully improve the environmental conditions of SMR, the following targets should be met:-

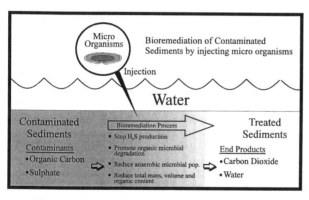

Figure 6. Bioremediation of Contaminated Sediment

- Promote degradation of organic content in sediment thereby improving its characteristics and texture.
- Reduce the high SOD of the surface sediment which has a resultant adverse impact on the DO level of the overlying water column; and
- Suppress H_2S and CH_3SH emission from the sediment surface which is accountable for the odour nuisance.

Bioremediation is a technique which can achieve these targets through the alteration and enhancement of the sediment's biological activities. In order to facilitate bioremediation for the SMR condition, the microbial population would be applied with supplementary oxygen releasing agent with a special carrier medium which could improve the porosity of the sediment. Once the targeted contaminants are degraded, the applied bacteria would die-off due to the lack of a metabolisable substrate. To ascertain the effectiveness of application of this technique in the local

Table 3. Results of the Bioremediation Pilot Trial

Sediment Quality	Result of Pilot Trial
Total Organic Carbon content	Reduction of up to 57%
Chemical Oxygen Demand	Reduction of up to 45% to 74%
Nitrogen	No change
Heavy metal concentration	No change
Average Acid Volatile Sulphide concentration	Reduction by up to 28%

condition of SMR, a bioremediation pilot trial was carried out at a selected section of about 300 m length at the intertidal zone upstream of the Lion Bridge over a period of about four months. The formulation was injected into the sediment via a steel lance by traversing over the river on a small boat (i.e. a boat-based method). A comparison of the sediment quality of the treated site before and after the trial was summarised in Table 3.

While the exact extent of improvement of the sediment and river water was not firmly defined under the 4-month trial period, it was found that bioremediation can significantly improve the sediment quality in SMR by reducing sediment organic content and inhibiting H_2S production. The trial also demonstrated that there was no adverse impact to the river water quality during application of the bioremediation process.

4.2 Dredging

It appears logical that the environmental conditions of SMR could be improved with removal of the contaminated sediment by dredging. A number of environmental benefits could be obtained in removing sediments from the intertidal zone where the most serious odour nuisance is encountered, and at the mid-stream of the main channel and the Fo Tan Nullah where the sediment is most heavily contaminated. A water quality model has been developed and the benefits to water quality of the river due to dredging was predicted and summarised in Table 4.

Although dredging can remove the contaminated sediment as a source of pollution and thereby improve the river water, significant environmental impacts could be resulted during the construction activities of dredging works. The heavy plants employed for the dredging works would likely create short-term noise nuisance, and the re-suspension of contaminated bed sediment during dredging would lead to serious deterioration of river water quality and odour problems. An empirical odour emission rate testing carried out by simulating the dredging operation has found that sediment disturbance could increase odour emissions by over 180%. Based on the estimated increase in emission rate, the odour modelling predicted that dredging would elevate odour levels by up to 25 ou m^{-3} over baseline conditions.

No matter either solution is adopted, due to re-sedimentation within SMR, supplementary improvement works would need to be carried out on a recurrent basis, which should be continuously monitored and reviewed by going through an Environmental Monitoring and Audit (EM&A) process.

5. THE ADOPTED IMPROVEMENT STRATEGY

In the light of bioremediation pilot trial results, an improvement strategy based on phased bioremediation of the SMR, with the flexibility of switching to a dredging-based strategy when necessary is adopted (Figure 7). The adopted improvement strategy includes two phases as follows:-

Phase 1	Phase 2
• Sediment bioremediation for priority areas of most severe sediment impacts, and • A bioremediation pilot trial in Fo Tan Nullah to determine whether the technique is able to improve the predominantly industrial contaminated sediments present in the nullah.	• Sediment bioremediation for remaining areas, • Sediment bioremediation at Fo Tan Nullah if the Phase 1 pilot trial is successful, and • Selective dredging at areas where level of improvement from bioremediation is found to be inadequate as indicated in the EM&A, and at Fo Tan Nullah if pilot trial under Phase 1 is unsuccessful.

Table 4. Predicted Benefits to Water Quality Achieved after Dredging

Parameter	Dry Season	Wet Season
Dissolved oxygen	Increases of up to 1.5 mg/l in areas upstream of Fo Tan Nullah	Increases of up to 1.5 mg/l in areas upstream of Fo Tan Nullah
BOD$_5$	Maximum reduction of 0.16 mg/l	Not significant
Ammoniacal nitrogen	Maximum reduction of 0.3 mg/l	Not significant
Organic nitrogen	Maximum reduction of 2 mg/l	Not significant

A schematic diagram showing the phased strategy for improving SMR is given in Figure 8. Based on a five years improvement programme, the cost for implementing the improvement strategy is estimated to be HK$44 million (US$5.6 million), which would be increased to a maximum of HK$64 million (US$8.2 million) if dredging option is to be adopted due to failure of bioremediation.

It should be noted that EM&A will play a key role during the implementation stage to define whether Phase 1 bioremediation is successful and whether Phase 2 or alternative strategies should be pursued. Apart from controlling the frequency and extent of bioremediation, EM&A will also define the environmental impacts during works and the requirements for any mitigation measures.

6. CONCLUSION

A robust yet flexible improvement strategy has been formulated for tackling the contaminated sediments within the SMR. Upon completion of the improvement works by about mid next decade and coupled with completion of other pollution source control measures such as the sewerage connection scheme, the designated Water Quality Objectives of the SMR will be achieved. In addition, environmental conditions of the river in terms of odour, ecology, visual appearance, etc would also be substantially enhanced rendered the use of the river for the intended secondary contact recreation purposes possible.

Figure 7. Phased Improvement Strategy of SMR

steering group and working group involved in managing the study.

7. ACKNOWLEDGEMENT

The authors wish to express thanks to the Director of Environmental Protection, Government of the Hong Kong Special Administrative Region, for permission to publish this paper. The views expressed are those of the authors and do not necessarily reflect in any way official government policy. Thanks are also expressed to the consultants responsible for the study and members of the study steering group and working group involved in managing the study.

8. REFERENCE

Balfours International (Asia) and Charles Haswell & Partners (Far East) (1988). *Sha Tin to Kai Tak effluent export scheme, Final Report*, Environmental Protection Department, Hong Kong Government.

Balfours International (Asia) (1990). *Tolo Harbour Catchment study on unsewered developments, Final Report*, Environmental Protection Department, Hong Kong Government.

Environmental Protection Department, Hong Kong (1997). *River water quality in Hong Kong for 1996*, Government Printer Hong Kong.

Lei, P.C.K., Wong, H.Y., Lui, P.H., Tang, D.S.W. (1996). *Tackling sewage pollution in the unsewered villages of Hong Kong*, Proceedings of the Third International Conference on Environmental Pollution, Budapest, Hungary, Volume 1, 334-341.

National Science and Technology Council (NSTC) (1995), *Biotechnology for the 21st century: new horizons*, http://www.nalusda.gov/bic/bio21/environ.htm

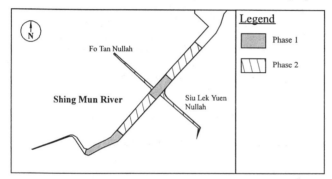

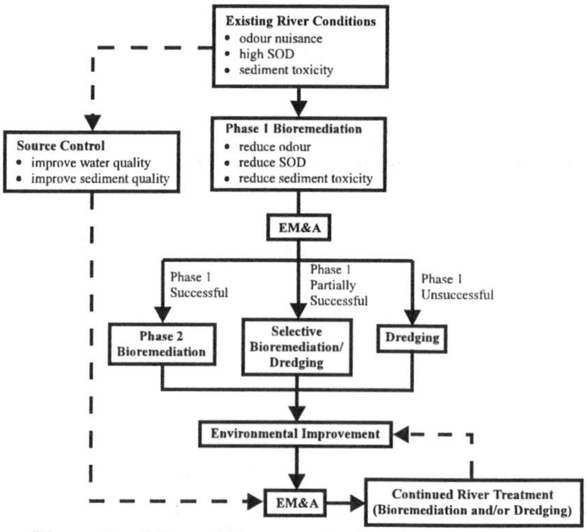

Figure 8. Adopted Strategy for Environmetal Improvement of SMR

USEPA (1991). *Biological remediation of contaminated sediments, with special emphasis on the great lake*, USEPA Publication no. NCEPI-EPA 600-9-91-001.

USEPA (1998). *Bioremediation – Keyterm product citations and abstracts*, http://www.epa.gov/gbwebdev/ged/publica/keycb7.htm

Wu, R.S.S. (1990). *A respirometer for continuous in-situ measurements of sediment oxygen demand*, Water Research, 24(3), 391-394.

Yong, R.N. (1995), *The fate of toxic pollutants in contaminated sediments*, Keynote paper in Dredging, Remediation, and Containment of Contaminated Sediments (Demars et al. Eds), 13-38, ASTM Publication.

4.4 Eco-hydraulics

Environmental Hydraulics, Lee, Jayawardena & Wang (eds)© 1999 Balkema, Rotterdam, ISBN 90 5809 035 3

Observations on characteristics of flow and production of attached algae at riffles and pools in gravel rivers

Y.Toda & S.Ikeda
Department of Civil Engineering, Tokyo Institute of Technology, Japan

ABSTRACT: Two field observations were performed in gravel rivers in Tokyo to study the hydraulic and ecological characteristics of riffles and pools. The hydraulic and ecological properties were measured by using various devices. It was found that the water surface texture of the riffles are affected by not only Froude number but also large gravels of channel bed. The activity of the production of the attached algae at riffles is found to be larger than that at pools. The primary production of the algae at riffles and pools are very large, yielding temporal variation of dissolved oxygen during day-time.

1. INTRODUCTION

There are many gravel-bed rivers in Japan, where two flow patterns are usually observed. One is riffle, which has shallow water depth and large flow velocity, and another one is pool, which shows deep water depth and small flow velocity. Recently, it has been recognized that the riffle and pool system provides good river eco-system, land-scape and/or sound-scape.

Many field surveys on fishes and insects in the system were performed in natural rivers (e.g., Shimatani et al., 1997). However, few quantitative data are available on the hydraulic characteristics and the ecological features, and, therefore, it is necessary to get more data on the system. This is the purpose of the present field observations.

In this study, two field observations were performed. The first field observation, which is termed observation A, was conducted to manifest the water surface texture at riffles. The second observation, which is termed observation B, was conducted to study the primary production at riffles and pools.

2. OBSERVATION A

2.1 Outline

The observation A was conducted at a riffle of Akikawa-river in Akiruno-city, Tokyo, on 11 and 12 Sept. 1996. In the observation area, the riffle extends with a length of about 60 m, and pools are generated at both upstream and downstream sides of the riffle.

Various types of riffles have been reported to exist in steep rivers (e.g., Furukawa, 1994). The following two types of riffles are observed in the present study: Chara-riffle: a flow field exists at the area with shallow water depth, where the spatial variation of water surface shows a wavy texture. The water surface does not break. Zara-riffle: a flow field almost same as chara-riffle. Water surface, however, breaks locally with entrainment of air.

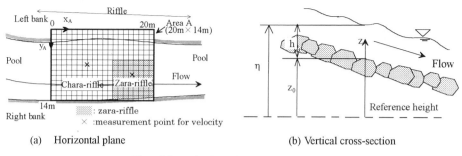

(a) Horizontal plane   (b) Vertical cross-section

Fig.1 Schematic view of the observation area

Figs.1(a) and (b) show the schematic view of the observation area.   The channel bed level and the water surface elevation were measured by using level meter (SOKIA Co.Ltd).   The area measured was taken to be 110 m length along the longitudinal direction.   An area termed A, which has dimensions of 20 m (in longitudinal direction) $\times$ 14 m (in transverse direction), was selected in the riffle, where the local water depth and the distribution of the channel bed gravels were measured in detail.   The local water depth was measured by using ruler, and the distribution of the bed gravel size was measured by using CCD digital video camera (SONY Co.Ltd.).   The video pictures of gravels were taken through a box, the base of which was made of transparent acrylic plate.   Since the flow field of riffle is much affected by large gravels (Toda and Ikeda, 1998), 5 largest gravels found in 1m $\times$ 1m area were sampled in a video picture.   These gravels were larger than d_{89}, and the diameters of the longest axis of the gravels were measured.   To study the hydraulic characteristics of riffles, 2 observation points in Fig.1(a) were chosen in the chara-riffles (at x_A=7 m and y_A=7 m) and zara-riffle (at x_A=16 m and y_A=9 m), respectively.   At each point of the observation, the vertical distribution of the longitudinal velocity component was measured by using electro-magnetic velocimetry (KENEK Co.Ltd.).   The flow velocity was measured with a frequency of 20 Hz for 60 seconds.

2.2 Results

Outline of the Geometry of Riffle and Pool: Fig.2 shows the vertical cross-section of the observation area along the longitudinal direction.   The riffle extends about 60 m longitudinally (between 20 m and 80 m), where the slope of the channel bed is fairly steep with an averaged slope of 1/162.   The averaged depth of the riffle is 38 cm.

Water Surface Texture and Channel Bed Gravels: The depth-averaged flow velocity, u, temporally-averaged local water depth, h, and local Froude number, Fr, are summarized in table 1.   Local Froude

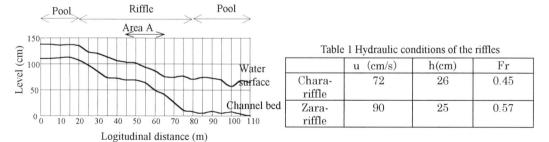

Fig.2 Vertical cross-section along longitudinal direction

Table 1 Hydraulic conditions of the riffles

	u (cm/s)	h(cm)	Fr
Chara-riffle	72	26	0.45
Zara-riffle	90	25	0.57

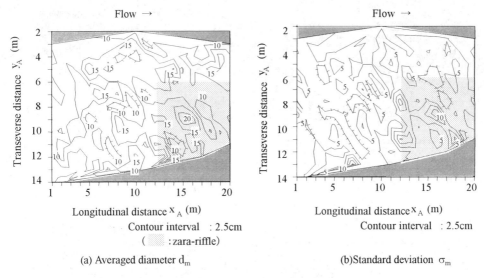

Flow →

Flow →

Longitudinal distance x_A (m)

Longitudinal distance x_A (m)

Contour interval : 2.5cm

Contour interval : 2.5cm

(: zara-riffle)

(a) Averaged diameter d_m

(b)Standard deviation σ_m

Fig.3 Spatial distribution of large gravels

number at zara-riffle shows a bit larger value than that at chara-riffle. However, there is not any significant difference between them. Figs.3(a) and (b) show the contour maps of the averaged diameter and the standard deviation of the 5 largest gravels sampled at each observation point in area A, respectively, in which the river banks are shown by oblique lines. Zara-riffle is shown by shade and the remainder part is chara-riffle. The area where the averaged diameter of gravels exceeds 15 cm is found to locate in zara-riffle. The result indicates that large gravels exist at zara-riffle. The standard deviation of the 5 largest gravels found in zara-riffle also take a larger value than that of chara-riffle. It is, therefore, concluded that the water surface texture at riffle is much affected by the size and the distribution of large gravels as well as Froude number.

3. OBSERVATION B

3.1 Outline

The observation B was conducted at Tamagawa-river in Oume-city of Tokyo during 24 Jul. and 11 Sep. 1997. The observation area was taken to be about 1.8 km length in longitudinal direction, where 3 riffles and 4 pools exist. The riffles and the pools are labeled as P1, R1, P2, R2 ... from the upstream side (see Fig.4). Since the flow velocity at R2 is too large to measure the flow properties accurately, the data on R2 are not included in this paper.

The imitated gravels made of concrete were settled in the riffles and the pools. Three imitated gravels were picked out at each location every week, and the algae attached to them was removed by toothbrush. The amount of chlorophyll-a of the algae was measured by using spectrophotometer(Central Science Co.Ltd.).

On 26 and 27 Aug. 1997, the hydraulic characteristics and the ecological features were measured in detail as described subsequently. The geometry of river was measured by using level meter and range finder (SOKIA Co.Ltd.). Flow velocity components were detected by electro-magnetic velocimetry at R3 riffle to know the flow rate of the river, in which they were sampled with a frequency of 20 Hz for 100 seconds. To observe the local distribution of attached algae around large gravels, the amounts of chlorophyll-a sampled at

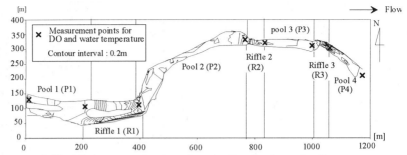

Fig.4 Contour map of water surface elevation for observation B

Table 2 Hydraulic properties at observed riffles and pools

	Riffle		Pool			
	R1	R3	P1	P2	P3	P4
U(cm/s)	122	96	71	44	55	44
H(cm)	27	36	31	40	59	76
Fr	0.69	0.49	0.41	0.22	0.23	0.16

the upstream side, the top and the downstream side of the large gravels were measured by using spectrophotometer. 5 largest gravels, the size of which are about 20 cm in the longest axis, were selected at R3 and P3, respectively. The area sampled the algae at each side of the gravels is 5cm×5 cm. Water temperature and dissolved oxygen were measured by using portable water quality meter (Horiba Co.Ltd.). The water temperature and the dissolved oxygen were sampled with an interval of 2 hours for 24 hours at the points depicted in Fig.4.

3.2 Results

Hydraulic Characteristics of Riffles and Pools: Spatially-averaged flow velocity, spatially-averaged water depth and Froude number of each riffle and pool are shown in Table 2. It is found that the Froude numbers of riffles are larger than that of pools.

Growth of Attached Algae: The temporal variations of the amount of chlorophyll-a of attached algae are depicted in Fig.5, in which the solid curves represent the logistic curves fitted to the observed data. The equation of logistic curve is represented by the following form:

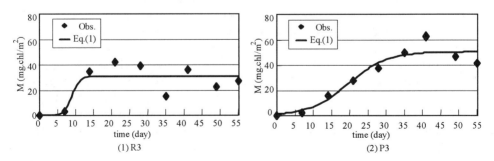

Fig.5 Variation of attached algae

$$M = \frac{M_0 \cdot S}{M_0 + (S - M_0) \cdot \exp(-\mu \cdot t)} \tag{1}$$

in which M is the amount of chlorophyll-a in mg.chl/m^2, M_0 is the amount of chlorophyll-a at t=0, μ is intrinsic growth rate in day^{-1} and S is saturation value in mg.chl/m^2. The intrinsic growth rate means the magnitude of growth speed in the early stage of growth, and therefore it shows the photosynthetic activity of attached algae. Table 3 shows the intrinsic growth rate and the saturation value of chlorophyll-a at riffles and pools. The intrinsic growth rates at riffles generally take larger values than those at pools, suggesting that the photosynthetic activity of the attached algae at riffles is larger than that at pools. The values of saturation of chlorophyll-a have considerable scatter, and the difference between the riffles and pools is not clear despite of the difference of the intrinsic growth rates.

Local distributions of the amount of attached algae around the large gravels are shown in table 4. At riffles, the amount of chlorophyll-a sampled at the upstream side of the large gravels is much less than those collected at the top and the downstream sides of the surface. However, chlorophyll-a of attached algae around gravels at pools does not varies locally. Since the flow velocity at riffle is large, the attached algae at upstream side of the gravels is probably removed by collision of the suspended particles such as sands or small gravels. In contrast, flow velocity at pool is small, and thus the small particles settle among the gravels calmly. Therefore, the algae at pool is expected to grow at all parts of the surface.

Temporal Variation of Dissolved Oxygen: Fig.6 shows the temporal variations of the concentration and the degree of saturation of dissolved oxygen. The concentration increases during the late morning, exceeding the supersaturation. It has been known that increase of dissolved oxygen in rivers is generally induced by reaeration and photosynthesis of attached algae. Since the dissolved oxygen is supersaturated during the late morning in this observation, the major reason for the increase is probably due to the active photosynthesis of the attached algae.

Production Rate of Oxygen: The dissolved oxygen at observation area is thus much affected by the photosynthesis of the attached algae. Therefore, the production rate of oxygen by the attached algae is crudely estimated subsequently. The photosynthesis of attached algae is approximately represented by the following chemical reaction formula (Tsuda, 1972):

$$106CO_2 + 90H_2O + 16NO_3 + PO_4 \rightarrow C_{106}H_{180}O_{45}N_{16}P + 154.5O_2 \tag{2}$$

Converting the intrinsic growth rate to the production rate of oxygen by using both Eq.(2) and the amount of chlorophyll-a included in the attached algae (Chl.a/$C_{106}H_{180}O_{45}N_{16}P$ =7.5$\times 10^{-3}$~1.3$\times 10^{-2}$ by weight, see Parsons et al., 1961), the primary production rate of oxygen per unit chlorophyll-a at each riffle and pool is estimated. The results are summarized in table 5. Despite such crude estimation, two features are remarkable. The production rate at observed riffles and pools is very large (e.g., the production rate of

Table 3 Characteristics of the growth

		Intrinsic growth rate in 1/day	Saturation value in mg.chl/m^2
Riffle	R1	0.72	45
	R2	0.96	31
Pool	P1	0.42	56
	P2	0.21	31
	P3	0.19	50
	P4	0.37	30

Table 4 Local distribution of the attached algae around large gravels (size of gravels: 20cm)

	Upstream side	Top	Downstream side
Riffle (R3)	7.0	35.7	33.6
Pool (P3)	20.3	19.9	19.1

in mg.chl/m^2

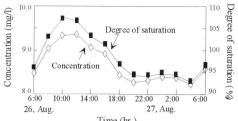

Fig.6 Temporal variations of dissolved oxygen
(at the downstream side of P4)

Table 5 Production rate of oxygen by attached algae

		Production rate per unit chlorophyll-a in mg.O_2/mg.chl/day	Production rate per unit area in g.O_2/m^2/day
Riffle	R1	110~190	5.2~8.7
	R3	150~260	4.8~8.0
Pool	P1	68~110	3.8~6.4
	P2	34~58	1.0~1.8
	P3	30~51	1.5~2.6
	P4	59~100	1.8~3.0

other water basin: 0.5 to 3.8 in g.O_2/m^2/day at Suwa lake and 0.2 to 1.7 in g.O_2/m^2/day at Biwa lake (Aruga, 1973)). Another feature is that the production rate of oxygen is large enough to induce the temporal fluctuation of dissolved oxygen.

4. CONCLUSIONS

The field observations conducted herein have revealed the followings:

1)The averaged diameter and the standard deviation of the largest bed gravels at zara-riffle are larger than those at chara-riffle, which primarily produce the difference of water surface textures between them.

2)The photosynthetic activity of attached algae at riffles is larger than that at pools.

3)The riffle and pool system has a large primary production.

4)The temporal variation of dissolved oxygen during day-time is mainly induced by the photosynthesis of attached algae.

ACKNOWLEDGEMENT

The present study has been conducted under the financial support of Grant-in-Aid for Scientific Research, the Ministry of Education and Culture of Japan (Grant No.08305017), and Tokyu Foundation for Better Environment. The observation was supported by the students of the Hydraulics Laboratory, Tokyo Institute of Technology. They are gratefully acknowledged.

REFERENCES

Aruga, Y.(1973) "Production of water plants", Kyoritsu Press (in Japanese).

Furukawa, T.(1994) "Fishing for ayu", Seito Press (in Japanese).

Parsons, T.R., Stephens, K. and Strickland, J.D.H.(1961) "On the chemical composition of eleven species of marine photoplankters", *Jour. Fish. Res. Bd.*, Canada, 18, 1001-1016.

Shimatani, Y. and Kayaba, Y. (1997) "Impacts of stream modification on habitat components and fish community in Tamagawa-river", *Jour. of Hydroscience and Hydraulic Engrg.*, JSCE, Vol.15, pp.49-58.

Tsuda, M (1972) "Ecology of water pollution", Association for Technology of Pollution Countermeasure (in Japanese).

Toda, Y., and Ikeda S. (1997) "Study on hydraulic characteristics of riffles by field observation", *Jour. of Hydroscience and Hydraulic Engrg.*, JSCE, Vol.15, pp.31-40.

Environmental Hydraulics, Lee, Jayawardena & Wang (eds) © 1999 Balkema, Rotterdam, ISBN 90 5809 035 3

A field observation on water purification system in reservoirs using wave energy

T. Komatsu, T. Okada, S. Nakashima & K. Fujita
Department of Civil Engineering, Kyushu University, Fukuoka, Japan

S. Marui
Kyushu Technical Office, Ministry of Construction, Kurume, Japan

Y. Matsunaga
Sasebo Heavy Industry Company, Japan

ABSTRACT: This study proposes a method which creates an unidirectional flow by making use of a natural wave energy. The unidirectional flow in a reservoir makes it possible to transport water of a surface layer including a large amount of dissolved oxygen (DO) to a bottom layer in which DO is not usually enough. The following important questions related to application of the method to practical use were taken into consideration. (1) how large are the scale and an occurring frequency of wind waves in real reservoirs? (2) how much is the volume of wave overtopping gained on a floating structure? (3) how much does the concentration of DO at the bottom layer change by sending the water of the surface layer to the bottom layer? The results suggest that the wave energy in reservoirs seems to be large enough to be used for water purification.

1. INTRODUCTION

It is the well known fact that if the mean residence time of water in a reservoir is long enough, the water quality goes toward deterioration. Generally speaking, the combination of nutrient release from the bottom with oxygen depletion in a profound zone is considered as a chief factor that causes deterioration of water quality in such stagnant waters. In order to give a contribution toward solving this issue, we have proposed a floating structure device which can produce an unidirectional flow by using the natural wave energy (see **FIG. 1**) (Komatsu *et al.*,1996b, 1997). The floating structure consists of a small tank with a slope in front and a pipe installed in the middle. Incident waves run up on the slope and pass over the crown of the structure. A water head difference between the water level in the tank and the mean water level of body water can be immediately obtained. The gained head difference makes it possible to generate an unidirectional downward flow through the pipe. Thus, the oxygen rich surface water due to wind mixing is transported to the profound zone where dissolved oxygen (DO) is almost depleted. Since the scale of wind waves which occur in reservoirs is much smaller than that of wind waves in the sea, the volume of water supply can not be expected to be enormous. However, this system can afford a high supply rate of DO and has great advantages of low construction cost, simplicity and use of only a natural energy source.

In this study, the following investigations are represented. (1) how large are the scale and an occurring frequency of wind waves in reservoirs? (2) how much is the volume of wave overtopping gained by prototypes of the floating structure in a real reservoir? (3) how much does the concentration of DO at the bottom layer change by sending the water of the surface layer to the bottom layer?

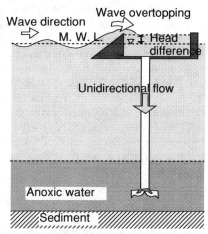

FIG. 1 System to send water of a surface layer into a bottom layer

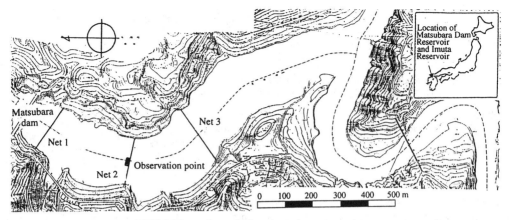

FIG. 2 Plane view of Matsubara dam reservoir

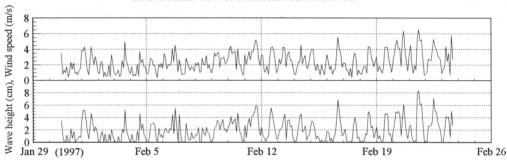

FIG. 3 Results of wind and wave observations

2. WAVES IN A RESERVOIR

Observations of winds and waves at Matsubara dam reservoir were carried out over five months, for two months from December (winter) in 1996 and for three months from June (summer) in 1997. Matsubara dam reservoir is located in the northern part of Kyushu, the west island of Japanese archipelago. **FIG. 2** shows the plane view of the reservoir. There is a driftwood protection system consisted of three vertical net sheets which were constructed across the reservoir at the distances of 100 m, 400 m and 700 m far from the dam, respectively. The reservoir has the width of 200 m and the depth of 40 m at the observation point.

2.1 Wind and Wave Observation Devices and Methods

Observations of winds and waves were carried out on a stage which is fixed on the second net and is located at the center of the reservoir (see **FIG.2**). A capacitance-type wave gage was used for the measurement of waves. A wind vane and an anemometer which were set up at 4.0 m high from the water surface were used for wind measurements. The data were obtained every 0.01 sec., for 10 min., every 2 hours.

2.2 Occurring Frequency of Wind Waves in a Reservoir

FIG. 3 shows the wind speeds and the wave heights from January 30 to February 23, 1997. The fluctuation of wave height is proportional to the wind speed. When the wind speed is large, the

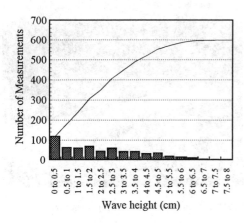

FIG. 4 Occurring frequency of wave heights in the winter

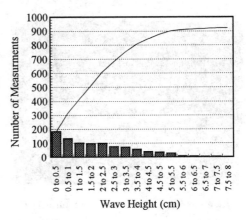

FIG. 5 Occurring frequency of wave heights in the summer

wave height is large too, and vice versa. The largest wave during the observation term was 9.2 cm high.

FIG. 4 and **5** show the occurring frequency of wave heights in the winter and the summer, respectively. The solid lines in the both figures denote accumulation of occurring frequency. One sees that the occurring frequency of wave more than 2.0 cm, 3.0 cm and 4.0 cm high is 49,1%, 32.3% and 18.6% of the waves in the winter and is 44.9%, 26.3% and 12.9% in the summer, respectively. Particularly in Japan, since the seasonal wind is stronger in winter than in summer, it seems that wave heights are larger in winter than in summer. However, the frequency of occurrence for wave heights in winter is almost the same as that in summer. Though the needed head difference to generate the unidirectional downward flow is strictly dependent on the density difference between the surface and the bottom water, even several centimeters head difference seems to be enough to operate the system. Therefore, if we assume that waves 3 cm high are enough to operate the system, we can anticipate that an annual net working rate of the system will be as much as 30% in Matsubara dam reservoir.

3. VOLUME OF WAVE OVERTOPPING GAINED BY THE FLOATING STRUCTURE AT FIELD WATER

3.1 Prototype Structure for Field Experiments

During field experiments, two kinds of the prototypes of floating structure with 4 m width were set at the center of the reservoir, though we expect that, in practical use, the plural floating structures with 10 m width will be set. Taking into consideration that in Matsubara dam reservoir the wind blows almost parallel with the main axis of the reservoir, Type 1 was fixed on the second net so as to point at the direction of the main axis (see **FIG. 6**). Therefore, Type 1 can gain the volume of wave overtopping from two directions, i.e., the upstream and the downstream. On the other hand, Type 2 is movable and rotatable to the windward by harnessing the wind energy (see **FIG. 7**). In addition, differing from Type 1 which was immersed when the tank was filled up with the water of wave overtopping, Type 2 was designed to be restrained from sinking by bigger floats.

3.2 Results of Field Experiments

The results of field experiments are shown in TABLE 1. Though the volume of wave overtopping depends on the type of prototype, the crown height and the wave condition, the averaged volume of

FIG. 6 Prototype of a floating structure, Type 1   **FIG. 7** Prototype of a floating structure, Type 2

TABLE 1 Results of field experiments

Prototype	Measurement Term	Number of Measurement days	Crown Height (cm)	Direction	Total Volume (m³)	Sum of Total Total Volume (m³)	Volume per day (m³/day)	Volume per unit time (l/s)	Volume per 10mwidth and per unit time (l/s)
Type 1	May 21 ~ May 28	7	4	Upstream	129	156	22.0	0.25	0.64
				Downstream	27				
	May 29 ~ Jun 3	5	6	Upstream	107	115	22.6	0.26	0.65
				Downstream	8				
	Jun 3 ~ Jun 17	14	2	Upstream	486	640	46.1	0.53	1.33
				Downstream	154				
	Jun 20 ~ Jun 26	6	4	Upstream	99	124	20.5	0.24	0.59
				Downstream	25				
	Jun 26 ~ July 3	7	2	Upstream	1057	1764	239.2	2.77	6.92
				Downstream	707				
	July 4 ~ July 9	5	3	Upstream	278	417	85.5	0.99	2.48
				Downstream	139				
Type 2	Aug. 7 ~ Aug. 21	14	3		896		64.4	0.75	1.86
	Aug. 27 ~ Sept. 4	8	2		551		68.5	0.79	1.98

wave overtopping per 4 m width and per unit time is 0.63 l/s; however the data obtained from June 26 to July 3 are excluded because of far larger value due to the heavy rain. If the structure of practical use is 10m wide, the volume of wave overtopping gained per unit time is about 1.6 l/s; 138 m³/day. It agrees well with the value which was estimated from the result of indoor experiments (Komatsu *et al.*, 1998).

FIG. 8 shows the relation for Type 2 between the dimensionless overtopping rate Q/HL and the relative crown height h_C/H. Q cm²/wave is

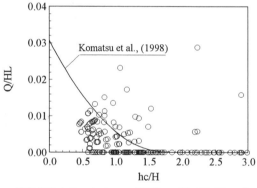

FIG. 8 Relation between Q/HL and h_C/H; Type 2

the overtopping quantity per a wave and per unit width. The solid line in the figure is obtained by the indoor experiments with irregular waves (Komatsu *et al.*,1996a). Though some extremely large values of Q/HL exist in the range of $h_C/H > 1$, one sees that the results of field experiments largely agree with those of the indoor experiments.

4. CHANGE OF CONCENTRATION OF DO BY SUPPLYING WATER OF SURFACE LAYER TO BOTTOM LAYER

4. 1 Experimental Methods

In order to estimate the effect of sending the surface water into the bottom layer on water

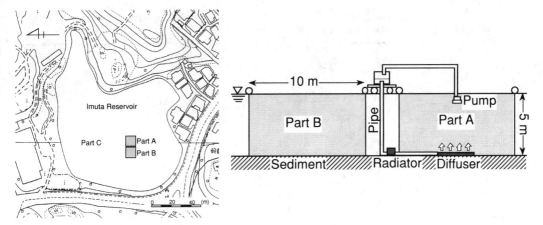

FIG. 9 Plane view of Imuta Reservoir   FIG. 10 Cross-section of isolation water structures

purification, an experiment using isolation water structures has been carried out since May of 1998 in Imuta reservoir. Imuta reservoir is located in the northern part of Kyushu (see **FIG. 2**). It is 5 m deep at the center and has the total water surface of 15,000 m^2 (see **FIG. 9**). Two isolation water structures; 10 m $\times$ 10 m, set up in the reservoir create two compartments, a controlled one; called Part A, and another one as a reference; called Part B. The unrestricted body water is called Part C. The side walls of the isolation waters are made of Tarpaulin, which is not permeable at all. At Part A, being pumped up from the surface, water is sent into the bottom layer through a pipe. The volume of water supply is 0.2 l/s. In order to adjust the temperature of the water sent to that of the water at the bottom layer, a radiator was installed in the middle of the pipe. A diffuser was installed in the outlet of the pipe, in order to mix the discharged water with the body water well.

4. 2 Results and Discussion

FIG. 11 shows the vertical distributions of the temperature of water and the concentration of DO three weeks later from starting the water supply. The water temperature in three parts are almost the same each other above 3.0 m of depth. The temperature between 3.0 m and 4.5 m of depth in Part A, however, is lower and below 4.5 m of the depth higher than the others. At the same time, the temperature of the water sent was 14.7 ℃ immediately before the outlet. Those indicate that the water at the bottom layer was warmed by mixture with the discharged water and that the temperature at the layer between 3.5 m and 4.5 m of depth decreases by the advection of colder water from the bottom layer. Having in regards that the temperature below 4.0m of depth at Part A is almost constant, it is considered that the circulation is generated below 4.0m of depth.
One sees that the concentrations of DO are about 1 mg/l below 4.0m of depth at Part A. **FIG. 12** shows the conductivity at each part. The conductivity at Part B and C increases monotonically with the increase of depth. On the other hand, the conductivity at Part A decreases with the increase of depth below 3.5m. Therefore, we can expect that the increase of only 1 mg/l of DO is enough to restrain nutrient release from the sediment. For the moment, this fact gives us confidence that this water purification system is capable to purify reservoirs in which the volume of wave overtopping can not be expected to be so enormous.

5. CONCLUSION

The followings were made clear from this the investigations.
 1) The scale and the occurring frequency of wind waves in summer are almost the same as those

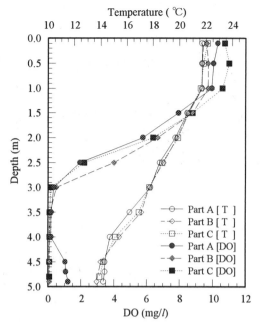

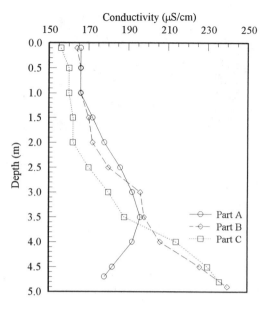

FIG. 11 Vertical distributions of temperature and concentration of DO; May 20 of 1998

FIG. 12 Vertical distributions of conductivity; May 20 of 1998

in winter in Matsubara dam reservoir.

2) The occurring frequency of wind wave heights which are bigger than 3.0 cm is about 30% in Matsubara dam reservoir, so we can anticipate that the annual rate of operation of this system will be as much as 30% in Matsubara dam reservoir.

3) The overtopping volume of about 138 m^3/day could be gained by one floating structure with 10 m width.

4) The increase of only 1mg/l of DO is large enough to restrain nutrient release from the sediment.

5) This water purification system is useful enough to purify water in reservoirs in which the volume of wave overtopping can not be expected to be so enormous.

ACKNOWLEDGEMENTS : This study was supported by the JSPS Research Fellowship for Young Scientists. We would like to thank Pro. Y. Nakamura and Mr. Y. Iseri for many valuable comments.

REFERENCES

Komatsu, T. *et al.* (1996a) : "An Experimental Study on the Effective Transformation of Wave Energy into Potential Energy", *J. Hydr., Coastal and Environ. Eng.*, *JSCE*, No.551/II-37, pp. 89-99 (in Japanese).

Komatsu, T. *et al.* (1996b) : "A Fundamental Study on Water Purification System in Man-made Lakes and Reservoirs Using Wave Energy", *J. Hydr. Eng.*, *JSCE*, 41, pp. 391-396 (in Japanese).

Komatsu, T. *et al.* (1997) : "An Experimental Study on Effective Conversion of Wave Energy into Potential Energy", *Pro. of the 27th Cong. of the IAHR*, D, pp. 571-576 .

Komatsu, T. *et al.* (1998) : "A Fundamental Study on Water Purification in Reservoirs Using Wave Energy",, *Pro. of the 11th Cong. of the APD-IAHR* (in press).

Environmental Hydraulics, Lee, Jayawardena & Wang (eds) © 1999 Balkema, Rotterdam, ISBN 90 5809 035 3

Basic concepts of urban disaster prevention channels into environmentally friendly channels

S. Shigeoka, K. Teshima & M. Oono
Pacific Consultants Company Limited, Tokyo, Japan

ABSTRACT: The objective of this project is to make a channel network by using the existing rivers, irrigation and drainage channels, as well as constructing numerous connecting channels, based on the concepts of urban disaster prevention channels of Koshigaya city. The improvement of the channel network involves the reclamation of waterside space in urban areas and prevention of natural disasters. The objectives are: 1)To provide clean waterside space to the people through friendly open channels, and to provide a feeling of prosperity and peace of mind. 2)To serve as water for fire-extinguishing purposes and household use during times of emergency such as great fires. 3)To discharge rainwater during floods or control floods. This thesis concerns the construction of the waterside network and planning for friendly open channels making up the network.

1. INTRODUCTION

The city of Koshigaya have an area of $60km^2$ with a population of 300,000 and a population density of 5,000 people $/km^2$. Five rivers flow through the city and along its boundaries. Numerous drainage and water supply channels are connected to the rivers thereby covering the city with the networks. The city has been conducting urban redevelopment projects while harmonizing nature and people by stressing. Safety, convenience, and comfortable environment as the basic concept. Historically speaking, this city is well known as the "riverside", "district" from olden times , surrounded by natural environment of water and greenery. However, due to the rapid industrial development and increase in population brought about by the high level of economic growth from the 1960s, this environment is gradually deteriorating, lowering the quality of the river and channel waters flowing through the city. Furthermore, the rate of discharge has greatly increased due to urbanization and increase in housing sites, leading to the increased possibility of floods. Measures have also become necessary in the case of large scale disasters such as earthquakes and fires. This paper delineates the basic concept and the basic development plan of the rivers and channels for disaster prevention channels as a part of a disaster prevention project in Koshigaya city. The plan was prepared based on the lesson leaned from the Great Hanshin Earthquake that hit Kobe in January 1995. The basic plan is to develop the rivers, drainage channels, and water supply channels into disaster prevention channels and environmentally friendly channels in order to utilize them for multipurpose use during ordinary and emergency periods.

2. PROJECT PLANNING AND DESIGN PROCEDURE

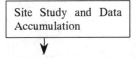 -Topography, geology, hydrology, and water quality.
-Urban planning and river study.
-Sewers, agricultural channels and underground facilities.
-Existing fire prevention facilities .

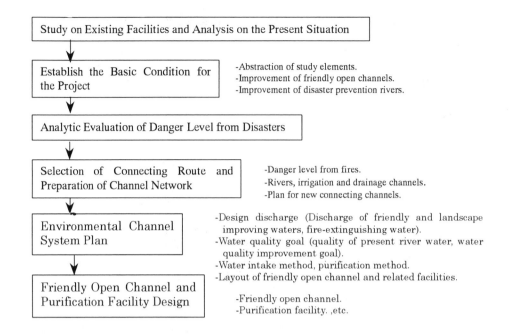

3. SELECTION OF NEW CONNECTING CHANNEL ROUTE AND IMPROVEMENT OF WATERSIDE NETWORK

In order to improve the waterside network using the existing rivers and channels, the channels for connecting the various channels need to be improved. In the case of Koshigaya city, the selection of the planned water connecting route will be a critical point in view of the water usage during disasters, based on the following: Land use in urban areas (commercial and housing areas), structure of buildings in the area, factories and facilities dealing in dangerous material, evacuation areas during disasters and availability of evacuation routes. The entire basin was evaluated in terms of the level of danger during fires, as shown below. This basic data was used in the selection evaluation of the new water connecting route.

3.1 Analysis on the Level of Danger During Disasters in the Basin and Planning of the Connecting Channel Route.

The evaluation items for the danger level is as follows:

1) Areas with no channels or rivers.
2) Density of wooden buildings (over 20 buildings/ha).
3) Areas with facilities dealing in dangerous material, such as gasoline stations.
4) Commercial and shopping areas.
5) Areas with no evacuation sites.
6) Areas with no fire-fighting facilities (fire stations, fire-fighting groups)
7) Areas with no water for fire-fighting purposes (earthquake proof fire tank).
8) Areas with a population density of over 80 people/ha.

The above eight items were categorized into degree of importance. The city was divided into a basic half mesh and the danger level (A,B,C) was evaluated.

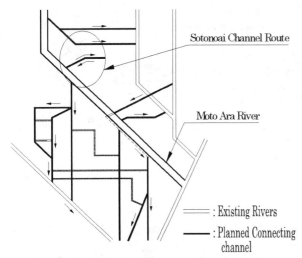

Sotonoai Channel Route

Moto Ara River

══ : Existing Rivers

── : Planned Connecting
      channel

Fig.1   Channel network

As a result, the area with a high danger level during fires was the old section of the city (commercial area and densely built-up area) located 10km by 3.0km along the railway line in where the Moto Ara River, a major river, runs through the city center. Thus, the plan for the connecting channel route was decided based on the danger level evaluation during fires. The newly planned connecting route consists of 11 routes as shown in Fig.1. The channel network will be completed by connecting these routes to the existing rivers and channels.

3.2   Planning the Environmental Friendly Channel System.

The Sotonoai rainwater main channel route is an example of a typical channel route of the project mentioned above.   As this route satisfied conditions such as site acquisition, adjustment with rainwater main channel projects and the conditions of the local people, it has been approved as a "Water and Recycled Sewage Project" by the Ministry of Construction. It is now a covered concrete channel, used as a footpath. However, foul odors are emitted from the gaps, creating an unpleasant environment.   By improving these channels into friendly open channels and rainwater drainage channels, the river water recycling project by water circulation was planned. The characteristic of this environmental friendly channel system project is to circulate the water from the major river Moto Ara, using the friendly and rainwater drainage channels, and return it to the starting point.   The water pumped from the Moto Ara River is purified and carried down the friendly open channel and discharged into the rainwater channel at the end.   Thus, the river water is recycled by returning the water to the Moto Ara River.   The outline of the facility is as follows:

Function of the Sotonoai Channel Route:

◆ Regular use 1) Purification of river water. 2) Renovation of the waterside and provide a place of recreation for the citizens.	◆Use during floods 1) Rainwater drainage. 2) Flood control.

◆Use during disasters such as fires and earthquakes
 1) Provide water for fire-extinguishing and household use.
 2) The friendly channel and promenade to serve as an evacuation route.
 3) The friendly channel and greenery to prevent the spreading of fire.

Major facilities:

    1) Water circulation and purification facility.
    2) Friendly open channel (channel, promenade, greenery).
    3) Rainwater main channel.

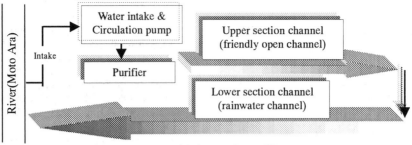

Fig.2   Sotonoai Rainwater Channel Route

4.   FRIENDLY OPEN CHANNEL PROJECT

4.1   Project Conditions

Water in the urban area will be secured for the friendly open channel.   In order to provide a feeling of prosperity and peace of mind to the citizens, the channel will be located in the upper section of the rainwater main channel.   The facilities to be attached are promenades, greenery and a water intake pit for fires.   When planning the friendly open channel in urban areas, it is important that they have other attachments such as greenery, and not just serve as a waterside space.   The following points were considered when planning the friendly open channel:

Scenic Effects of Water

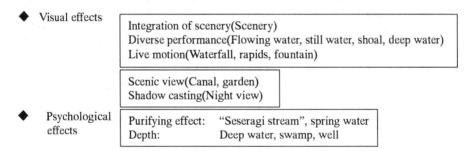

Preservation of Japanese Culture in the Waterside Space

The request for the "Seseragi Stream" by the citizens is greatly increasing, especially in the Sotonoai main line project area, where housing is increasing. The "Seseragi stream" is one of the most familiar type of friendly channel to the Japanese.   In order to realize this, suitable land must be secured and the stream should be suited to the other facilities on the route, preserving the environment.

4.2   Friendly Channel System

Friendly Open Channel

The overall image of the friendly open channel is as shown in the fig 3.

854

Promenade

The existing channel for the project is located in the center of a residential area. Although the path over the channel is narrow, it is used daily by the residents. Thus, the path running parallel to the channel was planned with the safety of the pedestrians and cyclists in mind, adapting to the scenery and surroundings so that it would be acceptable to the residents.

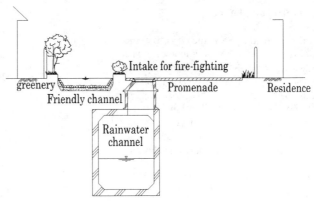

Fig.3   Cross Section of Friendly Channel

Greenery

Greenery by the "Seseragi stream" and footpaths naturally provide a tranquil atmosphere combined with the water scenery. The greenery according to the type and quantity will provide character and expression to the place, bringing a sense of season and change.

5.   PURIFICATION FACILITY PLAN

The purification facility assures that the water of the Moto Ara River is suitable as a "Seseragi stream" water.

5.1 Water to be Treated

In terms of extinguishing fires, the planned discharge for the friendly open channel should be adequate for one fire engine.   The treated water will be 0.15m³/s, to be supplied 24 hours.

5.2 Water Quality

Water quality of rivers in Koshigaya city

Present Condition:

Name of River	1/10 Drought	BOD(mg/l)	SS(mg/l)	DO(mg/l)	Coliform(100ml)
Oootosi-Furu Tone	—	6.8	34	8.0	—
Niigata	—	10.0	37	5.2	—
Moto Ara	1.22m³/s	5.1	20	7.9	29,000
Ayase	—	8.0	22	4.4	—

Planned Goal:
   The objective is to remove SS, purifying the present SS 20 to under SS 10.   The treated water will be for landscape improving water and friendly water.   However, the target quality will be basically for landscape water, assuming that water playing within the channel will not be conducted.

Goal Values of Basic Water Quality

Item	Landscape Water	Friendly Open Channel
Coliform	Under 1,000/100ml	Under 50/100ml
BOD	Under 10mg/l	Under 3mg/l
PH	5.8-5.6	5.8-8.6
Turbidity	Under 10 degrees	Under 5 degrees
Odor	Not unpleasant	Not unpleasant
Color	Under 40 degrees	Under 10 degrees

5.3 Treatment Method

As the removal of SS is the objective, the sand filter method was adopted. The following methods were considered: 1) Upward movement floor type continuous sand filtration. 2) Pressurized sand filtration. 3) Quick filtration method. However, considering construction costs, installation space and appearance, the pressurized filtration method was chosen.

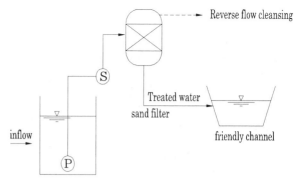

Fig.4　Sand filter method

Outline of Structure:	- It is a sealed structure.　Filtering is made by pressurization. - Filter material will be cleansed intermittently by the reverse flow.
Treatment Capacity:	- Filtering speed: 369m^3/m^2.day (high speed type) - Reverse flow cleansing volume:　Approx. 3.5% of original water volume.
Advantages:	- The height of the water tank does not have to be so high - By increasing the level of the water pump, the treated water　may be stored into elevated tanks.　Thus, a transfer pump will　not be necessary.
Disadvantages:	- Treated water is not available continuously.　(Reverse cleansing stage 15 minutes.) - Reverse cleansing unit necessary (reverse cleansing unit and treated water tank)

6. SUMMARY

An all-out effort was made by the River and Sewage Authority to get the land for the construction of friendly channels involved in water recycled sewage project from its' early planning stages to the detailed design state. Koshigaya city is densely populated and urbanization in this area is still rapidly progressing. Therefore, this kind of project will be brought about a great help in order to improve the waterside environment and establish the urban disaster prevention networks.　The biggest subject is to provide good enough water to the irrigation canals in all seasons, even in non-irrigation period to make it possible the daily use of friendly channels and disaster prevention channels.

Therefore, studies of water balance, provision of water and coordination with relevant organization is needed for this project to be succeeded.

Environmental Hydraulics, Lee, Jayawardena & Wang (eds) © 1999 Balkema, Rotterdam, ISBN 90 5809 035 3

Experimental studies on a system for catching *Oncomelania* in irrigation channels

Hechun Chen & Zhaohan Pan
Department of Architectural Engineering, The University of Hydrology and Electrical Engineering, Yichang, China

Changhao He
Department of Parasitology, Tongji Medical University, Wuhan, China

ABSTRACT: To prevent the spread of *Oncomelania* through irrigation channels, a new experimental method known as the *Oncomlania* Catching System (OCS) is presented. The proposed OCS imitates the hydraulic effects that are necessary both to separate snails from load materials and to deposit them in the collecting pool of the OCS. The results of experiments on live snails shows that the OCS possesses exceptionally high success rates in catching *Oncomelania* in channel flows.

1. INTRODUCTION

It is many years since experts from the World Health Organization stressed the need to use effective measures to prevent the spread of parasites after the building of large hydraulic projects and irrigation systems (WHO,1950). This particular plea was made in response to the conflict between economic development and ecological balance in Africa and Latin America, where such projects had often led to increased incidence of human *schistosomiasis*(hutt and Birkitt,1986;Mao, 1990). Many engineering projects and irrigation networks have recently been built in China, especially in Hubei Province, which contains the central basin of the Yangtze River. Here, the dominant species of blood fluke is *Schistosoma japonicum*, a parasite which often provokes extensive granulomatous reaction in the liver (Dunn,1969). In this region, invasion of irrigated areas by the parasite's major intermediate host, a water snail of the genus *Oncomelania,* poses an immediate threat to human health. One of the simplest ways to control the incidence of *schistosomiasis* is to intercept such snails before they enter the irrigated areas. However, the two methods used to catch the snails in China have either proved to be inefficient or expensive to implement (Xu, 1989-1990;Yang, 1989-1990). The nets used to filet load materials sometimes break under the weight of sediment and floating debris and, when they do stay intact, there are often problems with raised water levels and the extraction of the snails from the waste materials. The other method, draining the canals leaves some snails exposed on the bed, while others are deposited after passing through culvert pipes. However, the catching is often hampered by waste blocking the smaller pipes. A new experimental method known as the *Oncomelania* Catching System (OCS), combines the principles of fluid mechanics with those of snail biology. This system imitates the hydraulic effects that are necessary both to separate snails from load materials and to deposit them in collecting pool connected to the channel. The results of experiments on live snail shows that these tasks are achieved with exceptionally high success rates.

2. BIOMECHANICS CHARACTERS OF *ONCOMELANIA*

Oncomelania is a kind of water snail. Its body consists of two parts: the shell and the soft body. The shell is hard and approximately conical. Commonly, the shell has a height of less than 12mm and a diameter of less than 4.5mm. Although being a kind of amphibious animal, *oncomelania* has limited biological power.

*This project was supported by Natural Science Foundation of China (39270630)

Table 1. Biomechanics Characters of Oncomelania of Different Ages

Ages	1 week	3 week	5 week	7 week	9 week	mature
Diameter D(mm)	0.57 ± 0.04	0.80 ± 0.08	1.13 ± 0.11	1.97 ± 0.18	2.90 ± 0.16	3.16 ± 0.36
Height h(mm)	0.90 ± 0.09	1.37 ± 0.17	2.04 ± 0.38	4.49 ± 0.49	6.32 ± 0.54	8.29 ± 1.05
Weight(10^{-5}N)	0.19 ± 0.03	2.82 ± 0.55	4.47 ± 1.70	15.20 ± 4.64	20.10 ± 5.29	70.0 ± 12.0
Specific Weight (g/cm^3)	1.56	1.36	1.57	1.65	1.73	1.63
Suction Power* (10^{-5}N)	10.9 ± 1.70	111.2 ± 54.6	164.5 ± 66.4	421.1 ± 221.0	646.2 ± 331.8	1301 ± 571
Rate of Settling ω_0(10^{-2}m/s)	--	3.204	4.655	1.017	--	15.73

*for snail sticking to glass plate

Table 2. Suction and Threshold Velocity on Different Base Materials for Mature Snail

Base Materials	concrete	iron plate	wood plate	Glass plate	leaf of willow	leaf of poplar
Suction Power (10^{-5}N)	1112 ± 650	1272 ± 649	991 ± 401	1326 ± 588	731 ± 322	860 ± 473
Threshold velocity** （m/s）	$1.55\sim2.21$	$1.62\sim2.10$	$1.40\sim1.99$	$1.50\sim2.20$	$1.34\sim1.78$	$1.28\sim1.62$

** for snail sticking to various kinds of base material; while, for the snails just lying on the base material, the threshold velocity is only about 0.2m/s

On land, it can only crawl very slowly; in water, it can not swim like fish or frog and can not float on water surface too because its specific weight is greater than that of water. However, *oncomelania* has relatively strong adsorptive capability. It can stick to the surface of various kinds of objects no matter on land or in water. Precisely because of this reason, *oncomelania* is spread all around by various kinds of carriers materials floating in water flows. In order to find out an effective way of preventing the spread of the snail with water flows, we measured the biomechanics characters of the snail and studied their motion laws in static water and in water flows at the beginning of the work. Table 1. and Table 2 give some of the biomechanics characters of the snail we obtained.

3. THE ONCOMELANIA CATCHING SYSTEM AND EXPERIMENTAL METHOD

The *Oncomelania* Catching System, as shown in Figure 1, consists of a separation tank、a descending pool and a collecting box . It is housed in a glass channel 5m long, up to $1.0\sim1.5$m wide and 1.10m high. The separate tank take its shape by the gate 1、gate 2 and the side walls. By adjusting the opening of the two gates we can control the discharge、velocity and flow regimen in the separate tank. The designed flow regimen for the tank is submerged hydraulic jump .in which strong shear flow、fluctuating velocity 、pulsating pressure and noise are capable of inducing the snail to close its operculum and separate itself from the carrier materials such as small leaves. A array of baffle blocks are installed on the bed of the tank to strengthen the surface swirl and the turbulence of the submerged hydraulic jump .The descending pool is circled by the gate 2 、the downstream sill and side walls. It is a tapered divergent section where the flow is decelerated to ensure the snails deposit into the collecting box installed on the bottom of the pool. By adjusting he height of the downstream sill and the opening of gate 2 , the velocity distribution and flow regimen of the pool can be regulated.

Healthy *oncomelania* , $10\sim12$mm in length of variable age, collected from fields in Yinchen County , Hubei Province, were used in our tests. Leaves of willow and poplar of size of 2x3 cm^2 were taken as their carrier materials Experiments on the effect of the OCS were conducted in a sequence which was designed to identify optimum hydraulic conditions for both separating them from load material and catching them in the disturbed flow. This sequence also enabled comparison to be made between the catching rates for the OCS at different discharges and velocities. First, snails adhered to small leaves were placed onto the water surface in front of the gate 1 and their movements in the OCS afterward were observed and the rate of separation 、sinking behavior and the catch rate were recorded. To ensure the obtained rates are meaningful in statistical sense, the number of the test samples for every working case is greater than 60. The maximum model discharge is 0.33 m³/s and the maximum water level difference is 1.6m.

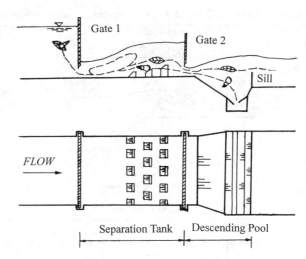

Figure 1.   Sketch of The Snail Catching System

4.   EXPERIMENTAL RESULTS

4.1 The Factors That affect The Catch Rate of The OCS

The results of our experiments shows there are many factors having an effect on the catch rate of the OCS. Simply, these factors can be classified into two categories: biological and hydraulic. The biological factors covers the size、shape、weight and the biological power of the snails; The hydraulic elements mainly includes discharge、water level、velocity and flow regimen etc. The former is natural and cannot be controlled. While, the latter could be regulated in light to our need by adjusting the opening of the gates and changing the structure of the OCS. Therefore, our experimental works were aimed at creating suitable hydraulic conditions , especially in finding optimum flow pattern for catching the snails in channel flows.

4.2 The Optimum Flow Pattern For Catching Snails with The OCS

We had tried a lot of flow patterns by adjusting the opening of gate1 and gate 2 、 the height of the downstream sill and the position 、 number and shape of the baffle blocks to find the optimum flow regimen of high catch rate. Table 3 shows the catch rates of four kinds of typical flow regimens we had tested. It is found that the submerged hydraulic jump is obviously of higher catch rate and is more stable than other flow patterns. Considering on this reason, we will focus our attention on the submerged hydraulic jump case in the flowing analysis.

4.3   The Way of The OCS Works

As Shown in Figure 1, when the snails adhered to small leaves enter into the separating tank, they have the alternative of swift running away with the main current near the bottom and circling one or more times in the surface swirl at first and then drifting downstream. The circling times is dependent on the strength of the surface swirl. It was observed that most of the snails adhered to carrier materials were separated from the carrier material at the interface of the surface swirl and the bottom main current, where high shear and strong turbulence characterize the flow, as shown in Figure 2, and the more the times the snails circle in the surface swirl the higher the separate rate is. As the snails escape from the swirl and drift downstream, some

859

Table 3   The Catch Rate of The Oncomelania Catching System at Different Flow Patterns

Flow Pattern	free hydraulic jump without baffle Blocks	submerged hydraulic jump without baffle Blocks
Sketch	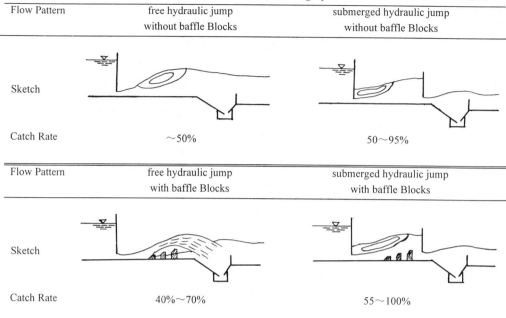	
Catch Rate	~50%	50~95%

Flow Pattern	free hydraulic jump with baffle Blocks	submerged hydraulic jump with baffle Blocks
Sketch		
Catch Rate	40%~70%	55~100%

of them will come into collision with the baffle blocks and lead to their separation from the small leaves. After the snails passing gate 2 with water flow, they will encounter a weak water jump just at the outlet of gate 2 , and a few of them are taken apart from the load materials. At last , the snails drift into the decelerating water flow in the divergent section. They descend quickly into the collecting box, while the carriers of small leaves float into the downstream channel. It was fond that 100% of the separated snails were deposited into the collecting box. This means the catch rate of the OCS are mainly determined by its separate rate.

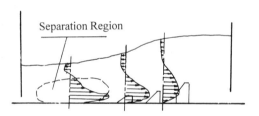

Figure 2. Separation Region and Velocity Distribution in The Separation Tank

5.   CONCLUSION

The results of the present study indicate that the *Oncomlania* Catching System would possess a number of advantages over the traditional netting and draining methods of *Oncomelania* control. The identification of the hydraulic conditions critical for efficient separation and catching should make it possible to trap most snails and then let the waste continue to flow downstream. Moreover, the OCS would require less land and smaller engineering than combined net and culvert systems and probably give a higher catch rate.

6. REFERENCES

Dunn, A. M.(1969)Veterinary Helminthology. London: Heinemann
Hutt, M. S. R. & Burkitt, D. P. (1986). The Geography of Non-infectious Disease. Oxford University Press

World Health Organization (1959), Man-made Water Resources and Their Impact on Schistosomiasis. Geneva:WHO

Xu, X. (1989-1990). Experiments for Preventing The Spread of Snils in Culverts and Slice Gates Using Catching Nets. Research report on blood fluke disease in Hubei Province. Wuhan: Annual Conferrence Procedings on Parasitology(in Chinses)

Yang, X. (1989-1990). A Study of the Deposition and Movements of Snails and their Ovum in static Water. Research report on blood fluke disease in Hubei Province. Wuhan: Annual Conferrence Procedings on Parasitology(in Chinses)

C. He etc., Simulation experimens for catching Oncomelania in irrigation canals. Annals of Tropical Medicine and Parasitology, Vol. 88, No.1, 103-106(1994)

C. Chen etc.,Experimental Studies on The Threshold Velocity of Oncomelania in Flume, Advances in Hydro-science and -Engineering, Volume II March 22-26,1995,Beijin, China

Environmental Hydraulics, Lee, Jayawardena & Wang (eds)© 1999 Balkema, Rotterdam, ISBN 90 5809 035 3

A low flow increase for improving river environment

Hiroe Sakata
ICE Japan Company Limited, Yokohama, Japan

ABSTRACT: Progress of urbanization is ever accelerating and causing various environmental impacts on mankind. Its impacts on river environment are no exception. Conspicuous impacts on urban rivers are increasing peak flood discharge and total runoff, shortening of the runoff period, decrease in the low flow rate, and the degradation of water quality. These phenomena are undesirable in view of urban environment. This paper introduces a low flow increase plan for improving the river and urban environment.

1. INTRODUCTION

Urbanization has been progressing steadily worldwide. In particular, this phenomenon is remarkable in Asian countries. Natural ground and farmland has been developed into housing areas and infrastructures. On the contrary, roofs and paved areas have increased thereby reducing infiltration capability. As a result, the runoff pattern in urban areas has changed. The increase in runoff volume and peak discharge in rivers and the deterioration of water quality are conspicuous.

Emphasis has been placed on improving flood-handling capabilities. Various measures have been taken for this purpose. Channel improvement work has been implemented and construction of diversion tunnels for making shortcuts of rivers is often seen in large cities to overcome flooding. Installation of flood detention and infiltration facilities is used to reduce the peak discharge of rivers. Reduction of infiltration capabilities in urban areas has caused the lowering of ground water tables and low flow rates of rivers and, as a result, the heat-island phenomenon. Streams without flow are often seen in urban areas thereby creating disagreeable scenery. To restore, improve, and maintain a river's natural environment is only possible by spending a large sum of money and exerting a great deal of effort by the river managing authority.

The Nameri River (hereafter referred to as the River) is a small size river whose length is approximately 5.6km with a catchment area of about 12km^2. It flows through the middle of Kamakura, one of the oldest cities in Japan, visited by nearly 20 million tourists annually.

Until 35 years ago, a large portion of the riverbanks was used for farmland, but has now developed into housing lots. As a result of this urbanization, the runoff pattern of the catchment area has changed: increased peak discharge and very low normal flow rate. In particular, only a small amount of water is seen during winter months.

As Kamakura is famous for being one of the most attractive tourist cities in Japan, effort is being made to maintain the historical landscape. In order to improve the landscape of the city, we have been promoting the installation of infiltration facilities to residential houses thereby recharging the ground water and, as a result, increasing the low flow of the River. As the project started only several years ago, it may take a long time to notice any visible results.

2. PROJECT RIVER AND BACKGROUND

The project area is in Kamakura City, which is approximately 43km south of central Tokyo. The features of the River and its catchment area are as follows:

Catchment Area:	11.9km^2
River Length:	5.6km
Channel Slope:	1/500 to 1/30
Developed Area:	5.0km^2 (42%)
Undeveloped Area:	6.9km^2 (58%)
Area Population:	47,000
No. of Houses and Buildings:	Approximately 11,800
Annual precipitation:	Approximately 1,500mm

Some 700 years ago, Kamakura was the political center of Japan and remained so for approximately 150 years. Many historical assets exist and millions of tourists visit the city yearly from all over the world. The city has elaborated on a City Master Plan to preserve the historical assets as well as to improve the urban environment.

A large portion of the plane area in the catchment area of the River was used as rice paddies and farmland until 35 years ago. But, now the area is fully redeveloped for housing, parking lots, buildings, stores, roads, etc. and, as a result, the area's infiltration capability has been reduced substantially and urban-type flooding started to occur. Most of the undeveloped area in the River's catchment area is mainly hills covered by vegetation. The geology of the hills is composed of weathered rock. As the hills are steep, the rainwater runs off quickly and very little infiltration takes place. Thus, the River's low flow often becomes only several liters per second during dry seasons. This condition must be remedied from an aesthetic viewpoint.

As a part of the City Master Plan, we proposed installing infiltration facilities in order to increase rainwater infiltration thereby recharging groundwater and, as a result, to increase the River's low flow rate as well as to reduce the peak flood discharge.

3. DESIRABLE FUNCTIONS OF THE NAMERI RIVER

Urbanization has greatly changed not only stormwater runoff patterns, but, also the water cycles and the natural environment, such as the living environment of wildlife and vegetation in river basins. While the increase in runoff volume and peak discharge and the shortening of the runoff period in urban rivers have become conspicuous, the habitat of wildlife has been reduced due to the decrease of rice fields, farmland, wooded areas, and wetlands as well as a decrease in the water bearing capacities of the city areas. Further, the decrease in the water bearing capacities of catchment areas has resulted in a heat-island phenomenon, i.e. excessive temperature rise. This condition may be improved by increasing the rivers' low flow rates.

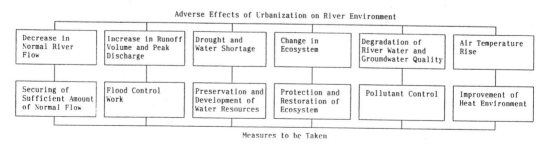

Fig. 1   Effects of Urbanization on Environment and Measures to be Taken to Remedy

Adverse effects of urbanization on environment and measures to be taken against the effects can be simplified as shown in Fig. 1. As seen in this figure, most of the adverse effects to river environment caused by urbanization may be alleviated by restoring the water cycle capability of the catchment area, i.e. increasing the infiltration capabilities.

A river in an urban area is a precious natural environment to which people can easily access and enjoy. In particular, a river's natural environment, such as greenery and clean water, is very precious resources that may enhance the amenity of the area and give us relaxed feelings. Greenery and clean water are very important factors for the landscape along the river. Further, as a river is very important when forming the city's landscape, it is vital to improve and maintain an agreeable urban river environment.

As a development strategy for the City Master Plan of Kamakura, the River's desirable functions were defined as follows:

a. To have sufficient flood conveying capacities
b. To maintain a sufficient amount of low flow
c. To maintain clean water quality
d. To have an aesthetically desirable landscape
e. To have a versatile biological system

It is of utmost importance for the River to have sufficient floodwater conveying capabilities. Thus, past river improvement work was undertaken by emphasizing flood protection capability for the safety of human life and protection of property. Those projects were mainly conducted in and adjacent to the river channel.

The items "b" through "e" above are all interrelated. Although the river water is no longer used for irrigation purpose in the area, a sufficient amount of water is necessary to maintain the biological system in the River as well as for aesthetic reason. A desirable biological system in the River could not be maintained without a sufficient amount of clean water even during dry seasons.

To make people aware of present environmental problems, environmental education for children, who shall be responsible for the future environment, is extremely important. Although the River is very small, children can closely observe fish, birds, insects, wildlife habitats, biological system as well as the water cycle. The River can provide children with precious learning material. Thus, we conduct a river walk program for children. From an educational viewpoint, it is also meaningful to maintain a sufficient amount of clean water in the River.

In view of the above background, we considered that the installation of infiltration facilities would be effective to recharge groundwater thereby increasing the low flow of the River. Thus, we have been promoting the installation of infiltration facilities in the city.

4. INFILTRATION FACILITIES

There are many types of infiltration facilities, such as infiltration inlets, trenches, pavements, ditches, wells, and ponds. The combination of rainwater infiltration inlets and a trench, as shown in Fig.2, was selected for our plan because of easy installation to private houses. Details are as shown in Fig. 3.

The capability of an infiltration facility varies greatly depending on such factors as the size and shape of the facility, the properties of the soil, the level of the groundwater table in the area, water depths in the facilities, etc. In particular, the infiltration capability of a specific infiltration facility is greatly related to the permeability of the saturated soil in which the facility is to be installed. Permeabilities of saturated soils having different grain sizes are as shown in Table 1.

Table 1 Permeability of Saturated Soil and Grain Size

Soil Type	Clay	Silt	Very Fine Sand	Fine Sand	Sand	Coarse Sand	Small Gravel
Grain Size (mm)	0~0.01	0.01~0.05	0.05~0.10	0.10~0.25	0.25~0.50	0.50~1.0	1.0~5.0
Permeability k (cm/sec)	3×10^{-6}	4.5×10^{-4}	3.5×10^{-3}	0.015	0.085	0.35	3.0

Source: Rainwater Storage and Infiltration Technologies Association (1995)

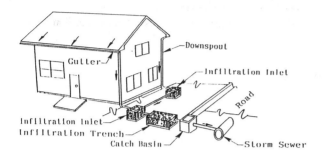

Fig. 2 Schematic View of Proposed Infiltration Inlets and Trench

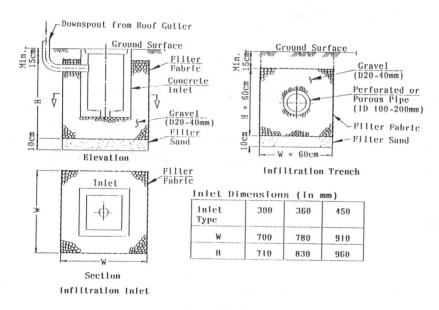

Inlet Dimensions (in mm)

Inlet Type	300	360	450
W	700	780	910
H	710	830	960

Fig. 3 Details of Proposed Infiltration Inlet and Trench

Numerical analysis methods have been developed to calculate the infiltration capabilities of an infiltration trench and inlet (Herath and Musiake, 1991). Also, various institutions have conducted in-situ tests to learn the infiltration capabilities of different types of facilities in different soil types. Thus, we utilize the results of those researches and tests for estimating the infiltration capabilities of the proposed types of infiltration facilities to be installed in different soil types. The following equations are proposed for the practical estimation of the infiltration rate of an infiltration inlet and trench (Rainwater Storage and Infiltration Technologies Association, 1995):

$$Q = k \, t \, K \tag{1}$$

where, k is the permeability of a saturated soil in cm/sec. K is the relative infiltration rate of a facility and can be expressed for an infiltration inlet as follows:

$$K = aH^2 + bH + c \tag{2}$$
$$a = 0.120W + 0.985 \tag{3}$$
$$b = 7.837W + 0.82 \tag{4}$$
$$c = 2.858W - 0.283 \tag{5}$$

where, W and H are the width and height of the inlet in meter respectively.

K for an infiltration trench is as follows:

$$K = aH + b \tag{6}$$
$$a = 3.093 \tag{7}$$
$$b = 1.34W + 0.677 \tag{8}$$

where W and H are the width and height of the trench in meter respectively.

By using the above equations, the infiltration capabilities of the proposed infiltration inlets and trench were calculated for different soil types as shown in Table 2.

The soils in the catchment area of the River exhibit spatial variety. But, in the plane area they are mainly loamy soils whose permeabilities are considered to be in the range of 1.5×10^{-3}cm/hour to 2.5×10^{-3}cm/sec. Thus, the infiltration capabilities of the proposed infiltration inlets are estimated to be 0.364m³/hour to 0.986m³/hour per each unit. The infiltration capability of the proposed infiltration trench is estimated as being 0.180m³/hour to 0.300m³/hour per one meter.

At a glance, these infiltration capabilities seem to be very small. But, they can be interpreted as follows: When rainfall having the intensity of 20mm/hour occurs, the amount of the rainfall on a house having a 100m² roof is $(20/1,000) \times 100 = 2.0$m³/hr. This amount of rainwater can be infiltrated into the ground through four Infiltration inlets each having the capacity of 0.500m³/hr or the combination of three inlets having the same capacity and two meters of infiltration trench having the capability of 0.250m³/hour. A rainfall having the intensity of 20mm/hr is a relatively strong rain. From this example, it is easy to understand how the proposed infiltration facilities are effective to reduce the direct runoff of rainwater from a housing area and how effective in restoring the water cycle affected by urbanization.

Table 2 Infiltration Capabilities of Proposed Infiltration Facilities

(Unit in m³/hr)

Soil Type	Permeability(cm/s)	Inlet Type 300	Inlet Type 360	Inlet Type 450	Trench (per 1m)
Silt (0.0~10.05)	4.5×10^{-4}	0.110	0.136	0.178	0.054
Sandy Silt (0.0~30.07)	2.5×10^{-3}	0.606	0.760	0.986	0.300
Very Fine Sand (0.05~0.10)	3.5×10^{-3}	0.848	1.064	1.381	0.421

The plane area along the River is fully developed. As it is difficult to secure sufficient open space for installing infiltration facilities, we request area residents to install infiltration facilities in the rainwater drainage system when they rebuild their homes. The extra cost required to install infiltration facilities are very small because they must install a rainwater drain system anyway. We also explain the importance and effects of infiltration facilities. As the residents consider the River to be their common asset, they understand the meaning of the facilities and are very cooperative.

5. CONCLUSION

Kamakura City started the above-mentioned project in 1996. We must emphasize that the project has been carried out by using only a small amount of public money, but in particular, with the strong support of the general public when they rebuild their homes. The installation of an infiltration facility to an individual's home is optional. During the 1996 fiscal year, 1,408 homes were rebuilt in Kamakura City. 31% of them, i.e. 437 homes, installed 2,319 infiltration inlet units. During 1997 fiscal year, 1,290 homes were rebuilt and 387 homes (30%) installed 2,075 infiltration inlet units.

It is said that more than 50% of homes in Japan are rebuilt within 50 years because the main building material is wood. If the above-mentioned rate of installing infiltration facilities continues, a substantial amount of rainwater infiltration will take place. We are confident that the project is an economical and cost-effective river environment improvement method. We can hardly wait to see an increase in the low-water flow of the River in the very near future.

6. REFERENCES

Herath, S., and Musiake K. (1991). "Design of infiltration facilities for urban flood control, " Proceedings of the International Symposium on Environmental Hydraulics, Hong Kong, Dec. 1991, 1425-1430.

Musiake, K., Herath, S., Hironaka, S., and Okamura, J. (1992). "Effects of Urban Storm Infiltration Systems and Their Evaluation," Proceedings of the International Water Resources Symposium, Tokyo, 1992, 591-596.

Rainwater Storage and Infiltration Technologies Association (1995). "Design and investigation of rainwater infiltration facilities technical guidance (Draft)."

4.5 River hydraulics/hydrology

Environmental Hydraulics, Lee, Jayawardena & Wang (eds) © 1999 Balkema, Rotterdam, ISBN 90 5809 035 3

A study on the numerical model of non-equilibrium sediment transport in unsteady flow

S.H. Lee

Rural Research Institute, Rural Development Corporation, An San City, Korea

ABSTRACT: A fully coupled one-dimensional mobile-bed river model for the condition of unsteady flow and non-equilibrium sediment transport with looped network system is created. The governing equations are as follows: continuity, motion, conservation of material in suspension, conservation of bed-material, sediment transport formula and roughness equation. The above equations are solved simultaneously using the Preissmann implicit scheme. Manning roughness coefficient with the bed form (ripple and dune) considered by van Rijn method is calculated at each time step. Applying this fully coupled sediment transport model to Belley reservoir of Rhone river, total trap efficiency (=0.4) gives reasonably good result comparing with the measurement (=0.49).

1. INTRODUCTION

Belley reservoir is located at the upstream part of Rhone river in France. The flushing of Verbois reservoir and Chancy-Pougny reservoir located at the upstream of Belley reservoir is operated every 3 years to evacuate the sediment deposits in these reservoirs. Because of this flushing, 600,000 m^3 of sediments are deposited at Belley reservoir after the measurement of river bed level. (result of flushing in 1990)

The length of Belley reservoir is about 18 km and the upstream reach of 4.5 km is natural river. Artificial canal with the length of 13.5 km is constructed to the downstream of hydraulic power station. (Fig. 1)

The flow is divided into two parts due to island and submerged dike in Cressin reservoir. Because of the velocity difference in these two canals, the sediment transport rates are quite different. The confluence and divergence problem is included in this research because of this reason. Taking into account the interaction of stream flow, sediment transport and bed forms, the effective roughness is used to compute the Chezy or Manning roughness coefficient by the method of van Rijn. (van Rijn, 1984)

As a result of measurement in 1987, 90 % of suspended material is in the range 0-200 µm, D_{50} is in the range 15-25 µm. Bed load, suspended load, settling, pick-up and effective roughness are considered. Van Rijn formula (van Rijn, 1984) is used for the sediment transport of bed load and suspended load.

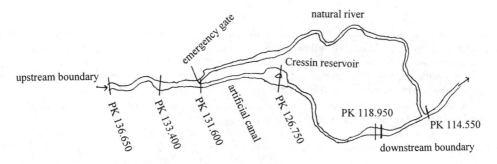

FIG. 1. Map of Belley Reservoir

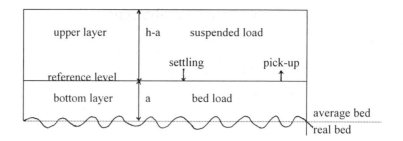

FIG. 2. Definition Sketch in the Sediment Transport Model

2. THEORY

2.1 Description of Sediment Transport Model

A phenomenon considered in this model is given in the following figure. (FIG. 2)

2.2 Governing Equations

The one-dimensional conservation equations for the sediment transport can be described by the following primary four equations:

Fluid continuity equation:

$$\frac{\partial A}{\partial t} + \frac{\partial Q}{\partial x} = B\frac{\partial z}{\partial t} + \frac{\partial Q}{\partial x} = q \tag{1}$$

Fluid motion equation:

$$\frac{\partial z}{\partial t} + \frac{1}{gA}\frac{\partial}{\partial x}(\beta\frac{Q^2}{A}) + \frac{Q^2}{K^2} + \frac{1}{gA}\frac{\partial Q}{\partial t} + \frac{Ke}{2g}\frac{\partial(Q/A)^2}{\partial x} = 0 \tag{2}$$

Suspended material conservation equation:

$$\frac{\partial CA}{\partial t} + \frac{\partial Qs}{\partial x} = S - qC \tag{3}$$

Bed material conservation equation:

$$(1-p)\frac{\partial Ab}{\partial t} + \frac{\partial Qb}{\partial x} + S = 0 \tag{4}$$

in which x = streamwise coordinate; t = time; A = wetted cross-sectional area; Q = discharge; Z = water-surface elevation above a datum; B = flow width; q = lateral flow; g = gravitational acceleration; β = momentum correction coefficient; K = conveyance; Ke = coefficient of expansion-contraction; C = average sediment concentration; Qs = volumetric suspended load; S = sediment flux between bottom layer and waterstream; Ab = bottom layer cross-section; Qb = bed load.

In addition to the above primary four equations, the van Rijn formula (part III, 1984) is used as sediment transport formula for suspended load and bed load as follows:

$$Qs = 0.012 \, [(V-Vcr)/\{(s-1)g \, D_{50}\}^{0.5}]^{2.4}D_{50}D_*^{-0.6} \, VBh^{-0.2} \tag{5}$$

$$Qb = 0.005 \, [(V-Vcr)/\{(s-1)g \, D_{50}\}^{0.5}]^{2.4}D_{50}^{1.2}VB \tag{6}$$

in which s = specific density; V = mean flow velocity; Vcr = critical mean flow velocity based on Shield's criterion; D_{50} = median grain size of bed material; D. = particle diameter; h = flow depth.

Roughness coefficient due to friction resistance uses Manning roughness coefficient derived from overall Chezy coefficient considering grain roughness and bed form roughness:

$$C' = 18 \log (12R / ks) \quad \text{or} \quad n = R^{1/6} / C' \tag{7}$$

in which C' = Chezy roughness coefficient; R = hydraulic radius; ks = effective roughness; n = Manning roughness coefficient.

The sediment flux between bottom layer and waterstream are given by Armanini and Di Silvio (1988) as follows:

$$S = (Qse - Qs) / L* \tag{8}$$

in which Qse = suspended load in equilibrium conditions; Qs = actual suspended load; L* = characteristic length which depends on the flow characteristic and sediment size.

In summary, a one-dimensional morphological system is described by the following seven equations:
   A. Fluid continuity equation   B. Fluid motion equation   C. Suspended material conservation equation
   D. Bed material conservation equation   E. Sediment transport equation (Bed load and Suspended load)
   F. Alluvial roughness equation   G. Sediment flux equation between bottom layer and water stream

2.3 Discretization

The above seven equations for fully coupled non-equilibrium sediment transport model are discretized using the Preissmann scheme of implicit finite difference. This scheme use the following approximations to the derivatives:

$$\frac{\partial f}{\partial t} = \frac{1}{\Delta t} [\psi \Delta f_{i+1} + (1-\psi) \Delta f_i] \tag{9}$$

$$\frac{\partial f}{\partial x} = \frac{1}{\Delta x} [(f_{i+1} + \theta \Delta f_{i+1}) - (f_i + \theta \Delta f_i)] \tag{10}$$

$$f(x,t) = \psi (f_{i+1} + \theta \Delta f_{i+1}) + (1-\psi)(f_i + \theta \Delta f_i) \tag{11}$$

in which i and n = gird point, θ = the weighting factors for time, ψ = the weighting factors for space.

In applications of the Preissmann scheme, it is supposed that all the function f(Z, Q, Zb, C) in the discretized algebraic equations are known at time level $n\Delta t$ and are differentiable with respect to Z, Zb, Q and C. Using a Taylor series expansion, the finite difference approximation leads to a system of four algebraic equations for every pair of points (i, i+1). One can obtain the linearized system for a pair of adjacent points (i, i+1).

2.4 Solution for Algebraic System and Boundary Condition

The system of four algebraic equations is solved for all computational points by the double sweep method which is often used to solve the St. Venant equations in fixed bed modeling.

Suppose that the discharge and concentration variations for upstream section i become the linear function of Z_i, Zb_i and C_i. It has the following relationship:

$$\Delta Q_i = F_i \Delta Z_i + G_i \Delta Zb_i + H_i \Delta C_i + K_i \tag{12}$$

$$\Delta C_i = F_j \Delta Z_i + G_j \Delta Zb_i + H_j \Delta Q_i + K_j \tag{13}$$

in which F_i, F_j, G_i, G_j, H_i, H_j, K_i and K_j are known coefficients for the given time.

873

FIG. 3. Looped Network

Using the above relationship, ΔQ_i, ΔZ_i, ΔC_i and ΔZb_i can be eliminated from the four algebraic equations. The following relationship is obtained:

$$\Delta Q_{i+1} = F_{i+1} \Delta Z_{i+1} + G_{i+1} \Delta Zb_{i+1} + H_{i+1} \Delta C_{i+1} + K_{i+1} \tag{14}$$

$$\Delta C_{i+1} = F_{j+1} \Delta Z_{i+1} + G_{j+1} \Delta Zb_{i+1} + H_{j+1} \Delta Q_{i+1} + K_{j+1} \tag{15}$$

The procedure can be applied from upstream to downstream for N computational points. A system of 4 (N-1) equations with 4N unknowns is obtained. If 4 boundary conditions are given, 2 for i = 1 and 2 for i = N, all coefficients $(F, G, H, K)_{i+1, j+1}$ can be computed by forward sweep. Unknown value $(\Delta Z_{i+1}, \Delta Zb_{i+1}, \Delta C_{i+1}, \Delta Q_{i+1})$ can be computed for all sections by backward sweep.

2.5 Looped Network with Internal Boundary Condition

The looped network has the points of confluence and divergence of tributaries or canals. Double sweep method is applied for the problem of confluence and divergence because of Cressin reservoir.
For divergence (A12), the following compatibility condition is used: (FIG. 3)

$$Z_A = Z_1 = Z_2, Zb_A = Zb_1 = Zb_2, C_A = C_1 = C_2, Q_A = Q_1 + Q_2 \tag{16}$$

For confluence (B34), the following compatibility condition is used:

$$Z_B = Z_3 = Z_4, Zb_B = Zb_3 = Zb_4, C_3 Q_3 + C_4 Q_4 = C_B Q_B, Q_B = Q_3 + Q_4 \tag{17}$$

3. APPLICATION

The simulated results of mobile bed unsteady flow model are compared with the measurement using the Manning roughness coefficient, n. To estimate n value for mobile bed, the soil conservation service (SCS) method is used as a basic n (=0.02) for initial value. After the model calculate n value with van Rijn formula as a basic n and additional n is added to basic n considering channel irregularity. (French, 1986) The variation of water level calculated by the model gives good result comparing with the measurement. (FIG.4)
The size distribution for bed material and suspended load are:

Bed material: $D_{16} = 0.034$ mm, $D_{50} = 0.150$ mm, $D_{84} = 0.288$ mm, $D_{90} = 0.375$ mm
Suspended load: $D_{16} = 3.4$ μm, $D_{50} = 11.5$ μm, $D_{84} = 28.0$ μm, $D_{90} = 36.0$ μm

Using $\Delta t = 5$ minute and D = 11.5 μm for the condition of Q = 700 m³/s, Z = 233.93 m, Zb = 255 m and varying C, the concentration variation along the Belley reservoir shows as follows: (FIG. 5)

The flushing operation in 1990 is lasted about 4 days. The discharge during 4 days changes with minimum Q = 512 m³/s and maximum Q = 990 m³/s. During flushing the concentration variation is observed with minimum C = 0.7 g/l and maximum C = 9 g/l. The concentration variation for 5 μm, 11.5 μm, 20 μm, and 30 μm are also given in the following figure. (FIG.6)

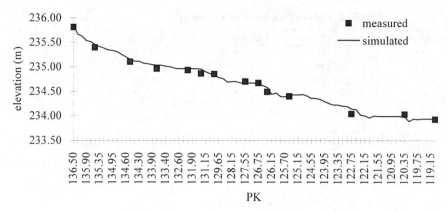

FIG. 4. Comparison of Water Surface Level

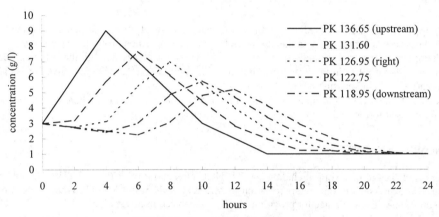

FIG. 5. Concentration Variation

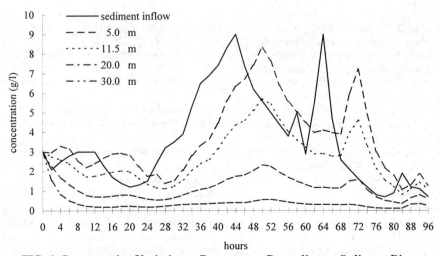

FIG. 6. Concentration Variation at Downstream Depending on Sediment Diameter

875

TABLE 1. Trap Efficiency of Sediment

Diameter(μm)	Inflow(m^3)	Deposit(m^3)	Outflow(m^3)	Trap(%)
5.0	295,000	3,000	292,000	1
11.5	295,000	84,000	211,000	28
20.0	295,000	201,000	94,000	68
30.0	295,000	284,000	31,000	89

The simulation result shows that D = 5 μm is almost transported to downstream, but D = 30 μm is almost deposited in reservoir. Suppose that sediment inflow is 295,000 m^3 and the size distribution is classified four classes such as D = 5 μm (30 %), D = 11.5 μm (30 %), D= 20 μm (20%) and D = 30 μm (20 %), the total trap efficiency is obtained as follows: (TABLE 1)

Trap efficiency = deposit/inflow = 119,000 / 295,000 = 40 % = 0.4

4. RESULT

A fully coupled one-dimensional sediment transport model for unsteady flow and non-equilibrium condition with looped network is created. It is applied to the Belley reservoir of Rhone river in France.
The results of simulation of this model show that D = 5 μm is almost pass through the reservoir and D = 30 μm is deposited about 89 % in the reservoir. The trap efficiency of simulation (=0.4) gives reasonably good result comparing with the measurement (=0.49) in 1990.
This model can be applicable for the non-equilibrium condition such as flushing where D_{50} of bed material is different from D_{50} of inflow sediment.

5. ACKNOWLEDGMENT

The writer is indebted to this work, including Dr. T.D.NGUYEN and Prof. P.BOIS, LTHE, Grenoble, France. The writer also expresses his sincere appreciation to Mr. E.TORMOS and Y.GIULIANI, CNR (Compagnie Nationale du Rhone), Lyon, France for providing the field data used in this study.

6. REFERENCES

Armanini, A., and Di Silvio, G. (1988). "A One-dimensional model for the transport of a sediment mixture in non-equilibrium condition.", *J. of Hydraulic Research*, Vol. 26, No. 3, pp. 276-286.
Cunge, J.A., Holly, F.M. Jr,Verwey, A. (1980). *Practical aspects of computational river hydraulics*. Pitman.
French, R.H. (1986). *Open-channel hydraulics*. McGraw-Hill.
Holly, F.M. Jr, and Rahuel, J.L. (1990). "New numerical/physical framework for mobile-bed modelling.", *J. of Hydraulic Research*, Vol. 28, No. 4, pp. 404-411.
Lee, S.H. (1993). *Modelisation numerique du transport solide en ecoulement non-permanent*. Thesis presented to Institut National Polytechnique de Grenoble (INPG) in partial fulfillment of the requirements for the Doctor degree (in French).
Van der Berg, J.A., and van Gelder, A. (1993). "Prediction of suspended bed material transport in flows over silt and very fine sand.", *Water Resources Research*, Vol. 29, No. 5, pp. 1401-1403.
Vanoni, V.A. (1977). *Sedimentation Engineering*. ASCE Task Committee.
Van Rijn, L.C. (1984). "Sediment transport: Part I. Bed load transport." *J. of Hyd. Div.*, ASCE, Vol. 110, No. 10, pp. 1431-1452.
Van Rijn, L.C. (1984). "Sediment Transport: Part II. Suspended load transport." *J. of Hyd. Div.*, ASCE, Vol. 110, No. 11, pp. 1613-1638.
Van Rijn, L.C. (1984). "Sediment transport: Part III. Bed forms and alluvial roughness" *J. of Hyd. Div.*, ASCE, Vol. 110, No. 12, pp. 1749-1754.
Van Rijn, L.C. (1988). "Application of sediment pick-up function." *J. of Hyd. Div.*, ASCE, Vol. 112, No. 9, pp. 1749-1754.

Environmental Hydraulics, Lee, Jayawardena & Wang (eds) © 1999 Balkema, Rotterdam, ISBN 90 5809 035 3

River junction design for urban flood control: A case study

J.H.W.Lee – *Department of Civil Engineering, The University of Hong Kong, China*

H.W.Tang – *College of Hydroelectric Engineering, Hohai University, Nanjing, China*

W.C.Chan – *Binnie Black and Veatch Hong Kong Limited, China*

G.Wilson – *Danish Hydraulic Institute, Denmark*

ABSTRACT: The Yuen Long Bypass Floodway is designed as the floodway to collect flows from the Sham Chung and the San Hui Channels, and also to serve as a diversion channel of the Yuen Long Main Nullah. Tests with a 1:50 undistorted Froude scale model were conducted to study water stages and flow characteristics in the floodway, and to investigate optimal design arrangements at the channel junctions and transitions. Special emphasis is placed on the complicated flow interactions at channel junctions, which cannot be resolved by 1-D mathematical models for unsteady free surface flows that are in commercial use. Five schemes were compared; the results indicate that the best choice can enhance the flood control capacity from that of a one in 10-year to a one in 200-year flood. The study serves as a good example of the complementary roles played by numerical and physical models in the solution of hydraulic engineering problems.

1. INTRODUCTION

Yuen Long is located in the northwestern part of Hong Kong. The geographical profile of the 30 km^2 catchment basin is basically a steep upward portion followed by an abrupt transition to a relatively flat lowland portion. The lowland portion is highly susceptible to severe flooding during the typhoon season; for example, during Typhoon Brenda in 1989 road links to the rest of the Territory were flooded, thus causing heavy losses due to disruption of economic activities. Less severe flooding occurs more frequently, for example during Typhoon Faye in 1992. In addition, rapid urbanisation of the area has significantly increased runoff rates by eliminating flood plain storage.

At present a system of natural and artificial channels (Main Nullah, San Hui Nullah, and Sham Chung River) carry the floodwaters through the town (Fig.1 and Fig.2a). For urban development the drainage system must be capable of handling a one in 200-year flood. Nevertheless, some sections of the present channels cannot accommodate a one in 10-year flood. The Yuen Long Bypass Floodway (YLBF) has been proposed to divert flood flows from the Sham Chung (SC) River, the San Hui (SH) channel, and part of the flow from the Yuen Long Main Nullah (MN) into the Kam Tin Floodway. Under a 200-year return period design condition, the floodway needs to i) divert approximately 15 percent of the main nullah flow (37.8 m^3/s); and ii) convey a total combined flow of 278 m^3/s to downstream within acceptable levels. The success of the design depends critically on complicated junction flow interactions which cannot be satisfactorily resolved by one- dimensional mathematical models for unsteady free surface flows that are in commercial use. A physical model study has been carried out to predict water levels and flow characteristics and study the optimal hydraulic design of the Bypass Floodway.

Although open channel junctions have been studied for some time (ASCE 1995), most of the previous work were concerned with rectangular channels of equal width and sharp-edged corners. Modi *et al.* (1981) investigated combining open channel flows using a conformal mapping approach and therefore did not account for energy losses. Best and Reid (1984) analyzed experimentally the geom-

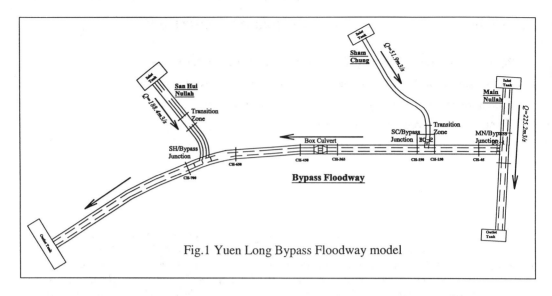

Fig.1 Yuen Long Bypass Floodway model

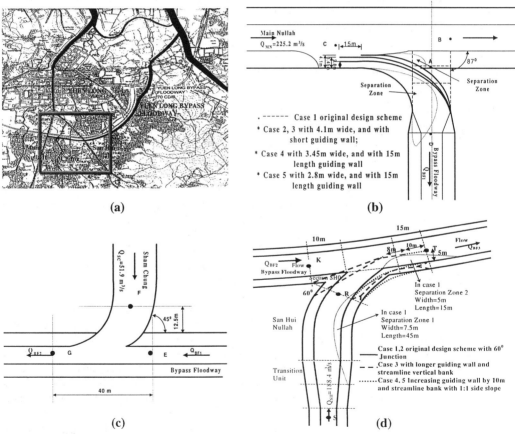

(a)

(b)

(c)

(d)

Fig. 2 Schematic diagram of Yuen Long Bypass Floodway junction details
a) Upper Yuen Long Catchment   b) Main Nullah/Bypass Floodway
c) Sham Chung/Bypass Floodway   d) San Hui/Bypass Floodway

878

etry of the separation zone at sharp-edged open channel junctions, while supercritical junction flows have also been studied (Hager 1989, Schwalt 1995). There appears to be no previous work for the problem at hand, which is characterised by i) a dividing flow - flow abstraction from the supercritical main nullah flow into a subcritical flow downstream; ii) a subcritical Sham Chung - Floodway junction; and iii) a supercritical-subcritical San Hui-Floodway junction.

2. PHYSICAL MODEL

Figure 2 shows the critical 1 km reach of the channel system under study. A 1:50 undistorted Froude scale model is designed and built in the Hydraulic Laboratory at the University of Hong Kong. The model, made of perspex, is composed of the Yuen Long Main Nullah, Sham Chung, San Hui, and the proposed Floodway. Confirmed by separate model tests, the channel roughness is correctly modelled with this choice of scale. The design upstream discharges in the channels and the stages at the downstream ends of the model limit in the MN and Floodway have been computed from the MIKE 11 hydrological model; these provide the boundary conditions for the physical model. The design parameters of the channels are summarized in Table 1.

In the model experiments, the upstream inflows were accurately controlled by specially designed and calibrated inlet tanks. The main nullah discharge at the model outlet and the model discharge at the floodway outlet were measured by V-notch and rectangular weirs respectively. Water depths were measured by point gauges; mean velocities were measured by calibrated propeller current meters. The flow was visualized with the aid of floats and dye injection and recorded on video.

3. EXPERIMENTAL RESULTS & DISCUSSION

Initial model tests with the original baseline design scheme (Case 1) have revealed that: a) only less than 5.27 m^3/s of the main nullah flow is diverted into the floodway, compared to the requirement of 37.8 m^3/s. The box culvert fails to extract effectively from the almost perpendicular supercritical MN flow; a large separation zone is observed at the entrance to floodway, which blocks the inflow from the MN (Fig.4, Case1). b) At the SC-Floodway junction only relatively weak separation is observed; there is little difference in stage upstream and downstream of the junction, and blocking effect is negligible. iii) With design flow of 188.4m^3/s, the supercritical flow (F=1.22) in the straight section of the SH is unstable. Fig.2d and Fig.4 show that separation zones are observed downstream of a hydraulic transition (with drop and expansion) designed to create sub-critical flow conditions before entering the floodway. The observed main SH flow is close to the left bank, with an average stage of 7.11 mPD;

Table 1: Design Parameters of the Channels in the Model Study

Parameters	Main Nullah	Sham Chung	San Hui	Bypass Floodway	
				SH Junction Upstream	SH Junction Downstream
Design $Q(m^3/s)$	225.2(upstream) 187.4(Downstream)	51.9	188.4	89.7 Between SC & SH	278.1 Combined
Bed Slope	1/320	1/440	1/280	1/1000	1/1000
Normal Depth(m)	3.06(upstream) 2.75(Downstream)	1.69	3.09	2.77	4.67
Critical Depth(m)	3.37(upstream) 3.01(Downstream)	1.50	3.48	2.13	3.76
Fr. Number	1.17(upstream) 1.15(Downstream)	0.84	1.22	0.77	0.69
Channel Section	⌐△⟍___╱ 9.7m	⌊___⌋ 9m	⌐△⟍___╱ 7.2m	⌐△⟍___╱ 10m	⌐△⟍___╱ 8m

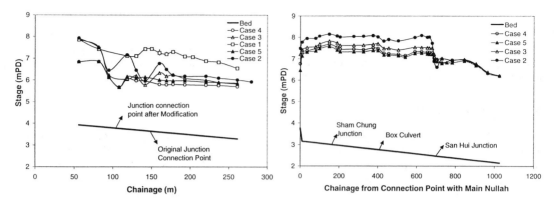

Fig. 3 Measured stages along Main Nullah & Bypass Floodway

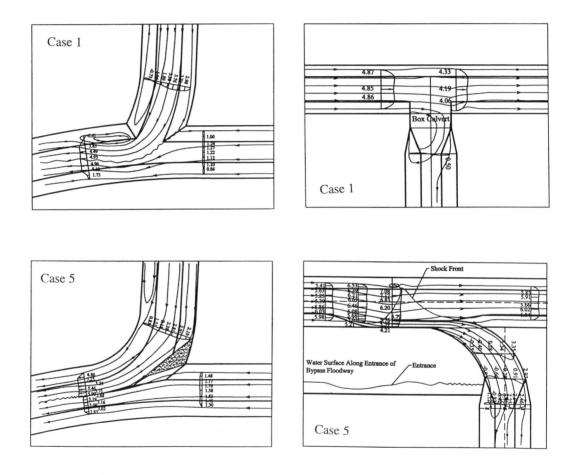

San Hui

Main Nullah

Fig. 4 Observed flow pattern of San Hui and Main Nullah junction in Case 1 & Case 5
Numbers refer to measured velocity (m/s)

the stage upstream and downstream of the junction are around 7.5 mPD and 6.7 mPD respectively - this large difference reflects the blocking effect from the lateral SH flow. Due to the backwater effect from the SH and SC junctions, the water stage in the Floodway is unacceptably high (Fig.3).

The design must be modified so that the flow diverted from the MN into the floodway can be increased and the stage controlled to within acceptable levels. Five schemes have been tested and summarized in Table 2.

In Table 2, Case 1 is the original baseline design scheme. In case 2, a fishmouth diversion channel (preceded by a trapezoidal to rectangular channel transition on the right bank) is provided at the MN junction, with unchanged SH junction. Case 3 is based on MN junction as in case 2, but with modified SH junction. For case 4, the width of diversion channel of MN junction is decreased while the length of guiding wall for both MN and SH junctions are increased. Case 5 is an improved version of Case 4. The various details are shown in Fig.2b) and Fig.2d). The measured diversion flows and water stage at key locations for the five cases are summarised in Table 3. The measured water surface profile along the MN and the Floodway are shown in Fig. 3.

Table 2 shows that, for Case 2, a discharge of 33.18 m^3/s can be diverted into the Floodway, but the water stage in all channels is unacceptably high due to both i) the increase in diverted flow; and ii) the significant backwater effect from SH junction. The stage upstream of the SH junction (WL_K) can be as high as 8 mPD and far exceeds bank levels. In Case 3, the diversion flow can be attained, but stages in all channels are still higher than desired. In Case 4, with increased length of guiding wall in both junctions, the flow conditions are much improved; a discharge of 51.9 m^3/s is diverted into Floodway. However, unstable waves are observed at the entrance to the MN guiding wall. In Case 5, the width of the cutwater channel is decreased to 2.8 m. A discharge of $42.25 m^3/s$ can be diverted into Floodway with acceptable stages and stable flow conditions. Our experiments indicate that Case 5 can meet all the flood control requirements and hydraulic constraints.

Fig.4 shows the observed flow pattern of MN and SH junction for Cases 1 and 5. It can be seen that the flow conditions at the two junctions are much improved for Case 5. The supercritical flow

Table 2: Summary of Test Schemes for Physical Model Study

Case#	MN Junction			SC Junction		SH Junction			
	Angle	Guide Wall	Inlet Width	Angle	Guide Wall	Angle	Guide Wall	Outlet Width	Transition (right bank)
Case 1	87^0	No	8.0m	45^0	No	60^0	No	Original	Sharp-edged
Case 2	0^0	Short	4.1m	45^0	No	60^0	No	Original	Sharp- edged
Case 3	0^0	Short	4.1m	45^0	No	12^0	8m	8.8m	Vertical
Case 4	0^0	15m	3.45m	45^0	No	0^0	18m	5.0m	Smooth
Case 5	0^0	15m	2.8m	45^0	No	0^0	18m	5.0m	Smooth

Table 3: Diverted Flow & Water Stages at Key Locations for Various Test Schemes

Case No.	Measured Diverted Flow Q_{BF1} (m^3/s)	Floodway Upstream X = 35m WL_D (mPD)	Sham Chung Upstream X = 207m WL_H (mPD)	Floodway Upstream of San Hui Junction X = 655m WL_K (mPD)	San Hui Upstream X = 17.5m WL_R (mPD)	Floodway Downstream of San Hui Junction X = 705m WL_T (mPD)
Case 1	5.27	7.363	7.495	7.450	7.065	6.420
Case 2	33.18	7.943	8.240	8.003	7.672	6.618
Case 3	38.78	7.528	7.528	7.865	7.520	8.300
Case 4	51.97	7.365	7.575	7.312	8.697	6.968
Case 5	42.25	7.310	7.470	7.232	8.598	6.872

X= distance from the junction.

881

from MN passes through a weak jump in the bend transition to the Floodway channel. In the beginning section of Floodway the maximum velocity is 2.94m/s, with an average stage of 7.31 mPD. These are lower than the observed values for Cases 2-4. The diverted flow is related to the oncoming flow and the width of the diversion channel. Tests showed that the diversion flow Q_1 varies as $Q_1/Q = K(b/B)$, where, K = coefficient related to the oncoming flow Q, with $K = 1.533Q^{-0.084}$, and b, B=bed widths of diversion channel and MN (=14.7 m at entrance to diversion channel) respectively. On the other hand, the action of the guidewall at the SH junction produces a critical or slightly supercritical SH at the junction. As pressure is continuous across a jet, the jet action effectively brings down the upstream water level in the floodway. The observed stage upstream of Floodway at location K (Fig.2d, Table 3) is 7.23 mPD. However, the reduction in stage at the junction will be accompanied by an increase in stage for a localised region in the SH, just upstream of the SH junction.

4. CONCLUDING REMARKS

Based on the measurements of and insights gained from the physical model study, the following points can be made: 1) The use of a diversion channel with appropriate length and width is successful in diverting the required flow from a supercritical flood stream into the Floodway. 2) The use of a hydraulic structure to create a local constriction and nearly critical flow conditions at the San Hui junction has the effect of significantly bringing down the stage in the Floodway upstream while still allowing the passage of the required flood flow. Experiments confirm that with contraction at the junction equal to 5.0m, the water depth decreases from 5.49 m of case 2 to 4.72m of case 5. Such design also facilitates flow diversion from the Main Nullah. 3) The modified Main Nullah and San Hui junctions with the original hydraulic transitions constitute an acceptable design for this urban flood control problem.

ACKNOWLEDGMENTS   The assistance of Feleke Arega, David Choi, NG Wai Kit, and Rowena Chan in the experimental work is gratefully acknowledged. This study is partially supported by the Territory Development Department.

REFERENCES

ASCE (1995). Hydraulic Design of Flood Control Channels *Technical Engineering and Design Guides - as Adapted from the US Army Corps of Engineers, No. 10* .

French, R.H. (1985). *Open-Channel Hydraulics*. McGraw-Hill. USA.

Best, J.L. and Reid,I.(1984) Separation Zone at Open-Channel Junctions. *J.Hydraulic Engineering* ASCE, Vol.110, No.11, 1588-1594.

Modi, P.N., Ariel, P.D., and Dandekar,M.M.(1981). Conformal Mapping for Channel Junction Flow. *J.Hdraulic Division,* ASCE, Vol.107, No.12, 1713-1733.

Schwalt, M. and Hager, W.H. (1995). Experiments to Supercritical Junction Flow. *Experiments in Fluids* 18 , 429-437.

Hager, W.H. (1989). Supercritical Flow in Channel Junctions. *J.Hydraulic Engineering*, ASCE, Vol.115, No.5, 595-616.

Environmental Hydraulics, Lee, Jayawardena & Wang (eds) © 1999 Balkema, Rotterdam, ISBN 90 5809 035 3

The influence of sea level rise upon flood/tidal level of the Pearl River Delta

Li Zi-hao, Huang Ben-sheng & Qiu Jing
Guangdong Provincial Research Institute of Water Conservancy and Hydro-Power, Guangzhou, China

ABSTRACT: In this paper, two dynamic factors affecting estuary water levels have been analyzed deeply. More than 190 representative flood series occurred in recent 40 years have been chosen and divided into 7 discharge grades. The corresponding tidal level correlation of 40 flood/tidal stations in the Pearl River Delta and Sanzao station at sea has been established, which can well reflect the physics mechanism influencing water level variation in estuary region. Under the different runoff, as sea level rising to different height, influence extent of sea level rise upon each flood/tidal water level station in the Pearl River delta has been calculated and 4 influence regions have been divided.

1. INTRODUCTION

Estuary water level is closely related to sea level, sea level rising leads water level to change directly. The water level in estuary delta networks is both influenced by oceanic tide and runoff. It differs not only from the variation in water level for inland rivers, which is controlled solely by runoff dynamics, but also from the tidal level variation caused by oceanic tide alone. Till now, not much research work has been done about the relationship between water levels in estuary network of the Pearl River Delta and sea level (Li Zi-hao,1985, 1994), not to say the systematic research work. In this paper, the factors affecting the water level of the estuary region are thoroughly analyzed. We selected more than 190 representative flood series occurred in recent 40 years, divided them into 7 discharge grades and established the corresponding tidal level correlation between 40 flood/tidal stations in the Pearl River Delta and the Sanzao station at sea (see Fig.1). The empirical formulae that can well reflect the relation between the estuary water level and sea level under the dual influence of the runoff and tide have been obtained, and used to predict the influence extent of sea level rise on tidal level of estuary delta.

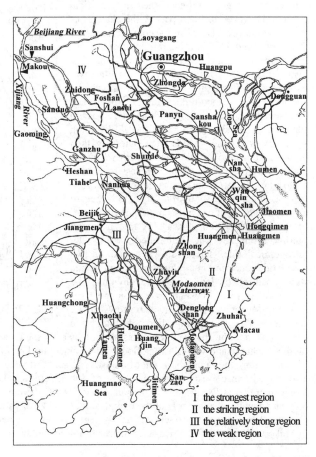

Fig.1 The position of flood/tidal level stations in the Pearl River Delta and four influence regions of sea level rise

2. THE RELATIONSHIP BETWEEN THE ESTUARY WATER LEVEL AND THE SEA LEVEL

The relationship between the flood and tidal level Z_2 in each water level station of the estuary delta and the sea tidal level Z_1 that represents sea level is

$$Z_2 = F(Z_1, Q) \tag{1}$$

For the inland river flow free from tide influence, the upstream and downstream water level or the water level and discharge have usually good correlation, that is,

$$Z_{down} = F(Z_{up})$$

or $$Z = F(Q) \tag{2}$$

Formula (2) is one of the hydrologic forecast methods commonly used in hydrology of inland. In the estuary region that is influenced both by the runoff and tide, the correlation between the upstream and downstream water levels is poor and can not satisfy the need of research and engineering.

2.1 Characteristics of Flood/Tidal Level in Estuary Region and Its Analytic Method

In low-water period, the tidal level in the delta of the estuary region reflects the regular tidal process. However, in flood period, the changes of the flood/tidal levels are more complicated. The flood wave and the tide wave influence each other, one increases while another decreases, or vice versa under the dual influences of the two dynamics of flood and tide. Fig. 2 shows the variation of flood/tidal levels of each representative stations, including the Sanzao station at sea, along the main flood diversion waterway of Xijiang—Modaomen waterway, then up to Makou station on the main river of Xijiang. It shows a typical flood change process from May 10 to May 30, 1975, the discharge in Makou rose from $12000m^3/s$ to $30700m^3/s$, then fell back to about $18000m^3/s$. It can be seen that the tidal level of the Sanzao station at sea was free from the runoff influence but controlled solely by astronomical tide. The tidal wave rule of two spring tides occurred in a month and irregular tide wave occurred on half a day are obvious. The tidal level at Denglongshan station on the river mouth of Modaomen waterway also shows such a variation. The connecting line of high tide

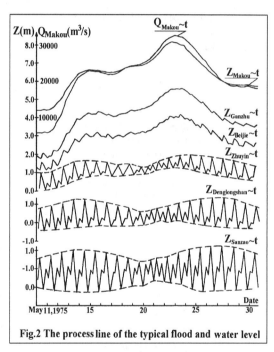

Fig.2 The process line of the typical flood and water level

corresponded to the change at Sanzao station. However, the overall waveform had been obviously influenced by the discharge and waveform of flood. The connecting line of low tide was distorted. Up to the upstream Zhuyin station, the fluctuation rule of the two rise-and-fall water level also occurred. The high and low tides also lagged and corresponded with those of Sanzao station. The flood dynamic influence was obviously larger than that at Denglongshan station. Up to Beijie, though the water level change caused by tide could only be vaguely seen, the rule of the overall flood/tidal level variation is the same as that at Makou station. In Ganzhu station, the tide influence is hardly seen in flood rise period, and is rather weak in flood fall period. In Makou station, the flood rise and fall are not influenced by tidal dynamics.

The tidal influence attenuates gradually from river outfall to upstream, while the influence of the runoff increases, as shown in Fig.2. Fig. 2 also shows the following features:
   ① Even if in larger runoff flood period, the high or low tide in each station near the offing of the Pearl

884

River Delta appears to lag behind the corresponding tidal change at Sanzao station. However, their tidal change is not only related to that of Sanzao station, but is also influenced by upstream runoff. Therefore, any attempt to find their correlation by simply plotting their respective tidal level without analysis will produce only a poor correlation or even no correlation at all.

② Upstream, the runoff influence increases and the tidal influence disappears gradually. But subject to the same flood discharge, the tidal influence at flood peak and during flood receding period is more apparent than during flood rising. Therefore, if only the runoff discharge is considered without considering the flood process, the correlation between the flood/tidal level upstream and downstream does not exist.

The tidal wave length is much larger than the wave height. It is a shallow water wave with wave velocity $C=(gH)^{1/2}$, much higher than the runoff or tide velocity. Therefore, the tidal wave at Sanzao can reach Sanshui and Makou within hours.

Assuming the roughness n and slope J remain unchanged, the flood wave velocity V_W can be obtained by Manning formula

$$V_w=V+(2SV/3bH) \tag{3}$$

where V_W is flood wave velocity; V is flow velocity; S is cross sectional area; b is river width and H is water depth. For the common channel cross section, $V_w>V$ such as parabolic cross section,

$$V_w=(13V/9) \tag{4}$$

From Makou to the offing, it is 130km long. Flood wave can travel to the offing from Makou within a day, When the flood in Makou station rises to peak discharge, all the delta networks had been filled with flood from the rising period, that is to say, the river channel storage process is finished. The change of flood/tidal level in each tide station is mainly controlled by the tidal dynamics, when water level rises to the relatively steady state that matches the upstream flood discharge, at the peak time and in one or three days after peak flood, and then the flood gradually subsides. From the analysis about the tidal wave propagation in the estuary region as show in Fig.2, we can consider that the high and low tidal level of a certain half tide on some days in each station can be related to the corresponding tidal at Sanzao station, hence the correlation can be set up in different flow regimes.

2.2 The Relationship between Estuary Water Level and Sea Level

For any engineering project, the flood peak and higher tidal levels are always of much concern. For rising of sea level, the most dangerous case is also decided by its influence extent on flood peak level and high tidal level. From the above analysis, we selected more than 190 typical major or small flood in the past 40 years for analysis. The discharge is divided into 7 grades. They are Q_{Makou} =5000, 10000, 15000, 20000, 25000, 30000, 37500m³/s (the flow difference should be kept within ±10%). For each flood, data are also selected from the relatively steady peak at Makou and the tidal level of each station in peak flood time or within one to three days after the peak and the corresponding high and low tide water level between each tide station and Sanzao station are plotted in Fig.3.

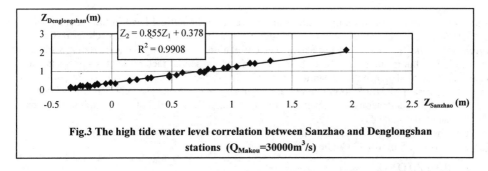

Fig.3 The high tide water level correlation between Sanzhao and Denglongshan stations (Q_{Makou}=30000m³/s)

Fig.3 shows that the correlation between the tidal level near river outfall and Sanzao station agree better.

The high tidal level correlation of Sanzao and Denglongshan stations can be written as

$$Z_2 = AZ_1 + B \qquad (5)$$

A is considered as a parameter mainly controlled by the tidal dynamics; B is a parameter purely controlled by the runoff. However, with the increase of discharge at Makou, those stations far away from the river outlet, such as Beijie, Tianhe, Ganzhu, Makou, Nanhua, Leliu, Lanshi, Zidong, Sando and Sanshui, when $Q_{Makou} > 15000 m^3/s$ (in Makou and Sanshui stations, when $Q_{Makou} > 10000 m^3/s$), the points appear scattered, indicating a poor correlation. There are two causes for it: The first one is when larger flood occurs, the tidal limit moves downward. Water levels in these stations are mainly influenced by the upstream flood discharge and less affected by the tidal wave. The second cause is that depending on the selected tidal level, each discharge regime may deviate to a certain degree from its corresponding upstream discharge and that furthermore its water level change is more influenced by the runoff than by tidal change, especially during a heavy flood. That is why the points appear rather scattered.

2.3 The Variation of Parameter A and B with Q

The variation of the values of A and B with discharge at each tidal level station is plotted. The correlation analysis indicates that A and B can be written as follow:

$$A = \alpha Q + C_1 \qquad (6)$$

$$B = \beta Q + C_2 \qquad (7)$$

where Q is the flood discharge at Makou, α, C_1 and β, C_2 are coefficients. For the same station, generally, A decreases with the increase of Q, while B increases with the increase of Q. For the upstream and downstream stations, with the increase of Q, the decrease of A at upstream station is large, that is to say, α is small, but the increase of B is larger, i.e. β is larger. It indicates the strength of the two dynamics and their interaction. Substitute formulae (6) and (7) into (5), we get

$$Z_2 = C_1 Z_1 \quad + \quad \beta Q \quad + \quad \alpha Z_1 Q \quad + \quad C_2 \qquad (8)$$

| tidal influence | runoff influence | flood and tide interaction influence | constant related to freezing basis level |
| ① | ② | ③ | ④ |

From formula (8), we can further conclude that the flood/tidal level of the estuary delta is influenced by both the oceanic tide and runoff, it is not the result of the linear superposition of the two effects, but shows complicated nonlinear interaction between the flood and tide. Formula (8) is a typical nonlinear relation, and the term ($\alpha Z_1 Q$) in it is just nonlinear one. The research carried out by Wu also proves the nonlinear characteristics in estuary water level variations (Wu, C. Y., 1997).

2.4 Discussion on Formula (8)

Formula (8) as above represents the relationship between the flood/tidal level in the estuary region and the sea level. Although the result is obtained through analysis of their correlation, each term, however, has a definite physical significance stated as above. The correlation is not a simple linear relation without time and spatial variation, but is one that includes both the time and the spatial factors. This is so because the selected data are from the tidal levels at each station corresponding to the tidal levels at Sanzao Station when the peak flow at the upstream Makou Station remains at a relatively stable level. The "corresponding" tidal level here refers to not any instantaneous tidal level, nor to any average tidal level, but to the high or low tidal level of the same tide on the same day.

When Q=constant, formula (8) becomes (5).

When Z_1=constant, formula (8) can be written as

$$Z_2 = (\beta + \alpha Z_1) Q + (C_1 Z_1 + C_2) \qquad (9)$$

$$Z_2 = \Upsilon Q + C_3 \qquad\qquad\qquad (10)$$

in which $\Upsilon = \beta + \alpha Z_1$, $C_3 = C_1 Z_1 + C_2$.

3. INFLUENCE OF SEA LEVEL RISE ON TIDAL LEVEL IN ESTUARY REGION

3.1 Influence Values of Sea Level Rise on Tidal Level in Estuary Region

Assuming that eustatic sea level rise is ΔZ_1, the influence value of sea level rise on high tide level of river networks ΔZ_2 can be written as follows by formula (8) :

$$\Delta Z_2 = (\alpha Q + C_1)\, \Delta Z_1 \qquad\qquad\qquad (11)$$

where, C_1 is a plus constant, generally, $\alpha < 0$, influence value of sea level rise decreases with increase of Q, but there is also special case $\alpha > 0$. At the waterways nearby Pearl River outlet and Yamen where tidal dynamic is strong, ΔZ_2 hardly varies, or may increase a little with increase of discharge Q.

3.2 The influence Extent of Sea Level Rise

The influence extent of sea level rise is actually a problem about influence of sea level rise on tidal limit in delta networks. By inference, sea level rise causes tidal range and tide wave energy increase, the tidal limit should move up. On the other hand, if the cross sectional area of the river keeps constant, rising of sea level may cause water depth increase, tidal wave velocity $C = (g\,(H + \Delta H))^{1/2}$ correspondingly increases and so does propagate power of tide wave, then tidal limit should move up too. However, rising of sea level is a slow process, it can cause sediment scouring and depositing and the relative variation of riverbed. So, it is very difficult to estimate the influence degree of sea level rise on tidal limit.
From formula (11), if $\alpha Q + C_1 = 0$, then $\Delta Z_2 = 0$, we can get:

$$Q_e = -C_1/\alpha \qquad\qquad\qquad (12)$$

where, the physical significance of Q_e is the limited discharge of some tidal level station affected by tide dynamic. That is to say, when upstream flood discharge $Q > Q_e$, the water level of this station is not affected by tidal dynamic already, so, the application range of formula (8) is $Q \leq Q_e$.

4. INFLUENCE OF SEA LEVEL RISE ON TIDAL LEVEL OF THE PEARL RIVER DELTA AND 4 INFLUENCE REGIONS

4.1 Influence Values of Sea Level Rise under the Different Discharge Grades upon High Tidal Level of the Pearl River Delta.

By the corresponding high tidal level correlation between each water level station and Sanzao station under the different discharge grades, as sea level rising to different height, using $\Delta Z_2 = A \Delta Z_1$ obtained from formula (5), the high tidal level increasing value ΔZ_2 of each water level station at the Pearl River delta can be calculated with the different discharge grades (Li Zi-hao, Huang Ben-sheng and Qiu Jing, 1996, 1997). By analyzing calculation results of 40 water level stations, it is shown that the influence of sea level rise being attenuated gradually from the offing to upstream. Generally, for the same station, influence value decreases with increase of runoff. But for Yamen and Humen waterway at the Pearl River Delta estuary and the other stations near river mouth where the ratio of runoff to tidal discharge is smaller and the tidal dynamic is strong, the influence values of sea level rise hardly varies with runoff, even increases a little with increase of runoff.

4.2 Four Influence Regions of Sea Level Rise in the Pearl River Delta

According to the enormous analysis and calculation as above, it is not difficult to see that the influence of sea level rise on delta tidal level is a concept of relative dynamic state, which varies with time and space. Firstly, that the influence degree is serious or not and the influence extent is wide or narrow varies with time. For example, influence degree and extent in flood period are very different from those in the low-water period, which shows the contrast and interaction degree between tide and runoff dynamic. Secondly, the influence of sea level rise on tide level has obviously characteristics of range and spatial too. It is strong nearby the river offing and weak in upstream of delta networks, this is also a contrast between tide and runoff dynamics. In addition, the influence of sea level rise is further closely related with waterway characteristic, the influence on the river mouth which the ratio of runoff and tidal discharge is small is stronger than that is the large one. All of these are due to the contrast of two dynamic factors in estuary delta.

According to the analysis and calculation done as before, influence of sea level rise on tide level in the Pearl River delta can be divided into 4 regions as follows: (Ⅰ) the strongest influence region, (Ⅱ) the striking influence region, (Ⅲ) the relatively strong influence region and (Ⅳ) the weak influence region (see Fig.1).

5. CONCLUSION

5.1 The empirical formula (8) has sufficient physical significance. It expresses the relation between the water levels at different stations and the sea level. Formula (11) can be used to calculate the influence values of sea level rising to different heights on the high tide water level of each tide station of the Pearl River Delta corresponding to the high tide water level at sea at different upstream flood discharges.

5.2 Generally, the influence values of sea level rise on water level of delta networks decreases gradually from the offing to upstream, and this decrease will be quicker for the strong runoff dynamic waterway and slower for the strong tide dynamic waterway.

5.3 Influence of sea level rise on water level of delta networks varies with runoff discharge of upstream. Usually, larger discharge is, smaller influence values is. so, either influence values or extent is the largest in the low-water period, and the influence can spread to Makou, Sanshui and upwards.

5.4 According to the influence values and extent of sea level rise on water level of estuary delta, 4 influence regions are divided as follows: (Ⅰ) the strongest influence region, (Ⅱ) the striking influence region, (Ⅲ) the relatively strong influence region and (Ⅳ) the weak influence region. Region Ⅰ includes Guangzhou, large areas of Panyu, Zhuhai, Yamen offing of Xinhui, Yinzhou lake and other regions nearby the river offing. The influence values on region Ⅰ is about the same as eustatic sea level rise values or a little larger, If eustatic sea level rise up 20 cm, during heavy flood period, the influence values on region Ⅰ is 19~24cm, on region Ⅱ is 10~18cm, and on region Ⅲ is 2~9cm. Region Ⅳ is hardly influenced by eustatic sea level rise, but in the low-water period, the influence values will be 11~16cm.

6.REFERENCES

Li, Z. H., (1985) "The effects of united embankments and gate dam projects on the flow and the channel changes in the Pearl River Delta", Tropical Geography, Vol.5, No.2, 99-107. (in Chinese).

Li, Z. H., (1994) "The effects of united embankments and gate dam projects on the flow in the Pearl River Delta", *Influence of sea level rise on Chinese delta regions and the countermeasures*, Academician Consultative Reports of The Chinese Academy of Sciences, Science Press, Peiking China, 129-137. (in Chinese).

Li, Z. H., Huang, B. S., and Qiu, J., (1996), "Preliminary study on influence of sea level rise upon flood/tidal level in the Pearl River Delta (Part 1 and 2)", Proc., 2th Congr. of Oceanic Technology, Kunming, China, 456-470. (in Chinese).

Li, Z. H., and Huang, B. S., (1997). "Research on relation between sea level rise and estuary water level", Technical Report, Guangdong Hydr. Res. Guangzhou, China. (in Chinese).

Wu, C. Y., (1997). "Application of artificial neural networks model in river networks region", Peal River, Vol. 98, 15-19. (in Chinese).

Environmental Hydraulics, Lee, Jayawardena & Wang (eds) © 1999 Balkema, Rotterdam, ISBN 90 5809 035 3

Different alignments of a new hydraulic structure using 2-D mathematical model

S.A.S. Ibrahim, M. B.A. Saad & A.A. El-Desoky
Hydraulics Research Institute (HRI), Delta Barrage, Egypt

ABSTRACT: Naga Hammahi Barrage extends across the River Nile width of 829 m. It has 100 gate openings. Due to the expiration of the life time, it was decided to replace it. Five alternatives were proposed to replace the old barrage. A two dimensional depth average flow model is employed to study the flow patterns created by passing the flow through hydraulic structure with different alignments. This was achieved by modeling a reach of 5.2 km at Naga Hammadi including the existing barrage. The flow patterns were investigated for the existing conditions covering minimum, dominant and maximum discharges. Also, the flow patterns were studied for each alternative for the same flow conditions. By comparing the results of each alternative with the existing conditions the most suitable alternative was selected. The best flow behavior was achieved up and down stream the structure which is located at a straight reach.

1. INTRODUCTION

Locations of main barrages constructed on the Nile river between Aswan and the Mediterranean sea in Egypt are shown in Fig. (1). Those main barrages provided complete control of the Nile River water and maintained satisfactory flow discharges through the irrigation system. High Aswan Dam (HAD) was constructed in 1969 at 6.5 Km upstream old Aswan Dam (OAD). As a result of controlling the flow discharge and releasing clear water from the reservoir, the Nile River started to degrade downstream each barrage and aggrade upstream the next. This generates a drop of water level, which has created negative effects for the barrages. Naga Hammadi Barrage was built during the period 1927-1930 on the River Nile 359 km downstream from Aswan Dam.

Because of the above mentioned reasons and due to the age of the barrages the Egyptian authority started to replace those old barrages by new ones. Replacement of Naga Hammadi Barrage is under investigation. The components of the proposed barrage are hydropower station, sluiceway with widths of 90 and 150 m respectively. The rest of the river width will be a closure dam. This means that some times the water will be released through a width of about one tenth of the river width. So, it was expected that a big morphological change will occur downstream the new structure.

This paper describes the 2-D mathematical model construction and calibration of Naga Hammadi barrage reach including the existing barrage. Flow patterns resulting from the passing flow through the new barrage which has different alignments from the old one are discussed.

2. THE MATHEMATICAL MODEL

The "Trisula" program for two dimensional horizontal (depth-averaged) flow computations developed by Delft Hydraulics is employed, Delft Hydraulics (1993). The mathematical formulation of the program based on the assumption that the vertical accelerations of water in rivers are small compared to the gravitational acceleration. The usual shallow water equations (with depth average velocities) are used. The set of equations are solved with a finite difference ADI (Alternating Direction Implicit) method on staggered grids.

3. MODEL CONSTRUCTION

A surveyed reach of about 5.2 km long (1.95 km upstream and 3.25 km downstream the existing barrage) and 0.5 km average width is schematized using curvilinear grids. The number of the grids are 84 and 54 in both flow and transverse directions. The length of the grids are selected to be small at the area of interest and close to the dimensions of the studied structure at its location.

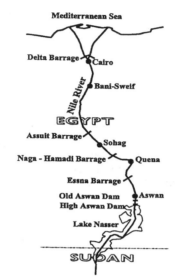

FIG. 1. Existing Main Hydraulic Structures

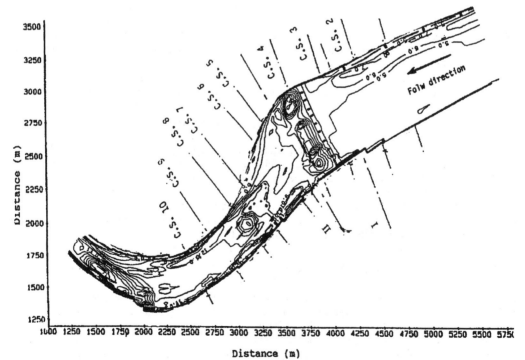

Distance (m)

FIG. 2. Naga Hammadi Depth Contour Map and Cross Sections Locations

The depth values were determined from the field contour map. Water levels (reference level) and velocity components in the flow and transverse directions were set to zero as initial conditions. The different discharges and their corresponding water levels were used as up and downstream boundary conditions respectively. Preliminary value for Manning coefficient (n) as initial value of the bed roughness for the whole reach was assumed. This value was changed during the calibration for each grid. A head difference at the barrage location was created by changing (n) value along the width of the reach at the barrage location. The time step of 0.5 minutes was selected which was applied for a total number of 800 steps to reach the steady state.

The simulation started with a discharge of 1114 m³/s in the prototype. The corresponding water levels were 65.3 and 59.68 m above mean sea level (AMSL) up and downstream of the barrage respectively. The resulting depth contour map in Fig. (2) was found very close to the measured depth map of the reach. The program results of the water slopes up and downstream of the barrage were compared with those of the prototype. The Manning coefficient (n) for the whole reach was changed to adjust the water slope to be similar to the field measurements. The program results for the velocity profiles at cross sections (I) and (II) located up and downstream the exiting barrage were compared with the prototype data. Cross section locations are shown in Fig. (2). Modifications in the roughness value are made to adjust the results to that of the prototype.

4. DESCRIPTION OF ALTERNATIVES AND MODEL RESULTS

Different alternative layouts were separately simulated. The flow behavior for all of them and the existing condition were investigated. In the following, description of the flow through the exiting barrage and the five different structures represented in the model:

1   The existing barrage where the flow passes through a width of about 800 m.
2   Insertion of a hydropower station in the barrage near the right abutment, while the rest of the existing barrage works as a sluiceway (alternative 1).
3   Insertion of a hydropower station and sluiceway near the right abutment, integration of the existing barrage into fill dam (alternative 2).
4   A new hydropower station, sluiceway and cofferdam at a distance of 150 m downstream the existing barrage near the left bank (alternative 3).

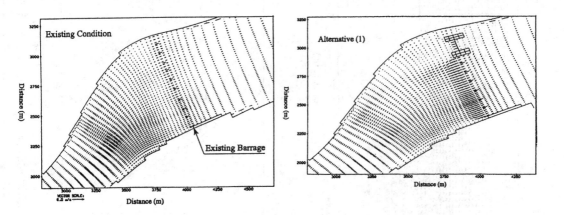

FIG. 3. Velocity Fields During Dominant Discharge

891

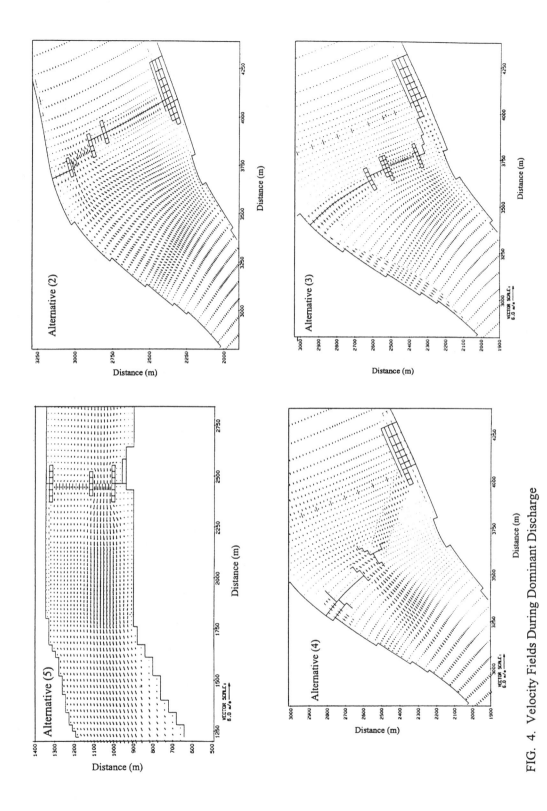

FIG. 4. Velocity Fields During Dominant Discharge

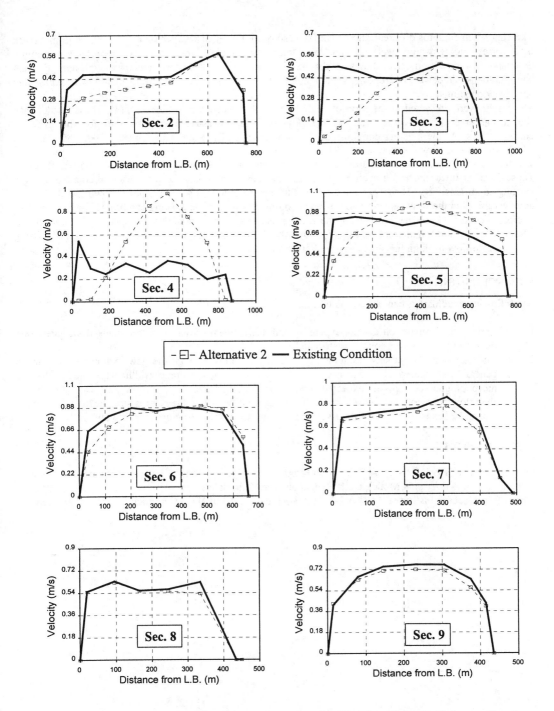

FIG. 5. Velocity Profiles Along 8 Cross Section During Passing The Flow Through Sluiceway Openings

5   The same components as the previous located near the right bank (alternative 4).
6   Constructing the same components like before but located at a distance of 3.5 km downstream the existing barrage (alternative 5).

The simulation runs for different alternative layouts covered river discharges of 810, 1600, 3000 m³/s with their corresponding water levels up and downstream the simulated reach. These represents minimum, dominant and maximum river flows in that reach respectively. The flow may passe through different configurations, sluiceway, hydropower station, half of the hydropower station and the hydropower station and sluiceway openings.

The results were presented in the form of velocity field plots as shown in Figs. (3), (4). These results have shown the presence of eddies and reverse flow up and downstream the structure for the different configurations. At the existing condition or during operation of the sluiceway (widths 800, 150 m), such eddies and reverse flow were not observed.

The results were presented in the form of velocity profiles at 10 cross sections covering the whole reach. These sections were chosen in such a way to facilitate the comparison between each alternative and the existing conditions. Fig. (2) shows the locations of the cross sections. The results have shown that there were large variations in the flow behaviors for different configurations and the existing conditions, until it vanished at cross section 9 (1550 m) downstream the existing barrage. These variations disappeared at a distance of 1300 m downstream last alternative. Fig. (5) shows the velocity profiles at 8 cross sections in case of passing the dominate flow through the sluiceway in alternative no. 2.

5. CONCLUSIONS

The following conclusions can be formulated:

1 Eddies, flow separation did not appear up and downstream of the existing barrage if all gates were open and the flow was equally distributed.

2 The variations in velocity resulting from passing the flow through the new hydraulics structure in general disappeared at a distance of 3-4 times the reach average width. This means that no morphological changes are expected downstream that distance.

3 Best flow behaviors is achieved in the the last alternative, where the river is straight for a distance of one and 2 times the width of the river up and downstream the structure respectively.

6. REFERENCES

Delft Hydraulics (1993) Trisula, A Simulation Program for Hydrodynamic Flow and Transport, Technical Manual for 2-D Trisula Model, Delft Hydr. Lab., the Netherlands.

Hydraulics Research Institute (1993) A 2-D Mathematical Model of Naga Hammadi Barrage, Final Report, Cairo.

Environmental Hydraulics, Lee, Jayawardena & Wang (eds) © 1999 Balkema, Rotterdam, ISBN 90 5809 035 3

Velocity distributions at channel bifurcations

M. Lutfor Rahman, M. R. Kabir & M. M. Hossain
Department of Water Resources Engineering, Bangladesh University of Engineering and Technology, Dhaka, Bangladesh

ABSTRACT: Two dimensional velocity distribution were measured in the small scale river bifurcation model constructed at the Hydraulics and River Engineering Laboratory of Bangladesh University of Engineering and Technology (BUET) , Dhaka. A series of experimental runs have been conducted to measure precisely two dimensional velocity distributions at channel bifurcation using Programmable Electromagnetic Velocity Meter (EMS). The result of the typical velocity distribution in the grid points near the bifurcation are discussed. It appears that the velocity around the nose were more unstable and two dimensional in nature than the other sections of the main and two downstream channels. The magnitude of the secondary velocity were varied from 2% to 20% of the primary velocity magnitude.

1. INTRODUCTION

Bifurcations are typical features in all alluvial rivers as well as estuaries. The morphological behaviour of the river at bifurcations is not yet a properly understood phenomena. The complexity of the bifurcation phenomena or problems lies in the determination of discharge distribution as well as sediment distribution of the downstream branches. ·

A bifurcation occurs when a river splits into two branches. This appears or happens when a middle bar forms in a channel or a distributary distributes water from the main river. At a bifurcation even if the upstream discharge Q_0 and sediment transport S_0 are known, then it would be difficult to predict the discharges Q_1 and Q_2 and sediment transport S_1 and S_2 in the two branches as it would depend on a number of factors such as the nose angle, cross-sectional areas, slopes of the downstream branches etc. This paper deals with the discharges i.e. velocity distribution only before going to experiments with the sediment transport distribution.

Dekker and van Voorthuizen (1994), Roosjen and Zwaneneberg (1995), Hannan (1995) and Islam (1996) have studied the morphology i.e. sediment distribution of a symmetrical river bifurcation using the same laboratory set up of the present study. This study using the same facilities, concentrated on the distribution of velocity over the downstream branches as well as main channel around the bifurcation area with discharge variation as a major variable. In order to collect data of velocity distribution near the bifurcation area, a series of experiments have been carried out using the physical model.

The velocity distribution in a channel section usually varies from one point to another. This is due to shear stress at the bottom and at the sides of the channel and due to the free surface. The velocity may have components in all three Cartesian coordinate directions. Most of the time, however, the components of the velocity in the vertical directions as well as transverse direction are small may be neglected (Chaudhury, M. H., 1994). Therefore , only the flow velocity distributions in the direction of flow needs to be considered. But this paper is only limited to the study of two dimensional velocity distribution at the physical laboratory bifurcation model.

2. EXPERIMENTAL SET-UP AND PROCEDURE

The model was a fixed bed with fixed banks (Fig.-1). The layout of the model comprised of three branches: a main branch (donated by B0) which bifurcated into two separate branches : branch B1 and branch B2. The main branch B0 was straight and its length and width were 4.55 m and 1.00 m respectively. The width, length and radius of curvature of branch B1 and branch B2 were as follows: branch B1: width =

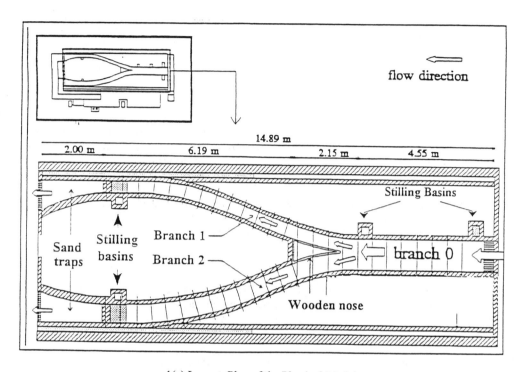

1(a) Layout Plan of the Physical Model

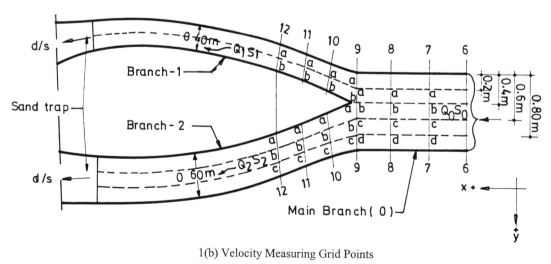

1(b) Velocity Measuring Grid Points

Fig. 1. Bifurcation Set-up with the Velocity Measuring Grid Points

0.40 m, length = 8.6 m and radius of curvature = 23.5 m ; branch B2: width = 0.60 m, length = 8.40 m and radius of curvature = 25.5 m. The detailed descriptions of the model may be seen elsewhere (Dekker and van Voorthuizen , 1994).

The model consists of two parts: permanent part and temporary part. The permanent part was the experimental facilities necessary for storage and regulation of water circulation through the model and guidance of this water to and from the temporary part. The permanent part could be divided into two elements: the water supply system and the regulating and measuring system.

The circulation of water within the model was a closed system. From the downstream reservoir the water is transported by means of a pipeline to the upstream reservoir. The regulating and measuring system consists of the tail gates, Rehbock weirs and stilling basins connected with Rehbock weirs. The tail gates were used to control the discharges flowing through the two downstream bifurcated branches. The Rehbock weirs were used to measure the discharges in the branch B1 and branch B2. The temporary part was the nose of the bifurcation. A symmetrical nose as shown in Fig. 1 was used to divide the inflow from the main branch. A symmetrical nose meant that the tip of the nose divided the inflow area according to the areas of the downstream branches.

2.1 Measurements

For data collection a symmetrical nose was taken up as explained earlier. Discharges in the main branch was selected considering its carrying capacity and the three discharges viz.: 40 l/s, 50 l/s and 60 l/s were used for the nose. Measurements around the bifurcated area include velocity in the main direction of flow (primary velocity) and the velocity in it's perpendicular direction (secondary velocity) in the x-direction and y-direction respectively and water depths. Two dimensional velocities were measured by the Electromagnetic Velocity Meter (EMS). The measurements of discharges on the downstream branches were taken by the Rehbock weirs.

2.1.1 Velocity measurements

Point velocities were measured along the verticals at each grid points around bifurcation. These grid points are shown in Fig. 1(b).

2.1.2 Discharge distribution ·

Along with the discharges ratio $Q_1/Q_2=1/2$, the water were distributed at the bifurcated channel. For this case tail gates were used to control the water flow. The distribution of discharges flowing through the main channels for different main channel discharges are; for 40 l/s water flowed in branch B1: 14 l/s and in branch B2: 26 l/s; for 50 l/s in branch B1: 16 l/s and in branch B2: 34 l/s and that for 60 l/s, in branch B1: 20 l/s and in branch B2: 40 l/s.

3. DATA ANALYSIS

In the following paragraphs, results of the physical model applied to the study of dividing water discharges through a bifurcation point are presented. Measured velocities are plotted in Fig.2 to know the hydrodynamic behavior at the bifurcated area.

It is seen that the primary velocity at section 7 of B0 is more than that of at section 9 at 'a' and 'c' . The regression coefficient 'R' of primary velocity at section 7 is more uniform than at section 9 just near the bifurcation point. It is proved that the velocity is more unstable near the bifurcation. The magnitude of the secondary velocity is greater and in positive in direction at the bifurcation point than the other area of the channel. This value is more near the tip of the bifurcation at sections 9. The variation in percentage of the secondary velocity is also more in the bifurcation about 15% to 20% of the primary velocity. So it is

897

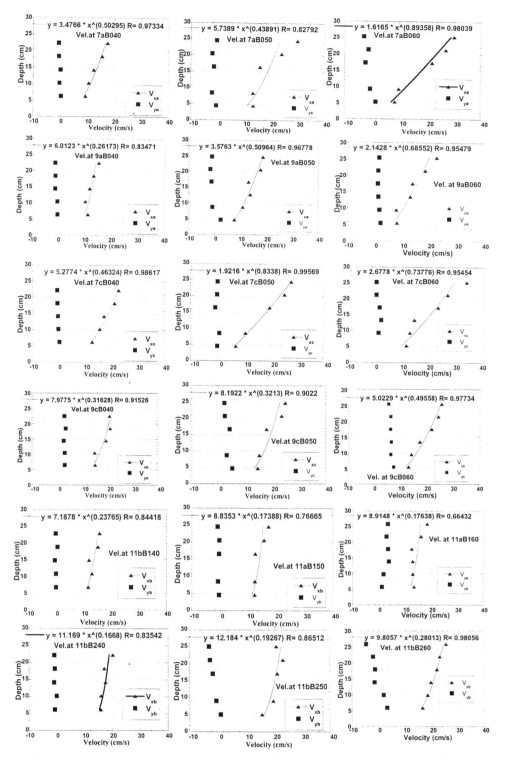

Fig. 2 Velocity profiles at sections 7 and 9 of branch B0 at 'a' and 'd' and that of at section 11 of branch B1 at 'a' and in branch B2 at 'c' for Q_0=40 l/s, 50 l/s and 60 l/s discharges

obvious that the turbulence in bifurcation has its guiding power to provide braiding system in the alluvial river. The both the primary and secondary velocities are more or less stable at the sections 11 of branches B1 and B2.

Normally, the shear stress is estimated with the formula applicable for steady, uniform flow. The emperical equations say van Rijn, Engelund-Hansen, Ackers-White etc. for sediment transport are associated with the shear stress parameter of the steady and uniform situation. From the velocity profiles at sections 7 and 9 of branch B0 and that of sections 11 of branches B1 and B2 , the shear stress are calculated. Taking the Manning's roughness coefficient for plastered cement rubble masionary 0.02, the Chezy's roughness coefficient is calculated from the relation $C=(1/n)R^{1/6}$ in which $R= (h*B)/(2h+B)$ where

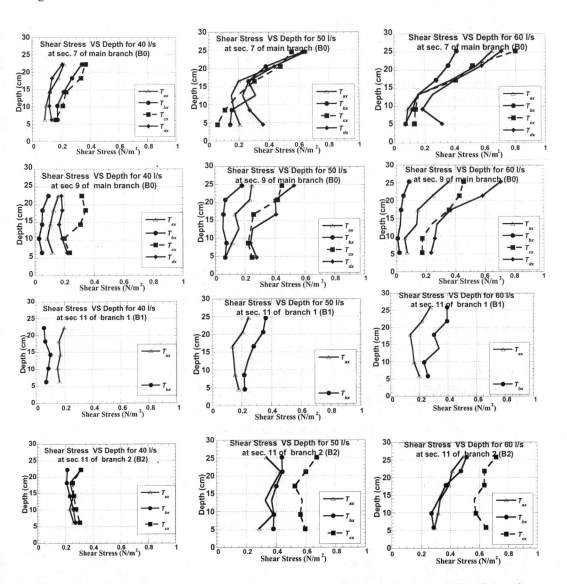

Fig. 3. Shear stress distributions at sections 7 and 9 of branch B0 at 'a', 'b', 'c', and 'd' and that of at section 11 of branch B1 at 'a', 'b' and in branch B2 at 'a', 'b', 'c' for Q_0=40 l/s, 50 l/s and 60 l/s discharges

h, the water depth from the bottom at the point of velocity measurement; B, width of the channels. From the known value of h, B and n=0.20, the value of C is calculated. Putting the value of C and measured velocity U in equation $T=\{(g^{1/2} *U)/C\}^2 * \rho$, one can get the shear stress T ; where T , Shear Stress; g, acceleration due to gravity; U, measured primary velocity and ρ , the density of water. The calculated shear stress are shown as in Fig. 3. From the figure it is clear that the shear stress variation near the bifurcation i.e. at section 9 is more non uniform than sections of the model. This shear stress will be used later on to calculate sediment transport flow in the bifurcation channels.

4. CONCLUSIONS

The study is based on the experimental data obtained from the small scale laboratory river bifurcation. The result have not been compared with the prototype data. Thus the conclusions from the study are valid only for the range covered by the experiments. The following conclusions are drawn from the study as follows :-

- The distribution of velocity near the bifurcations zone are unstable as seen in comparison with the sections 7 and 9 of main channel B0 as well as downstream branches B1 and B2 of sections 11.

- The instability is also highlighted by the best fit equations of primary velocity and the regression coefficients.

- As approaching the bifurcation nose from the main channel secondary flows no longer can be ignored which is about 20% of the primary flow.

5. REFERENCES

CHAUDHRY,M.H.(1994). "Open-Channel Flow", Reprinted in India by special arrangement with Prentice- Hall, Inc., Englewood Cliffs, N.J., U.S.A.

DEKKER, P.D. and J.M. van Voorthuizen (1994). "Research on the Morphological Behaviour of Bifurcations in Rivers", Delft University of Technology, The Netherlands

HANNAN,A. (1995). "A Laboratory Study of Sediment Distribution at Channel Bifurcation", M. Sc. Engineering Thesis, Department of Water Resources Engineering, BUET,Dhaka, Bangladesh.

ISLAM, G.M.T.(1996). "Laboratory Study of Sediment Distribution at Channel Bifurcation", M.Sc. Engg. Thesis, Department of Water Resources Engineering, Bangladesh University of Engineering and Technology, Dhaka, Bangladesh.

ROOSZEN,R. and C.Zwanenberg (1995). "Research on Bifurcations in Rivers", M.Sc.Thesis, Facuty of Civil Engineering, Delft University of Technology, The Netherlands.

Environmental Hydraulics, Lee, Jayawardena & Wang (eds) © 1999 Balkema, Rotterdam, ISBN 90 5809 035 3

Regulator with gravity flap gate

M.R.Kabir & A.K.M.Nazrul Islam Howlader
Department of Water Resources Engineering, Bangladesh University of Engineering and Technology, Dhaka, Bangladesh

J.J.Veldman
BUET-DUT University Linkage Project, Delft University of Technology, Netherlands

ABSTRACT: A scale model has been developed for a regulator with gravity flap gate. Series of experiments have been conducted for different opening angle of the flap. The head loss – discharge relation for different opening angle shows significantly different from the theoretical desk study. Based on model study necessary changes were introduced in the equations. Model study identified 20 - 70% variation in discharge estimated by the equations based on desk study. The number of vents in the regulator can be determined through this study with greater level of accuracy.

1. INTRODUCTION

Bangladesh is a land of rivers and can be divided into a tidal and non-tidal region. In tidal areas drainage regulator with invert level is well below the low tides level and a flap gate at the riverside are the common structures to control the water. For appropriate design of a regulator several aspects have to be considered amongst which hydraulic, operational and economic aspects are important. In the tidal areas various types of gates are used for automatic drainage such as radial, gravity flap gate etc. Therefore an understanding how gates interact with hydrodynamics is essential. Drainage capacity of the regulator with gravity flap is especially in situation with small head losses hindered by the weight of the gravity flap. In the coastal area both the tidal window for drainage and the discharge are reduced by the weight of the flap.

Khulna-Jessore Drainage Rehabilitation Project (KJDRP) is a large size flood control and drainage projects covering over 100,000 ha in south-west of Bangladesh. About twenty regulators having 2 to 21 vents are proposed to prevent the intrusion of saline water into the polder area and to drain out excess water into the river. The headloss – discharge relation of these regulators are mostly based on desk study and available literature and calculated discharges might be significantly different from the actual discharges. Therefore an attempt has been made to perform the present physical model study and recommend more accurate headloss - discharge relation which will be helpful in defining the behaviour and performance of the structure. A physical model was constructed for a regulator of KJDRP called 'Sholmari regulator' to investigate the effect of headloss, upstream water level, gate angle and gate weight on discharge. The research work has been divided into two parts. As a first paper, this will illustrate the effect of headloss; upstream water level and gate angle for fully submerged condition whereas a second paper will include the effect of gate weight. Furthermore the model results will be compared with the results obtained from desk study.

2. GOVERNING EQUATIONS

Flow through the regulator is governed by the basic equation like continuity, conservation of momentum and energy. Fig.1 shows a sketch of a regulator with flap gate in submerged flow conditions through a vent is fully submerged. In sections 1 to 2 and 3 to 4 the flow is accelerated. In these sections energy levels are assumed to be constant and energy equation can be applied. From 2 to 3 and from 4 to 5 the flow is

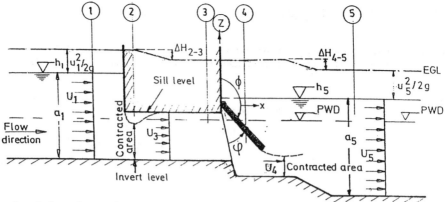

FIG.1. Sectional view of a regulator with flap gate

decelerated and energy loss is approximated by the well known Carnot formula as follows:

$$\Delta H_{i-j} = \left(u_i - u_j\right)^2 \Big/ 2g \tag{1}$$

2.1 Energy Losses

The following derivations were appeared from a desk study and literature research undertaken by the consultant of KJDRP (Matuszijk, 1996). These equations might give a first estimate, but for exact discharge head loss relations that include proper pressure distribution and energy losses can be determined through a physical model. Energy loss due to deceleration (2 - 3) at the entrance of regulator can be expressed as

$$\Delta H_{2-3} = \left(\left(\frac{1}{\mu_2} - 1\right)\frac{Q}{A_{vent}}\right)^2 \Big/ 2g \text{, with } \mu_2 = 0.5 - 0.7 \text{ and } A_{vent} = 2.1 \times 1.8 = 3.78 \text{ m}^2. \tag{2}$$

where, μ_2 is contraction coefficient and A_{vent} is cross-sectional area of vent. Energy loss due reduction of cross-section and contraction at the partly closed flap (3 - 4) and deceleration (4-5) at the outlet of regulator:

$$\Delta H_{4-5} = \left(\frac{Q}{A_4} - \frac{Q}{A_5}\right)^2 \Big/ 2g \tag{3}$$

Area under flap: $A_4 = \mu_4 A_{vent} \sin(\varphi)$ \hfill (4)

Contraction Coefficient at Flap: $\mu_4 = \mu_{4,max} - \left(1 - \mu_{4,min}\right)\cos(\varphi)$ with: $\mu_{4,max} = 1.0$, and $\mu_{4,min} = 0.65$ (5)

Area at downstream of regulator is assumed to be, $A_5 = 2 A_{vent}$ \hfill (6)

Therefore by substituting values from equation (4-6) in equation (3) head loss ΔH_{4-5} can be written as:

$$\Delta H_{4-5} = \left\{\left(\frac{1}{\left[\mu_{4,max} - \left(1 - \mu_{4,min}\right)\cos(\varphi)\right]\sin(\varphi)} - \frac{1}{2}\right)\frac{Q}{A_{vent}}\right\}^2 \Big/ 2g \tag{7}$$

Neglecting the energy loss due to friction inside the culvert (3-4) and combining above equations the headloss - discharge relation for the regulator can be written as follows:

902

$$Q = A_{vent} \sqrt{2g\Delta H \Bigg/ \left\{ \left(\frac{1}{\mu_2} - 1\right)^2 + \left(\frac{1}{\left[\mu_{4,max} - \left(1 - \mu_{4,min}\right)\cos(\varphi)\right]\sin(\varphi)} - \frac{1}{2}\right)^2 \right\}}$$ (8)

3. PHYSICAL MODEL

The physical model experiments were conducted in the 21 m long, 0.762 m wide and 0.762 m deep tilting flume at the Hydraulics Laboratory of Department of Water Resources Engineering, Bangladesh University of Engineering and Technology. KJDRP supplied the necessary data to carryout a scale model for Sholmari regulator: Number of Vents: 10; Vents Dimensions: 1.8 m x 2.1 m; Thickness of vent side wall: 0.6 m; Maximum design headloss: 1m; Assuming 1m head difference between upstream and downstream maximum velocity in a vent: 4.43 m/s; Contraction coefficient assumed as: 0.7; Design discharge was calculated as 11.74 m³/s per vent.

3.1 Model Construction

The high quality wooden model was constructed as per prototype data. The model dimensions are as below: Considering available flume width and maximum pump capacity 3-vent regulator was adopted through trial and error method where length scales, $n_L = 10$. To obtain reliable results from the model, it is necessary to ensure hydrodynamic similarity between the model and prototype. Therefore model was constructed on a non-distorted scale following the Froude scale relations. In order to minimise Reynolds effects the model was constructed on a large scale. The discharge scale, n_Q was calculated using the relation $n_Q = n_L^{5/2}$ (Huges, 1993). Using the discharge scale and the maximum available discharge at the flume = 0.19 m³/s, the maximum prototype discharge per vent, Q_p was calculated as 20.02 m³/s. The summary of the selected model parameters along with other trials and prototype data may be seen in Table 1.

3.2 Measurements

The model was run for four different discharges i.e. 15, 30, 60 and 135 l/s during eight different opening angle of flap gate i.e. 5, 10, 15, 20, 25 30, 40 and 85⁰. For each gate opening angle the downstream water depth was increased in steps of 2cm with the help of tail gate of the flume and continues until downstream water influence the upstream water level of the gate. Fig.2 shows the plot of measured headloss – discharge relation for different opening angle of flap. A family of curves with respect to the opening of flap can easily be seen from the figure.

3.3 Accuracy and error propagation

The error in the water level measurement from stilling basin is estimated as ± 0.5 mm (or 5mm in prototype).

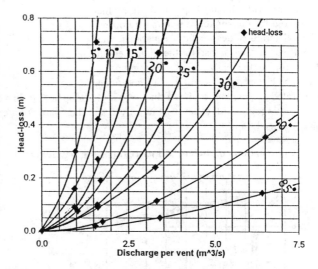

FIG. 2. Measured headloss – discharge relations for Different flap opening angle

Table 1. Summary of Prototype and Model Parameters

No. of Vent	Scale Ratio		Prototype Parameters		Model Parameters		
	n_L	n_Q	$W_{vent,p}$ (m)	Q_p per Vent (m^3/s)	Q_m per Vent (m^3/s)	$W_{vent,m}$ (m)	$a_{vent,m}$ (m)
2	6.3	99	1.8	9.46	0.095	0.285	0.333
3	10.0	316	1.8	20.02	0.063	0.180	0.210
4	12.6	563	1.8	26.76	0.048	0.143	0.166

However for small discharges when headlosses are small the maximum measuring error, $\Delta(h)$, using point gauges was considered to be ± 0.1 mm which means ± 1 mm on prototype scale. Considering the relation, $Q \cong \sqrt{\Delta h}$, where, $\Delta h = h_{u/s} - h_{d/s}$, the relative error in calculating discharge from the water level measurement will be as follows:

$$\frac{\Delta(Q)}{Q} = \frac{1}{2} \frac{\Delta(h) + \Delta(h)}{(h - h)} \qquad (9)$$

or simplified: $\dfrac{\Delta(Q)}{Q} = \dfrac{\Delta(h)}{\Delta h}$ (10)

4. ANALYSIS AND DISCUSSION

It has been observed (Fig.3) that the discharges computed with the equation from the previous desk study is higher than measured for the situation with open flap, and less then measured for the situations with partly opened flap. Therefore desk study equations were modified in two steps. In the first step the head-loss at the entrance of the regulator is discussed and the contraction coefficient calibrated against the results of the test with fully open flap. In a second step the geometrical description of the opening around the flap was improved and the contraction coefficient calibrated against the tests with partly closed flap.

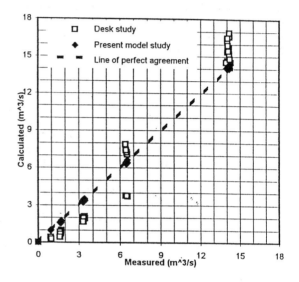

FIG. 3. Comparison of calculated and measured discharges

4.1 Contraction coefficient at upstream of the regulator (2-3)

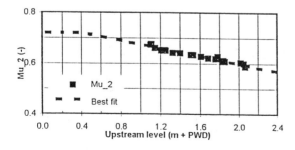

FIG. 4. Relation between upstream waterlevel and contraction coefficient μ_2

For the analysis of the entrance losses twenty two test were selected with fully submerged regulator and the flap in horizontal position (fully open). For these conditions the following cross sectional area can be applied: $A_2 = \mu_2 A_{vent}$, $A_3 = A_{vent}$, and $A_4 = A_{vent}$ The contraction coefficient μ_2 for the entrance of the regulator was determined from the measured discharges and head-losses by reverse calculation. The Fig.4 shows that μ_2 decreases with increasing water level. The following relation was proposed:

$$\mu_2 = \mu_{2,max} - Max\left\{0, \left(\mu_{2,max} - \mu_{2,min}\right) \frac{\left(h_1 - h_{sill}\right)}{h_{vent}}\right\}$$ (11)

with: $\mu_{2,min} = 0.56$ and $\mu_{2,max} = 0.72$

4.2 Contraction coefficient at downstream of flap (4-5)

When the flap is partly closed, the flow area is less than the opening of the vent. In this situation the water can pass under the bottom end of the flap, at both sides between the wall and the flap and through the narrow gap at the top of the flap near the hinges. The geometrical dimensions of these openings around the partly closed flap are:

$$A_{4,bottom} = \left(h_{flap} - h_{invert} + a_{step}\right)\left(W_{vent} + 2a_{step}\right)$$ (12)

$$A_{4,sides} = a_{flap}\left(W_{vent} + 2a_{step} - W_{flap}\right)$$ (13)

$$A_{4,hinge} = \left(W_{vent} + 2a_{step}\right)\left(h_{hinge} - h_{sill}\right)\sin(\varphi)$$ (14)

Now the total cross-section of the flow profile at section 4 is:

$$A_4 = \mu_4 \, Min\left(A_{vent}, A_{4,bottom} + A_{4,sides} + A_{4,hinge}\right)$$ (15)

The contraction coefficient μ_4 for the flow around the flap has been determined from measured discharges and head-losses applying the contraction coefficient μ_2 as proposed in the previous section. Fig.5 shows the correlation between μ_4 and the flap opening angle. From the graph the following relation was proposed:

$$\mu_4 = \mu_{4,max} - \left\{\left(1 - \mu_{4,min}\right)\sin(\varphi)\right\}$$ (16)

with $\mu_{4,min} = 0.46$ and $\mu_{4,max} = 1.00$

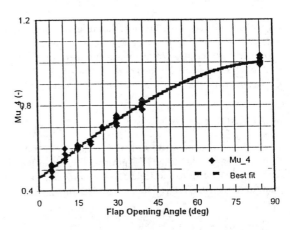

FIG. 5. Relation between flap opening angle and contraction coefficient μ_4

4.3 Total energy dissipation (1-5)

Discharge can be calculated considering the total energy dissipation at the regulator due to the contraction and expansion of flow and may be expressed as:

$$Q = \sqrt{2g\Delta H \Bigg/ \left(\left(\frac{1}{A_2} - \frac{1}{A_3}\right)^2 + \left(\frac{1}{A_4} - \frac{1}{A_5}\right)^2\right)}$$ (17)

Using equation 11 and 16 to compute the area A_2 and A_4, and consequently establish the headloss-discharge relation based on desk study Fig.3 shows very good matching between discharges calculated based on proposed modified equation and the measured discharges as compared to the previous desk study. Table 2 presents results of tests with minimum head-loss of 0.03 for fully open flap or 0.10 m for situation with partly open flap. These minimum headlosses ensure that the error in water level reading is not significantly

Table 2 Comparison of computed and measured discharges

Submerged (h_s >0.9 m)	μ_2		μ_4		Disch. per Vent Comp-measured		Discharge Comp/Measured		Selected tests	
	Min (-)	max (-)	Min (-)	max (-)	Aver (m^3/s)	St Dev (m^3/s)	Aver. (-)	St Dev (-)	Δh (m)	No. (-)
Desk study:										
Flap fully open	0.65	0.65	-	-	1.13	0.81	1.11	0.075	>0.03	22
Flap partly closed	0.65	0.65	0.65	1.00	-1.25	0.59	0.51	0.114	>0.10	33
Both	0.65	0.65	0.65	1.00	-0.34	1.38	0.73	0.306	>0.10	52
Present research:										
Flap fully open	0.56	0.72	-	-	0.03	0.11	1.00	0.011	>0.03	22
Flap partly closed	0.56	0.72	0.46	1.00	-0.01	0.08	1.00	0.032	>0.10	33
Both	0.56	0.72	0.46	1.00	0.01	0.10	1.00	0.026	>0.10	52

affecting computed discharges (Max error propagation in discharge: 3% for fully open and 1% for partly closed flap).

5. CONCLUSIONS

From the study following can be concluded:
- Discharges computed with the equation from the desk study appeared to be higher (upto 20%) than measured for open flap situation, and less (30 to 70%) than measured for the situation with partly opened flap.
- Some modifications were introduced in desk study equations.
- Accurate water level reading was utmost important especially at smaller head losses. Also accurate effective description of geometrical dimensions of flap opening is necessary.
- Model test appeared to be good tool for improving accuracy of headloss - discharge relation.
- Results from the model study can be utilised for Sholmari regulator in determining the number of vents.

6. ACKNOWLEDGEMENTS

The authors are grateful to SMEC International Pty Ltd., for providing field data and other related reports. Appreciation are due to SMEC project officials in Khulna-Jessore Drainage Rehabilitation Project, Khulna for arranging a trip to the field showing different regulators and holding discussions. The authors are also grateful to the BUET-DUT linkage project for financial support in carrying out the study.

7. REFERENCES

Huges, S. A. (1993), "Physical Models and Laboratory Techniques in Coastal Engineering", Advanced series on Ocean Engineering, Vol. 7, World Scientific publishing Co. Ltd., USA.

Lewin, J. (1995), "Hydraulic gates & valves in free surface flow and submerged outlets", Thomas Telford Publifications, London, UK.

Matuszyk, E. (1996), "Headlosses at flapgates for regulators hydraulics" SMEC and Associates, Khulna, Bangladesh.

4.6 Coastal hydraulics

Environmental Hydraulics, Lee, Jayawardena & Wang (eds) © 1999 Balkema, Rotterdam, ISBN 90 5809 035 3

A soft coastal defence system for the Belgium East Coast

M. Huygens, N. Van de Voorde & R. Verhoeven
Hydraulics Laboratory, University Gent, Belgium

ABSTRACT: Both a physical scale model and numerical simulations are used to validate the impact and efficiency of different sand supplement techniques to protect the coastal reach before Knokke at the Flemish east coast. A fundamental insight into the morphological behaviour of artificial beach nourishment can lead to a more reliable, scientifically based and from that economical coastal defence system. The experiments are conducted in a computer-controlled 2D-wave tank installation at Flanders Hydraulics Laboratory in Borgerhout. The morphological evolution of several representative bathymetries are recorded during characteristical tidal cycles. Comparison with existing field measurements reveals the similarities and shortcomings in the physical scale tests and the computer calculations. Some alternative suppletion forms are already suggested.

1. INTRODUCTION

Along the Flemish nearshore region, a combination of wave induced on- and offshore currents and a longshore flow provides a morphological equilibrium of natural beach replenishment. The port construction of Zeebrugge intersects the easterly longshore tidal drift. By that, the seaward harbour extension (with two breakwaters of 3.5 km) induces a persistent regression of the eastern coastline, forming an acute threat to the (touristical and ecologically valuable) beach of Knokke. A local tidal trench named "Appelzak" shifts ground and re-establishes itself just off the coast, in front of the groynes.

As a result, sandy beach material is transported offshore by the wave induced currents into the "Appelzak" trench from where the bottom material is removed by the predominant north eastern longshore tidal flow. Under successive storm action, the natural profile is gradually weakening and intervention is needed to ensure a sufficient safety level of shore protection. A rehabilitation of the natural sea-land environment, new technical possibilities and active integral coastal conservation actions, all-driven by political priorities, have made that recently preference is given to "soft", eco-friendly measures, i.e. beach nourishment, taking into account the natural dynamics of the shore profile. A first beach replenishment was executed in 1977-1978, a second beach nourishment over a limited stretch was executed in 1986. As a result of the intense local erosion at the beach of Knokke-Zoute after 5 years this sand was eroded and transported out of the region.

Therefore, to identify the potentials of different beach nourishment techniques an extended research program incorporating physical scale modelling and computer simulations is set up to explore a fundamental knowledge of the local beach morphology. From there, a proper coastal defence strategy for the whole Flemish coast is developed in the framework of a general protection plan "Coast 2002".

2. EXPERIMENTAL SET-UP

The complex nature of the hydrodynamics in coastal regions leads to an integrated approach of experimental work and numerical simulations. To investigate the impact and efficiency of several sand suppletion techniques a physical scale model is used to perform comparative tests. All experiments are conducted in a computer controlled 2D wave tank installation at Flanders Hydraulics Laboratory in Borgerhout. At the end of this rectangular, narrow wave tank (with a width of 0.7 m and a height of 0.85 m) a traditional wave generator of the paddle type is installed, so that perpendicular incident waves can be generated. To fill the model with a constant discharge a calibrated V-notch at the wave generator-end of the wave tank is used, while to empty it,

Figure 1. Experimental test flume

the valve at the bottom of the tank is manipulated. This way it is possible to simulate the real water level evolution over a tidal cycle. Traditional dynamic wave recorders register the local wave conditions in the wave tank (offshore / beach location). So, principally two hydraulic parameters are incorporated in this set-up: regular wave generation, be it only perpendicular incident and vertical tidal water level evolution. In this preliminar phase of the research program, the experiments in the 2D-wave tank only look for the cross-shore beach profile evolution, to identify the behaviour of different beach nourishment shapes under similar wave impact and tidal water level variations. The longshore tidal flows (ebb-flood) are not incorporated in the modelling. A global morphological investigation of the associated regions will be explored in a 3D-wave tank. Taking into account the practical limitations of the 2D wave tank, it is clear that two significant parameters of the in-situ situation cannot be simulated in the model: non-perpendicular incident waves and longshore flow. However in reality, all waves are nearly perpendicular incident due to refraction on the beach. The current investigation focusses only on the offshore directed sediment transport by the wave-induced undertow. Indeed, in this preliminary research stage one examines the possibility of intercepting the sand before it arrives in the tidal gully, from where the material is removed by the longshore tidal flow.

3. PHYSICAL SCALE TESTS

By using the traditional Froude scale law (linear scale 1/25) for all significant (horizontal and vertical) dimensions and for the modelling of the flow and wave characteristics, a reliable undistorted simulation of the stream pattern and the wave field can be achieved in the model. To achieve realistic sediment transport phenomena in the hydraulic model, the grain-Froude scale model is developped. A medium white sand of Mol (d_{50} = 170μm) as moveable sea bottom in the physical scale model ensures a proper qualitative simulation of the general transport phenomena. Indeed, the grain-Reynolds number Re. is nearly identical in both scale model and reality. Given the difference between the Shields parameter θ. (grain Froude number) in-situ and in scale model, the white sand of Mol appears not as moveable as the real sea bottom.

	In-situ sand	White sand of Mol
Mean grain diameter d_{50}	340 μm	170 μm
Critical shear stress τ_{cr}	0.234 N/m^2	0.210 N/m^2
Critical shear velocity u.	0.0153 m/s	0.0143 m/s
Dimensionless grain size D.	6.4216	3.1835
Grain Reynolds number Re.	3.31	1.55
Grain Froude number θ.	0.042	0.075

A typical serie of consecutive tidal cycles is simulated. Figure 2 shows the input data of the standard test serie (13 tidal cycles); significant wave height H, mean wave period T, high water level and low water level. This

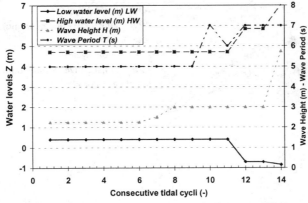

Figure 2. Input data - Hydrodynamical characteristics

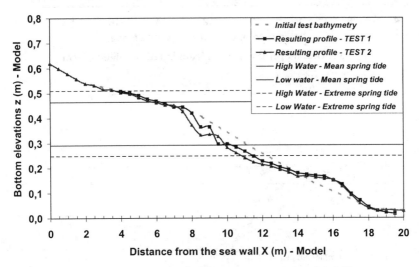

Figure 3. Resulting model bathymetries - Reproducability

input scheme is mainly based on the data from field records over the last years in the region and gives a representative image of the normal hydrodynamic on- /offshore action in the coastal reach.

Secondly, the quality and reproducability of the physical scale tests are controlled by performing identical hydrodynamic tidal cycles on an similar initial beach profile. Resulting bathymetries, as shown in figure 3, show a good agreement between both experiments. Typical features of both resulting beach profiles correspond quite well - an explicit berm just onshore the "Appelzak"-trench

- general erosion on foreshore with typical bar formation near wave breaking line (X_{model} = 8.5 m)

The development of the initial beach nourishment profile (suppletion 1986) as modelled in the physical scale experiments can be followed in figure 4. Again, one can clearly follow the formation of a stable berm at the onshore side of the tidal trench by eroding the relatively steep foreshore to a more flatten profile (equilibrium slope = 1/ 30 - 1/40). By that, this model tests already indicate the basic on- /offshore sediment transport mechanism in the region : bottom sand is moved offshore by extreme (breaking) wave action into the tidal gully "Appelzak" from where the sediments are carried away by the predominantly northeastern tidal (flood) current.

A detailed comparison between the model results and the field measurements in the critical coastal region under consideration reveals some interesting shortcomings of the current experiments. As indicated in figure 5, the recorded model beach profiles correspond quite well with the field registration in the offshore region, near the tidal gully. But the wet beach zone is much more eroded in the field, due to heavy storm impact where wave set-

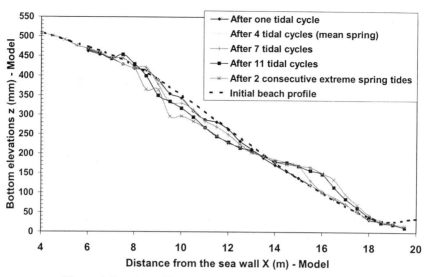

Figure 4. Beach profile evolution from initial suppletion form

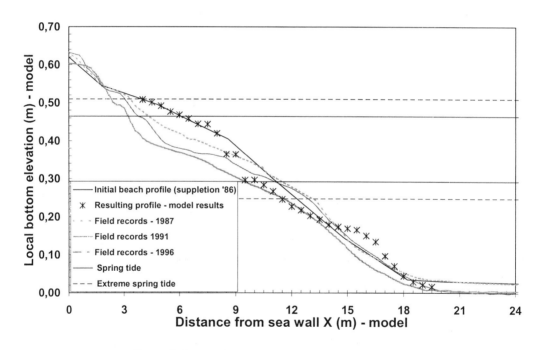

Figure 5. Comparison Model results - Field measurements

up rises the water level. From that, these high waterlevels together with extreme incident wave heights cause a substantial part of the erosion in that region. Together with a substantial eolian transport, storm events are identified as the main driving force to sediment transport. The resulting beach profile is mainly eroded during very limited time sequences of heavy storms; while the recovery stage (over a much longer period) isn't able to restore the beach. Another fact is the development of an equilibrium stage in the model, where no longshore removal appears when sand is transported from the foreshore to the tidal gully.

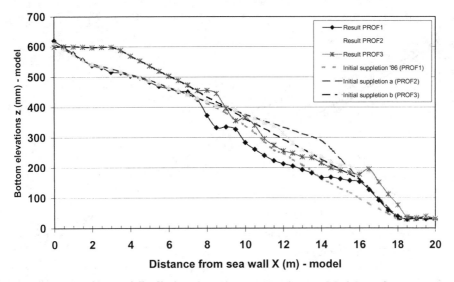

Figure 6. Preliminar beach suppletion forms - Model results

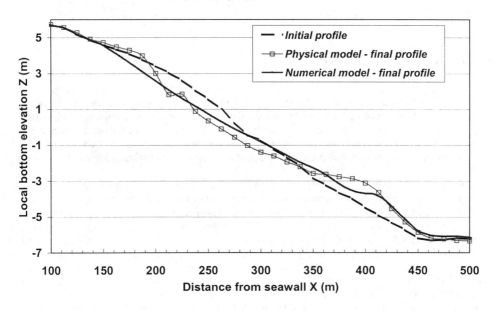

Figure 7. Comparison numerical-physical records

Finally, based on previous results with the initial beach nourishment profile (suppletion form 1986) some preliminar adaptions are suggested. Two proposed nourishment schemes, based on the previously analyzed model results, are illustrated in figure 6. The initial nourishment profile is chosen more closely to the equilibrium form; so a more stable shore profile is found (less offshore transport to the tidal gully). The suppletion form of *proposal a* is composed of the equilibrium berm slope on the foreshore and the resulting beach slope; while the *suppletion b* provides a touristical beach platform connected to the berm slope by a steeper shore line. Alternative cross-shore profiles with underwater feed berms are currently under investigation. More extended experiments, with explicit storm simulations and focus on wave set-up, will reveal more detailed information on the morphological development of the beach nourishment profile.

911

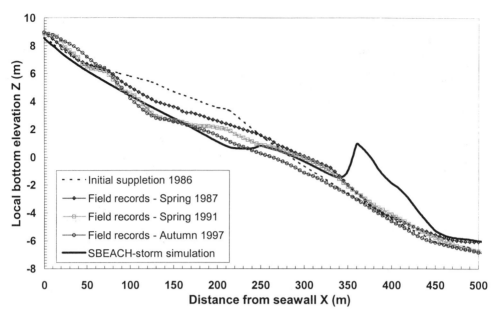

Figure 8. Numerical simulation of storm impact

4. MATHEMATICAL SIMULATIONS

Together with the physical experiments, numerical calculations with the "SBEACH"-software show a similar pattern under identical hydrodynamic conditions. Figure 7 indicates the qualitative correspondance between physical tests and computer records for the early mentioned cycle of 13 tides. Corresponding beach slopes, berm features and erosion spots prove the proper application of the software. The calculations confirm the scaling approach of the moveable sediments and the reliability of both simulation techniques. As a consequence, both physical and numerical results can be explored simultaneously, leading to an optimum interaction between both approaches.

Based on these results, several storm impacts are simulated in order to verify the field development. As mentioned before, storms are not yet incorporated in the physical model tests; while in-situ registrations show a clear storm erosion on the upper beach region Therefore, computer calculations look for the cross-shore beach profile development under several consecutive storm tides to check the field development with the numerical results. Figure 8 reveals the realistic development of the initial suppletion form (1986) under storm attack. The good correspondance with the in-situ bathymetries confirms the formerly formulated suggestion on the proportional impacts of storm events on the beach profile development. Especially the higher beach region is uniquely eroded under storm wave attack. The explicit formation of the foreshore berm, as calculated by the SBEACH-software, is not found in the field due to the longshore tidal flood flow. The contineous removal of the berm sand in the gully disturbs the equilibrium profile development as figured by the simulation.

5. CONCLUSION

By performing some initial scale model experiments, beach nourishment is identified as a possible remediation to the severe coastal erosion problems at the Flemish east coast. Its soft and eco-friendly development and application opens a wide spectrum of applications, while the fundamental knowledge on the typical behaviour of beach suppletion forms from the physical modelling will lead to a better understanding and economical implementation in the field. Together with the physical experiments, a proper application of the "SBEACH"-software is identified as an interactive, alternative engineering tool to develop a proper solution. Mathematical simulations identify storm events as the main erosive force for the cross-shore beach profile development.

6. REFERENCES

CUR (1997), "Manual on artificial beach nourishment", Report 130, Rijkswaterstaat Nederland.

Dean R.G. and Work P.A. (1995), "Assessment and prediction of beach-nourishment evolution", Journal of Waterway, Port, Coastal and Ocean Engineering, Vol. 121, No. 3 (May/June).

Dette H.H. (1983), "Technique of small scale beach fills", Proceedings COPEDEC, Colombo (Sri Lanka).

Hughes S.A. (1993), "Physical models and laboratory techniques in coastal engineering", World Scientific Publications Singapore.

Larson M. and Kraus N.C. (1989), "SBEACH: numerical model for simulating storm-induced beach change", US Army Corps of Engineers, CERC-89-9.

Roovers P.P.L., Kerckaert P., Burgers A., Noordam A. and De Candt P. (1981), "Beach protection as part of the harbour extension at Zeebrugge", Proceedings 25th PIANC Navigation Congress, Edingburgh, Scotland.

Vellinga P. (1986), "Beach and dune erosion during storm surges", Delft Hydraulics Communication No. 372.

Environmental Hydraulics, Lee, Jayawardena & Wang (eds) © 1999 Balkema, Rotterdam, ISBN 90 5809 035 3

Attenuation and phase delay of dynamic pressure propagated into reclaimed zone behind caisson-type seawall

T. Shigemura, K. Hayashi & K. Fujima
Department of Civil Engineering National Defense Academy, Yokosuka, Japan

M. Yokonuma
Aviation School of Ground Self Defense Force, Japan

ABSTRACT: Experimental study is carried out to investigate the attenuation and phase delay phenomena of the dynamic pressure which appear during propagation into the reclaimed zone of a caisson-type seawall. A series of model tests reveal that dynamic pressure propagates radially into the reclaimed zone and that the dynamic pressure propagates into the reclaimed zone attenuating the pressure and delaying its phase in a similar manner as it propagates into seabed vertically downward. Through these findings, some empirical formulae are driven which estimate the attenuation and phase delay of dynamic pressure during its propagation into the reclaimed zone.

1. INTRODUCTION

Recently, man-made islands have been constructed for various purposes where the water depth is relatively large and wave loads are considerably severe. In these projects, caisson-type seawalls have often been adopted to protect their shores directly against wave loads since it is difficult to build a conventional type breakwater there due to the depth. Unfortunately, settlement failures have often been generated in the zone of the reclaimed soil behind the caisson-type seawalls. For simplicity, the zone of the reclaimed soils will be called the reclaimed zone. In 1993, Port and Harbor Research Institute, Ministry of Transportation, Japan(PHRI) has initiated a comprehensive study on the mechanism of the settlement failures. Through their studies, it was revealed that the dynamic pressure propagated into the reclaimed zone through rubble mound and the zone of the backfill stones may play important role in causing the settlement failures(Takahashi et al, 1996). Here, the dynamic pressure means the wave pressure caused by the standing wave. The zone of the backfill stones will also be called the backfill zone in this paper. The authors have also conducted a series of model tests to examine the propagation properties of the dynamic pressure into the reclaimed zone to find similar results(Shigemura et al, 1998). However, the failure mechanism has not yet been clarified. This paper will mainly examine the attenuation and phase delay of the dynamic pressure while it propagates into the reclaimed zone, by conducting a series of model tests using the simplified seawall models.

2. EXPERIMENTAL SETUP AND PROCEDURES

A 20 m long, 0.3 m wide and 0.6 m deep wave flume was used to conduct the model tests. This flume has an absorbing type wave generator at one end and a model of a caisson-type seawall at the other end. The model tests were conducted with a scale of 1/30. The cross section of the standard seawall model was determined by referring to the cross sections of actual seawalls that have recently experienced the settlement failures. Further, cross section of the backfill zone was simplified to be like a vertical wall. Fig. 1 shows one of the cross section of a caisson-type seawall model determined in this way.

In this figure, installation points of the wave gauge, wave pressure and pore water pressure transducers are also shown together. Medium gravel was chosen as the materials to build both the rubble mound and backfill zone and sand was used as the materials to reclaim the space behind the backfill zone. The model tests were conducted by loading non-breaking standing waves. Six regular waves were chosen as the test waves that satisfy the criterion of non-breaking standing waves. Table 1 summarizes the characteristics of the test waves together with the physical properties of sand and gravel mentioned above.

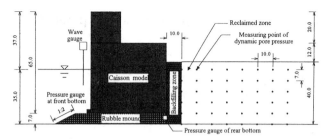

FIG. 1. Cross Section of Caisson-Type Seawall Model and Installation Points of Measuring Devices

Table 1 Characteristics of Test Waves and Physical Properties of Sand and Gravel

	WAVE-1	WAVE-2	WAVE-3	WAVE-4	WAVE-5	WAVE-6		Gravel	Sands
Period (s)	1.0	1.2	1.4	1.6	1.8	2.0	d(mm)	10~15	0.25
Water depth (cm)	35.0	35.0	35.0	35.0	35.0	35.0	S	2.64	2.59
Water height (cm)	12.0	12.0	12.0	12.0	12.0	12.0	k(m/s)	-----	0.39
H/L	0.25	0.19	0.15	0.13	0.11	0.11			
H/L	0.09	0.07	0.05	0.05	0.04	0.04			

Model tests were carried out for three cases (Case 1, Case 2 and Case 3) changing the thickness of the backfill zone, B to be 0.0, 5.0 and 10.0 cm, respectively. Case 1 has no backfill zone behind the caisson model. In each case, six test waves were loaded one by one on the seawall model in which eight pore pressure transducers were placed at a given height above the flume bottom horizontally, at an interval of 10 cm and signals from all measuring devices were recorded simultaneously for about 1 minute. This procedure was repeated six times by replacing eight pore pressure transducers from the bottom layer to the top layer, 7 cm apart from each other(see Fig. 1).

3. DATA PROCESSING AND RESULTS OF PRIMARY ANALYSIS

All of the recorded data were digitized at an interval of 0.02 seconds with the data noise removed using the FFT method. From this processed data, extreme values and phases of occurrence were determined respectively by applying the zero-up-cross method. From these determined values, significant wave height in front of the caisson model, $H_{1/3}$, significant amplitude of dynamic pressure at the bottom front face of the caisson model, $Po_{1/3}$ and the pore water pressure at eight points, $P_{1/3}$ were determined respectively. Further, $P_{1/3}$ was divided by $Po_{1/3}$ to convert $P_{1/3}$ into a dimensionless variable. Similarly, phase delays between the dynamic pressure at bottom front face of the caisson model and pore water pressure at given measuring points were determined by comparing the respective phases of the extreme values, and the significant value of the phase delay $\Delta t_{1/3}$ was determined similarly. $\Delta t_{1/3}$ was also divided by the period of the test wave, T to

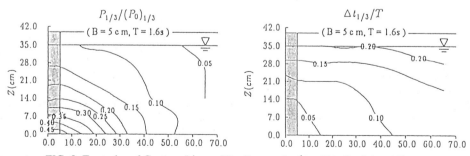

FIG. 2. Examples of Contour Lines of $P_{1/3}/Po_{1/3}$ and $\Delta t_{1/3}/T$ in Reclaimed Zone

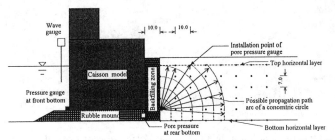

FIG. 3   Possible Propagation Paths of Dynamic Pressure in the Reclaimed Zone(Case 2)

convert $\Delta t_{1/3}$ into a dimensionless variable. The primary analysis was conducted on the data of $P_{1/3}/Po_{1/3}$ and $\Delta t_{1/3}/T$ to examine how the dynamic pressure propagates into the reclaimed zone after passing the rubble mound and backfill zone. Fig. 2 shows the examples of the contour lines of $P_{1/3}/Po_{1/3}$ and $\Delta t_{1/3}/T$ distributed in the reclaimed zone. In these figures, numerals on both axes indicate the distance measured from the line where the rear face of the caisson model intersects the flume bottom. As shown from these figures, contours of both $P_{1/3}/Po_{1/3}$ and $\Delta t_{1/3}/T$ distribute as if they run along the arcs of the concentric circles in the reclaimed zone. This clearly suggests that dynamic pressure would propagate almost radially in the reclaimed zone.

4. ANALYTICAL RESULTS AND DISCUSSION

In the previous chapter, it was revealed that dynamic pressure would propagate almost radially in the reclaimed zone. Thus, it was assumed that dynamic pressure would propagate into the reclaimed zone radially through a narrow horizontal slit provided on the boundary between backfill zone and reclaimed zone at 7.0 cm above the flume bottom. Fig. 3 shows the possible propagation paths of the dynamic pressure inside the reclaimed zone.

Based on this assumption, analysis will be made quantitatively on the attenuation and phase delay of the dynamic pressure during its propagation in the reclaimed zone.

4.1 Attenuation of Dynamic Pressure inside the Reclaimed Zone

To examine the attenuation of dynamic pressure during its propagation in the reclaimed zone, whole data of $P_{1/3}/Po_{1/3}$ were plotted against X/h on the semi-logarithmic paper where X is the shortest distance from the slit stated in the previous section and h is the water depth used for the model tests. These plots suggested that high correlation would exist between $P_{1/3}/Po_{1/3}$ and X/h. Regression analysis was conducted between $\log_{10}(P_{1/3}/Po_{1/3})$ and X/h for the classified data due to each period of the test waves. These analyses suggested that there should exist the following functional relationship between $P_{1/3}/Po_{1/3}$ and X/h:

$$P_{1/3}/Po_{1/3} = A \, exp(-\alpha (X/h)) \tag{1}$$

where A is a constant and α is the attenuation rate of dynamic pressure as it propagates into the reclaimed zone. However, as mentioned previously, this data is measured at six different layers in every case. Thus, ground condition of the reclaimed zone would possibly be different from each other each time the reclaimed zone was prepared. Further, the data include the dynamic pressure attenuated before the dynamic pressure reaches the reclaimed zone. Therefore, Eq. (1) can not evaluate the attenuation rate of dynamic pressure after the dynamic pressure reached in the reclaimed zone. So, the regression analysis was conducted again between $\log_{10}(P_{1/3}/Po_{1/3})$ and X/h for the data measured in each horizontal layer and some correction was made on the data of $P_{1/3}/Po_{1/3}$ so that the value of the coefficient A in Eq .(1) may become 1.0 at the point where X/h is 0.0. This correction will be called "initial value correction" from now on. The initial value correction was made on all data of $P_{1/3}/Po_{1/3}$ and the corrected data of $P_{1/3}/Po_{1/3}$ were plotted again against X/h. Fig. 4 shows some examples of these plots. These plots show the relationship between $P_{1/3}/Po_{1/3}$ and X/h inside the reclaimed zone. Regression analysis was conducted similarly between the corrected data of $P_{1/3}/Po_{1/3}$ and X/h. Table 2 summarizes the results of the regression analysis. In this table, r means the correlation coefficient

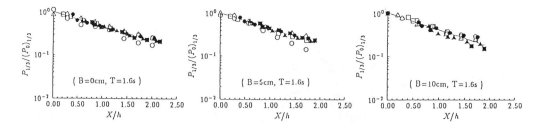

FIG. 4 Examples of the Plots of the Corrected Data of $P_{1/3}/Po_{1/3}$ against X/h

Table 2. Results of the Regression Analysis between the Corrected Data of $P_{1/3}/Po_{1/3}$ coefficient and X/h

Test Wave		WAVE-1	WAVE-2	WAVE-3	WAVE-4	WAVE-5	WAVE-6	Test case
T	(sec)	1.0	1.2	1.4	1.6	1.8	2.0	
H	(cm)	10.5	11.5	11.7	10.8	8.1	8.0	
	α	0.372	0.353	0.349	0.333	0.325	0.315	Case-1
	r	0.973	0.971	0.968	0.969	0.969	0.963	
H	(cm)	10.3	11.2	11.3	10.7	8.2	7.8	
	α	0.418	0.398	0.381	0.368	0.362	0.353	Case-2
	r	0.963	0.953	0.956	0.953	0.950	0.946	
H	(cm)	10.1	11.3	11.2	10.4	8.3	7.9	
	α	0.461	0.445	0.412	0.395	0.365	0.350	Case-3
	r	0.952	0.943	0.964	0.959	0.957	0.955	

As it can be seen from Fig. 4 and Table 2, quite reliable relationships were found between the corrected data of $P_{1/3}/Po_{1/3}$ and X/h which evaluate the attenuation of the dynamic pressure in the reclaimed zone.

Next, let's examine the physical features of the attenuation rate α of the dynamic pressure. Fig. 5 shows the relationship between the attenuation rate α summarized in Table 2 and the discharge coefficient given by Eq.(2).

$$C=(k/\gamma_w m_v) \cdot T/l^2 \tag{2}$$

where k is the permeability of the seabed sand, m_v is its volume compressibility, l is the depth of the permeable sea bed, γ_w is the unit weight of pore water and T is the wave period.

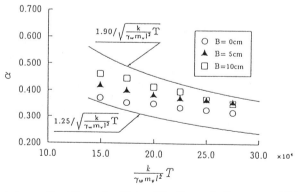

FIG. 5 Relationship between the Attenuation Rate of the Dynamic Pressure and Discharge Coefficient

The discharge coefficient C was proposed by Zen et al (1987) in their research of wave-induced liquefaction and densification in seabed. In their study, they indicated that C and the transmission coefficient a given by Eq.(3) were the influential variables to the propagation of wave pressure into seabed.

$$a=1.0+n \cdot m_w/m_v \tag{3}$$

where n is the porosity of the seabed sand and m_w is the compressibility of pore water. They also pointed out that T would be a single factor that affects propagation of the dynamic pressure if the physical properties of the seabed sand are constant. Sasaki et al(1993) derived the approximate solution for the governing equation derived by Zen et al and revealed that dynamic pressure would attenuate following the functional form shown below as it propagates downward into the seabed.

$$P/P_0 = (1-1/a)\exp(-\sqrt{\pi a /C} \ Z) \tag{4}$$

where Z is a dimensionless variable of the depth z divided by l. It should be noted however that these two works are studies on the propagation of dynamic pressure into the seabed vertically downward. Two rigid lines in Fig.5 are the curves determined by calculating the functions $b / \sqrt{C}$ which is quite similar to the attenuation rate of Eq.(4). Here, b is constant. The data of α follow the curves quite satisfactorily. These facts clearly indicate that dynamic pressure may propagate into the reclaimed zone in similarly as it propagates downward into seabed vertically.

4.2 Phase Delay of Dynamic Pressure during its Propagation into the Reclaimed Zone

To examine the phase delay of dynamic pressure during its propagation in the reclaimed zone, whole data of $\Delta t_{1/3}/T$ were plotted against X/h. As a result, it was revealed that there would exist high correlation between $\Delta t_{1/3}/T$ and X/h. Regression analysis was conducted between $\Delta t_{1/3}/T$ and X/h for the classified data due to each period of the test waves. These analyses indicated that there should exist the following linear relationship between $\Delta t_{1/3}/T$ and X/h.

$$dT1/3/T = \beta (X/h) + \gamma \tag{5}$$

where β is a rate of the phase delay and γ is the initial value of the phase delay at the point where X=0. However, as mentioned previously, the data used for the analysis is measured at six different layers in every case. So, the regression analysis was conducted again between $\Delta t_{1/3}/T$ and X/h for the data measured in each horizontal layer and some correction was made on the data of $\Delta t_{1/3}/T$ so that the value of γ in Eq.(5) may become 0.0 at the point where X/h is 0.0. This correction will also be called "initial value correction" from now on. This correction was made on all data of $\Delta t_{1/3}/T$ and the corrected data of $\Delta t_{1/3}/T$ were plotted again against X/h. Fig. 6 shows some examples of these plots. Regression analysis was conducted similarly between the corrected data of $\Delta t_{1/3}/T$ and X/h. Table 3 summarizes the results of the regression analysis. Fig.6 and Table 3 clearly indicate that linear relationship as shown below should exist between $\Delta t_{1/3}/T$ and X/h which evaluate the phase delay of the dynamic pressure in the reclaimed zone.

$$\Delta t_{1/3}/T = \beta (X/h) \tag{6}$$

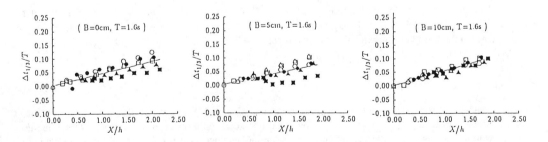

FIG. 6 Examples of Relationship between $\Delta t_{1/3}/T$ and X/h

Table 3. Results of the Regression Analysis between the Corrected Data of $\Delta t_{1/3}/T$

Test Wave		WAVE-1	WAVE-2	WAVE-3	WAVE-4	WAVE-5	WAVE-6	Test Case
T	(sec)	1.0	1.2	1.4	1.6	1.8	2.0	
H	(cm)	10.5	11.5	11.7	10.8	8.1	8.0	
β		0.051	0.044	0.046	0.044	0.051	0.054	Case-1
γ		0.862	0.773	0.796	0.847	0.877	0.923	
H	(cm)	10.3	11.2	11.3	10.7	8.2	7.8	
β		0.049	0.050	0.050	0.038	0.040	0.050	Case-2
γ		0.892	0.952	0.939	0.708	0.784	0.841	
H	(cm)	10.1	11.3	11.2	10.4	8.3	7.9	
β		0.058	0.051	0.051	0.052	0.051	0.059	Case-3
γ		0.832	0.800	0.883	0.943	0.910	0.924	

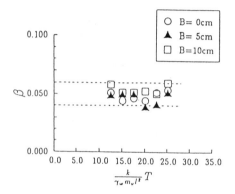

FIG. 7 Relationship between β and C

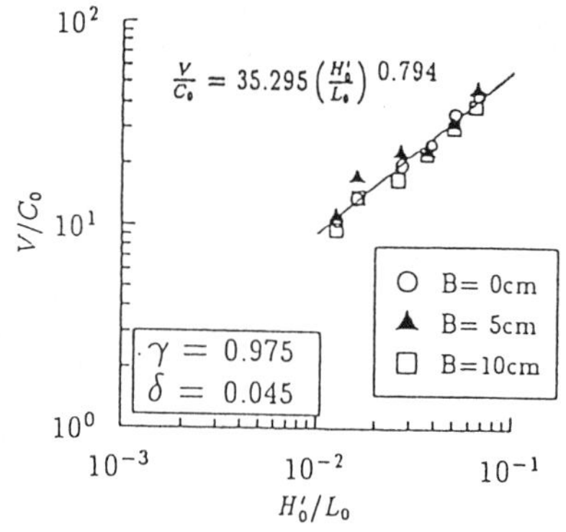

FIG. 8 Relationship between V/Co and Ho'/Lo

Next, let's examine the physical features of the rate of the phase delay of the dynamic pressure, β. Fig.7 shows the distribution of β plotted against the discharge coefficient, C.

As it can be seen from this figure, β does not varies noticeably but keeps constant value ranging from 0.04 to 0.06 even if the width of the backfill zone and the value of the discharge coefficient varied within the range in the present study. Finally, examination was made on the relationship between the phase velocity and characteristics of test waves. Transformation of Eq.(6) easily derives the following relationship that evaluates the phase velocity of dynamic pressure in the reclaimed zone.

$$X/\Delta t_{1/3} = h/\beta T = V \tag{7}$$

Fig. 8 shows the distribution of V/Co plotted against Ho'/Lo where Co is the phase velocity of deep water wave with period T and Ho'/Lo is the equivalent steepness of deep water wave with period T. Regression analysis derived the following functional relationship between both variables.

$$V/Lo = 35.30(Ho'/Lo)^{0.80} \tag{8}$$

5. CONCLUDING REMARKS

Experimental study was carried out to examine the attenuation and phase delay phenomena of dynamic pressure which appear during propagation into the reclaimed zone. As a result, it was revealed that dynamic pressure would propagate into the reclaimed zone in similar manner as it propagates vertically downward into the seabed. Further, some of the empirical formulae were obtained which would estimate the attenuation and phase delay of the dynamic pressure during its propagation into the reclaimed zone.

6. REFERENCES

Sakai, T.,Mase, H and Yamamoto, T(1993). "Fluctuation of pore water pressure around the surface zone of seabed due to the fluctuation of wave induced pressure", Proc. of Coastal Eng, JSCE, Vol.40,586-590.

Shigemura, T.,Yokonuma, M, Hayashi, K and Fujima, K(1998). "Propagation properties of dynamic pressure through reclaimed sand behind caisson-type seawalls", Proc. of Ports'98, Vol 1, ASCE,571-580.

Takahashi,S., Suzuki,K.,Tokubuchi,K. and Shimosako,K(1996)."Experimental analysis of the settlement failure mechanism shown by caisson-type seawalls", Proc. of 25th ICCE, ASCE, Chapter 148, 1902-1915.

Zen,K.,Yamazaki,H. and Watanabe,A.(1987). "Wave-induced liquefaction and densification in seabed", Report of Port and Harbor Research Institute, Vol.26,No.4, 125-180.

Environmental Hydraulics, Lee, Jayawardena & Wang (eds) © 1999 Balkema, Rotterdam, ISBN 90 5809 035 3

A constrained dual membrane wave barrier

Edmond Y.M. Lo
School of Civil and Structural Engineering, Nanyang Technological University, Singapore

ABSTRACT: The performance of a dual membrane wave barrier constrained by mooring lines attached along its sides is investigated. Previous studies have indicated that a single or dual membrane barrier system is effective when the tension is sufficiently large. The use of additional mooring lines provides further tension as well as modifying the coupling of the excited membrane modes to the propagating transmitted wave. These effects are investigated to show that the overall performance of the dual membrane wave barrier system can be improved.

1. INTRODUCTION

There has been recent interest in the use of flexible membranes (e.g. Kim & Kee, 1996; Lo, 1998) as a wave barrier for coastal applications. These applications are often driven by sedimentation and transport issues as such processes are strongly dependent on the local wave field. The possibility that the membrane barrier can be constructed rapidly in a cost effective manner has further added to its appeal.

In a typical arrangement, the membrane barrier spans the entire water depth and is fixed at both the sea bed and water surface, with tension in the membrane being provided by surface buoys. Past studies have indicated that good reflection of the incident wave energy over a wide range of wave frequencies can be achieved at a sufficiently large tension (Kim & Kee, 1996). Similar studies on flexible beam-like barriers (i.e. including structural stiffness) have also indicated that a sufficiently large beam stiffness is needed (Abul-Azm, 1994). An arrangement for improving barrier performance is the use of dual membranes (Cho et. al., 1998; Lo, 1998) or dual beams (Abu-Azm, 1994) spaced at appropriate distances apart. Then, the tension or stiffness requirement can be reduced without compromising barrier performance.

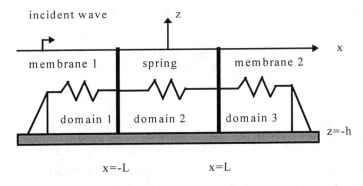

Figure 1: Definition sketch with one mooring line attachment at z=-h/2 shown

The magnitude of the transmitted wave past a membrane barrier depends on the degree of coupling between the membrane modes of motion and the propagating transmitted wave mode. For example, an optimal barrier is achieved if the membrane motion couples only to the evanescent modes on the transmission side. This coupling to the propagating or evanescent modes can be modified by constraining the membrane motion through additional mooring lines such as that shown in Figure 1 for a dual membrane wave barrier. The mooring lines also provide additional stiffness to the system. This paper reports on the results of such an investigation for the dual membrane arrangement shown in Figure 1.

2. THEORY

An eigenfunction approach is used to derive the motion of the membrane system in response to normally incident long crested monochromatic waves in water of constant depth h as shown in Figure 1. The derivation closely parallels that of Lo (1998) to which the reader is referred. In particular, the fluid is partitioned into three domains, that of the incident, the transmission, and the region in between. Incompressible, inviscid, and irrotational flow is assumed along with the usual kinematic and dynamic boundary conditions at the water surface, and outgoing waves for the transmitted and reflected waves at $\pm\infty$. With these, the velocity potentials ϕ_i in the three fluid domains are expanded as

$$\phi_1 = a_0 \cosh[k(z+h)]e^{ik(x+L)} + \alpha \cosh[k(z+h)]e^{-ik(x+L)} + \sum_{m=1}^{\infty} a_m \cos[\kappa_m(z+h)]e^{\kappa_m(x+L)} \quad (1)$$

$$\phi_2 = \cosh[k(z+h)](\gamma e^{ikx} + \delta e^{-ikx}) + \sum_{m=1}^{\infty} \cos[\kappa_m(z+h)](c_m e^{\kappa_m x} + d_m e^{-\kappa_m x}) \quad (2)$$

$$\phi_3 = \beta \cosh[k(z+h)]e^{ik(x-L)} + \sum_{m=1}^{\infty} b_m \cos[\kappa_m(z+h)]e^{-\kappa_m(x-L)} \quad (3)$$

where the propagating (i.e. the incident, reflected and transmitted wave) modes with wavenumber k, and the evanescent wave modes with wavenumbers κ_m at frequency ω are readily identified.

The membrane motion is modeled using a stretched membrane model. The dynamical equation for the membrane displacement ξ_1 at x=-L with tension T_1 and mass per unit area m_1 is

$$m_1\omega^2\xi_1 + T_1\frac{\partial^2\xi_1}{\partial z^2} - \sum_{j=1}^{J}\delta(z-z_j)[K\xi_1 + K(\xi_1-\xi_2)] = -i\omega\rho(\phi_1-\phi_2) \quad (4)$$

Here the pre-tensioned and taut mooring lines are modeled as simple horizontal springs attached to the membranes at points z_j, and δ is the Kronecker delta function. In practice, the horizontal attachment of the lines can be achieved by use of a rigging system as shown. If the lines are attached at an angle, the spring constants will have to be modified by the attachment angle, and the downward component of the mooring force will have to be accounted for. In the above, the total number of attachments points is J and all the spring constants per unit membrane width are assumed to be equal with magnitude K. A similar equation also holds for the membrane at x=+L. With both ends of the membranes being fixed, an eigenfunction expansion for the motion of the membranes is obtained as

$$\begin{Bmatrix} \xi_1(z) \\ \xi_2(z) \end{Bmatrix} = \sum_{n=1}^{\infty} \begin{Bmatrix} A_n /(\nu_1^2-\lambda_n^2) \\ B_n /(\nu_2^2-\lambda_n^2) \end{Bmatrix} \sin\lambda_n(z+h); \quad \nu_i = \omega\sqrt{(m_i/T_i)}; \quad \lambda_n = n\pi/h; \quad x = \mp L \quad (5)$$

Substituting equations (1) to (3) and (5) into equation (4) and applying the orthogonality of the membrane eigenfunctions over the depth h results in an equation relating A_n and B_n to ϕ_1 and ϕ_3 (through α, β, a_m, b_m), as well as ϕ_2 (through γ, δ, c_m and d_m). The dependence on ϕ_2 terms can be recast as a dependence on

ϕ_1 and ϕ_3 terms by applying the kinematic condition ($\partial\phi/\partial x$ being continuous) at $x=\pm L$ and utilizing the orthogonality of the wave eigenfunctions. This results in an equation at each membrane mode n of the form

$$A_n - \frac{2K}{hT_1}\sum_{j=1}^{J}\sum_{m=1}^{N}\left[\frac{2A_m}{v_1^2-\lambda_m^2}-\frac{B_m}{v_2^2-\lambda_m^2}\right]\sin\left[\lambda_n(z_j+h)\right]\sin\left[\lambda_m(z_j+h)\right]=$$

$$2\frac{i\omega\rho}{hT_1}\left\{\left[(a_o-\alpha)i\cot(2kL)-i\beta\csc(2kL)-(a_o+\alpha)\right]I_{n0}\right.$$

$$\left.-\sum_{m=1}^{M}\left[a_m(\coth(2\kappa_m L)+1)+b_m\operatorname{cosech}(2\kappa_m L)\right]I_{nm}\right\} \qquad n=1,2,\ldots N \qquad (6)$$

where $$\left\{\begin{matrix}I_{n0}\\I_{nm}\end{matrix}\right\}=\int_{-h}^{0}\left\{\begin{matrix}\cosh k(z+h)\\\cos\kappa_m(z+h)\end{matrix}\right\}\sin\lambda_n(z+h)dz \qquad (7)$$

Unlike the case of no mooring attachments along the membrane (Lo, 1998), each membrane mode is now coupled to all the others (with the number of membrane modes modeled being N), in addition to being dependent on the wave modes (with the number of evanescent modes modeled being M). A similar equation is obtained by considering the corresponding dynamical membrane equation at $x=+L$. The resulting set of 2N equations is further complemented by another 2M+2 equations by applying the equating the membrane velocity and fluid particle velocity at $x=\pm L$ as

$$\frac{\partial\phi_{1,3}}{\partial x}=-i\omega\xi_{1,2} \qquad x=\mp L \qquad (8)$$

and using the orthogonality of the wave eigenfunctions. The equations at $x=-L$ are given by

$$k(a_0-\alpha)J_0=-\omega\sum_{n=1}^{N}A_n I_{n0}/(v_1^2-\lambda_n^2) \qquad (9)$$

$$\kappa_m a_m J_m=-i\omega\sum_{n=1}^{N}A_n I_{nm}/(v_1^2-\lambda_n^2) \qquad m=1,2,..,M \qquad (10)$$

where $$\left\{\begin{matrix}J_0\\J_m\end{matrix}\right\}=\int_{-h}^{0}\left\{\begin{matrix}\cosh^2 k(z+h)\\\cos^2\kappa_m(z+h)\end{matrix}\right\}dz \qquad (11)$$

with a corresponding set at $x=+L$. The final matrix of 2N+2M+2 algebraic equations set are straight-forwardly solved for A_n, B_n, α, β, a_m, and b_m. Energy conservation of the sum of the reflected and transmitted wave energies being equal to the incident is used to verify the numerical accuracy of the solution. Convergence is very rapid, and 60 membrane modes and 20 evanescent are used are used for computing the results below where the energy conservation is satisfied to essentially computer accuracy.

3. RESULTS

The following results are presented using non-dimensional membrane parameters and spring constants as a function of non-dimensional wave frequency kh and membrane spacing L/h. The dimensionless values of membrane mass is $m'_i=m_i/\rho h$, membrane tension is $T'_i=T_i/\rho gh^2$ and spring constant is $K'=K/\rho gh$.

Computational runs were initially performed with the number and location of the mooring lines being varied. It was found that the first two (n=1,2) membrane modes and to a lesser extent, the third n=3 mode, coupled most strongly to the transmitted wave, with the transmission peaks coinciding with the dis-placement peaks of these modes. Thus the mooring lines should be placed to suppress the these membrane

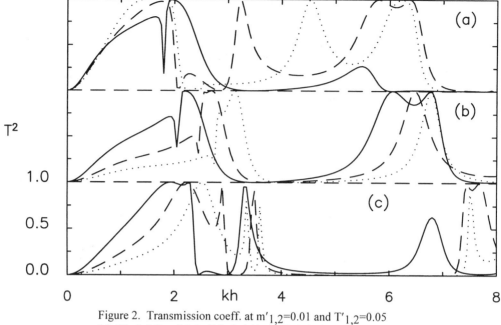

T^2

1.0

0.5

0.0

0　　　　　　2　　kh　　4　　　　　　6　　　　　　8

Figure 2. Transmission coeff. at $m'_{1,2}$=0.01 and $T'_{1,2}$=0.05
(a) K'=0, L/h = 0.2 (solid), 0.4 (dashed), 0.6 (dotted)
(b) L/h = 0.2, K'=0.05 (solid), 0.15 (dashed), 0.25 (dotted)
(c) L/h = 0.4, K'=0.05 (solid), 0.15 (dashed), 0.25 (dotted)

modes. The node of the first n=1 mode is at z=-h/2 while those of the second n=2 are at z=-h/4 and -3h/4. This implied that three (i.e. J=3 in equation 4) constraining points at these locations should be used, which was confirmed by the numerical trials. Such a mooring arrangement would also suppress the n=3 mode as well. Hence the subsequent results were computed using this mooring line arrangement. We had also computed cases where the number of mooring line attachments is increased. However, additional reductions in the transmission were minor.

The case of equal membrane mass $m'_{1,2}$=0.01 and tension $T'_{1,2}$=0.05 is presented in Figure 2 at various separation L/h of 0.2, 0.4, and 0.6. The energy transmission T^2 is given by the ratio $|\beta/a_0|^2$. In the absence of mooring lines (fig. 2a), the transmission is characterized by the a broad transmission peak centered around kh of 1.5 with additional peaks at larger kh depending on the spacing. This broad low frequency peak results from using too low a membrane tension (Lo, 1988) and renders the barrier ineffective.

The addition of mooring lines at L/h=0.2 (fig. 2b) significantly reduces the transmission of the broad peak There is a narrowing of the peak and a shift towards larger kh as the stiffness K' is increased from 0.05 to 0.25. The transmission peak is shifted to kh of 3.2 for K'=0.25, making the barrier effective over a much wider range of kh as compared to fig. 2a. We note that there is a continual shift of the transmission peak to higher kh with increasing K'. This is expected because if the membranes at the attachment points of -h/4, -h/2, and -3h/4 are fully constrained (i.e. infinite K'), the membrane modal expansion (equation 5) will have nonzero modes only at n=4 and its harmonics, with full suppression of the n=1,2, and 3 modes. However, such a large spring stiffness will be difficult to achieve in practice. At a larger membrane separation L/h=0.4 (fig. 2c), the effect of the mooring lines becomes less effective. While there is still a narrowing of the peak and a shift to higher kh, the overall reduction is less than at L/h=0.2, especially in at lower kh.

Earlier studies (Cho et.al., 1998; Lo, 1998) have indicated that unequal tension on the membranes when

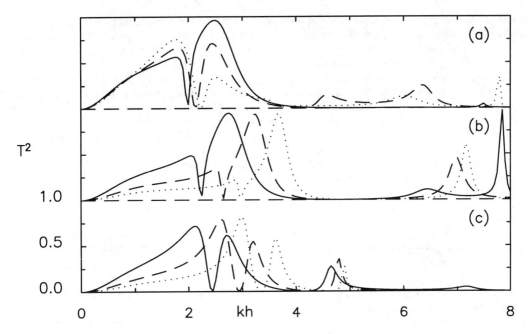

Figure 3. Transmission coeff. at $m'_{1,2}=0.01$ and $T'_1=0.05$, $T'_2=0.1$
(a) $K'=0$, $L/h = 0.2$ (solid), 0.3 (dashed), 0.4 (dotted)
(b) $L/h = 0.2$, $K'=0.05$ (solid), 0.15 (dashed), 0.25 (dotted)
(c) $L/h = 0.3$, $K'=0.05$ (solid), 0.15 (dashed), 0.25 (dotted)

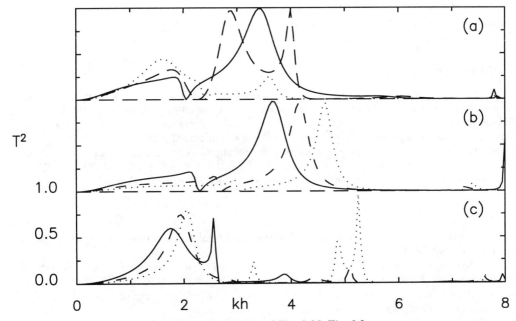

Figure 4. Transmission coeff. at $m'_{1,2}=0.01$ and $T'_1=0.05$, $T'_2=0.2$
(a) $K'=0$, $L/h = 0.2$ (solid), 0.4 (dashed), 0.6 (dotted)
(b) $L/h = 0.2$, $K'=0.05$ (solid), 0.15 (dashed), 0.25 (dotted)
(c) $L/h = 0.6$, $K'=0.05$ (solid), 0.15 (dashed), 0.25 (dotted)

combined with an appropriate separation can result in a rather effective wave barrier. We next show that this barrier arrangement can be further improved with mooring line restraints. Figure 3 shows the case of tension on the incident side T'_1=0.05 and tension on the transmission side T'_2=0.1 at various L/h. Without mooring lines (fig. 3a), the dual membranes reduces the width and peak of the broad transmission peak arising from a single membrane barrier, this peak being centered around kh of 2 (Lo, 1998). In particular, the single transmission peak of the single membrane becomes two distinct peaks with relative magnitudes depending on the membrane separation as shown in fig. 3a. With the inclusion of mooring lines (fig. 3b and 3c), one of the peaks is suppressed along with a general shift in the peaks to higher kh. At L/h=0.2 (fig. 3b), the lower peak is increasing suppressed with increasing stiffness K'. At K' of 0.25, the suppression of the lower peak results in an effective reflection of waves up to kh of 3. The upper peak which is also shifted to larger kh is further reduced in width with increasing K'. At a larger L/h of 0.3, the upper peak rather than the lower is suppressed. While there is also a shift to larger kh with increasing K', the presence of the lower peak implies that the barrier is not as effective as the case of smaller L/h of 0.2.

Another case of unequal membrane tension is shown in Figure 4 where the tension on the transmission side is increased to 0.2. In the absence of mooring lines, effective reflection over all kh is obtained at L/h of 0.6 (fig. 4a), with the largest transmission value of 0.4 at about kh of 1.6. However, this large membrane separation (actual separation is 2L/h) may be difficult to achieve in practice. With the addition of mooring lines, effective reflection is similarly achieved at much smaller separation of L/h of 0.2 as seen in fig. 4b. In fact, the transmission peak of the unconstrained case at kh of 1.6 is significantly reduced though at the expense of significantly enhancing the peak at larger kh. However, this peak which occurs at a higher kh of 3.6 (fig. 4a) is progressively shifted to even higher kh. Thus, at stiffness K' of 0.15 and 0.25, effective reflection is obtained up to kh of about 3.5 and 4 respectively. This is slightly better than the case shown earlier in fig. 3b at K' of 0.25, though the tension T'_2 there on the transmission side is less at 0.1. On comparing these two cases and also the case of T'_2=0.05 (fig. 2b), the best arrangement is likely to occur around L/h of 0.2 with K' of 0.15 to 0.25. Increasing T'_2 from 0.05 to 0.2 progressively shift the region of effective reflection to larger kh. The optimal setup will, of course, also depend on the exact shape of the incident wave spectrum.

We note that the mooring lines can also degrade membrane performance as shown in fig. 4c. Here L/h is 0.6 which correspond to the best barrier performance without mooring lines (fig. 4a). The addition of the mooring lines clearly results in increased wave transmission at all stiffness K' considered.

4. CONCLUSION

We had investigated the effect of mooring lines, modeled as linear springs, as additional constraints on membrane motions in a dual flexible membrane wave barrier. Three attachment locations at depths of -h/4, -h/2, and -3h/4 were needed to reduce the first three membrane modes, these being the modes which coupled to the transmitted wave. The barrier performance was improved with better reflection of the wave energy at the same or lower membrane tensions and over a wider range of wave frequency. For an effective barrier, a membrane separation of 2L/h of 0.4 and a spring stiffness K/ρgh of 0.15 to 0.25 should be used, though the barrier performance will continually improve with increasing spring stiffness.

5. REFERENCES

Abul-Azm, A.G. (1994). "Wave diffraction by double flexible breakwaters." Appl. Ocean Res., Elsevier, Great Britain, Vol. 16, 87-99.

Cho, I.H., Kee, S.T., and Kim M.H. (1998) "Performance of dual flexible membrane wave barriers in oblique waves", J. Wtrwy., Port, Coast., and Oc. Engrg., ASCE, 124(1), 21-30.

Kim, M.H., and Kee, S.T. (1996). "Flexible-membrane wave barrier. I: Analytic and numerical solutions." J. Wtrwy., Port, Coast., and Oc. Engrg., ASCE, 122(1), 46-53.

Lo, E.Y.M. (1998). "A flexible dual membrane wave barrier." To appear J. Wtrwy., Port, Coast., and Oc. Engrg., ASCE, Sept. '98 issue.

Williams, A.N. (1996). "Floating membrane breakwater." J. of Offshore Mechanics & Arctic Engrg, 118(1),46-52.

Environmental Hydraulics, Lee, Jayawardena & Wang (eds) © 1999 Balkema, Rotterdam, ISBN 90 5809 035 3

Effects of coral reefs on the nearshore coastal dynamics

Harshinie Karunarathna
Department of Civil Engineering, University of Moratuwa, Sri Lanka

ABSTRACT: The paper discusses hydraulic performance of coral reefs in the context of wave transformation, wave energy dissipation and associated mean water surface fluctuations through a set of experimental investigations. It is concluded through this study that a large portion of incident wave energy dissipates on the reef and a small fraction of wave energy propagate towards the beach. Therefore, the reef acts as a natural means of coast protection.

1. INTRODUCTION

Many tropical coastal regions in the Pacific, Atlantic and Indian oceans are fronted by coral reefs. Coral reefs can be of three different types: fringing reefs, barrier reefs and atolls. Fringing reefs especially have a very significant impact on the near shore wave climate due to substantial wave breaking and energy loss and further due to wave energy dissipation while traveling over the shallow and rough surface of the reef flat. Meanwhile, the wave induced mean currents generated at wave breaking is a prime factor governing sediment dynamics of the region.

The coral reef is one of the economically significant coastal ecosystems of tropical islands. The ability of a reef to provide goods and services is related directly to the presence or absence of certain organisms. The principal user groups who have a direct influence on the coral reefs, are the fishing community, tourists, developers (e.g. hoteliers) and those who exploit the reef for direct commercial benefits, such as for the production of materials for the construction industry.

From a coastal engineering view point the coral reef performs as an efficient natural submerged offshore energy dissipator with considerable energy dissipation taking place due to wave breaking on the surface of the reef. Any damage to the reef or deterioration would contribute to waves having higher levels of energy reaching the shoreline, thereby leading to increased erosion. Therefore, there is a need to understand the mechanics of wave structure interaction of coral reefs to evaluate the influence of key hydraulic parameters such as the energy dissipation characteristics over the length of the reef, the relative water depth, surface roughness and the overall geometry and porosity of the reef structure. In this context investigations of reef types which have a significant impact on nearshore wave dynamics are justified.

Recent predictions on global warming and sea level rise have identified the importance of the sea-defense functions of coral reefs and it is in this context that the use of artificial methods for rehabilitation, apart from restoring the natural resource base, should be given due consideration. There is a need to develop capabilities to rehabilitate degraded reefs as quickly as possible. A few research studies have been carried out to investigate the feasibility of the use of artificial concrete reef structures to promote the recovery of degraded reefs.

This paper focuses attention on understanding the role of coral reefs in coast protection. This subject has to be evaluated in the context of the need for engineering interventions for sustaining multiple uses of the coastal zone. Such interventions are considered very important in view of global warming and its impact on sea level rise.

2. REVIEW OF PREVIOUS STUDIES

A number of important studies have been carried out to study wave transformation on reefs. These studies include both experimental investigations and field measurements. Attention is focused on selected studies which refer to important aspects of hydraulics of wave-reef interaction.

Kono and Tsukayama (1980) conducted a pioneering study on wave transformation on reefs. A series of field measurements were done on a reef in Ryuko Islands, Japan. The study was supported by a supplementary set of experimental data collected on a model of the same reef. They found that wave attenuation on the reef depends on incident wave characteristics such as wave height and wave period, tidal range, mean water level, the surface friction and the bottom topography.

Gerritsen (1980) conducted field measurements on wave transformation at Ala Moana reef around Honolulu, Hawaii Islands. It was found that the energy dissipation on coral reefs occur due to two reasons namely, wave breaking on the slope of the reef and the bottom friction of the rough coral surfaces.

Nelson and Lesleighter (1985) presented non-dimensional design charts to estimate wave decay on platform coral reefs. But, their results were based on a set of data collected on a site-specific model reef. The transient wave set-up associated with breaking waves was neglected in their study considering the unconfined nature of the selected reef. Their results show that under extreme and severe wave conditions, around 50% of the incident wave energy dissipate within a very short distance along the reef.

Young (1989) did an extensive series of field measurements on wave transformation at the Great Barrier Reef, Australia. It was found that a large percentage of incident wave energy is dissipated on the reef due to breaking and bottom friction. According to the results, the front seaward slope of the reef plays a vital role in wave energy dissipation.

The above studies conclude that a significant portion of incident wave energy dissipates on the reef before they finally reach the beach. The primary sources of wave energy dissipation were found as wave breaking on the front slope of the beach and the bottom friction. The amount of wave attenuation on the reef depends on the parameters such as offshore wave characteristics, tidal range, bottom friction on the reef, etc.

3. INVESTIGATIONS ON FRINGING REEFS

3.1 Objective of the study

Fringing reefs among all other types of reef have a very significant impact on near shore wave dynamics and wave induced circulation. Most of the available information on wave transformation and associated hydrodynamic phenomena on fringing reefs to date are supported by either specific field investigations or experimental studies carried out using regular waves. Experimental data available on irregular waves have been collected on site-specific models reefs and they have not focused in detail on wave induced transient water surface fluctuations.

The objective of the present study is to investigate wave breaking, wave transformation and transient mean water levels induced by wave action on fringing reefs in general and thereby to analyse the role of fringing reefs in coast conservation in the context of global warming and associated sea level rise.

3.2. Description of the study

This section gives a brief description of the study presented in the paper. The methodology of the experimentation and the factors considered in designing the experiments are discussed together with incident wave parameters.

An extensive series of laboratory experiments were done to measure wave transformation and associated changes in mean water level on fringing reefs. The effects of deepwater wave characteristics and the steepness of the front slope of the reef on the transformation of waves were analysed. Finally, the change in mean water levels associated with wave motion on top of the reef was studied. A set of experimental data collected by Takayama et al. (1977) on fringing reefs was also used to reinforce the above results.

The experiments were done in a wave flume with irregular wave generation facilities. A physical model of a fringing reef with a steep seaward slope and a flat horizontal surface was used for measurements (Figure 1). Incident waves were generated with JONSWAP type short wave spectra. An array of capacitance type

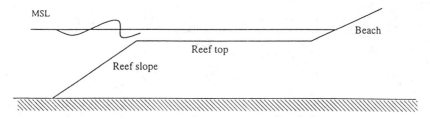

Figure 1. Schematic diagram of a fringing reef

wave probes placed across the reef profile was used to measure the incident waves and the accompanying mean water levels.

4. DISCUSSION OF RESULTS

This section of the paper is devoted to present and discuss a set of selected results on wave transformation, wave breaking and mean water surface fluctuations on fringing reefs.

Figures 2 (a) - (d) show wave height variation across the reef for four different deepwater wave steepnesses H_0/L_0 (H_0' is deepwater wave height and L_0 is deepwater wave length respectively) varying from 0.0095 to 0.0469. In these figures, the local significant wave heights ($H_{1/3}$) and the distance measured onshore (x) are normalised by the deepwater wave height. It can be seen from these figures that more than 50% of the incident wave energy on average is dissipated within a very short distance after the breaking point. Waves with high steepness lose more energy than waves with low steepness. Energy dissipation gradually diminishes as waves travel towards the shoreline along the flat reef but, only 10% - 20% of the wave energy on average remains when they finally reach the shore.

Figures 3(a) and (b) compare wave attenuation on fringing reefs with two different front slopes. Figure 3 (a) corresponds to a reef with a front slope of 1:2 while Figure 3 (b) corresponds to a slope of 1:10. Both wave shoaling and attenuation are more intense on the reef with the steeper slope.

This could be related to different patterns of wave breaking. On a highly steep slope, wave breaking is more rapid and violent thus, dissipating large amount of energy than on a less steep slope. Also, wave breaking process on a steep slope confines to a very narrow area.

The front slope of a natural reef in general is steep and may reach values as high as 1:2 or 1:1 which is very unlikely for a sandy beach. Therefore, wave energy dissipation on a reef could be several times larger than that in a normal sandy beach.

Figure 4(a) and (b) show the breaking wave height against the breaker depth on the slope of the reef for reef slopes 1:2 and 1:10. As can be seen from the figures, the breaking waves do not satisfy the depth limited breaking criterion $H_b= 0.78h_b$ or the criterion proposed by Goda, 1978 possibly due to the steepness of the front slope of the reef. Instead, they well agree with the equation

$$\frac{H_b}{L_0} = 1 - \exp\left[1.5\pi \frac{h_b}{L_0}(1 + \tan\theta)\right] \qquad (1)$$

where H_b is the breaking wave height, h_b is the water depth at breaking and $\tan\theta$ is the sea bottom slope.

Attention was also focused on transient mean water levels associated with the incident wave transformation on a reef as its significance is two fold. Firstly it interacts with the incident wave field. Secondly it increases the depth of submergence of the reef which may cause adverse impacts on the shoreline together with the effects of rise in sea level due to global warming.

Figures 5 (a) and (b) show the transient mean water levels associated with wave motions shown in Figures 2 (a) and (b) respectively. Irrespective of the tidal fluctuations, the increase in mean sea level on the top of the reef due to wave transformation can reach up to 5% - 15% of the deepwater wave height. This is a function of the deepwater wave steepness. At the time of high tide, positive addition of the tidal level and the

931

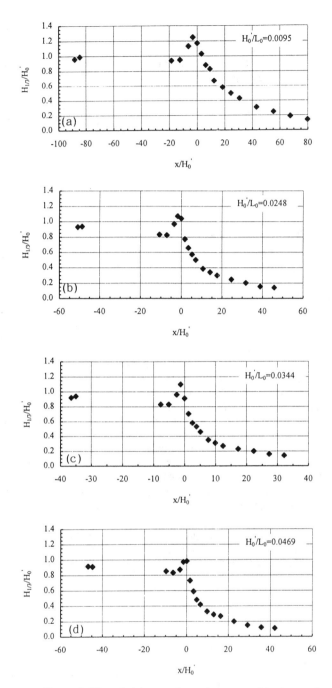

Figure 2. Wave height transformation across fringing reef

wave induced mean sea level represent the actual water level on the reef top. As a result, the depth of submergence increases and hence more land will be inundated.

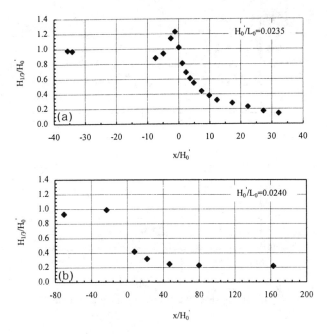

Figure 3. Wave height attenuation on fringing reefs with different slopes

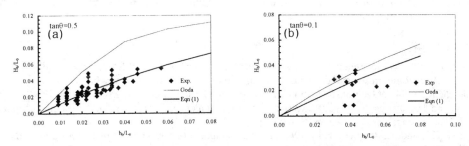

Figure 4. Breaking wave height on fringing reefs

5. CONCLUDING REMARKS

This paper has focused attention on the importance of coral reefs in coast protection. It has identified the need to have a clear understanding of the wave-structure interaction of coral reefs in order to appreciate the specific role played by coral reefs in coast protection. Such information can be of vital importance in planning strategies of the management of coastlines protected by coral reefs. The paper has presented selected results from important investigations relating to the hydraulic performance of fringing reefs. The results focus attention on important parameters, which govern wave-structure interaction of coral reefs. These parameters include deepwater wave characteristics, steepness of the front slope of the reef, transient mean water levels and depth of submergence of the reef.

It can be concluded through this study that the nearshore dynamics on a fringing reef is largely governed by the amount of wave breaking on the slope of the reef and hence energy loss. A considerable amount of incident wave energy dissipates on the slope and the surface of the reef and therefore, the shoreline is subjected to a very small percentage of the incident wave energy. In this context, a fringing reef can act as a natural means of coast protection.

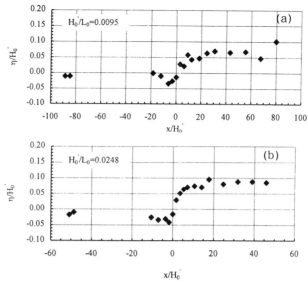

Figure 5. Mean water level on a fringing reef

Wave setup associated with wave attenuation increases the water level on the reef and at the same time generates mean currents which contribute to nearshore sediment dynamics in the region. In addition, the rough and uneven coral surface reduces the capacity of sediment transport and hence minimise sediment loss or shoreline erosion.

REFERENCES

Gerritsen, F., 1980, Wave attenuation and wave setup on a coastal reef, Proc. ICCE, pp. 444-461.

Goda, Y., 1985, Random sea and design of maritime structures, Tokyo University Press.

Gourlay, M.R., 1994, Wave transformation on a coral reef, Coastal Engineering, Vol.23, pp. 17-42.

Kono, T. and Tsukayama, S., 1980, Wave transformation on reefs and some considerations on its application to field, Coastal Engineering in Japan, Vol. 23, pp. 45-57

Nelson, R.C. and Lesleigther, E.J., 1985, Breaker height attenuation over platform coral reefs, Proc. Australian Conf. in Coastal Engineering, pp. 9-16.

Takayama, T., Kamiyama, Y. and Kikuchi, O., 1977, Wave transformation on reefs, Technical Note, Port and Harbor Research Institute, Japan (in Japanese).

Young, L.R., 1989, Wave transformation on coral reefs, J. Geophy. Res., Vol. 94, No.C7, pp. 9979-9789.

Environmental Hydraulics, Lee, Jayawardena & Wang (eds) © 1999 Balkema, Rotterdam, ISBN 90 5809 035 3

Modelling wave propagation in nearshore area

Liu Baiqiao
National Marine Data and Information Service, SOA, Tianjin, China

Zhao Zidan
Department of Water Resources and Harbor Engineering, Tianjin University, China

ABSTRACT: A practical model system for nearshore wave propagation is presented in the paper, which consists of directional wave spectrum propagation model and wave setup model . The results of numerical test done on a mild slope with a supposed elliptic bay show that the model system yields rather reasonable distribution of wave height and wave setup, and confirm that the interaction between wave and wave setup plays important role in nearshore wave computation.

1.INTRODUCTION

Many models for nearshore wave propagation successfully describe various wave transformation processes such as shoaling, refraction, diffraction, reflection and decay due to breaking and bottom friction. But few of them take the directional distribution of wave energy and interaction between wave and wave setup into account, which have great influence on wave energy distribution. This paper presents an attempt to supplement the usual computational model with consideration of directional wave spectrum and local water depth variation caused by wave setup, and, by means of numerical test, illustrates the effect of such a consideration on nearshore wave computation.

The model system presented in the paper consists of directional wave spectrum propagation model which is the extensive application of mild-slope equations for energy of simple harmonic wave, and wave setup model which uses the standard depth-average equations of fluid momentum and continuity with the inclusion of Radiation Stress(Longuet-Higgins and Stewart,1964). In the model system, wave and wave setup are coupled. Nearshore waves which induce the wave setup are simulated on the topography with local water depth composed of static depth and wave setup.

2.WAVE PROPAGATION MODEL

2.1 Combined Refraction and Diffraction Model of Directional Wave Spectrum

Combined refraction and diffraction model of directional wave spectrum bases on the Berkhoff's equations for wave amplitude and direction(Berkhoff, 1976):

$$\frac{1}{a}\left\{\frac{\partial^2 a}{\partial x^2} + \frac{\partial^2 a}{\partial y^2} + \frac{1}{cc_g}\left[\nabla a \cdot \nabla(cc_g)\right]\right\} + k^2 - |\nabla s|^2 = 0 \tag{1}$$

$$\nabla \cdot (a^2 cc_g \nabla s) = 0 \tag{2}$$

where a is the wave amplitude; ∇ is the spatial gradient operator($\partial / \partial x + \partial / \partial y$); x, y are longitude and lateral spatial Cartesian co-ordinates; k, s are the wave number and phase function; and c, c_g are the wave phase velocity and group velocity respectively. In the theory of linear wave, the irregular sea waves can be described by adding up many of component waves with different amplitude, frequencies, direction and random phases. It's energy, E, is the sum of that of all component waves:

$$E = \int_0^\infty \int_{-\pi}^\pi s(\omega,\theta)d\theta d\omega \tag{3}$$

or in the discrete form:

$$E = \lim_{M\to\infty} \lim_{N\to\infty} \sum_{m=1}^{m=M} \sum_{n=1}^{n=N} s(\omega_m,\theta_n)\delta\omega\delta\theta \tag{4}$$

$$= \lim_{M\to\infty} \lim_{N\to\infty} \sum_{m=1}^{m=M} \sum_{n=1}^{n=N} E_{m,n}$$

where $s(\omega,\theta)$ is the directional wave spectrum; $E_{m,n}$ is energy of the component wave with frequency in $\omega_m \sim \omega_m + \delta\omega$ and direction in $\theta_n \sim \theta_n + \delta\theta$. According to the relationship between energy and amplitude of the component wave:

$$E_{m,n} = \frac{1}{2}\rho g a_{m,n}^2 \tag{5}$$

where $a_{m,n}$ is the amplitude of corresponding component wave; ρ is the sea-water density; g is the acceleration of gravity. Substitution of equation (5) into equation (1) and (2) leads equations for energy and direction of the component wave:

$$|\nabla s_{m,n}|^2 = k_m^2 + \frac{1}{2E_{m,n}}(\frac{\partial^2 E_{m,n}}{\partial x^2} + \frac{\partial^2 E_{m,n}}{\partial y^2}) - \frac{1}{4E_{m,n}^2}[(\frac{\partial E_{m,n}}{\partial x})^2 + (\frac{\partial E_{m,n}}{\partial y})^2]$$

$$+ \frac{1}{2(cc_g)_{m,n}E_{m,n}}(\frac{\partial cc_g}{\partial x}\cdot\frac{\partial E_{m,n}}{\partial x} + \frac{\partial cc_g}{\partial y}\cdot\frac{\partial E_{m,n}}{\partial y}) \tag{6}$$

$$\nabla[E_{m,n}(cc_g)_{m,n}\nabla s_{m,n}] = 0 \tag{7}$$

Irrotationality of phase function of the component wave is also assumed, that is:

$$\nabla \times (\nabla s_{m,n}) = 0 \tag{8}$$

Equation(4), (6), (7) and(8) make up the combined refraction and diffraction mode of directional wave spectrum. It is the extensive application of mild-slop equation for energy of simple wave. If the directional distribution of wave energy is neglected, it will leads to combined refraction and diffraction mode of energy spectrum.

Bottom friction: Assume that the variation in wave energy with distance may be represented, locally, by

$$E_{m,n}(x_2) = E_{m,n}(x_1)\exp[-a_0*(x_2 - x_1)] \tag{9}$$

where a_0 is a wave energy attenuation coefficient attributable to energy dissipation in the boundary layer at the bed. The coefficient a_0 is approximated by:

$$a_0 = \frac{4k_m^2\sqrt{\dfrac{\upsilon}{2\omega_m}}}{2k_m(h+\bar{\zeta}) + \sinh 2k_m(h+\bar{\zeta})} \tag{10}$$

where υ is the kinematics viscosity; h is static water depth; $\bar{\zeta}$ is wave setup or setdown.

Breaking of the waves: Assume that energy of all the component waves is consumed in equal ratio when the irregular waves are in breaking, so:

$$\frac{(E_{m,n})_B}{(E_{m,n})_0} = \frac{E_B}{E_0} \tag{11}$$

The indexes " $_0$ " and " $_B$ " stand for deep water and breaking conditions respectively. According to the generalized *MICHE*-criteria in consideration of shallow water conditions (Kiyoshi Horikawa,1988), the breaking of the component wave can be checked using:

$$(E_{m,n})_B = \frac{1}{8}\rho g[0.14\frac{2\pi}{k}\tanh k(h+\bar{\zeta})]^2\frac{(E_{m,n})_0}{E_0} \tag{12}$$

where k is corresponding to mean wave period.

2.2 Wave Setup Model

Wave setup and wave-induced currents are accompaniments of wave field. Once details of the wave number and wave energy are known, it is possible to determine the wave setup and depth-average wave-induced current patterns in the test area. The wave setup, wave-induced currents are provided by solving the stand depth-average equations of fluid moment and continuity with the inclusion of Radiation Stress terms:

$$u\frac{\partial u}{\partial x} + v\frac{\partial u}{\partial y} = -g\frac{\partial\bar{\zeta}}{\partial x} + T_x + M_x - B_x \tag{13}$$

$$u\frac{\partial v}{\partial x} + v\frac{\partial v}{\partial y} = -g\frac{\partial\bar{\zeta}}{\partial y} + T_y + M_y - B_y \tag{14}$$

$$\frac{\partial}{\partial x}\left[u(\bar{\zeta}+h)\right] + \frac{\partial}{\partial y}\left[u(\bar{\zeta}+h)\right] = 0 \tag{15}$$

where u,v are depth-average velocities in the x,y co-ordinate directions; T_x and T_y, M_x and M_y, B_x and B_y are the Radiation Stress terms, lateral mixing terms, bottom friction terms respectively given by the expressions listed in many papers.

Wave setup, wave-induced current field are determined by wave condition in the test region. In the meantime, they rearrange the distribution of water depth, generate the water circulation, and therefor have great influence on wave deformation especially in the nearshore area. In the paper, equations above make up the coupled wave propagation model with a consideration of the interaction between wave and wave setup. Nearshore wave field which induces the wave setup is simulated on the topography with local water depth composed of static depth and wave setup. If the interaction terms between wave and wave setup is neglected, the model will leads to be uncoupled.

937

3.TEST CASE

The calculation of the wave propagation by the model gives good results for a lot of practical cases .In the following for an example the satisfactory performance of the model is demonstrated.

The test region, shown in Fig.1, is as large as 1.0km(X direction)× 1.28 km(Y direction), which is a system of topographies consisting of a mild slope and a elliptic bay.

The initial wave parameters ,the significant wave height and period, in deep water are listed as:

$$H_{1/3} = 2.0 \, \text{m} \tag{16}$$

$$T_{1/3} = 6.0 \, \text{s} \tag{17}$$

The directional wave spectrum is adopted as the form of :

$$S(\omega,\theta) = S(\omega)F(\theta) \tag{18}$$

where $S(\omega)$ is energy spectrum, given by Wen et al(1989):

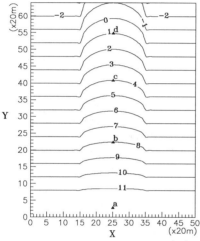

Fig.1. Bathymetric contour(m)

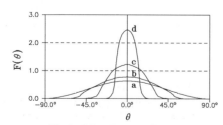

Fig.3. Directional functions

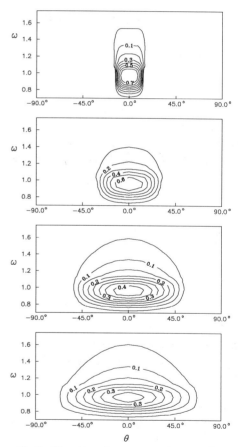

Fig.2. Directional wave spectra on the points labeled in Fig.1

938

$$S(\omega) = \frac{m_0 p}{\omega_0} \exp\left\{-95\left[\ln\frac{p(5.813-5.137\eta)}{(6.77+1.088p+0.013p^2)(1.307-1.426\eta)}\right]\cdot\left(\frac{\omega}{\omega_0}-1\right)^{12/5}\right\} \qquad (19)$$

$$\omega < 1.15\omega_0$$

$$S(\omega) = \frac{m_0(6.77+1.088p+0.013p^2)(1.307-1.426\eta)}{\omega_0(5.813-5.137\eta)}\left(\frac{1.15\omega_0}{\omega}\right)^m \qquad (20)$$

$$\omega \geq 1.15\omega_0$$

with:

$$m_0 = H_{1/3}^2 / 16 \qquad (21)$$

$$\omega_0 = 5.72 / T_{1/3} \qquad (22)$$

$$p = 96H_{1/3}^{1.35} / T_{1/3}^{2.7} \qquad (23)$$

$$\eta = 0.625 H_{1/3} / h \qquad (24)$$

$$m = 2(2-\eta) \qquad (25)$$

and $F(\theta)$ is the directional distribution function, given by the equation:

$$F(\theta) = K\cos^{ns}\theta \qquad (26)$$

$$K = \frac{1}{\int_{-\pi}^{\pi}\cos^{ns}\theta d\theta} \qquad (27)$$

The parameter ns controls the directional distribution pattern of wave energy. In the test, $ns = 2.0$.

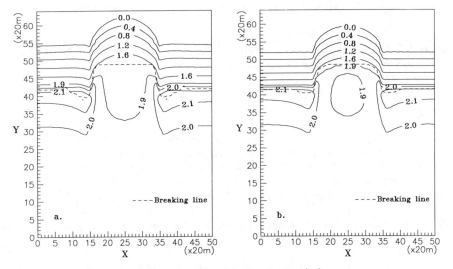

Fig.4. Wave height distribution(m)
a.The interaction between wave and wave set-up was considered
b.The interaction between wave and wave set-up was not considered

939

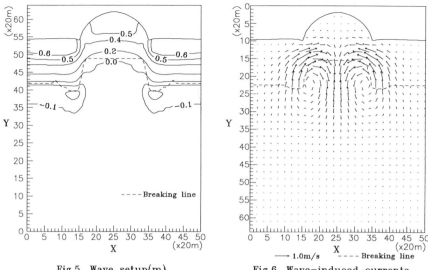

Fig.5. Wave setup(m)   Fig.6. Wave-induced currents

The directional wave spectra at the points of different water depth are given in Fig.2, and their corresponding directional distribution patterns are presented in Fig.3. From the figures, it can be found that the directional range of main wave energy becomes narrow in shallow water, and this change is more obvious in the component waves with low frequencies.

Fig.4a and Fig.4b present the distribution of wave height calculated using coupled , uncoupled model of directional wave spectrum. When the waves propagate from deep water to shoreline, the convergence of wave energy occurs at both sides of the bay and the divergence appears within the bay until the wave reaches the surf zone where the waves break and wave energy distributes along the contour lines. Because of wave setup and overbank surging ,the surf zone extends much to the land, the gradient of water depth becomes small in nearshore area, and the wave height declines weakly within the surf zone.

The mean water level, shown in Fig.5, falls firstly near the breaking line and then gradually rises in the surf zone. The wave setup reaches to it's maximum of 0.61m, about 31% of the height of deep water wave, at the shore.

Fig.6 presents the wave-induced currents calculated by the model. It is obvious that there are a pair of symmetric circulation at both sides of the bay. The maximum of current velocity is 1.23m/s, while that of offshore current is 0.61m/s.

4.REFERENCES

Berkhoff, J.G.W.(1976). Mathematical models for simple harmonic linear water waves, *Wave Diffraction and Refraction*, publication No.163, Delft Hydraulic Laboratory, Delft, the Netherlands.

Longuet-Higgins, M.S.and Stewart, T.W.(1964). Radiation stress and mass transport in gravity waves on steady non-uniform currents, *Journal of Fluid Mechanics*, 10, 529-549.

Kiyoshi Horikawa(1988). *Nearshore Dynamics and Coastal Processes*, University of Tokyo Press, Tokyo, Japan.

Wen et al.(1989). Improved form of wind wave frequency spectrum, *Acta Oceanologic Sinica*, 8(4), 467~483.

Environmental Hydraulics, Lee, Jayawardena & Wang (eds) © 1999 Balkema, Rotterdam, ISBN 90 5809 035 3

Study on water quality model for mild slope beach bay

Wang Zeliang & Huang Zhuchong
Department of Cooling Water, Institute of Water Resource and Hydroelectric Power Research, Beijing, China

ABSTRACT: 2-D depth integration models for predicating the pollutants distribution in the mild slope beach and shallow water sea-bay are studied. Slot method is used to treat moving boundary, The biodegradation equation for the organic pollutants is established. The water quality model is well verified by field data.

1. INTRODUCTION

For those sea-bays with mild slope beach, the shore line changes obviously following the tide rising and ebbing. In order to simulate the pollutant transport in this kind of sea-bay correctly, the moving boundary must be introduced. Recent years, a lot of study on the treatment of the moving boundary has been carried out, such as Water Level Judgment Method[Tao, 1983], the Weight Method[Shi, 1986], the Thin Layer Water Method[Zou, 1989], the Slot Method[Tao, 1984], The author used this method to simulate the tidal current by different difference schemes, it demonstrated that this method is available for different difference schemes. In order to simulate the pollutant transport on the shore area, the Slot Method is applied to the advection diffusion model which has been verified and tested in the paper. As we know, the ultimate fate of pollutants in the sea will not only be determined by the transport of the sea, but also by the bio-chemical reaction, in order to predict the actual distribution of pollutants , the bio-degradation must be considered in the model. Experimental study was done to determine the decay coefficient of organic pollutants by using the sea water from the Bohai Bay, the decay dynamic formula for COD was established. the models given in this paper have been used in the numerical study of pollutant transport in Bohai Bay, the comparison of the computation results and the field data is good.

2. MATHEMATICAL MODEL

2.1 Slot method

Slot method is a computational technique to treat the moving boundary condition. The basic idea is that supposing a set of slots exist in the beach, a closed boundary can be put in the slot to make the moving boundary problem to be a fixed boundary problem. Numerical experiments[Tao, 1984] showed that if the width of the slot is narrow enough, the existing of the slot will not affect the movement of water on the beach. The width of slot in unit width is defined as,

$$f(z) = \begin{cases} \varepsilon + (1-\varepsilon)e^{\alpha(z-z_b)} & z \le z_b \\ 1.0 & z > z_b \end{cases}, \quad z_b \text{ is bed}$$

elevation ; ε is coefficient of the slot; α is slot

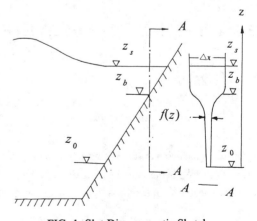

FIG. 1. Slot Diagrammatic Sketch

contraction coefficient, in this paper, $\alpha = 2.3$. In Fig. 1, z_0 is the slot bottom elevation, it is defined lower than the lowest water level, z_s is water surface elevation.

2.2 Hydrodynamic model

For a shallow water sea-bay with mild slope beach, the horizontal scales are much larger than vertical scales, then the 2-D depth averaged hydrodynamic models can be used. The continuity equation can be written as,

$$\frac{\partial P}{\partial x} + \frac{\partial Q}{\partial y} + F(\varsigma)\frac{\partial \varsigma}{\partial t} = 0, \quad F(\varsigma) = \begin{cases} 1 & \varsigma > z_b \\ 2f(\varsigma) - f^2(\varsigma) & \varsigma \le z_b \end{cases} \tag{1}$$

Assuming that $\Delta x = \Delta y$, then the momentum equations are as follows,

$$\frac{\partial P}{\partial t} + \frac{\partial}{\partial x}\left(\frac{P^2}{H}\right) + \frac{\partial}{\partial y}\left(\frac{PQ}{H}\right) + gh\frac{\partial \varsigma}{\partial x} = -g\frac{P\sqrt{P^2 + Q^2}}{C_z^2 H^2} + fQ + E\left(\frac{\partial^2 P}{\partial x^2} + \frac{\partial^2 P}{\partial y^2}\right) \tag{2}$$

$$\frac{\partial Q}{\partial t} + \frac{\partial}{\partial x}\left(\frac{PQ}{H}\right) + \frac{\partial}{\partial y}\left(\frac{Q^2}{H}\right) + gh\frac{\partial \varsigma}{\partial y} = -g\frac{Q\sqrt{P^2 + Q^2}}{C_z^2 H^2} - fP + E\left(\frac{\partial^2 Q}{\partial x^2} + \frac{\partial^2 Q}{\partial y^2}\right) \tag{3}$$

where

$$H = \begin{cases} \varepsilon(z_s - z_0) - \dfrac{1-\varepsilon}{\alpha}e^{\alpha(z_s - z_b)}(e^{\alpha(z_s - z_b)} - 1) & z_s \le z_b \\ \varepsilon(z_b - z_0) - \dfrac{1-\varepsilon}{\alpha}(1 - e^{\alpha(z_0 - z_b)}) + (z_s - z_b) & z_s > z_b \end{cases}$$

P, Q are the unit width discharges in the x and y directions respectively, $P = HU, Q = HV$, U, V are the velocity component in x and y direction respectively; ς is the surface elevation. H is the equivalent water depth including the slots; C_z is the Chezy coefficient; E is the eddy viscosity coefficient; and f is the Coriolis coefficient, where $f = 2\omega\sin\varphi$, and ω is the angular speed of the earth's rotation and φ is the latitude.

2.3 Advection-diffusion model

Assuming that $\Delta x = \Delta y$, 2-D depth averaged advection-diffusion model with slot is established as below,

$$\frac{\partial(PC)}{\partial x} + \frac{\partial(QC)}{\partial y} + \frac{\partial M}{\partial t} = \frac{\partial}{\partial x}\left(E_x H\frac{\partial C}{\partial x}\right) + \frac{\partial}{\partial y}\left(E_y H\frac{\partial C}{\partial y}\right) + HS_m - K_1 HC - K_2 HC, \tag{4}$$

$$M = \begin{cases} 2HC - \displaystyle\int_{z_0}^{z_s} Cf^2(z)dz \\ HC + H'C - \displaystyle\int_{z_0}^{z_b} Cf^2(z)dz \end{cases}; \quad H' = \varepsilon(z_b - z_0) + \frac{1-\varepsilon}{\alpha}(1 - e^{\alpha(z_0 - z_b)})$$

where C is the depth averaged concentration; E_x, E_y are the dispersion coefficients in the x and y directions respectively; S_m is the source term and K_1 is the bio-chemical degradation coefficient, it should be determined by experiment; K_2 rate of loss of pollutant concentration due to setting etc.

3 NUMERICAL VERIFICATION OF THE MOVING BOUNDARY MODEL

A semi-closed basin is shown in Fig 2. There is a slope in the left side and an open boundary in the right side. The open boundary condition is as follow: Water elevation condition: $\varsigma = \varsigma_0 \cos(\overline{\omega t + g})$, $\varsigma_0 = 0.95m$, $g = 60°$, ω is angular speed of M_2 tide component; Concentration condition:

$$\begin{cases} C = C_0 & \text{inflow} \\ \dfrac{\partial C}{\partial t} + V_n \dfrac{\partial C}{\partial n} = 0 & \text{outflow} \end{cases} ; \text{The dispersion coefficients are determined by the formula[Komatsu, 1994]}$$

as : $E_x = \alpha H|U|$, $E_y = \alpha H|V|$, α is parameter, it is supposed to be 300. in this paper, $\Delta x = \Delta y = 2000m$, $\Delta t = 934.045$ seconds, total grid points are $x \times y = 37 \times 12$.

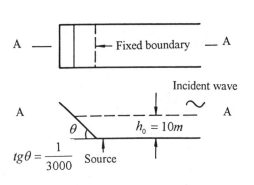

FIG. 2. The Sketch of 2-D Basin with Slope

There are five values of the parameters ε have been used in the computation. $\varepsilon = 0.08, 0.04, 0.02, 0.01$ and 0.005. Fig 3 gives the computation results of concentration distribution at $T = 970 \times \Delta t$ for different ε.

The computation result is convergent as the reduction of the width of ε, namely only if ε is small enough, the computation result is reliable. In this example, we can not use $\varepsilon = 0$ or without slots inside the beach, because the water surface changes obviously on the slope beach. In Fig 4, it gives the comparison of the results by using moving boundary condition and closed boundary condition at the lowest water level on the beach. The comparison shows that the results are much different, therefore the moving boundary has to be used to simulate the pollutant transport on a mild slope beach.

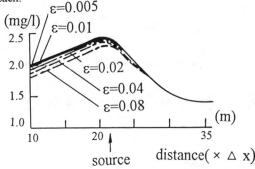

FIG. 3. The Computation Distribution for Different Slot Width

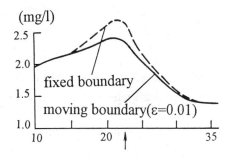

FIG. 4. Comparison of Concentration of Moving Boundary and That of Fixed Boundary

4 DETERMINATION OF THE DECAY COEFFICIENT FOR ORGANIC POLLUTANTS

The concentration of COD can be expressed in the form of first order reaction to describe the deoxygenation of ultimate carbonaceous COD. The COD function as in most models takes into account additional COD removal due to sedimentation, flocculation and scour[Falconer and Chen, 1996]:

$$\frac{DCH}{Dt} = -K_1 CH - K_2 CH \tag{5}$$

where C is concentration of ultimate carbonaceous $COD(mg/l)$, K_1 and K_2 have the same meaning as above.

In order to determine the bio-chemical degradation coefficient, laboratory studies of the degradation of COD have been undertaken by studying the sea water for the Bohai Bay. The experiments were undertaken for different temperatures, i.e. $10°C, 16°C, 20°C, 22°C, 24°C, 28°C$, and also different initial concentration was used in the experiment. After a detailed statistical analysis, the formula for the degradation of COD has been established, it is $C = C_0 e^{(K_1 t)}$. where K_1 is the degradation coefficient for COD; C_0 is the initial concentration of COD; t is the time; and $\ln K_1 = \dfrac{-5374}{T} + 15.35$, where T is the absolute temperature. K_1 can be expressed in the form of temperature(not the absolute temperature), it is $K_1 = K_{22} \times 1.064^{(t_c)}$, where K_{22} is the experimental result of K_1 when temperature is $22°C$ which is different from the normal temperature $20°C$ because of the special condition for Bohai Bay.

5 APPLICATION OF THE MILD SLOPE WATER QUALITY MODEL

Bohai Bay is a typical semi-closed bay in the western part of Bohai Sea. The main characters of the bay are very shallow water depth, wide mild slope beach, and heavy pollution loads. As described in the part 1, in the simulation of the transport of pollutants for this kind of sea-bay, the moving boundary must be considered in the models. In August 1993,a large scale measurement program was organized to measure the tidal current and the water quality of the Bohai Bay. Fig. 5 shows the location of different monitoring stations. It is possible to verify the mild slope water quality model with these field data.

FIG. 5. The Sketch of The Monitoring Stations

The main outfalls are Beitangkou, Dagukou and Canglangtsu. In this computation, ε in taken as 0.01, $\triangle x = \triangle y = 1000m$, $\triangle t = 360.0$ seconds, the bio-chemical degradation coefficient K_1 is given as the formula in part 3, the rate of loss of carbonaceous COD due to setting K_2 is omitted.

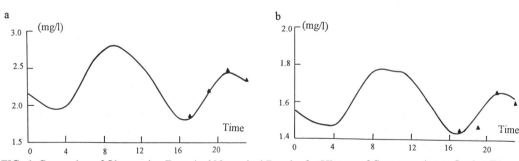

FIG. 6. Comparing of Observation Data And Numerical Results for History of Concentration at Station F1 and F2   a   F1, b   F2.   ▲ Observation; — Computation

944

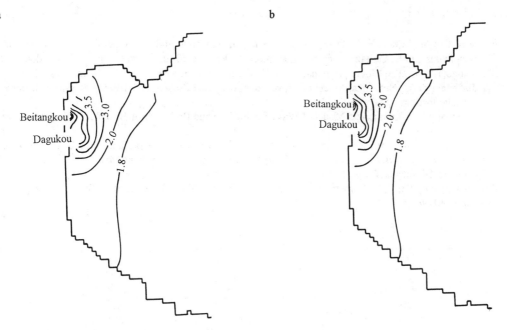

FIG. 7 Concentration Distribution of COD at Different Time (mg/l) (a flood tide; b ebb tide)

Fig. 6 gives the comparison of concentration of measurement with computation at station F1 and F2, the computation results are in good agreement with measurement data. Fig. 7 shows the computational concentration distribution of COD at 17:00 O'clock 2nd August(flood tide) and 21:00 O'clock (ebb tide), we can see that, the concentration of COD near the Beitangkou and Dagukou at flood tide is higher than that at ebb tide, the main reason is the effect of the direction of tidal current.

From the computation results above, it demonstrates that the present water quality model is available to simulate the pollutant distribution in the sea-bay with mild slope. Taking Bohai bay as an example, the mild slope beach is much widely, the width of the beach varies from several km to above 10 km, if the closed boundary condition and fixed source are used on the beach, it would bring much error in the computational results.

6   CONCLUSION

(1) To introduce Slot Method   into the advection-diffusion model to simulate the pollutant transport in the sea-bay with mild slope is necessary . Only if parameter ε is small enough, the accuracy of the computation results can be satisfied.

(2) The water quality model proposed in the paper has been well verified by field measured data, which can be used to simulate the water quality in the sea-bay with mild slope beach as well as normally sea bays and lakes.

(3) The water quality model presented in this paper was used successfully to investigate the environmental capacity, division of function area and oceanic disposal of wastewater of Bohai Bay.

(4) The release of pollutants from the mud of the sea-bed is not considered in the water quality presented in this paper, for those sea-bays with mud beach, the effect of release must be included in the model, further study will be needed for this kind of sea-bays.

7. REFERENCES

Falconer, R.A. and Chen, Y. (1996). "Modeling Sediment Transport and Water Quality Processes on Tidal
    Floodplains", in Floodplain Processes, Eds. M.G. Anderson et al., J. Wiley & Sons Ltd., London.
Shi Linbao(1986). "The Moving Boundary 2-D Simulation of the river mouth", *J. of Pear River*, 2, 4-8.
Tao Jianhua(1983). "Numerical Model of Wave Running up and Broken on A Beach", Report of Danish
    Hydraulic Institute.
Tao Jinahua(1984). "Numerical Model of Wave Running up and Broken on A Beach", *ACTA Oceanologica
    Sinica*, 6(5):692-700.
Toshimitsu Komatsu et al(1994). "Estimation of 1-D and 2-D Dispersion Coefficient in a Bay", *Proceeding
    of China-Japan Bilateral Symposium on Fluid Mechanics and Management tools for Environment*.
    Beijing, Hefei Science Press, 42-49.
Zou Guangyuan(1989). "A Numerical Study of the Motion of a Wave Running up A Beach", *ACTA
    Mechanica Sinic*a, 21 (1):1-8.

4.7 Miscellaneous

Environmental Hydraulics, Lee, Jayawardena & Wang (eds) © 1999 Balkema, Rotterdam, ISBN 90 5809 035 3

Stochastic analysis of Richards equation under uncertainty in unsaturated soil parameters

K.S. Hari Prasad, J.Q. Deng & M.S. Ghidaoui

Department of Civil Engineering, The Hong Kong University of Science and Technology, Kowloon, China

ABSTRACT: In the present study, stochastic analysis of Richards equation is performed under uncertainty in unsaturated soil parameters. The deviations in means and standard deviations of stochastic and deterministic models are studied considering the saturated hydraulic conductivity and Van Genuchten soil parameter 'n' as random variables. It is shown that the randomness in saturated hydraulic conductivity has a pronounced effect on the deviation in the means of the stochastic and deterministic models whereas the randomness in parameter 'n' has little effect. It is also shown that the cross correlation between the saturated hydraulic conductivity and the parameter 'n' has little effect on the deviations in means of stochastic and deterministic models. However, the cross correlation has significantly reduced the deviations in standard deviation between the stochastic and deterministic models.

1. INTRODUCTION

The movement of water in the unsaturated zone is important in many branches of hydrology such as agricultural engineering, soil science and environmental engineering. The flow in the unsaturated zone is usually modelled by solving the Richards equation which needs the relationship between the soil hydraulic properties and the transport mechanisms for its solution. A good description of the soil hydraulic properties is not easy to obtain in the field due to the spatial variability of field soils and the errors in the measurement and uncertainty exists in the hydraulic properties of soils. In such situations one has to resort to stochastic modelling of Richards equation for obtaining the expectations of the state variables (Yeh et al. 1985, Mantoglou and Gelhar, 1987). Bresler and Dagan (1983) considered the uncertainty in the saturated hydraulic conductivity and compared the stochastic solutions of Richards equation and Green Ampt equation. Govindaraju et al. (1992) studied the randomness in the parameter which governs the nonlinearity among the functional forms relating hydraulic conductivity, moisture content and pressure head to compare the Richards equation with simpler models. In the present study, stochastic simulation of Richards equation is performed to study the deviations between the stochastic and deterministic solutions of Richards equation under uncertainty in unsaturated soil parameters. The study is useful in quantifying the effect of uncertainty of each parameter on the deviations between stochastic and deterministic solutions. The study also includes the effect of cross correlation between the parameters on the deviations between the stochastic and deterministic models.

2. RICHARDS EQUATION

Richards equation for vertical unsaturated flow can be written as

$$\frac{\partial}{\partial z}\left\{ K(\theta)\left(\frac{\partial \psi}{\partial z} - 1 \right) \right\} = \frac{\partial \theta}{\partial t} \tag{1}$$

where θ is the moisture content, ψ is the pressure head, K is the hydraulic conductivity, t is the time coordinate and z is the vertical co-ordinate taken positive downwards. To solve Eqn.(1), one needs the constitutive relationships between θ - ψ and K - θ. Many empirical relationships exist in literature for these relationships. In the present study, Van Genuchten's relationship (1980) is used for θ - ψ relationship and Mualem's (1976) is used for K - θ relationship which are as follows.

θ - ψ Relationship (Van Genuchten, 1980)

$$S_e = \left[\frac{1}{1 + |\alpha\psi|^n} \right]^m \quad \psi < 0$$

$$= 1 \qquad \psi \geq 0$$

(2)

where α and n are the unsaturated soil parameters with m = 1-(1/n) and S_e is the effective saturation defined as

$$S_e = \frac{\theta - \theta_r}{\theta_s - \theta_r}$$

(3)

where θ_s is the saturated moisture content and θ_r is the residual moisture content.

K - θ Relationship (Mualem ,1976)

$$K = K_{sat} S_e^{1/2} \left[1 - \left(1 - S_e^{1/m} \right)^m \right]^2$$

(4)

where K_{sat} is the saturated hydraulic conductivity. It is clear from Eqn.(4) that the unsaturated hydraulic conductivity depends on K_{sat} and the parameter n. These parameters are usually estimated by fitting the constitutive relationships to field measured data and are usually subject to uncertainty. When the parameters are subject to uncertainty, stochastic analysis is the appropriate tool to determine the expectations of the state variables obtained from Richards equation. To carry out the stochastic simulation, one needs the probability distributions of the soil parameters and the cross correlation between the parameters. Carsel and Parrish (1988) developed statistical distributions for the saturated hydraulic conductivity and Van Genuchten's parameters. These distributions are very useful in performing stochastic simulations of Richards equation. In the following sections, stochastic simulation is performed to analyse the effect of randomness in the parameters K_{sat} and n and the correlation between these parameters on the deviations between the solutions of stochastic and deterministic models.

3. STOCHASTIC ANALYSIS OF RICHARDS EQUATION

During the stochastic simulation, the initial and boundary conditions are considered to be deterministic. The soil is initially assumed to be at a moisture content equal to θ_i and at time t=0, a soil moisture content θ equal to θ_s is applied on the ground surface while the lower boundary is held at a moisture content equal to θ_i. When a particular parameter P is considered as random, a set of N values of the parameter P is generated with a mean value of μ_p and standard deviation σ_p following the statistical distribution which represents the variation of parameter P, while the other parameters are treated as deterministic. Following Dagan and Bresler (1983), the mean and variance of the unsaturated hydraulic conductivity K are approximated as

$$|K(z,t)| = \sum_{i=1}^{N} K(z,t,P_i) \frac{1}{N}, \quad i = 1,2,\cdots\cdots,N$$

(5)

$$\|K(z,t)\| = \sum_{i=1}^{N} \left[K(z,t,P_i) - |K(z,t)| \right]^2 \frac{1}{N}, \quad i = 1,2,\cdots\cdots,N$$

(6)

where single and double vertical bars denote the mean and variance respectively and $K(z,t,P_i)$ is the hydraulic conductivity obtained by solving the Richards equation with the parameter value P_i. A value of $N = 20$ is used in the present study. For a comparative study of stochastic and deterministic models, one requires the differences in means and standard deviations of the stochastic and deterministic models. The time dependent normalized deviations in mean and standard deviation over the entire soil length are defined as

$$EM(t) = \left[\frac{\int_0^L \{|K(z,t)| - K_d(z,t)\}^2 \, dz}{L} \right]^{1/2} \bullet \left(\frac{100}{K_{sat,d}} \right) \%$$ (7)

$$ES(t) = \left[\frac{\int_0^L \|K(z,t)\| \, dz}{L} \right]^{1/2} \bullet \left(\frac{100}{K_{sat,d}} \right) \%$$ (8)

where L is the length of the soil profile and $K_d(z,t)$ are the deterministic hydraulic conductivity values obtained using the mean value μ_p of the parameter P. It should be noted here that the variance of the deterministic model is zero.

4. RESULTS AND DISCUSSION

Case 1. K_{sat} as a Random Parameter

The saturated hydraulic conductivity, K_{sat} varies by several orders in the field. The statistical distribution of K_{sat} is generally represented by lognormal distribution (Bresler and Dagan, 1983, Carsel and Parrish, 1988) to

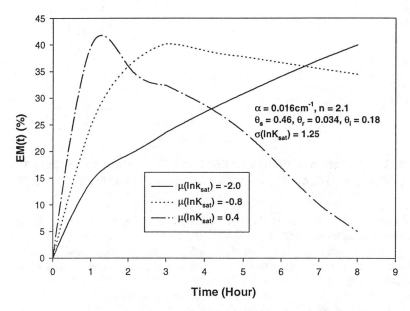

FIG. 1. Variation of EM(t) with time - Case 1

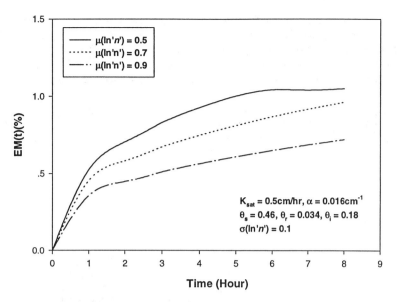

FIG. 2. Variation of EM(t) with time - Case 2

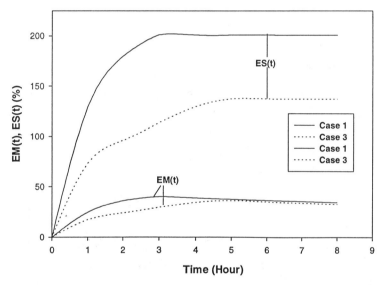

FIG. 3. Variation of EM(t) and ES(t) with time - Case 1 and Case 3
($\mu(\ln k_{sat}) = -0.8$, $\sigma(\ln K_{sat}) = 1.25$, $\mu(\ln'n') = 0.7$, $\sigma(\ln'n') = 0.1$ $\rho = 0.7$)

account for the wide spread in K_{sat} values observed in many field situations. Fig.1 shows the variation of EM(t) with time for three different $\mu(\ln K_{sat})$ values (-2.2, -0.8 and 0.4) with a $\sigma(\ln K_{sat})$ value of 1.25. It can be seen from Fig.1 that for $\mu(\ln K_{sat}) = 0.4$, (Indicating higher K_{sat} values) EM(t) increases rapidly with time, attains a maximum value and decreases gradually at large time as the system approaches steady state quickly where the difference between the stochastic and deterministic models approaches zero. In contrast, for lower $\mu(\ln K_{sat})$ values, rate of increase of EM(t) with time is slower. However, it is interesting to note that the maximum value of EM(t) is about 45% for all $\mu(\ln K_{sat})$ values.

Case 2. n as Random Parameter

The parameter n governs the nonlinearity in the soil moisture characteristic function (Eqn. (2)) and the hydraulic conductivity function (Eqn. (4)). The parameter n varies from 1.2 (for clay) to 2.7 (for sand) for most of the field soils. The variation in parameter n is assumed to be lognormal (Carsel and Parrish, 1988). In Fig. 2, variation of EM(t) is plotted against time for three different μ(ln 'n') values (0.5, 0.7 and 0.9) at σ(ln 'n') = 0.1. Fig. 2 suggests that as μ(ln 'n') increases (Indicating higher n values) EM(t) value decreases for all times. At high n values, the variation of nonlinearity in Mualem's equation (Eqn.(4)) with n decreases. Hence, the average value of K of all the realizations does not differ much with the deterministic K value which results in low EM(t) value for high n values. It can also be seen from Fig. 3 that the randomness in parameter n does not have an effect on EM(t) as the maximum value of EM(t) is about 1%.

Comparison of Case 1 and Case 2 indicates that the randomness in K_{sat} has a pronounced effect on EM(t) values, while the effect of randomness in n on EM(t) is insignificant. Case 1 and Case 2 consider the effect of randomness in parameters K_{sat} and n on EM(t), while treating the other parameter as deterministic. It has been observed in many field studies (Russo and Bresler, 1981, El-Kadi 1987) that the unsaturated soil parameters are not independent and are correlated to each other. Study of Carsel and Parrish (1988) has shown that in most field situations, the correlation between K_{sat} and n is quite significant and is around 0.7 in most of cases and suggested that an assumption of independence between the soil parameters is not plausible in stochastic simulation. In Case 3, the effect of correlation between K_{sat} and n on EM(t) and ES(t) is analysed.

Case 3. K_{sat} and n as Random Parameters - Effect of Cross Correlation

The parameters K_{sat} and n are assumed to be lognormally distributed and the correlation coefficient ρ between K_{sat} and n is taken as 0.7 with the following means and standard deviations; $\mu(\ln K_{sat})$ = -0.8, $\sigma(\ln K_{sat})$ = 1.25, μ(ln 'n') = 0.7, σ(ln 'n') = 0.1. Fig. 3 compares EM(t) and ES(t) for Case 1 and Case 3. It is clear from Fig.3 that the correlation between K_{sat} and n results in the decrease of EM(t) at all the times. However, the decrease in EM(t) in the presence of cross correlation is marginal. It can also be seen from Fig. 3 that ES(t) is quite sensitive to the parameter cross correlation. The parameter correlation decreases the value of ES(t) considerably at all times. The variation of ES(t) with time is similar to EM(t) , an increase in the value of ES(t) with time to reach a constant value at large times as both the stochastic and deterministic systems reach their respective steady states. The decrease in the value of ES(t) in the presence of parameter cross correlation can be explained as follows; for a fixed value of the parameter n (Case 1), variability of K_{sat} from its mean value is higher as compared to the variability of K_{sat} when cross correlation between K_{sat} and n (Case 3) is considered even though the average value of K_{sat} is the same in both the cases. When these values are used in the stochastic simulation, higher variability in conductivity values results for Case 1 as compared to Case 3.

5. CONCLUSIONS

In the present study, stochastic analysis of Richards equation is carried out to study the effect of uncertainty in soil parameters on the deviations in means and standard deviations of stochastic and deterministic solutions of Richards equations. It has been shown that the randomness in saturated hydraulic conductivity has significant effect on the deviations in means of stochastic and deterministic solutions of Richards equation. The randomness in Van Genuchten parameter 'n' has very little effect on the deviations in means. The correlation between saturated hydraulic conductivity and the parameter 'n' does not influence much the deviations between means of stochastic and deterministic models. However, it has significant effect on the deviations in standard deviation of stochastic and deterministic solutions. It can be concluded from the study the cross correlation between the parameters reduces the variability of stochastic solutions considerably.

6. ACKNOWLEDGEMENTS

Financial support for this work is provided by the Hong Kong University of Science and Technology, Hong Kong under the UPGC Research Infrastructure Grant Program, Project number RIG 94/95. EG14. The financial support is gratefully acknowledged.

7. REFERENCES

Bresler, E., and Dagan, G. (1983). Unsaturated Flow in Spatially Variable Field Soils, 2, Application of Water Flow Models to Various Fields, Water Resources Research, 19(2), 421-428.

Carsel, R. F., and Parrish, R. S. (1988). Developing Joint Probability Distributions of Soil Water Retention Characteristics, Water Resources Research, 24(5), 755-769.

Dagan G., and Bresler, E. (1983). Unsaturated Flow in Spatially Variable Field Soils, 1, Derivation of Models of Infiltration and Redistribution, Water Resources Research, 19(2), 413-420.

El-Kadi, A. I. (1987). Variability of Infiltration Under Uncertainty in Unsaturated Zone Parameters, Journal of Hydrology, 90, 61-80.

Govindaraju, R. S., Or, D., Kavvas, M. L., Rolston, D. E., and Biggar, J. (1992). Error Analyses of Simplified Unsaturated Flow Models Under Large Uncertainty in Hydraulic Properties, Water Resources Research, 28(11), 2913-2924.

Mantoglou, A., and Gelhar, L. W. (1987). Stochastic Modelling of Large Scale Transient Unsaturated Systems, Water Resources Research, 23(1), 37-46.

Mualem, Y. (1976). A new Model for Predicting the Unsaturated Hydraulic Conductivity of Unsaturated Porous Media, Water Resources Research, 12, 513-522.

Russo, D., and Bresler, E. (1981). Effect of Field Variability in Soil Hydraulic Properties on Solutions of Unsaturated Water and Salt Flows, Soil Science Society of America Journal, 45, 675-681.

Van Genuchten, M. T. (1980). A Closed Form Equation for Predicting the Hydraulic Conductivity of Unsaturated Soils, Soil Science Society of America Journal, 44, 892-898.

Yeh, T. C. J., Gelhar, L.W. and Gutjhar, A. L. (1985). Stochastic Analysis of Unsaturated Flow n Heterogeneous Soils, Statistically Isotropic Media, Water Resources Research, 21(4), 447-456.

Environmental Hydraulics, Lee, Jayawardena & Wang (eds) © 1999 Balkema, Rotterdam, ISBN 90 5809 035 3

Characteristics of ocean currents near Hong Kong by analysis of two ADCP data sets

Y.G.Wo
Montgomery Watson HK Limited, Hong Kong, China

J.H.W.Lee
University of Hong Kong, China

ABSTRACT: Hourly velocities measured by Acoustic Doppler Current Profiler (ADCP) at two locations L1 and L2 and tidal levels at L2 near Hong Kong are analyzed as a time series using the extended harmonic analysis method. The results indicate that the four major tidal constituents (M2, S2, K1 and O1) account for 81 and 40 percent of the power of the observed depth-averaged velocities at locations L1 and L2 respectively, and 79 percent of the observed tidal levels at L2. Surface velocities at both locations are not dominated by tidal effects. The results provide a benchmark that can assist in the interpretation of numerical tidal model predictions against field data near Hong Kong.

1. INTRODUCTION

Currents near Hong Kong are complicated and influenced by local topography, tides, winds and density stratification. In order to determine the contributions due to tides, ADCP (Acoustic Doppler Current Profiler) data at two locations in Hong Kong's coastal waters (L1 and L2 in Fig. 1), and water depth data at L2 are analyzed using an Extended Harmonic Analysis. The objective is to quantitatively investigate characteristics of currents near Hong Kong, with a view of improving our interpretation of numerical tidal model predictions.

2. DATA ANALYSIS

Fig. 1 shows the location of the measurement stations. L1 is located in the East Lamma Channel, while L2 lies further off shore near the Lema Channel, with a mean depth of about 28m and 29m respectively. The data analyzed are hourly velocities taken by ADCPs in a period of over 11 months at location L1 and over 6 months at location L2. At L2, the absolute sea water pressure was also measured on the sea bed for the same period covered by the velocity data. Variation of the atmospheric pressure at the water surface has already been taken into account. Missing values (less than 1% of total data) were linearly interpolated. Over the water column velocities were measured as speed and direction from the magnetic north at one meter depth intervals.

A time series of depth averaged velocities, velocities at 1m above the sea bed, at levels between bed and surface (22m and 21m above the sea bed at locations L1 and L2 respectively) and at the water surface was selected for analysis. These data can represent a vertical variation and a vertical mean of the currents. The vector currents were converted into two sets of scalar variables in an orthonormal basis as magnetic eastern and magnetic northern velocities for the analysis.

Given a sea water density, the bottom absolute sea water pressure at location L2 can be converted to a water depth. Based on historical data of salinity and temperature near location L2, the average density variation between wet and dry seasons is found to be less than 1%; a constant density $1023kg/m^3$ was then used for the depth calculation. The mean depth at L2 is computed to be 27.47m. Here the difference between the depth and its mean is defined as tidal level for further power analysis of the signal.

The discrete hourly time series of velocity/level is subjected to an extended harmonic analysis (e.g. Bowden 1983, Schuremann 1941, Zetler and Cummings 1965). The standard amplitudes and phases of selected tidal constituents with exactly known frequencies (from equilibrium tide theory) are determined by a least squares analysis. This method is superior to a standard Fourier analysis (e.g. Oppenheim and Shafer 1975) which suffers from signal leakage and other limitations. Based on the harmonic constants, the tidal velocity/level at the same location can be predicted for any time. The power or variance accounted for by the tidal prediction using the selected constituents can then be determined from the sum of the squares of the predicted tidal velocity/level (above the mean value). This value can be compared to the total power in the measured signal. The standard deviation between the tidal prediction and the data may also be used to indicate degree of tidal contribution in the signal.

Table 1. Amplitude contributions (%)

tidal Const.	L1 vel N. (%)	L1 vel E. (%)	L2 vel N. (%)	L2 vel E. (%)	L2 lev (%)
[2]Ssa	1.4	0.5	15.7	20.8	10.9
[2]Mm	0.9	0.6	8.4	12.5	1.9
[2]Msf	2.5	0.5	3.7	5.0	1.5
Mf	1.2	0.4	1.5	2.2	0.9
Q1	1.7	1.7	0.9	0.8	2.8
rho1	0.3	0.9	0.6	0.9	0.6
[1,2]O1	10.7	9.4	8.0	4.3	14.4
M1	0.3	0.3	0.1	0.8	0.8
[2]P1	3.8	3.4	2.2	1.9	5.7
[1,2]K1	13.0	13.2	10.2	7.1	18.3
J1	0.6	0.8	0.3	0.8	0.6
mu2	1.0	1.0	0.8	2.0	1.0
[2]N2	5.1	6.4	4.0	3.2	4.3
nu2	0.6	1.1	0.5	1.1	0.6
[1,2]M2	27.4	32.9	22.2	16.5	19.5
T2	0.6	0.5	0.3	1.4	0.5
[1,2]S2	11.1	12.8	9.8	6.4	8.3
[2]K2	2.8	3.6	2.3	1.9	2.2
MO3	2.0	0.7	0.1	0.5	0.4
M3	1.2	1.5	0.8	1.3	0.9
MK3	2.1	0.7	0.2	0.6	0.6
MN4	1.9	1.3	1.7	1.4	0.6
M4	4.8	3.8	3.5	4.3	1.5
MS4	2.9	2.0	2.2	2.6	1.1
sum set1	62.2	68.3	50.2	34.2	60.5
sum set2	78.7	83.3	86.5	79.4	87.0

As currents are converted to northern and eastern components, a weighted average of the power percentage of both components is used (Ave). The weight is the power of the corresponding northern or eastern component divided by the sum of the two components' power. For the standard deviation of both northern and eastern component (All), a square root of the sum of variance of eastern and northern components is applied under an assumption of two independent variables.

3. RESULTS

Based on previous analysis of tides in Hong Kong, the depth average velocity and tidal level data were initially analyzed using 24 constituents (the solar annual Sa was not included as the time series of the data is not longer than one year), which are Ssa, Mm, Msf, Mf, Q1, rho1, O1, M1, P1, K1, J1, mu2, N2, nu2, M2, T2, S2, K2, MO3, M3, MK3, MN4, M4, MS4.

Table 2 Power contributions (%) and std (m or m/s) from tidal level, magnetic eastern and northern velocities, their weighted average or all of std for given tidal constituents

ADCP current or tidal level	M2, S2, O1, K1 (power %)			M2, S2, O1, K1 (std m or m/s)			M2, S2, O1, K1, N2, K2, P1, Msf, Mm, Ssa (power %)			M2, S2, O1, K1, N2, K2, P1, Msf, Mm, Ssa (std m or m/s)		
	E vel	N vel	Ave	E Vel	N vel	All	E vel	N vel	Ave	E Vel	N vel	All
depth ave. at L1	79.2	82.8	81	0.08	0.07	0.11	83.1	86.4	85	0.07	0.07	0.10
surface adcp at L1	37.0	42.2	39	0.25	0.18	0.31	39.2	46.1	42	0.25	0.17	0.30
1m above bed at L1	57.6	76.7	69	0.11	0.08	0.13	60.4	80.0	72	0.10	0.08	0.13
22m above bed at L1	74.5	75.9	75	0.12	0.10	0.15	79.0	80.2	80	0.11	0.09	0.14
depth ave. at L2	41.5	23.5	40	0.17	0.07	0.18	54.6	39.4	53	0.15	0.06	0.16
surface at L2	39.7	43.7	40	0.35	0.14	0.38	58.0	54.5	58	0.29	0.12	0.32
1m above bed at L2	45.9	25.2	43	0.11	0.07	0.13	53.8	32.5	51	0.10	0.06	0.12
21m above bed at L2	38.5	15.7	37	0.20	0.09	0.22	50.5	29.2	49	0.18	0.08	0.20
tidal level at L2	79			0.16			92			0.10		

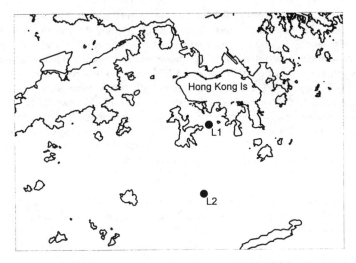

Fig. 1. Locations of two ADCPs L1 and L2

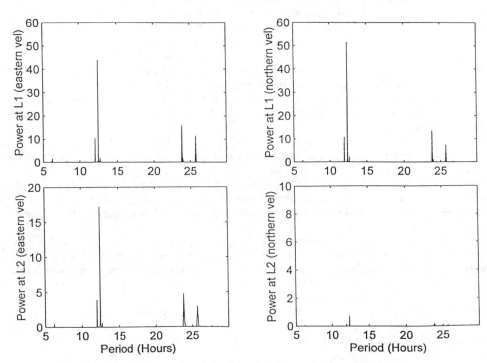

Fig. 2. Power spectrum of the signal from depth averaged
magnetic eastern and northern velocities at locations L1 (top) and L2 (bottom)

Table 1 lists the amplitude contribution of each tidal constituent in the percentage of the total amplitude from all 24 constituents considered. The results indicate that the four major tidal constituents M2, S2, O1 and K1, which are used in the Admiralty Method (Doodson 1941), are significant at both locations L1 and L2 for velocities and levels. At L2 located offshore but not at L1, Ssa, Mm, Msf are as important as the four major constituents for velocities, and for tidal level, however only Ssa is significant.

Based on the computed amplitude results in Table 1, two sets of tidal constituents were selected for analysis: the first set includes the four major tidal constituents M2, S2, O1, K1; the second set includes in addition N2, K2, P1, Mm, Msf, and Ssa which cover semi-diurnal, diurnal, monthly and semi-annual variations. Their contribution percentage in the total 24 tidal constituents are also included in the bottom of Table 1.

Fig. 2 shows the power spectrum of the time series for both magnetic eastern and northern depth-averaged velocities at locations L1 and L2. The four easily identifiable peaks correspond to the four major tidal constituents. At location L1, their contribution in percentage is 79.2% and 82.8% for the eastern and northern velocities respectively and the weighted average of the percentage is about 81% (Table 2). For the currents at 1m and 22m above the sea bed and at the water surface, their weighted average percentage were determined and also listed in Table 2. The percentage for the second set of tidal constituents can also be calculated (Table 2). Same exercises were applied to the velocity and tidal level data at location L2 and their results were included in Table 2.

Standard deviation between the signals from the tidal constituents and field data was calculated and also listed in Table 2. As indicated previously STD_{all} is defined as a square root of the sum of variance of both eastern and northern components.

4. DISCUSSION AND CONCLUSIONS

Based on the results of the analysis of ADCP data, the following points can be made:

1. At location L1 in East Lamma Channel, most of the observed current velocities can be explained by tidal variations, except at the surface where wind effects may dominate. In terms of depth-mean velocities, the four major tidal constituents M2,S2,O1,K1 account for 81% of the total variation; the corresponding value for a more extended set of 10 tidal constituents is 85%. This result suggests that a numerical tidal model driven by the four major constituents at the open boundary can be expected to simulate about 80 percent of the depth-averaged currents at L1. As more shallow water constituents can be incorporated through the computed water movement, this figure is probably also applicable to similar locations further onshore. This percentage may be used as a reference for checking performance of numerical models against field data.

2. At the more offshore location L2 near Lema Channel, only about 40 percent of the depth-mean velocity variations can be attributed to the four major constituents; the corresponding value for the second set of constituents is 53 percent.

3. At both locations L1 and L2, the four major tidal constituents account for about 40 percent of the surface velocity variation. This suggests that surface velocities are not mainly controlled by tides. They are probably influenced by local wind, although the results at L1 indicate these effects are limited to a thin surface layer.

4. The fact that tidal variations account for a significant percentage of the observed water level at L2 (79% for the four major constituents), but not the velocity suggest that oceanic circulation of a more regional or global nature may dominate the current at L2. It is thus necessary to take both tidal levels and currents into account in flow investigations and the use of currents in model calibration at locations around L2 and offshore in South China Sea.

5. The harmonic amplitude analysis in Table 1 indicates that the semi-annual and monthly constituents Ssa, Mm are as important as the four major constituents for velocities at L2, but not at L1. However for tidal levels at L2 only Ssa is significant.

5. REFERENCES

Bowden, K.F. (1983): *Physical oceanography of coastal waters*. Ellis Horwood.

Oppenheim, A.V. and Shafer R.W. (1975): *Digital signal processing*. Englewood Cliffs, NJ: Pretice-Hall

Shuremann, P (1941): *Manual of Harmonic Analysis and Prediction of Tides*, U.S. Coastal and Geodetic Survey, Washington.

Zetler,B.D., and Cummings,R.A.(1965): "A Harmonic Method for Predicting Shallow Water Tides", *Journal of Marine Research*, 42(2), pp.125-148.

Environmental Hydraulics, Lee, Jayawardena & Wang (eds) © 1999 Balkema, Rotterdam, ISBN 90 5809 035 3

Experimental study on the resistance of roll-wave flow

M. Miyajima
Department of Civil Engineering, Osaka Sangyo University, Japan

ABSTRACT: The periodic properties of the roll-wave flow over a smooth bed are investigated and the resistance properties of the flow on a rough bed are studied experimentally. The periodic properties are presumed to be similar to those of, as it is called, a simple pendulum. In addition, the resistance properties of the flow on a rough bed have proved to be considerably related to non-dimensional velocity gradient and the front-to-back depth ratio of the wave.

1. INTRODUCTION

Thin sheet flow on steep slope channel beds has periodical roll-wave trains. It is well known that this phenomenon happens when Froude number exceeds approximately 2. It is important to clarify the hydraulic characteristics and structures of roll-wave flow through fluid mechanics and engineering and environmental fields so as to gain basic information to better understand the flow mechanism, the natural process of erosion and movement of sediment and to further disaster prevention.

Experimentally derived data is scant with the exception of Ishihara (1954), Iwagaki and Iwasa (1955), Mayer (1959) and Murota and Miyajima (1993, 1995), due to the difficulty in measuring pertinent data since water depth is shallow, flow velocity is high and wave change is rapid. In this paper, the roll-wave flow is reproduced in the laboratory, using smooth and rough beds, so as to consider the velocity, period and friction of the roll-wave flow and to consider roll-wave front's effects for flow resistance. To pursue these experiments, non-intrusive equipment is used: a Laser-Doppler anemometer and a ultra sonic level meter.

A roll-wave flow running in a steep slope channel shows (Miyajima, 1997) that, because of very shallow and rapid flow, the mean velocity gradient is extremely large, so that various kinds of hydraulic quantities are arranged essentially by the velocity gradient. In the present study, the period and resistance of the roll-wave flow are investigated experimentally.

2. EXPERIMENTAL METHODS AND THE SCOPE OF THE CONDITIONS

By the use of an acrylic-made variable-slope channel with a length of 5 m and a width of B=20 cm, measuring points have been adopted near the downstream at 3.9 m from the upstream end of the channel. Periodic observations have been performed on a smooth bed surface, while resistance experiments on a rough bed surface. Experimental conditions and the scope are shown in Table 1. Including the experiments under additional conditions in the scope of the slopes of this channel, the sum of 44 cases for the smooth bed surfaces, and that of 25 cases for the rough bed surface

Table-1 Experimental Conditions

Run No.	Channel Slope S	Mean flow depth h m(mm)	Froude number Fr	Channel floor
1	1/5.97	2.5	6.5	smooth
2	1/5.97	3.5	7.2	smooth
3	1/5.97	4.4	7.6	smooth
4	1/8.25	3.1	6.0	smooth
5	1/8.25	4.2	6.5	smooth
6	1/8.25	4.7	6.7	smooth
7	1/12.7	4.0	4.6	smooth
8	1/12.7	4.7	4.7	smooth
9	1/12.7	5.1	5.2	smooth
10	1/5.86	4.8	2.8	rough
11	1/5.86	5.6	3.1	rough
12	1/5.86	6.1	3.4	rough
13	1/5.86	6.9	3.7	rough
14	1/5.90	2.9	5.6	rough
15	1/5.90	4.1	4.6	rough
16	1/5.90	5.3	3.9	rough
17	1/5.90	6.7	3.5	rough
18	1/6.90	4.8	5.0	rough
19	1/6.90	5.6	4.8	rough
20	1/7.67	4.9	3.2	rough
21	1/7.67	5.4	3.4	rough
22	1/7.67	6.1	3.3	rough
23	1/7.67	6.9	3.6	rough
24	1/9.79	6.8	2.3	rough
25	1/9.79	7.6	2.6	rough
26	1/9.79	8.8	2.6	rough

959

have been carried out to arrange the present experiments. Notations used in the experiments are indicated in Fig. 1. For the rough bed, a bed surface with D_{84}=1.9 mm, which is 84% grain diameter in the grain size distribution as grain diameter of roughness, was spread over the entire channel comparatively tightly. From a vertical distance of D_{84} (1.9mm) from the channel's bottom face, the area of its one-fourth (D_{84} / 4) below was assumed to be a virtual bed surface.

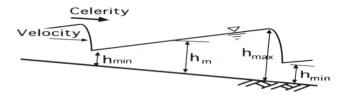

Fig. 1 Notations Used In The Experiments

Channel slope was set from 1/6 to 1/13 and the discharge rate was approximately 0.5 l/s to 1.3 l/s. This experiment was carried out in the condition of 2.3< Fr <7.6, 2000<Re<6000. In this experiment : h_m=$(h_{max}+h_{min})$ /2, Fr : Froude number; Fr=$U_m / \sqrt{gh_m}$, Re : Reynolds number; Re=$U_m \cdot h_m / \nu$, U_m=$Q/ (B \cdot h_m)$, B : channel width, g : gravitational acceleration, ν : coefficient of kinematic viscosity, Q : discharge rate.

3. DOMINATING FACTORS FOR ROLL-WAVE FLOW

This is considered to constitute one explanation of the fact that Dressler (1949) pointed out the existence of the roll-wave in connection with frictional resistance. Concerning the roll-wave flow, what is a control factor for flow in addition to the frictional resistance will be described here by using the dimensional analysis method. On studying, gravity effect g because of steep channel slope, a viscous force effect μ concerning shear force of flow because of mean velocity gradient being very steep, and further a period T of roll-wave are taken into consideration. Assuming that the description on the roll-wave flow for a turbulent flow is represented by equation (1), then equation (2) will be obtained. Here, ρ ; density.

$$F \ (g, \ \mu \ , \ T \ , \ h, \ U, \rho \ )= 0 \tag{1}$$

$$\Phi \ (Fr, \ Re, \ U/h \cdot T \ ) = 0 \tag{2}$$

Namely, a flow involving roll waves in turbulent flow is presumed to be related to not only the Froude number, but also Reynolds number and a factor peculiar to the roll-wave, a so-called non-dimensional velocity gradient.

It is important to recognize that roll-wave flow may be said to be wave motion, or rather to be the flow with very excellent fluidity because of fluid particle velocity equal to phase velocity. For example, in the case of the experiment Run No. 2, with a fluid velocity of 1.5 m/sec shown, this flow appears at a depth of 3.5 mm. It shows the flow with a very large velocity gradient corresponding to 430/sec in this case. Namely, with a property of a roll-wave flow having a very large velocity gradient, a non-dimensional quantity appears in the 3-rd term of Eq. (2) obtained from non-dimensionalizing the velocity gradient by the use of the period of the roll-wave flow, which is considered to express the property directly. Hereafter, this non-dimensional quantity will be called non-dimensional velocity gradient.

3.1 Froude Number And Non-Dimensional Velocity Gradient

In the previous section, three quantities, such as the Froude number, Reynolds number and non-dimensional velocity gradients, have been introduced as non-dimensional quantities dominating roll-wave flow. In this study, the Froude number and non-dimensional velocity gradient will be investigated additionally. It is obvious that the Froude number is dominant for the flow of a steep slope channel. Then, under the condition of this steep slope, roll-wave flow appears at very shallow depth. This means that rapid flow appearing at shallow depth results in large velocity gradient, as mentioned previously. Then, an example of velocity distribution for

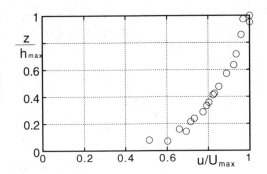

Fig. 2 Mean Velocity Distribution

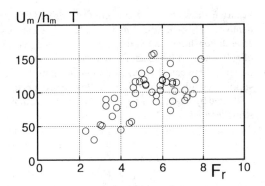

Fig. 3 Froude Number And Non-Dimensional
Velocity Gradient

roll-wave flow is shown in Fig. 2 (Hereafter in all Figures, $\bigcirc$; smooth, $\bullet$; rough bed). Here, non-dimensionalizing procedures are carried out by the maximum depth and maximum fluid velocity. Except positions near the surface of a bed, at any depth position it shows almost constant slopes distributed linearly. Therefore in this study, mean velocity gradient will be expressed by the ratio of mean velocity U_m to mean depth h_m.

3.2 Properties of Roll-Wave Flow

In Fig. 3, the axis of abscissas is indicated by the Froude number expressing flow with excellent gravitational effect, a property of the flow of a steep slope channel, while the axis of ordinates is indicated by the non-dimensional velocity gradient obtained from non-dimensionalizing large velocity gradient, representative of rapid fluid velocity and shallow depth, by the use of the period of roll-wave flow. It shows the tendency of the non-dimensional velocity gradient increasing together with the Froude number. This fact is considered in the great property of roll-wave flow.

3.3 Periodicity

Due to using the Froude number ($Fr = U_m/\sqrt{gh_m}$) in Fig. 3, rearranging axes of abscissa and ordinates leads to the roll-wave period T and term $\sqrt{(h_m/g)}$ appearing consequently. This relationship is shown in Fig. 4. Here, there is no clear relationship between the depth and period, which may result from very shallow depth. Then, in Fig. 5 the relationship between the depth and wave length of roll-wave is investigated. The axis of abscissas is indicated by the non-dimensional velocity gradient which is thought to characterize roll-wave flow, while the axis of ordinates by the ratio of water depth to wave length. It shows relatively strong relationship between the mean depth and wave length of roll-wave by interposing the non-dimensional velocity gradient.

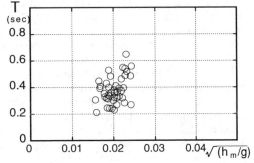

Fig. 4 Roll-Wave's Period And Water Depth

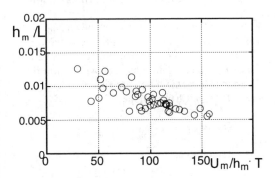

Fig. 5 Water Depth To Wave Length Ratio
And Non-Dim. Velocity Gradient

Figure 6 shows by the wave length L instead of the depth of roll-wave. Here, there is relatively clear linear-relationship between the wave length and period of roll-wave. It looks as if it reminds us of the relationship between the period and string length of a simple pendulum. Therefore, in this figure, period, evaluated by treating it as a simple pendulum with the string length L, is indicated by a straight line. It shows the tendency that the period obtained from experiments is equal to about 1/4 of that value derived from this line. Assuming that the analogy of roll-wave flow with the behavior of a simple pendulum in a gravitational field is true, it is presumed that it would become necessary to reconstruct recognition, concerning roll-wave flow, the friction of a bottom surface, viscosity and the non-linearity of flow.

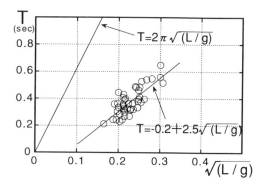

Fig. 6 Roll-Wave And A Simple Pendulum Period

4. DISCUSSIONS ON THE RESISTANCE OF ROLL-WAVE FLOW

Roll-wave flow down a steep slope channel can be considered to be a flow excellent in fluidity rather than wave motion because wave celerity c of the roll-wave flow substantially coincides with fluid velocity U in the vicinity of the wave crest area. By calculating Manning's roughness coefficient n_u from the mean water depth h_m of the experimental result, the result shown in Fig. 7 was obtained. Equations (3) and (4), to be described later, were used. It seems that as the Froude number increases, the roughness coefficient on the rough bed systematically decreases and gradually approaches the roughness coefficient result of the smooth bed at a location with a high Froude number. The smooth bed somewhat as the Froude number increases, but shows tends to decrease an approximately constant tendency. This result can be attributed to frictional resistance of the flow being affected by the existence of roll waves in some form or other.

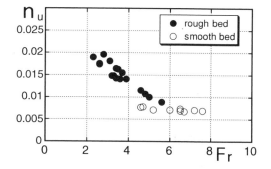

Fig. 7 Manning's Roughness Coefficient And Froude Number

4.1 The Roughness Coefficient of Uniform Flow

Roll-wave flow runs in a steep slope channel with very shallow depth. Therefore the behavior of Manning's roughness coefficient for this shallow depth will be recognized by shallow uniform flow. To arrange here, the following equations have been used. Here, S; Channel slope.

$$Q = B \cdot h_m \cdot U_m \qquad (3)$$

$$U_m = 1/n \cdot h_m^{2/3} \cdot S^{1/2} \qquad (4)$$

In general, the roughness coefficient of shallow flow is said to be strongly influenced not only by roughness but also by hydraulic water depth.

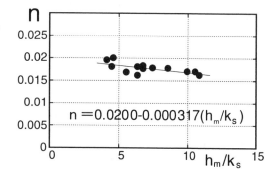

Fig. 8 Manning's Roughness Coefficient And Relative Water Depth

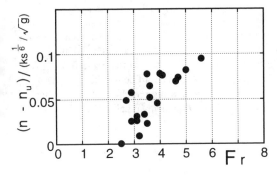

Fig. 9 Difference Of Manning's Coefficients
Between Uniform And Roll-Wave Flow

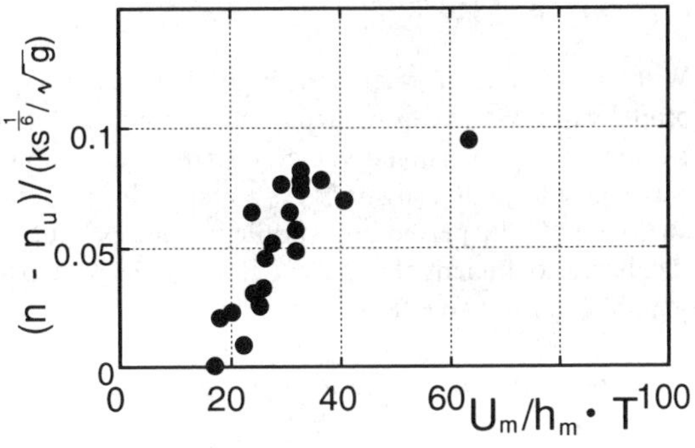

Fig. 10 Difference Of Manning's Coefficients
And Non-Dimensional Velocity Gradient

Then, the relationship between the roughness coefficient and relative depth for uniform flow in a channel used in this experiment is indicated in Fig. 8. It clearly shows the tendency that, with decreasing relative depth in the axis of abscissas, the roughness coefficient increases. Then, with the tendency between the relative depth and Manning's roughness coefficient at relatively shallow depth in this channel taken as a straight line, Eq. (5) has been set. ($k_s=D_{84}$)

$$\boldsymbol{n} = 0.0200-0.000317 \ (h_m \ / \ k_s) \tag{5}$$

4.2 The Resistance Property of Roll-Wave

Under these relationships, the difference between $\boldsymbol{n}$ evaluated from Eq. (5), Manning's roughness coefficient obtained from uniform flow, and the roughness coefficient $\boldsymbol{n}_u$ derived from mean depth for roll-wave, will be noticed. Assuming that this difference results from the presence of roll-wave, it should be related strongly with hydraulics quantities characterizing roll-wave flow. Therefore first, because roll-wave flow had a large Froude number, the difference of roughness coefficient between uniform and roll-wave flow is indicated in Fig. 9. It shows that, with the Froude number, the difference for the roughness coefficient increases. With the Froude number exceeding about 2, the apparent resistance of flow decreases.

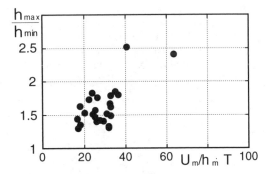

Fig. 11 The Front To Back Ratio Of Roll-Wave
Front And Non-Dim. Velocity Gradient

In addition, concerning roll-wave flow, mean velocity gradient or the non-dimensional velocity gradient is an important property, as mentioned previously, so that the difference of the roughness coefficient between uniform and roll-wave flow is indicated in Fig. 10. In this figure, with the non-dimensional velocity gradient, the tendency of the curve of the difference for the roughness coefficient is thought to become clear relatively. That is, the difference for the roughness coefficient is expressed by the function of the non-dimensional quantity, representative of roll-wave flow.

Next, Fig. 11 shows the relationship between the non-dimensional velocity gradient and front-to-back depth ratio h_{max}/h_{min} of the wave front of roll-wave, where h_{max} and h_{min} show the back maximum and front minimum depths, respectively. It also shows a strong relationship between the non-dimensional velocity gradient and roll-wave front.

963

5. CONCLUSION

With the non-dimensional velocity gradient set as a quantity characterizing roll-wave flow, with the periodic property of roll-wave examined on a smooth bed first, by resistance aspects of flow on a rough bed, the Froude number, non-dimensional velocity gradient and front-to-back depth ratio h_{max}/h_{min} of the wave front of roll-wave have been investigated. As a result, the periodic property of roll-wave could be presumed to have the analogy with the period of a simple pendulum. It has been also shown that the tendency of decreasing the roughness coefficient of roll-wave flow can be well expressed by the non-dimensional velocity gradient, representative of roll-wave flow.

In future, on the basis of these knowledge, general properties of roll-wave flow will be confirmed.

6. REFERENCES

Dressler, R.F. (1949): "Mathematical solution of the problem of roll-waves in inclined open channels", Communication on Pure and Applied Mathematics, Vol. 2, No.2/3, pp. 149-194.

Ishihara, T., And So On (1954): "Theory of The Roll-Wave Trains in Laminar Water Flow On a Steep Slope Surface", Trans., of JSCE, Vol.19, pp. 46-57. (In Japanese)

Iwagaki, Y., and Iwasa, Y. (1955): "On The Hydraulic Characteristics of The Roll-Wave", Trains., Proceeding of JSCE, Vol.40, No.1, pp. 5-12. (In Japanese)

Mayer, P.G. (1959): "Roll waves and slug flows in inclined open channels", Transactions of the ASCE, pp. 505-563.

Miyajima, M. (1997): "Experimental Study on Roll-wave Flow Characteristics", Proceedings, 2nd Sympo., on Environmental Fluid Mechanics, JSFM, pp. 445-446. (In Japanese)

Murota, A., and Miyajima, M. (1993): "Experimental Study on Internal Structures of Ultra-Rapid Flow (Mainly on Mean Flow and Wave Characteristics)", Proceedings of Hydraulic Engineering, JSCE, Vol. 37, pp. 563-568. (In Japanese)

Murota, A., and Miyajima, M. (1995): "Experimental Study on Internal Structures of Ultra-Rapid Flow (On 2 Dimensional Behaviors of Turbulence)", Annual Journal of Hydraulic Engineering, JSCE, Vol. 39, pp. 379-384. (In Japanese)

Murota, A., and Miyajima, M., and Muraoka, K. (1995): "Experimental Study on Internal Structures of Ultra-Rapid Flow (On Roll-Wave Characteristics)", Proc. of The 26th Congress of IAHR, London, HYDRA2000, Vol. 1, pp. 409-414.

Environmental Hydraulics, Lee, Jayawardena & Wang (eds) © 1999 Balkema, Rotterdam, ISBN 90 5809 035 3

Development of a non-contact method of measuring river depth from the air

Yoshiharu Okamoto
Civil Engineering Department, Niigata University, Japan

ABSTRACT: Using ordinary means, it is impossible to measure the depth of a rapidly flowing, wave tossed river. Therefore, a way was researched and developed to measure the depth of a river using a subsurface interface radar placed at a height of five to ten meters above the surface of the water.

1. INTRODUCTION

Traditionally, there have been two methods used to measure the depth of a river: sounding rods, and weighted lines. More recently a third method has been developed employing an echo sounder, which is now used almost exclusively.

A boat must be set afloat on a river in order to use an echo sounder to measure the water's depth. However, as it is difficult to set a boat down in rapidly flowing, choppy water, it is very difficult to measure the depth of such a river. In such a case, it is conceivable that the echo sounder's sensor may be lowered to the surface of the water from a position above the river. And yet, in reality, it would be virtually impossible to suspend the sensor on the surface of the water with any kind of stability due to the rushing, wavy nature of the river. And so, in fact, such a method is impossible. It is necessary that the sensor be able to measure the depth of the water without actually having to contact the surface of the water.

Therefore, a fourth method of measuring a river's depth may be considered: suspending an antenna above the water and projecting radio waves towards the water below. In other words, using radar.

The supersonic waves currently employed by echo sounders can only be sent short distances through the air, as they attenuate rapidly in the atmosphere. This renders the echo sounder impractical for use in non-contact means of measuring a river's depth.

2. THE PRINCIPLES OF A METHOD OF SENDING RADIO WAVES FROM THE AIR

This method involves transmitting a pulse of radio waves from an antenna above the river down to the surface of the water. The radio waves that hit the water's surface on a diagonal are fully reflected away. Of the radio waves that hit the water's surface on the perpendicular, some of the radio waves penetrate the water, while the remainder are reflected from the water's surface back to the antenna.

Most of the radio waves that penetrate the water will continue on into the ground. However, some will be reflected back up off the river's bottom and will return to the water's surface. The radio waves that return from the bottom on a diagonal will again reflect off the surface of the water back to the bottom of the river. Those that again reach the surface of the water on the perpendicular will also be mostly reflected back to the bottom of the river. However, some will go through the surface and reenter the atmosphere.

These radio waves which return up to the antenna are significantly weaker than those that were originally transmitted.

Using the information described above, we can calculate the depth of the river. This is done by comparing the time at which the radio waves that initially reflected off the surface of the water return to the antenna, to the time at which the radio waves that reflected off the bottom of the river return to the antenna.

This is very simple in principle, but in actuality it is not an easy matter as is explained in the next section.

3. PROBLEMS AND QUESTIONS REGARDING THE USE OF RADIO WAVES AS A METHOD OF MEASUREMENT

The following are problematic issues and questionable points regarding this method of measurement.

1. A certain amount is known regarding the characteristics of incidence and reflection of radio waves off the surface of sea water. However, it is not an exaggeration to say that nothing is known about these characteristics in regards to fresh water. Therefore at this point, we have no recourse but to analogize from the known characteristics of sea water.

2. The shorter the wavelength, the more difficult it becomes for radio waves to penetrate the water; the longer the wavelength, the easier it is for the radio waves to penetrate the water.

3. The composition of the soil at the bottom of rivers varies widely, each variety causing differences in the reflective characteristics of radio waves. That is, soil that readily reflects radio waves of one wavelength may well not equally reflect radio waves of another wavelength.

4. When there are waves on the river, radio waves that are sent out perpendicular to the river may not enter the water perpendicularly.

5. When the river is at high water, erosion occurs leaving silt and sand built up on the river bed. For this reason, the bottom of the river may not provide a distinct surface from which the radio waves can reflect.

6. During high water, the river's flow will contain large amounts of mud and foreign matter, causing a more pronounced attenuation of the radio waves in the water. Detectable radio waves may not be reflected from the river's bed.

7. It also appears that, depending on the situation, the speed at which the radio waves travel through muddy water may well differ from the speed at which they travel through clear water.

8. It is questionable whether or not the strong noise generated by the radio waves which strike the surface of the water on a diagonal could be removed.

4. UTILIZATION AND CHARACTERISTICS OF SUBSURFACE INTERFACE RADAR

A currently existing device known as a subsurface interface radar, (or, subsurface radar), is able to detect overlapping stratum in the ground (subsurface interface). Therefore, rather than developing a radar to be used exclusively to measure river depth, it was proposed that this existing tool be used.

Subsurface radar is placed on the flat ground, and uses two antenna; one antenna that transmits pulses of radio waves with weak directivity, and a second one for receiving the radio waves. It moves slowly rectilinearly, while its oscillograph shows the strength of the reflected waves, along with the time differences caused by the overlapping stratum under the antenna. Unlike ordinary radar which operates from a fixed point, it is necessary for the transmission and reception antenna of the subsurface radar to be moved.

The earth underground is composed of various stratum. Some stratum will reflect a certain length of radio wave, but no other wavelengths. Therefore, in order to detect the various stratum, the subsurface radar must project not just one wavelength of radio wave, but a wide range of wavelengths at the same time.

Transmitted radio waves are usually named by the center frequency of a broader band of waves, such as, for example, the "100 megahertz" band. Now as stated earlier, the longer the wavelength of the radio wave, the easier it penetrates the water. As the center frequency of the projected radio waves becomes lower, a point is reached at which it is possible to detect the radio waves that have been reflected back from the bottom of the river. So it is necessary that the subsurface radar be able to transmit radio waves using a freely designated central frequency.

Further, the subsurface radar is so made that a transmission of radio waves with a specific center frequency requires that a particular antenna must be used. A range of transmissions using a variety of center frequencies cannot be made with only one antenna. The device requires that a range of antennas be used, match with frequencies from the low to high end of the spectrum used.

Also, the radio waves transmitted from ordinary radar are directional. Yet the subsurface radar deals with a wide range of frequencies, which works against creating directional radio waves. Basically, the radio waves transmitted from the subsurface radar are non-directional.

Now in order to use the subsurface radar to measure the depth of a river, a helicopter or crane must be used to suspend the antenna in the air. Transmitting these non-directional radio waves from this position means that surrounding bodies cause spurious reflections of the radio waves. The problem now, being unable to make use of the directional radio waves provided by standard radar, is to prevent the radio waves from being transmitted in unnecessary directions.

5. DETERMINING WHAT VARIETY OF SUBSURFACE RADAR TO EMPLOY

The foremost requirement faced in selecting a suitable subsurface radar was to have a unit with a large number of available antennas, thus increasing the high/low range of frequencies the unit can produce. This is needed because a central frequency suitable for use in measuring a river's depth needed to be determined by trial and error, using the widest range of frequencies possible. The SIR SYSTEM-10A, produced by Geographical Survey Systems, Inc. of the United States, allows for the use of a variety of antennas and can transmit radio waves with a central frequency ranging from 20MHz to 2.5 GHz. This is the machine that has been utilized in this project.

6. SUITABLE TRANSMISSION FREQUENCY FOR THE MEASUREMENT OF RIVER DEPTH

Since the subsurface radar transmits radio waves of an extremely wide range of frequencies, virtually any central frequency chosen will result in the penetration of the water by the radio waves. The difference in the radio wave employed is that as the frequency of the radio waves goes higher, the reflection of the radio waves from the bottom of the river becomes weaker. There is a critical central frequency. By trial and error it was determined that good results were obtained by using an antenna that provides for a central frequency of 100 MHz. If a larger central frequency is used, the antenna becomes too large and unwieldy for practical use.

7. PROBLEMS AND COUNTERMEASURES REGARDING THE SUBSURFACE RADAR'S OMNIDIRECTIONAL RADIO WAVES

For all practical purposes, the radio waves transmitted by the underground radar are omnidirectional. This would not present a problem were the antenna floating in the air under its own power and if it were located on an endless expanse of water.

However, if the antenna is hanging from a helicopter, and the distance to the river's bank is not far, then spurious reflections occur as noise, and the radio waves returning from the bottom of the river cannot be distinguished from this noise. Unless this noise can be dealt with, the subsurface radar cannot be practically used to measure the depth of a river. As a countermeasure, making use of material that absorbs radio waves can be considered.

This material should absorb incoming radio waves, converting the energy to heat and allowing no reflection of the radio waves. With an ideal material, the radio waves would disappear immediately upon impact.

There are two different types of materials used to absorb radio waves, one that is lightweight and one that is heavy; each has basic differences in the manner in which they absorb radio waves. The light weight material can only dampen a narrow band of radio waves. This means that one type of light weight material would be inadequate. Many varieties of the light weight material would have to be used simultaneously to absorb the wide range of frequencies transmitted by the subsurface radar. But using many varieties of this material results in tremendous bulk and is ultimately impractical. If one variety of absorbing material is to be used, it must be the heavy type. The weight is still a problem, but if that problem is solved, then the use of the material becomes practical.

8. NECESSARY TRANSMISSION STRENGTH OF SUBSURFACE RADAR

The subsurface radar uses a pulse of radio waves, and here the transmission strength of the subsurface radar will be explained using the average strength of the radio waves transmitted during one pulse. When used in the investigation of thin stratum underground, the output strength of the subsurface radar is a few watts.

Also, there are two systems of antenna used by subsurface radar. In one, a single antenna is used for both the transmission of the radio waves and also the reception of the reflection of those waves. The second variety uses two antennas, one for transmitting and one for receiving. For this project the standard antenna used was the single antenna, with an output of five watts.

When the bottom of the standard antenna was placed on the water's surface, not suspended in the air, it was possible to measure the river's depth in any situation.

Therefore, if it can be determine how many meters high the antenna will be suspended in the air, the necessary transmission output strength can be calculated.

9. CONCRETE METHODS OF ELIMINATING STATIC-INDUCING RADIO WAVES USING RADIO WAVE ABSORBING MATERIAL

The way in which radio wave absorbing material is used to absorb the spurious radio waves that are the cause of static noise differs according to whether the subsurface radar uses a single antenna or a pair of antennas.

With a single antenna type of radar, the antenna is inserted in the bottom of a box constructed of radio wave absorbing material. The antenna is surrounded on the top and on four sides with material that absorbs radio waves. Further, a skirt of the material may be added around the open bottom of the box; the longer the skirt, the better. In this way, the static noise caused by the transmission and reflection of spurious radio waves can be prevented.

When separate transmission and reception antennas are used, each antenna is outfitted with the an identical box of absorption material covering it on the top and all four sides. They will be aligned together as one unit. In this case, a further skirt of absorption material around the open bottom of the boxes cannot be used.

10. SINGLE OR SEPARATED ANTENNAS: WHICH IS BETTER USED?

Here the problem presents itself as to which type of antenna should be used, a single antenna or a pair of transmitting/receiving antennas. It appears that only radio waves transmitted at a right angle to the river's surface, plus or minus 5 degrees, penetrate the water. This indicates that a single antenna would have good efficiency, whereas a pair of transmitting/receiving antennas would have remarkably worse efficiency, requiring a much higher transmission output. Assuming the height of the antenna to be 5 meters, a single antenna would require an output of about 100 watts. The two antenna system would require an output of several hundred watts.

Also, when separated antennas are used, a large portion of the radio waves transmitted strike the surface of the water at an angle, which causes static noise. It is reasonable to think that using a single antenna would remarkably reduce the amount of static noise produced.

Given the above, it seems obvious that a single antenna type of radar should be used. However, since high output, single antennas for subsurface radar are not mass produced, they must be special ordered by the user, resulting in very high costs. In this research project the use of high output single antennas was not possible. Therefore the following discussion is based on the assumption that a pair of high output separated antennas transmitting at 500 watts will be used.

The total distance from the point that the radio waves are transmitted out to the point that the radio waves audibly return to the antenna has to be over 40 meters.

11. POSSIBILITY OF MEASURING THE DEPTH OF A WAVE TOSSED RIVER

As radio waves will only penetrate the water's surface on, or very near, the perpendicular, it would seem to be a difficult task to use radio waves to measure the depth of a wave tossed river, as no flat surfaces of the water appear to be perpendicular to the antenna. However the reality of the situation is that at any given instant, there are a tremendous number of small flat surfaces on top of the river, among the waves. These innumerable small flat surfaces make it possible to measure the river's depth, in spite of the presence of waves.

12. CONCERNING ACCURACY IN MEASURING RIVER DEPTH

Using the radar to determine the depth of a river will essentially measure the shallowest water depth in an area slightly wider than the area in which the radio waves hit the water directly below the antenna. The measurement results are not necessarily from the spot exactly under the center of the antenna.

Because of this, the greater the water's depth, and the greater the degree of inclination of the river's bottom, the greater the margin of error becomes. If the river bed under the radar's transmission of radio waves is completely flat, then the margin of error becomes zero.

13. CONCLUSION

In the past it was impossible to measure the depth of a river that was flowing so quickly and roughly that a boat could not be safely launched upon it. However, with the use of the subsurface interface radar (subsurface radar) it has now become possible.

Environmental Hydraulics, Lee, Jayawardena & Wang (eds) © 1999 Balkema, Rotterdam, ISBN 90 5809 035 3

Effects of colloidal chemistry on fine particle deposition within a viscous porous flow

Zhishi Wang

Faculty of Science and Technology, University of Macau, Macau

ABSTRACT: Deposition of fine particles within the viscous porous flow is a common environmental hydraulic phenomenon occurring in deep bed filtration for water supply treatment, groundwater flow, wastewater disposal by land infiltration, estuary sedimentation as well as flow of the pore water in sediment. In this study, it has been also found that deposition of the fine particles on the model grain surface in the porous media depends upon the fine particle surface properties such as surface charge signs and magnitudes. There are two cases occurring with respect to the effects of the particle surface properties on the deposition in the viscous porous flow, i.e., favorable for the deposition for which the fine particle surface charge reversal occurs and unfavorable for the deposition. The zeta potential of the fine particles is the master variable to control these cases.

1. INTRODUCTION

In the water filtration treatment practice, one may usually deal with such cases that the treatment efficiency is closely related to the chemical control in the treatment process. For example, for water reuse a tertiary treatment has been developed for phosphate (PO_4^{3-}) removal from wastewater by coagulation and filtration (case 1) (Stumm 1978, Kraft, et. al 1990, Tanaka, et. al 1991). For the water direct filtration treatment, it is important how to control the inorganic coagulant Al(III), i.e., hydrolyzed aluminum ions) dosage and other chemical conditions for the better treatment result (case 2) (Wagner et. al 1982, Amirtharajah et. al 1984, Wiesner et. al 1989, Rebhum 1990). For the organic contaminated raw water, it must be considered to remove both suspended particles and the organic matter (COD) by coagulation and filtration as well as biodegradation (case 3) (Rittmann 1990, LeChevallier et. al 1992, Manem et. al 1992, Tobiason et. al 1992, Chandrakanth, et. al 1996). All these practical cases reflect one common focus - effects of zeta potentials of suspended particles on the deep bed filtration process. The traditional water filtration technology is focused on the physical aspect of the operation process such as controlling filtration velocity, filter media size and thickness, etc., which cannot meet the requirements for handling the cases mentioned above. The chemical aspect of the filtration operation becomes important, which involves how to control solution chemical conditions prior to the filtration, how to change the zeta potentials of the suspended particles in water for better filtration treatment efficiency. So it has been attempted to study: 1) variation of the suspended particle zeta potentials with the chemical conditions of water to be treated; and 2) effects of the zeta potential variation on the particle removal efficiency factors α of a well-defined filtration system.

2. METHODS AND MATERIALS

In this study, the experimental work was made by two subsequent phases, i.e., measuring the zeta potentials of suspended particles in water with various chemical conditions, using electrophoresis apparatus (phase 1) and measuring the particle removal efficiency factors α by water filtration in a defined filtration test system (phase 2). The suspended particle zeta potentials were measured using Zeta-Meter (III) with Pt electrodes and 1 cm x 1 cm cell, ZETA-METER Inc. USA. The turbidity of water was measured using UV-250 spectrophotometer with an integral sphere, Japan.

3. RESULTS AND DISCUSSIONS

3.1 Case 1 - Effects of inorganic phosphate anions PO_4^{3-}:

Figure 1 gives the zeta potential measurement for suspended hematite particles (d = 0.3 μm) with phosphate anions and without phosphates in solution, respectively. Figure 2 shows the measured particle removals by the well-defined filtration system. It indicates a close relation between the zeta potential variation and the removal efficient factor α. If the α value is near one, the chemical condition of filtered water is favorable for particle removal (i.e., high removal efficiency). And if the α value approaches zero, the chemical condition is unfavorable for particle removal by filtration (i.e., low filtration efficiency). Figure 2 shows that there is a special pH value at which the α value dramatically changes, which means that the filtration condition suddenly shifts from the favorable to unfavorable. The special pH value is defined as pH_α. In presence of phosphate anions in water, the zero point of charge of the hematite particle shifted from 7.2 to 4.8 while the pH_α shifted to the acidic side, i.e., from 7.0 to 4.5 accordingly.

3.2 Case 2 - Effects of inorganic coagulants Al(III):

Figure 3 gives the measurement of zeta potentials of suspended latex particles (d = 0.2 μm) with hydrolyzed aluminum ions Al(III) and without Al(III) in solution, respectively. Figure 4 shows the measured particle removal efficient factors α for the same case in the well-defined filtration system. In presence of Al(III), two pH_{pzc} values were observed for the latex particles, that is, pH_{pzc1} = 4.3; pH_{pzc2} = 8.0. And the pH_{pac2} value was observed to be near the pH_{pzc} of aluminum hydroxide precipitates (about 8.3). There were two pH_α values observed accordingly in the relevant filtration tests, i.e., $pH_{\alpha1}$ = 4.5 and $pH_{\alpha2}$ = 7.8.

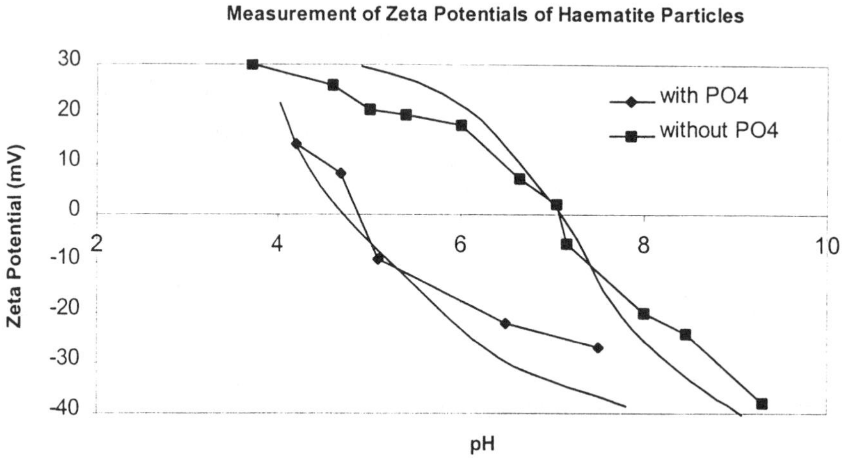

Measurement of Zeta Potentials of Haematite Particles

Figure 1. Measurement of zeta potentials of the hematite suspended particles in presence or absence of phosphate anions; and the solid lines are model-predicted ζ potentials in presence and absence of phosphate anions, respectively

972

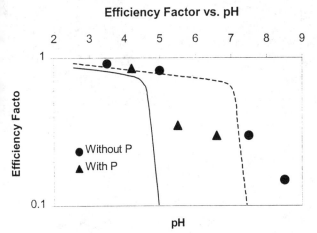

Efficiency Factor vs. pH

Figure 2. Measurement of particle removal efficiency factor α in presence or absence of phosphate anions; the solid line is model-predicted α in the case of presence of phosphate anions while the dash line is the case of absence.

3.3 Case 3 - Effects of organic matter (COD) and bio-film treatment:

The measured ζ potentials of suspended particles in raw water are quite different for the organic contaminated raw water from the relatively clean raw water (see Figures 5 and 6). In both cases, the ζ potential increases with adding aluminum coagulant dosage and so there occurs a zero point of charge ($\zeta = 0$) at a certain dosage, e.g., 40 mg/l in this test. However, when there exists organic matter (COD = 8 - 20 mg/l) in raw water, the ζ potential changes with the dosage with a complicated way. At a certain dosage range (in 35 - 125 mg/l in Figure 6), the ζ potential does not increase with the dosage and in stead, it keeps negative until a certain dosage (i.e., 125 mg/l) reaches. The jar tests of particle flocculation were also made accordingly.

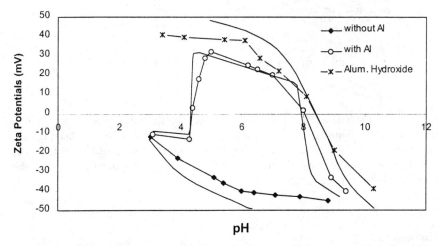

Zeta Potential Measurement of Latex Particles in Presence of Al (III)

Figure 3 Measurement of zeta potentials of the latex particles in presence of hydrolyzed aluminum ions

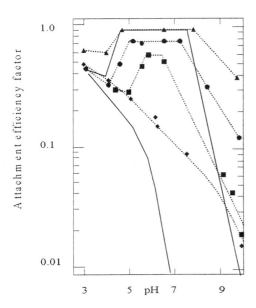

Figure 4 Measurement of latex particle removal effient factors α with and without hydrolyzed aluminum ions; and the solid lines are model-predicted α values.

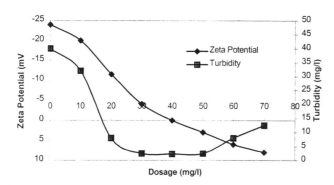

Figure 5 Measurement of zeta potentials of clay suspended particles in raw water without organic pollution

In order to remove both the suspended particles and organic matter from the contaminated raw water, a biofilm column treatment was designed prior to filtration pilot unit. Figure 7 shows that the zeta potential of suspended particles in the bio-treatment effluents at different ceramic bed depths reaches zero with increase of biomass of the column. It means that the biofilm pretreatment can reduce the absolute value of zeta potential to let it reach zero.

4. CONCLUSION

This study has revealed that:
1) Chemical conditions of raw water (pH, phosphate anions, hydrolyzed aluminum ions, etc.)

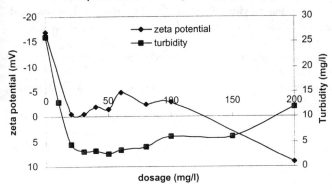

Figure 6 Measurement of zeta potentials of clay suspended particles in raw water with organic pollution

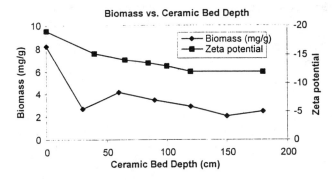

Figure 7: Measurement of both zeta potentials of suspended particles and biomass of the biofilm ceremic column

determine the zeta potentials of suspended particles. Particularly, the chemical conditions can shift the pH_{zpc} and reverse the surface charge signs of suspended particles so as to influence the filtration treatment efficiencies.

2) The biofilm column can not only be used to remove the biodegradable organic matter from raw water, but it can improve as well the follow-up operation of coagulation and filtration through reducing the absolute values of zeta potentials of suspended particles.

5. REFERENCES

Amirtharajah, A. and Suprenant, B. K. 1984. Direct filtration using the alum coagulation diagram. Proceedings of 1984 AWWA Annual Conference, Dallas.

Edzwald, J. K. et al. 1987. Organics, polymers, and performance in direct filtration. J. Environmental Engineering, ASCE, Vol. 113, No. 1, pp. 167.

Eriksson, L. et al. 1973. Desorption of hydrolyzed metal ions from hydrophobic interfaces: 1. latex-aluminum nitrate systems. J. Colloidal and Interface Science, Vol. 43, No. 3, pp.591.

Ives, K. J. 1969. Theory of filtration. Special Topic7. Proceedings of International Water Supply Congress & Exhibition, International Water Supply Association, London

Ives, K. J. 1982. Fundamentals of filtration. Proceedings of Symposium on Water Filtration, European Federation of Chemical Engineering. Antwerp.

Iwasaki, T. 1937. Some notes on sand filtration. J. AWWA, 29.

Kraft, A. and Seyfried, C. F. 1990. Ammonia and phosphate elimination by biologically intensified flocculation filtration processes. Chemical Water and Wastewater Treatment (H.H. Hahn and R. Klute, eds.). Springer-Verlag.

LeChevallier, M. W. and Becker, W. C. and Schorr, P. 1992. Evaluating the performance of biologically active rapid filters. J. AWWA, April.

Manem, J. A. and Rittmann, B. E. 1992. The effects of fluctuation in biodegradable organic matter on nitrification filters. J. AWWA, April, pp. 147.

Rebhum, M. 1990. Floc formation and breakup in continuous flow flocculation and contact filtration. Chemical Water and Wastewater Treatment (H.H. Hahn and R. Klute eds.) Springer-Verlag.

Rittmann, B. E. 1990. Analyzing biofilm processes used in biological filtration. J. AWWA, December, pp. 36.

Stumm, W. 1978. Chemical interactions in particle separation. Chemistry of Wastewater Technology (A.J. Rubin ed.) Ann Arbor Science Publishers, Inc.

Tanaka, K. and Aoki, M. and Takahashi, S. 1991. Study on development of phosphorus removal process by contact filtration. Water Science and Technology, Vol. 23, pp. 739.

Wagner, E. G. and Hudson, H. E. Jr, 1982. Low dosage high rate direct filtration. J. AWWA, 74(5) pp. 256.

Environmental Hydraulics, Lee, Jayawardena & Wang (eds) © 1999 Balkema, Rotterdam, ISBN 90 5809 035 3

Author index